Biocatalysis in the Pharmaceutical and Biotechnology Industries

Edited by
Ramesh N. Patel

CRC Press
Taylor & Francis Group
Boca Raton London New York

CRC Press is an imprint of the
Taylor & Francis Group, an informa business

CRC Press
Taylor & Francis Group
6000 Broken Sound Parkway NW, Suite 300
Boca Raton, FL 33487-2742

© 2007 by Taylor & Francis Group, LLC
CRC Press is an imprint of Taylor & Francis Group, an Informa business

First issued in paperback 2019

No claim to original U.S. Government works

ISBN 13: 978-0-367-44628-4 (pbk)
ISBN 13: 978-0-8493-3732-1 (hbk)

Library of Congress Cataloging-in-Publication Data

Biocatalysis in the pharmaceutical and biotechnology industries / edited by Ramesh N. Patel.
    p. ; cm.
  Includes bibliographical references and index.
  ISBN-13: 978-0-8493-3732-1 (hardcover : alk. paper)
  ISBN-10: 0-8493-3732-1 (hardcover : alk. paper)
  1. Enzymes--Biotechnology. 2. Biotechnology--Industrial applications. I. Patel, Ramesh N., 1942-
  [DNLM: 1. Catalysis. 2. Enzymes. 3. Biotechnology--methods. 4. Pharmaceutical
Preparations--chemical synthesis. QU 135 B6145 2006]

  TP248.65.E59B56 2006
  660.6'34--dc22
        2006008932

Visit the Taylor & Francis Web site at
http://www.taylorandfrancis.com

and the CRC Press Web site at
http://www.crcpress.com

Preface

There has been an increasing awareness of the enormous potential of microorganisms and enzymes for the transformation of synthetic chemicals in a highly chemo-, regio-, and enantioselective manner. Chiral intermediates are in high demand from pharmaceutical, agricultural, and other biotechnological industries for the preparation of bulk drug substances or fine chemicals. Bulk drug compounds or other fine chemicals can be produced by chemical or chemo-enzymatic synthesis. The advantages of biocatalysis over chemical synthesis are that enzyme-catalyzed reactions are often highly enantio- and regioselective. They can be carried out at ambient temperature and atmospheric pressure, thus avoiding the use of more extreme conditions which could cause problems with isomerization, racemization, epimerization, and rearrangement. Microbial cells and enzymes derived from microbial cells can be immobilized and reused for many cycles. Biocatalysis includes fermentation, biotransformation by whole cells or enzyme-catalyzed transformations, cloning and expression of enzymes, and directed evolution of enzymes to improve selectivity, substrate specificity, and stability. Various chapters in this book are contributed by internationally well-known scientists having many years of experience in different aspects of biocatalysis and biocatalytic applications in production of fine chemicals and chiral pharmaceutical intermediates.

This book contains 34 chapters with over 4000 references and more than 600 tables, equations, drawings, and micrographs. All the information cited in this book provides state-of-the-art knowledge and improves the ability of the reader to use different types of enzymatic reactions in synthesis of fine chemicals and chiral compounds and their application in biotechnological industries. Various chapters discuss the following important aspects in biocatalysis and its applications in various industries:

- Application of nitrilases and nitrile hydratases in synthesis of fine chemicals that describe cloning and expression of nitrilases and their use in production of chiral and achiral carboxylic acids, regioselective and chemoselective hydrolysis of nitriles, preparation of amides from nitriles, commercialized processes for preparation of nicotinamide, cyanovaleramide, acrylamide, and nitrile-containing polymers
- Biocatalytic deracemization processes that include dynamic kinetic resolution, stereoinversion processes, and enantioconvergent processes to prepare chiral compounds such as amino acids, amines, alcohols, diols, and epoxides in theoretical 100% yields
- Biocatalysis in pharmaceutical industries for synthesis of chiral intermediates and fine chemicals for chemoenzymatic synthesis of drugs such as anticancer, antiviral, antihypertensive, anticholesterol, anti-infective, anti-infammatory, antianxiety, and antipsychotic drugs
- Methods for directed evolution of lipases and esterases, assay development and screening of mutants for selection in esterification, transesterification, acylation and acyl hydrolytic reactions, and use of improved enzymes in organic synthesis
- Oxidative biocatalysis catalyzed by flavin-containing flavoprotein oxidases such as alcohol oxidases, amine oxidases, and sulfhydryl oxidases together with flavoprotein monooxygenases such as aromatic, heteroatom, and multicomponent monooxygenase in enzymatic oxygenation reactions

- Biotransformation (hydroxylation, dealkylation, N-oxide formation, and O-demet
  tion reactions) of natural and synthetic compounds for the generation of mole
  diversity and drug metabolites
- Enzymatic acylation of alcohols and amines in preparation of pharmaceuticals
  as anticonvulsant agents, anticancer agents, immunosuppressive compounds, ana
  drugs, antidepressants, anticholesterol drugs, antibiotics, β-adrenergic bloc
  calcium channel blockers, serotonin uptake inhibitors, antifungal agents,
  Alzheimer's agents, antiulcer agents, α-adrenoceptor agonists, and other drug
  stances
- Enzyme–metal complex-catalyzed asymmetric biotransformations and dynamic
  lution processes to prepare chiral alcohols, amines, and acetates
- Applications of aromatic hydrocarbon dioxygenases such as toluene dioxyge
  naphthalene dioxygenase, chlorobenzene dioxygenase in synthesis of fine chem
  and pharmaceuticals with continuous cofactor regeneration during biotransforma
- Investigation on baker's yeast reduction processes by genomic approach and pr
  ation of chiral alcohols and synthesis of anticancer, anticholesterol, and antihype
  sive drugs
- Techniques and applications of immobilization of enzymes as cross-linked en
  aggregates (CLEA) in synthesis of fine chemicals
- Application of C–C bond forming enzymes such as aldolases and transketolas
  synthesis of fine chemicals and pharmaceuticals
- Biocatalytic synthesis of nucleoside analogs by modification of base, sugar and o
  tion reactions, and applications of nucleoside analogs as antiviral agents
- Biocatalytic reduction of carboxylic acids by carbonyl reductases, mechanism o
  bonyl reductases, and cloning, and expression of enzyme with application in synthe
  fine chemicals
- Application of dehalogenases in biocatalysis and biodegradation with emphas
  haloalkane dehalogenases, haloacid dehalogenases, and halohydrin dehalogenase
  their application in preparation of chiral compounds
- Enzymatic synthesis of sugar esters and oligosaccharides from renewable resource
  includes regioselective synthesis of fatty acid sugar ester using lipases and proteases
  synthesis of oligosaccharides by transglycosidases and transglucosidases
- Efficient methodology and instrumentation for engineering custom enzymes
  directed evolution and solid-phase screening to maximize high throughput and sele
  of evolved enzymes and application of highly active enzymes in synthesis of inter
  ates for pharmaceuticals
- Biocatalytic enantioselective and diastereoselective deaminations for chemoenzy
  synthesis of antiviral agents using adenosine deaminase or adenylate deaminase
- Enzymatic resolution of lactones by lactonases and synthesis of chiral alcoho
  carbonyl reductases and effective cofactor regenerating systems for synthesis of pha
  ceutical intermediates
- Enzymatic acyloin condensations by decarboxylases and rational design of aryl
  nate decarboxylase for synthesis of fine chemicals and chiral intermediates
- Enantioselective biocatalysis for synthesis of pheromones and juvenile hormones
- Stereoselective and regioselective modifications of polyhydroxylated steroids by
  drogenases, lipases, and proteases to prepare novel steroidal compounds
- Industrial enzymatic processes for C–C, C–N, C–O bond formations by lyases su
  phenylalanine ammonia lyase, fumarase, malease, hydratases, and dehydratases
- State of the art and application in enantioselective synthesis of chiral cyanohydri
  hydroxynitrile lyases

- Chiral switches strategies, opportunities, and experiences, and biocatalysis in preparation of ibuprofen, ketoprofen, esomeprazole, methylphenidate, doxyzosin, levofloxacin, and other chiral molecules
- Cutting-edge methodology for gene shuffling, family of genes shuffling, directed evolution, and high-throughput screening of mutants to increase the selectivity, activity, and stability of enzymes
- Biocatalytic preparations of chiral amines by kinetic resolution, dynamic kinetic resolution; deracemization and stereoinversion processes using transaminases, amine oxidases, and lipases

Biocatalysis in the pharmaceutical and biotechnological applications is an indispensable resource for organic chemists, biochemists, microbiologists, biochemical engineers, biotechnologists, medicinal chemists, pharmacologists, and upper-level undergraduate and graduate students in these disciplines.

It is my pleasure to acknowledge sincere appreciation to all the authors for their contribution to this book. I would like to acknowledge continual support from David Fausel (production coordinator) and Anita Lekhwani (acquisition editor) and Richard Tressider (project editor) of the Taylor & Francis Group and Vinithan Sethumadhavan (project manager) of SPi. My interest in biocatalysis was developed and stimulated by David Gibson, Derek Hoare, Nicholas Ornston, Allen Laskin, Ching Hou, Laszlo Szarka, Christopher Cimarusti, John Scott, Richard Mueller, and many of my colleagues at the University of Texas, Yale University, Exxon Research and Engineering Company, and Bristol-Myers Squibb. I acknowledge their support and encouragement over the last 35 years. Finally, I would like to express my sincere thanks to my wife, Lekha, and my daughter, Sapana, for their support and encouragement while I worked on this book.

Editor

Ramesh N. Patel is a senior research fellow in charge of the Enzyme Technology Department in Process Research and Development at the Bristol-Myers Squibb Pharmaceutical Research Institute, New Brunswick, New Jersey. He was a National Institute of Health and American Chemical Society Postdoctoral Fellow (1971–1975) in the biology department at Yale University, New Haven, Connecticut, and a research scientist (1975–1987) in the biotechnology department at the Corporate Research Center, Exxon Research and Engineering Company, Clinton, New Jersey. Holder of over 70 U.S. and European patents, and author and coauthor of over 150 journal articles and 80 national and international conference presentations, he is a member of the American Association for the Advancement of Science, American Chemical Society, American Society for Microbiologists, Society for Industrial Microbiologists, and the American Oil Chemist's Society. He was awarded the 2004 Biotechnology Lifetime Achievement Award from the Biotechnology Division of the American Oil Chemist's Society. Dr. Patel received his bachelor's degree from Bombay University, Bombay, India; his master's degree from Maharaja Sayajirao University, Baroda, India; and his Ph.D. degree (1971) from the University of Texas, Austin.

Contributors

Yangsoo Ahn
Department of Chemistry
Pohang University of Science and
 Technology
Pohang, Korea

M. Alcalde
Department of Biocatalysis
Institute of Catalysis and Petrochemistry
CSIC
Madrid, Spain

Andrés R. Alcántara
Organic and Pharmaceutical Chemistry
 Department
Complutense University
Madrid, Spain

Laura Alessandrini
Department of Preclinical Sciences
LITA Vialba
University of Milan
Milan, Italy

Robert Azerad
Laboratoire de Chimie et Biochimie
 Pharmacologiques et Toxicologiques
Université René Descartes-Paris 5
Paris, France

A. Ballesteros
Department of Biocatalysis
Institute of Catalysis and Petrochemistry
CSIC
Madrid, Spain

Uwe T. Bornscheuer
Department of Biotechnology and Enzyme
 Catalysis
Institute of Biochemistry
Greifswald University
Greifswald, Germany

Stefan Buchholz
Creavis
Degussa AG
Hanau-Wolfgang, Germany

Reuben Carr
School of Chemistry
University of Manchester
Manchester, United Kingdom

Pierangela Ciuffreda
Department of Preclinical Sciences
LITA Vialba
University of Milan
Milan, Italy

William J. Coleman
KAIROS Scientific Inc.
San Diego, California

Luís A. Condezo
Organic and Pharmaceutical Chemistry
 Department
Complutense University
Madrid, Spain

Réne Csuk
Institute of Organic Chemistry
Martin Luther University Halle–Wittenberg
Halle (Saale), Germany

Lacy Daniels
College of Pharmacy
Texas A&H Health Science Center
Kingsville, Texas

Robert DiCosimo
DuPont Central Research and Development
Wilmington, Delaware

Susanne Dreyer
Department of Chemistry
University of Rostock
Rostock, Germany

Franz Effenberger
Institute of Organic Chemistry
University of Stuttgart
Stuttgart, Germany

Kurt Faber
Department of Chemistry
University of Graz
Graz, Austria

Jesús Fernández-Lucas
Organic and Pharmaceutical Chemistry
  Department
Complutense University
Madrid, Spain

M. Ferrer
Department of Biocatalysis
Institute of Catalysis and Petrochemistry
CSIC
Madrid, Spain

Wolf-Dieter Fessner
Institute of Organic Chemistry and
  Biochemistry
Technical University of Dornstadt
Dornstadt, Germany

Siegfried Förster
Institute of Organic Chemistry
University of Stuttgart
Stuttgart, Germany

Elena Fossati
Institute of Chemistry of Molecular Recognition
Milan, Italy

Marco W. Fraaije
Biochemical Laboratory
Groningen Biomolecular Sciences and
  Biotechnology Institute
University of Groningen
Groningen, The Netherlands

H. García-Arellano
Department of Organic and Inorganic
  Chemistry
University of Oviedo
CSIC
Madrid, Spain

Carlos A. García-Burgos
Organic and Pharmaceutical Chemistry
  Department
Complutense University
Madrid, Spain

I. Ghazi
Department of Organic and Inorganic
  Chemistry
University of Oviedo
CSIC
Madrid, Spain

Vicente Gotor
Department of Organic and Inorganic
  Chemistry
University of Oviedo
Oviedo, Spain

Vicente Gotor-Fernández
Department of Organic and Inorganic
  Chemistry
University of Oviedo
Oviedo, Spain

Harald Gröger
Service Center Biocatalysis
Degussa AG
Hanau-Wolfgang, Germany

Aurelio Hidalgo
Department of Biotechnology and Enzy
  Catalysis
Institute of Biochemistry
Greifswald University
Greifswald, Germany

Michael J. Homann
Schering-Plough Research Institute
Union, New Jersey

Kohsuke Honda
Division of Applied Life Sciences
Graduate School of Agriculture
Kyoto University
Kyoto, Japan

Gjalt W. Huisman
Codexis Inc.
Redwood City, California

Takeru Ishige
Division of Applied Life Sciences
Graduate School of Agriculture
Kyoto University
Kyoto, Japan

Dick B. Janssen
Groningen Biomolecular
  Sciences and Biotechnology
  Institute
University of Groningen
Groningen, The Netherlands

Stefan Jennewein
Institute of Organic Chemistry and
  Biochemistry
Technical University of Dornstadt
Dornstadt, Germany

Michihiko Kataoka
Division of Applied Life Sciences
Graduate School of Agriculture
Kyoto University
Kyoto, Japan

Mahn-Joo Kim
Department of Chemistry
Pohang University of Science and
  Technology
Pohang, Korea

Christoph Kobler
Institute of Organic Chemistry
University of Stuttgart
Stuttgart, Germany

Udo Kragl
Department of Chemistry
University of Rostock
Rostock, Germany

Wolfgang Kroutil
Department of Chemistry
University of Graz
Graz, Austria

James J. Lalonde
Codexis Inc.
Redwood City, California

Julia Lembrecht
Department of Chemistry
University of Rostock
Rostock, Germany

Andreas Liese
Institute of Technical Biocatalysis
Technical University of
  Hamburg-Harburg
Hamburg, Germany

Mahmoud Mahmoudian
Research and Development
Process Chemicals
Rohm & Haas
Spring House, Pennsylvania

Akinobu Matsuyama
Corporate Development Center
Daicel Chemical Industries Ltd.
Tsukuba, Japan

Kenji Miyamoto
Department of Biosciences and
  Informatics
Keio University
Yokohama, Japan

Kenji Mori
RIKEN Research Center for Allergy and
  Immunology
Wako, Japan

Hiromichi Ohta
Department of Biosciences and
  Informatics
Keio University
Yokohama, Japan

Rebecca E. Parales
Section of Microbiology
University of California
Davis, California

Jaiwook Park
Department of Chemistry
Pohang University of Science and
  Technology
Pohang, Korea

Ramesh N. Patel
Process Research and Development
Bristol-Myers Squibb
New Brunswick, New Jersey

F.J. Plou
Department of Biocatalysis
Institute of Catalysis and Petrochemistry
CSIC
Madrid, Spain

Martina Pohl
Institute of Molecular Enzyme
    Technology
Heinrich Heine University of
    Düsseldorf
Jülich, Germany

Francisca Rebolledo
Department of Organic and Inorganic
    Chemistry
University of Oviedo
Oviedo, Spain

Sol M. Resnick
Biotechnology Core Research &
    Development
Dow Chemical Company
San Diego, California

D. Reyes-Duarte
Department of Biocatalysis
Institute of Catalysis and
    Petrochemistry
CSIC
Madrid, Spain

Sergio Riva
Institute of Chemistry of Molecular
    Recognition
Milan, Italy

Steven J. Robles
KAIROS Scientific Inc.
San Diego, California

John P.N. Rosazza
Center for Biocatalysis and Bioprocessing
University of Iowa
Iowa City, Iowa

Enzo Santaniello
Department of Preclinical Sciences
LITA Vialba
University of Milan
Milan, Italy

Jan Schumacher
Department of Chemistry
University of Rostock
Rostock, Germany

Roger A. Sheldon
Department of Biotechnology
Delft University of Technology
Delft, The Netherlands

Sakayu Shimizu
Division of Applied Life Sciences
Graduate School of Agriculture
Kyoto University
Kyoto, Japan

Yolanda Simeó
Department of Chemistry
University of Graz
Graz, Austria

Ajay Singh
Department of Biology
University of Waterloo
Waterloo, Ontario, Canada

José V. Sinisterra
Organic and Pharmaceutical Chemistry
    Department
Complutense University
Madrid, Spain

Jon D. Stewart
Department of Chemistry
University of Florida
Gainesville, Florida

Wen-Chen Suen
Schering-Plough Research
    Institute
Union, New Jersey

Nicholas J. Turner
School of Chemistry
University of Manchester
Manchester, United Kingdom

Willem J.H. van Berkel
Laboratory of Biochemistry
Wageningen University
Wageningen, The Netherlands

Padmesh Venkitasubramanian
Center for Biocatalysis and Bioprocessing
University of Iowa
Iowa City, Iowa

Owen Ward
Department of Biology
University of Waterloo
Waterloo, Ontario, Canada

Hiroaki Yamamoto
Lifescience Development Center
CPI Company
Daicel Chemical Industries Ltd.
Tsukuba, Japan

Mary M. Yang
KAIROS Scientific Inc.
San Diego, California

Aleksey Zaks
Schering-Plough Research Institute
Union, New Jersey

Ningyan Zhang
Merck Research Laboratories
West Point, Pennsylvania

Table of Contents

1 Nitrilases and Nitrile Hydratases

Robert DiCosimo

CONTENTS

1.1 INTRODUCTION

Nitrilase and nitrile hydratase (NHase) are two classes of enzymes that are finding increasing use as catalysts for the conversion of nitriles to carboxylic acids and amides, respectively; in addition, NHases are often used in combination with amidases to produce carboxylic acids. These reactions can be enantioselective, chemoselective, and/or regioselective, and there are often no equivalent chemical catalysts that can produce the desired product with the same selectivity afforded by the enzyme-catalyzed reactions. A large number of publications, patent applications, and patents describe the preparation and use of these enzyme catalysts, and numerous reviews of nitrilase- and NHase-catalyzed reactions have been published. The most recent work in this area has been summarized here, with a focus on the application of these enzymes as catalysts in synthesis, as well as in process development or in commercial processes.

1.2 NITRILASES

Nitrilases have been the subject of a number of recent reviews [1–7]. All known nitrilases share a highly conserved region of amino acid sequence which includes a cysteine that is responsible for the catalytic activity of the enzyme [1,8]. The mechanism that has been proposed for conversion

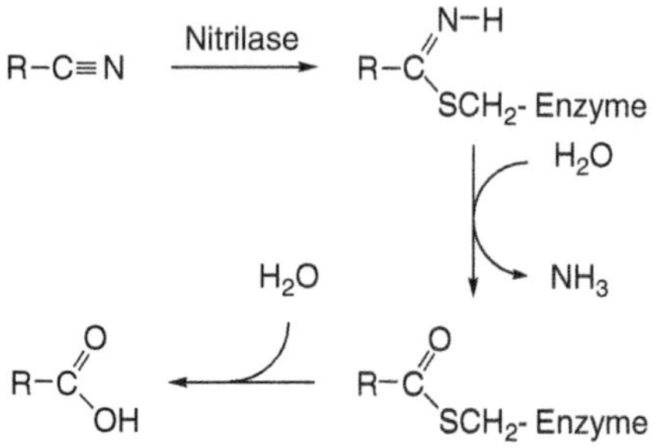

$$R-C\equiv N \xrightarrow{\text{Nitrilase}} R-C\underset{SCH_2\text{- Enzyme}}{\overset{N-H}{\parallel}}$$

$$\xrightarrow{-H_2O,\ NH_3} R-C\underset{SCH_2\text{- Enzyme}}{\overset{O}{\parallel}} \xrightarrow{H_2O} R-C\underset{OH}{\overset{O}{\parallel}}$$

SCHEME 1.1 Proposed mechanism for hydrolysis of nitrile to carboxylic acid by nitrilase. (Re
from Kobayashi, M., Goda, M, and Shimizu, S., *Biochem. Biophys. Res. Commun.*, 253, 662, 199

of a nitrile to a carboxylic acid by nitrilase is depicted in Scheme 1.1, where, after bindi
the enzyme active site, the nitrile reacts with a cysteine sulfhydryl residue to produ
intermediate thioimidate; subsequent hydrolysis of this thioimidate produces the ammo
salt of the carboxylic acid [9]. A crystal structure has not yet been reported for nitrilase

1.2.1 HETEROLOGOUS NITRILASE EXPRESSION

A number of recent patent applications describe the preparation of nitrilase catalysts, w
either the nitrilase gene was isolated from a wild-type cell and expressed in a transformant su
Escherichia coli, or variants of the wild-type nitrilase gene were created through directed e
tion techniques and heterologously expressed. The nitA gene encoding an enantiosele
nitrilase from *Rhodococcus rhodochrous* NCIMB 11216 was cloned and expressed in *E.*
and the resulting transformants screened for activity against a variety of aliphatic and ar
phatic nitriles [10]. A "PnitA-NitR" system for regulatory gene expression in *Streptomycete*
been developed, based on the expression mechanism of *R. rhodochrous* J1 nitrilase, wh
highly induced by ε-caprolactam [11]; heterologous protein expression yielded nitrilase lev
as high as 40% of soluble protein. The nitrilase from *Acidovorax facilis* 72W has been clone
expressed in *E. coli*, where the amount of nitrilase protein produced (active and inactive) wa
of total soluble protein, and active nitrilase comprised 12% of total soluble protein [12].

The nitrilase gene from the photosynthetic cyanobacterium *Synechocystis* sp. strain PCC
has been expressed in *E. coli*, and the purified nitrilase isolated from this recombinant strai
characterized [13]. The observed substrate specificity of the purified nitrilase most cl
resembled that of previously described aliphatic nitrilases, and the temperature optima (
45°C) and pH optima (pH 7 to 7.5) were similar to the nitrilases of the mesophilic bacter
rhodochrous J1 or *Alcaligenes faecalis* JM3. The purified enzyme was active in the presenc
wide range of organic solvents; for example, after incubation of the nitrilase for 10 min
mixture of 50 mM phosphate buffer and solvent, the rate of hydrolysis of benzonitrile to be
acid was not significantly affected by 40% dimethyl sulfoxide or *n*-heptane, 20% methan
10% ethanol or dimethyl formamide. The turnover rates of substrates with poor water solub
e.g., dodecanoic acid nitrile and naphthalenecarbonitrile, were increased in the presence of
water-soluble and water-immiscible solvents.

Two nitrilase genes, ZmNIT1 and ZmNIT2, have been isolated from maize (*Zea m*
and heterologously expressed in *E. coli*. [14]. ZmNIT2 and *Arabidopsis* NIT4 have a relat
high homology (69.3%) but ZmNIT2 had no activity toward β-cyanoalanine, the substra
Arabidopsis NIT4, and instead hydrolyzed indole-3-acetonitrile (IAN) to indole-3-acetic
(IAA), where AtNIT4 had no activity for hydrolysis of IAN.

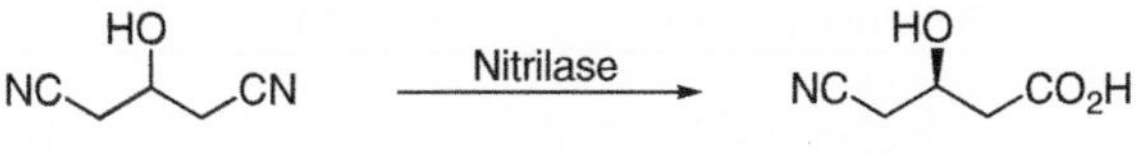

Nitrilase	% ee (0.100 M)	% ee (2.25 M)
Wild-type	94.5	87.8
Ala190His	97.9	98.1

SCHEME 1.2 Desymmetrization of prochiral 3-hydroxyglutaronitrile to (R)-3-hydroxy-4-cyanobutyric acid.

1.2.2 PRODUCTION OF CHIRAL CARBOXYLIC ACIDS

A method for the desymmetrization of prochiral 3-hydroxyglutaronitrile using a nitrilase has been demonstrated (Scheme 1.2) [15,16]; esterification of the resulting (R)-3-hydroxy-4-cyanobutyric acid produced an intermediate useful for the manufacture of the cholesterol-lowering drug Lipitor (atorvastatin calcium). Nitrilases identified in genomic libraries created by extraction of DNA directly from environmental samples were expressed in *E. coli*, then the resulting library was screened for nitrilases that were highly enantioselective for this reaction [17]. Using one of these (R)-specific nitrilases, (R)-3-hydroxy-4-cyanobutyric acid was produced using a 100 mM initial nitrile concentration in 98% yield and 94.5% ee. The enantioselectivity of this wild-type nitrilase decreased with increasing nitrile concentration, where only 87.8% ee was obtained using a more industrially relevant substrate concentration of 2.25 M. Mutagenesis of the nitrilase using a technique that combinatorially saturated each amino acid in the protein to each of the other 19 amino acids resulted in an improved variant (Ala190His), which was expressed in *E. coli*; this variant gave an enantiomeric excess of 98.1% and 98.5% at 2.25 M and 3 M substrate concentration, respectively, with a volumetric productivity of 619 g/L/d at 3 M final product concentration. Nitrilases from this library have been used to produce a range of (R)-mandelic acid derivatives and analogs, and (S)-phenyllactic acid, with high yields and enantioselectivities [16], and a nitrilase from this collection has been expressed in *Pseudomonas fluorescens* [18].

P. putida, *Microbacterium paraoxydans*, and *M. liquefaciens* possess enantioselective nitrilase activity for hydrolysis of (RS)-mandelonitrile to (R)-(−)-mandelic acid with good specific activity (0.33 to 0.50 U/mg), and high ee (>93%) and E values (Scheme 1.3) [19]. P. putida was preferred, as it demonstrated a higher reaction rate, lower K_m, good yield and ee values, and higher stability compared with the other two microorganisms.

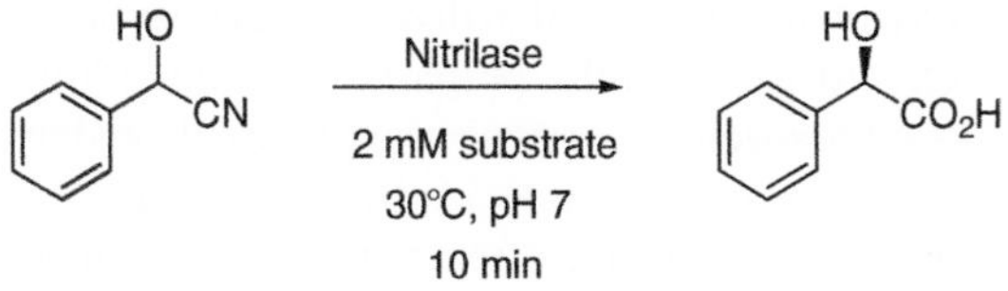

Nitrilase	% Conversion	% ee
Pseudomonas putida	29.2	99.98
Microbacterium paraoxydans	15.3	99.89
Microbacterium liquefaciens	16.6	93.81

SCHEME 1.3 Conversion of (RS)-mandelonitrile to (R)-mandelic acid.

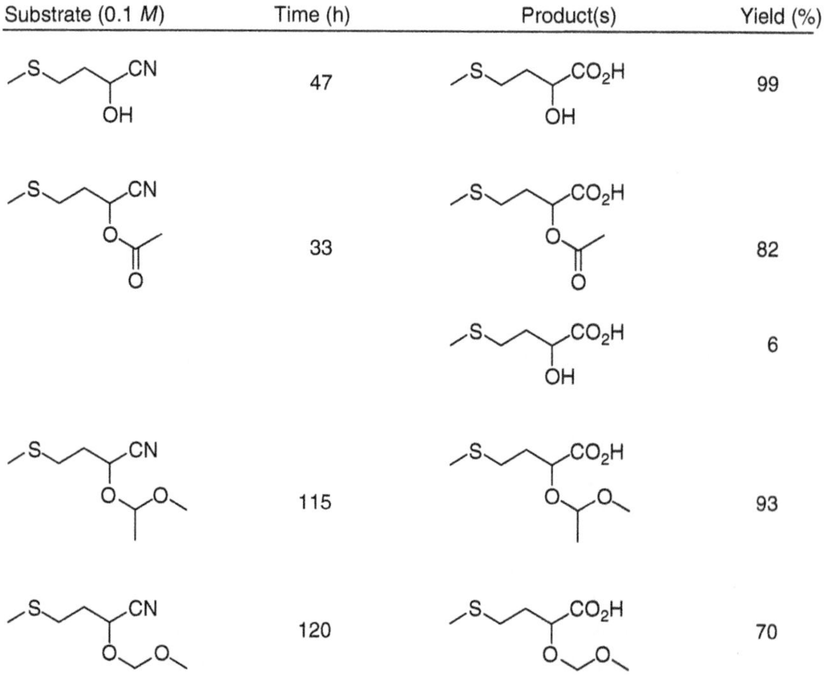

Substrate (0.1 M)	Time (h)	Product(s)	Yield (%)
	47		99
	33		82
			6
	115		93
	120		70

SCHEME 1.4 Nitrilase-catalyzed hydrolysis of nitrile precursors of methionine and methionine an

An immobilized *E. coli* transformant expressing the nitrilase from *A. faecalis* (A 8750) [20] was employed in the synthesis of hydroxy analogs of methionine that are use nutritional additives in cattle feed (Scheme 1.4) [21,22]. There was no enantioselec for hydrolysis of 2-hydroxy-4-methylthiobutanenitrile, 2-acetoxy-4-methylthiobutanen 2-(1-ethoxyethoxy)-4-methylthiobutanenitrile, or 2-methoxymethoxy-4-methylthiobu nitrile when compared with the hydrolysis of mandelonitrile, which produced *R*-(−)-ma acid in 99% ee at pH 8.5 (presumably by re-equilibration of the cyanohydrin during the c of the conversion). The absence of enantioselectivity for the production of methionine a hydroxyl analogs was not critical, as the enantiomers are nutritionally equivalent. Mo nitrilases with improved activity for hydrolysis of 2-amino-4-(methylthio)butyronitrile prepared by making an *A. faecalis* C162N nitrilase mutant and a *Comamonas testos* Q162C nitrilase mutant [23]. 2-Hydroxy-4-methylthiobutyric acid has also been pre using *R. rhodochrous* B24-1 (FERM P-17515) nitrilase [24].

As an alternative to using an enantioselective nitrilase for the hydrolysis of ra α-hydroxynitriles to chiral α-hydroxycarboxylic acids, it has been proposed that α-hydroxynitriles (readily prepared by the addition of HCN to aldehydes or ke using *R*- or *S*-specific oxynitrilases) [25] or their derivatives can be hydrolyzed to the d chiral α-hydroxycarboxylic acids by nitrilase, or a combination of NHase and amidase The α-hydroxynitriles are often not stable in aqueous solution, where the cyanohydri re-equilibrate with HCN and aldehyde/ketone, resulting in a loss of chirality. Prior conve of α-hydroxynitriles to the corresponding 2-acetoxynitriles prevented cyanohydrin de position or re-equilibration, and the hydrolysis of 2-acetoxybutenenitrile, 2-acetoxyhep nitrile, 2-acetoxy-2-(2-furyl)acetonitrile, and 2-acetoxy-2,3,3-trimethylbutanenitrile (A7 with several *P. fluorescens* nitrilases was examined. No hydrolysis of the sterically hin ATMB was observed, and all wild-type strains were found to contain esterases that conv the 2-acetoxynitriles to 2-hydroxynitriles, which then spontaneously decomposed to alde

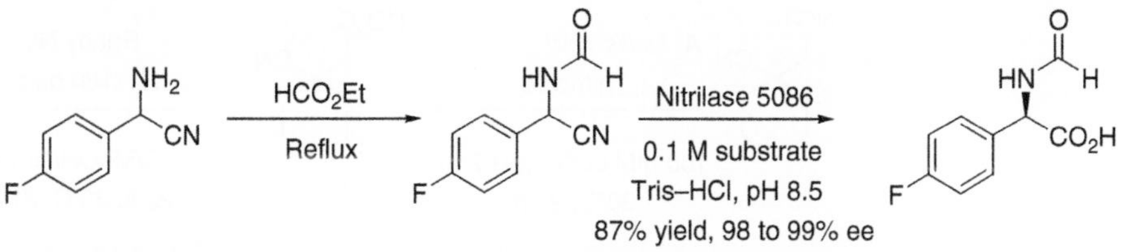

SCHEME 1.5 Dynamic kinetic resolution of *N*-formyl-4-fluorophenylglycinonitrile to produce (*R*)-*N*-formyl-4-fluorophenylglycine.

and cyanide. Chemoselective hydrolysis to the desired 2-acetoxycarboxylic acids was ultimately achieved using either isolated nitrilase or recombinant *E. coli* strains that heterologously expressed nitrilase activities originating from *P. fluorescens* EBC191, *R. rhodochrous* NCIMB 111216, or *Synechocystis* spp. PCC6803.

The dynamic kinetic asymmetric hydrolysis of aromatic aminonitriles, phenyl-glycinonitrile, and 4-fluorophenylglycinonitrile was performed at high pH to produce the corresponding amino acids in high ee [27]. *N*-Acylation of aromatic aminonitriles at pH 8 resulted in spontaneous racemization, allowing enantioselective nitrilase-catalyzed hydrolysis of the (*R*)-enantiomer to produce the corresponding *N*-acylamino acids in up to 99% ee (Scheme 1.5).

New nitrilases with modified substrate/activity profiles were obtained by making a single amino acid change in an *A. faecalis* nitrilase at position 296 such that the amino acid was not tyrosine [28]. Two mutants of *A. faecalis* nitrilase (Y296 → C and Y296 → A) were used for the preparation of substituted chiral carboxylic acids from racemic nitriles, including 2-chloromandelonitrile. The enantioselective preparation of hydroxycarboxylic acids from racemic nitriles, particularly for conversion of 2-chloromandelonitrile to (*R*)-2-chloromandelic acid, was also demonstrated using mutants of *A. faecalis* NitB nitrilase [29].

1.2.3 REGIOSELECTIVE AND CHEMOSELECTIVE NITRILE HYDROLYSIS

A. facilis 72W or *E. coli* SW91 (a transformant expressing the 72W nitrilase) cells having a regioselective microbial nitrilase were utilized for conversion of 2-methylglutaronitrile to 4-cyanopentanoic acid (Scheme 1.6), an intermediate in the preparation of 1,5-dimethyl-2-piperidone (an industrial solvent with chemical properties similar to *N*-methylpyrrolidinone) [30–32]. Whole cells were immobilized in alginate beads, and the beads were chemically cross-linked with glutaraldehyde and polyethylenimine before use. An average volumetric productivity of 79 g 4-cyanopentanoic acid/L/h (216 g/L final concentration of the ammonium salt, 98.5% yield at 100% conversion) was achieved over the course of 195 consecutive batch reactions with catalyst recycle using the immobilized *E. coli* SW91 transformant, where the remaining catalyst activity was 67% of initial activity, and catalyst productivity was 3500 g 4-cyanopentanoic acid/g dry cell weight.

1-Cyanocyclohexaneacetic acid is an intermediate in the synthesis of gabapentin [33] (1-aminomethyl-1-cyclohexaneacetic acid), which has been used for treatment of cerebral

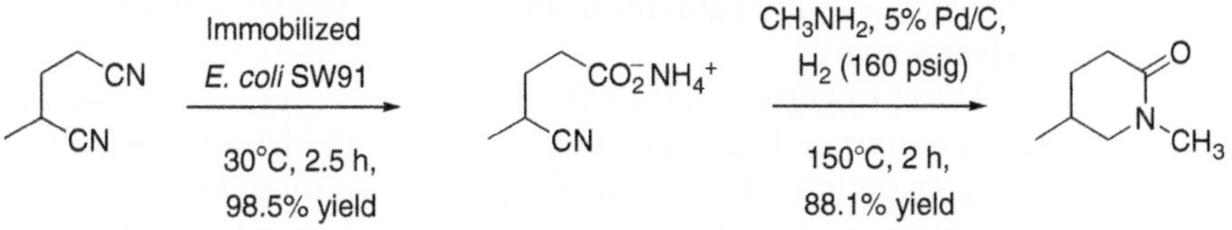

SCHEME 1.6 Regioselective hydrolysis of 2-methylglutaronitrile to 4-cyanopentanoic acid.

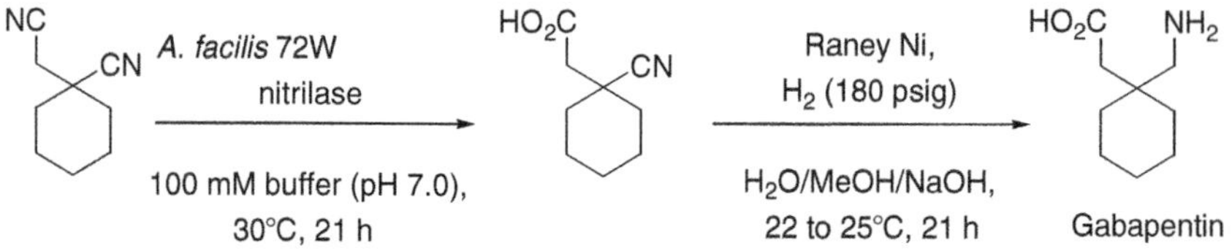

SCHEME 1.7 Regioselective hydrolysis of 1-cyanocyclohexaneacetonitrile to 1-cyanocyclohex
cetic acid.

diseases. Microbial nitrilase catalysts such as *A. facilis* 72W and *E. coli* SS1001 (a
formant expressing the *A. facilis* 72W nitrilase) have been used for the preparatie
1-cyanocyclohexaneacetic acid by regioselective hydrolysis of 1-cyanocyclohexaneaceto
(Scheme 1.7) [34]. Reactions were performed using either unimmobilized cells or ca
alginate–immobilized cells as catalysts, where quantitative conversion of the dinitrile
100% regioselectivity to the desired product was obtained. For example, immobilized *I*
SS1001 cells were used in 35 consecutive batch reactions to prepare 1-cyanocyclohexane.
acid with a catalyst productivity of 77 g product/g dry cell weight.

An immobilized *E. coli* transformant expressing an *A. faecalis* nitrilase was
regioselective for hydrolysis of a series of C_3–C_5 aliphatic dinitriles, producing the
sponding cyano acids, and no diacid, at complete conversion of dinitrile [21]. The
selective hydrolysis of α,ω-dicyanoalkanes and β-substituted glutaronitriles by *Arabic
thaliana* AtNIT1 was examined [35], and selectivity of hydrolysis of dinitriles to ω-c
carboxylic acids depended on chain length; at complete conversion of dinitrile the c
carboxylic acid was the sole product for chain length of up to six carbon a
Glutaronitriles were hydrolyzed to the corresponding cyanobutanoic acids with
selectivity. Aromatic poly-nitriles such as orthophthalonitrile, isophthalonitrile,
terephthalonitrile were converted to the corresponding cyanocarboxylic acids using
crobial catalyst prepared by the recombinant expression of the *Rhodococcus* sp. A
89484 nitrilase in *E. coli* [36,37].

R. rhodochrous LL100-21 expressed different nitrile-hydrolyzing enzymes, depe
on the mononitrile used as nitrogen source during growth [38]. Cell suspensions g
on propionitrile or benzonitrile converted only the aliphatic group of 2-(cyanome
benzonitrile, producing 2-(cyanophenyl)acetic acid as the sole reaction product, purpo
by the action of a regioselective nitrilase. In contrast, 3-(cyanomethyl)benzonitrile
4-(cyanomethyl)benzonitrile were converted to 3- and 4-(cyanomethyl)benzoic acid, res
ively, as the major hydrolysis product, although small amounts of the corresponding di
were also observed. The regioselectivity of the nitrilase for 3- and 4-(cyanome
benzonitriles differs markedly from the 2-cyano derivative, where the aliphatic side
of the 2-cyano derivative may have prevented hydrolysis of the aromatic cyano grou
to steric hindrance. The aliphatic cyano group of 3- and 4-(cyanomethyl)benzoic acid wa
hydrolyzed, suggesting that the nitrilase enzyme could only hydrolyze this cyano group
there is a cyano group (or other substituent) in the *ortho* position. The same cells, when g
on acetonitrile, converted 2-(cyanomethyl)benzonitrile to the amides 2-(cyanophenyl)a
mide and 2-(cyanomethyl)benzamide in low yields; in this case, hydrolysis was the res
NHase activity.

AtNIT1 hydrolyzed the (*E*)-isomers of stereoisomeric α,β-unsaturated nitriles exclu
to the corresponding (*E*)-carboxylic acids with high specificity, thereby also making po
the preparation of isomerically pure (*Z*)-nitriles (Scheme 1.8) [39]. (*E*)-Selectivity
not obtained when the location of the double bond was changed from α,β-unsaturati

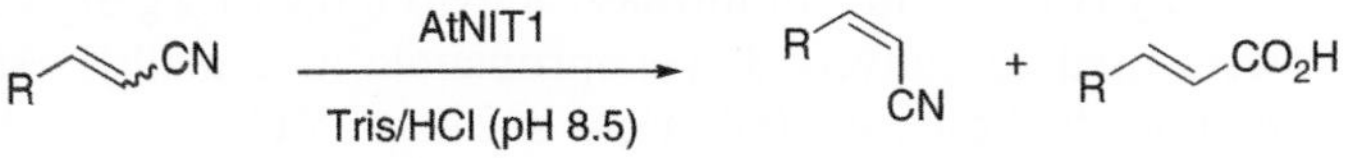

$R=CH_3-, Ph-, PhCH(CH_3)CH_2-, CH_2=CHCH_2-, CH_3O-$

SCHEME 1.8 Regioselective hydrolysis of stereoisomeric α,β-unsaturated nitriles to the corresponding (*E*)-carboxylic acids.

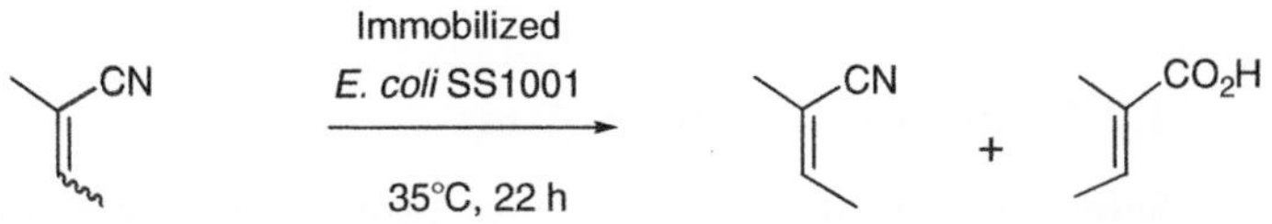

$E:Z = 72:28$ (1.34 *M*)   100%   100%

SCHEME 1.9 Regioselective hydrolysis of (*E,Z*)-2-methyl-2-butenenitrile to (*E*)-2-methyl-2-butenoic acid.

β,γ- unsaturation (e.g., for hydrolysis of (*E,Z*)-3-heptenenitrile). *A. facilis* 72W nitrilase (expressed in *E. coli* SS1001) exhibited a similar stereoisomeric preference, where for the regioselective hydrolysis of (*E,Z*)-2-methyl-2-butenenitrile, only (*E*)-2-methyl-2-butenoic acid was produced at complete conversion of the (*E*)-nitrile (Scheme 1.9) [40]. Solvent extraction of the unreacted (*Z*)-nitrile from the aqueous product mixture containing the (*E*)-acid ammonium salt, followed by acidification of the aqueous phase and solvent extraction of the resulting (*E*)-acid, allowed for the simple separation and isolation of both reaction products in high yield and purity.

Nitrilases can also be highly chemoselective; *A. facilis* 72W nitrilase converted aliphatic or aromatic cyanocarboxylic acid esters to the corresponding dicarboxylic acid monoesters at 100% conversion with little or no hydrolysis of the ester functionality (Table 1.1) [41]. The microbial cells were first heat-treated at 50°C for 1 h to inactivate a co-expressed NHase

TABLE 1.1

Chemoselective Hydrolysis of Cyanocarboxylic Acid Esters Using *A. facilis* 72W Microbial Nitrilase

Cyanocarboxylic Acid Ester	Concentration (*M*)	Dicarboxylic Acid Monoester	Yield (%)
Cyanoacetic acid Ethyl ester	0.40	Propanedioic acid Monoethyl ester	100
Cyanoacetic acid Propyl ester	0.40	Propanedioic acid Monopropyl ester	100
3-Cyanopropanoic Acid methyl ester	2.00	Butanedioic acid Monomethyl ester	100
5-Cyanopentanoic Acid methyl ester	1.00	Hexanedioic acid Monomethyl ester	97
4-Cyanobenzoic acid Methyl ester	0.052	1,4-Benzenedicarboxylic Acid monomethyl ester	94

activity (with no loss of nitrilase activity), then 1.2 g dry cell weight/L of unimmobilized were used to convert 3-cyanopropanoic acid methyl ester (226 g/L) to butanedioic monomethyl ester in 100% yield in 2 h at 25°C.

A general survey of a variety of arylacetonitrilases has been performed using a of aromatic and arylaliphatic nitrile substrates [42]. Each biocatalyst exhibited a u substrate-conversion profile, except that α-substituted arylaliphatic nitriles were gen not readily converted to the corresponding carboxylic acids. Chemoselective hydroly the nitrile functionality of substrates having both nitrile and ester groups was de strated, as was the regioselective hydrolysis of single nitrile groups in dinitrile subst The biocatalysts also mediated the synthesis of a range of α-hydroxycarboxylic acids aldehydes in the presence of cyanide.

1.2.4 ACHIRAL CARBOXYLIC ACIDS

Several microbial cell catalysts for production of ammonium acrylate have been rep A Rhodococcal isolate having only a combination of NHase and amidase was com to *R. ruber* NCIMB 40757, which has only an aliphatic nitrilase [43]. The microbial nit catalyst was preferred, where the nitrilase was much more stable in the presence of acrylo and ammonium acrylate; a fed-batch reaction produced ~500 g product/L. Polyacryla immobilized *Acinetobacter* sp. AK226 was used to produce 300 g/L of ammonium ac from acrylonitrile with a catalyst productivity of 4000 g product/g dry cell weight; the rea was run in the absence of buffer, thereby producing a very pure aqueous reaction produc *R. rhodochrous* NCIMB 40757 and NCIMB 40833 each had a K_m for acrylonitrile l 500 μM and a Ki for ammonium acrylate above 100,000 μM; immobilized cells were us either continuous or fed-batch reactions to produce 400 g/L product with less than acrylonitrile remaining [45]. Screening isolates from environmental samples for tolerar acrylonitrile identified *A. denitrificans* C-32 as a nitrilase-containing strain with high ac for hydrolysis of acrylonitrile [46].

The nitrilase activity of *A. facilis* 72W whole cells, or *E. coli* transformants expre the *A. facilis* 72W nitrilase, have been used as catalysts to hydrolyze glycolonitrile to colic acid at high concentration and in >99% yield [47]. Conversion of acetone cyanohyd 2-hydroxyisobutyric acid was also demonstrated using *A. facilis* 72W nitrilase, where s quent dehydration of 2-hydroxyisobutyric acid produced methacrylic acid [48].

1.2.5 AMIDES VIA NITRILASE CATALYSIS

Nitrilases normally catalyze the addition of two equivalents of water to a nitrile to di produce the corresponding carboxylic acid and ammonia, but in some instances pro other than carboxylic acids have been generated [9]. Inclusion of hydroxylamine (375 in phosphate buffer (pH 7.3) containing benzonitrile (50 mM) and whole cel *R. rhodochrous* LL100-21 (grown on benzonitrile to induce nitrilase) at 30°C produ 3% yield of benzohydroxamic acid in addition to benzoic acid; no production o hydroxamic acid was observed in the absence of the nitrilase catalyst [49]. An en found in the genome of *P. fluorescens* was cloned in *E. coli* and found to have lase activity, and also NHase activity when hydroxycinnamonitrile was used as the strate [50]; hydrolysis of this substrate was not due to a combination of NHase amidase, and a homologous nitrilase from a different strain of *P. fluorescens* was sl to produce only the corresponding acid.

The nitrilase AtNIT1 from the plant *A. thaliana* (overexpressed in *E. coli*) preferred alip nitriles as substrates [51]. Except for fluoro-substituents, substitution at the carbon adjacent nitrile function completely inhibited hydrolysis. The resolution of (RS)-2-fluoroarylacetonitriles by AtNIT1 produced the corresponding (R)-amides, and not the expected carboxylic acids, as the major products in 88 to 92% ee at ca. 50% conversion, and recrystallization of the product improved R-amide ee to >99%; acid-catalyzed hydrolysis of the (R)-amide produced the desired (R)-carboxylic acid in 88 to 91% yields without racemization (Scheme 1.10) [52].

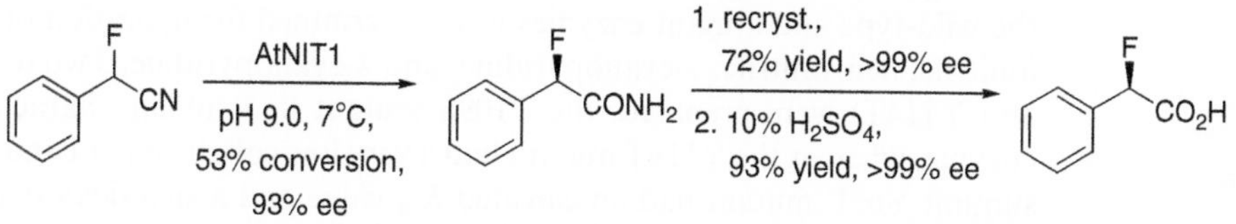

SCHEME 1.10 Conversion of 2-fluorobenzyl cyanide to (R)-2-fluoro-2-phenylacetamide by AtNIT1 nitrilase.

1.3 NHase and NHase/AMIDASE

There have been several recent reviews on synthetic applications of NHase and NHase/amidase for the conversion of nitriles to amides and carboxylic acids, respectively [53–56], and additional reviews that address the mechanism by which Fe- and Co-containing NHase enzymes convert nitriles to amides [57–62]. The metal ion is believed to function as a Lewis acid, and three reaction mechanisms (Scheme 1.11) have been proposed [57]. For the nonheme Fe-NHase of *Rhodococcus* sp. N771, enzyme activity was found to be regulated by nitrosylation and photo-induced de-nitrosylation of the iron center, and posttranslational modification of two cysteine ligands to cysteine sulphenic acid (Cys-SOH) and cysteine sulphinic acid (Cys-SO$_2$H) has been proposed as a requirement for catalytic activity [59,60,62,63].

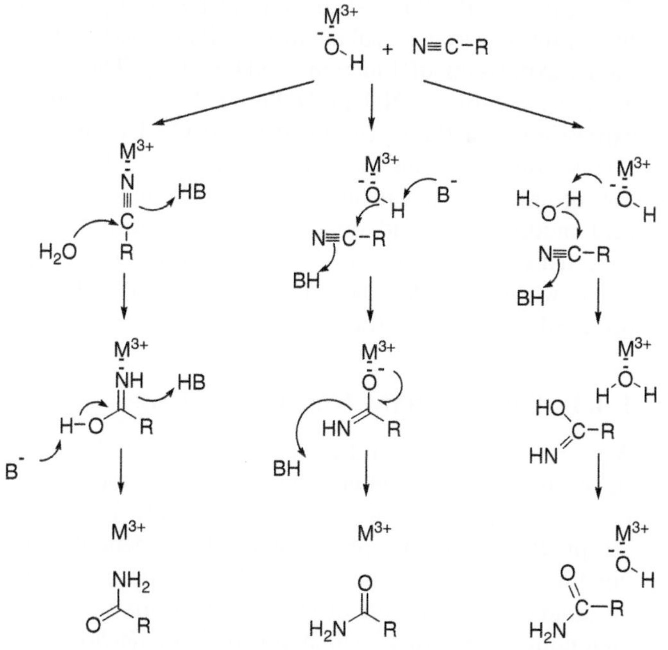

SCHEME 1.11 Proposed mechanism for hydrolysis of nitrile to amide by nitrile hydratase. (Redrawn from Huang, W., Jia, J., Cummings, J., Nelson, M., Schneider, G., and Lindqvist, Y., *Structure*, 5, 691, 1997.)

1.3.1 CRYSTAL STRUCTURES

The crystal structures of an Fe-NHase from *Rhodococcus* sp. N-771 [57,64], and a Co-N from *Pseudonocardia thermophila* [65] have been reported. Mutants of *P. thermophila* 3095 NHase that were targeted to produce changes in substrate binding, catalysis, and f tion of the active center were prepared, and mutational and structural analyses of the sul binding and metal specificity of these mutants have also been reported [66]. NHase acti the wild-type and mutant enzymes were determined for hydration of acrylonitrile, met lonitrile, benzonitrile, 3-cyanopyridine, and 4-cyanopyridine. Two α-subunit mutants (' and Y114T) were prepared; the T109S mutant had similar characteristics to the wil enzyme, whereas the Y114T mutant had a very low cobalt content and catalytic activity. ' subunit Y68F mutant had an elevated K_m value and a significantly decreased kcat val acrylonitrile, methacrylonitrile, and benzonitrile.

The crystal structure of the cobalt NHase of *Bacillus smithii* SC-J05-1 has been deter [67]. The amino acid sequence identity between the NHase from *B. smithii* and *Rhodo* sp. N-771 was 53.2% for the α subunit and 33.9% for the β subunit, and sequence id between the *B. smithii* and *P. thermophila* NHase was 64.0% for the α subunit and 42.2 the β subunit. The metal center is located in a central cavity formed at the interface be the *B. smithii* α and β subunits. The ligands to the cobalt atom are three sulfur ato Cys120, Cys123, and Cys125, and two main chain amide nitrogen atoms of Ser12 Cys123; all ligands are in the α subunit. The Phe52 in the β subunit partially cove metal center and narrows the active site cleft, and it was suggested that this structural f might contribute to the substrate selectivity of this NHase, where hydrolysis of ali nitriles is preferred to aromatic nitriles.

1.3.2 HETEROLOGOUS PROTEIN EXPRESSION

The sequences flanking NHase genes have been proposed to encode for protein "activ that transport iron or cobalt to the NHase active site, and are often required for he ogous expression of functional NHase [68]. The heterologous expression of cataly active *C. testosteroni* NI1 Fe-NHase in *E. coli* as an α2β2 tetramer was accomplished i expression with the *E. coli* GroES and GroEL chaperones [69]. The purified recom NHase was highly similar to the enzyme purified from *C. testosteroni*. The mass spectr the recombinant enzyme indicated that the majority of the α subunits contained one s acid modification, and observation of a second sulfinic modification indicated tha expected sulfenic modification was also produced. The production of these different ei forms when expressed recombinantly suggested that these posttranslational modific occurred by auto-oxidation, or were assisted by chaperones or other proteins in *E. c*

1.3.3 ENANTIOSELECTIVE NITRILE HYDROLYSIS

A series of α,α-disubstituted malononitriles and related substrates were used to stu substrate specificity and enantioselectivity of the NHase and amidase from *R. rhodo* IFO 15564 [70]. The amidase preferentially hydrolyzed the pro-(*R*) carbamyl gro the prochiral diamides produced by the nonenantioselective hydration of the corres ing dinitrile. The introduction of a fluorine atom at the α position caused an inhibitory on amidase. A precursor of (*S*)-α-methyldopa was prepared (Scheme 1.12); the subst malononitrile was first incubated with *R. rhodochrous* cells to produce the (*R*)-amide/ca ylic acid in 95% yield and 98.2% ee, and subsequent methylation of the resulting a carboxylic acid with diazomethane, followed by Hofmann rearrangement of the fi

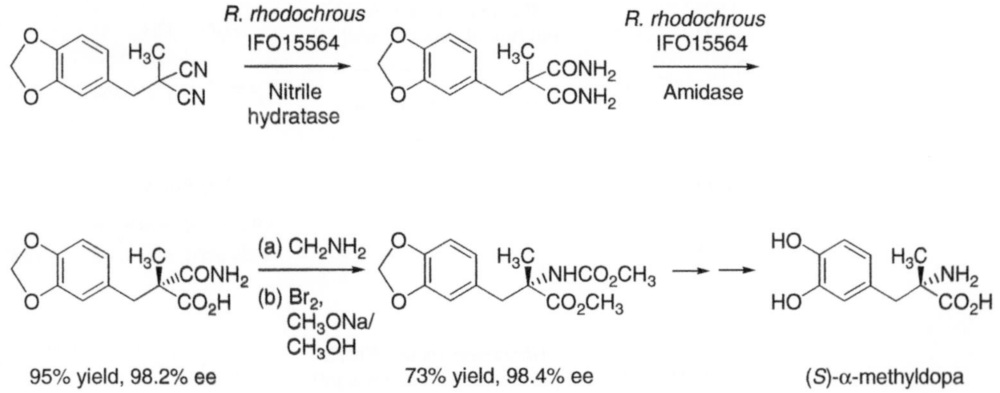

SCHEME 1.12 Enantioselective hydrolysis of the pro-(R) carbamyl group of a prochiral diamide produced by the nonenantioselective hydration of a substituted malononitrile.

recrystallized amide/ester (99.8% ee) and trapping with methanol, provided the desired precursor in 73% yield and 98.4% ee. The preparation of (±)-α-cyano-α-fluoro-α-phenylacetic acid (CFPA) from diethyl-α-fluoro-α-phenylmalonate was also demonstrated. α,α-Disubstituted malonamides RCH$_2$CMe(CONH$_2$)$_2$ (R = Ph, 2-ClC$_6$H$_4$, 3-ClC$_6$H$_4$, 4-ClC$_6$H$_4$, 4-MeC$_6$H$_4$, 4-MeOC$_6$H$_4$, 4-FC$_6$H$_4$, 4-BrC$_6$H$_4$, PhCH$_2$, 1-propyl) were enantioselectively hydrolyzed by *Rhodococcus* sp. CGMCC 0497 whole cells to produce the corresponding (R)-malonamic acids in 92 to 98% yields and 91 to 99% ee [71]; the products could also be converted to (R)- or (S)-α,α-dialkylated amino acids.

The enantioselective biotransformation of geminally dihalogenated cyclopropanecarbonitriles and amides by *Rhodococcus* sp. AJ270 microbial cells has been reported [72,73]. Both reaction rate and enantioselectivity of the NHase and amidase were strongly controlled by the nature of gem-disubstituents on the cyclopropane ring; the amidase generally exhibited steric dependence on the substituents, whereas both steric and electronic factors affected the NHase. An effective synthesis of optically active 2,2-disubstituted-3-phenylcyclopropane-carboxylic acid and amide in both enantiomeric forms was demonstrated, including a multi-gram scale synthesis of enantiopure 2,2-dichloro-3-phenylcyclopropanecarboxylic acid and amide in both enantiomeric forms. This same microbial catalyst was used to examine the hydrolysis of 2,2-dimethyl-3-substituted cyclopropanecarbonitriles [74]. *Cis*-3-aryl-2,2-dimethylcyclopropanecarbonitriles were not hydrolyzed, but racemic *trans*-isomers were hydrolyzed in high yields and with high enantioselectivity (in most instances) to produce (+)-(1R,3R)-3-aryl-2,2-dimethylcyclopropanecarboxylic acids and (−)-(1S,3S)-3-aryl-2,2-dimethylcyclopropanecarboxamides; optically pure geminally dimethyl-substituted cyclopropanecarboxylic acids and amides, including chrysanthemic acids, were produced in both enantiomeric forms (Scheme 1.13).

Chiral α-substituted α-amino acids have found use in the preparation of pharmaceuticals, and numerous methods have been reported for their preparation. *Rhodococcus.* sp. AJ270 cells catalyzed the enantioselective hydrolysis of α-substituted DL-glycine nitriles to produce D-(−)-α-amino acid amides and L-(+)-α-amino acids in high yields and ee (Scheme 1.14) [75]. The enantioselective synthesis of (R)-(+)-α-arylalanine amides and (S)-(−)-α-arylalanines was also accomplished by the kinetic resolution of racemic amides using the amidase activity of *Rhodococcus.* sp. AJ270 whole cells [76]. Both the reaction rate and S-enantioselectivity of the biocatalytic kinetic resolution were dependent upon the nature and the substitution pattern of the aryl substituent. Racemic α-ethyl phenylglycine amide was

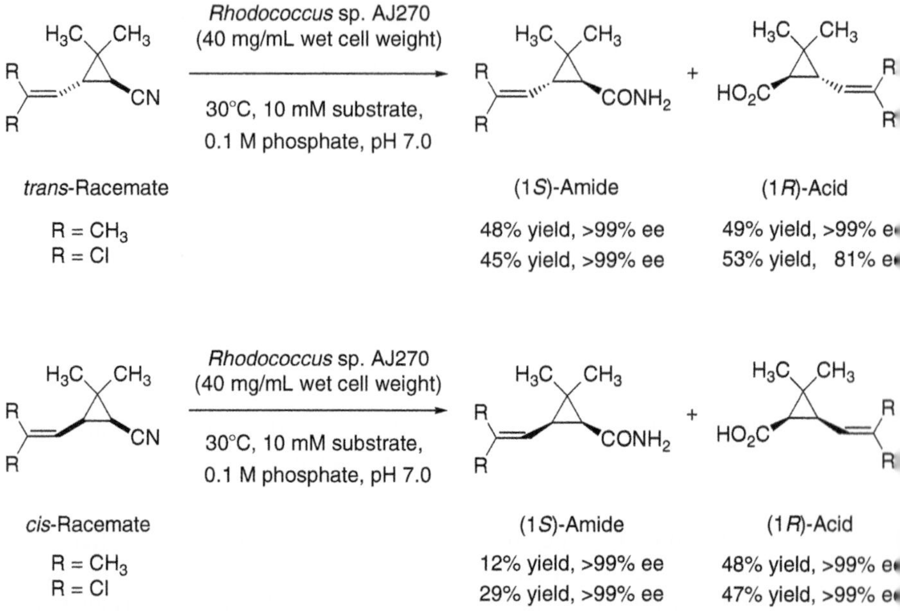

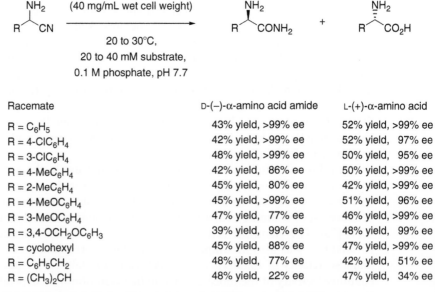

SCHEME 1.13 Biocatalytic hydration/hydrolysis of racemic chrysanthemic nitriles.

efficiently resolved into (R)-$(+)$-α-ethyl phenylglycine amide (96% ee) and (S)-$(-)$-α phenylglycine (99% ee), and (R)-$(+)$-1-amino-1-carbamoyl-1,2,3,4-tetrahydronaphth and (S)-$(-)$-1-amino-1-carboxy-1,2,3,4-tetrahydronaphthalene were prepared in ex yields at >99.5% ee (Scheme 1.15). The chemical hydrolysis of the remaining chiral could also be performed with no measurable loss of chirality, making both enantid

Racemate	D-(–)-α-amino acid amide	L-(+)-α-amino acid
R = C₆H₅	43% yield, >99% ee	52% yield, >99% ee
R = 4-ClC₆H₄	42% yield, >99% ee	52% yield, 97% ee
R = 3-ClC₆H₄	48% yield, >99% ee	50% yield, 95% ee
R = 4-MeC₆H₄	42% yield, 86% ee	50% yield, >99% ee
R = 2-MeC₆H₄	45% yield, 80% ee	42% yield, >99% ee
R = 4-MeOC₆H₄	45% yield, >99% ee	51% yield, 96% ee
R = 3-MeOC₆H₄	47% yield, 77% ee	46% yield, >99% ee
R = 3,4-OCH₂OC₆H₃	39% yield, 99% ee	48% yield, 99% ee
R = cyclohexyl	45% yield, 88% ee	47% yield, >99% ee
R = C₆H₅CH₂	48% yield, 77% ee	42% yield, 51% ee
R = (CH₃)₂CH	48% yield, 22% ee	47% yield, 34% ee

SCHEME 1.14 Enantioselective biotransformations of DL-α-amino nitriles.

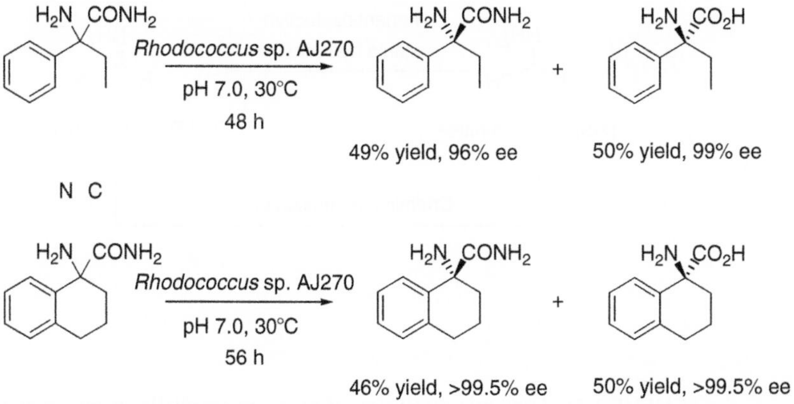

SCHEME 1.15 Biocatalytic kinetic resolution of racemic α-ethyl phenylglycine amide and 1-amino-1-carbamoyl-1,2,3,4-tetrahydronaphthalene.

forms of the arylalanines readily available from the racemic amides. *Rhodococcus*. sp. AJ270 was used to prepare enantiopure S-(+)-2-aryl-3-methylbutyric acids and R-(+)-2-aryl-3-methylbutyramides from the hydrolysis of 2-aryl-3-methylbutyronitriles [77]; in this application, the NHase displayed a low S-enantioselectivity for the nitriles, whereas the amidase had a strict S-enantioselectivity for the 2-aryl-3-methylbutyramides.

Five new bacterial isolates, *R. erythropolis* 11.1, *Rhodococcus* sp. 27.1, *Pantoea endophytica* 26.2.2, *Pantoea* sp. 17.3.1, and *Nocardioides* sp. 29.3, were each found to have an enantioselective NHase for hydration of (RS)-2-phenylpropionitrile and (RS)-phenylglycine nitrile [78]. The NHase enantioselectivities were generally low and poorly (S)-selective. The amidases were either (S)- or (R)-enantioselective, and could be used in kinetically controlled reactions for synthesis of pure (S)- or pure (R)-phenylglycine, respectively. *P. endophytica* produced (S)-phenylglycine in >99% ee through hydrolysis of (S)-phenylglycine amide by an (S)-specific amidase, and (R)-phenylglycine (>99% ee) was produced using the (R)-selective amidase of *Pantoea* sp.

A recent patent application disclosed a method for the production of artificially evolved R- or S-enantioselective NHases by recombining two or more genes encoding a nonenantioselective NHase and/or by mutating one or more nonenantioselective NHase genes in one or more cycles of recombination or mutation. The recombination of two or more NHase genes was typically performed by recursive combination, whole genome recombination, synthetic recombination, or *in silico* recombination [79]. This method could also be used to produce enantioselective nitrilases. An enantioselective method for the conversion of a racemic mixture of aminonitriles to R-amino acids (Scheme 1.16) or S-amino acids was described.

DL-Methionine has been prepared from 2-amino-4-methylthiobutyronitrile or the corresponding amide at 35 to 60°C with a *P. thermophila* JC M3095 or JC M3032 catalyst that produced both NHase and amidase [80]; in this case the combination of NHase and amidase activities afforded no enantioselectivity, and the product was produced as the racemate. The combination of NHase and amidase activities of *R. rhodochrous* DSM 43198 was also used to convert 2-hydroxy-4-methylthiobutanenitrile to DL-methionine at 5 to 40°C [81].

Chiral α-hydroxycarboxylic acids have been prepared by hydrolysis of (R)- or (S)-cyanohydrins to the corresponding (R)- or (S)-α-hydroxycarboxylic acids using whole cells of *R. erythropolis* NCIMB 11540 [82,83] that have a NHase/amidase enzyme combination with high specific activity for hydrolysis of the nitrile group of cyanohydrins. Enantiopure cyanohydrins

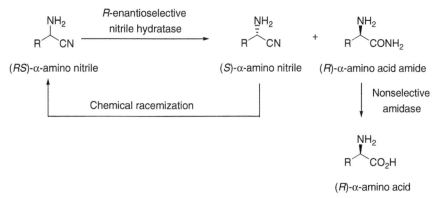

SCHEME 1.16 Dynamic resolution of DL-α-amino nitriles using an enantioselective nitrile hydra

were readily prepared using (R)- or (S)-hydroxynitrile lyases (HNL) [84,85], and the pr
were obtained in high yield without racemization, decomposition, or side reaction
pharmaceutical intermediate (R)-2-hydroxy-4-phenylbutyric acid was prepared using
of the corresponding nitrile in 98% yield (ee > 98%) (Scheme 1.17). The NHase and ar
from *R. erythropolis* NCIMB 11540 were separately cloned and expressed in *E. coli* [86], a
extracts were used for the hydrolysis of different aromatic cyanohydrins. The product
either α-hydroxy amides or acids was demonstrated with retention of enantiopurity, a
production of NHase and amidase in separate transformants allowed for the use of addi
amidase catalyst in instances where amide hydrolysis was rate-limiting.

The effect of the *N*-protecting group on the biotransformation of *N*-tolylsulfonyl- a
butyloxycarbonyl-protected β-amino nitriles to the corresponding β-amino amides and
was examined using whole cells of *Rhodococcus* sp. R312 or *R. erythropolis* NCIMB
as catalysts [87,88]. The bioconversion products of five-membered carbocyclic nitrile
mainly the respective acids, whereas the six-membered carbocyclic nitriles produced a
as the major reaction product. The type of protecting group (sulfonamide or carbama
not affect the nature of the products produced, but yields were higher usin
carbamate-protected derivatives. The enantioselectivity of the NHase/amidase activi
R. equi A4, *R. erythropolis* NCIMB 11540, and *Rhodococcus* sp. R312 was evaluat
conversion of *cis*- and *trans*-2-aminocyclopentane/cyclohexane nitriles to the correspo
β-amino amides and acids (Scheme 1.18) [89]. Five-membered alicyclic 2-amino nitrile
converted significantly faster than the six-membered ring compounds, and the produ
trans-2-amino nitriles (amides and acids) were produced considerably faster than t
counterparts (which produced only amides). The ees of the *cis* isomers were consistently
than that of the *trans* isomers.

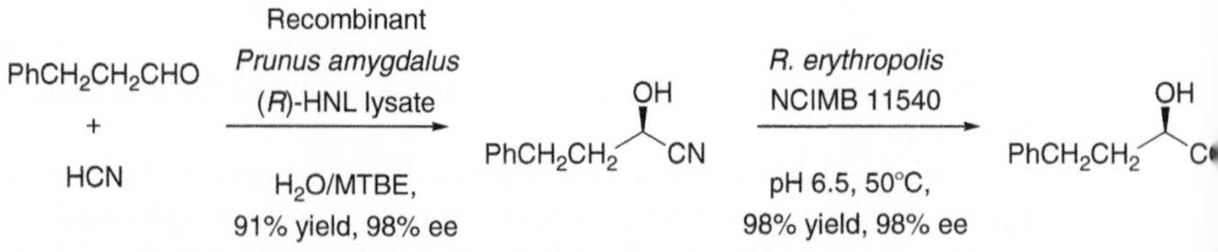

PhCH₂CH₂CHO   Recombinant
               Prunus amygdalus   *R. erythropolis*
    +   (R)-HNL lysate   NCIMB 11540
                                        PhCH₂CH₂ $\overset{OH}{\underset{}{\diagup}}$ CN   PhCH₂CH₂ $\overset{OH}{\underset{}{\diagup}}$ C
   HCN   H₂O/MTBE,   pH 6.5, 50°C,
               91% yield, 98% ee   98% yield, 98% ee

SCHEME 1.17 Preparation of (R)-2-hydroxy-4-phenylbutyric acid using a combination of a
hydroxynitrile lyase, nitrile hydratase, and amidase.

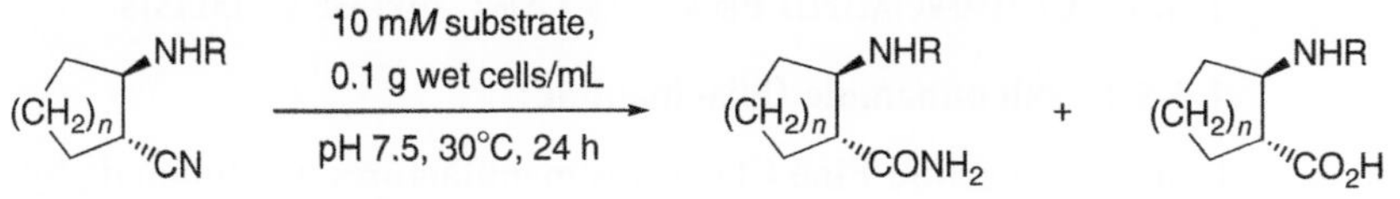

$n = 1$, R = Bz	R. equi A4	40% yield, 94% ee	55% yield, 75% ee
	R. erythropolis 11540	30% yield, >99% ee	63% yield, 48% ee
	R. sp. R 312	7% yield, >99% ee	87% yield, 15% ee
$n = 1$, R = Ts	R. equi A4	14% yield, >99% ee	44% yield, 2% ee
	R. erythropolis 11540	13% yield, >99% ee	86% yield, 5% ee
	R. sp. R 312	10% yield, >99% ee	34% yield, 14% ee
$n = 2$, R = Bz	R. equi A4	22% yield, 56% ee	36% yield, >95% ee
	R. erythropolis 11540	16% yield, 67% ee	15% yield, >95% ee
	R. sp. R 312	14% yield, 38% ee	7% yield, >95% ee
$n = 2$, R = Ts	R. equi A4	54% yield, 65% ee	13% yield, >99% ee
	R. erythropolis 11540	56% yield, 59% ee	15% yield, 97% ee
	R. sp. R 312	42% yield, 77% ee	16% yield, 87% ee

SCHEME 1.18 Enantioselective conversion of *cis*- and *trans*-2-aminocyclopentane/cyclohexane nitriles to the corresponding β-amino amides and acids.

1.3.4 NHase/Amidase Catalysts for Production of Achiral Carboxylic Acids

Microbial catalysts having a combination of NHase and amidase activities had a significantly higher specific activity than the microbial nitrilases that were screened for hydrolysis of 3-hydroxyalkanenitriles to the corresponding 3-hydroxyalkanoic acids (Scheme 1.19) [90]. *C. testosteroni* 5-MGAM-4D cells were immobilized in alginate beads, and the resulting NHase/amidase biocatalyst hydrolyzed 3-hydroxyvaleronitrile to 3-hydroxyvaleric acid in 99 to 100% yields in a series of 85 consecutive batch reactions with biocatalyst recycle for the production of 118 g/L 3-hydroxyvaleric acid. The catalyst productivity for this series of reactions was 670 g 3-hydroxyvaleric acid/g dry cell weight, the initial volumetric productivity was 44 g 3-HVA/L/h, and the recovered NHase and amidase activities in the final reaction were 29% and 40%, respectively, of their initial activities. Similar results were obtained for hydrolysis of 3-hydroxybutryonitrile and 3-hydroxypropionitrile. The combined NHase and amidase activities of *C. testosteroni* 5-MGAM-4D have also been employed as catalyst for the hydrolysis of acrylonitrile and methacrylonitrile to acrylic acid and methacrylic acid, respectively [91], and for hydrolysis of acetone cyanohydrin to 2-hydroxyisobutyric acid [92].

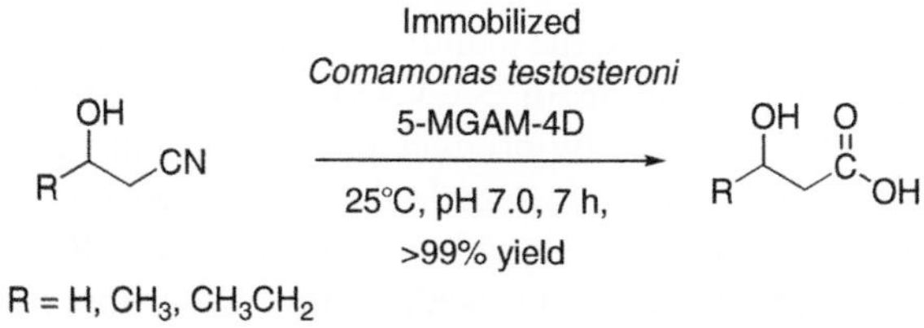

SCHEME 1.19 Two-step conversion of 3-hydroxyalkanenitriles to 3-hydroxyalkanoic acids using nitrile hydratase/amidase.

1.3.5 COMMERCIALIZED PROCESSES USING NHASE CATALYSTS

1.3.5.1 Nicotinamide (Niacinamide)

Lonza Guangzhou Fine Chemicals manufactures nicotinamide by a process where 2-n 1,5-diaminopentane (a by-product of nylon-6,6 manufacture) is first converted to 3-c pyridine in a series of three chemically catalyzed reactions, then the nitrile is hydra nicotinamide using *R. rhodochrous* J1 cells immobilized in polyacrylamide particles (S 1.20) [93–95]. A continuous feed of 3-cyanopyridine at concentrations of between I 20 wt.% is supplied in the direction of process flow, with a countercurrent feed of bioca in a series of stirred-tank batch reactors. The process generates >3500 MT of nicotina with >99.3% selectivity at 100% conversion of 3-cyanopyridine. An additional pl planned, with an initial 6000 MT/y capacity that could be expanded to 9000 MT/y [96

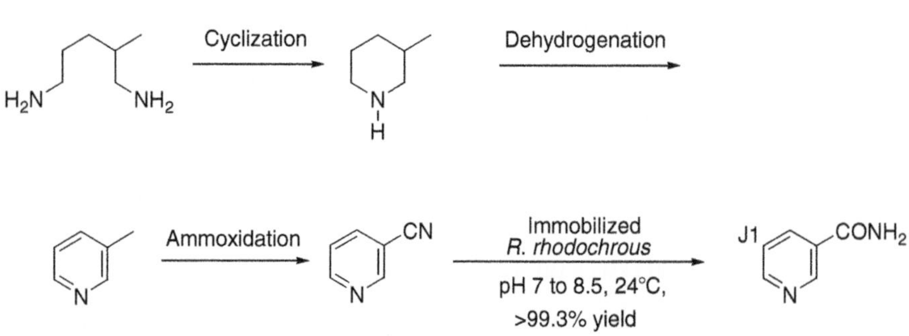

SCHEME 1.20 Process for production of nicotinamide employing biocatalytic hydrat 3-cyanopyridine.

R. *rhodochrous* J1 has a high K_m (200 mM) and low tolerance for 3-cyanopyridin contains a red pigment that requires decolorizing the product solution. Recent work by has focused on developing alternate biocatalysts for nicotinamide production, inc *Amycolatopsis*, *Actinomadura*, and *Rhodococcus* [97–99]. *Rhodococcus* sp. FZ4 13597) was isolated from soil samples using 3-pyridinaldoxime as inducer of NHase a [96]; it expressed a cobalt-dependent NHase with improved properties for producti nicotinamide relative to J1 (Table 1.2). Using unimmobilized FZ4 cells, a total of nicotinamide/g FZ4 dry cell weight was obtained in a single batch reaction to produce (w/v) nicotinamide in 0.1 M potassium phosphate buffer (pH 6.0).

1.3.5.2 5-Cyanovaleramide

A biocatalytic process was commercialized for production of 5-cyanovaleramide (5-CV a starting material in the manufacture of the herbicide azafenidin (Scheme 1.21) [100]. N catalyzed production of 5-CVAM resulted in higher yields of 5-CVAM, higher ca productivity, lower by-product production, and generated significantly less process than alternative chemical methods. The process was first piloted using *R. erythropolis* A [101–103] immobilized in calcium alginate beads. Reactions were run at 5°C, as the stabi the enzyme decreased markedly above this temperature. Sixty consecutive batch rea converted a total of 1.1 MT of adiponitrile to produce a 9.0 wt.% solution of 5-CVAM i yield, with a catalyst productivity of greater than 1000 kg 5-CVAM/kg dry cell w

TABLE 1.2
Comparison of *Rhodococcus* sp. FZ4 and *R. rhodochrous* J1 for Hydration of 3-Cyanopyridine

	Rhodococcus sp. FZ4	*R. rhodochrous* J1
K_m (mM 3-cyanopyridine)	160	200
Relative activity (%) in 10% (w/v) 3-cyanopyridine	100	63
Relative activity (%) in 30% (w/v) nicotinamide	100	n.d.
Relative activity after incubation at 60°C for 15 min	93	80
Pigmentation	None	red

n.d., not determined.

5-CVAM was isolated by distillation of water, dissolution of the resulting solids in boiling methanol, and filtration of insoluble adipamide and salts; the resulting methanolic 5-CVAM solution was used directly in the subsequent azafenidin process step.

Commercial-scale fermentation of *R. erythropolis* A4 required a light-activation step of the microbial NHase [104,105] as an additional cost for catalyst manufacture. Further screening of microbial NHases identified *P. chlororaphis* B23 as a more effective biocatalyst for 5-CVAM production. *P. chlororaphis* B23 was first isolated and characterized by H. Yamada and coworkers [106–108], and was later used by the Nitto Chemical Industry Co. for the manufacture of acrylamide from acrylonitrile [109–112]. In the initial commercial-scale production run, *P. chlororaphis* B23 microbial cells were immobilized in calcium alginate beads, and 58 consecutive batch reactions were run at 5°C with biocatalyst recycle to convert a total of 12.7 MT of adiponitrile to 13.6 MT of 5-CVAM (97% conversion, 96% selectivity, 93% yield) as a 19.2 wt.% solution in water [113,114]. The immobilized-cell catalyst was highly regioselective, with less than 5% selectivity to by-product adipamide. The catalyst productivity in this first commercial-scale run was 3150 kg 5-CVAM/kg dry cell weight; a total of ~150 MT of 5-CVAM were prepared by this process for azafenidin manufacture.

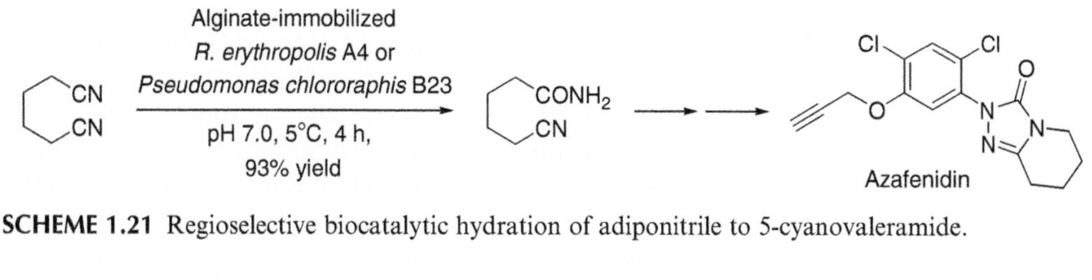

SCHEME 1.21 Regioselective biocatalytic hydration of adiponitrile to 5-cyanovaleramide.

1.3.5.3 Acrylamide

New NHase catalysts continue to be developed and compared with the existing wild-type whole-cell catalysts used in the commercial production of acrylamide from acrylonitrile. Mitsubishi Rayon has produced mutant enzymes of *R. rhodochrous* J1 NHase with significantly improved thermal stability and catalytic activity at 50 to 70°C [115,116]. SNF Floerger (Saint-Etienne, France) has licensed and commercialized the manufacture of acrylamide using Mitsubishi Rayon's immobilized *R. rhodochrous* J1 [117], and has independently developed a *R. pyridinovorans* whole-cell catalyst for this process [118]. A thermostable NHase has been cloned from *Geobacillus thermoglucosidasius* strain Q-6 and expressed in *E. coli* [119 catalyst had good performance at high acrylonitrile (6 wt.%) and acrylamide (35 concentrations, and the enzyme retained >35% initial activity after 30 min at 70°C. mutants of *P. thermophila* JCM3095 NHase have been produced, where substitutions a positions in the α subunit and 15 positions in the β subunit resulted in mutant enzyme improved specific activity or substrate selectivity [120].

An *E. coli* transformant that expressed a thermally stable NHase produced by *C. teroni* 5-MGAM-4D was immobilized in calcium alginate and evaluated for product acrylamide in batch reactions with catalyst recycle [121,122]. Catalyst productivity dec with increasing acrylonitrile concentration or reaction temperature, but was relative sensitive to acrylamide concentration. A total of 975 g acrylamide/g dry cell weight an acrylamide/L were produced at 5°C in 206 consecutive batch reactions with catalyst r with initial and final volumetric productivities of 142 and 76 g acrylamide/L/h, respec This catalyst productivity was comparable to that of commercial polyacrylamide-imi ized *Rhodococcus* sp. 774 and *P. putida* B23 catalysts (500 and 850 g acrylamide/g d weight, respectively), but less than that of the currently employed polyacrylamide-imi ized *R. rhodochrous* J1 (>7000 g acrylamide/g dry cell weight) [123]. Although the 5-MC 4D NHase had good thermal stability at 35°C, catalyst productivity at this temperatu significantly less than that at 5°C. The major determinant of biocatalyst productivity w concentration of acrylonitrile in the reaction mixture, as was the case in previous studie thermally stable microbial NHases, where the nucleophilic reaction of protein func groups with acrylonitrile was proposed as a mechanism for enzyme inactivation [124, 1

In addition to the commercial production of >30,000 MT/y of acrylamide usi *rhodochrous* J1, >15,000 MT acrylamide/y are currently produced in the Peoples Re of China (PRC) using *Nocardia* sp. microbial NHase biocatalysts. In an initial study suitability of *Nocardia* sp. 9112-118 for commercial production of acrylamide, the cell immobilized in alginate and used to produce 250 g/L acrylamide at 20 to 23°C in a : batch reactor [126]. For the fermentation of *Nocardia* sp. RS, optimization of pH regu and addition of glucose-Co^{2+} in fed-batch fermentations increased NHase activity to U/mL fermentation broth, among the highest volumetric productivities for all rep NHase fermentations [127]. The acrylamide tolerance of *Nocardia* sp. RS was improv feeding acrylonitrile periodically into shake flasks containing growing cells and screen cells surviving in the high concentration of the resulting acrylamide, thereby producii acrylamide-tolerant *Nocardia* sp. strain RS-1 [128,129]. The final acrylamide concent and percent conversion of acrylonitrile catalyzed by the RS-1 strain were 587 g/L and 99 respectively, a 30.6% higher final acrylamide concentration than that produced by the p RS strain; further optimization of reaction conditions using RS-1 produced a final acryl concentration of 641 g/L. Recently, two NHase genes, NHBA and NHBAX, from *No* sp. YS-2002 have been cloned, and expressed in *E. coli* [130].

Alternatives to the polyacrylamide immobilization of microbial cell catalysts for ind acrylamide production have also been evaluated. A hollow-fiber membrane bioreacto evaluated for the conversion of acrylonitrile to acrylamide using unimmobilized ce *Nocardia* sp., a microbial cell catalyst that is used for commercial production of acryl by the Nantian Corporation (Jiangsu, PRC) [131]. Initial studies at laboratory scale v 10% (v/v) fermentation broth as a catalyst charge at 20°C. Different membrane material evaluated, including polysulfone (PS), polyvinylidene difluoride (PVDF), and polyacr trile (PAN). The PS hollow-fiber membrane was scaled up, having the advantages o flux, high resistance to fouling, operation in a cross-flow filtration mode, and easy clear 50,000-Da nominal molecule weight cutoff was preferred for protein filtration. An decay of flux, almost 35% of the initial filtration rate, occurred during the first 2 to 3 h filtration, after which the filtration rate remained unchanged; the decay of the membrane flux was attributed to absorption and compaction of cells and protein on the inner surface of the hollow fibers. The conversion rate of acrylonitrile was 99.9%, and the productivity of acrylamide was 20.67 g/g cell/h over 5 h of operation. The reaction was scaled up in a 6000 L hollow-fiber membrane reactor to produce 300 g/L acrylamide over 17 h. The conversion rate of acrylonitrile was 99.9%, and the efficiency of enzyme activity and acrylamide productivity were 10.96 g/g cell/h and 187.2 g/g cell, respectively, compared to 5.22 g/g cell/h and 112 g/g cell, respectively, when using the immobilized-cell catalyst in a fed-batch mode over 20 h.

A process has been described for producing amides, including acrylamide, that employs unimmobilized microbial cells at a reaction temperature of 10°C, where the cells have an NHase specific activity of at least 50 U/mg dry cell weight [132]. This process claims the advantage of efficiently producing acrylamide without experiencing a loss of catalyst specific activity as a result of immobilization, and increased reaction rates and volumetric productivity were observed when comparing unimmobilized cells with an equivalent amount (on a dry cell weight basis) of immobilized cells.

In a study comparing free-cell and immobilized-cell catalysts, *R. rhodochrous* M33 cells were immobilized in acrylamide-based polymer gels, and compared to unimmobilized cells for the production of acrylamide [133]. The optimum pH and temperature for NHase activity in both free and immobilized cells were 7.4 and 45°C, respectively, but the optimum temperature for acrylamide production was 20°C. The immobilized-cell NHase was more stable than the free-cell NHase in the presence of high concentrations of acrylamide. Under optimal conditions, a final acrylamide concentration of ~400 g/L was achieved, with a conversion yield of almost 100% after 8 h of reaction when using 150 g/L of immobilized cells (corresponding to a 1.91 g dry cell weight/L); the immobilized cell enzyme activity decreased rapidly with repeated use. Comparing the quality of the acrylamide produced by immobilized cells with that produced by free cells, the aqueous acrylamide solution produced by the former had lower color, salt content, turbidity, and foam formation than that produced by the latter; this aqueous acrylamide solution could be used directly in commercial applications, whereas the free-cell product required further downstream purification.

1.3.6 Modification of Nitrile-Containing Polymers

Enzymatic treatment of PAN polymers by NHase, or a combination of NHase and amidase, can be used to hydrate and/or hydrolyze surface nitrile groups; the treated polymers or fibers were more hydrophilic, had improved antistatic properties, and readily reacted with acid dyes under conditions where untreated polymers were inert. *R. rhodochrous* NCIMB 11216 produced NHase and amidase activities that hydrolyzed nitrile groups of both granular PAN and acrylic fibers [134]; 1.8% of PAN40 (molecular mass, 40 kDa) total available nitrile groups and 1.0% of PAN190 (molecular mass, 190 kDa) total available nitrile groups were converted to the corresponding carboxylic acids, and a maximum of 16% of surface nitrile groups of acrylic fibers were converted only to the corresponding amides. Similar improvements in acid dyeing of acrylic fibers was achieved by treatment with the NHase of *Arthrobacter* sp. Ecu1101 [135], or *Brevibacterium imperiale* and *Corynebacterium nitrilophilus* [136].

1.3.7 Nitrilase or NHase?

Bacterial nitrile hydrolysis has been widely documented, whereas there are relatively few reports of fungal nitrile hydrolysis. A nitrile-converting enzyme activity was induced in *Aspergillus niger* K10 by 3-cyanopyridine [137], and the resulting microbial biocatalyst hydrolyzed the cyano group of benzonitrile, 3- and 4-substituted benzonitrile, cyanopyridines,

2-phenylacetonitrile, and thiophen-2-acetonitrile into acid and/or amide groups. Amide significant reaction products for hydrolysis of 2- and 4-cyanopyridine, 4-chlorobenzo 4-tolunitrile, and 1,4-dicyanobenzene, while α-substituted acrylonitriles were converte to amides. This pattern of nitrile conversion is novel for filamentous fungi that hyd nitriles mostly through the nitrilase pathway, therefore the nitrile-converting enzyme is a nitrilase affording extremely large amounts of some amides or a NHase.

1.4   CONCLUSIONS

Both nitrilases and NHases have found increasing use in the production of agrochemica pharmaceuticals, as well as commodity chemicals. The advent of methods for di evolution of enzymes has made it possible to move beyond the screening of microbial c collections for the desired catalyst activity; enzymes can be evolved for improvem enantioselectivity and regioselectivity, temperature stability, substrate K_m, and sul and product inhibition or inactivation. The expression of heterologous nitrilases and N in suitable hosts often results in significant increases in microbial specific activity compared with that of the wild-type microbe from which the gene was isolated. It is in ing to note that for both nitrilases and NHases, microbial cell catalysts are utilized much frequently than isolated enzymes, as there are rarely any microbially catalyzed reactio compete with these enzymes for substrate, or produce undesired reaction by-products.

Many of the recent examples cited above demonstrate the use of nitrilase and I as catalysts to generate product concentrations that are required for economical co cial production, and report volumetric productivity (g product/L/h) and biocatalyst uctivity (g product/g microbial biocatalyst dry cell weight) in these reactions. It i generally recognized that catalyst cost and overall process economics must be cons in the design of any industrial process employing a biocatalyst, where the cost of facture must favorably compete with alternative chemical methods. The design o processes must take into account the expected capital costs and downstream proc costs (e.g., conversion of the ammonium salt of a carboxylic acid to the free separation and recovery of product from reaction by-products in high yield and p disposal of reaction by-products and salts generated in product isolation/purification a number of recent publications and patents address the engineering aspects of sc along with biocatalyst development.

REFERENCES

1. O'Reilly, C. and Turner, P.D., The nitrilase family of CN hydrolyzing enzymes — a comp study, *J. Appl. Microbiol.*, 95, 1161, 2003.
2. Martinkova, L. and Mylerova, V., Synthetic applications of nitrile-converting enzymes, *Cur Chem.*, 7, 1279, 2003.
3. Martinkova, L. and Kren, V., Nitrile- and amide-converting microbial enzymes: stereo-, reg chemoselectivity, *Biocatal. Biotransform.*, 20, 73, 2002.
4. Banerjee, A., Sharma, R., and Banerjee, U.C., The nitrile-degrading enzymes: current stat future prospects, *Appl. Microbiol. Biotechnol.*, 60, 33, 2002.
5. Brenner, C., Catalysis in the nitrilase superfamily, *Curr. Opin. Struct. Biol.*, 12, 775, 2002.
6. Groger, H., Enzymatic routes to enantiomerically pure aromatic α-hydroxy carboxylic a further example for the diversity of biocatalysis, *Adv. Synth. Catal.*, 343, 547, 2001.
7. Wieser, M. and Nagasawa, T., Stereoselective nitrile-converting enzymes, in *Stereoselective talysis*, Patel, R.N., Ed., Marcel Dekker, New York, 2000, chap. 17.

8. Pace, H.C. and Brenner, C., The nitrilase superfamily: classification, structure and function, *Genome Biol.* (online computer file), 2, reviews 0001.1, 2001.
9. Kobayashi, M., Goda, M., and Shimizu, S., Nitrilase catalyzes amide hydrolysis as well as nitrile hydrolysis. *Biochem. Biophys. Res. Commun.*, 253, 662, 1998.
10. Ress-Loeschke, M., Hauer, B., Mattes, R., and Engels, D., U.S. Patent Appl. 2003157672 A1, 2003.
11. Herai, S., Hashimoto, Y., Higashibata, H., Maseda, H., Ikeda, H., Omura, S., and Kobayashi, M., Hyper-inducible expression system for *Streptomycetes*. *Proc. Natl. Acad. Sci. USA*, 101, 14031, 2004.
12. Chauhan, S., DiCosimo, R., Fallon, R.D., Gavagan, J.E., and Payne, M.S., U.S. Patent 6,870,038 B2, 2005.
13. Heinemann, U., Engels, D., Buerger, S., Kiziak, C., Mattes, R., and Stolz, A., Cloning of a nitrilase gene from the cyanobacterium *Synechocystis* sp. strain PCC6803 and heterologous expression and characterization of the encoded protein, *Appl. Environ. Microbiol.*, 69, 4359, 2003.
14. Park, W.J., Kriechbaumer, V., Mueller, A., Piotrowski, M., Meeley, R.B., Gierl, A., and Glawischnig, E., The nitrilase ZmNIT2 converts indole-3-acetonitrile to indole-3-acetic acid, *Plant Physiol.*, 133, 794, 2003.
15. DeSantis, G., Wong, K., Farwell, B., Chatman, K., Zhu, Z., Tomlinson, G., Huang, H., Tan, X., Bibbs, L., Chen, P., Kretz, K., and Burk, M.J., Creation of a productive, highly enantioselective nitrilase through gene site saturation mutagenesis (GSSM), *J. Am. Chem. Soc.*, 125, 11476, 2003.
16. DeSantis, G., Zhu, Z., Greenberg, W.A., Wong, K., Chaplin, J., Hanson, S.R., Farwell, B., Nicholson, L.W., Rand, C.L., Weiner, D.P., Robertson, D.E., and Burk, M.J., An enzyme library approach to biocatalysis: development of nitrilases for enantioselective production of carboxylic acid derivatives, *J. Am. Chem. Soc.*, 124, 9024, 2002.
17. Robertson, D.E., Chaplin, J.A., DeSantis, G., Podar, M., Madden, M., Chi, E., Richardson, T., Milan, A., Miller, M., Weiner, D.P., Wong, K., McQuaid, J., Farwell, B., Preston, L.A., Tan, X., Snead, M.A., Keller, M., Mathur, E., Kretz, P.L., Burk, M.J., and Short, J.M., Exploring nitrilase sequence space for enantioselective catalysis, *Appl. Environ. Microbiol.*, 70, 2429, 2004.
18. Squires, C. and Talbot, H., *Pseudomonas fluorescens*: a robust manufacturing platform, *Speciality Chemicals Magazine*, 24, 25, 2004.
19. Kaul, P., Banerjee, A., Mayilraj, S., and Banerjee, U.C., Screening for enantioselective nitrilases: kinetic resolution of racemic mandelonitrile to (*R*)-(−)-mandelic acid by new bacterial isolates, *Tetrahedron Asymmetry*, 15, 207, 2004.
20. Favre-Bulle, O., Pierrard, J., David, C., Morel, P., and Horbez, D., U.S. Patent 6180359 B1, 2001.
21. Rey, P., Rossi, J.-C., Taillades, J., Gros, G., and Nore, O., Hydrolysis of nitriles using an immobilized nitrilase: applications to the synthesis of methionine hydroxy analogue derivatives, *J. Agric. Food Chem.*, 52, 8155, 2004.
22. Favre-Bulle, O. and Le Thiesse, J.-C., U.S. Patent Appl. 2004082046 A1, 2004.
23. Pierrard, J., Favre-Bulle, O., and Jourdat, C., PCT Int. Appl. WO 2001034786 A1, 2001.
24. Nagasawa, T. and Matsuyama, A., Jpn. Kokai Tokkyo Koho JP 2001054380 A2, 2001.
25. Effenberger, F., Forster, S., and Wajant, H., Hydroxymitrile lyases in stereoselective catalysis, *Curr. Opin. Biotechnol.*, 11, 532, 2000.
26. Heinemann, U., Kiziak, C., Zibek, S., Layh, N., Schmidt, M., Griengl, H., and Stolz, A., Conversion of aliphatic 2-acetoxynitriles by nitrile-hydrolysing bacteria, *Appl. Microbiol. Biotechnol.*, 63, 274, 2003.
27. Chaplin, J.A., Levin, M.D., Morgan, B., Farid, N., Li, J., Zhu, Z., McQuaid, J., Nicholson, L.W., Rand, C.A., and Burk, M.J., Chemoenzymatic approaches to the dynamic kinetic asymmetric synthesis of aromatic amino acids, *Tetrahedron Asymmetry*, 15, 2793, 2004.
28. Zelinski, T., Kesseler, M., Hauer, B., and Friedrich, T., PCT Int. Appl. WO 2004076655 A1, 2004.
29. Bensoussan, C., Bontoux, M.C., Gelo, P.M., and Foray, F., Fr. Demande 2822460 A1, 2002.
30. Cooling, F.B., Fager, S.K., Fallon, R.D., Folsom, P.W., Gallagher, F.G., Gavagan, J.E., Hann, E.C., Herkes, F.E., Phillips, R.L., Sigmund, A., Wagner, L.W., Wu, W., and DiCosimo, R., Chemoenzymatic production of 1,5-dimethyl-2-piperidone, *J. Mol. Catal. B Enzym.*, 11, 295, 2001.
31. Hann, E.C., Sigmund, A.E., Hennessey, S.M., Gavagan, J.E., Short, D.R., Ben-Bassat, A., Chauhan, S., Fallon, R.D., Payne, M.S., and DiCosimo, R., Optimization of an immobilized-cell biocatalyst for production of 4-cyanopentanoic acid, *Org. Process Res. Dev.*, 6, 492, 2002.

32. Chauhan, S., Wu, S., Blumerman, S., Fallon, R.D., Gavagan, J.E., DiCosimo, R., and Payne Purification, cloning, sequencing and over-expression in *Escherichia coli* of a regioselective al nitrilase from *Acidovorax facilis* 72W, *Appl. Microbiol. Biotechnol.*, 61, 118, 2003.

33. Jennings, R.A., Johnson, D.R., Seamans, R.E., and Zeller, J.R., U.S. Patent 5,362,883, 19

34. Burns, M.P. and Wong, J.W., U.S. Patent Appl. 2005009154 A1, 2005.

35. Effenberger, F. and Osswald, S., Enzyme-catalyzed reactions. 41. Selective hydrolysis of al dinitriles to monocarboxylic acids by a nitrilase from *Arabidopsis thaliana*, *Synthesis*, 12, 1866

36. Kamachi, H. and Aoki, H., Jpn. Kokai Tokkyo Koho JP 2001136977 A2, 2001.

37. Fujita, I., Aoki, H., and Wada, K., Jpn. Kokai Tokkyo Koho JP 2003325194 A2, 2003.

38. Dadd, M.R., Claridge, T.D.W., Walton, R., Pettman, A.J., and Knowles, C.J., Regiose biotransformation of the dinitrile compounds 2-, 3- and 4-(cyanomethyl)benzonitrile by t bacterium *Rhodococcus rhodochrous* LL100-21, *Enzyme Microb. Technol.*, 29, 20, 2001.

39. Effenberger, F. and Osswald, S., (*E*)-Selective hydrolysis of (*E,Z*)-α,β-unsaturated nitr the recombinant nitrilase AtNIT1 from *Arabidopsis thaliana*, *Tetrahedron Asymmetry*, 12, 2001.

40. Hann, E.C., Sigmund, A.E., Fager, S.K., Cooling, F.B., Gavagan, J.E., Ben-Bassat, A., Chauh Payne, M.S., and DiCosimo, R., Regioselective biocatalytic hydrolysis of (*E,Z*)-2-methyl-2-b nitrile for production of (*E*)-2-methyl-2-butenoic acid, *Tetrahedron*, 60, 577, 2004.

41. Chauhan, S., DiCosimo, R., Fallon, R.D., Gavagan, J.E., and Payne, M.S., U.S. Patent 6 B1, 2002.

42. Brady, D., Beeton, A., Zeevaart, J., Kgaje, C., van Rantwijk, F., and Sheldon, R.A., Charac tion of nitrilase and nitrile hydratase biocatalytic systems, *Appl. Microbiol. Biotechnol.*, 64, 76

43. Webster, N.A., Ramsden, D.K., and Hughes, J., Comparative characterisation of two *Rhodc* species as potential biocatalysts for ammonium acrylate production, *Biotechnol. Lett.*, 23, 95

44. Okamoto, M. and Nagano, O., Jpn. Kokai Tokkyo Koho JP 2004305062 A2, 2004.

45. Symes, K.C. and Hughes, J., U.S. Patent 6,162,624, 2000.

46. Poltavskaya, S.V., Kozulina, T.N., Singirtsev, I.N., Kozulin, S.V., Shub, G.M., and Voroni Development and implementation of biocatalytic method for acrylic acid manufactur Isolation of *Alcaligenes denitrificans* strain transforming acrylonitrile into ammonium ac Optimization of the culture medium, *Biotekhnologiya*, 1, 62, 2004.

47. Chauhan, S., DiCosimo, R., Fallon, R.D., Gavagan, J.E., and Payne, M.S., U.S. Patent 6 B1, 2002.

48. Chauhan, S, DiCosimo, R., Fallon, R., Gavagan, J., Manzer, L.E., and Payne, M.S., U.S. 6,582,943 B1, 2003.

49. Dadd, M.R., Claridge, T.D.W., Pettman, A.J., and Knowles, C.J., Biotransformation of be trile to benzohydroxamic acid by *Rhodococcus rhodochrous* in the presence of hydroxyl *Biotechnol. Lett.*, 23, 221, 2001.

50. Subramanian, V., Christenson, C.P., Conboy, C.B., Li, K., Redwine, O.D., Chopin, L. Miller, B.A., Morabito, P.L., Winterton, R.C., and Rosner, B.M., PCT Int. Appl. WO 2004 A2, 2004.

51. Osswald, S., Wajant, H., and Effenberger, F., Characterization and synthetic applicati recombinant AtNIT1 from *Arabidopsis thaliana*, *Eur. J. Biochem.*, 269, 680, 2002.

52. Effenberger, F. and Osswald, S., Enantioselective hydrolysis of (*RS*)-2-fluoroarylaceto using nitrilase from *Arabidopsis thaliana*, *Tetrahedron Asymmetry*, 12, 279, 2001.

53. Yamada, H., Shimizu, S., and Kobayashi, M., Hydratases involved in nitrile conversion: scr characterization and application, *Chem. Rec.*, 1, 152, 2001.

54. Martinkova, L. and Mylerova, V., Synthetic applications of nitrile-converting enzymes, *Cur Chem.*, 7, 1279, 2003.

55. Martinkova, L. and Kren, V., Nitrile- and amide-converting microbial enzymes: stereo-, regi chemoselectivity, *Biocatal. Biotransform.*, 20, 73, 2002.

56. Banerjee, A., Sharma, R., and Banerjee, U.C., The nitrile-degrading enzymes: current stat future prospects, *Appl. Microbiol. Biotechnol.*, 60, 33, 2002.

57. Huang, W., Jia, J., Cummings, J., Nelson, M., Schneider, G., and Lindqvist, Y., Crystal str of nitrile hydratase reveals a novel center in a novel fold, *Structure*, 5, 691, 1997.

58. Cowan, D.A., Cameron, R.A., and Tsekoa, T.L., Comparative biology of mesophilic and thermophilic nitrile hydratases, *Adv. Appl. Microbiol.*, 52, 123, 2003.
59. Endo, I. and Odaka, M., Studies on photoreactive enzyme — nitrile hydratase, *Prog. Biotechnol.*, 22, 159, 2002.
60. Endo, I., Nojiri, M., Tsujimura, M., Nakasako, M., Nagashima, S., Yohda, M., and Odaka, M., Fe-type nitrile hydratase, *J. Inorg. Biochem.*, 83, 247, 2001.
61. Radu, S.D., Nitrile hydration by the cobalt-containing nitrile hydratase. DFT investigation of the mechanism, *Revista de Chimie* (Bucharest, Romania), 56, 359, 2005.
62. Kobayashi, M. and Shimizu, S., Nitrile hydrolases, *Curr. Opin. Chem. Biol.*, 4, 95, 2000.
63. Tsujimura, M., Odaka, M., Nakayama, H., Dohmae, N., Koshino, H., Asami, T., Hoshino, M., Takio, K., Yoshida, S., Maeda, M., and Endo, I., A novel inhibitor for Fe-type nitrile hydratase: 2-cyano-2-propyl hydroperoxide. *J. Am. Chem. Soc.*, 125, 11532, 2003.
64. Nagashima, S., Nakasako, M., Dohmae, N., Tsujishima, M., Takio, K., Odaka, M., Yohda, M., Kamiya, N., and Endo, I., Novel nonheme iron center of nitrile hydratase with a claw setting of oxygen atoms, *Nat. Struct. Biol.*, 5, 347, 1998.
65. Miyanaga, A., Fushinobu, S., Ito, K., and Wakagi, T., Crystal structure of cobalt-containing nitrile hydratase, *Biochem. Biophys. Res. Commun.*, 288, 1169, 2001.
66. Miyanaga, A., Fushinobu, S., Ito, K., Shoun, H., and Wakagi, T., Mutational and structural analysis of cobalt-containing nitrile hydratase on substrate and metal binding, *Eur. J. Biochem.*, 271, 429, 2004.
67. Hourai, S., Miki, M., Takashima, Y., Mitsuda, S., and Yanagi, K., Crystal structure of nitrile hydratase from a thermophilic *Bacillus smithii*, *Biochem. Biophys. Res. Comm.*, 312, 340, 2003.
68. Komeda, H., Kobayashi, M., and Shimizu, S., Characterization of the gene cluster of high-molecular-mass nitrile hydratase (H-NHase) induced by its reaction product in *Rhodococcus rhodochrous* J1, *Proc. Natl. Acad. Sci. USA*, 93, 4267, 1996.
69. Stevens, J.M., Rao Saroja, N., Jaouen, M., Belghazi, M., Schmitter, J.-M., Mansuy, D., Artaud, I., and Sari, M.-A., Chaperone-assisted expression, purification, and characterization of recombinant nitrile hydratase NI1 from *Comamonas testosteroni*, *Protein Expression Purif.*, 29, 70, 2003.
70. Yokoyama, M., Kashiwagi, M., Iwasaki, M., Fuhshuku, K., Ohta, H., and Sugai, T., Realization of the synthesis of α,α-disubstituted carbamylacetates and cyanoacetates by either enzymatic or chemical functional group transformation, depending upon the substrate specificity of *Rhodococcus amidase*, *Tetrahedron Asymmetry*, 15, 2817, 2004.
71. Wu, Z.-L. and Li, Z.-Y., Practical synthesis of optically active α,α-disubstituted malonamic acids through asymmetric hydrolysis of malonamide derivatives with *Rhodococcus* sp. CGMCC 0497, *J. Org. Chem.*, 68, 2479, 2003.
72. Wang, M.-X., Feng, G.-Q., and Zheng, Q.-Y., Synthesis of high enantiomeric pure gem-dihalocyclopropane derivatives from biotransformations of nitriles and amides, *Tetrahedron Asymmetry*, 15, 347, 2004.
73. Wang, M.-X., Feng, G.-Q., and Zheng, Q.-Y., Nitrile and amide biotransformations for efficient synthesis of enantiopure gem-dihalocyclopropane derivatives, *Adv. Synth. Catal.*, 345, 695, 2003.
74. Wang, M.-X. and Feng, G.-Q., Nitrile biotransformation for highly enantioselective synthesis of 3-substituted 2,2-dimethylcyclopropanecarboxylic acids and amides, *J. Org. Chem.*, 68, 621, 2003.
75. Wang, M.-X. and Lin, S.-J., Practical and convenient enzymatic synthesis of enantiopure α-amino acids and amides, *J. Org. Chem.*, 67, 6542, 2002.
76. Wang, M.-X., Lin, S.-J., Liu, J., and Zheng, Q.-Y., Efficient biocatalytic synthesis of highly enantiopure α-alkylated arylglycines and amides, *Adv. Synth. Catal.*, 346, 439, 2004.
77. Wang, M.-X., Li, J.-J., Ji, G.-J., and Li, J.-S., Enantioselective biotransformations of racemic 2-aryl-3-methylbutyronitriles using *Rhodococcus* sp. AJ270, *J. Mol. Catal. B Enzym.*, 14, 77, 2001.
78. Hensel, M., Lutz-Wahl, S., and Fischer, L., Stereoselective hydration of (*RS*)-phenylglycine nitrile by new whole cell biocatalysts, *Tetrahedron Asymmetry*, 13, 2629, 2002.
79. Ramer, S.W., Huisman, G., Millis, J., Sheldon, R., Del Cardayre, S., Tobin, M., Cox, A., and Davis, S.C., U.S. Patent Appl. 20020137153 A1, 2002.

80. Ishikawa, T. and Hayakawa, K., Jpn. Kokai Tokkyo Koho JP 2004254690 A2, 2004.

81. Nagasawa, T. and Matsuyama, A., Jpn. Kokai Tokkyo Koho JP 2004081169 A2, 2004.

82. Osprian, I., Fechter, M.H., and Griengl, H., Biocatalytic hydrolysis of cyanohydrins: an ε approach to enantiopure α-hydroxy carboxylic acids, *J. Mol. Catal. B Enzym.*, 24–25, 89, 200.

83. Griengl, H., Osprian, I., Schoemaker, H.E., Reisinger, C., and Schwab, H., PCT Int. App 2004076385 A2, 2004.

84. Poechlauer, P., Skranc, W., and Wubbolts, M., The large-scale biocatalytic synthesis of e pure cyanohydrins, in *Asymmetric Catalysis on Industrial Scale*, Blaser, H.-U. and Schm Eds., Wiley-VCH, Weinheim, Germany, 2004, p. 151.

85. Schwab, H., Glieder, A., Kratky, C., Dreveny, I., Poechlauer, P., Skranc, W., Mayrho Wirth, I., Neuhofer, R., and Bona, R., U.S. Patent 6,861,243 B2, 2005.

86. Reisinger, C., Osprian, I., Glieder, A., Schoemaker, H.E., Griengl, H., and Schwab, H., Enz hydrolysis of cyanohydrins with recombinant nitrile hydratase and amidase from *Rhod erythropolis*, *Biotechnol. Lett.*, 26, 1675, 2004.

87. Preiml, M., Hillmayer, K., and Klempier, N., A new approach to β-amino acids: biotrans tion of *N*-protected β-amino nitriles, *Tetrahedron Lett.*, 44, 5057, 2003.

88. Preiml, M., Hönig, H., and Klempier, N., Biotransformation of β-amino nitriles: the role *N*-protecting group, *J. Mol. Catal. B Enzym.*, 29, 115, 2004.

89. Winkler, M., Martinkova, L., Knall, A.C., Krahulec, S., and Klempier, N., Synthesis and bial transformation of β-amino nitriles, *Tetrahedron*, 61, 4249, 2005.

90. Hann, E.C., Sigmund, A.E., Fager, S.K., Cooling, F.B., Gavagan, J.E., Ben-Bassat, A., Chau Payne, M.S., Hennessey, S.M., and DiCosimo, R., Biocatalytic hydrolysis of 3-hydroxyalkane to 3-hydroxyalkanoic acids, *Adv. Synth. Catal.*, 345, 775, 2003.

91. DiCosimo, R., Fallon, R., Gavagan, J.E., and Manzer, L., U.S. Patent Appl. 20031484 2003.

92. Chauhan, S., DiCosimo, R., Fallon, R., Gavagan, J., Manzer, L.E., and Payne, M.S., U.S. 6,582,943 B1, 2003.

93. Chassin, C., A biotechnological process for the production of nicotinamide, *Chim. Oggi*, 14, 9

94. Shaw, N.M., Robins, K.T., and Kiener, A., Lonza: 20 years of biotransformations, *Adv. Catal.*, 345, 425, 2003.

95. Heveling, J., Armbruster, E., Utiger, L., Rohner, M., Dettwiler, H.-R., and Chuck, R.J Patent 5,719,045, 1998.

96. Anonymous *Chem. Eng. News*, 81(23),12, 2003.

97. Robins, K.T. and Nagasawa, T., U.S. Patent 6,444,451 B1, 2002.

98. Robins, K.T. and Nagasawa, T., U.S. Patent Appl. 20030148478 A1, 2003.

99. Robins, K.T. and Nagasawa, T., U.S. Patent Appl. 20040142447 A1, 2004.

100. Shapiro, R., DiCosimo, R., Hennessey, S.M., Stieglitz, B., Campopiano, O., and Chiang Discovery and development of a manufacturing-scale synthesis of azafenidin, *Org. Proce Dev.*, 5, 593, 2001.

101. Bernet, N., Arnaud, A., and Galzy, P., Optimization of culture conditions of *Brevibacterium* for the production of nitrile hydratase, *Biocatalysis*, 3, 259, 1990.

102. Ingvorsen, K., Godtfredsen, S.E., and Tsuchiya, R.T., Microbial hydrolysis of organic nitri amides, *Ciba Found. Symp.*, 140, 16, 1988.

103. Andresen, O. and Godtfredsen, S. E., European Patent EP 178106 B1, 1993.

104. Nagamune, T., Kurata, H., Hirata, M., Honda, J., Hirata, A., and Endo, I., Photose phenomena of nitrile hydratase of *Rhodococcus* sp. N-771, *Photochem. Photobiol.*, 51, 87, 1

105. Honda, J., Kandori, H., Okada, T., Nagamune, T., Shichida, Y., Sasabe, H., and En Spectroscopic observation of the intramolecular electron transfer in the photoactivation pr of nitrile hydratase, *Biochemistry*, 33, 3577, 1994.

106. Nagasawa, T., Nanba, H., Ryuno, K., Takeuchi, K., and Yamada, H., Nitrile hydra *Pseudomonas chlororaphis* B23: purification and characterization, *Eur. J. Biochem.*, 162, 691

107. Yamada, H., Ryuno, K., Nagasawa, T., Enomoto, K., and Watanabe, I., Optimum conditions for production by *Pseudomonas chlororaphis* B23 of nitrile hydratase, *Agric Chem.*, 50, 2859, 1986.

108. Asano, Y., Yasuda, T., Tani, Y., and Yamada, H., Microbial degradation of nitrile compounds. Part VII. A new enzymic method of acrylamide production, *Agric. Biol. Chem.*, 46, 1183, 1982.
109. Ashina, Y. and Suto, M., Development of an enzymic process for manufacturing acrylamide and recent progress, *Bioprocess Technol.*, 16, 91, 1993.
110. Kobayashi, M., Nagasawa, T., and Yamada, H., Enzymic synthesis of acrylamide: a success story not yet over, *Trends Biotechnol.*, 10, 402, 1992.
111. Nagasawa, T., Ryuno, K., and Yamada, H., Superiority of *Pseudomonas chlororaphis* B23 nitrile hydratase as a catalyst for the enzymic production of acrylamide, *Experientia*, 45, 1066, 1989.
112. Ryuno, K., Nagasawa, T., and Yamada, H., Isolation of advantageous mutants of *Pseudomonas chlororaphis* B23 for the enzymic production of acrylamide, *Agric. Biol. Chem.*, 52, 1813, 1988.
113. DiCosimo, R., Hann, E.C., Eisenberg, A., Fager, S.K., Perkins, N.E., Gallagher, F.G., Cooper, S.M., Gavagan, J.E., Stieglitz, B., and Hennessey, S.M., Biocatalytic production of 5-cyanovaleramide from adiponitrile, *ACS Symp. Ser.*, 767, 114, 2000.
114. Hann, E.C., Eisenberg, A., Fager, S.K., Perkins, N.E., Gallagher, F.G., Cooper, S.M., Gavagan, J.E., Stieglitz, B., Hennessey, S.M., and DiCosimo, R., 5-Cyanovaleramide production using immobilized *Pseudomonas chlororaphis* B23, *Bioorg. Med. Chem.*, 7, 2239, 1999.
115. Watanabe, F., Jpn. Kokai Tokkyo Koho JP 2004215513 A2, 2004.
116. Watanabe, F., Jpn. Kokai Tokkyo Koho JP 2004222538 A2, 2004.
117. Ramey, S. and Tichenor, G., Enzymatically produced acrylamide: improving the environment, an additional benefit of building a better product, WEFTEC.02 Conference Proceedings, Chicago, September 28, 2002.
118. Wieser, M. and Pommares, P., PCT Int. Appl. WO 2003066800 A3, 2004.
119. Furuya, K., Tamaki, A., Nagasawa, S., and Suzuki, A., PCT Int. Appl. WO 2004108942 A1,2004.
120. Yamaki, T., Banba, S., Matoishi, K., Ito, K., Kobayashi, H., Tanaka, E., and Oikawa, T., PCT Int. Appl. WO 2004056990 A1, 2004.
121. Petrillo, K.L., Wu, S., Hann, E.C., Cooling, F.B., Ben-Bassat, A., Gavagan, J.E., DiCosimo, R., and Payne, M.S., Over-expression in *Escherichia coli* of a thermostable nitrile hydratase from *Comamonas testosteroni* 5-MGAM-4D, *App. Microbiol. Biotechnol.*, 67, 664, 2005.
122. Mersinger, L.J., Hann, E.C., Cooling, F.B., Ben-Bassat, A., Gavagan, J.E., Wu, S., Petrillo, K.L., Payne, M.S., and DiCosimo, R., Production of acrylamide using alginate-immobilized *E. coli* expressing *Comamonas testosteroni* 5-MGAM-4D nitrile hydratase, *Adv. Synth. Catal.*, 347, 1125, 2005.
123. Nagasawa, T., Shimizu, H., and Yamada, H., The superiority of the third-generation catalyst, *Rhodococcus rhodochrous* J1 nitrile hydratase, for industrial production of acrylamide, *Appl. Microbiol. Biotechnol.*, 40, 189, 1993.
124. Padmakumar, R. and Oriel, P., Bioconversion of acrylonitrile to acrylamide using a thermostable nitrile hydratase, *Appl. Biochem. Biotechnol.*, 77–79, 671, 1999.
125. Graham, D., Pereira, R., Barfield, D., and Cowan, D., Nitrile biotransformations using free and immobilized cells of a thermophilic *Bacillus* spp., *Enz. Microb. Technol.*, 26, 368, 2000.
126. Han, J., Zhang, Y., Xue, J., Huang, J., and Chen, X., Research on manufacture of acrylamide on industrial scale by microbial method, *Shanghai Huagong*, 20, 7, 1995.
127. Liu, M., Li, C., Gao, Y., Wang, Y., and Cao, Z., Optimization of glucose-Co^{2+} coupling fed batch fermentation for production of nitrile hydratase of high activity by *Nocardia* sp. RS, *Guocheng Gongcheng Xuebao*, 3, 555, 2003.
128. Liu, M., Li, C., Gao, Y., Huang, Y., and Cao, Z., Improved process for high acrylamide accumulation with acclimated microorganism *Nocardia* sp., *Huagong Xuebao*, 55, 1678, 2004.
129. Liu, M., Li, C., Huang, Y., Gao, Y., and Cao, Z., Improving the acrylamide-tolerance of nitrile hydratase in *Nocardia* sp. by extreme cultivation, *Guocheng Gongcheng Xuebao*, 4, 250, 2004.
130. Shi, Y., Yu, H., Sun, X., Tian, Z., and Shen, Z., Cloning of the nitrile hydratase gene from *Nocardia* sp. in *Escherichia coli* and *Pichia pastoris* and its functional expression using site-directed mutagenesis, *Enzyme Microb. Technol.*, 35, 557, 2004.
131. Sun, X., Shi, Y., Yu, H., and Shen, Z., Bioconversion of acrylonitrile to acrylamide using hollow-fiber membrane bioreactor system, *Biochem. Eng. J.*, 18, 239, 2004.
132. Murao, K. and Ishii, K., U.S. Patent Appl. 2004048348 A1, 2004.

133. Kim, B.-Y. and Hyun, H.-H., Production of acrylamide using immobilized cells of *Rhode rhodochrous* M33. *Biotechnol. Bioprocess Eng.*, 7, 194, 2002.
134. Tauber, M.M., Cavaco-Paulo, A., Robra, K.H., and Gubitz, G.M., Nitrile hydratase and a from *Rhodococcus rhodochrous* hydrolyze acrylic fibers and granular polyacylonitriles *Environ. Microbiol.*, 66, 1634, 2000.
135. Wang, N., Xu, Y., Lu, D.-N., and Xu, J.-H., Enzymatic surface modification of acrylic *AATCC Review*, 4, 28, 2004.
136. Battistel, E., Morra, M., and Marinetti, M., Enzymatic surface modification of acrylonitrile *Appl. Surf. Sci.*, 177, 32, 2001.
137. Snajdrova, R., Kristova-Mylerova, V., Crestia, D., Nikolaou, K., Kuzma, M., Lemai Gallienne, E., Bolte, J., Bezouska, K., Kren, V., and Martinkova, L., Nitrile biotransforma *Aspergillus niger, J. Mol. Catal. B Enzym.*, 29, 227, 2004.

2 Biocatalytic Deracemization: Dynamic Resolution, Stereoinversion, Enantioconvergent Processes, and Cyclic Deracemization

Yolanda Simeó, Wolfgang Kroutil, and Kurt Faber

CONTENTS

2.1 INTRODUCTION

Driven by the increased demand for chiral drugs in enantiomerically pure form, following the release of Food and Drug Administration's (FDA) marketing guidelines, the search for novel catalytic methods to access enantiopure compounds is a major topic in contemporary organic synthesis [1–3]. In this context, biocatalysis has been applied in preparative organic chemistry at an increasing pace over the last two decades [4–7]. From the two principles of biocatalytic reactions where chiral substrates are involved—(i) desymmetrization of *meso*- and prochiral compounds [8,9]; and (ii) kinetic resolution of racemates [10–13]—the latter is remarkably dominant in number of applications (~1:4) [14]. This is because (out of combinatorial considerations) there are more racemic molecules possible than *meso*- and prochiral compounds, which bear an element of symmetry within the molecule. Consequently, racemates always have been (and will be) an indispensable starting point for the synthesis of chiral materials in nonracemic form. In principle, kinetic resolut racemates is based on the difference in reaction rates of enantiomers with a chiral (bio)ca In an ideal case, the reactive "fitting" enantiomer is quickly converted, while the "w enantiomer remains untouched. Thus, the reaction comes to a standstill at 50% conve where both enantiomeric substrate and product can be separated by physical means.

Despite its widespread application, kinetic resolution is impeded by several inherent vantages, especially on an industrial scale. The most obvious drawbacks of kinetic resoluti

1. The theoretical yield of each enantiomer can never exceed a limit of 50%.
2. Separation of the formed product from the remaining substrate may be labori particular for cases in which simple extraction or distillation fails and chromatog methods are required [15].
3. In the majority of processes, only one stereoisomer is desired and there is little or for the other. In some rare cases, the unwanted isomer may be used through a se synthetic pathway in an enantioconvergent fashion, but this requires a highly f synthetic strategy [16].
4. For kinetic reasons, the optical purity of the substrate and the product is depleted point, where separation of product and substrate is most desirable from a prepa point of view, i.e., at 50% conversion [17].

After all, it should be kept in mind that an ideal production process leads to a enantiomeric product in 100% yield. So alternatives to kinetic resolution techniques th provide a single stereoisomer from a racemate are highly advantageous as they pro double theoretical yield of the desired stereoisomeric product [18]. These latter proces encompassed under the general term "deracemization" [19–21] (Scheme 2.1). All of techniques deal with a common stereochemical phenomenon, i.e., both substrate enanti have to be processed through two different pathways. In this chapter, general strategie lead to the formation of a single enantiomeric product in 100% theoretical yield f racemate are reviewed.

2.2   IMPROVING KINETIC RESOLUTION

2.2.1   RE-RACEMIZATION AND REPEATED RESOLUTION

To avoid the loss of half of the material in kinetic resolution, it has been a co practice to racemize the unwanted stereoisomer after separation from the desired duct and to subject it again to kinetic resolution in a subsequent cycle until virtually the racemic material has been converted into a single stereoisomer [22,23]. On a supe look, repeated resolution appears to be less than optimal and certainly lacks thetic elegance, bearing in mind that an infinite number of cycles are theore required to transform all of the racemic starting material into a single stereois However, it is a viable option for resolutions on an industrial scale, in particul continuous processes, where the racemized material is simply fed back into the subse batch of the resolution process.

Racemization, in general, is an energetic "downhill" reaction due to an incre entropy [24], and thus has been considered more often as an undesired side re rather than as a synthetically useful transformation. As a consequence, the con racemization of organic compounds has been scarcely studied deliberately, and a s cant part of the data available to date stems from industrial research predomi reported in the patent literature. It was only recently that the importance of sy

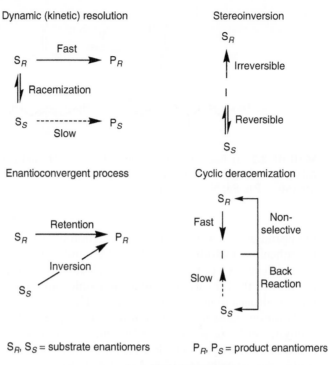

Dynamic (kinetic) resolution

$$S_R \xrightarrow{\text{Fast}} P_R$$

Racemization

$$S_S \dashrightarrow[\text{Slow}] P_S$$

Stereoinversion

S_R

Irreversible

I

Reversible

S_S

Enantioconvergent process

Retention

$$S_R \longrightarrow P_R$$

Inversion

S_S

Cyclic deracemization

S_R

Fast

Non-selective

I

Slow

Back Reaction

S_S

S_R, S_S = substrate enantiomers P_R, P_S = product enantiomers

I = achiral intermediate

SCHEME 2.1 General strategies for the transformation of a racemate into a single stereoisomeric product.

protocols for the controlled racemization of organic compounds has been recognized, as highlighted by the review of Zwanenburg et al. [25]. A detailed investigation of the data available reveals that chemical racemization techniques largely depend on harsh reaction conditions, predominantly (i) thermal racemization, as well as (ii) strong acid catalysis or (iii) base catalysis [26], and (iv) through chirally labile intermediates [25]. Overall, ~75% of all racemizations fall under these categories. As a consequence of the harsh reaction conditions, the possibility for process control in chemical racemization is very limited and undesired side reactions, such as elimination, condensation, rearrangement, and/or decomposition, set a low ceiling on the preparative utility of these processes. However, milder methods for chemical racemization have been recently developed [27]. For instance, chiral *sec*-alcohols and amines can be racemized through a transition metal–catalyzed oxidation–reduction sequence [28] or by π-allyl formation [29] (Scheme 2.2). The mechanism of transition metal–catalyzed racemization by hydrogen transfer has been extensively investigated [30], and a recent study indicates that two different hydridic pathways can be involved in these reactions [31]: a metal monohydride mechanism and a metal dihydride mechanism. Whereas the first mechanism applies to rhodium, iridium, and most nonhalide ruthenium complexes, the second mechanism operates for ruthenium dihalide catalyst precursors.

   An example of a ruthenium catalyst is shown in Scheme 2.3, which is one of the few Ru-complexes that are applicable to a broad substrate spectrum. A relevant feature of this complex is that no external base is needed as a co-catalyst, since one of the oxygens of

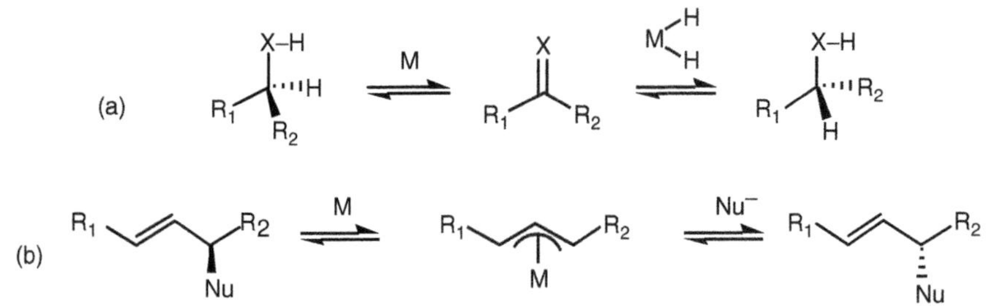

SCHEME 2.2 (a) Racemization of alcohols (X = O) and amines (X = NH) by transition catalyzed hydrogen transfer (M = Rh, Ir, Ru). (b) Racemization of allylic alcohols by π-allyl m ism (M = Ru, Pd; Nu = acetate).

the ligand acts as a base. Thus, the reaction of the basic oxygen with an alcohol giv to ruthenium hydride intermediate "A" and the ketone, which is reduced by ruth intermediate "A" to form the racemic alcohol and ruthenium species "B." A similar anism with the formation of ruthenium amine intermediates has been recently propos the dehydrogenation of amines to imines [32,33].

However, most hydrogen transfer catalysts need the addition of an external base catalyst, which can affect the performance of the enzyme and can also cause side reacti substrates and/or products. For instance, the presence of base in the resolution of β-h cohols can generate epoxides. As a consequence, it is advisable to use a catalyst tha without external base.

On the other hand, racemization of allylic esters has been accomplished with (π-all ladium complexes using Pd(0) catalysis. This isomerization can proceed through di mechanisms [34]. Interconversion of enantiomers of sec-amines is also performed by catalyzed dehydrogenation–hydrogenation (Scheme 2.7c).

Another strategy for racemization is to (chemically) convert compounds bearing a stereocenter into a chirally unstable intermediate. For instance, amino acid amides or can be racemized via Schiff-base derivatives of aromatic aldehydes involving the α- group [35] or, alternatively, for free amino acids, mixed acid anhydrides are used. I sense, several racemic unsaturated amino acids have been resolved by hydroly

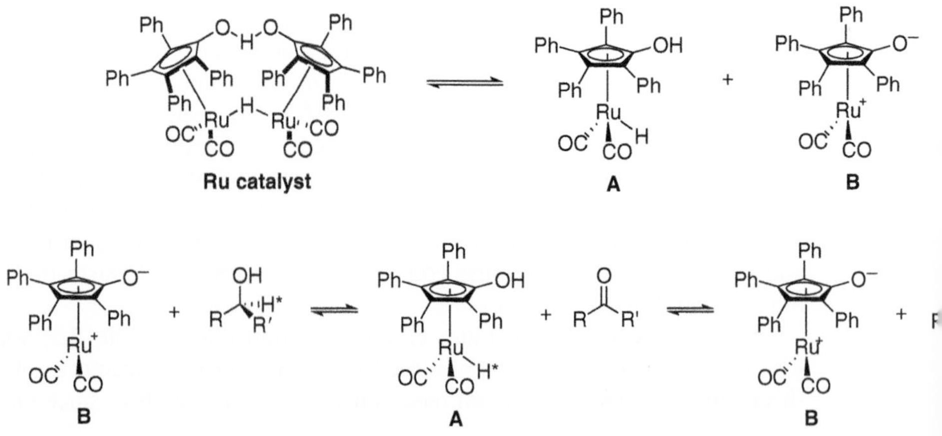

SCHEME 2.3 Racemization of alcohols catalyzed by an Ru catalyst.

the corresponding amino acid amide racemic mixture with an aminopeptidase from *Pseudomonas putida* ATCC 12633 and subsequent Schiff-base formation of the remaining amide enantiomer [36]. However, after acid/base-catalyzed racemization of the respective intermediates, the starting compound has to be liberated again, which overall turns this technique into a rather tedious multistep procedure.

The general disadvantage of most of these techniques based on chemocatalysis—i.e., the requirement of harsh reaction conditions—can be avoided by the employment of biocatalytic racemization methods, which proceed under mild reaction conditions, e.g., room temperature, atmospheric pressure, and physiological pH. Under these conditions, side reactions are largely suppressed [37]. Unfortunately, Nature does not rely on racemates to a large extent and, as a consequence, biochemical racemization is a rather scarce feature, making racemases a small group of enzymes that can only be found in certain biological niches. One of the major targets for biochemical racemization involves stereogenic centers in carbohydrates, i.e., *sec*-alcohol groups. However, since both stereoisomers of these reactions represent diastereomers rather than enantiomers, "epimerization" would be a more correct term. Various enzymes—epimerases—are involved in the racemization of *sec*-hydroxyl groups. These enzymes are usually NAD^+-dependent and of very little interest for practical applications.

Among noncarbohydrate "true" racemases, base catalysis seems to be the general scheme of biochemical racemization, and two major groups can be classified according to their reaction mechanism [38]: (i) racemases employing a one-base mechanism, i.e., the proton at the chiral center is abstracted by the same base functionality of the enzyme, which re-adds it—and (ii) those employing a two-base mechanism, i.e., one base is capable of *abstracting* the proton at the chiral center and another base *puts it back* from the opposite side, thus resembling a ping-pong mechanism.

Mandelate racemase [EC 5.1.2.2] belongs to the latter group employing a two-base mechanism [39]. The Mg^{2+}-dependent enzyme is capable of racemizing a remarkably broad substrate spectrum, which opens up large possibilities for the deracemization of various kinds of α-hydroxyacids. Substrates that meet the following constraints are accepted by mandelate racemase [40]:

1. The α-hydroxyacid moiety is (almost strictly) required, as the only exception to this rule seems to be an α-hydroxy carboxamide functionality [41].
2. A π-electron system has to be present in the β,γ-position. The latter can be freely varied, including the corresponding α-hydroxy-β,γ-alkenoic and α-hydroxy-α-aryl carboxylic acids. Even heteroaromatic systems are accepted at reasonable rates [42]. In general, electron-withdrawing groups attached to the β,γ-unsaturated (or aromatic) system that help to stabilize the (anionic) transition intermediate through resonance, increase the reaction rates significantly [43].

An example of mandelate racemase application, based on a two-enzyme system consisting of (i) a lipase-catalyzed enantioselective acylation followed by (ii) mandelate racemase-catalyzed racemization of the remaining nonreacted substrate enantiomer, is shown in Scheme 2.4 [44]. Thus, in the first step, (±)-mandelic acid is subjected to lipase-catalyzed *O*-acylation in an organic solvent producing (*S*)-*O*-acetyl mandelate, leaving the (*R*)-enantiomer behind. Due to the high selectivity ($E > 200$), the reaction comes to a standstill at 50% conversion. In the second step, the organic solvent is switched to aqueous buffer [45], and the remaining unreacted (*R*)-mandelic acid is racemized in the presence of (*S*)-*O*-acetyl mandelate, which is a nonsubstrate. When this two-step process is repeated four times, (*S*)-*O*-acetyl mandelate is obtained in ~80% chemical yield and >99% enantiomeric

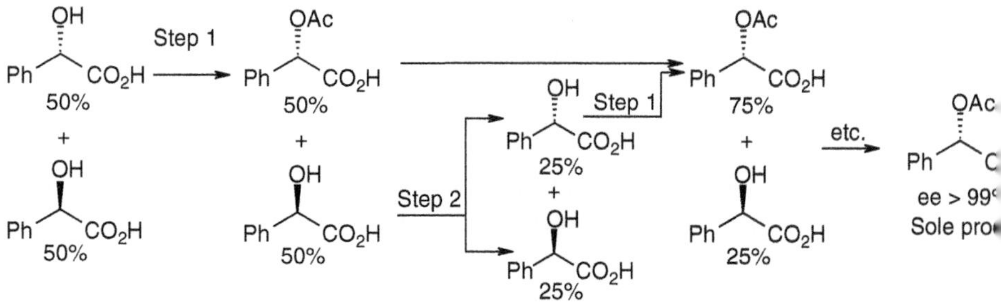

Step 1: *Pseudomonas* sp. lipase, *i*-Pr$_2$O, vinylacetate ($E > 200$)
Step 2: Mandelate racemase, aqueous buffer

SCHEME 2.4 Deracemization of mandelate by repeated resolution, employing a lipase–racema enzyme process.

excess as the sole product. It should be emphasized that separation of the formed p from the remaining starting material is not required due to the high specificity racemase employed.

A relevant group of racemases are amino acid racemases, which are widely u industry because of the high commercial importance of amino acids and their deriv Although the biosynthesis of α-amino acids is highly stereospecific with respect L-stereoisomers, an impressive number of D-analogs have been found in various bio sources, usually as components of highly potent natural products [46,47]. Instead of a a biosynthesis through "mirror-image" metabolic pathways, D-amino acid isomers are ally obtained by biocatalytic racemization and kinetic resolution; thus it is not surprisi racemases acting on α-amino acids and derivatives thereof are quite common.

Another remarkable group of isomerase enzymes are hydantoin racemases. Mono tuted hydantoins bearing an aromatic substituent in position 5, e.g., phenylhyd (Scheme 2.5, R = Ph), spontaneously racemize under weak alkaline conditions [has been discussed that the velocity of chemical racemization is influenced by the elec nature of the substituent, such as its electronegativity and the presence of ad π-systems, which facilitates enolate stabilization due to participation of mesomeric stru [49,50]. As a consequence, 5-alkyl- or 5-aryl-alkyl-substituted hydantoins are chirally and spontaneous racemization is too slow to allow dynamic kinetic resolution. For types of substrates, hydantoin racemases were found in *Arthrobacter* and *Pseudo* species, which enzymatically catalyze the interconversion of enantiomers. Hydantoir mases find their application in the so-called hydantoinase process for the industria production of either D- (or L-)amino acids, depending on the follow-up "hydant enzymes" used (the recommended name of these enzymes is "dihydropyrimid [EC 3.5.2.2]).

2.2.2 DYNAMIC RESOLUTION

The disadvantages of kinetic resolution can largely be avoided by employing a so dynamic resolution [51–53] (Scheme 2.6). Such a process comprises kinetic resolutio an additional feature—*in situ* racemization of the starting material—which is u achieved using chemocatalysis. Ultimately, both the substrate enantiomers S_R+ $ transformed into a single product enantiomer P_R in 100% theoretical yield. In cont kinetic resolution, where the reaction slows down at 50% conversion (or comes eve standstill, if the enantioselectivity is sufficiently high), when the fast-reacting enantiome

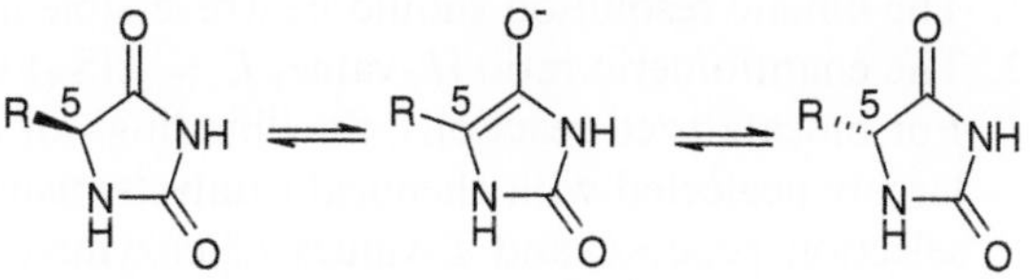

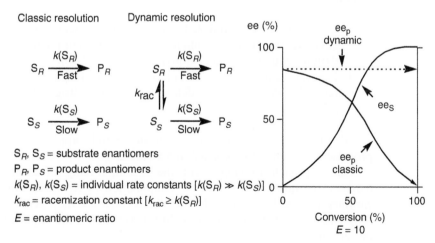

SCHEME 2.5 Chemical racemization of 5-substituted hydantoins via enolates.

consumed and only the slow-reacting counterpart S_S remains, substrate racemization ensures the continuous formation of S_R from S_S during the course of the reaction, and thus avoids the depletion of S_R. Therefore, the reaction does not come to a standstill and it can be run to completion by gradually converting all the racemic starting material into product P_R. In order to indicate the nonstatic behavior of such a process, the term "dynamic resolution" has been aptly coined. The major advantages of this methodology are twofold. The ee of the product is higher compared to classic resolutions and the separation of the substrate–product mixture can be omitted, due to the absence of an unwanted remaining substrate enantiomer. This concept has been applied to a variety of compounds such as α-amino acids, hemithioacetal esters, α-(hetero)arylcarboxylic acid esters, α-substituted nitriles, cyanohydrins, α-substituted thioesters, as well as 4-substituted oxazolin-5-ones and thiazolin-5-ones.

The following properties are typical for dynamic resolution processes [54,55]. From Scheme 2.6 it can be seen that (for kinetic resolutions) the ee of the product (ee_P) is at its maximum at the onset of the reaction and gradually begins to decline as the slower reacting enantiomer is accumulated in the reaction mixture, in particular around halfway through the reaction. This depletion does not occur if the substrate is constantly racemized during the resolution process and, thus, in a dynamic resolution the ee_P is *not* a function of the conversion but remains constant throughout the reaction. Since the catalyst always faces a racemic starting material ([S_R] always equals [S_S]), it is understandable that the selection of the faster-reacting enantiomer from the substrate remains a simple task, as opposed to kinetic resolution where depletion of S_R occurs.

In order to design a successful dynamic resolution process, both parallel reactions— kinetic resolution [$k(S_R) \gg k(S_S)$] and *in situ* racemization (k_{rac})—have to be carefully tuned, taking into account the following aspects:

Classic resolution   Dynamic resolution

$$S_R \xrightarrow[\text{Fast}]{k(S_R)} P_R \qquad S_R \xrightarrow[\text{Fast}]{k(S_R)} P_R$$

$$k_{rac} \updownarrow$$

$$S_S \xrightarrow[\text{Slow}]{k(S_S)} P_S \qquad S_S \xrightarrow[\text{Slow}]{k(S_S)} P_S$$

S_R, S_S = substrate enantiomers
P_R, P_S = product enantiomers
$k(S_R)$, $k(S_S)$ = individual rate constants [$k(S_R) \gg k(S_S)$]
k_{rac} = racemization constant [$k_{rac} \geq k(S_R)$]
E = enantiomeric ratio

ee (%)

ee_P dynamic

ee_S

ee_P classic

Conversion (%)
$E = 10$

SCHEME 2.6 Kinetic principles of kinetic and dynamic resolution.

1. The kinetic resolution should be irreversible in order to ensure high enantiosele
2. The enantiomeric ratio [E-value, $E = k(S_R)/k(S_S)$] should be at least greater th For biocatalyzed reactions, the "binding" of the substrate enantiomers (which largely neglected with chemical catalysts) usually plays an important role in the selection process, and E-values of enzyme-catalyzed reactions are therefore d through Michaelis–Menten kinetics: $E = (k_{cat}/K_m)_{S_R}/(k_{cat}/K_m)_{S_S}$.
3. To avoid depletion of S_R, racemization (k_{rac}) should be at least equal to, or g than, the reaction rate of the fast enantiomer $k(S_R)$.
4. In case the selectivities are only moderate, k_{rac} should be greater than $k(S_R)$ by a of about 10.
5. For obvious reasons, any spontaneous side reactions involving the substrate e omers as well as racemization of the product should be absent.
6. Dynamic resolution is generally limited to compounds possessing one stereoc However, under certain circumstances, multicenter compounds can be proces long as the stereocenters are (stereo)chemically very similar. In such a case, the re proceeds through several diastereomeric intermediates.

A common scenario for dynamic resolution processes based on enantioselective bi lysis makes use of a combination of an enzyme-catalyzed kinetic resolution coupled to racemization of the remaining substrate enantiomer through chemocatalysis. Four pr situations can be encountered:

1. Compounds with a chirality center bearing an acidic proton—adjacent to an acti carbonyl group, such as an ester or ketone—usually undergo facile racemi through the formation of an achiral enolate species through base-catalyzed abstraction [56,57].
2. When such racemization is impossible, e.g., in case of a secondary alcohol, race tion can be achieved by a (reversible) decomposition reaction, such as the cleav hemi(thio)acetals and cyanohydrins [58,59].

If chemocatalysis relies on acid or base, its combination with the biocatalyst may b difficult due to the incompatibility of enzymes with strong basic or acidic media. C quently, biocompatible *in situ* racemization techniques are of high value. Typical exa include:

3. Transition metal–catalyzed racemization of *sec*-alcohols, based on Pd^{II}-cat. allylic rearrangement of allylic acetate esters [60] (Scheme 2.7a) or Ru-catalyze dation–reduction sequences [61] (see Scheme 2.7b).
4. Benzylic amines can be racemized under biocompatible conditions us reversible Pd/C-catalyzed dehydrogenation–hydrogenation reaction (see Scheme [62].

All of these techniques have been successfully applied to dynamic resolution. prominent examples are depicted in Scheme 2.7.

One approach to circumvent the incompatibility of chemical and biocatalys dynamic resolution consists in the combination of two biocatalysts, bearing in min enzymes are easily compatible with each other as they generally work under the (physiological) reaction conditions. In this viewpoint, the application of racemases se be very promising.

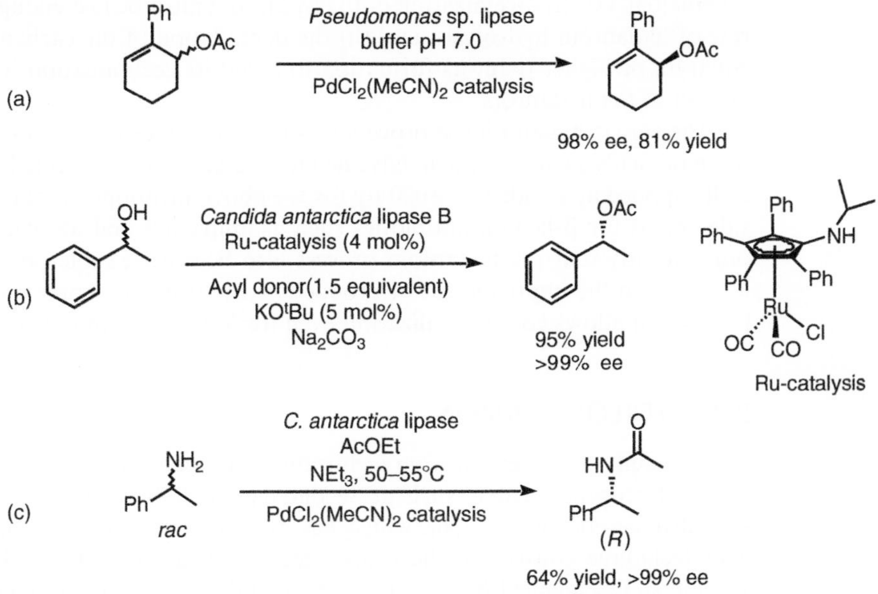

SCHEME 2.7 (a) Dynamic resolution of an allylic *sec*-alcohol through combined Pd^{II}- and lipase catalysis. (b) Dynamic resolution of *sec*-alcohols through combined Ru- and lipase catalysis. (c) Dynamic resolution of a benzylic amine through Pd/C- and lipase catalysis.

One example of this strategy is the dynamic kinetic resolution of racemic hydantoins to either the D- (or L-)amino acid. This process consists of three steps (Scheme 2.8): (i) ring-opening hydrolysis of the hydantoin is performed by a D- (or L-)specific hydantoinase; (ii) carbamoylase-catalyzed hydrolysis of the resulting *N*-carbamoylamino acid; which shifts the equilibrium towards (iii) completion. For complete transformation of the *rac*-starting material in a dynamic kinetic resolution, substrate racemization is performed by chemical and/or enzymatic catalysis using a hydantoin racemase. As three enzymes are involved in this process, the relative activity (or expression level) of the enzymes must match the kinetic

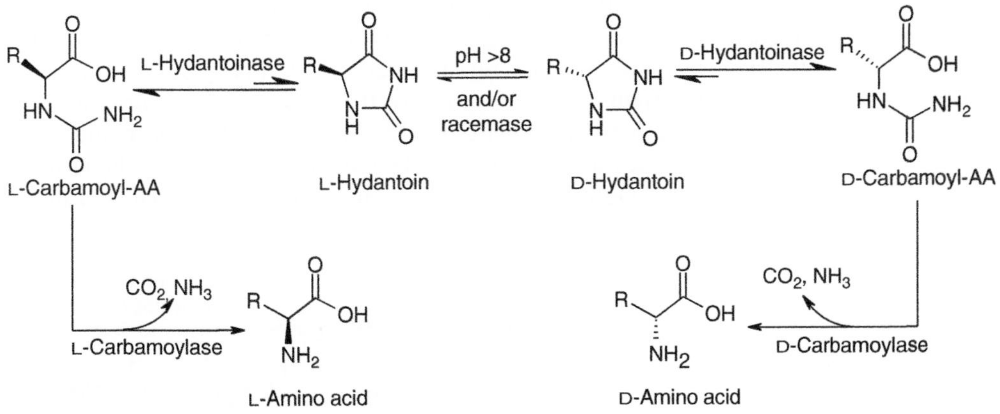

SCHEME 2.8 Hydantoinase process for production of D- (or L-)amino acids by dynamic kinetic resolution.

criteria [63], i.e., (i) racemization of the hydantoin must be fast enough to meet (or excee rate of hydantoin hydrolysis; and (ii) the degradation of the carbamoyl amino acid m equal to, or higher than, its formation to avoid its accumulation, which would cause mation of the hydantoin.

The classic hydantoinase process was first introduced in the 1970s for the product D-amino acids such as D-phenylglycine and D-p-OH-phenylglycine [64]. Today, it is co cially applied at a scale of >1000 t/y for the above-mentioned amino acids, which are u side chains for β-lactam antibiotics such as ampicillin and amoxicillin, respectively. only recently that the hydantoin process also became feasible on a large scale for L- acids due to the improvement of productivity by using a recombinant whole-cell bioc [65], which allows commercialization even for low-priced amino acids such as L-methi

2.3 STEREOINVERSION

The difficulty to achieve *in situ* racemization with compounds possessing a configurati stable stereogenic center, such as secondary alcohols, may be overcome by emplo so-called stereoinversion [66], which has only been achieved through biocatalytic me This technique consists in the transformation of a racemate S_R/S_S involving the en specific stereochemical inversion of one enantiomer (S_S) by a chemically stable achir prochiral) intermediate I (Scheme 2.1).

For instance, stereoinversion of *sec*-alcohols was achieved by employing whol [67–76] (Scheme 2.9). Although no definite proof exists, it was proposed by various g that microbial stereoinversion occurs by an oxidation–reduction sequence. Thus, one tiomer of a racemic mixture is selectively oxidized to the corresponding ketone catalysis of a dehydrogenase, while the mirror-image counterpart remains unaffected. the ketone is reduced again in a subsequent step by a different enzyme displaying op stereochemical preference. Overall, this two-step oxidation–reduction sequence cons formally a deracemization process. Due to the involvement of a consecutive oxid reduction reaction, the net redox balance of the process is zero and (in an ideal ca external cofactor recycling is necessary since the redox equivalents, such as NAD(P)H be recycled internally between both steps, e.g., by using whole-cell systems.

The success of a biocatalytic stereoinversion by a redox process is determined following crucial point: for the entropy balance of the process, which is required to ach high optical purity of the product, at least one of the two redox reactions has irreversible [77,78]. The origin of this irreversibility is currently under investigatio the data available to date reveal a rather puzzling picture. For instance, deracemi of various terminal (±)-1,2-diols by the yeast *Candida parapsilosis* has been clain operate through an (R)-specific NAD$^+$-linked dehydrogenase and an (S)-specific NA dependent reductase. Although no detailed data were given, the latter step was propo be irreversible. Observations on the fungus *Geotrichum candidum* prove the requirem

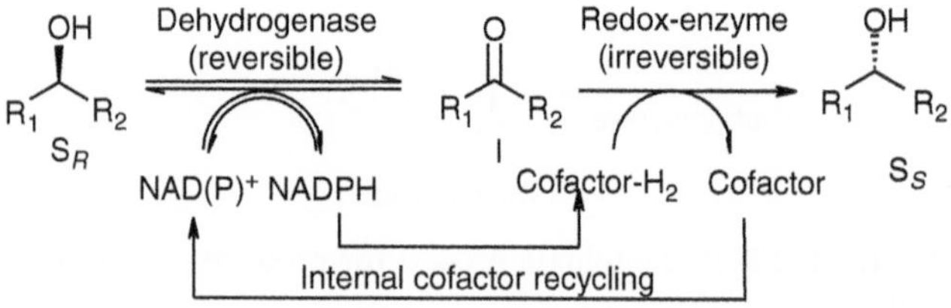

SCHEME 2.9 Deracemization of *sec*-alcohols based on a biocatalytic oxidation–reduction seque

molecular oxygen, which would suggest the involvement of an alcohol oxidase rather than an alcohol dehydrogenase.

Table 2.1 gives an overview of secondary alcohols that have been successfully deracemized by microbial stereoinversion employing various organisms. It is noteworthy that biocatalytic deracemization by stereoinversion of secondary alcohols was only accomplished by using whole cells.

Molecules possessing two chiral alcohol moieties were deracemized by one-step stereo-inversion employing *Corynesporium cassiicola* and *Candida boidinii*: *rac*-1,2-indanediol and *rac*-1,2-cyclohexanediol were deracemized employing *C. cassiicola* to give the (*S,S*)-diol in >99% ee at up to 83% isolated yield [72,79]; employing *C. boidinii*, (*R,R*)-2,4-pentanediol was obtained from the racemate [73].

Steroinversion was also observed for more complex molecules: one hydroxyl group of the two alcohol moieties of debenzylbutanolides, which possesses four chiral centers, was inverted by plant cells of *Catharanthus roseus* (periwinkle) to afford a product showing >99% diastereomeric excess [80].

Another example of a biocatalytic stereoinversion is the deracemization of α-chiral aryl- or aryloxy-carboxylic acids by isomerases [81]. Employing this methodology, deracemization of 2-aryl- and 2-aryloxy-propanoic acids [82,83] was performed using fermenting or resting cells of *Nocardia diaphanozonaria* JCM3208: 2-aryl- and *para*-substituted 2-aryloxy-propanoic acids were obtained in high yields with excellent ee from the corresponding racemates (Scheme 2.10) [84].

Mechanistically, this stereoinversion process was proposed to occur in the following sequence: (i) formation of an "activated" acyl-CoA-derivative of the (*S*)-acid; (ii) epimerization of the latter to yield the (*R*)-isomer; and finally (iii) hydrolysis of the (*R*)-acyl-CoA-ester. To eliminate the competing β-oxidation of the substrate, inhibitors of acyl-CoA-dehydrogenase, the first enzyme in the β-oxidation pathway, can be added [85].

Moreover, α-chiral carboxylic acids can also be stereoinverted employing an oxidase, namely glycolate oxidase, which oxidizes the (*S*)-enantiomer of the α-hydroxy acid to the keto-acid, which in turn is reduced in a second step by lactate dehydrogenase to the (*R*)-enantiomer, allowing to obtain enantiopure (*R*)-α-hydroxy acids (Scheme 2.11) [86,87]. A similar strategy with crude enzyme preparation as well as whole cells was employed for the conversion of racemic mandelic acid into its (*R*)-enantiomer [88,89].

An analogous combination of an oxidase (in this case a D-amino acid oxidase) and an amino acid dehydrogenase (leucine dehydrogenase) has been used to produce L-methionine from the racemate by enzymatic stereoinversion [90].

2.4 ENANTIOCONVERGENT PROCESSES

Deracemization may be achieved through so-called enantioconvergent processes, in such a way that each of the enantiomers is converted into the same product enantiomer P_R through two independent pathways (Scheme 2.1). Thus, whereas enantiomer S_R is reacted to product P_R through *retention* of configuration, its counterpart S_S is transformed with *inversion* of configuration. In general, both reactions are conducted in a stepwise fashion. The crucial prerequisites for the proper functioning of such systems are:

1. The first step must combine excellent enantiospecificity *and* stereospecificity with regard to retention or inversion. In other words, the requirements of the chiral recognition *and* stereochemistry of the respective transformation are high, and it is not surprising that these specificities are usually only achieved by enzymes.

TABLE 2.1
Deracemization of *rac-sec*-Alcohols Bearing a Single Stereocenter by Microbial Stereoinversion

Product	Organism	ee max (%)	Conversion /yield (%)	Re
Ph⌒OH (S) CO₂Et	*Candida parapsilosis*	>99	85–90% yield	
(lactone, OH, H, D)	*Rhodococcus erythropolis*	94	70% yield	
(R) OH CO₂Et	*Geotrichum candidum*	96	75% yield	
R″–C₆H₄ OH O O OR″ R′ = Me, Et; R″ = H, p-MeO, p-NO₂, p-Me, o-Me	*Candida parapsilosis*	99	62–75% yield	
OH R₁ R₂ (R) R₁ = Ar, Het-Ar, Ar–CH₂–; R₂ = Me, Et, c-hexyl, vinyl	*Sphingomonas paucimobilis*	99	up to 90% yield	
Ph OH (R) HN O Ph	*Cunninghamella echinulata*	92	57% yield	
F OH (R)	*Aspergillus terreus* CCT 3320	>99	59% yield	

TABLE 2.1 (continued)
Deracemization of *rac-sec*-Alcohols Bearing a Single Stereocenter by Microbial Stereoinversion

Product	Organism	ee max (%)	Conversion/yield (%)	Reference
OH, (S), R = NO$_2$, Br	*Aspergillus terreus* CCT 4083	>99	Up to 86% conversion	139
OH, (R), X = H, Cl, OMe	*Geotrichum candidum*	Up to 99	55–100% conversion	68, 140
OH, (R) Ph, N, *m,p*	*Catharanthus roseus*	Up to >99	92–100% yield	70, 141
O, Ph, Ph, OH, (S) for pH = 4–5, (R) for pH 7.5–8	*Rhizopus oryzae*	95–97 (R)	73–76% yield for (R)	142
		85 (S)	71% yield for (S)	
OH, OH, (S)	*Candida boidinii*	Up to 100	Up to 72% conversion	143
	Pichia methanolica	Up to 100	Up to 88% conversion	
	Hansenula polymorpha	Up to 60	Up to 100% conversion	
OH, HO, (S) ()$_n$	*C. parapsilosis*	100, n = 2,3	Up to 100% yield	77
		79, n = 1		
OH, (R) or (S)	(R)-alcohol: *Nocardia fusca, N. globerula, N. erythropolis*;	100 (R)	83% conversion for (R)	76, 144
	(S): *Nocardia pseudosporangifera*	98 (S)	70% conversion for (S)	

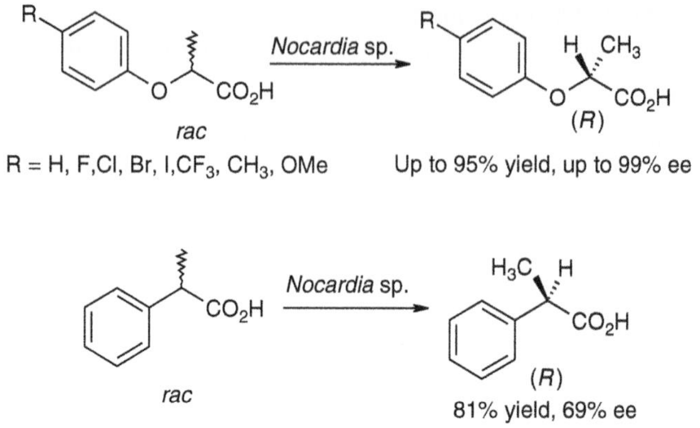

SCHEME 2.10 Deracemization of α-phenyl- and p-substituted α-aryloxypropanoic acids by enz
Nocardia diaphanozonaria JCM3208.

2. Since the starting material for the second step (i.e., S$_S$) is enantiomerically enric
   even pure), only high stereospecificity is required with respect to inversion or re
   of configuration. Hence, this step may also be performed by using a chemocata
3. An important factor for the economy of the whole process is the compatibility
   reaction conditions of both steps. If, for instance, the conditions are incom
   separation of product P$_R$ (formed during step 1 from S$_R$) from the remaining e
   meric starting material S$_S$ is required, which usually accompanies loss of material
   all, it appears anachronistic to separate materials from each other that are
   combined at the end of the process. Thus, successful enantioconvergent pr
   should always be performed in a one-pot fashion.

2.4.1 ENANTIOCONVERGENCE USING TWO BIOCATALYSTS

The only enzymes that may transform non-natural compounds with concomitant *in*
of configuration during catalysis are: (i) glycosidases [91]; (ii) dehalogenases [92]; (iii)
tases [93]; and (iv) epoxide hydrolases [94]. Whereas glycosidases cannot be employ

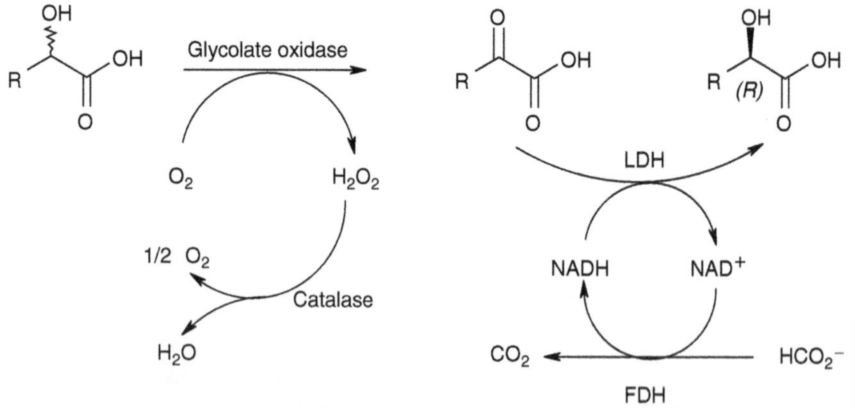

SCHEME 2.11 Glycolate oxidase/LDH-catalyzed stepwise deracemization of *rac*-2-hydroxy
LDH, lactate dehydrogenase; FDH, formate dehydrogenase.

deracemization, because their substrates are diastereomers rather than enantiomers, dehalogenases are not widely distributed in Nature and they exhibit a limited substrate tolerance. Sulfatases are under investigation. The application of aryl-sulfatases in preparative biotransformations is very limited [95], but alkyl sulfatases seem to be quite flexible in catalyzing the enantioselective hydrolysis of sulfate esters through inversion of configuration [96].

Epoxide hydrolases from microbial sources, such as bacteria, fungi, and (red) yeasts, have been shown to possess a great potential for the stereoselective hydrolysis of epoxides to furnish the corresponding vicinal diols [97–99]. In contrast to ester hydrolysis catalyzed by lipases, esterases, or proteases, where the absolute configuration at the stereogenic center(s) always remains the same throughout the reaction, enzymatic hydrolysis of epoxides may take place through an attack on either carbon atom of the oxirane ring, and it is the structure of the substrate and of the enzyme that determines the regioselectivity of the attack [100]. This is exemplified in Scheme 2.12. If the (*S*)-enantiomer is preferentially hydrolysed from the racemate with *retention* of configuration (pathway a), kinetic resolution furnishes a mixture of (*S*)-diol and unreacted (*R*)-epoxide. On the contrary, the corresponding (*R*)-diol is produced from the (*S*)-oxirane, if the enzyme acts with *inversion* of configuration (pathway b). Therefore, enantioconvergent hydrolysis of epoxides should be feasible when appropriate enzymes are available.

An elegant deracemization of (±)-styrene oxide was developed by making use of two epoxide hydrolase activities from fungal sources [101]. Whereas *Aspergillus niger* preferentially hydrolyzed the (*R*)-enantiomer with retention of configuration by producing (*R*)-phenylethan-1,2-diol (Scheme 2.12, epoxide hydrolase 1), *Beauveria bassiana* (formally denoted as *B. sulfurescens*) showed opposite enantiopreference, i.e., (*S*), with matching

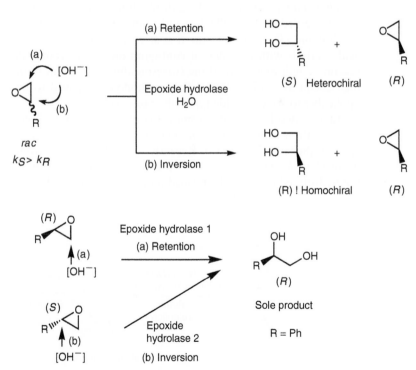

SCHEME 2.12 Enzymatic hydrolysis of epoxides proceeding with retention or inversion of configuration.

opposite regioselectivity, causing inversion of configuration (epoxide hydrolase 2 combination of both bicoatalysts in a single reactor led to almost complete deracemiz

The same methodology has been employed to prepare an enantiopure building bl Eliprodil, a promising neuroprotective agent, using high substrate concentrations in a sic reactor and low temperatures, to minimize significant spontaneous hydrolysis [102

2.4.2 ENANTIOCONVERGENCE USING COMBINED BIO- AND CHEMOCATALYSIS

In order to transform both enantiomers obtained from lipase-catalyzed kinetic resolu *sec*-alcohols by acyl-transfer or ester hydrolysis, the following one-pot two-step techniq employed [103–107]. Without separation, the mixture of *sec*-alcohol and the correspo pseudo-enantiomeric carboxylic acid ester was subjected to chemical inversion of the a by treatment with mesyl chloride or (for large-scale reactions) with fuming nitric acid carefully controlled reaction conditions, yielding a mixture of enantiomeric activate nonactivated esters. For small-scale reactions, Mitsunobu conditions may be li employed [108]. Both compounds were hydrolyzed by strong base with concurrent *in* and *retention* of configuration for the activated and nonactivated ester, respectively consequence, a single enantiomeric *sec*-alcohol was formed as the sole product.

Bacterial epoxide hydrolases have been shown to be the biocatalysts of choice f enantioselective hydrolysis of 2,2-disubstituted oxiranes, showing virtually absolute er selectivities ($E > 200$) [109]. In this case, the reaction proved to proceed invariably th *retention* of configuration. As the existence of enzymes that attack a quaternary carbor with *inversion* of configuration is rather unlikely, deracemization by a two-enzyme sys described above is impossible (see Scheme 2.12). However, the combination of bi chemocatalysis proved to be very efficient (Scheme 2.13) [110].

Thus, kinetic resolution of 2,2-disubstituted epoxides using a *Nocardia* sp. epoxide I lase proceeded with excellent enantio- and regioselectivity by furnishing the correspo (*S*)-diol and (*R*)-epoxide in the first step. Then, the remaining epoxide was transform acid catalysis with *inversion* of configuration in the second step under carefully con reaction conditions to yield the corresponding (*S*)-diol in virtually enantiopure form high chemical yields (>90%). This methodology proved to be highly flexible and wa applicable to styrene-oxide type substrates [111].

In a related fashion, deracemization of *rac*-1-methyl-1,2-epoxycyclohexane was ac by incubation with *Corynebacterium* sp. C12 and subsequent acid catalysis of the rem epoxide (Scheme 2.14) [112]. Recently, a similar chemo-enzymatic procedure was re using *Methylobacterium* sp. as biocatalyst [113]. Instead of the chemical transformation nonhydrolysed oxirane, the formed diol can be converted into the remaining epoxide

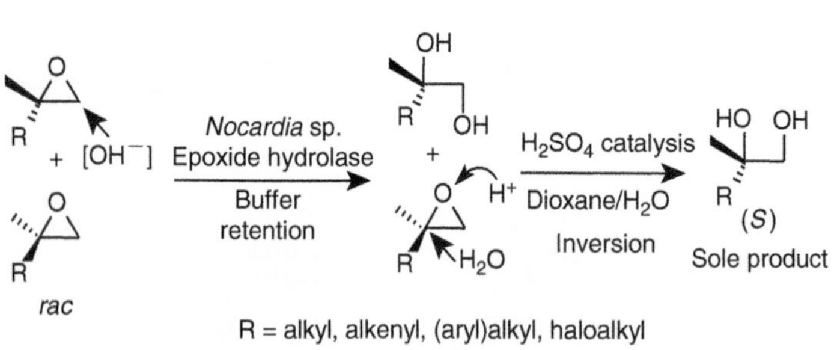

R = alkyl, alkenyl, (aryl)alkyl, haloalkyl

SCHEME 2.13 Enantioconvergent hydrolysis of 2,2-disubstituted epoxides by combination of b chemocatalysis.

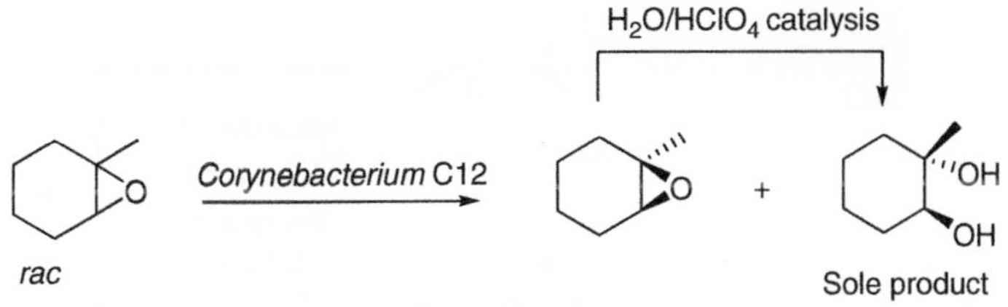

SCHEME 2.14 Deracemization of 1-methyl-1,2-epoxycyclohexane.

Enantioconvergent processes using a single biocatalyst can also be accomplished with certain sulfatases, since the stereochemical course of sulfate ester hydrolysis (i.e., *retention vs. inversion*) can be controlled by the choice of the appropriate subtype of this enzyme class. Thus, whereas aryl-sulfatases generally act through *retention* of configuration, some alkyl-sulfatases lead to *inversion*. Consequently, when a racemic sulfate ester is used as the substrate of these alkyl-sulfatases, both the formed product *sec*-alcohol and the remaining noncon-verted sulfate ester possess the same absolute configuration, and hence constitute a homo-chiral product mixture (Scheme 2.15) [115]. Removal of the sulfate ester group from the remaining nonhydrolyzed (S)-sulfate ester with retention of configuration was achieved under acid catalysis [116] and led to the formation of the corresponding (S)-*sec*-alcohol as the sole product.

Although the *stereo*selectivitites of a *sec*-alkyl sulfatase RS2 from *Rhodococcus* sp. were absolute with respect to inversion, *enantio*selectivities ranged from low to moderate (*E*-values up to 21). The latter values could be improved by using low concentrations of Fe^{3+} acting as enantioselective inhibitor [117]. Highly enantioselective sulfatases were recently identified in hyperthermophilic *Archeaea*, such as *Sulfolobus* sp. [118].

2.4.3 ENANTIOCONVERGENCE USING A SINGLE BIOCATALYST

Processes depending on more than one catalyst are generally sensitive with respect to the kinetic tuning of both reactions and, therefore, enantioconvergent reactions that depend on a *single* catalyst would be more reliable in practice. However, the requirements for this single catalyst are extremely high, i.e., it has to exhibit not only high *enantioselectivity*, but also show at the same time *opposite regioselectivity* for the transformation of each enantiomer in order

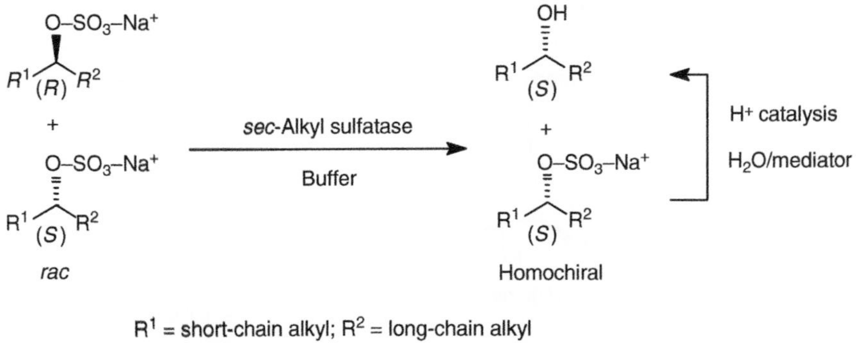

R^1 = short-chain alkyl; R^2 = long-chain alkyl

SCHEME 2.15 Chemo-enzymatic deracemization of *sec*-alcohols by their corresponding sulfate esters using inverting *sec*-alkyl sulfatases.

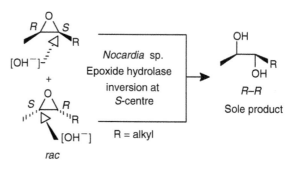

SCHEME 2.16 Enantioconvergent hydrolysis of 2,3-disubstituted epoxides using a single biocata

to make the overall process enantioconvergent. Therefore, such processes catalyzed
single (bio)catalyst are very rare (Scheme 2.16) [119,120]. For instance, an epoxide hyd
from *Nocardia* sp. hydrolyzed both enantiomers of *cis*-2,3-disubstituted epoxides
opposite *regio*selectivity by attack at their respective (S)-oxirane carbon atom with
inversion of configuration, yielding the corresponding (R,R)-diol as the sole product in
92% ee and 85% chemical yield [121].

Detailed analysis of this system using a range of 2,3-disubstituted *rac*-oxiranes re
that the relative *cis*- or *trans*-configuration was of critical importance for the enantioco
gent microbial biohydrolysis (Scheme 2.17) [122]. Due to the fact that each enantiome
cis-configurated *rac*-epoxide possesses a single (S)-configurated carbon atom (match)
enantiomers are hydrolyzed at comparable rates and with opposite regioselectivity, lead
the same stereoisomeric product. In contrast, in the *trans*-series, one enantiomer (poss
two "matches") is hydrolyzed very fast with low regioselectivity, whereas the mirror-

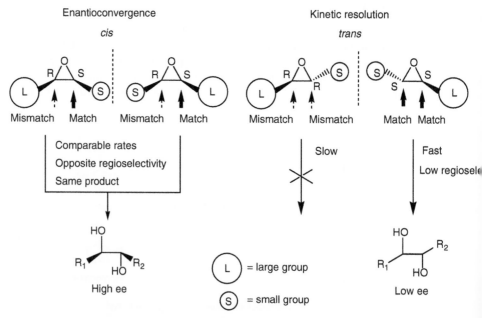

SCHEME 2.17 Relative stereochemistry of 2,3-disubstituted oxiranes determines enantioconverge
kinetic resolution.

counterpart remains unaffected as the result of two "mismatches." Consequently, kinetic resolution leads to the formation of the corresponding diol in low ee

2.5   DERACEMIZATION BY A CYCLIC OXIDATION–REDUCTION SEQUENCE

Deracemization of compounds bearing a chiral *sec*-hydroxyl or amino group can be achieved by a novel process based on a cyclic oxidation–reduction sequence [123,124]. The system consists of two independent reactions outlined in Scheme 2.18.

First, one enantiomer of the secondary alcohol or amine (S_R) is selectively oxidized from the starting racemate ($S_R + S_S$) to yield the achiral intermediate I, i.e., the corresponding ketone or imine, respectively. Then, the product is chemically reduced in a nonselective fashion to yield again a racemic mixture. Both reactions alone are of limited use for the preparation of enantiopure material, since step 1 (i.e., a kinetic resolution by enantioselective oxidation) is limited to a 50% theoretical yield of chiral nonreacting substrate enantiomer S_S and achiral intermediate I, and step 2 does not show any chiral induction at all. However, combination of both steps in a cyclic mode leads to a highly versatile deracemization technique.

The functioning of this system is explained by the following example. If the selectivity of step 1 is assumed to be absolute, only S_R is selectively oxidized to form achiral I in 50% yield by leaving S_S untouched. In the second step, intermediate I is nonselectively reduced to furnish $S_R + S_S$ in equal amounts of 25% each. As a consequence, the enantiomeric composition of S_R/S_S after a single cycle now equals 25/75. The diagram shown in Scheme 2.18 reveals that further cycles lead to a gradual increase of enantiomer S_S at the expense of S_R, and that the enantiomeric excess of the substrate is already well above 90% after only four cycles, assuming absolute enantioselectivity. Overall, if the cyclic process is driven in the forward direction, enantiomer S_S represents the "sink" of material in the whole system.

For practical applications, however, enantioselectivities in step 1 often range below E-values of 100. For these cases, the enantioselectivity determines two crucial factors of the

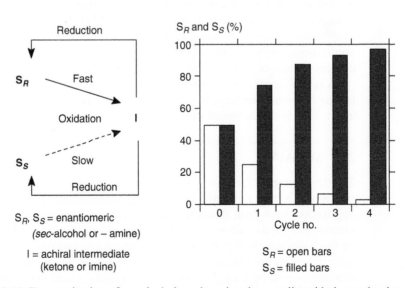

SCHEME 2.18 Deracemization of *sec*-alcohols and -amines by a cyclic oxidation–reduction sequence.

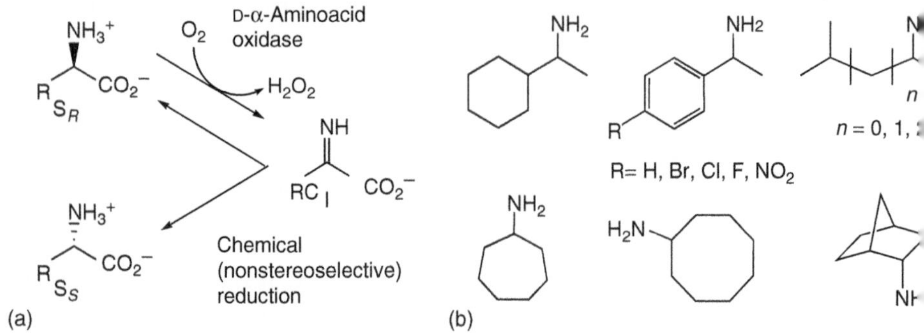

SCHEME 2.19 (a) Deracemization of *rac*-α-amino acids using enantioselective bio-oxidation cou nonselective chemical reduction of the imino acid intermediate I. (b) Selection of chiral amines th be oxidized by an Asn336Ser variant of the amine oxidase from *Aspergillus niger*.

system: (i) the maximum obtainable ee at equilibrium; and (ii) the number of cycles th required to reach this value. The merits and limits of cyclic deracemization systems hav described based on the underlying kinetics [123,124].

This procedure has been employed for the deracemization of α-amino acids by (i) co an enantioselective oxidation (catalyzed by a D-specific amino acid oxidase, producin corresponding nonchiral imino acid as a short-lived intermediate) to a (ii) chemica selective reduction step, e.g., with sodium borohydride or amine borane (Scheme 2.19) 128]. Overall, this process led to the formation of L-amino acids from the racemate [12 and has been used for the resolution of racemic pipecolic acid, an L-proline homolog [

In order to extend the applicability of this elegant process to chiral amines (which a oxidized by amino acid oxidases), an amine oxidase from *A. niger* (predominantly acti α-methylbenzyl amine) was identified as a suitable biocatalyst. Directed evolution o enzyme resulted in an amine oxidase possessing not only a wide substrate spectrum bu good enantioselectivity [130–132]. The Asn336Ser variant of the amine oxidase sh highest activity toward substrates bearing a methyl substituent and a bulky alkyl/aryl adjacent to the amino-carbon atom (Scheme 2.18). In all cases examined so far, the en variant was enantioselective for the (*S*)-isomer of the *rac*-amine substrate. For large applications, improvements of enzyme stability, activity, and selectivity are still require deracemization of chiral amines should now be feasible by using this enzyme.

A related approach is based on the electro- and biochemical oxidation and/or reduct lactate involving the recycling of NADH [133]. In this way, complete inversion of L-l into D-lactate was achieved in a model reactor (Scheme 2.20).

2.6 SUMMARY

The development of methods for the preparation of chiral compounds in 100% chemic optical yields from racemates is one of the current challenges in asymmetric synthesis. S approaches have been described so far, which are either based on modifications of kinetic resolution, such as re-racemization and repeated resolution or dynamic resoluti on the transformation of enantiomers through enantioconvergent pathways, which is u achieved by combination of chemo- and/or biocatalysts in sequential reactions or— elegantly—even by a single (bio)catalyst. Finally, cyclic deracemization and stereoinv through an oxidation–reduction sequence are feasible options. It has to be empha

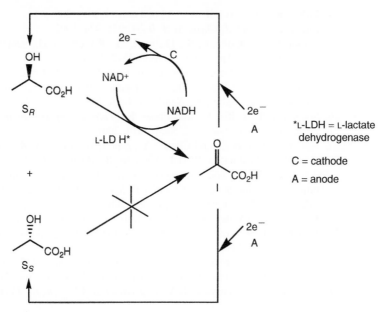

SCHEME 2.20 Deracemization of α-hydroxyacid through a bio- and electrochemical cyclic oxidation–reduction sequence.

however, that each of the above-described approaches offers a solution for only certain types of stereochemical problems and the corresponding substrate classes, but none of the methods can be employed as a general solution.

ACKNOWLEDGMENT

Financial support by Fundación Ramón Areces is gratefully acknowledged.

REFERENCES

1. Stinson, S.C., *Chem. Eng. News.*, Sept 28, 46–78, 1992.
2. Sheldon, R.A., *Chirotechnology*, Marcel Dekker, New York, 1993.
3. Collins, A.N., Sheldrake, G.N., and Crosby, J., Eds., *Chirality in Industry*, vols. I and II, Wiley, Chichester, 1992 and 1997.
4. Faber, K., *Biotransformations in Organic Chemistry*, 5th ed., Springer, Heidelberg, 2004.
5. Drauz, K. and Waldmann, H., *Enzyme Catalysis in Organic Synthesis*, 2nd ed., vols. I and II, VCH, Weinheim, Germany, 2002.
6. Liese, A., Seelbach, K., and Wandrey, C., *Industrial Biotransformations*, VCH, Weinheim, Germany, 2000.
7. Bornscheuer, U.T. and Kazlauskas, R.J., *Hydrolases in Organic Synthesis*, VCH, Weinheim, Germany, 1999.
8. Schoffers, E., Golebiowski, A., and Johnson, C.R., *Tetrahedron*, 52, 3769–3826, 1996.
9. Kagan, H.B. and Fiaud, J.C., *Topics Stereochem.*, 18, 249–330, 1988.
10. Horeau, A., *Tetrahedron*, 31, 1307–1309, 1975.
11. Sih, C.J. and Wu, S.H., *Topics Stereochem.*, 19, 63–125, 1989 (for biocatalyzed reactions).
12. Martin, V.S., Woodard, S.S., Katsuki, T., Yamada, Y., Ikeda, M., and Sharpless, K.B., *J. Am. Chem. Soc.*, 103, 6237–6240, 1981 (for nonbiocatalyzed reactions).
13. Balavoine, G., Moradpour, A., and Kagan, H.B., *J. Am. Chem. Soc.*, 96, 5152–5158, 1974 (for prebiotic reactions).

14. Faber, K., and Kroutil, W., ~13.000 entries, May 2006.
15. Faber, K., *Indian J. Chem. Sect. B*, 31, 921–924, 1992.
16. Cotterill, I.C., Jaouhari, R., Dorman, G., Roberts, S.M., Scheinmann, F., and Wakefiel *J. Chem. Soc. Perkin Trans.,* 1, 2505–2512, 1991.
17. Chen, C.S., Fujimoto, Y., Girdaukas, G., Sih, C.J., *J. Am. Chem. Soc.*, 104, 7294–7299, 1
18. Azerad, A.R. and Buison, D., *Curr. Opinion. Biotechnol.*, 11, 565–571, 2000.
19. Stecher, H. and Faber, K., *Synthesis*, 1–16, 1997.
20. Strauss, U.T., Felfer, U., and Faber, K., *Tetrahedron Asymmetry*, 10, 107–117, 1999.
21. Faber, K., *Chem. Eur. J.*, 7, 5004–5010, 2001.
22. Xie, Y.C., Liu, H.Z., and Chen, J.Y., *Biotechnol. Lett.*, 20, 455–458, 1998.
23. Kamphuis, J., Boesten, W.H.J., Kaptein, B., Hermes, H.F.M., Sonke, T., Broxterman, Q. den Tweel, W.J.J., and Schoemaker, H.E., The production and uses of optically pure natu unnatural amino acids, in *Chirality in Industry*, Collins, A.N., Sheldrake, G.N., and Cro Eds., Wiley, New York, 1992, pp. 187–208.
24. Eliel, E.L. and Wilen, S.H., *Stereochemistry of Organic Compounds*, Wiley, New York, 19
25. Ebbers, E.J., Ariaans, G.J.A., Houbiers, J.P.M., Bruggink, A., and Zwanenburg, B., *Tetra* 53, 9417–9476, 1997.
26. Um, P.J. and Drueckhammer, D.G., *J. Am. Chem. Soc.*, 120, 5605–5610, 1998.
27. Huerta, F.F., Minidis, A.B.E., and Bäckvall, J.E., *Chem. Soc. Rev.*, 30, 321–331, 2001.
28. Larsson, A.L.E., Persson, B.A., and Bäckvall, J.E., *Angew. Chem. Int. Ed. Engl.*, 36, 121 1997.
29. Allen, J.V. and Williams, J.M.J., *Tetrahedron Lett.*, 37, 1859–1862, 1996.
30. Palmer, M.J. and Wills, M., *Tetrahedron Asymmetry*, 10, 2045–2061, 1999.
31. Pàmies, O. and Bäckvall, J.E., *Chem. Eur. J.*, 7, 5052–5058, 2001.
32. Samec, J.S.M. and Bäckvall, J.E., *Chem. Eur. J.*, 8, 2955–2961, 2002.
33. Éll, A.H., Samec, J.S.M., Brasse, C., and Bäckvall, J.E., *Chem. Commun.*, 1144–1145, 200
34. Granberg, K.L. and Bäckvall, J.E., *J. Am. Chem. Soc.*, 114, 6858–6863, 1992.
35. Broxterman, Q.B., Boesten, W.H.J., Kamphuis, J., Kloosterman, M., Meijer, E.M Schoemaker, H.E., Chemo-enzymatic production methods for optically pure intermediates tential commercial interest, in *Opportunities in Biotransformations*, Copping, L.G., Martir Pickett, J.A., Bucke, C., and Bunch, A.W., Eds., Elsevier, London, 1990, pp. 148–169.
36. Wolf, L.B., Sonke, T., Tjen, K.C.M.F., Kaptein, B., Broxterman, Q.B., Schoemaker, H. Rutjes F.P.J.T., *Adv. Synth. Catal.*, 343, 662–674, 2001.
37. Adams, E., *Adv. Enzymol. Relat. Areas. Mol. Biol.*, 44, 69–138, 1976.
38. Gallo, K.A., Tanner, M.E., and Knowles, J.R., *Biochemistry*, 32, 3991–3997, 1993.
39. Kenyon, G.L., Gerlt, J.A., and Kozarich, J.W., *Acc. Chem. Res.*, 28, 178–186, 1995.
40. Felfer, U., Goriup, M., Koegl, M.F., Wagner, U., Larissegger-Schnell, B., Faber, K., and I W., *Adv. Synth. Catal.*, 347, 951–961, 2005.
41. Goriup, M., Strauss, U.T., and Faber, K., *J. Mol. Catal. B Enzym.*, 15, 207–211, 2001.
42. Felfer, U., Strauss, U.T., Kroutil, W., Fabian, W.M.F., and Faber, K., *J. Mol. Catal. B I* 15, 213–222, 2001.
43. Hegeman, G.D., Rosenberg, E.Y., and Kenyon, G.L., *Biochemistry*, 9, 4029–4035, 1970.
44. Strauss, U.T., and Faber, K., *Tetrahedron Asymmetry*, 10, 4079–4081, 1999.
45. *Pogorevc, M., Stecher, H., and Faber, K., *Biotechnol. Lett.*, 24, 857–860, 2002.
46. Schleifer, K.H. and Kandler, O., *Bacteriol. Rev.*, 36, 407–477, 1972.
47. Bycroft, B.W., *Dictionary of Antibiotics and Related Substrates*, Chapman & Hall, New York
48. Yamada, H., Shimizu, S., Shimada, H., Tani, Y., Takahashi, S., and Ohashi, T., *Biochir* 395–399, 1980.
49. Syldatk, C., Müller, R., Pietzsch, M., and Wagner, F., Microbial and enzymatic produc L-amino acids from DL-monosubstituted hydantoins, in *Biocatalytic Production of Amino Ac Derivatives*, Rozell, J.D. and Wagner, F., Eds., Hanser Publishers, New York, 1992, pp. 1

---

*Mandelate racemase was found to be inactive in various organic solvent systems.

50. Pietzsch, M., Syldatk, C., and Wagner, F., *Ann. N. Y. Acad. Sci.*, 672, 478–483, 1992.
51. Ward, R.S., *Tetrahedron Asymmetry*, 6, 1475–1490, 1995.
52. Caddick, S. and Jenkins, K., *Chem. Soc. Rev.*, 25, 447–456, 1996.
53. Noyori, R., Tokunaga, M., and Kitamura, M., *Bull. Chem. Soc. Jpn.*, 68, 36–56, 1995.
54. Kitamura, M., Tokunaga, M., and Noyori, R., *J. Am. Chem. Soc.*, 115, 144–152, 1993.
55. Kitamura, M., Tokunaga, M., and Noyori, R., *Tetrahedron*, 49, 1853–1860, 1993.
56. Um, P.J. and Drueckhammer, D.G., *J. Am. Chem. Soc.*, 120, 5605–5610, 1998.
57. Fülling, G. and Sih, C.J., *J. Am. Chem. Soc.*, 109, 2845–2846, 1987.
58. Inagaki, M., Hiratake, J., Nishioka, T., and Oda, J., *J. Am. Chem. Soc.*, 113, 9360–9361, 1991.
59. Brinksma, J., van der Deen, H., van Oeveren, A., and Feringa, B.L., *J. Chem. Soc. Perkin. Trans.*, 1, 4159–4163, 1998.
60. Allen, J.V., and Williams, J.M.J., *Tetrahedron Lett.*, 37, 1859–1862, 1996.
61. Larsson, A.L.E., Persson, B.A., and Bäckvall, J.E., *Angew. Chem.*, 109, 1256–1258, 1997.
62. Reetz, M.T. and Schimossek, K., *Chimia*, 50, 668–669, 1996.
63. Kitamura, M., Tokunaga, M., and Noyori, R., *J. Am. Chem. Soc.*, 115, 144–152, 1993.
64. Cecere, F., Galli, G., Della Penna, G., and Rappuoli, B., Ger Offen 2621076, addition to Ger Offen 2, 422,737, 1976; *Chem. Abst.*, 86, 107042t, 1976.
65. May, O., Verseck, S., Bommarius, A., and Drauz, K., *Org. Process Res. Dev.*, 6, 452–457, 2002.
66. Carnell, A.J., *Adv. Biochem. Eng. Biotechnol.*, 63, 57–72, 1999.
67. Buisson, D., Azerad, R., Sanner, C., and Larcheveque, M., *Biocatalysis*, 5, 249–265, 1992.
68. Nakamura, K., Inoue, Y., Matsuda, T., and Ohno, A., *Tetrahedron Lett.*, 36, 6263–6266, 1995.
69. Fantin, G., Fogagnolo, M., Giovannini, P.P., Medici, A., and Pedrini, P., *Tetrahedron Asymmetry*, 6, 3047–3053, 1995.
70. Takemoto, M. and Achiwa, K., *Tetrahedron Asymmetry*, 6, 2925–2928, 1995.
71. Tsuchiya, S., Miyamoto, K., and Ohta, H., *Biotechnol. Lett.*, 14, 1137–1142, 1992.
72. Carnell, A.J., Iacazio, G., Roberts, S.M., and Willetts, A.J., *Tetrahedron Lett.*, 35, 331–334, 1994.
73. Matsumura, S., Kawai, Y., Takahashi, Y., and Toshima, K., *Biotechnol. Lett.*, 16, 485–490, 1994.
74. Shimizu, S., Hattori, S., Hata, H., and Yamada, H., *Enzyme. Microb. Technol.*, 9, 411–416, 1987.
75. Chadha, A. and Baskar, B., *Tetrahedron Asymmetry*, 13, 1461–1464, 2002.
76. Ogawa, A.J., Xie, S.-X., and Shimizu, S., *Biotechnol. Lett.*, 21, 331–335, 1999.
77. Hasegawa, J., Ogura, M., Tsuda, S., Maemoto, S., Kutsuki, H., and Ohashi, T., *Agric. Biol. Chem.*, 54, 1819–1827, 1990.
78. Azerad, R. and Buisson, D., Stereocontrolled reduction of β-ketoesters with *Geotrichum candidum*, in *Microbial Reagents in Organic Synthesis*, S. Servi, Ed., NATO ASI Series C, Kluwer, Dordrecht, vol. 381, 1992, pp. 421–440.
79. Bulman Page P.C., Carnell, A.J., and McKenzie, M.J., *Synlett*, 774–776, 1998.
80. Takemoto, M., Matsuoka, Y., Achiwa, K., and Kutney, JP., *Tetrahedron Lett.*, 41, 499–502, 2000.
81. Faber, K. and Kroutil, W., *Curr. Opin. Chem. Biol.*, 9, 181–187, 2005.
82. Kato, D., Mitsuda, S., and Ohta, H., *Org. Lett.*, 4, 371–373, 2002.
83. Mitsukura, K., Yoshida, T., and Nagasawa, T., *Biotechnol. Lett.*, 24, 1615–1621, 2002.
84. Kato, D., Mitsuda, S., and Ohta, H., *J. Org. Chem.*, 68, 7234–7242, 2003.
85. Kato, D., Miyamaoto, K., and Ohta, H., *Tetrahedron Asymmetry*, 15, 2965–2973, 2004.
86. Adam, W., Lazarus, M., Boss, B., Saha-Möller, C.R., Humpf, H.U., and Schreier, P., *J. Org. Chem.*, 62, 7841–7843, 1997.
87. Adam, W., Lazarus, M., Saha-Möller, C.R., and Schreier, P., *Tetrahedron Asymmetry*, 9, 351–355, 1998.
88. Takahashi, E., Nakamichi, K., and Furui, M., *J. Ferment. Bioeng.*, 80, 247–250, 1995.
89. Tsuchiya, S., Miyamoto, K., and Ohta, H., *Biotechnol. Lett.*, 14, 1137–1142, 1992.
90. Nakajima, N., Esaki, N., and Soda, K., *J. Chem. Soc. Chem. Commun.*, 947–948, 1990.
91. Sinnott, M.L., *Chem. Rev.*, 90, 1171–1202, 1990.
92. Leisinger, T. and Bader, R., *Chimia*, 47, 116–121, 1993.
93. Roy, A.B., *The Enzymes*, 5, 1–19, 1971.
94. Orru, R.V.A., Archelas, A., Furstoss, R., and Faber, K., *Adv. Biochem. Eng. Biotechnol.*, 63, 145–167, 1999.

95. Pelsy, G. and Klibanov, A.M., *Biotechnol. Bioeng.*, 25, 919–928, 1983.
96. Wallner, S.R., Pogorevc, M., Trauthwein, H., and Faber, K., *Eng. Life Sci.*, 4, 512–516, 2(
97. Faber, K., Mischitz, M., and Kroutil, W., *Acta Chem. Scand.*, 50, 249–258, 1996.
98. Archelas, A. and Furstoss, R., *Annu. Rev. Microbiol.*, 51, 491–525, 1997.
99. Archer, I.V.J., *Tetrahedron*, 53, 15617–15662, 1997.
100. Mischitz, M., Mirtl, C., Saf, R., and Faber, K., *Tetrahedron Asymmetry*, 7, 2041–2046, 19(
101. Pedragosa-Moreau, S., Archelas, A., and Furstoss, R., *J. Org. Chem.*, 58, 5533–5536, 1993
102. Manoj, K.M., Archelas, A., Baratti, J., and Furstoss, R., *Tetrahedron*, 57, 695–701, 2001.
103. Danda, H., Nagatomi, T., Maehara, A., and Umemura, T., *Tetrahedron*, 47, 8701–8716, 1(
104. Lemke, K., Ballschuh, S., Kunath, A., and Theil, F., *Tetrahedron Asymmetry*, 8, 2051–2055
105. Vänttinen, E. and Kanerva, L.T., *Tetrahedron Asymmetry*, 6, 1779–1786, 1995.
106. Takano, S., Suzuki, M., and Ogasawara, K., *Tetrahedron Asymmetry*, 4, 1043–1046, 1993.
107. Mitsuda, S., Umemura, T., and Hirohara, H., *Appl. Microbiol. Biotechnol.*, 29, 310–315, 1(
108. Steinreiber, A., Stadler, A., Mayer, S.F., Faber, K., and Kappe, C.O., *Tetrahedron Lett.*, 42,
6286, 2001.
109. Mischitz, M., Kroutil, W., Wandel, U., and Faber, K., *Tetrahedron Asymmetry*, 6, 1261–1272
110. Orru, R.V.A., Mayer, S.F., Kroutil, W., and Faber, K., *Tetrahedron*, 54, 859–874, 1998.
111. Pedragosa-Moreau, S., Morisseau, C., Baratti, J., Zylber, J., Archelas, A., and Fursto
Tetrahedron, 53, 9707–9714, 1997.
112. Archer, I.V.J., Leak, D.J., and Widdowson, D.A., *Tetrahedron Lett.*, 37, 8819–8822, 1996.
113. Überbacher, B.J., Osprian I., Mayer S.F., and Faber, K., *Eur. J. Org. Chem.*, 7, 1266–1270
114. Monfort, N., Archelas, A., and Furstoss, R., *Tetrahedron*, 60, 601–605, 2004.
115. Pogorevc, M., Kroutil, W., Wallner, S.M., and Faber, K., *Angew. Chem. Int. Ed. Engl.*, 41,
4054, 2002.
116. Wallner, S.R., Nestl, B., and Faber, K., *Tetrahedron*, 61, 1517–1521, 2005.
117. Pogorevc, M., Strauss, U.T., Riermeier, T., and Faber, K., *Tetrahedron Asymmetry*, 13,
1447, 2002.
118. Wallner, S.R., Nestl, B.M., and Faber, K., *Org. Lett.*, 6, 5009–5010, 2004.
119. Pedragosa-Moreau, S., Archelas, A., and Furstoss, R., *Tetrahedron*, 52, 4593–4606, 1996.
120. Bellucci, G., Chiappe, C., and Cordoni, A., *Tetrahedron Asymmetry*, 7, 197–202, 1996.
121. Kroutil, W., Mischitz, M., and Faber, K., *J. Chem. Soc. Perkin. Trans.*, 1, 3629–3636, 199
122. Mayer, S.F., Steinreiber, A., Orru, R.V.A., and Faber, K., *Eur. J. Org. Chem.*, 4537–4542,
123. Kroutil, W. and Faber, K., *Tetrahedron Asymmetry*, 9, 2901–2913, 1998.
124. Free shareware programs (Cyclo) running under Windows and Macintosh are available at <
borgc185.kfunigraz.ac.at> or directly from the authors (W Kroutil, A Kleewein, K Faber ©
A description of how to use the program is given in the help-file that accompanies the pro
125. Huh, J.W., Yokoigawa, K., Esaki, N., and Soda, K., *J. Ferment. Bioeng.*, 74, 189–190, 199
126. Soda, K., Oikawa, T., and Yokoigawa, K., *J. Mol. Catal. B Enzym.*, 11, 149–153, 2001.
127. Alexandre, F.R., Pantaleone, D.P., Taylor, P.P., Fotheringham, I.G., Ager, D.J., and Turne
Tetrahedron Lett., 43, 707–710, 2002.
128. Turner, N.J., *Curr. Opin. Chem. Biol.*, 8, 114–119, 2004.
129. Huh, J.W., Yokoigawa, K., Esaki, N., and Soda, K., *Biosci. Biotech. Biochem.*, 56, 2081–2082
130. Alexeeva, M., Enright, A., Dawson, M.J., Mahmoudian, M., and Turner, N.J., *Angew. Che
Ed. Engl.*, 41, 3177–3180, 2002.
131. Carr, R., Alexeeva, M., Enright, A., Eve, T.S.C., Dawson, M.J., and Turner, N.J., *Angew.
Int. Ed. Engl.*, 42, 4807–4810, 2003.
132. Alexeeva, M., Carr, R., and Turner, N.J., *Org. Biomol. Chem.*, 1, 4133–4137, 2003.
133. Biade, A.E., Bourdillon, C., Laval, J.M., Mairesse, G., and Moiroux, J., *J. Am. Chem. So(
893–899, 1992.
134. Shimizu, S., Hattori, S., Hata, H., Yamada, H., *Appl. Environ. Microbiol.*, 53, 519–522, 19
135. Padhi, S.K., Pandian, N.G., and Chadha, A., *J. Mol. Catal. B Enzym.*, 29, 25–29, 2004.
136. Allan, G.R. and Carnell, A.J., *J. Org. Chem.*, 66, 6495–6497, 2001.
137. Cardus, G.J., Carnell, A.J., Trauthwein, H., and Riermeir, T., *Tetrahedron Asymmetry*, 1(
243, 2004.

138. Comasseto, J.V., Omori, A.T., Andrade, L.H., and Porto, A.L.M., *Tetrahedron Asymmetry*, 14, 711–715, 2003.
139. Comasseto, J.V., Andrade, L.H., Omori, A.Z., Assis, L.F., and Porto, A.L.M., *J. Mol. Catal. B Enzym.*, 29, 55–61, 2004.
140. Nakamura, K., Fujii, M., and Ida, Y., *Tetrahedron Asymmetry*, 12, 3147–3153, 2001.
141. Takemoto, M. and Achiwa, K., *Chem. Pharm. Bull.*, 46, 577–580, 1998.
142. Demir, A.S., Hamamci, H., Sesenoglu, O., Neslihanoglu, R., Asikoglu, B., and Capanoglu, D., *Tetrahedron Lett.*, 43, 6447–6449, 2002.
143. Goswami, A., Mirfakhrae, K.D., and Patel, R.N., *Tetrahedron Asymmetry*, 10, 4239–4244, 1999.
144. Xie, S.-X., Ogawa, J., and Shimizu, S., *Appl. Microbiol. Biotechnol.*, 52, 327–331, 1999.

3 A Decade of Biocatalysis at Glaxo Wellcome

Mahmoud Mahmoudian

CONTENTS

3.1 INTRODUCTION

The biotech industry has evolved into three major areas of application: (1) "red biotechnology" (pharmaceutical, therapeutic/medical applications); (2) "green biotechnology" (agricultural applications); and (3) "white biotechnology" (industrial applications). White biotechnology is at a relatively early stage in the chemical industry, but is seen as a potential key driver to the industry's future. Industrial applications of biotechnology today include biofeedstocks that replace fossil fuel with sugars and starch, bioprocesses such as fermentation for vitamin production, biocatalysis in active pharmaceutical ingredient production, and other applications in textiles and leather, animal feed, pulp and paper, energy, metals, minerals and waste processing. McKinsey & Company [1] estimates that by 2010 the chemical industry could generate 10 to 20% of its sales revenue from chemicals derived from biotechnology. The largest component would be fine chemicals at 30 to 60% of sales from biotech, specialty chemicals (15%), polymers (6%), and bulk chemicals (12%). The most likely area of increased penetration

for biotechnology is in fine chemical/pharmaceutical manufacture, where it is estimate
$30 billion to $60 billion in additional value could be generated by 2010 as evident by th
in demands toward biologics, chiral products, and natural products.

Increased petrochemical raw material cost (doubling of natural gas prices since 200
competition from low-cost producers such as China and India have eroded profit margins
chemical industry. This has therefore encouraged ways to find the more cost-effective ro
existing and new chemical entities. Biotechnology capabilities are likely to be a techn
differentiator for western companies as they face increased competition and commodi
pressures. In polymers, biotechnology has been used in some applications, but it rem
largely niche area today. Biotech routes can be used in the production of certain mon
such as acrylamide, adipic acid, caprolactam, and dicarboxylic acids, and can be u
polymerization for making polyester. New polymers using biotech routes are DuPont's S
(1,3-propanediol), polylactic acid polymers (PLAs from Cargill Dow), polyhydroxyalka
(PHAs from Metabolix), and Nexia's Spider Silk. However, the key, but still elusive, tar
white biotechnology is yet to penetrate bulk chemicals. Acetic acid, succinic acid, and
oleochemicals might be among the next bulk chemicals to become "biotech." The key t
cost production here is the shift toward waste biomass as a feedstock for fermentation,
would substitute for petroleum-based feedstocks. Although biomass as a feedstock is
tially cost-competitive to petroleum-based feedstocks, there are still some challenges
A switch to biomass would require building an entirely new value chain, with massive
investments and collaboration among different players. Furthermore, technological ad
such as in the enzymic conversion of cellulose are still needed for commercial applicatio

Red biotechnology, on the other hand, has profoundly affected the pharmac
industry. Not only are there numerous new entrants that are developing new protein t
ies, but the conventional, "small-molecule" drugs have also become more complex
newer drugs are increasingly difficult to synthesize chemically and in many cases are
Most regulatory authorities often require development of a single enantiomeric route
poses huge synthetic challenges where the potential of biotech approaches for regiose
and enantioselective catalysis has become crucial. Furthermore, environmental implic
in using undesired solvents and reagents in synthetic chemistry have prompted an inc
use of biotech tools in this sector. Ideally, from an industrial viewpoint, it is desired to
close integration between biotech and chemical catalysis at various stages of drug dis
and development to maximize its full potential. In the 1980s and 1990s several big ph
ceutical companies (Glaxo Wellcome, Merck, Schering-Plough, Pfizer, and Bristol-
Squibb) set up specialized biotech groups to complement internal R&D programs. B
enabled these giants to secure additional patent coverage and use it effectively as levera
price negotiations with outsourcing partners.

The pharmaceutical industry is one of the biggest and most lucrative in the world
annual sales of around $400 billion. Pfizer rivals Microsoft in market capitalization, a
two are exceeded in size only by General Electric. Pfizer and other giants such as Glaxo
Kline, Bristol Myer-Squibb, and Merck routinely report multibillion-dollar profits. But
its outward strength, the industry is ailing. The "pipelines" of forthcoming drugs on wh
future depends have been drying up for some time. In this competitive landscape the ph
ceutical sector is facing mounting pressures on number of fronts: (1) delivering new dr
market rapidly and efficiently; (2) patents expirations and stiff competition from generic
(3) difficulty to maintain its market share and competitive edge; (4) the need to impr
product pipelines; (5) a gradual but steady shift of innovation base from big pharma
biotech sector; (6) the realization that it is no longer cost-effective to do the entire R&D u
single roof; (7) tackling and understanding complex disease mechanisms; (8) navi
through complex regulatory hurdles; and (9) meeting shareholders' expectations. Alt

globally research funding has doubled since 1991, the number of new drugs emerging each year has fallen by half. In 2003, America's Food and Drug Administration (FDA) approved only 21 "new chemical entities"—down from 53 in 1996. The more pharma spends on research, the less it seems to have to show for it. This is probably due to a combination of factors. One is the industry's obsession with producing "blockbusters" (drugs with annual sales of >$1 billion). As the search for such bestsellers continues, firms may mistakenly be passing up smaller, but still profitable, opportunities, and as patents expire and revenues from old drugs dry up, costs are rising inexorably, not just in research but also in sales and marketing. So far, the industry has responded with cost-cutting, organizational changes, mergers, and acquisitions. The underlying challenge, however, is to address the *innovation deficit*.

There has been a technological explosion in the last decade. Rapid advances in technology have made significant impact on drug discovery. Achievements in human genomics, proteomics, and bioinformatics present huge possibilities to unravel new biological targets to understand disease states. The two much-hyped technologies, combinatorial chemistry and high-throughput screening (HTS), which appeared on the scene in the 1990s, promised to speed up the development of new drugs by exploiting automation, and generated lots of hits. It was hoped that testing many new compounds quickly would increase the rate at which new leads were produced. While these approaches looked promising, and the quantity improved, the quality did not. The number of new leads going into clinical testing did not increase, and enthusiasm for the new technologies waned. Similarly, the sequencing of the human genome was expected to revolutionize the process of drug discovery, and while it is undoubtedly a remarkable achievement, the link between genes and disease state is still obscure. Furthermore, the human genome project provided thousands of potential targets for new drugs that we must painfully sift through. This "information overload" has created a huge hurdle that industry must overcome. The genome is estimated to contain around 5000 pharmaceutically relevant genes; however, the 100 best-selling drugs target 43 genes between them, and the top 200, just 47—the whole industry is therefore, running on less than 50 genes. As a result, much effort is now being focused on using combinatorial chemistry and HTS more appropriately than in the past, and finding new ways to identify targets early, to determine the structure of proteins, and to test compounds for activity and behavior. The key aim is to distinguish winners from losers as early as possible. Failure can occur at any point in the discovery process, and the later the failure, the more costly the loss. A target may be important but not chemically tractable; drug compounds generated may not work, or not work well enough; a promising lead may turn out to be toxic—"fail early, fail cheap."

The enthusiasm for combinatorial chemistry also diverted attention away from compounds derived from natural products. Once a mainstay of pharmaceutical research, and the source of antibiotics and anticancer drugs, natural products were left behind in the rush to automation because they are complex and hard to make. But now they are making a comeback in the form of novel scaffolds for chemical libraries using combinatorial techniques. Despite the abundance of tools, most drug discovery and development efforts still fail because of a lack of understanding of how the drugs work, and an inability to predict reliably how the human body will handle them. To be successful the pharmaceutical industry must embrace and combine old and new techniques, to have the right blend of physiology, pharmacology, and target-oriented chemistry on one hand, and genomics, molecular modeling, and structural biology on the other. To be competitive pharma must also accelerate and improve efficiency in all aspects of drug discovery and development—the entire process typically costs $900 million and takes 10 to 15 y (Figure 3.1). Only 1 in 1000 compounds tested makes it into human trials, and only 1 in 5 of those emerges as a drug. Outsourcing, strategic alliances, and partnership with the academic sector and biotech sector are now the key components of pharmaceutical R&D.

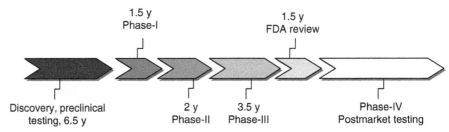

FIGURE 3.1 A typical drug discovery path in the pharmaceutical sector.

During 1989–2002 several major bioprocesses were developed at Glaxo Wellcom
were crucial to the market launch of many blockbusters including Epivir (anti-HIV),
vudine (anti-hepatitis), abacavir (Ziagen, anti-HIV), and Zanamavir (Relenza, anti-f
2002, however, the biotech group at the newly merged GlaxoSmithKline was sadly disba
This chapter is a tribute to a decade of biocatalysis at Glaxo Wellcome and attempts to
nostalgic but concise overview of some of the major bioprocesses developed in this era
generation of antibiotics, antileukemic, anti-inflammatory, and antiviral agents, as
several novel enabling technology platforms for production of optically pure amino aci
amines, and high-throughput experimentations.

3.2   β-LACTAM ANTIBIOTICS

The trinems **1** and **2**, bearing the tricyclic skeleton, are members of a novel class of
synthetic β-lactam antibiotics discovered at Glaxo Wellcome, Italy (Figure 3.2). The
prepared by a multistep synthesis starting from the commercially available acetoxya
none **3** and the chiral alcohol **4** [3]. For large-scale production of **1** and **2**, an efficient sy
of the chiral alcohol **4** was needed. We developed both enzymic and chemical reso
approaches for the preparation of multi-kg quantities of (+)-(1S, 2S)-*trans*-2-methoxy
hexanol **4** from the corresponding racemate (±) *trans*-2-methoxycycloheaxol **5** [4]: fir
lipase-catalyzed transesterification of the unwanted (1R, 2R) enantiomer, followed by
ation of the required alcohol **4** from the acylated species **6** by a simple partition metho
secondly, the synthesis of diastereomeric L-valine ester derivatives followed by separat
the required diastereomer **7** by fractional crystallization and hydrolysis back to the alc
(Figure 3.3).

For enzymic resolution, previous work had shown that the racemic acetate of *t*
methoxycyclohexanol may be hydrolyzed to yield (−)-(1R, 2R)-2-methoxycyclohexa
high enantiomeric excess (ee) using SAM-II lipase from *Pseudomonas* spp. [5], or pi
acetone powder [6]. Similarly, the racemic butanoate of *trans*-2-methoxycyclohexan

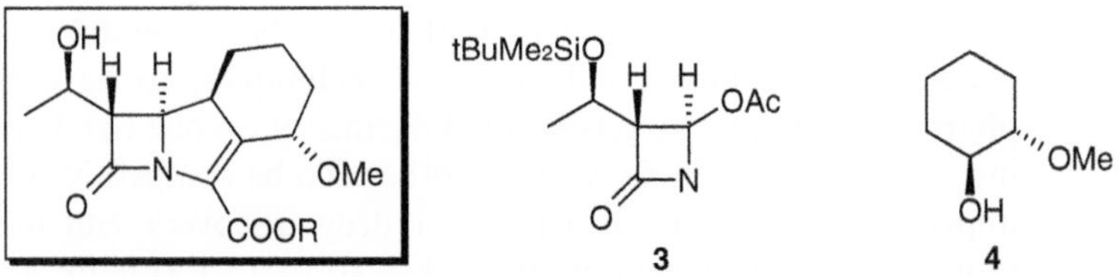

1 R = Na
2 R = CH(CH$_3$)OC(O)OC$_6$H$_{11}$

FIGURE 3.2 Structure of tricyclic β-lactam antibiotics and key chiral intermediates for their sy

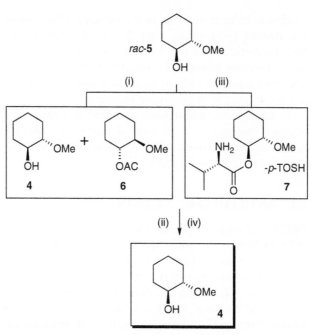

FIGURE 3.3 Chemoenzymic strategy for the resolution of racemic *trans*-2-methoxycyclohexanol **5**: (i) lipase, triethylamine, cyclohexane, vinyl acetate; (ii) aqueous extraction; (iii) L-valine, *p*-toluenesulfonic acid (-*p*-TOSH), toluene, reflux, fractional crystallization; (iv) NaOH.

hydrolyzed by lipases from *Pseudomonas* and *Candida* spp. to yield the (−)-enantiomer of *trans*-2-methoxycyclohexanol [7]. Since we required the (+)-(1*S*, 2*S*) enantiomer **4**, we considered enantioselective acylation of the racemic alcohol **5** to provide a more direct route to our target compound without the need for subsequent ester hydrolysis. Amongst the lipases screened, enzymes from *Candida antarctica* and *Pseudomonas fluorescence* gave the highest ee for the enantioselective acylation of racemic **5**. Reactions were carried out in cyclohexane using vinyl acetate as acyl donor, and were monitored by chiral gas chromatography (GC) and high-performance liquid chromatography (HPLC). The immobilized *C. antarctica* lipase (CAL, Novozyme) exhibited excellent stability retaining over half of its initial activity after nine cycles of reuse. Process optimization showed the fasted bioconversions with cyclohexane as solvent. Acetaldehyde, resulting from transesterification of vinyl acetate, caused reaction inhibition, but this was reversible by simply washing the acetaldehyde-treated enzyme with cyclohexane, restoring its initial activity. Under optimized conditions CAL (37.4 g/L), vinyl acetate (1.7 M), racemic **5** (1.4 M) were used, with small quantities of triethylamine (0.16 M) to neutralize any acetic acid formed by hydrolysis of vinyl acetate; ee >98%, conversion 55% by GC. Cyclohexane was a particularly suitable solvent both for bioconversion and, due to its very low polarity, for the selective extraction of the required alcohol **4** into water at the end of a reaction cycle. The optically pure alcohol **4** was isolated in >99% ee (36% yield) by extraction into ethyl acetate followed by evaporation of organic solvent.

For L-valine resolution, we used chemical esterification of racemic **5** with L-valine in the presence of *p*-toluenesulfonic acid [8]. The formation of crystalline ester derivatives and the ability to optimize the efficiency of the resolution, either by varying the amino acid or by using alternative acids, made this approach very attractive to us for providing a large-scale resolution of the racemic *trans*-2-methoxyclohexanol **5**. In practice, we found that a remarkably simple and efficient resolution could be achieved by heating the racemic alcohol and

L-valine in toluene at reflux in the presence of *p*-toluenesulfonic acid (1.3 molar equiva
The desired (+)-L-valine ester **7** was obtained in good yield (typically >27%) an
diastereomeric excess (de 96% by proportionate mortality ratio, PMR) on filtration
cooled reaction mixture. Treatment of the (+)-L-valine ester **7** in a mixture of *t*-butyl
ether and aqueous sodium hydroxide under phase-transfer conditions then gave the
alcohol **4** in the 95th percentile recovery.

In conclusion, whilst both chemical and enzymic methods were simple to opera
afforded material of good optical purity, the enzymic approach was advantageous
economically and environmentally, and was the method of choice for production of the
alcohol **4** on a multi-kg scale.

3.3   ANTILEUKEMIC AGENTS

506U78 (2-amino-9-β-D-arabinofuranoyl-6-methoxy-9H-purine) was developed i
1970s by Burroughs Wellcome as a potential antiviral agent but was dropped due to t
issues. After the merger with Glaxo, Glaxo Wellcome developed this compound f
treatment of leukemia [9,10]. 506U78 is a prodrug of *ara*-G (9-β-D-arabinofuranosyl
ine) (Figure 3.4). Given the difficulty in synthesizing *ara*-G using traditional chemica
niques, and its poor water solubility, we turned to 506U78, which is several times
soluble than *ara*-G and can be synthesized relatively easily using enzyme technology. *A*
506U78 is rapidly demethoxylated by adenosine deaminase to *ara*-G [11]; thus, many
obstacles to using *ara*-G were circumvented by developing 506U78. Our colleagues
former Burroughs Wellcome used uridine phosphorylase (UPase) and purine nucl
phosphorylase (PNP), catalyzing net transfer of arabinose from a pyrimidine to a
base with retention of the β-D configuration, for the synthesis of 506U78 (Figure 3.5)
enzyme was cloned and overexpressed in independent *Escherichia coli* strains containi
corresponding genes on a multicopy plasmid.

In the UK, we developed, optimized, and scaled up a fermentation process for the e
production, and to simplify regulatory approval, all preparations of organism ban
fermentation processes were carried out in animal-free media: e.g., tryptone could be re
with soya peptone without compromising cell yields and expression levels of each e
Seed cultures (250 mL) of recombinant *E. coli* strains, for production UPase and PNP (
Wellcome collection), were grown at 37°C in a medium containing soya peptone (1

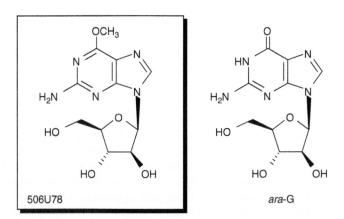

FIGURE 3.4 Structures of the antileukemic agent 506U78 and its prodrug *ara*-G.

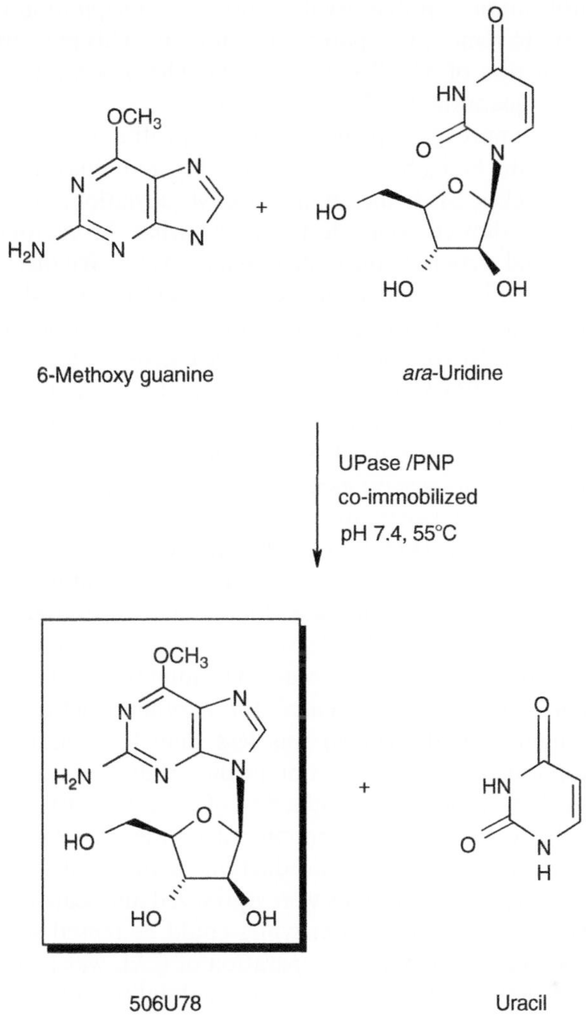

6-Methoxy guanine   ara-Uridine

UPase /PNP
co-immobilized
pH 7.4, 55°C

506U78   Uracil

FIGURE 3.5 Enzyme-catalyzed formation of 506U78. UPase, uridine phosphorylase; PNP, purine nucleoside phosphorylase.

yeast extract (5 g/L), and NaCl (5 g/L). All media were supplemented with tetracycline (20 mg/L final concentration) for UPase, or kanamycin (50 mg/L) for PNP strains. Cultures were transferred to a production medium—soya peptone (10 g/L), yeast extract (5 g/L), potassium phosphate (13.1 g/L), sodium ammonium phosphate (3.5 g/L), citric acid (2 g/L), $MgSO_4 \cdot 7H_2O$ (0.2 g/L), glycerol (6.3 g/L) postautoclave—when growth was in early to mid-log phase. Production stage flasks (2 L) were used to inoculate fermenters. Fermentation conditions were optimized to attain best enzyme production. Crude lysates of these enzymes were also found not to be stable at the high reaction temperature (55°C), but could be stabilized by direct co-immobilization of the enzymes onto an ion-exchange support (DEAE-52)—this also facilitated removal and recovery of the enzyme. Typically, bioconversions were carried out with up to 200 g/L substrate input and the co-immobilized enzymes could be reused several times in bioreactors. Upon completion of each cycle, beads were filtered and washed with the reaction buffer (10 mM potassium phosphate, pH 7.4). The crude product was crystallized from hot aqueous base (pH 11); this was followed by a final

purification step that involved product precipitation from hot water, after a charcoa
ment, to remove any potential endotoxin. This has formed the basis of a scalable proc
production of 506U78 (Figure 3.5). The process was transferred to the factory to p
larger quantities for further evaluation.

To improve its pharmacokinetic profile, we found that esters of 506U78, in par
5'-acetate, had a better water solubility and bioavailability [12]. All of the chemical sy
approaches used, including selective acylation, or deacylation of the correspondi
acetate, showed poor selectivity, requiring chromatography to remove other undesire
ates, and were not amenable to scale-up [13]. We therefore embarked on a program to
enzymes that would regioselectively acylate the 5'-hydroxyl position of 506U78, an
mized conditions for its production. Several enzymes (CAL, Novozyme; *Bacillus lichen*
protease, Novozyme; *Mucor miehei* lipase, Novozyme; *B. licheniformis* protease,
Savinase, Novozyme; *Alcaligenes* sp. lipase, Altus; Lipolase, Novozyme) were fou
acylate 506U78 regioselectively but with up to 20% of the other related acylated pr
impurities. Novozyme-435, an immobilized preparation of CAL, was selected for p
optimization—several parameters (solvents, acylating agents, temperature, substrate c
tration) were systematically investigated and reactions were optimized to minimi
impurity levels (Figure 3.6). Preparative reactions were carried out under optimized
tions. Typically, bioconversions were carried out in anhydrous 1,4-dioxane, with vinyl a
(20 to 50% v/v) as the acyl donor, and up to 100 g/L of 506U78. On completion
reaction, the enzyme beads were removed by filtration and, after washing with neat met
were stored at 4°C before reuse. The immobilized enzyme was found to be stable when
at 4°C and could be reused for another reaction cycle. The level of related imp
(3'-mono-and diacetates) was less than 0.5% and the reaction mixture could sim
evaporated to dryness to obtain the product in acceptable purity. This formed the b
a scalable process for production of 5'-acetyl 506U78 (Figure 3.6) [14].

In summary, a co-immobilized UPase and PNP preparation from recombinant
strains was used in the production of the antileukemic agent 506U78. Fermentatic
bioconversion conditions were optimized and scaled up with up to 200 g/L of substrat
and the co-immobilized enzymes could be reused several times in bioreactors. In a p
process an immobilized preparation of CAL was used to produce 5'-esters of 506U78,
were shown to have a better water solubility and bioavailability. Both processes were
up to produce much larger quantities for further evaluation.

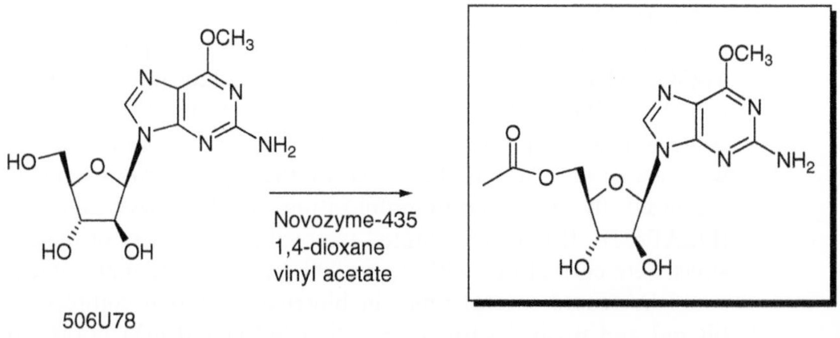

FIGURE 3.6 *Candida antarctica*–catalyzed acylation of 506U78. CAL, *C. antarctica* lipase: Novozy

3.4 ANTI-INFLAMMATORY AGENTS

In the early 1990s Glaxo Wellcome was developing a novel group of compounds with broad anti-inflammatory properties [15]. A key intermediate in the synthetic route is the nucleoside analog **9**, (1-[2-chloro-6-[(2,2-diphenylethyl)amino]-9H-purin-9-yl]-1-deoxy-β-D-ribofuranuronic acid) (Figure 3.7). Although chemical oxidation of the 5′-hydroxyl of precursor **8**, (2-chloro-N-(2,2-diphenylethyl)-adenosine), using transition metal oxidants (such as $KMnO_4$), was carried out in good chemical yields, this posed considerable problems for scale-up due to the heterogeneous nature of the reaction, and due to environmental and handling implications. The lack of regioselectivity of the oxidation reaction also necessitates the use of a protection–de-protection sequence to protect the 2′/3′-hydroxyl groups, thus introducing two extra steps into the route (Figure 3.7). We envisaged that an enzymic oxidation of a nucleoside precursor would offer potential for an improved synthesis—it is environmentally clean and obviates the need for protection of other functional groups.

Nucleoside oxidase is produced by *Pseudomonas* spp. and related Gram-negative bacteria [16–18]. Crude extracts of *Stenotrophomonas* (*Pseudomonas*) *maltophilia* exhibiting nucleoside oxidase activity were shown to convert the 5′-hydroxyl groups of natural purine and pyrimidine nucleosides to their corresponding carboxylic acids at analytical scale [19]—the purified enzyme catalyzes a two-step oxidation of a nucleoside by an aldehyde intermediate consuming one molecule of molecular oxygen [17,18]. Small-scale cultures of *S. maltophilia* (FERM BP-2252) were grown routinely at 25°C in 50 mL volumes of medium—yeast extract (25 g/L), glucose (30 g/L), K_2HPO_4 (1 g/L), KCl (1 g/L), $MgSO_4$ (0.5 g/L)—in shake flasks (250 rev/min, 5 cm throw). For larger-scale cultivations, 450 to 500 L fermentations were inoculated from two 8-h-old 5 L seed fermenters grown as above (2% v/v). Crude extracts of *S. maltophilia* were initially found to oxidize selectively the 5′-hydroxyl of **8** to generate **9** on mg scale; the substrate was also cleaved slowly to release the purine base, presumably by a

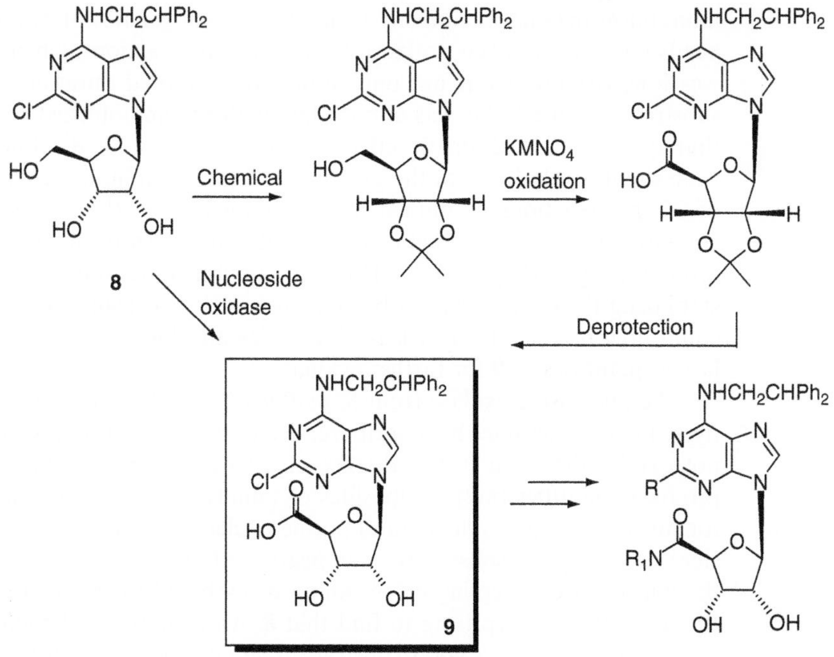

FIGURE 3.7 Nucleoside oxidase-catalyzed oxidation of **8**.

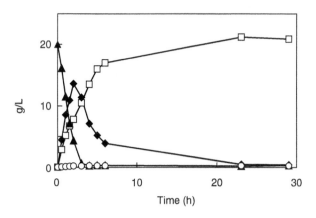

FIGURE 3.8 Production of **9** by crude extracts of *Stenotrophomonas maltophilia*. The reacti
carried out at room temperature in a magnetically stirred flask containing 3.5 mL of a clarifie
lysate (pH 6). **8** was added as a solid (70 mg) and stirred to obtain a homogeneous susp
Periodically, samples were removed, diluted into the mobile phase, and the clarified solution a
by HPLC. **8** (▲); **9** (□); aldehyde (◆); base (○).

phosphorylase activity [20] (Figure 3.8). This was, however, not considered to be sign
when cells with a high nucleoside oxidase activity were used and crude extracts were fo
be satisfactory for use in reactions without further purification. Crude extracts cont
nucleoside oxidase were used in bioconversions with up to 20 g/L input with no evide
inhibition at high substrate concentrations; **8** was quickly oxidized to **9** and reaction
completed within 24 h. Oxidation of **8** was shown to go through a transient formation
corresponding aldehyde intermediate to produce the carboxylic acid in high chemical
(Figure 3.8).

To simplify downstream processing, and to allow the enzyme to be reused, we ch
immobilize the enzyme directly from crude homogenates of *S. maltophilia* onto Eup￼
beads (10 g dry beads/g protein). In free enzyme reactions, when the enzyme/substrat
was low, oxidation was not only slow but of limited duration and stopped before
substrate was used; this was also evident with the immobilized enzyme. We found, ho
that quinol has a dramatic effect on the oxidation of **8**; although initial reactior
were significantly lower in the presence of 1 g/L of quinol, even at a substrate concen￼
of 20 g/L reactions continued up to completion in the presence of quinol, where
corresponding control reactions (without quinol) stopped at a product concentrat
around 15 g/L (Figure 3.9). This was attributed to quinol having a protective r￼
stabilizing the enzyme during bioconversions. It was found that the same batch of er
could now be reused for at least five cycles and bioconversions were scaled up to p￼
larger quantities of **9** for further evaluation.

The nucleoside oxidase from *S. maltophilia* has been found to accept natural puri
pyrimidine nucleosides having ribose, deoxyribose, or arabinose as a sugar moiety, bu
not oxidize the sugar in the absence of a base [16–18]. We envisaged using this enzy
produce quantities (mg) of ribosides of unnatural purine bases and carbocyclic nucle
for further testing. As the natural purine nucleosides (adenosine, guanosine, inosine, x￼
sine) are all good substrates, it appears that the enzyme is quite tolerant of chan
the purine base, accepting either amino or carbonyl functionality at both positions 2
It was, however, surprising to find that **8**, with the bulky diphenylethylamino group
6-position, was also a good substrate. The enzyme is tolerant of different functiona
the 2-position including chloro **10** or phenyl ethyl amino **11** (Figure 3.10). Furthermo

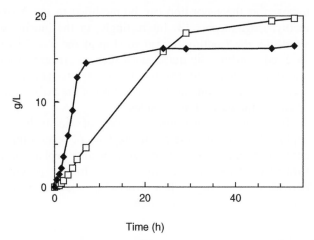

Time (h)

FIGURE 3.9 Reaction profile for oxidation of **8** using immobilized nucleoside oxidase from *Stenotrophomonas maltophilia* with quinol. The reaction (5 mL) was carried out in 50 mM potassium phosphate buffer (pH 6). **8** (100 mg) was added as a solid and stirred at room temperature to obtain a homogeneous suspension. The reaction was started by the addition of washed immobilized beads at 40% (w/v). At intervals, samples were removed, cleared of enzyme beads, and analyzed by HPLC. Reactions were carried out either in the presence (□) or absence (◆) of quinol at 1 g/L.

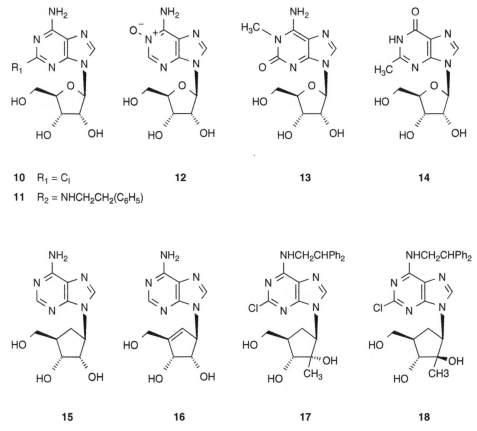

FIGURE 3.10 Substrate tolerance of the nucleoside oxidase from *Stenotrophomonas maltophilia*.

nitrogen at the 1-position can be modified to the N-oxide **12** or methylated in the c methyl isoguanosine **13**. Interestingly, in the inosine series, with 2-methylinosine **1** substrate, the reaction tended to stop at the 5'-aldehyde intermediate and was not f oxidized. Carbocyclic nucleosides such as aristeromycin **15** and neplanocin-A **16** wes selectively oxidized to give the 5'-carboxylates (Figure 3.10). The enzyme did not, ho accept a methyl group at the 2'-position of the carbocyclic moiety **17–18** or the 2',3' acet of any of the natural nucleosides.

In summary, the synthetic utility of the nucleoside oxidase from *S. maltophil.* exploited to produce the 5'-carboxylates of several purine nucleoside analogs, includi carbocyclic nucleosides aristeromycin and neplanocin-A on a preparative scale. Th formed the basis of a scalable process for the generation of nucleoside 5'-carboxyli derivatives. The enzyme has surprisingly wide substrate specificity toward unnatural n sides, especially in the base moiety [20].

3.5   ANTIVIRAL AGENTS

During the late 1980s and early 1990s a range of compounds were under evaluation at Wellcome as potential antiviral agents. These were oxathiolane nucleoside Epivir (La dine **19**), 2-aminopurine nucleoside Abacavir (Ziagen **20**), Zanamavir (Relenza **21**), cyclic nucleosides *c*-BVdU **22**, *c*-dG **23**, and Carbovir **24** (Figure 3.11). We went develop scalable biological routes, which were crucial to the market launch of sev these blockbusters (sales >\$1 billion), to complement our chemical synthesis program

3.5.1   EPIVIR (LAMIVUDINE)

(2'*R*-*cis*)-2'-deoxy-3-thiacytidine (3TC, Epivir, **19**) (Figure 3.11) has been approved FDA, and is marketed for the therapy of human immunodeficiency virus (HIV), in cc ation with Retrovir (Zidovudine, AZT) and Abacavir (Ziagen **20**), in adults and ch HIV is the causative agent of acquired immunodeficiency syndrome (AIDS), which i acterized by a chronic suppression of many immune functions and a concomitant incr susceptibility to opportunistic infections [21,22]. Epivir is a potent and selective inhib the reverse transcriptase enzyme, which is required for the replication of the HIV ge catalyzing the conversion of the HIV RNA to a double-stranded DNA copy. Epivir active against hepatitis B virus (HBV) and is sold as Lamivudine in China. In the early when we began developing the racemic BCH189 ($\pm$**19**, Figure 3.12), licensed from Bi Pharma, Canada, we found that in contrast to the majority of nucleoside analogs, display antiviral activity primarily or exclusively residing in the "natural" β-D-isom enantiomers of $\pm$**19** are equipotent *in vitro* against HIV-1 and HIV-2, but the "unna β-L-(−)-**19** isomer (Epivir) (Figure 3.12) is substantially less cytotoxic than its correspo "natural" β-D-(+)-isomer [23,24]. Clinical studies have shown Epivir to be well toler: high doses, with a good bioavailability and pharmacokinetic profile compared with nucleoside analogs, and it does not appear to cause bone marrow toxicities *in vitro* [2.

Early supplies of the individual enantiomers for biological evaluation were made preparative chiral HPLC; this method was, however, not amenable to scale up to prov larger quantities of Epivir required for further evaluation. One of the initial routes cons was the enzymic resolution using 5'-nucleotidase and alkaline phosphatase that allow access to both enantiomers of **19**. The chemically synthesized monophosphate (tive ($\pm$)-**27** was resolved using 5'-nucleotidase from *Crotalus atrox* venom, and the re

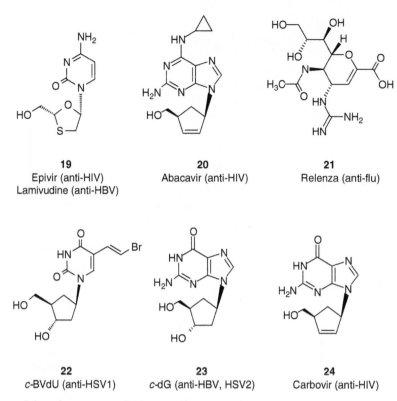

19
Epivir (anti-HIV)
Lamivudine (anti-HBV)

20
Abacavir (anti-HIV)

21
Relenza (anti-flu)

22
c-BVdU (anti-HSV1)

23
c-dG (anti-HBV, HSV2)

24
Carbovir (anti-HIV)

FIGURE 3.11 Selected structures of Glaxo Wellcome's antiviral agents.

mixture was separated by chromatography and purified on silica gel, to give (+)-**19** (ee > 99%, Figure 3.12). Hydrolysis of the remaining monophosphate (−)-**27** with alkaline phosphatase from *E. coli* (EC 3.1.3.1) afforded (−)-**19** (Epivir) in an optically pure form (Figure 3.12). To establish a chiral synthesis for Epivir, we needed to determine the absolute stereochemistry of the individual enantiomers of (+) and (−)-**19**. Epivir was treated with 4-bromobenzoyl chloride to afford the corresponding N^4-amide; the presence of the heavy bromine atom allowed the absolute stereochemistry to be determined by x-ray crystallography, confirming the "unnatural" configuration of Epivir [27]. To produce much larger quantities of Epivir, the 5′-nucleotidase route was inefficient; the most expeditious way forward was to investigate a scalable end-stage resolution, by enantioselective deamination of (±)-**19** with cytidine deaminase [28] (EC 3.5.4.5), to support the development of Epivir (Figure 3.12). At that time, cytidine deaminase was not commercially available and had not been widely used for preparative transformations. The *E. coli* cytidine deaminase seemed a promising source as it was known to be inducible to reasonably high levels in enteric bacteria (*E. coli, Salmonella typhimurium*), allowing these organisms to grow rapidly with cytidine as sole nitrogen source [29]. Initially, small-scale reactions confirmed that (±)-**19** was a substrate and that the enantioselectivity was exquisite. The enzyme was partially purified by ammonium sulfate fractionation (55 to 75% saturation) from a clarified cell extract of a wild-type *E. coli* B. The enzyme (0.3 unit) was incubated with (±)-**19**, 0.33 mg in a 1 mL reaction volume, for 24 h; then, chiral HPLC analysis indicated that the (+)-**19** had been deaminated to give the uridine analog (+)-**25**, leaving Epivir, essentially optically pure [28] (Figure 3.12). We therefore chose to investigate *E. coli* as a source of large quantities of enzyme for use in production of Epivir, and developed a scalable route involving cloning and

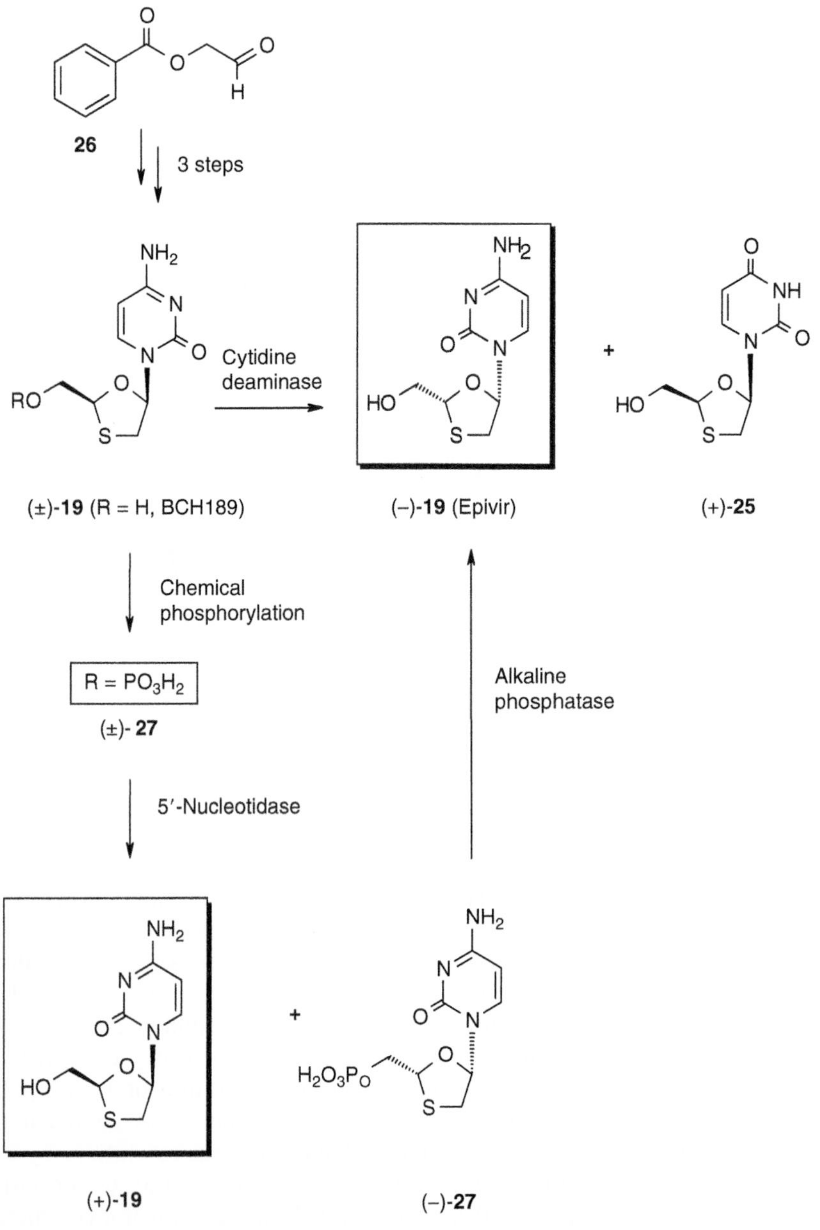

FIGURE 3.12 Synthesis of Epivir.

overexpression, fermentation and immobilization of cytidine deaminase for reuse, and
opment of an efficient isolation process that could be used in the factory to produce
tonne quantities of Epivir [28].

   After the initial demonstration of activity in a clarified cell extract from *E. coli*
turned our attention to developing a robust procedure for isolation and immobilization
enzyme. Cytidine deaminase was partially purified by column chromatography at a
scale; a 20-fold purification, with good recovery of the enzyme, was obtained by ion-exc
and hydrophobic interaction chromatography followed by ammonium sulfate precipi
As the scale of operation was increased, the procedure was modified to include a

adsorption step rather than column chromatography; after cell disruption, the clarified lysate was adsorbed onto cellulose DE-52 and the enzyme was eluted with 0.3 to 0.5 M NaCl (50 to 70% recovery). The significant improvement in the production of cytidine deaminase with the constitutive mutant and especially the recombinant strain prompted us to investigate the possibility of simplifying enzyme preparation further by using crude extracts without enzyme purification. The batch adsorption step was, therefore, omitted from the isolation procedure. The cell extracts were found to be satisfactory for use in the biotransformation. There was no evidence of competing side reactions and the enzyme activity was found to be remarkably stable. Typically, 4 kg (wet weight) of *E. coli* 3732E cell paste was suspended in 20 L of lysis buffer—KH_2PO_4 (50 mM), EDTA (1 mM), DTT (1 mM), ρ-hydroxybenzoic acid ethyl ester (500 ppm), pH (7.5)—and disrupted by three passages through a Manton–Gaulin homogenizer. The extract was clarified using continuous centrifugation and/or microfiltration and concentrated approximately fivefold. The solution was further clarified by centrifugation (20,000*g* for 60 min) prior to immobilization. A 4-kg batch of the mutant strain (3732E) typically yielded 150,000 units of cytidine deaminase, whereas 15,000,000 units could be obtained from 4 kg of the clone (3804E); one unit of activity was defined as the amount of enzyme required to deaminate 1 μmole cytidine/min at 25°C. For larger-scale operations, we found that even the clarification step could be omitted and the crude extract was directly immobilized onto Eupergit-C (Rhom Pharma, Darmstadt, Germany) without centrifugation, to give a stable enzyme preparation. At Glaxo's Greenford pilot plant in the United Kingdom, a 20-kg campaign was carried out as a series of 3-kg (500 L working volume) batches of racemic **19** (Figure 3.12) using immobilized cytidine deaminase from the constitutive mutant 3732E. Deamination of ($\pm$)-**19** was initially rapid, but approached completion through a very gentle asymptote. Careful chiral HPLC was needed to judge when the residual Epivir had reached sufficiently high optical purity (ee > 99.5%). In practice, cessation of alkalinization of the medium by ammonia release was a particularly useful indicator that the reaction was "completed." The same batch of enzyme was used for at least 15 cycles. Although the reaction time increased from 35 h to more than 70 h, this was attributed primarily to physical loss of the beads during collection and washing. To improve volumetric productivity, the effect of increased substrate concentration was investigated; no inhibition was noted up to 30 g/L input. These higher concentrations were therefore used when the process was transferred to the factory [28].

Having successfully scaled up the bioconversion stage of the process, we focused on developing a simple and efficient strategy for isolation of Epivir. Initial work involved small-scale isolation procedures using ion-exchange chromatography (QAE Sephadex), followed by desalting of Epivir on a polystyrene divinyl benzene resin (XAD16) column and recovery by freeze drying. Neither the ion-exchange step, due to poor flow characteristics of the column, nor freeze drying were suitable for operation at larger scale. A series of resins were therefore tested and Duolite A113, a strongly basic polystyrene quaternary ammonium resin with a good capacity for the uridine analog [(+)-**25**, Figure 3.12], was selected and used for larger-scale operations. The freeze-drying step was also replaced by direct crystallization of product from the concentrated solution. At the completion of each cycle of reaction (500 L, 3-kg substrate input), the pH was adjusted to 10.5 with concentrated ammonia solution, and the mixture was applied to a column of Duolite A113 super resin in the OH⁻ cycle. The uridine analog (+)-**25** adsorbed to the resin, whereas Epivir was not retained. The column was washed with 0.04% ammonia solution. This wash and the column spent, which contained the Epivir solution, were combined, the pH adjusted to 7.5 with concentrated H_2SO_4, and the mixture applied to an XAD16 resin column. The column was washed with distilled water and Epivir was eluted with 33% (v/v) acetone in water. Fractions containing Epivir were combined and concentrated approximately fourfold on a Balfour wiped film evaporator.

The concentrate was warmed to 50°C to dissolve any crystals and the solution was va filtered through Whatman-54 filter paper before further concentration to a slurry (3 L a rotary evaporator. After cooling, the crystalline product was recovered by filt Typically, 1.15 kg of highly pure Epivir (average recovery, 76%) was recovered from 3-kg batch. Purity was better than 97% by HPLC and the ee was at least 99.8%. Usin approach, 20 kg of optically pure Epivir was isolated to support the initial develo program. For further scale-up, the process was transferred to our production site at ston, Cumbria, UK, where several tonnes of Epivir were produced, using immo cytidine deaminase from the recombinant strain.

One major drawback with the routes—(i) enzymic resolution of (±)-**19** using 5′-n tidase and alkaline phosphatase and (ii) enantioselective deamination of (±)-**19** wit dine deaminase—discussed earlier is late-stage resolution of the racemic drug (Figure In practice, 50% of the valuable starting material was unused and, as the scale of ope increased, alternative routes for the manufacture of Epivir were sought. Nevert resolution using immobilized cytidine deaminase was the only scalable route *initially* able that enabled us to rapidly produce multitonne quantities of optically pure Epi clinical trials; this resulted in a 2-y clinical lead before a chemical asymmetric route w in place.

As an alternative to end-stage resolution, we also investigated enzymic methods to resolution prior to the addition of the cytosine base. One such route considered w enantioselective hydrolysis of racemic *trans*-5-propionyloxy-1,3-oxathiolane-2-methanol ate **28**, to obtain the key intermediate (−)-**28** [30] (Figure 3.13). The *trans*-oxath benzoate **28**, synthesized from benzoyloxyacetaldehyde **26** (3 steps), was an attractiv strate because it could easily be accommodated into the existing synthetic route. Interes **26** was also a common starting material for the synthesis of (±)-**19**, BCH189 (3 Figure 3.12). A number of commercially available lipases and proteases were screen the ability to hydrolyse racemic oxathiolane **28** enantioselectively. *M. miehei* lipas identified as the most efficient biocatalyst. All other enzymes tested were less enantiose than *M. miehei* lipase, but it is interesting to note that *C. cylindracae* and *Chromobac viscosum* lipases and subtilisin showed the opposite enantioselectivity to *M. miehei* albeit in very low ee [5 to 15% ee (+)-2S isomer] [30]. Bioconversion of **28** with *M.* lipase afforded enantiomerically enriched residual ester of the correct absolute stereocher (−)-2R, for the subsequent synthesis of Epivir (2 steps, Figure 3.13).

In summary, our finding that *E. coli* cytidine deaminase can deaminate racer (BCH189, Figure 3.12) enantioselectively extends the understanding of the specificity enzyme, and it has become a generally useful reagent for resolution of unnatural c nucleosides. The cloning and overexpression of the enzyme, under the control of a hig inducible promotor, was essential to the development of a scalable process. Also cruc developing a large-scale process was the demonstration that cytidine deaminase can be bilized to give a stable enzyme preparation that can be reused many times in the reactio absence of substrate inhibition allowed high volumetric productivities to be achieve assisted downstream processing. A robust scalable product isolation process was dev to yield crystalline Epivir. Here a simple two-column process, with adsorption–desc rather than chromatographic steps, was used. Overall, yields through the resolution pro 76% were obtained. The overall process proved remarkably robust in a factory settin batches of enzyme surviving entire production campaigns. The resolution with cytidir minase was the only scalable route available to us for synthesis of Epivir through much preclinical and clinical development of the drug. However, end-stage resolution was likely to be the optimal economic route in the longer term, and chemical resolution of a: stage intermediate now forms the basis of the manufacturing process for Epivir [31].

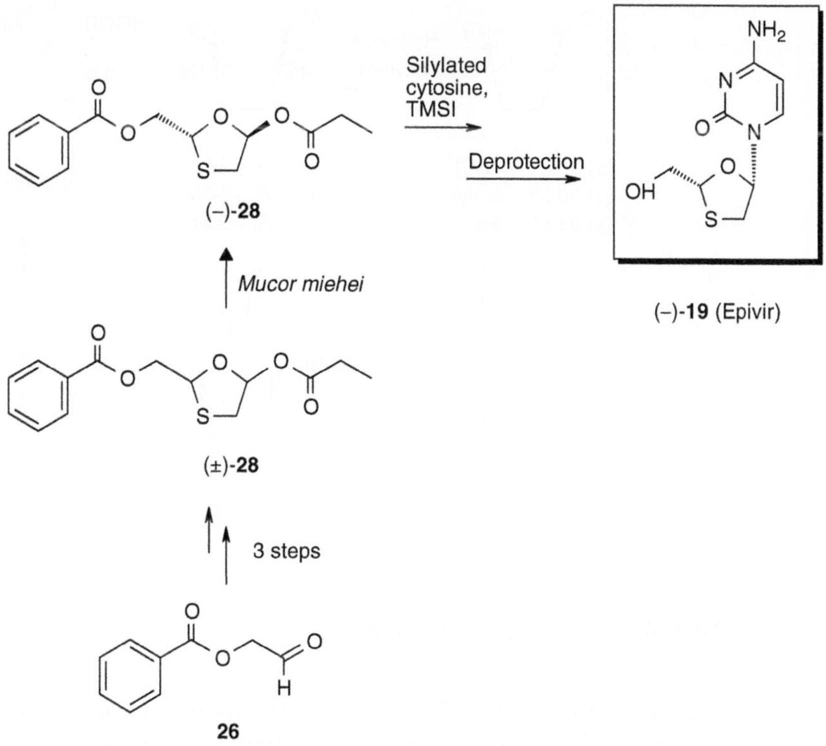

FIGURE 3.13 An alternative chemoenzymic route for synthesis of Epivir.

3.5.2 ABACAVIR (ZIAGEN)

Abacavir **20** (Ziagen) has been approved by the FDA for the treatment of HIV and HBV infections in adults and children—it is a selective and potent reverse transcriptase inhibitor [32] (Figure 3.14). This novel 2-aminopurine nucleoside analog is lipophilic and water-soluble, synergistic *in vivo* with protease and other reverse transcriptase inhibitors, as well as being well tolerated and orally absorbed with significant CNS penetration [33]. The triple combination regimens, involving the various reverse transcriptase and protease inhibitors, are often difficult to adhere to; patients may have to take between 10 to 20 pills/d at different times and, depending on the drugs involved, either with or without food and drink. In 2000 the FDA approved Trizivir, a combination therapy of a fixed dose of Abacavir (Ziagen), Retrovir (AZT), and Epivir (3TC)—the three nucleoside reverse transcriptase inhibitors (NRTIs) already approved by the FDA. Trizivir provides a well-tolerated and compact dosing regimen (one tablet twice a day, with no dietary restrictions), which can be particularly important in improving patient adherence. This new simple dosing regimen therefore addresses one of the most significant clinical challenges (patient adherence) faced in HIV treatment today.

The unsubstituted γ-lactam **31** (2-azabicyclo[2.2.1]hept-5-en-3-one) is a potential intermediate for the synthesis of Abacavir (Figure 3.14). The ChiroTech group (now part of Dow Chemical) in Cambridge, United Kingdom, had developed a process for the resolution of racemic **31** using γ-lactamase containing microorganisms such as *P. solanacearum* NCIMB 40249 and *Rhodococcus* sp. NCIMB 40213 [34,35]; these enzymes are, however, proprietary and not commercially available for general use. To circumvent ChiroTech's patent, we at Glaxo Wellcome envisaged that by activating the lactam ring with acyl-protecting groups

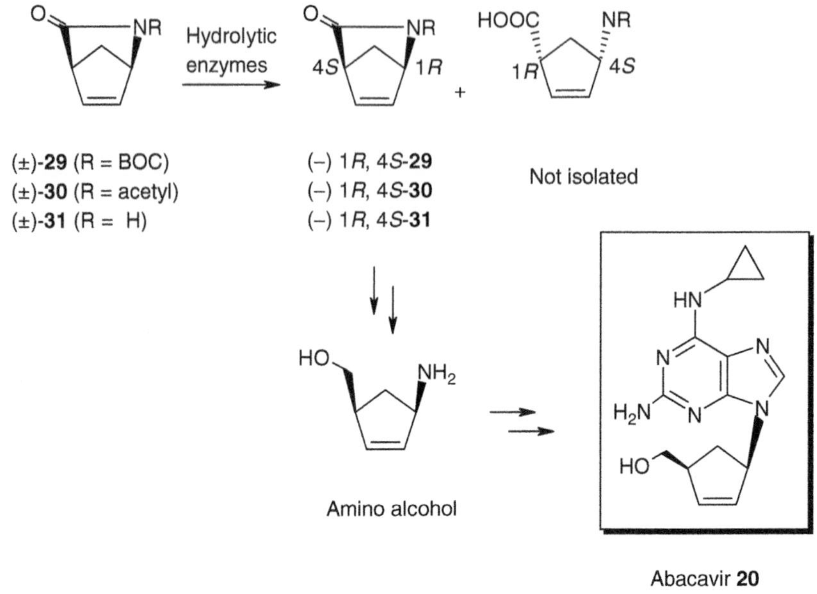

FIGURE 3.14 Production of Abacavir.

such as butyloxycarbonyl (BOC) or acetyl, we may be able to find a conventional hyd
enzyme, rather than needing a specialized γ-lactamase, that would hydrolyse the
bond of **29** and **30** enantioselectively. We therefore embarked on a program to prod
N-BOC-substituted γ-lactam **29** in an optically pure form for the synthesis of Abacav

A number of commercially available hydrolytic enzymes were screened for the ab
hydrolyse the lactam bond of racemic **29** [(±)-*tert* butyl 3-oxo-2-azabicyclo(2.2.1)hept-
2-carboxylate], enantioselectively (Figure 3.14). There was substantial chemical hydrol
the N-BOC protecting group under aqueous conditions, but this could be minim
reactions contained up to 50% (v/v) of organic solvents such as tetrahydrofuran. We
surprisingly that several esterases, proteases, and lipases hydrolyzed (+) 1S, 4R-**29**
corresponding N-acyl amino acid, leaving behind the residual (−) 1R, 4S-**29** of the
absolute configuration for the synthesis of Abacavir (Figure 3.14). Savinase was
promising and reactions were found to be highly enantioselective. Savinase (subtilis
3.4.21.62) is a serine-type protease, manufactured by Novozyme, and is produced b
merged fermentation of recombinant alkalophilic *Bacillus* sp. It is cheaply available i
for use in detergent industry and is an active ingredient of washing powders [37]. Typ
reactions were carried out at 30°C in phosphate buffer (pH 8.0) containing up to 50%
organic solvent such as tetrahydrofuran and up to 100 g/L of racemic **29**. Reaction m
were monitored by reverse phase and chiral HPLC. Upon completion of rea
(50% conversion) the ee of (−)-**29** was better than 99%. The reaction mixture was f
and (−)-**29** was isolated in good chemical yield (84% theoretical) by extraction into cyc
ane followed by evaporation of organic solvent [36]. Similarly, savinase hydrolyzed the
bond of racemic **30** (*cis*-2-acetyl-2-azabicyclo[2.2.1]hept-5-en-3-one) enantioselectiv
afford (−) 1R, 4S-**30** ee > 99% (Figure 3.14); interestingly, savinase did not hydroly
unactivated racemic **31**. This was the first example of the use of detergent enzymes i
large-scale production of pharmaceutically important active intermediates. This has f
the basis for the development of a simple and scalable process for preparation of op
pure N-substituted γ-lactams for the production of Abacavir [36,38].

In summary, we have developed a simple, efficient, cost-effective, and practical enzymic procedure for the preparation of *N*-substituted γ-lactams, using a readily bulk-available commercial protease, in very high optical (ee > 99%) and chemical yields (84% theoretical). This process was scaled up and integrated into the chemical route, which resulted in a significant overall cost saving to manufacture Abacavir. We subsequently used our strong patent coverage for price negotiations with outsourcing partners [38].

3.5.3 ZANAMAVIR (RELENZA)

Relenza **21** (2,3-didehydro-2,4-dideoxy-4-guanidinyl-*N*-acetylneuraminic acid, Figure 3.15) is a potent and selective inhibitor of influenza virus sialidase (neuraminidase) and has been approved by the FDA for the treatment of types A and B influenza—the two types most responsible for flu epidemics. The neuraminidase removes sialic acids from glycoconjugates aiding virus penetration into respiratory tract and releases mature viroins from the infected cells. Relenza works by preventing the spread of influenza from one cell to another within the respiratory tract. To target the site of infection directly, Relenza is inhaled (using a Diskhaler device) into the airways, where the flu virus replicates, and begins to function to destroy the

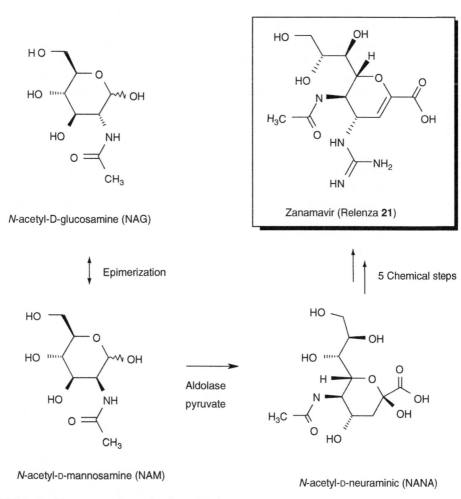

FIGURE 3.15 Chemoenzymic production of Relenza.

virus. Relenza was licensed from Biota Australia after a collaborative effort that le rational design from the x-ray crystal structure of sialic acid bound to influenza sialida the aid of molecular modeling and computational chemistry techniques [39].

N-acetyl-D-neuraminic acid (NANA, Figure 3.15) is the key intermediate for syntl Relenza. NANA is the most prominent of the sialic acids, a group of 9-carbon amino incorporated at the terminal positions of glycoproteins and glycolipids, which p important role in a variety of biological recognition processes [40]. The chemical sy of NANA is lengthy, requiring complex protection and deprotection steps, and does n much potential for economic large-scale production [41]. NANA can be isolate biological materials such as milk, eggs, edible birds' nests [40,42], or bacterial ce polymer colominic acid [43–45], but levels are modest and purification difficult. Th promising option for long-term supply appeared to be the use of the NANA aldolase e Several bacteria contain this enzyme, which appears to have a role in the catabo NANA that is cleaved to pyruvate and N-acetyl-D-mannosamine (NAM). Under appr conditions, the enzyme can be used in reverse to synthesize NANA (Figure 3.15).

There are, however, a number of hurdles to be overcome for the development efficient biochemical process for production of NANA: (i) NAM is expensive and n produced *in situ* by epimerization of N-acetyl-D-glucosamine (NAG) at C_2 [46], t equilibrium lies in favour of NAG; (ii) the K_m for NAM is high (0.7 M) [47]; (iii) N inhibitory [47]; and (iv) pyruvate is usually used in excess (10 molar excess) to pu equilibrium over to NANA, which necessitates the removal of large amounts of r pyruvate, but pyruvate and NANA are difficult to separate. The synthesis of NANA the aldolase enzyme either from *E. coli* or *Clostridium perfringens* had been pre reported by several groups [48–50]. These groups developed batch processes for the p tion of NANA from NAM and pyruvate using free or immobilized NANA aldolase. I to drive the equilibrium toward NANA, the cheaper pyruvate was generally used i excess, making the downstream processing rather difficult.

The Wandrey group at Juelich developed an elegant continuous process for NAN thesis introducing the NANA-2-epimerase enzyme for epimerization of NAG and inte the epimerization with NANA synthesis in an enzyme-membrane reactor [47]. Althou process has an excellent space time yield, the epimerase enzyme is not freely availab product stream is very dilute and gradient chromatography is required for purification, r scale-up difficult. The Marukin Company [51] also achieved "simultaneous" epimerizati aldol condensation by operating the enzyme process at high pH (e.g., pH 10.5), but this significant compromises of both enzyme activity and epimerization rate, and again the p isolation issue was not adequately addressed. None of the reported procedures, ho offered a fully effective and integrated process solution for the large-scale produc NANA using NANA aldolase, and in particular methods for isolation of NANA the reaction mixture were inadequate. We therefore embarked on the development approaches for NANA production that surmounted most of these problems [52,53].

It was clear that any approach to NANA synthesis would require a plentiful so enzyme. We cloned the enzyme from *E. coli* and overexpressed it in an inducible syste promoter) [53]. As very high expression levels were achieved (10 units/mg protein, rep ing 30% of cell protein), we hoped that we would not need to purify the enzyme f Indeed, reactions with crude extracts were clean, with no observed side reactions. To mi the processing required we simply homogenized the bacterial cells and directly im ized the enzyme from crude extracts onto Eupergit-C beads without any clarificatio debris did not interfere with immobilization of NANA aldolase and could be washe after the immobilization step. The immobilized enzyme was stable for at least ten reaction cycles.

Our initial strategy for production of NANA was to carry out the aldolase reaction on equilibrium mixtures of NAG and NAM (4:1 ratio), resulting from base-catalyzed epimerization of NAG, using a 5 to 7 molar excess of pyruvate to drive the equilibrium toward the NANA product (Figure 3.15). This resulted in more than 90% of NAM being converted into NANA; at completion, the reaction thus comprised NANA (21 g/L), NAM (2g/L), NAG (65 g/L), and pyruvate (56 g/L). To isolate NANA we removed pyruvate initially by complexation with bisulphite rather than resorting to the (low-capacity) ion-exchange separation of NANA and pyruvate. Indeed, after removing sodium ions with an IR120(H^+) column, pyruvate was complexed very effectively using an A113 column in the HSO_3^- form. The column could be regenerated by simply using hot water to strip pyruvate and recharging with sodium metabisulphite—in principle the pyruvate could be reutilized. As the only remaining acid, NANA was selectively bound onto an anion-exchange resin, Duolite A113 PLUS(OAc^-), eluted and crystallized by addition of acetonitrile. To improve the volumetric productivity for NANA, we studied the effect of increasing NAG/NAM concentration in equilibrium mixtures (4:1 ratio of NAG/NAM) and found out that high levels of NAG are grossly inhibitory to the aldolase reaction. To avoid the inhibition, we therefore investigated the possibility of enriching NAG/NAM mixtures for NAM.

In the second approach, a selective precipitation of NAG using isopropanol was developed to produce an NAM-enriched mixture (1:5 ratio of NAG/NAM). This was used in the reaction at a very high NAM concentration (up to 20% w/v) so that NAM itself drives the reaction and it is only necessary to add a small molar excess of pyruvate (e.g., 1.5-fold) (Figure 3.16). At the "conclusion" of the reaction, the mixture comprised NANA (155 g/L), NAM (13 g/L), NAG (24 g/L), and pyruvate (29 g/L) (Figure 3.17). Under these conditions, NANA could now be crystallized directly from the reaction mixture simply by the addition of acetic acid. Furthermore, we found that upon completion of the reaction, after filtering off

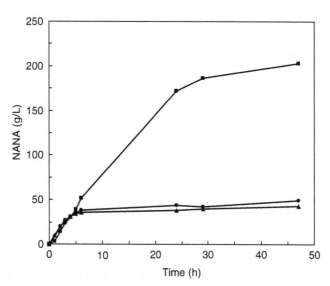

FIGURE 3.16 Comparison of production of NANA from NAM-enriched and equilibrium mixtures. Reactions (20 mL) were carried out at 20°C using either NAM-enriched (■) or equilibrium mixtures (●, ▲). The molar ratio of pyruvate/NAM was either 2:1 (■, ▲) or 5:1 (●). Reactions were started by the addition of 8 g wet weight of washed immobilized enzyme beads. The NAG/NAM concentrations were as follows: 66:224 g/L (■),169:43 g/L (●, ▲).

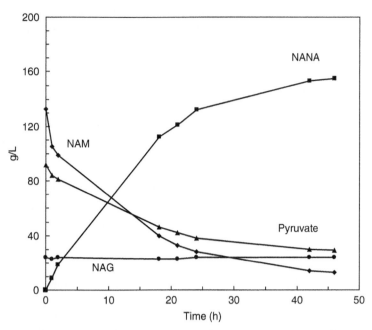

FIGURE 3.17 Production of NANA from an NAM-enriched mixture. The reaction was carried
20°C in a 7 L LH-fermentor containing 3.5 L of an NAM-enriched mixture. Sodium pyruvate
was added in a 1.5 molar ratio with respect to NAM. After adjusting the pH to 7.5 with NaOH, r
was started by addition of 1.7-kg wet weight of washed immobilized enzyme beads. NAN
pyruvate (▲), NAG (●), NAM (◆).

the immobilized enzyme beads and concentrating the reaction mixture, we could ach
NANA concentration of approximately 200 g/L by adding six volumes of acetic acid
crystalline NANA in excellent purity [53].

In order to understand the contribution to sialidase binding made by each of the
on the dihydropyran ring of Relenza, we used the aldolase technology to produce r
quantities of various NANA analogs, for structure activity relationship (SAR) studies.
such study, using 2-deoxy-glucose as substrate in the aldolase reaction, **32** was produc
used to synthesize **33** and **34** in a one-pot process that lacked a substituent at the 5-posi
the dihydropyran ring [54] (Figure 3.18). Compounds **33** and **34** were evaluated as inh
of influenza A and B sialidase and were shown to display markedly reduced affinity
viral enzymes when compared to Relenza [55]. Subsequent evaluation of these comp
against influenza virus sialidase has established the critical importance of the 5-acet
group for good binding affinity of Relenza.

In summary, we developed two processes for NANA production, which have bot
operated at substantial scale [52,53]. In the first process, the use of a large molar
of pyruvate (five- to sevenfold) to drive the equilibrium to NANA has been made f
by development of a bisulphite complexation method for pyruvate removal from the re
mixture. In the second process, the development of a method for enrichment of
in NAG/NAM mixtures has allowed NAM to be used at very high concentratio
this obviates the need to use a large molar excess of pyruvate. The NAG residue rec
from the enrichment procedure can of course be recycled through the epimerization
dure. Under these circumstances it has been possible to develop a method for re
of NANA from the reaction mixture by a simple crystallization: [NANA] > 150 g/L,

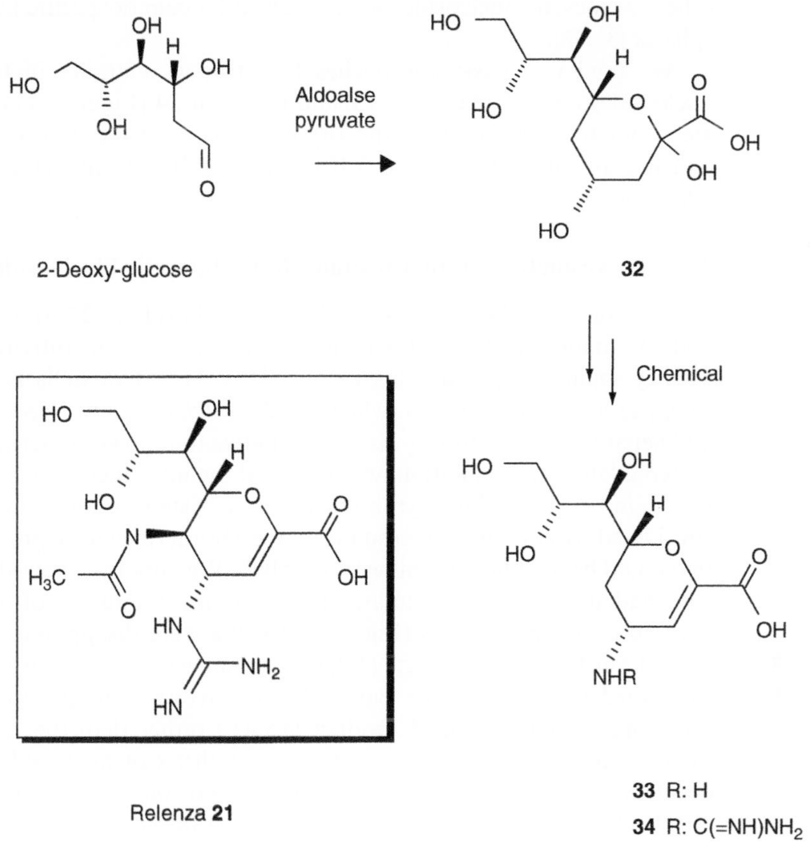

FIGURE 3.18 Aldolase-catalyzed bioconversion of 2-deoxy-glucose.

process yield 75% from NAM, purity of crystalline NANA >99%. In the pilot plant multi-kg quantities of NANA were produced using a recycled stable immobilized aldolase preparation. In the factory setting the same batch of enzymes was reused for more than 2000 cycles in batch column reactors, without any significant loss of activity, to produce multitonne quantities of NANA. This was the first example of an integrated process for industrial scale application of NANA aldolase, which was crucial for the market launch of Relenza.

3.5.4 CARBOCYCLIC NUCLEOSIDES

We at Glaxo Wellcome had spent a number of years investigating the potential of carbocyclic nucleosides as antiviral agents. Carbocyclic analogs of purine and pyrimidine nucleosides have anti-HIV and antiherpetic properties [56–58]; traditionally, however, these compounds have been synthesized and tested initially in racemic form. Where the enantiomers had been examined, biological activity shown by the racemate was found to reside primarily in the "natural" enantiomer, whereas the corresponding "unnatural" isomer was found to be inactive or showed substantially reduced potency [58–61]. Enzymes have frequently been used to resolve the enantiomers of carbocyclic and 2',3'-dideoxynucleoside analogs for further biological testing [62–66]. For example, adenosine deaminase was used to prepare optical isomers of a wide range of carbocyclic analogs of purine nucleosides [31,59,64,65], whereas

other carbocyclic nucleosides were resolved by enantiospecific hydrolysis of their mo⋯
sphates [31,58].

We used two basic approaches to introduce chirality in the synthesis of carb⋯
nucleosides c-BVdU **22**, c-dG **23**, and Carbovir **24** (Figure 3.11): (i) resolution of inte⋯
ates prior to base addition; and (ii) synthesis from the natural antibiotic aristeromy⋯
and neplanocin-A **16** (Figure 3.10) produced by the filamentous bacterium *Strept*⋯
citricolor.

3.5.4.1 Resolution of Intermediates to Carbocyclic Nucleosides c-BVdU and c-dG

(+)-Carbocyclic 2′-deoxy-5-[(*E*)-2-bromovinyl]uridine **22** (c-BVdU) and (+)-carb⋯
2′-deoxyguanosine **23** (c-dG) (Figure 3.11) are potent antiviral agents; (+)-**22** po⋯
activity against herpes simplex virus 1 (HSV-1) and varicella zoster virus (chicken p⋯
shingles) *in vitro* and *in vivo*, while (+)-**23** is active against HSV-2, human cytomegal⋯
and hepatitis B virus [67,68]. As an alternative to total asymmetric synthesis of **22** and ⋯
to avoid late-stage resolution, we envisaged the introduction of an enzymic or microbi⋯
to obtain key homochiral synthons, at an early stage in the synthetic strategy. One ap⋯
considered was the enzymic resolution of ester intermediates, prior to the addition of th⋯
that could be used in the synthesis of **22** [62]. We initially followed the work of Sicsic et ⋯
who had used pig liver esterase (PLE) for the resolution of 4-*cis*-acetamido-cyclo⋯
enecarboxylic acid (±)-**35** (Figure 3.19). We were disappointed, however, to find t⋯
optical purity of the products (1*R*,4*S* residual ester, ee 28%) was very modest and ha⋯
overstated (ee 87%) in the paper due to an error in the optical rotation value ⋯
homochiral ester. The acid product (1*S*,4*R*) rather than the residual ester (1*R*,4*S*) ⋯
the correct absolute stereochemistry for synthesis of nucleoside analogs (Figure 3.1⋯
decided, therefore, to look at the benzoyl amino compound (±)-**36**, which was resolve⋯
effectively using PLE and to our surprise with the correct opposite enantioselectivity (⋯
3.19). This highlights that the effect of substitution remote from the center of hydroly⋯
profoundly affect the enantioselectivity of the enzymic hydrolysis of 4-amino-cyclope⋯
carboxylic acid derivatives [62]. Although the enantioselectivity was still relatively mod⋯

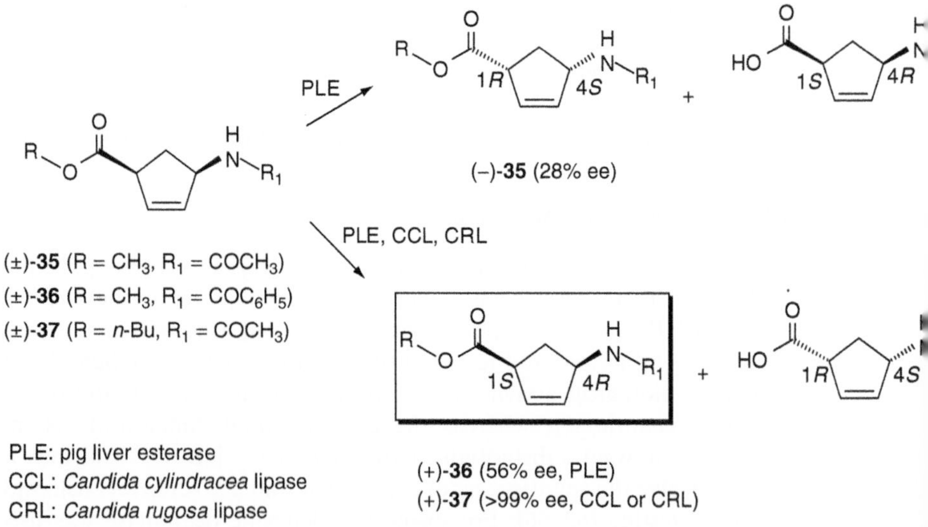

PLE: pig liver esterase
CCL: *Candida cylindracea* lipase
CRL: *Candida rugosa* lipase

FIGURE 3.19 Enzymic resolution of carbocyclic nucleoside precursors.

residual ester product [(+)-1*S*,4*R* **36**, ee 56%] was of the correct absolute stereochemistry for synthesis (8 steps) of (+)-*c*-BVdU **22** (Figure 3.20). We were nevertheless able to improve the enantioselectivity by careful attention to the reaction conditions, particularly the cosolvent used; for example, methanol at 5% (v/v) increased the enantioselectivity of reactions by at least twofold. Interestingly with dimethyl sulphoxide (DMSO) at 50% (v/v) the enantioselectivity was reversed such that the residual ester was enriched in the wrong (−)-1*R*,4*S* enantiomer [62]. Although the enantioselectivity was not complete, the residual ester product could be obtained in good optical purity by allowing the reaction to proceed beyond 50%; for example, the residual ester could be obtained in 90% ee (65% yield) by allowing the reaction to proceed to 65% conversion or 95% ee (58% yield) at 69% conversion [62]. We were also interested to note that *C. cylindracae* or *C. rugosa* lipases were reported some years later to resolve higher ester of the acetamido compound *n*-butyl (±)-**37** (Figure 3.19) or *n*-hexyl esters; the residual esters (+)-1*S*,4*R* were obtained in excellent optical purity (ee > 99%) [70].

In an attempt to shorten synthetic routes to **22** and **23**, we also investigated an alternative method using the stereoselective reduction of **38** by growing cells of *M. circinelloides* [71]. Optimization of fermentation conditions, pH, and temperature resulted in the production of the desired intermediate (+)-**38**, in an optically pure form (ee > 98%; yield, 62%), which was used for the convergent syntheses of (+)-**22** (3 steps) and (+)-**23** (4 steps) (Figure 3.20).

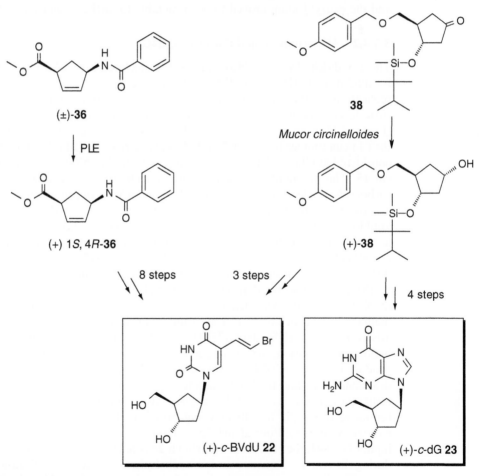

FIGURE 3.20 Chemoenzymic synthesis of (+)-*c*-BVdU **22** and (+)-*c*-dG **23**.

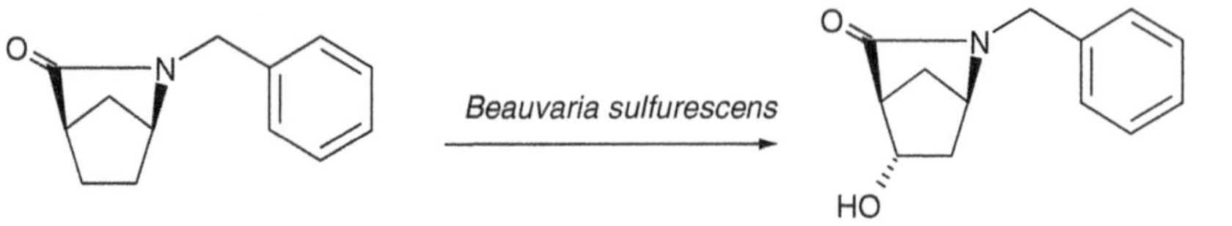

FIGURE 3.21 Microbial hydroxylation of the bicyclic lactam **39** as a potential intermediate for cyclic nucleoside synthesis.

Furthermore, the α-hydroxy (+)-**38** has been shown to be a versatile intermediate that used for the convergent synthesis of a variety of chiral purine and pyrimidine carb nucleoside analogs [71].

There was potential for a still more attractive route by the enantioselective, regios and stereospecific hydroxylation of the bicyclic lactam (±)-**39** (Figure 3.21). Althou 5-*exo*-hydroxylation of (±)-**39** was reported previously using *Beauvaria sulfuresce* optical purity of the product **40** was not measured quantitatively [72]. Although we ob similar results, we were disappointed to find that the desired hydroxylated product had of only 11%. Screening of other fungi increased this beyond 50%, but yields were quite r and the benzyl lactam proved to be intractable to further chemistry [73] (Figure 3.21)

3.5.4.2 Aristeromycin and Carbovir (*c*-d4G)

The 2′,3′-didehydro-2′,3′-dideoxycarbocyclic nucleoside, (±)-Carbovir, is a potent and ive inhibitor of HIV *in vitro* [74]. Its hydrolytic stability and ability to inhibit infectio replication of the virus in human T-cell lines at concentrations 200- to 400-fold toxic levels has made Carbovir a potentially useful antiretroviral agent. (−)-Carbo (Figure 3.11) is approximately twofold more active than the corresponding racemate, w the (+)-enantiomer is at least 75-fold less active than (−)-**24** [75]. One of the routes to considered at Glaxo Wellcome was synthesis from the chiral natural product (−)-ariste cin **15**. The potential of aristeromycin as a chiral pool starting material for synth carbocyclic nucleosides was recognized at Glaxo Wellcome in the mid-1980s and we in a substantial effort in the development of fermentation and isolation conditions fo antibiotic. (−)-**15** was a very attractive starting material because it was readily avail a secondary metabolite of *S. citricolor* and would afford (−)-**24** without the need to re an optical resolution. Synthesis of (−)-**24** from (−)-**15** involved nine steps requiri distinct transformations: (i) an adenine to guanine base interconversion; and (ii) introd of the 2′,3′-double bond from the 2′,3′-diol [76]. One approach for the base conv involved the hydrolytic deamination of *cis*-4-[2,6-diamino-9*H*-purin-9-yl]-2-cyclope methanol dihydrochloride (−)-**41** using adenosine deaminase; (−)-**41** was prepared aristeromycin in eight steps [76] (Figure 3.22). To obtain sufficient material for t biological evaluation, the chemoenzymic route was investigated and a process develo produce (−)-Carbovir **24** on a kg scale [77]. We used commercial preparations of ade deaminase (calf intestinal mucosa and *Aspergillus* sp.), and following standard purifi techniques directly adsorbed the enzyme onto an anion-exchanger to provide a rapid m for its recovery (50 to 340 units/mg protein). Bioconversions with the partially p enzyme were carried out at pH 7.5 in fermenters (up to 70 L working volume). At input of (−)-**41**, the product (−)-Carbovir **24** (Figure 3.22) was much less soluble, cryst during the course of reaction and was subsequently recovered by filtration (>90% thec yield). To allow enzyme reuse, adenosine deaminase was immobilized onto Eupergit-C

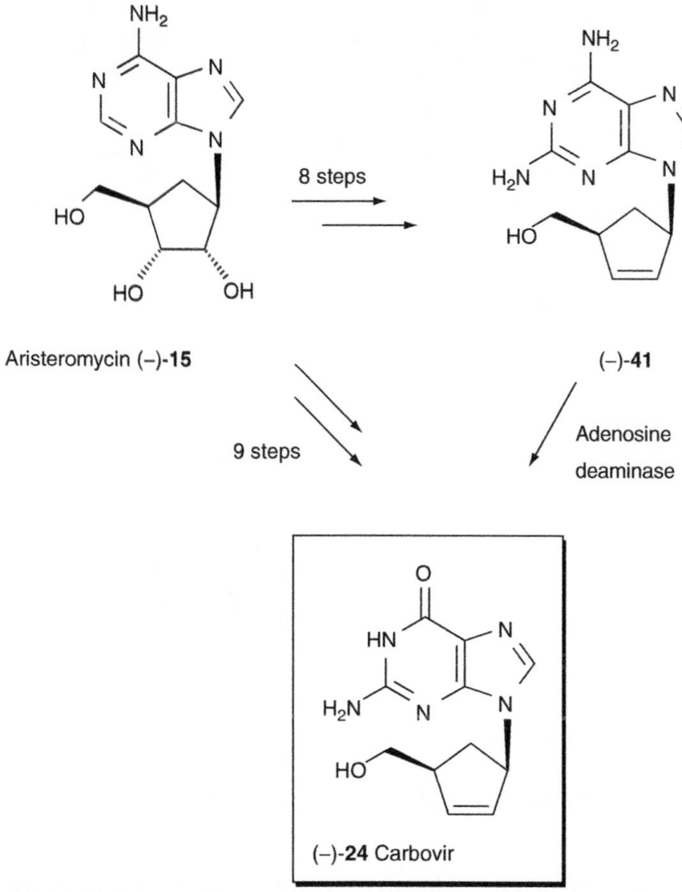

FIGURE 3.22 Chemoenzymic synthesis of Carbovir (−)-**24** from aristeromycin.

a stable enzyme preparation. Given the low solubility of (−)-Carbovir, the concentration of (−)-**41** was reduced to 2.5 g/L in bioconversions with immobilized enzyme so that the beads could easily be recovered from the reaction mixture without interference from the product. The enzyme was reused up to 10 cycles without any significant loss of activity. This work demonstrated the potential of adenosine deaminase as a catalyst for large-scale production of optically pure (−)-Carbovir **24** [31]. In contrast to Abacavir (−)-**20**, however, Carbovir suffers form poor oral absorption, limited brain penetration, low aqueous solubility, and potential for renal and cardiac failure [31]. Despite the structural similarity between Abacavir and Carbovir (Figure 3.11), it is remarkable that an apparently small change in the molecule can have such a profound effect on its biological activity.

Another approach to base interconversion arose through an observation during isolation of biosynthetic mutants of the aristeromycin production strain. Neplanocin-A **16** (Figure 3.10) was fed to block mutants looking for conversion of aristeromycin; one mutant was found not to produce aristeromycin, but to modify the base moiety of the added neplanocin-A, especially by deamination. On feeding aristeromycin, this was converted to carbocyclic inosine, carbocyclic xanthosine, and carbocyclic guanosine (Figure 3.23). The conversion of aristeromycin to carbocyclic inosine by *Streptomyces* sp. from the "mutant" culture was rapid and the ability to do this was widespread amongst the organisms we investigated. Further conversion of carbocyclic xanthosine or carbocyclic guanosine was much slower. Carbocyclic

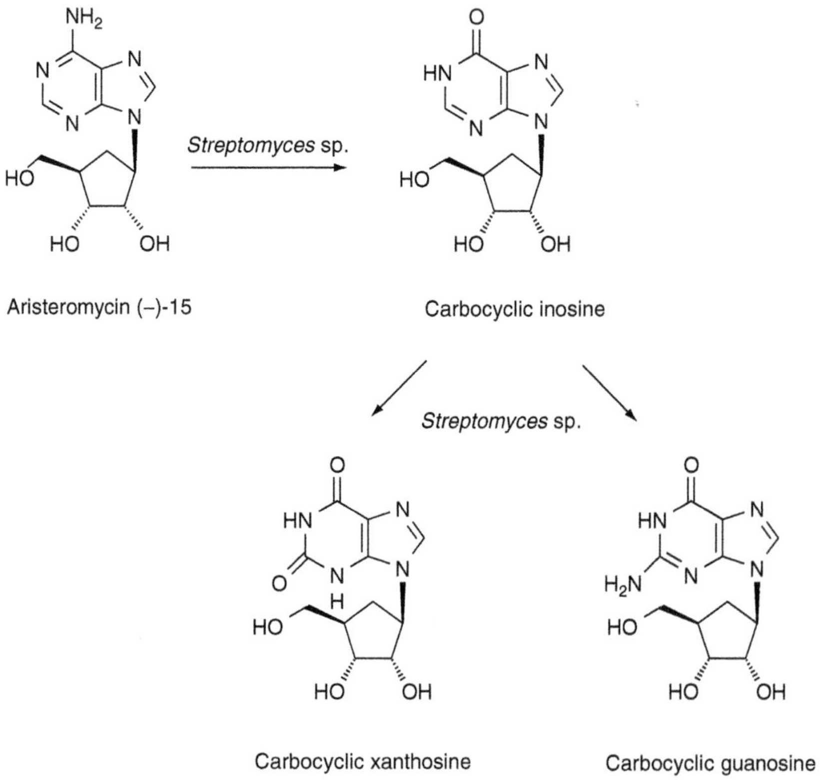

FIGURE 3.23 Microbial base interconversion of adenine to guanine.

inosine could be added to the cultures and was transformed to carbocyclic xanthosine
carbocyclic guanosine, but neither of these was further converted. Conditions co
manipulated to favor either carbocyclic xanthosine or carbocyclic guanosine as final pr
[73] (Figure 3.23).

3.6   ENABLING TECHNOLOGIES

3.6.1   CHIRAL AMINES

Chiral amines are valuable synthons that feature prominently in many agrochemica
pharmaceutical drug pipelines. A recent survey of the top 400 drugs revealed the wide
prevalence of compounds containing amino groups attached to asymmetric centers [7
combined 2002 sales of only 12 blockbusters containing chiral amines was in excess
billion, covering a variety of indications including depression, bacterial infection, ather
osis, prostatic hypertrophy, growth failure, migraine, and Alzheimer's [78] (Figure
Chiral amines have great utilities as catalysts for asymmetric synthesis [79], resolving
[80], and chiral auxiliaries/chiral bases [81]. Methods for the preparation of chiral ami
largely based on resolution of racemates (limited to 50% yield), either by recrystalliza
diastereomeric salts [82–83] or by kinetic resolutions of racemates using hydrolases [
To improve reaction yields beyond 50%, asymmetric approaches have been develope
(i) transaminases for the conversion of ketones into chiral amines [89–92]; (ii) mi
reduction of oximes into chiral amines [93]; and (iii) asymmetric hydrogenation of

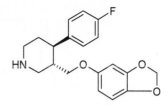

Roche — Tamiflu (anti-flu)
Sales > $1 billion

GSK — Paxil (asthma, depression)
Sales ca. $3 billion

Merck — Trusopt (elevated
intraocular pressure)
Sales $455 million

Sanofi, BMS — Plavix (atherosclerosis)
Sales $3 billion

Pfizer — Zoloft (depression)
Sales $2.8 billion

Syngenta — Piperophos (herbicide)
Sales $24 million

BayerAG — Iprovalicarb (fungicide)
Sales > $20 million

FIGURE 3.24 Structures of selected agrochemicals and pharmaceutical drugs containing chiral amino groups.

using chiral ruthenium catalysts [94–96]. Alternative strategies are based on dynamic kinetic resolution (DKR) of amines that employ enzymes in combination with transition-metal catalysts for *in situ* racemizations [97–100]. For example, in the DKR approach, Reetz and Schimossek [98] developed a chemoenzymic route for the synthesis of α-methylbenzylamine derivatives using CAL in combination with a palladium (Pd/C) catalyst to racemize the

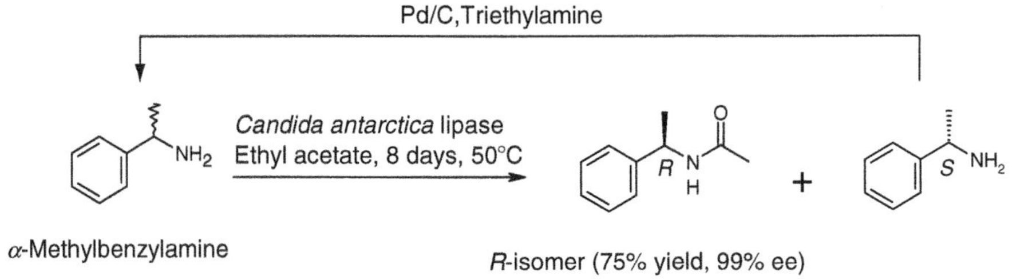

Pd/C,Triethylamine

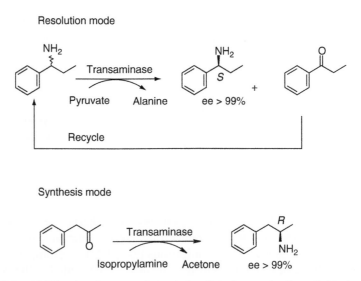

Candida antarctica lipase
Ethyl acetate, 8 days, 50°C

α-Methylbenzylamine

R-isomer (75% yield, 99% ee)

FIGURE 3.25 Dynamic kinetic resolution using palladium for *in situ* racemization.

undesired *S*-amine isomer (Figure 3.25). Although yields greater than 50% were obtair
> 99%, *R*-amide), there were compatibility issues between the enzymic and the chemic
due to the harsh conditions required for racemizations. This, together with the ver
racemizations rates (up to 8 days), has unfortunately limited the general utility c
approach. Current commercial processes for production chiral amines include:

1. BASF's ChiPros technology [97] based on lipase-catalyzed resolutions; this is, ho
    a multistep process requiring acylation, deprotection, and subsequent larg
    separation of the amine product.
2. Cambrex, using Celgene's transaminase technology [89,90], can be operated ei
    "resolution" or "synthesis" mode with yields >50%; however, the *R/S*-transan
    are not commercially available, the substrate range is rather limited to primary a
    and a very limited range of amino acceptors can be used in the resolution mode (
    3.26).
3. Biokatalyse, using a laborious multistep route to produce α-methylbenzylamin
    (ee > 99%, 87% yield) (Figure 3.27).

In the 1990s we were interested to develop biocatalytic routes to chiral amines and
acids to complement our chemical development programs. In collaboration with Univ

Resolution mode

NH₂

Transaminase

Pyruvate   Alanine

NH₂

S

ee > 99%

+

Recycle

Synthesis mode

Transaminase

Isopropylamine   Acetone

R

NH₂

ee > 99%

FIGURE 3.26 Celgene's (Cambrex) transaminase technology for production of chiral amines.

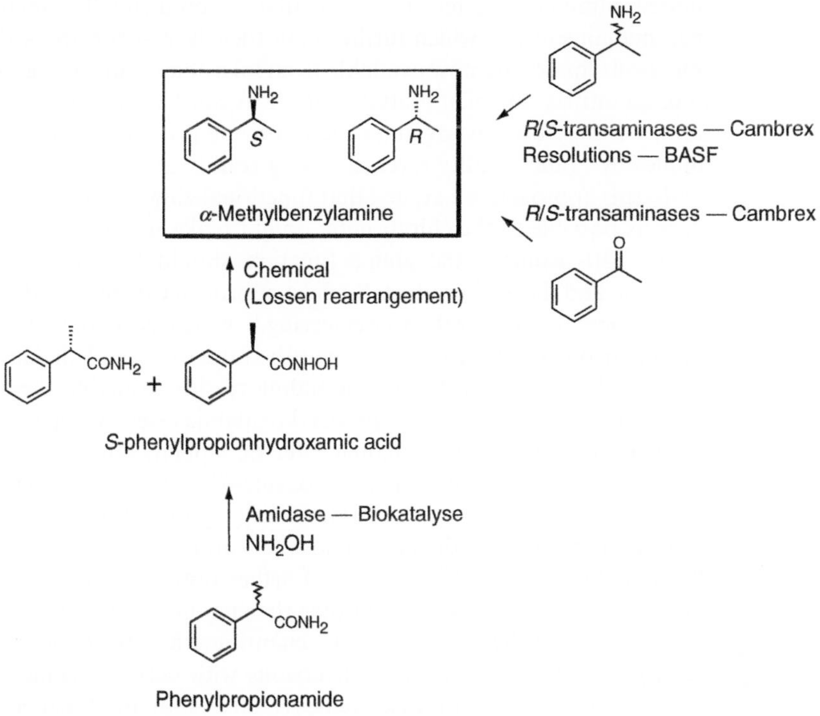

FIGURE 3.27 Commercial production of chiral amines.

of Warwick and Edinburgh, we at Glaxo Wellcome developed two elegant processes: (i) resolution of amines by "substrate engineering" [102]; and (ii) deracemization of amino acids and amines, in high optical and chemical yields, using enantioselective oxidases coupled with nonselective chemical reducing agents [103–105]. After the merger with SmithKline Beecham, however, our deracemization technology was "transferred" to Ingenza (a start-up company from Edinburgh University), who now exploits this approach for the synthesis of chiral amines and amino acids.

3.6.1.1 Resolution of Amines by Substrate Engineering

As a general rule of thumb, kinetics resolutions of esters of chiral acids **42** are more likely to succeed using esterases (e.g., porcine liver esterase) than lipases, whereas the resolution of esters of chiral alcohols **43** are more likely to succeed using lipases than esterase (Figure 3.28). This "functionality reversal," therefore, greatly extends the use of lipases in resolution processes, and is an innovative concept of considerable practical potential [102]. Only a

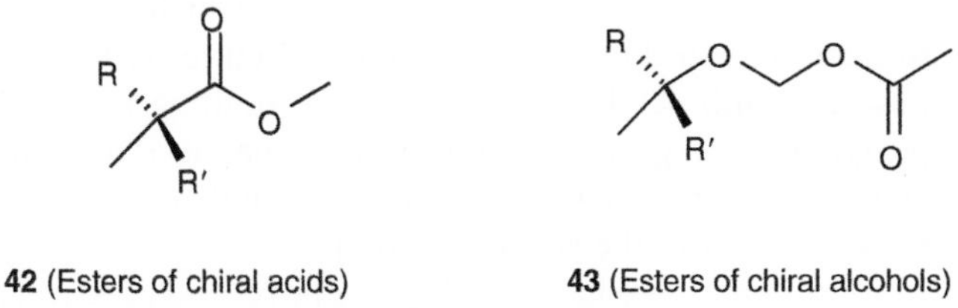

42 (Esters of chiral acids)   **43** (Esters of chiral alcohols)

FIGURE 3.28 Enzymic resolutions by "functionality reversal."

limited number of esterases are available commercially—these tend to be mainly mammalian sources, which further limit their large-scale applicability. In contrast, ho numerous lipases are now available from microbial sources (e.g., *P. fluorescence* and C bulk quantities. In collaboration with Professor Crout at Warwick University, United dom, we therefore envisaged a process for the production of either *R*- or *S*-isomers of amines by "functionality reversal" using readily available lipases [102].

In this approach, we argued that functionalizing the amino group, with an ester link close as possible to the chiral center, might make it more amenable to lipase resolutio access both isomers, the amine function should be regenerated from both the re substrate and the acid product. Such an approach can be regarded as "substrate enginee complementary to "protein engineering," which is used to tailor the enzyme to m particular substrate. Several racemic-substituted 1-phenylethylamines (α-methylbenzyl were synthesized from the corresponding readily available acetophenones. These wer used in the synthesis of ethyl and octyl oxalamic esters **44a–g** in good chemical yields (3.29). Initially the ethyl oxalamic ester of 1-phenylethylamine **44a** was screened ag: panel of lipases; CAL B (CAL, Novozyme-435) showed the highest enantioselectivity for the hydrolysis of the *R*-oxalamic ester (Figure 3.29). Both the residual substrate a corresponding acid product could now be readily hydrolyzed chemically to generate eit. *R*- or *S*-chiral amine (Figure 3.29). Further optimization through modification of th chain length (C_2 to C_8) revealed that the enantioselectivity was significantly improve the extension of the alkyl chain; the enantiomeric ratio *E* increased from 11 to 104 w octyl group (Figure 3.30). Bioconversions with octyl oxalamic esters **44b–g** proceede very high degree of enantioselectivities (*E* 30 to 100); the *R*-oxalamic acids and the *S*-ox esters were conveniently hydrolyzed chemically to the corresponding optically pure ; (Figure 3.29). Interestingly, in contrast to CAL, lipases from porcine pancreas (PPl *P. fluorescence* (PFL) showed opposite *S*-selectivity hydrolyzing the *S*-oxalamic (Figure 3.30). Furthermore, in sharp contrast to CAL, increasing the ester chain from methyl to octyl had very little effect on the enantiomeric ratio with PPL an (Figure 3.30). We also analyzed the selectivity of CAL and the influence of increasin chain length using published x-ray crystal structure of CAL [106]. Using comput; modeling (minimization and dynamics programs), with mapping and calculations likely binding sites for both enantiomers with varying chain length, we found a likely "; pocket" in the active site of CAL that preferentially accommodates the methyl grou oxalamic ester of 1-phenylbenzyl amine resulting in *R*-oxalamic acid product. The act: seems to be narrow and U-shaped, allowing entry of the benzyl group—there is hydrophobic binding region that is involved in the binding of the ester chain up t length, and any further increase beyond C_8 does not appear to have any significant ef enantioselectivity of CAL [107].

Current approaches for commercial production of chiral amines are limited o primary amines (Figure 3.27). We therefore wondered if our process could be used to p chiral secondary amines. We were surprised to find that under nonoptimized conditic octyl oxalamic ester **45** was kinetically resolved with moderate selectivity (*E* 4) by CAL *R*-oxalamic acid and *S*-oxalamic ester, which, after chemical hydrolysis, afforded isomers of *N*-α-dimethylbenzylamine (Figure 3.31). Similarly modest selectivity (*E* observed with octyl oxalamic ester **46** resulting in the *R*- and *S*-isomers of 5-(3-pyri pentylamine (Figure 3.31). Given the apparent structural similarities between **44b** (*E* and **45** (*E* 4), this suggests that the additional methyl group might interfere with binding methyl pocket in the active site [107].

In summary, we have developed a novel and practical method for the resolution of ; of readily synthesized amines by "substrate engineering." This simple process is cap:

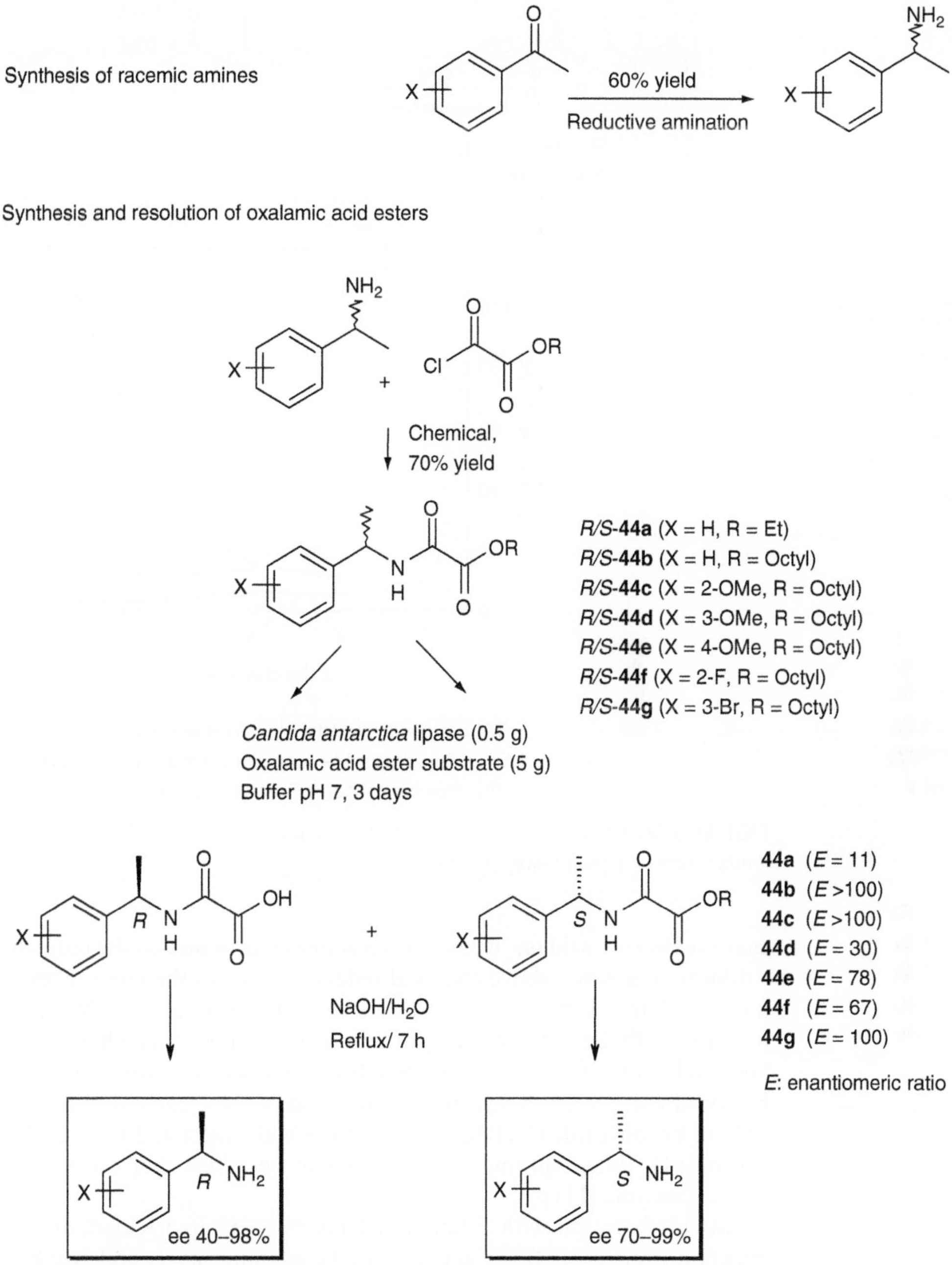

FIGURE 3.29 Resolution of amines by substrate engineering.

accessing both *R*- and *S*-isomers of primary and secondary amines in very high optical purities [102] (ee up to 99%).

3.6.1.2 Deracemization of Amino Acids and Amines

Based on the original reports by Hafner and Wellner [108] and Soda and coworkers [109,110] we envisaged a deracemization approach where a racemic amino acid is oxidized by an

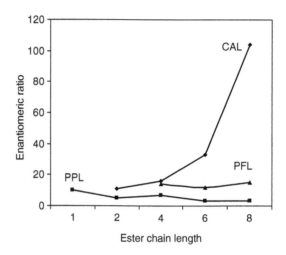

R' = Me, Et, butyl,
hexyl, octyl

CAL: *Candida antarctica* lipase (*R*-selective)
PFL: *Pseudomonase fluorescence* lipase (*S*-selective)
PPL: Porcine panceatic lipase (*S*-selective)

FIGURE 3.30 Effect of ester chain length on enantioselectivity of lipase-catalyzed hydrolysis
lamic esters of 1-phenylethylamine.

enantioselective oxidase, which converts one enantiomer to the corresponding achiral
Addition of a nonselective chemical reducing agent to the mixture results in the reduc
imine back to 1:1 mixture of the amino acid. If this process is allowed to undergo a su
number of cycles, the resulting amino acid eventually reaches optical purity—yiel
approach 100% if the imine can be efficiently reduced before it undergoes hydrolysis
ketone (Figure 3.32). A significant limitation of these early studies, however, was the
sodium borohydride (NaBH$_4$), which reacts with water and is easily decomposed at l
neutral pH, thus requiring large excesses to be added that can lead to deactivation
oxidase enzyme [111].

In collaboration with Edinburgh University, United Kingdom, we developed a
mization process [103] for acyclic/alicyclic α-amino acids and β-substituted α-amin
using a combination of *R*-selective D-amino acid oxidases (D-AAO: porcine kidney,
nopsis variabilis [112]), and L-AAO (*Proteus myxofaciens* [113]; *Cellulomonas cellulan*
(Figure 3.32 and Figure 3.33). In this approach amino acids (prepared in racemic for
as a mixture of diastereomers from the corresponding racemic aldehydes by the S
reaction [115]) in buffer solutions, containing the appropriate chemical reducing agent
incubated with D- or L-AAO and reactions were monitored by chiral HPLC. As an alte
to NaBH$_4$, we used NaCNBH$_3$, Pd/C-HCO$_2$NH$_4$ catalytic transfer hydrogenation [11
amine boranes [117] (NH$_3$:NH$_3$) in combination with either commercially available
or microbial L-AAO grown in fermenters; for example, *P. myxofaciens* L-AAO, wh

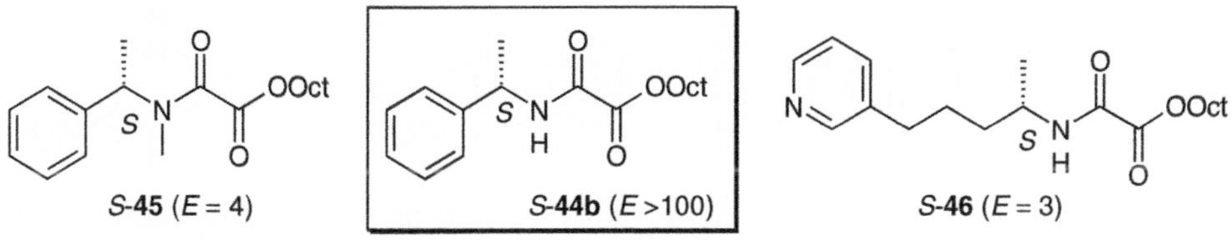

FIGURE 3.31 Effect of amino group substitution on enantioselectivity of *Candida antarctica* lipase (CAL)-catalyzed hydrolysis of octyl oxalamic ester derivatives.

been overexpressed in *E. coli* [113], was used either as a whole cell biocatalyst or as partially disrupted cells. A variety of acyclic (L-2-amino-4-Z-aminobutyric acid, L-lysine, D-α-aminobutyric acid, D-allylglycine, D-methionine, D-leucine), and alicyclic (D- and L-phenylalanine, L-piperazine-2-carboxylic acid, D-cyclopentyl glycine, L-proline) α-amino acids, as well as β-substituted α-amino acids (L-2-amino-3-methylhexanoic acid, L-2-β-methyl phenylalanine) were produced in very high chemical and optical yields (ee > 99%, yield 60 to 95%) (Figure 3.33).

Similarly, the same deracemization principle was applied to produce chiral amines, in high optical and chemical yields, using an enantioselective monoamine oxidase (MAO) coupled with nonselective chemical reducing agents [104,105] (Figure 3.34). Amine oxidases are classified mechanistically into two distinct groups. Type-I are copper-dependent and require topa quinone as a co-factor, whereas type-II are flavin-dependent. With type-I enzymes, the imine remains enzyme-bound and is not released as a free intermediate; these enzymes are therefore unlikely to be suitable for deracemizations. Type-II oxidases are more closely related to the D- and L-amino acid oxidases, in that they have a bound flavin unit and generate "free" imine [118]; known sources include bovine liver [119] and *Aspergillus niger* [120]. Schilling and Lerch [120,121] reported the cloning and expression of a type-II monoamine oxidase from *A. niger* (MAO-N), and subsequently Sablin *et al.* [122] purified the enzyme to homogeneity for substrate specificity and kinetic studies. The enzyme was very active toward simple aliphatic amines (amylamine, butylamine) but was also active, albeit at a much lower rate, toward benzylamine. In our experiments MAO-N showed very low but detectable activity toward L-(*S*)-α-methylbenzylamine with even slower oxidation of the D-(*R*)-enantiomer; the enzyme was only partially enantioselective. We therefore embarked on a program to improve both the catalytic activity and enantioselectivity using *in vitro* evolution methods [104]. A library of variants was generated by randomly mutating the plasmid containing MAO-N gene by using the *E. coli* XL1-Red mutator strain [105].

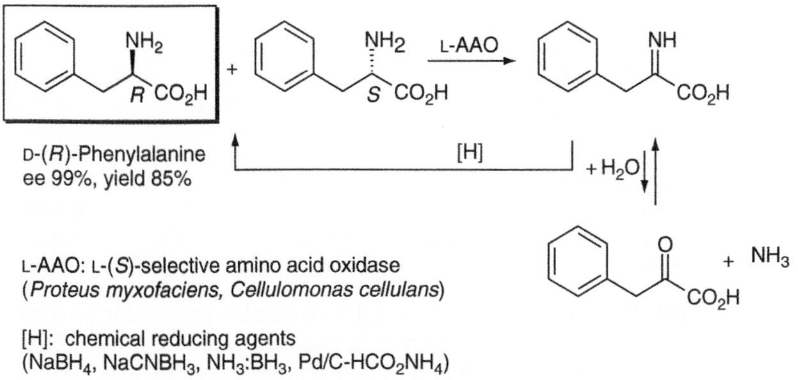

FIGURE 3.32 Deracemization of (±)-α-amino acids using *S*-selective L-amino acid oxidases coupled with chemical reducing agents.

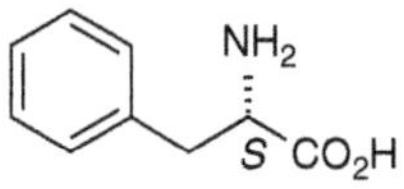

L-(S)-Phenylalanine
ee > 99%, yield 80%
Porcine kidney D-AAO
NaCNBH$_3$

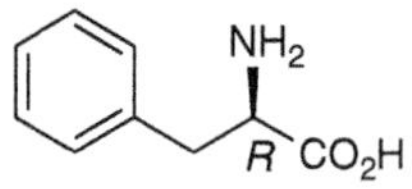

D-(R)-Phenylalanine
ee 99%, yield 85%
Proteus myxofaciens L-AAO
NH$_3$:BH$_3$

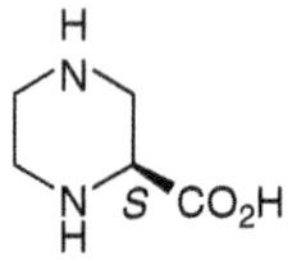

L-(S)-Piperazine-2-carboxylic acid
ee 99%, yield 95%
Porcine kidney, Trigonopsis variabilis [
Pd/C-HCO$_2$NH$_4$

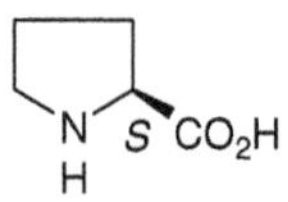

L-(S)-Proline
ee > 99%, yield 94%
Porcine Kidney D-AAO
NaCNBH$_3$

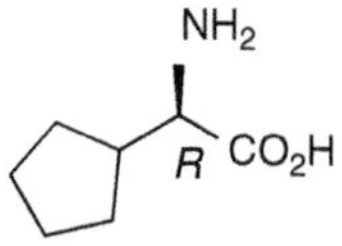

D-(R)-Cyclopentyl glycine
ee 94%, yield 80%
Proteus myxofaciens L-AAO
NH$_3$:BH$_3$

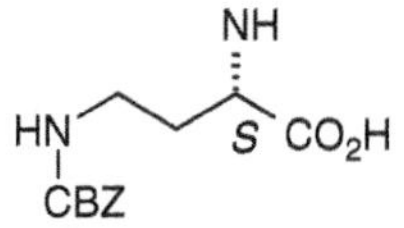

L-(S)-2-Amino-4-Z-aminobutyric acid
ee > 99%, yield 85%
Trigonopsis variabilis D-AAO
NaCNBH$_3$

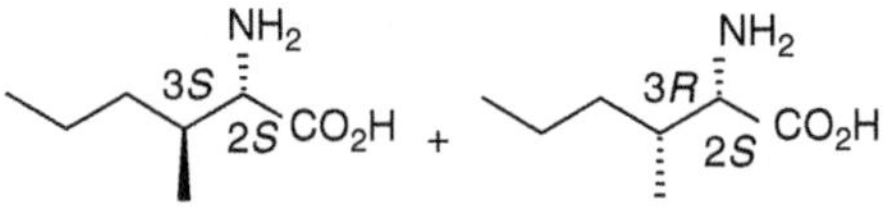

Isomers of L-(2S)-2-amino-3-methylhexanoic acid
ee 99%, yield 90%
Trigonopsis variabilis D-AAO, NaBH$_4$, NaCNBH$_3$

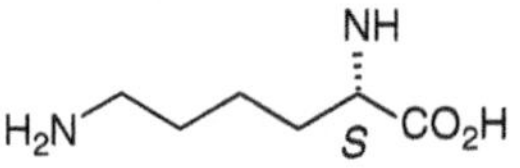

L-(S)-lysine
ee 98%, yield 95%
Trigonopsis variabilis D-AAO, NaCNB

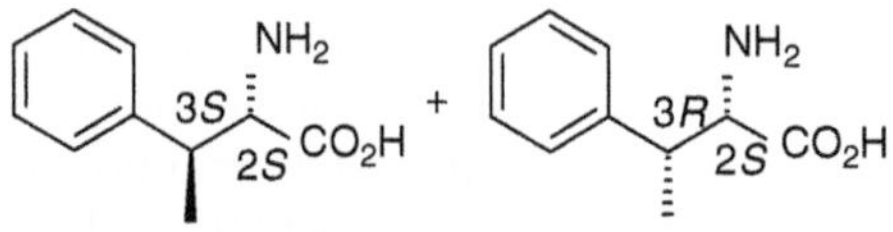

Isomers of L-(2S)-β-methyl phenylalanine
ee 99%, yield 85%
Trigonopsis variabilis D-AAO, NaCNBH$_3$

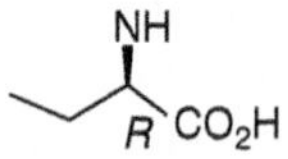

D-(R)-alpha-aminobutyric acid
ee 90%, yield 60%
Proteus myxofaciens L-AAO, NH$_3$:BH

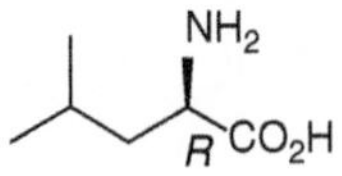

D-(R)-leucine
ee > 99%, yield 80%
Proteus myxofaciens L-AAO
NH$_3$:BH$_3$

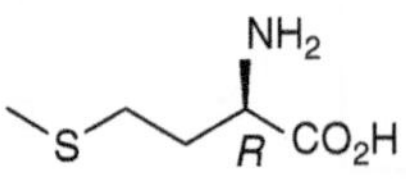

D-(R)-methionine
ee > 99%, yield 84%
Proteus myxofaciens L-AAO
NaCNBH$_3$, NH$_3$:BH$_3$

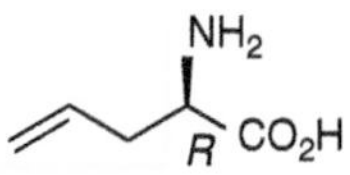

D-(R)-allylglycine
ee > 99%, yield 79%
Proteus myxofaciens L-AAO
NaCNBH$_3$, NH$_3$:BH$_3$

FIGURE 3.33 Glaxo Wellcome's deracemization process for production of optically pure natu
unnatural amino acids using D- or L-amino acid oxidases (AAO). (From Enright, A. and Mahm
M., Glaxo Wellcome R&D, UK, unpublished work, 2001.)

An S-selective MAO-N mutant (Asn336Ser) was identified which showed significantly improved catalytic activity (47-fold) and enantioselectivity (sixfold) toward L-(S)-α-methylbenzylamine compared with the wild-type enzyme; enantioselectivity of the mutant (L- vs. D-isomer) was 100:1 compared with 17:1 for the wild-type enzyme [104]. This was confirmed by chiral HPLC where complete oxidation of L-(S)-α-methylbenzylamine was observed after 24 h, whereas no detectable conversion of the corresponding D-isomer was evident. Several reducing agents were screened (NaBH$_4$, NaCNBH$_3$, Pd/C-HCO$_2$NH$_4$, NH$_3$:NH$_3$) with 1 mM DL-α-methylbenzylamine using the mutant MAO-N; best conditions were observed with NH$_3$:NH$_3$, which afforded D-(R)-α-methylbenzylamine (ee 93%, yield 77%) (Figure 3.34). Using the mutant MAO-N we studied its substrate specificity toward a variety of alicyclic and acyclic primary and secondary amines using a high-throughput plate format; rate of oxidations were monitored by measuring H$_2$O$_2$ formation using a colorimetric enzyme assay [104] (Figure 3.34 and Figure 3.35). The wild-type MAO-N was inactive in most parts toward the amines studied but was most active toward simple straight chain amines, and showed very poor activity for the more sterically demanding branched amines. In sharp contrast, however, the mutant MAO-N (Asn336Ser) showed good activity toward branched acyclic amines, substituted α-methylbenzylamines (X = OMe, Me, Br, NO$_2$), alicyclic amines, and the secondary amine 1-methyltetrahydroisoquinoline [104] (Figure 3.35). In addition to showing a broad substrate specificity, the mutant MAO-N was highly S-selective giving the D-(R)-amine products in high optical and chemical yields [104] (Figure 3.34). Working with the wild-type MAO-N we found that, using site-directed mutagenesis, the identity of mutation at position 348 greatly influenced the efficiency of expression. We therefore used the mutant MAO-N (Asn336Ser) to introduce a second mutation (Met348Lys), by site-directed mutagenesis, which combined the catalytic enhancement of the Asn336Ser mutation with an increase in expression [104]. The activity of the double mutant enzyme (Asn336Ser, Met348Lys) was then examined with a variety of amine

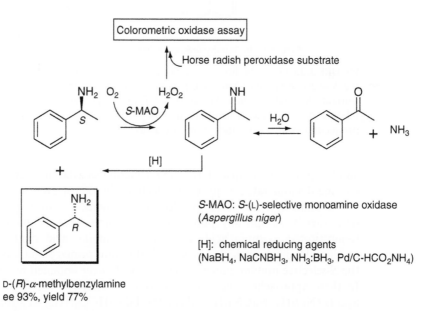

D-(R)-α-methylbenzylamine
ee 93%, yield 77%

FIGURE 3.34 Deracemization of (±)-amines using an S-selective oxidase coupled with chemical reducing agents.

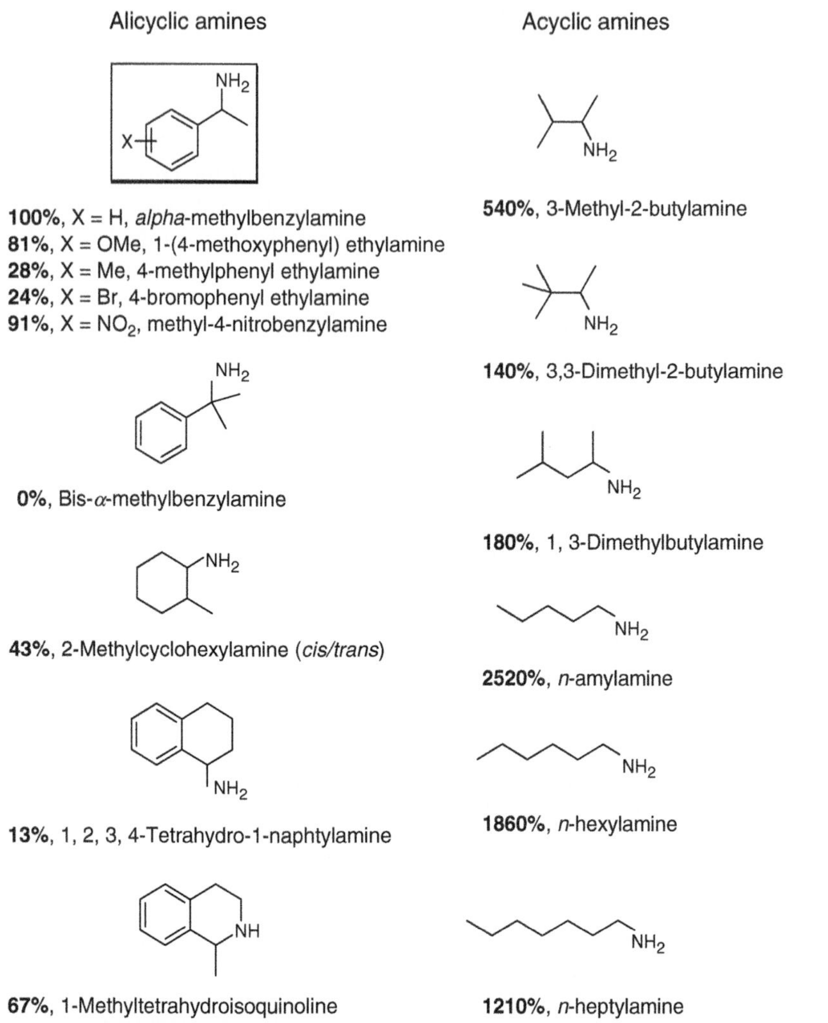

Alicyclic amines	Acyclic amines

100%, X = H, *alpha*-methylbenzylamine
81%, X = OMe, 1-(4-methoxyphenyl) ethylamine
28%, X = Me, 4-methylphenyl ethylamine
24%, X = Br, 4-bromophenyl ethylamine
91%, X = NO₂, methyl-4-nitrobenzylamine

540%, 3-Methyl-2-butylamine

0%, Bis-α-methylbenzylamine

140%, 3,3-Dimethyl-2-butylamine

43%, 2-Methylcyclohexylamine (*cis/trans*)

180%, 1, 3-Dimethylbutylamine

2520%, *n*-amylamine

13%, 1, 2, 3, 4-Tetrahydro-1-naphtylamine

1860%, *n*-hexylamine

67%, 1-Methyltetrahydroisoquinoline

1210%, *n*-heptylamine

FIGURE 3.35 Glaxo Wellcome's deracemization process for production of optically pure amin the *S*-selective mutant (Asn336Ser) monoamine oxidase from *Aspergillus niger*. (From Alexec Enright, A., Mahmoudian, M., et al., PCT Int. Appl. WO 03/080855, 2003.) All rates are m relative to L-(*S*)-α-methylbenzylamine (100%). Rate of oxidation was monitored by measurin production using a colorimetric oxidase assay.

substrates (rates compared to L-(*S*)-α-methylbenzylamine set at 100%, Figure 3.3 enzyme demonstrated exquisite *S*-selectivity in almost every case (ee 90 to 99%) [10⁴ In summary, we have developed two deracemization processes for product (i) a variety of optically pure natural and unnatural L- and D-α-amino acids as β-substituted α-amino acids using D- or L-amino acid oxidases [103]; and (ii) a of alicyclic and branched acyclic optically pure primary and secondary D-(*R*)-amine the *S*-selective mutant amine oxidase from *A. niger* obtained by directed evolution [10 In these approaches, the oxidase enzyme is coupled with nonselective chemical re agents (NaBH₄, NaCNBH₃, NH₃:BH₃, Pd/C-HCO₂NH₄) to afford amino acids and in high chemical and optical purities (ee > 99%, yield 70 to 99%) [103–105].

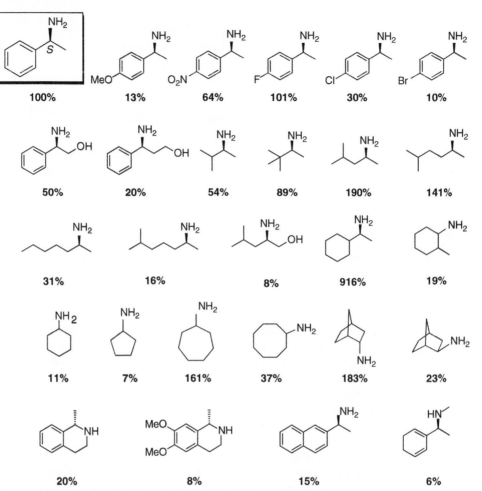

FIGURE 3.36 Glaxo Wellcome's process for production of optically pure amines using the *S*-selective double mutant (Asn336Ser, Met348Lys) monoamine oxidase. (From Alexeeva, M., Enright, A., Mahmoudian, M., et al., PCT Int. Appl. WO 03/080855, 2003.) All rates are measured relative to L-(*S*)-α-methylbenzylamine (100%). Rate of oxidation was monitored by measuring H_2O_2 production using a colorimetric oxidase assay.

3.6.2 HIGH-THROUGHPUT SCREENING USING INFRARED THERMOGRAPHY

At Glaxo Wellcome we often screened isolated enzymes, recombinant microorganisms, and chiral ligands for enantioselective transformations. For HTS, the bottleneck is usually in the speed of chiral analyses involving laborious development of chiral assays. We were therefore particularly interested in developing new approaches for rapid determination of ee. This is, however, frequently hampered by the availability of a suitable chiral assay; chiral HPLC/GC and chiral shift NMR are the most preferred techniques but these methods can suffer from being too time-consuming to develop. Other approaches used by several groups include fluorescence [123–125], mass spectroscopy [126], capillary electrophoresis [127], and circular dichroism [128]. Among these methods infrared thermography (IRT) has

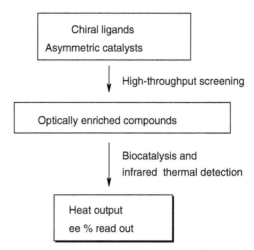

FIGURE 3.37 Infrared thermography (IRT) for high-throughput screening (HTS) of catal asymmetric transformations.

attracted considerable interest in recent years because noninvasive thermal imag chemical reactions can be performed through detection of emitted infrared rad [129]. Acceleration of drug screening is one of the potential applications of IRT pharmaceutical industry. Reetz et al. [130–131] highlighted the use of thermal imag screening of enantioselective biocatalysts and observed a differential heat output dur thermal imaging of lipase-catalyzed acylation of chirally pure *R*- and *S*-alcohols [13 therefore envisaged a high-throughput chiral screening approach based on thermal de of enzyme-catalyzed kinetic resolutions of racemic mixtures and went on to devel further [132] (Figure 3.37).

In this approach immobilized CAL was used in 96-well microtitre plates con solutions of racemic 1-phenylethanol *Rac*-**47** (ee 1%, prepared by sodium borol reduction of acetophenone) and *R*-**47** (ee 86%, prepared by Corey–Bakshi–Shibata reduction of acetophenone) (Figure 3.38a, second row). Standard solutions of the ▪ mixtures of optically pure *R*- and *S*-**47** (0 to 100% *R*- or *S*-isomer) were used for deter ee values (Figure 3.38a, first row). Temperature changes were monitored periodically ▪ infrared camera (thermal imaging) and reactions were started by adding solutions c acetate in toluene to each well. Before start of reactions, cold spots (blue) were evident the endothermic process of solvent evaporation (Figure 3.38a). Subsequently, we ob differential heat outputs (yellow) corresponding to various concentrations of the *R*-e mer (Figure 3.38b, first row). Similarly, wells containing either *Rac*-**47** or *R*-**47** (prepa chemical reductions) showed elevated heat outputs (Figure 3.38b, second row). CAL tion of *Rac*-**47** was therefore shown to be *R*-enantioselective (the *R*-isomer is prefer acylated leaving behind the corresponding *S*-isomer, Figure 3.38). This suggested t excellent *R*-enantioselectivity was also obtained after CBS reduction [133] of acetopl (ee 86% by chiral HPLC) (Figure 3.38). The temperature changes from enzymic re were quantified by measuring heat output as a function of time, and the areas und temperature curve could now be used to calculate ee values of *Rac*-**47** (ee 8%) and o enriched *R*-**47** (ee 94%). These corresponded very well with ee values measured by HPLC (1 vs. 86% respectively) [132]. Similarly we applied IRT determination of ee t substrates such as 2-hexanol **48** and 1-phenylethylamine **49** (Figure 3.39). The time-r

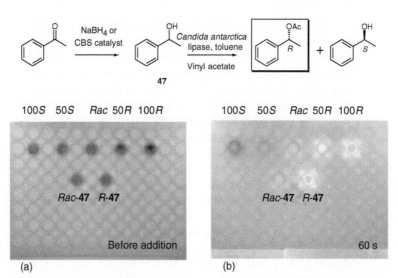

FIGURE 3.38 (See color insert following page 526) Time-resolved thermal imaging of *Candida antarctica*–catalyzed acylation of 1-phenylethanol **47**. NaBH$_4$ reduction of acetophenone afforded *Rac*-**47** (ee 1%, chiral HPLC), whereas Corey–Bakshi–Shibata (CBS) reduction gave *R*-**47** (ee 86%, chiral HPLC). Colors correspond to the following temperatures: (a) at time zero—black (12.5 to 13.5°C), purple (15 to 16°C), pink (17 to 18°C), orange (18.5 to 19.5°C); (b) after 60 s—orange (18.5 to 19°C), yellow (19.5 to 20°C), white (20.5°C), gray (21.4 to 21.6°C).

IRT temperature profile (standard curves) for *C. antarctica* at various mixtures of optically pure **48** or **49** (100*S*–100*R*) was obtained in microtitre plates, which gave good linear correlations with the observed ee values. The data were analyzed using the statistical software package Minitab [132].

FIGURE 3.39 *Candida antarctica*–catalyzed acylation of 2-hexanol **48** and 1-phenylethylamine **49** using infrared thermography (IRT) for determination of enantiomeric excess (ee). Either vinyl acetate or ethyl methoxyacetate was used as acylating agents. Standard curves for ee determination by IRT were obtained. Temperature areas (s.°K), second x Kelvin, representing various mixtures of either **48** or **48** isomers (100*S*–100*R*) were obtained in microtitre plates, which gave good linear correlations with the observed ee values. The data were analyzed using the statistical software package Minitab. (From Millot, N., Borman, P., Mahmoudian, M., et al., *Org. Process R&D.*, 6, 463–470, 2002.)

Having these encouraging results in hand, we decided to extend the application
to rapid screening of a variety of racemic secondary alcohols in a 96-well plate usin
(Table 3.1). However, an important criterion of the HTS campaign is speed, i.e., the
to rapidly identify the substrates of interest for further study. Although desirab
unrealistic to expect an HTS process to deliver completely reliable and reproducib
for every substrate. With this is mind we proceeded with analysis of thermal i
pictures of the microtitre plate, taken at regular time intervals, which showed that st
heat output was immediately observed for alcohol G9 having a tertiary amino grou
position [132] (Table 3.1). Significant temperature rises were also seen after 1
substrates with a tertiary amino group at α-position (E9 and F9). For compariso
outputs from reactions of similar alcohols not containing any basic functionality (e
B4, D4, B12, and G11) were detected much later. This may suggest that kinetics of ac
of alcohols E9–G9 could be substantially different than for reactions with alcohols
no basic functionality. Attempts to obtain a more accurate readout for each well pr
be more tricky; interestingly, despite showing small temperature rises, no reaction at
detected in 11 wells (B1, H3, F6, A7, B7, C9, D9, B11, C11, E11, D12) as judged
highlighting a potential limitation of IRT. For example, the highest temperature r
3.1°K (Kelvin) for a well with no reaction. This was therefore used as background r
46 wells gave a temperature rise under 3.1°K, whereas the remaining 50 wells sh
temperature rise above 3.1°K. This detection rate is typical for HTS, where the emp
on rapid detection of interesting wells for further study rather than an exhaustive asse
of each substrate. Furthermore, although chiral analyses (chiral GC) of selecte
showed that in some cases acylations were indeed enantioselective, this did not corr
well with the generated heat outputs (Table 3.2). These findings on the substrate spe
of CAL for acylations of alcohols are also consistent with previously reported data
active site [134].

In summary, IRT is a novel technique that can be used to screen a potentially large i
of asymmetric catalysts or substrates in a high-throughput fashion, and also for rapi
mination of ee. Here we used IRT as a fast, simple, and practical method for initial scree
the substrate specificity of CAL for the generation of chiral secondary alcohols and
using a 96-well microtitre plate format. The potential drawback of IRT was also high
when detecting the heat output in a 96-well format, and the apparent lack of cor
between the observed heat output and enantioselectivity in some cases. However, we
that IRT can provide a crude but practical initial method, when screening large num
catalysts and/or substrates, to rapidly identify a particular structural class that can be
accurately analyzed by conventional methods. For example, the data generated by IRT
whole plate (96 substrates) took only 15 min—several orders of magnitude quicke
analysis by GC or HPLC.

3.7 SUMMARY

The use of biotech tools to discover new molecules, and to develop and improve e
mentally friendly processes for their production, is widespread in pharmaceutical, agr
ical, chemical, and biotech sectors as well academia. During 1989–2002, we at
Wellcome had an active biotech program, routinely utilizing enzymes and recon
microorganisms in most aspects of our drug discovery and development pipeline
generate complex templates, which would otherwise be difficult to synthesize che
for use in combinatorial libraries; (ii) produce natural products and secondary met
through fermentation processes; (iii) functionalize novel molecules for structure
relationship (SAR) to improve activity, potency, and/or water solubility; (iv) use mi

TABLE 3.1
Rapid Screening of Racemic Secondary Alcohols in 96-Well Plate Format Using *Candida antarctica* Lipase (CAL) and Infrared Thermography (IRT)

H	G	F	E	D	C	B	A	
3.5; 54	5.1; 100	3.4; 28	2.1; 52	1.9; 2	2.4; 12	2.8; 0	6.4; 57	1
2.5; 60	4.8; 62	3.9; 44	2.9; 40	3.9; 41	3.1; 72	3.8; 44	7.1; 43	2
1.9; 0	4.2; 99	4.3; 51	4.5; 60	5.2; 60	4.2; 30	4.5; 63	85; 7	3
4.9; 54	4.6; 60	4.9; 56	5.2; 62	5.4; 55	6.2; 60	6.6; 58	5.7; 59	4
1.8; 32	1.7; 29	2.8; 30	2.8; 7	4.3; 59	4.4; 47	4.0; 57	4.0; 50	5
3.1; 31	2.7; 55	3.1; 0	2.8; 33	3.2; 54	2.8; 7	3.5; 58	4.0; 100	6
1.6; 6	3.0; 60	4.4; 61	4.0; 83	3.1; 20	2.5; 80	2.5; 0	2.6; 0	7
2.1; 53	2.1; 30	3.0; 43	3.5; 50	3.4; 53	2.9; 50	2.5; 93	3.2; 67	8
2.3; 60	6.8; 74	3.4; 56	3.1; 59	2.8; 0	2.9; 0	3.1; 43	3.6; 59	9
3.1; 40	3.2; 40	4.5; 63	2.0; 62	2.6; 65	4.1; 68	5.1; 55	5.0; 100	10
2.7; 36	3.4; 46	6.0; 53	2.4; 0	2.2; 20	1.8; 0	2.7; 0	3.4; 7	11
3.6; 57	2.4; 51	4.2; 78	2.5; 46	1.7; 0	2.6; 46	2.6; 46	2.8; 40	12

Note: Numbers are related to the temperature rise (°K) measured by IRT and conversions (%) measured by GC after 90 min. A–H and 1–12 represent rows of microtitre plate.

TABLE 3.2
Correlation between Enantiomeric Ratio (*E*) and the Observed
Temperature Rise (Kelvin, °K) by Thermal Imaging for Lipase-Catalyzed
Acylation of Selected Wells in Table 3.1

Wells	Conversion (%)	ee (%)	E	IRT (°K)
A8	67	0	1.0	3.2
C2	72	33	4.8	3.1
D2	41	24	1.9	3.9
D7	20	72	7.3	3.1
G8	30	7	1.2	2.1
G10	40	30	2.2	3.2
H10	40	52	4.4	3.1

Note: Conversions and enantiomeric excess (ee) determinations were carried out using chiral gas chromatography (GC).

IRT, infrared thermography.

models to generate metabolites that were used as standards in mammalian meta (impurity profiling); and (v) produce gram to tonne quantities of key enantiomericall intermediates for the synthesis of development compounds. Biocatalytic approache often considered at the very inception of a chemical route, for example, to byp environmentally "unfriendly" step (e.g., use of transitional metal oxidants such as pot permanganate) or even to replace several chemical stages with a single enzymic step studies reviewed in this chapter highlight the importance of close integration of approaches at various stages of drug discovery and chemical development proce pharma. The potential benefits of biotech tools will be fully realized as more com adopt a similar strategy.

Biotechnology will grow in many sectors over the next 10 y and can no longer be i in certain product areas and market segments. A diverse set of tools will be needed wit new technologies being assembled at interface of traditional disciplines. In sharp cont the pharmaceutical and biotech sectors, surprisingly there are only a relatively few ch products made through biotech approaches in the chemical sector, with some notable tions being acrylamide production and some commodity chemicals such as high-fructos syrup; overall this is proving to be a challenging space indeed. There are, however, a n of companies actively trying to penetrate the bio-based products market: (i) DuPont/C cor/Staley (1,3-propane-diol biopolymer); (ii) Cargill Dow (polylactic acid biodegr plastic); and (iii) Eastman/Genencor (vitamin-C process). At Rohm and Haas we are oping key capabilities and multiple business platforms at the intersection of bioscienc polymer sciences to explore, expand, and develop new market opportunities.

ACKNOWLEDGMENTS

This chapter is dedicated to Helen, Laura, and Yasmin who have supported me unfa I gratefully acknowledge contributions of many colleagues from Glaxo's Natural Pr Discovery Department (1989–2002), and Glaxo Wellcome Operations for the scale these bioprocesses.

REFERENCES

1. McKinsey & Company, Biotechnology in chemical industry, *The McKinsey Quarterly*, 2002.
2. Van Arnum, F., White biotechnology: the chemical industry's next challenge, *Focus*, 2003; fine chemicals/biopharmaceuticals, *Chemical Market Reporter*, 263: i24 pFR6–1, 2003.
3. Tamburini, B., Perboni, A., Rossi, T., et al., Preparation of 10-(1-hydroxyethyl)-11-oxo-1-azatri-cyclo[7.2.0.0.3,8]undec-2-ene-2-carboxylic acid derivatives as antibacterials, Eur. Patent Appl., 0416953, 1991.
4. Stead, P., Marley, H., Mahmoudian, M., et al., Efficient procedures for the large-scale preparation of (+)-(1S, 2S)-*trans*-2-methoxycyclohexanol, a key chiral intermediate in the synthesis of tricyclic β-lactam antibiotics, *Tetrahedron Asymmetry*, 7(8), 2247–2250, 1996.
5. Laumen, K., Breitgoff, D., Seemayer, R., et al., Enantiomerically pure cyclohexanols and cyclo-hexane-1,2-diol derivatives; chiral auxiliaries and substitutes for (−)-8-phenylmenthol: a facile enzymatic route, *J. Chem. Soc. Chem. Commun.*, 148–150, 1989.
6. Basavaiah, D. and Krishna, P.R., Pig liver powder as biocatalyst: enantioselective synthesis of *trans*-2-alkoxycyclohexan-1-ols. *Tetrahedron*, 50, 10521–10530, 1994.
7. Honig, H. and Seufer-Wasserthal, P., A general method for the separation of enantiomeric *trans*-2-substituted cyclohexanols, *Synthesis*, 1137–1140, 1990.
8. Halpern, B. and Westley, J.W., Chemical resolution of secondary alcohols, *Aust. J. Chem.*, 19, 1533–1538, 1966.
9. Averett, D.R., Koszalka, G.W., Fyfe, J.A., et al., 6-Methoxypurine arabinoside as a selective and potent inhibitor of varicella-zoster virus, *Antimicrob. Agents Chemother.*, 35, 851–857, 1991.
10. Lambe, C.U., Averett, D.R., Paff, M.T., et al., 2-Amino-6-methoxypurine arabinoside: an agent for T-cell malignancies, *Cancer Res.*, 55, 3352–3356, 1995.
11. Krenitsky, T.A., Averett, D.R., Moorman, A.R., et al., Method of treating T-Cell lymphoblastic leukaemia with *ara*-G nucleoside derivatives, U.S. Patent Appl., 5492897, 1996.
12. Moorman, A.R., Chamberlain, S.D., Jones, L.A., et al., 5′-ester prodrugs of the varicella-zoster antiviral agent, 6-methoxypurine arabinoside, *Antiviral Chem. Chemother.*, 3, 141–146, 1992.
13. Eaddy, J., Glaxo Wellcome R&D, USA, unpublished work, 1997.
14. Mahmoudian, M., Eaddy, J., and Dawson M.J., Enzymatic acylation of 506U78: a powerful new anti-leukaemic agent, *Biotechnol. Appl. Biochem.*, 29, 229–233, 1999.
15. Gregson, M., Ayres, B.E., Ewan, G.B., et al., 2,6-Diaminopurine derivatives, PCT Int. Appl., WO 9417090, 1994.
16. Isono, Y. and Hoshino, M., Nucleoside oxidase and assay method utilizing same, U.S. Patent Appl., 5156955, 1992.
17. Isono, Y., Sudo, T., and Hoshino, M., Purification and reaction of a new enzyme, nucleoside oxidase, *Agric. Biol. Chem.*, 53, 1663–1669, 1989.
18. Isono, Y., Sudo, T., and Hoshino, M., Properties of a new enzyme, nucleoside oxidase, from *Pseudomonas maltophilia* LB-84, *Agric. Biol. Chem.*, 53, 1671–1677, 1989.
19. Misaki, H., Ikuta, S., and Matsunia, K., Nucleoside oxidase and process for making same, and process and kit for using same, U.S. Patent Appl., 4385112, 1983.
20. Mahmoudian, M., Rudd, B.A.M., Cox, B., et al., A versatile procedure for the generation of nucleoside 5′-carboxylic acids using nucleoside oxidase, *Tetrahedron*, 54, 8171–8182, 1998.
21. Gallo, R.C., Sarin, P.S., Gelman, E.P., et al., Isolation of a human T-cell leukaemia virus in AIDS, *Science*, 220, 865–867, 1983.
22. Fauchi, A.S., Masur, H., Gelman, E.P., et al., NIH conference: the AIDS update, *Ann. Intern. Med.*, 102, 800–813, 1985.
23. Coates, J.A.V., Cammack, N., Jenkinson, H.J., et al., (−)-2′-Deoxy-3′-thiacytidine is a potent, highly selective inhibitor of human immunodeficiency virus Type 1 and Type 2 replication *in vitro*, *Antimicrob. Agents Chemother.*, 36, 733–739, 1992.
24. Coates, J.A.V., Cammack, N., Jenkinson, H.J., et al., The separated enantiomers of 2′-deoxy-3′-thiacytidine (BCH189) both inhibit human immunodeficiency virus replication *in vitro*, *Antimicrob. Agents Chemother.*, 36, 202–205, 1992.
25. Cameron, J.M., Collis, P., Daniel, M., et al., Lamivudine, *Drugs of the Future*, 18, 319–323, 1993.

26. Francis, C.E., Hunter, A., Berney, J.J., et al., *In vitro* myelotoxicity studies with 3T(International conference on AIDS and 3rd STD world congress (Amsterdam), Abs. n 6027, 1992.

27. Storer, R., Clemens, I.R., Lamont, B., et al., The resolution and absolute stereochemistr enantiomers of *cis*-1-[2-(hydroxymethyl)-1,3-oxathiolan-5-yl]cytosine (BCH189): equipote HIV agents, *Nucleosides Nucleotides*, 12, 225–236, 1993.

28. Mahmoudian, M., Baines, B.S., Drake, C.S., et al., Enzymatic production of optically pur (Lamivudine): a potent anti-HIV agent. *Enzyme Microb. Technol.*, 15, 749–755, 1993.

29. Hammer-Jespersen, K., Nucleoside metabolism, in *Metabolism of nucleotides, nucleosi nucleobases in microorganisms*, Munch-Petersen, A., Ed., Academic Press, London pp. 203–258.

30. Cousins, R.P.C., Mahmoudian, M., and Youds, P.M., Enzymatic resolution of oxa intermediates: an alternative approach to the anti-viral agent lamivudine (3TC$^{®}$), *Tetr Asymmetry*, 6, 393–396, 1995.

31. Mahmoudian, M. and Dawson, M.J., Chemoenzymatic production of the antiviral agent E *Biotechnology of antibiotics*, 2nd Ed., Strohl, W.R., Ed., Marcel Dekker Inc., New Yor pp. 753–777.

32. Daluge, S.M., Therapeutic nucleosides, Eur. Patent Appl., 0434450, 1991.

33. Daluge, S.M., Good, S.S., Faletto, M.B., et al., 1592U89 a novel carbocyclic nucleoside a with potent selective anti-HIV activity, *Antimicrob. Agents Chemother.*, 41, 1082–1093, 19

34. Evans, C.T. and Roberts, S.M., Chiral azabicycloheptanone and a process for their prep Eur. Patent Appl., 0424 064 B1, 1991.

35. Evans, C.T., Roberts, S.M., Shoberu, K.A., et al., Potential use of carbocyclic nucleoside treatment of AIDS: chemoenzymatic synthesis of the enantiomers of carbovir, *J. Che Perkin Trans.*, 1, 589–592, 1992.

36. Mahmoudian, M., Lowdon, A., Jones, M.F., et al., A practical enzymatic procedure resolution of *N*-substituted 2-azabicyclo[2.2.1]hept-en-3-one. *Tetrahedron Asymmetry*, 1(1206, 1999.

37. Novozyme: Savinase 6T, Lipolase 100T, Novo Nordisk enzyme data sheet, 1996.

38. Dawson, M.J., Mahmoudian, M., and Wallis, C., Process for the enantioselective hydr *N*-derivatised lactam, Int. Appl. No. PCT/EP98/05291, 1999.

39. Von Itzstein, M., Wu, W.Y., Kok, G.B., et al., Rational design of potent sialidase-based in of influenza virus replication, *Nature*, 363, 418–423, 1993.

40. Schauer, R., Chemistry metabolism and biological function of sialic acids, *Adv. Carbohydr Biochem.*, 40, 131–234, 1982.

41. De Ninno, M.P., The synthesis and glycosidation of *N*-acetyl-D-neuraminic acid, *Synt. 583–593, 1991.

42. Juneja, L.R., Koketsu, M., Nishimoto, K., et al., Large-scale preparation of sialic ac chalaza and egg-yolk membrane, *Carbohydr. Res.*, 214, 179–186, 1991.

43. McGuire, E.J. and Binkley, S.B., The structure and chemistry of colominic acid, *Biochem 247–251, 1964.

44. Uchida, Y., Tsukada, Y., and Sugimori, T., Improved microbial production of colomir a homopolymer of *N*-acetyl-D-neuraminic acid, *Agric. Biol. Chem.*, 37, 2105–2110, 1973.

45. Marugane-Shoyu, K.K., Preparation of colominic acid, Jap. Patent Appl. 144989, 1989.

46. Spivak, C.T. and Roseman, S., Preparation of *N*-acetyl-D-mannosamine (2-acetamido-2-d mannose) and D-mannosamine hydrochloride (2-amino-2-deoxy-D-mannose), *J. Am. Che* 81, 2403–2404, 1959.

47. Kragl, U., Gygax, D., Ghisalba, O., et al., Enzymatic two-step synthesis of *N*-acetyl-D- neu acid in the enzyme membrane reactor, *Angew Chem. Int. Ed. Engl.*, 30, 827–828, 1991.

48. Auge, C., David, S., and Gautheron, C., Synthesis with immobilized enzyme of the most im sialic acid, *Tetrahedron Lett.*, 25, 4663–4664, 1984.

49. Bednarski, M.D., Chenault, H.K., Simon, E.S., et al., Membrane-enclosed enzymatic ((MEEC): a useful, practical new method for the manipulation of enzymes in organic sy *J. Am. Chem. Soc.*, 109, 1283–1285, 1987.

50. Kim, M.J., Hennen, W.J., Sweers, H.M., et al., Enzymes in carbohydrate synthesis: N-acetyl-D-neuraminic acid aldolase catalyzed reactions and preparations of N-acetyl-2-deoxy-D-neuraminic acid derivatives, *J. Am. Chem. Soc.*, 110, 6481–6486, 1988.

51. Tsukada, Y. and Ohta, Y., Process for producing N-acetyl-D-neuraminic acid, Eur. Patent Appl. 0578825A1, 1992.

52. Dawson, M.J., Noble, D., and Mahmoudian, M., Process for the preparation of N-acetyl-D-neuraminic acid, PCT W0 9429476, 1994.

53. Mahmoudian, M., Noble, D., Drake, C.S., et al., An efficient process for production of N-acetylneuraminic acid using N-acetylneuraminic acid aldolase, *Enzyme Microb. Technol.*, 20, 393–400, 1997.

54. Starkey, I.D., Mahmoudian, M., Noble, D., et al., Synthesis and influenza virus sialidase inhibitory activity of the 5-desacetoamido analogue of 2,3-didehydro-2,4-dideoxy-4-guanidinyl-N-acetylneuraminic acid (GG167), *Tetrahedron Lett.*, 36, 299–302, 1995.

55. Woods, J.M., Bethell, R.C., Coates, J.A.V., et al., 4-Guanidino-2,4-dideoxy-2,3-dehydro-N-acetylneuraminic acid is a highly effective inhibitor both of the sialidase (neuraminidase) and of growth of a wide range of influenza A and B viruses *in vitro*, *Antimicrob. Agents Chemother.*, 37, 1473–1479, 1993.

56. Cermak, R.C. and Vince, R., ($\pm$)-4β-Amino-2α,3α-dihydroxy-1β-cyclopentanemethanol hydrochloride: carbocyclic ribofuranosylamine for the synthesis of carbocyclic nucleosides, *Tetrahedron Lett.*, 22, 2331–2332, 1981.

57. Marquez, V.E. and Lim, M., Carbocyclic nucleosides, *Med. Res. Rev.*, 6, 1–40, 1986.

58. Borthwick, A.D., Butt, S., Biggadike, K., et al., Synthesis and enzymatic resolution of carbocyclic 2'-*ara*-fluoro-guanosine: a potent new anti-herpetic agent, *J. Chem. Soc. Chem. Commun.*, pp. 656–658, 1988.

59. Vince, R. and Brownell, J., Resolution of racemic carbovir and selective inhibition of human immunodeficiency virus by the ($-$)-enantiomer, *Biochem. Biophys. Res. Commun.*, 168, 912–916, 1990.

60. Balzarini, J., Baumgartner, H., Bodenteich, M., et al., Synthesis and anti-viral activity of the enantiomeric forms of carba-5-iodo-2'-deoxyuridine and carba-(E)-5-(2-bromovinyl)-2'-deoxyuridine, *J. Med. Chem.*, 32, 1861–1865, 1989.

61. Balzarini, J., Metabolism and mechanism of anti-retroviral action of purine and pyrimidine derivatives, *Pharm. World Sci.*, 16, 113–126, 1994.

62. Mahmoudian, M., Baines, B.S., Dawson, M.J., et al., Resolution of 4-amino-cyclopentanecarboxylic acid methyl esters using hydrolytic enzymes, *Enzyme Microb. Technol.*, 14, 911–916, 1992.

63. Krenitsky, T.A., Koszalka, G.W., Tuttle, J.V., et al., An enzymatic synthesis of purine D-arabinonucleosides, *Carbohydrate Res.*, 97, 139–146, 1981.

64. Herdewijn, P., Balzarini, J., de Clercq, E.A., et al., Resolution of aristeromycin enantiomers, *J. Med. Chem.*, 28, 1385–1386, 1985.

65. Secrist, J.A., Montgomery, J.A., Shealy Y.F., et al., Resolution of racemic carbocyclic analogues of purine nucleosides through the action of adenosine deaminase: antiviral activity of the carbocyclic 2'-deoxyguanosine enantiomers, *J. Med. Chem.*, 30, 746–749, 1987.

66. Hoong, L.K., Strange, L.E., Liotta, D.C., et al., Enzyme-mediated enantioselective preparation of pure enantiomers of the anti-viral agent 2',3'-dideoxy-5-fluoro-3'-thiacytidine (FTC) and related compounds, *J. Org. Chem.*, 57, 5563–5565, 1992.

67. Cameron, J.M., New anti-herpes drugs in development, *Rev. Med. Virol.*, 3, 225–236, 1993.

68. Boehme, R.E., Bereford, A., Hart, G.J., et al., GR95168X (chiral carbocyclic BVdU) is highly effective against simian *Varicella* virus infections of African green monkeys, *Antiviral Res.*, 23, 97–98, 1994.

69. Sicsic, S., Ikbal, M., and Le Goffic, F., Chemoenzymatic approach to carbocyclic analogues of ribonucleotides and nicotinamide ribose, *Tetrahedron Lett.*, 28, 1887–1888, 1987.

70. Csuk, R. and Doerr, P., Bioanalytical transformations: the 4-acetamido-cyclopent-2-ene carboxylate route revisited; synthesis of ($+$)- and ($-$)-aristeromycin, *Tetrahedron*, 51, 5789–5798, 1995.

71. Borthwick, A.D., Crame, A.J., Mahmoudian, M., et al., A short, convergent synthesis of two chiral anti-viral agents ($+$)-carbocyclic 2'-deoxy-5-[(E)-2-bromovinyl]uridine and ($+$)-carbocyclic 2'-deoxyguanosine, *Tetrahedron Lett.*, 36, 6929–6932, 1995.

72. Archelas, A. and Morin, C., Biohydroxylation of a 2-azabicyclo[2.2.1]heptane, *Tetrahedro* 25, 1277–1278, 1984.

73. Dawson, M.J., Mahmoudian, M., Rudd, B.A.M., et al., Biosynthesis and biotransformatic preparation of nucleoside analogues, in *Chiral Europe*, 1996, pp. 35–38.

74. White, E.L., Parker, W.B., Macy, L.J., et al., Comparison of the effect of carbovir, A2 dideoxynucleoside triphosphates on the activity of human immunodeficiency virus rever scriptase and selected human polymerases, *Biochem. Biophys. Res. Commun.*, 161, 393–39

75. Carter, S.G., Kessler, J.A., and Rankin, C.D., Activities of (−)-carbovir and 3′-azido-3′-de midine against human immunodeficiency virus *in vitro*, *Antimicrob. Agents Chemother.*, 3-1300, 1990.

76. Exall, A.M., Jones, M.F., Mo, C.L., et al., Synthesis of (−)-aristeromycin and X-ray stru (−)-carbovir, *J. Chem. Soc. Perkin Trans.*, 1, 2467–2477, 1991.

77. Pun, K.T., Baines, B.S., and Lawrence, G.C., Isolation, immobilization and use of ad deaminase in the synthesis of (−)-carbovir, a purine nucleoside analogue, 5th European c on biotechnology, Copenhagen, Abs. no. TUP 111, 1990, p. 292.

78. Mahmoudian, M., Top 400 prescription drugs, *MedAd News*, 3–20, Eastman Chemical Cc USA, unpublished work. 2003.

79. Adamo, M.F.A., Aggarwal, V.K., and Sage, M.A., Epoxidation of alkenes by amine precursors: implication of aminium ion and radical cation intermediates, *J. Am. Chem. Sc* 8317–8318, 2000.

80. Nieuwenhuijzen, J.W., Grimbergen, R.F.P., Koopman, C., et al., The role of nucleation in in optical resolutions with families of resolving agents, *Angew. Chem. Int. Ed.*, 41, 4281–428

81. Henderson, K.W., Kerr, W.J., and Moir, J.H., Enantioselective deprotonation reactions novel homochiral magnesium amide base, *Chem. Commun.*, 479–480, 2000.

82. Balint, G., Egri, M., and Czugler, J., Resolution of 1-phenylethylamine by its acidic deri *Tetrahedron Asymmetry*, 12, 1511–1518, 2001.

83. Varies, T.R., Wynberg, H., van Echten, E., et al., The family approach to the resol racemates, *Angew. Chem. Int. Ed.*, 37, 2349–2354, 1998.

84. Messina, F., Botta, M., Corelli, M.P., et al., Stereoselective synthesis of aryl-2-benzofura namines and aryl-1H-indole-2-methanamines through palladium-mediated annulation o arylpropargylamines, *Tetrahedron Asymmetry*, 11, 1681–1685, 2000.

85. Luna, A., Alfonso, I., and Gotor, V., Biocatalytic approaches toward the synthesis enantiomers of *trans*-cyclopentane-1,2-diamine, *Org. Lett.*, 4, 3627–3629, 2002.

86. Takayama, S., Lee, S.T., Hung, S.C., et al., Designing enzymatic resolution of amines *Commun.*, 127–128, 1999.

87. Gutman, A.L., Meyer, E., Kalerin, E., et al., Enzymatic resolution of racemic amines in cor reactor in organic solvents, *Biotechnol. Bioeng.*, 40, 760–767, 1992.

88. Jaeger, K., Liebeton, K., Zonta, A., et al., Biotechnological application of *Pseudomonas aer* lipase: efficient kinetic resolution of amines and alcohols, *Appl. Microbiol. Biotechnol.*, 46, 1996.

89. Stirling, D.I., The use of aminotransferases for the production of chiral amino acids and an *Chirality in Industry*, Collins, A.N., et al., Eds., Wiley, New York, 1992, pp. 209–222.

90. Matcham, G., Bhatia, M., Lang, W., et al., Enzyme and reaction engineering in bioca synthesis of (*S*)-methoxyisopropylamine, *Chimia*, 53, 584–589, 1999.

91. Shin, J.S., Kim, B.G., Liese, A., et al., Kinetic resolution of chiral amines using enzyme me reactors, *Biotechnol. Bioeng.*, 73, 179–187, 2001.

92. Shin, J.S. and Kim, B.G., Comparison of the ω-transaminase from different microorganis application to production of chiral amines, *Biosci. Biotechnol. Biochem.*, 65, 1782–1788, 2

93. Takaaki, Y., Hiroshi, T., and Takeo Y., Production of optically active α-methylbenzylami Patent Appl. 4234993A, 1991.

94. Uematsu, N., Fujii, A., Hashiguchi, S., et al., Asymmetric transfer hydrogenation of imines *Chem. Soc.*, 118, 4916–4917, 1996.

95. Okuda, J., Verch, S., Sturmer, R., et al., Chiral complexes of titanium containing a linke cyclopentadienyl ligand: synthesis, structure, and asymmetric imine hydrogenation ca *J. Organomet. Chem.*, 605, 55–67, 2000.

96. Kainz, S., Brinkmann, A., Leitner, W., et al., Iridium-catalyzed enantioselective hydrogenation of imines in supercritical carbon dioxide, *J. Am. Chem. Soc.*, 121, 6421–6429, 1999.
97. Hieber, G. and Ditrich, K., Introducing ChiPros biocatalytic production of chiral intermediates on a commercial scale, *Chimia Oggi*, 19, 16–20, 2001.
98. Reetz, M.T. and Schimossek, K., Lipase-catalyzed dynamic kinetic resolution of chiral amines: use of palladium as the racemisation catalyst, *Chimia*, 50, 668–689, 1996.
99. Choi, Y.K., Kim, M.J., Ahn, Y., et al., Lipase palladium-catalyzed asymmetric transformations of ketoximes to optically active amines, *Org. Lett.*, 3, 4099–4101, 2001.
100. Pamies, O., Ell, A.H., Samec, J.S.M., et al., An efficient and mild ruthenium-catalyzed racemisation of amines: application to the synthesis of enantiomerically pure amines, *Tetrahedron Lett.*, 43, 4699–4702, 2002.
101. Hirrlinger, B. and Stolz, A., Formation of a chiral hydroxamic acid with an amidase from *Rhodococcus erythropolis* MP50 and subsequent chemical Losen rearrangement to a chiral amine, *Appl. Environ. Microbiol.*, 63, 3390–3393, 1997.
102. Chapman, D.T., Crout, D.H.G., Mahmoudian, M., et al., Enantiomerically pure amines by a new method: biotransformation of oxalamic esters using the lipase from *Candida antarctica*, *Chem. Commun.*, 2415–2416, 1996.
103. Enright, A. and Mahmoudian, M., Deracemisation of chiral amines and amino acids, Glaxo Wellcome R&D, UK, unpublished work, 2001.
104. Alexeeva, M., Enright, A., Mahmoudian, M., et al., Deracemisation of amines, PCT Int. Appl. WO 030-80855, 2003.
105. Alexeeva, M., Enright, A., Mahmoudian, M., et al., Deracemisation of α-methylbenzylamine using an enzyme obtained by *in vivo* evolution, *Angew. Chem. Int. Ed.*, 41, 3177–3180, 2002.
106. Uppenberg, J., Hansen, M.T., Patkar, S., et al., The sequence, crystal structure determination and refinement of two crystal forms of lipase B from *Candida antarctica*, *Structure*, 2, 293–308, 1994.
107. Chapman, D.T., Mahmoudian, M., and Crout, D.H.G., Glaxo Wellcome R&D, UK, unpublished work, 1995.
108. Hafner, E.W. and Wellner, D., Demonstration of imino acids as products of the reactions catalyzed by D- and L-amino acid oxidases, *Proc. Natl. Acad. Sci. USA*, 68, 987–991, 1971.
109. Huh, J.W., Yokoigawa, K., Soda, K.J., et al., Synthesis of L-proline from the racemate by coupling of enzymic enantiospecific oxidation and chemical non-enantiospecific reduction, *J. Ferment. Bioeng.*, 74, 189–190, 1992.
110. Huh, J.W., Yokoigawa, K., Soda, K.J., et al., Total conversion of racemic pipecolic acid into the L-enantiomer by a combination of enantiospecific oxidation with D-amino acid oxidase and reduction with sodium borohydride, *Biosci. Biotech. Biochem.*, 56, 2081–2082, 1992.
111. Quay, S. and Massey, V., Effect of pH on the interaction of benzoate and D-amino acid oxidase, *Biochemistry*, 16, 3348–3354, 1977.
112. Pollegioni, L., Buto, S., Tischer, W., et al., Characterization of D-amino acid oxidase from *Trigonopsis variabilis*, *Biochem. Mol. Biol. Int.*, 31, 709–717, 1993.
113. Pantaleone, D.P., Geller, A.M., and Taylor, P.P., Purification and characterization of an L-amino acid deaminase used to prepare unnatural amino acids, *J. Mol. Cat. B Enzyme*, 11, 795–803, 2001.
114. Braun, M., Kim, J.M., and Schmid, R.D., Purification and some properties of an extracellular L-amino acid oxidase from *Cellulomonas cellulans* AM8 isolated from soil, *Appl. Microbiol. Biotechnol.*, 37, 594–598, 1992.
115. Truong, M., Lecornue, F., and Fadel, A., First synthesis of (1*S*,2*S*)- and (1*R*,2*R*)-1-amino-2-isopropylcyclobutanecarboxylic acids by asymmetric Strecker reaction from 2-substituted cyclobutanones, *Tetrahedron Asymmetry*, 14, 1063–1072, 2003.
116. Johnstone, R.A.W., Wilby, A.H., and Entwistle, I.D., Heterogeneous catalytic transfer hydrogenation and its relation to other methods for reduction of organic compounds, *Chem. Rev.*, 85, 129–170, 1985.
117. Hutchins, R.O., Learn, K., Nazer, B., et al., Amine boranes as selective reducing and hydroborating agents: a review, *Org. Prep. Proc. Int.*, 16, 335–372, 1984.
118. Frebort, I., Matsushita, K., and Adachi, O., Involvement of multiple copper/topa containing and flavin containing amine oxidases and NADP aldehyde dehydrogenases in amine degradation by filamentous fungi, *J. Ferment. Bioeng.*, 84, 200–212, 1997.

119. Silverman, R.B., Cesarone, J.M., and Liu, X., Stereoselective ring opening of 1-phenylcy pylamine catalyzed by monoamine oxidase, *J. Am. Chem. Soc.*, 115, 4955–4961, 1993.
120. Schilling, B. and Lerch, K., Amine oxidases from *Aspergillus niger*: identification of a nove dependent enzyme, *Biochim. Biophys. Acta.*, 1243, 529–539, 1995.
121. Schilling, B. and Lerch, K., Cloning, sequencing and heterologous expression of the mor oxidase gene from *Aspergillus niger*, *Mol. Gen. Genet.*, 247, 430–438, 1995.
122. Sablin, S.O., Yankovskaya, V., Bernard, S., et al., Isolation and characterization of an ev ary precursor of human monoamine oxidases A and B, *Eur. J. Biochem.*, 253, 270–279, 1
123. Korbel, G.A., Lalic, G., and Shair, M.D., Reaction microarrays: a method for rapidly dete the enantiomeric excess of thousands of samples, *J. Am. Chem. Soc.*, 123, 361–362, 2001.
124. Badalassi, F., Wahler, D., Klein, G., et al., A versatile periodate-coupled fluorogenic a hydrolytic enzymes, *Angew. Chem. Int. Ed.*, 39, 4067–4070, 2000.
125. Reetz, M.T. and Sostmann S., 2,15-Dihydroxyhexahelicene (HELIXOL): synthesis and u enantioselective fluorescent sensor, *Tetrahedron*, 57, 2515–2520, 2001.
126. Reetz, M.T., Becker, M.H., Klein, H.W., et al., A method for high-throughput scree enantioselective catalysts, *Angew. Chem. Int. Ed.*, 38, 1758–1759, 1999.
127. Reetz, M.T., Kühling, K.M., Deege, A., et al., Super high throughput screening of enantio catalysts by using capillary array electrophoresis, *Angew. Chem. Int. Ed.*, 39, 3891–3893,
128. Ding, K., Ishii, A., and Mikami, K., Super high-throughput screening of chiral liga activators: asymmetric activation of chiral diol zinc catalysts by chiral nitrogen activator enantioselective addition of diethylzinc to aldehydes, *Angew. Chem. Int. Ed.*, 38, 497–501,
129. Taylor, S.J. and Morken, J.P., Thermographic selection of effective catalysts from an polymer-bound library, *Science*, 280, 267–270, 1998.
130. Reetz, M.T., Becker, M.H., Kühling, K.M., et al., Time-resolved IR-thermographic detec screening of enantioselectivity in catalytic reactions, *Angew. Chem. Int. Ed.*, 37, 2647–264
131. Reetz, M.T., Becker, M.H., Liebl, M., et al., IR-thermographic screening of thermone endothermic transformations: the ring closing olefin metathesis reaction, *Angew. Chem.* 39, 1236–1238, 2000.
132. Millot, N., Borman, P., Mahmoudian, M., et al., Rapid determination of enantiomeric exce infrared thermography, *Org. Process R&D*, 6, 463–470, 2002.
133. Corey, E.J. and Helal, C.J., Reduction of carbonyl compounds with chiral oxazabo catalysts: a new paradigm for enantioselective catalysis and a powerful new synthetic *Angew. Chem. Int. Ed.*, 37, 1986–2012, 1998.
134. Anderson, E.M., Larsson, K.M., and Kirk, O., One biocatalyst many applications: the *Candida antarctica* B lipase in organic synthesis, *Biocatal. Biotrans.*, 16, 181–185, 1998.

4 Biocatalysis for Synthesis for Chiral Pharmaceutical Intermediates

Ramesh N. Patel

CONTENTS

4.1 INTRODUCTION

Chirality is a key factor in the efficacy of many drug products and agrochemicals, and thus the production of single enantiomers of chiral intermediates has become increasingly important in the pharmaceutical industry [1]. Single enantiomers can be produced by chemical or chemo-enzymatic synthesis. The advantages of biocatalysis over chemical synthesis are that enzyme-catalyzed reactions are often highly enantioselective and regioselective. They can be carried out at ambient temperature and atmospheric pressure, thus avoiding the use of more extreme conditions that could cause problems with isomerization, racemization, epimerization, and rearrangement. Microbial cells and enzymes derived therefrom can be immobilized and reused for many cycles. In addition, enzymes can be overexpressed to make biocatalytic processes economically efficient, and enzymes with modified activity can be tailor-made. The preparation of thermostable and pH-stable enzymes by random and site-directed mutagenesis has led to the production of novel biocatalysts. A number of review articles [2–10] have been published on the use of enzymes in organic synthesis. This chapter provides examples of the use of enzymes for the synthesis of single enantiomers of key intermediates for drugs.

4.2 ANTIVIRAL DRUGS

4.2.1 Enzymatic Preparation of (1S,2S)-[3-Chloro-2-Hydroxy-1-(Phenylmethyl)propyl]carbamic Acid, 1,1-Dimethylethyl Ester

An essential step in the life cycle of the human immunodeficiency virus (HIV-1) is the proteolytic processing of its precursor proteins by HIV-1 protease, a virally encoded enzyme. Inhibition of HIV-1 protease arrests the replication of HIV *in vitro*, and thus, HIV-1 protease is an attractive target for chemotherapeutic intervention. Barrish et al. [11] reported the discovery of a new class of selective HIV-protease inhibitors, which incorporate a C2 symmetric aminodiol core as its key structural feature. Members of this class, particularly

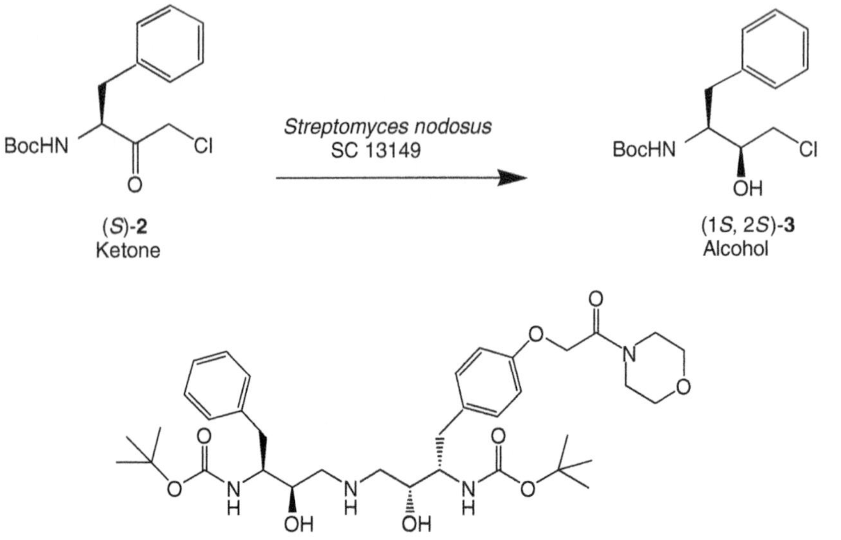

FIGURE 4.1 Synthesis of chiral intermediates for HIV-protease inhibitor 1. Enantioselective en reduction of (1S)-[3-chloro-2-oxo-1(phenylmethyl)propyl]carbamic acid, 1,1-dimethylethyl este to the corresponding (1S,2S)-3 by *Streptomyces nodosus* SC 13149.

compound **1** (Figure 4.1), display potent anti-HIV activity in cell culture. The diastere tive microbial reduction of (1S)-[3-chloro-2-oxo-1-(phenylmethyl)propyl]carbamic 1,1-dimethylethyl ester (**2**) to **3**, a key intermediate in the total chemical synthesis c pound **1**, has been demonstrated [12]. *Streptomyces nodosus* SC 13149 converted keto the corresponding chiral alcohol **3** in a reaction yield of 67% with an enantiomeric exc of 99.9% and a diastereomeric purity of >99%. *Mortierella ramanniana* SC 13850 reaction yield of 54% with an ee of 99.9% and a diastereomeric purity of 92% for **3**. A stage fermentation–biotransformation process was developed using cells of *S. nodo* 13149. A reaction yield of 80% with a diasteromeric purity of >99% and an ee of 99.8 obtained.

4.2.2 ENZYMATIC PREPARATION OF (1S,2R)-[3-CHLORO-2-HYDROXY-1-(PHENYLMETHYL)PROPYL]CARBAMIC ACID, 1,1-DIMETHYLETHYL ESTER

Atazanavir (**4**, Figure 4.2a) is an acyclic aza-peptidomimetic, a potent HIV-protease in [13,14]. An enzymatic process has been developed for the preparation of (1S,2R)-[3-ch hydroxy-1-(phenylmethyl)propyl]carbamic acid, 1,1-dimethylethyl ester (**5**) for th synthesis of the HIV-protease inhibitor, atazanavir (Figure 4.2b). The diastereose reduction of (1S)-[3-chloro-2-oxo-1-(phenylmethyl)propyl]carbamic acid, 1,1-dimeth ester (**2**) was carried out using *Rhodococcus*, *Brevibacterium*, and *Hansenula* strains to ˛ **5**. Three strains of *Rhodococcus* gave >90% yield with a diastereomeric purity of >98 an ee of 99.4% [15]. An efficient single-stage fermentation–biotransformation proce developed for the reduction of ketone **2** with cells of *Rhodococcus erythropolis* SC to yield **5** in 95% with a diasteromeric purity of 98.2% and an ee of 99.4%. Ch reduction of chloroketone **2** using NaBH$_4$ produces primarily the undesired chloro diastereomer [16].

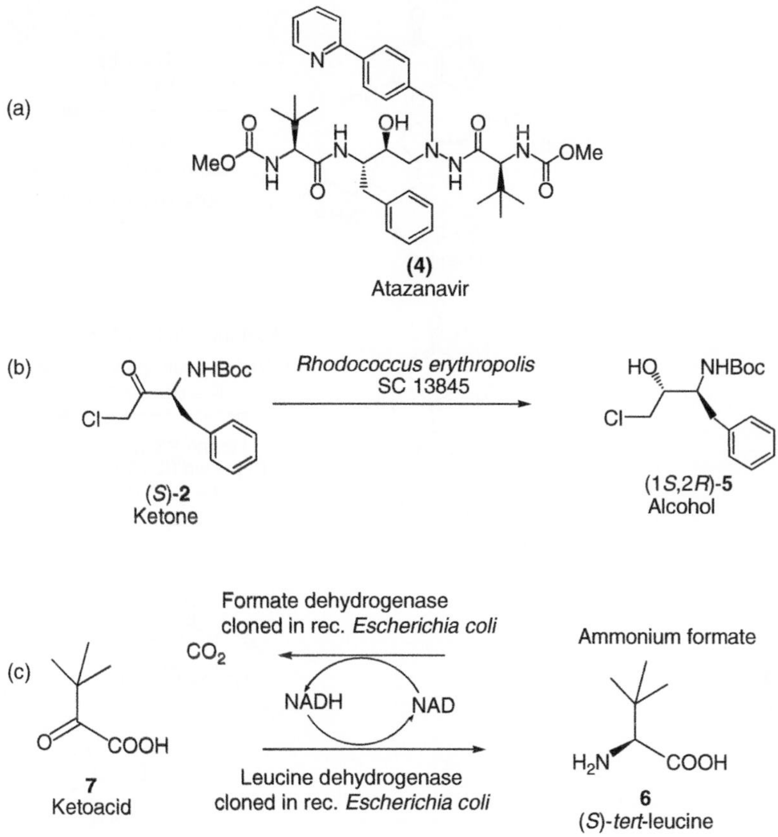

FIGURE 4.2 (a) Synthesis of chiral intermediates for antiviral agent, atazanavir **4**. (b) Enantioselective enzymatic reduction of (1*S*)-[3-chloro-2-oxo-1-(phenylmethyl)propyl]carbamic acid, 1,1-dimethylethyl ester (**2**) to the corresponding (1*S*,2*R*)-**5** by *Rhodococcus erythropolis* SC 13845. (c) Enzymatic reductive amination of ketoacid **7** to (*S*)-*tert*-leucine (**6**) by leucine dehydrogenase.

4.2.3 ENZYMATIC SYNTHESIS OF (*S*)-*TERT*-LEUCINE

Synthesis of atazanavir also required the (*S*)-*tert*-leucine **6** (Figure 4.2c). An enzymatic reductive amination of ketoacid **7** to amino acid **6** by recombinant *Escherichia coli* expressing leucine dehydrogenase from *Thermoactinimyces intermedius* was developed. The reaction required ammonia and NADH as a cofactor. Nicotinamide adenine dinucleotide (NAD) produced during the reaction was regenerated back to NADH using recombinant *E. coli* expressing formate dehydrogenase (FDH) from *Pichia pastoris*. A reaction yield of >95% with an ee of >99.5% was obtained for **6** at 100 g/L substrate input (R. Hanson, S. Goldberg, and R. Patel, unpublished results).

4.2.4 REGIOSELECTIVE ENZYMATIC AMINOACYLATION

Lobucavir (**12**, Figure 4.3) is a cyclobutyl guanine nucleoside analog under development as an antiviral agent for the treatment of herpes virus and hepatitis B [17]. A prodrug in which one of the two hydroxyls is coupled to valine, **14**, has also been considered for development. Regioselective aminoacylation is difficut to achieve by chemical procedures, but appeared to be suitable for an enzymatic approach [18]. Synthesis of the lobucavir L-valine prodrug **14**

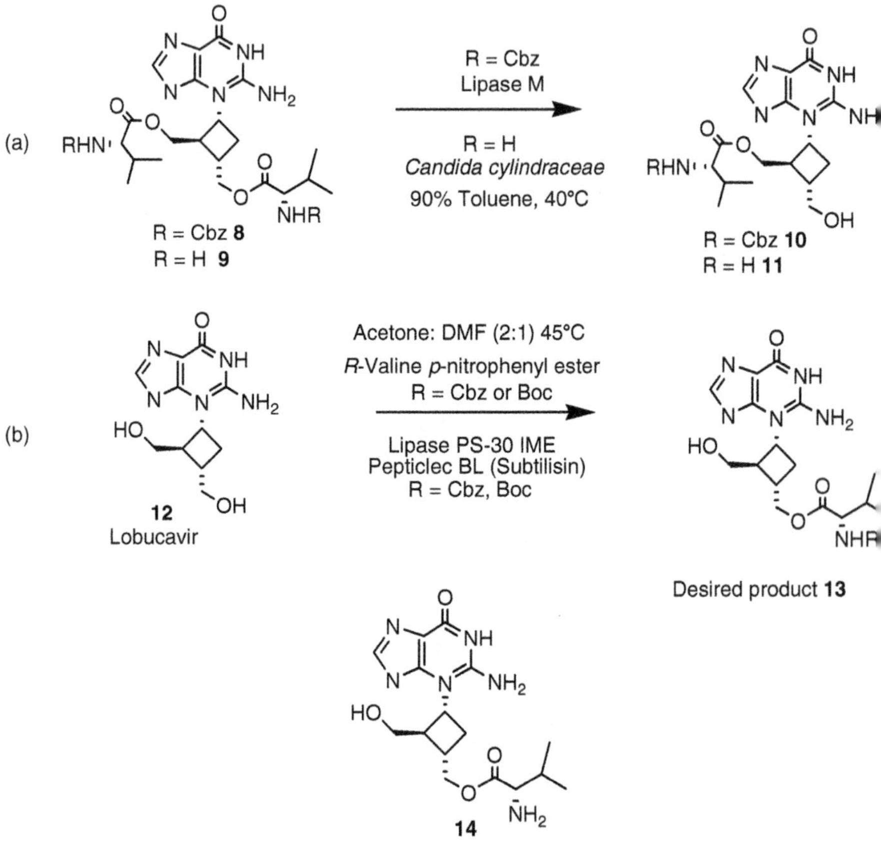

FIGURE 4.3 Synthesis of chiral intermediates for the lobucavir prodrug **14**. (a) Regioselective en hydrolysis of **8** and **9**. (b) Regioselective enzymatic aminoacylation of lobucavir (**12**).

requires regioselective coupling of one of the two hydroxyl groups of lobucavir (**1** valine. Enzymatic processes were developed for aminoacylation of either hydroxyl g lobucavir [18]. The selective hydrolysis of the methyl ester **8** with lipase M gave **10** yield. When the methyl ester of **9** dihydrochloride was hydrolyzed with lipase from C cylindraceae, **11** was obtained in 87% yield. The final intermediates for lobucavir prod methyl ester of **13**, could be obtained by transesterification of lobucavir using ChiroCL (61% yield) or more selectively by using lipase from *Pseudomonas cepacia* (84% yield)

4.2.5 CRIXIVAN (HIV-PROTEASE INHIBITOR)

4.2.5.1 Enzymatic Preparation of *cis*-(1*S*,2*R*)-Indandiol and *trans*-(1*R*,2*R*)-Indandio

Cis-(1*S*,2*R*)-indandiol **16** or *trans*-(1*R*,2*R*)-indandiol **16** (Figure 4.4a) are both p precursors to *cis*-(1*S*,2*R*)-1-aminoindan-2-ol (**19**), a key chiral synthon for Crixivan (vir, **20**), a HIV-protease inhibitor. Enrichment and isolation of microbial cultures yield strains, *Rhodococcus* sp. B 264-1 (MB 5655) and I-24 (MA 7205), capable of biotransf indene **15** to *cis*-(1*S*,2*R*)-indandiol and *trans*-(1*R*,2*R*)-indandiol, respectively [19]. Isol. 5655 was found to have a toluene dioxygenase, while isolate MA 7205 was found to toluene and naphthalene dioxygenases as well as a naphthalene monooxygenase tha lyzes the above biotransformation. When scaled up in a 14 L fermentor, MB 5655 pr

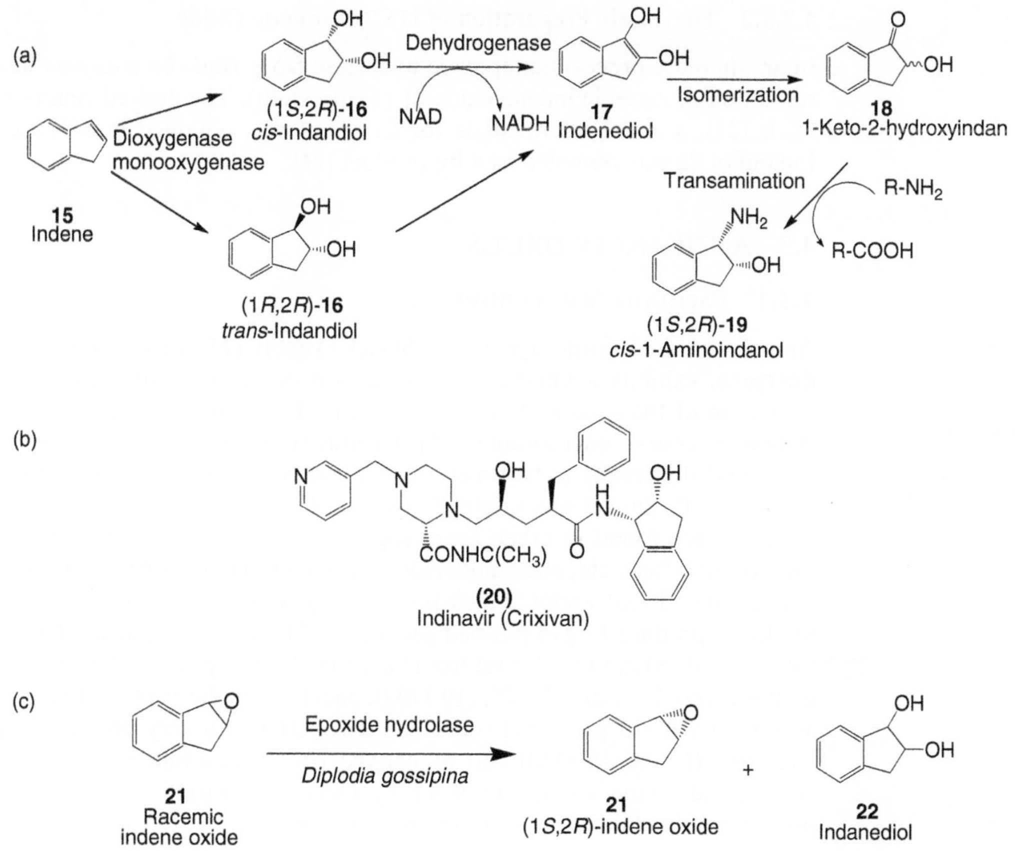

FIGURE 4.4 Synthesis of chiral intermediates for Crixivan (**20**). (a) Microbial oxygenation of indene **15** to *cis*-indandiol **16** and *trans*-indandiol **16**. (b) Resolution of racemic indene oxide **21** to (1*S*,2*R*)-indene oxide **21** by epoxide hydrolase from *Diplodia gossipina*.

up to 2.0 g/L of *cis*-(1*S*,2*R*)-indandiol **16** with an ee of >99%. *Rhodococcus* sp. MA 7205 cultivated under similar conditions produced up to 1.4 g/L of *trans*-(1*R*,2*R*)-indandiol **16** with an ee of >98%. Process development studies yielded titers of >4.0 g/L of *trans*-(1*R*,2*R*)-indandiol [20]. A metabolic engineering approach [21] and a directed evolution technique [22] were evaluated to avoid side reactions, block degradative pathways, and enhance the key reaction to convert indene to *cis*-aminoindanol **19** or *cis*-indanediol. The application of multiparameter flow cytometry was employed for the measurement of indene toxicity to the strain, and it was found that concentrations up to 0.25 g/L of indene (0.037 g indene/g dry cell weight) in batch bioconversions did not influence cell physiology. Using this information, the implementation of a single-phase indene fed-batch bioconversion was carried out. Cytoplasmic membrane integrity and membrane polarization of a large number of cells were measured during bioconversions and compared to a control in order to assess any toxic effects of indene feeding. The results indicated that indene supply at a rate of 0.1 g/L/h was feasible without any deleterious effects. *Cis*-(1*S*,2*R*)-indandiol **16** production rate was enhanced from 20 mg/L/h in a previously reported silicone oil two-liquid phase system up to 200 mg/L/h by a combination of suitable indene feeding rates in the stationary phase and operating with a high biomass concentration.

4.2.5.2　Enzymatic Preparation of (1S,2R)-Indene Oxide

In an alternate process, an epoxide hydrolase from *Diplodia gossipina* has been used resolution of racemic indene oxide 21 (Figure 4.4b). The desired enantiomer (1S,2R)-oxide (21), a chiral intermediate for Crixivan, was obtained in 14% yield and >99 Indandiol 22 was obtained as a by-product [24].

4.3　ANTICANCER DRUGS

4.3.1　PACLITAXEL SEMISYNTHESIS

Among the antimitotic agents, paclitaxel (Taxol) (23, Figure 4.5), a complex, pol diterpene, exhibits a unique mode of action on microtubule proteins responsible formation of the spindle during cell division. Paclitaxel is known to inhibit the depc ization process of microtubulin [25]. Various types of cancers have been treated witl taxel, and the results in treatment of ovarian cancer and metastatic breast cancer a promising. Paclitaxel was originally isolated from the bark of the yew *Taxus brevifo* has also been found in other *Taxus* spp. Paclitaxel was obtained from *T. brevifolia* very low (0.07%) yield, and cumbersome purification from other related taxanes was re It is estimated that about 20,000 pounds of yew bark (equivalent to about 3,000 tre needed to produce 1 kg of purified paclitaxel [27]. The development of a semisynthetic for the production of paclitaxel from baccatin III (24) (paclitaxel without the C-13 side or 10-deacetylbaccatin III (25) (10-DAB, paclitaxel without the C-13 side chain and th acetate) and C-13 paclitaxel side chain 27 or 31 was a very promising approach. T baccatin III, and 10-DAB can be derived from renewable resources such as the r shoots, and young *Taxus* cultivars [28]. Thus, preparation of paclitaxel by a semisy process would eliminate the cutting of yew trees.

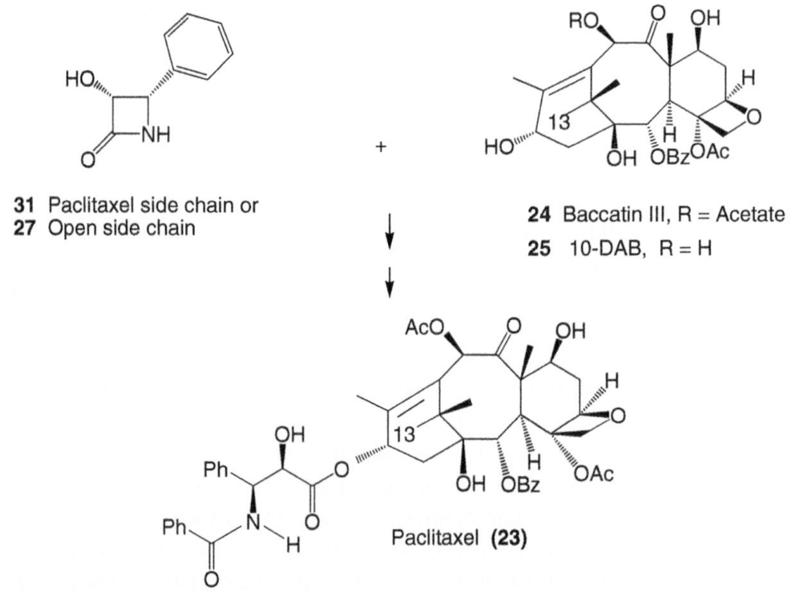

31 Paclitaxel side chain or
27 Open side chain

24 Baccatin III, R = Acetate
25　10-DAB, R = H

Paclitaxel (23)

FIGURE 4.5　Semisynthesis of paclitaxel (23), an anticancer agent. Coupling of baccatin III (C-13 paclitaxel side-chain synthons 27 or 31.

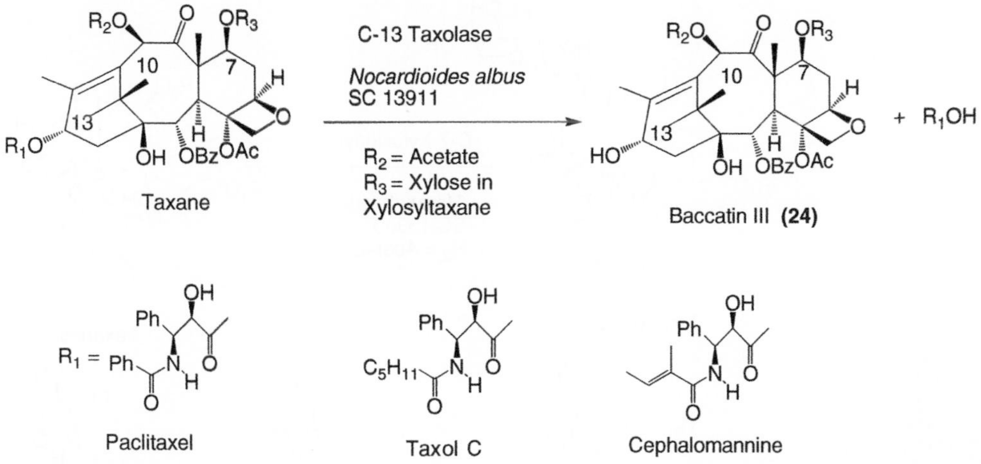

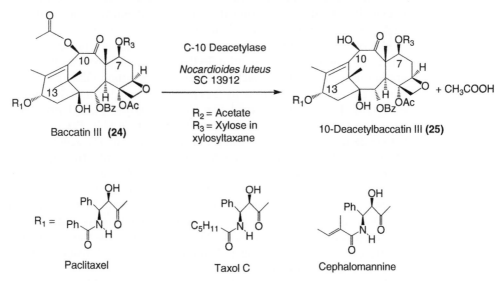

FIGURE 4.6 Enzymatic hydrolysis of C-13 side chain of taxanes by C-13 taxolase from *Nocardioides albus* SC 13911.

Using selective enrichment techniques, two strains of *Nocardioides* were isolated from soil samples that contained novel enzymes C-13 taxolase and C-10 deacetylase [29,30]. The extracellular C-13 taxolase derived from the filtrate of the fermentation broth of *Nocardioides albus* SC 13911 catalyzed the cleavage of the C-13 side chain from paclitaxel and related taxanes such as taxol C, cephalomannine, 7-β-xylosyltaxol, 7-β-xylosyl-10-deacetyltaxol, and 10-deacetyltaxol (Figure 4.6). The intracellular C-10 deacetylase derived from fermentation of *Nocardioides luteus* SC 13912 catalyzed the cleavage of the C-10 acetate from paclitaxel, related taxanes, and baccatin III to yield 10-DAB (Figure 4.7). The C-7 xylosidase derived from fermentation of *Moraxella* sp. (Figure 4.8) catalyzed the cleavage of the C-7 xylosyl group [31] from various taxanes. Fermentation processes were developed for growth of *N. albus* SC

FIGURE 4.7 Enzymatic hydrolysis of C-10 acetate of taxanes and baccatin III **24** by C-10 deacetylase from *Nocardioides luteus* SC 13912.

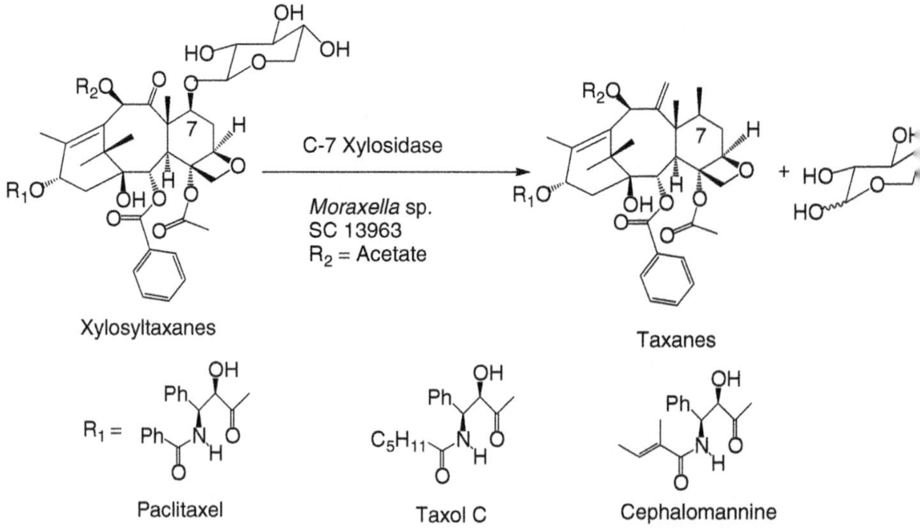

FIGURE 4.8 Enzymatic hydrolysis of C-7 xylose by C-7 xylosidase from *Moraxella* sp. 13963.

13911 and *N. luteus* SC 13912 to produce C-13 taxolase and C-10 deacetylase, respecti
5000 L batches, and a bioconversion process was demonstrated for the conversion of pac
and related taxanes in extrtacts of *Taxus* cultivars to the single compound 10-DAB **2**
both enzymes. In the bioconversion process, ethanolic extracts of the whole young plane
different cultivars of *Taxus* were first treated with a crude preparation of the C-13 taxe
give complete conversion of measured taxanes to baccatin III **24** and 10-DAB **25** in 6 h. *M*
SC 13192 whole cells were then added to the reaction mixture to give complete conver
baccatin III **24** to 10-DAB **25**. The concentration of 10-DAB **25** was increased by 5.5 to
in the extracts from various *Taxus* cultivars by treatment with the two enzymes. The **b**
version process was also applied to extracts of the bark of *T. bravifolia* to give a 12-fold i
in 10-DAB **25** concentration. Enhancement of the 10-DAB concentration in yew extra
potentially useful in increasing the amount and purification of this key precursor
paclitaxel semisynthetic process using renewable resources.

Another key precursor for the paclitaxel semisynthetic process is the chiral C-13 pa
side chain. Two different enantioselective enzymatic processes were developed for the
aration of the chiral C-13 paclitaxel side-chain synthon [32,33]. In one process, the e
selective microbial reduction of 2-keto-3-(N-benzoylamino)-3-phenyl propionic aci
ester (**26**) (Figure 4.9) to yield (2R,3S)-N-benzoyl-3-phenyl isoserine ethyl ester (.
been demonstrated using two strains of *Hansenula* [32]. Preparative-scale bioreduc
ketone **26** was demonstrated using cell suspensions of *Hansenula polymorpha* SC 138
Hansenula fabianii SC 13894 in independent experiments. In both batches, reaction y
>80% and ees of >94% were obtained for **27**. A 20% yield of the undesired antidiaster
was obtained with *H. polymorpha* SC 13865 compared with a 10% yield with *H. fabi*
13894. A 99% ee was obtained with *H. polymorpha* SC 13865 compared with a 94%
H. fabianii SC 13894. In a single-stage bioreduction process, cells of *H. fabianii* were g
a 15 L fermentor for 48 h; then the bioreduction process was initiated by addition of
substrate and 250 g of glucose and continued for 72 h. A reaction yield of 88% with a
95% was obtained for **27**.

In an alternate process for the preparation of the C-13 paclitaxel side chain, the e
selective enzymatic hydrolysis of racemic acetate *cis*-3-(acetyloxy)-4-phenyl-2-azetidinc

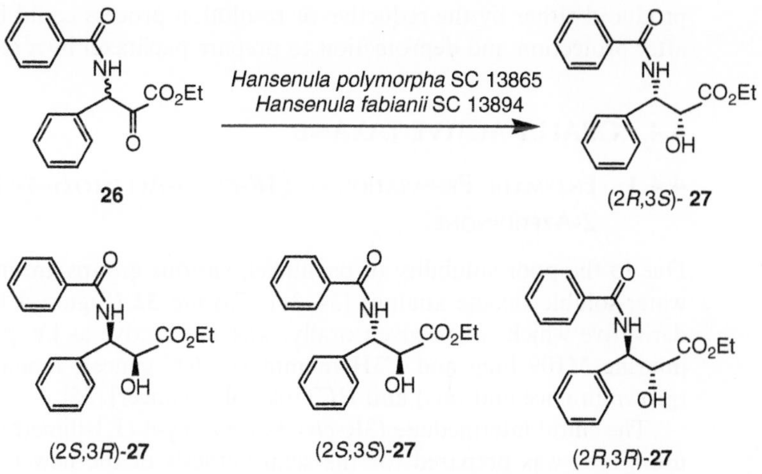

26   **(2R,3S)- 27**

Hansenula polymorpha SC 13865
Hansenula fabianii SC 13894

(2S,3R)-27   **(2S,3S)-27**   **(2R,3R)-27**

FIGURE 4.9 Enzymatic synthesis of C-13 side chain of paclitaxel **23**: enantioselective microbial reduction of 2-keto-3-(*N*-benzoylamino)-3-phenyl propionic acid, ethyl ester (**26**).

(Figure 4.10) to the corresponding (*S*)-alcohol **29** and the unreacted desired (*R*)-acetate **30** was demonstrated [33] using lipase PS-30 from *P. cepacia* (Amano International Enzyme Company) and BMS lipase (extracellular lipase derived from the fermentation of *Pseudomonas* sp. SC 13856). Reaction yields of >48% (theoretical maximum yield 50%) with ees of >99.5% were obtained for the (*R*)-acetate. BMS lipase and lipase PS-30 were immobilized on Accurel polypropylene (PP), and the immobilized lipases were reused (ten cycles) without loss of enzyme activity, productivity, or the ee of the product **30** in the resolution process. The enzymatic process was scaled up to 250 L (2.5 kg substrate input) using immobilized BMS lipase and lipase PS-30, respectively. From each reaction batch, *R*-acetate **30** was isolated in 45% yield (theoretical maximum yield 50%) and 99.5% ee. The (*R*)-acetate was chemically converted to (*R*)-alcohol **31**. The C-13 paclitaxel side-chain synthon (2*R*,3*S*-**27** or **31**)

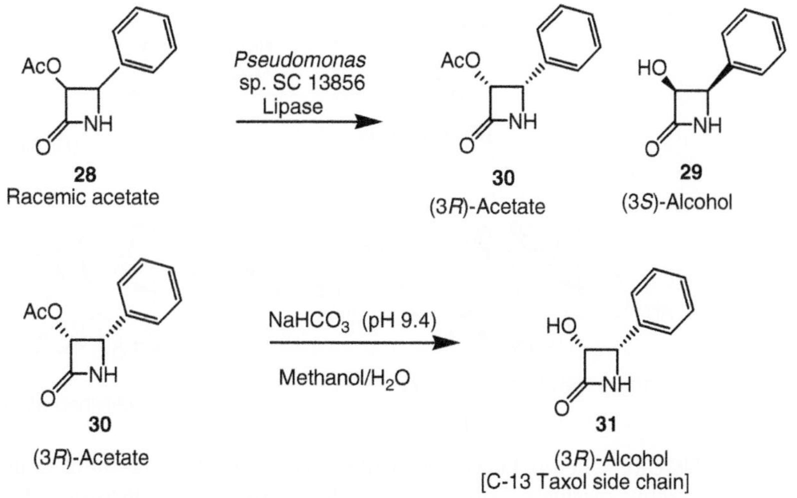

28   **30**   **29**
Racemic acetate   (3*R*)-Acetate   (3*S*)-Alcohol

Pseudomonas sp. SC 13856 Lipase

30   **31**
(3*R*)-Acetate   (3*R*)-Alcohol
                                               [C-13 Taxol side chain]

NaHCO$_3$ (pH 9.4)
Methanol/H$_2$O

FIGURE 4.10 Enantioselective enzymatic hydrolysis of *cis*-3-(acetyloxy)-4-phenyl-2-azetidinone **28**.

produced either by the reductive or resolution process could be coupled to bacattin III after protection and deprotection to prepare paclitaxel by a semisynthetic process [28].

4.4   ORALLY ACTIVE TAXANE

4.4.1   Enzymatic Preparation of (3R-cis)-3-Acetyloxy-4-(1,1-Dimethylethyl)-2-Azetidinone

Due to the poor solubility of paclitaxel, various groups are involved in the developme water-soluble taxane analogs [34–36]. Taxane **32** (Figure 4.11a) is a water-soluble ta derivative which, when given orally, was as effective as i.v. paclitaxel in five tumor m [murine M109 lung and C3H mammary 16/C cancer, human A2780 ovarian cancer (grown in mice and rats) and HCT/pk colon cancer] [35].

The chiral intermediate (3R-cis)-3-acetyloxy-4-(1,1-dimethylethyl)-2-azetidinone (**35** ure 4.11b) was prepared for the semisynthesis of the new taxane **32**. The enantioseld enzymatic hydrolysis of cis-3-acetyloxy-4-(1,1-dimethylethyl)-2-azetidinone (**33**) to

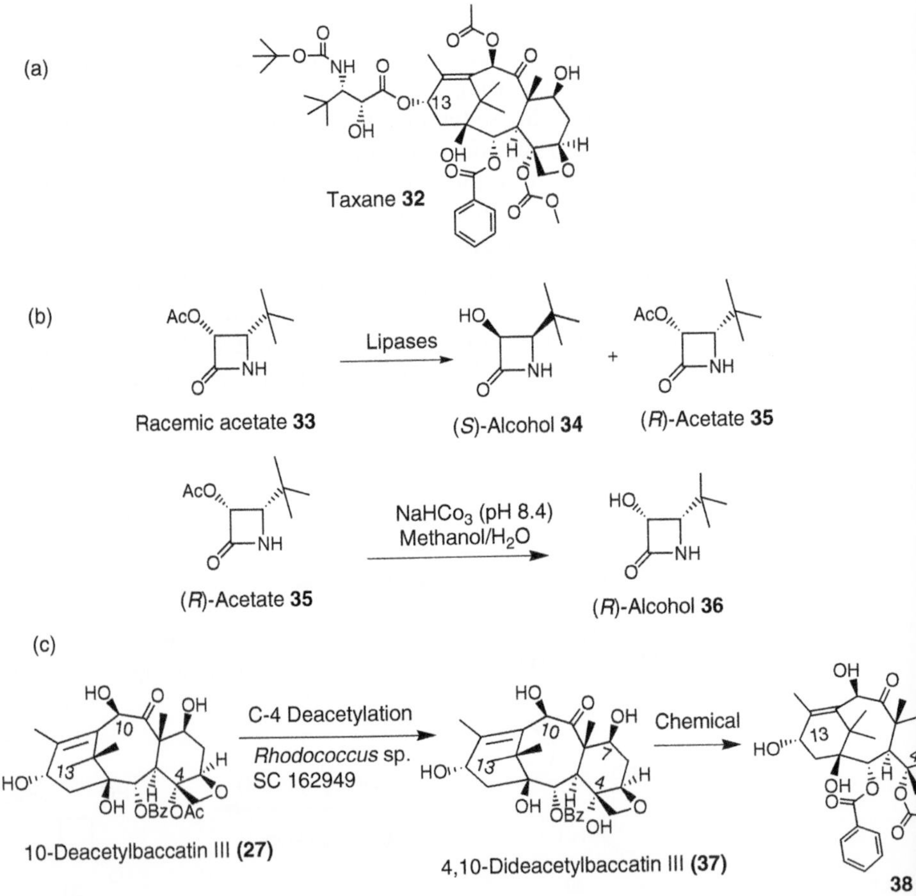

FIGURE 4.11 (a) Enzymatic synthesis of C-13 side chain **36** of orally active taxane **32**. (b) E selective enzymatic hydrolysis of cis-3-acetyloxy-4-(1,1-dimethylethyl)-2 azetidinone (**33**). (c) En C-4 deacetylation of 10-deacetylbaccatin III (**27**).

corresponding undesired (S)-alcohol **34** and unreacted desired (R)-acetate **35** was carried out using immobilized lipase PS-30 (Amano International Enzyme Company) or BMS lipase (extracellular lipase derived from the fermentation of *Pseudomonas* sp. SC 13856). Reaction yields of >48% (theoretical maximum yield 50%) with ees of >99% were obtained for the (R)-acetate **35**. Acetoxy β-lactam **35** was converted to hydroxy β-lactam **36** for use in the semisynthesis of **32** [37].

The synthesis of oral taxane **32** also required 4,10-dideacetylbaccatin **37** (Figure 4.11c) as starting material for the synthesis of the C-4 methylcarbonate derivative of 10-dideacetylbaccatin **38**. A microbial process was developed for deacetylation of 10-deacetylbaccatin III (**27**) to 4,10-dideacetylbaccatin III (**37**) using a *Rhodococcus* sp. SC 162949 isolated from soil using culture enrichment techniques [38].

4.5 EPOTHILONES

4.5.1 MICROBIAL HYDROXYLATION OF EPOTHIOLONE B

The tubulin-polymerizing chemotherapeutic agents, such as taxanes, have shown to be one of the most effective agents in the treatment of ovarian cancer. The clinical success of paclitaxel has stimulated research into compounds with similar modes of activity in an effort to emulate its antineoplastic efficacy while minimizing its less desirable aspects, which include nonwater solubility, difficult synthesis, and emerging resistance. The epothilones are a novel class of natural product cytotoxic compounds derived from the fermentation of the myxobacterium *Sorangium cellulosum* that are nontaxane microtubule-stabilizing compounds, triggering apoptosis [39,40]. The natural product epothilone B **39** (Figure 4.12) has demonstrated broad spectrum *in vitro* and *in vivo* antitumor activity, including tumors with paclitaxel resistance mediated by overexpression of P-glycoprotein or β-tubulin mutation [41]. The role of **39** as a potential paclitaxel successor has initiated interest in its synthesis, resulting in several total syntheses of **39** and various derivatives thereof [42–44]. The epothilone analogs were synthesized in an effort to optimize the water solubility, *in vivo* metabolic stability, and antitumor efficacy of this class of antineoplasic agents [45,46].

A fermentation process was developed for the production of epothilone B and titer of epothilone B was increased by continuous feed of sodium propionate during fermentation. The inclusion of XAD-16 resin during fermentation to adsorb epothilone B and to carry out volume reduction made the recovery of the product very simple [40]. A microbial

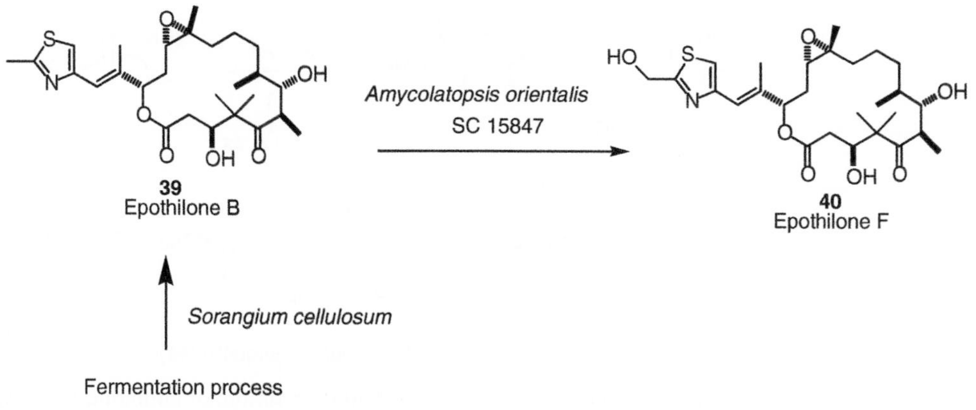

FIGURE 4.12 Microbial hydroxylation of epothilone B to epothilone F.

hydroxylation process was developed for conversion of epothilone B **39** to epothilone **
Amycolatopsis orientalis SC 15847 [47].

4.6 DEOXYSPERGUALIN

4.6.1 Enantioselective Enzymatic Acetylation of Racemic
7-[N,N'-*bis*(Benzyloxycarbonyl)-N-(Guanidinoheptanoyl)]-α-Hydroxy-C

An antitumor antibiotic spergualin was discovered in the culture filtrate of a bacteria
and its structure was determined to be 15S-1-amino-10-guanidino-11,15-dihydro
12-triazanonadecane-10,13-dione [48]. The total synthesis was accomplished by th
catalyzed condensation of 11-amino-1,1-dihydroxy-3,8-diazaundecane-2-one wit
7-guanidino-3-hydroxy-heptanamide followed by the separation of the 11-epimeric m
Antibacterial or antitumor activity of the racemic spergualin was about half of that
natural spergualin [49] indicating the importance of the configuration at C-11 for ant
activity.

The lipase-catalyzed enantioselective acetylation of racemic 7-[N,N'-*bis*(benzyloxy
nyl)-N-(guanidinoheptanoyl)]-α-hydroxy-glycine (**41**, Figure 4.13) to the corresp
(S)-acetate **42** and unreacted alcohol (R)-**43** has been developed [50]. (S)-acetate **42** i
intermediate in the chemical synthesis of (−)-15-deoxyspergualin (**44**), a related immu
pressive agent and antitumor antibiotic [51]. The reaction was carried out in methy
ketone (MEK) using lipase from *Pseudomonas* sp. (lipase AK) with vinyl acetate
acylating agent. A reaction yield of 48% (theoretical maximum yield 50%) with an
98% was obtained for (S)-acetate **42**. The unreacted alcohol (R)-**43** was obtained in 41
and 94% ee.

4.7 ENZYMATIC PREPARATION OF (S)-2-CHLORO-1-
(3-CHLOROPHENYL)ETHANOL

The synthesis of the leading candidate compound in an anticancer program [52,53] re
(S)-2-chloro-1-(3-chlorophenyl)ethanol (**45**, Figure 4.14) as an intermediate. Other p
candidate compounds used analogs of the (S)-alcohol. About 100 microbial culture
screened for reduction of the corresponding ketone **46** to the (S)-alcohol **45**, and *H
morpha* SC 13824 (73.8% ee) and *R. globerulus* SC SC16305 (71.8% ee) had the

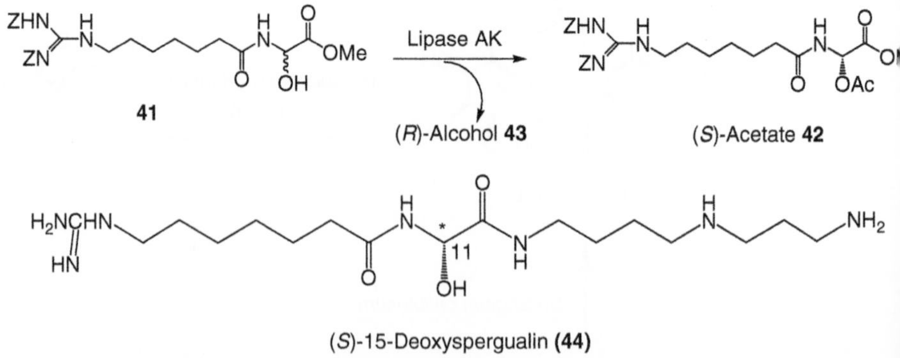

FIGURE 4.13 Synthesis of chiral intermediates for antitumor antibiotic 15-deoxyspergualin (**44**
tioselective enzymatic acylation of racemic **41** to yield (S)-acetate **42**.

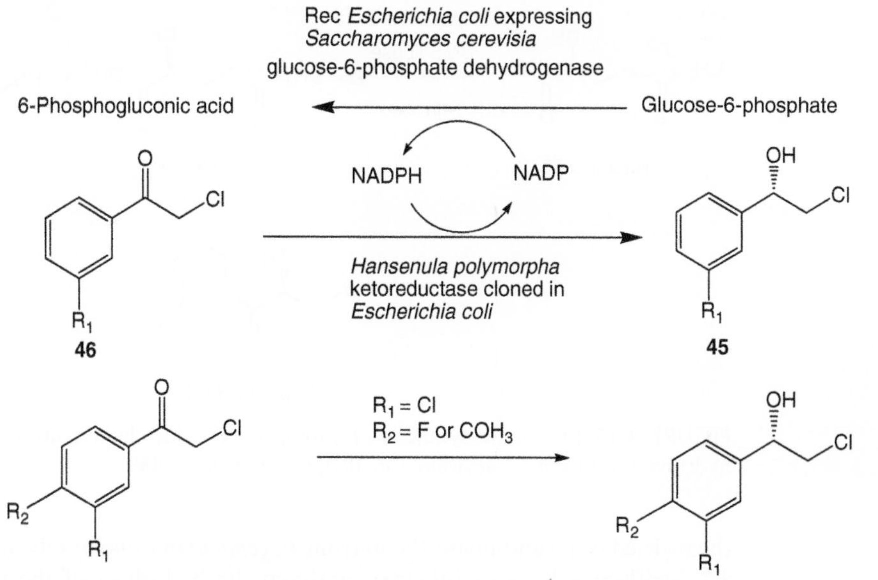

FIGURE 4.14 Enantioselective enzymatic reduction of **46** to **45**, an intermediate in anticancer drug.

enantioselectivity. A ketoreductase from *H. polymorpha*, after purification to homogeneity, gave (*S*)-alcohol **45** with 100% ee. Amino acid sequences from the purified enzyme were used to design PCR primers for cloning the ketoreductase. The cloned ketoreductase required NADP(H), had a subunit molecular weight of 29,220, and a native molecular weight of 88,000. The cloned ketoreductase was expressed in *E. coli* together with a cloned glucose-6-phosphate dehydrogenase from *Saccharomyces cerevisiae* to allow regeneration of the NADPH required by the ketoreductase. An extract of *E. coli* containing the two recombinant enzymes was used to reduce 2-chloro-1-(3-chloro-4-fluorophenyl)-ethanone (**46**) and two related ketones to the corresponding (*S*)-alcohols. Intact *E. coli* cells provided with glucose were used to prepare (*S*)-**45** in 89% yield with 100% ee [54].

4.8 ANTIHYPERTENSIVE DRUGS

4.8.1 ANGIOTENSIN-CONVERTING ENZYME INHIBITORS

4.8.1.1 Captopril: Enzymatic Preparation of (*S*)-3-Acetylthio-2-Methylpropanoic Acid

Captopril is designated chemically as 1-[(2*S*)-3-mercapto-2-methylpropionyl]-L-proline (**47**, Figure 4.15). It is used as an antihypertensive agent through suppression of the renin-angiotensin-aldosterone system [55,56]. Captopril prevents the conversion of angiotensin I to angiotensin II (AII) by inhibition of angiotensin-converting enzyme (ACE). The potency of captopril **47** as an inhibitor of ACE depends critically on the configuration of the mercaptoalkanoyl moiety; the compound with the *S*-configuration is about 100 times more active than its corresponding *R*-isomer [57]. The required 3-mercapto-(2*S*)-methylpropionic acid moiety has been prepared from the microbially derived chiral 3-hydroxy-(2*R*)-methylpropionic acid, which is obtained by the hydroxylation of isobutyric acid [58]. The synthesis of the (*S*)-side chain of captopril by the lipase-catalyzed enantioselective hydrolysis of the thioester bond of racemic 3-acetylthio-2-methylpropanoic acid (**48**) to yield (*S*)-**49** has been demonstrated [59]. Among various lipases evaluated, the lipase from *Rhizopus oryzae* ATCC 24563

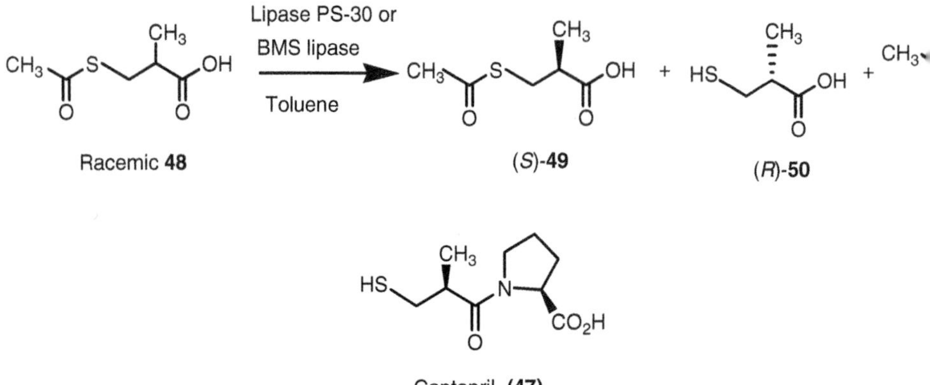

FIGURE 4.15 Enzymatic synthesis of captopril (47) side-chain synthon: enantioselective em hydrolysis of racemic 3-acylthio-2-methylpropanoic acid (48).

(heat-dried cells) and lipase PS-30 from *P. cepacia* in organic solvent systems (1,1,2-tri 1,2,2-trifluoroethane or toluene) catalyzed the hydrolysis of the thioester bond of desired enantiomer of racemic 48 to yield desired (*S*)-49, (*R*)-3-mercapto-2-methylpro acid (50), and acetic acid (Figure 4.15). Reaction yields of >24% (maximum theoretic is 50%) with ees of >95% were obtained for (*S*)-49 using each lipase.

4.8.1.2   Zofenopril: Enzymatic Preparation of (*S*)-3-Benzoylthio-2-Methylpropanoi

In an alternative approach to prepare the chiral side chain of captopril (47), and zof (51), the lipase-catalyzed enantioselective esterification of racemic 3-benzoylthio-2-propanoic acid (52, Figure 4.16) in an organic solvent was demonstrated to yield methyl ester 53 and unreacted acid enriched in the desired (*S*)-54 [60]. Using lipase with toluene as solvent and methanol as nucleophile, the desired (*S*)-54 was obtained yield (maximum theoretical yield is 50%) with 97% ee. The amount of water a concentration of methanol supplied in the reaction mixture were very critical. Wa used at 0.1% concentration in the reaction mixture. Higher than 1% water led

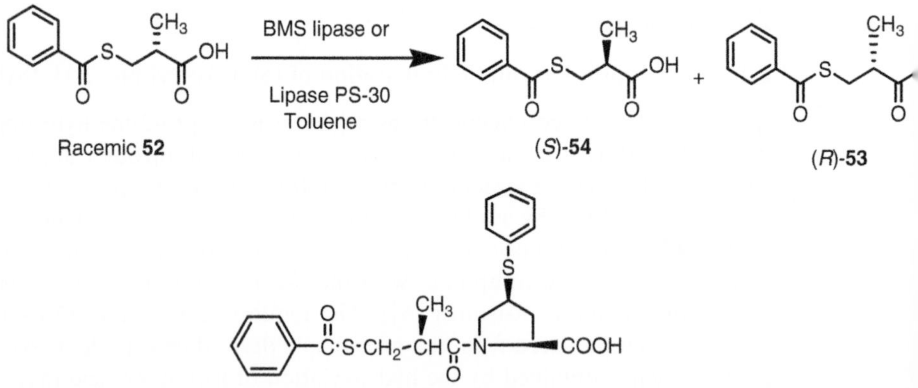

FIGURE 4.16 Synthesis of zofenopril (51) side-chain (*S*)-54: enantioselective enzymatic esterific racemic 3-benzylthio-2-methylpropanoic acid (52).

aggregation of enzyme in the organic solvent with a decrease in the rate of reaction due to mass transfer limitation. The rate of esterification decreased as the methanol/substrate ratio was increased from 1:1 to 4:1. Higher methanol concentrations probably inhibited the esterification reaction by stripping the essential water from the enzyme. Crude lipase PS-30 was immobilized on Accurel polypropylene (PP) in absorption efficiencies of 98.5%. The immobilized lipase efficiently catalyzed the esterification reaction, giving 45% reaction yield with 97.7% ee of (S)-54. The immobilized enzyme under identical conditions gave a similar ee and yield of product in 23 additional reaction cycles without any loss of activity and productivity. (S)-54 is a key chiral intermediate for the synthesis of captopril [61] or zofenopril [62].

4.8.1.3 Monopril: Enzymatic Preparation of (S)-2-Cyclohexyl- and (S)-2-Phenyl-1,3-Propanediol Monoacetates

(S)-2-Cyclohexyl-1,3-propanediol monoacetate (55) and (S)-2-phenyl-1,3-propanediol monoacetate (56) are key chiral intermediates for the chemo-enzymatic synthesis of Monopril (57, Figure 4.17), an antihypertensive drug that acts as an ACE inhibitor. The asymmetric hydrolysis of 2-cyclohexyl-1,3-propanediol diacetate (58) and 2-phenyl-1,3-propanediol diacetate (59) to the corresponding (S)-monoacetate 55 and (S)-monoacetate 56 by porcine pancreatic lipase (PPL) and *Chromobacterium viscosum* lipase has been demonstrated [63]. In a biphasic system using 10% toluene, reaction yields of >65% with ees of 99% were obtained for (S)-55 using each enzyme. (S)-56 was obtained in 90% reaction yield with 99.8% ee using *C. viscosum* lipase under similar conditions.

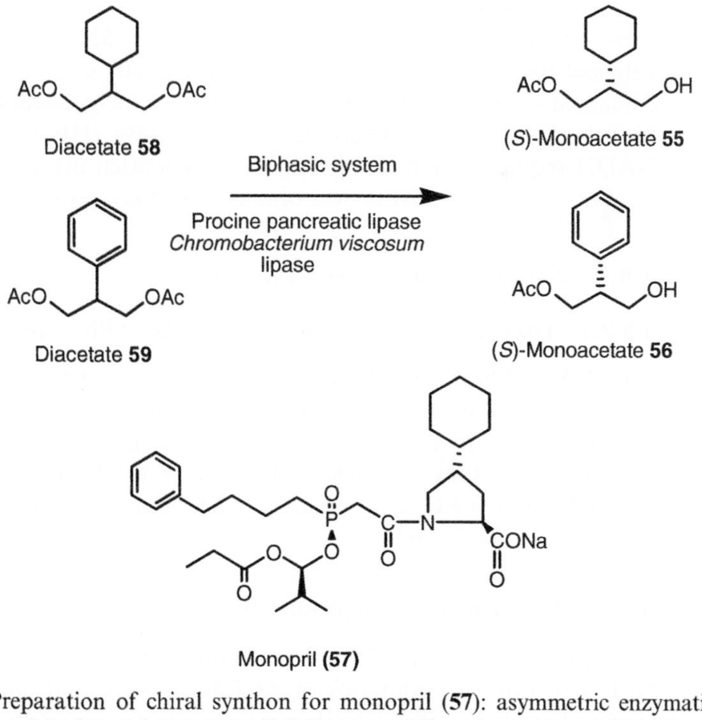

FIGURE 4.17 Preparation of chiral synthon for monopril (57): asymmetric enzymatic hydrolysis of 2-cyclohexyl-58 and 2-phenyl-1,3-propanediol diacetate (59) to the corresponding (S)-monoacetates 55 and 56.

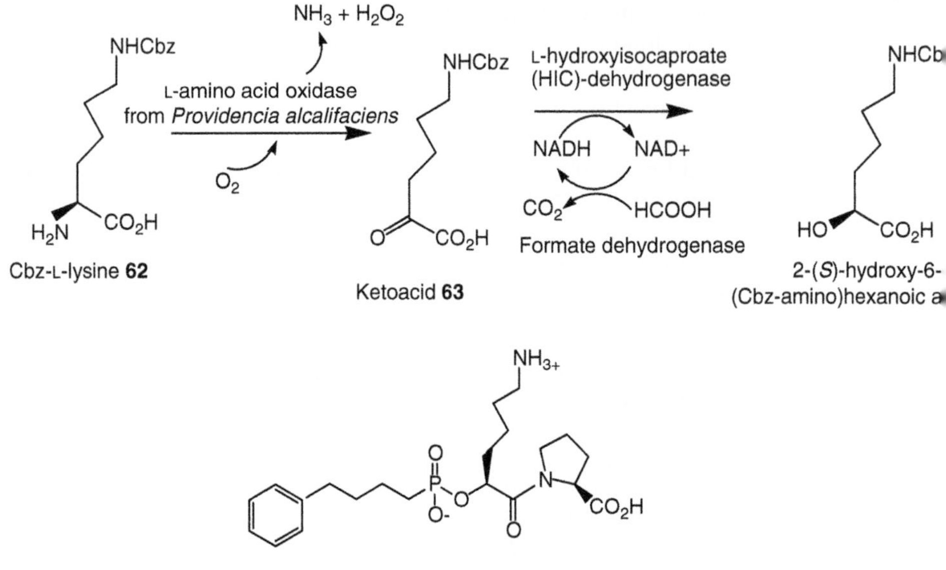

FIGURE 4.18 Synthesis of chiral synthon for ceranopril (**60**): enzymatic conversion of Cbz-L-lysi Cbz-L-oxylysine **61**.

4.8.1.4   Ceranopril: Enzymatic Preparation of *N*-ε-Carbobenzoxy-L-Oxylysine

Ceranopril (**60**) is another ACE inhibitor [64] that requires chiral intermediate carbobe L-oxylysine **61** (Figure 4.18). A biotransformation process has been developed to prep Cbz-L-oxylysine [65]. *N*-ε-carbobenzoxy-L-lysine (**62**) was first converted to the corresp ketoacid **63** by oxidative deamination using cells of *Providencia alcalifaciens* SC 90. contained L-amino acid oxidase and catalase. The ketoacid **63** was subsequently conve **61** using L-hydroxy isocaproate dehydrogenase (HIC) from *Lactobacillus confusu* NADH required for this reaction was regenerated using FDH from *C. boidinii*. A re yield of 95% with 98.5% ee was obtained for the overall process.

4.8.2   Neutral Endopeptidase Inhibitors

4.8.2.1   Enzymatic Preparation of (*S*)-α-[(Acetylthio)methyl]benzenepropanoic Aci

(*S*)-α-[(Acetylthio)methyl]benzenepropanoic acid (**64**, Figure 4.19) is a key chiral mediate for the neutral endopeptidase inhibitor (NEP) **65** [66]. The lipase PS-30 ca enantioselective hydrolysis of the thioester bond of racemic α-[(acetylthio)methyl]be propanoic acid (**66**) in organic solvent to yield (*R*)-α-[(mercapto)methyl]benzenepro acid (**67**) and (*S*)-**64**. A 40% reaction yield (maximum theoretical yield is 50%) with ' was obtained for (*S*)-**64** [59].

4.8.3   Angiotensin-Converting Enzyme and Neutral Endopepsidase Inhibitors

Omapatrilat (**68**, Figure 4.20) is an antihypertensive drug that acts by inhibiting AC NEP [67]. Effective inhibitors of ACE have been used not only in the treatment of tension but also in the clinical management of congestive heart failure. NEP, like A a zinc metalloprotease and is highly efficient in degrading atrial natriuretic peptide (A

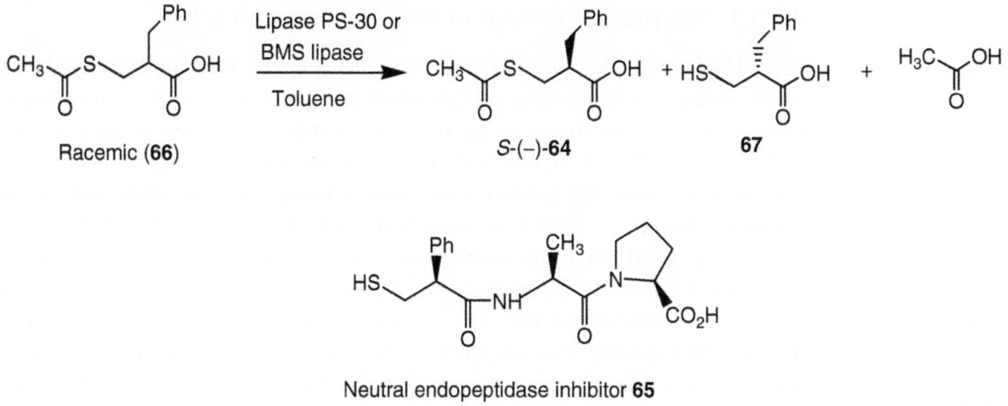

FIGURE 4.19 Preparation of chiral synthon for neutral endopeptidase inhibitor 65: enantioselective enzymatic hydrolysis of racemic α-[(acetylthio)methyl]benzenepropanoic acid (66).

28-amino acid peptide secreted by the heart in response to atrial distension. By interaction with its receptor, ANP promotes the generation of cyclic guanosine monophosphate (cGMP) through guanylate cyclase activation, thus resulting in vasodilatation, natriuresis, diuresis, and inhibition of aldosterone. Therefore, simultaneous potentiation of ANP by NEP inhibition and attenuation of AII by ACE inhibition should lead to complementary effects in the management of hypertension and congestive heart failure [68].

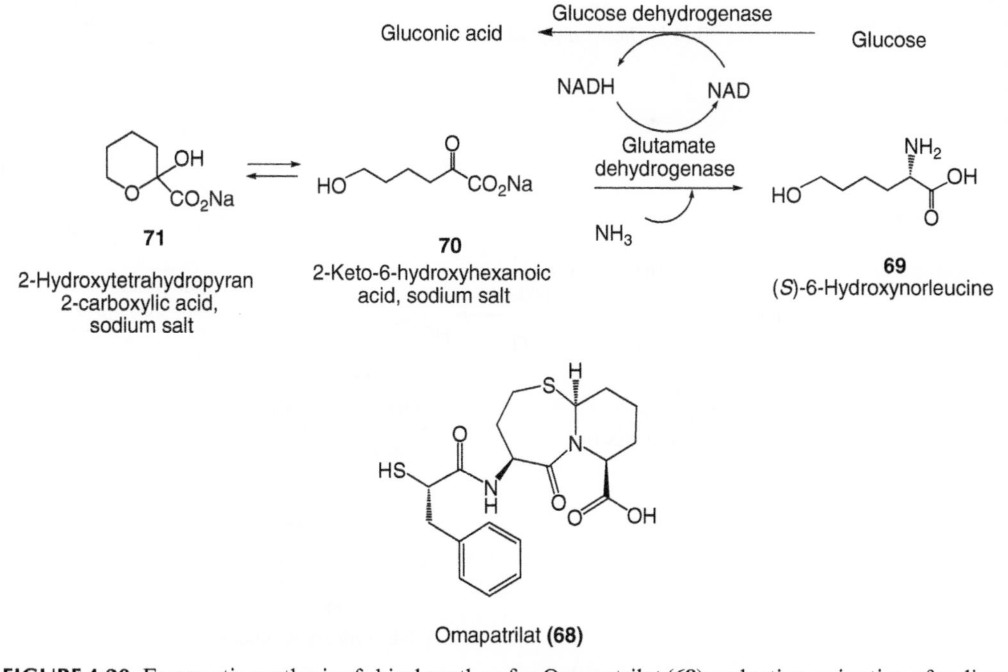

FIGURE 4.20 Enzymatic synthesis of chiral synthon for Omapatrilat (68): reductive amination of sodium 2-keto-6-hydroxyhexanoic acid (70) to (S)-6-hydroxynorleucine (69) by glutamate dehydrogenase.

4.8.3.1 Enzymatic Synthesis of (S)-6-Hydroxynorleucine

(S)-6-Hydroxynorleucine (**69**, Figure 4.20) is a key intermediate in the synthesis of ⊲ trilat. Reductive amination of ketoacids using amino acid dehydrogenases has been ⊰ method for the synthesis of natural and unnatural amino acids [69,70]. The synthe complete conversion of 2-keto-6-hydroxyhexanoic acid (**70**) to (S)-6-hydroxynorleuc was demonstrated by reductive amination using phenylalanine dehydrogenase (PDF *Sporosarcina* sp. or beef liver glutamate dehydrogenase [71]. Beef liver glutamate d⊲ genase was used for preparative reactions at 100 g/L substrate concentration. As d⊲ compound **70**, in equilibrium with 2-hydroxytetrahydropyran-2-carboxylic acid sodi (**71**), was converted to **69**. The reaction requires ammonia and NADH. NAD p⊩ during the reaction was recycled to NADH by the oxidation of glucose to gluco⊮ using glucose dehydrogenase from *Bacillus megaterium*. The reaction was complete i⊪ 3 h, with reaction yields of 92% and ees of >99% for (S)-6-hydroxynorleucine.

The synthesis and isolation of ketoacid **70** required several steps. In a secon⊂ convenient, process the ketoacid was prepared by treatment of racemic 6-hydroxyno⊪ (**72**) (produced by hydrolysis of 5-(4-hydroxybutyl) hydantoin (**73**)) with D-amino acid and catalase (Figure 4.21). After the ee of the remaining (S)-6-hydroxynorleuci⊪

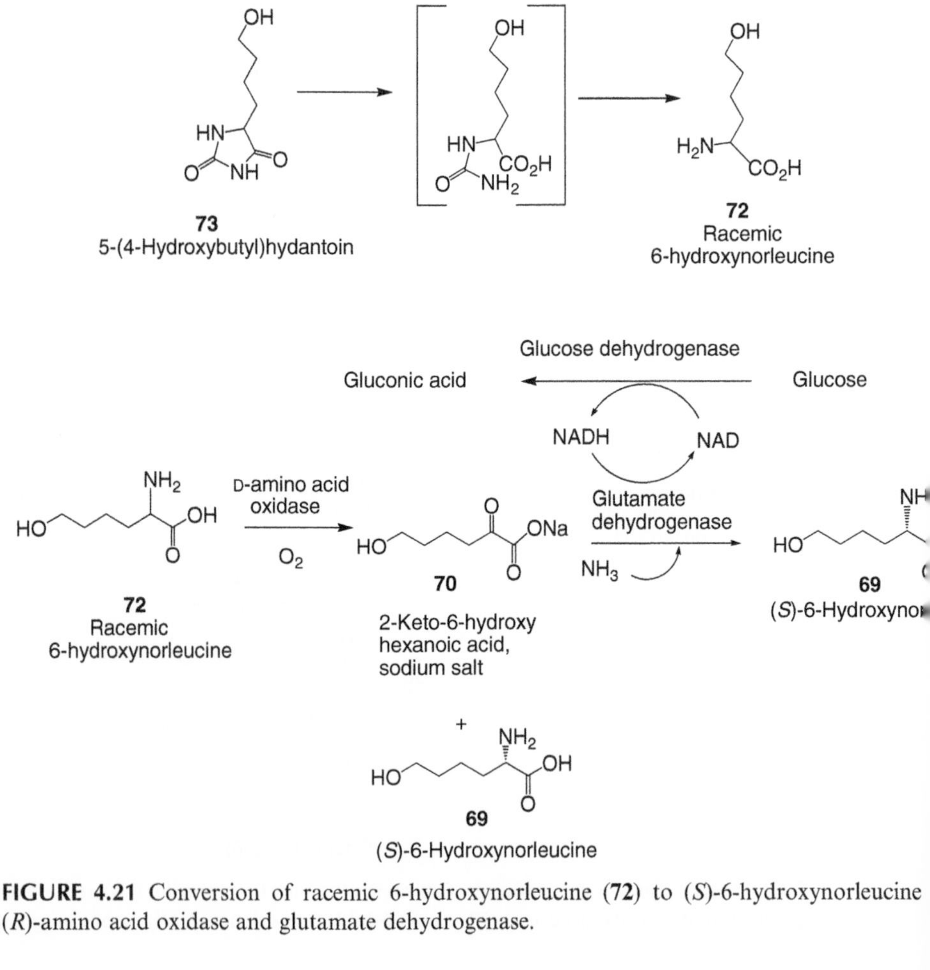

73
5-(4-Hydroxybutyl)hydantoin

72
Racemic
6-hydroxynorleucine

72
Racemic
6-hydroxynorleucine

70
2-Keto-6-hydroxy
hexanoic acid,
sodium salt

69
(S)-6-Hydroxynor⊪

69
(S)-6-Hydroxynorleucine

FIGURE 4.21 Conversion of racemic 6-hydroxynorleucine (**72**) to (S)-6-hydroxynorleucine (R)-amino acid oxidase and glutamate dehydrogenase.

increased to >99%, the reductive amination procedure was used to convert the mixture containing the 2-keto-6-hydroxyhexanoic acid entirely to (S)-6-hydroxynorleucine in 97% yield with 98% ee from racemic 6-hydroxynorleucine at 100 g/L. Porcine kidney D-amino acid oxidase and beef liver catalase or *T. variabilis* whole cells (source of both the oxidase and catalase) were used successfully for this transformation [71]. The (S)-6-hydroxynorleucine prepared by the enzymatic process was converted chemically to omapatrilat (**68**) as described previously [72].

4.8.3.2 Enzymatic Synthesis of Allysine Ethylene Acetal

(S)-2-Amino-5-(1,3-dioxolan-2-yl)-pentanoic acid [(S)-allysine ethylene acetal] (**74**, Figure 4.22) is one of the three building blocks used in an alternative synthesis of omapatrilat (**68**). It had been previously prepared following an eight-step chemical synthesis from 3,4-dihydro[2H]pyran [73]. An alternate synthesis of **74** was demonstrated by reductive amination of ketoacid acetal **75** using PDH from *T. intermedius* [74]. The reaction required ammonia and NADH; NAD produced during the reaction was recycled to NADH by the oxidation of formate to CO_2 using FDH. *T. intermedius* PDH was cloned and expressed in *E. coli*, inducible by β-D-isopropylthio-galactoside (IPTG). Fermentation of *T. intermedius* yielded 184 units of PDH activity per liter of whole broth in 6 h. In contrast, *E. coli* BL21 (DE3) (pPDH155K) produced over 19,000 units per liter of whole broth in about 14 h. *C. boidinii* [75] or *P. pastoris* [76] grown on methanol are useful sources of FDH. Expression of *T. intermedius* PDH in *P. pastoris*, inducible by methanol, allowed obtaining both enzymes from a single fermentation.

A procedure using heat-dried cells of *E. coli* containing cloned PDH and heat-dried *C. boidinii* was scaled up. A total of 197 kg of **74** was produced in three 1,600 L batches using a 5% concentration of substrate **75** with an average yield of 91% and an ee of >98%. A second-generation procedure, using dried recombinant *P. pastoris* containing *T. intermedius* PDH inducible with methanol and endogenous FDH, induced when *P. pastoris* was grown in a medium containing methanol, allowed both enzymes to be produced during a single fermentation. The procedure with *P. pastoris* was also scaled up to produce 15.5 kg of **74** in a maximum yield of 97% and >98% ee in a 180 L batch using 10% ketoacid **75** concentration. The (S)-allysine ethylene acetal (**74**) produced by the enzymatic process was converted to omapatrilat (**68**) [72].

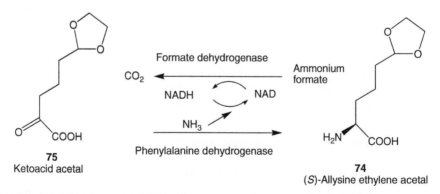

FIGURE 4.22 Enzymatic synthesis of chiral synthon for Omapatrilat (**68**): reductive amination of keto-acid acetal **75** to amino acid acetal **74** by phenylalanine dehydrogenase. Regeneration of NADH was carried out using formate dehydrogenase (FDH).

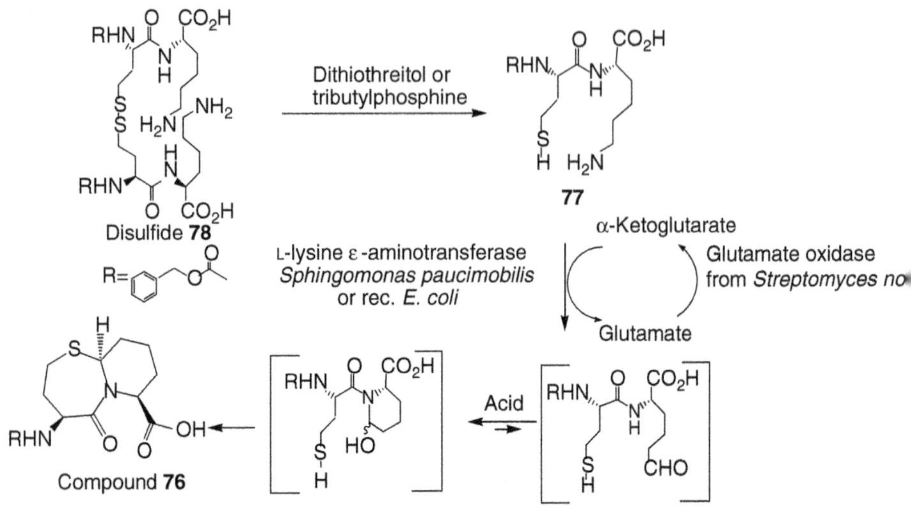

FIGURE 4.23 Enzymatic synthesis of chiral synthon for omapatrilat (**68**): conversion of disulfie thiazepine **76** by L-lysine ε-aminotransferase.

4.8.3.3 Enzymatic Synthesis of Thiazepine

[4S-(4a,7a,10ab)]-1-Octahydro-5-oxo-4-[[(phenylmethoxy)carbonyl]amino]-7H-pyrido [1,3]thiazepine-7-carboxylic acid (**76**, Figure 4.23) is a key intermediate in the synth omapatrilat (**68**) [67]. An enzymatic process was developed for the preparation of com **76**. A selective culture technique was used to isolate eight different types of microbial c able to utilize N-α-Cbz-S-lysine as the sole source of nitrogen. Cell extracts were evalua oxidation of the ε-amino group of (S)-lysine in the thiol substrate **77** generated in situ fr disulfide N^2-[N[[(phenylmethoxy) carbonyl]-L-homocysteinyl]-L-lysine]-1,1-disulfide (treatment with dithiothreitol (DTT). Product **76** formation was observed with four c One of the cultures, Z-2, later identified as *Sphingomonas paucimobilis* SC 16113 was u process development. Due to the low activity of enzyme [L-lysine ε-aminotransferase in *S. paucimobilis* SC 16113, and to minimize **77** hydrolysis, LAT was overexpressed in strain GI724(pAL781-LAT) and a biotransformation process was developed [77 aminotransferase reaction required α-ketoglutarate as the amine acceptor. Glutamate during this reaction was recycled back to α-ketoglutarate by glutamate oxidase (GOX *S. noursei* SC 6007. The extracellular GOX was cloned and expressed in *S. lividans*.

Biotransformation of compound **78** to compound **76** was carried out using LA *E. coli* GI724[pal781-LAT] in the presence of α-ketoglutarate and DTT (or tribu sphine) and GOX. Maximum reaction yields of 65 to 67% were obtained. The r yield in the absence of GOX averaged only about 33 to 35%. However, the reactio increased to a maximum of 70%, by increasing the α-ketoglutarate to 40 mg/ml (10 × i in concentration) and conducting the reaction at 40°C [77].

4.9 ANTICHOLESTROL DRUGS

4.9.1 Microbial Production of (S)-4-Chloro-3-Oxobutanoate Esters

Chiral β-hydroxy esters are versatile synthons in organic synthesis, specifically in th aration of natural products [78,79]. The reduction of 4-chloro-3-oxobutanoic acid ester (**79**) to (S)-4-chloro-3-hydroxybutanoic acid methyl ester (**80**, Figure 4.24)

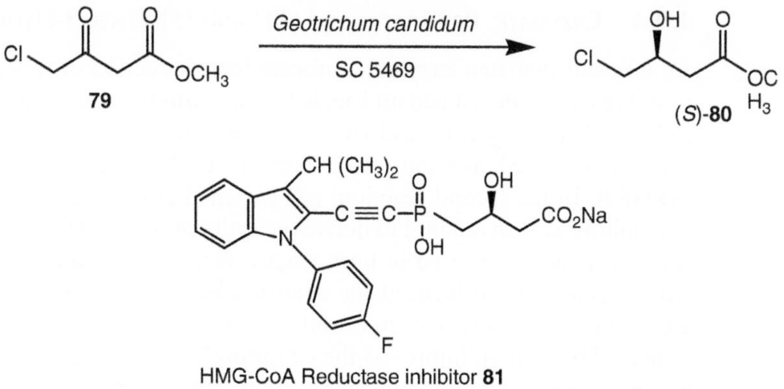

FIGURE 4.24 Synthesis of a chiral synthon for the cholesterol-lowering drug **81**: enantioselective microbial reduction of 4-chloro-3-oxobutanoic acid methyl ester **79** to (S)-4-chloro-3-hydroxybutanoic acid ester **80**.

suspensions of *Geotrichum candidum* SC 5469. S-(−)-**80** is a key chiral intermediate in the total chemical synthesis of **81**, a cholesterol antagonist that acts by inhibiting 3-hydroxy-3-methylglutaryl (HMG) coenzyme A (CoA) reductase [80]. In the biotransformation process, a reaction yield of 95% and ee of 96% were obtained for S-(−)-**80** by glucose-, acetate- or glycerol-grown cells (10% w/v) of *G. candidum* SC 5469 at 10 g/L substrate input. The ee of S-(−)-**80** was increased to 98% by heat treatment of cell suspensions (55°C for 30 min) prior to conducting the bioreduction of **79**.

4.9.2 Enzymatic Preparation of (S)-4-Chloro-3-Hydroxybutanoate

In an alternate approach, the asymmetric reduction of ethyl 4-chloroacetoacetate to (S)-4-chloro-3-hydroxybutonoate was demonstrated by a secondary alcohol dehydrogenase (PfODH) from *P. finlandica*. The gene encoding PfODH was cloned from *P. finlandica* and overexpressed in *E. coli*. FDH was used to regenerate the cofactor NADH required for this reaction. With recombinant *E. coli* coexpressing both PfODH and FDH from *Mycobacetrium* sp., (S)-4-chloro-3-hydroxybutonoate was produced in 98.5% yield and 99% ee at 32 g/L substrate input [81].

4.9.3 Enzymatic Preparation of (R)-4-Cyano-3-Hydroxybutyrate

An enzymatic process was developed for the preparation of 4-halo-3-hydroxybutyric acid derivatives by ketoreductase-catalyzed conversion of 4-halo-3-ketobutyric acid derivatives. Thus, the genes encoding halohydrin dehalogenase from *Agrobacterium tumefaciens*, ketoreductase from *C. magnoliae*, glucose dehydrogenase from *B. subtilis*, and FDH from *C. boidinii* were separately cloned into *E. coli* BL21. Each enzyme was then produced by fermentation, isolated, and characterized. Ethyl (R)-4-cyano-3-hydroxybutyrate was prepared from ethyl 4-chloroacetoacetate by the following procedure. Ethyl 4-chloroacetoacetate was incubated at pH 7.0 with ketoreductase, glucose dehydrogenase, and NADP for 40 h to produce ethyl (S)-chloro-3-hydroxybutyrate. The ethyl (S)-chloro-3-hydroxybutyrate was extracted with ethyl acetate, dried, filtered, and concentrated to yield a 97% ester. The dried ethyl (S)-chloro-3-hydroxybutyrate was dissolved in phosphate buffer and mixed with halohydrin dehalogenase and sodium cyanide at pH 8.0. After 57 h, (R)-4-cyano-3-hydroxybutyrate was recovered, an intermediate used in many HMG CoA reductase inhibitor syntheses [82].

4.9.4 ENZYMATIC PREPARATION OF (*R*)- AND (*S*)-ETHYL-3-HYDROXYBUTYRATE

An efficient two-step enzymatic process for production of (*R*)- and (*S*)-ethyl-3-hydro:
rate (HEB) was developed and scaled up to a multi-kg scale. Both enantiomers were ol
at 99% chemical purity and over 96% ee, with an overall process yield of 73%. T
reaction involved acetylation of racemic HEB with vinyl acetate for the produc
(*S*)-HEB. In the second reaction, (*R*)-enriched ethyl-3-acetoxybutyrate (AEB) was su
to alcoholysis with ethanol to derive optically pure (*R*)-HEB. Immobilized *C. antarctic*
B (CALB) was employed in both stages, with high productivity and selectivity. The
butyric acid ester influenced the enantioselectivity of the enzyme. Thus, extending tl
alkyl chain from ethyl to octyl resulted in a decrease in ee, whereas using bulky grour
as benzyl or *t*-butyl, improved the enantioselectivity of the enzyme. The immobilized (
was packed in a column and the reactants were circulated through the enzyme bed u
targeted conversion was reached. The desired products were separated from the r
mixture in each of the two stages by fractional distillation. The main features of the
are the exclusion of solvent (thus ensuring high process throughput), and the use of tr
enzyme for both the acetylation and the alcoholysis steps to prepare kilogram quant
(*S*)-HEB and (*R*)-HEB [83].

4.9.5 ENZYMATIC SYNTHESIS OF ETHYL (3*R*,5*S*)-DIHYDROXY-6-(BENZYLOXY) HEXANOA

The diol ethyl (3*R*,5*S*)-dihydroxy-6-(benzyloxy) hexanoate (**83a**, Figure 4.25) is
intermediate in the synthesis of [4-[4α,6β(*E*)]]-6-[4,4-bis[4-fluorophenyl]3-(1-metl
tetrazol-5-yl)-1,3-butadienyl]tetrahydro-4-hydroxy-2H-pyren-2-one (**84**), a potenti:

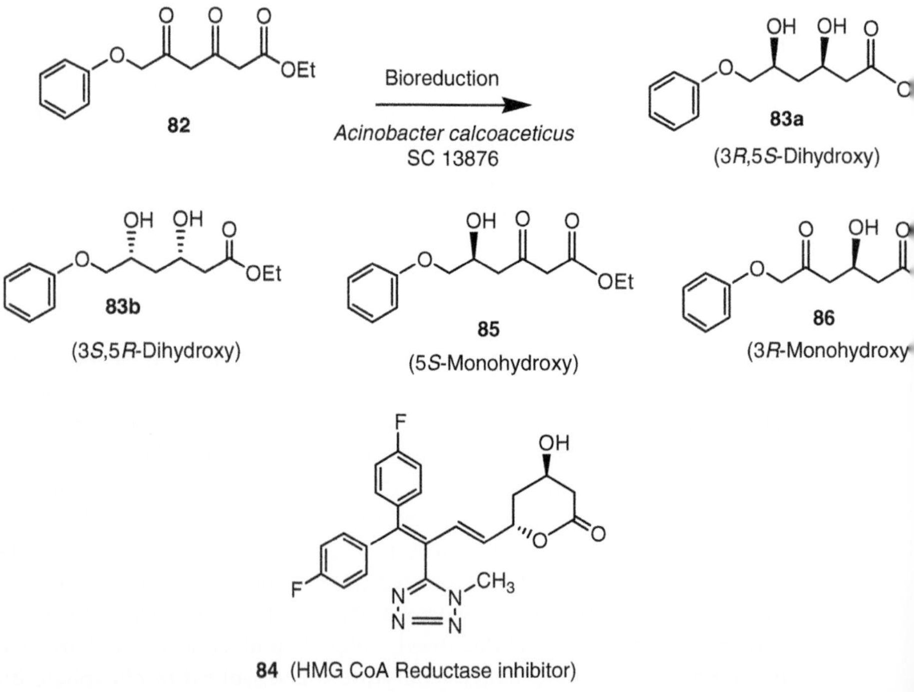

84 (HMG CoA Reductase inhibitor)

FIGURE 4.25 Synthesis of a chiral synthon for the cholesterol-lowering drug (5*R*,3*R*)-**84**: enan
tive microbial reduction of 3,5-dioxo-6-(benzyloxy)hexanoic acid, ethyl ester (**82**) to (3*S*,5*R*)-dih
6-(benzyloxy)hexanoic acid, ethyl ester (**83a**).

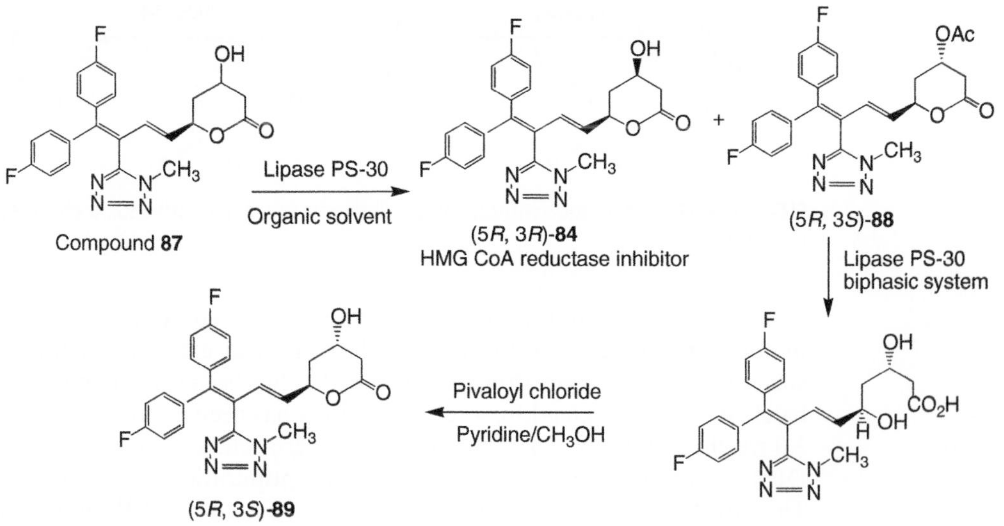

FIGURE 4.26 Synthesis of anticholesterol drug (5R,3R)-**84**: diastereoselective enzymatic acetylation of **87**.

anticholesterol drug that acts by inhibition of HMG CoA reductase [84]. The enantioselective reduction of the diketone ethyl 3,5-dioxo-6-(benzyloxy)hexanoate (**82**) to the diol ethyl (3R,5S)-dihydroxy-6-(benzyloxy)hexanoate (**83a**) [85] has been demonstrated by *Acinetobacter calcoaceticus* SC 13876 in a yield of 85% and diastereoselectivity of 97%.

Cell extracts of *A. calcoaceticus* SC 13876 in the presence of NAD$^+$, glucose, and glucose dehydrogenase reduced **82** to the corresponding isomeric monohydroxy compounds **85** and **86**, which were further reduced to the compound **83a**. A reaction yield of 92% and a diastereomeric purity of 98% were obtained when the reaction was carried out at 10 g/L in a 1 L batch.

Using an enzymatic diastereoselective acetylation process, the (5R,3R)-alcohol **83a** (Figure 4.26) was prepared from **87** [86]. Lipase PS-30 and BMS lipase (produced by fermentation of *Pseudomonas* strain SC 13856) efficiently catalyzed the acetylation of the **87** (4 g/L) to yield the (5R,3S)-acetate **88** and unreacted desired (5R,3R)-alcohol **83a**. A maximum reaction yield of 49% and an ee of 98.5% were obtained for (5R,3R)-alcohol **83a** when the reaction was conducted in toluene in the presence of isopropenyl acetate as an acyl donor. In MEK at 50 g/L substrate concentration, a maximum reaction yield of 46% and an ee of 96% were obtained for **83a**. The enzymatic process was scaled up to a 640 L preparative batch using immobilized lipase PS-30. From the reaction mixture (5R,3R)-alcohol **83a** was isolated in 35% overall yield (theoretical maximum yield 50%) with 98.5% ee and 99.5% chemical purity. The (5R,3S)-acetate **88** produced by this process was enzymatically hydrolyzed by lipase PS-30 in a biphasic system to prepare the corresponding (5R,3S)-alcohol **89**.

4.9.6 ENZYMATIC PREPARATION OF A 2,4-DIDEOXYHEXOSE DERIVATIVE

The chiral 2,4-dideoxyhexose derivative required for the HMG CoA reductase inhibitors has also been prepared using 2-deoxyribose-5-phosphate aldolase (DERA). The reactions start with a stereospecific addition of acetaldehyde **90** (Figure 4.27) to a substituted acetaldehyde to form a 3-hydroxy-4-substituted butyraldehyde **91**, which reacts subsequently with another acetaldehyde to form a 2,4-dideoxyhexose derivative **92**. DERA has been expressed in *E. coli* [87].

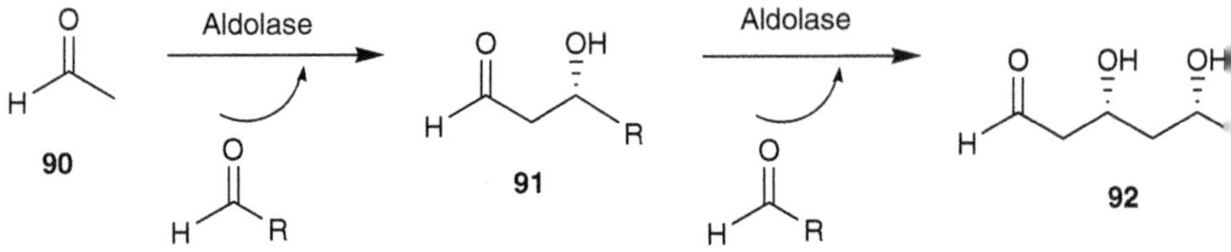

FIGURE 4.27 Enzymatic synthesis of 2,4-dideoxyhexose derivative **92**, a chiral synthon for ant
trol drugs.

The above process has been improved and optimized. An improvement of almost 4
in volumetric productivity relative to the published enzymic reaction conditions ha
achieved, resulting in an attractive process that has been run on up to a 100 g scale in a
batch at a rate of 30.6 g/L/h. The catalyst load has been improved tenfold as well, from
2.0 weight % DERA. These improvements were achieved by a combination of disco
DERA with improved activity and reaction optimization to overcome substrate inh
The two stereogenic centers are set by DERA with ee at >99.9% and diastereomeric ex
96.6%. In addition, downstream chemical processes have been developed to conv
enzymic product efficiently to versatile intermediates applicable to preparation of ato
tin and rosuvastatin [88].

4.9.7 Enzymatic Synthesis of S-[1-(Acetoxyl)-4-(3-Phenyl)butyl]phosphonic Ac Diethyl Ester

Squalene synthase is the first pathway-specific enzyme in the biosynthesis of choleste
catalyzes the head-to-head condensation of two molecules of farnesyl pyrophosphate
to form squalene. It has been implicated in the transformation of FPP into pres
pyrophosphate (PPP). FPP analogs are a major class of inhibitors of squalene s
[89]. However, this class of compounds lacks specificity and is a group of potential inl
of other FPP-consuming transferases such as geranyl geranyl pyrophosphate syntha
increase enzyme specificity, analogs of PPP and other mechanism-based enzyme inh
such as **94**, have been synthesized [90].

S-[1-(Acetoxyl)-4-(3-phenyl) butyl]phosphonic acid, diethyl ester (**93**, Figure 4.28)
chiral intermediate required for the total chemical synthesis of **94**. The enantios
acetylation of racemic [1-(hydroxy)-4-(3-phenyl)butyl]phosphonic acid, diethyl (9
been demonstrated with *G. candidum* lipase in toluene using isopropenyl acetate as t

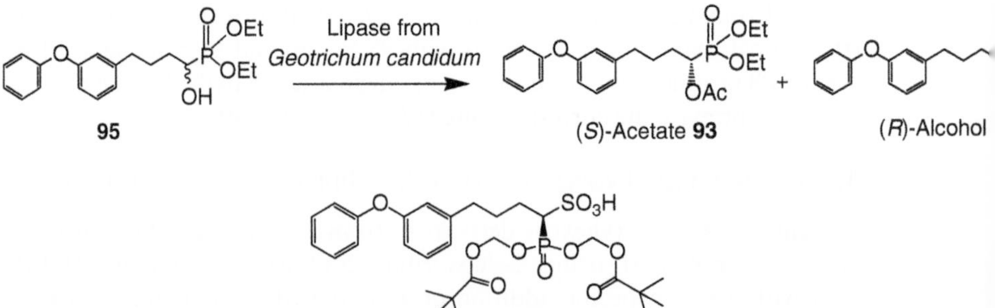

FIGURE 4.28 Enzymatic synthesis of a chiral synthon for the squalene synthase inhibitor **94**:
selective enzymatic acetylation of racemic **95** to (*S*)-acetate **93**.

donor [91]. A reaction yield of 38% (theoretical maximum yield 50%) and an ee of 95% were obtained for chiral **93**.

4.10 THROMBOXANE A2 ANTAGONIST

4.10.1 ENZYMATIC PREPARATION OF LACTOL [3AS-(3Aα,4α,7α,7Aα)]-4, 7-EPOXYISOBENZOFURAN-1-(3H)-OL

Thromboxane A2 (TxA2) is an exceptionally potent vasoconstrictor substance produced by the metabolism of arachidonic acid in blood platelets and other tissues. Together with its potent antiaggregatory and vasodilator activities, TxA2 plays an important role in the maintenance of vascular homeostasis, and contributes to the pathogenesis of a variety of vascular disorders. Approaches towards limiting the effect of TxA2 have focused on either inhibiting its synthesis or blocking its action at its receptor sites by means of an antagonist [92,93]. The lactol [3aS-(3aα,4α,7α,7aα)]-hexahydro-4,7-epoxyisobenzo-furan-1-(3H)-ol (**96**, Figure 4.29a) or the corresponding chiral lactone **97** are key intermediates in the total synthesis of [1S-[1α,2α(Z),3α,4α [[-7-[3-[[[[1-oxoheptyl)-amine]acetyl]methyl]-7-oxabicyclo-[2.2.1] hept-2-yl]-5-heptanoic acid (**98**), a new cardiovascular agent of potential use in the treatment of thrombotic disease [94].

The enantioselective oxidation of (exo,exo)-7-oxabicyclo[2.2.1]heptane-2,3-dimethanol (**99**) to the corresponding (S)-lactol **96** and (S)-lactone **97** has been demonstrated by *Nocardia globerula* ATCC 21505 and *Rhodococcus* sp. ATCC 15592 [95]. Lactone **97** was obtained in a maximum yield of 70% and 96% ee using cell suspensions of *N. globerula* ATCC 21505. An overall reaction yield of 46% (lactol and lactone combined) and ees of 96.7% and 98.4% were obtained for lactol **96** and lactone **97**, respectively, using cell suspensions of *Rhodococcus* sp. ATCC 15592.

The enantioselective hydrolysis of the diacetate (exo,exo)-7-oxabicyclo[2.2.1]heptane-2,3-dimethanol (**100**) to the corresponding S-monoacetate ester **101** (Figure 4.29b) has been demonstrated using lipase PS-30 from *P. cepacia* [96]. A maximum reaction yield of 75% and ee of >99% were obtained when the reaction was conducted in a biphasic system with 10% toluene. Lipase PS-30 was immobilized on Accurel polypropylene (PP) and the immobilized enzyme was reused (five cycles) without loss of enzymic activity, productivity, or ee of product **101**. The reaction process was scaled up to 80 L (400 g of substrate) and the product **101** was isolated in a maximum yield of 80% with 99.3% ee. The S-monoacetate was oxidized to its corresponding aldehyde, which was hydrolyzed to the (S)-lactol **96** used in the chemo-enzymatic synthesis of thromboxane A2 antagonist **98**.

4.11 CALCIUM CHANNEL BLOCKER

4.11.1 ENZYMATIC PREPARATION OF [(3R-CIS)-1,3,4,5-TETRAHYDRO-3-HYDROXY- 4-(4-METHOXYPHENYL)-6-(TRIFLUROMETHYL)-2H-1-BENZAZEPIN-2-ONE]

Diltiazem **102** (Figure 4.30), a benzothiazepinone calcium channel-blocking agent that inhibits influx of extracellular calcium through L-type voltage-operated calcium channels, has been widely used clinically in the treatment of hypertension and angina [97]. Since diltiazem has a relatively short duration of action [98], an 8-chloro derivative has recently been introduced into the clinic as a more potent analog [99]. Lack of extended duration of action and little information on structure–activity relationships in this class of compounds led Floyd et al. [100] to prepare isosteric 1-benzazepin-2-ones; this led to the identification of (*cis*)-3-(acetoxy)-1-[2-(dimethylamino)ethyl]-1,3,4,5-tetrahydro-4-(4-methoxyphenyl)-6-trifluoromethyl-2H-1-benzazepin-2-one (**103**) as a

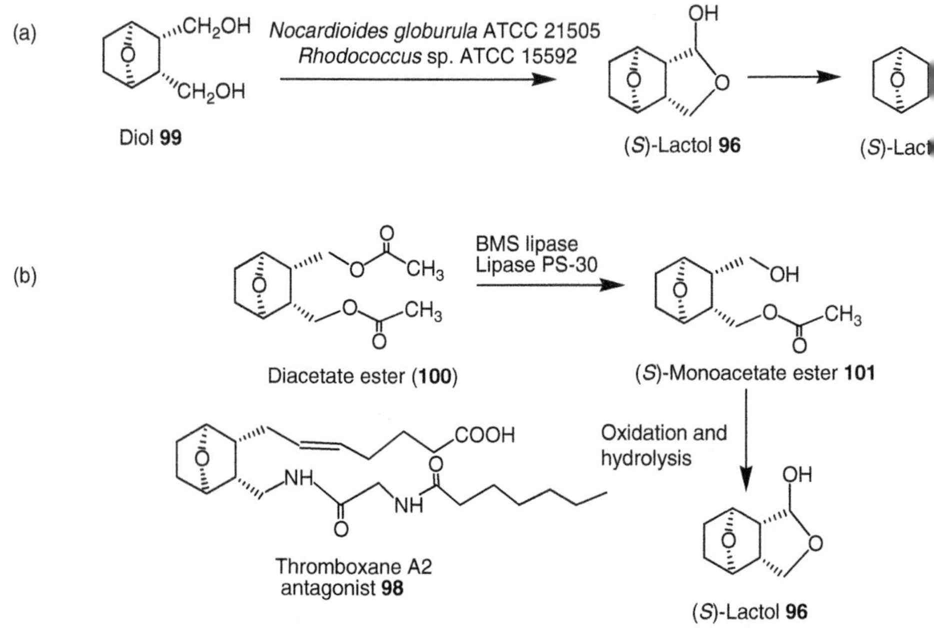

FIGURE 4.29 Synthesis of chiral synthon for thromboxane A2 antagonist 98. (a) Stereos microbial oxidation of (exo,exo)-7-oxabicyclo[2.2.1]heptane-2,3-dimethanol (99) to the corresp lactol 96 and lactone 97. (b) Asymmetric enzymatic hydrolysis of (exo,exo)-7-oxabicyclo[2.2.1]h 2,3-dimethanol, diacetate ester (100) to the corresponding (S)-monoacetate ester 101.

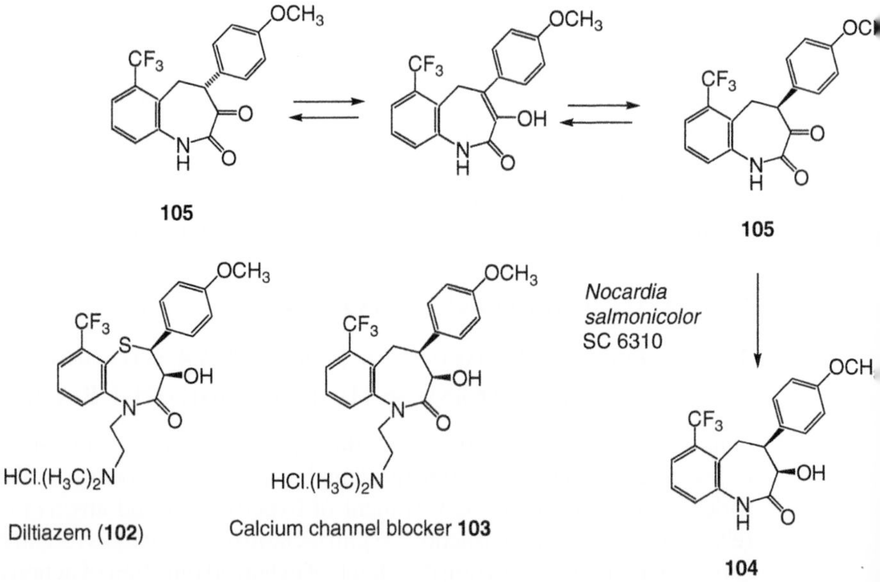

FIGURE 4.30 Synthesis of chiral synthon for calcium channel blocker 103: microbial redu 4,5-dihydro-4-(4-methoxyphenyl)-6-(trifluoromethyl)-1H-benzazepin-2,3-dione (105).

longer-lasting and more potent antihypertensive agent. A key intermediate in the synthesis of this compound was (3*R-cis*)-1,3,4,5-tetrahydro-3-hydroxy-4-(4-methoxyphenyl)-6-(trifluromethyl)-2H-1-benzazepin-2-one (**104**). An enantioselective process was developed for the reduction of 4,5-dihydro-4-(4-methoxyphenyl)-6-(trifluoromethyl)-1H-1-benzazepin-2,3-dione (**105**) to **104** using *N. salmonicolor* SC 6310, in 96% reaction yield with 99.8%ee [101].

4.12 POTASSIUM CHANNEL OPENER

4.12.1 MICROBIAL OXYGENATION TO PREPARE CHIRAL EPOXIDE AND DIOL

It has long been known that K channels play a major role in neuronal excitability and a critical role in the basic electrical and mechanical function of a wide variety of tissues, including smooth and cardiac muscle [102,103]. A new class of highly specific compounds that either open or block K channels has been developed [104]. The synthesis and antihypertensive activity of K-channel openers based on monosubstituted *trans*-4-amino-3,4-dihydro-2,2-dimethyl-2H-1-benzopyran-3-ol (**106**, Figure 4.31) have been demonstrated [105,106]. Chiral epoxide **107** and diol **108** are potential intermediates for the synthesis of **106**. The enantioselective microbial oxygenation of 6-cyano-2,2-dimethyl-2H-1-benzopyran (**109**) to the corresponding chiral epoxide **107** and chiral diol **108** has been demonstrated [107]. *M. ramanniana* SC 13840 and *Corynebacterium* sp. SC 13876 gave maximum yields of 67.5% and 32% and ees of 96% and 89%, respectively, for (+)-*trans* diol **108**. *Corynebacterium* sp. SC 13876 also gave chiral epoxide **107** in a maximum yield of 17% and 88% ee.

A single-stage process (fermentation/epoxidation) for the biotransformation of **109** was developed using *M. ramanniana* SC 13840. In a 25 L fermentor, (+)-*trans* diol **108** was obtained in a maximum yield of 61% and ee of 92.5%. In a two-stage process using a cell suspension (10% w/v, wet cells) of *M. ramanniana* SC 13840, the (+)-*trans* diol **108** was obtained in a maxmum yield of 76% with an ee of 96% when the reaction was carried out in a 5 L Bioflo fermentor. Glucose was supplied to regenerate NADH required for this reaction. From the reaction mixture, (+)-*trans* diol **108** was isolated in 65% overall yield with 97% ee and 98% chemical purity.

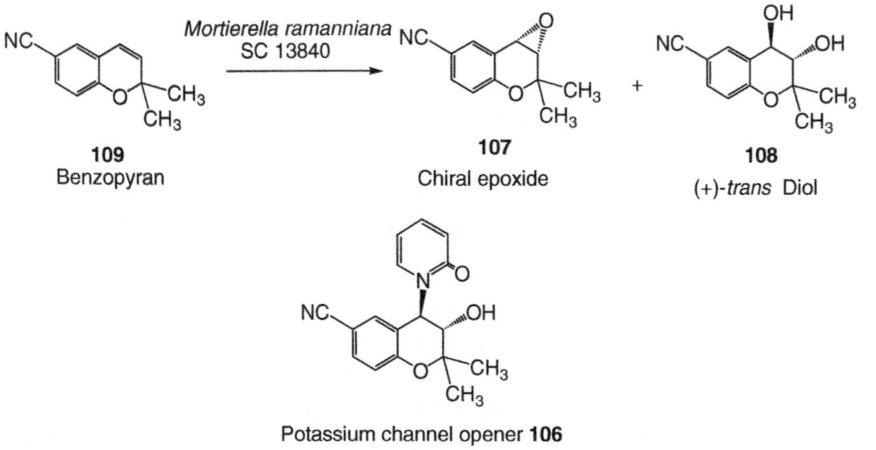

FIGURE 4.31 Preparation of chiral synthons for potassium channel openers **106**: oxygenation of 2,2-dimethyl-2H-1-benzopyran-6-carbonitrile (**109**) to the corresponding chiral epoxide **107** and (+)-*trans* diol **108** by *Mortierella ramanniana* SC 13840.

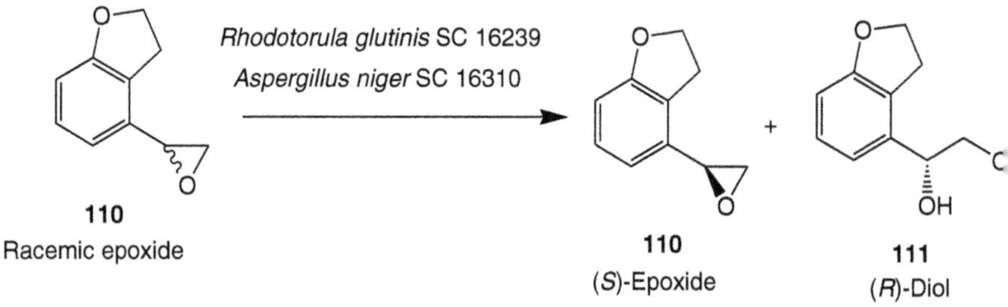

FIGURE 4.32 Synthesis of chiral intermediates for a melatonin receptor agonist: enantio microbial hydrolysis of racemic epoxide **110** to the corresponding (R)-diol **111** and unreac epoxide **110**.

In an enzymatic resolution approach, (+)-*trans* diol **108** was prepared by the en lective acetylation of racemic diol with lipases from *C. cylindraceae* and *P. cepaci* enzymes catalyzed the acetylation of the undesired enantiomer of the racemic diol to y monoacetylated product and unreacted (+)-*trans* diol **108**. A reaction yield of 40% (t ical maximum yield 50%) and an ee of >90% were obtained with each lipase [108].

4.13　MELATONIN RECEPTOR AGONIST

4.13.1　ENANTIOSELECTIVE ENZYMATIC HYDROLYSIS OF RACEMIC 1-{2′,3′-DIHYDRO BENZO[B]FURAN-4′-YL}-1,2-OXIRANE

Epoxide hydrolase catalyzes the enantioselective hydrolysis of an epoxide to the corre ing enantiomerically enriched diol and unreacted epoxide [109,110]. The (S)-epox (Figure 4.32) is a key intermediate in the synthesis of a number of prospective drug can [111]. The enantiospecific hydrolysis of the racemic 1-{2′, 3′-dihydro benzo[b]furan-4′- oxirane (**110**) to the corresponding (R)-diol **111** and unreacted S-epoxide **110** ha demonstrated [112]. Two *A. niger* strains (SC 16310 and SC 16311) and *Rhodotorula* SC 16293 selectively hydrolyzed the (R)-epoxide, leaving behind the (S)-epoxide **110** i ee and 45% yield (theoretical maximum yield 50%). Several solvents at 10% v/v were ev in an attempt to improve the ee and yield. Solvents had significant effects on both the of hydrolysis and the ee of unreacted (S)-epoxide **110**. Most solvents gave a lower ee and slower reaction rate than that of reactions without any solvent supplement, al MTBE gave a reaction yield of 45% (theoretical maximum yield 50%) and an ee of 99 unreacted (S)-epoxide **110**.

4.13.2　BIOCATALYTIC DYNAMIC KINETIC RESOLUTION OF (R,S)-1-{2′,3′-DIHYDROBENZO[B]FURAN-4′-YL}-ETHANE-1,2-DIOL

Most commonly used biocatalytic kinetic resolution of racemates often provide com with high ee, but the maximum theoretical yield of product is only 50%. The reaction contains approximately 50:50 mixture of reactant and product that possesses onl differences in physical properties (e.g., a hydrophobic alcohol and its acetate), a separation may be very difficult. These issues with kinetic resolutions can be addre employing a "dynamic kinetic resolution" process involving a biocatalyst or biocataly metal-catalyzed *in situ* racemization [113,114].

S-1-{2′,3′-Dihydrobenzo[b]furan-4′-yl}ethane-1,2-diol (**111**, Figure 4.33) is a p precursor of S-epoxide **110** [112]. The dynamic kinetic resolution of the racemic diol

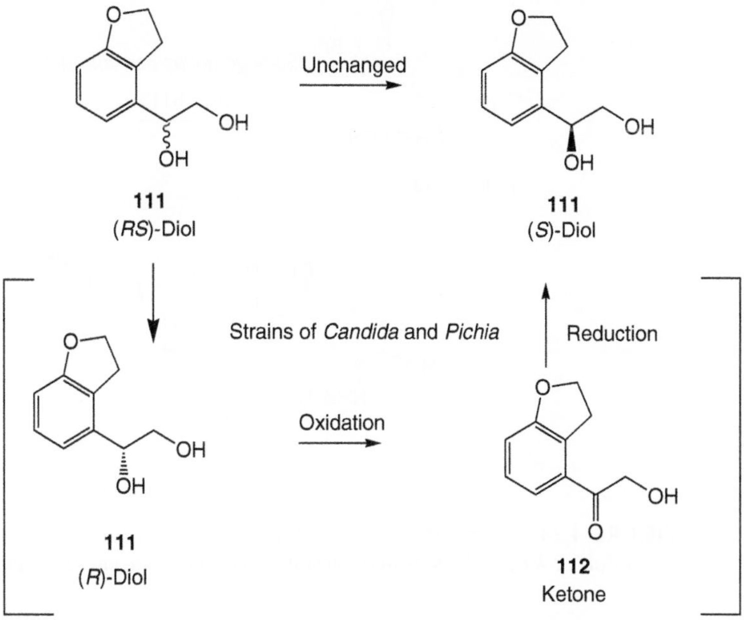

FIGURE 4.33 Synthesis of chiral intermediates for a melatonin receptor agonist: stereoinversion of racemic diol **111** to (*S*)-diol **111** by *Candida boidinii* and *Pichia methanolica*.

the (*S*)-enantiomer **111** has been demonstrated [115]. Seven cultures [*C. boidinii* SC 13821, SC 13822, SC 16115, *P. methanolica* SC 13825, SC 13860, and *H. polymorpha* SC 13895, SC 13896] were found to be promising, providing (*S*)-diol **111** in 87 to 100% ees and 60 to 75% yields. A new compound was formed during these biotransformations and was identified as the hydroxy ketone **112** from an LC-MS. The area of the high-performance liquid chromatography (HPLC) peak for hydroxy ketone **112** first increased with time, reached a maximum, and then decreased, as expected for the proposed dynamic kinetic resolution pathway. *C. boidinii* SC 13822, *C. boidinii* SC 16115, and *P. methanolica* SC 13860 transformed the racemic diol **111** in 3 to 4 days to (*S*)-diol **111** in >70% yield and 90 to 100% ee.

4.14 β-3-RECEPTOR AGONIST

β-3-Adrenergic receptors are found on the cell surfaces of both white and brown adipocytes and are responsible for lipolysis, thermogenesis, and relaxation of intestinal smooth muscle [116]. Consequently, several research groups are engaged in developing selective β-3 agonists for the treatment of gastrointestinal disorders, type II diabetes, and obesity [117,118]. Three different biocatalytic syntheses of chiral intermediates required for the total synthesis of β-3-receptor agonists **113** (Figure 4.34) have been demonstrated [119].

4.14.1 MICROBIAL REDUCTION OF 4-BENZYLOXY-3-METHANESULFONYLAMINO-2′-BROMOACETOPHENONE

The microbial reduction of 4-benzyloxy-3-methanesulfonylamino-2′-bromoacetophenone (**114**, Figure 4.34) to the corresponding (*R*)-alcohol **115** has been demonstrated [119] using *S. paucimobilis* SC 16113. The growth of *S. paucimobilis* SC 16113 was carried out in a 750 L fermentor and cells (60 kg) harvested from the fermentor were used to conduct the biotransformation in 10 L and 200 L preparative batches using 20% (w/v, wet cells).

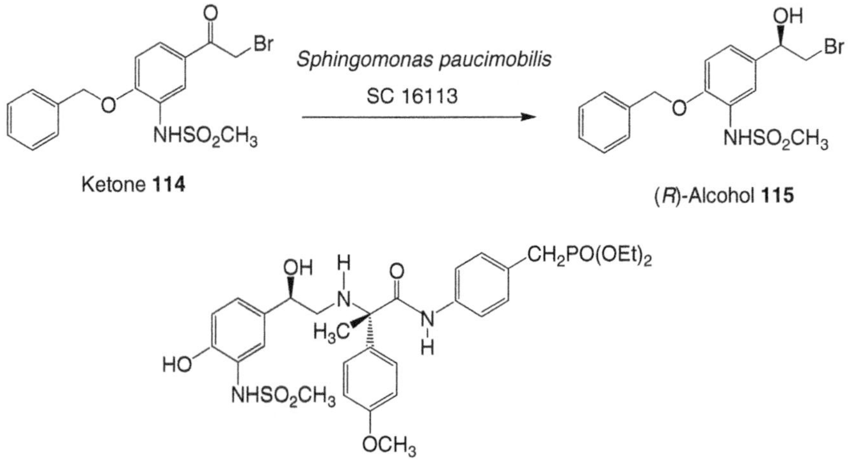

FIGURE 4.34 Enzymatic synthesis of chiral synthon for β-3-receptor agonist 113: enantio
reduction of 4-benzyloxy-3-methanesulfonylamino-2′-bromo-acetophenone (114) to (R)-alcoho

In some batches, the fermentation broth was concentrated threefold by microfiltrati
subsequently washed with buffer by diafiltration and used directly in the biored
process. In all the batches, reaction yields of >85% and ees of >98% were obtaine
isolation of alcohol 115 from the 200 L batch gave 320 g (80% yield) of product v
ee of 99.5%.

In an alternate process, frozen cells of *S. paucimobilis* SC 16113 were used with X
hydrophobic resin (50 g/L) adsorbed substrate at 10 g/L concentration. In this proc
average reaction yield of 85% and an ee of >99% were obtained for alcohol 115. At the
the biotransformation, the reaction mixture was filtered on a 100 mesh (150 μ) stainle
screen, and the resin retained by the screen was washed with water. The product w
desorbed from the resin with acetonitrile and crystallized in a 75% overall yield and 99

4.14.2 Enzymatic Resolution of Racemic α-Methylphenylalanine Amides

The chiral amino acids 116 and 117 (Figure 4.35) are intermediates for the synth
β-3-receptor agonists [117,118]. These may be obtained by the enzymatic resolution of r
α-methylphenylalanine amide 118 and α-methyl-4-methoxyphenylalanine amide 119, r
ively, by an amidase from *Mycobacterium neoaurum* ATCC 25795 [119]. Wet cells (1(
completed resolution of amide 118 in 75 min with a yield of 48% (theoretical maximu
50%) and an ee of 95% for the desired (S)-amino acid 116. Alternatively, freeze-dried ce
suspended in 100 mM potassium phosphate buffer (pH 7.0) at 1% concentration
complete resolution in 60 min with a yield of 49.5% (theoretical maximum yield 50%)
ee of 99% for the (S)-amino acid 116.

Freeze-dried cells of *M. neoaurum* ATCC 25795 and partially purified amidase (a
activity in cell extracts purified fivefold by diethyl aminoethyl celluose column chr
graphy) were used for the biotransformation of compound 119. A reaction yield of 4*
an ee of 78% were obtained for the desired product 117 using freeze-dried cells. The res
was completed in 50 h. Using partially purified amidase, a reaction yield of 49% and a
94% were obtained after 70 h.

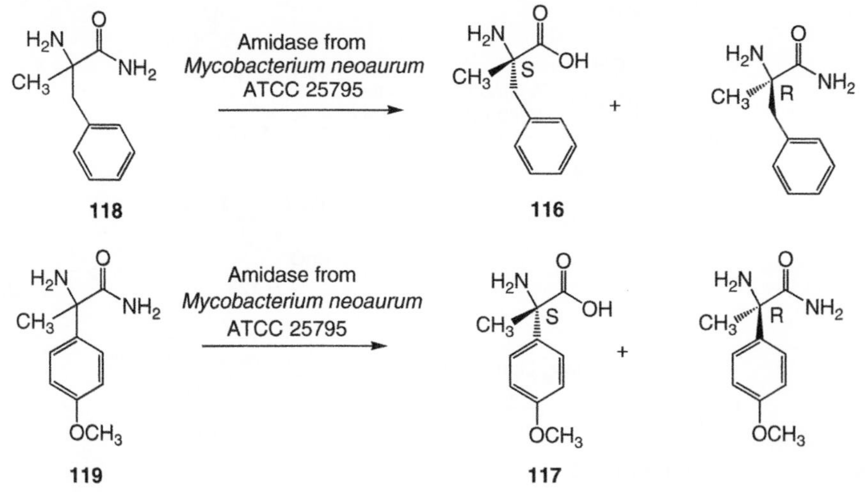

FIGURE 4.35 Enantioselective enzymatic hydrolysis of α-methyl phenylalanine amide (**118**) and α-methyl-4-hydroxyphenylalanine amide (**119**) by amidase.

4.14.3 ENANTIOSELECTIVE HYDROLYSIS OF DIETHYL METHYL-(4-METHOXYPHENYL)-PROPANEDIOATE

The (S)-monoester **120** (Figure 4.36) is a key intermediate for the synthesis of β-3-receptor agonists. The enantioselective enzymatic hydrolysis of diester **121** to the desired acid ester **120** by pig liver esterase [119] has been demonstrated. In various organic solvents the reaction yields and ees of monoester **120** were dependent upon the solvent used. High ees (>91%) were obtained with methanol, ethanol, and toluene as a cosolvent. Ethanol gave the highest reaction yield (96.7%) and ee (96%) for the desired acid ester **120**. It was observed that the ee of the (S)-monoester **120** was increased by decreasing the temperature from 25 to 10°C, when biotransformation was conducted in a biphasic system using ethanol as a cosolvent. A semipreparative 30 g scale hydrolysis was carried out using 10% ethanol as a cosolvent in a 3 L reaction mixture (pH 7.2) at 10°C for 11 h. A maximum reaction yield of 96% and an ee of 96.9% were obtained.

4.15 β-2-RECEPTOR AGONIST

4.15.1 ENANTIOSELECTIVE ENZYMATIC ACYLATION

A potent β-2-receptor agonist formoterol **122** (Figure 4.37) is marketed as a diastereomeric mixture in spite of the varying efficacy of its stereoisomer. The preparation of the (R,R)-

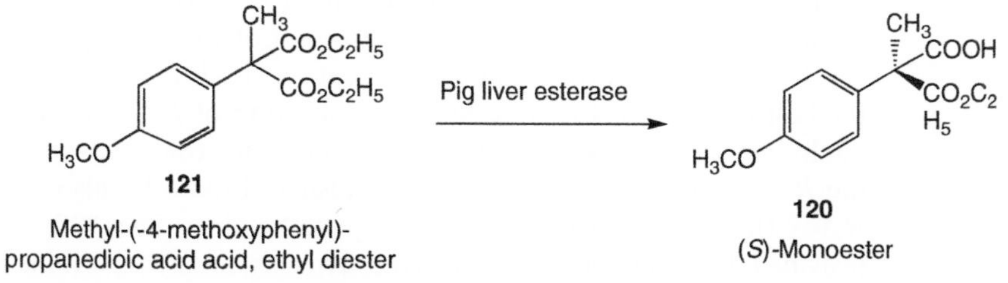

FIGURE 4.36 Enantioselective enzymatic hydrolysis of methyl-(4-methoxyphenyl)-propanedioic acid, ethyl diester (**121**) to (S)-monoester **120**.

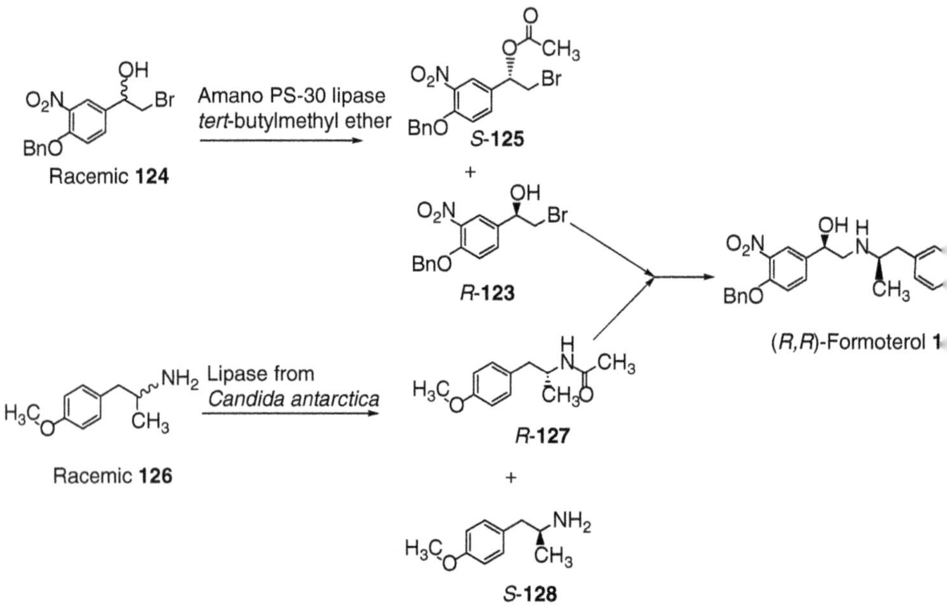

FIGURE 4.37 Enzymatic synthesis of chiral synthon for β-2-receptor-agonist **122**: enantios enzymatic acylation of racemic **124** and **126** to (*R*)-**123** and (*R*)-**127**.

stereoisomer was achieved by an enzymatic resolution process [120]. The *R*-bromohyd was prepared by enzymatic acylation of racemic alcohol **124** to yield acetylated produc and unreacted desired *R*-**123** (46% isolated yield) using lipase PS-30 from *P. c* The resolution of **126** was achieved by enzymatic acylation to yield desired *R*-**1,** unreacted *S*-**128** using *C. antarctic* lipase. An overall reaction yield of only 11% an of 96% were obtained for *R*-**127**. Remarkably, the addition of 0.15 equivalent of triethy led to a 42% conversion to *R*-**127** in 94% ee in 4 h. Following hydrolysis and chromatog *R*-**127** was isolated in 21% yield with 94% ee. Subsequently, coupling of the two intermediates led to synthesis of (*R*,*R*)-formoterol.

4.16   TRYPTASE INHIBITOR

4.16.1   ENZYMATIC PREPARATION OF *S-N*(*TERT*-BUTOXYCARBONYL)- 3-HYDROXYMETHYLPIPERIDINE

S-N(*tert*-butoxycarbonyl)-3-hydroxymethylpiperidine (**129**, Figure 4.38) is a key intern in the synthesis of a potent tryptase inhibitor [121]. *S*-**129** was made from *R*,*S*-3-h methylpiperidine by fractional crystallization of the corresponding L(−)-dibenzoyl ta salt followed by hydrolysis and esterification [122]. Lipase from *P. cepacia* was foun the best enzyme for the stereospecific resolution of *R*,*S-N*-(*tert*-butoxycarbonyl)-3-hy methylpiperidine (**130**). *S*-**129** was obtained in 16% yield and >95% ee by hydrol the *R*,*S*-acetate by lipase PS from *P. cepacia*. Lipase PS also catalyzed esterification *R*,*S-N*-(*tert*-butoxycarbonyl)-3-hydroxy methylpiperidine (**130**) with succinic anhydrid provided *R-N*-(*tert*-butoxycarbonyl)-3-hydroxy methylpiperidine (**132**) and the (*S*)-he cinate ester **133**, which could be easily separated and hydrolyzed by base to the (*S*)-**1**

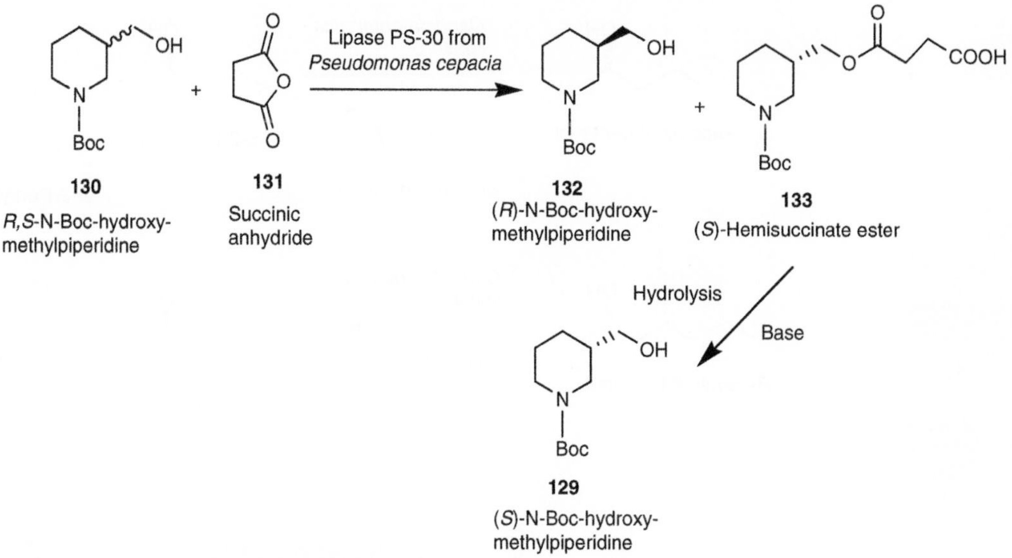

FIGURE 4.38 Preparation of a chiral synthon for a tryptase inhibitor: enzymatic resolution of racemic (*tert*-butoxycarbonyl)-3-hydroxymethylpiperidine (**130**) to (*S*)-**129**.

yield and ee could be improved greatly by repetition of the process. Using the repeated esterification/resolution procedure, (*S*)-**129** was obtained in 32% yield (maximum theoretical yield 50%) and 98.9% ee [123].

4.17 ANTI-ALZHEIMER'S DRUGS

4.17.1 ENZYMATIC PREPARATION OF (*S*)-2-PENTANOL AND (*S*)-2-HEPTANOL

(*S*)-2-Pentanol (Figure 4.39) is an intermediate in the synthesis of several potential anti-Alzheimer's drugs that inhibit β-amyloid peptide release and/or its synthesis [124,125]. The enzymatic resolution of racemic 2-pentanol and 2-heptanol by lipase B from *C. antarctica* has been demonstrated [126].

Commercially available lipases were screened for the enantioselective acetylation of racemic 2-pentanol in an organic solvent (hexane) in the presence of vinyl acetate as an acyl donor. *C. antarctica* lipase B gave a reaction yield of 49% (theoretical maximum yield 50%) and 99% ee for (*S*)-2-pentanol at 100 g/L substrate input. Among acylating agents tested, succinic anhydride was found to be of choice due to easy recovery of the (*S*)-2-pentanol at the end of the reaction. Reactions were carried out using racemic 2-pentanol as solvent as well as substrate. Using 0.68 mol equivalent of succinic anhydride and 13 g of lipase B per kg of racemic 2-pentanol, a maximum reaction yield of 43% and an ee of >98% were obtained for (*S*)-2-pentanol, isolated in 38% overall yield. The resolution of 2-heptanol was also carried out using lipase B under similar conditions to give a maximum reaction yield of 44% and an ee of >99% of (*S*)-2-heptanol, isolated in 40% overall yield.

In an alternate approach, the enantioselective reduction of 2-pentanone to the corresponding (*S*)-2-pentanol (Figure 4.39) has been demonstrated by *Gluconobacter oxydans*. Using triton X-100 treated cells of *G. oxydans*, preparative scale reduction of 2-pentanone was carried out and 1.06 kg of (*S*)-2-pentanol was prepared [127].

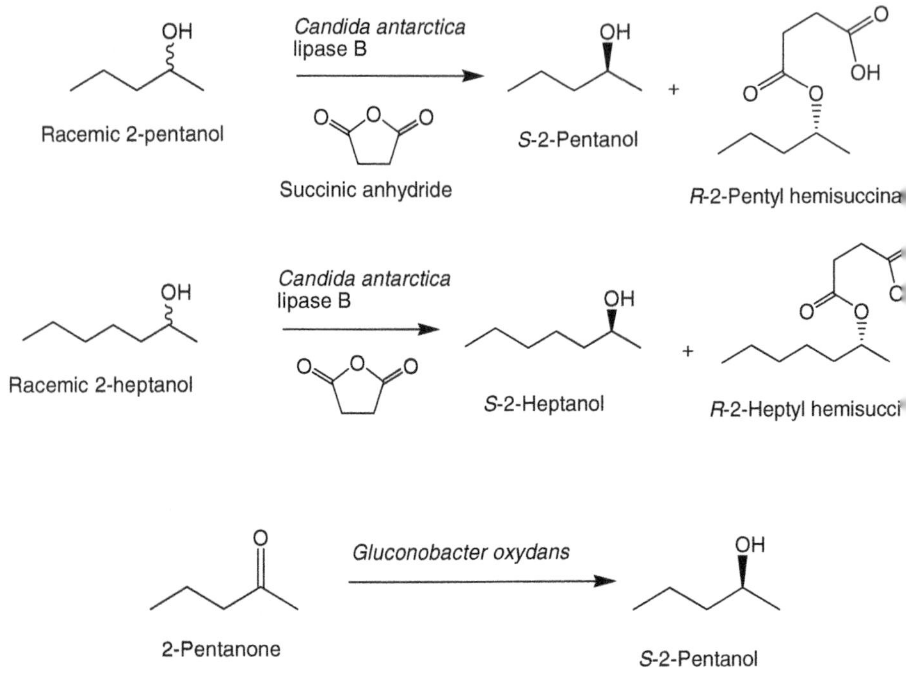

FIGURE 4.39 Synthesis of chiral intermediates for anti-Alzheimer's drugs: enzymatic resol racemic 2-pentanol and 2-heptanol by *Candida antarctica* lipase and enantioselective microbia tion of 2-pentanone to (*S*)-2-pentanol.

4.17.2 ENANTIOSELECTIVE ENZYMATIC REDUCTION OF 5-OXOHEXANOATE AND 5-OXOHEXANENITRILE

Ethyl-(*S*)-5-hydroxyhexanoate (**134**) and (*S*)-5-hydroxyhexanenitrile (**135**, Figure 4. key chiral intermediates in the synthesis of pharmaceuticals. Both chiral compound been prepared by enantioselective reduction of ethyl-5-oxohexanoate (**136**) and 5-ox nenitrile (**137**) by *P. methanolica* SC 16116. Reaction yields of 80 to 90% and >95% e obtained for each chiral compound. In an alternate approach, the enzymatic resolute racemic 5-hydroxyhexanenitrile (**138**) by enzymatic succinylation was demonstrate immobilized lipase PS-30 to obtain (*S*)-5-hydroxyhexanenitrile (**135**) in 35% yield (the maximum yield 50%). (*S*)-5-Acetoxyhexanenitrile (**139**) was prepared by enantios enzymatic hydrolysis of racemic 5-acetoxyhexanenitrile (**140**) by *C. antarctica* lip reaction yield of 42% and an ee of >99% were obtained [128].

4.17.3 ENANTIOSELECTIVE MICROBIAL REDUCTION OF SUBSTITUTED ACETOPHENONE

The chiral intermediate (*S*)-1-(2′-bromo-4′-fluoro phenyl)ethanol (**142**, Figure 4.4 prepared by the enantioselective microbial reduction of 2-bromo-4-fluoro acetop (**141**) [129]. Organisms from genus *Candida, Hansenula, Pichia, Rhodotorula, Sacchar Sphingomonas*, and Baker's yeast reduced **141** to **142** in >90% yield and 99% ee.

In an alternate approach, the enantioselective microbial reduction of methyl-, eth *tert*-butyl-4-(2′-acetyl-5′-fluorophenyl) butanoates **143**, **145**, and **147**, respectively, wa onstrated using strains of *Candida* and *Pichia*. Reaction yields of 40 to 53% and ees c 99% were obtained for the corresponding (*S*)-hydroxy esters **144**, **146**, and **148**. The re

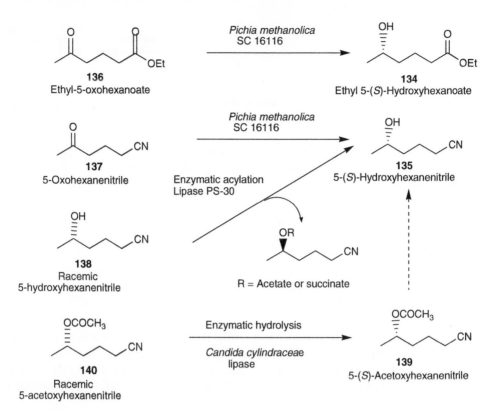

FIGURE 4.40 Synthesis of chiral intermediates for anti-Alzheimer's drugs: enantioselective microbial reduction of ethyl-5-oxohexanoate (**136**) and 5-oxohexanenitrile (**137**) and enzymatic resolution of 5-hydroxyhexanenitrile (**138**) and 5-acetoxyhexanenitrile (**140**).

that catalyzed the enantioselective reduction of ketoesters was purified to homogeneity from cell extracts of *P. methanolica* SC 13825. It was cloned and expessed in *E. coli* and recombinant cultures were used for the enantioselective reduction of the keto-methyl ester **143** to the corresponding (*S*)-hydroxy methyl ester **144**. On a preparative 300 L scale, a reaction yield of 98% with an ee of 99% was obtained [129].

4.18 RETINOID RECEPTOR GAMMA-SPECIFIC AGONISTS

4.18.1 ENZYMATIC PREPARATION OF 2-(*R*)-HYDROXY-2-(1′,2′,3′,4′-TETRAHYDRO-1′,1′,4′, 4′-TETRAMETHYL-6′-NAPHTHALENYL) ACETATE

A number of studies have demonstrated that retinoids (vitamin A derivatives) are essential for normal growth, vision, tissue homeostasis, and reproduction [130]. Retinoic acid and its natural and synthetic analogs (retinoids) exert a wide variety of biological effects by binding to, or activating, a specific receptor or sets of receptors [131]. They have been shown to effect cellular growth and differentiation, and are promising drugs for the treatment of cancers [132]. A few retinoids are already in clinical use for the treatment of dermatological diseases such as acne and psoriasis [133]. (*R*)-3-Fluoro-4-[[hydroxy-(5,6,7,8-tetrahydro-5,5,8,8-tetramethyl-2-naphthalenyl)-acetyl]amino]benzoic acid **149** (Figure 4.42) is a retinoic acid receptor gamma-specific agonist potentially useful as a dermatological and anticancer drug [134].

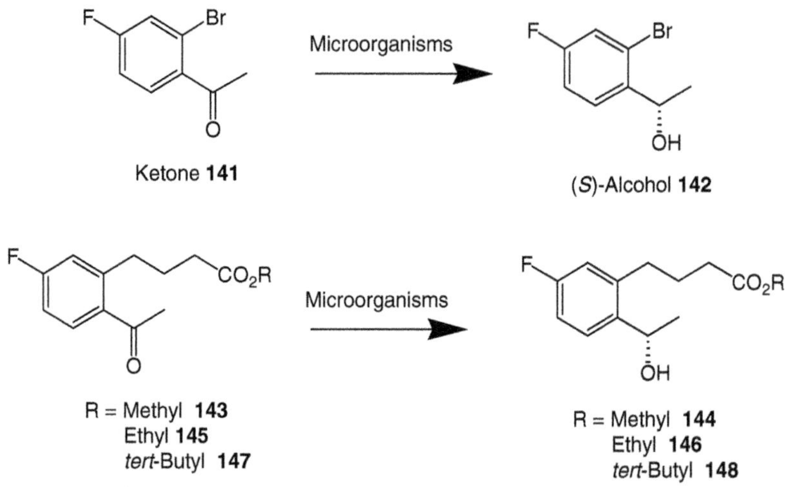

FIGURE 4.41 Synthesis of chiral intermediates for anti-Alzheimer's drugs: enantioselective m reduction of 2-bromo-4-fluoro acetophenone (**141**) and methyl 4-(2′-acetyl-5′-fluoro butanoate (**143**).

Ethyl 2-(*R*)-hydroxy-2-(1′,2′,3′,4′-tetrahydro-1′,1′,4′,4′-tetramethyl-6′-naphthalenyl) (**150**) and the corresponding acid **151** were prepared as intermediates in the synth the retinoic acid receptor gamma-specific agonist **149** [135]. Enantioselective mi reduction of ethyl 2-oxo-2-(1′,2′,3′,4′-tetrahydro-1′,1′,4′,4′-tetramethyl-6-naphthalenyl)

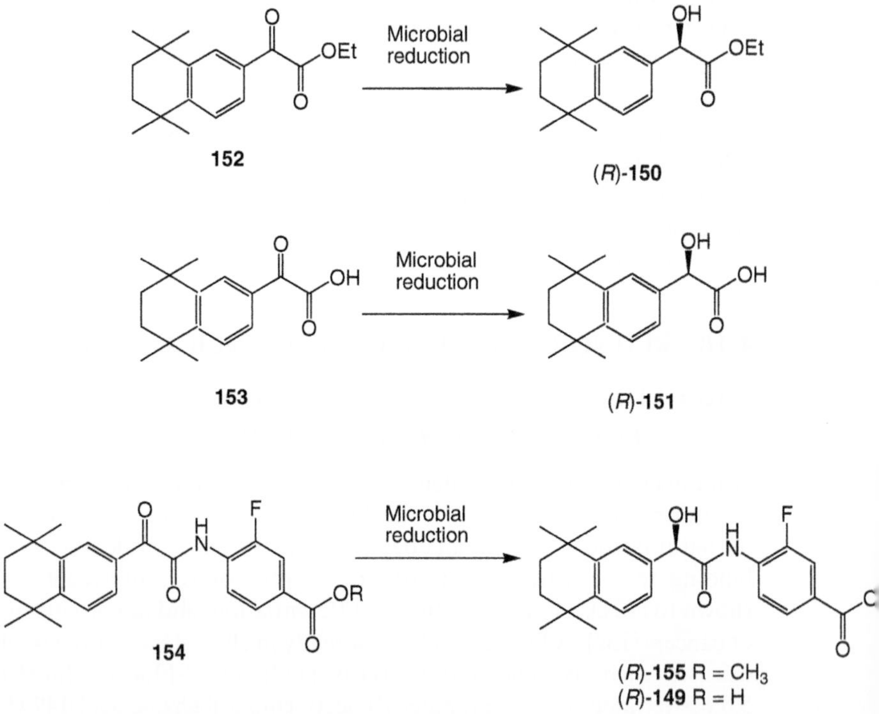

FIGURE 4.42 Enzymatic synthesis of chiral synthons for the retinoid receptor gamma-specific **149**: enantioselective microbial reduction of ketoester **152**, ketoacid **153**, and ketoamide **154**.

(**152**) to alcohol **150** was carried out using *Aureobasidium pullulans* SC 13849 in 98% yield with an ee of 96%. At the end of the reaction, hydroxyester **150** was adsorbed onto XAD-16 resin and, after filtration, recovered in 94% yield from the resin with acetonitrile extraction. The recovered (*R*)-hydroxyester **150** was treated with Chirazyme L-2 or pig liver esterase to convert it to the corresponding (*R*)-hydroxyacid **151** in quantitative yield.

Among microorganisms screened for the reduction of 2-oxo-2-(1′,2′,3′,4′-tetrahydro-1′,1′,4′,4′-tetramethyl-6′naphthalenyl)acetic acid (**153**) to hydroxy acid **151**, *C. maltosa* SC 16112 and two strains of *C. utilis* (SC 13983, SC 13984) gave >53% reaction yields with >96% ee. The enantioselective microbial reduction of ketoamide **154** to the corresponding (*R*)-hydroxyamide **155** by *A. pullulans* SC 13849 was also demonstrated [135].

4.19 ANTI-INFECTIVE DRUGS

4.19.1 MICROBIAL HYDROXYLATION OF PLEUROMUTILIN OR MUTILIN

Pleuromutilin (**156**, Figure 4.43) is an antibiotic from *Pleurotus* or *Clitopilus* basidiomycetes strains that kills mainly gram-positive bacteria and mycoplasms. A more active semisynthetic analog, tiamulin, has been developed for the treatment of animals and poultry infection and has been shown to bind to prokaryotic ribosomes and inhibit protein synthesis [136]. Metabolism of pleuromutilin derivatives results in hydroxylation by microsomal cytochrome P-450 at the 2- or 8-position and inactivates the antibiotics [137]. Modification of the 8-position of pleuromutilin and its analogs is of interest as a means of preventing the metabolic hydroxylation. Microbial hydroxylation of pleuromutilin **156** or mutilin **157** would provide a functional group at this position to allow further modification. The target analogs would maintain the biological activity of the parent compounds but not be susceptible to metabolic inactivation.

Biotransformation of mutilin and pleuromutilin by microbial cultures has been investigated to provide a source of 8-hydroxymutilin or 8-hydroxypleuromutilin [138]. *S. griseus* strains SC 1754 and SC 13971 (ATCC 13273) hydroxylated mutilin to 2-(*S*)-hydroxymutilin (**158**), 7-(*S*)-hydroxymutilin (**159**), and 8-(*S*)-hydroxymutilin (**160**, Figure 4.43). *Cunninghamella echinulata* SC 16162 (NRRL 3655) gave (2*S*)-hydroxymutilin and (2*R*)-hydroxypleuromutilin (**161**) from biotransformation of mutilin and pleuromutilin, respectively. The biotransformation of mutilin by the *S. griseus* strain SC 1754 was scaled up in 15, 60, and 100 L fermentations to produce a total of 49 g of (8*S*)-hydroxymutilin, 17 g of (7*S*)-hydroxymutilin, and 13 g of (2*S*)-hydroxymutilin from 162 g of mutilin [138].

A C-8 ketopleuromutilin **162** derivative has been synthesized from the biotransformation product 8-hydroxymutilin [139]. A key step in the process was the selective oxidation at C-8 of 8-hydroxymutilin using tetrapropylammonium perruthenate. The presence of the C-8 keto group precipitated interesting intramolecular chemistry to afford **163** with a novel pleuromutilin-derived ring system by acid catalyzed conversion of C-8 ketopleuromutilin.

4.19.2 ENZYMATIC PREPARATION OF (*R*)-1,3-BUTANEDIOL AND (*R*)-4-CHLORO-3-HYDROXYBUTONOATE

(*R*)-1,3-butanediol (**164**, Figure 4.44) is a key starting material of azetidinone derivatives **165**, which are key chiral intermediates for the syntheses of penem **166** and carbapenem antibiotics [140]. From a microbial screen the *C. parapsilosis* strain IFO 1396 was identified which produced (*R*)-1,3-butanediol from the racemate. The (*S*)-1,3-butanediol oxidizing enzyme (CpSADH), which produced (*R*)-1,3-butanediol from the racemate, was cloned in *E. coli*. The recombinant culture catalyzed the enantioselective oxidation of secondary alcohols and also

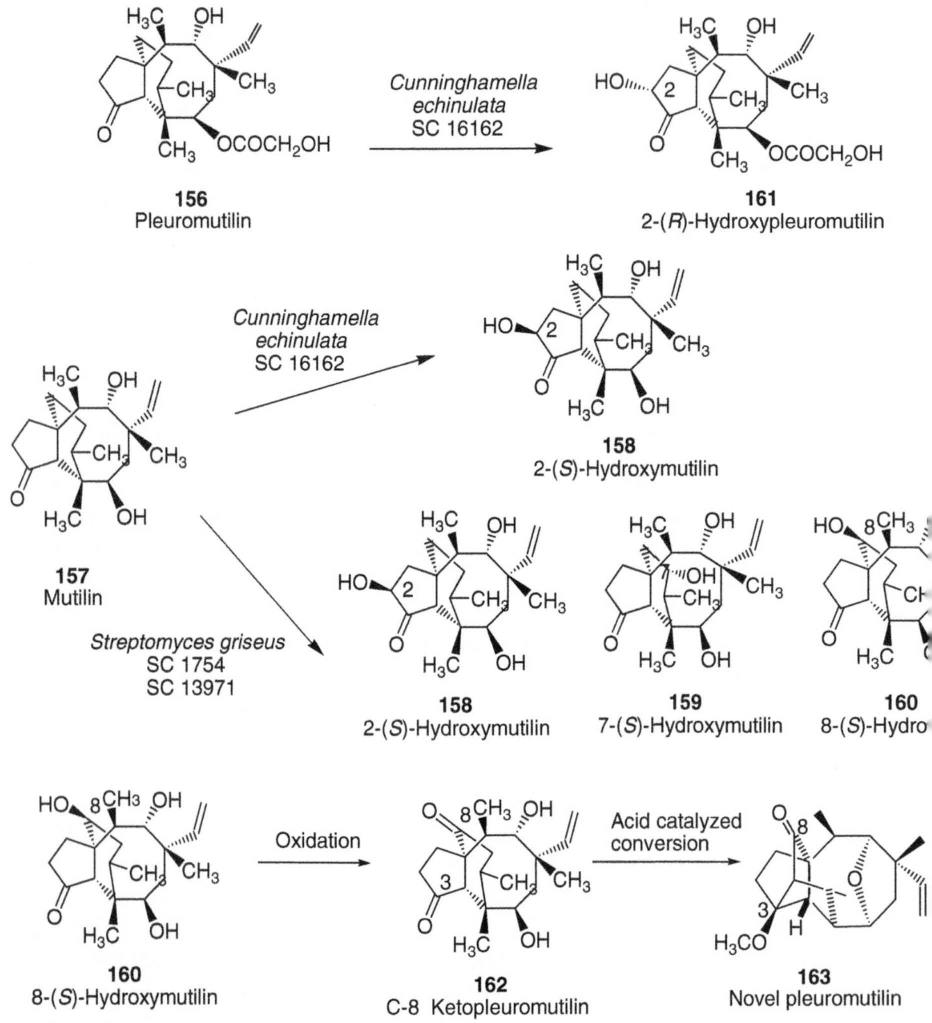

FIGURE 4.43 Microbial hydroxylation of pleuromutilin (**156**) and mutilin (**157**).

catalyzed the asymmetric reduction of aromatic and aliphatic ketones to their corresp (S)-secondary alcohols. Using the recombinant enzyme, (R)-1,3-butanediol was prod 97% yield and 95% ee using 150 g/L input of the racemate. Recombinant enzyme (Cp was also used for reduction of ethyl 4-chloroacetoacetate (**167**) to produce (R)-4-cl hydroxybutonoate (**168**) in 95% yield and 99% ee using 36 g/L substrate input. Isop was used to regenerate the NADH required for this reduction. (R)-4-Chloro-3-hyd tonoate is useful for the synthesis of L-carnitine (**169**) and (R)-4-hydroxypyrr (**170**) [141].

4.19.3   ENZYMATIC SYNTHESIS OF L-β-HYDROXYVALINE

The asymmetric synthesis of β-hydroxy-α-amino acids by various methods has been strated [142–144] because of their utility as starting materials for the total synth monobactam antiobiotics. L-β-hydroxyvaline **171** is a key chiral intermediate requ the total synthesis of orally active monobactam [145], Tigemonam **172** (Figure 4.4

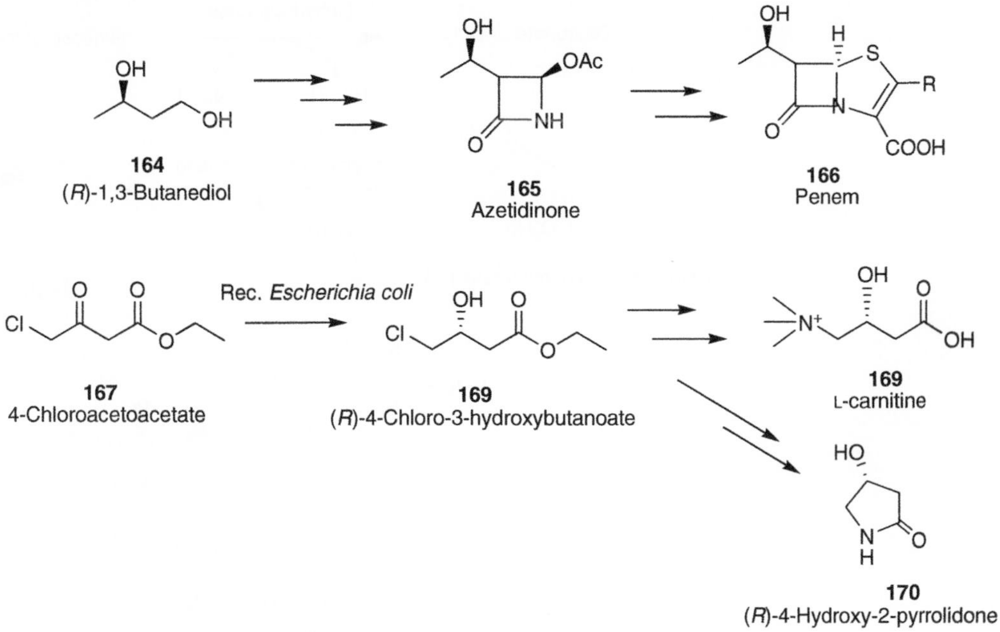

FIGURE 4.44 Enzymatic resolution of 1,3-butanediol and enantioselective enzymatic reduction of ethyl 4-chloroacetoacetate (**167**).

resolution of Cbz-β-hydroxyvaline by chemical methods has been demonstrated [146]. The synthesis of L-β-hydroxyvaline **171** from α-keto-β-hydroxyisovalerate **173** by reductive amination using leucine dehydrogenase from *B. sphaericus* ATCC 4525 has been demonstrated (Figure 4.45) [147]. NADH required for this reaction was regenerated by either FDH from *C. boidinii* or glucose dehydrogenase from *B. megaterium*. The immobilized cofactors such as polyethylene glycol-NADH and dextrans-NAD were effective in the biocatalytic process. The required substrate **173** was generated either from α-keto-β-bromoisovalerate or its ethyl esters by hydrolysis with sodium hydroxide *in situ*. In an alternate approach, the substrate **173** was also generated from methyl-2-chloro-3,3-dimethyloxiran carboxylate and the corresponding isopropyl and 1,1-dimethylethyl ester. These glycidic esters are converted to substrate **173** by treatment with sodium bicarbonate and sodium hydroxide. In this process, an overall reaction yield of 98% and an ee of 99.8% were obtained for the L-β-hydroxyvaline **171**.

4.20   α1-ADRENORECEPTOR ANTAGONIST

4.20.1   Enzymatic Esterification and Ammonolysis

Afuzosin (**174**, Figure 4.46), a quinozoline derivative, acts as a potent and selective antagonist of α1-adrenoreceptor-mediated contaction of the prostate and the prostatic capsule, thereby reducing the symptoms associated with benign prostatic hypertrophy [148]. Several routes have been reported for the chemical synthesis of Afuzosin, with tetrahydro-*N*-[3-(methylamino)-propyl]-2-furancarboxamide (**175**) as a widely used intermediate [149]. Its synthesis from 2-tetrahydrofuroic acid is difficult, involving toxic reagents and drastic reaction conditions [150]. Lipase-catalyzed ammonolysis reactions using ammonia as the nucleophile have been demonstrated with esters [151]. A lipase-catalyzed process has been described for the

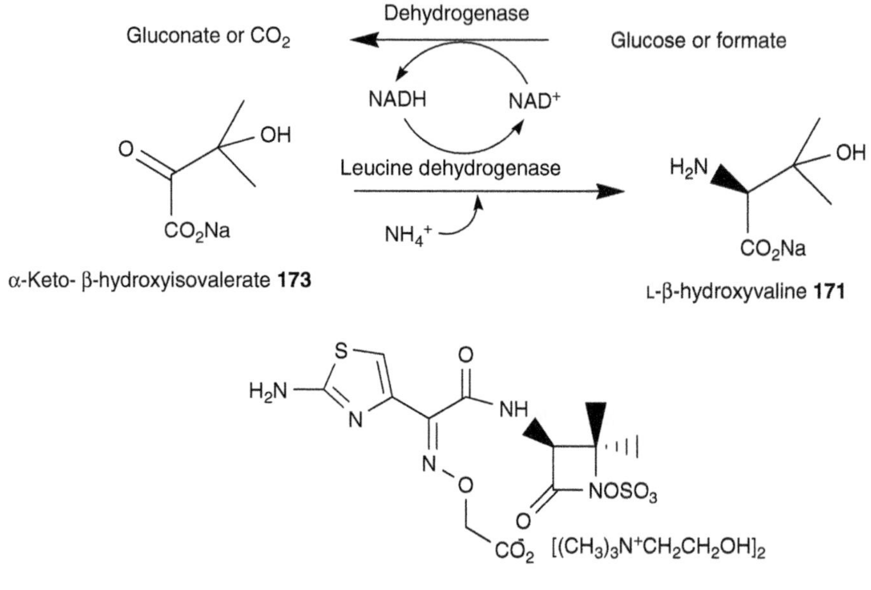

FIGURE 4.45 Enzymatic preparation of L-β-hydroxyvaline **171**, an intermediate for synth Tigemonam **172**.

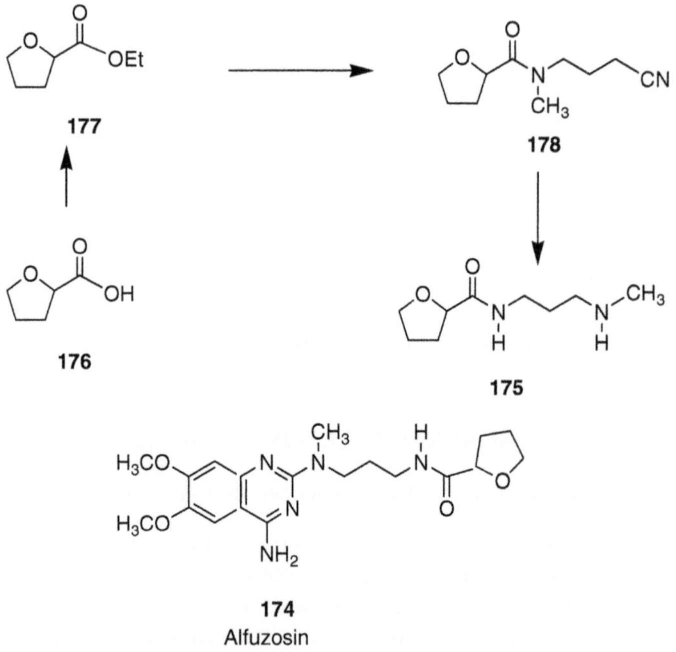

FIGURE 4.46 Preparation of a chiral synthon for an α1-adrenoreceptor antagonist **174**: atic synthesis of tetrahydro-*N*-[3-(methylamino)-propyl]-2-furancarboxamide (**175**) from car acid **176**.

one-pot conversion of carboxylic acids into substituted amides through *in situ* formation of the ethyl ester and subsequent ammonolysis [152]. The procedure was optimized for the preparation of **175** and involved the treatment of the corresponding carboxylic acid **176** with ethyl alcohol to prepare ester **177** in the presence of immobilized *C. antarctica* lipase followed by addition of *N*-methyl-1,3-propanediamine. The amide **175** was obtained in 72% yields. Immobilized enzyme was reused over eight cycles in this process.

This process was proven to be general and can be applied to open-chain, cyclic, hydroxy-, amino-, dicarboxylic, and unsaturated acids [152]. The enzyme shows regioselective behavior in relation to primary and secondary amino groups.

4.21   ENDOTHELIN RECEPTOR ANTAGONIST

4.21.1   ENANTIOSELECTIVE MICROBIAL REDUCTION OF KETO ESTER AND CHLOROKETONE

Endothelin is present in elevated levels in the blood of patients with hypertension, acute myocardial infraction, and pulmonary hypertension. Two endothelin receptor subtypes have been identified that bind endothelin, thus causing vasoconstriction [153,154]. Endothelin antagonists such as compound **179** (Figure 4.47) have potential therapeutic value. Enantioselective microbial reduction of a ketoester **180** and a chlorinated ketone **181** to their corresponding (*S*)-alcohols **182** and **183** was demonstrated by *P. delftensis* MY 1569 and *R. piliminae* ATCC 32762 with ees of >98% and >99%, respectively [155]. Reductions were scaled up to 23 L to produce the desired (*S*)-alcohols in 88% and 97% yields, respectively.

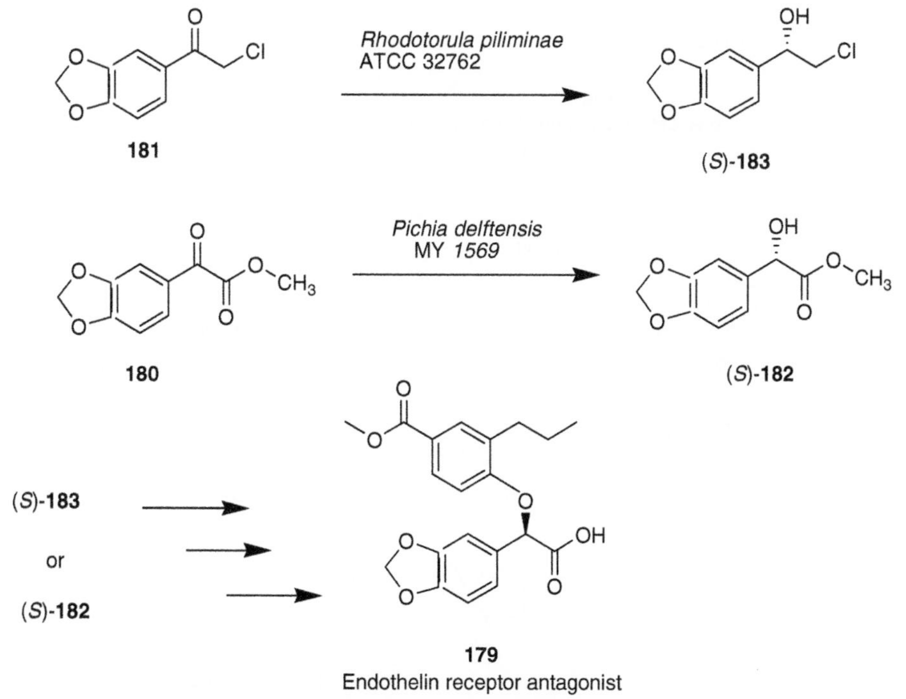

FIGURE 4.47 Preparation of a chiral synthon for an endothelin receptor antagonist **179**: enantioselective microbial reduction of ketoester **180** and chloroketone **181**.

4.22   ANTIANXIETY DRUG

4.22.1   ENZYMATIC PREPARATION OF 6-HYDROXYBUSPIRONE

Buspirone (Buspar, **184**, Figure 4.48) is a drug used for the treatment of anxie
depression that is thought to produce its effects by binding to the serotonin 5HT1A r
[156]. Mainly as a result of hydroxylation reactions, it is extensively converted to
metabolites [157], and blood concentrations return to low levels a few hours after do
major metabolite, 6-hydroxybuspirone **185**, produced by the action of liver cytochron
CYP3A4, is present at much higher concentrations in human blood than buspiron
This metabolite has anxiolytic effects in an anxiety model using rat pups and binds
human 5HT1A receptor [158]. Although the metabolite has only about a third of the
for the human 5HT1A receptor as buspirone, it is present in human blood at 30 to 4
higher concentration than buspirone following a dose of buspirone, and therefore
responsible for much of the effectiveness of the drug [159]. For the developm
6-hydroxybuspirone as a potential antianxiety drug, preparation and testing of t
enantiomers as well as the racemate was of interest. Both the *R*- and *S*-enantiomers, i
by chiral HPLC, were effective in tests using a rat model of anxiety [160]. Wher
R-enantiomer showed somewhat tighter binding and specificity for the 5HT1A re
the *S*-enantiomer had the advantage of being cleared more slowly from the blo
enzymatic process was developed for resolution of 6-acetoxybuspirone **186**. L-amin

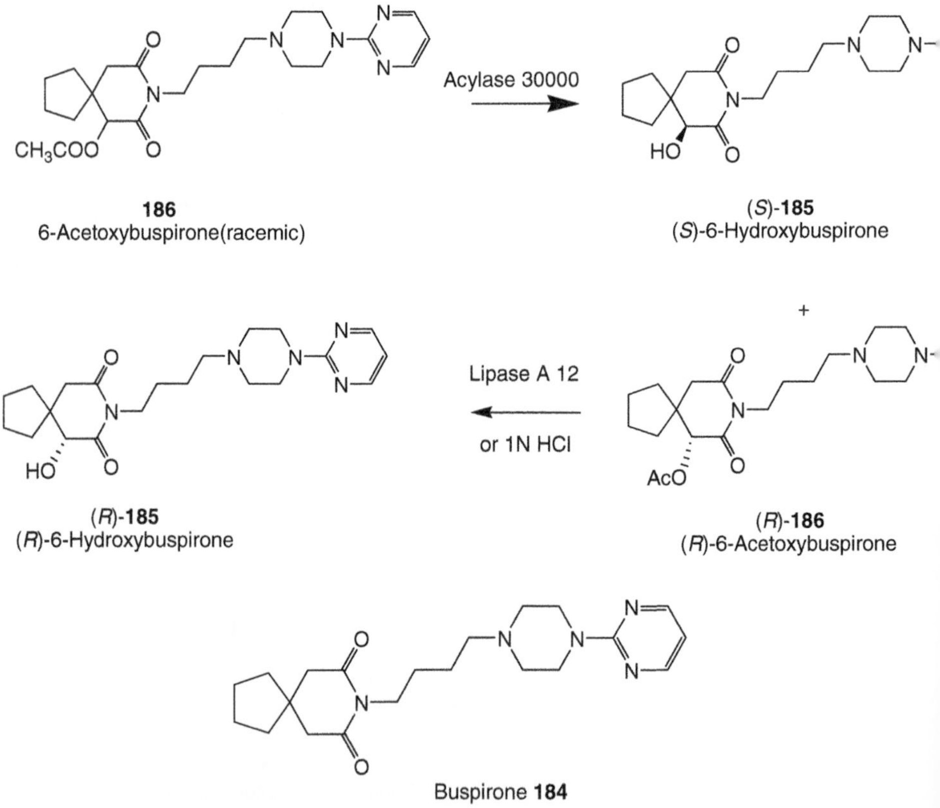

FIGURE 4.48 Lipase-catalyzed preparation of (*S*)-**185** and (*R*)-6-hydroxybuspirone **185** by a re
process.

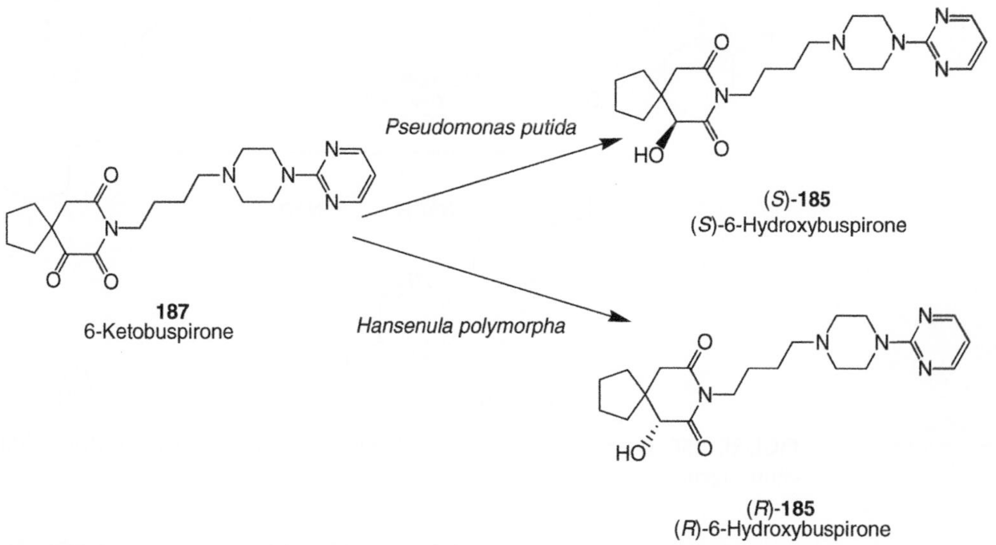

FIGURE 4.49 Enzymatic preparation of (S)-**185** and (R)-6-hydroxybuspirone **185** by reduction of 6-ketobuspirone **187**.

acylase from *Aspergillus melleus* (Amano acylase 30000) was used to hydrolyze racemic 6-acetoxybuspirone to (S)-6-hydroxybuspirone in 96% ee after 46% conversion. The remaining (R)-6-acetoxybuspirone with 84% ee was converted to (R)-6-hydroxybuspirone by acid hydrolysis. The ee of both enantiomers could be improved to >99% by crystallization as a metastable polymorph [161]. Direct hydroxylation of buspirone to (S)-6-hydroxbuspirone by *S. antibioticus* ATCC 14980 has also been described [161]. In an alternate process, enantioselective microbial reduction of 6-ketobuspirone **187** (Figure 4.49) to either (R)- or (S)-6-hydroxybuspirone was described. About 150 microorganisms were screened for the enantioselective reduction of **187**. *R. stolonifer* SC 13898, *R. stolonifer* SC 16199, *Neurospora crassa* SC 13816, *Mucor racemosus* SC 16198, and *P. putida* SC 13817 gave >50% reaction yields and >95% ees of (S)-6-hydroxybuspirone. The yeast strains *H. polymorpha* SC 13845 and *C. maltosa* SC 16112 gave (R)-6-hydroxybuspirone in >60% reaction yield and >97% ee [162].

4.23 ANTIPSYCHOTIC AGENT

4.23.1 ENZYMATIC REDUCTION OF 1-(4-FLUOROPHENYL)-4-[4-(5-FLUORO-2-PYRIMIDINYL)-1-PIPERAZINYL]-1-BUTANONE

During the past few years, much effort has been directed toward the understanding of the sigma receptor system in the brain and endocrine tissues. This effort has been motivated by the hope that the sigma site may be a target of a new class of antipsychotic drugs [163,164]. The characterization of the sigma system helped to clarify the biochemical properties of the distinct haloperidol-sensitive sigma binding site, the pharamacological effects of sigma drugs in several assay systems, and the transmitter properties of a putative endogenous ligand for the sigma site [165,166]. R-(+) compound **188** [BMY-14802] is a sigma ligand and has a high affinity for sigma binding sites and antipsychotic efficacy [167,168]. The stereoselective microbial reduction of keto compound 1-(4-fluorophenyl)-4-[4-(5-fluoro-2-pyrimidinyl)-1-piperazinyl]-1-butanone **189** to yield the corresponding hydroxy compound R-(+)-BMY-14802 **188** (Figure 4.50) has been developed by Patel et al. [169]. Among various microorganisms

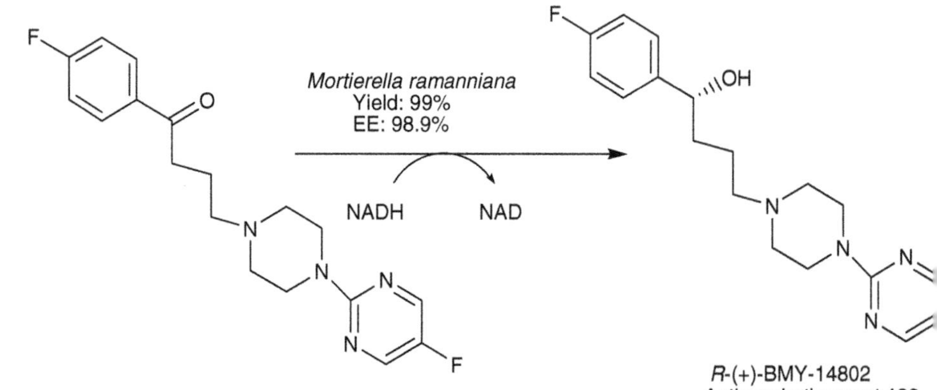

FIGURE 4.50 Enantioselective microbial reduction of **189** to prepare (*R*)-(+)-BMY-14802, an chotic agent.

evaluated for the reduction of ketone **189**, *M. ramanniana* ATCC 38191 predom reduced compound **189** to *R*-(+)-BMY-14802, and *Pullularia pullulans* ATCC 16623 r compound **189** to *S*-(−)-BMY-14802. An ee of >98% was obtained in each reaction.

In a two-stage process for the reduction of compound **189**, cells of *M. ramanniana* 38191 were grown in a 380 L fermentor, and cells harvested after 31 h growth were u the reduction in a 15 L fermentor using 20% cell suspensions (20% w/v, wet cells). Ketc was used at 2 g/L concentration and glucose was supplemented at 20 g/L concentration the biotransformation process to generate NADH required for the reduction. After biotransformation period, about 90% yield and 99.0% ee of *R*-(+)-BMY-1480 obtained. The *R*-(+)-BMY-14802 was isolated from the fermentation broth in overa yield, 99.5% ee, and 99% chemical purity.

A single-stage fermentation–biotransformation process was demonstrated for the tion of ketone **189** to *R*-(+)-BMY-14802 by the cells of *M. ramanniana* ATCC 3819 were grown in a 20 L fermentor containing 15 L of medium. After 40 h of grow biotransformation process was initiated by addition of 30 g of ketone **189** and 3C glucose, and was completed in 24 h, with a reaction yield of 100% and an ee of 98.9% (+)-BMY-14802. At the end of the biotransformation process, cells were removed by tion and product was recovered from the filtrate in overall 80% recovery.

4.24   ENZYMATIC ACYLOIN CONDENSATION

Asymmetric α-hydroxyketones (acyloins) are important classes of intermediates in c synthesis due to their bifunctional aspect, especially having one chiral center amen further modification. Enzyme-mediated acyloin formation could provide an advant environment-friendly method to prepare optically active asymmetric acyloins [170]. *A* formation mediated by yeast pyruvate decarboxylase and bacterial benzoylformate deca lase [171] has been reported. Although phenylpyruvate decarboxylase (PPD) [1 decarboxylaton of phenylpyruvic acid has been known for a long time, recently we re the acyloin condensation catalyzed by PPD [173,174]. *Achromobacter eyrydice* PPD w to catalyze the asymmetric acyloin condensation of phenylpyruvate (**190**, Figure 4.5 various aldehydes **191** to produce optically active acyloins PhCH₂COCH(OH)R **19**

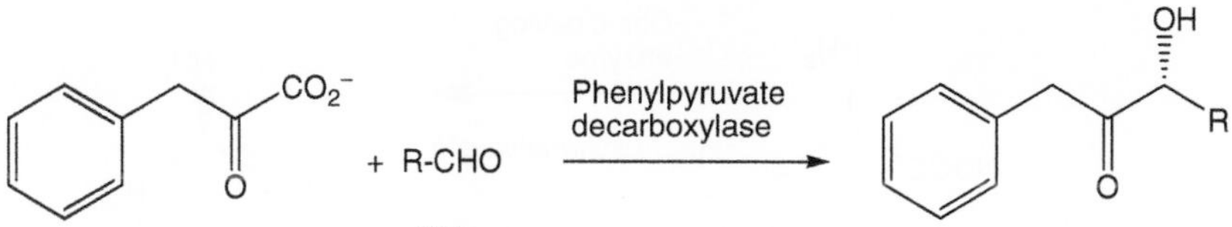

190

Phenylpyruvate

191a–s

192a–s

(R): a (Me), b (Et), c (n-Pr), d (n-Bu), e (PhCH$_2$), f (Ph), g (H),
h (Me(CH$_2$)$_6$), i (Me(CH$_2$)$_8$), j (Me(CH$_2$)$_{10}$), k (PhCH=CH), l (Br$_3$C), m (Me$_2$CH),
n (Me$_3$CCH$_2$), o (BrCH$_2$), p (BrCH$_2$CH$_2$), q (CH$_2$=CH), r (ClCH$_2$), s (HOCH$_2$)

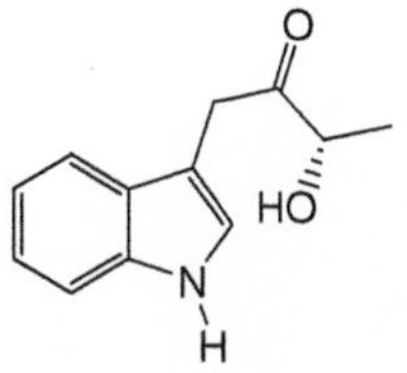

193

FIGURE 4.51 Phenylpyruvate decarboxylase catalyzed acyloin condensation reactions.

acyloin condensation yield decreased with increasing chain length for straight-chain aliphatic aldehydes from 76% for acetaldehyde to 24% for valeraldehyde. The ees of the acyloin products were of 87 to 98%. Low yields of acyloin products were obtained with chloroacetaldehyde (13%) and glycoaldehyde (16%). Indole-3-pyruvate was a substrate of the enzyme and provided acyloin condensation product 3-hydroxy-1-(3-indolyl)-2-butanone (**193**) with acetaldehyde in 19% yield.

4.25 ENANTIOSELECTIVE ENZYMATIC CLEAVAGE OF CARBOBENZYLOXY GROUPS

Amino groups often require protection during synthetic transformations elsewhere in the molecule; at some point, the protecting group must be removed. Enzymatic protection and deprotection under mild conditions have been demonstrated previously. Penicillin G amidase and phathalyl amidase have been used for the enzymatic deprotection of the phenylacetyl and phthaloyl groups from the corresponding amido or imido compounds [175,176]. Acylases have been used widely in the enantioselective deprotection of N-acetyl-DL-amino acids [177]. Enzymatic deprotection of N-carbamoyl L-amino acids and N-carbamoyl D-amino acids has been demonstrated by microbial L-carbamoylases and D-carbamoylases, respectively [178,179].

The carbobenzyloxy (Cbz) group is commonly used to protect amino and hydroxyl groups during organic synthesis. Chemical deprotection is usually achieved by hydrogenation with a palladium catalyst [180,181]. However, during chemical deprotection some groups are reactive (e.g., carbon–carbon double bonds) under, or may interfere (e.g., thiols or sulfides) with, the hydrogenolysis conditions. An enantioselective enzymatic deprotection process has been developed that can be performed under mild conditions without damaging any otherwise susceptible groups in the molecule. A microbial culture was isolated from soil and identified as *S. paucimobilis* strain SC 16113; this culture catalyzed the enantioselective cleavage of Cbz groups (Figure 4.52) from various Cbz-protected amino acids [182]. Only Cbz-L-amino acids were deprotected, giving complete conversion to the corresponding L-amino acid. Cbz-D-amino acids gave <2% reaction yield.

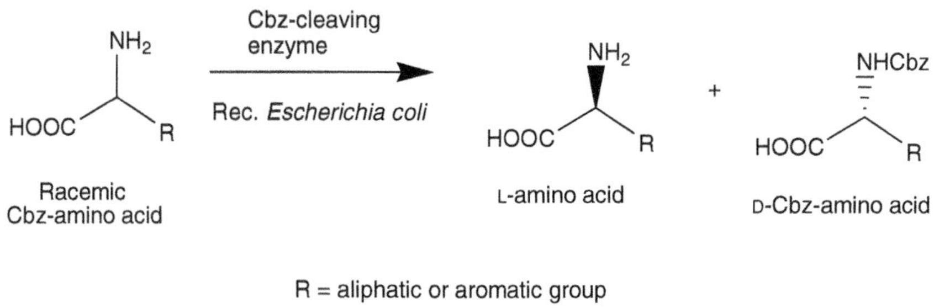

R = aliphatic or aromatic group

FIGURE 4.52 Enantioselective enzymatic cleavage of carbobenzyloxy (Cbz) groups from Cb₂ acids.

Racemic Cbz-amino acids were also evaluated as substrates for hydrolysis by cell extra▶ *paucimobilic* SC 16113. As anticipated, only the L-enantiomer was hydrolyzed, giving the ᵣ acids in >48% yields and >99% ees. The unreacted Cbz-D-amino acids were recovered iᵣ yield and >98% ee [182]. This enzyme has been cloned and overexpressed in *E. coli* [183].

4.26 CONCLUSION

The production of single enantiomers of drug intermediates is increasingly importan▶ pharmaceutical industry. Organic synthesis is one approach to the synthesis of sing▶ tiomers, and biocatalysis provides an alternate opportunity to prepare pharmace▶ useful chiral compounds. The advantages of biocatalysis over chemical catalysis a▶ enzyme-catalyzed reactions are stereoselective and regioselective and can be carried▶ ambient temperature and atmospheric pressure. The use of different classes of enzy▶ the catalysis of many different types of chemical reactions is capable of generating variety of chiral compounds. This includes the use of hydrolytic enzymes such as esterases, proteases, dehalogenases, acylases, amidases, nitrilases, lyases, epoxide hyd▶ decarboxylases, and hydantoinases in the resolution of a variety of racemic compound▶ the asymmetric synthesis of enantiomerically enriched chiral compounds. Oxido-red▶ and aminotransferases have been used in the synthesis of chiral alcohols, aminoal▶ amino acids, and amines. Aldolases and decarboxylases have been effectively used iᵣ metric synthesis by aldol condensation and acyloin condensation reactions. Monoxy▶ have been used in enantioselective and regioselective hydroxylation, epoxidatio▶ Baeyer–Villiger reactions. Dioxygenases have been used in the chemo-enzymatic sy▶ of chiral diols. During the last decade, progress in biochemistry, protein chemistry, m▶ cloning, and random and site-directed mutagenesis, directed evolution of biocatalys▶ fermentation technology has opened up unlimited access to a variety of enzymes and▶ bial cultures as tools in organic synthesis.

ACKNOWLEDGMENT

The author would like to acknowledge Ronald Hanson, Animesh Goswami, Amit B▶ Venkata Nanduri, Jeffrey Howell, Steven Goldeberg, Robert Johnston, Mary-Jo D▶ Dana Cazzulino, Thomas Tully, Thomas LaPorte, Shankar Swaminathan, Lawrence John Venit, John Wasylyk, Michael Montana, Ronald Eiring, Sushil Srivatava, A▶ Singh, Rapheal Ko, Linda Chu, Clyde McNamee, John Thottathil, David Kronenth▶ Richard Mueller for research collaboration and Dr. Richard Mueller for reviewi▶ manuscript and making valuable suggestions.

REFERENCES

1. Food & Drug Administration, FDA's statement for the development of new stereoisomeric drugs, *Chirality*, 4, 338, 1992.
2. Buckland, B., Robinson D., and Chartrain, M., Biocatalysis for pharmaceuticals: status and prospects for a key technology, *Metabol. Eng.*, 2, 42, 2000.
3. Pesti, J.A. and DiCosimo, R., Recent progress in enzymatic resolution and desymmetrization of pharmaceuticals and their intermediates, *Curr. Opin. Drug Discovery Dev.*, 6, 884, 2003.
4. Patel, R.N., Biocatalytic synthesis of chiral pharmaceutical intermediates, *Food Technol. Biotechnol.*, 42, 305, 2004.
5. Faber, K. and Kroutil, W., New enzymes for biotransformations, *Curr. Opin. Chem. Biol.*, 9(2), 181, 2005.
6. Sheldon, R.A. et al., Biocatalysis in ionic liquids, *Green Chem.*, 4(2), 147, 2002.
7. Turner, N.J., Enzyme catalyzed deracemization and dynamic kinetic resolution reactions, *Curr. Opin. Chem. Biol.*, 8(2), 114, 2004.
8. Mahmoudian, M., Development of bioprocesses for the generation of anti-inflammatory, anti-viral and anti-leukaemic agents, *Focus Biotechnol.*, 1(Novel Frontiers in the Production of Compounds for Biomedical Use), 249, 2001.
9. Zaks, A., Industrial biocatalysis, *Curr. Opin. Chem. Biol.*, 5(2), 130, 2001.
10. Patel, R.N., Stereoselective biocatalysis for synthesis of some chiral pharmaceutical intermediates, in *Stereoselective Biocatalysis*, Patel, R.N., Ed., Marcel Dekker, New York, 2000, pp. 87–130.
11. Barrish, J. et al., Amino diol HIV-protease inhibitors, 1: design, synthesis, and preliminary SAR, *J. Med. Chem.*, 37, 1758, 1994.
12. Patel, R.N. et al., Preparation of chiral synthon for HIV-protease inhibitor: stereoselective microbial reduction of *N*-protected α-aminochloroketone, *Tetrahedron Asymmetry*, 8, 2547, 1997.
13. Bold, G. et al., New aza-dipeptide analogs as potent and orally absorbed HIV-1 protease inhibitors: candidates for clinical development, *J. Med. Chem.*, 41, 3387, 1998.
14. Robinson, B. et al., BMS-232632, a highly potent human immunodeficiency virus protease inhibitor that can be used in combination with other available antiretroviral agents, *Antimicrob. Agents Chemother.*, 44, 2093, 2000.
15. Patel, R.N., Chu, L., and Mueller, R., Diastereoselective microbial reduction of (*S*)-[3-chloro-2-oxo-1-(phenylmethyl)propyl]carbamic acid, 1,1-dimethylethyl ester, *Tetrahedron Asymmetry*, 14, 3105, 2003.
16. Xu, Z. et al., Process research and development for an efficient synthesis of the HIV-protease inhibitor BMS-232632, *Organic Process R&D*, 6, 323, 2002.
17. Ireland, C., Leeson, P., and Castaner, J., Lobucavir: antiviral, *Drugs Future*, 22, 359, 1997.
18. Hanson, R. et al., Regioselective enzymatic aminoacylation of lobucavir to give an intermediate for lobucavir prodrug, *Bioorg. Med. Chem.*, 8, 2681, 2000.
19. Chartrain, M. et al., Bioconversion of indene to *cis*-(1*S*,2*R*)-indandiol and *trans*-(1*R*,2*R*)-indandiol by *Rhodococcus* sp., *J. Ferment Technol.*, 86, 550, 1998.
20. Buckland, B.C., Robinson, D.K., and Chartrain, M., Biocatalysis for pharmaceuticals: status and prospects for a key technology, *Met. Eng.*, 2, 42, 2000.
21. Chartrain, M. et al., Metabolic engineering and directed evolution for the production of pharmaceuticals, *Curr. Opin. Biotechnol.*, 11, 209, 2000.
22. Zhang, N. et al., Directed evolution of toluene dioxygense from *Pseudomonas putida* for improved selectivity toward *cis*-(1*S*,2*R*)-indanediol during indane bioconversion, *Metabol. Eng.*, 2, 339, 2000.
23. Amanullah, A. et al., Measurement of strain-dependent toxicity in the indene bioconversion using multiparameter flow cytometry, *Biotechnol. Bioeng.*, 80, 239, 2002.
24. Zhang, J. et al., Bioconversion of indene to *trans*-2*S*,1*S*-bromoindanol and 1*S*,2*S*-indene oxide by a bromoperoxidase/dehydrogenase preparation from *Curvularia protuberata* MF5400, *J. Ferment. Bioeng.*, 80, 244, 1995.
25. Holton, R., Biediger, R., and Joatman, P., Semisynthesis of taxol and taxotere, in *Taxol: Science and Application*, Suffness, M., Ed., CRC Press, New York, 1995.

26. Kingston, D., Natural taxoids: structure and chemistry, in *Taxol: Science and Applicatic* ness, M., Ed., CRC Press, New York, 1995.

27. Baldini, E. et al., Multicenter randomized phase III trial of Epirubicin plus Paclitaxel v ubicin followed by Paclitaxel in metastatic breast cancer patients: focus on cardiac sat *J. Cancer*, 91, 45, 2004.

28. Patel R.N., Tour de paclitaxel: biocatalysis for semisynthesis, *Ann. Rev. Microbiol.*, 98, 36

29. Hanson R.L. et al., Site-specific enzymatic hydrolysis of taxanes at C-10 and C-13, *J. Biol.* 269, 22145, 1994.

30. Nanduri, B. et al., Fermentation and isolation of C-10 deacetylase for the production of from baccatin III, *Biotech. Bioeng.*, 48, 547, 1995.

31. Hanson, R. et al., Enzymic hydrolysis of 7-xylosyltaxanes by xylosidase from *Morax Biotechnol. Appl. Biochem.*, 26, 153, 1997.

32. Patel, R.N. et al., Stereoselective microbial reduction of 2-keto-3-(N-benzoylamino)-3 propionic acid ethyl ester: synthesis of taxol side-chain synthon, *Tetrahedron Asymm* 2069, 1993.

33. Patel, R.N. et al., Enzymic preparation of (3R-cis)-3-(acetyloxy)-4-phenyl-2-azetidinone: side-chain synthon, *Biotech. Appl. Biochem.*, 20, 23, 1994.

34. Baloglu, E. and Kingston, D., The taxane diterpenoids, *J.Nat. Prod.*, 62, 1068, 1999.

35. Rose, W., Preclinical pharmacology of BMS-275183, an orally active taxane, *Clin. Cancer* 2016, 2001.

36. Holton, R., Nadizadeh, H., and Beidiger, R., Preparation of substituted taxanes as an agents, Eur. Patent Appl. 17 pp. EP 534708 A1 19930331 CAN 119:49693 AN 1993:44969

37. Patel, R.N. et al., Enzymatic preparation of (3R)-cis-3-acetyloxy-4-(1,1-dimethylethyl)-2 none: a side-chain synthon for an orally active taxane, *Tetrahedron Asymmetry*, 14, 3673,

38. Hanson, R. and Patel, R., Process for the preparation of C-4 deacetyltaxanes, PCT Int. Ap 2000010989 A1 20000302 CAN 132:194525 AN 2000:144872, 2000.

39. Goodin, S., Kane, M.P., and Rubin, E.H., Epothilones: mechanism of action and biologic *J. Clin. Oncol.*, 22(10), 2015, 2004.

40. Nicolaou, K.C., Roschangar, F., and Vourloumis, D., Chemical biology of epothilone *wandte Chemie*, International Edition, 37(15), 2014, 1998.

41. Altmann, K-H., The merger of natural product synthesis and medicinal chemistry: on th istry and chemical biology of epothilones, *Org. Biomol. Chem.*, 2(15), 2137, 2004.

42. Mulzer, J., Mantoulidis, A., and Ohler, E., Total syntheses of epothilones B and D, *J. Org.* 65(22), 7456, 2000.

43. Altmann, K-H. et al., The natural products epothilones A and B as lead structures for an drug discovery: chemistry, biology, and SAR studies, *Progr. Med. Chem.*, 42, 171, 2004.

44. Lin, N., Brakora, K., and Seiden, M., BMS-247550 (Bristol-Myers Squibb/GBF), *Cur Investig. Drugs* (Thomson Current Drugs), 4(6), 746, 2003.

45. Gerth, K. et al., Studies on the biosynthesis of epothilones: hydroxylation of epo A a epothilones E and F, *J. Antibiot.*, 55(1), 41, 2002.

46. Li, W. et al., Microbial transformation method for the preparation of an epothilone, P Appl., WO 2000039276 A2 20000706 CAN 133:88294 AN 2000:457191, 2000.

47. Basch, J.D. et al., The epothilone B hydroxylase and ferredoxin genes of Amycolatopsis a use in the development of strains for the manufacture of hydroxylated epothilones, PCT In WO 2004078978 A1 20040916 CAN 141:255532 AN 2004:756881, 2004.

48. Umezawa, H. et al., Structure of an antitumor antibiotic, spergualin, *J. Antibiot.*, 34, 162

49. Umeda, Y. et al., Synthesis and antitumor activity of spergualin analogues, III: novel met synthesis of optically active 15-deoxyspergualin and 15-deoxy-11-O-methylspergualin, *J. A* 40, 1316, 1987.

50. Patel, R.N., Banerjee, A., and Szarka, L.J., Stereoselective acetylation of racemic 7-[N benzyloxycarbonyl)-N-(guanidinoheptanoyl)]-α-hydroxyglycine, *Tetrahedron Asymmetry*, 1997.

51. Nemoto, K. et al., Effect of 15-deoxyspergualin on graft-v-host disease in mice, *Trans. Proc* 3985, 1987.

52. Carboni, J.M., Hurlburt, W.W., and Gottardis, M.M., Synergistic methods and compositions using IGF-1 receptor inhibitors and EGF receptor inhibitors for treating cancer, PCT Int. Appl., WO 2004030625 A2 20040415 CAN 140:315050 AN 2004:308360, 2004.

53. Beaulieu, F. et al., Preparation of isopurines and related compounds as tyrosine kinase inhibitors for the treatment of cancer, PCT Int. Appl., WO 2004063151 A2 20040729 CAN 141:140461 AN 2004:606437, 2004.

54. Hanson, R.L. et al., Purification and cloning of a ketoreductase used for the preparation of chiral alcohols, *Adv. Synth. Cat.*, 347(7&8), 1073, 2005.

55. Ondetti, M.A. and Cushman, D.W., Inhibition of renin–angiotensin system: a new approach to the theory of hypertension, *J. Med. Chem.*, 24, 355, 1981.

56. Ondetti, M.A., Rubin, B., and Cushman, D.W., Design of specific inhibitors of angiotensin-converting enzyme: new class of orally active antihypertensive agents, *Science*, 196, 441, 1977.

57. Cushman, D.W. and Ondetti, M.A., Inhibitors of angiotensin-converting enzyme for treatment of hypertension, *Biochem. Pharm.*, 29, 1871, 1980.

58. Goodhue, C.T. and Schaeffer, J.R., Preparation of L-(+)-β-hydroxyisobutyric acid by bacetrial oxidation of isobutyric acid, *Biotechnol. Bioeng.*, 13, 203, 1971.

59. Patel, R.N. et al., Stereoselective enzymatic hydrolysis of α-[acetylthio)methyl]benzenepropanoic acid and 3-acetylthio-2-methylpropanoic acid, *Biotechnol. Appl. Biochem.*, 16, 34, 1992.

60. Patel, R.N. et al., Stereoselective enzymatic esterification of 3-benzoylthio-2-methylpropanoic acid, *Appl. Microbiol. Biotechnol.*, 36, 29, 1991.

61. Moniot, J.L., Preparation of N-[2-(mercaptomethyl)propionyl]-L-prolines, U.S. Patent Appl., CN 88-100862, 1988.

62. Ondetti, M.A., Miguel, A., and Krapcho, J., Mercaptoacyl derivatives of substituted prolines, U.S. Patent, 4316906, 1982.

63. Patel, R.N., Robison, R.S., and Szarka, L.J., Stereoselective enzymic hydrolysis of 2-cyclohexyl- and 2-phenyl-1,3-propanediol diacetate in biphasic systems, *Appl. Microbiol. Biotechnol.*, 34, 10, 1990.

64. Karenewsky, D.S. et al., (Phosphinyloxy)acyl amino acid inhibitors of angiotensin converting enzyme (ACE), 1: discovery of (S)-1-[6-amino-2-[[hydroxy(4-phenylbutyl)phosphinyl]oxy]-1-oxo-hexyl]-L-proline, a novel orally active inhibitor of ACE, *J. Med. Chem.*, 31, 204, 1988.

65. Hanson, R.L. et al., Transformation of N-ε-CBZ-L-lysine to CBZ-L-oxylysine using L-amino acid oxidase from *Providencia alcalifaciens* and L-2-hydroxy-isocaproate dehydrogenase from *Lactobacillus confusus*, *Appl. Microbiol. Biotechnol.*, 37, 599, 1992.

66. Delaney, N.G. et al., Amino acid and peptide derivatives as inhibitors of neutral endopeptidase and their use as antihypertensives and diuretics, Eur. Patent, EP361365, 1988.

67. Robl, J.A. et al., Dual metalloprotease inhibitors: mercaptoacetyl-based fused heterocyclic dipeptide mimetics as inhibitors of angiotensin-converting enzyme and neutral endopeptidase, *J. Med. Chem.*, 40, 1570, 1997.

68. Seymour, A.A., Swerdel, J.N., and Abboa-Offei, B.E., Antihypertensive activity during inhibition of neutral endopeptidase and angiotensin converting enzyme, *J. Cardiovasc. Pharmacol.*, 17, 456, 1991.

69. Wichman, R. et al., Continuous enzymic transformation in an enzyme membrane reactor with simultaneous NAD(H) regeneration, *Biotechnol. Bioeng.*, 23, 2789, 1981.

70. Kragl, U., Vasic-Racki, D., and Wandrey, C., Continuous processes with soluble enzymes, *Chem. Ing. Tech.*, 64, 499, 1992.

71. Hanson, R.L. et al., Enzymatic synthesis of L-6-hydroxynorleucine, *Bioorg. Med. Chem*, 7, 2247, 1999.

72. Patel, R.N., Enzymatic synthesis of chiral intermediates for Omapatrilat, an antihypertensive drug, *Biomol. Eng.*, 17(6), 167, 2001.

73. Rumbero, A. et al., Chemical synthesis of allysine ethylene acetal and conversion *in situ* into 1-piperideine-6-carboxylic acid: key intermediate of the α-aminoadipic acid for β-lactam antibiotics biosynthesis, *Bioorg. Med. Chem*, 3, 1237, 1995.

74. Hanson, R. et al., Enzymatic synthesis of allysine ethylene acetal, *Enzyme Microb.Technol.*, 26, 348, 2000.

75. Schütte, H. et al., Purification and properties of formaldehyde dehydrogenase and fo dehydrogenase from *Candida boidinii*, *Eur. J. Biochem*, 62, 151, 1976.

76. Hou, C.T. et al., NAD-linked formate dehydrogenase from methanol-grown *Pichia pa* NRRL-Y-7556, *Arch. Biochem. Biophys.*, 216, 296, 1982.

77. Patel, R.N., Biocatalytic preparation of a chiral synthon for a vasopeptidase inhibitor: enzy conversion of N2-[N-[(phenylmethoxy)carbonyl]L-homocysteinyl]-L-lysine (1>1′)-disulfic [4S-(4a,7a,10ab)]1-octahydro-5-oxo-4-[(phenylmethoxy) carbonyl]amino]-7H-pyrido-[2,1-b][1,: zepin-7-carboxylic acid methyl ester by a novel L-lysine ε-aminotransferase, *Enzyme M Technol.*, 27, 376, 2000.

78. Ward, O. and Young, C., Reductive biotransformations of organic compounds by cells or en of yeast, *Enzyme Microb. Technol.*, 12, 482, 1990.

79. Csuk, R. and Glanzer, B., Yeast mediated stereoselective biocatalysis, in *Stereoselective Bi lysis*, Patel, R.N., Ed., Marcel Dekker, New York, 2000.

80. Patel, R.N. et al., Stereoselective reduction of β-ketoesters by *Geotrichum candidum*, *E. Microb. Technol.*, 14, 731, 1992.

81. Matsuyama, A., Yamamoto, H., and Kobayashi, Y., Practical application of recombinant ` cell biocatalysts for the manufacturing of pharmaceutical intermediates such as chiral alc *Org. Process R&D*, 6, 558, 2002.

82. Davis, S. et al., Halohydrin dehalogenases and method for production of 4-cyano-3-hydro tyric acid esters and amides, PCT Int. Appl., WO 2004015132, A2 20040219, CAN 140:19819 2004:143313, 2004.

83. Fishman, A. et al., Enzymatic resolution as a convenient method for the production of resolved materials on an industrial scale, *Biotechnol. Bioeng.*, 74, 256, 2001.

84. Sit, S. et al., Synthesis, biological profile, and quantitative structure–activity relationship of a of novel 3-hydroxy-3-methylglutaryl coenzyme A reductase inhibitors, *J. Med. Chem.*, 33, 1990.

85. Patel, R.N. et al., Enantioselective microbial reduction of 3,5-dioxo-6-(benzyloxy)hexanoi ethyl ester, *Enzyme Microb. Technol.*, 15, 1014, 1993.

86. Patel, R.N., McNamee, C., and Szarka, L., Enantioselective enzymic acetylation of r. [4-[4α,6β(E)]]-6-[4,4-bis(4-fluorophenyl)-3-(1-methyl-1H-tetrazol-5-yl)-1,3-butadienyl]tetrahyc hydroxy-2H-pyran-2-one, *Appl. Microbiol. Biotechnol.*, 38, 56, 1992.

87. Gijsen, H. and Wong, C-H., Sequential three- and four-substrate aldol reactions cataly aldolases, *J. Amer. Chem. Soc.*, 116, 8422, 1994.

88. Greenberg, W. et al., Development of an efficient, scalable, aldolase-catalyzed process fo tioselective synthesis of statin intermediates, *Proc. Natl. Acad. Sci. USA*, 101, 5788, 2004.

89. Steiger, A., Pyun, H., and Coates, R., Synthesis and characterization of aza analog inhibi squalene and geranylgeranyl diphosphate synthases, *J. Org. Chem.*, 57, 3444, 1992.

90. Lawrence, M., Biller, S., and Fryszman, O., Preparation of α-phosphonosulfinic squalene : tase inhibitors, U.S. Patent 5447922, 1995.

91. Patel, R.N., Banerjee, A., and Szarka, L., Stereoselective acetylation of [1-(hydroxy)-4-(3-ph phenyl)butyl]phosphonic acid diethyl ester, *Tetrahedron Asymmetry*, 8, 1055, 1997.

92. Nakane, M., Preparation and formulation of 7-oxabicycloheptane substituted sulfonamid taglandin analogs useful in treatment of thrombolic disease, U.S. Patent 4,663,336, 1987.

93. Ford-Hutchinson, A.W., Innovations in drug research: inhibitors of thromboxane and trienes, *Clin. Exper. Allergy.*, 21, 272, 1991.

94. Das, J. et al., Synthesis of optically 7-oxabicyclo[2.2.1]heptanes and assignment of a configuration, *Synthesis*, 12, 1100, 1987.

95. Patel, R.N. et al., Stereoselective microbial/enzymatic oxidation of (exo,exo)-7-oxabicyclc heptane-2,3-dimethanol to the corresponding chiral lactol and lactone, *Enzyme Microb. T* 14, 778, 1992.

96. Patel, R.N. et al., Stereoselective enzymic hydrolysis of (exo,exo)-7-oxabicyclo[2.2.1]hept: dimethanol diacetate ester in a biphasic system, *Appl. Microbiol. Biotechnol.*, 37, 180, 199:

97. Chaffman, M. and Brogden, R.N., Diltiazem: a review of its pharmacological proper therapeutic efficacy, *Drugs*, 29, 387, 1985.

98. Kawai, C. et al., Comparative effects of three calcium antagonists, diatiazem, verapamil and nifedipine, on the sinoatrial and atrioventricular nodes, *Exp. Cir.*, 63, 1035, 1981.
99. Isshiki, T., Pegram, B., and Frohlich, E., Immediate and prolonged hemodynamic effects of TA-3090 on spontaneously hypertensive (SHR) and normal Wistar–Kyoto (WKY) rats, *Cardiovasc. Drug Ther.*, 2, 539, 1988.
100. Floyd, D.M. et al., Synthesis of benzazepinone and 3-methylbenzothiazepinone analogs of diltiazem, *J. Org. Chem.*, 55, 5572, 1990.
101. Patel, R.N. et al., Stereospecific microbial reduction of 4,5-dihydro-4-(4-methoxyphenyl)-6-(trifluoromethyl-1H-1)-benzazepin-2-one, *Enzyme Microb. Technol.*, 13, 906, 1991.
102. Edwards, G. and Weston, A.H., Structure–activity relationships of K+ channel openers, *Trends Pharmacol. Sci.* 11, 417–422, 1990.
103. Robertson, D.W. and Steinberg, M.I., Potassium channel openers: new biological probes, *Ann. Med. Chem.*, 24, 91, 1989.
104. Hamilton, T.C. and Weston, A.H., Cromakalim, nicorandil and pinacidil: novel drugs which open potassium channels in smooth muscle, *Gen. Pharmac.*, 20, 1, 1989.
105. Ashwood, V.A. et al., Synthesis and antihypertensive activity of 4-(cyclic amido)-2H-1-benzopyrans, *J. Med. Chem.*, 29, 2194, 1986.
106. Bergmann, R., Eiermann, V., and Gericke, R.J., 4-Heterocyclyloxy-2H-1-benzopyran potassium channel activators, *J. Med. Chem.*, 33, 2759, 1990.
107. Patel, R.N. et al., Stereoselective epoxidation of 2,2-dimethyl-2H-1-benzopyran-6-carbonitrile, *Bioorg. Med. Chem.*, 2, 535, 1994.
108. Patel, R.N. et al., Stereoselective acetylation of 3,4-dihydro-3,4-dihydroxy-2,2-dimethyl-2H-1-benzopyran-6-carbonitrile, *Tetrahedron Asymmetry*, 6, 123, 1995.
109. Archelas, A. and Furstoss, R., Biocatalytic approaches for the synthesis of enantiopure epoxides, *Top. Curr. Chem.*, 200, 159, 1999.
110. Mischitz, M. et al., Asymmetric microbial hydrolysis of epoxides, *Tetrahedron Asymmetry*, 6, 1261, 1995.
111. Catt, J.D. et al., Preparation of benzofuran and dihydrobenzofuran melatonergic agents, U.S. 5856529, CAN 130:110151, 1999.
112. Goswami, A. et al., Stereospecific enzymatic hydrolysis of racemic epoxide: a process for making chiral epoxide, *Tetrahedron Asymmetry*, 10, 3167, 1999.
113. Stecher, H. and Faber, K., Biocatalytic deracemization techniques: dynamic resolutions and stereoinversions, *Synthesis*, 1, 1, 1997.
114. Fantin, G. et al., Kinetic resolution of racemic secondary alcohols via oxidation with *Yarrowia lipolytica* strains, *Tetrahedron Asymmetry*, 11, 2367, 2000.
115. Goswami, A., Mirfakhrae, K.D., and Patel, R.N., Deracemization of racemic 1,2-diol by biocatalytic stereoinversion, *Tetrahedron Asymmetry*, 10(21), 4239, 1999.
116. Arch, J.R.S., β3-adrenoceptors and other putative atypical β-adrenoceptors, *Pharmacol. Rev. Commun.*, 9, 141, 1997.
117. Bloom, J.D. et al., Disodium (*R,R*)5-[2-(3-chlorophenyl)-2-hydroxyethyl]amino]propyl]-1,3-benzodioxole-2,2-dicarboxylate: a potent β-adrenergic agonist virtually specific for β-3-receptors, *J. Med. Chem.*, 35, 3081, 1989.
118. Fisher, L.G. et al., BMS-187257, a potent, selective, and novel heterocyclic β-3 adrenergic receptor agonist, *Bioorg. Med. Chem. Lett.*, 6, 2253, 1994.
119. Patel, R.N. et al., Microbial synthesis of chiral intermediates for β-3-receptor agonists, *J. Am. Oil Chem. Soc.*, 75, 1473, 1998.
120. Campos, F., Bosch, M., and Guerrero, A., An efficient enantioselective synthesis of (*R,R*)-formoterol, a potent bronchodilator, using lipases, *Tetrahedron Asymmetry*, 11, 2705, 2000.
121. Bisacchi, G., Preparation of amidino and guanidino azetidinone compounds as tryptase inhibitors, WO 9967215, A1 19991229, CAN 132:64103, AN 1999:819347, 1999.
122. Wirz, B. and Walther, W., Enzymic preparation of chiral 3-(hydroxymethyl)piperidine derivatives, *Tetrahedron Asymmetry*, 3, 1049, 1992.
123. Goswami, A. et al., Chemical and enzymatic resolution of (*R,S*)-*N*-(*tert*-butoxycarbonyl)-3-hydroxymethylpiperidine, *Organic Process R&D*, 5, 415, 2001.

124. Schenk, D., Games, D., and Seubert, P., Potential treatment opportunities for Alzheimer's di through inhibition of secretases and Aβ immunization, *J. Mol. Neurosci.*, 17, 259, 2001.

125. Audia, J. et al., Preparation of *N*-(phenylacetyl)di- and tripeptide derivatives for inhib β-amyloid peptide release, PCT Int. Appl., WO 9822494, A2 19980528, CAN 129:41414, 12

126. Patel, R.N. et al., Enzymatic resolution of racemic secondary alcohols by lipase B from *Ca antarctica*, *J. Am. Oil Chem. Soc.*, 77, 1015, 2000.

127. Nanduri, V., Purification of a stereospecific 2-ketoreductase from *Gluconobacter oxydans*, Microbiol. Biotechnol.*, 25, 171, 2000.

128. Nanduri, V.B. et al., Biochemical approaches to the synthesis of ethyl 5-(*S*)-hydroxyl hexa and 5-(*S*)-hydroxyhexanenitrile, *Enzyme Microb. Technol.*, 28(7&8), 632, 2001.

129. Patel, R.N., Enantioselective microbial reduction of substituted acetophenones, *Tetrah Asymmetry*, 15, 1247, 2004.

130. Kagechika, H. et al., Retinobenzoic acids, 1: structure–activity relationships of aromatic a with retinoidal activity, *J. Med. Chem.*, 31, 2182, 1988.

131. Shudo, K. and Kagechika, H., Structure–activity relationships of a new series of syntheti noids, in *Chemistry and Biology of Synthetic Retinoids*, Dawson, M.I. and Okamura, W.H., CRC Press, Boca Raton, FL, 1989.

132. Moon, RC. and ad Itri, L.M., Retinoids and cancer, *Retinoids*, 2, 327, 1984.

133. Kagechika, H. et al., Retinobenzoic acids, 1: structure–activity relationships of aromatic a with retinoidal activity, *J. Med. Chem.*, 31, 2182, 1988.

134. Belema, M., Zusi, F.C., and Tramposch, K.M., Active enantiomer of RAR γ-specific agonist Int. Appl., WO 0016769 A1 20000330 CAN 132:246379 AN 2000:209902, 2000.

135 Patel, R.N. et al., Enantioselective microbial reduction of 2-oxo-2-(1′,2′,3′,4′-tetrahydro-1′,1 tetramethyl-6′-naphthalenyl)acetic acid and its ethyl ester, *Tetrahedron Asymmetry*, 13, 349,

136. Hoegenauer, G., Mechanism of action of antibacterial agents: tiamulin and pleuromutilin, otics, 5, 344, 1979.

137. Berner, H. et al., Chemistry of pleuromutilin, IV: synthesis of 14-*O*-acetyl-8α-hydroxym *Tetrahedron*, 39, 1317, 1983.

138. Hanson, R.L. et al., Hydroxylation of mutilin by *Streptomyces griseus* and *Cunningh echinulata*, *Organic Process R&D*, 6, 482, 2002.

139. Springer, D.M. et al., Synthesis and activity of a C-8 ketopleuromutilin derivative, *Bioorg. Chem. Lett.*, 13, 1751, 2003.

140. Iwata, H., Tanaka, R., and Ishiguro, M., Structures of the alkaline hydrolysis products of antibiotic, SUN5555, *J. Antibiot.* 43, 901, 1990.

141. Matsuyama, A., Yamamoto, H., and Kobayashi, Y., Practical application of recombinant cell biocatalysts for the manufacturing of pharmaceutical intermediates such as chiral alc *Organic Process R&D*, 6, 558, 2002.

142. O'Donnell, M.J., Bennett, D.W., and Wu, S., The stereoselective synthesis of α-amino ac phase-transfer catalysis, *J. Am. Chem. Soc.*, 111, 2353, 1989.

143. Schmidt U. et al., Amino acids and peptides, 71: total synthesis of the didemnins, III: synth protected (2*R*,3*S*)-alloisoleucine and (3*S*,4*R*,5*S*)-isostatine derivatives—amino acids from h acids, *Synthesis*, 28, 256, 1989.

144. Bold, G., Duthaler, R.O., and Riediker, M., Enantioselective synthesis with titanium carboh complexes, Part 3: enantioselective synthesis of (D)-threo-β-hydroxy-α-amino acids with ti carbohydrate complexes, *Angew. Chem. Int. Ed. Engl.*, 28, 497, 1989.

145. Gordon, E.M. et al., *O*-Sulfated β-lactam hydroxamic acids (monosulfactams): novel mon β-lactam antibiotics of synthetic origin, *J. Amer. Chem. Soc.*, 104, 6053, 1982.

146. Godfrey, J.D., Mueller, R.H., and van Langen, D.J., β-Lactam synthesis: cyclization 1,2-acyl migration-cyclization: the mechanism of the 1,2-acyl migration-cyclization, *Tetr Lett.*, 27, 2793, 1986.

147. Hanson, R.L. et al., Synthesis of L-β-hydroxyvaline from α-keto-β-hydroxyisovalerat leucine dehydrogenase from *Bacillus* species, *Bioorg. Chem.*, 18, 116, 1990.

148. Langer, S.Z. and Hicks, P.E., Alpha-adrenoceptor subtypes in blood vessels: physiolo pharmacology, *J. Cardiovasc. Pharmacol.*, 6, 547, 1984.

149. Manoury, P.M. et al., Synthesis and antihypertensive activity of a series of 4-amino-6,7-dimethoxy-quinazoline derivatives, *J. Med. Chem.*, 29, 19, 1986.
150. Manoury, P., Binet, J., and Cavero, I., Pyridine derivatives and their use in therapeutic, Eur. Pat. Appl., EP 61379 A1 19820929 CAN 98:89179 AN 1983:89179, 1982.
151. Sanchez, V.M., Rebolledo, F., and Gotor, V., *Candida antarctica* lipase catalyzed resolution of ethyl ($\pm$)-3-aminobutyrate, *Tetrahedron Asymmetry*, 8, 37–40, 1997.
152. Baldessari, A. and Mangone, C.P., One-pot biocatalyzed preparation of substituted amides as intermediates of pharmaceuticals, *J. Mol. Catalysis B: Enzymatic.*, 11, 335, 2001.
153. Fukuroda, T. and Nishikibe, M., Enhancement of pulmonary artery contraction induced by endothelin-β-receptor antagonism, *J. Cardiovasc. Pharmacol.*, 31(Suppl. 1, Endothelin V), S169–S171, 1989.
154. Sumner, M.J. et al., Endothelin ETA and ETB receptors mediate vascular smooth muscle contraction, *Br. J. Pharmacol.*, 107, 858, 1992.
155. Krulewicz, B. et al., Asymmetric biosynthesis of key aromatic intermediates in the synthesis of an endothelin receptor antagonist, *Biocatalysis Biotransform.*, 19, 267, 2001.
156. Jajoo, H.K. et al., Metabolism of the antianxiety drug buspirone in human subjects, *Drug Met. Disposition*, 17(6), 634, 1989.
157. Mayol, R.F., Buspirone metabolite for the alleviation of anxiety, Cont.-in-part of U.S. Ser. No. 484,161 US 6150365 A 20001121 CAN 133:359250 AN 2000:819478, 2000.
158. Yevich, J.P. et al., Antianxiety composition using BMY 28674, and preparation thereof, U.S. Pat. Appl. Publ., Cont.-in-part of U.S. Ser. No. 588,220. U.S. 2003069251 A1 20030410 CAN 138:281137 AN 2003:282120, 2003.
159. Yevich, J.P. et al., *S*-6-Hydroxy-buspirone for treatment of anxiety, depression and related disorders, U.S. Pat. Appl. Publ., 2003.
160. Yevich, J.P. et al., *R*-6-hydroxybuspirone for the treatment of anxiety, depression, and other psychogenic disorders. PCT Int. Appl., WO 2003009851 A1 20030206 CAN 138:131155 AN 2003:97305, 2003.
161. Hanson, R.L. et al., Preparation of (*R*)- and (*S*)-hydroxybuspirone by enzymatic resolution and hydroxylation, *Tetrahedron Asymmetry*, 16(16), 2711, 2005.
162. Patel, R.N. et al., Enantioselective microbial reduction of 6-oxo-8-[4-[4-(2-pyrimidinyl)-1-piperazinyl]butyl]-8-azaspiro[4.5]decane-7,9-dione, *Tetrahedron Asymmetry*, 16(16), 2778, 2005.
163. Ferris, C.D. et al., 3H]opipramol labels a novel binding site and sigma receptors in rat brain membranes, *J. Neurochem.*, 57, 729, 1991.
164. Junien, J.L. and Leonard, B.E., Drugs acting on sigma and phencyclidine receptors: a review of their nature, function, and possible therapeutic importance, *Clin. Neuropharmacol.*, 12, 374, 1989.
165. Steinfels, G.F., Tam, S.W., and Cook, L., Discriminative stimulus properties of a σ receptor agonist in the rat: role of monoamine systems, *Neuropsychopharmcology*, 2, 201, 1989.
166. Massamiri, T. and Duckles, S.P., Sigma receptor ligands inhibit rat tail artery contractile responses by multiple mechanisms, *J. Pharmacol. Exp. Ther.*, 253, 124, 1990.
167. Yevich, J.P. et al., Synthesis and evaluation of *N*-substituted 1-(5-fluoro-2-pyrimidinyl)piperazine derivatives as potential anti-ischemic agents, *Bioorg. Med. Chem. Lett.*, 16(4), 1941, 1994.
168. Yevich, J.P. et al., Synthesis and biological characterization of α-(4-fluorophenyl)-4-(5-fluoro-2-pyrimidinyl)-1-piperazinebutanol and analogs as potential atypical antipsychotic agents, *J. Med. Chem.*, 35(24), 4516, 1992.
169. Patel, R.N. et al., Microbial reduction of 1-(4-fluorophenyl)-4-[4-(5-fluoro-2-pyrimidinyl)-1-piperazinyl]butan-1-one, *Biotechnol. Appl. Biochem.*, 17, 139, 1993.
170. Ward, O.P. and Baev, M.V., Decarboxylases in stereoselective catalysis, in *Stereoselective Biocatalysis*, Patel, R.N., Ed., Marcel Dekker, New York, 2000.
171. Iding, H. et al., Application of α-keto acid decarboxylases in biotransformations, *Biochim. Biophys. Acta*, 1385, 307, 1998.
172. Iding, H. et al., Benzoylformate decarboxylase from *Pseudomonas putida* as stable catalyst for the synthesis of chiral 2-hydroxy ketones, *Chem. Eur. J.*, 6, 1483, 2000.
173. Guo, Z. et al., Asymmetric acyloin condensation catalyzed by phenylpyruvate decarboxylase, *Tetrahedron Asymmetry*, 12, 571, 2001.

174. Guo, Z. et al., Asymmetric acyloin condensation catalyzed by phenylpyruvate decarboxyl. 2: substrate specificity and purification of the enzyme, *Tetrahedron Asymmetry*, 12, 571, :

175. Waldmann, H., Heuser, A., and Schulze S., Selective enzymic removal of protecting grc phenylacetamide as amino protecting group in phosphopeptide synthesis, *Tetrahedron I* 8725, 1996.

176. Costello, C.A., Kreuzman, A.J., and Zmijewski, M.J., Selective deprotection of phthalyl p amines, *Tetrahedron Lett.*, 37, 7469, 1996.

177. Bommarius, A.S. et al., L-methionine related L-amino acids by acylase cleavage of the sponding *N*-acetyl-DL-derivatives, *Tetrahedron Asymmetry*, 8, 3197, 1997.

178. Ogawa, J. and Shimizu, S., Stereoselective synthesis using hydantoinases and carbamoy *Stereoselective Biocatalysis*, Patel, R.N., Ed., Marcel Dekker, New York, 2000.

179. Ogawa, J., Miyake, H., and Shimizu S., Purification and characterization of *N*-carbamoyl acid amidohydrolase with broad substrate specificity from *Alcaligenes xylosoxidans*, *App* *biol. Biotechnol.*, 43, 1039, 1995.

180. Sajiki, H., Hattori, K., and Hirota K., The formation of a novel Pd/C-ethylenediamine catalyst: chemoselective hydrogenation without deprotection of the *O*-benzyl and *N*-Cbz *J. Org. Chem.*, 63, 7990, 1998.

181. Royer, G.P., Chow, W., and Hatton, K.S., Palladium/polyethylenimine catalysts, *J. Mo*, 31, 1, 1985.

182. Patel, R.N. et al., Enantioselective enzymatic cleavage of *N*-benzyloxycarbonyl grou *Synth. Catalysis*, 345, 830, 2003.

183. Nanduri, V.B. et al., Cloning and expression of a novel enantioselective *N*-carbobe cleaving enzyme, *Enzyme Microb. Technol.*, 34(3&4), 304, 2004.

5 Directed Evolution of Lipases and Esterases for Organic Synthesis

Aurelio Hidalgo and Uwe T. Bornscheuer

CONTENTS

5.1 LIPASES AND ESTERASES

5.1.1 DEFINITION, STRUCTURAL FEATURES, PROPERTIES

Lipases and esterases are, undoubtedly, enzymes with great biotechnological importance. Approximately 55% of all enzymes used in biocatalysis are hydrolases, lipases (~30%), and esterases (~8%) being the most predominant enzymes besides proteases, nitrilases, and others [1,2]. Both of these enzymes belong to the structural superfamily of α/β-hydrolases. This family groups several hydrolytic enzymes of widely differing phylogenetic origin and catalytic function. The common structural motif comprises eight beta-sheets connected by alpha-helices. These enzymes seem to diverge from a common ancestor in order to preserve the

arrangement of the catalytic residues. They all have a catalytic triad, the elements of wh
borne on loops, which are the best-conserved structural features in the fold. Only the h
in the nucleophile–histidine–acid catalytic triad is completely conserved. In the case of
and esterases the catalytic triad is composed of Ser-Asp-His (Ser-Glu-His for some l
the Ser residue being in the consensus sequence Gly-X-Ser-X-Gly. The unique topologi
sequential arrangement of the triad residues produces a catalytic triad which is, in a s
mirror image of the serine protease catalytic triad [3].

Lipases (triacylglycerol hydrolases, EC 3.1.1.3) are the most frequently used hydro
organic synthesis [4,5]. They are quite widespread in nature and a considerable nur
them are commercially available. In nature they catalyze the hydrolysis and transest
tion of triglycerides present in fats and oils. The reaction mechanism comprises fou
detailed in Figure 5.1:

1. The substrate reacts with the active site serine, yielding a tetrahedral interr
   stabilized by the catalytic His and Asp (Glu) residues.
2. The alcohol is released and a covalent acyl–enzyme complex is formed.
3. Attack of a nucleophile (water in hydrolysis, alcohol in transesterification) form
   a tetrahedral intermediate.
4. The intermediate collapses to yield the product and regenerates the free enzym

Lipases are commercially used in laundry detergents, in the food industry for
making, for modification of fats and oils, for synthesis of sugar esters, in the co
industry and personal care for the synthesis of emollient esters. A major area is the
lipases in fine chemicals industry, as they often exhibit excellent enantio- and stereosele
and, therefore, allow for the production of a range of optically active building bloc.
instance, Novozymes produces the lipases from *Candida antarctica B* (CAL-B) and *RN
cor miehei* (RML) for applications in organic synthesis and lipid modification, respecti

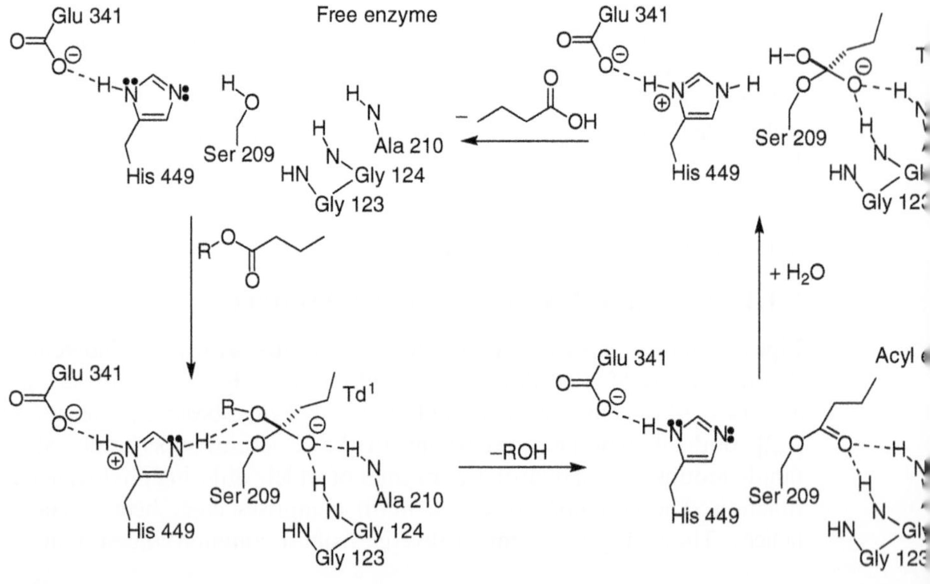

FIGURE 5.1 Hydrolysis of a butyric acid ester catalyzed by lipase or esterase. The amino acid n
ing corresponds to the active site of the *Candida rugosa* lipase (CRL).

Lipases can be classified according to sequence alignment into three major groups. The mammalian lipases form one group, from which the most relevant member is the porcine pancreatic lipase (PPL). This commercial lipase preparation is an impure acetone extract from the animal source. The different proteins in the preparation exhibit different behavior in biocatalytic reactions [7] and, in fact, many of the biotransformations usually attributed to this lipase are now suspected to be originated by the contaminants in the preparation [8]. Nevertheless, the commercial preparation of PPL is currently employed in the industrial resolution of some compounds, e.g., the resolution of glycidyl butyrate developed by Ladner and Whitesides [9].

The fungal lipases can be subdivided into two groups. The first family is the *Candida rugosa* family, and it includes the lipases from *C. rugosa, Geotrichum candidum,* and the pancreatic cholesterol esterase (although it is not from fungal origin, the sequence similarity places it in this group). The second subgroup is the *Rhizomucor* family, which includes, amongst others, small lipases such as CAL-B, RML, and those from *Thermomyces lanuginosus* (TLL) and *Rhizopus oryzae* (ROL). For more information on the latter group, readers are referred to the review by Bornscheuer et al. [10].

Finally, the bacterial lipases can also be subdivided into two groups: the *Staphylococcus* family and the *Pseudomonas* family, which includes lipases such as *Burkholderia cepacia* lipase (BCL), *Pseudomonas fluorescens* lipase (PFL), and *Chromobacterium viscosum* lipase (CVL). Readers are also encouraged to consult the Lipase Engineering Database for further information about lipase classifications and alignments (http://www.led.uni-stuttgart.de).

Lipases prefer substrates containing long-chain fatty acids. Furthermore, since the natural substrate of lipases, i.e., triglycerides, are insoluble in water, lipases are able to function at the hydrophilic–hydrophobic interface. At this interface, lipases exhibit a phenomenon termed interfacial activation, which causes a remarkable increase in activity upon contact with a hydrophobic surface. This effect is due to the presence of a flexible peptide (the "flap" or "lid") covering the active site. Lipases undergo a drastic conformational change in the presence of hydrophobic surfaces, causing the lid to displace and expose the hydrophobic active site to the interface. The catalytic process involves a series of differentiated stages: contact with the interface, conformational change, penetration in the interface, and finally the catalysis itself. The principle of interfacial activation applies not only to liquid–liquid interfaces, such as two-phase aqueous systems, but also to solid–liquid interfaces, such as in hydrophobic chromatographic supports [11,12].

Not every lipase of biotechnological interest can be produced by overexpressing the gene of interest in a heterologous host. For instance, the *Bacillus* lipases can be generally overproduced in *Escherichia coli,* but other enzymes, e.g., some from *Pseudomonas,* can hardly be overexpressed in active form in this host, and homologous expression in *Pseudomonas* sp. is preferred. The expression of fungal lipases requires careful control of codon usage and folding issues if overexpression in bacterial hosts is targeted [13]. The commercial lipases CAL-B, RML, and TLL are all expressed in *Aspergillus* sp. [14,15].

Esterases (carboxyl esterases, EC 3.1.1.1) have the same reaction mechanism as lipases, but differ from them by their substrate specificity, since they prefer short-chain fatty acids whereas lipases prefer long-chain fatty acids. Another difference lies in the "interfacial activation" phenomenon, which has been observed only for lipases. In contrast to lipases, only a few esterases have practical uses in organic synthesis, because lipases are generally more enantioselective and resistant to organic solvents, although certain examples of thermophilic esterases resistant to organic solvents have been described [16,17]. The most widely used esterase is the preparation isolated from pig liver [6] but there are several reports of kinetic resolutions of racemates catalyzed by other esterases [18–23]. Other applications of esterases include the deprotection of labile functional groups [24], the use of feruloyl esterases in pulp bleaching [25] or in vanillin production [26,27].

According to sequence similarities, Arpigny and Jaeger grouped lipases and este
eight families. Among them, the most remarkable groups are group I, which contair
lipases," mostly from Pseudomonads; group II, which comprises the GDSL-motif con
lipases and esterases; group IV, similar to the human hormone-sensitive lipase; gr
similar to epoxide hydrolases, dehalogenases, and haloperoxidase; group VI, containir
of the smallest esterases known; and group VII, which contains larger esterases [28] (
[29]). However, such a classification based on sequence similarity cannot always exp
differences found in substrate specificity.

5.1.2   Uses in Biocatalysis

Enzyme catalyzed reactions have a number of advantages that make them ideal for
synthetic reactions. However, certain requirements should ideally be fulfilled by the b
lysts in order to make them viable alternatives to conventional chemical process
biocatalyst used in such reactions should exhibit the following properties (always dep
on the particular reaction):

- Chemo- and regioselectivity
- Stereoselectivity
- Desired thermostability
- Stability against organic solvents
- Easy recovery and reutilization
- Broad substrate range

Despite being overexpressed in suitable hosts, only few esterases have been used
synthesis of optically pure compounds, because they often exhibit moderate or no e
selectivity. Nevertheless, there are several examples described in which either wild-
mutated esterases are used as biocatalysts to profit from their chemo-, regio-, or e
selectivity. These have been gathered in a review about microbial carboxyl ester
Bornscheuer [29].

Biotechnology requires enzymes that are functional and stable under a wide ra
unnatural conditions [30], but enzymes do not always fulfill these requirements, and
fore, adequate solutions have to be implemented. On one hand, reaction and
engineering were used to change typical variables of the reaction, such as pH, tempe
type of organic cosolvent and its concentration, and immobilization support. The
through a combinatorial approach, the ideal conditions can be found for every r
studied. On the other hand, the enzyme itself can be altered to accept nonnatural su
[31], evolved to become less prone to inactivation under harsh operating conditions [24,
and even mutated to direct the immobilization toward a predetermined region of the
[38]. The strategies for such alterations can either be through structural studies,
modeling to decide which residues are replaced (site-directed mutagenesis), or a
approach in which the enzyme is mutated without previous knowledge of the three-
sional structure and the variants generated are subsequently screened for the desired p
i.e., by the directed evolution approach.

5.2   DIRECTED EVOLUTION OF LIPASES AND ESTERASES

Directed evolution is a technique of enzyme alteration and selection to produce
biocatalysts, based on the creation of random mutants that does not require prior kn
of structure–function relationship.

A directed evolution strategy requires:

1. A mutagenesis method, as unbiased as possible (when no other constraints or requirements are preimposed)
2. A high-throughput selection method that specifically measures the property of interest, as close to the original reaction as possible

The evolution process begins with the selection of a known enzyme and of the property or activity that needs to be evolved (e.g., increased enzyme activity, improved thermal stability, higher tolerance to the presence of organic solvents, modified optimal working temperature or pH, altered specificity, and enantioselectivity). A large mutant library—usually within the range of 10^5 to 10^{10} mutants—is then prepared by mutagenesis of the entire gene, according to the techniques detailed below. After cloning and medium- or high-throughput expression, a collection of enzyme molecules is generated. This library is subjected to screening in terms of the evolved property to select the mutants exhibiting the desired characteristics. Isolated mutants might subsequently serve as improved starting points for additional rounds of mutagenesis to accumulate beneficial mutations for the best result (Figure 5.2) [39].

5.2.1 CREATION OF DIVERSITY

5.2.1.1 Mutagenesis Methods

Methods to create mutant libraries can be divided into two major categories: nonrecombining mutagenesis, in which a parent gene is subjected to random mutagenesis to yield variants with

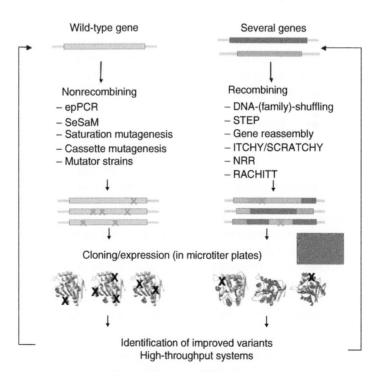

FIGURE 5.2 The process of directed evolution. For abbreviations, see text.

point mutations; and, when several parental genes (generally homologous or with degree of similarity) are available, they can be randomly fragmented, shuffled, and structed to create a library of recombined offspring. Although an overview of some m is presented here, readers are referred to the reviews by Neylon [40] and Kurtzman [4 more detailed survey.

5.2.1.1.1   Creation of New Mutations

Error-prone polymerase chain reaction (epPCR) is the most widespread method to ge mutations quickly and efficiently. It consists of an error-prone version of the well-amplification reaction using DNA polymerase from thermophilic microorganisms. controlled conditions (addition of extra Mg^{2+} and Mn^{2+} and usage of unbalanced concentrations) a reproducible error rate is achieved. The error rate is kept low to ge adaptative mutations, since a higher error rate will lead to deleterious mutations or i variants. Nevertheless, a homogeneous mutational spectrum cannot be achieved usin DNA polymerase with Mn^{2+} and unbalanced nucleotides. Such a fact, known as poly bias, will result in a higher tendency to exchange the desoxynucleotides A and T than G Improved DNA polymerases that exhibit a more homogeneous mutational spectri now commercially available (Strategene) for this purpose. The bottleneck introduced ligation of the mutated epPCR product into a cloning or expression vector is partially by the "bringer" technique developed by Bichet et al., based on amplifying the whole p under mutagenic conditions; but many controls are needed in order to ensure transfor of only mutated plasmids, especially of those mutated only in the target gene [42].

To overcome the positional and mutational bias connected to epPCR, a more c approach (sequence saturation mutagenesis [SeSaM]) was developed by Wong et al. th introduce any mutation on any position in the gene by using universal bases [43].

An easier alternative to create random mutations involves the use of mutator strain strains are deficient in three of the primary DNA repair pathways and exhibit a mutati approximately 5000 times that of the wild-type host strain. On the one hand, wi approach, the ligation step is avoided, but on the other, there is no control on the le of the mutation, which can affect the promotor present in the plasmid, its copy numb other features of the vector. Nevertheless, there are several references that describe suc creation of mutants using this technique [44–46].

Once a key position to enzyme function has been identified through point mutage is a usual strategy to saturate that position, i.e., to replace the mutated amino acid remaining 18 proteinogenic amino acids. Furthermore, by random mutagenesis the spots" can be identified, i.e., those amino acid positions that are key to the desired pr even if they are far away from the active site, and would seem to be, at first glance, irre

However, another approach was recently developed by Reetz et al. [47] that co randomization with site-directed mutagenesis. First, certain residues were chosen that interact with both the acyl and alcohol in the binding pocket of *P. aeruginosa* lipase. Thes amino acids were randomized in pairs, and such pairs were chosen according to str criteria, namely, the structural motif they were part of, i.e., a loop, a sheet, or a helix. domize a loop, two contiguous amino acids were randomized: for a sheet, amino acids at p n and $(n + 2)$; for a 3_{10} helix, n and $(n + 3)$; and for an α-helix, n and $(n + 4)$. In this f not only was the size of the library generated drastically reduced to 3000 clones per libra the strategy also proved adequate to expand the range of substrates accepted by this e

5.2.1.1.2   Recombination of Variants

When beginning from already selected variants, mutagenesis methods are preferred th combine the best features of both parental donors. These are also known as "s mutagenesis methods. The first example was developed by Stemmer [48] and termed "DNA shuffling". It consists of a DNAse-dependent degradation and subsequent recombination of the fragments without primers (self-priming PCR) followed by a final PCR with primers. An alternative to avoid bias introduced by DNAse I digestion [49] is the staggered extension process (StEP) developed by Zhao et al. [50]. It consists of the amplification of short fragments of the parental genes, so that in subsequent cycles, the resulting short fragments can anneal on any other parental genes. A further improvement on the classic shuffling approach was developed by Coco et al., who devised an alternative DNA-shuffling method, random chimeragenesis on transient templates (RACHITT), based on the ordering, trimming, and joining of randomly cleaved single-stranded parental gene fragments annealed onto a transient full-length single-stranded template [51]. This method exhibited higher recombination frequencies and 100% chimerical products. Two alternative methods that focus on increasing or controlling the recombination frequency and the fraction of chimerical products over the unshuffled parental strains are the "combinatorial libraries enhanced by recombination in yeast" (CLERY) [52], which combines *in vitro* DNA shuffling with *in vivo* recombination in yeast and the degenerate oligonucleotide gene shuffling (DOGS) [53].

However, similarity of sequence exhibited by the parental genes was a necessary constraint to apply recombination techniques; several methodologies have been introduced to recombine several parental sequences without regard to homology but with control over the demarcation points, such as the Gene Reassembly method [54], the nonhomologous random recombination (NRR) [55], the incremental truncation for the creation of hybrid enzymes (ITCHY), and its variation, the thio-ITCHY [56].

As a general trend, several mutagenesis methods can be combined in order to cover the distance in the sequence space that separates the starting point (often the wild-type enzyme, but not necessarily) from the optimal variant for a given property (activity, thermostability, enantioselectivity, etc.). Thus, point mutation methods can be used in conjunction with recombination methods in order to obtain the best variant possible, for instance, by shuffling mutants containing point mutations in different regions of the gene.

5.2.1.2 Environmental Diversity

An alternative approach to create enzyme diversity is to screen libraries of environmental DNA (eDNA), also termed the "metagenomic approach" [57]. It is generally assumed that only 0.001 to 1% of all microorganisms are culturable, depending on their origin [58]. Therefore, rather than trying to isolate and culture microorganisms from diverse environments that may exhibit the desired enzyme activities, only their DNA is isolated, genomic libraries are created, and subsequently screened. All these procedures are carried out in high-throughput format [58–60]. Lipases and esterases have been isolated from environmental samples following this protocol and screened for hydrolytic activity against triolein or tributyrin [61] or directly amplified analyzing conserved regions using carefully designing primers [62]. A powerful example is the discovery of 200 nitrilases (approximately ten times more than the reported number of nitrilases up to the time of publication) from different biotopes around the globe by Diversa Corp. using high-throughput cloning and protein expression coupled with *in vivo* selection [63].

5.2.2 Library Screening and Selection

Following creation of the mutants, these are most often cultured in solid media or in microtiter plates. The microorganisms are then incubated until enough biomass is obtained to allow for accurate and sensitive detection of enzyme activity.

A screening assay is based on the development of a detectable signal to identify a by a catalytic activity that stands out from that of the background of clones in the This process may be directed either toward the identification and isolation of v highly active mutants from an essentially inactive background population, or tow quantitative measurement of the activity exhibited by each cell or colony for the identi of the most active mutants out of a moderately active background population [39] screening methods must be sensitive to small functional changes that arise from amino acid substitutions, and also reliable enough to distinguish one single variant thousands [64].

Ideally, the screening assay should be based on a signal generated by the inte between a chromogenic or fluorogenic substrate and the enzyme of interest, although measurements, e.g., through coupled enzymatic reactions, are also possible [39].

Last but not least, screening methods should resemble the reaction of interest as cl possible, to avoid the risk of selecting an enzyme that is very active against the sc (surrogate) substrate but not so active against the target substrate. This is often stated well-known axiom of directed evolution: *you only get what you screen for.*

In the following sections, some screening assays and selection techniques for lipa esterases are reviewed. The assays for enantioselectivity screening in principle are speci of methods given in Section 5.2.2.1 and Section 5.2.2.2. However, as the altera enantioselectivity is one of the most challenging targets in directed evolution, met determine enantioselectivity are summarized separately in Section 5.2.2.3.

5.2.2.1 Screening for Hydrolytic Activity

Hydrolytic activity of esters can be determined using a wide variety of substrate natural and some surrogate, i.e., nonnatural substrates designed to provide an detectable signal when they are converted by the enzyme. Nevertheless, not all activit are susceptible to being implemented in the high-throughput format required for the ing of the vast libraries created by the mutagenesis protocols used in directed evolut example is the simple pH-stat assay using tributyrin or triolein emulsions as sut Nevertheless, lipolytic activity can still be screened in a high-throughput format, o with triolein or tributyrin-agar through halo formation.

Colorimetric and fluorometric assays are undoubtedly the most widespread as hydrolytic activities. They involve the cleavage of an ester to yield a chromophore phore that is detected. The most commonly used chromophores/fluorophores are phenol, fluorescein, resorufin, and coumarin. Some important disadvantages of hy activity assays of lipases and esterases are the rather poor solubility in aqueous media risk of strong autohydrolysis at extreme pH or elevated temperature using chromo fluorogenic substrates. To circumvent this problem, two strategies have been describe literature. In the first, the "traditional" esters of *p*-nitrophenol or coumarin are repl the corresponding acyloxymethylethers, or diacylglycerol analogs (Figure 5.3). This the substrate much more stable, as the ester susceptible to enzymatic cleavage is se from the chromophore (or fluorophore), avoiding autohydrolysis since the alcohol n now a worse leaving group than the coumarin or the *p*-nitrophenoxide ion. Dependin particular structure, the cleaved alcohol is then directly decarboxylated or first oxidi periodate and then subjected to BSA-catalyzed β-elimination in order to release th mophore/fluorophore. Alternatively, the diols can be separated and quantified b performance liquid chromatography (HPLC), without elimination [65–67]. This meth is also applicable to the screening and characterization of enantioselective enzym Disadvantages are the need for synthesis of the specifically designed substrates a

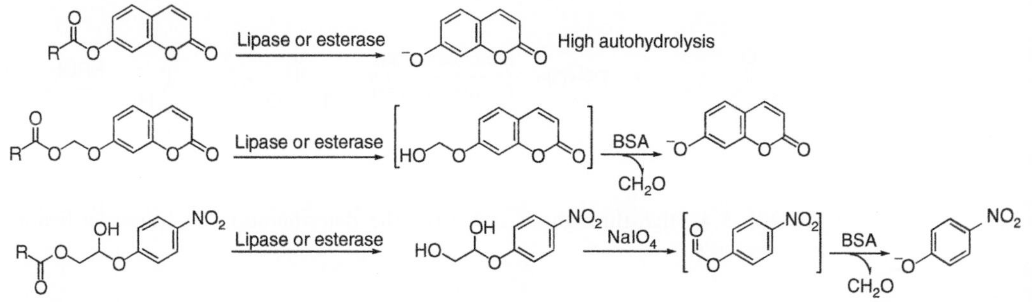

FIGURE 5.3 Fluorogenic and chromogenic substrates for lipases and esterases. The separation of the enzymatic and chromogenic reactions enables the reduction of background signal due to autohydrolysis.

only end-point measurements are possible rather than quantification of enzyme kinetics. The second strategy uses the method of back-titration with adrenaline of the sodium periodate consumed in the oxidation of the diol generated by enzymatic cleavage [69,70]. Alternatively, a high-throughput assay in solid phase was recently developed by Babiak and Reymond using esters of coumarin [71]. All these methodologies practically suppress any background reactivity of the substrates in the absence of enzyme.

A remarkably sensitive assay was recently developed by Moore et al. based on surface-enhanced resonance Raman scattering (SERRS). Although specific instrumentation for the Raman measurements is needed, the assay works in the picoliter and femtoliter range (similar to *in vivo* enzyme concentration). The principle is based on the use of a benzotriazole dye, which is masked and joined by a linker to the substrate. Lipase-catalyzed turnover of the substrate releases the dye, which complexes to dispersed silver nanoparticles and thus generates an SERRS response proportional to the enzymatic reaction. Fourteen different lipases were tested and the enantioselectivity values determined for the hydrolysis of the model compound 3-phenylbutyric acid, invariable with the dye group used. Thus, despite the use of "surrogate" substrates, not only activity but also enantioselectivity could be quantified at extremely low concentrations [72].

Very recently, an ultra-high-throughput screening method was reported by Schuster et al., based on the coexpression of pHluorin (a pH-sensitive modified green fluorescent protein) and an esterase. As the recombinant esterase catalyzes the hydrolysis of an ester, an acid is released that causes changes in the emission spectrum of pHluorin, which can be detected. However, not all the esters tested as substrate provided a significant fluorescent signal (compared to the control), and, therefore, this technique is limited to substrates not hydro-lyzed by the host cell, and which can enter the cell either alone or with the help of a cosolvent. The assay in microtiter plate format was successfully implemented to a flow cytometer, in order to separate those cells expressing esterase activity from the control cells expressing an inactive esterase [73].

5.2.2.2 Screening for Transesterification

Transesterification is usually measured by gas chromatography (GC) or HPLC, and, there-fore, is not susceptible to scaling up to high-throughput screening, although some examples in medium-throughput screening have been reported for other reactions [74].

A fluorometric method to determine transesterification by lipases and esterases in organic solvents was described by Konarzycka-Bessler and Bornscheuer [75]. Using vinyl acetate, the transesterification reaction is made irreversible as the vinyl alcohol undergoes keto-enol tautomerization to acetaldehyde, which reacts with 7-hydrazino-4-nitrobenzafurazane (NBDH) to yield a fluorescent hydrazone (Figure 5.4).

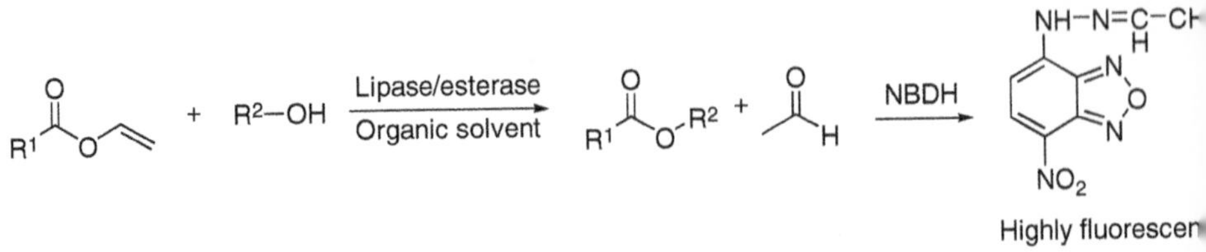

Highly fluorescen

FIGURE 5.4 High-throughput assay for the determination of transesterification activity of li and esterases.

5.2.2.3 Screening for Enantioselectivity

Although it is well known that each enantiomer of a chiral compound may exhibit diff biological activities, a number of commodity products and fine chemicals, such as c agrochemicals, flavors, fragrances, and pheromones, are still used in the form of ra mixtures. Therefore, the general trend is to replace these mixtures with single enantic that can be obtained directly, either by an asymmetric synthesis or by the resolution of racen In addition, for newly designed drugs, it may be necessary to obtain both enantiome pharmacological and toxicological studies [76]. Thus, enantioselectivity is often a property for directed evolution of enzymes involved in organic synthesis of pharmace compounds or synthesis intermediates. One of the main disadvantages of screenin enantioselectivity is that it depends on the reaction catalyzed by the enzymes, and n reactions are susceptible of being implemented in a high-throughput screening setup.

In 1997, Janes et al. described the first high-throughput screening method for ena electivity based on the separate hydrolysis of the enantiomers of p-nitrophenyl esters chiral acid. The "Quick E" method determined the ratio of initial rates for each enant corrected for competitive binding using simultaneously a nonchiral resorufin ester (Figui [77]. The Quick E method has been applied successfully in order to test the sub selectivity of a series of hydrolases [46,78,79].

A variation of the Quick E method for the determination of apparent enantiosele (E_{app}, based on the determination of rates using both enantiomers separately; E_{true}-valu based on the kinetic resolution of racemates, and, therefore, include competition o enantiomers for the active site) was developed shortly afterwards using a pH indicator with the enantiopure ester of interest. Thus, not only the need to use an unnatural ch genic substrate but also the bias introduced by the presence of a bulky chromop fluorophore moiety [80] was eliminated. This principle has also been implemented c solid phase by Copeland and Miller [81], using a fluorescent pH indicator attached to a

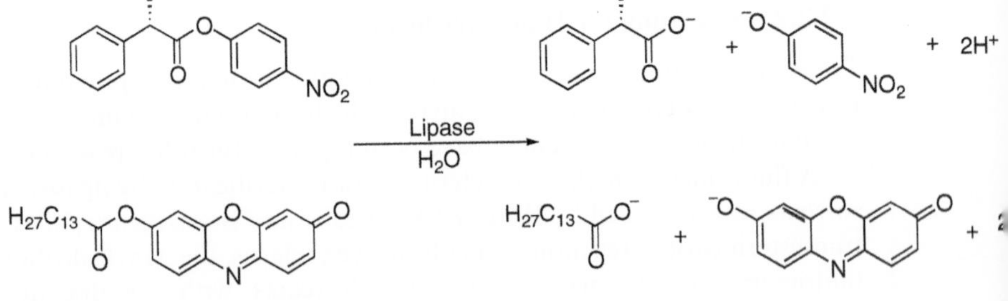

FIGURE 5.5 The "Quick E" method.

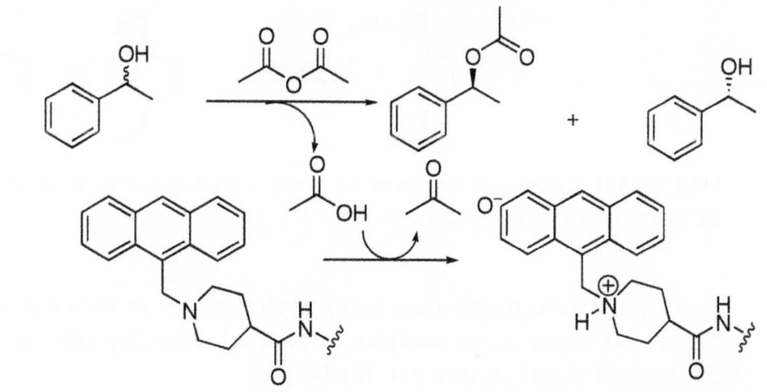

FIGURE 5.6 Screening for the enantioselective acylation of α-phenethyl alcohol with acetic anhydride, using a fluorescent indicator on solid support.

support. This assay was used to screen a catalytic peptide for the enantioselective acylation of secondary alcohols, such as α-phenethyl alcohol with acetic anhydride (Figure 5.6).

An alternative method is the commercially available "acetic acid" test (R-Biopharm GmbH, Darmstadt, Germany), which couples the hydrolysis of enantiopure chiral acetates with an acetate-dependent enzymatic cascade, leading to the formation of NADH (Figure 5.7) [82].

However, the acetic acid method can only be applied to chiral alcohols. If the chiral center is in the acyl group of an ester, alternative methods have to be used. This could be the use of *p*-nitrophenyl or resorufin esters of a chiral acid [45,83]. Nevertheless, there is a chance of selecting mutants that are optimal for the hydrolysis of these chromogenic/fluorogenic esters and not the real substrates.

All methods mentioned above make use of pure enantiomeric compounds, which are not always available. In order to use the racemate as substrate, a separation technique is needed prior to quantification of each enantiomer. This separation can be according to chirality or, in the case of isotopically labeled substrates, according to molecular mass. HPLC and GC have been adapted to high-throughput screening in order to be used [74]. With this setup, about 700 measurements were carried out per day to screen a mutant library from *P. aeruginosa* lipase toward the enantioselective esterification of 2-phenylpropanol.

Mass spectrometry (MS) is used with one isotopically labeled compound in an enantiomer pair in kinetic resolutions [84] or as shown in Figure 5.8, in the biotransformation of a *meso* compound.

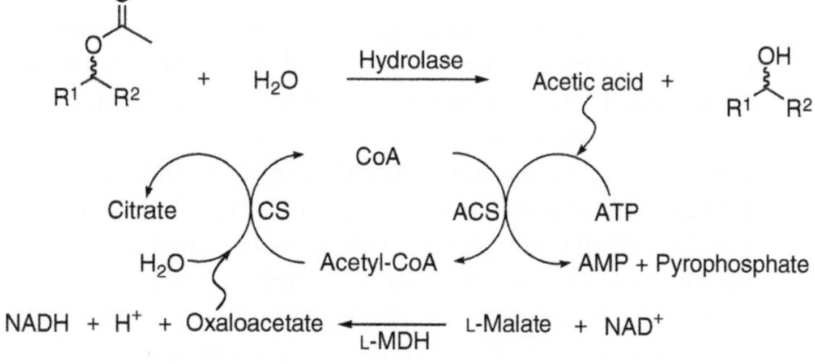

FIGURE 5.7 The "acetic acid" test.

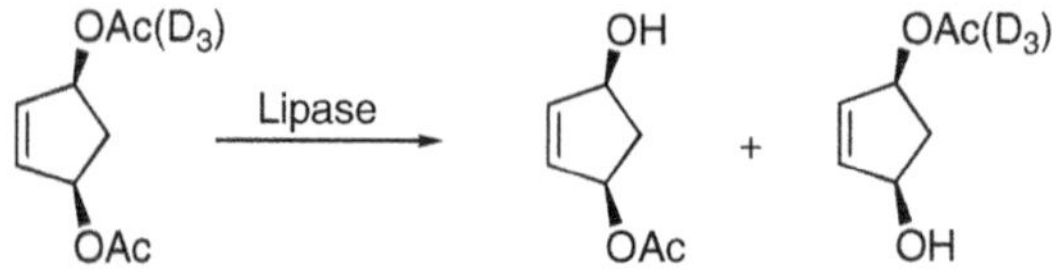

FIGURE 5.8 Kinetic resolution of an isotopically labeled *meso* compound, amenable of monite by mass spectrometry (MS).

Capillary electrophoresis using cyclodextrins in the electrolyte has also been ada process as many as 96 samples in parallel, allowing the determination of 7000 sam derivatized chiral amines per day [85].

An elegant approach involves the use of an ELISA assay in microtiter plate forma monoclonal antibodies raised against the (*S*)-enantiomer of mandelic acid, with less t cross-reactivity to the (*R*)-enantiomer [86]. Although there have not been any ex reported for lipase screening, the technique seems promising and cost-effective.

5.2.2.4   Screening and Selection in Solid Media

High-throughput screening can be performed directly on the colonies expressing variants in a solid culture (agar plate) or in membranes when cell lysis is required screening in solid and not liquid medium, the formation of insoluble, colored pro preferred, e.g., X-gal or α-naphthyl acetate and Fast Blue overlay agar assay for activity detection [87]. An example of the above-mentioned strategy is the selec mutants of *P. fluorescens* esterase I (PFEI) created with the mutator strain XL-1 the resolution of an intermediate in the synthesis of epothilones. In this case the hydro the adequate ester was evidenced by the presence of a pH indicator in the solid coupled with growth enhancement of the colonies producing an esterase with the a selectivity [44].

5.2.2.5   Selection as a Tool for Screening

Mutagenesis techniques create libraries often ranging between 10^5 and 10^{10} individ order to screen a significant amount of sequence space, the size of the library can be by discarding inactive or uninteresting mutants. Such selection can be carried out *in in vivo*.

Phage display is one of the most common techniques for *in vitro* selection of mutants in a large library (Figure 5.9). It involves cloning of the gene of interest (in t every individual in a library of mutants) in fusion with a gene encoding a coat protei virion. When the phage is assembled, the foreign protein is displayed on its surface. Th a physical linkage between gene and expression product is achieved by means of particle. The phages are then captured by affinity interaction of the displayed enzyme immobilized ligand. The nature of this binding depends on the enzyme, e.g., the tag c substrate, a suicide-substrate coupled to biotin (which is subsequently captured on vidin beads), or an immobilized transition state analog. The selected phages are replicated, and amplified by simple infection [88]. Danielsen et al. illustrated the sele Lipolase displaying phages with a biotinylated phosphorylating inhibitor that ena enrich the library by 180-fold in a single round [89].

Protein libraries can be displayed not only on the surface of bacteriophages but bacteria and yeast. Bacterial display presents certain advantages over the much mo spread phage display. First, only one host is needed to propagate the library, compar

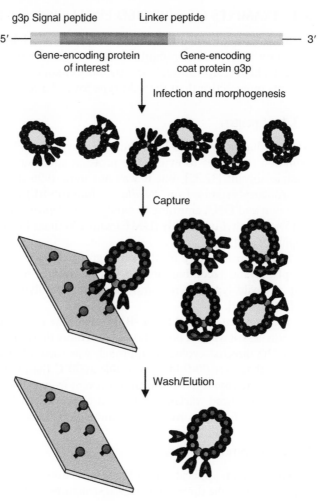

FIGURE 5.9 The procedure of phage display.

two—the bacteriophage and bacterium—in phage display. Second, the selected variants can be directly amplified without further transfer of the genetic material to another host. Third, the risk of affinity artifacts due to avidity effects might be less pronounced. Fourth, direct screening using fluorescence-activated cell sorter (FACS) is possible [90]. Bacterial display of an esterase from *B. gladioli* was achieved by Schultheiss et al. using an artificial gene composed of the esterase gene and the essential autotransporter domains in *E. coli*. The esterase activity was successfully directed to the outer-membrane fraction as confirmed by different techniques [91].

In vivo selection can be postulated when the target activity is essential for viability, and growth (e.g., overcoming increasing concentrations of antibiotics or providing an essential nutrient) selection might be applicable [92].

The principle is illustrated by Reetz and Rüggeberg with an example in which survival is coupled to the hydrolysis of a certain enantiomer that releases a growth-inhibiting compound [93]. Thus, microorganisms expressing the lipase variant with the adequate enantioselectivity will not cleave this compound, and, therefore, survive, promoting an effective enrichment of the culture in the enzyme variant with the desired enantioselectivity.

5.3   EXAMPLES OF DIRECTED EVOLUTION OF LIPASES AND ESTERASES

In this section, we will detail some examples in which lipases and esterases have been e
to obtain more stable variants under certain denaturing conditions (presence of
solvents, high temperature), variants more enantioselective in a certain reaction, and e
variants more active than the wild type toward a new substrate.

5.3.1   STABILITY

Already during 1991–1993 the pioneering work of Chen and Arnold applied basic d
evolution principles to evolve the protease subtilisin toward activity in the presence
organic solvent [32,33], but it was not until 1996 that the first esterase was evolve
p-nitrobenzyl esterase from *Bacillus subtillis* (pNBE) was altered by several rounds of r
mutation and DNA shuffling to catalyze the deprotection of the antibiotic nucleus lora
in 15% dimethylformamide (DMF) with 150 times higher activity compared to the wi
(Figure 5.10) [24].

In a separate study by the same group, pNBE was evolved by means of eight
of epPCR and recombination to yield a variant with a melting temperature 17°C
than the wild-type enzyme, and increased hydrolytic activity toward p-nitrophenyl
(pNPA) [34,37].

C. antarctica lipase B (CAL-B) is one of the most used biocatalysts in organic che
Nevertheless, it was not possible to obtain more thermostable variants by rational mutag
When the directed evolution approach was used, after two rounds of mutation by e
variants that were 20-fold more stable at 70°C than the wild type were found. Positic
and 281 were found to be critical to prevent irreversible inactivation and protein aggre
in these enhanced variants, which also proved to be more active against p-nitro
butyrate and 6,8-difluoro-4-methylumbelliferyl octanoate [35].

5.3.2   ENANTIOSELECTIVITY

One of the main applications of lipases and esterases in organic synthesis is the resolu
racemates since the three-dimensional structures of the active site and binding sites of er
provide a natural scaffold for enantiorecognition.

Shortly after the report of the directed evolution of pNBE, the first example of d
evolution of a lipase (from *P. aeruginosa*) was reported by Reetz et al. for the
resolution of the racemic ester detailed in Figure 5.11 [94].

The initial enantioselectivity factor, E, for this transformation was 1.1 (in favor
(S)-acid), and after four rounds an E of 11.3 was obtained. Further mutants were crea
combining mutations on the positions identified to be critical in the generation of tl
variants for every round, which led to the identification of a more enantioselective
($E = 21$) [95]. For the same reaction, a DNA shuffling approach proved effective, yie

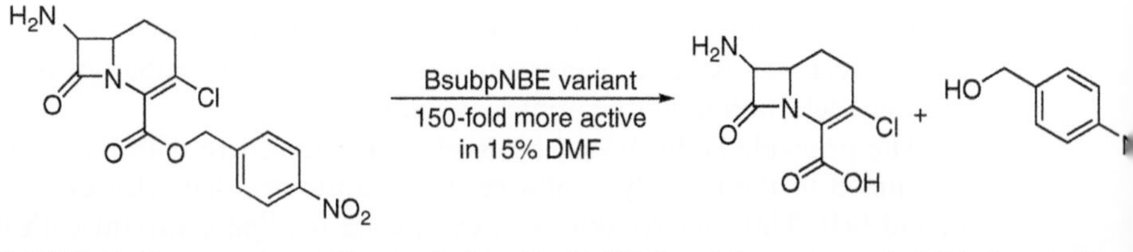

FIGURE 5.10 Deprotection of loracarbef by BsubpNBE and its variants in 15% dimethylforn
(DMF).

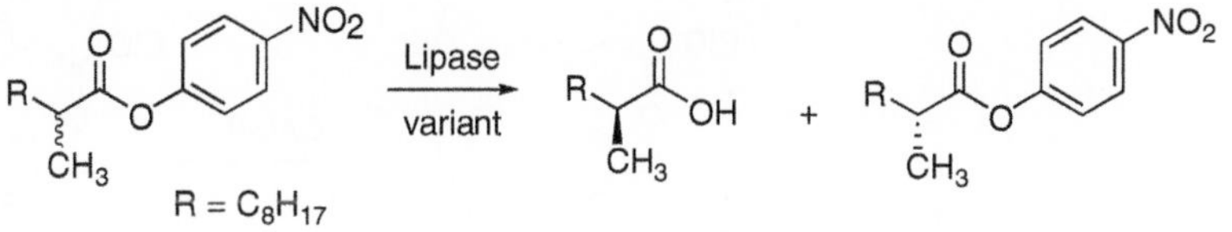

FIGURE 5.11 Resolution of a racemic ester by an evolved lipase from *Pseudomonas aeruginosa*.

variant that exhibited $E = 32$. Furthermore, a modified version of Stemmer's combinatorial multiple-cassette mutagenesis was applied to two of the obtained mutants and a mutagenic oligocassette that allowed simultaneous randomization at previously determined "hot spots." This resulted in the most enantioselective variant, displaying a selectivity factor of $E = 51$. In addition, variants with good (R)-selectivity ($E = 30$) were also identified [96].

A simple and elegant example is the generation of enantioselective mutants for the hydrolytic kinetic resolution illustrated in Figure 5.12 using the mutator strain *E. coli* XL-1 Red [44].

An attempt to rationalize the enantiopreference toward primary and secondary alcohols was developed by Kazlauskas et al. and is commonly known as the "Kazlauskas Rule" [97]. This rule is mainly based on the size of substituents attached to the stereogenic center. High E-values are achieved only when a small (i.e., methyl-) and a large (i.e., phenyl-) substituent are present. Thus, secondary alcohols with residues having only small differences in size are "difficult to resolve."

Attempts to resolve racemic mixtures of esters of secondary alcohols with mutants of the *P. fluorescens* esterase I have been reported. Henke and Bornscheuer evolved this enzyme using the mutator strain *E. coli* XL-1 Red, to develop enantiopreference in the hydrolysis of methyl 3-phenylbutyrate [45]. The same enzyme was evolved by a single round of epPCR to obtain a variant containing three point mutations that could be used in the kinetic resolution of the acetic acid 1-methyl-prop-2-ynyl ester. Subsequently, the role of each mutation on the enantioselectivity, the reaction rate, and the solubility of the mutant was studied [98].

Regarding esters of tertiary alcohols, a specific amino acid motif (GGGX) is needed in the oxyanion pocket of lipases and esterases in order for them to be accepted as substrates. Nevertheless, their natural enantioselectivity is still low but can be optimized through rational protein design, as exemplified by Henke et al. to increase the enantioselectivity of an esterase from *B. subtilis* from $E = 3$ to an E-value of 19 [99].

CAL-B was also engineered by shuffling its gene with those of lipases from *Hyphozyma* sp. CBS 648.91 and *Cryptococcus tsukubaensis* ATCC 24555 in order to create a lipase B variant with increased activity against the hydrolysis of diethyl 3-(3',4'-dichlorophenyl)gluta-rate, which yields a chiral synthon for the preparation of an NK1/NK2 dual antagonist (Figure 5.13).

The work of Koga et al. illustrates that directed evolution could also be carried out without the need for an *in vivo* expression system. Using a novel technique for the construc-tion and screening of a protein library by single-molecule DNA amplification by PCR

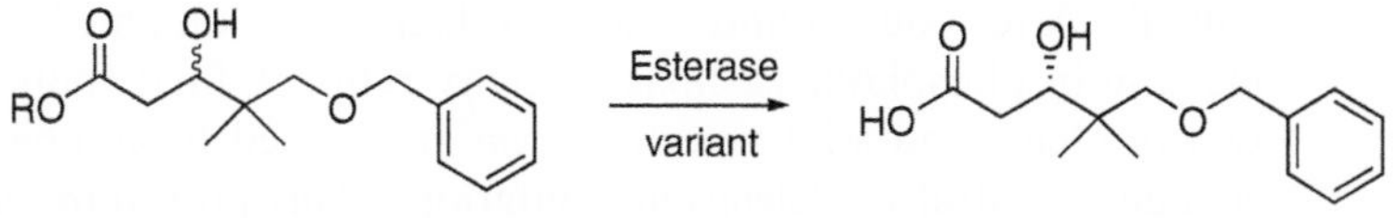

FIGURE 5.12 Kinetic resolution of a precursor ester for epothilone synthesis by *Pseudomonas fluor-escens* esterase I evolved using a mutator strain.

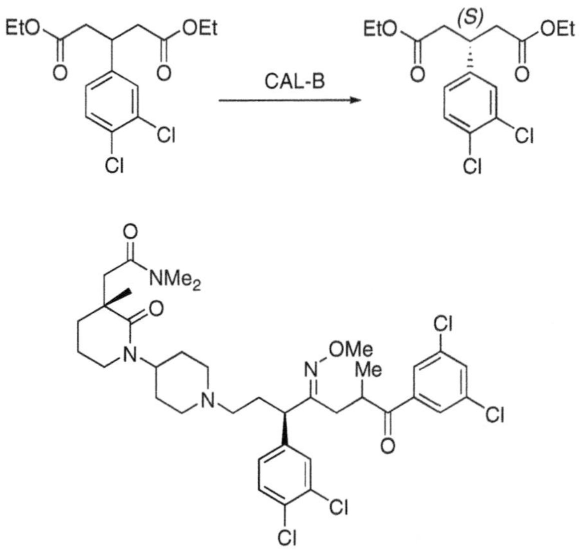

FIGURE 5.13 Hydrolysis of diethyl 3-(3′,4′-dichlorophenyl)glutarate by a lipase CAL-B vari
increased activity, to obtain a chiral synthon for the preparation of an NK1/NK2 dual antago

followed by *in vitro* coupled transcription/translation system termed single-molecul
linked *in vitro* expression (SIMPLEX), the enantioselectivity of *B. cepacia* KWI-56 lip
evolved toward (*R*)-enantioselectivity in the hydrolysis of 3-phenylbutyric acid *p*-nitr
ester. The library was generated saturating four positions (L17, F119, L167, and L26
then diluted until only five molecules of DNA were present per well in the microtite
These molecules were amplified using a single-molecule PCR product and expressed
since each gene fragment already carries a T7 promoter, a ribosome-binding site, a
terminator. The DNA corresponding to active wells showing the desired enantiosel
was once again diluted to give one molecule per well, reamplified, and rechecked. T
mutant exhibited a selectivity factor of $E = 38$ toward the (*R*)-enantiomer, whereas
type exhibited $E = 33$ for the opposite enantiomer [100].

5.3.3 CHEMOSELECTIVITY

Directed evolution can be used to alter the regiospecificity of a reaction, having the
accept other functional groups different than the one which is its natural substrate
enzymes already catalyze reactions on alternative functional groups, but at a very sl
compared to their main catalytic function. Such behavior, termed "catalytic prom
could be the basis for a directed evolution strategy aimed at altering the chemoselect
an enzyme. For further information on this concept, readers are referred to the w
Bornscheuer and Kazlauskas [101] and Aharoni et al. [102].

Phospholipase A$_1$ activity—already present in many lipases as promiscuous activit
substantially enhanced in the *Staphylococcus aureus* lipase (SAL) by sequential ro
epPCR. After four rounds, two products were obtained, displaying a 5.9- and
increase in phospholipase/lipase activity ratio. A final round of DNA shuffling wi
two products and wild-type SAL was performed to combine beneficial mutations
eliminate neutral or deleterious mutations. This procedure yielded a variant contai
amino acid mutations displaying a 11.6-fold increase in absolute phospholipase acti
a 11.5-fold increase in phospholipase/lipase ratio compared to the starting point [10

similar way, a single round of epPCR yielded a 17-fold increase in the phospholipase/lipase ratio of the thermoalkalophilic lipase of *B. thermocatenulatus* [104].

Recently, Fujii et al. reported the enhancement of amidase activity of *P. aeruginosa* lipase after one single round of random mutagenesis. Mutant libraries were screened for hydrolytic activity against oleyl-naphthylamide compared to the hydrolysis rate of the corresponding ester. Three mutational sites were identified to enhance amidase activity, and the double mutant F207S/A213D was found to have the highest amidase activity, twofold that of the wild type. These mutations were located near the calcium binding site, far from the active site [31].

The chain length selectivity of lipases can be altered by site-directed mutagenesis, such as exemplified by Joerger and Haas for the *R. oryzae* (formerly *R. delemar*) lipase (RDL) [105], and also by directed evolution [89]. While trying to isolate new enzyme variants of the extracellular lipase from *T. lanuginosa* with enhanced activity in the presence of detergent, Danielsen et al. randomized nine amino acids in two regions flanking the flexible α-helical lid. A S83T mutation was found in six of the seven most active variants, which in the homologous RDL had been proven to alter the chain length preference.

5.4 CONCLUSION

Considerable progress has been made in the past few years with respect to discovery and improvement of lipases and esterases for biocatalysis. Directed evolution and the necessary methods for library creation and high-throughput screening stimulated the research for even better hydrolases. The examples covered in this chapter demonstrate that indeed many significant achievements were made leading to enzymes with often substantially improved properties—especially with respect to enantioselectivity—further extending the possibility for applying them in biocatalytic processes.

ACKNOWLEDGMENTS

We acknowledge financial support provided through the European Community's Human Potential Programme under contract HPRN-CT2002-00239.

REFERENCES

1. Faber, K., *Biotransformations in Organic Chemistry*, 5th ed., Springer, Berlin, 2004.
2. Lloyd, R.C. et al., Use of hydrolases for the synthesis of cyclic amino acids, *Tetrahedron*, 60, 714, 2004.
3. Ollis, D. et al., The alpha/beta hydrolase fold, *Protein Eng.*, 5, 197, 1992.
4. Kazlauskas, R.J. and Bornscheuer, U.T., *Biotransformation with Lipases*, in *Biotechnology Series*, vol. 8a, Kelly, D.R., Ed., Wiley-VCH, Weinheim, Germany, 1998, p. 37.
5. Schmid, R.D. and Verger, R., Lipases: interfacial enzymes with attractive applications, *Angew. Chem. Int. Ed.*, 37, 1608, 1998.
6. Bornscheuer, U.T. and Kazlauskas, R.J., *Hydrolases in Organic Synthesis: Regio- and Stereoselective Biotransformations*, 2nd ed., Wiley-VCH, Weinheim, Germany, 2005.
7. Segura, R.L. et al., Different properties of the lipases contained in porcine pancreatic lipase extracts as enantioselective biocatalysts, *Biotechnol. Prog.*, 20, 825, 2004.
8. Segura, R.D.L. et al., Purification and identification of the enzyme responsible of most esterasic activity in commercial porcine pancreatic lipase preparations, *Enzyme Microb. Technol.*, 39, 817, 2006.
9. Ladner, W.E. and Whitesides, G., Lipase-catalyzed hydrolysis as a route to esters of chiral epoxy alcohols, *J. Am. Chem. Soc.*, 106, 7250, 1984.

10. Bornscheuer, U.T. et al., Lipases from *Rhizopus* species: genetics, structures and applica
    Protein Engineering in Industrial Biotechnology, Alberghina L., Ed., Harwood Acaden
    lishers, Amsterdam, 2000.

11. Bastida, A. et al., A single step purification, immobilization, and hyperactication of lip
    interfacial adsorption on strongly hydrophobic supports, *Biotechnol. Bioeng.*, 58, 486, 199

12. Palomo, J.M. et al., Purification, immobilization, and stabilization of a lipase from
    thermocatenulatus by interfacial adsorption on hydrophobic supports, *Biotechnol. Prog.*,
    2004.

13. Jaeger, K.-E. and Eggert, T., Lipases for biotechnology, *Curr. Opin. Biotechnol.*, 13, 390,

14. Huge-Jensen, B. et al., *Rhizomucor miehei* triglyceride lipase is processed and secrete
    transformed *Aspergillus oryzae*, *Lipids*, 24, 781, 1989.

15. Hoegh, I. et al., Two lipases from *Candida antarctica*—cloning and expression in *As*
    oryzae, *Can. J. Botan.*, 73, S869, 1995.

16. Sobek, H. and Gorisch, H., Purification and characterization of a heat stable esterase f
    themoacidophilic archaeobacterium *Sulfolobus acidocaldarius*, *Appl. Environ. Microbiol.*,
    1988.

17. Luthi, E., Jasmat, N.B., and Bergquist, P.L., Overproduction of an acetylxylan esterase f
    extreme thermophilic bacterium *Caldocellum saccharolyticum*, *Appl. Microbiol. Biotech*
    2677, 1990.

18. Nishizawa, M. et al., Stereoselective production of (+)-*trans*-chrysanthemic acid by a m
    esterase: cloning, nucleotide sequence, and overexpression of the esterase gene of *Arth*
    globiformis in *Escherichia coli*, *Appl. Environ. Microbiol.*, 61, 3208, 1995.

19. Quax, W.J. and Broekhuizen, C.P., Development of a new *Bacillus* carboxyl esterase for u
    resolution of chiral drugs, *Appl. Microbiol. Biotechnol.*, 41, 425, 1994.

20. Krebsfänger, N., Schierholz, K., and Bornscheuer, U.T., Enantioselectivity of a reco
    esterase from *Pseudomonas fluorescens* towards alcohols and carboxylic acids, *J. Biotech.*
    105, 1998.

21. Krebsfänger, N. et al., Characterization and enantioselectivity of a recombinant estera
    Pseudomonas fluorescens, *Enzyme Microb. Technol.*, 22, 641, 1998.

22. Baumann, M., Hauer, B.H., and Bornscheuer, U.T., Rapid screening of hydrolases
    enantioselective conversion of "difficult-to-resolve" substrates, *Tetrahedron Asymme*
    4781, 2000.

23. Hou, C.T., Screening of microbial esterases for asymmetric hydrolysis of 2-ethylhexylb
    J. Ind. Microbiol., 11, 73, 1993.

24. Moore, J.C. and Arnold, F.H., Directed evolution of a *para*-nitrobenzyl esterase for a
    organic solvents, *Nat. Biotechnol.*, 14, 458, 1996.

25. Record, E. et al., Overproduction of the *Aspergillus niger* feruloyl esterase for pulp b
    application, *Appl. Microbiol. Biotechnol.*, 62, 349, 2003.

26. Priefert, H., Rabenhorst, J., and Steinbüchel, A., Biotechnological production of vanilli
    Microbiol. Biotechnol., 56, 296, 2001.

27. Cheetham, P.S.J., Gradley, M.L., and Sime, J.T., Flavour/aroma materials and their prep
    PCT Patent, WO0050622, 2000.

28. Arpigny, J.L. and Jaeger, K.-E., Bacterial lipolytic enzymes: classification and properties, *B*
    J., 343, 177, 1999.

29. Bornscheuer, U.T., Microbial carboxyl esterases: classification, properties and applica
    biocatalysis, *FEMS Microbiol. Rev.*, 26, 73, 2002.

30. Eijsink, V.G.H. et al., Directed evolution of enzyme stability, *Biomol. Eng.*, 22, 21, 2005.

31. Fujii, R. et al., Directed evolution of *Pseudomonas aeruginosa* lipase for improved amide
    lyzing activity, *Prot. Eng. Des. Sel.*, 18, 93, 2005.

32. You, L. and Arnold, F.H., Directed evolution of subtilisin E in *Bacillus subtilis* to enhan
    activity in aqueous dimethylformamide, *Protein Eng.*, 9, 77, 1993.

33. Chen, K. and Arnold, F.H., Tuning the activity of an enzyme for unusual environments: se
    random mutagenesis of subtilisin E for catalysis in dimethylformamide, *Proc. Natl. Acad. S*
    90, 5618, 1993.

34. Gershenson, A. et al., Tryptophan phosphorescence study of enzyme flexibility and unfolding in laboratory-evolved thermostable esterases, *Biochemistry*, 39, 4658, 2000.
35. Zhang, N. et al., Improving tolerance of *Candida antarctica* lipase B towards irreversible thermal inactivation through directed evolution, *Protein Eng.*, 16, 599, 2003.
36. Spiller, B. et al., A structural view of evolutionary divergence, *Proc. Natl. Acad. Sci. USA*, 96, 12305, 1999.
37. Giver, L. et al., Directed evolution of a thermostable esterase, *Proc. Natl. Acad. Sci. USA*, 95, 12809, 1998.
38. Abián, O. et al., Stabilization of penicillin G acylase from *Escherichia coli*: site-directed mutagenesis of the protein surface to increase multipoint covalent attachment, *Appl. Environ. Microbiol.*, 70, 1249, 2004.
39. Cohen, N. et al., *In vitro* enzyme evolution: the screening challenge of isolating the one in a million, *Trends Biotechnol.*, 19, 507, 2001.
40. Neylon, C., Chemical and biochemical strategies for the randomization of protein encoding DNA sequences: library construction methods for directed evolution, *Nucleic Acids Res.*, 32, 1448, 2004.
41. Kurtzman, A.L. et al., Advances in directed protein evolution by recursive genetic recombination: applications to therapeutic proteins, *Curr. Opin. Biotechnol.*, 12, 361, 2001.
42. Bichet, A. et al., The "Bringer" strategy: a very fast and highly efficient method for construction of mutant libraries by error-prone polymerase chain reaction of ring-closed plasmids, *Appl. Biochem. Biotechnol.*, 117, 115, 2004.
43. Wong, T.S. et al., Sequence saturation mutagenesis (SeSaM): a novel method for directed evolution, *Nucleic Acids Res.*, 32, e26, 2004.
44. Bornscheuer, U.T., Altenbuchner, J., and Meyer, H.H., Directed evolution of an esterase for the stereoselective resolution of a key intermediate in the synthesis of epothilones, *Biotechnol. Bioeng.*, 58, 554, 1998.
45. Henke, E. and Bornscheuer, U.T., Directed evolution of an esterase from *Pseudomonas fluorescens*: random mutagenesis by error-prone PCR or a mutator strain and identification of mutants showing enhanced enantioselectivity by a resorufin-based fluorescence assay, *Biol. Chem.*, 380, 1029, 1999.
46. Horsman, G.P. et al., Mutations in distant residues moderately increase the enantioselectivity of *Pseudomonas fluorescens* esterase towards methyl 3-bromo-2-methyl propanoate and ethyl 3-phenylbutyrate, *Chem. Eur. J.*, 9, 1933, 2003.
47. Reetz, M.T. et al., Expanding the range of substrate acceptance of enzymes: combinatorial active-site saturation test, *Angew. Chem. Int. Ed.*, 44, 4192, 2005.
48. Stemmer, W.P.C., Rapid evolution of a protein by *in vitro* DNA shuffling, *Nat. Biotechnol.*, 370, 389, 1994.
49. Joern, J.M., Meinhold, P., and Arnold, F.H., Analysis of shuffled gene libraries, *J. Mol. Biol.*, 316, 643, 2002.
50. Zhao, H., Molecular evolution by staggered extension process (StEP) *in vitro* recombination, *Nat. Biotechnol.*, 16, 258, 1998.
51. Coco, W.M. et al., DNA shuffling methods for generating highly recombined genes and evolved enzymes, *Nat. Biotechnol.*, 19, 354, 2001.
52. Abecassis, V., Pompon, D., and Truan, G., High efficiency family shuffling based on multi-step PCR and *in vivo* DNA recombination in yeast: statistical and functional analysis of a combinatorial library between human cytochrome P450 1A1 and 1A2, *Nucleic Acids Res.*, 28, e88, 2000.
53. Bergquist, P.L., Reeves, R.A., and Gibbs, M.D., Degenerate oligonucleotide gene shuffling (DOGS) and random drift mutagenesis (RNDM): two complementary techniques for enzyme evolution, *Biomol. Eng.*, 22, 63, 2005.
54. Short, J.M., Synthetic ligation reassembly in directed evolution, U.S. Patent, US6537776, 2003.
55. Bittker, J.A. et al., Directed evolution of protein enzymes using nonhomologous random recombination, *Proc. Natl. Acad. Sci. USA*, 101, 7011, 2004.
56. Lutz, S., Ostermeier, M., and Benkovic, S.J., Rapid generation of incremental truncation libraries for protein engineering using alpha phosphothioate nucleotides, *Nucleic Acids Res.*, 29, e16, 2001.
57. Lorenz, P. and Eck, J., Metagenomics and industrial applications, *Nature*, 3, 510, 2005.

58. Lorenz, P. et al., The impact of non-cultivated biodiversity on enzyme discovery and ev *Biocat. Biotransform.*, 21, 87, 2003.

59. Uchiyama, T. et al., Substrate-induced gene-expression screening of environmental meta libraries for isolation of catabolic genes, *Nat. Biotechnol.*, 23, 88, 2005.

60. Handelsman, J., Sorting out metagenomes, *Nat. Biotechnol.*, 23, 38, 2005.

61. Henne, A. et al., Screening of environmental DNA libraries for the presence of genes cc lipolytic activity on *Escherichia coli*, *Appl. Environ. Microbiol.*, 66, 3113, 2000.

62. Bell, P.J.L. et al., Prospecting for novel lipase genes using PCR, *Microbiology*, 148, 2283,

63. Robertson, D.E. et al., Exploring nitrilase sequence space for enantioselective catalys *Environ. Microbiol.*, 70, 2429, 2004.

64. Arnold, F.H. et al., How enzymes adapt: lessons from directed evolution, *Trends Biochem.* 100, 2001.

65. Leroy, E., Bensel, N., and Reymond, J.-L., A low background high-throughput screenin fluorescence assay for lipases and esterases using acyloxymethylethers of umbelliferone, *Med. Chem. Lett.*, 13, 2105, 2003.

66. Klein, G. and Reymond, J.-L., An enantioselective fluorimetric assay for alcohol dehydr using albumin-catalyzed β-elimination of umbelliferone, *Bioorg. Med. Chem. Lett.*, 8, 11

67. Grognux, J. et al., Universal chromogenic substrates for lipases and esterases, *Tetrahedro metry*, 15, 2981, 2004.

68. Grognux, J. and Reymond, J.-L., Classifying enzymes from selectivity fingerprints, *Ch chem.*, 5, 826, 2004.

69. Wahler, D. et al., Adrenaline profiling of lipases and esterases with 1,2-diol and carbc acetates, *Tetrahedron*, 60, 703, 2004.

70. Wahler, D. and Reymond, J.-L., The adrenaline test for enzymes, *Angew. Chem. Int. Ed.*, ⁴ 2002.

71. Babiak, P. and Reymond, J.L., A high-throughput, low-volume enzyme assay on solid *Anal. Chem.*, 77, 373, 2005.

72. Moore, B.D. et al., Rapid and ultra-sensitive determination of enzyme activities using enhanced resonance Raman scattering, *Nat. Biotechnol.*, 22, 1133, 2004.

73. Schuster, S. et al., pHluorin-based *in vivo* assay for hydrolase screening, *Anal. Chem.*, 7 2005.

74. Reetz, M.T. et al., A GC-based method for high-throughput screening of enantioselecti lysts, *Catal. Today*, 67, 389, 2001.

75. Konarzycka-Bessler, M. and Bornscheuer, U.T., A high-throughput-screening method fc mining the synthetic activity of hydrolases, *Angew. Chem. Int. Ed.*, 42, 1418, 2003.

76. Murer, P. et al., Combinatorial "library on bead" approach to polymeric materials wit enhanced chiral recognition, *Chem. Commun.*, 2559, 1998.

77. Janes, L.E., Kazlauskas, R.J., and Quick, E., A fast spectroscopic method to mea enantioselectivity of hydrolases, *J. Org. Chem.*, 62, 4560, 1997.

78. Liu, A.M.F. et al., Mapping the substrate selectivity of new hydrolases using colorimetri ing: lipases from *Bacillus thermocatenulatus* and *Phiostoma piliferum*, esterases from *Pseue fluorescens* and *Streptomyces diastatochromogenes*, *Tetrahedron Asymmetry*, 12, 545, 200

79. Somers, N.A. and Kazlauskas, R.J., Mapping the substrate selectivity and enantioselec esterases from thermophiles, *Tetrahedron Asymmetry*, 15, 2991, 2004.

80. Janes, L.E., Löwendahl, C., and Kazlauskas, R.J., Quantitative screening of hydrolase using pH indicators: identifying active and enantioselective hydrolases, *Chem. Eur. J.*, 1998.

81. Copeland, G.T. and Miller, S.J., Selection of enantioselective acyl transfer catalysts from ʒ peptide library through a fluorescence-based activity assay: an approach to kinetic resol secondary alcohols of broad structural scope, *J. Am. Chem. Soc.*, 123, 6496, 2001.

82. Baumann, M., Stürmer, R., and Bornscheuer, U.T., A high-throughput-screening methoc identification of active and enantioselective hydrolases, *Angew. Chem. Int. Ed.*, 40, 4201,

83. Reetz, M.T. et al., Creation of enantioselective biocatalysts for organic chemistry by evolution, *Angew. Chem. Int. Ed.*, 36, 2830, 1997.

84. Reetz, M.T. et al., A method for high-throughput screening of enantioselective catalysts, *Angew. Chem. Int. Ed.*, 38, 1758, 1999.
85. Reetz, M.T. et al., Super-high-throughput screening of enantioselective catalysts by using capillary array electrophoresis, *Angew. Chem. Int. Ed.*, 39, 3891, 2000.
86. Taran, F. et al., High-throughput screening of enantioselective catalysts by immunoassay, *Angew. Chem. Int. Ed.*, 41, 124, 2002.
87. Higerd, T.B. and Spizizen, J., Isolation of two acetyl esterases from extracts of *Bacillus subtilis*, *J. Bacteriol.*, 114, 1184, 1973.
88. Soumillon, P., Selection of phage-displayed enzymes, in *Evolutionary Methods in Biotechnology: Clever Tricks for Directed Evolution*, Brakmann, S. and Schwienhorst, A., Eds., Wiley-VCH, Weinheim, Germany, 2004.
89. Danielsen, S. et al., *In vitro* selection of enzymatically active lipase variants from phage libraries using a mechanism-based inhibitor, *Gene*, 272, 267, 2001.
90. Samuelson, P. et al., Display of proteins on bacteria, *J. Biotechnol.*, 96, 129, 2002.
91. Schultheiss, E. et al., Functional esterase surface display by the autotransporter pathway in *Escherichia coli*, *J. Mol. Catal. B: Enzym.*, 18, 89, 2002.
92. Lorenz, P. and Eck, J., Screening for novel industrial biocatalysts, *Eng. Life Sci.*, 4, 501, 2004.
93. Reetz, M.T. and Rüggeberg, C.J., A screening system for enantioselective enzymes based on differential cell growth, *Chem. Commun.*, 7, 1428, 2002.
94. Reetz, M.T., An overview of high-throughput screening systems for enantioselective enzymatic transformations, in *Directed Enzyme Evolution: Screening and Selection Methods*, Arnold, F.H. and Georgiou, G., Eds., Humana Press, Totowa, NJ, p. 259.
95. Liebeton, K. et al., Directed evolution of an enantioselective lipase, *Chem. Biol.*, 7, 709, 2000.
96. Reetz, M.T. et al., Directed evolution of an enantioselective enzyme through combinatorial multiple-cassette mutagenesis, *Angew. Chem. Int. Ed.*, 40, 3589, 2001.
97. Kazlauskas, R.J. et al., A rule to predict which enantiomer of a secondary alcohol reacts faster in reactions catalyzed by cholesterol esterase, lipase from *Pseudomonas cepacia*, and lipase from *Candida rugosa*, *J. Org. Chem.*, 56, 2656, 1991.
98. Schmidt, M. et al., Directed evolution of an esterase from *Pseudomonas fluorescens* yields a mutant with excellent enantioselectivity and activity for the kinetic resolution of a chiral building block. *ChemBioChem*, 7, 805–809, 2006.
99. Henke, E. et al., A molecular mechanism of enantiorecognition of tertiary alcohols by carboxylesterases, *ChemBioChem*, 4, 485, 2003.
100. Koga, Y. et al., Inverting enantioselectivity of *Burkholderia cepacia* KWI-56 lipase by combinatorial mutation and high-throughput screening using single-molecule PCR and *in vitro* expression, *J. Mol. Biol.*, 331, 585, 2003.
101. Bornscheuer, U.T. and Kazlauskas, R.J., Catalytic promiscuity in biocatalysis: using old enzymes to form new bonds and follow new pathways, *Angew. Chem. Int. Ed.*, 43, 6032, 2004.
102. Aharoni, A. et al., The "evolvability" of promiscuous protein functions, *Nat. Genet.*, 37, 73, 2004.
103. van Kampen, M.D. and Egmond, M.R., Directed evolution: from a staphylococcal lipase to a phospholipase, *Eur. J. Lipid Sci. Technol.*, 102, 717, 2000.
104. Kauffmann, I. and Schmidt-Dannert, C., Conversion of *Bacillus thermocatenulatus* lipase into an efficient phospholipase with increased activity towards long-chain fatty acyl substrates by directed evolution and rational design, *Protein Eng.*, 14, 919, 2001.
105. Joerger, R.D. and Haas, M.J., Alteration of chain length selectivity of a *Rhizopus delemar* lipase through site-directed mutagenesis, *Lipids*, 29, 377, 1994.

6 Flavin-Containing Oxidative Biocatalysts

Marco W. Fraaije and Willem J.H. van Berkel

CONTENTS

6.1 INTRODUCTION

Flavoenzymes are able to catalyze a remarkably wide variety of oxidative reactions such as regio- and enantioselective monooxygenations and highly regiospecific oxidations. Such reactions are often difficult, if not impossible, to be achieved using chemical approaches. Due to their regio- and enantioselectivity and catalytic efficiency, these enzymes have been shown to be highly valuable biocatalysts for the synthesis of a variety of fine chemicals. Until a decade ago, only a scarce number of flavin-dependent enzymes had been cloned and over-expressed, which limited the number of flavoenzyme-based biocatalytic applications. However, in recent years a large number of novel flavin-containing biocatalysts have been discovered and new biocatalytic processes have been developed based on oxidative flavoenzymes.

6.2 GENERAL PROPERTIES OF FLAVIN-CONTAINING OXIDATIVE BIOCATALYSTS

In nature, many oxidative reactions are carried out by flavin-dependent oxidoreductases [1]. These enzymes typically harbour a flavin mononucleotide (FMN) or flavin adenine dinucleotide (FAD) molecule as organic cofactor (Figure 6.1). By varying the direct protein environment around the flavin, thereby modulating the redox properties of the organic cofactor, evolution has created an immense set of redox and nonredox active flavoenzymes. Well-known examples are flavin-containing dehydrogenases and oxidases that are able to oxidize organic substrates.

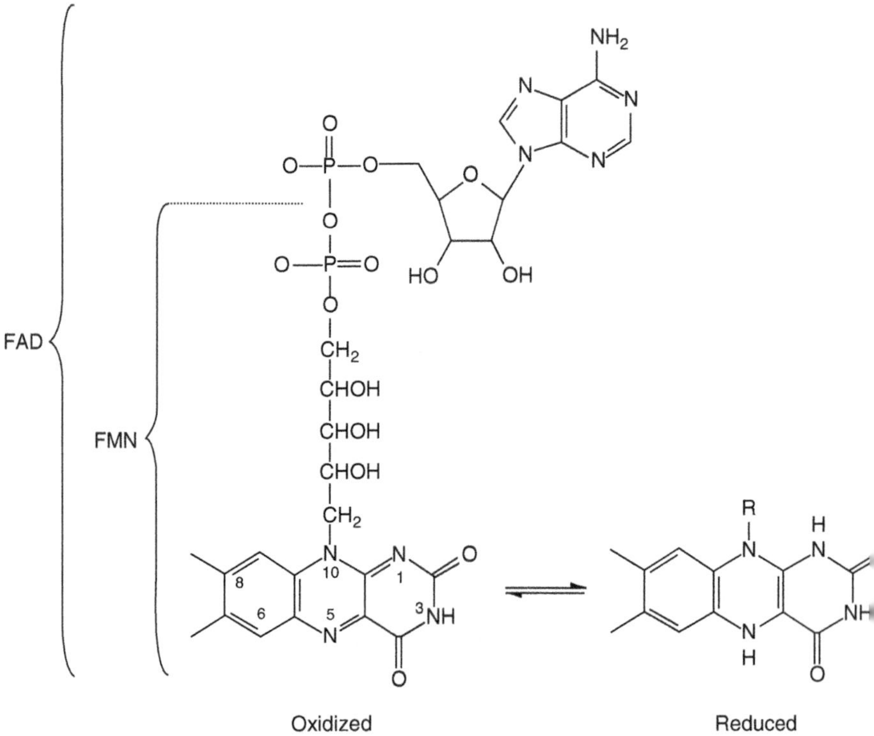

FIGURE 6.1 Structural formulas of the flavin cofactors flavin adenine dinucleotide (FAD) as mononucleotide (FMN).

Electron transfer is also mediated by flavoproteins and light production can be cata flavoenzymes. A large number of flavoenzymes are also able to use molecular ox substrate, enabling oxyfunctionalization of hydrocarbons through, for example, Villiger oxidations or epoxidations. In fact, flavins are the only organic cofactors able to utilize molecular oxygen for oxygenation reactions. Other known oxygenating c always depend on a metal ion for their reactivity. Studies on oxidative flavoenzymes du past decades have shown that these enzymes display an astonishing variability in the reactions catalyzed (Figure 6.2). Such a flexibility of using molecular oxygen for differe of reactions while retaining a high degree of enantio- and/or regioselectivity has so far encountered in any other enzyme family [2]. The flexibility in reactivity is also reflecte diverse physiological processes in which these enzymes are involved: regulating protein [3], xenobiotics metabolism in humans [4], biosynthesis of toxins [5], drug activation [growth regulation [7], and pollutant degradation by microorganisms [8].

Enantio- and regioselective oxygenations and oxidations are difficult to achieve b ical means, while these reaction types can lead to valuable fine chemicals. Chemical s routes that catalyze selective oxygen insertion or oxidation reactions often involve tedi costly blocking and deblocking steps and are catalyzed by heavy metals. Therefore dependent monooxygenases and oxidases represent highly attractive alternatives a biocatalysts are able to catalyze a huge variety of monooxygenation and oxidation re while exhibiting a remarkable selectivity. Exploitation of these biocatalysts affords e environment-friendly synthesis routes ("green chemistry") that have several adv

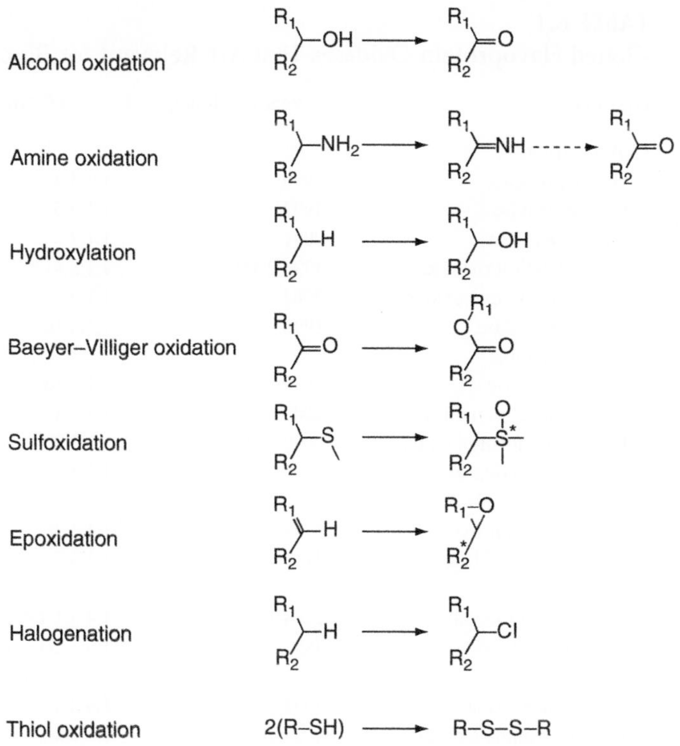

FIGURE 6.2 Example reaction types catalyzed by flavin-dependent oxidoreductases.

over chemical routes. This includes the renewability of the catalyst, reduced formation of by-products thanks to fewer side reactions, and the possibility to use mild process conditions. Except for synthesis purposes, flavin-containing enzymes can also be applied in sensing applications.

This review will give an overview on the availability and biocatalytic potential of several types of flavoenzymes of two classes: oxidases and monooxygenases.

6.3 FLAVOPROTEIN OXIDASES

Flavoprotein oxidases are enzymes that have the potential to be highly valuable for biotechnological processes as they are able to use molecular oxygen as a clean and cheap oxidant while they do not require expensive coenzymes. An increasing number of oxidases have been identified, cloned, and overexpressed in the past few years (see Table 6.1). The discovery of these novel oxidases has also revealed a number of new oxidase-linked reactivities. Not only are relatively simple alcohol or amine oxidations carried out by flavin-containing oxidases, but more chemically demanding reactions are also mediated by these enzymes. For example, vanillyl-alcohol oxidase is able to cleave an ether bond [9], while prenylcysteine oxidase (EC 1.8.3.5) is able to break a thioether bond [10]. Another striking example of the delicate catalytic power of flavoprotein oxidases is the plant enzyme (*S*)-reticuline oxidase (EC 1.21.3.3), which catalyzes the regio- and stereospecific oxidative cyclization of benzophenanthrodine alkaloids [11]. Also the biosynthesis of cannabinoids was recently shown to depend on a similar plant oxidase [12]. The common denominator for oxidases is the fact that upon

TABLE 6.1
Cloned Flavoprotein Oxidases That Are Relevant for Biocatalysis

Enzymes	Year of Cloning	EC Number	Structure/Cofactor
Carbohydrates			
Glucose oxidase	1990	1.1.3.4	+/−
Hexose oxidase	1997	1.1.3.5	−/+
Lactose oxidase	2001	1.1.3.x	−/+
Sorbitol/xylitol oxidase	1998/2000	1.1.3.41	−/+
D-Gluconolactone oxidase	2004	1.1.3.x	−/+
Pyranose oxidase	1996	1.1.3.10	+/+
Aliphatic alcohols			
Alcohol oxidase	1985	1.1.3.13	−/−
Isoamyl alcohol oxidase	2000	1.1.3.x	−/+
Long-chain alcohol oxidase	2000/2004	1.1.3.20	−/−
Cholesterol oxidase	1986	1.1.3.6	+/−, +
Aromatic alcohols			
Vanillyl-alcohol oxidase	1998	1.1.3.38	+/+
Aryl-alcohol oxidase	1999	1.1.3.7	−/−
Amines			
D-Amino acid oxidase	2004/1997/2002	1.4.3.1/1.4.3.3 1.4.3.19	+/−
L-Amino acid oxidase	1990/2003/2003/1988	1.4.3.2/1.4.3.11 1.4.3.14/1.4.3.16	+/−
Monoamine oxidase	1995	1.4.3.4	+/+
Fructosyl amine oxidase	1997/2002/2005	1.5.3.x	−/+
Others			
Sulfhydryl oxidase	2000	1.8.3.2	+/−
NADH oxidase (H$_2$O forming)	1982	1.6.3.x	−/−

a, algae; b, bacterial; f, fungal; y, yeast; p, plant; +, crystal structure; −, no crystal structure; +, covalent fl
noncovalent flavin.

oxidation of the organic substrate, the reduced flavin cofactor transfers the electr
molecular oxygen, yielding hydrogen peroxide:

$$S + O_2 \rightarrow P + H_2O_2$$

An increasing number of oxidases is produced and successfully applied for industri
poses. In the past, only three specific oxidases have been extensively used: glucose o
(e.g., for biosensing and gluconic acid production), cholesterol oxidase (e.g., for
assays), and D-amino acid oxidase (e.g., for synthesis of antibiotics). The biocatalytic e
ation of oxidases was mainly restricted by availability. Only in recent years has the nur
available (recombinant) oxidases increased significantly.

Many of the newly discovered oxidases contain a covalently bound FAD as c
(Table 6.1). The physiological rationale for covalent cofactor binding is still not well
stood. Nevertheless, the advantage of covalent coupling of a cofactor is multifold. One
and obvious advantage is that cofactor dissociation is prevented, which would lead to
activity. In fact, efficient commercial usage of D-amino acid oxidase and glucose oxi
frequently hampered by the fact that these enzymes can lose the flavin cofactor unde
conditions. Except for saturating the enzyme active site, formation of a covalent bo

also have beneficial effects on the reactivity of the cofactor. For vanillyl-alcohol oxidase it was shown that covalent flavinylation is an autocatalytic process [13,14], and that the covalent histidyl–FAD bond increases the redox potential of the flavin cofactor, resulting in a tenfold increase in enzyme activity [15]. Furthermore, covalent linkage may well add significantly to the protein stability as the additional covalent bond has a similar stabilizing effect to that found in intramolecular disulfide bridges [16].

Another striking observation is the fact that most discovered oxidases are of fungal origin. Only a limited number of oxidases have been found in bacteria while archaebacterial representatives are really scarce. The sporadic occurrence of oxidases in (archae)bacterial genomes prevents effective mining of sequenced genomes. The low abundance of oxidases in nature might reflect that they are energy-wasting enzymes as they shuttle the generated electrons directly to molecular oxygen without exploiting the respiratory chain. As a result, most oxidoreductases in bacteria have evolved in such a way that they are able to utilize alternative electron acceptors. In fact, it is thought that many fungal oxidases serve a competitive role, instead of being part of catabolic routes. By producing hydrogen peroxide from molecular oxygen, microbes competing for the same nutrients are eliminated. Furthermore, several fungal oxidases are involved in an extracellular cascade catalytic machinery, producing hydrogen peroxide needed for the degradation of lignin by peroxidases. The low abundance of oxidases in archaebacteria appears logical as these microbes typically grow in anoxic conditions. In fact, a known archaeal oxidase, NADH oxidase, appears to fulfill a detoxifying role by scavenging molecular oxygen at the expense of NADH. This oxidase might have biocatalytic relevance as it can be used to regenerate NAD^+ from NADH [17].

6.4 CARBOHYDRATE OXIDASES

Glucose oxidase from *Aspergillus niger* is the most widely employed oxidase. Since decades it has been used for biosensing, bleaching, bread improvement, and as oxygen scavenger [18]. A typical biocatalytic application is the use of (isolated) enzyme to produce gluconic acid. Illustrative for its wide applicability is the number of patents, which include the use of glucose oxidase, which is well above 5000. As glucose oxidase has been the subject of biochemical studies for a long time, many of its molecular properties are well described, including its crystal structure (Figure 6.3) [19,20] and catalytic mechanism [21].

Glucose oxidase from *A. niger* is a homodimeric FAD-containing glycoprotein. It catalyzes the oxidation of β-D-glucose by molecular oxygen to D-glucono-1,5-lactone, which subsequently hydrolyzes spontaneously to gluconic acid. The enzyme obeys a ping-pong bi-bi kinetic mechanism, which implies that the glucono-δ-lactone product generated in the reductive half-reaction dissociates before the reduced enzyme reacts with oxygen to regenerate the oxidized enzyme. It has been proposed that β-D-glucose is oxidized by a concerted transfer of a proton from its C1-hydroxyl to an active site histidine and a direct hydride transfer from its C1 position to flavin-N5 [22].

Glucose oxidase has a narrow specificity with electron-donor substrates. Numerous derivatives of β-D-glucose are converted, but the physiological substrate is by far the best. In contrast, a large number of substrates can replace molecular oxygen in the oxidative half-reaction. In addition to benzoquinones and naphthaquinones, which act as efficient two-electron acceptors, the enzyme is active with many one-electron oxidants including metal–ion complexes. These mediators are widely applied in glucose oxidase electrodes [21].

Only recently several alternative flavin-dependent oxidases acting on carbohydrates have been discovered. One of these oxidases, hexose oxidase from seaweed, is presently produced at a large scale by Danisco and has shown to be a valuable biocatalyst. Production of this

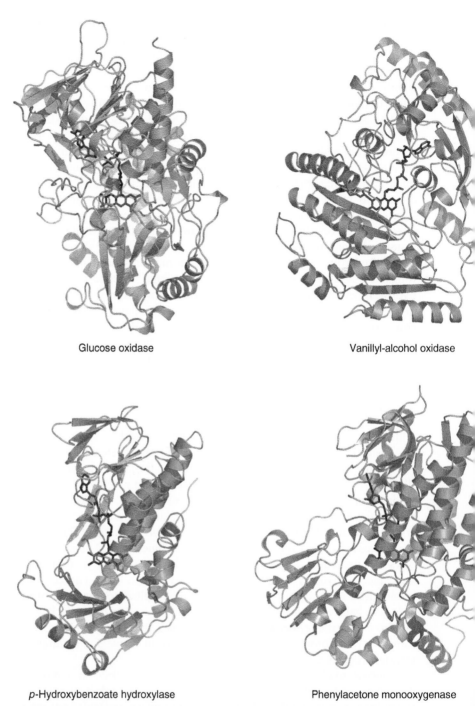

Glucose oxidase

Vanillyl-alcohol oxidase

p-Hydroxybenzoate hydroxylase

Phenylacetone monooxygenase

FIGURE 6.3 Crystal structures of glucose oxidase, vanillyl-alcohol oxidase, p-hydroxybenzoa
xylase, and phenylacetone monooxygenase (PDB files 1GAL, 1VAO, 1PHH, and 1W4X, resp
The flavin cofactor is shown in black and is shown in a similar orientation in each figure. No
vanillyl-alcohol oxidase also the covalent histidyl linkage is shown.

oxidase was only feasible after developing an efficient microbial expression system for this plant enzyme [23]. Hexose oxidase is able to oxidize a variety of carbohydrates including D-glucose, D-galactose, maltose, cellobiose, and lactose. By this, it contrasts with the restricted substrate specificity of glucose oxidase. Another striking difference is the mode of flavin cofactor binding. While in glucose oxidase the FAD cofactor is dissociable, hexose oxidase contains a covalently bound FAD cofactor, attached to a histidine residue [23].

Recently, yet another carbohydrate oxidase has become available for biocatalytic applications: pyranose oxidase. This oxidase originates from fungi and, like hexose oxidase, contains a covalently bound FAD cofactor. It has been shown that this fungal enzyme can be overexpressed in *Escherichia coli* [24]. In 2004 the crystal structure of pyranose oxidase was elucidated showing some structural resemblance with glucose oxidase [25,26]. Nevertheless, the active sites of these fungal oxidases differ to a great extent. This agrees with the fact that pyranose oxidase selectively oxidizes D-glucose and related carbohydrates at the C2 position, while glucose oxidase and hexose oxidase oxidize exclusively the C1-hydroxyl moiety. As a consequence, a totally novel product spectrum can be obtained using pyranose oxidase. For example, the oxidation of D-glucose by pyranose oxidase yields 2-keto-D-glucose, which can serve as precursor for the synthesis of, for example, D-fructose [27] or the antibiotic cortalcerone [28]. To get an impression of the biocatalytic potential of pyranose oxidase, the reader is referred to a review by Giffhorn [29]. The thermostability and catalytic efficiency of *Peniophora gigantea* pyranose oxidase has been improved by a dual site-directed mutagenesis and molecular evolution approach [30].

The above-mentioned carbohydrate oxidases preferably act on monosaccharides. Recent reports on newly discovered oxidases have changed this situation. A fungal lactose oxidase and some plant oxidases have been found to accept di- and polysaccharides [31,32]. Lactose oxidase catalyzes the production of lactobionic acid, which can be used as food additive and in a range of other applications [33]. In addition, bacterial oxidases acting on alditols (xylitol and sorbitol oxidase) and a fungal oxidase acting on aldonolactones (D-gluconolactone oxidase) were identified and expressed in recent years [34–36].

6.5 ALCOHOL OXIDASES

Several flavoprotein oxidases have been described that are active with short-chain aliphatic alcohols. The most well-known enzymes are the methanol oxidases from yeast [37]. For oxidation of long-chain alcohols (C_4–C_{22}), other yeast and also plant oxidases have been identified and expressed as recombinant enzymes [38–40]. Recently, a fungal oxidase has been described that is primarily active on a branched-chain alcohol: isoamyl alcohol [41].

FMN-dependent alcohol oxidases that are active with 2-hydroxy acids are also widespread [42,43]. Like other secondary alcohol oxidases [44], these enzymes can be applied in a coupled enzyme approach for the development of deracemization processes. Thus, glycolate oxidase from spinach was used for the enantioselective oxidation of various *rac*-2-hydroxy acids, yielding the corresponding 2-keto acid and the nonconverted (R)-α-hydroxy acid. The 2-keto acid was then further converted using D-lactate dehydrogenase and formate dehydrogenase (for NADH regeneration), yielding the (R)-α-hydroxy acid in high enantiomeric excess and high yield [45]. In another chemoenzymatic approach (see also below), sodium borohydride was used to deracemize DL-lactate using L-lactate oxidase as the enantioselective enzyme [46].

Lactate is a substrate of great interest in clinical and sport medicine. Therefore, lactate oxidase–based biosensors are being developed for continuous monitoring of the condition of patients and athletes. These miniaturized devices are based on the action of immobilized lactate oxidase and electrochemical detection [47,48].

Alcohol oxidases that have evolved to oxidize cholesterol and related steroids ha٭ found in several bacteria. These 3β-hydroxysteroid oxidases, more commonly referre cholesterol oxidases, are bifunctional enzymes because they also catalyze the isomeriza the initially formed 5-cholesten-3-one into 4-cholesten-3-one.

Two types of cholesterol oxidases have been described containing a dissociable covalently bound (II) FAD cofactor [49]. The type I form of the enzyme contains a I Asn that are critically involved in substrate oxidation and a Glu that acts as a base isomerization reaction. The type II cholesterol oxidase has a more hydrophilic active s contains an Arg that replaces the His–Asn pair in the type I reaction. The type II enzy a higher redox potential and thus a stronger oxidation power.

Cholesterol oxidase can be exploited as clinical reagent [50] and as insecticide [5 clinical use in cholesterol serum determinations is based on the colorimetric detection enzymatically produced hydrogen peroxide. The enzyme is also active with other ster several structurally unrelated alcohols [52]. With regard to the substrate specificity, be stressed that the catalytic efficiency of cholesterol oxidase is sensitive to the activity substrate in the mixed phase. Truncation of an active site loop appeared to be import movement of cholesterol from the lipid bilayer [53]. Mutation of a wide variety of amino acid residues altered the catalytic efficiency of the type I enzyme without chan substrate specificity [54].

Two flavoprotein oxidases that efficiently convert aromatic alcohols are avail recombinant form. Studies on vanillyl-alcohol oxidase revealed that this enzyme specific toward phenolic substrates, while it displays an exceptional flexibility in ox reactivity. In addition to alcohol oxidations, amine oxidation, dehydrogenation, dem tion, and hydroxylation reactions can be catalyzed [9,55]. Elucidation of the vanillyl- oxidase structure (Figure 6.3) confirmed that the phenolic moiety of the substra prerequisite for binding and activation of the substrate [56]. It was also shown that v alcohol oxidase catalysis involves the formation of quinone methide product inte ates [57]. These electrophilic compounds are attacked by water in the enzyme acti leading, for instance, to the production of chiral 1-(4′-hydroxyphenyl)alcohols [58]. By the water activity during turnover, the relative amount of hydroxylated or alkenic could be modulated [59]. Furthermore, the crystallographic data enabled the rational of an enzyme variant that displays an inverse enantioselectivity toward the hydroxyla 4-ethylphenols [60]. Vanillyl-alcohol oxidase can also be used for the production of vanillin [61]. Using molecular evolution it was possible to turn the suicide inhibitor into a vanillin precursor [62].

Aryl-alcohol oxidase from *Pleurotus eryngii* was cloned and expressed recently [6 monomeric enzyme is related in structure to glucose oxidase. Substrate profiling revealed that aryl-alcohol oxidase is complementary to vanillyl-alcohol oxidase by ac nonphenolic aromatic substrates [64]. The enzyme also oxidizes unsaturated aliphatic alcohols [65].

6.6   AMINE OXIDASES

The ability of certain flavoprotein oxidases to oxidize amines has also attracted bioc interest. A classic example concerns D-amino acid oxidase, which catalyzes the stereo deamination of D-amino acids to the corresponding α-keto acids along with the proc of ammonia and hydrogen peroxide through an imino acid intermediate (Figur D-amino acid oxidase has a broad substrate specificity [66] and is used on an in scale for the synthesis of glutaryl-7-aminocephalosporinic acid from the natural an cephalosporin C [67]. This reaction is part of a two-enzyme process toward the produc

7-aminocephalosporinic acid, a key intermediate in the production of semisynthetic cephalosporin antibiotics. The enzyme is also used in biosensors and for the production of α-keto acids [68].

The relationships between structure and function of pig kidney and yeast D-amino acid oxidase have been studied in great detail. Both enzymes are homodimers, with each subunit containing a relatively weakly bound FAD [69]. The mammalian enzyme is more thermostable, but is a less efficient biocatalyst. For industrial applications, the enzyme is used in immobilized form, resulting in improved biocatalyst stability [70]. To protect the enzyme from hydrogen peroxide inactivation, methionine residues susceptible to oxidation were removed by site-directed mutagenesis [71].

L-amino acid oxidase is the enantiomerically opposite of D-amino acid oxidase. It is a homodimeric FAD-dependent glycoprotein and present at significantly high concentrations in snake venoms [72,73]. Snake venom L-amino acid oxidase has a more narrow substrate specificity than D-amino acid oxidase, exhibiting a marked preference for bulky hydrophobic amino acids. Comparison of the crystal structures of L- and D-amino acid oxidase revealed a mirror-symmetrical relationship between the two substrate binding sites, which facilitates the enantiomeric selectivity while preserving a common arrangement of the atoms involved in catalysis [72]. Several recombinant forms of bacterial L-amino acid oxidases have recently been described [74,75]. These enzymes differ strongly in substrate preference. The enzyme from *Rhodococcus opacus* exhibits a much broader substrate specificity than the ophidian L-amino acid oxidase and is an attractive catalyst for enzymatic synthesis.

Recently, it was shown that D- and L-amino acid oxidase are valuable biocatalysts for deracemization of α-amino acids [76,77]. By performing *in situ* chemical reduction of the formed imine product, one of the two amino acid enantiomers was converted in the other enantiomer. This elegant and effective combination of an enzymatic oxidation and chemical reduction demonstrates the versatile applicability of these types of enzymes. The chemoenzymatic system can also be used to interconvert diastereomeric amino acids bearing more than one stereocenter [78].

Monoamine oxidase is a flavoenzyme that is closely related in structure to L-amino acid oxidase [79]. In vertebrates, this 8α-*S*-cysteinyl-FAD-containing enzyme is bound to the mitochondrial outer membrane where it catalyzes the oxidative deamination of neurotransmitters and biogenic amines. The human enzyme, which occurs in two isoforms (A and B), is involved in a large number of diseases, and is an important target for antidepressant and neuroprotective drugs. Phylogenetic analysis has indicated that monoamine oxidase was already present in early eukaryotes, but that its gene was lost in worm, fly, plant, and yeast before gene duplication in terrestrial vertebrates occurred [80]. A single monoamine oxidase gene seems to be present in aquatic vertebrates, whereas a different situation was found with amphibians [81].

Another eukaryotic monoamine oxidase was discovered in *A. niger* [82]. This enzyme has a considerably higher turnover number on many aliphatic and aromatic amines than the mammalian isoforms but is not active with biogenic amines [83]. *A. niger* monoamine oxidase has been exploited for the deracemization of chiral primary amines. To enhance the efficiency of this process, *in vitro* evolution was used to convert the fungal enzyme into a more competent biocatalyst [84]. Through this approach, it was also possible to enhance the activity with secondary amines [85].

The development of high-level expression systems for producing recombinant human liver monoamine oxidases in the methylotrophic yeast *Pichia pastoris* [86,87] has facilitated structure–function relationship studies and provided much insight into the mode of binding of clinically used drug inhibitors [88–90]. Despite this, the mechanism of substrate oxidation remains controversial [91]. Based on the active site topology and previous structure activity studies, a concerted polar nucleophilic mechanism for monoamine oxidase catalysis, ing substrate–flavin adduct formation, was proposed [92]. Very recently, however, a radical, suggested to be the key missing link in support of the single electron mechanism, was identified in monoamine oxidase A [93].

Monoamine oxidases have a broad substrate specificity and can be used in applications. As noted above, these enzymes are useful for the deracemization c amines. Another application concerns the immobilization of different monoamine isoforms to create online immobilized enzyme reactors. These systems can be used rapid identification of monoamine oxidase inhibitors in complex chemical and bi mixtures [94]. Because altered levels of monoamine oxidase activity are associated wit neurological and psychiatric diseases, there is also a need for imaging the enzyme i cells and tissues. Toward this goal, a fluorogenic probe for monoamine oxidase wa oped that can act as an irreversible redox switch [95]. In short, this switch is based enzymatic oxidation of the ethylamino group of an aminocoumarin substrate, affore aldehyde product, which subsequently undergoes spontaneous intramolecular conde furnishing an indole moiety that alters the fluorescence profile.

Relatively newly discovered flavoprotein oxidases that act on an amine moiety fructosyl amine oxidases [96]. These fungal enzymes are also called amadoriases as on Amadori products: glycated proteins. As for some of the above-mentioned o these were again found to contain a covalently bound FAD cofactor. Their fungal physiology is still obscure. It has been shown that these enzymes facilitate when using fructosyl-amino acids as nitrogen source [97]. Furthermore, a report suggests a role in recycling aged (i.e., glycated) amino acids in fungal cells [⁹ fructosyl amine oxidases have been studied for their putative value for d glycated proteins in blood samples. This would enable efficient clinical diagnosis of mellitus. Amadoriases may also be useful to eliminate Maillard reaction products. To i the catalytic properties of these enzymes, molecular evolution has successfully been resulting in thermostable variants with improved affinity for glycated amino acids [9

6.7   SULFHYDRYL OXIDASES

Sulfhydryl oxidases represent a relatively new group of flavoprotein oxidases [100 enzymes catalyze the insertion of disulfide bonds into proteins and/or the oxidation c molecular weight thiols. Flavin-containing sulfhydryl oxidases are present in many eu species, where they are targeted to different subcellular compartments and specific o reactions. Most sulfhydryl oxidases contain a small FAD-binding domain with a unio which can be found as a single-domain protein (Erv/ALR family) or fused to an N-t thioredoxin domain in the QSOX family. Some of the single-domain sulfhydryl oxid *in vivo* linked to the function of protein disulfide isomerase and a source of disulfid for oxidative protein folding in the cell [101]. *In vitro*, however, single-domain su oxidases are only active with small–molecular weight thiols such as dithiothreitol and glutathione.

Chicken sulfhydryl oxidase is presently the most well-documented flavin-de biocatalyst to introduce disulfide bridges directly into a wide range of unfolded proteins and peptides [100]. This QSOX-type enzyme is a heavily glycosylated hom with each subunit containing a noncovalently bound FAD and a redox-active ide. The catalytic efficiency of chicken sulfhydryl oxidase is not strongly dependen size or charge of the protein substrate. However, with RNAse as a protein su the inclusion of protein disulfide isomerase is required for the efficient recovery o disulfide bonds.

Sulfhydryl oxidases may be applied in breadmaking by improving the strength and handling properties of the dough [102]. Removal of low–molecular weight thiols such as reduced glutathione during mixing may prevent their participation in the thiol–disulfide exchange reactions, which result in the depolymerization of the gluten proteins and thereby reduce the dough elasticity and increase its extensibility. Moreover, the hydrogen peroxide generated may be used in a peroxidase-mediated reaction to catalyze the oxidative gelation of the wheat flour pentosans. A similar mechanism of peroxidase activation can account for the improving effect in breadmaking by application of different carbohydrate oxidases [103,104].

Apart from all the above-mentioned oxidases, other flavin-containing oxidases are also available in recombinant form. Most of these enzymes, like sarcosine oxidase [105], polyamine oxidase [106,107], nitroalkane oxidase [108,109], and γ-N-methylaminobutyrate oxidase [110], have a limited substrate specificity. Nevertheless, knowledge about their structure and function will contribute to the future development of flavin-containing oxidative biocatalysts.

6.8   FLAVOPROTEIN MONOOXYGENASES

Flavoprotein monooxygenases catalyze the insertion of one atom of dioxygen into the substrate according to the following scheme:

$$S + NAD(P)H + O_2 \rightarrow P + NAD(P)^+ + H_2O \tag{6.2}$$

Several types of flavin-dependent monooxygenases have been discovered during the past few decades. While some of these enzymes have been isolated from eukaryotes (e.g., fungi and yeasts), the best-studied flavin-dependent monooxygenases are from bacterial origin. In fact, almost all cloned flavoprotein monooxygenases are from bacteria. This is in contrast with the situation for oxidases where many eukaryotic representatives have been cloned and overexpressed. Also different from the oxidases is the observation that all characterized monooxygenases contain a flavin cofactor that is dissociable. This might indicate that for catalytic functioning the flavin should not be restricted in mobility. This is in line with the fact that for some monooxygenases the flavin has been found to adopt multiple orientations within the protein structure during the catalytic cycle [111,112]. This is in contrast with the oxidases in which the active site is typically buried in the protein structure and can be regarded as relatively rigid.

Roughly, three classes of flavoprotein monooxygenases can be identified (Table 6.2). The first entails hydroxylases that typically act on aromatic compounds; the prototype for these monooxygenases is p-hydroxybenzoate hydroxylase (PHBH). The second comprises monooxygenases that can catalyze Baeyer–Villiger oxidations and heteroatom oxidations. In this review we refer to this class as heteroatom monooxygenases. The first structure of a member of this class of monooxygenases has just been elucidated [113]. The third can be defined as a group of monooxygenases that depend on two different subunits: the two-component monooxygenases. This review only aims at illustrating the biocatalytic potential of flavoprotein monooxygenases. The following sections describe the catalytic versatility of some specific monooxygenases that have been cloned and described in the literature. However, this only reflects part of the industrial applicability of such oxidative enzymes. For example, at Lonza several industrial processes rely on the use of isolated bacterial strains that exhibit specific hydroxylation activities [114]. While molecular details of these biocatalysts might not be known, several of them might well represent flavoproteins.

TABLE 6.2
Cloned Flavoprotein Monooxygenases That Are Relevant for Biocatalysis

Enzymes	Year of Cloning	EC Number	Structure
Aromatic hydroxylases			
4-Hydroxybenzoate 3-hydroxylase	1988	1.14.13.2	+
Phenol hydroxylase	1992	1.14.13.7	+
2-Hydroxybiphenyl 3-monooxygenase	1997	1.14.13.44	−
Baeyer–Villiger monooxygenases			
Phenylacetone monooxygenase	2004	1.14.13.92	+
Cyclohexanone monooxygenase	1988	1.14.13.22	−
Cyclopentanone monooxygenase	2002	1.14.13.16	−
Cyclododecanone monooxygenase	2001	1.14.13.x	−
Steroid monooxygenase	1999	1.14.13.54	−
4-Hydroxyacetophenone monooxygenase	2001	1.14.13.84	−
Two-component monooxygenases			
Tryptophan 7-halogenase	2000	1.14.14.x	+m
Phenol 2-monooxygenase	2003	1.14.14.x	+r
Styrene monooxygenase	1996	1.14.14.x	−
Phenazine 1-carboxylic acid monooxygenase	2001	1.14.14.x	−
Dibenzothiophene monooxygenase	1994	1.14.14.x	−

b, bacterial; y, yeast; m, monooxygenase component; r, reductase component.

6.9 AROMATIC HYDROXYLASES

The most extensively studied class of flavoprotein monooxygenases is the group of ar
hydroxylases that shows sequence homology with PHBH. The three-dimensional struc
this prototype monooxygenase was already elucidated in 1979 (Figure 6.3) [115,11
triggered several groups to study its molecular details [111,117]. The folding topo
PHBH is shared by several flavin-containing biocatalysts including glucose oxidase,
terol oxidase, D-amino acid oxidase, monoamine oxidase, and phenol hydroxylase [1

Flavoprotein aromatic hydroxylases contain a noncovalently bound FAD cofact
depend on NADH or NADPH for activity. Most representatives have many catalyti
erties in common and their overall reactions can be divided into two half-reactions [2]
reductive half-reaction, the FAD cofactor becomes reduced and the NADP$^+$ coenz
released. In the oxidative half-reaction, the reduced flavin reacts with molecular (
yielding the flavinhydroperoxide oxygenation species. Protonation of the distal ox
the peroxide moiety increases the electrophilic reactivity of the flavin peroxide and fa
its attack on the nucleophilic carbon center of the substrate. Rapid reaction studi
substrate analogs have indicated that the aromatic product is initially formed in i
isomeric form, which isomerizes to give the energetically favored dihydroxy isomer [2

Flavoprotein aromatic hydroxylases display a subtle mechanism of substrate reco
[119]. For efficient turnover, the substrate should act as an effector to induce flavin rec
by NAD(P)H. Moreover, upon binding to the enzyme active site, the substrate must I
activated to allow hydroxylation and to prevent the unproductive decompostion
flavinhydroperoxide [111]. Studies from PHBH variants, created by site-directed muta
have shown that a great number of strictly conserved amino acid residues are invo
substrate binding and activation [117]. As a result, these monooxygenases typically di
narrow substrate specificity and perform highly regioselective hydroxylations [120–12

There are a few flavin-dependent aromatic hydroxylases with a rather relaxed substrate specicifity. One of these enzymes, phenol hydroxylase, was originally obtained from the yeast *Trichosporon cutaneum*, but is now available in recombinant form [123]. Another aromatic hydroxylase with a rather broad substrate spectrum concerns 2-hydroxybiphenyl 3-monooxygenase. This enzyme is induced in *Pseudomonas azelaica* HBP1, and is active with a wide range of 2'-substituted phenols [124]. The 2-hydroxybiphenyl 3-monooxygenase gene has been cloned in *E. coli* and the recombinant strain was used as a whole-cell biocatalyst for the production of 2-substituted catechols [125,126]. Molecular evolution has been used to improve the catalytic scope and performance of 2-hydroxybiphenyl 3-monooxygenase. In this way it was possible to increase the hydroxylation efficiency [127] and extend the substrate spectrum [128,129].

6.10 HETEROATOM MONOOXYGENASES

The second flavoprotein monooxygenase class entails enzymes that share sequence homology with phenylacetone monooxygenase. The crystal structure of this enzyme (Figure 6.3) was recently solved and represents the first structure of a Baeyer–Villiger monooxygenase [113]. Most heteroatom monooxygenases contain a tightly bound FAD cofactor and are typically dependent on NADPH as electon donor [130]. Representatives of this class have been found to catalyze a relatively broad range of asymmetric oxygenation reactions with high enantio- or enantiotoposelectivity. In addition to Baeyer–Villiger reactions, a range of other oxidation reactions can be catalyzed including sulfoxidations, amine hydroxylations, organoboron oxidations, and epoxidations. Based on genome analysis it has been found that many bacteria and fungi contain heteroatom monooxygenases that preferentially act as Baeyer–Villiger monooxygenases [131]. Plant and vertebrate genomes are rich in heteroatom monooxygenases that preferentially act on soft nucleophilic heteroatoms (e.g., S and N) and are known as flavin-containing monooxygenases (FMOs). In humans these FMOs are involved in detoxification. Mutations in FMO genes have been shown to result in serious health problems [132].

Microbial Baeyer–Villiger monooxygenases appear to be mainly involved in the degradation of natural ketones. A number of bacterial Baeyer–Villiger monooxygenases have been extensively studied for their biocatalytic potential. Since the 1970s, cyclohexanone monooxygenase from *Acinetobacter* NCIB 9871 has been explored to a great extent with respect to its catalytic properties [133–136]. The enzyme has also been used in dynamic kinetic resolution reactions, which combine *in situ* racemization with a kinetic resolution step [137]. So far, over 100 different substrates have been described for cyclohexanone monooxygenase including ketones, aldehydes, sulfides, amines, olefins, seleno compounds, and organoboronic compounds. Also for other Baeyer–Villiger monooxygenases quite a number of substrates have been identified [130,138]. Nevertheless, each specific Baeyer–Villiger monooxygenase appears to prefer a certain class of substrates. For example, 4-hydroxyacetophenone monooxygenase efficiently converts acetophenones and benzaldehydes while it is poorly active with cyclic aliphatic ketones [130,139,140]. Therefore, it is unlikely that such enzyme activity will be detected in high-throughput fluorescence assays based on the conversion of 2-coumaryloxyketones [141]. The typical relaxed substrate acceptance of Baeyer–Villiger monooxygenases probably reflects a common mechanistic feature of these types of enzymes. Kinetic studies have revealed that Baeyer–Villiger monooxygenases, independent from substrate binding, will form and stabilize a peroxyflavin when NADPH and oxygen are available [142]. Consequently, these enzymes will oxidize substrates that are reactive enough and are able to reach the activated flavin. Depending on the shape of the substrate binding pocket, regio- and/or enantioselective oxygenation reactions are catalyzed. This concept of selectivity tuning by preferential binding of the substrate was recently demonstrated in an enzyme redesign study

using phenylacetone monooxygenase [143]. By partial deletion of a loop that sha
substrate binding pocket, the substrate acceptance and enantioselectivity could be
cantly altered. The resulting mutant could accept substrates that are not converted
wild-type enzyme while displaying altered enantioselectivity.

It has been found that bacterial Baeyer–Villiger monooxygenases also represent i
ing targets for treating infections. Many pathogenic bacteria carry one or more
Villiger monooxygenase genes in their genome [131]. In fact, Baeyer–Villiger monooxy
have already been exploited as drug target for several decades to treat tubercula
leprosy. The corresponding mycobacteria, *Mycobacterium tuberculosis* and *M. lepr*
duce a Baeyer–Villiger monooxygenase that converts a number of commonly used an
into toxic bactericidal products [144].

6.11   TWO-COMPONENT MONOOXYGENASES

The third class of flavoprotein monooxygenases consists of multicomponent enzyn
variety in this group of monooxygenases is quite large concerning cofactor use and cc
specificity. Generally, these enzymes comprise two polypeptide chains: a reductase c
ent carrying out flavin reduction and an oxygenase component oxidizing the subs
molecular oxygen [145,146]. The reducing equivalents needed for the oxygenation ar
ferred from NAD(P)H to a flavin bound to the reductase component. With some red
a tightly bound flavin acts as extra serving-hatch of electrons [147]. Subsequently, the
flavin is transferred to the oxygenase component where the activation of molecular
and the reaction with substrate occur. A well-known example of this monooxygenase
4-hydroxyphenylacetate-3-monooxygenase from *E. coli* W, consisting of a large ox
(59 kDa) and a small reductase (19 kDa) [148]. Another example of a two-component
oxygenase is styrene monooxygenase that has been found in a number of pseudo
[149–152]. This enzyme is of value for biotechnological applications as it catalyze
enantioselective epoxidation reactions [153,154].

In addition to hydroxylation and epoxidation reactions, desulfurization reactions
be carried out by two-component flavoprotein monooxygenases. Dibenzothiophene
desulfurization has been studied in a variety of microorganisms [155]. The genes resp
for the degradation pathway have been cloned from *Rhodococcus* sp. IGTS8 [156–1.
removal of sulfur from the substrate requires action of two monooxygenase protein;
and DszC) that are dependent on the availability of reduced FMN. Reduced I
generated by a reductase (DszD) that uses NADH as electron source and is related
reductase components of 4-hydroxyphenylacetate-3-monooxygenase and styrene me
genase, discussed above. Biocatalytic processes using desulfurizing flavoprotein
oxygenases may find application in the desulfurization of fossil fuels [155]. C
approaches have been used to engineer strains with improved desulfurization activit
expression of DszD or another flavin reductase in *E. coli* and *P. putida* enhanced the
rate of desulfurization [159,160]. Furthermore, activities toward (highly) alkylate
have been improved using a chemostat approach [161] and gene shuffling [162].

A special member of the two-component flavin-dependent monooxygenase fa
tryptophan 7-halogenase [163–165]. Several bacterial enzymes have been found
able to insert a halogen atom into an organic substrate [166]. This reaction is perfo
the expense of a reduced flavin cofactor and molecular oxygen. The catalyzed halog
reactions are typically regioselective and, as a result, these enzymes bear great p
for biocatalytic applications [167]. Recently, the crystal structure of the monoox
component of tryptophan 7-halogenase was elucidated [168]. Based on the bindin
of the flavin, the tryptophan substrate, and a chloride ion in the active site, an o

halogenation mechanism, involving the formation and channeling of hypochlorous acid, is proposed.

In view of the need of expensive coenzymes, flavin-dependent monooxygenases are mostly applied in whole-cell sytems. However, with the continuous advancements in heterologous protein expression and coenzyme regeneration, even (partially) purified monooxygenases will become increasingly attractive for the environmentally benign and cost-competitive production of high-value chemicals.

REFERENCES

1. Massey, V., The chemical and biological versatility of riboflavin, *Biochem. Soc. Trans.*, 28(4), 283–296, 2000.
2. Massey, V., Activation of molecular oxygen by flavins and flavoproteins, *J. Biol. Chem.*, 269(36), 22459–22462, 1994.
3. Suh, J.K., Poulsen, L.L., Ziegler, D.M., et al., Yeast flavin-containing monooxygenase generates oxidizing equivalents that control protein folding in the endoplasmic reticulum, *Proc. Natl. Acad. Sci. USA*, 96(6), 2687–2691, 1999.
4. Ziegler, D.M., Flavin-containing monooxygenases: enzymes adapted for multisubstrate specificity, *Trends Pharmacol. Sci.*, 11(8), 321–324, 1990.
5. Keller, N.P., Watanabe, C.M., Kelkar, H.S., et al., Requirement of monooxygenase-mediated steps for sterigmatocystin biosynthesis by *Aspergillus nidulans*, *Appl. Environ. Microbiol.*, 66(1), 359–362, 2000.
6. DeBarber, A.E., Mdluli, K., Bosman, M., et al., Ethionamide activation and sensitivity in multidrug-resistant *Mycobacterium tuberculosis*, *Proc. Natl. Acad. Sci. USA*, 97(17), 9677–9682, 2000.
7. Zhao, Y., Christensen, S.K., Fankhauser, C., et al., A role for flavin monooxygenase-like enzymes in auxin biosynthesis, *Science*, 291(5502), 306–309, 2001.
8. Wieser, M., Wagner, B., Eberspacher, J., et al., Purification and characterization of 2,4,6-trichlorophenol-4-monooxygenase, a dehalogenating enzyme from *Azotobacter* sp. strain GP1, *J. Bacteriol.*, 179(1), 202–208, 1997.
9. Fraaije, M.W., Veeger, C., and van Berkel, W.J.H., Substrate specificity of flavin-dependent vanillyl-alcohol oxidase from *Penicillium simplicissimum*: evidence for the production of 4-hydroxycinnamyl alcohols from 4-allylphenols, *Eur. J. Biochem.*, 234(1), 271–277, 1995.
10. Digits, J.A., Pyun, H.J., Coates, R.M., et al., Stereospecificity and kinetic mechanism of human prenylcysteine lyase, an unusual thioether oxidase, *J. Biol. Chem.*, 277(43), 41086–41093, 2002.
11. Kutchan, T.M. and Dittrich, H., Characterization and mechanism of the berberine bridge enzyme, a covalently flavinylated oxidase of benzophenanthridine alkaloid biosynthesis in plants, *J. Biol. Chem.*, 270(41), 24475–24481, 1995.
12. Sirikantaramas, S., Morimoto, S., Shoyama, Y., et al., The gene controlling marijuana psychoactivity: molecular cloning and heterologous expression of delta1-tetrahydrocannabinolic acid synthase from *Cannabis sativa* L, *J. Biol. Chem.*, 279(38), 39767–39774, 2004.
13. Fraaije, M.W., van den Heuvel, R.H.H., van Berkel, W.J.H., et al., Structural analysis of flavinylation in vanillyl-alcohol oxidase, *J. Biol. Chem.*, 275(49), 38654–38658, 2000.
14. Hassan-Abdallah, A., Bruckner, R.C., Zhao, G., et al., Biosynthesis of covalently bound flavin: isolation and *in vitro* flavinylation of the monomeric sarcosine oxidase apoprotein, *Biochemistry*, 44(17), 6452–6462, 2005.
15. Fraaije, M.W., van den Heuvel, R.H.H., van Berkel, W.J.H., et al., Covalent flavinylation is essential for efficient redox catalysis in vanillyl-alcohol oxidase, *J. Biol. Chem.*, 274(50), 35514–35520, 1999.
16. Caldinelli, L., Iametti, S., Barbiroli, A., et al., Dissecting the structural determinants of the stability of cholesterol oxidase containing covalently bound flavin, *J. Biol. Chem.*, 280(24), 22572–22581, 2005.
17. Ödman, P., Wellborn, W.B., and Bommarius, A.S., An enzymatic process to α-ketoglutarate from L-glutamate: the coupled system L-glutamate dehydrogenase/NADH oxidase, *Tetrahedron Asymmetry*, 15(18), 2933–2937, 2004.

18. Wilson, R. and Turner, A.P.F., Glucose oxidase: an ideal enzyme, *Biosens. Bioelectr.*, 7(185, 1992.

19. Hecht, H.J., Kalisz, H.M., Hendle, J., et al., Crystal structure of glucose oxidase from *Asg niger* refined at 2.3 Å resolution, *J. Mol. Biol.*, 229(1), 153–172, 1993.

20. Wohlfahrt, G., Witt, S., Hendle, J., et al., 1.8 and 1.9 Å resolution structures of the *Per amagasakiense* and *Aspergillus niger* glucose oxidases as a basis for modelling substrate co* *Acta Crystall. Section D*, 55(5), 969–977, 1999.

21. Leskovac, V., Trivic, S., Wohlfahrt, G., et al., Glucose oxidase from *Aspergillus ni* mechanism of action with molecular oxygen, quinones, and one-electron accepto *J. Biochem. Cell. Biol.*, 37(4), 731–750, 2005.

22. Wohlfahrt, G., Triviæ, S., Zeremski, J., et al., The chemical mechanism of action of oxidase from *Aspergillus niger*, *Mol. Cell. Biochem.*, 260(1), 69–83, 2004.

23. Wolff, A.M., Hansen, O.C., Poulsen, U., et al., Optimization of the production of *Chondru. hexose oxidase in *Pichia pastoris*, *Protein Expr. Purif.*, 22(2), 189–199, 2001.

24. Vecerek, B., Maresova, H., Kocanova, M., et al., Molecular cloning and expression pyranose 2-oxidase cDNA from *Trametes ochracea* MB49 in *Escherichia coli*, *Appl. M Biotechnol.*, 64(4), 525–530, 2004.

25. Hallberg, B.M., Leitner, C., Haltrich, D., et al., Crystal structure of the 270 kDa homote* lignin-degrading enzyme pyranose 2-oxidase, *J. Mol. Biol.*, 341(3), 781–796, 2004.

26. Bannwarth, M., Bastian, S., Heckmann-Pohl, D., et al., Crystal structure of pyranose 2- from the white-rot fungus *Peniophora* sp., *Biochemistry*, 43(37), 11683–11690, 2004.

27. Liu, T.E., Wolf, B., Geigert, J., et al., Convenient laboratory procedure for producing *arabino*-hexos-2-ulose (D-glucosone), *Carbohydrate Res.*, 113(1), 151–157, 1983.

28. Koths, K., Halenbeck, R., and Moreland, M., Synthesis of the antibiotic cortalceron D-glucose using pyranose 2-oxidase and a novel fungal enzyme, aldos-2-ulose dehy *Carbohydrate Res.*, 232(1), 59–75, 1992.

29. Giffhorn, F., Fungal pyranose oxidases: occurrence, properties and biotechnical applica carbohydrate chemistry, *Appl. Microbiol. Biotechnol.*, 54(6), 727–740, 2000.

30. Bastian, S., Rekowski, M.J., Witte, K., et al., Engineering of pyranose 2-oxidase from *Per gigantea* towards improved thermostability and catalytic efficiency, *Appl. Microbiol. Bio* 67(5), 654–663, 2005.

31. Xu, F., Golightly, E.J., Fuglsang, C.C., et al., A novel carbohydrate: acceptor oxidoreducta *Microdochium nivale*, *Eur. J. Biochem.*, 268(4), 1136–1142, 2001.

32. Custers, J.H., Harrison, S.J., Sela-Buurlage, M.B., et al., Isolation and characterisation o: of carbohydrate oxidases from higher plants, with a role in active defence, *Plant J.*, 39(2), 1 2004.

33. Xu, F., Applications of oxidoreductases: recent progress, *Ind. Biotechnol.*, 1(1), 38–50, 20*

34. Hiraga, K., Eto, T., Yoshioka, I., et al., Molecular cloning and expression of a gene enc novel sorbitol oxidase from *Streptomyces* sp. H-7775, *Biosci. Biotechnol. Biochem.*, 62(2), 3 1998.

35. Yamashita, M., Omura, H., Okamoto, E., et al., Isolation, characterization, and molecular of a thermostable xylitol oxidase from *Streptomyces* sp. IKD472, *J. Biosci. Bioeng.*, 89(4), 3 2000.

36. Salusjarvi, T., Kalkkinen, N., and Miasnikov, A.N., Cloning and characterization of glucono oxidase of *Penicillium cyaneo-fulvum* ATCC 10431 and evaluation of its use for produ D-erythorbic acid in recombinant *Pichia pastoris*, *Appl. Environ. Microbiol.*, 70(9), 5503–551

37. van der Klei, I.J., Harder, W., and Veenhuis, M., Biosynthesis and assembly of alcohol * a peroxisomal matrix protein in methylotrophic yeasts: a review, *Yeast*, 7(3), 195–209, 19*

38. Vanhanen, S., West, M., Kroon, J.T., et al., A consensus sequence for long-chain fatty-acid oxidases from *Candida* identifies a family of genes involved in lipid ω-oxidation in ye* homologues in plants and bacteria, *J. Biol. Chem.*, 275(6), 4445–4452, 2000.

39. Eirich, L.D., Craft, D.L., Steinberg, L., et al., Cloning and characterization of thr* alcohol oxidase genes from *Candida tropicalis* strain ATCC 20336, *Appl. Environ. Mi* 70(8), 4872–4879, 2004.

40. Cheng, Q., Liu, H.T., Bombelli, P., et al., Functional identification of AtFao3, a membrane bound long chain alcohol oxidase in *Arabidopsis thaliana*, *FEBS Lett.*, 574(1–3), 62–68, 2004.
41. Yamashita, N., Motoyoshi, T., and Nishimura, A., Molecular cloning of the isoamyl alcohol oxidase-encoding gene (*mre*A) from *Aspergillus oryzae*, *J. Biosci. Bioeng.*, 89(3), 522–527, 2000.
42. Amar, D., North, P., Miskiniene, V., et al., Hydroxamates as substrates and inhibitors for FMN-dependent 2-hydroxy acid dehydrogenases, *Bioorg. Chem.*, 30(3), 145–162, 2002.
43. Yorita, K., Misaki, H., Palfey, B.A., et al., On the interpretation of quantitative structure–function activity relationship data for lactate oxidase, *Proc. Natl. Acad. Sci. USA*, 97(6), 2480–2485, 2000.
44. Kroutil, W., Mang, H., Edegger, K., et al., Biocatalytic oxidation of primary and secondary alcohols, *Adv. Synth. Catal.*, 346(2–3), 125–142, 2004.
45. Adam, W., Lazarus, M., Boss, B., et al., Enzymatic resolution of chiral 2-hydroxy carboxylic acids by enantioselective oxidation with molecular oxygen catalyzed by the glycolate oxidase from spinach (*Spinacia oleracea*), *J. Org. Chem.*, 62(22), 7841–7843, 1997.
46. Oikawa, T., Mukoyama, S., and Soda, K., Chemo-enzymatic D-enantiomerization of DL-lactate, *Biotechnol. Bioeng.*, 73(1), 80–82, 2001.
47. Leegsma-Vogt, G., Rhemrev-Boom, M.M., Tiessen, R.G., et al., The potential of biosensor technology in clinical monitoring and experimental research, *Biomed. Mater. Eng.*, 14(4), 455–464, 2004.
48. Poscia, A., Messeri, D., Moscone, D., et al., A novel continuous subcutaneous lactate monitoring system, *Biosens. Bioelectron.*, 20(11), 2244–2250, 2005.
49. Sampson, N.S. and Vrielink, A., Cholesterol oxidases: a study of nature's approach to protein design, *Acc. Chem. Res.*, 36(9), 713–722, 2003.
50. MacLachlan, J., Wotherspoon, A.T., Ansell, R.O., et al., Cholesterol oxidase: sources, physical properties and analytical applications, *J. Steroid Biochem. Mol. Biol.*, 72(5), 169–195, 2000.
51. Corbin, D.R., Grebenok, R.J., Ohnmeiss, T.E., et al., Expression and chloroplast targeting of cholesterol oxidase in transgenic tobacco plants, *Plant. Physiol.*, 126(3), 1116–1128, 2001.
52. Ahn, K.W. and Sampson, N.S., Cholesterol oxidase senses subtle changes in lipid bilayer structure, *Biochemistry*, 43(3), 827–836, 2004.
53. Sampson, N.S., Kass, I.J., and Ghoshroy, K.B., Assessment of the role of an omega loop of cholesterol oxidase: a truncated loop mutant has altered substrate specificity, *Biochemistry*, 37(16), 5770–5778, 1998.
54. Xiang, J. and Sampson, N.S., Library screening studies to investigate substrate specificity in the reaction catalyzed by cholesterol oxidase, *Protein Eng. Des. Sel.*, 17(4), 341–348, 2004.
55. van den Heuvel, R.H.H., Fraaije, M.W., Laane, C., et al., Regio- and stereospecific conversion of 4-alkylphenols by the covalent flavoprotein vanillyl-alcohol oxidase, *J. Bacteriol.*, 180(21), 5646–5651, 1998.
56. Mattevi, A., Fraaije, M.W., Mozzarelli, A., et al., Crystal structures and inhibitor binding in the octameric flavoenzyme vanillyl-alcohol oxidase: the shape of the active-site cavity controls substrate specificity, *Structure*, 5(7), 907–920, 1997.
57. Fraaije, M.W. and van Berkel, W.J.H., Catalytic mechanism of the oxidative demethylation of 4-(methoxymethyl)phenol by vanillyl-alcohol oxidase: evidence for formation of a *p*-quinone methide intermediate, *J. Biol. Chem.*, 272(29), 18111–18116, 1997.
58. Drijfhout, F.P., Fraaije, M.W., Jongejan, H., et al., Enantioselective hydroxylation of 4-alkylphenols by vanillyl alcohol oxidase, *Biotechnol. Bioeng.*, 59(2), 171–177, 1998.
59. van den Heuvel, R.H.H., Partridge, J., Laane, C., et al., Tuning of the product spectrum of vanillyl-alcohol oxidase by medium engineering, *FEBS Lett.*, 503(2–3), 213–216, 2001.
60. van den Heuvel, R.H.H., Fraaije, M.W., Ferrer, M., et al., Inversion of stereospecificity of vanillyl-alcohol oxidase, *Proc. Natl. Acad. Sci. USA*, 97(17), 9455–9460, 2000.
61. van den Heuvel, R.H., Fraaije, M.W., Laane, C., et al., Enzymatic synthesis of vanillin, *J. Agric. Food Chem.*, 49(6), 2954–2958, 2001.
62. van den Heuvel, R.H., van den Berg, W.A., Rovida, S., et al., Laboratory-evolved vanillyl-alcohol oxidase produces natural vanillin, *J. Biol. Chem.*, 279(32), 33492–33500, 2004.
63. Varela, E., Guillen, F., Martinez, A.T., et al., Expression of *Pleurotus eryngii* aryl-alcohol oxidase in *Aspergillus nidulans*: purification and characterization of the recombinant enzyme, *Biochim. Biophys. Acta.*, 1546(1), 107–113, 2001.

64. Ferreira, P., Medina, M., Guillen, F., et al., Spectral and catalytic properties of aryl-alcohol a fungal flavoenzyme acting on polyunsaturated alcohols, *Biochem. J.*, 38g(3), 731–738, 2

65. Guillén, F., Martínez, Á.T., and Martínez, M.J., Substrate specificity and properties of alcohol oxidase from the ligninolytic fungus *Pleurotus eryngii*, *Eur. J. Biochem.*, 209(2), 1992.

66. Pilone, M.S. and Pollegioni, L., D-Amino acid oxidase as an industrial biocatalyst, *Bio transform.*, 20(3), 145–159, 2002.

67. Pollegioni, L., Caldinelli, L., Molla, G., et al., Catalytic properties of D-amino acid o cephalosporin C bioconversion: a comparison between proteins from different sources, *Bi Prog.*, 20(2), 467–473, 2004.

68. Tishkov, V.I. and Khoronenkova, S.V., D-Amino acid oxidase: structure, catalytic mechan practical application, *Biochemistry* (Moscow), 70(1), 40–54, 2005.

69. Pilone, M.S., D-Amino acid oxidase: new findings, *Cell. Mol. Life Sci.*, 57(12), 1732–174

70. Betancor, L., Hidalgo, A., Fernandez-Lorente, G., et al., Use of physicochemical tools mine the choice of optimal enzyme: stabilization of D-amino acid oxidase, *Biotechnol. Pro* 784–788, 2003.

71. Ju, S.S., Lin, L.L., Chien, H.R., et al., Substitution of the critical methionine residues *nopsis variabilis* D-amino acid oxidase with leucine enhances its resistance to hydrogen *FEMS Microbiol. Lett.*, 186(2), 215–219, 2000.

72. Pawelek, P.D., Cheah, J., Coulombe, R., et al., The structure of L-amino acid oxidase re substrate trajectory into an enantiomerically conserved active site, *EMBO J.*, 19(16), 42 2000.

73. Geyer, A., Fitzpatrick, T.B., Pawelek, P.D., et al., Structure and characterization of th moiety of L-amino-acid oxidase from the Malayan pit viper *Calloselasma rhodostor J. Biochem.*, 268(14), 4044–4053, 2001.

74. Geueke, B. and Hummel, W., Heterologous expression of *Rhodococcus opacus* L-am oxidase in *Streptomyces lividans*, *Protein Expr. Purif.*, 28(2), 303–309, 2003.

75. Nishizawa, T., Aldrich, C.C. and Sherman, D.H., Molecular analysis of the rebeccamycin acid oxidase from *Lechevalieria aerocolonigenes* ATCC 39243, *J. Bacteriol.*, 187(6), 2084–20

76. Soda, K., Oikawa, T., and Yokoigawa, K., One-pot chemo-enzymatic enantiomeriz racemates, *J. Mol. Catal. B: Enzym.*, 11(4–6), 149–153, 2001.

77. Turner, N.J., Controlling chirality, *Curr. Opin. Biotechnol.*, 14(4), 401–406, 2003.

78. Enright, A., Alexandre, F.R., Roff, G., et al., Stereoinversion of beta- and gamma-su alpha-amino acids using a chemo-enzymatic oxidation–reduction procedure, *Chem. C* 2636–2637, 2003.

79. Binda, C., Newton-Vinson, P., Hubalek, F., et al., Structure of human monoamine ox a drug target for the treatment of neurological disorders, *Nat. Struct. Biol.*, 9(1), 22–26,

80. Roelofs, J. and Van Haastert, P.J., Genes lost during evolution, *Nature*, 411(6841), 10 2001.

81. Setini, A., Pierucci, F., Senatori, O., et al., Molecular characterization of monoamine o zebrafish (*Danio rerio*), *Comp. Biochem. Physiol. B Biochem. Mol. Biol.*, 140(1), 153–161,

82. Schilling, B. and Lerch, K., Cloning, sequencing and heterologous expression of the mo oxidase gene from *Aspergillus niger*, *Mol. Gen. Genet.*, 247(4), 430–438, 1995.

83. Sablin, S.O., Yankovskaya, V., Bernard, S., et al., Isolation and characterization of an e ary precursor of human monoamine oxidases A and B, *Eur. J. Biochem.*, 253(1), 270–27

84. Alexeeva, M., Carr, R., and Turner, N.J., Directed evolution of enzymes: new biocata asymmetric synthesis, *Org. Biomol. Chem.*, 1(23), 4133–4137, 2003.

85. Carr, R., Alexeeva, M., Dawson, M.J., et al., Directed evolution of an amine oxidas preparative deracemisation of cyclic secondary amines, *Chem. Biochem.*, 6(4), 637–639, 2

86. Newton-Vinson, P., Hubalek, F., and Edmondson, D.E., High-level expression of hun monoamine oxidase B in *Pichia pastoris*, *Protein Express. Purif.*, 20(2), 334–345, 2000.

87. Li, M., Hubalek, F., Newton-Vinson, P., et al., High-level expression of human liver mo oxidase A in *Pichia pastoris*: comparison with the enzyme expressed in *Saccharomyces c Protein Express. Purif.*, 24(1), 152–162, 2002.

88. Edmondson, D.E., Binda, C., and Mattevi, A., The FAD binding sites of human monoamine oxidases A and B, *Neurotoxicology*, 25(1–2), 63–72, 2004.
89. Binda, C., Li, M., Hubalek, F., et al., Insights into the mode of inhibition of human mitochondrial monoamine oxidase B from high-resolution crystal structures, *Proc. Natl. Acad. Sci. USA*, 100(17), 9750–9755, 2003.
90. Hubalek, F., Binda, C., Khalil, A., et al., Demonstration of isoleucine 199 as a structural determinant for the selective inhibition of human monoamine oxidase B by specific reversible inhibitors, *J. Biol. Chem.*, 280(16), 15761–15766, 2005.
91. Scrutton, N.S., Chemical aspects of amine oxidation by flavoprotein enzymes, *Nat. Prod. Rep.*, 21(6), 722–730, 2004.
92. Edmondson, D.E., Mattevi, A., Binda, C., et al., Structure and mechanism of monoamine oxidase, *Curr. Med. Chem.*, 11(15), 1983–1993, 2004.
93. Rigby, S.E., Hynson, R.M., Ramsay, R.R., et al., A stable tyrosyl radical in monoamine oxidase A, *J. Biol. Chem.*, 280(6), 4627–4631, 2005.
94. Markoglou, N., Hsuesh, R., and Wainer, I.W., Immobilized enzyme reactors based upon the flavoenzymes monoamine oxidase A and B, *J. Chromatogr. B*, 804(2), 295–302, 2004.
95. Chen, G., Yee, D.J., Gubernator, N.G., et al., Design of optical switches as metabolic indicators: new fluorogenic probes for monoamine oxidases (MAO A and B), *J. Am. Chem. Soc.*, 127(13), 4544–4545, 2005.
96. Yoshida, N., Sakai, Y., Isogai, A., et al., Primary structures of fungal fructosyl amino acid oxidases and their application to the measurement of glycated proteins, *Eur. J. Biochem.*, 242(3), 499–505, 1996.
97. Akazawa, S., Karino, T., Yoshida, N., et al., Functional analysis of fructosyl-amino acid oxidases of *Aspergillus oryzae*, *Appl. Environ. Microbiol.*, 70(10), 5882–5890, 2004.
98. Yoshida, N., Takatsuka, K., Katsuragi, T., et al., Occurrence of fructosyl-amino acid oxidase-reactive compounds in fungal cells, *Biosci. Biotechnol. Biochem.*, 69(1), 258–260, 2005.
99. Sakaue, R. and Kajiyama, N., Thermostabilization of bacterial fructosyl-amino acid oxidase by directed evolution, *Appl. Environ. Microbiol.*, 69(1), 139–145, 2003.
100. Thorpe, C., Hoober, K.L., Raje, S., et al., Sulfhydryl oxidases: emerging catalysts of protein disulfide bond formation in eukaryotes, *Arch. Biochem. Biophys.*, 405(1), 1–12, 2002.
101. Gross, E., Kastner, D.B., Kaiser, C.A., et al., Structure of Ero1p, source of disulfide bonds for oxidative protein folding in the cell, *Cell*, 117(5), 601–610, 2004.
102. Vignaud, C., Kaid, N., Rakotozafy, S., et al., Partial purification and characterization of sulfhydryl oxidase from *Aspergillus niger*, *J. Food Sci.*, 67, 2016–2022, 2002.
103. Vemulapulli, V., Miller, R., and Hoseney, R.C., Glucose oxidase in breadmaking systems, *Cereal Chem.*, 75(2), 439–442, 1998.
104. Poulsen, C. and Bak Hostrup, P., Purification and characterization of a hexose oxidase with excellent strengthening effects in bread, *Cereal Chem.*, 75(1), 51–57, 1998.
105. Wagner, M.A., Trickey, P., Chen, Z.W., et al., Monomeric sarcosine oxidase, 1: flavin reactivity and active site binding determinants, *Biochemistry*, 39(30), 8813–8824, 2000.
106. Binda, C., Coda, A., Angelini, R., et al., A 30-angstrom-long U-shaped catalytic tunnel in the crystal structure of polyamine oxidase, *Structure Fold. Des.*, 7(3), 265–276, 1999.
107. Huang, Q., Liu, Q., and Hao, Q., Crystal structures of Fms1 and its complex with spermine reveal substrate specificity, *J. Mol. Biol.*, 348(4), 951–959, 2005.
108. Fitzpatrick, P.F., Orville, A.M., Nagpal, A., et al., Nitroalkane oxidase, a carbanion-forming flavoprotein homologous to acyl-CoA dehydrogenase, *Arch. Biochem. Biophys.*, 433(1), 157–165, 2005.
109. Valley, M.P., Tichy, S.E., and Fitzpatrick, P.F., Establishing the kinetic competency of the cationic imine intermediate in nitroalkane oxidase, *J. Am. Chem. Soc.*, 127(7), 2062–2066, 2005.
110. Chiribau, C.B., Sandu, C., Fraaije, M., et al., A novel gamma-*N*-methylaminobutyrate demethylating oxidase involved in catabolism of the tobacco alkaloid nicotine by *Arthrobacter nicotinovorans* pAO1, *Eur. J. Biochem.*, 271(23–24), 4677–4684, 2004.
111. Entsch, B. and van Berkel, W.J., Structure and mechanism of *para*-hydroxybenzoate hydroxylase, *FASEB J.*, 9(7), 476–483, 1995.

112. Enroth, C., Neujahr, H., Schneider, G., et al., The crystal structure of phenol hydrox complex with FAD and phenol provides evidence for a concerted conformational chang enzyme and its cofactor during catalysis, *Structure*, 6(5), 605–617, 1998.

113. Malito, E., Alfieri, A., Fraaije, M.W., et al., Crystal structure of a Baeyer–Villiger monooxy *Proc. Natl. Acad. Sci. USA*, 101(36), 13157–13162, 2004.

114. Schmid, A., Dordick, J.S., Hauer, B., et al., Industrial biocatalysis today and tomorrow, 409(6817), 258–268, 2001.

115. Wierenga, R.K., de Jong, R.J., Kalk, K.H., et al., Crystal structure of *p*-hydroxybenzoate xylase, *J. Mol. Biol.*, 131(1), 55–73, 1979.

116. Schreuder, H.A., Prick, P.A., Wierenga, R.K., et al., Crystal structure of the *p*-hydroxyb hydroxylase–substrate complex refined at 1.9 Å resolution: analysis of the enzyme–substr enzyme–product complexes, *J. Mol. Biol.*, 208(4), 679–696, 1989.

117. Entsch, B., Cole, L.J., and Ballou, D.P., Protein dynamics and electrostatics in the fun *p*-hydroxybenzoate hydroxylase, *Arch. Biochem. Biophys.*, 433(1), 297–311, 2005.

118. Mattevi, A., The PHBH fold: not only flavoenzymes, *Biophys. Chem.*, 70(3), 217–222, 19!

119. Moonen, M.J.H., Fraaije, M.W., Rietjens, I.M.C.M., et al., Flavoenzyme-catalyzed oxyge and oxidations of phenolic compounds, *Adv. Synth. Catal.*, 344(10), 1023–1035, 2002.

120. Peelen, S., Rietjens, I.M.C.M., Boersma, M.G., et al., Conversion of phenol derivatives tc xylated products by phenol hydroxylase from *T. cutaneum*: a comparison of regioselectiv rate of conversion with calculated molecular orbital substrate characteristics, *Eur. J. B.* 227(1–2), 284–291, 1995.

121. van der Bolt, F.J.T., van den Heuvel, R.H.H., Vervoort, J., et al., ^{19}F NMR study regiospecificity of hydroxylation of tetrafluoro-4-hydroxybenzoate by wild-type and *p*-hydroxybenzoate hydroxylase: evidence for a consecutive oxygenolytic dehalogenation r ism, *Biochemistry*, 36(46), 14192–14201, 1997.

122. Jadan, A.P., Moonen, M.J.H., Boeren, S.A., et al., Biocatalytic potential of *p*-hydroxyb hydroxylase from *Rhodococcus rhodnii* 135 and *Rhodococcus opacus* 557, *Adv. Synth.* 346(2–3), 367–375, 2004.

123. Kälin, M., Neujahr, H.Y., Weissmahr, R.N., et al., Phenol hydroxylase from *Trichospor neum*: gene cloning, sequence analysis, and functional expression in *Escherichia coli*, *J. Ba* 174(22), 7112–7120, 1992.

124. Held, M., Suske, W., Schmid, A., et al., Preparative scale production of 3-substituted c using a novel monooxygenase from *Pseudomonas azelaica* HBP 1, *J. Mol. Catal. B: Enzym.* 87–93, 1998.

125. Held, M., Schmid, A., Kohler, H.P., et al., An integrated process for the production of toxic c from toxic phenols based on a designer biocatalyst, *Biotechnol. Bioeng.*, 62(6), 641–648, 199

126. Schmid, A., Vereyken, I., Held, M., et al., Preparative regio- and chemoselective functiona of hydrocarbons catalyzed by cell free preparations of 2-hydroxybiphenyl 3-monooxy *J. Mol. Catal. B: Enzym.*, 11(4–6), 455–462, 2001.

127. Meyer, A., Schmid, A., Held, M., et al., Changing the substrate reactivity of 2-hydroxyb 3-monooxygenase from *Pseudomonas azelaica* HBP1 by directed evolution, *J. Biol. Chem.* 5575–5582, 2002.

128. Meyer, A., Wursten, M., Schmid, A., et al., Hydroxylation of indole by laboratory 2-hydroxybiphenyl 3-monooxygenase, *J. Biol. Chem.*, 277(37), 34161–34167, 2002.

129. Meyer, A., Held, M., Schmid, A., et al., Synthesis of 3-*tert*-butylcatechol by an eng monooxygenase, *Biotechnol. Bioeng.*, 81(5), 518–524, 2003.

130. Kamerbeek, N.M., Janssen, D.B., van Berkel, W.J.H., et al., Baeyer–Villiger monooxygen emerging family of flavin-dependent biocatalysts, *Adv. Synth. Catal.*, 345(6–7), 667–678, 2

131. Fraaije, M.W., Kamerbeek, N.M., van Berkel, W.J., et al., Identification of a Baeyer– monooxygenase sequence motif, *FEBS Lett.*, 518(1–3), 43–47, 2002.

132. Cashman, J.R., The implications of polymorphisms in mammalian flavin-containing me genases in drug discovery and development, *Drug Discov. Today*, 9(13), 574–581, 2004.

133. Walsh, C.T. and Chen, Y.-C.J., Enzymatic Baeyer–Villiger-oxidations by flavin-dependen oxygenases, *Angew. Chem. Int. Ed. Engl.*, 100(2), 342–352, 1988.

134. Mihovilovic, M.D., Müller, B., and Stanetty, P., Monooxygenase-mediated Baeyer–Villiger oxidations, *Eur. J. Org. Chem.*, 2002(22), 3711–3730, 2002.
135. Alphand, V., Carrea, G., Wohlgemuth, R., et al., Towards large-scale synthetic applications of Baeyer–Villiger monooxygenases, *Trends Biotechnol.*, 21(7), 318–323, 2003.
136. Ottolina, G., de Gonzalo, G., Carrea, G., et al., Enzymatic Baeyer–Villiger oxidation of bicyclic diketones, *Adv. Synth. Catal.*, 347(7–8), 1035–1040, 2005.
137. Gutierrez, M.-C., Furstoss, R., and Alphand, V., Microbial transformations 60: enantioconvergent Baeyer–Villiger oxidation via a combined whole cells and ionic exchange resin-catalysed dynamic resolution process, *Adv. Synth. Catal.*, 347(7–8), 1051–1059, 2005.
138. Fraaije, M.W., Wu, J., Heuts, D.P., et al., Discovery of a thermostable Baeyer–Villiger monooxygenase by genome mining, *Appl. Microbiol. Biotechnol.*, 66(4), 393–400, 2005.
139. Moonen, M.J.H., Westphal, A.H., Rietjens, I.M.C.M., et al., Enzymatic Baeyer–Villiger oxidation of benzaldehydes, *Adv. Synth. Catal.*, 347(7–8), 1027–1034, 2005.
140. Mihovilovic, M.D., Kapitan, P., Rydz, J., et al., Biooxidation of ketones with a cyclobutanone structural motif by recombinant whole-cells expressing 4-hydroxyacetophenone monooxygenase, *J. Mol. Catal. B Enzym.*, 32(1), 135–140, 2005.
141. Sicard, R., Chen, L.S., Marsaioli, A.J., et al., A fluorescence-based assay for Baeyer–Villiger monooxygenases, hydroxylases and lactonases, *Adv. Synth. Catal.*, 347(7–8), 1041–1050, 2005.
142. Sheng, D., Ballou, D.P., and Massey, V., Mechanistic studies of cyclohexanone monooxygenase: chemical properties of intermediates involved in catalysis, *Biochemistry*, 40(37), 11156–11167, 2001.
143. Bocola, M., Schulz, F., Leca, F., et al., Converting phenylacetone monooxygenase into phenylcyclohexanone monooxygenase by rational design: towards practical Baeyer–Villiger monooxygenases, *Adv. Synth. Catal.*, 347(7–8), 979–986, 2005.
144. Fraaije, M.W., Kamerbeek, N.M., Heidekamp, A.J., et al., The prodrug activator EtaA from *Mycobacterium tuberculosis* is a Baeyer–Villiger monooxygenase, *J. Biol. Chem.*, 279(5), 3354–3360, 2004.
145. Kirchner, U., Westphal, A.H., Muller, R., et al., Phenol hydroxylase from *Bacillus thermoglucosidasius* A7, a two-protein component monooxygenase with a dual role for FAD, *J. Biol. Chem.*, 278(48), 47545–47553, 2003.
146. Thotsaporn, K., Sucharitakul, J., Wongratana, J., et al., Cloning and expression of *p*-hydroxyphenylacetate 3-hydroxylase from *Acinetobacter baumannii*: evidence of the divergence of enzymes in the class of two-protein component aromatic hydroxylases, *Biochim. Biophys. Acta*, 1680(1), 60–66, 2004.
147. van den Heuvel, R.H., Westphal, A.H., Heck, A.J., et al., Structural studies on flavin reductase PheA2 reveal binding of NAD in an unusual folded conformation and support novel mechanism of action, *J. Biol. Chem.*, 279(13), 12860–12867, 2004.
148. Prieto, M.A. and Garcia, J.L., Molecular characterization of 4-hydroxyphenylacetate 3-hydroxylase of *Escherichia coli*: a two-protein component enzyme, *J. Biol. Chem.*, 269(36), 22823–22829, 1994.
149. Hartmans, S., van der Werf, M.J., and de Bont, J.A., Bacterial degradation of styrene involving a novel flavin adenine dinucleotide-dependent styrene monooxygenase, *Appl. Environ. Microbiol.*, 56(5), 1347–1351, 1990.
150. Panke, S., Witholt, B., Schmid, A., et al., Towards a biocatalyst for (*S*)-styrene oxide production: characterization of the styrene degradation pathway of *Pseudomonas* sp. strain VLB120, *Appl. Environ. Microbiol.*, 64(6), 2032–2043, 1998.
151. Di Gennaro, P., Colmegna, A., Galli, E., et al., A new biocatalyst for production of optically pure aryl epoxides by styrene monooxygenase from *Pseudomonas fluorescens* ST, *Appl. Environ. Microbiol.*, 65(6), 2794–2797, 1999.
152. O'Leary, N.D., Duetz, W.A., Dobson, A.D., et al., Induction and repression of the sty operon in *Pseudomonas putida* CA-3 during growth on phenylacetic acid under organic and inorganic nutrient-limiting continuous culture conditions, *FEMS Microbiol. Lett.*, 208(2), 263–268, 2002.
153. Otto, K., Hofstetter, K., Rothlisberger, M., et al., Biochemical characterization of StyAB from *Pseudomonas* sp. strain VLB120 as a two-component flavin-diffusible monooxygenase, *J. Bacteriol.*, 186(16), 5292–5302, 2004.

154. Hollmann, F., Hofstetter, K., Habicher, T., et al., Direct electrochemical regeneration ⟨ oxygenase subunits for biocatalytic asymmetric epoxidation, *J. Am. Chem. Soc.*, 127(1⟩ 6541, 2005.

155. Gray, K.A., Mrachko, G.T., and Squires, C.H., Biodesulfurization of fossil fuels, *Cu⟨ Microbiol.*, 6(3), 229–235, 2003.

156. Denome, S.A., Oldfield, C., Nash, L.J., et al., Characterization of the desulfurization ge⟨ *Rhodococcus* sp. strain IGTS8, *J. Bacteriol.*, 176(21), 6707–6716, 1994.

157. Piddington, C.S., Kovacevich, B.R., and Rambosek, J., Sequence and molecular charact⟨ of a DNA region encoding the dibenzothiophene desulfurization operon of *Rhodococcus* ⟨ IGTS8, *Appl. Environ. Microbiol.*, 61(2), 468–475, 1995.

158. Matsubara, T., Ohshiro, T., Nishina, Y., et al., Purification, characterization, and overe⟨ of flavin reductase involved in dibenzothiophene desulfurization by *Rhodococcus erythroṕ Appl. Environ. Microbiol.*, 67(3), 1179–1184, 2001.

159. Galan, B., Diaz, E., and Garcia, J.L., Enhancing desulphurization by engineering a flav⟨ tase-encoding gene cassette in recombinant biocatalysts, *Environ. Microbiol.*, 2(6), 687–6⟨

160. Reichmuth, D.S., Hittle, J.L., Blanch, H.W., et al., Biodesulfurization of dibenzothio⟨ *Escherichia coli* is enhanced by expression of a *Vibrio harveyi* oxidoreductase gene, *B⟨ Bioeng.*, 67(1), 72–79, 2000.

161. Arensdorf, J.J., Loomis, A.K., DiGrazia, P.M., et al., Chemostat approach for the evolution of biodesulfurization gain-of-function mutants, *Appl. Environ. Microbiol.*, 68 698, 2002.

162. Coco, W.M., Levinson, W.E., Crist, M.J., et al., DNA shuffling method for generati⟨ recombined genes and evolved enzymes, *Nat. Biotechnol.*, 19(4), 354–359, 2001.

163. Keller, S., Wage, T., Hohaus, K., et al., Purification and partial characterization of tr⟨ 7-halogenase (PrnA) from *Pseudomonas fluorescens*, *Angew. Chem. Int. Ed. Engl.* 2300–2302, 2000.

164. Burd, V.N. and van Pee, K.H., Cloning and sequencing of the gene of tryptophan-7-ha⟨ from *Pseudomonas fluorescens* strain CHA0, *Biochemistry* (Moscow), 69(6), 674–677, 20⟨

165. Yeh, E., Garneau, S., and Walsh, C.T., Robust *in vitro* activity of RebF and RebH component reductase/halogenase, generating 7-chlorotryptophan during rebeccamycin⟨ thesis, *Proc. Natl. Acad. Sci. USA*, 102(11), 3960–3965, 2005.

166. van Pee, K.H. and Unversucht, S., Biological dehalogenation and halogenation reactions *sphere*, 52(2), 299–312, 2003.

167. Unversucht, S., Hollmann, F., Schmid, A., et al., FADH$_2$-dependence of tryptophan 7-ha⟨ *Adv. Synth. Catal.*, 347(7–8), 1163–1167, 2005.

168. Dong, C., Kotzsch, A., Dorward, M., et al., Crystallization and X-ray diffraction of a h⟨ ing enzyme, tryptophan 7-halogenase, from *Pseudomonas fluorescens*, *Acta Crystall. S⟨ 60(8), 1438–1440, 2004.

7 Preparation of Chiral Pharmaceuticals through Enzymatic Acylation of Alcohols and Amines

Vicente Gotor-Fernández, Francisca Rebolledo, and Vicente Gotor

CONTENTS

7.1   INTRODUCTION

Nowadays, biocatalysis is a standard methodology for the production of chemicals. talytic steps are already in use to produce a wide range of products, including agric chemicals, drugs, and important commodity chemicals such as acrylamide [1]. These talytic reactions can be carried out in organic solvents or in aqueous media. H biocatalysis in nonconventional media has shown to be an excellent strategy for chemists in the production of chemicals that are difficult to obtain by chemical conve procedures [2].

The use of enzymes in organic solvents is currently of special relevance for the preparation of products of fine chemicals. Recently, tremendous efforts have been made to establish enantio-selective routes for the preparation of enantiomerically pure compounds due to their importance in the pharmaceutical, agricultural, and food industries [3]. Among the enzymes tested, lipases are the biocatalysts that have shown the greatest utility, especially through enzymatic transesterification reactions. In the last few years, there has been an ever-increasing trend for chiral drug substances to focus on single enantiomers instead of racemic mixtures. The most important reason for developing stereochemically pure and defined compounds is the difference of biological activity, displayed in many cases by each enantiomer of a chiral compound [4].

In this context, biocatalysis has been well recognized as an excellent strategy for the preparation of chiral pharmaceuticals [5]. Biocatalytic processes are environment-friendly in contrast to conventional chemical catalysis, especially when these make use of heavy metal catalysis. The role of biotransformations in the pharmaceutical, fine chemicals, or food industries is clearly expanding, as the pharmaceutical companies largely use this methodology [6]. In addition, the application of enzyme technology in the chemical industry has recently been well documented [7].

Both isolated enzymes and whole cells, either in soluble or immobilized form, have been used successfully in the synthesis of pharmacologically valuable materials [8,9]. However, hydrolytic enzymes, especially lipases, are widely used in organic synthesis as environment-friendly catalysts that posses broad substrate specificities, display high stereoselectivity, are commercially available, and do not require the use of cofactors [10].

These biocatalysts have been exploited for asymmetric synthesis transformations, led by the growing demand for enantiopure pharmaceuticals. Furthermore, lipase-catalyzed reactions are normally carried out under mild conditions, and can be used in organic solvents. In addition, biocatalysis in nonaqueous media has been widely used for the resolution of alcohols, acids, or lactones through enzymatic transesterification reactions using different lipases [11]. Moreover, other processes, such as the enzymatic acylation of amines or ammonia, have shown themselves to be of great utility for the resolution of amines and the preparation of chiral amides [12].

The main difference between enzymatic acylation of alcohols and amines is the use of the corresponding acyl donor, because activated esters that are of utility in acylation of alcohols react with amines in the absence of a biocatalyst, and nonactivated esters must be used to carry out an enzymatic aminolysis or ammonolysis reaction. In addition, lipases are the most efficient hydrolases to catalyze the acylation of amines and ammonia, because these hydrolytic enzymes have very low amidase activity, although in some cases the hydrolysis or alcoholysis of amides can be useful to achieve chiral amines [13].

Stereoselective biotransformations can be grouped into two different classes: asymmetric synthesis and kinetic resolution (KR) of racemic mixtures. Conceptually, they differ from each other in the fact that the asymmetric synthesis implies the formation of one or more chirality elements in a substrate, a KR is based on a transformation which, subsequently, makes easier the separation of the two enantiomers of the racemic substrate. This fact involves a practical difference: in a KR only half of the starting material is used (Scheme 7.1). When only one enantiomer of a substrate is required this constitutes a disadvantage of KRs and different approaches have been developed to overcome this limitation [14]. More attention has been recently paid in the dynamic kinetic resolution (DKR) [15] and consists in carrying out an *in situ* continuous racemization of the substrate, so that, theoretically, all of the racemic starting material can be used for transformation into one enantiomer. This new strategy has appeared in asymmetric catalysis during the last decade and the most common methodology involves a lipase as biocatalyst and a metal-organic complex as chemical catalyst [16]. Although, DKR of alcohols has been widely used for their resolution by enzymatic hydrolysis or transesterification, the application to the resolution of amines has been much less investigated [17]. A new

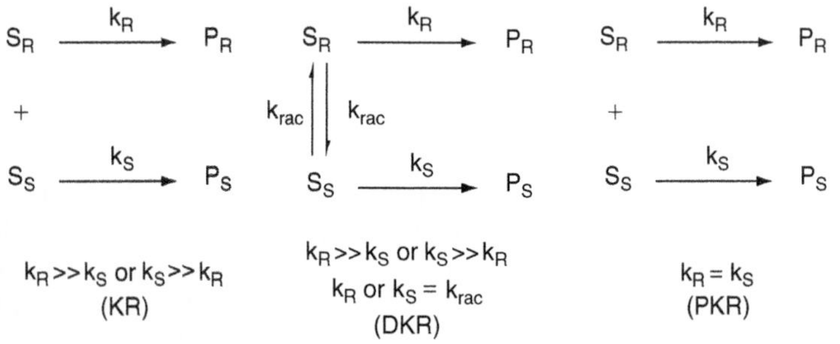

$$S_R \xrightarrow{k_R} P_R \qquad S_R \xrightarrow{k_R} P_R \qquad S_R \xrightarrow{k_R} P_R$$

$$+ \qquad k_{rac} \updownarrow k_{rac} \qquad +$$

$$S_S \xrightarrow{k_S} P_S \qquad S_S \xrightarrow{k_S} P_S \qquad S_S \xrightarrow{k_S} P_S$$

$$k_R \gg k_S \text{ or } k_S \gg k_R$$
(KR)

$$k_R \gg k_S \text{ or } k_S \gg k_R$$
$$k_R \text{ or } k_S = k_{rac}$$
(DKR)

$$k_R = k_S$$
(PKR)

SCHEME 7.1 Kinetic resolution (KR), dynamic kinetic resolution (DKR), and paralle resolution (PKR).

concept has been demonstrated in the resolution of racemic mixtures in which both enan of a substrate react with similar rate to give two different products with high enant excess, this strategy was called by Vedejs and Chen as parallel kinetic resolution (PK This concept in stereoselective biocatalysis has been recently reviewed [19].

As mentioned above there is other possibility, of great interest is the desymmetriz proquiral or *meso* compounds using hydrolytic enzymes. The desymmetrization of sy compounds consists of a modification that eliminates one or more elements of symr the substrate. If the symmetry elements that preclude chirality are eliminated, enant tivity can be achieved [20]. Enantioselective enzymatic desymmetrizations (EEDs) b the field of asymmetric synthesis and, accordingly, a maximum yield of 100% can be ; [21]. For this reason, they constitute a very interesting alternative to KRs for the prep of optically active compounds, which is reflected in the increasing number of en desymmetrizations applied to synthesis published in the literature during the recent ye

To date, the enzymatic acylation of alcohols in organic solvents catalyzed by lipases KR is still one of the most popular methods for the resolution of different hydroxyl com although other types of resolutions such as DKR are increasing with time. Among the acyl donors used in these resolutions, the most efficient are vinyl esters [23]. These es also be employed in EEDs for the desymmetrization of proquiral or *meso*-diol compo an enzymatic acylation of alcohols in organic solvents. Depending on the substrate, en resolution and desymmetrization processes have shown their utility in the prepar pharmaceutical intermediates [24].

This chapter covers a wide range of resolutions of racemates and desymmetriz proquiral or *meso* compounds through enzymatic acylation of alcohols and amine lipases for the preparation of a number of pharmaceuticals and chiral building block synthesis of chiral drugs that has been described in the literature. The enzymatic resol pharmaceuticals through the acylation of the hydroxyl group or hydrolysis of esters been reviewed in the last decade for many authors [1,5,6,8,16,22–34]. The enantios processes have been considered without including examples where biocatalyst has be for the regioselective modification of different derivatives.

In this chapter, we have classified the enzymatic processes depending on the material in the reaction: alcohol, aminoalcohol, and amine derivatives.

7.2 ENZYMATIC ACYLATION OF ALCOHOLS

Among the alcohol derivatives, we have divided the section depending on the nu hydroxyl groups present in the molecule, in this manner we have three different d monoalcohol, diol, and polyalcohol derivatives.

7.2.1 Resolution of Monoalcohols

7.2.1.1 Aminoglutethimide (1)

It was originally developed as an anticonvulsant and now this itself and other derivatives are effective drugs against breast cancer for postmenopausal patients. The enzymatic transesterification of racemic 2-ethyl-5-hydroxy-2-phenylpentamonitrile (**2**) was done with different lipases (Scheme 7.2) obtaining the best results with *Pseudomonas cepacia* lipase (PSL) and *Pseudomonas fluorecens* lipase (LAK) [35]. Enzymatic acylation of **2** by PSL in hexane provided (*R*)-alcohol with 99% ee at 66% conversion after 8 h, while using LAK in diisopropyl ether (DIPE) (*S*)-alcohol is obtained with 96% ee at 86% conversion after 6 h.

The cholesterol derivative **5** is a cholesterol metabolite that has been used as the (25*RS*)-epimeric mixture in studies concerning the inhibitory effects of oxysterols on the hydroxymethylglutarate coenzyme (HMG CoA) reductase activity. The 3β-*O*-silyl ether of (25*RS*)-**5** was subjected to an acylation process with vinyl acetate catalyzed by PSL in chloroform affording the (25*R*)-acetate **6** in 30% conversion after 24 h (Scheme 7.3) [36].

7.2.1.2 Clavularin A and Clavularin B (7 and 8)

These anticancer agents were synthesized from (±)-**9** by transesterification reaction catalyzed by lipase LIP (from *Pseudomonas* sp.) using vinyl acetate in *tert*-butyl methyl ether (TBME) [37]. After 12 d at room temperature, (−)-acetate **10** was obtained in 36% yield and 95% ee, while (+)-alcohol **9** was recovered in 43% yield and 55% ee (Scheme 7.4).

7.2.1.3 Coronafacic Acid (11)

Compound (+)-**11** is a common chiral building block of natural products present in diverse pharmaceuticals. The resolution of racemic **12** in THF catalyzed by PSL supported on celite using vinyl acetate as acyl donor at room temperature for 4 h (Scheme 7.5) afforded allylic acetate (−)-**13** (>98% ee and 44% yield) and (+)-alcohol **12** (>99% ee and 46% yield) [38].

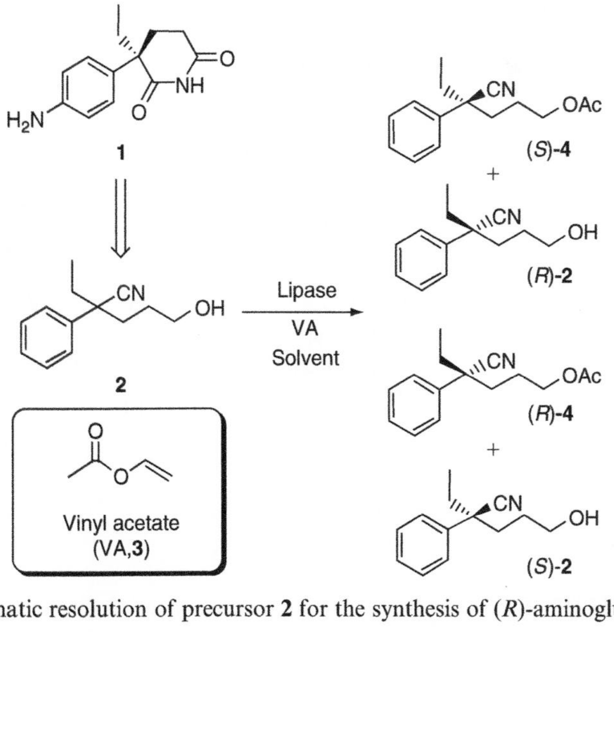

SCHEME 7.2 Enzymatic resolution of precursor **2** for the synthesis of (*R*)-aminoglutethimide **1**.

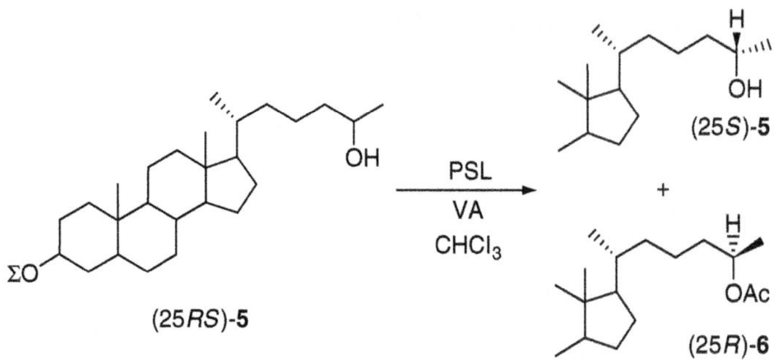

SCHEME 7.3 Enzymatic resolution of **5**.

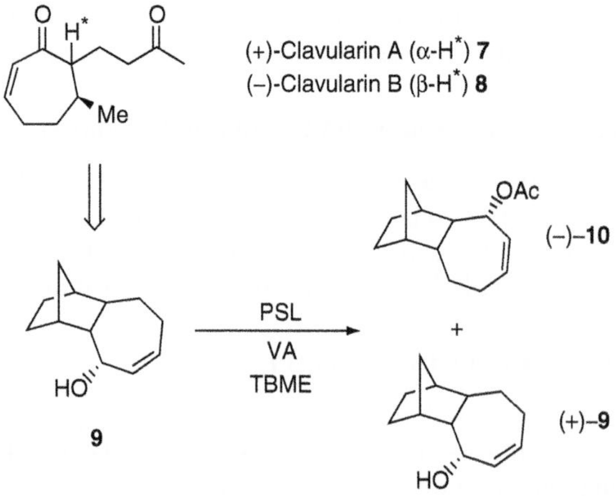

SCHEME 7.4 Enzymatic resolution of precursor **9** for the synthesis of clavularin A and clavula

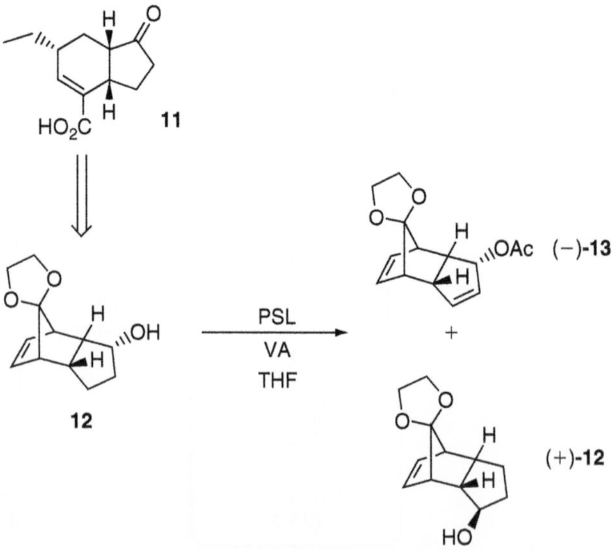

SCHEME 7.5 Enzymatic resolution of precursor **12** for the synthesis of coronafacic acid.

7.2.1.4 (+)-Docosa-4,15-dien-1-yn-3-ol (14)

Compound **22** shows *in vitro* immunosuppressive and antitumor activities. The lipase-catalyzed biotransformation of racemic **14Z** was carried out with CAL-B (Novozyme 435) and vinyl acetate in TBME obtaining the acylated product (4*E*,15*Z*)-acetate **15Z** and the unreacted alcohol **14Z** being formed in 95 and 65% ee, respectively, after 47% conversion in 2 h at 30°C (Scheme 7.6) [39]. Similarly, racemic **14E** was acylated with CAL-B at 30°C in 2 h, obtaining the remaining alcohol with 81% ee and the acetate **15E** with 94% ee for a 45% conversion.

7.2.1.5 E-4018 (16)

Compound **16** is a potent analgesic, the (*R*)-enantiomer being more active than the (*S*)-isomer. The alcohol **17** is an effective building block for the synthesis of **16** and its enzymatic resolution was achieved by a lipase-catalyzed transesterification using PSL, molecular sieves, and vinyl acetate as solvent and acyl donor, in this way the monoacetate (*S*)-**18** was obtained in 55% yield and 96% ee at 60°C (Scheme 7.7) [40].

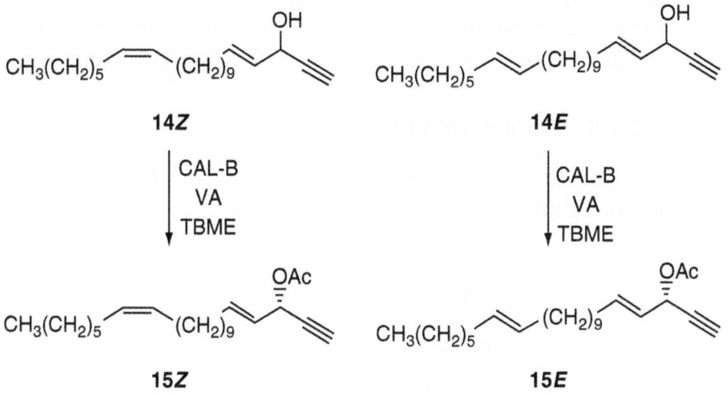

SCHEME 7.6 Enzymatic resolution of precursors **15Z** and **15E** for the synthesis of a docosa derivative.

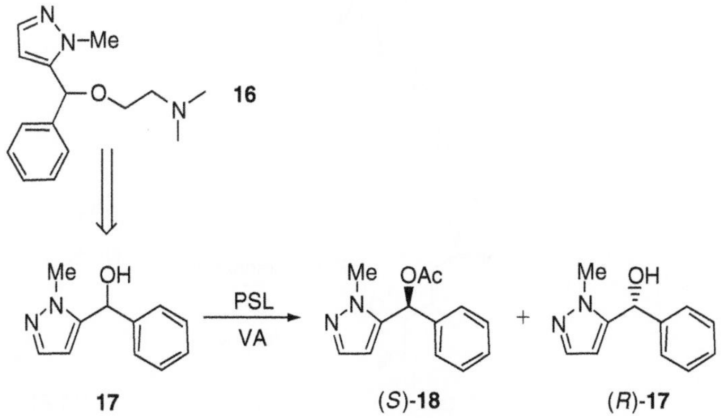

SCHEME 7.7 Enzymatic resolution of precursor **17** for the synthesis of **16**.

7.2.1.6   Epothilone A (19)

Pseudomonas AK lipase was used to synthesize the chiral intermediate **20** in th
synthesis of the potent antitumor agent epothilone A [41]. The process was carr
at room temperature using vinyl acetate as acyl donor and hexane as solvent, pr
after 72 h the enantiomerically enriched alcohol (*S*)-**20** in 48% isolated yield and
(Scheme 7.8).

7.2.1.7   Fluoxetine (22), Tomoxetine (23), and Nisoxetine (24)

Compounds **22**, **23**, and **24** are the group of nontricyclic antidepressants that act by in
the uptake of norepinephrine and serotonin. The synthesis of these three compoun
carried out from racemic 3-chloro-1-phenylpropan-1-ol (**25**) by enzymatic transesteri
in hexane using vinyl butanoate as acyl donor, and CAL-B as biocatalyst at 30°C (
7.9) [42]. Alcohol (*S*)-**25** was recovered with 96% ee and 33% isolated yield and acetat
with 97% ee and 31% yield.
   In a different chemoenzymatic approach, these pharmaceutical derivatives were
sized using 3-hydroxy-3-phenylpropanenitrile **28** as the starting material (Scheme 7.
Transesterification of **28** was achieved using both immobilized and nonimmobilized
biocatalyst at 40°C in DIPE. Using immobilized enzyme after 16 h, the (*S*)-alcohol
furnished in 46% yield and 99% ee, same values were obtained for the (*R*)-acetate 29
the free enzyme, the reaction was slower and after 180 h the alcohol was obtained in 4
the acetate in 44% yields, with ee > 99% for both compounds.

7.2.1.8   Formoterol (30)

The potent β2-receptor agonist formoterol (*R*,*R*)-**30** is on the market as a diaste
mixture despite its varying efficacy of stereoisomer. The preparation of the (*R*,*R*)-stere

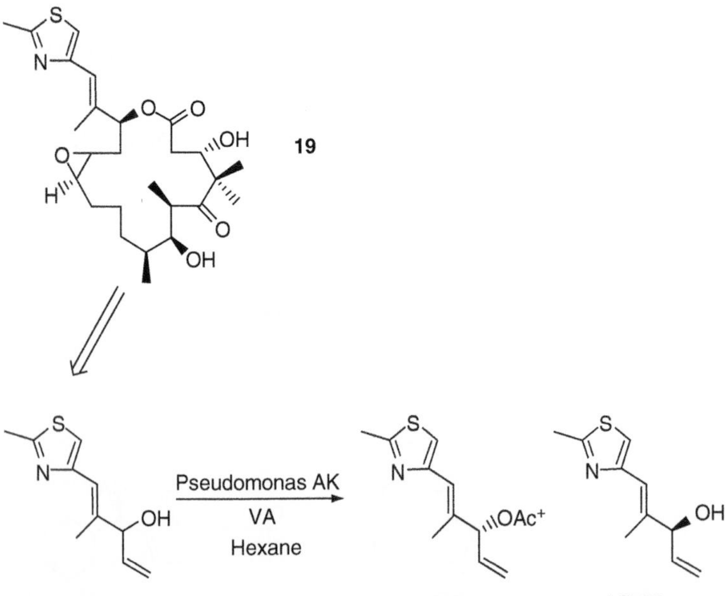

SCHEME 7.8 Enzymatic resolution of precursor **20** for the synthesis of **19**.

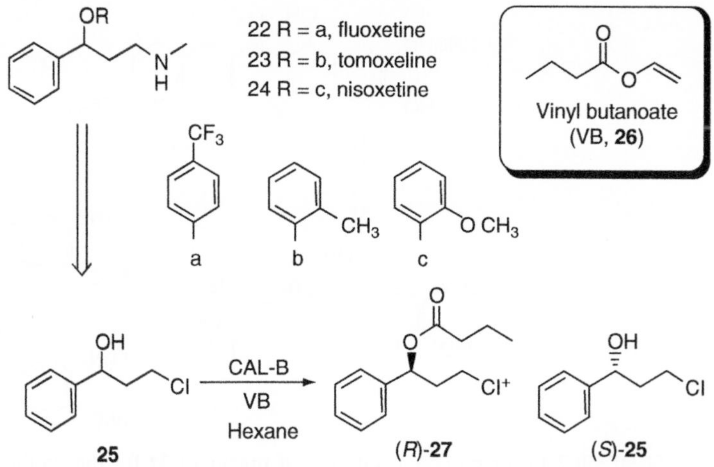

SCHEME 7.9 Enzymatic resolution of precursor **25** for the synthesis of **22, 23,** and **24.**

was achieved by enzymatic resolution of (*R*)-bromohydrin **31** (Scheme 7.11) [44]. Immobilized lipase PS-30 from *Pseudomonas cepacia*, vinyl acetate, and TBME were used as solvent, and after 50% conversion at 37°C and 99 h reaction, the acetylated product (*S*)-**32** (48% yield and 86% ee) and the unreacted desired (*R*)-**31** (46% isolated yield and 96% ee) were obtained.

7.2.1.9 HMG CoA Inhibitors

Compound [4-[4α,6β(E)]]-6[4,4-bis(4-fluorophenyl)-3-(1-methyl-1H-tetrazol-5-yl)-1,3-butadienyl]-tetrahydro-4-hydroxy-2H-pyren-2-one, (*R*)-(+)-**33**, is an anticholesterol drug that acts by inhibition of HMG CoA reductase. It was resolved by the lipase-catalyzed stereoselective acetylation of the racemic alcohol using PS-30 lipase in toluene and iso-propenyl acetate as acyl donor obtaining the (*R*)-alcohol with ee > 98% and 49% yield (Scheme 7.12) [45]. The industrial scale resolution process was also developed in methyl ethyl ketone at 50 g/L substrate concentration, a reaction yield of 46% and 96% ee was obtained. Furthemore, lipase was immobilized on Accurel PP and the enzyme was reused five times without loss of activity. The enzymatic process was scaled up to a 640 L preparative batch using immobilized PS-30 lipase at 4 g/L of racemic substrate in toluene as a solvent, isolating (*R*)-alcohol in 40% yield with 98.5% ee.

7.2.1.10 β-Lactams

β-Lactam antibiotics have been obtained using an one-step enzymatic process from their corresponding building blocks. Nagai and coworkers developed the enzymatic synthesis

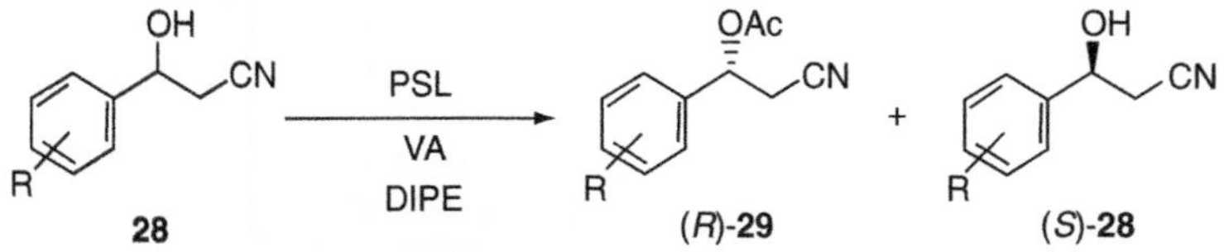

SCHEME 7.10 Enzymatic resolution of precursor **28** for the synthesis of **22, 23,** and **24.**

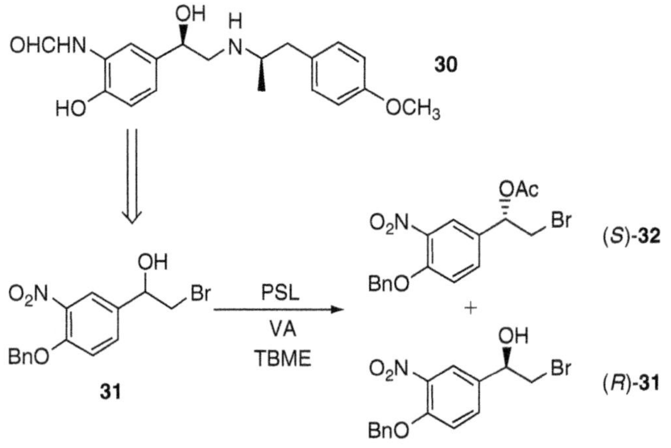

SCHEME 7.11 Enzymatic resolution of precursor **31** for the synthesis of formoterol.

of optically active β-lactams by lipase-catalyzed KR using the transesterification
hydroxymethyl β-lactam **36** in dichloromethane in the presence of vinyl acetate
donor (Scheme 7.13) [46]. The reaction yields of 35 to 50% and ee of 93 to to 99
obtained depending on the specific substrate used in this process. Lipase B from *Pseud*
fragi and lipase PS-30 were used in these reactions.

7.2.1.11 3-Methylnonacosanol (38)

Compound **38** is traditionally used for the treatment of jaundice, enlargement of the liv
spleen, leprosy, and other skin diseases. The (*S*)-**38** was prepared through the acetat
mediate (*S*)-**40** obtained by the enzymatic resolution of racemic alcohol **38** (Schem
[47,48]. CRL catalyzed the transesterification of **39** using vinyl acetate in DIPE a
temperature. After 5 h and 40% conversion, the acetate (*S*)-**40** (ee > 96%) and the r

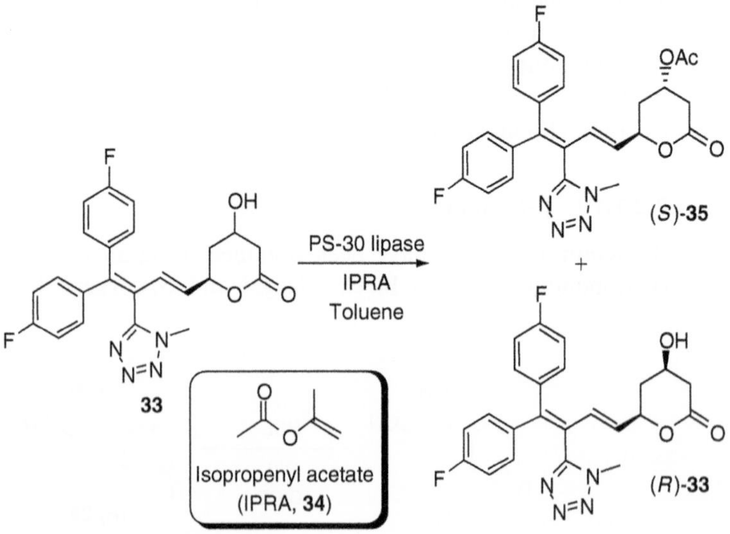

SCHEME 7.12 Enzymatic resolution of **33**.

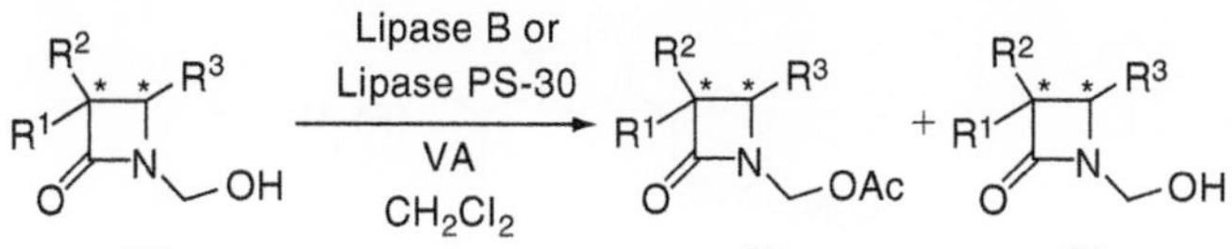

R^1 = H, Me, Bu

R^2 = H, H, Me

R^3 = Ph, PhCH=CH, PhCH$_2$CH$_2$, C$_6$H$_4$OMe

SCHEME 7.13 Enzymatic resolution of **36**.

(R)-**39** (71% ee) were produced. A second acetylation of the partially resolved **39** led to the (R)-alcohol with 98% ee.

7.2.1.12 Nifenalol (41) and Sotalol (42)

Drugs bearing a structured unit of 2-amino-1-arylethanol such as **41** and **42** are of great importance as β-adrenergic blockers. They are also used in the therapy of asthma, bronchitis, and congestive heart failure. Among their enantiomers, only (R)-(−)-**41** and (S)-(+)-**42** are β-adrenergic blockers and are effective in the treatment of cardiovascular diseases. Both **41** and **42** have a common precursor, i.e., 2-bromo-1-(4-nitrophenyl)ethanol (**43**), while their final products have opposite configurations at C-1 (Scheme 7.15).

The transesterification reaction of **43** using PS-C-II as biocatalyst, vinyl acetate as acyl donor, and toluene as solvent led to the formation of (S)-acetate **44** in 57% yield and (R)-alcohol **43** in 42% yield, both compounds obtained with ee > 99% [49].

7.2.1.13 Nilvadipine (45)

Nilvadipine is a calcium antagonist of the dihydropyridine group. Owing to its high receptor affinity, nilvadipine blocks L-type calcium channels in vascular muscle cells. This leads to prolonged vascular relaxation and lowering of blood pressure. The transesterification of racemic isopropyl 2-hydroxymethyl-1,4-dihydro-6-methyl-4-(3-nitrophenyl)-3-methoxycarbonyl-5-pyridinecarboxylate (**46**) was carried out in acetone using vinyl acetate as acyl donor and PSL as biocatalyst at 40°C obtaining the (S)-acetate **47** in 72% ee and 55% yield and the remaining unreacted (R)-alcohol in 97% ee and 42% yield after 44 h (Scheme 7.16) [50].

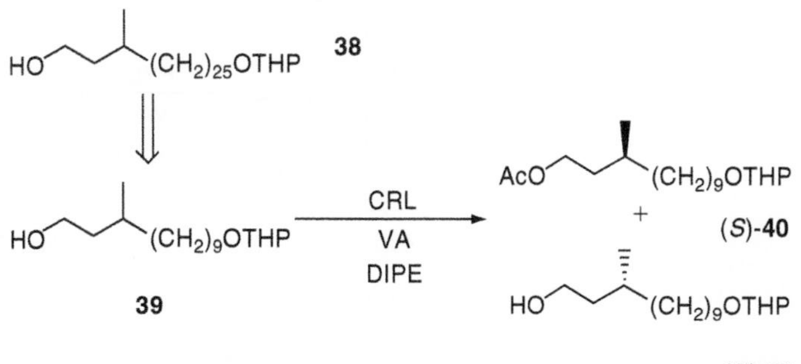

SCHEME 7.14 Enzymatic resolution of **39** for the production of 3-methylnonacosanol.

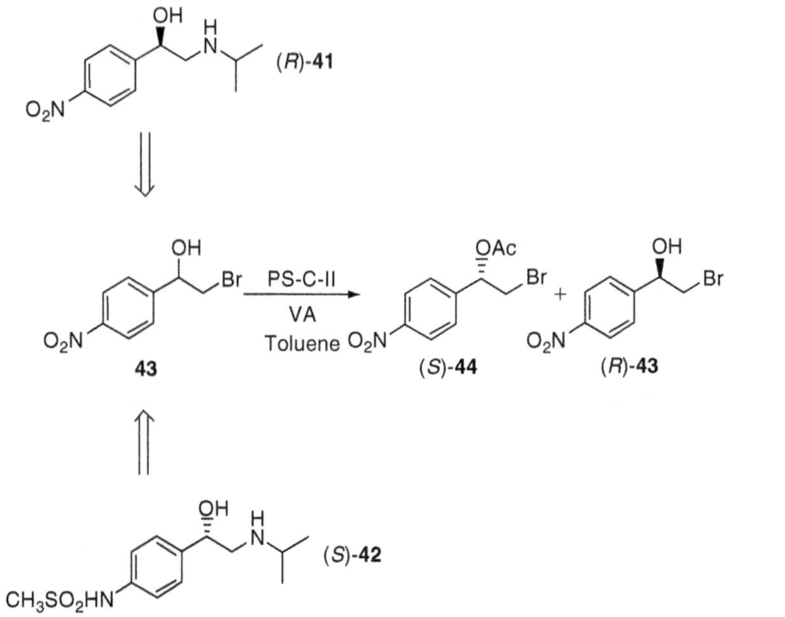

SCHEME 7.15 Enzymatic resolution of **43** for the production of (*R*)-nifenalol and (*S*)-sotalol.

7.2.1.14   OPC-29030 (48)

Compound   (*S*)-3,4-dihydro-6-[3-(1-*o*-tolyl-2-imidazolyl)sulfinylpropoxy]-2(1*H*)-quir
(OPC-29030, **48**) is a platelet adhesion inhibitor that suppresses the production of
hydroxyeicosatetraenoic acid in plateletes and has been under clinical trials. Metabolit

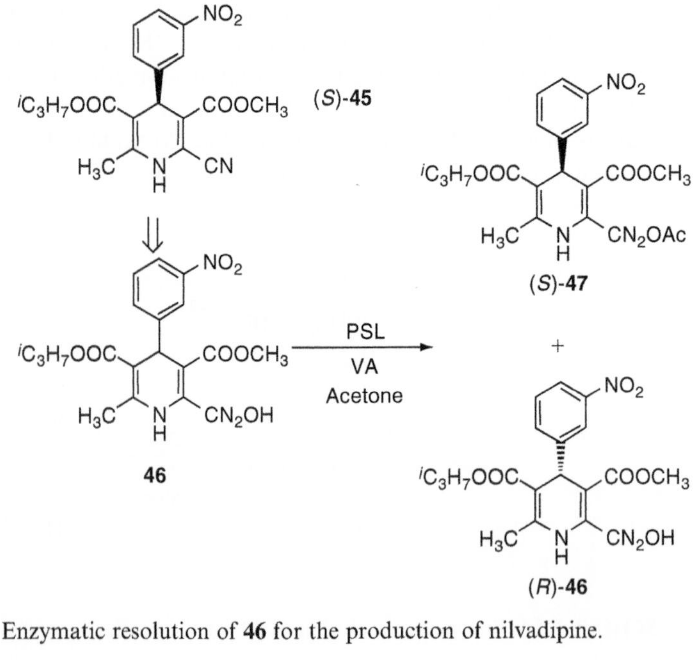

SCHEME 7.16 Enzymatic resolution of **46** for the production of nilvadipine.

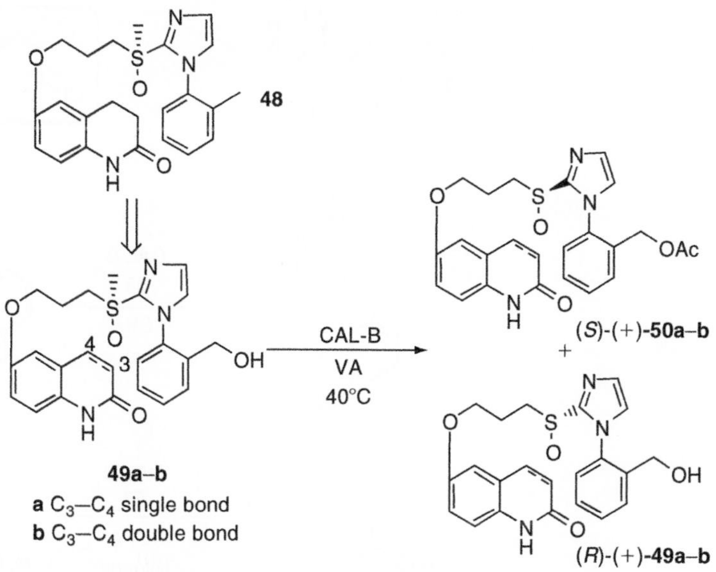

SCHEME 7.17 Enzymatic resolution of **49a–b** for the production of OPC-29030.

were resolved at room temperature using lipase-catalyzed transesterification with CAL-B (Novozyme 435) and vinyl acetate in CH_2Cl_2 (Scheme 7.17) [51]. (*S*)-Acetate **50a** was obtained in 93% ee and (*R*)-alcohol **49a** with 82% after 47% conversion, while (*S*)-acetate **50b** was recovered in 40% isolated yield and 97% ee and (*R*)-alcohol **49b** in 89% ee and 46% isolated yield.

7.2.1.15   Propanolol (51)

Propanolol **51** belongs to the group of β-adrenergic blocking agents (β-blockers) of the general structure $ArOCH_2CH(OH)CH_2NHR$, where Ar is an aryl and R is an alkyl. These compounds are used for the treatment of hypertension and angina pectoris. It has been shown that β-adrenergic receptor blocking activity resides mostly in (*S*)-enantiomers. Synthesis of (*S*)-**51** was achieved by the DKR of (±)-**52** using a combination of the ruthenium complex **53** and CAL-B (Novozyme-435) in toluene at 80°C, and *p*-chlorophenyl acetate (**54**) as acyl donor (Scheme 7.18) [52]. The (*R*)-acetate **55** was produced in ee > 99% and 86% isolated yield after 1 d reaction. The enzyme was recycled and used again for another cycle without any loss of activity. Using the same procedure, the racemic **56** was subjected to DKR to produce the (*R*)-acetate **57** (ee > 99%, conversion 92%, and isolated yield 84%) precursor of (*R*)-denopamine (**58**), a potent orally active β₁ receptor agonist for the treatment of heart failure.

The stereoselective preparation of chiral building blocks in the synthesis of **51** can be achieved via acylation of alcohols **59** [53] or **61** [54] in organic solvents using the lipase from *Pseudomonas cepacia* (Scheme 7.19). Using the cyanohydrin **59**, the (*S*)-alcohol was obtained in 96% ee after 56% conversion, while using the chloroalcohol **61** the acetate precursor (*R*)-**62** reached ee > 95% after 47% conversion.

7.2.1.16   Serotonin Antagonism and Uptake Inhibitors

The indanamine derivative MDL 27777A (**63**) is a serotonin uptake inhibitor, and studies showed that (+)-**63** is at least 10 times more active in inhibiting serotonin uptake both *in vitro* and *in vivo* than (−)-**63**. The inhibitor (+)-**63** was synthesized in both "cold" form and

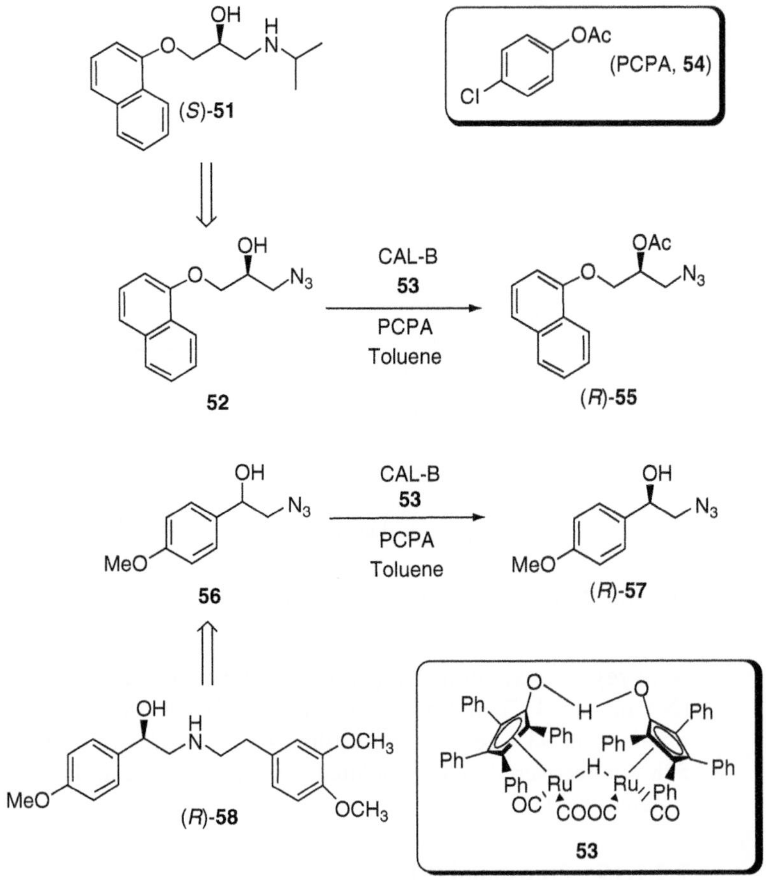

SCHEME 7.18 Dynamic kinetic resolution of **52** and **56**, precursors of (S)-propanolol and (R)-deno

[14]C-labeled. Preparation of [14]C-**63** is the first example of the use of stereoselective enz
acylation in organic solvents for the resolution of radiolabeled compounds [55]. Th
achieved using vinyl acetate, TBME, and lipase P as biocatalyst (Scheme 7.20). Af
at 22°C, the acetate (−)-**65** (43% yield, 97% ee) and the alcohol (+)-**64** (46% yield, 9
were recovered.

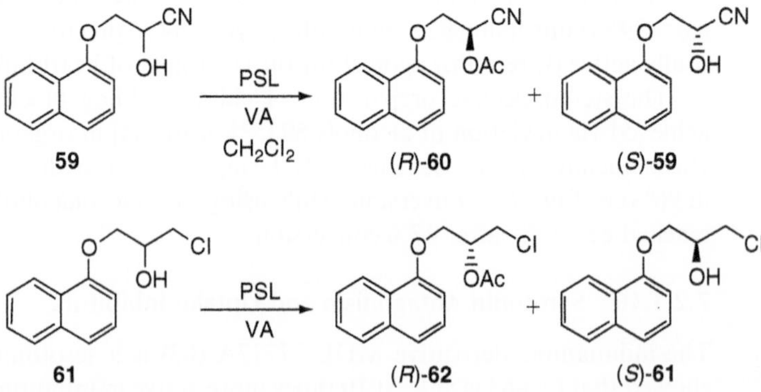

SCHEME 7.19 Enzymatic resolution of **59** and **61**, precursors of (S)-propanolol.

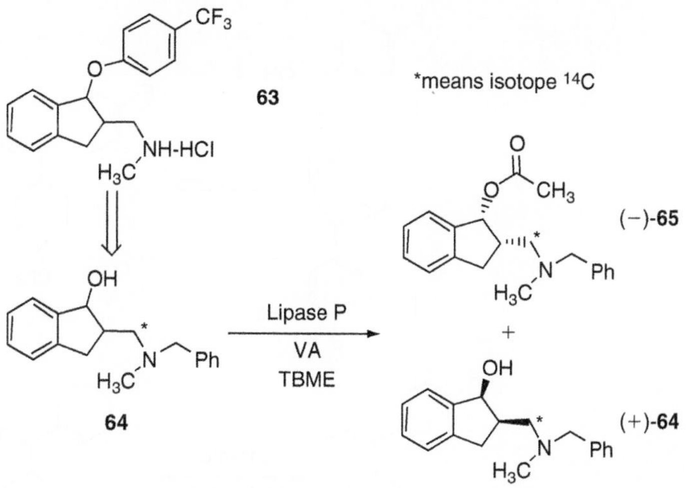

SCHEME 7.20 Enzymatic resolution of **64**, precursor of MDL 28618A, [(+)-**63**].

Serotonin antagonist **66**, which was identified as a nonnarcotic analgesic and muscle relaxant, has two chiral centers and therefore exists in the form of four stereoisomers with different biological properties. Resolution of compounds with several chiral centers presents a significant challenge, so **66** was divided into two fragments, the secondary alcohols **67** and **68**, were resolved by enzymatic acylation obtaining four chiral building blocks (*R*)-**69**, (*S*)-**67**, (*R*)-**70**, and (*S*)-**68**, in good yield and excellent optical purity (Scheme 7.21). The simple combination of these blocks gave four different enantiomers of **66** [56].

7.2.2 RESOLUTION OF DIOLS

7.2.2.1 Baclofen (71)

Compound 4-amino-3-(4-chlorophenyl)butanoic acid (**71**) is an analog of γ-aminobutyric acid, which is an inhibitory neurotransmitter that regulates the control of neuronal activity in the central nervous system to regulate several other physiological mechanisms. Unlike GABA, baclofen can cross the blood–brain barrier, and racemic baclofen is widely used as antispatic agent. The enantiomers of this compound differ in the pharmacodynamic and toxilogical properties; the (−)-enantiomer is not only more active but also more toxic than the (+)-enantiomer.

Esterification of 2-(4-chlorophenyl)-1,3-propanediol (**72**) by acetic anhydride in benzene, using porcine pancreas lipase (PPL), gave the hemiester (*S*)-**73** in high ee (>96%) and 93% yield (Scheme 7.22).

7.2.2.2 Camptosar (74)

Camptosar is used in the treatment of ovarian cancer. Sih developed the enzymatic resolution of diol **75** using PSL supported on celite. Using isopropenyl acetate as acyl donor in TBME, (*S*)-**75** was obtained in 40% isolated yield and 99% ee after 1 d reaction at 25°C (Scheme 7.23) [57].

7.2.2.3 Cispentacin (77)

The β-amino acid cispentacin, (1*R*,2*S*)-2-aminocyclopentanecarboxylic acid (**77**), is an antifungal antibiotic. Diol (1*RS*,2*SR*)-**78** was first monosilylated and later the silyloxy alcohol

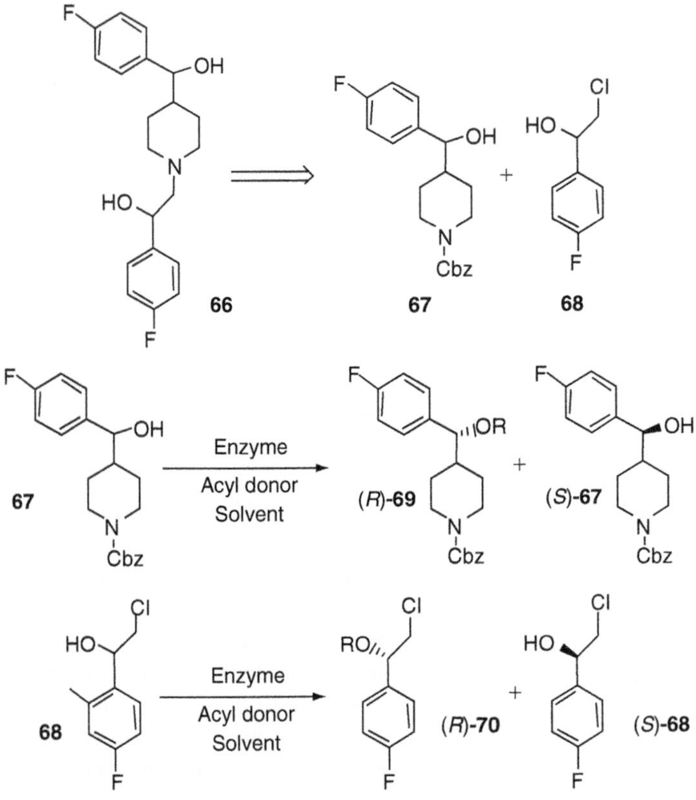

SCHEME 7.21 Enzymatic resolution of **67** and **68**, precursors of **66**.

(1*RS*,2*SR*)-**79** selectively acetylated with vinyl acetate using PSL in TBME affordi
40% conversion the acetate (1*R*,2*S*)-**80** with an ee of 94% (Scheme 7.24). The re
enantiomerically enriched silyloxy alcohol (1*S*,2*R*)-**79** with an ee of 51% was subjec
second transesterification that was terminated at 20% conversion (60% from sta
yielded a further fraction of (1*R*,2*S*)-**80** with an ee of 98% and the remaining
(1*S*,2*R*)-**79** with 99% ee [58].

7.2.2.4 ε-Hydroxytramadol (81) and δ-Hydroxytramadol (82)

Tramadol (**83**) and its modified analogs ε-hydroxytramadol (**81**) and δ-hydroxytr
(**82**) exhibit interesting pharmacological properties as analgesics. The *Candida*

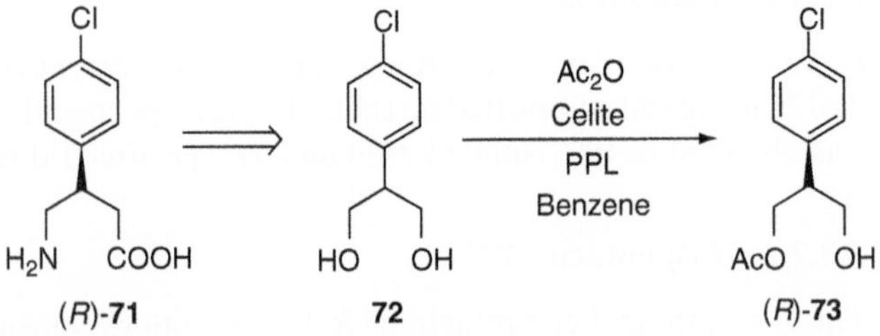

SCHEME 7.22 Enzymatic resolution of **72**, precursors of baclofen.

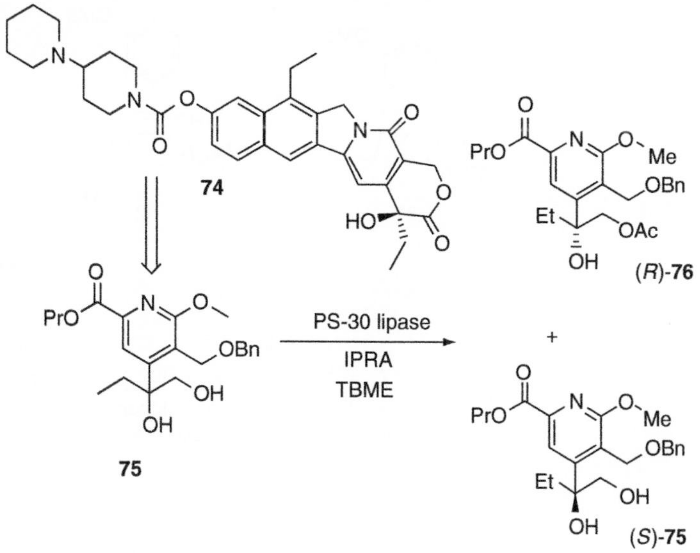

SCHEME 7.23 Enzymatic resolution of **75** for the synthesis of camptosar.

lipase (CRL)-catalyzed transesterification of racemic **81** was studied using vinyl acetate as acyl donor and solvent or isopropenyl acetate as acyl donor and toluene as solvent (Scheme 7.25) [59]. In the first case, after 7 d at room temperature and 56% conversion, alcohol **81** was obtained with 97% ee and ester **84** with 91% ee. When isopropenyl acetate was used as acyl donor, alcohol was obtained in 60% ee and ester **77** in 99% ee after 5 d of reaction at 34% conversion. Having obtained favorable results with the transesterification of **81**, the CRL-catalyzed acylation of *rac*-**82** was investigated. However, the acylation of **82** was not as selective as that of *rac*-**81**. Using isopropenyl acetate (IPRA) and toluene at room temperature the conversion reached 25% after 16 d affording the ester **85** in 68% ee and the alcohol **82** in 87% ee.

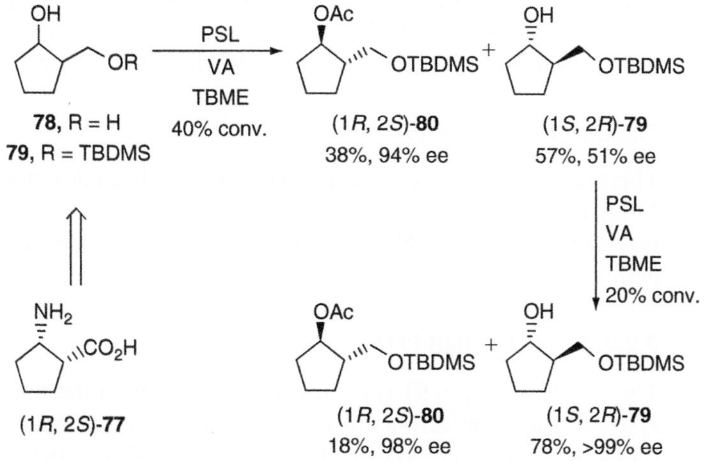

SCHEME 7.24 Enzymatic resolution of **79** for the synthesis of cispentacin.

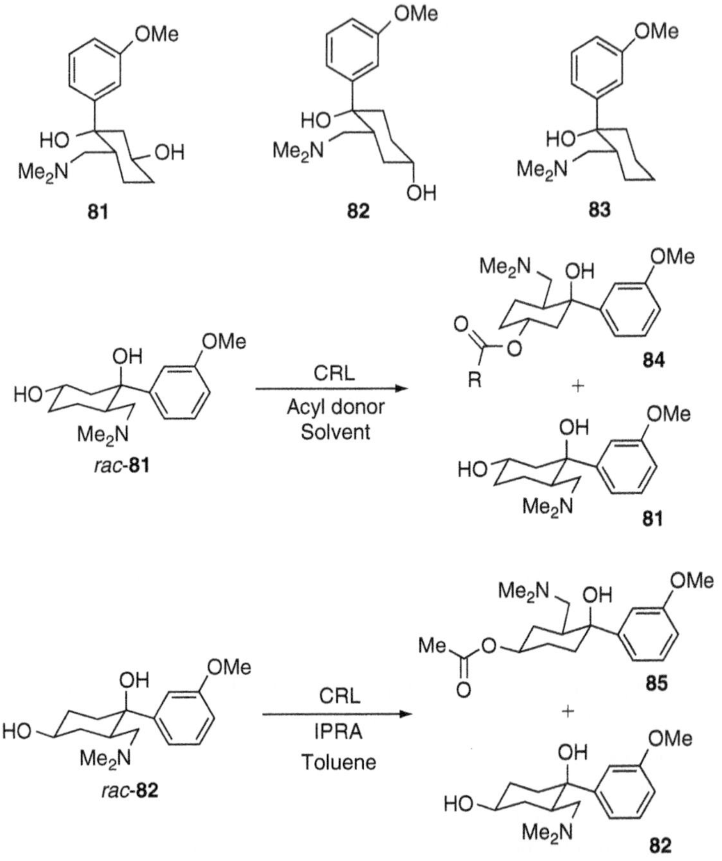

SCHEME 7.25 Enzymatic resolution of ε-hydroxytramadol and δ-hydroxytramadol.

7.2.2.5 Leustroduscin B (86)

Leustroduscin B is a potent colony-stimulating factor inducer, and it induces cyt
productions via NF-κB activation at the transcription level as well as at the posttra
tional level. Desymmetrization of *meso*-diol **87** at room temperature with lipase AK in
using vinyl acetate as acyl donor gave the corresponding (R)-acetate that was imme
protected to form **88** in an overall yield of 86 and 90% ee (Scheme 7.26) [60].

7.2.2.6 Propanolol (51)

Therapeutic properties of (S)-propanolol have been described in Section 7.2.1 "Resolu
Monoalcohols." Lipase P from *Pseudomonas fluorecens* catalyzed the stereoselective
tion of the *meso*-diol **89** using vinyl acetate. The (S)-**90** was recovered in 92% isolate
with 94% ee, after 4 h reaction at 8°C (Scheme 7.27) [61].

7.2.2.7 SCH 51048 (91)

The tetrahydrofuran **91** is an antifungal agent with therapeutic potential against a va
systemic fungal infections in normal and immunocompromised infection models. T
thesis of **91** was achieved using an enzymatic desymmetrization of the homoallylic

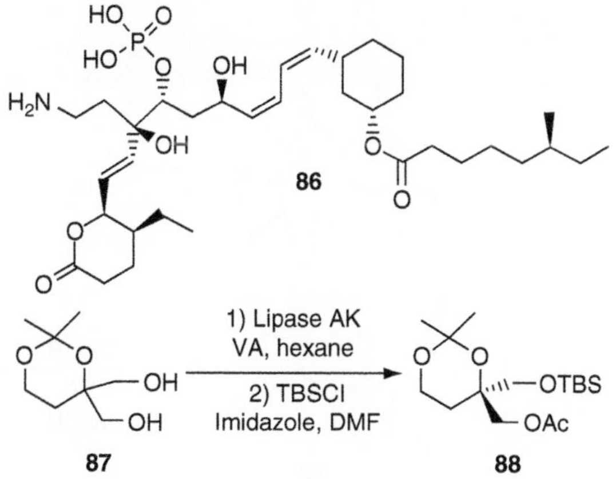

SCHEME 7.26 Enzymatic resolution of **87** for the synthesis of leustroduscin B.

with CAL-B (Novozyme 435) to provide the desired (*S*)-monoacetate **93** in 71% yield and 98% ee using vinyl acetate and MeCN at 0°C (Scheme 7.28) [62]. These conditions allowed to prepare up to 30 kg batches of (*S*)-**93**, and the catalyst reusability was demonstrated over six cycles without major loss of activity.

7.2.2.8 Sobrerol (94)

Compound (±)-*trans*-5-(1-hydroxy-1-methylethyl)-2-methyl-2-cyclohexen-1-ol [(±)-*trans*-sobrerol, **94**] is a mucolytic drug, and in spite of its differences in the pharmacological activity between the (+) and (−) forms, it is produced and marketed as racemate. Immobilized PSL selectively acetylated **94** using *tert*-amyl alcohol as solvent and vinyl acetate as acylating agent, the monoacetate (−)-**95** and the diol (+)-**94** were obtained in 99% ee after 50% conversion in 2 h reaction at 45°C (Scheme 7.29) [63].

7.2.2.9 Spyrotryprostatins A (96) and B (97), (−)-Physostigime (98), and (−)-Esermethole (99)

Oxindole and indoline skeletons having a stereogenic quaternary carbon center at the C-3 position are found in many biologically important indole alkaloids, such as spyrotryprosta-tins A and B, (−)-physostigime, and (−)-esermethole. Kita and coworkers reported the desymmetrization of diols **100a–d** using lipase OF and 1-ethoxyvinyl-2-furoate as acyl donor in mixtures of *i*Pr$_2$O:THF at 30°C obtaining the corresponding monoacetates **102a–d** in good yields and high ee (Table 7.1 and Scheme 7.30) [64,65].

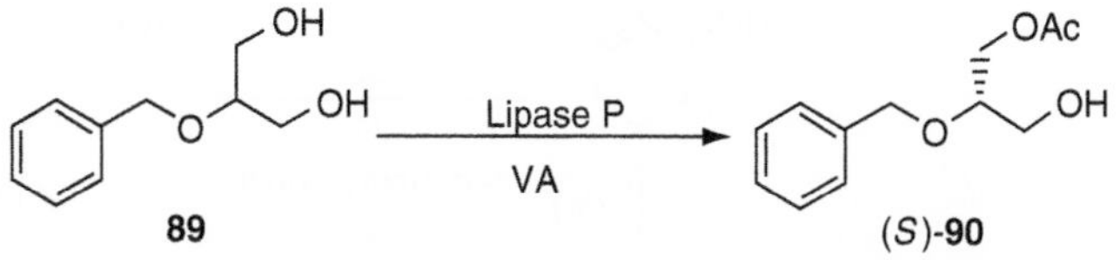

SCHEME 7.27 Enzymatic resolution of **89** for the synthesis of (*S*)-propanolol.

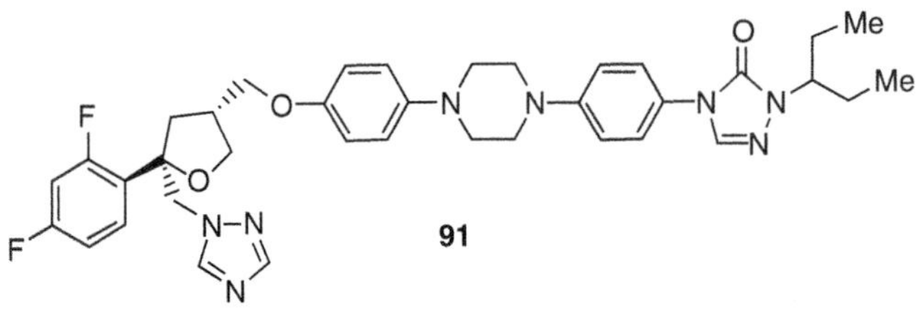

SCHEME 7.28 Enzymatic desymmetrization of **92** for the synthesis of retiferol.

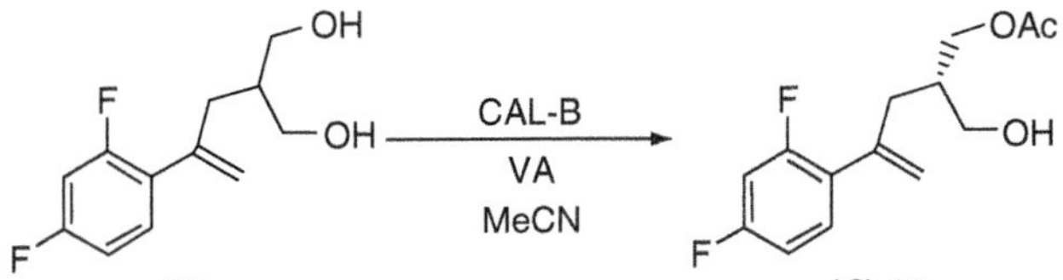

7.2.3 RESOLUTION OF POLYALCOHOLS

7.2.3.1 Glycerol Derivatives

Tumor inhibitory activity of glycosylglycerols and glycoglycerolipids is demonstrate▮
basis of their *in vitro* and *in vivo* antitumor promoting effect on Epstein–Barr viru▮
antigen activation induced by the tumor promoter 12-*O*-tetradecanoylphorbol-13-▮
PSL catalyzed the regioselective transesterification of the primary hydroxyl group o▮
THF using different trifluoroethyl esters obtaining the 1,3-di-*O*-benzyl-2-*O*-(6-*O*-a▮
glucopyranosyl)-glycerols **105** in good yields (Scheme 7.31) [66]. To obtain the 3-est▮
CAL-B was used in THF using the appropiate trifluoroethyl ester affording 3-▮
2-*O*-(2,3,4,6-tetra-*O*-chloroacetyl-β-D-glucopyranosyl)-glycerols in good diastereose▮
and yields.

7.2.3.2 Inositol Phosphates

A significant number of physiological processes in differentiated higher cells are closel▮
with inositol metabolism. Important examples are the activation of thrombocytes in th▮
clotting process, hormonal signal transduction, signal transformation, contrac▮
muscles, control of cell proliferation, etc. D-*myo*-inositol phosphate **108** was synthe▮
a short route from D-1-acetoxy-4,6-di-*O*-benzyl-*myo*-inositol (**110**), which was easily o▮
by an initial highly regio- and enantioselective acylation of 4,6-di-*O*-benzyl-*myo*-inosit▮
catalyzed by PSL (Scheme 7.32) [67]. CAL-B showed good selectivity toward racemi▮
O-benzyl-*myo*-inositol (**111**) using vinyl acetate as acyl donor yielding 49% of the unco▮
inositol derivative (−)-2,6-di-*O*-benzyl-*myo*-inositol with 99% ee and monoacetylated ▮
112 with 49% yield, which was easily converted to (+)-2,6-di-*O*-benzyl-*myo*-ino.▮

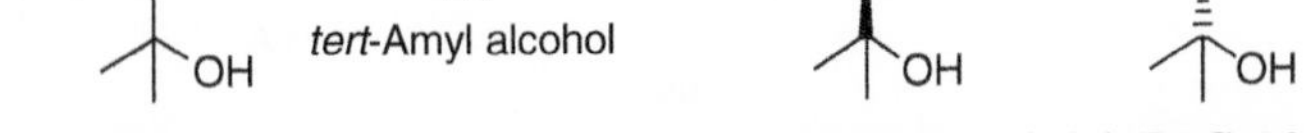

SCHEME 7.29 Enzymatic resolution of (±)-*trans*-sobrerol (**94**).

TABLE 7.1
Enzymatic Desymmetrization of Diols 100a–d

Entry	Compound	R^1	R^2	iPr$_2$O:THF	t (h)	102 (ee [%])	102 (Yield [%])
1	100a	Me	H	5:1	20	79	60
2	100b	Boc	H	100:0	20	97	68
3	100c	Boc	5-OMe	5:1	3	98	77
4	100d	Boc	6-OMe	5:1	19	91	79

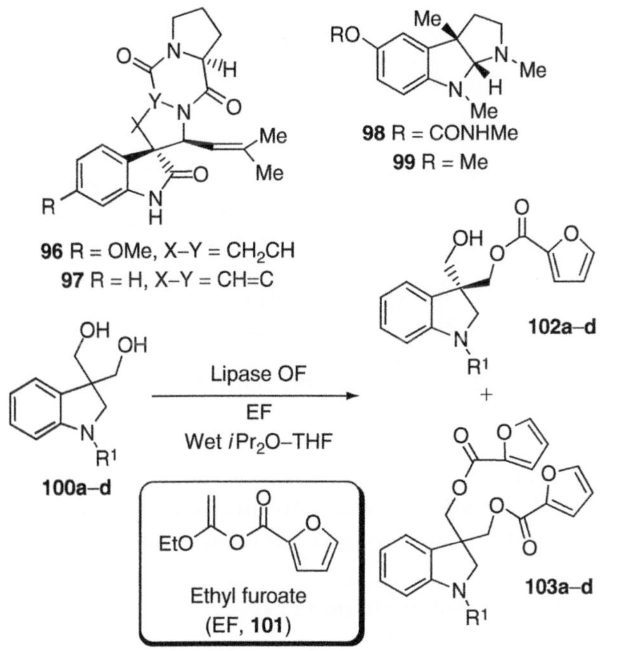

SCHEME 7.30 Enzymatic desymmetrization of diols **100a–d** for the synthesis of **96–99**.

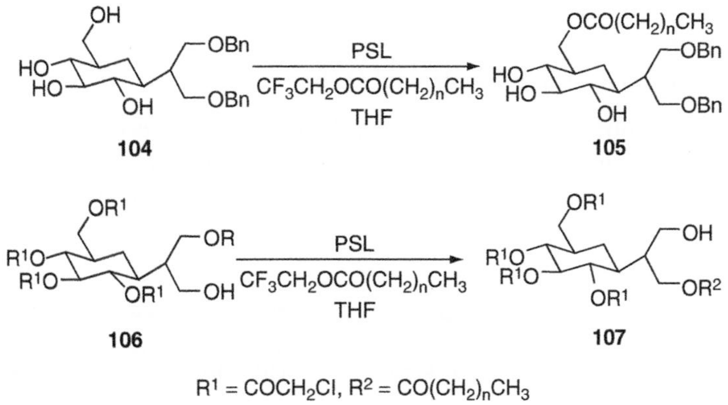

$R^1 = COCH_2Cl$, $R^2 = CO(CH_2)_nCH_3$

SCHEME 7.31 Selective modifications of glycerol derivatives.

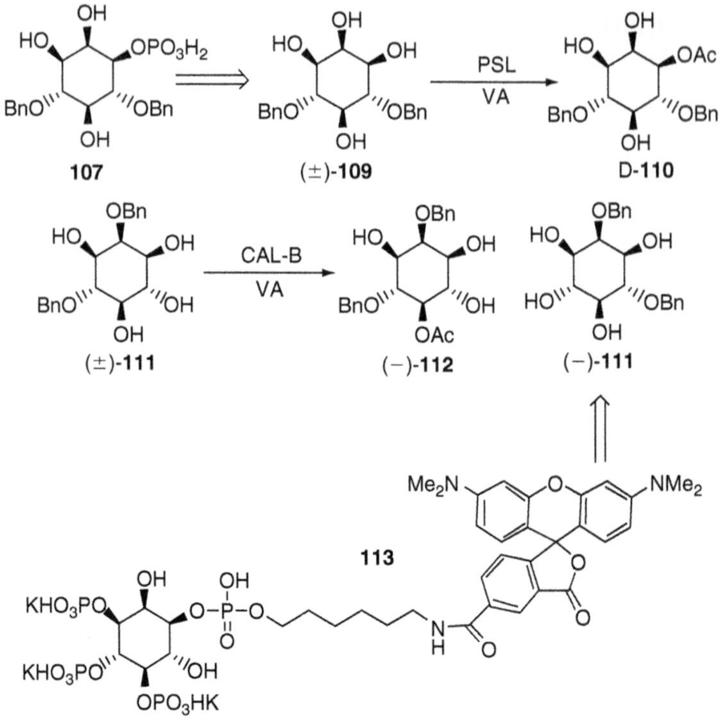

SCHEME 7.32 Resolution of **109** and **111**, precursors of inositol derivatives.

chemical hydrolysis [68]. Pure (−)-2,6-di-O-benzyl-*myo*-inositol was also used for the s
sis of 5′-TAMRA-labeled (P1-tethered)-D-*myo*-inositol-3,4,5-triphosphate (**113**), a bic
ical agent for binding assays to identify and characterize pharmaceutically relevant pr

7.2.3.3 Phosphonotrixin (114)

Phosphonotrixin, a secondary metabolite, has attracted interest due to its antibiot
herbicidal properties. The synthesis of (S)-(−)-**114** was achieved through desymmetriza
triol **115** occurred in vinyl acetate and catalyzed by PPL [69]. After 5 h at room temperat
monoacetate **116** was obtained in 88% isolated yield and the diacetate **117** in 7% yield (S
7.33).

7.2.3.4 (S)-α-Tocotrienol (118)

Tocotrienols inhibit cholesterol biosynthesis by posttranscriptional suppression of β-hy
β-methyl-glutaryl-coenzyme A reductase activity. Also, tocotrienols are effective antio:
and inhibit the oxidation of low density lipoproteins associated with coronary heart c

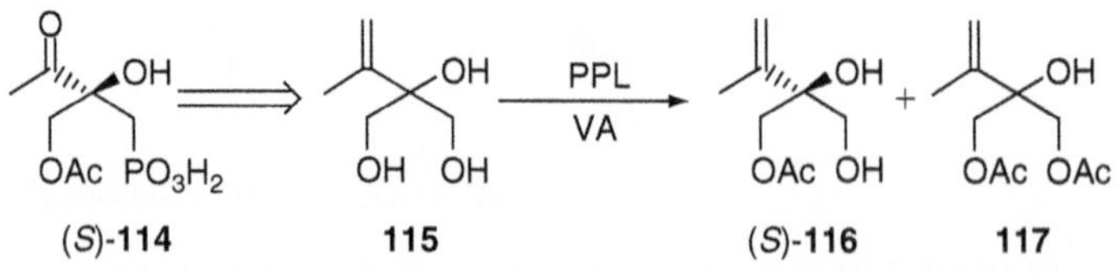

SCHEME 7.33 Desymmetrization of **115**, precursor of (S)-phosphonotrixin.

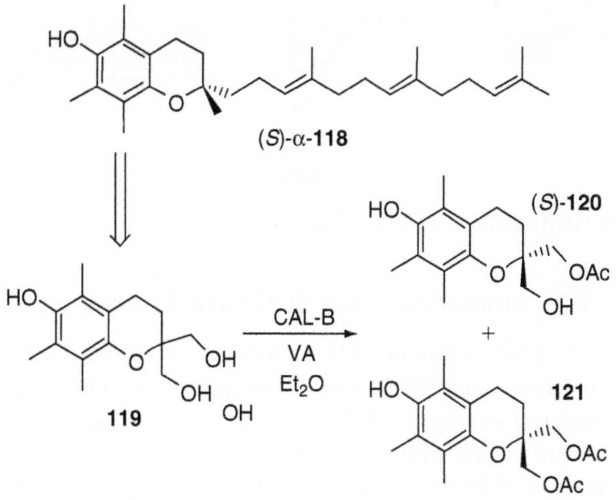

SCHEME 7.34 Enzymatic desymmetrization of **119**, precursor of (S)-α-tocotrienol.

In addition, they have shown anticarcenogic and neuroprotective properties. The desymmetrization of triol **119** with vinyl acetate in the presence of CAL in Et$_2$O gave enantiomerically pure (ee > 98%) monoester **120** in 60% yield and the corresponding achiral diester **121** in 27% isolated yield after 2 h reaction (Scheme 7.34) [70].

7.3 ENZYMATIC ACYLATION OF AMINOALCOHOLS

7.3.1 (5S,9S)-(+)-INDOLIZIDINE 209D (122)

All four possible stereoisomers of indolizidine 209D are potent blockers for nicotinic acetycholine receptors. Chênevert and coworkers reported the enzyme-catalyzed transeterification of diol **123** using CAL in vinyl acetate at room temperature to give optically active monoester **124** in 95% ee and 80% yield after 3 h reaction (Scheme 7.35) [71].

7.3.2 JATROPHAM (125)

(R)-(−)-5-hydroxy-3-methyl-3-pyrrolin-2-one, (R)-**125** is an antitumor alkaloid that has been synthesized through the KR of the racemate using lipase PL as biocatalyst and vinyl acetate as acyl donor and solvent at 25°C, isolating the (R)-alcohol in 35% yield with 98% ee along with the acetate (S)-**126** in 53% yield with 50% ee (Scheme 7.36) [72].

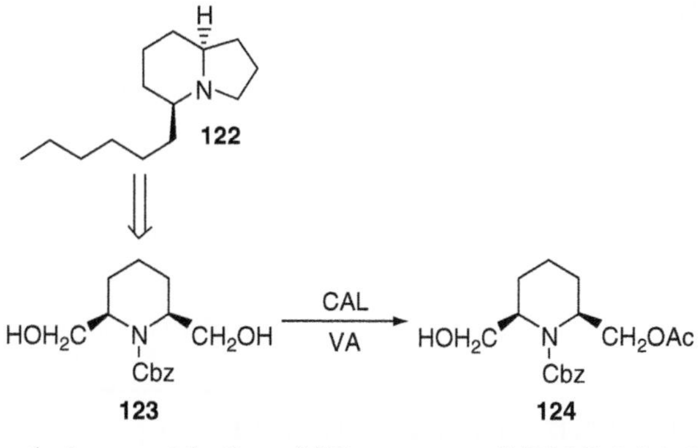

SCHEME 7.35 Enzymatic desymmetrization of **123**, precursor of (5S,9S)-(+)-indolizidine 209D.

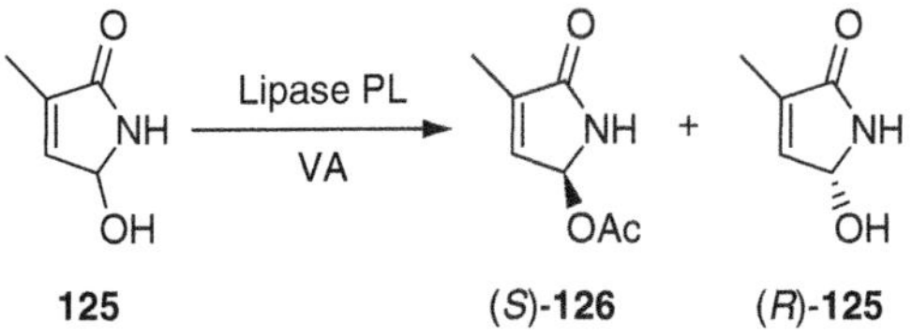

SCHEME 7.36 Kinetic resolution of jatropham.

7.3.3 NORPHENYLEPHRINE (127) AND OCTOPAMINE (128)

(*R*)-(−)-Noradrenaline **129** is a natural cathecolamine that acts as an α-adrenoceptor ago
also shows activity toward β-receptors. Norphenylephrine (**127**) and octopamine (**1**
structural analogs of **129** that have similar properties. The synthesis of enantiomerica
127 and **128** were possible through the resolution of derivatives **130a–b** using butanoic ac
presence of PSL in a mixture of toluene/THF as solvent (Scheme 7.37) [73]. *N*-Boc and *
derivatives of octopamine were also resolved using butanoic anhydride instead of butan

7.4 ENZYMATIC ACYLATION OF AMINES

In this section, we have included separately the resolution of amines, mainly primary
and 1,2-diamines.

7.4.1 RESOLUTION OF AMINES

7.4.1.1 Cinacalcet (133)

Cinacalcet·HCl is an oral calcimimetic drug, which has received the FDA approval in
is used in the treatment of hyperparathyroidism and for the preservation of bone de
patients with kidney failure or hypercalcemia due to cancer [74]. A precursor of **13**

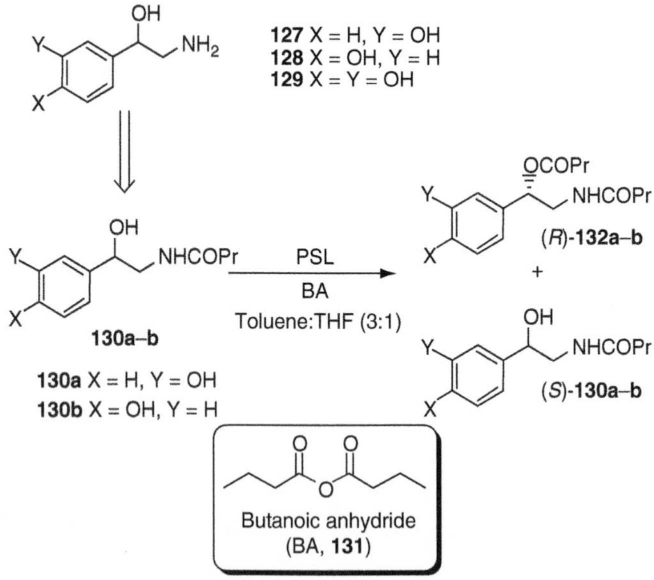

SCHEME 7.37 Enzymatic resolution of **130a–b**, precursors of norphenylephrine and octopami

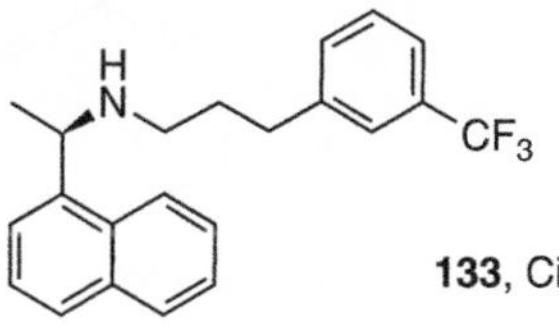

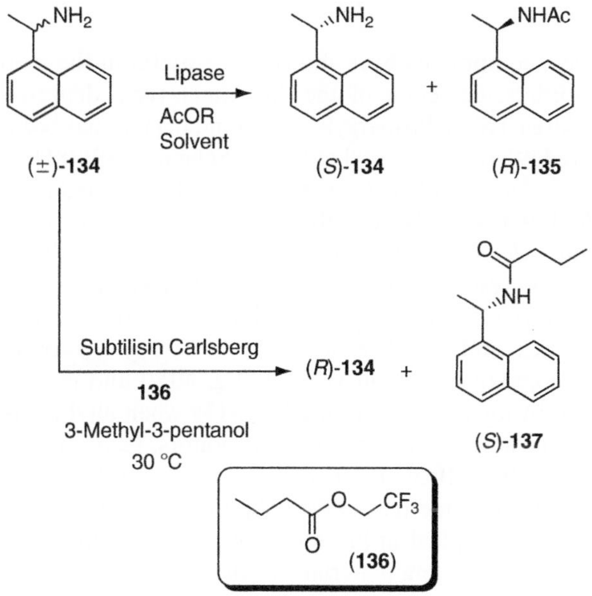

SCHEME 7.38 Resolution of ($\pm$)-1-(1-naphthyl)ethylamine (**134**).

optically active (R)-1-(1-naphthyl)ethylamine [(R)-**134**] [75] or its amide derivative (R)-**135**, which has been obtained by enzymatic resolution of ($\pm$)-**134** (Scheme 7.38). When CAL-B was used as catalyst, results were greatly dependent on the acyl donor and the solvent used. Low and moderate enantiomeric ratios were obtained using ethyl acetate as the acyl donor ($E = 5$ and 24 with ethyl acetate and diethyl ether as solvents, respectively) [76,77]. However, excellent results were obtained with isopropyl acetate [78] in 1,2-dimethoxyethane and ($\pm$)-1-phenylethyl acetate [79] in 1,4-dioxane ($E > 200$ in both cases), allowing the isolation of the acetamide (R)-**135** with ee > 99%. In addition, when the acetylation of ($\pm$)-**134** was carried out with *Pseudomonas aeruginosa* lipase, ethyl acetate as acyl donor, and TBME as solvent, (R)-**135** was obtained with ee > 99% [80].

The protease subtilisin Carlsberg has also been used to catalyze the resolution of ($\pm$)-**134** with different activated esters such as 2,2,2-trifluoroethyl butyrate (**136**) [81] and methacrylate [82], or cianomethyl pent-4-enoate [83] and 3-methyl-3-pentanol as solvent. In all the cases, the enzyme exhibited a high enantioselectivity ($E > 100$) toward the (S) enantiomer, the enantiopreference being opposite to that shown by the lipases. The resolution of ($\pm$)-**134** with 2,2,2-trifluoroethyl butyrate (Scheme 7.38) was scaled up to 1.6 kg of substrate using a continuous-flow column bioreactor containing subtilisin immobilized on glass beads [84]. Thus, 560 g of pure (R)-1-(1-naphthyl)ethylamine with 99% ee was prepared.

7.4.1.2 (*R*)-*N*-Propargyl-1-Aminoindan: Rasagiline (138)

Amine **138** and some of its derivatives have shown to be highly selective and potent inhibitors of the B form of monoamine oxidase, in contrast the levorotatory enantiomer is inactive.

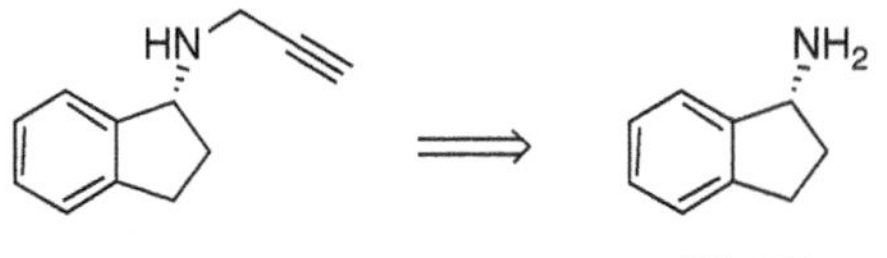

138, Rasagiline   (*R*)-**139**

SCHEME 7.39 Retrosynthesis of rasagiline **138**.

These compounds have shown to be useful in the treatment of Parkinson's disease, m
disorders, dementia of the Alzheimer type, depression, and the hyperactive syndr
children. (*R*)-*N*-Propargyl-1-aminoindan (**138**) has been easily prepared from (*R*)-1-amin
[(*R*)-**139**] and propargyl chloride (Scheme 7.39) [85].

Different hydrolytic enzymes have been tested in the resolution of (±)-**139**. Amida
(*R*)-**139** was catalyzed with aminoacylase I from *Aspergillus melleus* [86] and penicillin a
from *Alcaligenes faecalis* [87] using methyl methoxyacetate and phenylacetamide a
donors, respectively, but with very low enantioselectivities ($E \leq 9$). In contrast, su
was very efficient and, as described above for (*R*)-**134**, the aminolysis of 2,2,2-trifluo
butyrate and (±)-**139**, scaled up to 330 g of racemic amine, yielded the unreacted
aminoindan with high yield (132 g, 40%) and ee > 98% [84]. CAL-B also was an ex
catalyst for the resolution of (±)-**139** when allyl pent-4-enoate and dibenzyl carbona
used as acyl and alkoxycarbonyl donors, respectively. In both cases, the lipase was s
toward the (*R*) enantiomer of the amine catalyzing the formation of the correspondi
pent-4-enoamide (ee > 99%) [88] and (*R*)-benzyl carbamate (ee = 95%) [89]. This lip
also been applied in an interesting process combining enzyme and metal catalysis (S
7.40). The starting material was the readily available ketoxime **140** and the catalysts w
and CAL-B. Thus, the coupling of the Pd-catalyzed reduction of the ketoxime **140** a
subsequent Pd and CAL-B catalyzed DKR of the resulting racemic amine, afforde
amide (*R*)-**141** with very high chemical and optical yields [90]. In this process, an additi
as *N*-ethyldiisopropylamine was required to suppress the reductive deamination of the

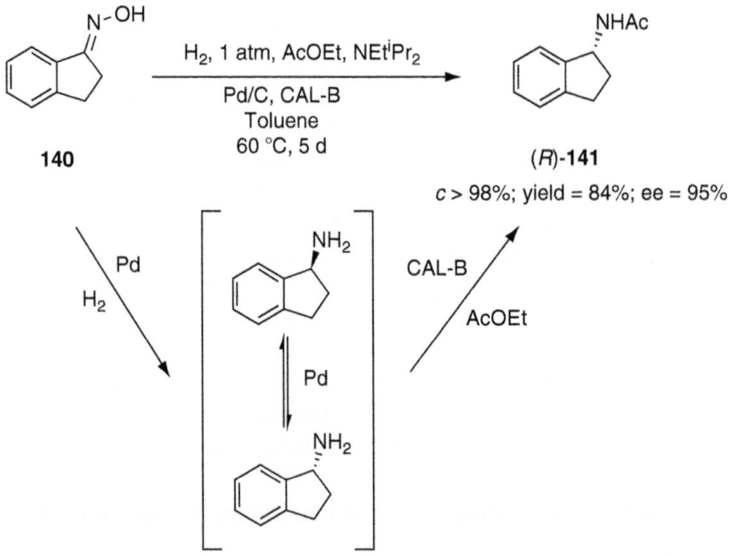

c > 98%; yield = 84%; ee = 95%

SCHEME 7.40 Asymmetric transformation of **140** into (*R*)-**141**.

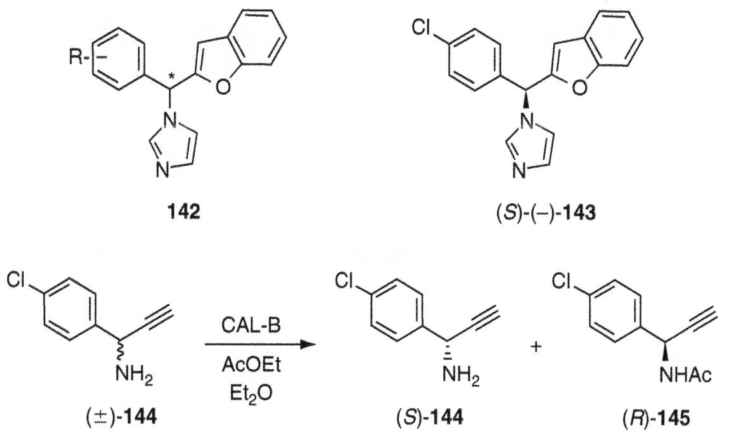

142   **(S)-(−)-143**

(±)-**144**   (S)-**144**   (R)-**145**

SCHEME 7.41 Resolution of (±)-**144**.

7.4.1.3 1-[(2-Benzofuranyl)phenylmethyl]imidazols 142

Imidazols **142** are a new class of potent aromatase inhibitors that show promise as chemotherapeutic agents for the treatment of estrogen dependent tumors. In particular, the 4-chloro analog **143** has shown activity either *in vitro* or *in vivo*, the (+) form being 15-fold more active than its counterpart. A chemoenzymatic synthesis of both enantiomers of **143** has been developed starting from racemic 1-(4-chlorophenyl)-2-propynylamine [(±)-**144**] [91]. Resolution of (±)-**144** was carried out by CAL-B catalyzed acetylation using ethyl acetate as the acyl donor and diethyl ether as the solvent (Scheme 7.41). After 25 h reaction at room temperature, the remaining amine (S)-**144** and the converted acetamide (R)-**145** were isolated with ee of 98% and >98%, respectively (E > 200). Conventional hydrolysis of (R)-**145** yielded the amine (R)-**144** that was transformed into (S)-(−)-**143** (ee = 97.5%) by a three-step synthesis. Analogously, (S)-**144** was converted into (R)-(+)-**143** in 40 to 45% overall yield in three steps.

The enzymatic resolution has also been efficiently applied to other (±)-1-aryl-2-propynylamines (R = H, 4-F, 3-F, and 3-Me) [92], precursors of imidazols **142**.

7.4.1.4 1-Arylpropan-2-Amines (146): Amphetamine and Derivatives

Amines **146** possess central and peripheral stimulant activity and suppress appetite. The presence of methyl groups on the nitrogen atom and methoxy groups on the aromatic ring is known to increase the effects of the drug. In addition, the stereochemistry of these amines has a great influence on their pharmacological properties. Thus, (S)-(+)-1-phenylpropan-2-amine [(S)-(+)-amphetamine, dexedrine] has greater pharmacological activity as stimulant [93] and hyperthermic [94] agent than its (R)-(−)-enantiomer.

Both enantiomers of the amphetamine (**146a**) and the isomeric *o*-, *m*-, and *p*-methoxyamphetamines (**146b–d**) have been prepared with very high ee by CAL-B catalyzed resolution of the corresponding racemic amines using ethyl acetate as acyl donor and solvent [95]. The unreacted amines (S)-**146a–d** and the converted acetamides (R)-**147a–d** were easily separated by selective extraction. Enantioselectivity of these reactions was from moderate to high (Scheme 7.42), allowing (S)-amines to be obtained with very high ee at 50% conversion. (R)-Acetamides, isolated from the enzymatic reactions with ee 82% to 89%, were finally obtained with ee > 98% after recrystallization from hexane–chloroform. The use of a catalytic amount of triethylamine in the reaction of (±)-**146d** slightly improved the enantioselectivity

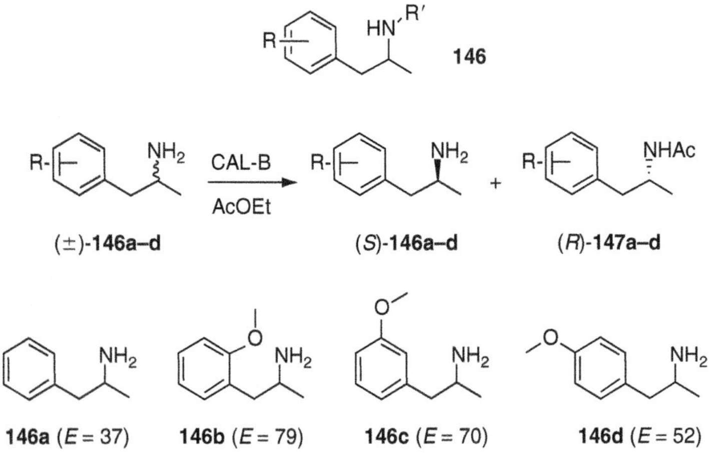

SCHEME 7.42 Resolution of amphetamine and the isomeric methoxyamphetamines.

$(E = 66)$ and the reaction rate [44]. The enantiomerically pure acetamide (R)-14 hydrolyzed and the resulting (R)-**146d** was used in the synthesis of the most active st mer of formoterol (**30**) (Scheme 7.43). The epoxide (R)-**148** was obtained from brom (R)-**31**, prepared by enzymatic resolution (Scheme 7.11).

Fenfluramine **146f** (Scheme 7.44) is also an amphetamine derivative with anorect erties but without stimulant effects. Pharmacological studies have revealed that (S)-14 active isomer and (R)-**146f** is responsible for the adverse effects [96]. In addition, nor amine **146e** is the main metabolite of **146f** and also its synthetic precursor [97]. Resol $(\pm)$-**146e** has been efficiently accomplished by enantioselective CAL-B catalyzed am reaction using $(\pm)$-1-phenylethyl acetate [$(\pm)$-**149**] as acylating agent and 1,4-dio solvent (Scheme 7.44). Very high enantioselectivities were obtained for amine (E_a) as ester (E_e) [79].

7.4.1.5 Mexiletine (151)

1-(2,6-Dimethylphenoxy)propan-2-amine (**151**, mexiletine) is an antiarrhythmic ag (R)-enantiomer is more potent for experimental arrhythmias and in binding stu

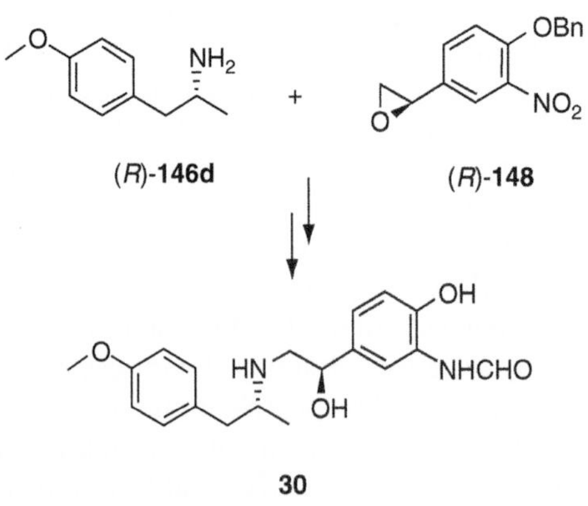

SCHEME 7.43 Synthesis of (R,R)-formoterol (**30**).

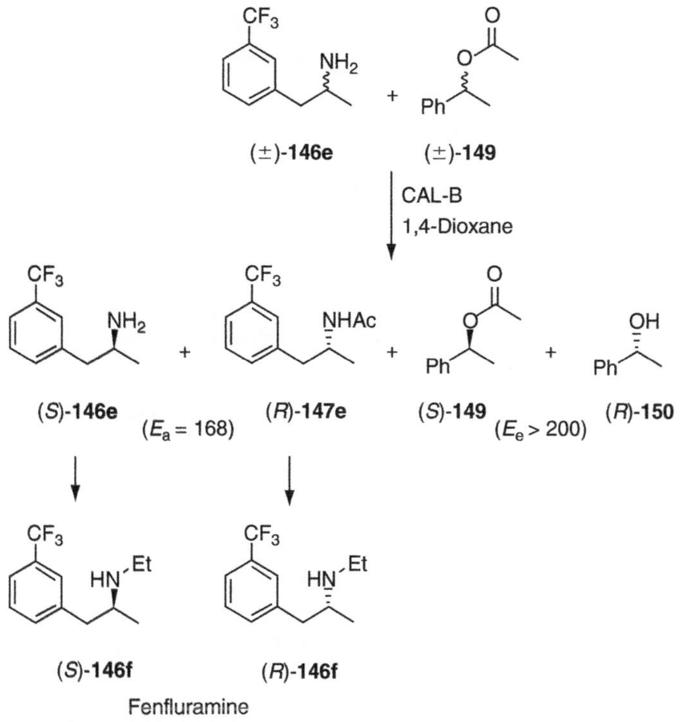

SCHEME 7.44 Chemoenzymatic preparation of fenfluramine **(146f)**.

cardiac sodium channels than its (*S*)-counterpart [98]. Using the enzymatic method described for amphetamines (Scheme 7.42), (*S*)-**151** (ee = 99%) and (*R*)-acetamide **152** (ee = 92%) were prepared (Scheme 7.45). A simple recrystallization of the (*R*)-acetamide allowed its isolation in enantiopure form [95].

7.4.1.6 Labetalol (153)

Labetalol (Scheme 7.46) is an antihypertensive agent [99], the most active stereoisomer being the divelalol with *R,R* configuration. A building block of this compound is (*R*)-4-phenylbutan-2-amine [(*R*)-**154**], which has been prepared by CAL-B catalyzed amidation of (±)-**154** with ethyl acetate. The lipase catalyzed the formation of the corresponding (*R*)-acetamide (*c* = 51%, ee = 86%) with moderate enantioselectivity (*E* = 41), but its recrystallization from hexane–chloroform increased the ee up to 99% [95]. The resolution of (±)-**154** was also carried out by acylation reaction catalyzed by penicillin acylase from *Alcaligenes faecalis* in

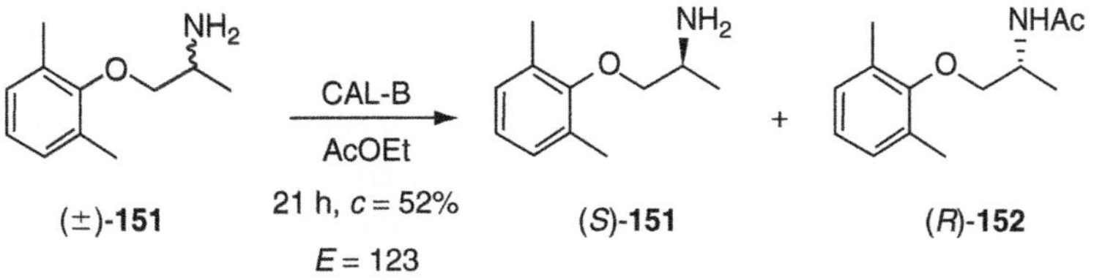

SCHEME 7.45 Enzymatic resolution of mexiletine.

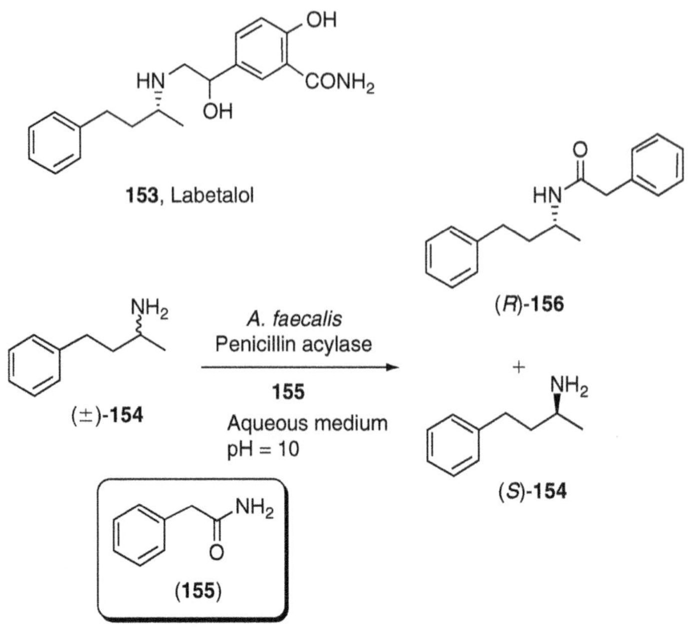

SCHEME 7.46 Resolution of ($\pm$)-**154**, precursor of labetalol.

aqueous medium using phenylacetamide (**155**) as the acyl donor [100]. From this p phenylacetamide (*R*)-**156** was obtained with ee $= 96\%$ ($c = 45\%$, $E = 120$).

7.4.1.7   YH1885 (157)

The pyrimidinic compound **157** has showed utility for the treatment of gastroeso reflux disease and duodenal ulcers. Synthesis of racemic **157** has been carried out from methyl-1,2,3,4-tetrahydroisoquinoline (Scheme 7.47) [101], but recently this secondary has been efficiently resolved by lipase-catalyzed alkoxycarbonylation reaction. Th treatment of ($\pm$)-**158** with allyl *m*-methoxyphenyl carbonate (**159**) in toluene con 0.05% w/w water in the presence of chiroCLEC-CR, a cross-linked enzyme crystal o afforded (*S*)-**158** (yield $= 46\%$; ee $= 99.6\%$) and (*R*)-**160** (yield $= 47\%$; ee $= 98.4\%$) afte reaction [102].

7.4.1.8   SCH66336 (161)

SCH66336 [(*R*)-**161**] is a selective, nonpeptide, nonsulfhydryl farnesyl protein transfer tor that is currently undergoing phase II clinical trials for the treatment of solid tumor In the synthesis of (*R*)-**161** from compound **162** (Scheme 7.48), two racemic interm ($\pm$)-**163** and ($\pm$)-**164** are required to be resolved by enzymatic processes. Compou does not contain chiral center, but it exists as a pair of enantiomers due to atropiso about the exocyclic double bond. Both ($\pm$)-**163** and ($\pm$)-**164** have been efficiently reso enzymatic aminolysis reactions [104].

    After an exhaustive screening of reaction conditions (233 commercially available en 12 organic solvents, and a wide variety of esters and carbonates as acylating agent results for the resolution of ($\pm$)-**163** were obtained with Toyobo LIP-300 (a lipoprotei from *Pseudomonas aeruginosa*), TBME as solvent, and 2,2,2-trifluoroethyl isobutyra as acyl donor (Scheme 7.49). The acylation of ($\pm$)-**163** yielded (+)-**166** (ee $= 97$

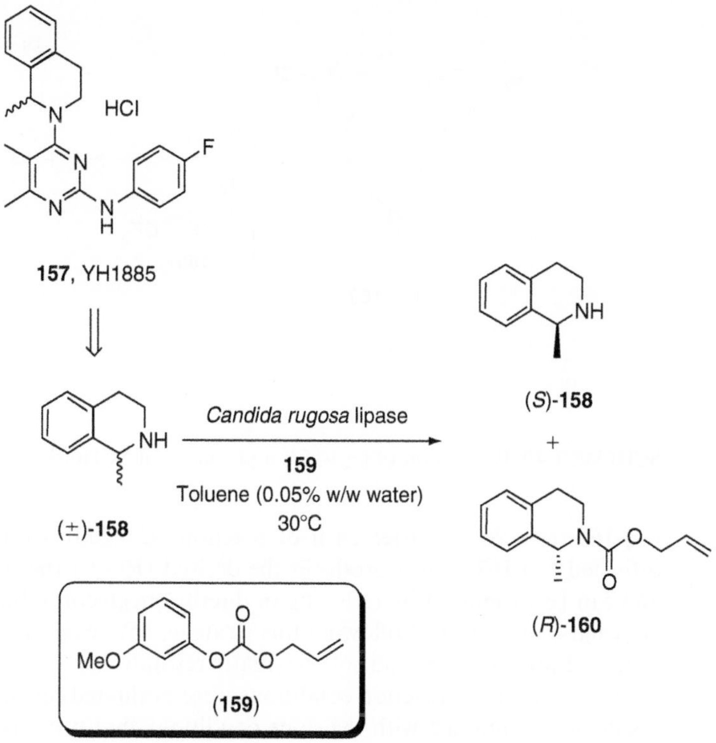

SCHEME 7.47 Resolution of (±)-**158**, precursor of YH1885.

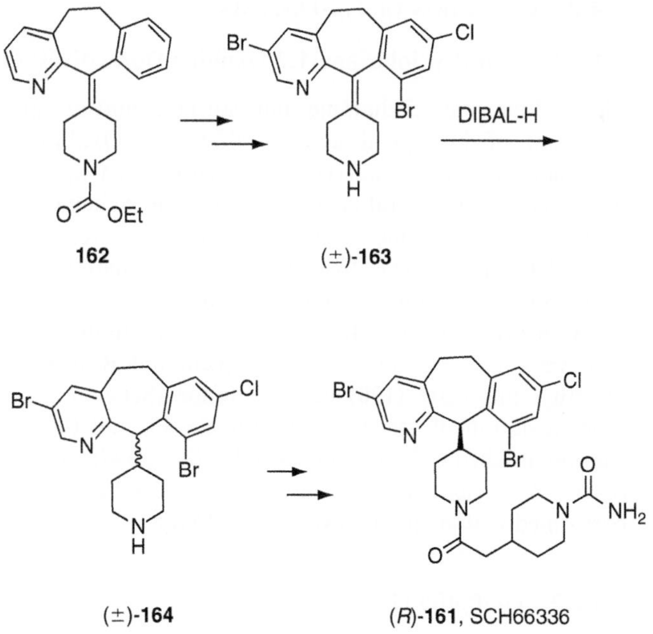

SCHEME 7.48 Synthesis of (*R*)-**161**.

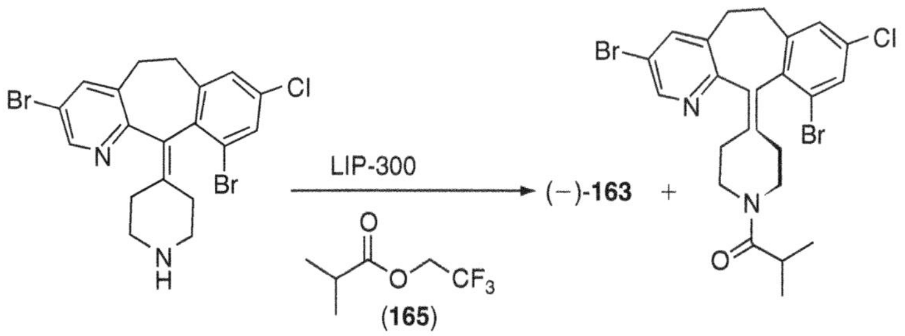

SCHEME 7.49 Resolution of (±)-**163**, a precursor of SCH66336.

(−)-**163** (ee = 96.3%) after 24 h of reaction (E > 200). Conventional hydrolysis of afforded (+)-**163**, which produces the desired (R)-(+)-**164** on reduction. The unwan **163** can be racemized by refluxing in di(ethyleneglycol) dibutyl ether and again subj enzymatic resolution. Following this strategy, 65% overall yield of (+)-**163** (ee = 98. obtained after three rounds of enzymatic resolution.

Wide variety of reaction conditions were evaluated for the resolution of (±)-**164**, results were obtained with the same conditions that were used for (±)-**163**. In this c reaction also proceeded with high enantioselectivity (E > 100) and either (S)-**164** produced (R)-isobutyramide was obtained with very high ee (98.5% and 94.3%, respe

7.4.2 RESOLUTION OF 1,2-DIAMINES

7.4.2.1 *trans*-Cyclohexane-1,2-Diamine Derivatives: Oxaliplatin (167) and NDD

The cyclohexane-1,2-diamine unit can be found in various compounds displaying spectrum of biological activity [105,106]. Oxaliplatin, [*trans*-(1R,2R)-cyclohexa diamine]oxalatoplatinum(II), has recently been approved for combination chemo of metastatic colorectal cancer. Oxaliplatin (Scheme 7.50) is significantly more acti its *trans*-(1S,2S) isomer and the racemic mixture. In addition, another platinum derived from *trans*-(1R,2R)-cyclohexane-1,2-diamine, NDDP (**168**), is also an anti agent currently in phase II clinical trials.

An excellent method to prepare both enantiomers of *trans*-cyclohexane-1,2-diami with very high ee consisted the sequential KR of (±)-**169** using CAL-B as catal dimethyl malonate (**170**) as acyl donor (Scheme 7.50) [107]. The enzyme cataly monoacylation of the diamine with E = 45, and the subsequent acylation of the r enantioenriched (1R,2R)-monoamide (E = 68) afforded the diamide (1R,2R)-**171** wit ee and 45% yield. Enantiopure diamine (1S,2S)-**169** was obtained when the react performed with a slightly excess acyl donor.

7.4.2.2 U-50,488 (172)

Compound U-50,488 (**172**) and other structural analogs have been reported to be selective κ-opioid agonists, free from the adverse side effects of μ agonists like morph majority of these pharmacologically active compounds have similar configuration

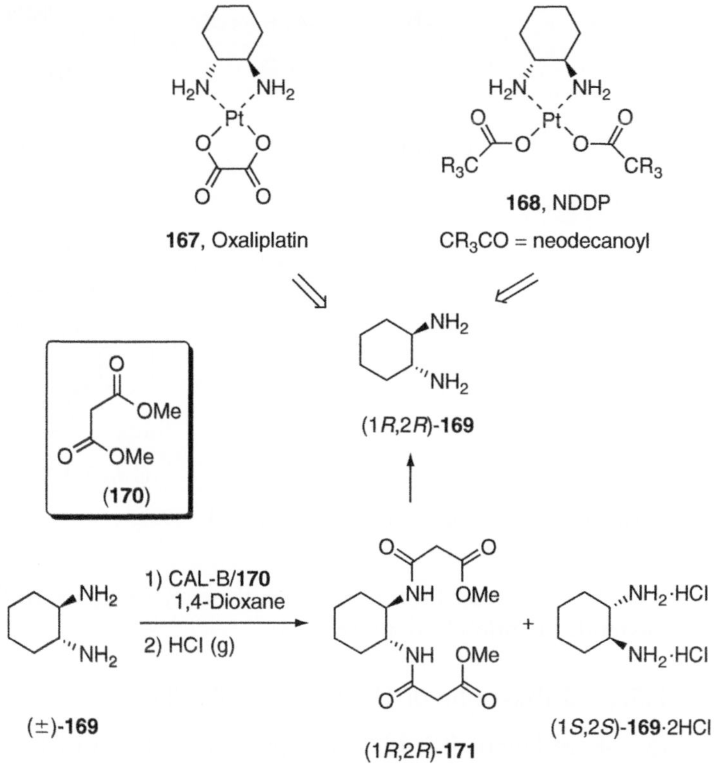

SCHEME 7.50 Enzymatic preparation of (1*R*,2*R*)-**169**, precursor of the antitumoral agents **167** and **168**.

stereogenic centers. Thus, (1*S*,2*S*)-(−)-**172** (Scheme 7.51) exhibits greater κ agonist activity than its enantiomer, and the *cis* diastereomer of U-50,488 has practically no affinity for κ receptors [108].

(1*S*,2*S*)-**172** has been obtained by a chemoenzymatic route starting from the inexpensive cyclohexene oxide [109]. The key steps involved in this route were the stereospecific synthesis of (±)-*trans*-2-(pyrrolidin-1-yl)cyclohexanamine **173** and its subsequent enzymatic resolution by CAL-B catalyzed aminolysis of ethyl acetate. The enzymatic process gave very high enantioselectivity (*E* = 170), CAL-B preferentially catalyzed the acetylation of the (1*R*,2*R*) enantiomer of **173**. To facilitate the isolation of the amide (1*R*,2*R*)-**175** (ee = 94%) and the unreacted diamine (1*S*,2*S*)-**173** (ee = 99%), the reaction mixture was treated with benzylchloroformate. Thus, (1*S*,2*S*)-**173** was transformed into the benzylcarbamate (1*S*,2*S*)-**174** and the mixture formed by **174** and **175** was easily separated by flash chromatography. Finally, carbamate (1*S*,2*S*)-**174** was submitted to reduction with lithium aluminum hydride (LAH) and subsequent amidation with 3,4-dichlorophenylacetyl chloride to yield the analgesic (1*S*,2*S*)-**172**.

7.5   RESOLUTION OF ESTERS BY AMINOLYSIS OR AMMONOLYSIS

7.5.1   IBUPROFEN

A derivative of the anti-inflammatory ibuprofen, its 2-chloroethyl ester (*S*)-**176**, has been obtained with ee = 96% by CAL-B catalyzed ammonolysis of (±)-**176** (Scheme 7.52) [110]. Reaction was performed in *tert*-butyl alcohol as solvent and ammonia was bubbled through

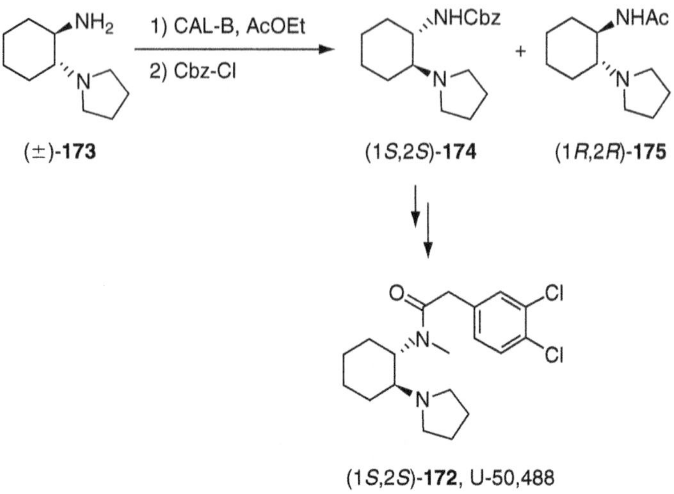

SCHEME 7.51 Chemoenzymatic synthesis of U–50,488.

the solution. Under these conditions, the enzyme catalyzed the ammonolysis of (R)-1'
moderate enantioselectivity ($E = 28$).

7.5.2   3-Pyrrolidinol Derivatives 178–181

(S)-3-Pyrrolidinol [(S)-**182**] is a precursor of a variety of compounds with acti
κ-receptor agonists. Scheme 7.53 demonstrates some of these compounds: **178** [11
[112], **180** [113], and **181** [114].

   A chemoenzymatic synthesis of (S)-**182** has been carried out from ethyl ($\pm$)-4-ch
hydroxybutanoate (**183**), the key step being the resolution of ($\pm$)-**183** by ammo
reaction catalyzed by CAL-B in 1,4-dioxane as a solvent. CAL-B catalyzed the am
lysis of (S)-**183** with moderate enantioselectivity, the corresponding amide (S)-**18**
isolated with 93% ee after 1.5 h reaction (Scheme 7.54). The amide (S)-**184** was
transformed into (S)-**182** (87% of overall yield) [115].

7.5.3 (R)-GABOB (185)

(R)-4-Amino-3-hydroxybutanoic acid [(R)-GABOB] is a compound of great imp
because of its biological function as a neuromodulator in the mammalian central r
system. Moreover, (R)-GABOB (**185**) is a precursor of (R)-carnitine (**186**, vitami
a therapeutic agent for the treatment of myocardial ischemia. (R)-GABOB has be
thesized by a chemoenzymatic route involving the asymmetric ammonolysis of di

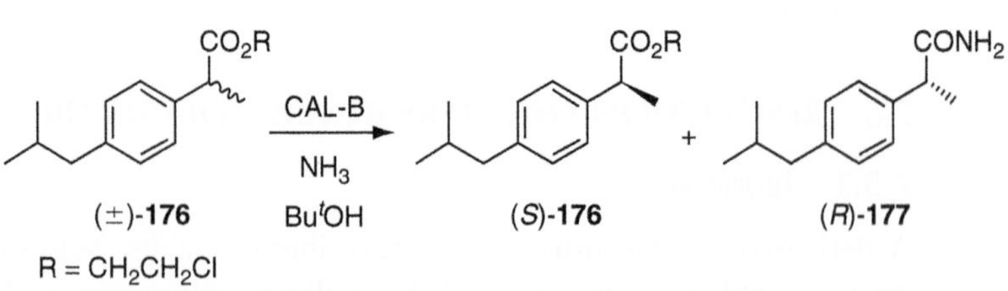

SCHEME 7.52 Resolution of ($\pm$)-**176**.

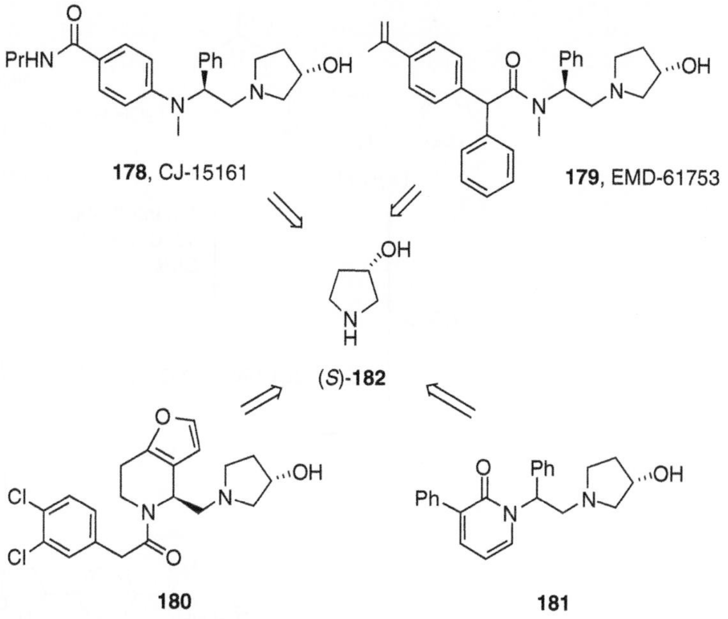

SCHEME 7.53 3-Pyrrolidinol derivatives.

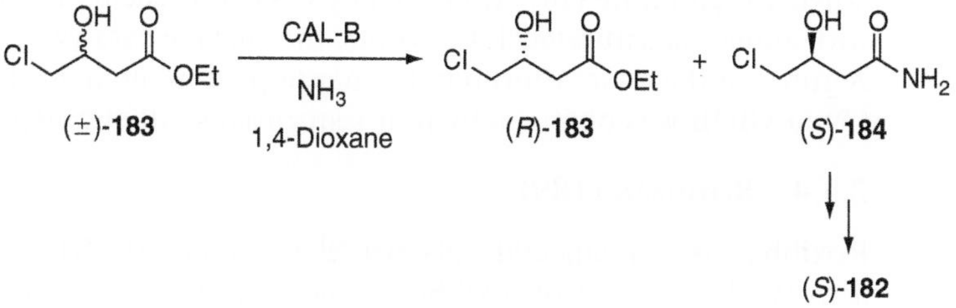

SCHEME 7.54 Chemoenzymatic synthesis of (S)-3-pyrrolidinol.

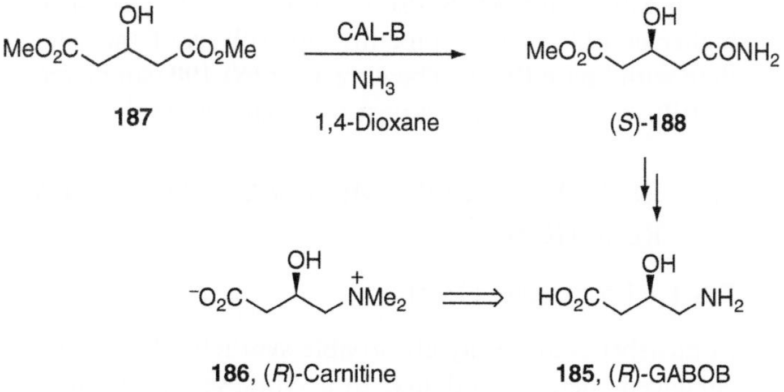

SCHEME 7.55 Chemoenzymatic synthesis of (R)-GABOB.

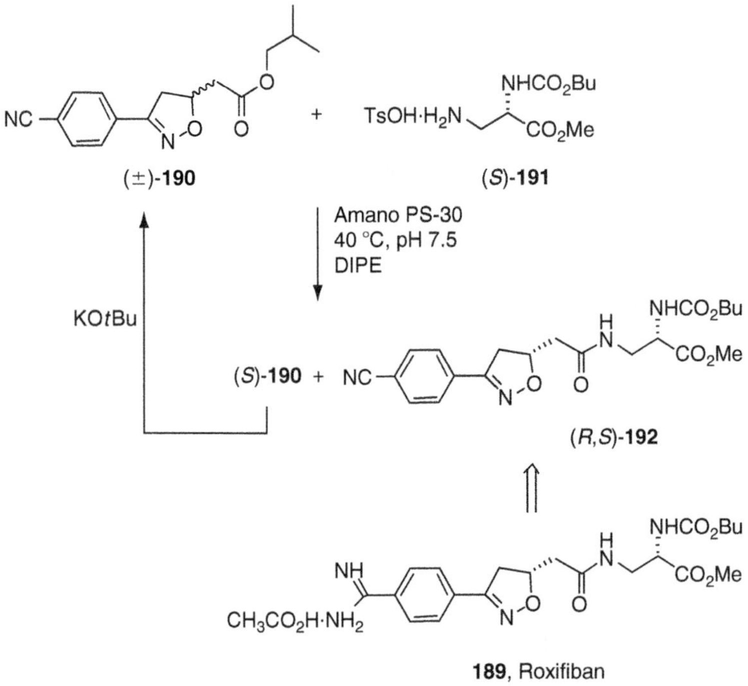

SCHEME 7.56 Enzymatic preparation of (R,S)-**192**, precursor of roxifiban.

3-hydroxyglutarate (**187**) catalyzed by CAL-B (Scheme 7.55). The reaction was car█
with ammonia saturated 1,4-dioxane, the enzyme catalyzing the transformation of █
R group of the ester, affording the enantiopure monoamide (S)-**188**. From this mon█
(R)-GABOB was obtained by four sequential synthetic steps [116].

7.5.4 ROXIFIBAN (189)

Roxifiban is a nonpeptide platelet glycoprotein IIb/IIIa antagonist with antithr█
activity. A precursor of roxifiban is the optically active isoxazol derivative (R,S)-**19**
which has been prepared by aminolysis of the racemic isobutyl 2-isoxazolylacetate (█
using a pretreated lipase Amano PS-30 [118]. As shown in Scheme 7.56, nucleophile █
enantiopure diaminoester derivative (S)-**191** that was used as its p-toluenesulfon█
because the free amine was unstable. When enzymatic reaction was carried out in █
the formation of (R,S)-**192** occurs with high diastereoselectivity [diastereomeric exces█
86%]. Although CAL-B also catalyzed this aminolysis reaction yielding (R,S)-**192** w█
higher de (92%), the competitive hydrolysis of the ester (±)-**190** took place at grea█
than with lipase PS-30. The unreacted (S)-**190** can be recovered, racemized by treatm█
KOtBu, and recycled in a next enzymatic aminolysis reaction.

7.6 RESOLUTION OF AMINO ACID DERIVATIVES BY N-ACYLATION REACTIONS

7.6.1 LORACARBEF (193)

Loracarbef is an orally absorbable synthetic β-lactam antibiotic that is characterize█
enhanced chemical stability. In a synthesis of **193** the racemic azetidinone cis-(±)-█
used as key intermediate, the (2R,3S)-isomer being used in subsequent chemistry█

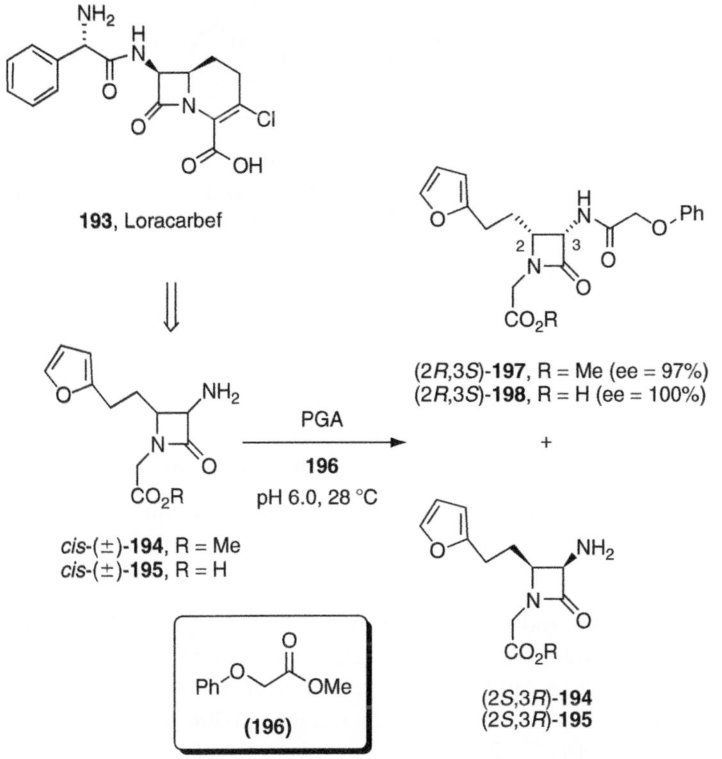

SCHEME 7.57 Resolution of azetidinones **194** and **195**, precursors of loracarbef.

antibiotic (Scheme 7.57). The required (2*R*,3*S*)-azetidinone was obtained by the enzymatic resolution of either *cis*-(±)-**194** or *cis*-(±)-**195**, using immobilized penicillin G amidase (PGA) from *E. coli*. [119]. Reactions were carried out in water, at pH 6.0, using methyl phenylacetate and methyl phenoxyacetate (**196**) as acyl donors. With both acyl donors, the enzyme catalyzed the formation of the amide with the required (2*R*,3*S*) configuration, with very high ee (≥96% starting from **194** and 100% starting from **195**). Scheme 7.57 demonstrates the results obtained with methyl phenoxyacetate (**196**).

The loracarbef also contains in its structure a L-phenylglycine unit. This α-amino acid can be obtained by enzymatic acylation of the corresponding racemic phenylglycine methyl ester using the PGA and methyl 4-hydroxyphenylacetate as the acyl donor in toluene or dichloromethane. The corresponding 4-hydroxyphenylacetamide derived from (*S*)-phenylglycine was obtained with a very high enantioselectivity [120].

7.6.2 Proline and Pipecolic Acid Derivatives 199–202

L-Proline and L-pipecolic acid have properties as anticonvulsants and they have been used as starting materials of many pharmacologically active compounds [121]. For instance, L-proline is a constituent of the antihypertensive enalapril (**199**), enalaprilat (**200**), and captopril (**201**) (Scheme 7.58). In addition, both enantiomers of pipecolic acid (**203**) have been used to obtain AF-DX 384 (**202**), a selective antagonist of muscarinic M$_2$ receptors, the (*R*)-isomer exhibiting higher affinity than (*S*)-enantiomer [122].

An efficient approach to obtain these cyclic α-amino acids with high enantiomeric purity has been demonstrated by the resolution of their corresponding racemic methyl esters

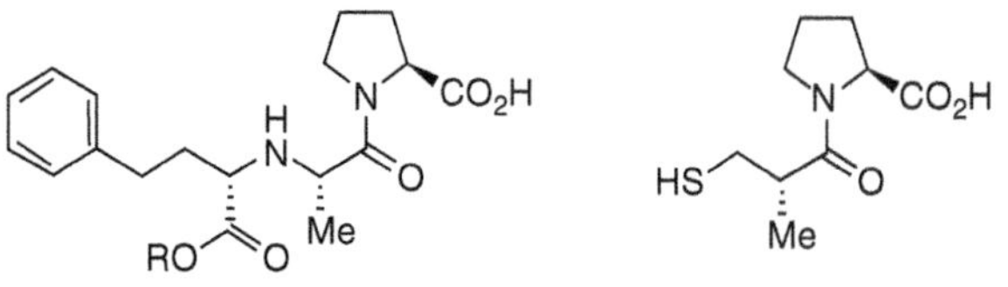

199, R = Et: Enalapril
200, R = H: Enalaprilat

201, Captopril

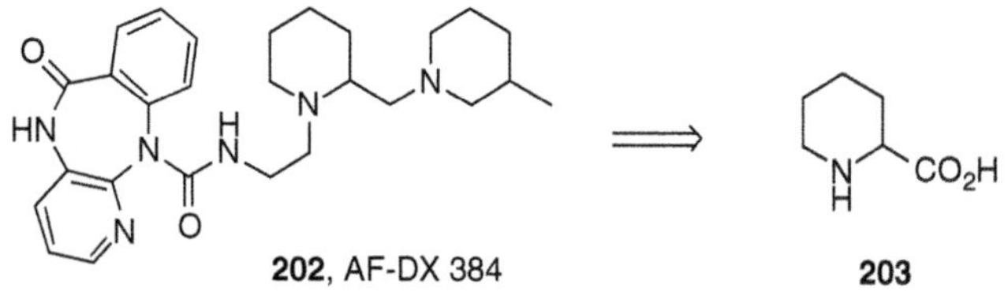

202, AF-DX 384

203

SCHEME 7.58 L-Proline and pipecolic acid derivatives.

($\pm$)-**204** and ($\pm$)-**205** by aminolysis of 2,2,2-trifluoroethyl butanoate (**136**) using C (immobilized on celite) as catalyst and TBME as solvent. In both cases, the enzyme wa efficient in catalyzing the acylation of the secondary amino group (Scheme 7.59) lead the formation of the butanamides (S)-**206** and (S)-**207** with very high enantioselec ($E > 100$) [17,123].

By an adequate selection of the reaction conditions the DKR of these substrat achieved. Thus, using vinyl butanoate as the acyl donor and triethylamine as an additi **206** and (S)-**207** were obtained with very high ee (97%) and yields (90% and 70%, respe [17]. The use of vinyl butanoate allows the release of acetaldehyde, which *in situ* racemi remaining (R)-**204** and (R)-**205**.

7.6.3 MOIRAMIDE B (208) AND ANDRIMID (209)

The pseudopeptide natural products moiramide B and andrimid represent a new c antibiotics that target bacterial fatty acid biosynthesis [124]. One of the starting mater the synthesis of **208** and **209** is the optically active D-β-phenylalanine or some of its deri [125]. Similar to the resolution of proline and pipecolic acid derivatives, CAL-A cataly resolution of racemic β-phenylalanine ethyl ester [($\pm$)-**210**] in DIPE as solvent. After reaction with 2,2,2-trifluoroethyl butanoate (**136**), the unreacted (R)-**210** and the butai (S)-**211** produced by enzymatic reaction were isolated in 90% ee and 98% ee, respe (Scheme 7.60) [126].

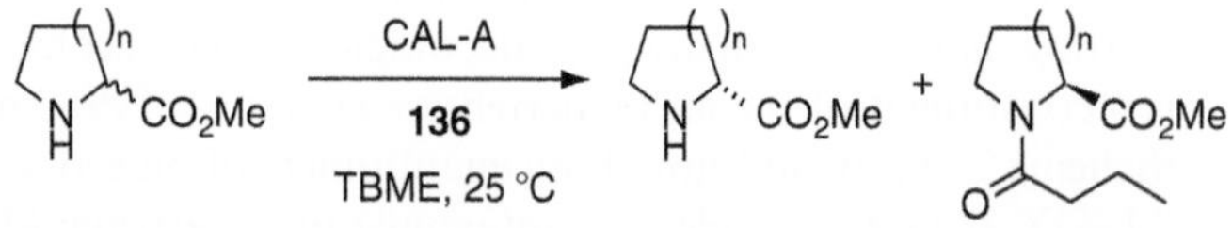

($\pm$)-**204**, n = 1	0.5 h, c = 50%	(R)-**204**	(S)-**206**
($\pm$)-**205**, n = 2	9 h, c = 49%	(R)-**205**	(S)-**207**

SCHEME 7.59 Resolution of ($\pm$)-**204** and ($\pm$)-**205**.

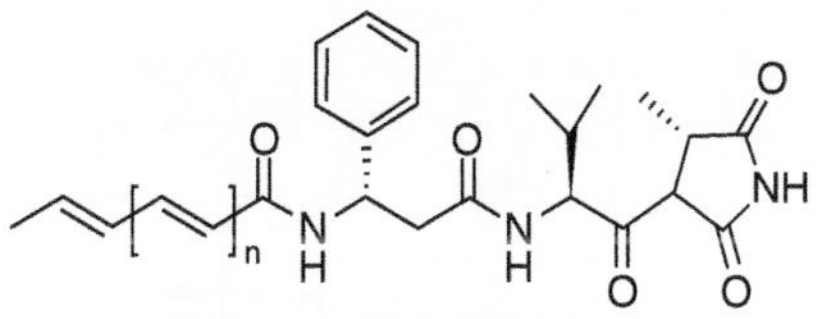

208, n = 1: Moriamide B
209, n = 2: Andrimid

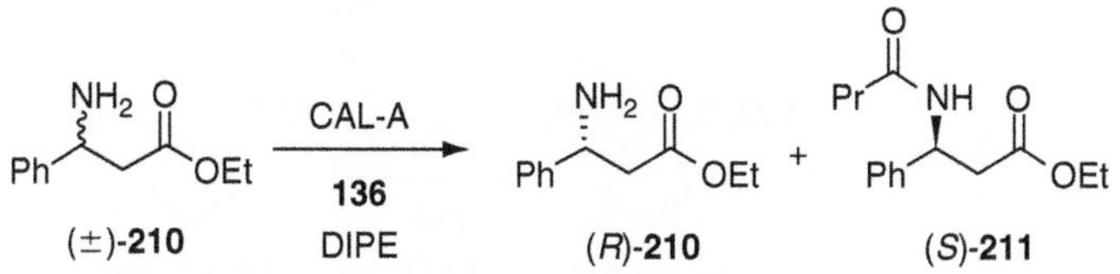

SCHEME 7.60 Resolution of ($\pm$)-β-phenylalanine ethyl ester **(210)**.

7.6.4   XEMILOFIBAN **(212)** AND **FR-184764 (213)**

These compounds are peptido-mimetics based on the RGD sequence of fibrinogen, which effectively disrupts the platelet–fibrinogen interaction thus inhibiting platelet aggregation and preventing thrombus formation.

The optically active β-amino ester (*S*)-**214** (Scheme 7.61) has been used in the syntheses of xemilofiban **(212)** [127] and its analogous FR-184764 **(213)** [128]. The resolution of ($\pm$)-**214** has been efficiently achieved by enzymatic acylation using phenylacetic acid **(215)** as acyl donor and an immobilized PGA-450 as catalyst [129]. The reaction was carried out at pH 5.7 and 28°C. Under these conditions, the enzyme showed a very high enantioselectivity and both the phenylacetamide (*R*)-**216** and the unreacted β-amino ester (S)-**214** were isolated with

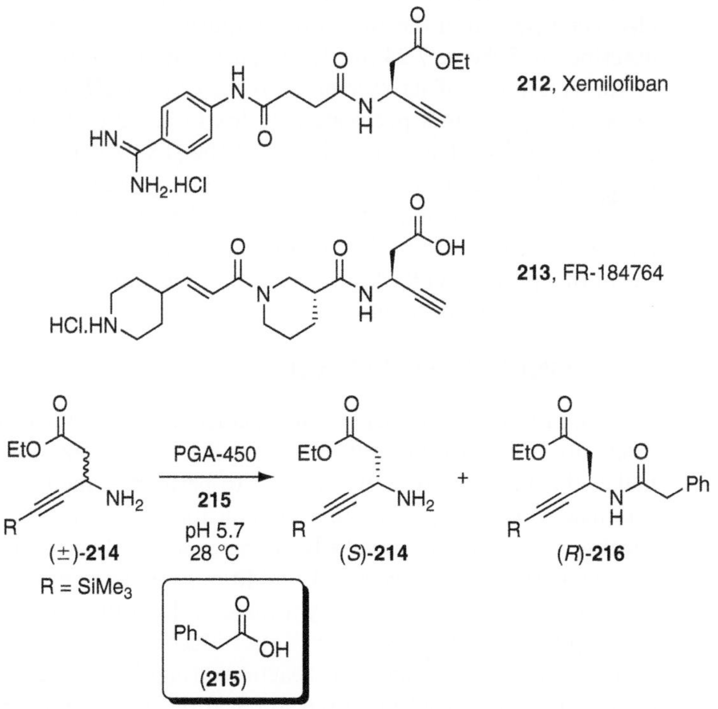

212, Xemilofiban

213, FR-184764

SCHEME 7.61 Resolution of ($\pm$)-**214**, precursor of xemilofiban and FR-184764.

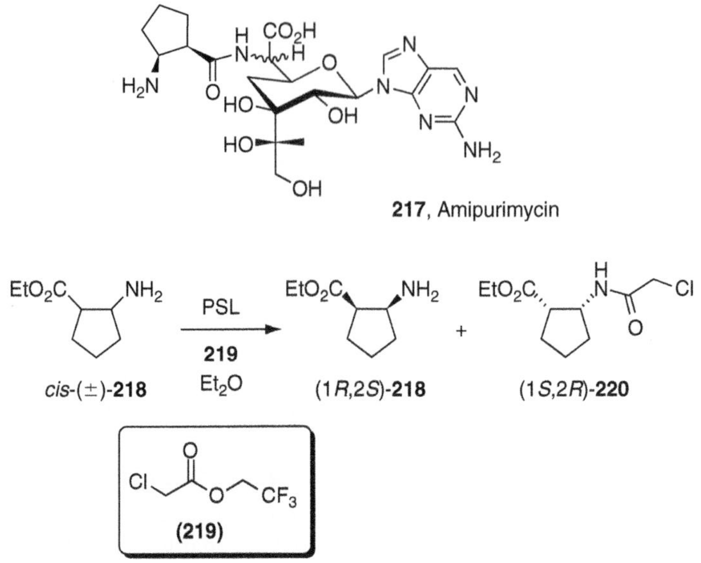

217, Amipurimycin

SCHEME 7.62 Resolution of (±)-**218**.

>99% ee and 98% ee, respectively. This bioconversion was scaled to resolve 6.1 kg of a
three runs, from which the required (S)-**214** was isolated in 96 to 98% ee and 43 to 46

7.6.5 AMIPURIMYCIN (217)

In addition to the already described pharmacological properties of (1,
aminocyclopentane carboxylic acid (**77**: cispentacin, Scheme 7.24), this β-amino
also a component of the antibiotic amipurimycin (**217**, Scheme 7.62). An alternative
described in Scheme 7.24 for the preparation of enantiomerically pure cispentancin
of the resolution of its racemic ethyl ester (±)-**218** by N-acylation reaction catal
lipases [130]. Using a pretreated PS lipase, 2,2,2-trifluoroethyl acetate as acyl don
diethyl ether as solvent, acetylation of the (1S,2R) enantiomer of the β-amino ester
with very high enantioselectivity (E >100), giving 49% of conversion after 1 h re
When a more activated ester such as 2,2,2-trifluoroethyl chloroacetate (**219**) was u
reaction rate increased, 51% of conversion being achieved only after 5 min of the r
From this process, (1R,2S)-**218** and chloroacetamide (1S,2R)-**220** were isolated witl
ee and 95% ee, respectively.

7.7 CONCLUDING REMARKS

The use of biocatalysis has become a conventional tool for organic and bioorganic c
Their utility in carrying out very selective transformations under mild reaction co
makes them a very attractive catalyst to perform some transformations in a syntheti
Pharmaceutical companies are using biocatalytic processes for the preparation of chir
because they offer enormous advantages, such as their environmental friendly proper
the low cost of the processes. In addition, we strongly believe that the genetic eng
techniques will continue to play a major role in the future research in biocatalysis a
open up new possibilities to carry out industrial processes in several sectors. In this
we have shown how enzymatic reactions using hydrolases, mainly acylation of alcoh
amines, have provided potential application for the preparation of chiral pharmac
during the past few years.

REFERENCES

1. Zaks, A., Industrial biocatalysis, *Curr. Opin. Chem. Biol.*, 5(2), 130–136, 2001.
2. Roberts, S.M., Biocatalysis in synthetic organic chemistry, *Tetrahedron*, 60(3), 499–500, 2004.
3. Thomas, S.M., DiCosimo, R., and Nagarajan, V., Biocatalysis: applications and potentials for the chemical industry, *Trends Biotechnol.*, 20(6), 238–242, 2002.
4. Federsel, H.J., Drug chirality-scale-up manufacturing and control, *CHEMTECH*, 23(12), 24–33, 1993.
5. Gotor, V., Biocatalysis applied to the preparation of pharmaceuticals, *Org. Proc. Res. Dev.*, 6(4), 420–426, 2002.
6. Straathof, A.J.J., Panke, S., and Schmid, A., The production of fine chemicals by biotransformations, *Curr. Opin. Biotecnol.*, 13(6), 548–556, 2002.
7. Huisman, G.W. and Gray, D., Towards novel processes for the fine chemical and pharmaceutical industries, *Curr. Opin. Biotechnol.*, 13(4), 352–358, 2002.
8. Patel, R.N., Microbial/enzymatic synthesis of chiral intermediates for pharmaceuticals, *Enzyme Microb. Tech.*, 31(6), 804–826, 2002.
9. Patel, R.N., Microbial/enzymatic synthesis of chiral pharmaceutical intermediates, *Curr. Opin. Drug. Disc. Dev.*, 6(6), 902–920, 2003.
10. Reetz, M.T., Lipases as practical biocatalysts, *Curr. Opin. Chem. Biol.*, 6(2), 145–150, 2002.
11. Ghanem, A. and Aboul-Enein, H.Y., Lipase mediated chiral resolution of racemates in organic solvents, *Tetrahedron Asymmetry*, 15(21), 3331–3351, 2004.
12. Alfonso, I. and Gotor, V., Biocatalytic and biomimetic aminolysis reactions: useful tools for selective transformations on polifunctional substrates, *Chem. Soc. Rev.*, 33(4), 201–209, 2004.
13. Kato, K., Goug, Y., Saito, T., et al., Enzymatic resolution of 2,2,2-trifluoro-1-arylethylamine derivatives by *Pseudomonas fluorescens* lipase in organic solvents, *J. Mol. Catal., B Enzym.*, 30(2), 61–68, 2004.
14. Faber, K., Non-sequential processes for the transformation of a racemate into a single stereoisomer product. Proposal for stereochemical classification, *Chem. Eur. J.*, 7(23), 5005–5010, 2001.
15. Pellissier, H., Dynamic kinetic resolution, *Tetrahedron*, 59(42), 8291–8327, 2003.
16. Pàmies, O. and Bäckvall, J.E., Combination of enzymes and metal catalysis. A powerful approach in asymmetric catalysis, *Chem. Rev.*, 103(8), 3247–3262, 2003.
17. Liljeblad, A., Kiviniemi, A., and Kanerva, L.T., Aldehyde-based racemization in the dynamic kinetic resolution of *N*-heterocyclic α-aminoesters using *Candida antarctica* lipase A, *Tetrahedron*, 60(3), 671–677, 2004.
18. Vedejs, E. and Chen, X., Parallel kinetic resolution, *J. Am. Chem. Soc.*, 119(10), 2584–2585, 1997.
19. Dehli, J.R. and Gotor, V., Parallel kinetic resolution of racemic mixtures: a new strategy for the preparation of enantiopure compounds? *Chem. Soc. Rev.*, 31(6), 365–370, 2002.
20. Wills, M.C.J., Enantioselective desymmetrization, *J. Chem. Soc. Perkin Trans.*, 1(13), 1765–1784, 1999.
21. Schoffers, E., Gelebiowski, A., and Johnson C.R., Enantioselective synthesis through enzymatic asymmetrization, *Tetrahedron*, 52(11), 3769–3826, 1996.
22. García-Urdiales, E., Alfonso, I., and Gotor, V., Enantioselective desymmetrization in organic synthesis, *Chem. Rev.*, 105(1), 313–354, 2005.
23. Gotor, V., Pharmaceuticals through enzymatic transesterification and aminolysis reactions, *Biocatal. Biotransformation*, 18(2), 87–103, 2000.
24. Pesti, J.A. and DiCosimo, R., Recent progress in enzymatic resolution and desymmetrization of pharmaceuticals and their intermediates, *Curr. Opin. Drug. Discov. Devel.*, 6(6), 884–901, 2003.
25. Margolin, A.L., Enzymes in the synthesis of chiral drugs, *Enzyme Microb. Technol.*, 15(4), 266–280, 1993.
26. Schulze, B. and Wubbolts, M.G., Biocatalysis for industrial production of fine chemicals, *Curr. Opin. Biotechnol.*, 10(6), 609–615, 1999.
27. Patel, R.N. (Ed.), *Stereoselective Biocatalysis*, Marcel Dekker, New York, 2000, pp. 87–130.
28. Patel, R.N., Enzymatic synthesis of chiral intermediates for drug development, *Adv. Synth. Catal.*, 343(6–7), 527–546, 2001.

29. Patel, R.N., Biocatalytic synthesis of intermediates for the synthesis of chiral drugs sub *Curr. Opin. Biotechnol.*, 12(6), 587–604, 2001.
30. Roberts, S.M., Preparative biotransformations, *J. Chem. Soc. Perkin Trans.*, 1, 1475–149§
31. Jaeger, K.-E. and Eggert, T., Lipases for biotechnology, *Curr. Opin. Biotechnol.*, 13(4), 3 2002.
32. Laumen, K., Kittelmann, M., and Ghisalba, O., Chemo-enzymatic approaches for the cre; novel chiral building blocks and reagents for pharmaceutical applications, *J. Mol. C; Enzym.*, 19–20, 55–66, 2002.
33. Patel, R.N., Enzymatic preparation of chiral pharmaceutical intermediates by lipases *Biotechnol.*, 527–561, 2002.
34. Park, H.G., Do, J.H., and Chang, N.C., Regioselective enzymatic acylation of multi-h compounds in organic synthesis, *Biotechnol. Bioprocess Eng.*, 8(1), 1–8, 2003.
35. Im, D.S., Cheong, C.S., Lee, S.H., et al., Chemo-enzymatic synthesis of (*R*)-(+)-aminog mide by kinetic resolution of (±)-4-cyano-4-phenyl-1-hexanol, *J. Mol. Catal., B Enzym.*, ' 185–191, 2003.
36. Ferraboschi, P., Pecora, F., Reza-Elahi, S., et al., Chemoenzymatic syntheses of (25*R*)- an(25-hydroxy-27-nor-cholesterol, a steroid bearing a secondary hydroxy group in the sid(*Tetrahedron Asymmetry*, 10(13), 2497–2500, 1999.
37. Hiroya, K., Zhang, H., and Ogasawara, K., Preparation of the synthetic equivalents c cyclohexadienone and cycloheptadienone. The enantio- and diastereo-controlled synthesis clavularin B, *Synlett*, (5), 529–532, 1999.
38. Mehta, G. and Reddy, D.S., Enzymatic resolution of dioxygenated dicyclopentadienes: ε pure hydrindanes, hydroisoquinolones, diquinanes and application to a synthesis of (+)- facic acid, *Tetrahedron Lett.*, 40(5), 991–994, 1999.
39. Morishita, K., Kamezawa, M., Ohtani, T., et al., Chemoenzymic synthesis of (+)-doco dien-1-yn-3-ol, a component of the marine sponge *Cribrochalina vasculum*, confirmatior structure and absolute configuration of the acetylenic alcohol, by lipase-catalyzed biotran tions, *J. Chem. Soc. Perkin Trans.*, 1(4), 513–518, 1999.
40. Frigola, J., Berrocal, J.M., Cuberes, M.R., et al., Procede de séparation de carbinols, Patent 235497 D16049-FA, 1995.
41. Zhu, B. and Panek, J.S., Methodology based on chiral silanes in the synthesis of polypro derived natural products—total synthesis of epothiolone A, *Eur. J. Org. Chem.*, (9), 170(2001.
42. Liu, H.-L., Helge, B., and Anthonsen, T., Chemoenzymatic synthesis of the non-tricyclic ant sants fluoxetine, tomoxetine and nisoxetine, *J. Chem. Soc. Perkin Trans.*, 1(11), 1767–1769, ;
43. Kamal, A., Khanna, G.B.R., and Ramu, R., Chemoenzymatic synthesis of both enantio fluoxetine, tomoxetine and nisoxetine: lipase catalyzed resolution of 3-aryl-3-hydroxypropane *Tetrahedron Asymmetry*, 13(18), 2039–2051, 2002.
44. Campos, F., Bosch, M.P., and Guerrero, A., An efficient enantioselective synthesis of formoterol, a potent bronchodilator, using lipases, *Tetrahedron Asymmetry*, 11(13), 2705–271
45. Patel, R.N., McNamee, C.G., and Szarka, L.J., Enantioselective enzymatic acetylation of 4-4-α,6-β(*E*)-6,4,4-bis(4-fluorophenyl)-3-(1-methyl-1-H-tetrazol-5-yl)-1,3-butadienyl-1 dro-4-hydroxy-2H-pyran-2-one, *Appl. Microbiol. Biotechnol.*, 38(1), 56–60, 1992.
46. Nagai, H., Shizawa, T., and Achiwa, K., Convenient syntheses of optically active β-lact enzymic resolution, *Chem. Pharm. Bull.*, 41(11), 1933–1938, 1993.
47. Sankaranarayanan, S., Sharma, A., Kulkarni, B.A., et al., Preparation of the versatile chir and (*S*)-12-(tetrahydropyranyloxy)-3-methyldodecanol: application to the syntheses of branched insect pheromones, *J. Org. Chem.*, 60(12), 4251–4253, 1995.
48. Kulkarni B.A., Sankaranarayanan S., Subbaraman A.S., et al., Synthesis of racemic a; enantiomer of 3-methylnonacosanol, a new plant growth regulator from *Lowsonia inermis* *hedron Asymmetry*, 10(8), 1571–1577, 1999.
49. Kapoor, M., Anand, N., Ahmad, K., et al., Synthesis of β-adrenergic blockers (*R*)-(−)-n and (*S*)-(+)-sotalol via a highly resolution of a bromohydrin precursor, *Tetrahedron Asy* 16(3), 717–725, 2005.

50. Ebiike, H., Ozawa, Y., Achiwa, K., et al., Enzyme-catalysed synthesis of biologically active (*S*)-nilvadipine, *Heterocycles*, 35(2), 603–606, 1993.
51. Kitano, K., Matsubara, J., Ohtani, T., et al., An efficient synthesis of optically active metabolites of platelet adhesion inhibitor OPC-29030 by lipase-catalyzed enantioselective transesterification, *Tetrahedron Lett.*, 40(28), 5235–5238, 1999.
52. Pàmies, O. and Bäckvall, J.-E., Dynamic kinetic resolution of β-azido alcohols. An efficient route to chiral aziridines and β-aminoalcohols, *J. Org. Chem.*, 66(11), 4022–4025, 2001.
53. Wang, Y.-F., Chen, S.-T., Liu, K.-C., et al., Lipase-catalyzed irreversible transesterification using enol esters: resolution of cyanohydrins and syntheses of ethyl (*R*)-2-hydroxy-4-phenylbutyrate and (*S*)-propanolol, *Tetrahedron Lett.*, 30(15), 1917–1920, 1989.
54. Bevinakatti, H.S. and Banerji, A.A., Practical chemoenzymatic synthesis of both enantiomers of propanolol, *J. Org. Chem.*, 56(18), 5372–5375, 1991.
55. Cregge, R.J., Wagner, E.R., Freedman, J., et al., Lipase-catalyzed transesterification in the synthesis of a new chiral unlabeled and carbon-14 labeled serotonin uptake inhibitor, *J. Org. Chem.*, 55(14), 4237–4238, 1990.
56. Carr, A.A., Nieduzak, T.R., and Miller, F.P., 1,4-Disubstituted piperidine derivatives and their preparation, pharmaceutical compositions, and use in analgesics and muscle relaxants, EP 317997, CAN 111:232581, 1989.
57. Sih, J.C., Application of immobilized lipase in production of camptosar (CPT-11), *J. Am. Oil Chem. Soc.*, 73(11), 1377–1378, 1996.
58. Theil, F. and Ballschuh, S., Chemoenzymatic synthesis of both enantiomers of cispentacin, *Tetrahedron Asymmetry*, 7(12), 3565–3572, 1996.
59. Gais, H.-J., Griebel, C., and Buschmann, H., Enzymatic resolution of analgesics: δ-hydroxytramadol, ε-tramadol and *O*-desmethyltramadol, *Tetrahedron Asymmetry*, 11(4), 917–928, 2000.
60. Shimada, K., Kaburagi, Y., and Fukuyama, T., Total synthesis of leustroduscin B., *J. Am. Chem. Soc.*, 125(14), 4048–4049, 2003.
61. Terao, Y., Murata, M., Achiwa, K., et al., Highly efficient lipase-catalyzed asymmetric synthesis of chiral glycerol derivatives leading to practical synthesis of *s*-propranolol, *Tetrahedron Lett.*, 29(40), 5173–5176, 1988.
62. Saksena, A.K., Girijavallabhan, V.M., Lovey, R.G., et al., Highly stereoselective access to novel 2,2,4-trisubstituted tetrahydrofurans by halocyclization-practical chemoenzymatic synthesis of Sch-51048, a broad spectrum orally active antifungal agent, *Tetrahedron Lett.*, 36(11), 1787–1790, 1995.
63. Bovara, R., Carrea, G., Ferrara, L., et al., Resolution of (±)-*trans*-sobrerol by lipase PS-catalyzed transesterification and effects of organic solvents on enantioselectivity, *Tetrahedron Asymmetry*, 2(9), 931–938, 1991.
64. Akai, S., Tsujino, T., Naka, T., et al., Lipase-catalysed enantioselective desymmetrization of prochiral 3,3-*bis*(hydroxymethyl)oxindoles, *Tetrahedron Lett.*, 42(41), 7315–7317, 2001.
65. Akai, S., Tsujino, T., Akiyama, E., et al., Enantiodivergent preparation of optically active oxindoles having a stereogenic quaternary carbon center at the C3 position via the lipase-catalyzed desymmetrization protocol: effective use of 2-furoates for either enzymatic esterification or hydrolysis, *J. Org. Chem.*, 69(7), 2478–2486, 2004.
66. Colombo, D., Compostella, F., Ronchetti, F., et al., Chemoenzymatic synthesis and antitumor promoting activity of 6′- and 3-esters of 2-*O*-β-D-glucosyglycerol, *Bioorg. Med. Chem.*, 7(9), 1867–1871, 1999.
67. Laumen, K. and Ghisalba O., Preparative-scale chemo-enzymic synthesis of optically pure D-*myo*-inositol 1-phosphate, *Biosci. Biotechol. Biochem.*, 58(11), 2046–2049, 1994.
68. Laumen, K. and Ghisalba O., Chemo-enzymatic synthesis of both enantiomers of *myo*-inositol 1,3,4,5-tetrakisphosphate, *Biosci. Biotechol. Biochem.*, 63(8), 1374–1377, 1999.
69. Chênevert, R., Simard, M., Bergeron, J., et al., Chemoenzymatic formal synthesis of (*S*)-(-)-phosphonotrixin, *Tetrahedron Asymmetry*, 15(12), 1889–1892, 2004.
70. Chênevert, R. and Courchesne, G., Synthesis of (*S*)-α-tocotrienol via an enzymatic desymmetrization of an achiral chroman derivative, *Tetrahedron Lett.*, 43(44), 7971–7973, 2002.

71. Chênevert, R., Ziarani, G.M., Morin, M.P., et al., Enzymatic desymmetrization of *meso* and *cis, cis*-2,4,6-substituted piperidines. Chemoenzymatic synthesis of (5S,9S)-(+)-indc 209D, *Tetrahedron Asymmetry*, 10(16), 3117–3122, 1999.

72. Mase, N., Nishi, T., Takamori, Y., et al., First synthesis of (R)-(−)-5-hydroxy-3-methyl-3-p 2-one (jatropham) by lipase-catalyzed kinetic resolution, *Tetrahedron Asymmetry*, 10(23) 4471, 1999.

73. Lundell, K., Katainen, E., Kiviniemi, A., et al., Enantiomers of adrenaline-type aminoalcc *Burkholderia cepacia* lipase-catalyzed asymmetric acylation, *Tetrahedron Asymmetry*, 3723–3729, 2004.

74. Franceschini, N., Joy, M.S., and Kshirsagar, A., Cinacalcet HCl: a calcimimetic agent management of primary and secondary hyperparathyroidism, *Expert Opin. Investig. Drug* 1413–1421, 2003.

75. Van Wagenen, B.C., Moe, S.T., Balandrin, M.F., et al., Preparation of 1-arylethylan calcium receptor ligands, US 6211244, CAN 134:280612, 2001.

76. Skupinska, K.A., McEachern, E.J., Baird, I.R., et al., Enzymatic resolution of bicyclic 1 arylamines using *Candida antarctica* lipase B., *J. Org. Chem.*, 68(9), 3546–3551, 2003.

77. Reetz, M.T. and Dreisbach, C., Highly efficient lipase-catalyzed kinetic resolution of chiral *Chimia*, 48(12), 570, 1994.

78. Reeve, C.D., Resolution of chiral amines, WO 9931264, CAN 131:43668, 1999.

79. García-Urdiales, E., Rebolledo, F., and Gotor, V., Enzymatic one-pot resolution of two philes: alcohol and amine, *Tetrahedron Asymmetry*, 11(7), 1459–1463, 2000.

80. Jaeger, K.-E., Schneidinger, B., Rosenau, F., et al., Bacterial lipases for biotechnological tions, *J. Mol. Catal., B Enzym.*, 3(1–4), 3–12, 1997.

81. Kitaguchi, H., Fitzpatrick, P.A., Huber, J.E., et al., Enzymic resolution of racemic amines role of the solvent, *J. Am. Chem. Soc.*, 111(8), 3094–3095, 1989.

82. Margolin, A.L., Fitzpatrick, P.A., Dubin, P.L., et al., Chemoenzymic synthesis of opticall (meth)acrylic polymers, *J. Am. Chem. Soc.*, 113(12), 4693–4694, 1991.

83. Takayama, S., Moree, W.J., and Wong, C.-H., Enzymic resolution of amines and amino ε using pent-4-enoyl derivatives, *Tetrahedron Lett.*, 37(35), 6287–6290, 1996.

84. Gutman, A.L., Meyer, E., Kalerin, E., et al., Enzymatic resolution of racemic amir continuous reactor in organic solvents, *Biotechnol. Bioeng.*, 40(7), 760–767, 1992.

85. Youdim, M.B.H., Finberg, J.P.M., Levy, R., Sterling, J., Lerner, D., and Berger-Paskin enantiomers of *N*-propargyl-1-aminoindan compounds as inhibitors of B-form of mor oxidase enzyme, EP 436492, CAN 115:158747, 1991.

86. Youshko, M.I., van Rantwijk, F., and Sheldon, R.A., Enantioselective acylation of chiral catalyzed by aminoacylase I, *Tetrahedron Asymmetry*, 12(23), 3267–3271, 2001.

87. van Langen, L.M., Oosthoek, N.H.P., Guranda, D.T., et al., Penicillin acylase-ca resolution of amines in aqueous organic solvents, *Tetrahedron Asymmetry*, 11(22), 459 2000.

88. Takayama, S., Lee, S.T., Hung, S.-C., et al., Designing enzymatic resolution of amines *Commun.*, (2), 127–128, 1999.

89. Hacking, M.A.P.J., van Rantwijk, F., and Sheldon, R.A., Lipase catalyzed reactions of ε and arylaliphatic carbonic acid esters, *J. Mol. Catal., B Enzym.*, 9(4–6), 201–208, 2000.

90. Choi, Y.K., Kim, M.J., Ahn, Y., et al., Lipase/palladium-catalyzed asymmetric transforma ketoximes to optically active amines, *Org. Lett.*, 3(25), 4099–4101, 2001.

91. Messina, F., Botta, M., Corelli, F., et al., Chiral azole derivatives, 3. Synthesis of the enar of the potent aromatase inhibitor 1-[2-benzofuranyl(4-chlorophenyl)methyl]-1*H*-imidazolε *hedron Lett.*, 40(40), 7289–7292, 1999.

92. Messina, F., Botta, M., Corelli, F., et al., Resolution of (±)-1-aryl-2-propynylamines transfer catalyzed by *Candida antarctica* lipase, *J. Org. Chem.*, 64(10), 3767–3769, 1999.

93. Taylor K.M. and Snyder S.H., Amphetamine: differentiation by *d* and *l* isomers of k involving brain norepinephrine or dopamine, *Science*, 168(938), 1487–1489, 1970.

94. Hajos, G.T. and Garattini, S., A note on the effect of (+)- and (−)-amphetamine metabolism, *J. Pharm. Pharmacol.*, 25(5), 418–419, 1973.

95. González-Sabín, J., Gotor, V., and Rebolledo, F., CAL-B-catalyzed resolution of some pharmacologically interesting β-substituted isopropylamines, *Tetrahedron Asymmetry*, 13(12), 1315–1320, 2002.

96. Goument, B., Duhamel, L., and Mauge, R., Asymmetric syntheses of (*S*)-fenfluramine using sharpless epoxidation methods, *Tetrahedron*, 50(1), 171–188, 1994.

97. Fogagnolo, M., Giovannini, P.P., Guerrini, A., et al., Homochiral (*R*)- and (*S*)-1-heteroaryl- and 1-aryl-2-propanols via microbial redox, *Tetrahedron Asymmetry*, 9(13), 2317–2327, 1998.

98. Turgeon, J., Uprichard, A.C., Belanger, P.M., et al., Resolution and electrophysiological effects of mexiletine enantiomers, *J. Pharm. Pharmacol.*, 43(9), 630–635, 1991.

99. Gold, E.H., Chang, W., Cohen, M., et al., Synthesis and comparison of some cardiovascular properties of the stereoisomers of labetalol, *J. Med. Chem.*, 25(11), 1363–1370, 1982.

100. Guranda, D.T., van Langen, L.M., van Rantwijk, F., et al., Highly efficient and enantioselective enzymatic acylation of amines in aqueous medium, *Tetrahedron Asymmetry*, 12(11), 1645–1650, 2001.

101. Breen, G.F., Forth, M.A., and Popkin, M.E., Process for preparing 1-methyl-1,2,3,4-tetrahydroisoquinoline or a salt thereof, WO 2002088088, CAN 137:352906, 2002.

102. Breen, G.F., Enzymatic resolution of a secondary amine using novel acylating reagents, *Tetrahedron Asymmetry*, 15(9), 1427–1430, 2004.

103. Njoroge, F.G., Taveras, A.G., Kelly, J., et al., (+)-4-[2-[4-(8-Chloro-3,10-dibromo-6,11-dihydro-5H-benzo[5,6]cyclohepta[1,2-*b*]- pyridin-11(*R*)-yl)-1-piperidinyl]-2-oxo-ethyl]-1-piperidinecarboxamide (SCH-66336): a very potent Farnesyl protein transferase inhibitor as a novel antitumor agent, *J. Med. Chem.*, 41(24), 4890–4902, 1998.

104. Morgan, B., Zaks, A., Dodds, D.R., et al., Enzymatic kinetic resolution of piperidine atropisomers: synthesis of a key intermediate of the farnesyl protein transferase inhibitor, SCH66336, *J. Org. Chem.*, 65(18), 5451–5459, 2000.

105. Lucet, D., Le Gall, T., and Mioskowski, C., The chemistry of vicinal diamines, *Angew. Chem. Int. Ed. Engl.*, 37(19), 2580–2627, 1998.

106. Michalson, E.T. and Szmuszkovicz, J., Medicinal agents incorporating the 1,2-diamine functionality, *Prog. Drug Res.*, 33, 135–149, 1989.

107. Alfonso, I., Astorga, C., Rebolledo, F., et al., Sequential biocatalytic resolution of (±)-*trans*-cyclohexane-1,2-diamine. Chemoenzymic synthesis of an optically active polyamine, *Chem. Commun.*, (21), 2471–2472, 1996.

108. De Costa, B.R., Bowen, W.D., Hellewell, S.B., et al., Alterations in the stereochemistry of the κ-selective opioid agonist U50,488 result in high-affinity σ ligands, *J. Med. Chem.*, 32(8), 1996–2002, 1989.

109. González-Sabín, J., Gotor, V., and Rebolledo, F., Chemoenzymatic preparation of optically active *trans*-cyclohexane-1,2-diamine derivatives: an efficient synthesis of the analgesic U-(−)-50,488, *Chem. Eur. J.*, 10(22), 5788–5794, 2004.

110. de Zoete, M.C., Kock-van Dalen, A.C., van Rantwijk, F., et al., Ester ammonolysis: a new enzymic reaction, *J. Chem. Soc. Chem. Commun.*, (24), 1831–1832, 1993.

111. Ghosh, A., Sieser, J.E., Caron, S., et al., Synthesis of the kappa-agonist CJ-15,161 via a palladium-catalyzed cross-coupling reaction, *Chem. Commun.*, (15), 1644–1645, 2002.

112. Gottschlich, R., Ackermann, K.A., Barber, A., et al., EMD 61753 as a favorable representative of structurally novel arylacetamido-type κ opiate receptor agonists, *Bioorg. Med. Chem. Lett.*, 4(5), 677–682, 1994.

113. Naylor, A., Judd, D.B., Scopes, D.I.C., et al., 4-[(Alkylamino)methyl]furo[3,2-*c*]pyridines: a new series of selective κ-receptor agonists, *J. Med. Chem.*, 37(14), 2138–2144, 1994.

114. Semple, G., Andersson, B.-M., Chhajlani, V., et al., Synthesis and biological activity of kappa opioid receptor agonists. Part 2: preparation of 3-aryl-2-pyridone analogues generated by solution- and solid-phase parallel synthesis methods, *Bioorg. Med. Chem. Lett.*, 13(6), 1141–1145, 2003.

115. García-Urdiales, E., Rebolledo, F., and Gotor, V., Enzymatic ammonolysis of ethyl (±)-4-chloro-3-hydroxybutanoate. Chemoenzymatic syntheses of both enantiomers of pyrrolidin-3-ol and 5-(chloromethyl)-1,3-oxazolidin-2-one, *Tetrahedron Asymmetry*, 10(4), 721–726, 1999.

116. Puertas, S., Rebolledo, F., and Gotor, V., Enantioselective enzymic aminolysis and amme of dimethyl 3-hydroxyglutarate. Synthesis of (R)-4-amino-3-hydroxybutanoic acid, *J. Org.* 61(17), 6024–6027, 1996.

117. Zhang, L.-h., Chung, J.C., Costello, T.D., et al., The enantiospecific synthesis of an isoxaze RGD mimic platelet GPIIb/IIIa antagonist, *J. Org. Chem.*, 62(8), 2466–2470, 1997.

118. Sigmund, A.E., McNulty, K.C., Nguyen, D., et al., Enantioselective enzymatic aminoly racemic 2-isoxazolylacetate alkyl ester, *Can. J. Chem.*, 80(6), 608–612, 2002.

119. Zmijewski, M.J., Briggs, B.S., Thompson, A.R., et al., Enantioselective acylation of a beta intermediate in the synthesis of loracarbef using penicillin G amidase, *Tetrahedron Lett.*, 1621–1622, 1991.

120. Basso, A., Braiuca, P., De Martin, L., et al., D-Phenylglycine and D-4-hydroxyphenylglycine esters via penicillin G acylase catalysed resolution in organic solvents, *Tetrahedron Asyn* 11(8), 1789–1796, 2000.

121. Petersen, M. and Sauter, M., Biotechnology in the fine-chemicals industry. Cyclic amino a enantioselective biocatalysis, *Chimia*, 53(12), 608–612, 1999.

122. Martin, J., Deagostino, A., Perrio, C., et al., Syntheses of R and S isomers of AF-L a selective antagonist of muscarinic M$_2$ receptors, *Bioorg. Med. Chem.*, 8(3), 591–600, 200

123. Liljeblad, A., Lindborg, J., Kanerva, A., et al., Enantioselective lipase-catalyzed react methyl pipecolinate: transesterification and *N*-acylation, *Tetrahedron Lett.*, 43(13), 247 2002.

124. Pohlmann, J., Lampe, T., Shimada, M., et al., Pyrrolidinedione derivatives as antibacteria with a novel mode of action, *Bioorg. Med. Chem. Lett.*, 15(4), 1189–1192, 2005.

125. Davies, S.G. and Dixon, D.J., Asymmetric syntheses of moiramide B and andrimid, *J. Che Perkin Trans.*, 1(17), 2635–2644, 1998.

126. Gedey, S., Liljeblad, A., Lazar, L., et al., Preparation of highly enantiopure β-amino es *Candida antarctica* lipase A, *Tetrahedron Asymmetry*, 12(1), 105–110, 2001.

127. Cossy, J., Schmitt, A, Cinquin, C., et al., A very short, efficient and inexpensive synthesi prodrug form of SC-54701A a platelet aggregation inhibitor, *Bioorg. Med. Chem. Lett* 1699–1700, 1997.

128. Yamanaka, T., Ohkubo, M., Takahashi, F., et al., An efficient synthesis of the orally-active IIIa antagonist FR184764, *Tetrahedron Lett.*, 45(13), 2843–2845, 2004.

129. Landis, B.H., Mullins, P.B., Mullins, K.E., et al., Kinetic resolution of β-amino esters by a using immobilized penicillin amidohydrolase, *Org. Proc. Res. Dev.*, 6(4), 539–546, 2002.

130. Kanerva, L.T., Csomós, P., Sundholm, O., et al., Approach to highly enantiopure β-ami esters by using lipase catalysis in organic media, *Tetrahedron Asymmetry*, 7(6), 1705–1716

8 Dynamic Kinetic Resolution and Asymmetric Transformations by Enzyme–Metal Combinations

Mahn-Joo Kim, Yangsoo Ahn, and Jaiwook Park

CONTENTS

8.1 INTRODUCTION

The methods for the preparation of optically active compounds are of great importance in pharmaceutical and fine chemical industries [1]. One popular method is the resolution of racemic mixtures by enzymes such as lipases and esterases [2–16]. This method has been widely used for the preparation of optically active alcohols, acids, and their esters. However, the enzymatic resolution method has an intrinsic limitation; the yield cannot exceed 50% for a single enantiomer. Thus, the resolution is usually accompanied by additional processes such as separation, racemization, and recycling of unwanted enantiomers. The limitation, however, can be overcome if kinetic resolution could be transformed to dynamic kinetic resolution

(DKR) by coupling with a racemization reaction for the *in situ* conversion of un enantiomers to products [17].

Recently, several groups have reported the use of a metal complex as the racemizing along with an enzyme in the DKR [18–26]. Allen and Williams employed palladium racemizing catalyst together with lipase in the DKR of allyl acetates [27] and rhodiun DKR of secondary alcohols [28]. Reetz and Schimossek used palladium for the racemiz 1-phenylethylamine in its DKR with lipase [29]. Later, the Bäckvall group reported the of secondary alcohols employing a ruthenium complex for the racemization. Recen developed a new Ru-based catalyst for the efficient racemization of secondary alcohols temperature [30]. The room temperature DKRs of secondary alcohols with the Ru were successfully performed with both lipase and subtilisin to provide (*R*)- and (*S*)-p respectively, in good yields. The enzyme–metal combination strategy has also been ap the asymmetric transformations of prochiral ketones and ketoximes.

This chapter covers the recent developments in the DKR and the asymmetric tra ations by enzyme–metal combinations.

8.2 DYNAMIC KINETIC RESOLUTION BY ENZYME–METAL COMBINATIONS

8.2.1 DYNAMIC KINETIC RESOLUTION BY LIPASE–RUTHENIUM COMBINATION

8.2.1.1 Dynamic Kinetic Resolution of Secondary Alcohols

The first DKR of secondary alcohols, reported by the Williams group in 1996 [2 performed by the combination of lipase as the resolving catalyst and rhodium c as the racemizing catalyst (Scheme 8.1). The lipase-Rh catalyzed DKR of 1-pheny using vinyl acetate as an acyl donor gave a modest result of 76% conversion and 80% e Bäckvall et al. reported a significantly improved procedure using a diruthenium comple an immobilized and thermally stable lipase (*Candida antarctica* lipase B, CALB; N 435). The reactions performed at elevated temperature (70°C) gave excellent result presence of a special acyl donor *p*-chlorophenyl acetate (PCPA). Popular acyl donors vinyl acetate and isopropenyl acetate were incompatible with the racemizing catalyst The DKR of 1-phenylethanol by the procedure gave optically pure (*R*)-α-phenylethyl (>99.5% ee) in a high yield (100% conversion, 92% isolated yield) (Table 8.1). Howe DKR required a stoichiometric amount of acetophenone that acts as a hydrogen acce racemization, otherwise a large amount (>20%) of acetophenone was formed by decrea yield of (*R*)-α-phenylethyl acetate [32].

We found that an indenylruthenium complex **2** racemized secondary alcohols with aid of ketones during DKR [34]. The DKR with **2**, however, required a catalytic am triethylamine and molecular oxygen to activate **2**. The combination of an imm *Pseudomonas cepacia* lipase (PCL, Lipase PS-C of Amano Enzyme Co., Japan) a 60°C was effective for the DKR of benzylic alcohols but was less effective for a alcohols (Table 8.2).

We then found a commercially available cymene–ruthenium complex **3** and its a hydride form **4** as good racemizing catalysts. The activated complex **4**, in particular, satisfactory performances in the DKR of aliphatic alcohols as well as benzylic alcohol 8.3). A noticeable feature of the catalyst is the high racemizing activity toward allylic [35], so that their DKR can be performed at room temperature to provide high y products with excellent optical purities (Table 8.4). Additional feature is its good ac ionic liquids such as [EMIm]BF$_4$ and [BMIm]PF$_6$ ([EMIm] = 1-ethyl-3-methylimida [BMIm] = 1-butyl-3-methylimidazolium) [36–43]. The DKRs in the ionic liquids w

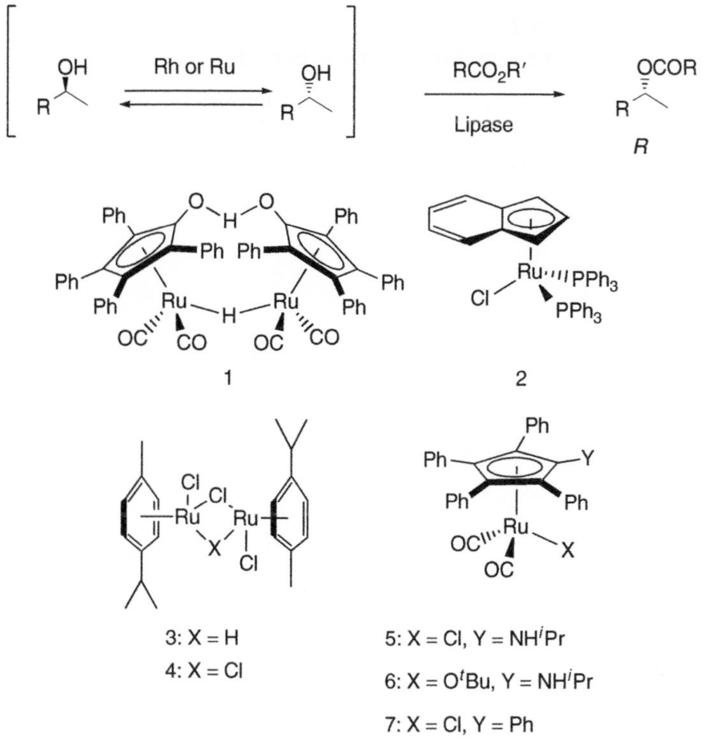

SCHEME 8.1 DKR of secondary alcohols.

possible at room temperature wherein the racemizing catalyst and enzyme were reusable after the products were extracted with ether (Table 8.5) [44].

In an effort to develop racemizing catalysts that are active at room temperature, we synthesized a novel aminocyclopentadienyl ruthenium chloride complex **5** that transforms to its active form **6** upon *in situ* treatment with potassium *t*-butoxide [30]. The *in situ* generated species **6** displayed high activity in the DKR of aliphatic alcohols, as well as aromatic alcohols at room temperature (Table 8.6). Interestingly, it can be used with isopropenyl acetate that is incompatible with other ruthenium catalysts **1–4**. Used in the previous DKRs, isopropenyl acetate is more practical than PCPA; it is readily available, more active than PCPA, and is easily separable from the DKR products [45]. Although the

TABLE 8.1
DKR of 1-Phenylethanol with 1

Acyl Donor	ee (%)	Yield[a] (%)
Vinyl acetate	>99	50
Isopropenyl acetate	>99	72
4-Chlorophenyl acetate	>99	100

[a] Determined by ^{1}H NMR and GC.

Note: The reactions were performed on a 2-mmol scale in 5 mL of *t*-BuOH with 2 mol% of complex **1**, 1 equivalent of acetophenone, 50 mg of Novozym 435, and acyl donor at 70°C under Ar atmosphere.

TABLE 8.2
DKR of Secondary Alcohols with 2

Substrate	Product	ee[a] (%)	Yield[b] (%)
		96	86
		99	82
		99	98
		82	88
		97	60

[a] Measured by HPLC equipped with a chiral column.
[b] Determined by ^{1}H NMR.
Note: The reactions were performed on a 0.25-mmol scale in 2-mL of CH_2Cl_2 with 5 mol% of complex **2**, 5 mol% of O_2, 3 equivalent of triethylamine, 3 equivalent of PCPA, and 40 mg of PCL at 60°C for 43 h.

mechanism for the catalytic racemization is not clear yet, according to our mech studies, the amino group in **5** or **6** seems to be crucial for the racemization, while report by the Bäckvall group suggests a different pathway [46–48]. The Bäckvall group modified complex **7** as the efficient racemizing catalyst, which is similar to **5** but with amino group.

A new catalyst system of [TosN$(CH_2)_2NH_2$] RuCl(p-cymene) and 2,2,6,6-tetrame piperidinyloxy was reported by Sheldon et al. for the DKR of alcohols [49]. However, tested only for 1-phenylethanol to afford 1-phenylethyl acetate in 76% yield.

8.2.1.2 Dynamic Kinetic Resolution of Functionalized Alcohols

The DKR of functionalized alcohols such as diols, hydroxy esters, hydroxy aldehydes, alcohols, and hydroxy nitriles were well performed by lipase–ruthenium bicatalysi DKR of diols was achieved with diruthenium catalyst **1** and CALB in the prese. PCPA to give the corresponding diacetates of (R,R)-configuration from the mixture and *meso*-isomers (Table 8.7) [50,51]. The DKR of rigid benzylic diols with **1** gave results in terms of *de* compared to those of more flexible aliphatic diols, reflecting that displays higher stereoselectivity toward benzylic diols than aliphatic diols.

TABLE 8.3
DKR of Secondary Alcohols with 4

Substrate	Product	ee (%)	Yield (%)
		94	95
		99	93
		99	93
		>99	85

Note: The reactions were performed in CH_2Cl_2 with 4 mol% of complex **4**, 1 equivalent of triethylamine, 1.5 equivalent of PCPA, and PCL at 40°C under Ar atmosphere.

The DKR of hydroxy esters were also accomplished with PCL and **1** at 60°C ~70°C [52–55]. In most cases the enantioselectivities were good, but the yields were moderate (Table 8.8 and Table 8.9). The use of H_2 was necessary in the DKR of γ- and δ-hydroxy esters to suppress the formation of ketones.

The DKR of small functionalized alcohols such as 2-hydroxybutanoic acid, 2-hydroxypropanal, and 1,2-propanediol was carried out after the protection of the terminal groups with a bulky group, because the bulky protecting groups enhanced the enantioselectivity of enzyme [56]. In the DKR of hydroxy acids, *t*-butyl group was the best protecting group for the carboxylic acid functionality (Table 8.10). The trityl group was a proper choice for the protection of primary alcohols in diols such as 1,2-propanediol, 1,2-butanediol, and 1,3-butanediol [56]. For example, 1,2-benzenedimethanol was used for protecting the formyl groups of α- and β-hydroxy aldehydes (Table 8.11) [56]. High enantiomeric excesses (95% and more) were obtained in the DKRs of the protected diols and hydroxy aldehydes. Similarly, 2,6-dimethyl-4-heptanol was used as a hydrogen source to suppress the formation of the oxidized side products.

The DKR of β-azidoalcohols [55] and β-hydroxynitriles [57–59] were also accomplished by employing **1** and CALB, with PCPA as the acyl donor. The DKR of β-azidoalcohols were performed at 60°C while that of β-hydroxynitriles required a higher temperature (100°C) to increase the racemization rate. The optical purities of products were satisfactory in all cases. In the case of β-hydroxynitriles, dehydrogenation lowered the yield.

TABLE 8.4
DKR of Allyl Alcohols with 4

Substrate	Product	ee[a] (%)	Yield[b] (%)
		>99	84
		99	91
		99	85
		99	92
		95	90
		>99	85

[a] Measured by HPLC or GC equipped with a chiral column.
[b] Determined by ^{1}H NMR.
Note: The reactions were performed on a 0.5-mmol scale in 2 mL of CH_2Cl_2 with 4 mol% of complex **4**, 1 equivalent of triethylamine, 1.5 equivalent of PCPA, and 75 mg of PCL at room temperature for 2 d.

8.2.2 DYNAMIC KINETIC RESOLUTION BY SUBTILISIN–RUTHENIUM COMBINATION

8.2.2.1 Dynamic Kinetic Resolution of Secondary Alcohols

The lipase-catalyzed DKRs provide only (*R*)-products. To obtain (*S*)-products, we r enzyme with a complementary (*S*)-stereoselectivity. We surveyed (*S*)-selective enzyme patible with the racemizing catalyst **6**. Though subtilisin was suitable, its commercia was not applicable to DKR due to its low enzymatic activity and instability in nona medium. However, we succeeded in enhancing its activity and stability by treating it surfactant before use. Thus, the DKR with surfactant-treated subtilisin, **6**, and trifluo butanoate as an acylating agent gave the (*S*)-products in good yields with high optical at room temperature (Scheme 8.2 and Table 8.12) [60].

The (*S*)-selective DKR of alcohols with subtilisin was also possible in ionic liquid a temperature (Table 8.13) [44]. In this case, the cymene–ruthenium complex **4** was used racemization catalyst. In general, the optical purities of (*S*)-esters were lower than t (*R*)-esters as described in Table 8.5.

TABLE 8.5
DKR of Secondary Alcohols with 4 in [BMIm]PF$_6$

Substrate	Product	ee[a] (%)	Yield[b] (%)
		99	85
		99	85
		98	87
		99	87
		99	85
		99	87
		99	92
		99	85
		99 (de 99%)	87
		99 (de 97%)	86

[a] Measured by HPLC equipped with a chiral column.
[b] Determined by ^{1}H NMR.
Note: The reactions were performed on a 0.3-mmol scale in 1 mL of [BMIm]PF$_6$ with 8 mol% of complex **4**, 1 equivalent of triethylamine, 3 equivalent of 1,1,1-trifluoroethyl acetate, and 45 mg of LPS-TN-M at room temperature for 2–4 d under Ar atmosphere.

TABLE 8.6
DKR of Secondary Alcohols with 6

Substrate	Product	ee[a] (%)	Yield[b] (%)
		>99	95
		>99	94
		>99	90
		95	89
		>99	86[c]
		>99	97
		>99	95
		>99	90
		91	89[c]
		98	93
		81	62
		>99	90
		99	97

[a] Measured by HPLC equipped with a chiral column.
[b] Isolated yields.
[c] Measured by GC equipped with a chiral column.
Note: The reactions were performed on a 1.0-mmol scale in 3.2 mL of toluene with isopropenyl acetate (1.5 equivalent), Novozym 435 (3 mg), Na_2CO_3 (1.0 equivalent), **5** (4 mol%), and potassium *tert*-butoxide (5 mol%) at 25°C for 2–6 d under Ar atmosphere.

TABLE 8.7
DKR of Diols

Substrate	Catalyst	Product	e^a (%)	(R,R)/meso^a	Yield^b (%)
(diol structure)	1	(diacetate structure)	>99	86/14	63
(diol structure)	1	(diacetate structure)	>99	38/62	90
(diol structure)	1	(diacetate structure)	>97	90/10	63
(diol structure)	1	(diacetate structure)	>99	74/26	43
(diol structure)	1	(diacetate structure)	>99	98/2	76
	6		>99	99/1	95
(diol structure)	1	(diacetate structure)	>99	98/2	77
	6		>99	98/2	94
(diol structure)	1	(diacetate structure)	>99	100/0	78
(diol structure)	1	(diacetate structure)	>96	89/11	64

^{a} Measured by HPLC or GC equipped with a chiral column.
^{b} Isolated yields.
Note: The reactions were performed on a 0.5-mmol scale in 1 mL of toluene with PCPA (3 equivalent), Novozym 435 (30 mg), and **1** (4 mol %) at 70°C for 1–2 d under Ar atmosphere.

8.2.3 Dynamic Kinetic Resolution by Lipase–Palladium Combination

8.2.3.1 Dynamic Kinetic Resolution of Allyl Acetates

The first DKR of allyl acetates was accomplished through coupled Pd-catalyzed racemization and enzymatic hydrolysis of allyl acetates in a buffer solution [27]. However, the DKR was limited to cyclohexenyl acetates to give symmetrical palladium–allyl intermediates. Among them, only 2-phenyl-2-cyclohexenyl acetate was resolved with good results (96% conversion, 81% yield, and 96% ee) (Scheme 8.3).

TABLE 8.8
DKR of α-Hydroxy Esters with 1

Substrate	Product	ee^a (%)	Yield^b (%)
OH / CO$_2$Me (phenyl)	OAc / CO$_2$Me (phenyl)	94	80
OH / CO$_2$Me (H$_3$CO-phenyl)	OAc / CO$_2$Me (H$_3$CO-phenyl)	94	76
OH / CO$_2$Me (Br-phenyl)	OAc / CO$_2$Me (Br-phenyl)	98	69
OH / CO$_2$Me (cyclohexyl)	OAc / CO$_2$Me (cyclohexyl)	98	80
OH / CO$_2$Me (phenethyl)	OAc / CO$_2$Me (phenethyl)	30	62
OH / CO$_2$Me (butyl)	OAc / CO$_2$Me (butyl)	80	60

^{a} Measured by HPLC or GC equipped with a chiral column.
^{b} Isolated yields.
Note: The reactions were performed on a 0.25-mmol scale in 1.25 mL of cyclohexane with PCPA (2 equivalent), LPS-C (15 mg), and **1** (2 mol%) at 60°C for 2–3 d under Ar atmosphere.

We improved the DKR of allyl acetates by replacing enzymatic hydrolytic reacti⬚ enzymatic transesterification reaction and by employing Pd(PPh$_3$)$_4$ and 1,1′-*bis*(diphe⬚ sphino)ferrocene (dppf) as the racemizing catalyst system (Scheme 8.4). The DKR re⬚ were performed with 2-propanol as an acyl acceptor in tetrahydrofuran (THF) [61]. ⬚ of the chelating ligand (dppf) decreased the formation of byproducts (1,3-dienes). ⬚ acyclic allylic acetates were transformed to their corresponding allylic alcohols a⬚ temperature with good yields and excellent enantioselectivities (Table 8.14).

8.2.3.2   Dynamic Kinetic Resolution of Amines

Reetz et al. reported for the first time the DKR of 1-phenylethylamine by em⬚ palladium on carbon and CALB [29]. However, the DKR required a very long ⬚ time (8 d) at 50°C ~55°C and provided a poor isolated yield (60%) (Scheme 8.5). R⬚ Bäckvall et al. reported that diruthenium complex **1** racemizes aromatic amines a⬚ in toluene, but the racemization conditions were not applicable to the DKR in⬚ enzymes [62].

TABLE 8.9
DKR of β-Hydroxy Esters with 1

Substrate	Product	ee[a] (%)	Yield[b] (%)
		95	76
		99	74
		96	80
		70	82

[a] Measured by GC equipped with a chiral column.
[b] Isolated yields.
Note: The reactions were performed on a 0.8-mmol scale in 8 mL of TBME with PCPA (3 equivalent), LPS-C (48 mg), and **1** (6 mol%) at 60°C for 6 d under Ar atmosphere.

TABLE 8.10
DKR of 2-Hydroxybutanoic Acid Esters with 1

Substrate	Product	ee[a] (%)	Yield[b] (%)
		86	88
		93	91
		94	92
		>99	88

[a] Measured by HPLC equipped with a chiral column.
[b] Isolated yields.
Note: The reactions were performed on a 0.3-mmol scale in 1 mL of toluene with PCPA (6.5 equivalent), PCL (24 mg), and **1** (10 mol%) at 70°C for 4 d under Ar atmosphere.

TABLE 8.11
DKR of Protected Diols and Hydroxy Aldehydes

Substrate	Product	ee^a (%)	Yield^b (%)
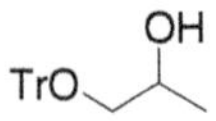 (structure: TrO, OH)	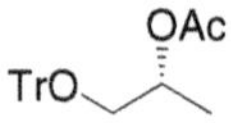 (structure: TrO, OAc)	>99	96
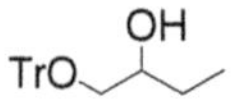 (structure: TrO, OH)	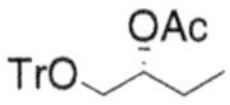 (structure: TrO, OAc)	99	91
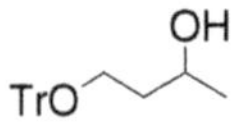 (structure: TrO, OH)	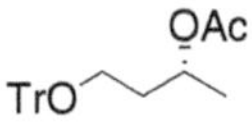 (structure: TrO, OAc)	95	97
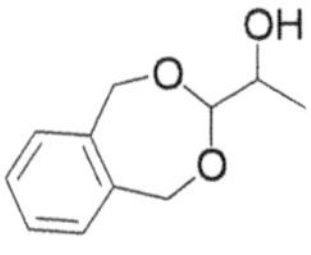 (structure: benzo-dioxepine, OH)	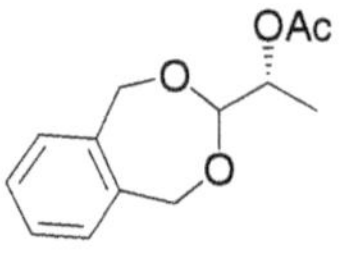 (structure: benzo-dioxepine, OAc)	98	95
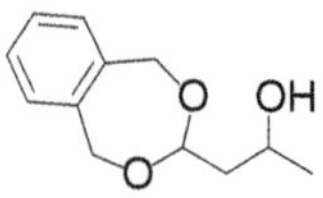 (structure: benzo-dioxepine, OH)	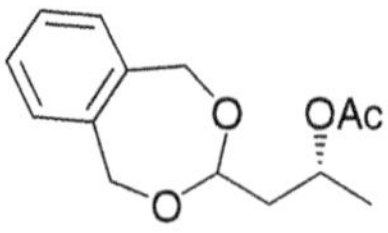 (structure: benzo-dioxepine, OAc)	96	90

^{a} Measured by HPLC equipped with a chiral column.
^{b} Isolated yields.
Note: The reactions were performed on a 0.3-mmol scale in 1 mL of toluene with PCPA (6.5 equivalent), PCL (24 mg), and **1** (10 mol%) at 70°C for 4 d under Ar atmosphere.

8.3 ASYMMETRIC TRANSFORMATIONS BY ENZYME–METAL COMBO-CATALYSIS

8.3.1 ASYMMETRIC TRANSFORMATION BY LIPASE–RUTHENIUM COMBINATION

8.3.1.1 Asymmetric Reductive Acetylation of Ketones

The catalytic alcohol racemization with diruthenium catalyst **1** is based on a rev
transfer hydrogenation mechanism including ketone as an oxidized intermediate. The
the formation of ketone as a major side product is inevitable in the DKR of sec
alcohols with **1**. To circumvent the problem of ketone formation, we envisioned a
DKR process starting from ketone (Scheme 8.6). A key factor of this process w
selection of hydrogen donors compatible with the DKR conditions. Finally, 2,6-dir
heptan-4-ol, which cannot be acylated by lipases, and molecular hydrogen was
as hydrogen donors [63,64]. Asymmetric reductive acetylation of ketones under 1
hydrogen was performed in ethyl acetate as an acyl donor and solvent, which needed

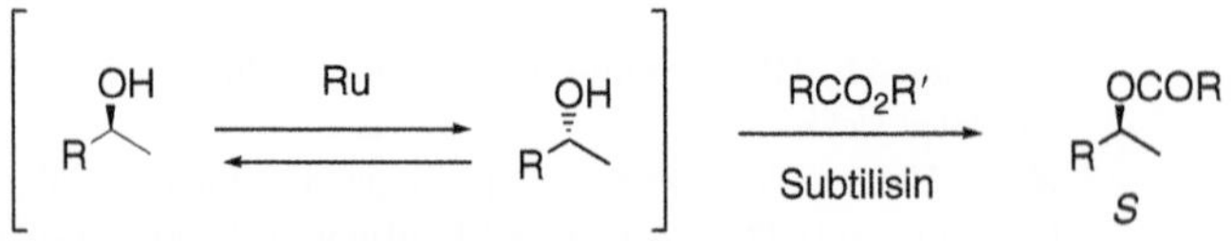

SCHEME 8.2 (*S*)-Selective DKR of secondary alcohols.

TABLE 8.12
(S)-Selective DKR of Secondary Alcohols by Subtilisin–Ru Combination

Substrate	Product	ee[a] (%)	Yield[b] (%)
		92	95
		99	92 (90)
		94	93 (91)
		92	77 (76)
		98	80 (78)
		98	80 (74)
		98	77 (67)
		95	90 (90)

[a] Measured by HPLC or GC equipped with a chiral column.
[b] Determined by ^{1}H NMR (isolated yields in parentheses).
Note: The reactions were performed on a 0.3-mmol scale in 1 mL of THF with 2,2,2-trifluoroethyl butyrate (1.7 equivalent), the subtilisin (7.5 mg), Na$_2$CO$_3$ (63.6 mg), **5** (4 mol%), and potassium *tert*-butoxide (5 mol%) at 25°C for 3 d under Ar atmosphere.

TABLE 8.13

(S)-Selective DKR of Secondary Alcohols in [BMIm]PF$_6$

Substrate	Product	ee[a] (%)	Yield[b] (%)
		97	89
		97	90
		85	90
		85	90
		99	80
		87	92
		82	91
		86	84
		86 (de 52%)	78
		96 (de 63%)	83

[a] Measured by HPLC equipped with a chiral column.

[b] Determined by ^{1}H NMR.

Note: The reactions were performed on a 0.3-mmol scale in 1 mL of [BMIm]PF$_6$ with 8 mol% of complex **4,** 1 equivalent of triethylamine, 3 equivalent of 1,1,1-trifluoroethyl butyrate, and 20 mg of subtilisin-CLEC at room temperature for 6 d under Ar atmosphere.

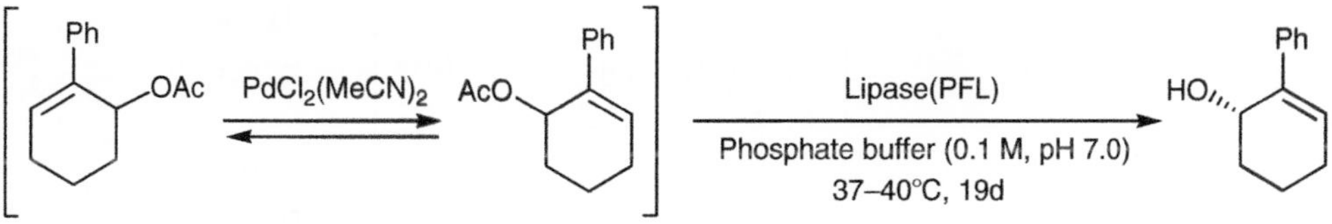

SCHEME 8.3 DKR of 2-phenyl-2-cyclohexenyl acetate.

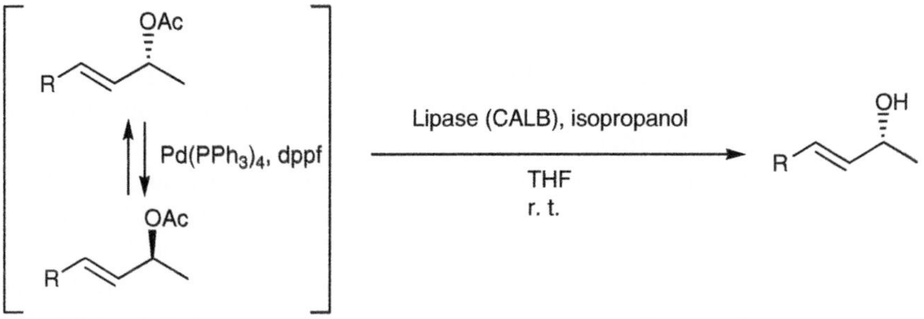

SCHEME 8.4 DKR of allyl acetates.

TABLE 8.14
DKR of Allyl Acetates by Lipase–Pd Combination

Substrate	Product	ee[a] (%)	Yield[b] (%)
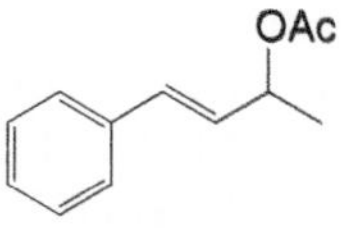	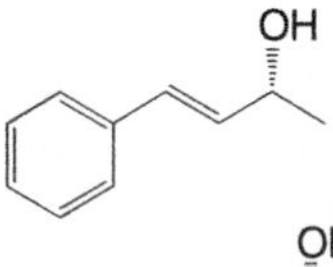	98	83
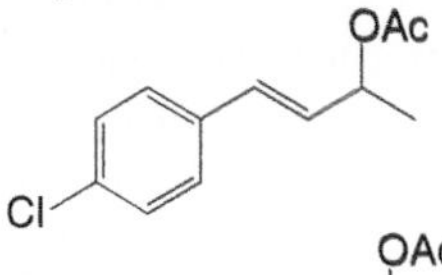	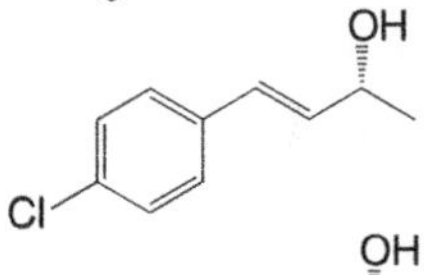	97	77
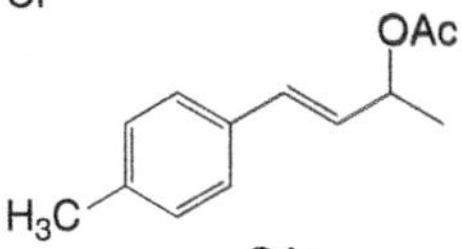	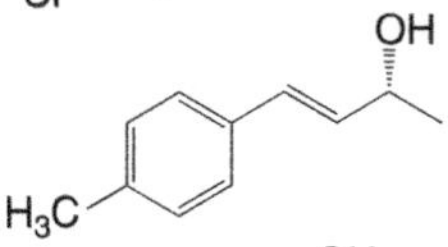	98	82
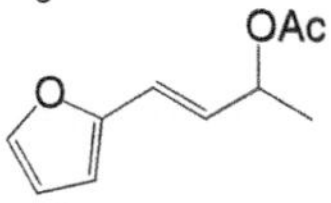	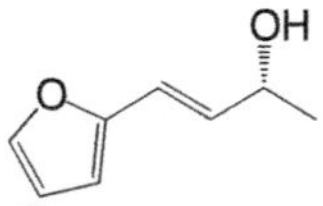	>99	87
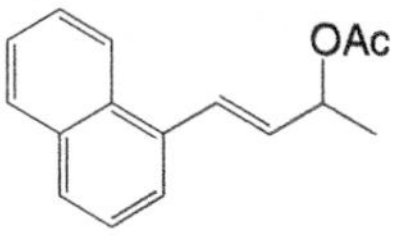	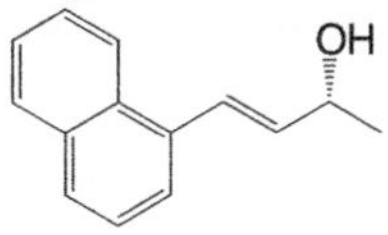	98	70

[a] Measured by HPLC equipped with a chiral column.
[b] Determined on the basis of HPLC results.
Note: The reactions were performed on a 0.53-mmol scale in 2 mL of THF with 5 mol% of Pd(PPh₃)₄, 15 mol% equivalent of dppf, 0.4 mL of i-PrOH, and 200 mg of CAL at room temperature for 2 d under Ar atmosphere.

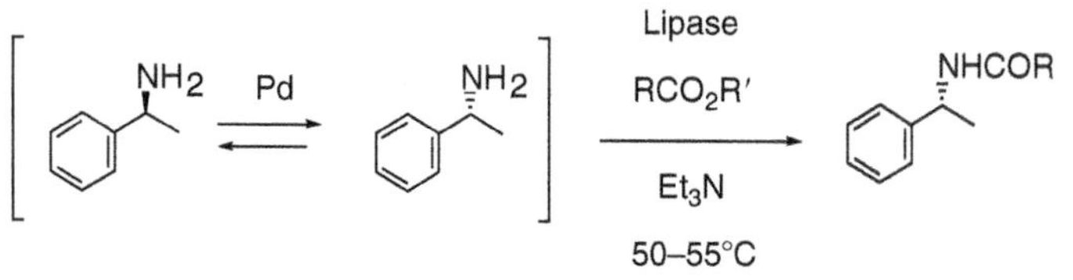

SCHEME 8.5 DKR of 1-phenylethylamine.

reaction time (96 h). Ethanol formation from ethyl acetate did not cause critical pr
and various ketones were transformed successfully into the corresponding chiral a
(Table 8.15) [64].

Asymmetric reductive acetylation process was also applicable to acetoxyaryl keton
For example, 3′-acetoxyacetophenone was transformed to (R)-1-(3-hydroxyphenyl)eth
ate under 1 atm of H_2 in 95% yield. The overall reaction seems to be a simple asym
reductive intramolecular acyl migration. In fact, however, it is the result from nine ca
steps: two ruthenium-catalyzed reductions, two ruthenium-catalyzed epimerizations
lipase-catalyzed deacylations, and two lipase-catalyzed acylations (Scheme 8.7). This p
was applicable to a wide range of acyloxyphenyl ketones (Table 8.16). In most case
yields and excellent optical purities were obtained.

8.3.1.2 Asymmetric Hydrogenation of Enol Acetates

After succeeding in the asymmetric reductive acylation of ketones, we studied to see
acetates can be used as acyl donors and precursors of ketones at the same time th
deacylation and keto-enol tautomerization (Scheme 8.8). The overall reaction thus
ponds to the asymmetric reduction of enol acetate. For example, 1-phenylvinyl
was transformed to (R)-1-phenylethyl acetate by CALB and diruthenium complex 1
presence of 2,6-dimethyl-4-heptanol in 89% yield (98% ee) [63]. Molecular hydrogen (
was almost equally effective for the transformation (86% yield, 96% ee) [64]. A broad ra
enol acetates were prepared from ketones and were successfully transformed to the
sponding (R)-acetates under 1 atm of H_2 (Table 8.17). From unsymmetrical aliphatic ke
enol acetates were obtained as the mixtures of regio- and geometrical isomers. N
however, the efficiency of the process was little affected by the isomeric composition
enol acetates.

8.3.2 ASYMMETRIC TRANSFORMATIONS BY LIPASE–PALLADIUM COMBINATION

8.3.2.1 Asymmetric Reductive Acetylation of Ketoximes

The strategy for the asymmetric reductive acylation of ketones was extended to ketoxi
coupling the reduction of ketoximes to the DKR of amines (Scheme 8.9). The asym

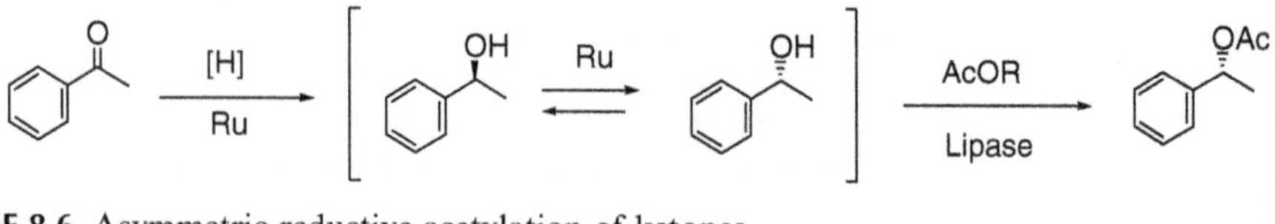

SCHEME 8.6 Asymmetric reductive acetylation of ketones.

TABLE 8.15
Asymmetric Reductive Acetylation of Ketones

Substrate	Product	ee[a] (%)	Yield[b] (%)
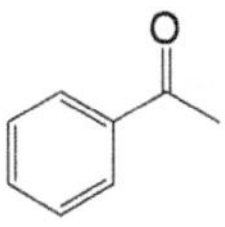	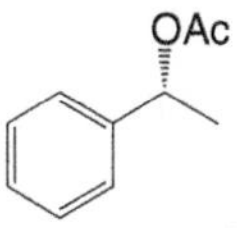	96	81
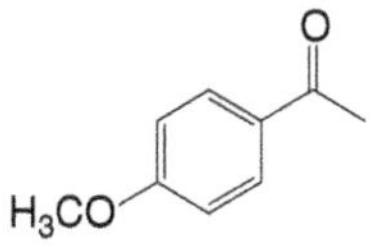	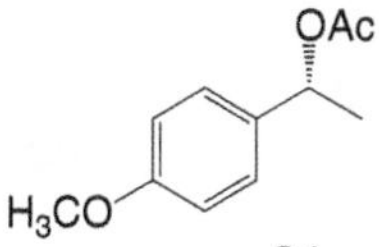	99	85
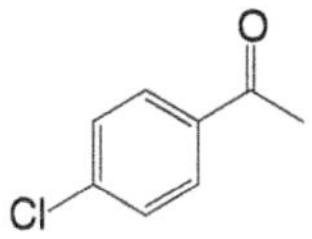	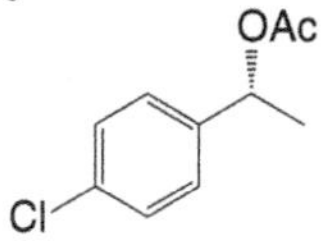	97	72
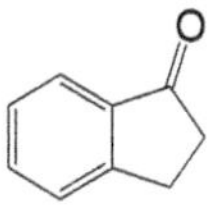	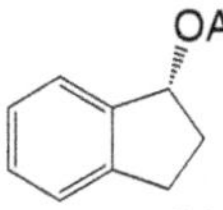	99	89
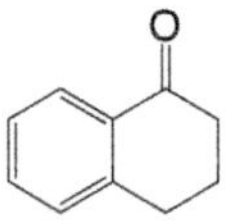	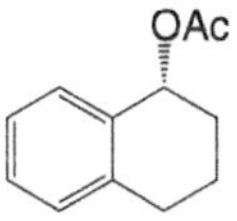	99	87
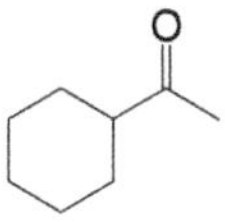	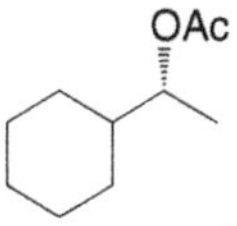	90	87
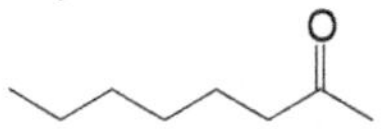	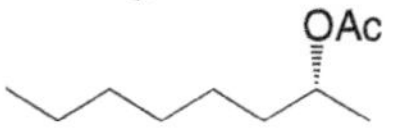	91	87
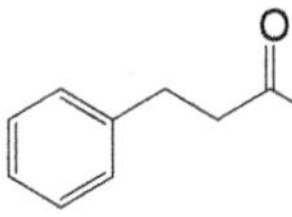	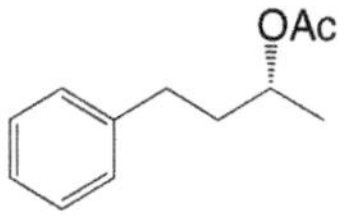	72	83

[a] Measured by HPLC or GC equipped with a chiral column.
[b] Isolated yields.
Note: The reactions were performed on a 1.0-mmol scale in 3 mL of ethyl acetate with Novozym 435 (84 mg) and **1** (2 mol%) at 70°C for 2–4 d under 1 atm of H_2.

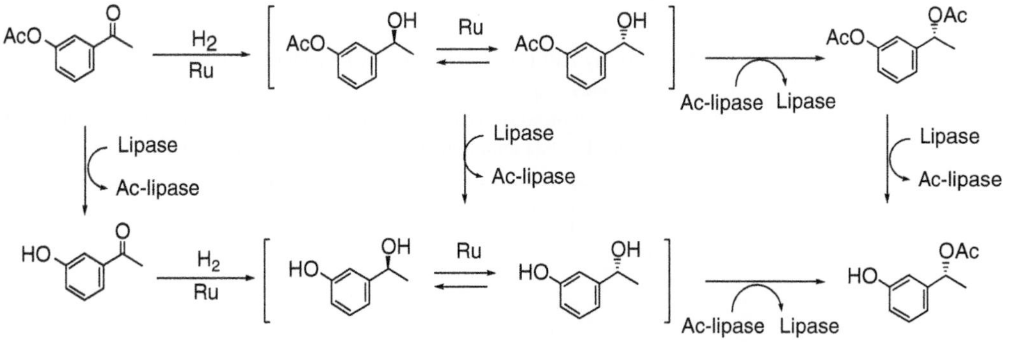

SCHEME 8.7 Reaction pathway for the asymmetric transformation of 3′-acetoxyacetophenone.

TABLE 8.16
Asymmetric Transformations of Acyloxyphenyl Ketones

Substrate	Product	ee[a] (%)	Yield[b] (%)
		98	95
		98	96
		93	94
		96	93
		89	88
		98	88
		96	92
		98	89

[a] Measured by HPLC or GC equipped with a chiral column.
[b] Isolated yields.
Note: The reaction were performed on a 0.56-mmol scale in 1.5 mL of tolune with LPS-D (25 mg), and **1** (0.022 mmol) at 70°C for 3d under Ar atmosphere.

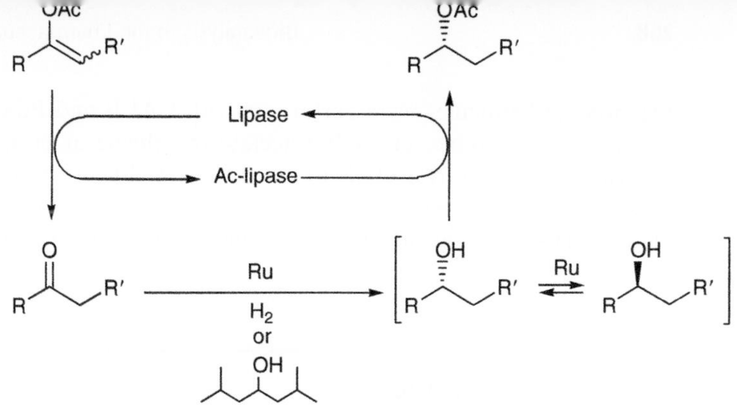

SCHEME 8.8 Asymmetric hydrogenation of enol acetates.

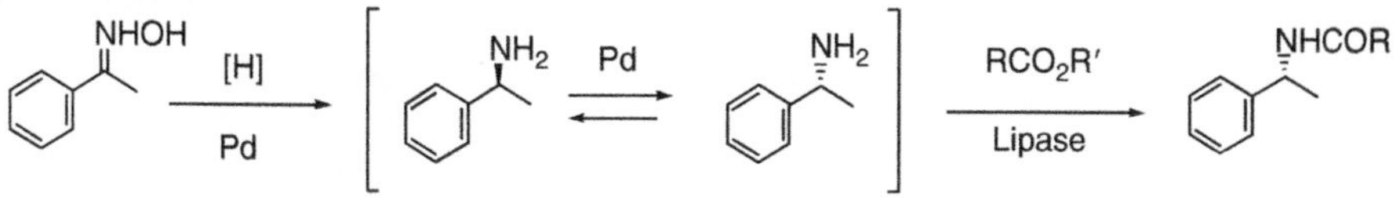

SCHEME 8.9 Asymmetric reductive acetylation of ketoximes.

TABLE 8.17
Asymmetric Hydrogenation of Enol Acetates

Substrate	Product	ee[a] (%)	Yield[b] (%)
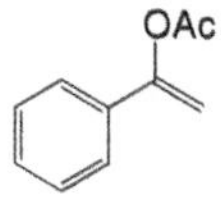	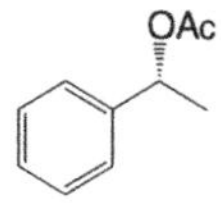	98	89
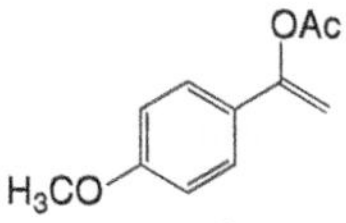	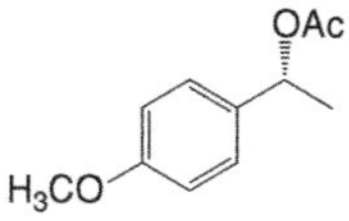	98	80
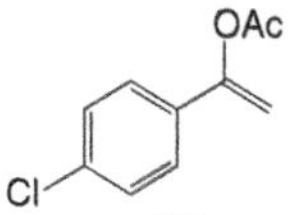	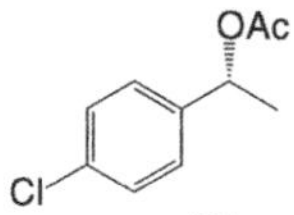	97	91
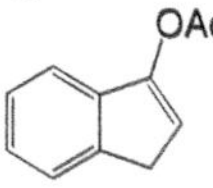	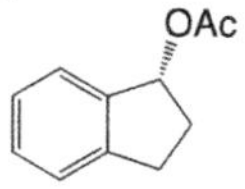	99	87
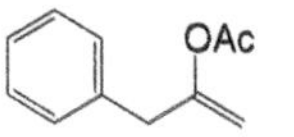	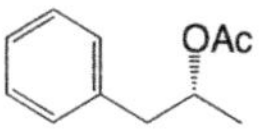	79	90
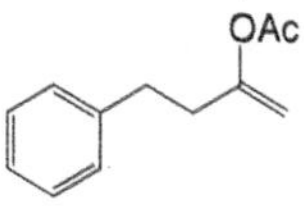	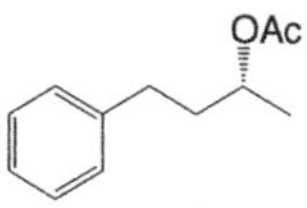	94	92
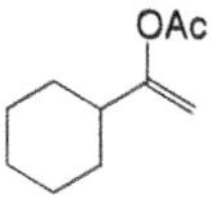	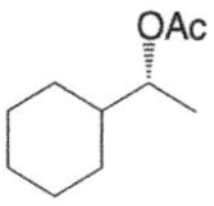	99	94
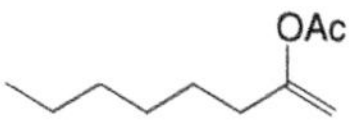	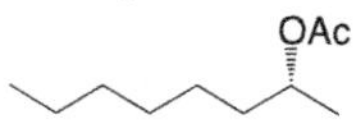	91	95

[a] Measured by HPLC or GC equipped with a chiral column.
[b] Isolated yields.
Note: The reaction were performed on a 1.0-mmol scale in 3 mL of toluene with Novozym 435(30 mg) and **1** (2 mol%) at 70°C for 2–4 d under 1 atm of H$_2$.

reactions of ketoximes were performed with CALB and Pd/C in the presence of hyd
diisopropylethylamine, and ethyl acetate in toluene at 60°C for 5 d (Table 8.18) [
comparison to the direct DKR of amines, the yields of chiral amides increased signifi
The use of diisopropylethylamine to keep the reaction condition basic was a factor
increase. However, the major factor would be the slow generation of amines, which ma

TABLE 8.18
Asymmetric Reductive Acetylation of Ketoximes

Substrate	Product	ee[a] (%)	Yield[b] (%)
		98	80
		97	84
		94	81
		96	82
		98	76
		95	84
		97	70
		99	89

[a] Measured by HPLC equipped with a chiral column.
[b] Isolated yields.
Note: The reactions were performed on a 0.37-mmol scale in 3.7 mL of toluene
with Novozym 435 (200 wt%), diisopropylethylamine (3 equivalent), ethyl
acetate (2 equivalent), and Pd (66 wt%, 5% Pd/C) at 60°C for 5 d under 1 atm
of H_2.

the amine concentration low enough to suppress side reactions including the reductive deamination. Disappointingly, this process is limited to benzylic amines. Low turnover frequencies also need to be overcome.

8.4   CONCLUSION

This chapter describes that enzyme–metal combo-catalysis provides a novel approach for the conversion of racemic substrates to single enantiomeric products. The key feature of this methodology is DKR by the combination of metal-catalyzed racemization with enzymatic resolution. It has been demonstrated that racemic alcohols, esters, and amines can be efficiently converted to the corresponding enantiomeric products through the enzyme–metal catalyzed DKR. In the DKR of alcohols, a pair of complementary procedures are now available for the preparation of both (*R*)- and (*S*)-products. The DKR, performed at room temperature with commercially available enzymes and metal catalyst, can be applicable to a wide range of substrates. However, the DKR of amines is limited to benzylic amines and requires high temperature. Accordingly, for the efficient DKR of amines, further efforts will be directed toward developing practical racemizing catalysts with high activity and broad specificity at room temperature [67].

REFERENCES

1. Sheldon, R.A., *Chirotechnology, Industrial Synthesis of Optically Active Compounds*, Marcel Dekker, New York, 1993.
2. Wong, C.-H. and Whitesides, G.M., *Enzymes in Synthetic Organic Chemistry*, Pergamon Press, Oxford, UK, 1994.
3. Koskinen, A.M.P. and Klibanov, A.M., Eds., *Enzymatic Reactions in Organic Media*, Blackie Academic & Professional, Glasgow, Scotland, 1996.
4. Faber, K., *Biotransformations in Organic Chemistry*, 3rd ed., Springer, Berlin, Germany, 1997.
5. Bornscheuer, U.T. and Kazlauskas, R.J., *Hydrolases in Organic Synthesis*, Wiley-VCH, Weiheim, Germany, 1999.
6. Drauz, K. and Waldmann, H., *Enzyme Catalysis in Organic Synthesis: A Comprehensive Handbook*, 2nd ed., vols. I–III, Wiley-VCH, Weinheim, Germany, 2002. .
7. Kim, M.-J., Choi, G.-B., Kim, J.-J., et al., Lipase-catalyzed transesterification as a practical route to homochiral acyclic *anti*-1,2-diol. A new synthesis of (+)- and (−)-*endo*-brevicomin, *Tetrahedron Lett.*, 36, 6253, 1995.
8. Kim, M.-J. and Lim, I.-T., Synthesis of unsaturated C8–C9 sugars by enzymatic chain elongation, *Synlett*, 2, 138, 1996.
9. Kim, M.-J., Lim, I.-T., Choi, G.-B., et al., The efficient resolution of protected diols and hydroxy aldehydes by lipases: steric auxiliary approach and synthetic applications, *Bioorg. Med. Chem. Lett.*, 6, 71, 1996.
10. Kim, M.-J., Lim, I.-T., Kim, H.-J., et al., Enzymatic single aldol reactions of remote dialdehydes, *Tetrahedron: Asymmetry*, 8, 1507, 1997.
11. Lee, D. and Kim, M.-J., Lipase-catalyzed transesterification as a practical route to homochiral *syn*-1,2-diols. The synthesis of the taxol side chain, *Tetrahedron Lett.*, 39, 2163, 1998.
12. Chung, S.-K., Chang, Y.-T., Lee, E.J., et al., Synthesis of two enantiomeric pairs of myo-inositol-(1,2,4,5,6) and myo-inositol-(1,2,3,4,5)pentakisphosphate, *Bioorg. Med. Chem. Lett.*, 8, 1503, 1998.
13. Lee, D., Kim, K.C., and Kim, M.-J., Selective enzymatic acylation of 10-deacetylbaccatin III, *Tetrahedron Lett.*, 39, 9039, 1998.
14. Lee, D. and Kim, M.-J., Enzymatic selective dehydration and skeleton rearrangement of paclitaxel precursors, *Org. Lett.*, 1, 925, 1999.

15. Im, A.S., Cheong, C.S., and Lee, S.H., Lipase-catalyzed remote kinetic resolution of qu carbon-containing alcohols and determination of their absolute configuration, *Bull. Korea Soc.*, 24, 1269, 2003.
16. Kang, H.-Y., Ji, Y., Yu, Y.-K., et al., Synthesis of α-ketobutyrolactones and γ-hydrox acids, *Bull. Korean Chem. Soc.*, 24, 1819, 2003.
17. Ward, R.S., Dynamic kinetic resolution, *Tetrahedron: Asymmetry*, 6, 1475, 1995.
18. Stürmer, R., Enzymes and transition metal complexes in tandem—a new concept for kinetic resolution, *Angew. Chem. Int. Ed. Engl.*, 36, 1173, 1997.
19. El Gihani, M.T. and Williams, J.M.J., Dynamic kinetic resolution, *Curr. Opin. Biotechno* 1999.
20. Azerad, R. and Buisson. D., Dynamic resolution and stereoinversion of secondary alcc chemo-enzymatic processes, *Curr. Opin. Biotechnol.*, 11, 565, 2000.
21. Huerta, F.F., Minidis, A.B.E., and Bäckvall, J.-E., Racemisation in asymmetric synthesis. I kinetic resolution and related processes in enzyme and metal catalysis, *Chem. Soc. Rev.*, 2001.
22. Kim, M.-J., Ahn, Y., and Park, J., Dynamic kinetic resolutions and asymmetric transforma enzymes coupled with metal catalysis, *J. Curr. Opin. Biotechnol.*, 13, 578, 2002.
23. Pellissier, H., Dynamic kinetic resolution, *Tetrahedron*, 59, 8291, 2003.
24. Pàmies, O. and Bäckvall, J.-E., Combination of enzymes and metal catalysts. A powerful a in asymmetric catalysis, *Chem. Rev.*, 103, 3247, 2003.
25. Pàmies, O. and Bäckvall, J.-E., Combined metal catalysis and biocatalysis for an efficien mization process, *Curr. Opin. Biotechnol.*, 14, 407, 2003.
26. Turner, N.J., Enzyme catalysed deracemisation and dynamic kinetic resolution reaction *Opin. Chem. Biol.*, 8, 114, 2004.
27. Allen, J.V. and Williams, J.M.J., Dynamic kinetic resolution with enzyme and palladium ations, *Tetrahedron Lett.*, 37, 1859, 1996.
28. Dinh, P.M., Howarth, J.A., Hudnott, A.R., et al., Catalytic racemization of alcohols: app to enzymatic resolution reactions, *Tetrahedron Lett.*, 37, 7623, 1996.
29. Reetz, M.T. and Schimossek, K., Lipase-catalyzed dynamic kinetic resolution of chiral am of palladium as the racemization catalyst, *Chimia*, 50, 668, 1996.
30. Choi, J.H., Kim, Y.H., Nam, S.H., et al., Aminocyclopentadienyl ruthenium chloride: racemization and dynamic kinetic resolution of alcohols at ambient temperature, *Angew. C Ed. Engl.*, 41, 2373, 2002.
31. Larsson, A.L.E., Persson, B.A., and Bäckvall, J.-E., Enzymatic resolution of alcohols coup ruthenium-catalyzed racemization of the substrate alcohol, *Angew. Chem. Int. Ed. Engl.*, 1997.
32. Persson, B.A., Larsson, A.L.E., Ray, M.L., and Bäckvall, J.-E., Ruthenium- and enzyme-c dynamic kinetic resolution of secondary alcohols, *J. Am. Chem. Soc.*, 121, 1645, 1999.
33. Lee, H.K. and Ahn, Y., Lipase/ruthenium-catalyzed dynamic kinetic resolution of β-hyd ylferrocene derivatives, *Bull. Korean Chem. Soc.*, 25, 1471, 2004.
34. Koh, J.H., Jeong, H.M., Kim, M.-J., et al., Enzymatic resolution of secondary alcohols with ruthenium-catalyzed racemization without hydrogen, *Tetrahedron Lett.*, 40, 6281, 19'
35. Lee, D., Huh, E.A., Kim, M.-J., et al., Dynamic kinetic resolution of allylic alcohols med ruthenium- and lipase-based catalysts, *Org. Lett.*, 2, 2377, 2000.
36. Kim, K.W., Song, B., Choi, M.Y., et al., Biocatalysis in ionic liquids: markedly enhanced e electivity of lipase, *Org. Lett.*, 3, 1507, 2001.
37. Lee, J.K. and Kim, M.-J., Ionic liquid-coated enzyme for biocatalysis in organic solvent *Chem.*, 67, 6845, 2002.
38. Kim, M.-J., Choi, M.Y., Lee, J.K., et al., Enzymatic selective acylation of glycosides liquids: significantly enhanced reactivity and regioselectivity, *J. Mol. Catal. B. Enzym.*, 2003.
39. Erbeldinger, M., Mesiano, A.J., and Russel, A., Enzymatic catalysis of formation of Z-aspa ionic liquid—an alternative to enzymatic catalysis in organic solvents, *Biotechnol. Prog.*, 2000.

40. Lau, R.M., van Rantwijk, F., Seddon, K.R., et al., Lipase-catalyzed reactions in ionic liquids, *Org. Lett.*, 2, 4189, 2000.

41. Itoh, T., Akasaki, E., Kudo, K., et al., Lipase-catalyzed enantioselective acylation in the ionic liquid solvent system: reaction of enzyme anchored to the solvent, *Chem. Lett.*, 262, 2001.

42. Schoefer, S.H., Kraftzik, N., Wasserscheid, P., et al., Enzyme catalysis in ionic liquids: lipase catalysed kinetic resolution of 1-phenylethanol with improved enantioselectivity, *Chem. Commun.*, 425, 2001.

43. Park, S. and Kazlauskas, R., Improved preparation and use of room-temperature ionic liquids in lipase-catalyzed enantio- and regioselective acylations, *J. Org. Chem.*, 66, 8395, 2001.

44. Kim, M.-J., Kim, H.M., Kim, D., et al., Dynamic kinetic resolution of secondary alcohols by enzyme–metal combinations in ionic liquid, *Green Chem.*, 6, 471, 2004.

45. Choi, J.H., Choi, Y.K., Kim, Y.H., et al., Aminocyclopentadienyl ruthenium complexes as racemization catalysts for dynamic kinetic resolution of secondary alcohols at ambient temperature, *J. Org. Chem.*, 69, 1972, 2004.

46. Csjernyik, G., Bogár, K., Bäckvall, J.-E., New efficient ruthenium catalysts for racemization of alcohols at room temperature, *Tetrahedron Lett.*, 45, 6799, 2004.

47. Martin-Matute, B., Edin, M., Bogár, K., et al., Highly compatible metal and enzyme catalysts for efficient dynamic kinetic resolution of alcohols at ambient temperature, *Angew. Chem. Int. Ed. Engl.*, 43, 6535, 2004.

48. Martin-Matute, B., Edin, M., Bogár, K., et al., Combined ruthenium(II) and lipase catalysis for efficient dynamic kinetic resolution of secondary alcohols. Insight into the racemization mechanism, *J. Am. Chem. Soc.*, 127, 8817, 2005.

49. Dijksman, A., Elzinga, J.M., Li, Y.X., et al., Efficient ruthenium-catalyzed racemization of secondary alcohols: application to dynamic kinetic resolution, *Tetrahedron: Asymmetry*, 13, 879, 2002.

50. Persson, B.A., Huerta, F.F., and Bäckvall, J.-E., Dynamic kinetic resolution of secondary diols via coupled ruthenium and enzyme catalysis, *J. Org. Chem.*, 64, 5237, 1999.

51. Edin, M. and Bäckvall, J.-E., On the mechanism of the unexpected facile formation of *meso*-diacetate products in enzymatic acetylation of alkanediols, *J. Org. Chem.*, 68, 2216, 2003.

52. Huerta, F.F., Laxmi, S.Y.R., and Bäckvall, J.-E., Dynamic kinetic resolution of α-hydroxy acid esters, *Org. Lett.*, 2, 1037, 2000.

53. Runmo, A.B.L., Pàmies, O., Faber, K., et al., Dynamic kinetic resolution of γ-hydroxy acid derivatives, *Tetrahedron Lett.*, 43, 2983, 2002.

54. Pàmies, O. and Bäckvall, J.-E., Enzymatic kinetic resolution and chemoenzymatic dynamic kinetic resolution of δ-hydroxy esters. An efficient route to chiral δ-lactones, *J. Org. Chem.*, 67, 1261, 2002.

55. Pàmies, O. and Bäckvall, J.-E., Dynamic kinetic resolution of β-azido alcohols. An efficient route to chiral aziridines and β-amino alcohols, *J. Org. Chem.*, 66, 4022, 2001.

56. Kim, M.-J., Choi, Y.K., Choi, M.Y., et al., Lipase/ruthenium-catalyzed dynamic kinetic resolution of hydroxy acids, diols, and hydroxy aldehydes protected with a bulky group, *J. Org. Chem.*, 66, 4736, 2001.

57. Pàmies, O. and Bäckvall, J.-E., Efficient lipase-catalyzed kinetic resolution and dynamic kinetic resolution of β-hydroxy nitriles. A route to useful precursors for γ-amino alcohols, *Adv. Synth. Catal.*, 343, 726, 2001.

58. Pàmies, O. and Bäckvall, J.-E., Chemoenzymatic dynamic kinetic resolution of β-halo alcohols. An efficient route to chiral epoxides, *J. Org. Chem.*, 67, 9006, 2002.

59. Pàmies, O. and Bäckvall, J.-E., An efficient route to chiral α- and β-hydroxyalkanephosphonates, *J. Org. Chem.*, 68, 4815, 2003.

60. Kim, M.-J., Chung, Y.I., Choi, Y.K., et al., (*S*)-selective dynamic kinetic resolution of secondary alcohols by the combination of subtilisin and an aminocyclopentadienylruthenium complex as the catalysts, *J. Am. Chem. Soc.*, 125, 11494, 2003.

61. Choi, Y.K., Suh, J.H., Lee, D., et al., Dynamic kinetic resolution of acyclic allylic acetates using lipase and palladium, *J. Org. Chem.*, 64, 8423, 1999.

62. Pàmies, O., Ell, A.H., Samec, J.S.M., et al., An efficient and mild ruthenium-catalyzed racemization of amines: application to the synthesis of enantiomerically pure amines, *Tetrahedron Lett.*, 43, 4699, 2002.

63. Jung, H.M., Koh, J.H., Kim, M.-J., et al., Concerted catalytic reactions for conversion of keton enol acetates to chiral acetates, *Org. Lett.*, 2, 409, 2000.

64. Jung, H.M., Koh, J.H., Kim, M.-J., et al., Practical ruthenium/lipase-catalyzed asymmetric tr formations of ketones and enol acetates to chiral acetates, *Org. Lett.*, 2, 2487, 2000.

65. Kim, M.-J., Choi, M.Y., Han, M.Y., et al., Asymmetric transformations of acyloxyphenyl ket by enzyme–metal multicatalysis, *J. Org. Chem.*, 67, 9481, 2002.

66. Choi, Y.K., Kim, M.-J., Ahn, Y., et al., Lipase/palladium-catalyzed asymmetric transformatio ketoximes to optically active amines, *Org. Lett.*, 3, 4099, 2001.

67. Kim, W.-H., Karvembu, R., and Park, J., Alumina-supported ruthenium catalysts for the racer tion of secondary alcohols, *Bull. Korean Chem. Soc.*, 25, 931, 2004.

9 Biotransformation of Natural or Synthetic Compounds for the Generation of Molecular Diversity

Robert Azerad

CONTENTS

9.1 INTRODUCTION

In the last decade, combinatorial chemistry has become a versatile and powerful tool for generating libraries of new chemical entities for the pharmaceutical industry. Meanwhile, biocatalytic methods have developed and provided complementary approaches and in some cases powerful alternatives to conventional synthetic chemical techniques, due to their high chemo-, regio-, and stereoselectivity. Moreover, the ability of enzyme or microbial catalysis to perform difficult chemical reactions, such as hydroxylations of nonactivated carbons, on structurally complex molecules, without the need of protection/deprotection steps of reactive functional groups, was recognized. That promoted new strategies for the use of biocatalysis as an additional tool for the multiple modification of synthetic or natural products [1], and the development of a biocombinatorial chemistry ("combinatorial biocatalysis") for generating molecular diversity, as a novel approach in drug discovery and development [2–4].

True combinatorial methods with associated deconvolution strategies have been described using enzymes either in the preparation of combinatorial libraries or in the postsynthetic modification of combinatorial libraries to amplify their structural diversity [5,6].

Several classes of enzymic reactions have been used, sometimes in combined rounds [4] or in a one-pot reaction cascade: acylations with lipases [4,5,7], glycosylations [8], oxido-reductions by dehydrogenases [9,10], aldolase reactions [11], etc.

More examples are related to parallel syntheses, operating with a collection of microorganisms to produce either from a single substrate or a library of related substrates, a number of different derivatives, making use of the versatility of the microorganism-catalyzed biotransformations [12—14].

This chapter focuses on biooxidative transformations [15,16] (hydroxylations an droxylations) of structurally complex natural or synthetic compounds of biological int generate in a parallel or combinatorial approach, known animal metabolites or novel tives exhibiting new or modified pharmacological or pharmacokinetic properties. T accomplishment of this methodology came from the use of microorganism models tc the numerous animal or human detoxication reactions of drugs and xenobiotics [1 shown to generate new bioactive molecules [17]. This concept has been widely applie fields of steroids [18–20], terpenoids [13,21], alkaloids [22], and synthetic chemicals [1 functionalize these compounds and to generate new oxidized chemical entities. Cloned P450 monooxygenases [23–25] and cloned bacterial dioxygenases [26,27] have als successfully employed to this end.

In contrast to combinatorial biocatalysis, "combinatorial biosynthesis" [28–32] ma of genetic contribution and recent improvements in the understanding of natural biosy pathways [33–35] to produce mutated or recombinant microorganisms and enzymes generate modified natural product analogs, in the same way as those structures have over time in nature [36–38]. This has been demonstrated in the field of polyketide mac e.g., with erythromycin, where a wide diversity of modified products have been gener engineering every step of the modular biosynthetic megasynthase and associated re [39], including the introduction of a functional complete pathway in a heterologous overproduce these complex analogs [40–42].

Combinatorial biosynthesis (including mutational biosynthesis, precursor-direct synthesis, and cyclization pattern modifications) constitutes an important source future development of novel secondary metabolite structures and for the genera structural diversity in natural product libraries. Several examples of this promising ap will be presented in addition to the use of classical biotransformation studies, betwe classes of natural polyketides: FK-506/FK-520 and avermectin/milbemycin.

9.2 FLAVONOID DERIVATIVES

Flavonoids represent a very interesting group of ubiquitous plant secondary meta They have been shown to have a wide range of beneficial pharmacological and effects generally related to their antioxidant properties. Based on their backbone st (Figure 9.1), they are grouped into several major classes that include substituted flav flavanones 2, isoflavones 3, chalcones 4, etc. from which an enormous variety of struc derived by modification of the backbone substituents.

Although 6500 flavonoids have so far been identified in the *Handbook of Natura noids* [43], biological transformation of these compounds has been a matter of relativ studies. In addition to chemical synthesis, it may be a valuable approach to increa

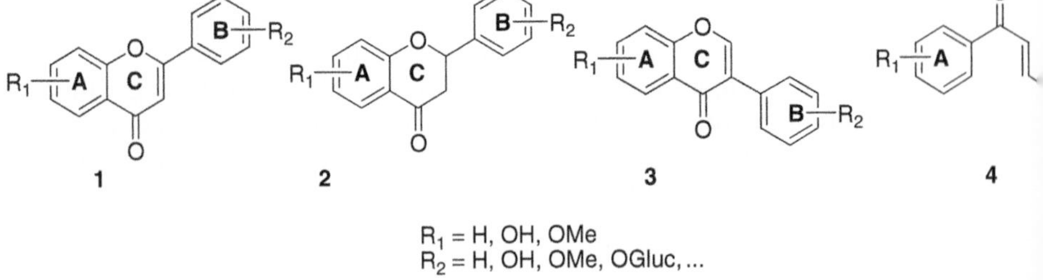

R₁ = H, OH, OMe
R₂ = H, OH, OMe, OGluc, ...

FIGURE 9.1 Main backbones and families of flavonoid compounds.

natural diversity and produce unnatural and potentially active derivatives by enzymatic or microbial conversion.

Enzymatic transformations of flavonoids have been essentially directed to the regioselective glycosylation of the phenolic groups and the subsequent acylation of the resulting glycosides. Several glycosidases readily accept flavonoids as substrates to prepare regioselectively glycosylated derivatives in moderate yields [44]. The resulting flavonoid glycosides were then subsequently modified, using classical lipase or esterase reactions with various activated aromatic or aliphatic ester groups [45–47] to obtain regioselectively acylated derivatives mimicking natural compounds of particular interest due to their intrinsic biological activities.

Other transformations are mainly concerned with oxidative reactions of flavonoid compounds, either oxidative coupling or additional introduction of hydroxyl groups.

For example, theaflavins are formed by fermentative oxidation of catechin in black tea and share with other polyphenols of tea a considerable interest because of their potential benefits for human health and reduction in the risk of cardiovascular diseases and cancer. In an attempt to prepare significant amounts of similar theaflavin derivatives, 18 different related compounds have been obtained using coupling reactions of selected pairs of flavonoid reactants catalyzed by the horseradish peroxidase/hydrogen peroxide system [48], as shown in Figure 9.2. The reaction involves the oxidation of the B-rings to quinones, followed by the

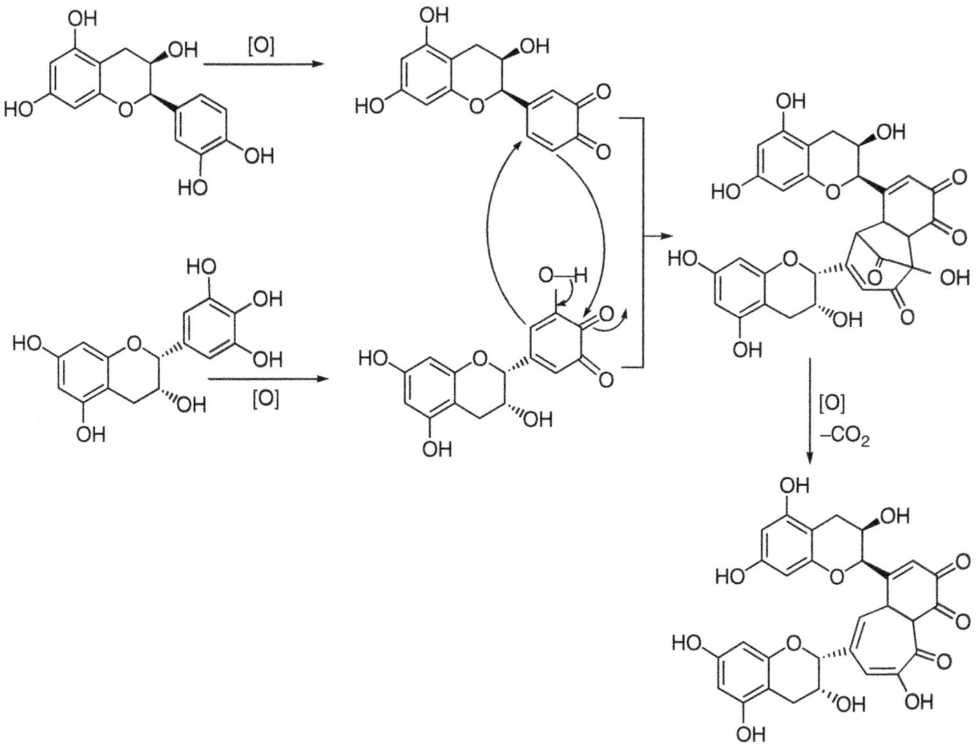

FIGURE 9.2 Hydrogen peroxide/horseradish peroxidase-catalyzed oxidation of a couple of flavanoid moieties to a theaflavin-like derivative. (Adapted from Sang, S., Lambert, J.D., Tian, S., et al., *Bioorg. Med. Chem.*, 12, 459–467, 2004.)

Michael addition of the gallocatechin quinone to the catechin quinone prior to the ca addition across the ring and subsequent decarboxylation.

Most of the microbial (whole-cell) transformations of flavonoids have been found to in rings A and B, particularly at C-5 to C-7, C-4' and to a lesser extent at C-8 or C-3' [49,50] transformations essentially include monohydroxylation [51–54], O-demethylation [52, or O-methylation [58,59], and less frequently C-methylation [59] or conjugation [55, Characteristic examples of these reactions using various fungi or *Streptomyces* strai shown in Figure 9.3 and Figure 9.4, respectively. Some of them are exact mimics of the or human metabolism of dietary flavonoids by dedicated liver CYP450s [63–67]. Ba strains isolated from intestinal tract [68,69] and fermented foods or drinks [70] have als used for similar transformations.

A peculiar reaction has been described for prenylated isoflavone compounds, s 7-O-methyl luteone [71–73], which is cyclized by *Botrytis cinerea* to the correspo dihydrofurano- and dihydropyrano-fused derivatives upon an initial monooxyg catalyzed epoxidation of the prenyl side chain (Figure 9.5).

There have been only a few reports concerning the microbial transformation of cha **4**. In addition to hydroxylation or O-demethylation reactions, a strain of *Aspergillus a*

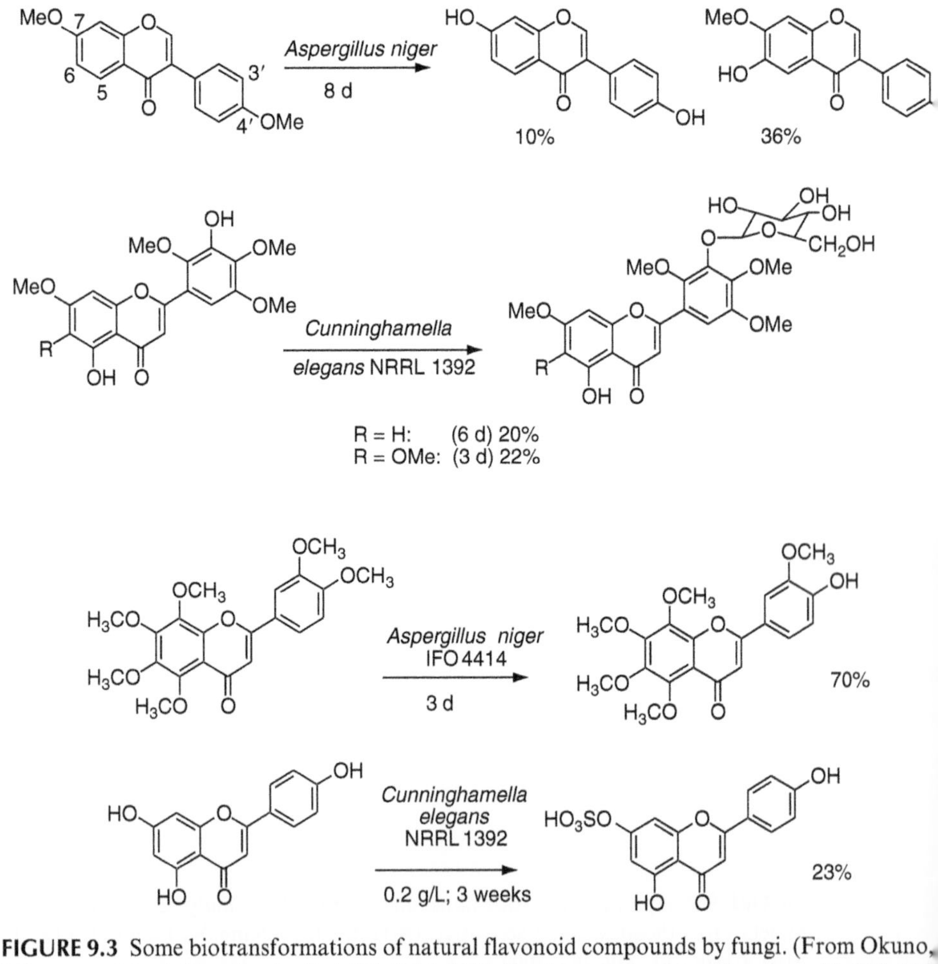

FIGURE 9.3 Some biotransformations of natural flavonoid compounds by fungi. (From Okuno, Miyazawa, M., *J. Nat. Prod.*, 67, 1876–1878, 2004; Ibrahim, A.R., *Phytochemistry*, 53, 209–212

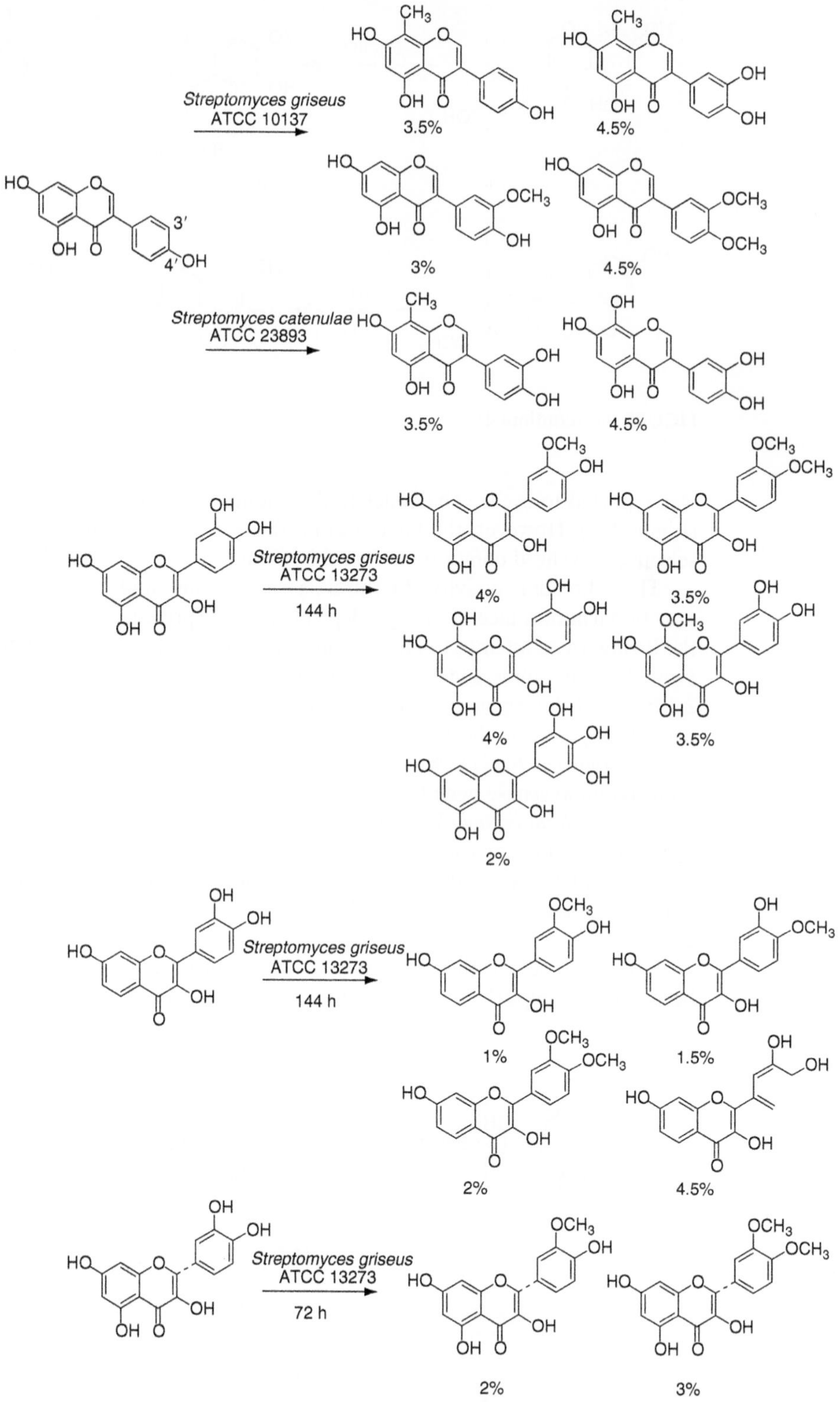

FIGURE 9.4 Some biotransformations of natural flavonoid compounds by bacteria. (From Hosny, M., Dhar, K., and Rosazza, J.P.N., *J. Nat. Prod.*, 64, 462–465, 2001; Hosny, M. and Rosazza, J.P.N., *J. Nat. Prod.*, 62, 1609–1612, 1999; Klus, K. and Barz, W., *Phytochemistry*, 47, 1045–1048, 1998.)

(*continued*)

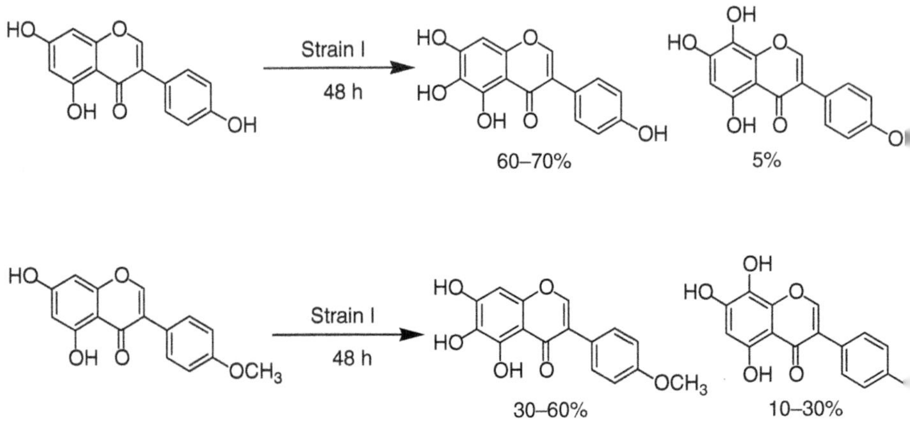

FIGURE 9.4 (continued)

was reported to cyclize chalcones to flavanones [74], mimicking a plant biosynthetic (Figure 9.6). However, the flavonoid products are racemic, unlike the corresponding selectively cyclized compounds found in plants.

The antioxidant activity of natural flavonoid compounds, considered to be of medici nutritional importance, is strongly dependent on the presence of an *o*-dihydroxylated stru the B-ring. In an attempt to increase molecular diversity and antioxidative activity of na synthetic compounds, genetically engineered dioxygenase systems such as recombinant m 2,3-biphenyl dioxygenases have been recently developed to generate *o*-dihydroxylated derivatives [75] from the corresponding natural or synthetic flavonoids.

Streptomyces lividans whole-cells expressing a shuffled broad-specificity large uni biphenyl dioxygenase gene *bphA1* (*2072*) derived from *Pseudomonas pseudoalcaligenes* and *Burkholderia cepacia* LB400, in combination with *bphA2* (small subunit of the iror protein), *bphA3* (ferredoxin), and *bphA4* (ferredoxin reductase) transform in m yields flavone, flavanone, 6-hydroxyflavanone, and 6-hydroxyflavone into corresp *o*-dihydroxylated or monohydroxylated derivatives in various positions of the phenyl group [76].

Similar results, with a prominent conversion to 2′,3′- or 3′,4′-catechol derivativ been recently reported using an *Escherichia coli* strain expressing the same diox

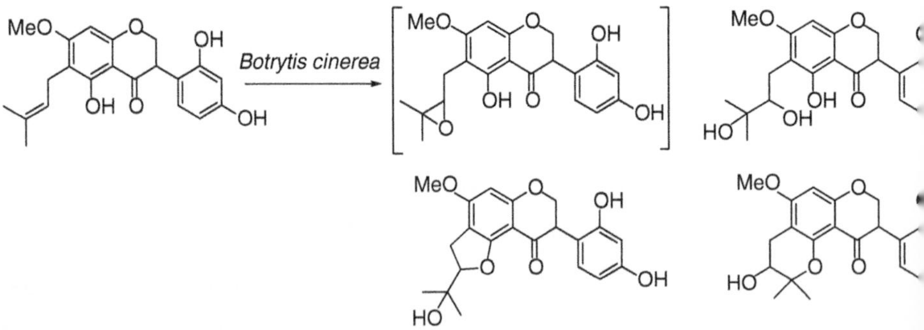

FIGURE 9.5 Biotransformation of 7-*O*-methyl luteone by whole-cells or enzymic extract of *cinerea*.

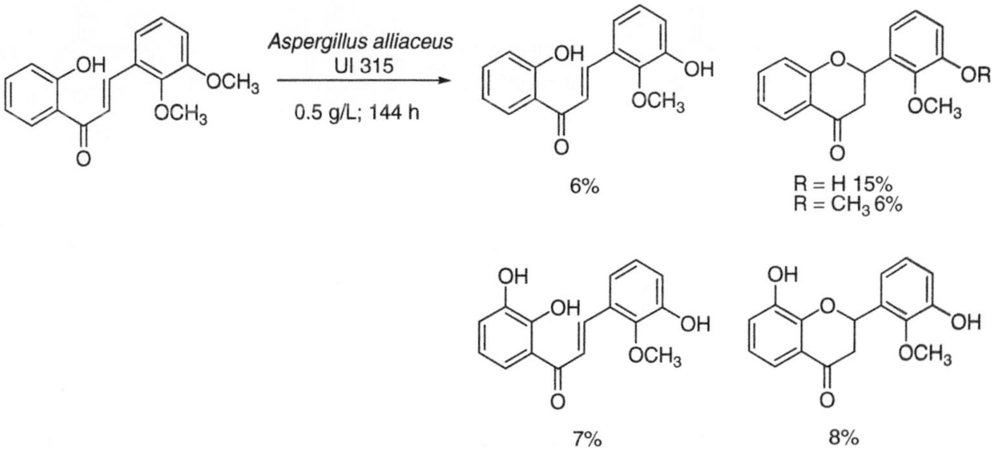

FIGURE 9.6 Biotransformation of a chalcone by a strain of *Aspergillus alliaceus*. (From Sanchez-Gonzalez, M. and Rosazza, J.P.N., *J. Nat. Prod.*, 67, 553–558, 2004.)

system, in addition to the dihydrodiol dehydrogenase gene *bhpB* from *P. pseudoalcaligenes* KF707 [77]. By this method, "unnatural" flavonoids bearing a catechol structure in unusual positions were generated (Figure 9.7), which have not been previously detected in nature and which showed stronger antioxidant activities than their flavonoid natural precursors.

Other results dealing with the introduction of new hydroxy groups into isoflavonoid compounds (including halogenated compounds) using recombinant microbial enzymes have been independently reported by a Chilean group [78]. Recombinant whole-cells of *E. coli* similarly expressing biphenyl-2,3-dioxygenase (BphA) with or without the associated biphenyl-2,3-dihydrodiol 2,3–dehydrogenase (BphB) of *Burkholderia* sp. LB400 were used to generate, from fourteen synthetic isoflavonoids, 2′,3′- and 3′-4′-dihydrodiols, plus the corresponding monohydroxylated compounds at the 2′-, 3′- and/or 4′-positions, resulting from a nonenzymatic dehydration in the absence of the dehydrogenase. Recombinant cells containing both BphA and BphB produced, sometimes in high yields, the corresponding catechol compounds, besides some monohydroxylated derivatives (Figure 9.8).

Another example of the diversification of flavonoid structures by recombinant enzymic methods has been described [79], using several plant flavonoid-modifying enzymes with different substrate specificities, namely an *O*-glucosyltransferase (GT) from *Arabidopsis thaliana* and several *O*-methyltransferases (OMTs) from *Mentha piperita*, expressed in *E. coli*. Single or mixed recombinant cultures were tested for *O*-glucosidation and *O*-methylation of quercetin **5** as a model substrate for the generation of regioselective derivatives.

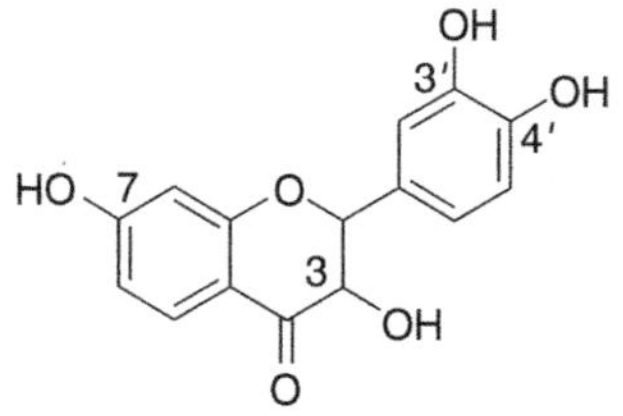

5

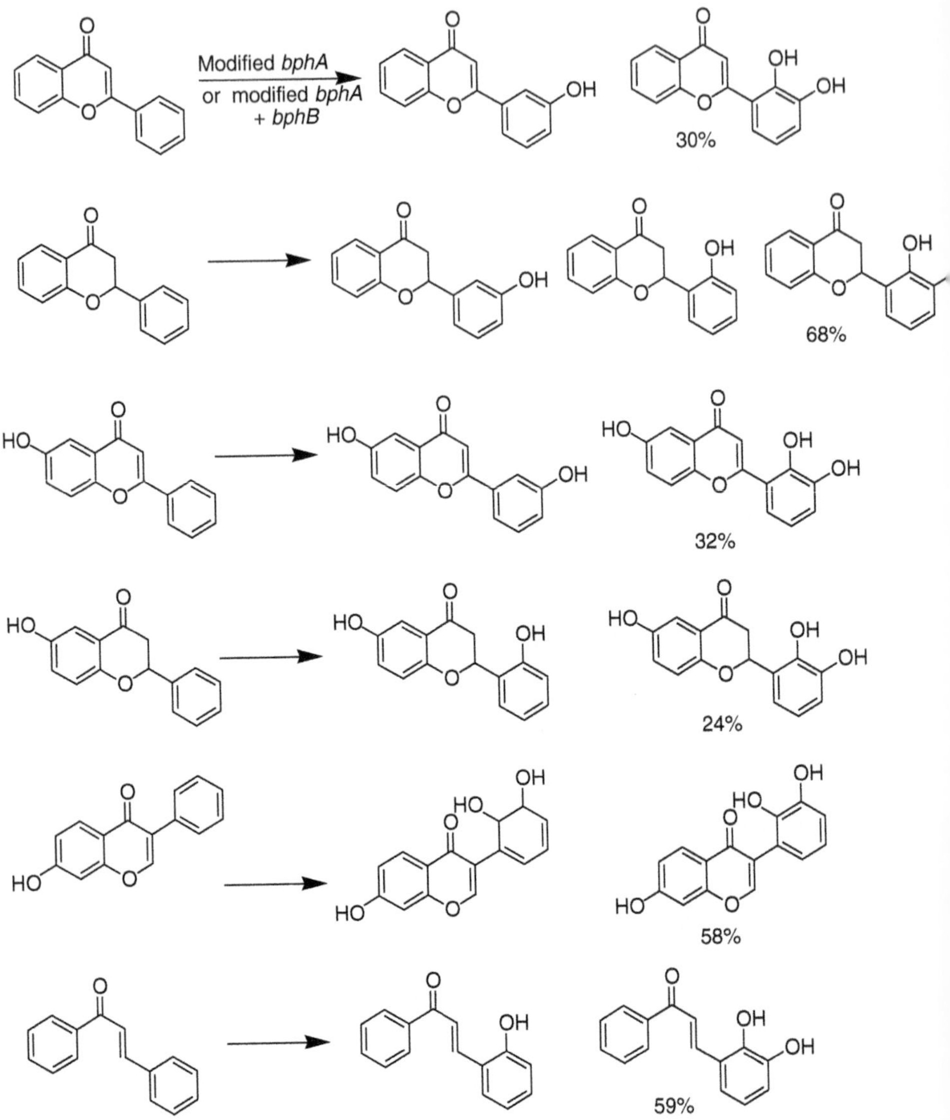

FIGURE 9.7 Hydroxylated derivatives generated from simple flavonoids by whole-cells of expressing shuffled BphA (2,3-biphenyl dioxygenase) and BphB (2,3-dihydrodiol dehydrogenase

When induced cultures expressing each of the individual genes were supplemented quercetin (up to 0.3 mM), regiospecific 7-, 3'-, or 4'-O-methylated and 7-O-glucos derivatives were formed in 25 to 50% yields, in agreement with previous *in vitro* s with purified enzymes. When combinatorial mixed-culture experiments were perf involving several OMTs and/or GT, a variety of the expected partially and comp modified flavonoids were obtained, indicating the feasibility of a simultaneous expr of the different enzyme classes.

9.3   FK-506 AND FK-520

FK-506 (**6**, tacrolimus, L-679,934) and FK-520 (**7**, ascomycin, immunomycin, L-68 two immunosuppressive macrolide compounds, respectively isolated from *S. tsuku*

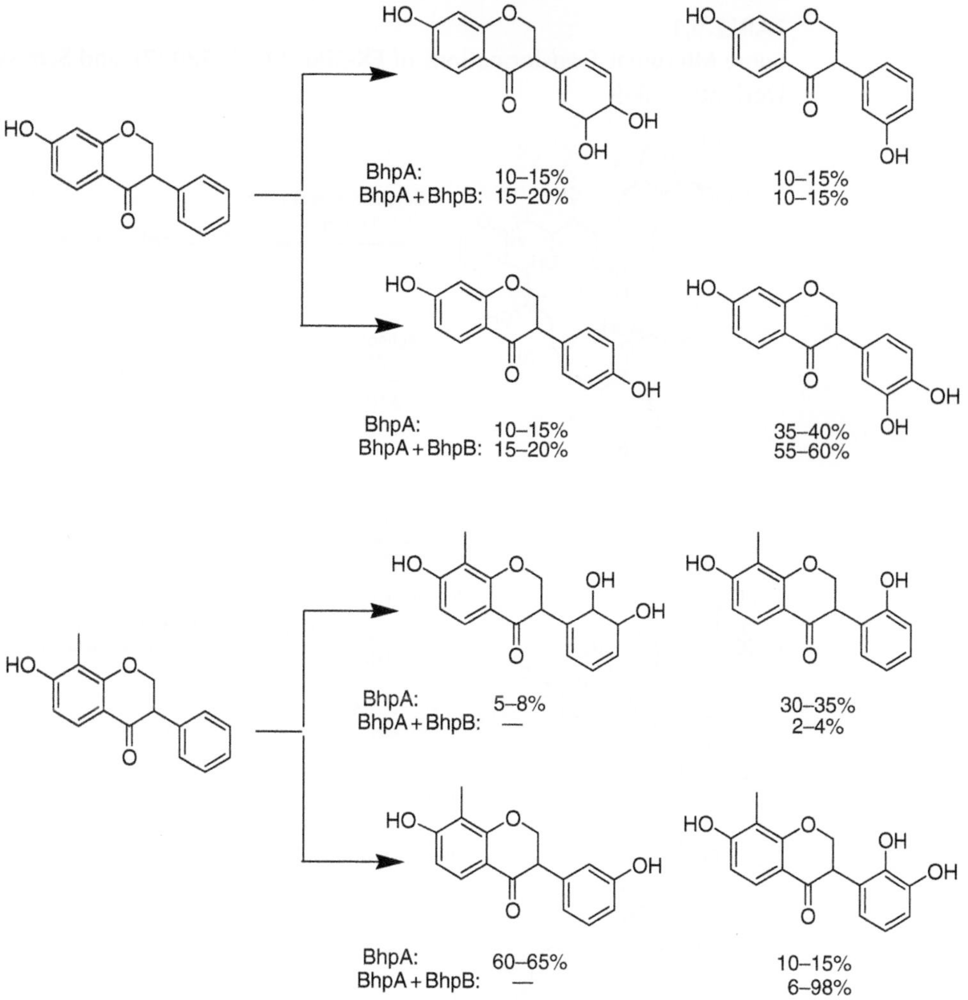

FIGURE 9.8 Hydroxylated derivatives generated from two synthetic isoflavonoids by whole-cells of *E. coli* expressing BphA (2,3-biphenyl dioxygenase) and BphB (2,3-dihydrodiol dehydrogenase).

[80,81] and *S. hygroscopicus* subsp. *yakushimaensis* No. 7238 [82] or var. *ascomyceticus* [83], have been developed as highly effective therapeutic agents in human organ transplants. Due to undesirable side effects and limited bioavailability, both molecules have been the subject of an extensive screening program at the Merck Research Laboratories for their structural modification and for the generation of new active derivatives. Due to the complexity of these molecules, mainly biological approaches were contemplated and a screening for microbial transformations was initiated [84]. A series of actinomycetes were isolated, which were able to perform regioselective hydroxylation or *O*-demethylation reactions of the natural compounds **6–7** or the semisynthetic derivates **8–9**. A summary of the main biotransformations observed and effectively used for the preparation of their respective derivatives is shown in Table 9.1.

Regioselective single or combined *O*-demethylations in positions −13, −15, or −31 were prominent reactions, sometimes accompanied by rearrangement or cyclization reactions [85–94]. The 13-*O*-desmethyl derivative of **6** was previously isolated as its major metabolite in liver microsomal incubations [95,96].

TABLE 9.1

Some Microbial Transformations of FK-506 (6), FK-520 (7), and Semisynthetic Derivatives 8–9

Streptomyces rimosus
(MA187)
——————→ 31-desmethyl FK-506

Actinomycete sp.
(MA6474)
——————→ 13-desmethyl rearranged (= L-683,519)
ATCC 53828

Actinoplanes sp.
(MA6474) 13-desmethyl rearranged
——————→ (= L-685,487)
ATCC 53828

Streptomyces sp.
(MA6870) Hydroxylation at C-19 methyl group and
——————→ hemiketal formation with C-22 carbonyl
ATCC 55281

Actinoplanes sp.
(MA6559) 31-desmethylimmunomycin (= L-683,742)
——————→
ATCC 53771 31- and 13-desmethyl rearranged (= L-683,756)

Streptomyces lavandulae 31-desmethylation + hydroxylation at C-19 methyl
——————→ group and hemiketal formation with C-22 carbonyl
ATCC 55209

Streptomyces lydicus (?)
(MA6890)
——————→ 13-desmethyl rearranged
ATCC 55387

Actinoplanes sp.
——————→ 6-alkoxy-15-desmethyl ascomycin
ATCC 53771

9, R = Et, Pr

However, none of the microorganisms screened was able to produce the 15-*O*-desmethylated derivative of **7**. An elegant solution for this problem was found by using the *Actinoplanes* sp. strain ATCC 53771 to produce the 15,31-*O,O*-bisdesmethyl derivative, then transforming it with the 31-*O*-methyltransferase isolated from *S. ascomyceticus* [90,97], which catalyzes the specific methylation of the hydroxyl group at position −31 of **7** [98]

Another reaction catalyzed by a different *Streptomyces* strain was an allylic hydroxylation of the methyl group at C-19, followed by a cyclic hemiketal formation with the C-22 carbonyl group [94,99].

Using modified culture media, the conversion of FK-520 to its 31-*O*-desmethyl derivative by *Actinoplanes* sp. ATCC 53771 could be improved from 15% to 35–45% yield to produce the desmethyl derivative in a 900 L-scale fermentation [94]. Besides the expected *O*-desmethyl derivative (18 g from 90 g FK-520), small amounts of the 31-*O*-desmethyl *epi* derivative (0.08 g) and of the 31-*O*-desmethyl-19(22)-hydroxymethyl ketal (0.64 g) were isolated.

Alternatively, the bioavailability of these highly lipophilic molecules could be increased by *O*-glucosylation at position −24, using a bioconversion with a *Bacillus subtilis* strain [100], or by *O*-phosphorylation at position −32, carried out by *Rhizopus oryzae* ATCC 11145 [84].

Recently, different approaches to the synthesis of new modified derivatives of FK-506 or FK-520 were undertaken by genetic engineering methods, following the nearly complete elucidation of the biosynthetic pathway of these molecules catalyzed by polyketide synthase (PKS) clusters in combination with post-PKS reactions [101,102].

One of these approaches was the disruption of a few genes, particularly those responsible for the post-PKS reactions, such the one involved in the hydroxylation of position −9, which led to the 9-deoxo or the 9-deoxo-31-*O*-desmethyl derivatives [103].

Another more sophisticated approach ("combinatorial biosynthesis") was the replacement of selected acetyltransferase genes of the specific PKS clusters, by heterologous genes controlling the incorporation of new different acyl units in the sequential elaboration of the FK-506 or FK-520 skeleton, to synthetize novel unnatural polyketides [28,29]. The FK-506 or the FK-520 modular polyketide synthases (Type I) contain 10 modules [35,102] that catalyze sequential additions of two-, three- or four-carbon units to a shikimate-derived starter unit to form the polyketide chain (Figure 9.9). The chain is cyclized after a subsequent addition of pipecolic acid. Each module contains an acyltranferase that selects the particular acylCoA precursor for each round of chain extension.

The replacement of the ascomycin acyltransferase domains (AT7 and AT8) specific for the incorporation of putative methoxymalonate units [104] resulting in the formation of the 13- and 15-methoxy groups, by a malonyl, methylmalonyl, or ethylmalonyl heterologous acyltransferase domains from the PKS clusters of FK-520 or Rapamycin (a closely related structural analog), allowed the production of 13-desmethoxy-, 13,15-bisdesmethoxy- and various analogs methyl- or ethyl-substituted at positions −13 or −15 [104,105]. Other chimeric PKS with preprogrammed substrate specificity, in which biosynthetic modules are truncated, exchanged, or repositioned, should lead to the preparation of novel modified polyketide compounds, with potentially different biological profiles.

9.4 AVERMECTINS AND MILBEMYCINS

Avermectins (**10**) are a mixture of eight closely related 16-membered macrocyclic lactones isolated as fermentation products from *S. avermitilis* ATCC 31267 cultures (Figure 9.10). They contain at C-13, a disaccharide of the unusual methylated deoxysugar L-oleandrose and their diversity results from structural differences at C-5, C-22,23 and a variable side chain at C-25. Avermectins, the semisynthetic 22,23-dihydroavermectins B called ivermectins (**11**), and several of their derivatives are widely used for the treatment of animal diseases caused by

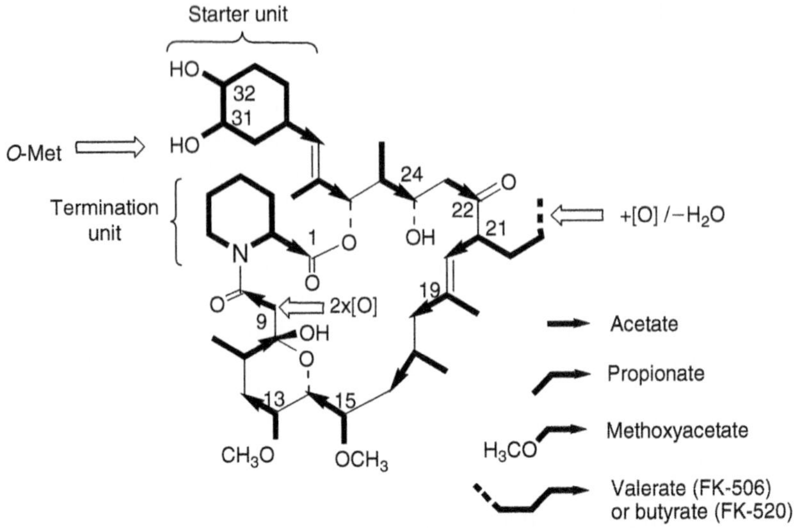

FIGURE 9.9 The biosynthetic pattern of FK-506/FK-520. The biosynthetic units resulting from sequential acylCoA additions are indicated by bold arrows. Post-PKS modifications and their t units are indicated with empty arrows.

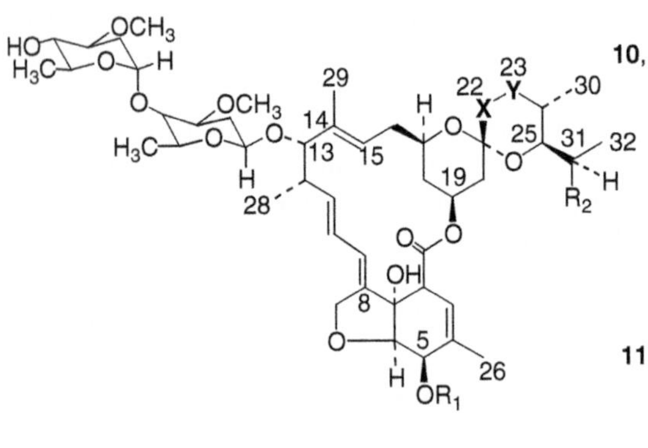

	R$_1$	R$_2$	X-Y(22,23)
10, Avermectin A1a	CH$_3$	C$_2$H$_5$	CH=CH
A1b	CH$_3$	CH$_3$	CH=CH
A2a	CH$_3$	C$_2$H$_5$	CH$_2$—CHO
A2b	CH$_3$	CH$_3$	CH$_2$—CHO
B1a	H	C$_2$H$_5$	CH=CH
B1b	H	CH$_3$	CH=CH
B2a	H	C$_2$H$_5$	CH$_2$—CHO
B2b	H	CH$_3$	CH$_2$—CHO
11, Ivermectin B1a	H	C$_2$H$_5$	CH$_2$—CH$_2$
B1b	H	CH$_3$	CH$_2$—CH$_2$

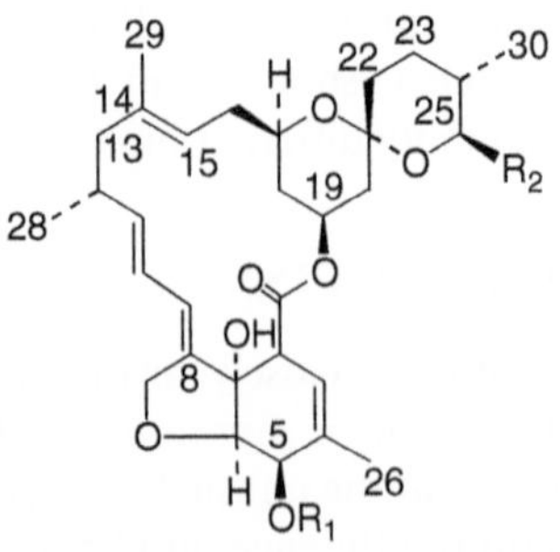

	R$_1$	R$_2$
12, Milbemycin A$_3$	H	CH$_3$
A$_4$	H	C$_2$H$_5$
D	H	CH(CH$_3$)$_2$
G	CH$_3$	CH(CH$_3$)$_2$
13, Nemadectin	H	C(CH$_3$)=CH—CH

FIGURE 9.10 The main natural or semisynthetic components of the avermectin and milb families.

parasitic nematodes and arthropods, and also for crop protection since 1980–1985. Ivermectins have recently been shown to be highly effective in treating human tropical parasitic diseases caused by microfilariae, such as onchocerciasis (river blindness) and strongyloidiasis.

Milbemycins (12), a closely related family of compounds produced by various *S. hygroscopicus* strains, lack the disaccharide unit at C-13 and differ from the 22,23-dihydroavermectin aglycones by their substituents at C-25 (Figure 9.10). Another related compound, Nemadectin (13) is produced by *S. thermoachaensis* or *S. cyaneogriseus*. They all similarly exhibit a broad spectrum of antiparasitic (anthelminthic), acaricidal, and insecticidal activities.

To obtain new derivatives for subsequent use as metabolite reference standards in animal metabolism studies, and to functionalize intermediates for subsequent site-directed chemical modifications, the microbial transformation of some of these natural products was extensively investigated by several industrial groups (Ramos Tombo et al. from CIBA-Geigy, Basel; Nakagawa et al. from Sankyo Ltd., Tokyo; and Arison et al. from Merck & Co, Rahway).

Results have been obtained using incubations of milbemycin compounds with various actinomycetes or fungi cultures, as shown with Milbemycin A4 (Figure 9.11). The isolated derivatives were formed by epoxidation at C-14,15 [106]; by hydroxylation at methyl groups located in external (mainly) allylic positions, C-26 [107], -28 [108,109], -29 [106,110], -30 [109,111,112], or in the C-25 side chain (C-31, C-32 [112]); and by hydroxylation at methylene groups involved in the carbon ring, such as C-24 [109] or C-13 [106,109,113]. The last one has been particularly useful in giving access to the more active avermectin (or ivermectin) family through various enzymatic, microbial, or chemical glycosylation reactions [114,115].

Acaricidal activity of milbemycins depends on the substituents at C-25 position. Oxidations on this side chain are difficult to obtain by chemical transformations, and the described microbial oxidations of the side chain at this position have been used as a preliminary functionalization method to prepare novel C-25 substituted Milbemycin A4 derivatives, with noticeably increased acaricidal activity [116].

Similarly, biotransformations have been described on avermectins, ivermectins or their aglycones to give reactive sites for directed chemical modifications. Diastereomeric side chain-hydroxylated derivatives have been obtained in a scaled-up process from Avermectin A1a [117–119] or Ivermectin B1a aglycone [120,121] (Figure 9.12).

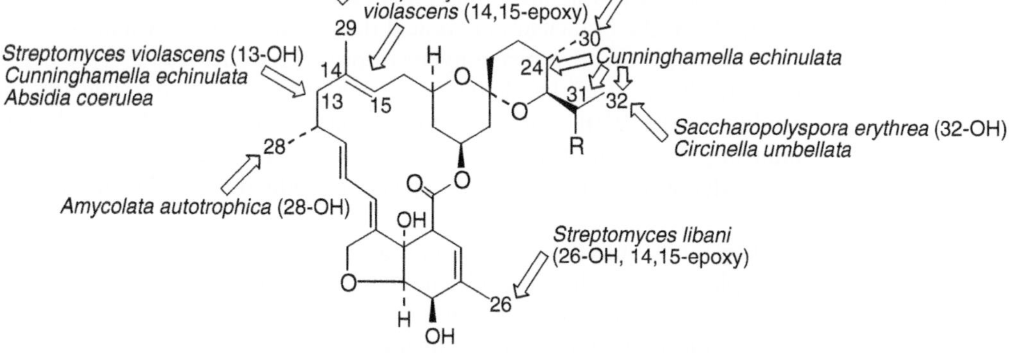

FIGURE 9.11 Some typical oxidative microbial transformations of milbemycins (milbemycin A4, R = H).

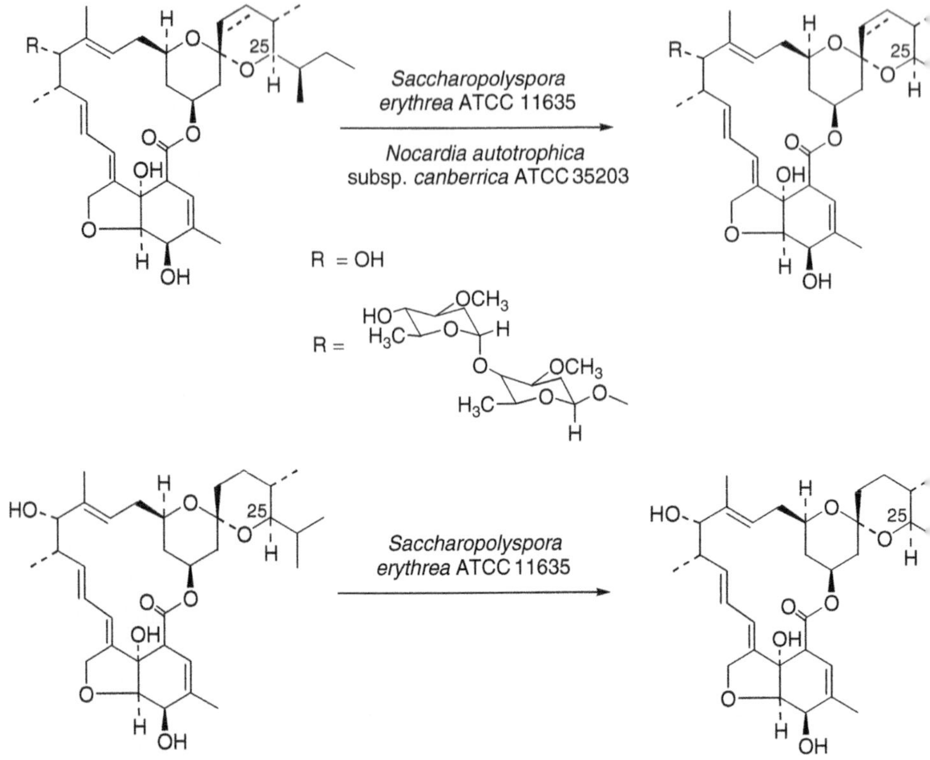

FIGURE 9.12 Microbial conversions of Avermectin A1a or ivermectin aglycones leading to hy tions on the C-25 side chain.

The cloning of the gene cluster for avermectin biosynthesis has been reported ind ently by the Merck group [122] and the Kitasato Institute group [123–125]. The know the complete genome sequence of *S. avermitilis* [126,127] and the progressive elucidatio biosynthetic pathway of avermectins, still not achieved, have led to rational approa the selective production of desired components, or the production of novel ave derivatives, by "engineered biosynthesis," through mutagenesis or recombinant DN nologies.

The avermectin polyketide backbone is derived by sequential elongation from acetate and five propionate units (Figure 9.13) added to an α-branched chain fatty acid arising from the branched amino acid degradation pathway and corresponding to th substituent (*S*-α-methyl propionyl or isopropyl residues for "a" type or "b" type aver respectively). Postpolyketide synthase reactions are directed to dehydration at C-22,2 modulates the B1:B2 ratio, monooxygenase reactions at C-6 and at C-8 methyl inducing the furane ring closure followed by the C-5 carbonyl reduction, *O*-meth by *S*-adenosylmethionine, and stepwise introduction of the oleandrose disacchari [128].

The main objects of these engineered biosynthetic approaches have been dire the replacement of the C-25 side chain, to the modification of the C-5 substitution, an modification of the ratio in the biosynthetized product mixtures to favor the most acti

Avermectin homologs with longer side chains at C-25 are produced by *S. avermitil* cultures are supplied with high concentrations of 2-methylpentanoate or 2-methylhex

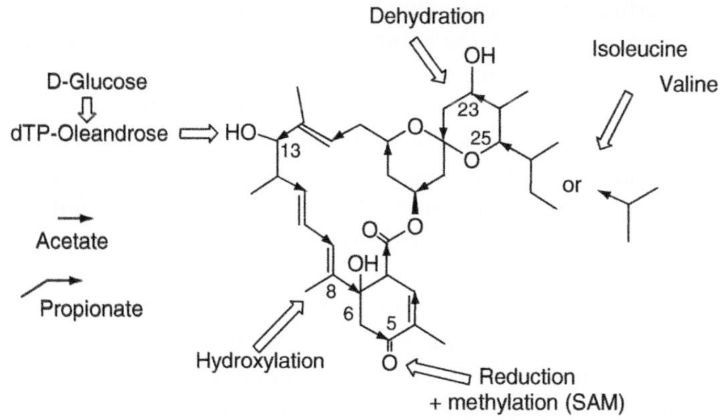

FIGURE 9.13 Incorporation of acetate and propionate units into the avermectin backbone. Post-PKS reactions are indicated with their target sites. (Adapted from Ikeda, H., Ishikawa, J., Hanamoto, A., et al., *Nat. Biotechnol.*, 21, 526–531, 2003.)

[129], indicating the ability of this species to incorporate new fatty acidCoAs as starter units for the polyketide synthase. Alternatively, *S. avermitilis* mutants that contain no functional branched chain 2-oxoacid dehydrogenase and are unable to grow with isoleucine, valine, and leucine as sole carbon sources have been isolated [130]. These mutants could synthetize avermectins when supplemented with exogenous α-branched chain fatty acids. Moreover, such mutants have been shown by the Pfizer group to form a corresponding series of novel avermectins when supplemented with a wide variety of fatty acids [131,132]. Noticeably, supplementation with cyclohexanecarboxylic acid (CHC) or introduction of a pathway that provided CHCCoA into the mutant [133] led to a C-25 cyclohexyl-substituted avermectin mixture CHC-B1 (Doramectin) and CHC-B2 (Figure 9.14). Engineering of *AveC*, a gene with undetermined mechanistic function, by DNA shuffling and extensive variant screening, has led to a considerably improved ratio of the more active Doramectin (CHC-B1) over the undesirable CHC-B2 component (1:0.07 vs. 1:1 for the parent strain), thus allowing a large scale development for the production of Doramectin [134].

The 5-keto compounds produced by *AveF* mutants or engineered strains result from a complete synthesis, lacking the non-PKS dependent reduction at this position [123,135]. As the yield of the corresponding selective chemical oxidation of the 5-hydroxyl group of

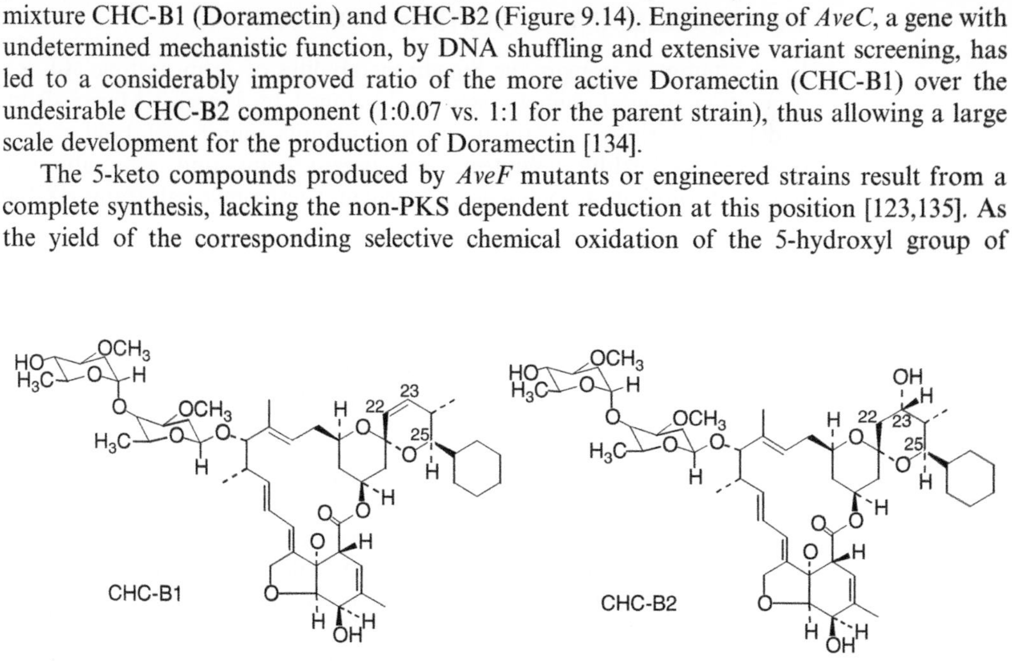

FIGURE 9.14 The main 25-cyclohexyl analogs of avermectins: CHC-B1 (Doramectin) and CHC-B2.

avermectin B components is low, the biosynthetic 5-oxo compounds thus obtained p
important role in the combinatorial biosynthesis of avermectins (and milbemycins), th
their chemical conversion to very active 5-oxime derivatives [136].

On the other hand, conversion of avermectins having a free hydroxyl group at C-
component) to their O-methylated counterpart ("A" component) is catalyzed by a 5-m
O-transferase depending on the aveD gene [124,135]. Since the anthelminthic activity c
components is superior to that of "A," mutations that cause the inactivation or the
lation of the methyltransferase are important for the production of more active avern
mixtures.

As expected, the pentacyclic ring of milbemycins is also derived from a poly
biosynthesis involving seven acetate and five propionate units. However, the milbe
(and Nemadectin) PKS should contain a functional dehydratase and an additional en
ductase in their tenth synthetic module to generate a fully saturated chain at C-
Similarly, the presence of a saturated carbon chain at C-22,23, which is absent in n
avermectins, is explained by fully active enoyl reductase and dehydratase domains
second synthetic module [123].

Exchange or complementation of PKS domains between S. avermitilis and milbemy
other polyketide producing strains, if normally processed, should provide new opport
for combinatorial biosynthesis of novel components, including the direct production of
dihydroavermectins [137]. As an example, the replacement of the original dehydrat
module 2 of avermectin synthetase in S. avermitilis by the functional dehydratase fro
erythromycin eryAII module 4 by homologous recombination resulted in a recom
strain producing only C-22,23-unsaturated avermectin compounds [138]. Concernir
post-PKS reactions, it has been demonstrated that the gene product corresponding tc
mectin 5-O-methyl transferase from S. avermitilis is able to methylate at C-5 a milbe
substrate with high efficiency [139].

9.5 MISCELLEANOUS BIOTRANSFORMATIONS

In an attempt to prepare the human metabolites of Trimegestone (14, RU27987), a se
generation progestomimetic steroid developed by Hoechst-Marion-Roussel in 1990
and particularly, to obtain in sizable amounts minor metabolites of unknown str
observed in healthy volunteers, the parent compound was submitted to an ext
biotransformation screening using fungal and actinomycetal strains [18,140].

14

Nine oxidized derivatives (Table 9.2) were obtained in substantial yields, some of t
a single or a major product depending on the microorganism, allowing an easy prepara
each derivative in the 20 to 100 mg scale, a sufficient amount for complete identificatic
in vitro and in vivo assays. Among the new compounds obtained, two of them (1β-hy
and 6β-hydroxy derivatives) were identified as the unknown human metabolites prev

TABLE 9.2
Biotransformation of Trimegestone by Microorganisms

Metabolite Structure	Strain	Incubation Time (h)	Yield (%)
	Absidia cylindrospora LCP57.1569 Curvularia lunata NRRL2380 Mucor hiemalis	24 24 72	20 70 63
	Absidia cylindrospora LCP57.1569 Circinella minor	24 72	10 23
	Absidia corymbifera LCP63.1800 Cunninghamella elegans ATCC 26269 Absidia cylindrospora LCP57.1569	168 48 24	34 20 32
	Fusarium roseum ATCC 14717	48	45
	Mortierella isabellina NRRL 1757 Fusarium roseum ATCC 14717	48 48	15 12
	Cunninghamella bainieri ATCC 9244	72	10
	Cunninghamella bainieri ATCC 9244	72	6
	Absidia corymbifera LCP63.1800 Fusarium roseum ATCC 14717	168 168	10 8
	Streptomyces rimosus NRRL 2234	96	24

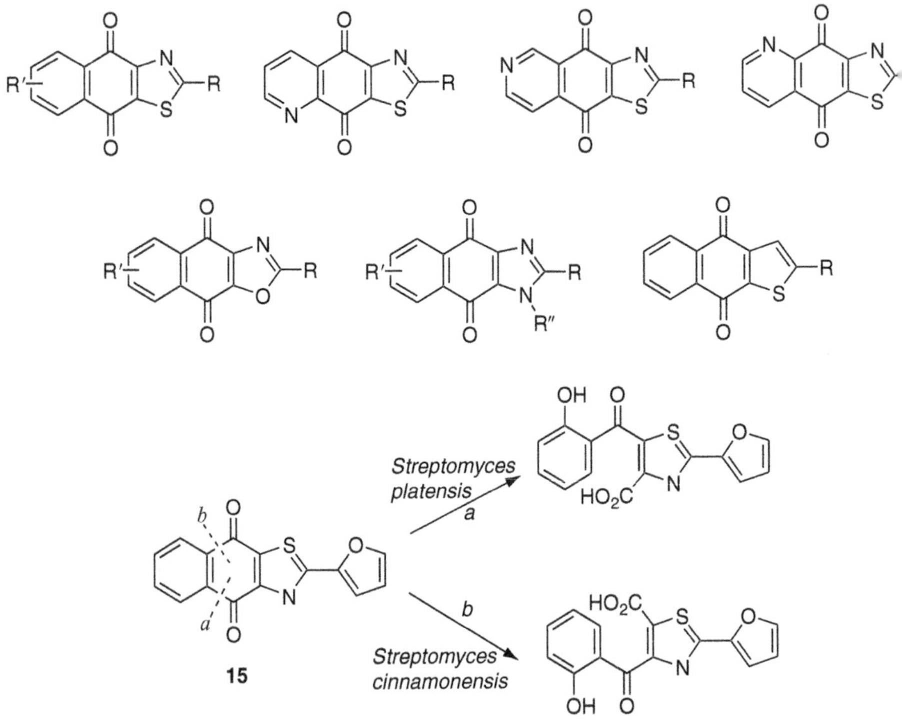

FIGURE 9.15 A library of heterocyclic-substituted naphthoquinones, quinolidione, and isoqu dione analogues, and the oxidative cleavage reactions of INO5042 (**15**) carried out by two *Strep* strains.

detected. In addition, the 1β-hydroxy derivative exhibited a selective progestomimetic comparable to that of the parent compound.

A library of substituted heterocyclic tricyclic 1,4-dioxo-1,4-dihydronapthalene der and their quinoline and isoquinoline analogs (Figure 9.15) were prepared by Labc Innothera (Arcueil, France) for assaying their pharmacological activities in vein disea inflammatory edemas [141]. However, their solubility and biodisponibility were very l as a possibility to increase their hydrophilicity, biotransformation methods were cons Surprisingly, a biotransformation study with microorganisms, using naphthothiazole tive INO5042 (**15**) as a model compound, indicated that several *Streptomyces* strain able to oxidatively cleave this compound in a dioxygenase-like mechanism [14 Depending on the strain used, two isomeric trifunctional derivatives that might interesting pharmacological properties were produced in good yields (Figure 9.15) approach related to parallel and combinatorial syntheses, the whole library was sub to biotransformation by the two selected strains, in miniaturized assays, allowi production, in most cases, of two sublibraries of isomeric oxidized derivatives [14 thus became available for pharmacological studies.

9.6 CONCLUSION

From the examples, it is clear that microbial transformations (particularly, oxidative tra ations related to detoxication mechanisms) may constitute a choice method for the prep of new derivatives of complex molecules, without any need for protection or deprotection techniques. Making use of the high activity and versatility of microorganism-catalyzed reactions, in a pseudo-combinatorial approach, a number of novel compounds with potentially new or modified pharmacological or pharmacokinetic activities can be obtained, which becomes available for biological assays. The miniaturization and automation of the microorganism screening, using microwell plates for high throughput culture, incubation, and product recovery and analysis [9,145–147], will certainly alleviate the corresponding labor that represented the main bottleneck of this method. In future, the development of genetic recombinant techniques exploiting the expression of metagenome capabilities [32], associated with combinatorial biosynthesis methods will probably be used for the discovery and preparation of new chemical entities.

REFERENCES

1. Rich, J.O., Michels, P.C., and Khmelnitsky, Y.L., Combinatorial biocatalysis, *Curr. Opin. Chem. Biol.*, 6, 161–167, 2002.
2. Alreuter, D.H. and Clatk, D.S., Combinatorial biocatalysis: taking the lead from nature, *Curr. Opin. Biotechnol.*, 10, 130–136, 1999.
3. Müller, M., Chemical diversity through biotransformations, *Curr. Opin. Biotechnol.*, 15, 591–598, 2004.
4. Michels, P.C., Khmelnitsky, Y.L., Dordick, J.S., et al., Combinatorial biocatalysis: a natural approach to drug discovery, *TIBTECH*, 16, 210–215, 1998.
5. Mozahaev, V.V., Budde, C.L., Rich, J.O., et al., Regioselective enzymatic acylation as a tool for producing solution-phase combinatorial libraries, *Tetrahedron*, 54, 3971–3982, 1998.
6. Krstenansky, J.L. and Khmelnitsky, Y.L., Biocatalytic combinatorial synthesis, *Bioorg. Med. Chem.*, 7, 2157–2162, 1999.
7. Tricand de la Goutte, J., Khan, J.A., and Vulfson, E.N., Identification of novel polyphenol oxidase inhibitors by enzymatic one-pot synthesis and deconvolution of combinatorial libraries, *Biotechnol. Bioeng.*, 75, 93–99, 2001.
8. Gebhardt, S., Bihler, S., Shubert-Zsilavecz, M., et al., Biocatalytic generation of molecular diversity: modification of ginsenoside Rb1 by β-1,4-galactosyltransferase and *Candida antartica* lipase, Part 4, *Helv. Chem. Acta*, 85, 1943–1959, 2002.
9. Chartrain, M., Greasham, R., Moore, J., et al., Asymmetric bioreduction: application to the synthesis of pharmaceuticals, *J. Mol. Catal., B Enzym.*, 11, 503–512, 2001.
10. Secundo, F., Carrea, G., De Amici, M., et al., A combinatorial biocatalysis approach to an array of cholic acid derivatives, *Biotechnol. Bioeng.*, 81, 391–396, 2003.
11. Lins, R.J., Flitsch, S., Turner, N.J., et al., Generation of a dynamic combinatorial library using sialic acid aldolase and *in situ* screening against wheat germ agglutinin, *Tetrahedron*, 60, 771–780, 2004.
12. Azerad, R., Microbial models for drug metabolism, in *Adv. Biochem. Eng./Biotechnol. (Biotransformations)*, vol. 63, Faber, K. and Scheper, T., Eds., Springer, Heidelberg, 1999, pp. 169–218.
13. Azerad, R., Microbial hydroxylation of terpenoid compounds, in *Stereoselective Biocatalysis*, Patel, R., Ed., Marcel Dekker, New York, 2000, pp. 153–180.
14. Azerad, R., Oxidations using a biocatalyst. Hydroxylation at a saturated carbon, in *Asymmetric Oxidation Reactions: A Practical Approach in Chemistry*, Katsuki, T., Ed., Oxford University Press, New York, 2001, pp. 181–200.
15. Li, Z., Van Beilen, J.B., Duetz, W.A., et al., Oxidative biotransformations using oxygenases, *Curr. Opin. Chem. Biol.*, 6, 136–144, 2002.
16. Urlacher, V., Lutz-Wahl, S., and Schmid, R.D., Microbial P450 enzymes in biotechnology, *Appl. Microbiol. Biotechnol.*, 64, 317–325, 2004.
17. Fura, A., Shu, Y.-Z., Zhu, M., et al., Discovering drugs through biological transformation: role of pharmacologically active metabolites in drug discovery, *J. Med. Chem.*, 47, 4339–4351, 2004.

18. Lacroix, I., Biton, J., and Azerad, R., Microbial models of mammalian metabolism. Mi transformations of Trimegestone (RU 27987), a 3-keto-$\Delta^{4,9(10)}$-norsteroid drug, *Bioorg Chem.*, 7, 2329–2341, 1999.

19. Mahato, S.B., Banerjee, S., and Podder, S., Steroid transformations by microorganisn *Phytochemistry*, 28, 7–40, 1989.

20. Fernandes, P., Cruz, A., Angelova, B., et al., Microbial conversion of steroid compounds: developments, *Enzyme Microb. Technol.*, 32, 688–705, 2003.

21. Lamare, V. and Furstoss, R., Bioconversion of sesquiterpenes, *Tetrahedron*, 46, 4109–413

22. Abraham, W.R. and Spassov, G., Biotransformations of alkaloids: a challenge, *Heterocyc* 711–741, 2002.

23. Vail, R.B., Homann, M.J., Hanna, I., et al., Preparative synthesis of drug metabolites using cytochrome P450s 3A4, 2C9 and 1A2 with NADPH-P450 reductase expressed in *E. coli*, *Microbiol. Biotechnol.*, 32, 67–74, 2005.

24. Guengerich, F.P., Cytochrome P450 enzymes in the generation of commercial products *Discovery*, 1, 359–366, 2002.

25. Rushmore, T.H., Reider, P.J., Slaughter, D., et al., Bioreactor systems in drug metal synthesis of cytochrome P450-generated metabolites, *Metabol. Eng.*, 2, 115–125, 2000.

26. Hudlicky, T., Gonzales, D., and Gibson, D.T., Enzymatic dihydroxylation of aroma enantioselective synthesis: expanding asymmetric methodology, *Aldrichim. Acta.*, 32, 1999.

27. Boyd, D.R. and Sheldrake, G.N., The dioxygenase-catalysed formation of vicinal *cis*-diol *Prod. Rep.*, 15, 309–324, 1998.

28. Hutchinson, C.R., Combinatorial biosynthesis for new drug discovery, *Curr. Opin. Micro* 319–329, 1998.

29. Khosla, C. and Zawada, R.J.X., Generation of polyketide libraries via combinatorial biosy *TIBTECH*, 14, 335–341, 1996.

30. Rodriguez, E. and McDaniel, R., Combinatorial biosynthesis of antimicrobials and other products, *Curr. Opin. Microbiol.*, 4, 526–534, 2001.

31. Weist, S. and Süssmuth, R.D., Mutational biosynthesis—a tool for the generation of str diversity in the biosynthesis of antibiotics, *Appl. Microbiol. Biotechnol.*, 68, 141–150, 2005.

32. Courtois, S., Cappellano, C.M., Ball, M., et al., Recombinant environmental libraries access to microbial diversity for drug discovery from natural products, *Appl. Environ. Mic* 69, 49–55, 2003.

33. Hopwood, D.A., Genetic contributions to understanding polyketide synthases, *Chem. R* 2465–2497, 1997.

34. Rawlings, B.J., Type I polyketide biosynthesis in bacteria (Part A-erythromycin biosynthesi *Prod. Rep.*, 18, 190–227, 2001.

35. Rawlings, B.J., Type I polyketide biosynthesis in bacteria (Part B), *Nat. Prod. Rep.*, 18, 2: 2001.

36. Walsh, C.T., Polyketide and nonribosomal peptide antibiotics: modularity and versatility, S 303, 1805–1810, 2004.

37. Cane, D.E. and Walsh, C.T., The parallel and convergent universes of polyketide syntha nonribosomal peptide synthetases, *Chem. Biol.*, 6, R319–R325, 1999.

38. Sieber, S.A. and Marahiel, M.A., Learning from Nature's drug factories: nonribosomal sy of macrocyclic peptides, *J. Bacteriol.*, 185, 7036–7043, 2003.

39. Pohl, N.L., Non-natural substrates for polyketide synthases and their associated mo enzymes, *Curr. Opin. Chem. Biol.*, 6, 773–778, 2002.

40. Pfeifer, B.A. and Khosla, C., Biosynthesis of polyketides in heterologous hosts, *Microbio Biol. Rev.*, 65, 106–118, 2001.

41. Pfeifer, B.A., Admiraal, S.J., Gramajo, H., et al., Biosynthesis of complex polyketid metabolically engineered strain of *E. coli*, *Science*, 291, 1790–1792, 2001.

42. Pfeifer, B., Hu, Z., Licari, P., et al., Process and metabolic strategies for improved prod of *Escherichia coli*-derived 6-deoxyerythronolide B, *Appl. Environ. Microbiol.*, 68, 328 2002.

43. Harborne, J.B. and Baxter, H., *Handbook of Natural Flavonoids*, vols. 1–2, Wiley, Chichester, 1999.
44. Gao, C., Mayon, P., Macmanus, D.A., et al., Novel enzymatic approach to the synthesis of flavonoid glycosides and their esters, *Biotechnol. Bioeng.*, 71, 235–243, 2001.
45. Danieli, B. and Bertario, A., Chemo-enzymatic synthesis of 6″-*O*-(3-arylprop-2-enoyl) derivatives of the flavonol glucoside isoquercitin, *Helv. Chim. Acta*, 76, 2981–2991, 1993.
46. Ardhaoui, M., Falcimagne, A., Engasser, J.M., et al., Enzymatic synthesis of new aromatic and aliphatic esters of flavonoids using *Candida antartica* lipase as biocatalyst, *Biocatal. Biotransformation.*, 22, 253–259, 2004.
47. Riva, S., Enzymatic modification of the sugar moieties of natural glycosides, *J. Mol. Catal., B Enzym.*, 19–20, 43–54, 2002.
48. Sang, S., Lambert, J.D., Tian, S., et al., Enzymatic synthesis of tea theaflavin derivatives and their anti-inflammatory and cytotoxic activities, *Bioorg. Med. Chem.*, 12, 459–467, 2004.
49. Ibrahim, A.R. and Abul-Hajj, Y.J., Microbiological transformations of (+/−) flavanone and (+/−) isoflavanone, *J. Nat. Prod.*, 53, 644–656, 1990.
50. Ibrahim, A.R. and Habul-Hajj, Y.J., Microbiological transformations of flavone and isoflavone, *Xenobiotica*, 20, 363–373, 1990.
51. Abulhajj, Y.J., Ghaffari, M.A., and Mehrotra, S., Importance of oxygen functions in the biological hydroxylation of flavonoids by *Absidia blackesleeana*, *Xenobiotica*, 21, 1171–1177, 1991.
52. Miyazawa, M., Ando, H., Okuno, Y., et al., Biotransformation of isoflavones by *Aspergillus niger*, as biocatalyst, *J. Mol. Catal., B Enzym.*, 27, 91–95, 2004.
53. Ibrahim, A.R. and Abulhajj, Y.J., Microbiological transformation of chromone, chromanone, and ring A hydroxyflavones, *J. Nat. Prod.*, 53, 1471–1478, 1990.
54. Kostrzewa-Suslow, E., Dmochowka-Gladysz, J., Bialonska, A., et al., Microbial transformations of flavanone and 6-hydroxyflavanone by *Aspergillus niger* cultures, 7th International Symposium on Biocatalysis and Biotransformations (Biotrans' 2005), Delft, The Netherlands, July 3–8, 2005, 190.
55. Ibrahim, A.R., Galal, A.M., Ahmed, M.S., et al., *O*-demethylation and sulfation of 7-methoxylated flavanones by *Cunninghamella elegans*, *Chem. Pharm. Bull.*, 51, 203–206, 2003.
56. Farooq, A. and Tahara, S., Fungal metabolism of flavonoids and phytoalexins, *Curr. Top. Phytochem.*, 2, 1–33, 1999.
57. Okuno, Y. and Miyazawa, M., Biotransformation of nobiletin by *Aspergillus niger* and the antimutagenic activity of a metabolite, 4′-hydroxy-5,6,7,8,3′-pentamethoxyflavone, *J. Nat. Prod.*, 67, 1876–1878, 2004.
58. Hosny, M., Dhar, K., and Rosazza, J.P.N., Hydroxylations and methylations of quercetin, fisetin, and catechin by *Streptomyces griseus*, *J. Nat. Prod.*, 64, 462–465, 2001.
59. Hosny, M. and Rosazza, J.P.N., Microbial hydroxylations and methylations of genistein by *Streptomyces griseus*, *J. Nat. Prod.*, 62, 1609–1612, 1999.
60. Ibrahim, A.R. and Abulhajj, Y.J., Aromatic hydroxylation and sulfation of 5-hydroxyflavone by *Streptomyces fulvissimus*, *Appl. Environ. Microbiol.*, 55, 3140–3142, 1989.
61. Ibrahim, A.R., Galal, A.M., Mossa, J.S., et al., Glucose-conjugation of the flavones of *Psiadia arabica* by *Cunninghamella elegans*, *Phytochemistry*, 46, 1193–1195, 1997.
62. Ibrahim, A.R., Sulfation of naringenin by *Cunninghamella elegans*, *Phytochemistry* 53, 209–212, 2000.
63. Bayer, T., Colnot, T., and Dekant, W., Disposition and biotransformation of the estrogenic isoflavone daidzein in rats, *Toxicol. Sci.*, 62, 205–211, 2001.
64. Nikolic, D., Li, Y., Chadwick, L.R., et al., Metabolism of 8-prenylnaringenin, a potent phytoestrogen from hops (*Humulus lupulus*) by human liver microsomes, *Drug Metab. Dispos.*, 32, 272–279, 2004.
65. Breinholt, V.M., Rasmussen, S.E., Brosen, K., et al., *In vitro* metabolism of genistein and tangeretin by human and murine cytochrome P450s, *Pharmacol. Toxicol.*, 93, 14–22, 2003.
66. Manach, C. and Donovan, J.L., Pharmacokinetics and metabolism of dietary flavonoids in humans, *Free Radic. Res.*, 38, 771–785, 2004.

67. Nikolic, D. and van Breemen, R.B., New metabolic pathways for flavanones catalyzed by microsomes, *Drug Metab. Dispos.*, 32, 387–397, 2004.
68. Hur, H.G., Lay, J.O., Beger, R.D., et al., Isolation of human intestinal bacteria metaboli natural isoflavone glycosides daidzin and genistin, *Arch. Microbiol.*, 174, 422–428, 2000.
69. Wang, X.-L., Hur, H.-G., Lee, J.H., et al., Enantioselective synthesis of *S*-equol from daidzein by a newly isolated anaerobic human intestinal bacterium, *Appl. Environ. Micro* 214–219, 2005.
70. Klus, K. and Barz, W., Formation of polyhydroxylated isoflavones from the isoflavones g and biochanin A by bacteria isolated from tempe, *Phytochemistry*, 47, 1045–1048, 1998.
71. Tahara, S., Saitoh, F., and Mizutani, J., 7-*O*-Methyl-luteone metabolism in *Botrytis c* identification of the epoxy-intermediate and absolute configuration of the pyrano-iso metabolite, *Z. Naturforsch. C*, 48, 16–21, 1993.
72. Tanaka, M., Mizutani, J., and Tahara, S., Cyclization of a prenylated isoflavone via en epoxidation, *Biosci. Biotechnol. Biochem.*, 60, 171–172, 1996.
73. Tahara, S., Tanaka, M., and Barz, W., Fungal metabolism of prenylated flavonoids, *Phy istry*, 44, 1031–1036, 1997.
74. Sanchez-Gonzalez, M. and Rosazza, J.P.N., Microbial transformations of chalcones: hy tion, *O*-demethylation, and cyclization to flavanones, *J. Nat. Prod.*, 67, 553–558, 2004.
75. Kim, S.-Y., Jung, J., Lim, Y., et al., Cis 2′,3′-dihydrodiol production on flavone B-ring by dioxygenase from *Pseudomonas pseudoalcalinogenes* KF707 expressed in *E. coli*, *Anto Leeuwenhoek*, 84, 261–268, 2003.
76. Chun, H.-K., Ohnishi, Y., Shindo, K., et al., Biotransformation of flavone and flavar Streptomyces lividans cells carrying shuffled biphenyl dioxygenase genes, *J. Mol. Catal., B* 21, 113–121, 2003.
77. Shindo, K., Kagiyama, Y., Nakamura, R., et al., Enzymatic synthesis of novel anti flavonoids by Escherichia coli cells expressing modified metabolic genes involved in t catabolism, *J. Mol. Catal., B Enzym.*, 23, 9–16, 2003.
78. Seeger, M., Gonzalez, M., Camara, B., et al., Biotranformation of natural and s isoflavonoids by two recombinant microbial enzymes, *Appl. Environ. Microbiol.*, 69, 504 2003.
79. Willis, M.G., Giovanni, M., Prata, R.T.N., et al., Bio-fermentation of modified fla an example of *in vivo* diversification of secondary metabolites, *Phytochemistry*, 65, 2004.
80. Tanaka, H., Kuroda, A., Marusawa, A., et al., Structure of FK506: a novel immunosup isolated from *Streptomyces*, *J. Am. Chem. Soc.*, 109, 5031–5033, 1987.
81. Kino, T., Hatanaka, H., Hashimoto, M., et al., FK506, a novel immunosuppressant isolat a *Streptomyces*, *J. Antibiot.*, 40, 1249–1255, 1987.
82. Hatanaka, H., Kino, T., Miyata, S., et al., FR-900520 and FR-920523, novel immunosupp isolated from a *Streptomyces*. I. Taxonomy of the producing strain, *J. Antibiot.*, 41, 159 1988.
83. Morisaki, M. and Arai, T., Identity of immunosuppressant FR-900520 with ascomycin, *biot.*, 45, 126–128, 1992.
84. Chen, T.S., Li, X., Petuch, B., et al., Structural modification of the immunosupressants and ascomycin using a biological approach., *Chimia*, 53, 596–600, 1999.
85. Inamine, E.S., Wicker, L.S., Chen, S.S.T., et al., Immunosuppressant agent, Eur. Patent 1990; *Chem. Abstr.*, 113, 76606z, 1990.
86. Chen, T.S., Arison, B.H., Wicker, L.S., et al., Microbial transformation of immunosup compounds. 1. Desmethylation of FK-506 and immunomycin (FR-900520) by *Actinopl* ATCC-53771, *J. Antibiot.*, 45, 118–123, 1992.
87. Chen, T.S., Arison, B.H., Wicker, L.S., et al., Microbial transformation of immunosup compounds. 2. Specific desmethylation of 13-methoxy group of FK-506 and FR-90 Actinomycete sp. ATCC-53828, *J. Antibiot.*, 45, 577–580, 1992.
88. Shafiee, A., Arison, B.H., Chen, S.S.T., et al., Microbial transformation product having i suppressive activity, U.S. Patent 5,352,783, 1994; *Chem. Abst.*, 122, 8154, 1995.

89. Shafiee, A., Chen, T., and Cameron, P., Microbial demethylation of immunosuppressant FK-506: isolation of 31-*O*-FK-506-specific demethylase showing cytochrome P-450 characteristics from *Streptomyces rimosus* MA187, *Appl. Environ. Microbiol.*, 61, 3544–3548, 1995.

90. Shafiee, A., Chen, T., Arison, B.H., et al., Enzymatic synthesis and immunosuppressive activity of novel desmethylated immunomycins (ascomycins), *J.Antibiot.*, 46, 1397–1405, 1993.

91. Arison, B.H., Inamine, E., Chen, S.S.T., et al., A novel immunosuppressant prepared by microbial transformation of L-683,590, Eur. Patent 378,321, 1990; *Chem. Abst.*, 117, 46696, 1992.

92. Arison, B.H., Inamine, E., Chen, S.S.T., et al., A novel immunosupressant L-68376 prepared by microbial transformation of L-683,590, Eur. Patent 378,320, 1990; *Chem. Abst.*, 114, 99961, 1991.

93. Arison, B.H., Inamine, E., and Chen, S.S.T., A novel immunosuppressant prepared by microbial transformation of L-679,934, Eur. Patent 378,317, 1990; *Chem. Abstr.* 114, 99960v, 1991.

94. Haag, M., Baumann, K., Billich, A., et al., Bioconversion of ascomycin and its 6-alkoxyderivatives, *J. Mol. Cat., B Enzym.*, 5, 389–394, 1998.

95. Christians, U., Radeke, H.H., Kownatzki, R., et al., Isolation of an immunosuppressive metabolite of FK506 generated by human microsome preparations, *Clin. Biochem.*, 24, 271–275, 1991.

96. Iwasaki, K., Shiraga, T., Nagase, K., et al., Isolation, identification and biological activities of oxidative metabolites of FK506, a potent immunosuppressive macrolide lactone, *Drug Metab. Disp.*, 21, 971–977, 1993.

97. Motamedi, H., Shafiee, A., Cai, S.J., et al., Characterization of methyltransferase and hydroxylase genes involved in the biosynthesis of the immunosuppressants FK506 and FK520., *J. Bacteriol.*, 178, 5243–5248, 1996.

98. Shafiee, A., Motamedi, H., and Chen, T.S., Enzymology of FK506 biosynthesis: purification and characterization of 31-desmethyl FK506-*O*-methyltransferase from *Streptomyces* sp. MA6858, *Eur. J. Biochem.*, 225, 764–769, 1994.

99. Chen, T.S., Petuch, B., White, R., et al., Microbial transformation of immunosupressive compounds. IV. Hydroxylation and hemiketal formation of ascomycin (immunomycin) by *Streptomyces* sp. MA6970 (ATCC no. 55281), *J. Antibiot.*, 47, 1557–1559, 1994.

100. Petuch, B.R., Arison, B.H., Hsu, A., et al., Microbial transformation of immunosuppressive compounds. III. Glucosylation of immunomycin (FR900520) and FK506 by *Bacillus subtilis*, *J. Ind. Microbiol.*, 13, 131–135, 1994.

101. Motamedi, H., Cai, S.J., Shafiee, A., et al., Structural organization of a multifunctional polyketide synthetase involved in the biosynthesis of the macrolide immunosuppressant FK506, *Eur. J. Biochem.*, 244, 74–80, 1997.

102. Motamedi, H. and Shafiee, A., The biosynthetic gene cluster for the macrolactone ring of the immunosuppressant FK506, *Eur. J. Biochem.*, 256, 528–534, 1998.

103. Shafiee, A., Motamedi, H., Dumont, F.J., et al., Chemical and biological characterization of two FK506 analogs produced by targeted gene disruption in *Streptomyces* sp. MA6548, *J. Antibiot.*, 50, 418–423, 1997.

104. Wu, K., Chung, L., Revill, W.P., et al., The FK520 gene cluster of *Streptomyces hygroscopicus* var. *ascomycetinus* (ATCC 14891) contains genes for biosynthesis of unusual polyketide extender units, *Gene*, 251, 81–90, 2000.

105. Revill, W.P., Voda, J., Reeves, C.D., et al., Genetically engineered analogs of ascomycin for nerve regeneration, *J. Pharmacol. Experiment. Therapeut.*, 302, 1278–1285, 2002.

106. Nakagawa, K., Sato, K., Okazaki, T., et al., Microbial conversion of milbemycins—13β, 29-dihydroxylation of milbemycins by soil isolate *Streptomyces cavourensis*, *J. Antibiot.*, 44, 803–805, 1991.

107. Nakagawa, K., Sato, K., Tsukamoto, Y., et al., Microbial conversion of milbemycins—microbial conversion of milbemycins A4 and A3 by *Streptomyces libani*, *J. Antibiot.*, 47, 502–506, 1994.

108. Nakagawa, K., Sato, K., Tsukamoto, Y., et al., Microbial conversion of milbemycins: 28-hydroxylation of milbemycins by *Amycolata autotrophica*, *J. Antibiot.*, 46, 518–519, 1993.

109. Nakagawa, K., Miyakoshi, S., Torikata, A., et al., Microbial conversion of milbemycins—hydroxylation of milbemycin-A4 and related compounds by *Cunninghamella echinulata* ATCC 9244, *J. Antibiot.*, 44, 232–240, 1991.

110. Nakagawa, K., Sato, K., Tsukamoto, Y., et al., Microbial conversion of milbemyci
     hydroxylation of milbemycins by genus *Syncephalastrum*, *J. Antibiot.*, 45, 802–805,

111. Nakagawa, K., Torikata, A., Sato, K., et al., Microbial conversion of milbemycins: 30-oxid
     milbemycin-A4 and related compounds by *Amycolata autotrophica* and *Amycolatopsis me
     nei*, *J. Antibiot.*, 43, 1321–1328, 1990.

112. Nakagawa, K., Sato, K., Tsukamoto, Y., et al., Microbial conversion of milbemycins: oxid
     milbemycin A4 and related compounds at the C-25 ethyl group by *Circinella umbellata* and
     cylindrospora, *J. Antibiot.*, 48, 831–837, 1995.

113. Ramos Tombo, G.M., Ghisalba, O., Schär, H.-P., et al., Diasteroselective microbial hydrox
     of milbemycin derivatives, *Agric. Biol. Chem.*, 53, 1531–1535, 1989.

114. Schulman, M., Doherty, P., and Arison, B., Microbial conversion of avermectins by *Saccharopc
     erythrea*—glycosylation at C-4′ and C-4″, *Antimicrob. Agents Chemother.*, 37, 1737–1741, 199

115. Wei, L., Wei, G., Zhang, H., et al., Synthesis of new, potent avermectin-like insecticidal
     Carbohydr. Res., 340, 1583–1590, 2005.

116. Tsukamoto, Y., Nakagawa, K., Kajino, H., et al., Synthesis of novel 25-substituted milbem
     derivatives and their acaricidal activity against *Tetranychus urticae*, *Biosci. Biotechnol. B*
     61, 1650–1657, 1997.

117. Schulman, M., Doherty, P., Zink, D., et al., Microbial conversion of avermectins by *Saccha*
     pora erythrea: hydroxylation at C-27, *J. Antibiot.*, 47, 372–375, 1994.

118. Chartrain, M., White, R., Goegelman, R., et al., Bioconversion of avermectin into
     avermectin, *J. Ind. Microbiol.*, 6, 279–284, 1990.

119. Arison, B.H., Doherty, P.J., and Schulman, M.D., Bioconversion products of 27-h
     avermectin, Brit. Patent 2,281,911, 1995; *Chem. Abstr.*, 123, 54274u, 1995.

120. Schulman, M., Doherty, P., Zink, D., et al., Microbial conversion of avermectins by *Saccha*
     pora erythrea: hydroxylation at C28, *J.Antibiot.*, 46, 1016–1019, 1993.

121. Arison, B.H., Doherty, P.J., and Schulman, M.D., Fermentation of 28-hydroxy ive
     aglycone with *Saccharopolyspora*, U.S. Patent 5,124,258, 1992; *Chem. Abstr.*, 117, 169566

122. McNeil, D.J., Occi, J.L., Gewain, K.M., et al., Correlation of the avermectin polyketide s
     genes to the avermectin structure. Implications for designing novel avermectins, *Ann. N. Y*
     Sci., 721, 123–132, 1994.

123. Ikeda, H. and Omura, S., Avermectin biosynthesis, *Chem. Rev.*, 97, 2591–2609, 1997.

124. Ikeda, H., Tonomiya, T., and Omura, S., Organization of biosynthetic gene cluster
     polyketide anthelmintic macrolide avermectin in *Streptomyces avermitilis*: analysis of en
     domains in four polyketide synthases, *J. Ind. Microbiol. Biotechnol.*, 27, 170–176, 2001.

125. Ikeda, H., Tonomiya, T., Usami, M., et al., Organization of the biosynthetic gene cluster
     polyketide anthelmintic macrolide avermectin in *Streptomyces avermitilis*, *Proc. Natl. Ac*
     USA, 96, 9509–9514, 1999.

126. Omura, S., Ikeda, H., Ishikawa, J., et al., Genome sequence of an industrial microoi
     Streptomyces avermitilis: deducing the ability of producing secondary metabolites, *Pro*
     Acad. Sci. USA, 98, 12215–12220, 2001.

127. Ikeda, H., Ishikawa, J., Hanamoto, A., et al., Complete genome sequence and comp
     analysis of the industrial microorganism *Streptomyces avermitilis*, *Nat. Biotechnol.*, 21, 5
     2003.

128. Yoon, Y.J., Kim, E.-S., Hwang, Y.-S., et al., Avermectin: biochemical and molecular basi
     biosynthesis and regulation, *Appl. Microbiol. Biotechnol.*, 63, 626–634, 2004.

129. Chen, T.S., Inamine, E.S., Hensens, O.D., et al., Directed biosynthesis of avermectins
     Biochem. Biophys., 269, 544–547, 1989.

130. Hafner, E.W., Holley, B.W., Holdom, K.S., et al., Branched-chain fatty acid requirem
     avermectin production by a mutant of *Streptomyces avermitilis* lacking branched-chain 2-c
     dehydrogenase activity, *J. Antibiot.*, 44, 349–356, 1991.

131. Dutton, C.J., Gibson, S.P., Goudie, A.C., et al., Novel avermectins produced by mu
     biosynthesis, *J. Antibiot.*, 44, 357–365, 1991.

132. Denoya, C.D., Fedechko, R., Hafner, E.W., et al., A second branched-chain α-keto aci
     drogenase gene cluster (*bkdFGH*) from *Streptomyces avermitilis*: its relationship to ave

biosynthesis and the construction of a *bkdF* mutant suitable for the production of novel antiparasitic avermectins, *J. Bacteriol.*, 177, 3504–3511, 1995.

133. Cropp, T.A., Wilson, D.J., and Reynolds, K.A., Identification of a cyclohexycarbonyl CoA biosynthetic gene cluster and application in the production of doramectin, *Nat. Biotechnol.*, 18, 980–983, 2000.

134. Stutzman-Engwall, K., Conlon, S., Fedechko, R., et al., Semi-synthetic DNA shuffling of *aveC* leads to improved industrial scale production of doramectin by *Streptomyces avermitilis*, *Metabol. Eng.*, 7, 27–37, 2005.

135. Ikeda, H., Wang, L.-R., Ohta, T., et al., Cloning of the gene encoding avermectin B 5-*O*-methyltransferase in avermectin-producing *Streptomyces avermitilis*, *Gene*, 206, 175–180, 1998.

136. Tsukamoto, Y., Sato, K., Mio, S., et al., Synthesis of 5-keto-5-oxime derivatives of milbemycins and their activities against microfilariae, *Agric. Biol. Chem.*, 55, 2615–2621, 1991.

137. Gaisser, S., Kellenberger, L., Kaja, A.L., et al., Direct production of ivermectin-like drugs after domain exchange in the avermectin polyketide synthase of *Streptomyces avermitilis* ATCC 31272, *Org. Biomol. Chem.*, 1, 2840–2847, 2003.

138. Yong, J.H. and Byeon, W.H., Alternative production of avermectin components in *Streptomyces avermitilis* by gene replacement, *J. Microbiol.*, 43, 277–284, 2005.

139. Hong, Y.-S., Hwang, B.Y., Kim, H.S., et al., Biotransformation of Milbemycin D to Milbemycin G by *Streptomyces lividans* conferring avermectin *O*-methyltranferase activity, *Biotechnol. Lett.*, 20, 991–995, 1998.

140. Biton, J., Azerad, R., Marchandeau, J.P., et al., Process for the preparation of new 1- or 6-hydroxylated steroids and their use as medicaments, Eur. Patent 808,845, 1997; *Chem. Abst.*, 128, 34924, 1997.

141. Boutherin-Falson, O., Desquand-Billiald, S., Favrou, A., et al., Preparation of heterocyclic tricyclic 1,4-dihydro-1,4-dioxonaphthalene vasoconstrictors, PCT Appl. WO 9721684, 1997; *Chem. Abst.*, 127, 121–716, 1997.

142. Le Texier, L., Roy, S., Fosse, C., et al., A biosynthetic microbial ability applied to the oxidative ring cleavage of non-natural heterocyclic quinones, *Tetrahedron Lett.*, 42, 4135–4137, 2001.

143. Fosse, C., Le Texier, L., Roy, S., et al., Parameters and mechanistic studies on the oxidative ring cleavage of synthetic heterocyclic naphthoquinones by *Streptomyces* strains, *Appl. Microbiol. Biotechnol.*, 65, 446–456, 2004.

144. Fosse, C., Le Texier, L., Roy, S., et al., unpublished results.

145. Duetz, W.A., Minas, W., Kühner, M., et al., Miniaturized microbial growth systems in screening, *Bio World*, 2, 8–10, 2001.

146. Clerval, R., Kühner, M., Li, Z., et al., The come-back of high throughput screening of wild-type microbial strains through the use of miniaturised growth systems and LC-MS, *Bio World*, 6, 24–26, 2001.

147. Lye, G.J., Ayazi-Shamlou, P., Baganz, F., et al., Accelerated design of bioconversion processes using automated microscale processing techniques, *TIBTECH*, 21, 29–37, 2003.

148. Arison, B.H., Inamine, E., Chen, S.S.T., et al., Microbial transformation product of L-679,934, Eur. Patent 378,317, 1990.

149. Chen, S.S.T., White, R.F., Dezeny, G., et al., New C-31 desmethyl FR-900520 cyclic hemiketal immunosuppressant agent, Eur. Patent 526,934, 1993; *Chem. Abstr.*, 118, 190103, 1993.

150. Shafiee, A., Kaplan, L., Chen, S.S.T., et al., C-31-methylated FR-900520 cyclic hemiketal immunosuppressant agents obtained by fermentation, U.S. Patent 5,149,701 A, 1992; *Chem. Abst.*, 118, 32932, 1993.

151. Shafiee, A., Kaplan, L., Chen, S.S.T., et al., Manufacture of derivatives of FR-520 for use as immunosuppressants with *Streptomyces lavendulae*, Eur. Patent 526,935 A1, 1993; *Chem. Abst.*, 118, 190104, 1993.

10 Applications of Aromatic Hydrocarbon Dioxygenases

Rebecca E. Parales and Sol M. Resnick

CONTENTS

10.1 INTRODUCTION

Aromatic hydrocarbon dioxygenases refer specifically to enzymes that catalyze the oxidation of the aromatic ring of aromatic hydrocarbons, compounds containing only carbon and hydrogen. This group of enzymes represents an important subset of a large family of "aromatic-ring-hydroxylating dioxygenases" or "Rieske nonheme iron oxygenases" as they have been variously designated, all with similar mechanism and structure [1,2]. Aromatic-ring-hydroxylating dioxygenases initiate bacterial pathways for the degradation of a wide range of aromatic compounds, including polycyclic aromatic hydrocarbons, nitroaromatic and

chlorinated aromatic compounds, aromatic acids, and heterocyclic aromatic comp
Toluene dioxygenase (TDO) from *Pseudomonas putida* F1 was the first characterized ar
hydrocarbon dioxygenase. Identified by Gibson and coworkers, TDO [3,4] was sho
catalyze *cis*-dihydroxylation of benzene and toluene [5–7]. Our understanding of the str
function, and substrate specificity of aromatic hydrocarbon dioxygenases has come fr
studies of TDO and naphthalene dioxygenase (NDO). To date, more than 100 Rieske no
iron oxygenases have been identified on the basis of biological activity or nucleotide se
identity. These multicomponent enzyme systems are cofactor requiring proteins (EC 1.)
that catalyze the addition of molecular oxygen to the aromatic ring through re
dihydroxylation. The initial oxidation of the aromatic ring is the most difficult step
degradation of this class of compounds. This addition of hydroxyl groups to the
stable aromatic ring activates the molecule for further oxidation and eventual ring cle
Aromatic-ring-hydroxylating dioxygenases are unrelated in both structure and fu
to aromatic-ring-cleavage (or ring-fission) dioxygenases (EC 1.13.11.-) that oxidize and
the catecholic intermediates produced during the catabolism of aromatic compounds.

10.2 DISTRIBUTION AND DIVERSITY

Many of the substrates for aromatic-ring-hydroxylating dioxygenases are toxic envirc
tal pollutants. Some, such as certain chlorinated and nitroaromatic compounds, are syr
while others are naturally occurring biological or pyrolysis products, or compone
petroleum [8]. Bacterial degradation of these chemicals is critical for recycling of carl
earth as well as removal of toxic pollutants at contaminated sites.

10.2.1 DISTRIBUTION AND DETECTION OF AROMATIC-RING-HYDROXYLATING DIOXYGEN
IN BACTERIA

Most aromatic-ring-hydroxylating dioxygenases have been identified from bacterial i
capable of growth on specific aromatic compounds. These strains are typically isola
selective enrichment and subsequent plating on minimal medium containing the ar
substrate as the sole source of carbon. Depending on their vapor pressures and solut
various substrates have been incorporated into the agar media, provided in the vapor
[9], or sprayed as an insoluble layer onto the plate surface [10]. Bacteria capable of deg
many naturally occurring aromatic compounds can be easily isolated from most soil sa
but other good sources of inocula include samples from sites with a history of expo
environmental pollutants such as creosote, gasoline, refined petroleum products, o
aromatic pollutants. A wide range of both Gram-positive and Gram-negative bacteri
been found to harbor aromatic compound degradation pathways that are initiated b
hydroxylating dioxygenases, although many of the isolates are pseudomonads or
proteobacteria. This isolation bias is due to the rapid doubling times and minimal g
factor requirements exhibited by this group of bacteria that enable them to do
enrichment cultures. Recently, there have been more efforts to characterize the dive
organisms capable of degrading aromatic compounds, and a wider range of genera a
known to utilize dioxygenase-mediated degradation pathways.

Several colorimetric indicators are available for detecting dioxygenase activity or a
of downstream enzymes required for aromatic compound degradation. These inclu
well-known conversion of indole to indigo [11], which is catalyzed by many aromatic
carbon dioxygenases, and the conversion of indole carboxylic acids to indigo by
aromatic acid dioxygenases [12]. The detection of ring-cleavage products formed fr
catecholic intermediates of aromatic compound degradation has been used to i

coupled activities of aromatic-ring-hydroxylating dioxygenases, *cis*-dihydrodiol dehydrogenases (DDH), and *meta* ring-cleavage dioxygenases [13]. Gene-based discovery approaches for the detection of aromatic hydrocarbon dioxygenases can be employed for screening libraries prepared directly from environmental samples. However, several challenges exist for the application of these techniques to multicomponent enzyme systems. Activity screens require that each gene is present and sufficiently expressed for detection of activity. Homology-based hybridization and PCR amplification typically do not allow the identification of high-genetic diversity, and hybridization of multiple components can be complicated by the "mosaic" organization of the encoding genes [14]. One approach identified novel polycyclic aromatic hydrocarbon degradation genes by screening the genomic DNA from aromatic hydrocarbon degrading isolates for lack of hybridization to standard dioxygenase gene probes [15].

10.2.2 TYPES OF REACTIONS CATALYZED

The primary reaction catalyzed by aromatic-ring-hydroxylating dioxygenases on aromatic substrates is a *cis*-dihydroxylation of the carbon–carbon double bond of adjacent unsubstituted carbon atoms. In most cases, this reaction results in the formation of a stable chiral *cis*-dihydrodiol (Reaction A, Table 10.1). Similarly, oxidation of aromatic acids such as benzoate occurs at the carboxylated carbon and an adjacent unsubstituted carbon, and a chiral *cis*-dihydroxylated cyclohexadiene carboxylic acid is formed (Reaction B, Table 10.1). Dioxygenase-catalyzed dechlorination can occur with chlorinated benzoates, benzenes, and biphenyls as substrates; dioxygenation at a chlorine-substituted carbon results in the elimination of chloride and formation of a catechol (Reactions C and D, Table 10.1) [16–19]. Nitrite, ammonia, or sulfite elimination has been demonstrated with certain enzymes when provided with nitroaromatic, aminoaromatic, or sulfoaromatic substrates (Reactions E–H, Table 10.1) [20–25]. With these substrates, the displacement reactions also result in the formation of catecholic products. Certain members of this large family of multicomponent enzymes catalyze angular dioxygenation of certain multiring substrates [26,27] (Reaction I, Table 10.1). Substrates that are oxidized by angular dioxygenation include carbazole, diphenylethers, dibenzofuran, and dibenzo-*p*-dioxin [28–32]. In addition, a small subset of ring hydroxylating dioxygenases function as monooxygenases with their native substrates. Such enzymes include salicylate 5-hydroxylase [33] and salicylate 1-hydroxylase [34–36] (Reaction P, Table 10.1), 2-oxo-1,2-dihydroquinoline-8-monooxygenase [37] and toluene sulfonate methyl-monooxygenase [38] (Reaction K, Table 10.1), methoxybenzoate monooxygenase [39], vanillate demethylase [40–44], and 5,5′-dehydrodivanillic acid *O*-demethylase [45], the latter three reactions involving an *O*-demethylation (Reaction N, Table 10.1). Three other reaction types have been documented primarily for NDO, TDO, and in some cases, carbazole dioxygenase. These include benzylic hydroxylation [46–54] (Reaction J, Table 10.1), oxygen-dependent desaturation [46,51,52,54,55] (Reaction L, Table 10.1), sulfoxidation [13,49,50,56–58] (Reaction M, Table 10.1), and *N*-dealkylation [59,60] (Reaction O, Table 10.1). Finally, dioxygen-dependent alcohol oxidation (Reaction Q, Table 10.1) has been reported for purified NDO [46].

10.3 CHARACTERISTICS OF AROMATIC-RING-HYDROXYLATING DIOXYGENASES

Aromatic-ring-hydroxylating dioxygenases are capable of oxidizing a wide range of substrates [60,61]. Specific enzymes demonstrate a remarkable diversity both in the number of substrates oxidized and in the types of reactions catalyzed. Our understanding of the substrate

TABLE 10.1
Reactions Catalyzed by Aromatic-Ring-Hydroxylating Dioxygenases

Reaction Type	Substrate	Enzyme	Product
A. cis-Dihydroxylation	Naphthalene	Naphthalene dioxygenase / O_2	Naphthalene cis-1,2-dihydrodiol
B. cis-Dihydroxylation	Benzoate	Benzoate dioxygenase / O_2	Benzoate cis-1,2-dihydrodiol
C. cis-Dihydroxylation and dehalogenation	Chlorobenzene	Chlorobenzene dioxygenase / O_2	Catechol
D. cis-Dihydroxylation, dehalogenation and decarboxylation	2-Chlorobenzoate	Chlorobenzoate dioxygenase / O_2	Catechol
E. cis-Dihydroxylation and nitrite elimination	Nitrobenzene	Nitrobenzene dioxygenase / O_2	Catechol
F. cis-Dihydroxylation and deamination	Aniline	Aniline dioxygenase / O_2	Catechol
G. cis-Dihydroxylation, deamination and decarboxylation	Anthranilate	Anthranilate dioxygenase / O_2	Catechol
H. cis-Dihydroxylation, and desulfonation	p-Sulfobenzoate	p-Sulfobenzoate dioxygenase / O_2	Protocatechua
I. Angular dihydroxylation	Dibenzofuran	Dibenzofuran dioxygenase / O_2	2,2',3-Trihydroxy biph

TABLE 10.1 (continued)
Reactions Catalyzed by Aromatic-Ring-Hydroxylating Dioxygenases

Reaction Type	Substrate	Enzyme	Product
J. Benzylic hydroxylation	Indene	Naphthalene dioxygenase / O_2	Indenol
K. Methyl group hydroxylation	2-Nitrotoluene	Naphthalene dioxygenase / O_2	2-Nitrobenzyl alcohol
L. Oxygen-dependent desaturation	Indan	Naphthalene dioxygenase / O_2	Indene
M. Sulfoxidation	Methyl phenyl sulfide	Naphthalene dioxygenase / O_2	Methyl phenylsulfoxide
N. O-Dealkylation	4-Methoxybenzoate	Naphthalene dioxygenase / O_2	4-Hydroxybenzoate
O. N-Dealkylation	N-Methylaniline	Naphthalene dioxygenase / O_2	Aniline
P. Net aromatic-ring hydroxylation	Salicylate	Salicylate 5-hydroxylase / O_2	Gentisate
Q. Dioxygen-dependent alcohol oxidation	(S)-1-Phenethyl alcohol	Naphthalene dioxygenase / O_2	Acetophenone

Source: Adapted from Parales, R.E. and Resnick, S.M. in *Pseudomonas Volume 4: Molecular Biology of Emerging Issues*, Academic/Plenum, New York, 2006. With permission.

specificities of various aromatic hydrocarbon dioxygenases has been based on the studi
wild-type strains, blocked mutant strains that accumulate the products of dioxy
catalyzed reactions, recombinant strains expressing cloned dioxygenase genes, and in
cases with purified enzymes. While over 100 aromatic-ring-hydroxylating dioxygenase
been reported, much of our knowledge on substrate specificity has been obtained t
detailed studies with relatively few enzymes that include TDOs, NDOs, biphenyl
genases (BPDOs), chlorobenzene dioxygenase, and carbazole dioxygenase. The ra
substrates and selectivity of reactions catalyzed make these aromatic hydrocarbon
genases particularly interesting for a variety of applications.

10.3.1 MULTICOMPONENT NATURE

Rieske nonheme iron oxygenases are two- or three-component enzyme systems. A
either one or two electron transport proteins that transfer electrons from NAD(P)H
catalytic oxygenase component. Therefore, to set up an *in vitro* biotransformation sys
least two proteins must be purified and recombined in an optimized ratio. In ad
a source of reductant is continuously required. To date, protein components from
aromatic hydrocarbon dioxygenase systems have been purified and studied in detail,
ing those of the NDO system from *Pseudomonas* sp. NCIB 9816-4 [62–64]. Dur
reaction catalyzed by NDO, electrons are transferred sequentially from NAD(P)H
reductase, to the ferredoxin, to the Rieske center of the oxygenase and finally to the
the active site (Figure 10.1). Two electrons are necessary to complete the reaction cyc
The 35 kD reductase contains one molecule of FAD and a plant-type iron–sulfur center
accept electrons from either NADH or NADPH [63,66]. The Rieske iron–sulfur
containing ferredoxin is a monomer of approximately 11 kD [62,66]. The oxygenas
$\alpha_3\beta_3$ hexamer [67] in which each α subunit contains a Rieske [2Fe-2S] center and
nuclear Fe^{2+} at the active site. When individual subunits of the oxygenase were purifi
reconstituted [68,69], both subunits were found to be essential for activity; similar resul
seen with TDO [70] and BPDO [71].

10.3.2 BROAD SUBSTRATE RANGE

The substrate specificities of several aromatic-ring-hydroxylating dioxygenases hav
investigated in detail. NDO from *Pseudomonas* sp. NCIB 9816-4 catalyzes the oxida
more than 75 different substrates [60] by not only *cis*-dihydroxylation (dioxyge
reactions, but also benzylic monohydroxylation (monooxygenation [53,72]), desatr
[52,54], *O*- and *N*-dealkylation [55,59], and sulfur oxidation (sulfoxidation [58]) re
(Table 10.1). TDO has the ability to catalyze the same range of reaction types as

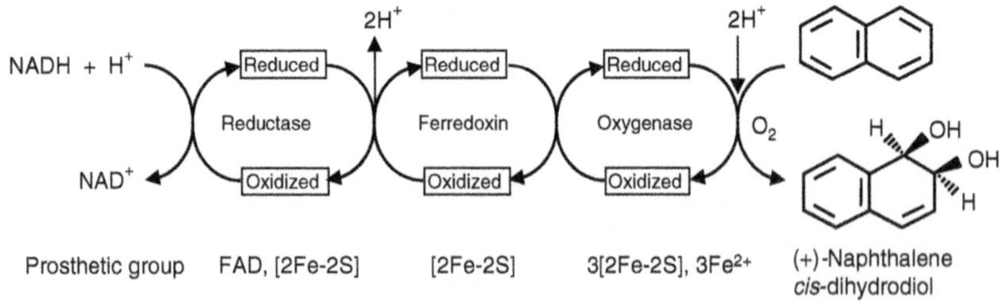

FIGURE 10.1 Reaction catalyzed by the three-component NDO enzyme system from *Pseudom*
NCIB 9816–4.

(Table 10.1). TDOs from *P. putida* F1 and *P. putida* UV4 have equally impressive substrate ranges, with the ability to catalyze the oxidation of more than 100 substrates, including monocyclic aromatic compounds, polycyclic and heterocyclic aromatic compounds, substituted aromatics, conjugated mono- and polyalkenes, and a variety of halogenated and nonhalogenated aliphatic olefins [73–76]. Table 10.2 shows an extensive list of substrates oxidized by aromatic-ring-hydroxylating dioxygenases, the reaction types catalyzed (Table 10.1), and relevant references, which indicate the enzyme(s) known to catalyze the reactions. The specificities of NDO and TDO often overlap, but in general, TDO prefers smaller (1–3–ring) substrates, while NDO prefers larger (2–4–ring) substrates. The BPDO from *Sphingomonas yanoikuyae* strain B1 (isolated for its ability to grow on biphenyl) has been shown to catalyze *cis*-dihydroxylation of a wide range of polycyclic and heterocyclic aromatic compounds including not only naphthalene, phenanthrene, anthracene, acenaphthene, but also high molecular weight polycyclics such as benzo[a]pyrene, benzo[a]anthracene, chrysene and benzo[b]-naphtha[2,1-d]thiophene [77–82]. In fact, *cis*-dihydrodiols from the latter two substrates as well as from acridine and phenazine undergo sequential *cis*-dihydroxylation to yield *bis-cis*-diol metabolites [81,83,84]. The substrate specificity of carbazole dioxygenase from *P. resinovorans* CA10 has also been investigated. This enzyme, which catalyzes angular dioxygenation (Reaction I, Table 10.1) on its native substrate, is capable of both *cis*-dihydroxylation and angular dioxygenation with several aromatic hydrocarbons and heterocycles [56].

10.3.3 REGIOSELECTIVITY AND ENANTIOSELECTIVITY

As a group, aromatic-ring-hydroxylating dioxygenases exhibit relaxed specificity as evidenced by their vast and structurally diverse substrate range. Yet, the substrates that are accepted are generally oxidized with both high regioselectivity and high enantioselectivity. Studies with different dioxygenases and a series of benzocyclic aromatic compounds suggest that the types of reactions catalyzed are determined by the fit of the given molecule within the active site. Examples of this point are illustrated by detailed characterization of oxidation products formed by TDO and NDO from indan, indene and related compounds. TDO oxidizes indan to (1*R*)-indanol and converts indene to *cis*-(1*S*, 2*R*)-indandiol and (1*R*)-indenol [53]. In contrast, NDO oxidizes indan to (1*S*)-indanol and oxidizes indene to *cis*-(1*R*, 2*S*)-indandiol and (1*S*)-indenol [52]. Such studies also illustrate the trends observed in product enantioselectivity for specific reaction types that is often a conserved intrinsic characteristic for a particular dioxygenase.

Some of the trends in enantioselectivity have been well characterized for the *cis*-dihydroxylation, benzylic monohydroxylation and sulfoxidation reactions catalyzed by NDO [60], and other aromatic-ring-hydroxylating dioxygenases. For example, NDO typically catalyzes benzylic hydroxylation and sulfoxidation to yield benzylic alcohols and sulfoxides of (*S*)-configuration [60,76]. In contrast, TDO typically catalyzes benzylic hydroxylation and sulfoxidation to generate products of (*R*)-configuration [57,58,85]. This enantiocomplimentarity for TDO and NDO has been documented for benzylic hydroxylation of 2-indanol [86] and chiral indanols [87]. Other organisms have been identified that are capable of generating enantiopure *cis*-diols of opposite chirality to that of previously identified metabolites. For example, a carbazole-utilizing strain oxidized biphenyl, biphenylene, and 9-fluorenone to previously unobserved *cis*-dihydrodiol or angular monohydrodiol enantiomers [88,89].

Recent reports showed that strains expressing TDO and NDO catalyze dihydroxylation of conjugated monoalkene and polyalkenes to yield the corresponding monols and enantiopure cyclic *cis*-diols [76]. Both the dioxygenase and the alkene substrates were important factors in determining the preference for 1,2-dihydroxylation of conjugated alkene or arene groups, and monohydroxylation at benzylic or allylic centers. These factors are illustrated with styrene and a

TABLE 10.2
Reactions Catalyzed by Rieske Nonheme Iron Oxygenases for Specific Substrates

Substrate	Type of Oxidation Reaction[a]																
	Dihydroxylations											Other Oxidations					
	A	B	C	D	E	F	G	H	I	J	K	L	M	N	O	P	Q
Aromatic hydrocarbons																	
Naphthalene [66,168–173]	✓																
Benzene [174–176]	✓																
Toluene [4,46,177,178]	✓										✓						
Ethylbenzene [46,179]	✓											✓					
Styrene [90,180–183]	✓									✓							
Isopropylbenzene [184–186]	✓																
o-Xylene [46,187]	✓										✓						
Biphenyl [188–194]	✓																
Anthracene [56,195,196]	✓																
Phenanthrene [196–198]	✓																
9,10-Dihydrophenanthrene [47,48]	✓									✓							
Chrysene [77,81,83]	✓																
Pyrene [199–201]	✓																
Fluoranthene [56,202]	✓																
Acenaphthene [49]	✓									✓							
Fluorene [50,203,204]	✓									✓							
Substituted aromatic compounds																	
Aniline [22,205–207]						✓											
Anthramilate [23,208]							✓										
Benzoate [209–211]		✓															
Salicylate [33–36]																✓	
o-Halobenzoates [212–215]		✓	✓														
Chlorobenzoates [19,215,216]	✓	✓	✓	✓													
Isopropylbenzoate [217]																	

Dihalobenzenes (F, Cl, Br, I) [222]

Halotoluenes (F, Cl, Br, I) [222]

(poly)Chlorobiphenyls [223–227]

Phthalate (*o*-) [228–232]

Isophthalate [233]

Terephthalate [233–235]

Toluates (*o*-, *m*-) [236]

4-Methoxybenzoate [237]

Vanillate [41,42,44]

(*S*)-1-Phenethyl alcohol [46]

p-Toluene sulfonate [38,238]

Aminobenzene sulfonate [239]

p-Sulfobenzoate [25]

Nitrobenzene [240]

2-Nitrotoluene [20,241–243]

3-Nitrotoluene [240,241]

4-Nitrotoluene [240,241]

2,4-Dinitrotoluene [240,244]

2,6-Dinitrotoluene [127,245]

2,2'-Dinitrobiphenyl [246]

Benzonitrile [94,183]

1,2,4-Trimethylbenzene [49]

1-,2-Substituted naphthalenes [247–251]

Dimethylnaphthalenes [49,247]

7-Oxodehydroabietic acid [252]

5,5'-Dihydrodivanillic acid [45]

Heterocyclic aromatic compounds

Carbazole [28,56,229,253–255]

Dibenzofuran [27,50,246,256–258]

Dibenzodioxin [246,259–262]

Dibenzothiophene [13,50,56]

Indole [11]

Fluorenone [49,204]

Quinolines [61,141]

continued

TABLE 10.2 (contiued)
Reactions Catalyzed by Rieske Nonheme Iron Oxygenases for Specific Substrates

Substrate	Type of Oxidation Reaction[a]																
	Dihydroxylations											Other Oxidations					
	A	B	C	D	E	F	G	H	I	J	K	L	M	N	O	P	Q
2-Oxo-1,2-dihydroquinoline [37,263]																	√
Pyrazon [264]	√																
3-Methyl benzothiophene [49]											√		√				
2-Methylbenzo-1,3-thiole [57]													√				
Carbocyclic, alkyl-aryl ether, thioether, or N-alkyl substrates																	
Indan [51–53]										√		√					
Indene [52,144,265]	√									√		√					
1,2-Dihydronaphthalene [54,266]	√									√		√					
Tetralin [267,268]	√									√							
Methyl phenyl sulfide [57,58]													√				
Ethyl phenyl sulfide [57,58]													√				
Methyl p-tolyl sulfide [58]													√				
p-Methoxyphenyl methyl sulfide [58]													√				
Methyl p-nitrophenyl sulfide [58]													√				
Anisole [55]	√																
Phenetole [55]	√																
Carboxydiphenylethers [269]									√								
N-Methylindole [60]															√		
N-Methylaniline [59]														√	√		
N,N-Dimethylaniline [59]														√	√		

[a]The types of reactions are shown in Table 10.1: A, *cis*-Dihydroxylation (C=C); B, *cis*-Dihydroxylation (at and adjacent to a carboxyl bearing carbon); C, *cis*-Dihydroxylation and dehalogenation; D, *cis*-Dihydroxylation, decarboxylation, and dehalogenation; E, *cis*-Dihydroxylation and nitrite elimination; F, *cis*-Dihydroxylation and deamination; G, *cis*-Dihydroxylation, deamination, and decarboxylation; H, Dihydroxylation and desulfonation; I, Angular dihydroxylation; J, Benzylic hydroxylation; K, Methyl group hydroxylation; L, Oxygen-dependent desaturation; M, Sulfoxidation; N, O-Dealkylation; O, N-Dealkylation; P, Net aromatic-ring monohydroxylation; Q, Dioxygen-

series of *meta*-substituted styrenes that were substrates for TDO-catalyzed arene ring *cis*-dihydroxylation, which yielded dihydrodiols of >98% ee. In contrast, NDO catalyzed exclusive formation of alkenadiols of (1*R*)-absolute configuration. With four of the seven substrates, the alkanediols were observed to accompany the ring dihydroxylation products formed by TDO [76]. The conversion of styrene to (*R*)-1-phenyl-1-,2-ethanediol has been demonstrated with purified NDO [46,90]. In a separate example, TDO-catalyzed dihydroxylation and benzylic monohydroxylation of 2-methylindene yielding enantiopure (1*S*, 2*R*)-dihydroxy-2*S*-methylindan (26% yield), small amounts of (1*R*)-2-methylinden-1-ol (~1%) and the isomerization product 2-methyl-1-indanone (~57%), while NDO preferentially catalyzed monohydroxylation at the benzylic center yielding (1*S*)-2-methylinden-1-ol (67%, >98% ee) without isomerization to the ketone. Allylic hydroxylation yielded 2-hydroxy-methylindene as a minor product (5%). These trends observed for the absolute configuration of the mono- and dihydroxylation products formed by TDO and NDO were consistent with those previously reported for the enzymes with related benzocyclic substrates [47,53,54,87,91,92].

Chlorobenzene dioxygenase from *Pseudomonas* sp. strain P51 was initially noted for its ability to oxidize 1,2-dichlorobenzene, naphthalene, and a wide range of substituted benzenes, toluenes, biphenyls, and 3-ring aromatic *O*-heterocyclic compounds [17,93]. Chlorobenzene dioxygenase has also been developed as an efficient recombinant biocatalyst for *cis*-dihydroxylation of benzyl cyanide, cinnamonitrile and a range of substituted benzonitriles; the enantiomeric purities of the *cis*-dihydroxylated products ranged from 43 to 97% ee [94].

Although crystal structures of several aromatic hydrocarbon dioxygenases are now available [67,95–99], the prediction of enzyme substrate specificity based on sequence or structure data is not yet reliable. The most useful approaches to identify an enzyme capable of catalyzing a desired reaction still remain (1) characterization of dioxygenase specificity profiles using diagnostic substrates, and (2) screening different enzymes on particular substrates for the desired target reaction or products.

10.4   BIOCATALYSIS STRATEGIES

Several properties associated with aromatic-ring-hydroxylating dioxygenases pose challenges for their application to the targeted preparation of oxidation products. These include their cofactor requirements, multicomponent nature, and low-specific activity and stability. The requirement for reduced nicotinamide cofactor(s) (NAD(P)H) combined with the multicomponent nature of the iron–sulfur containing subunits demand that an intricate balance be met to maintain enzyme stability and sustained productivity for the oxygenase-catalyzed reactions. Many monooxygenase and dioxygenase reactions have been demonstrated using purified oxygenase components with reduced cofactor supplied either through direct addition (stochiometric amounts) or via cofactor regeneration systems. In these cases, the specific activities of purified enzymes are typically considered low (nmol/min/mg) when compared with other single component, cofactor-independent enzymes such as hydrolases (μmol/min/mg), which have been widely applied for bioresolution of pharmaceutical intermediates. As a result, applications based on aromatic-ring-hydroxylating dioxygenases have been based largely on whole-cell biotransformations typically with enzymes overexpressed in recombinant hosts such as *Escherichia coli*.

10.4.1   WHOLE-CELL BIOPROCESSES

Due to the cofactor requirements and increased efforts associated with the use of purified enzyme components or cell-free preparations, biocatalytic applications employing aromatic-ring-hydroxylating dioxygenases have been predominantly developed using whole-cell

systems. These whole-cell biotransformation systems are typically facilitated by the i
overexpression of multicomponent dioxygenases in recombinant host strains, such a
or *Pseudomonas* sp., which can be grown to high cell densities. These approaches ha
employed for the application of oxygenases in bioprocesses attaining reasonably hi
metric yields [74,100,101].

10.4.2   COFACTOR REGENERATION

The whole-cell or resting-cell biotransformation approach has facilitated the produ
multikilogram quantities of chiral metabolites and relies on the integrity of multicon
dioxygenase activity, typically expressed in recombinant hosts with reduced c
supplied through the metabolism of exogenous carbon substrates (e.g., glucose,
pyruvate). A number of elegant approaches have been developed to enable bio
with oxygenases in cell-free systems. The use of purified oxygenases can be facili
including systems for enzymatic [102] or electrochemical [103] regeneration of
NAD(P)H cofactor. The direct regeneration of flavin-dependent monooxygenases
been demonstrated through the use of an organometallic redox complex [104]. T
common methods for enzymatic regeneration of nicotinamide coenzyme include
dehydrogenase (FDH) for NAD(P)H [102,105], alcohol dehydrogenase [106], and
6-phosphate dehydrogenase for NADPH [107]. FDH has been the most commonly
enzymatic regeneration system owing to its formation of carbon dioxide and speci
either NADH or NADPH recycle. Examples of oxidoreductases utilized with
coupled cofactor recycle include 2-hydroxybiphenyl monooxygenase [102,108], styren
oxygenase [109], and cyclohexanone monooxygenase [106].

Indirect electrochemical regeneration of NADH was employed with the flavin-de
2-hydroxybiphenyl-3-monooxygenase to give unoptimized rates that were ~50% of th
process [103]. Recently, direct chemical regeneration was achieved for the $FADH_2$-de
styrene monooxygenase (StyA) utilizing formate and a nonnative organometallic redox
[104]. Initial productivities for styrene monooxygenase-catalyzed epoxidation of ind
compared with fully enzymatic (FDH) and chemoenzymatic regeneration of FADH
regeneration systems were also demonstrated to replace NAD(P)H for several heme-co
P450 monooxygenases. Although not demonstrated for ring hydroxylating dioxygen
direct (electro)chemical regeneration offers potential for "cofactor-free" chemoer
biocatalysis using isolated oxygenase enzymes for chemical synthesis.

10.4.3   REACTION AND PROCESS ENGINEERING APPROACHES

The oxygen-dependent alcohol oxidation reaction catalyzed by purified NDO was
demonstrated by the oxidation of benzyl alcohol to benzaldehyde and 1-phenylethyl
to acetophenone (the corresponding aldehydes were also observed as produc
nitrotoluenes and xylenes) [46]. This reaction was also catalyzed with benzyl alco
3,4-dimethylbenzyl alcohol using recombinant xylene monooxygenase, XylMA, exp
E. coli [110]. Xylene monooxygenase catalyzed the multistep oxidation of one methyl
toluenes, xylenes, and pseudocumene to corresponding alcohols, aldehydes, and ac
has been used to develop aqueous-organic two-liquid phase (ATP) systems to exploit
and control product formation in multistep biooxidations [111,112]. The applic
recombinant XylMA-based whole-cell catalysis in ATP systems with fed–batch cu
using *bis*-(2-ethylhexyl)phthalate as carrier solvent, has been scaled to 30 L for the pre
conversion of pseudocumene to 3,4-dimethylbenzaldehyde at a concentration of 37
product was isolated in 97% purity and 65% overall yield [113]. The ATP system has

advantages for the regulation of concentrations (and associated toxicity) of apolar substrates and products in the aqueous microenvironment for the control of product formation and for facilitating product recovery [112]. Many aspects critical to the successful implementation of oxidative biocatalysis (e.g., high enzyme activity and specificity, product degradation, cofactor recycling, reactant toxicity, and substrate and oxygen mass transfer) can be addressed through biocatalyst and bioprocess engineering approaches, which have been recently reviewed [100].

10.4.4 BIOCATALYST IMPROVEMENT

Enzymes with new or improved activity or stability have been developed by a variety of laboratory engineering approaches, including "rational design" based on available or modeled crystal structures, saturation mutagenesis, error-prone PCR, and directed evolution of enzymes using DNA shuffling [114–122]. A promising new mutagenesis method, "combinatorial active-site saturations test" (CAST), involves saturation mutagenesis of spatially close pairs of amino acids at the active site, to allow more extensive active site reshaping with relatively small enzyme libraries [123]. Examples of the use of some of these methods to modify oxygenases follow.

Site-directed mutagenesis of amino acids located near the hydrophobic active site pocket of NDO from *Pseudomonas* sp. NCIB 9816-4 demonstrated that the enzyme was able to tolerate a wide range of single amino acid substitutions near the active site [124,125]. Enzymes with substitutions at position 352 of the oxygenase α subunit (phenylalanine in the wild-type enzyme) had altered enantioselectivity with naphthalene, biphenyl, phenanthrene and anthracene, and changes in the regioselectivity with biphenyl and phenanthrene [124,125]. For example, replacement of Phe352 with smaller amino acids (Gly, Ala, Val, Ile, Leu, and Thr) resulted in enzymes that produced increased amounts of biphenyl *cis*-3,4-dihydrodiol relative to biphenyl *cis*-2,3-dihydrodiol. In addition, the NDO-F352V and NDO-F352T enzymes formed significant amounts of (-)-biphenyl *cis*-(3S, 4R)-dihydrodiol, a compound not produced by wild-type NDO. Enzymes with substitutions at position 206 (NDO-A206I and NDO-A206I/L253T) formed significantly more phenanthrene *cis*-1,2-dihydrodiol than wild type, and several of the enzymes formed phenanthrene *cis*-9,10-dihydrodiol, a product not formed by the wild type [126].

In addition to the reported role of Phe352 in NDO from *Pseudomonas* sp. NCIB 9816-4 [124,125], substitution of the corresponding residue (Phe350) in NDO from *Ralstonia* sp. U2 with a threonine resulted in an enzyme with the ability to remove nitrite from 2,3- and 2,6-dinitrotoluene, but not from 2,4-dinitrotoluene [127]. Similarly, substitution of Val350 with either Phe or Met in the 2,4-dinitrotoluene dioxygenase from *Burkholderia cepacia* R34 resulted in enzymes with improved activities with substituted phenols [128]. Changing the corresponding residues in *B. xenovorans* LB400 BPDO and *Ralstonia* sp. PC12 tetrachlorobenzene dioxygenase (TecA) α subunits (Phe378 and Phe366, respectively) resulted in a reduction in overall activity, a change in the regiospecificity with chlorobiphenyl substrates for BPDO [129], and an increased preference for methyl group monooxygenation with mono- and dichlorotoluenes by TecA [130]. TecA also had a slightly increased preference for ring attack on 2,4,5-trichlorotoluene [130]. Substitution of tyrosine or tryptophan at this position either in TecA or in NDO resulted in inactive enzymes [125,130].

In another site-directed mutagenesis study, Phe227, Ile335, Thr376, and Phe377 in the α subunit of BPDO from *P. pseudoalcaligenes* KF707 were found to be important for determining the position of oxidation with various polychlorinated biphenyl (PCB) congeners [131]. Previous studies also demonstrated that Thr376 played an important role in determining regiospecificity with PCBs [132,133]. A variant large subunit biphenyl dioxygenase gene, *bphA*, was evolved by DNA shuffling of the corresponding genes from *P. pseudoalcc* KF707 and *B. xenovorans* LB400 [134]. The amino acid sequence of the evolved differed by 4 and 15 amino acids when compared with those of KF707 and LB400, r ively [134]. When expressed in *E. coli*, this variant showed extended substrate specificit ability to dihydroxylate several phenyl-substituted heterocyclic compounds that we substrates for the parent strains; the new substrates included 1-phenylpyrazole, 2-phen idine, 4-phenylpyrimidine, 1-benzylimidazole, 4-phenylisothiazole, 2-benzylpyridine, a: eral other heterocyclic aromatic compounds [134].

Directed evolution was conducted with TDO to obtain variants that produce amounts of the indene by-products (1-indenol and 1-indenone), while maintaining dih ylation activity to yield (-)-*cis*-(1*S*, 2*R*)-indandiol in high ee [135]. Three rounds of m esis yielded variants producing more *cis*-indandiol relative to the indenol by-produ enantioselectivity was altered to favor the production of the undesired (+)-*cis*-inc enantiomer [135] (see Section 5.3).

10.5   APPLICATIONS

Aromatic hydrocarbon dioxygenases have proven useful in a number of biotech applications. While they catalyze an impressive array of different reaction types, the selective *cis*-dihydroxylation of nonactivated aromatic compounds is unique to this e class and has been the basis for most applications. Examples include dioxygenase-ca synthesis of chiral intermediates for the preparation of natural products, polyfunctio metabolites, and pharmaceutical intermediates; expression of recombinant NDO in a neered bacterial strain for the production of indigo from glucose; and target-specific l radation of environmental pollutants.

10.5.1   Enantioselective Synthesis Using Dioxygenase-Derived Synthons

Dioxygenase-derived chiral metabolites have been utilized in a variety of multistep synthe the preparation of fine chemicals, natural products, pharmaceutical intermediates, a logically active compounds (Figure 10.2). The use of enzymatically formed *cis*-diols in e selective synthesis has been the subject of a comprehensive review [74] detailing synthetic rationales, listing *cis*-diols accessible via dioxygenase-catalyzed *cis*-dihydroxation an application in asymmetric methodology for the synthesis of a wide range of cyclitols, c itols, conduramines, inositols, heteroatom carbohydrates, alkaloids, and a variety of products. The *cis*-dihydrodiols of dictamnine and 4-chlorofuroquinoline yielded p derivatives from which a range of furoquinoline alkaloids were synthesized [136]. A su of recent progress in the synthesis of morphine alkaloids included the use of several meta derived via dioxygenase biocatalysis. Cyclohexadiene *cis*-dihydrodiols of phenethyl b and bromobenzene, as well as 3-bromocatechol (produced by a strain overexpressing TI DDH), have been employed as synthons in two separated synthetic strategies to p advanced intermediates for the synthesis of morphine alkaloids [137]. The biooxida 4-bromoanisole by recombinant *E. coli* expressing TDO and DDH yielded *p*-methoxyl catechol. This functionalized catechol was coupled with trimethoxyphenylacetyl convergent syntheses of combretastatins A-1 and B-1, members of a class of oxygenated products with potent cytotoxic activity [138]. An efficient chemoenzymatic synth strawberry furanone (4-hydroxy-2,5-dimethyl-2,3-dihydrofuran-3-one), a naturally occ flavor compound, was enabled through directed evolution of TDO and tetrachlorob dioxygenase operons that yielded improved enzymes for the conversion of *p*-xylene requisite diol synthon, *cis*-1,2-dihydroxy-3,6-dimethyl-3,5-cyclohexadiene (Figure 10.3a

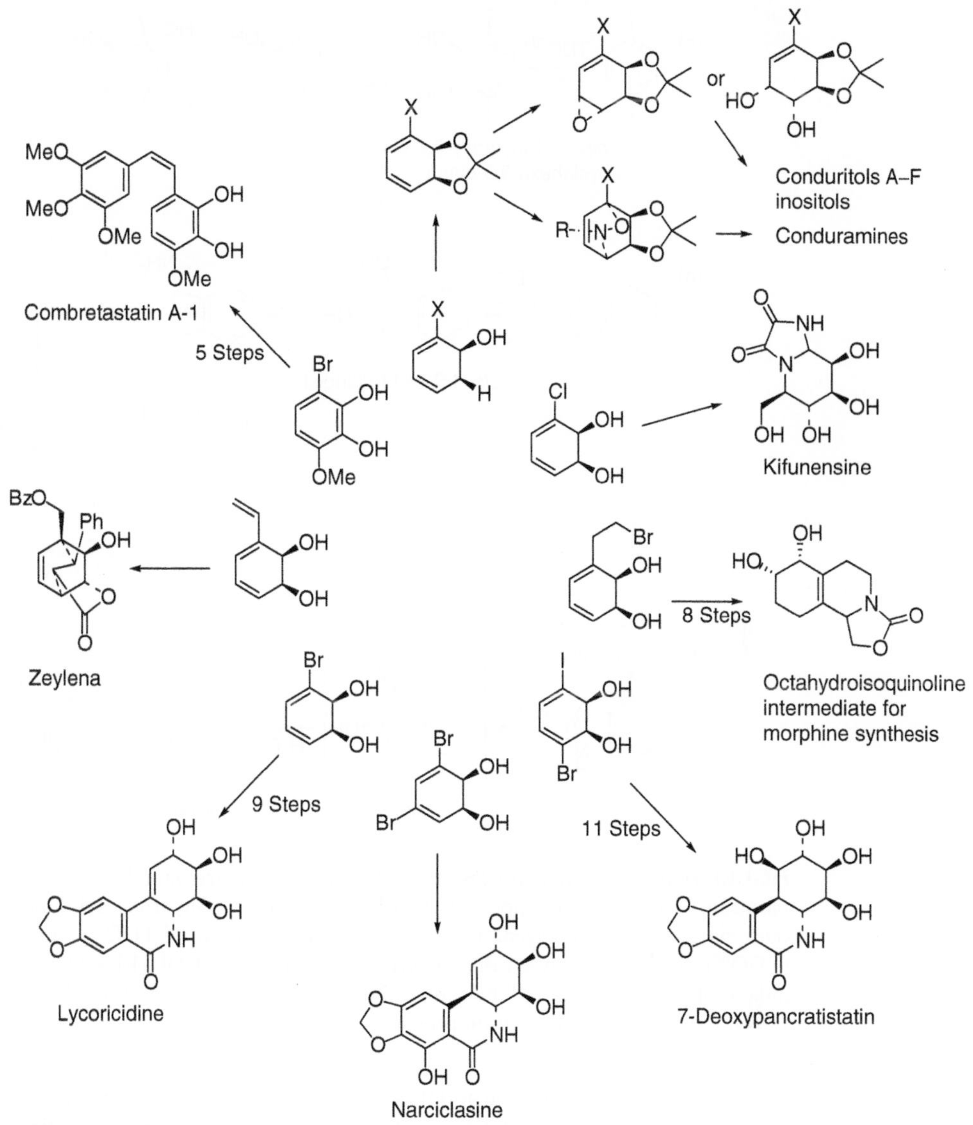

FIGURE 10.2 Examples of natural products, natural product intermediates, and analogs prepared via chemoenzymatic synthesis from the aromatic hydrocarbon dioxygenase-derived *cis*-dihydrodiols or catechols shown in the center. Hudlicky and coworkers provide an extensive review of specific synthetic applications to products from *cis*-dihydroarenediols. (From Hudlicky, T., Gonzalez, D., and Gibson, D.T., *Aldrichimica Acta*, 32, 35–62, 1999.)

10.5.2 DIOXYGENASE-CATALYZED REACTIONS FOR LEAD DEVELOPMENT AND NATURAL PRODUCT MODIFICATION

While *cis*-dihydrodiol metabolites, and in some cases *bis-cis*-diols [81,83,84], have been isolated from PAHs, benzo- and carbocyclic alkenes and polyenes, azaarenes, quinolines, and other heterocyclic aromatic compounds, the bacterial *cis*-dihydroxylation of an alkaloid natural product was reported only recently. Biotransformation of the parent furoquinoline alkaloid dictamnine by BPDO in *S. yanoikuyae* strain B8/36 yielded the first isolable alkaloid *cis*-dihydrodiol

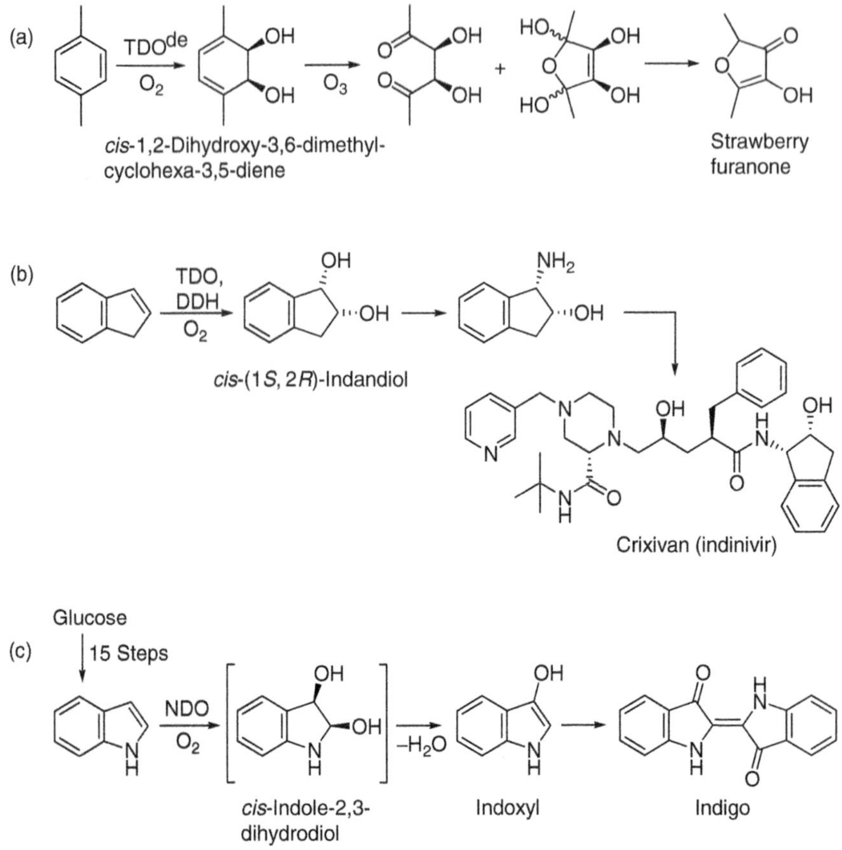

(a) ... TDO^{de} / O_2 → cis-1,2-Dihydroxy-3,6-dimethyl-cyclohexa-3,5-diene → (O_3) → + → Strawberry furanone

(b) ... TDO, DDH / O_2 → cis-(1S, 2R)-Indandiol → ... → Crixivan (indinivir)

(c) Glucose / 15 Steps ... NDO / O_2 → [cis-Indole-2,3-dihydrodiol] → (−H_2O) Indoxyl → Indigo

FIGURE 10.3 Biocatalytic processes developed based on aromatic hydrocarbon dioxygenases. paration of the flavor compound strawberry furanone (4-hydroxy-2,5-dimethyl-2,3-dihydro one) via enzymatic dioxygenation of *p*-xylene by a variant of TDO obtained by directed e (TDO^{de}). (b) TDO-catalyzed dihydroxylation of indene and DDH-based resolution to (*cis*)-(indandiol preparation of indinivir. (c) NDO-catalyzed conversion of indole to indigo in a reco *E. coli* (glucose-fed) fermentation process.

metabolites [136]; dictamnine and its synthetic precursor, 4-chlorofuroquinoline, were su for asymmetric dihydroxylation each yielding two enantiopure *cis*-dihydrodiol regio Absolute configuration of the (7S, 8R)- and (5R, 6S)-*cis*-dihydrodiols, correlated ^{1}H-NMR analysis of diastereomeric boronates prepared with (R)- and (S)-2-(1-methox phenylboronic acid [140] and comparison of CD spectra, was consistent with those establi the *cis*-dihydrodiols formed from quinoline and acridines [84,141]. Small quantities of diols of variable enantiopurity were also observed and presumed to result from BPDO-ca dihydroxylation of the dictamnine furan ring yielding a dihydrodiol undergoing spon reversible ring-opening to the aldehyde that is enzymatically reduced to the observed similar metabolic sequence [furan → furan *cis*-diol ↔ aldehyde↔ furan *trans*-diol → acy observed during dihydroxylation of benzofuran [142] and benzothiophene [143] inv equilibrium mixture of *cis/trans* diols and would result in the variable enantiopurity of the acyclic diol. This approach illustrates the potential use of dioxygenase-catalyzed *cis*-dihy tion to access and expand the chemical diversity of not only plant alkaloids but also other products; such metabolites may be of considerable value for lead modification and oxy nalization to afford synthons useful in the preparation of complex molecules with int biological activities.

10.5.3 Indinavir Production

Biocatalytic production of enantiopure (-)-*cis*-(1*S*, 2*R*)-indandiol is of interest due to the availability of a reaction allowing direct conversion to *cis*-(1*S*)-amino-(2*R*)-indanol (Figure 10.3B), which is a key intermediate in the chemical synthesis of Merck's HIV-1 protease inhibitor Indinivir Sulfate (Crixivan) [144]. Whole-cell biotransformations conducted with toluene-induced *P. putida* F39/D (a mutant strain lacking DDH activity) or a recombinant TDO expressed in *E. coli* demonstrated that wild-type TDO oxidized indene to (-)-*cis*-(1*S*, 2*R*)-indandiol (~30% ee) and (1*R*)-indenol as the main products, with traces of 1-indenone formed [53,145]. However, the (-)-*cis*-(1*S*, 2*R*)-indandiol was obtained in >98% ee in the late stages of indene conversion with wild-type *P. putida* F1 [144] or by coexpression of DDH together with TDO in *E. coli* [145]. The increased enantiopurity of (-)-*cis*-(1*S*, 2*R*)-indandiol was found to occur at the expense of total indandiol yield as a result of kinetic resolution catalyzed by DDH that is selective for the undesired (+)-*cis*-(1*R*, 2*S*)-indandiol [145]. Directed evolution was used to select the variants of TDO that produced reduced amounts of the indene by-products 1-indenol and 1-indenone, while maintaining high (-)-*cis*-(1*S*, 2*R*)-indandiol enantiopurity [135]. After three rounds of mutagenesis, variants that produced significantly more *cis*-indandiol relative to the undesired by-product indenol were obtained. However, stereoselectivity was altered to favor the production of the undesired (+)-*cis*-indandiol enantiomer [135]. Neither these strategies nor the application of oxygenases from *Rhodococcus* strains [146] eliminated formation of indene by-products, and the maximum yields were limited to <60% (-)-*cis*-(1*S*, 2*R*)-indandiol. An alternative route to the vicinal aminoindanol involved the TDO-catalyzed enantioselective monohydroxylation of 2-indanol to (-)-*cis*-(1*S*, 2*R*)-indandiol. This reaction served as the basis for a process to prepare chiral 1-hydroxy-2-substituted indan intermediates [147]. *P. putida* strains UV-4 [148] and F39/D expressing TDO oxidized 2-indenol to (-)-*cis*-(1*S*, 2*R*)-indandiol in >98% ee and >85% yield; minor products included *trans*-1,2-indandiol (<15%) and 2-hydroxy-1-indanone (<2%) [149].

10.5.4 Indigo Production

The oxidation of indole to indigo was first shown in recombinant *E. coli* strains expressing NDO from *Pseudomonas* sp. NCIB 9816-4 [11]. NDO oxidized indole to an unstable *cis*-dihydroindolediol that dehydrates to indoxyl, and subsequently undergoes spontaneous oxidation to indigo. This reaction is catalyzed by many related dioxygenases, and this simple colorimetric test has been widely utilized for detection and isolation of strains expressing mono- and dioxygenases, and in screening for mutants of these strains. Commercial interest in the reaction led Genencor International to genetically engineer a cost-competitive, multistep pathway for the production of indigo from glucose in *E. coli* (Figure 10.3C) [101]. The process for indigo production was based on a recombinant *E. coli* strain in which the tryptophan pathway was modified to allow a high level of indole production and cloned NDO from *P. putida* was expressed [101]. Numerous modifications were made to the strain to improve metabolite flux, eliminate the formation of the by-product isatin, and ultimately increase the production of indigo to levels exceeding 18 g/L. Despite the technical success of the process, the commercial production of indigo has not been implemented at an industrial scale.

10.5.5 Bioremediation of TCE Contamination

Trichloroethylene (TCE) is widely used as a solvent and is classified as a priority pollutant by the U.S. Environmental Protection Agency. TCE is difficult to degrade and some of its breakdown products are actually more toxic than TCE itself. TDO from *P. putida* F1 and other di- and monooxygenases are capable of oxidizing TCE [150–152], but at this time, no

bacteria are known to utilize TCE as a carbon and energy source. TDO converts TCE i
nontoxic products formate and glyoxylate [153], and a hybrid TDO–BPDO was fo
have enhanced TCE-degrading activity [154]. The purified hybrid protein was shown
higher catalytic efficiency and a lower K_m for TCE than wild-type TDO [152]. Althoug
functions as an inducer of the genes encoding TDO in *P. putida* F1 [155], resting c
exposed to TCE rapidly lose TCE oxidation activity [150]. A recent study demonstrat
the addition of benzene or toluene restored TCE-degrading activity to *P. putida* F1 cell
suggesting that TCE degradation could be improved with this strategy. A field trial m
ing TCE cooxidation suggested that this type of *in situ* bioremediation strategy m
feasible. In this study, toluene or phenol, and oxygen or hydrogen peroxide were ac
cosubstrates to stimulate TCE cooxidation by indigenous bacteria, and >90% of th
cis-dichloroethylene, and vinyl chloride were removed [157].

10.5.6   CONSTRUCTION OF NEW BIODEGRADATION PATHWAYS

Aromatic hydrocarbon dioxygenases, such as TDO, have been used to engineer new pa
for the degradation of recalcitrant compounds. Genes encoding TDO from *P. putida*
cytochrome P450cam monooxygenase were used to construct a *Pseudomonas* strain
metabolize polyhalogenated compounds by sequential reductive and oxidative reacti
this engineered pathway, cytochrome P450cam catalyzes the conversion of highly chlo
alkanes, such as pentachloroethane, to TCE under low oxygen tension, and TDO o
TCE to glyoxalate and formate [158]. This type of hybrid pathway might find applica
the clean up of sites contaminated with mixtures of polyhalogenated ethanes.

The *tod* genes encoding TDO from *P. putida* F1 were also cloned into the gene
Deinococcus radiodurans, an extremely radiation resistant bacterium. This recombinan
was able to degrade toluene and related aromatic hydrocarbons in the presence of higl
of radiation [159]. Expression of the mercury resistance gene (*merA*) together with th
genes in *Deinococcus radiodurans* resulted in an engineered strain with the ability to rer
mixed radioactive waste containing aromatic hydrocarbon pollutants and the heavy
mercury [160].

Many bacterial isolates are capable of degrading only a limited number of ar
hydrocarbon pollutants, and genetic engineering has been used to increase the su
range of specific microorganisms. When a constructed cassette, carrying genes for tl
version of styrene to phenylacetate, was introduced into *P. putida* F1 carrying th
plasmid, the engineered strain was capable of growth on an expanded range of ar
hydrocarbons, including benzene, toluene, ethylbenzene, *m*-xylene, *p*-xylene, and
[161]. To reduce the undesirable possibility of horizontal transfer of genetically engi
DNA among bacteria in the environment, the cassette was introduced to a minitran
carrying an engineered gene containment system [162]. Such a strain could prove usefu
bioremediation of low molecular weight aromatic hydrocarbon pollution.

10.6   OUTLOOK

We have tried to emphasize the versatility of aromatic-ring-hydroxylating dioxygenas
respect to the range of substrates attacked and the types of reactions catalyzed. Equally
more important, is the ability of many of these enzymes to form chiral products i
enantiomeric purity. To identify an enzyme that oxidizes a particular target and/or pro
specific product, one can attempt to isolate a new bacterial strain with the desired ab
screen the large number of well-characterized dioxygenases that are currently availal
alternative is to modify an available enzyme known to have a weak or similar acti

random mutagenesis methods [116,163–165] or by rational design, taking advantage of the growing number of available dioxygenase crystal structures [67,95–99]. Finally, screening of metagenomic libraries [166,167], especially with samples from diverse environments, may allow the identification of new dioxygenases with useful activities without the need to isolate and characterize the host bacterium.

Once an enzyme with the desired selectivity is obtained, numerous steps are still required to develop viable and economical commercial processes. Efficiency can be realized through high protein expression coupled with process development. Overall productivity may also be increased by improving the activity or thermostability of the enzyme, or by reducing product inhibition. Process design must take into consideration the multicomponent nature of the enzyme and the NADH requirement when determining whether to use purified enzymes or whole-cells. In spite of the multiple challenges involved in using multicomponent dioxygenases, this class of enzymes holds significant promise for the development of environmentally friendly processes for reactions that are difficult, inefficient, or otherwise challenging to carry out by standard chemical methods.

ACKNOWLEDGMENTS

We thank David Gibson for critical review of the manuscript and Juan Parales for preparing Table 10.1. Research in the Parales laboratory is supported by the Army Research Office, the Strategic Environmental Research and Development Program, and the University of California Toxic Substances Research and Teaching Program.

REFERENCES

1. Gibson, D.T. and Parales, R.E., Aromatic hydrocarbon dioxygenases in environmental biotechnology, *Curr. Opin. Biotechnol.*, 11, 236–243, 2000.
2. Butler, C.S. and Mason, J.R., Structure–function analysis of the bacterial aromatic ring-hydroxylating dioxygenases, *Adv. Microbial. Physiol.*, 38, 47–84, 1997.
3. Yeh, W.-K., Gibson, D.T., and Liu, T.-N., Toluene dioxygenase: a multicomponent enzyme system, *Biochem. Biophys. Res. Commun.*, 78, 401–410, 1977.
4. Subramanian, V., Liu, T.-N., Yeh, W.-K., and Gibson, D.T., Toluene dioxygenase: purification of an iron-sulfur protein by affinity chromatography, *Biochem. Biophys. Res. Commun.*, 91, 1131–1139, 1979.
5. Gibson, D.T., Koch, J.R., and Kallio, R.E., Oxidative degradation of aromatic hydrocarbons by microorganisms I. Enzymatic formation of catechol from benzene, *Biochemistry*, 7, 2653–2661, 1968.
6. Kobal, V.M., Gibson, D.T., Davis, R.E., and Garza, A., X-ray determination of the absolute stereochemistry of the initial oxidation product formed from toluene by *Pseudomonas putida* 39/D, *J. Am. Chem. Soc.*, 95, 4420–4421, 1973.
7. Gibson, D.T., Hensley, M., Yoshioka, H., and Mabry, T.J., Formation of (+)-*cis*-2,3-dihydroxy-1-methylcyclohexa-4,6-diene from toluene by *Pseudomonas putida*, *Biochemistry*, 9, 1626–1630, 1970.
8. Blumer, M., Polycyclic aromatic compounds in nature, *Sci. Am.*, 234, 35–41, 1976.
9. Gibson, D.T., Initial reactions to the bacterial degradation of aromatic hydrocarbons, *Zbl. Bakt. Hyg. I Abt. Orig. B.*, 162, 157–168, 1976.
10. Kiyohara, H., Nagao, K., and Yana, K., Rapid screen for bacteria degrading water-insoluble, solid hydrocarbons on agar plates, *Appl. Environ. Microbiol.*, 43, 454–457, 1982.
11. Ensley, B.D., Ratzkin, B.J., Osslund, T.D., Simon, M.J., Wackett, L.P., and Gibson, D.T., Expression of naphthalene oxidation genes in *Escherichia coli* results in the biosynthesis of indigo, *Science*, 222, 167–169, 1983.
12. Eaton, R.W. and Chapman, P.J., Formation of indigo and related compounds from indolecarboxylic acids by aromatic acid-degrading bacteria: chromogenic reactions for cloning genes encoding dioxygenases that act on aromatic acids, *J. Bacteriol.*, 177, 6983–6988, 1995.

13. Kodama, K., Umehara, K., Shimizu, K., Nakatani, S., Minoda, Y., and Yamada, K., I◖ tion of microbial products from dibenzothiophene and its proposed oxidation pathwa◖ *Biol. Chem.*, 37, 45–50, 1973.

14. Kim, E. and Zylstra, G.J., Functional analysis of genes involved in biphenyl, nap phenanthrene, and *m*-xylene degradation by *Sphingomonas yanoikuyae* B1., *J. Ind. ↳ Biotechnol.*, 23, 294–302, 1999.

15. Zylstra, G.J., Kim, E., and Goyal, A.K., Comparative molecular analysis of genes for p aromatic hydrocarbon degradation, *Genet. Eng.*, 19, 257–269, 1997.

16. Haddock, J.D., Horton, J.R., and Gibson, D.T., Dihydroxylation and dechlorination o◖ ated biphenyls by purified biphenyl 2,3-dioxygenase from *Pseudomonas* sp. strain *J. Bacteriol.*, 177, 20–26, 1995.

17. Werlen, C., Kohler, H.-P., and van der Meer, J.R., The broad substrate chlorobenzene d ase and *cis*-chlorobenzene dihydrodiol dehydrogenase of *Pseudomonas* sp. strain P51 a evolutionarily to the enzymes for benzene and toluene degradation, *J. Biol. Chem.*, 271, 40◖ 1996.

18. Fetzner, S., Müller, R., and Lingens, F., Purification and some properties of 2-halo◖ 1,2-dioxygenase, a two-component enzyme system from *Pseudomonas cepacia* 2CBS, *J. B* 174, 279–290, 1992.

19. Nakatsu, C.H. and Wyndham, C., Cloning and expression of the transposable chlorob◖ 3,4-dioxygenase genes of *Alcaligenes* sp. strain BR60, *Appl. Environ. Microbiol.*, 59, 36◖ 1993.

20. An, D., Gibson, D.T., and Spain, J.C., Oxidative release of nitrite from 2-nitrotoluene by component enzyme system from *Pseudomonas* sp. strain JS42, *J. Bacteriol.*, 176, 7462–74◖

21. Spanggord, R.J., Spain, J.C., Nishino, S.F., and Mortelmans, K.E., Biodegrad◖ 2,4-dinitrotoluene by a *Pseudomonas* sp., *Appl. Environ. Microbiol.*, 57, 3200–3205, 1991◖

22. Fukumori, F. and Saint, C., Nucleotide sequences and regulational analysis of genes in◖ conversion of aniline to catechol in *Pseudomonas putida* UCC22(pTDN1), *J. Bacteriol.*, ▮ 408, 1997.

23. Bundy, B.M., Campbell, A.L., and Neidle, E.L., Similarities between the *antABC*-encode◖ nilate dioxygenase and the *benABC*-encoded benzoate dioxygenase of *Acinetobacter* s◖ ADP1, *J. Bacteriol.*, 180, 4466–4474, 1998.

24. Nishino, S.F. and Spain, J.C., Oxidative pathway for the biodegradation of nitrobe◖ *Comamonas* sp. strain JS765, *Appl. Environ. Microbiol.*, 61, 2308–2313, 1995.

25. Locher, H.H., Leisinger, T., and Cook, A.M., 4-Sulphobenzoate 3,4-dioxygenase: purific◖ properties of a desulphonative two-component enzyme system from *Comamonas testoste* *Biochem. J.*, 274, 833–842, 1991.

26. Engesser, K.H., Strubel, V., Christoglou, K., Fischer, P., and Rast, H.G., Dioxygenolytic of aryl ether bonds: 1,10-dihydro-1,10-dihydroxyfluoren-9-one, a novel arene dihydr◖ evidence for angular dioxygenation of dibenzofuran, *FEMS Microbiol. Lett.*, 65, 205–21◖

27. Bünz, P.V. and Cook, A.M., Dibenzofuran 4,4a-dioxygenase from *Sphingomonas* sp. str◖ angular dioxygenation by a three-component enzyme system, *J. Bacteriol.*, 175, 6467–64◖

28. Sato, S.-I., Nam, J.-W., Kasuga, K., Nojiri, H., Yamane, H., and Omori, T., Identifica◖ characterization of genes encoding carbazole 1,9a-dioxygenase in *Pseudomonas* sp. stra◖ *J. Bacteriol.*, 179(15), 4850–4858, 1997.

29. Engesser, K.H., Fietz, W., Fischer, P., Schulte, P., and Knackmuss, H.-J., Dioxygenolytic of aryl ether bonds: 1,2-dihydro-1,2-dihydroxy-4-carboxybenzophenone as evidence f◖ 1,2-dioxygenation in 3- and 4-carboxy biphenyl ether degradation, *FEMS Microbiol. ↳* 317–322, 1990.

30. Wittich, R.-M., Wilkes, H., Sinnwell, V., Francke, W., and Fortnagel, P., Metabolism of *p*-dioxin by *Sphingomonas* sp. strain RW1, *Appl. Environ. Microbiol.*, 58, 1005–1010, 199◖

31. Nojiri, H. and Omori, T., Molecular bases of aerobic bacterial degradation of dioxins: inv◖ of angular dioxygenation, *Biosci. Biotechnol. Biochem.*, 66, 2001–2016, 2002.

32. Nojiri, H., Habe, H., and Omori, T., Bacterial degradation of aromatic compounds via dioxygenation, *J. Gen. Appl. Microbiol.*, 47, 279–305, 2001.

33. Zhou, N.-Y., Al-Dulayymi, J., Baird, M.S., and Williams, P.A., Salicylate 5-hydroxylase from *Ralstonia* sp. strain U2: a monooxygenase with close relationships to and shared electron transport proteins with naphthalene dioxygenase, *J. Bacteriol.*, 184, 1547–1555, 2002.

34. Demaneche, S., Meyer, C., Micoud, J., Louwagie, M., Willison, J.C., and Jouanneau, Y., Identification and functional analysis of two aromatic ring-hydroxylating dioxygenases from a *Sphingomonas* strain that degrades various polycyclic aromatic hydrocarbons, *Appl. Environ. Microbiol.*, 70, 6714–6725, 2004.

35. Pinyakong, O., Habe, H., Yoshida, T., Nojiri, H., and Omori, T., Identification of three novel salicylate 1-hydroxylases involved in the phenanthrene degradation of *Sphingobium* sp. strain P2, *Biochem. Biophys. Res. Commun.*, 301, 350–357, 2003.

36. Cho, O., Choi, K.Y., Zylstra, G.J., Kim, Y.-S., Kim, S.-K., Lee, J.H., Sohn, H.-Y., Kwon, G.-S., Kim, Y.M., and Kim, E., Catabolic role of a three-component salicylate oxygenase from *Sphingomonas yanoikuyae* B1 in polycyclic aromatic hydrocarbon degradation, *Biochem. Biophys. Res. Commun.*, 327, 656–662, 2005.

37. Rosche, B., Tshisuaka, B., Fetzner, S., and Lingens, F., 2-Oxo-1,2-dihydroquinoline 8-monooxygenase, a two-component enzyme system from *Pseudomonas putida* 86, *J. Biol. Chem.*, 270, 17836–17842, 1995.

38. Junker, F., Kiewitz, R., and Cook, A.M., Characterization of the *p*-toluenesulfonate operon *tsaMBCD* and *tsaR* in *Comamonas testosteroni* T-2, *J. Bacteriol.*, 179(3), 919–927, 1997.

39. Bernhardt, F.-H., Erdin, N., Staudinger, H., and Ullrich, V., Interactions of substrates with a purified 4-methoxybenzoate monooxygenase system (*O*-demethylating) from *Pseudomonas putida*, *Eur. J. Biochem.*, 35, 126–134, 1973.

40. Venturi, V., Zennaro, F., Degrassi, G., Okeke, B.C., and Bruschi, C.V., Genetics of ferulic acid bioconversion to protocatechuic acid in plant-growth-promoting *Pseudomonas putida* WCS358, *Microbiology*, 144, 965–973, 1998.

41. Segura, A., Bunz, P.V., D'Argenio, D.A., and Ornston, L.N., Genetic analyis of a chromosomal region containing *vanA* and *vanB*, genes required for conversion of either ferulate or vanillate to protocatechuate in *Acinetobacter*, *J. Bacteriol.*, 181, 3439–3504, 1999.

42. Brunel, F. and Davison, J., Cloning and sequencing of *Pseudomonas* genes encoding vanillate demethylase, *J. Bacteriol.*, 170, 4924–4930, 1988.

43. Civolani, C., Barghini, P., Roncetti, A.R., Ruzzi, M., and Schiesser, A., Bioconversion of ferulic acid into vanillic acid by means of a vanillate-negative mutant of *Pseudomonas fluorescens* strain BF13, *Appl. Environ. Microbiol.*, 66, 2311–2317, 2000.

44. Priefert, H., Rabenhorst, J., and Steinbüchel, A., Molecular characterization of genes of *Pseudomonas* sp. strain HR199 involved in bioconversion of vanillin to protocatechuate, *J. Bacteriol.*, 179(8), 2595–2607, 1997.

45. Sonoki, T., Obi, T., Kubota, S., Higashi, M., Masai, E., and Katayama, Y., Coexistence of two different *O* demethylation systems in lignin metabolism by *Sphingomonas paucimobilis* SYK-6: cloning and sequencing of the lignin biphenyl-specific *O*-demethylase (LigX) gene., *Appl. Environ. Microbiol.*, 66, 2125–2132, 2000.

46. Lee, K. and Gibson, D.T., Toluene and ethylbenzene oxidation by purified naphthalene dioxygenase from *Pseudomonas* sp. strain NCIB 9816-4, *Appl. Environ. Microbiol.*, 62, 3101–3106, 1996.

47. Boyd, D.R., Sharma, N.D., Kerley, N.A., McMordie, R.A.S., Sheldrake, G.N., Williams, P., and Dalton, H., Dioxygenase-catalyzed oxidation of dihydronaphthalenes to yield arene hydrate and *cis*-dihydronaphthalenediols, *J. Chem. Soc. Perkin Trans.* 1, 67–74, 1996.

48. Resnick, S.M. and Gibson, D.T., Regio- and stereospecific oxidation of 9,10-dihydroanthracene and 9,10-dihydrophenanthrene by naphthalene dioxygenase: structure and absolute stereochemistry of metabolites, *Appl. Environ. Microbiol.*, 62, 3355–3359, 1996.

49. Selifonov, S.A., Grifoll, M., Eaton, R.W., and Chapman, P.J., Oxidation of naphthenoaromatic and methyl-substituted aromatic compounds by naphthalene 1,2-dioxygenase, *Appl. Environ. Microbiol.*, 62, 507–514, 1996.

50. Resnick, S.M. and Gibson, D.T., Regio- and stereospecific oxidation of fluorene, dibenzofuran, and dibenzothiophene by naphthalene dioxygenase from *Pseudomonas* sp. strain NCIB 9816-4, *Appl. Environ. Microbiol.*, 62, 4073–4080, 1996.

51. Brand, J.M., Cruden, D.L., Zylstra, G.J., and Gibson, D.T., Stereospecific hydroxylation c by *E. coli* containing the cloned toluene dioxygenase genes from *Pseudomonas putida* F *Environ. Microbiol.*, 58, 3407–3409, 1992.

52. Gibson, D.T., Resnick, S.M., Lee, K., Brand, J.M., Torok, D.S., Wackett, L.P., Schocke and Haigler, B.E., Desaturation, dioxygenation and monooxygenation reactions ca by naphthalene dioxygenase from *Pseudomonas* sp. strain 9816-4, *J. Bacteriol.*, 177, 261 1995.

53. Wackett, L.P., Kwart, L.D., and Gibson, D.T., Benzylic monooxygenation catalyzed by dioxygenase from *Pseudomonas putida*, *Biochemistry*, 27, 1360–1367, 1988.

54. Torok, D.S., Resnick, S.M., Brand, J.M., Cruden, D.L., and Gibson, D.T., Desaturati oxygenation of 1,2-dihydronaphthalene by toluene and naphthalene dioxygenase, *J. Ba* 177, 5799–5805, 1995.

55. Resnick, S.M. and Gibson, D.T., Biotransformation of anisole and phenetole by hydrocarbon-oxidizing bacteria, *Biodegradation*, 4, 195–203, 1993.

56. Nojiri, H., Nam, J.-W., Kosaka, M., Morii, K., Takemura, T., Furihata, K., Yamane, Omori, T., Diverse oxygenations catalyzed by carbazole 1,9a-dioxygenase from *Pseudom* strain CA10, *J. Bacteriol.*, 181, 3105–3113, 1999.

57. Allen, C.C.R., Boyd, D.R., Dalton, H., Sharma, N.D., Haughey, S.A., McMordie, McMurray, B.T., Sheldrake, G.N., and Sproule, K., Sulfoxides of high enantiopurit bacterial dioxygenase-catalyzed oxidation, *J. Chem. Soc., Chem. Commun.*, 119–120, 1995

58. Lee, K., Brand, J.M., and Gibson, D.T., Stereospecific sulfoxidation by toluene and naph dioxygenases, *Biochem. Biophys. Res. Commun.*, 212, 9–15, 1995.

59. Lee, K., Biochemical studies on toluene and naphthalene dioxygenases, Ph.D. disse University of Iowa, Iowa City, Iowa, 1995.

60. Resnick, S.M., Lee, K., and Gibson, D.T., Diverse reactions catalyzed by naphthalene diox from *Pseudomonas* sp. strain NCIB 9816, *J. Ind. Microbiol.*, 17, 438–457, 1996.

61. Boyd, D.R., Sharma, N.D., Modyanova, L.V., Carroll, J.G., Malone, J.F., Allen, Hamilton, J.T.G., Gibson, D.T., Parales, R.E., and Dalton, H., Dioxygenase-ca *cis*-dihydroxylation of pyridine-ring systems, *Can. J. Chem.*, 80, 589–600, 2002.

62. Haigler, B.E. and Gibson, D.T., Purification and properties of ferredoxin$_{NAP}$, a compo naphthalene dioxygenase from *Pseudomonas* sp. strain NCIB 9816, *J. Bacteriol.*, 172, 4 1990.

63. Haigler, B.E. and Gibson, D.T., Purification and properties of NADH-ferredoxin$_{NAP}$ rec a component of naphthalene dioxygenase from *Pseudomonas* sp. strain NCIB 9816, *J. Ba* 172, 457–464, 1990.

64. Ensley, B.D. and Gibson, D.T., Naphthalene dioxygenase: purification and properti terminal oxygenase component, *J. Bacteriol.*, 155, 505–511, 1983.

65. Karlsson, A., Parales, J.V., Parales, R.E., Gibson, D.T., Eklund, H., and Ramaswamy, S., structure of naphthalene dioxygenase: side-on binding of dioxygen to iron, *Science*, 299 1043, 2003.

66. Simon, M.J., Osslund, T.D., Saunders, R., Ensley, B.D., Suggs, S., Harcourt, A., Suen, Cruden, D.L., Gibson, D.T., and Zylstra, G.J., Sequences of genes encoding naphthalene genase in *Pseudomonas putida* strains G7 and NCIB 9816-4, *Gene*, 127, 31–37, 1993.

67. Kauppi, B., Lee, K., Carredano, E., Parales, R.E., Gibson, D.T., Eklund, H., and Ramasw Structure of an aromatic ring-hydroxylating dioxygenase-naphthalene 1,2-dioxygenase, *St* 6, 571–586, 1998.

68. Suen, W.-C. and Gibson, D.T., Isolation and preliminary characterization of the subunit terminal component of naphthalene dioxygenase from *Pseudomonas putida* NCIB *J. Bacteriol.*, 175, 5877–5881, 1993.

69. Suen, W.-C. and Gibson, D.T., Recombinant *Escherichia coli* strains synthesize active f naphthalene dioxygenase and its individual α and β subunits, *Gene*, 143, 67–71, 1994.

70. Jiang, H., Parales, R.E., and Gibson, D.T., The α subunit of toluene dioxygenase from *monas putida* F1 can accept electrons from reduced ferredoxin$_{TOL}$ but is catalytically inacti absence of the β subunit, *Appl. Environ. Microbiol.*, 65, 315–318, 1999.

71. Hurtubise, Y., Barriault, D., and Sylvestre, M., Characterization of active recombinant His-tagged oxygenase component of *Comamonas testosteroni* B-356 biphenyl dioxygenase, *J. Biol. Chem.*, 271, 8152–8156, 1996.

72. Resnick, S.M., Torok, D.S., Lee, K., Brand, J.M., and Gibson, D.T., Regiospecific and stereoselective hydroxylation of 1-indanone and 2-indanone by naphthalene dioxygenase and toluene dioxygenase, *Appl. Environ. Microbiol.*, 60, 3323–3328, 1994.

73. Lange, C.C. and Wackett, L.P., Oxidation of aliphatic olefins by toluene dioxygenase: enzyme rates and product identification, *J. Bacteriol.*, 179, 3858–3865, 1997.

74. Hudlicky, T., Gonzalez, D., and Gibson, D.T., Enzymatic dihydroxylation of aromatics in enantioselective synthesis: expanding asymmetric methodology, *Aldrichimica Acta*, 32, 35–62, 1999.

75. Boyd, D.R. and Sheldrake, G.N., The dioxygenase-catalysed formation of vicinal *cis*-diols, *Nat. Prod. Rep.*, 15, 309–324, 1998.

76. Boyd, D.R., Sharma, N.D., Bowers, N.I., Brannigan, I.N., Groocock, M.R., Malone, J.F., McConville, G., and Allen, C.C.R., Biocatalytic asymmetric dihydroxylation of conjugated mono- and poly-alkenes to yield enantiopure cyclic *cis*-diols, *Adv. Synth. Catal.*, 347, 1081–1089, 2005.

77. Gibson, D.T., *Beijerinckia* sp. strain B1: a strain by any other name, *J. Ind. Microbiol. Biotechnol.*, 23, 284–293, 1999.

78. Schocken, M.J. and Gibson, D.T., Bacterial oxidation of the polycyclic aromatic hydrocarbons acenaphthene and acenaphthylene, *Appl. Environ. Microbiol.*, 48, 10–16, 1984.

79. Mahaffey, W.R., Gibson, D.T., and Cerniglia, C.E., Bacterial oxidation of chemical carcinogens: formation of polycyclic aromatic acids from benz[a]anthracene, *Appl. Environ. Microbiol.*, 54, 2415–2423, 1988.

80. Gibson, D.T., Roberts, R.L., Wells, M.C., and Kobal, V.M., Oxidation of biphenyl by a *Beijerinckia* species, *Biochem. Biophys. Res. Commun.*, 50, 211–219, 1973.

81. Boyd, D.R., Sharma, N.D., Agarwal, R., Resnick, S.M., Schocken, M.J., Gibson, D.T., Sayer, J.M., Yagi, H., and Jerina, D.M., Bacterial dioxygenase-catalyzed dihydroxylation and chemical resolution routes to enantiopure *cis*-dihydrodiols of chrysene, *J. Chem. Soc. Perkin Trans.* 1, 1715–1723, 1997.

82. Jerina, D.M., Van Bladeren, P.J., Yagi, H., Gibson, D.T., Mahadevan, V., Neese, A.S., Koreeda, M., Sharma, N.D., and Boyd, D.R., Synthesis and absolute configuration of the bacterial *cis*-1,2-, *cis*-8,9-, and *cis*-10,11-dihydrodiol metabolites of benz[a]anthracene formed by a strain of *Beijerinckia*, *J. Org. Chem.*, 49, 3621–3628, 1984.

83. Boyd, D.R., Sharma, N.D., Hempenstall, F., Kennedy, M.A., Malone, J.F., Allen, C.C.R., Resnick, S.M., and Gibson, D.T., *bis-cis*-Dihydrodiols: a new class of metabolites resulting from biphenyl dioxygenase-catalyzed sequential asymmetric *cis*-dihydroxylation of polycyclic arenes and heteroarenes, *J. Org. Chem.*, 64, 4005–4011, 1999.

84. Boyd, D.R., Sharma, N.D., Carroll, J.G., Allen, C.C.R., Clarke, D.A., and Gibson, D.T., Multiple site dioxygenase-catalysed *cis*-dihydroxylation of polycyclic azaarenes to yield a new class of *bis-cis*-diol metabolites, *Chem. Commun.*, 1201–1202, 1999.

85. Finn, K.J., Cankafi, P., Jones, T.R.B., and Hudlicky, T., Enzymatic oxidation of thioanisoles: isolation and absolute configuration of metabolites, *Tetrahedron Asymmetry*, 15, 2833–2836, 2004.

86. Bowers, N.I., Boyd, D.R., Sharma, N.D., Goodrich, P.A., Groocock, M.R., Blacker, A.J., Goode, P., and Dalton, H., Stereoselective benzylic hydroxylation of 2-substituted indanes using toluene dioxygenase as biocatalysts, *J. Chem. Soc. Perkin Trans.* 1, 1453–1461, 1999.

87. Lee, K., Resnick, S.M., and Gibson, D.T., Stereospecific oxidation of (*R*)- and (*S*)-indanol by naphthalene dioxygenase from *Pseudomonas* sp. strain NCIB 9816-4, *Appl. Environ. Microbiol.*, 63, 2067–2070, 1997.

88. Resnick, S., Diverse reactions catalyzed by multicomponent aromatic hydrocarbon dioxygenases, Ph.D. thesis, University of Iowa, Iowa City, 1997.

89. Resnick, S.M. and Gibson, D.T., Oxidation of polycyclic aromatic compounds and formation of novel *cis*-diol enantiomers by carbazole utilizing pseudomonads, in *Abstract of the 96th General Meeting of the American Society for Microbiology*, New Orleans, LA, O-11, 1996.

90. Lee, K. and Gibson, D.T., Stereospecific dihydroxylation of the styrene vinyl group by naphthalene dioxygenase from *Pseudomonas* sp. strain 9816-4, *J. Bacteriol.*, 178, 3353–33

91. Resnick, S.M. and Gibson, D.T., Oxidation of 6,7-dihydro-5H-benzocycloheptene by strains expressing naphthalene dioxygenase, biphenyl dioxygenase, and toluene dioxygena homochiral monol or *cis*-diol enantiomers as major products, *Appl. Environ. Microbiol.*, (1368, 1996.

92. Boyd, D.R., Dorrity, M.R.J., Malone, J.F., McMordie, R.A.S., Sharma, N.D., Dalton, Williams, P., Bacterial metabolism of 6,7-dihydro-5H-benzocycloheptene by *Pseudomona* Synthesis and absolute configuration of benzylic alcohol and *cis*-diol metabolites of 6,7-5H-benzocycloheptene, *J. Chem. Soc. Perkin. Trans. I*, 1990, 489–494, 1990.

93. Raschke, H., Meier, M., Burken, J.G., Hany, R., Muller, M.D., van der Meer, J.R., and Kohl Biotransformation of various substituted aromatic compounds to chiral dihydrodihydrox tives, *Appl. Environ. Microbiol.*, 67, 3333–3339, 2001.

94. Yildirim, S., Franco, T.T., Wohlgemuth, R., Kohler, H.-P.E., Witholt, B., and Sch Recombinant chlorobenzene dioxygenase from *Pseudomonas* sp. P51: a biocatalyst for re tive oxidation of aromatic nitriles, *Adv. Synth. Catal.*, 347, 1060–1072, 2005.

95. Furusawa, Y., Nagarajan, V., Tanokura, M., Masai, E., Fukuda, M., and Senda, T. structure of the terminal oxygenase of biphenyl dioxygenase from *Rhodococcus* sp. strai *J. Mol. Biol.*, 342, 1041–1052, 2004.

96. Friemann, R., Ivkovic-Jensen, M.M., Lessner, D.J., Yu, C.-L., Gibson, D.T., Paral Eklund, H., and Ramaswamy, S., Structural insights into the dioxygenation of nitroar pounds: the crystal structure of the nitrobenzene dioxygenase, *J. Mol. Biol.*, 348, 1139–11

97. Nojiri, H., Ashikawa, Y., Noguchi, H., Nam, J.-W., Urata, M., Fujimoto, Z., Uchir Terada, T., Nakamura, S., Shimizu, K., Yoshida, T., Habe, H., and Omori, T., Structu terminal oxygenase component of angular dioxygenase, carbazole 1,9a-dioxygenase, *J. M* 351, 355–370, 2005.

98. Dong, X., Fushinobu, S., Fukuda, E., Terada, T., Nakamura, S., Shimizu, K., Nojiri, H., O Shoun, H., and Wakagi, T., Crystal structure of the terminal oxygenase component of dioxygenase from *Pseudomonas fluorescens* IP01, *J. Bacteriol.*, 187, 2483–2490, 2005.

99. Martins, B.M., Svetlitchnaia, T., and Dobbek, H., 2-Oxoquinoline 8-monooxygenase o component: active site modulation by Rieske-[2Fe-2S] center oxidation/reduction, *Struc* 817–824, 2005.

100. Buhler, B. and Schmid, A., Process implementation aspects for biocatalytic hydrocarbon tionalization, *J. Biotechnol.*, 113, 183–210, 2004.

101. Berry, A., Dodge, T.C., Pepsin, M., and Weyler, W., Application of metabolic engineering to both the production and use of biotech indigo, *J. Ind. Microbiol. Biotechnol.*, 28, 127–133,

102. Schmid, A., Vereyken, I., Held, M., and Witholt, B., Preparative regio- and chemc functionalization of hydrocarbons catalyzed by cell free preparations of 2-hydroxy 3-monooxygenase, *J. Mol. Catal., B Enzym.*, 11, 455–462, 2001.

103. Hollmann, F., Schmid, A., and Steckhan, E., First synthetic application of a monoo employing indirect electrochemical NADH regeneration, *Angew. Chem. Int. Ed.*, 40, 2001.

104. Hollmann, F., Lin, P.-C., Witholt, B., and Schmid, A., Stereospecific biocatalytic epoxida first example of direct regeneration of a FAD-dependent monooxygenase for catalysis *Chem. Soc.*, 125, 8209–8217, 2003.

105. Rissoma, S., Schwarz-Linekb, U., Vogelb, M., Tishkovc, V.I., and Kragl, U., Synthesis epsilon-lactones in a two-enzyme system of cyclohexanone mono-oxygenase and forma drogenase with integrated bubble-free aeration, *Tetrahedron Asymmetry*, 8, 2523–2526, 1

106. Zambianchi, F., Pasta, P., Carrea, G., Colonna, S., Gaggero, N., and Woodley, J.M. isolated cyclohexanone monooxygenase from recombinant *Escherichia coli* as a biocat Baeyer–Villiger and sulfide oxidations, *Biotechnol. Bioeng.*, 78, 489–496, 2002.

107. Wong, C.-H. and Whitesides, G.M., Enzyme-catalyzed organic synthesis: NAD(P)H regeneration by using glucose-6-phosphate and the glucose-5-phosphate dehydrogena *Leuconostoc mesenteroides*, *J. Am. Chem. Soc.*, 103, 4890–4899, 1981.

108. Lutz, J., Mozhaev, V.V., Khmelnitsky, Y.L., Witholt, B., and Schmid, A., Preparative application of 2-hydroxybiphenyl 3-monooxygenase with enzymatic cofactor regeneration in organic-aqueous reaction media, *J. Mol. Catal., B Enzym.*, 19/20, 177–187, 2002.

109. Hofstetter, K., Lutz, J., Lang, I., Witholt, B., and Schmid, A., Biocatalytic asymmetric epoxidation with NADH regeneration in organic-aqueous emulsions, *Angew. Chem. Int. Ed. Engl.*, 43, 2163–2166, 2004.

110. Buhler, B., Schmid, A., Hauer, B., and Witholt, B., Xylene monooxygenase catalyzes the multistep oxygenation of toluene and pseudocumene to corresponding alcohols, aldehydes, and acids in *Escherichia coli* JM101, *J. Biol. Chem.*, 275, 10085–10092, 2000.

111. Buhler, B., Witholt, B., Hauer, B., and Schmid, A., Characterization and application of xylene monooxygenase for multistep biocatalysis, *Appl. Environ. Microbiol.*, 68, 560–568, 2002.

112. Buhler, B., Bollhalder, I., Hauer, B., Witholt, B., and Schmid, A., Use of the two-liquid phase concept to exploit kinetically controlled multistep biocatalysis, *Biotechnol. Bioeng.*, 81, 683–694, 2003.

113. Buhler, B., Bollhalder, I., Hauer, B., Witholt, B., and Schmid, A., Chemical biotechnology for the specific oxyfunctionalization of hydrocarbons on a technical scale, *Biotechnol. Bioeng.*, 82, 833–842, 2003.

114. Harayama, S., Artificial evolution by DNA shuffling, *Trends Biotechnol.*, 16, 76–82, 1998.

115. Crameri, A., Raillard, S.-A., Bermudez, E., and Stemmer, W.P.C., DNA shuffling of a family of genes from diverse species accelerates directed evolution, *Nature*, 391, 288–291, 1998.

116. Cirino, P.C. and Arnold, F.H., Protein engineering of oxygenases for biocatalysis, *Curr. Opin. Chem. Biol.*, 6, 130–135, 2002.

117. Zhao, H. and Arnold, F.H., Optimization of DNA shuffling for high fidelity recombination, *Nucleic Acids Res.*, 6, 1307–1308, 1997.

118. Lutz, S. and Ostermeier, M., Preparation of SCRATCHY hybrid protein libraries: size- and in-frame selection of nucleic acid sequences, *Methods Mol. Biol.*, 231, 143–151, 2003.

119. Ostermeier, M. and Lutz, S., The creation of ITCHY hybrid protein libraries, *Methods Mol. Biol.*, 231, 129–141, 2003.

120. Cirino, P.C., Mayer, K.M., and Umeno, D., Generating mutant libraries using error-prone PCR, *Methods Mol. Biol.*, 231, 3–9, 2003.

121. Georgescu, R., Bandara, G., and Sun, L., Saturation mutagenesis, *Methods Mol. Biol.*, 231, 75–83, 2003.

122. Coco, W.M., RACHITT: gene family shuffling by random chimeragenesis on transient templates, *Methods Mol. Biol.*, 231, 111–127, 2003.

123. Reetz, M.T., Bocola, M., Carballeira, J.D., Zha, D., and Vogel, A., Expanding the range of substrate acceptance of enzymes: combinatorial active-site saturation test, *Angew. Chem. Int. Ed. Engl*, 44, 4192–4196, 2005.

124. Parales, R.E., Lee, K., Resnick, S.M., Jiang, H., Lessner, D.J., and Gibson, D.T., Substrate specificity of naphthalene dioxygenase: effect of specific amino acids at the active site of the enzyme, *J. Bacteriol.*, 182, 1641–1649, 2000.

125. Parales, R.E., Resnick, S.M., Yu, C.L., Boyd, D.R., Sharma, N.D., and Gibson, D.T., Regioselectivity and enantioselectivity of naphthalene dioxygenase during arene *cis*-dihydroxylation: control by phenylalanine 352 in the α subunit, *J. Bacteriol.*, 182, 5495–5504, 2000.

126. Yu, C.-L., Parales, R.E., and Gibson, D.T., Multiple mutations at the active site of naphthalene dioxygenase affect regioselectivity and enantioselectivity, *J. Ind. Microbiol. Biotechnol.*, 27, 94–103, 2001.

127. Keenan, B.G., Leungsakul, T., Smets, B.F., Mori, M.A., Henderson, D.E., and Wood, T.K., Protein engineering of the archetypal nitroarene dioxygenase of *Ralstonia* sp. strain U2 for activity on aminonitrotoluenes and dinitrotoluenes through alpha-subunit residues leucine 225, phenylalanine 350, and glycine 407, *J. Bacteriol.*, 187, 3302–3310, 2005.

128. Keenan, B.G., Leungsakul, T., Smets, B.F., and Wood, T.K., Saturation mutagenesis of *Burkholderia cepacia* R34 2,4-dinitrotoluene dioxygenase at DntAc valine 350 for synthesizing nitrohydroquinone, methylhydroquinone, and methoxyhydroquinone, *Appl. Environ. Microbiol.*, 70, 3222–3231, 2004.

129. Zielinski, M., Kahl, S., Hecht, H.J., and Hofer, B., Pinpointing biphenyl dioxygenase residu are crucial for substrate interaction, *J. Bacteriol.*, 185, 6976–6980, 2003.

130. Pollmann, K., Wray, V., Hecht, H.-J., and Pieper, D.H., Rational engineering of the reg tivity of TecA tetrachlorobenzene dioxygenase for the transformation of chlorinated to *Microbiology*, 149, 903–913, 2003.

131. Suenaga, H., Watanabe, T., Sato, M., Ngadiman, and Furukawa, K., Alteration of regiospe in biphenyl dioxygenase by active-site engineering, *J. Bacteriol.*, 184, 3682–3688, 2002.

132. Mondello, F.J., Turcich, M.P., Lobos, J.H., and Erickson, B.D., Identification and modifica biphenyl dioxygenase sequences that determine the specificity of polychlorinated biphenyl ation, *Appl. Environ. Microbiol.*, 63, 3096–3103, 1997.

133. Kimura, N., Nishi, A., Goto, M., and Furukawa, K., Functional analyses of a variety of c dioxygenases constructed from two biphenyl dioxygenases that are similar structurally bu ent functionally, *J. Bacteriol.*, 179, 3936–3943, 1997.

134. Misawa, N., Shindo, K., Takahashi, H., Suenaga, H., Iguchi, K., Okazaki, H., Harayama, Furukawa, K., Hydroxylation of various molecules including heterocyclic aromatics using binant *Escherichia coli* cells expressing modified biphenyl dioxygenase genes, *Tetrahedr* 9605–9612, 2002.

135. Zhang, N., Stewart, B.G., Moore, J.C., Greasham, R.L., Robinson, D.K., Buckland, B. Lee, C., Directed evolution of toluene dioxygenase from *Pseudomonas putida* for in selectivity toward *cis*-indandiol during indene bioconversion, *Metab. Eng.*, 2, 339–348, 20

136. Boyd, D.R., Sharma, N.D., O'Dowd, C.R., Carroll, J.G., Loke, P.L., and Allen, C.C. Dihydrodiol, arene oxide and phenol metabolites of dictamnine: key intermediates in the radation and biosynthesis of furoquinoline alkaloids, *Chem. Commun.*, 31, 3989–3991, 20

137. Zezula, J. and Hudlicky, T., Recent progress in the synthesis of morphine alkaloids, *Sy* 388–405, 2005.

138. Bui, V.P., Hudlicky, T., Hansen, T.V., and Stenstrom, Y., Direct biooxidation of ar corresponding catechols with *E. coli* JM109(pDTG602). Application to synthesis of comb tins A-1 and B-1, *Tetrahedron Lett.*, 43, 2839–2841, 2002.

139. Newman, L.M., Garcia, H., Hudlicky, T., and Selifonov, S.A., Directed evolution of the genase complex for the synthesis of furanone flavor compounds, *Tetrahedron*, 60, 7 2004.

140. Resnick, S.M., Torok, D.S., and Gibson, D.T., Chemoenzymatic synthesis of chiral boron the ^{1}H NMR determination of the absolute configuration and enantiomeric excess of bacte synthetic *cis*-diols, *J. Org. Chem.*, 60, 3546–3549, 1995.

141. Boyd, D.R., Sharma, N.D., Dorrity, M.R.J., Hand, M.V., McMordie, R.A.S., Malon Porter, H.P., Chima, J., Dalton, H., and Sheldrake, G.N., Structure and stereochemistry dihydrodiol and phenol metabolites of bicyclic azaarenes from *Pseudomonas putida* UV4, *J Soc. Perkin Trans.* 1, 1065–1071, 1993.

142. Boyd, D.R., Sharma, N.D., Brannigan, I.N., Haughey, S.A., Malone, J.F., Clarke and Dalton, H., Dioxygenase-catalysed formation of *cis/trans*-dihydrodiol metabolites of and bi-cyclic heteroarenes, *Chem. Commun.*, 2361–2362, 1996.

143. Eaton, R.W. and Nitterauer, J.D., Biotransformation of benzothiophene by isopropylb degrading bacteria, *J. Bacteriol.*, 176, 3992–4002, 1994.

144. Connors, N., Prevoznak, R., Chartrain, M., Reddy, J., Singhvi, R., Patel, Z., Olewin Salmon, P., Wilson, J., and Greasham, R., Conversion of indene to *cis*-(1*S*, 2*R*)-indan mutants of *Pseudomonas putida* F1, *J. Ind. Microbiol. Biotechnol.*, 18, 353–359, 1997.

145. Reddy, J., Lee, C., Neeper, M., Greasham, R., and Zhang, J., Development of a biocon process for production of *cis*-1*S*, 2*R*-indandiol from indene by recombinant *Escheric* constructs, *Appl. Microbiol. Biotechnol.*, 51, 614–620, 1999.

146. O'Brien, X.M., Parker, J.A., Lessard, P.A., and Sinskey, A.J., Engineering an indene bio sion process for the production of *cis*-aminoindanol: a model system for the production c synthons, *Appl. Microbiol. Biotechnol.*, 59, 389–399, 2002.

147. Blacker, A.J., Boyd, D.R., Dalton, H., and Bowers, N., WO 96/37628, Preparation of h compounds by conversion with dioxygenase. May 20, 1996.

148. Boyd, D.R., Sharma, N.D., Bowers, N.I., Goodrich, P.A., Groocock, M.R., Blaker, A.J., Clarke, D.A., Howard, T., and Dalton, H., Stereoselective dioxygenase-catalyzed benzylic hydroxylation at prochiral methylene groups in the chemoenzymatic synthesis of enantiopure vicinal aminoindanols, *Tetrahedron Asymmetry*, 7, 1559–1562, 1996.

149. Lakshman, M.K., Chaturvedi, S., Zajc, B., Gibson, D.T., and Resnick, S.M., A general chemoenzymatic synthesis of enantiopure *cis* β-amino alcohols from microbially derived *cis*-glycols, *Synthesis*, 9, 1352–1356, 1998.

150. Wackett, L.P. and Gibson, D.T., Degradation of trichloroethylene by toluene dioxygenase in whole cell studies with *Pseudomonas putida* F1, *Appl. Environ. Microbiol.*, 54, 1703–1708, 1988.

151. Kesseler, M., Dabbs, E.R., Averhoff, B., and Gottschalk, G., Studies on the isopropylbenzene 2,3-dioxygenase and 3-isopropylcatechol 2,3-dioxygenase genes encoding by the linear plasmid of *Rhodococcus erythropolis* BD2, *Microbiology*, 142, 3241–3251, 1996.

152. Maeda, T., Takahashi, Y., Suenaga, H., Suyama, A., Goto, M., and Furukawa, K., Functional analyses of Bph-Tod hybrid dioxygenase, which exhibits high degradation activity toward trichloroethylene, *J. Biol. Chem.*, 276, 29833–29838, 2001.

153. Li, S. and Wackett, L.P., Trichloroethylene oxidation by toluene dioxygenase, *Biochem. Biophys. Res. Commun.*, 185, 443–451, 1992.

154. Furukawa, K., Hirose, J., Hayashida, S., and Nakamura, K., Efficient degradation of trichloroethylene by a hybrid aromatic ring dioxygenase, *J. Bacteriol.*, 176(7), 2121–2123, 1994.

155. Shingleton, J.T., Applegate, B.M., Nagel, A.C., Bienkowski, P.R., and Sayler, G.S., Induction of the *tod* operon by trichloroethylene in *Pseudomonas putida* TVA8, *Appl. Environ. Microbiol.*, 64, 5049–5052, 1998.

156. Morono, Y., Unno, H., Tanji, Y., and Hori, K., Addition of aromatic substrates restores trichloroethylene degradation activity in *Pseudomonas putida* F1, *Appl. Environ. Microbiol.*, 70, 2830–2835, 2004.

157. Hopkins, G.D. and McCarty, P.L., Field evaluation of *in-situ* aerobic cometabolism of trichloroethene and three dichloroethene isomers using phenol and toluene as the primary substrates, *Environ. Sci. Technol.*, 29, 1628–1637, 1995.

158. Wackett, L.P., Sadowsky, M.J., Newman, L.M., Hur, H.-G., and Li, S., Metabolism of polyhalogenated compounds by a genetically engineered bacterium, *Nature*, 368, 627–629, 1994.

159. Lange, C.C., Wackett, L.P., Minton, K.W., and Daly, M.J., Engineering a recombinant *Deinococcus radiodurans* for organopollutant degradation in radioactive mixed waste environments, *Nat. Biotechnol.*, 16, 929–933, 1998.

160. Brim, H., McFarlan, S.C., Fredrickson, J.K., Minton, K.W., Zhai, M., Wackett, L.P., and Daly, M.J., Engineering *Deinococcus radiodurans* for metal remediation in radioactive mixed waste environments, *Nat. Biotechnol.*, 18, 85–90, 2000.

161. Lorenzo, P., Alonso, S., Velasco, A., Diaz, E., Garcia, J.L., and Perera, J., Design of catabolic cassettes for styrene biodegradation, *Antonie Van Leeuwenhoek.*, 84, 17–24, 2003.

162. Díaz, E., Munthali, M., de Lorenzo, V., and Timmis, K.N., Universal barrier to lateral spread of specific genes among microorganisms, *Mol. Microbiol.*, 13, 855–861, 1994.

163. Jaeger, K.E. and Eggert, T., Enantioselective biocatalysis optimized by directed evolution, *Curr. Opin. Biotechnol.*, 15, 305–313, 2004.

164. Otten, L.G. and Quax, W.J., Directed evolution: selecting today's biocatalysts, *Biomol. Eng.*, 22, 1–9, 2005.

165. Zhao, H., Chockalingam, K., and Chen, Z., Directed evolution of enzymes and pathways for industrial biocatalysis, *Curr. Opin. Biotechnol.*, 13, 104–110, 2002.

166. Daniel, R., The soil metagenome—a rich resource for the discovery of novel natural products, *Curr. Opin. Biotechnol.*, 15, 199–204, 2004.

167. Lorenz, P., Liebeton, K., Niehaus, F., and Eck, J., Screening for novel enzymes for biocatalytic processes: accessing the metagenome as a resource of novel functional sequence space, *Curr. Opin. Biotechnol.*, 13, 572–577, 2002.

168. Jeon, C.O., Park, W., Padmanabhan, P., DeRito, C., Snape, J.R., and Madsen, E.L., Discovery of a bacterium, with distinctive dioxygenase, that is responsible for *in situ* biodegradation in contaminated sediment, *Proc. Natl. Acad. Sci. USA*, 100, 13591–13596, 2003.

169. Larkin, M.J., Allen, C.C.R., Kulakov, L.A., and Lipscomb, D.A., Purification and cha⋅ tion of a novel naphthalene dioxygenase from *Rhodococcus* sp. strain NCIMB12038, *J. E* 181, 6200–6204, 1999.

170. Denome, S.A., Stanley, D.C., Olsen, E.S., and Young, K.D., Metabolism of dibenzothiop naphthalene in *Pseudomonas* strains: complete DNA sequence of an upper naphthalene pathway, *J. Bacteriol.*, 175, 6890–6901, 1993.

171. Takizawa, N., Kaida, N., Torigoe, S., Moritani, T., Sawada, T., Satoh, S., and Kiyol Identification and characterization of genes encoding polycyclic aromatic hydrocarbon c̓ ase and polycyclic aromatic hydrocarbon dihydrodiol dehydrogenase in *Pseudomon⋅* OUS82, *J. Bacteriol.*, 176, 2444–2449, 1994.

172. Bosch, R., Garcia-Valdes, E., and Moore, E.R.B., Complete nucleotide sequence and evo⋅ significance of a chromosomally encoded naphthalene-degradation lower pathway from *monas stutzeri* AN10, *Gene*, 245, 65–74, 2000.

173. Ensley, B.D., Gibson, D.T., and Laborde, A.L., Oxidation of naphthalene by a multico enzyme system from *Pseudomonas* sp. strain NCIB 9816, *J. Bacteriol*, 149, 948–954, 198

174. Irie, S., Doi, S., Yorifuji, T., Takagi, M., and Yano, K., Nucleotide sequencing and cha⋅ tion of the genes encoding benzene oxidation enzymes of *Pseudomonas putida*, *J. Bacter* 5174–5179, 1987.

175. Geary, P.J., Mason, J.R., and Joannou, C.L., Benzene dioxygenase from *Pseudomonas pu⋅* (NCIB 12190), *Methods Enzymol.*, 188, 52–60, 1990.

176. Tan, H.-M., Tang, H.-Y., Joannou, C.L., Abdel-Wahab, N.H., and Mason, J.R., The *P̓ nas putida* ML2 plasmid-encoded genes for benzene dioxygenase are unusual in codon u̓ low in G+C content, *Gene*, 130, 33–39, 1993.

177. Zylstra, G.J. and Gibson, D.T., Toluene degradation by *Pseudomonas putida* F1: n̓ sequence of the *todC1C2BADE* genes and their expression in *E. coli*, *J. Biol. Chem.*, 26⋅ 14946, 1989.

178. Mosqueda, G., Ramos-Gonzalez, M.I., and Ramos, J.L., Toluene metabolism by the tolerant *Pseudomonas putida* DOT-T1 strain, and its role in solvent impermeabilization, *G̓ 69–76, 1999.

179. Corkery, D.M. and Dobson, A.D., Reverse transcription-PCR analysis of the regu⋅ ethylbenzene dioxygenase gene expression in *Pseudomonas fluorescens* CA-4, *FEMS M Lett.*, 166, 171–176, 1998.

180. Warhurst, A.M., Clarke, K.F., Hill, R.A., Holt, R.A., and Fewson, C.A., Metabolism ⋅ by *Rhodococcus rhodochrous* NCIMB 13259, *Appl. Environ. Microbiol.*, 60, 1137–1145, 1

181. Junker, L. and Hay, A.G., Preliminary evidence for bacterial growth on styrene via pathway, in *Abstracts of the 102nd General Meeting of the American Society for Mic⋅ 2002*. American Society for Microbiology, 2002.

182. Junker, L. and Hay, A.G., Bacterial growth on styrene using a mosaic pathway, in *Abstr⋅ 103rd General Meeting of the American Society for Microbiology 2003*, American S⋅ Microbiology, 2003.

183. Boyd, D.R., Sharma, N.D., Byrne, B., Hand, M.V., Malone, J.F., Sheldrake, G.N., Bl̓ and Dalton, H., Enzymatic and chemoenzymatic synthesis and stereochemical assignme̓ dihydrodiol derivatives of monosubstituted benzenes, *J. Chem. Soc. Perkin Trans.*, 1, 1935–1̓

184. Eaton, R.W. and Timmis, K.N., Characterization of a plasmid-specified pathway for ca̓ of isopropylbenzene in *Pseudomonas putida* RE204, *J. Bacteriol.*, 168, 123–131, 1986.

185. Aoki, H., Kimura, T., Habe, H., Yamane, H.T.K., and Omori, T., Cloning, nucleotide s̓ and characterization of the genes encoding enzymes involved in the degradation of c̓ 2-hydroxy-6-oxo-7-methylocta-2,4-dienoic acid in *Pseudomonas fluorescens* IP01, *J. Bioeng.*, 81, 187–196, 1996.

186. Dabrock, B., Riedel, J., Bertram, J., and Gottschalk, G., Isopropylbenzene (cumene̓ substrate for the isolation of trichloroethene-degrading bacteria, *Arch. Microbiol.*, 158, 9–

187. Kim, D., Kim, Y.S., Kim, S.K., Kim, S.W., Zylstra, G.J., Kim, Y.M., and Kim, E., M⋅ aromatic hydrocarbon degradation by *Rhodococcus* sp. strain DK17, *Appl. Environ. Micr⋅ 3270–3278, 2002.

188. Furukawa, K., Arimura, N., and Miyazaki, T., Nucleotide sequence of the 2,3-dihydroxybiphenyl dioxygenase gene of *Pseudomonas pseudoalcaligenes*, *J. Bacteriol.*, 169, 427–429, 1987.
189. Erickson, B.D. and Mondello, F.J., Nucleotide sequencing and transcriptional mapping of the genes encoding biphenyl dioxygenase, a multicomponent polychlorinated biphenyl-degrading enzyme in *Pseudomonas* strain LB400, *J. Bacteriol.*, 174, 2903–2912, 1992.
190. Sylvestre, M., Sirois, M., Hurtubise, Y., Bergeron, J., Ahmad, D., Shareck, F., Barriault, D., Guillemette, I., and Juteau, J.M., Sequencing of *Comamonas testosteroni* strain B-356-biphenyl/ chlorobiphenyl dioxygenase genes: evolutionary relationships among Gram-negative bacterial biphenyl dioxygenases, *Gene*, 174, 195–202, 1996.
191. Masai, E., Yamada, A., Healy, J., Hatta, T., Kimbara, K., Fukuda, M., and Yano, K., Characterization of biphenyl catabolic genes of gram-positive polychlorinated biphenyl degrader *Rhodococcus* sp. strain RHA1, *Appl. Environ. Microbiol.*, 61, 2079–2085, 1995.
192. Asturias, J.A., Díaz, E., and Timmis, K.N., The evolutionary relationship of biphenyl dioxygenase from Gram-positive *Rhodococcus globerulus* P6 to multicomponent dioxygenases from Gram-negative bacteria, *Gene*, 156, 11–18, 1995.
193. Arai, H., Kosono, S., Taguchi, K., Maeda, M., Song, E., Fuji, F., Chung, S.-Y., and Kudo, T., Two sets of biphenyl and PCB degradation genes on a linear plasmid in *Rhodococcus erythropolis* A421, *J. Ferment. Bioeng.*, 86, 595–599, 1998.
194. Hayase, N., Taira, K., and Furukawa, K., *Pseudomonas putida* KF715 *bphABCD* operon encoding biphenyl and polychlorinated biphenyl degradation: cloning, analysis, and expression in soil bacteria, *J. Bacteriol.*, 172, 1160–1164, 1990.
195. Akhtar, M.N., Boyd, D.R., Thompson, N.J., Koreeda, M., Gibson, D.T., Mahadevan, V., and Jerina, D.M., Absolute stereochemistry of the dihydroanthracene-*cis*- and -*trans*-1,2-diols produced from anthracene by mammals and bacteria, *J. Chem. Soc. Perkin Trans.*, 1, 2506–2511, 1975.
196. Jerina, D.M., Selander, H., Yagi, H., Wells, M.C., Davey, J.F., Mahadevan, V., and Gibson, D.T., Dihydrodiols from anthracene and phenanthrene, *J. Am. Chem. Soc.*, 98, 5988–5996, 1976.
197. Saito, A., Iwabuchi, T., and Harayama, S., A novel phenanthrene dioxygenase from *Nocardioides* sp. strain KP7: expression in *Escherichia coli*, *J. Bacteriol.*, 182, 2134–2141, 2000.
198. Laurie, A.D. and Lloyd-Jones, G., The *phn* genes of *Burkholderia* sp. strain RP007 constitute a divergent gene cluster for polycyclic aromatic hydrocarbon catabolism, *J. Bacteriol.*, 181, 531–540, 1999.
199. Krivobok, S., Kuony, S., Meyer, C., Louwagie, M., Willison, J.C., and Jouanneau, Y., Identification of pyrene-induced proteins in *Mycobacterium* sp. strain 6PY1: evidence for two ring-hydroxylating dioxygenases, *J. Bacteriol.*, 185, 3828–3841, 2003.
200. Khan, A.A., Wang, R.F., Cao, W.W., Doerge, D.R., Wennerstrom, D., and Cerniglia, C.E., Molecular cloning, nucleotide sequence, and expression of genes encoding a polycyclic aromatic ring dioxygenase from *Mycobacterium* sp. strain PYR-1, *Appl. Environ. Microbiol.*, 67, 3577–3585, 2001.
201. Walter, U., Beyer, M., Klein, J., and Rehm, H.J., Degradation of pyrene by *Rhodococcus* sp. UW1, *Appl. Microbiol. Biotechnol.*, 34, 671–676, 1991.
202. Story, S.P., Parker, S.H., Hayasaka, S.S., Riley, M.B., and Kline, E.L., Convergent and divergent points in catabolic pathways involved in utilization of fluoranthene, naphthalene, anthracene, and phenanthrene by *Sphingomonas paucimobilis* var. EPA505, *J. Ind. Microbiol. Biotechnol.*, 26, 369–382, 2001.
203. Grifoll, M., Selifonov, S.A., Gatlin, C.V., and Chapman, P.J., Actions of a versatile fluorene-degrading bacterial isolate on polycyclic aromatic compounds, *Appl. Environ. Microbiol.*, 61, 3711–3723, 1995.
204. Selifonov, S.A., Grifoll, M., Gurst, J.E., and Chapman, P.J., Isolation and characterization of (+)-1,1a-dihydroxy-1-hydrofluoren-9-one formed by angular dioxygenation in the bacterial catabolism of fluorene, *Biochem. Biophys. Res. Commun.*, 193, 67–76, 1993.
205. Fujii, T., Takeo, M., and Maeda, Y., Plasmid-encoded genes specifying aniline oxidation from *Acinetobacter* sp. strain YAA, *Microbiology*, 143, 93–99, 1997.
206. Urata, M., Uchida, E., Nojiri, H., Omori, T., Obo, R., Miyaura, N., and Ouchiyama, N., Genes involved in aniline degradation by *Delftia acidovorans* strain 7N and its distribution in the natural environment, *Biosci. Biotechnol. Biochem.*, 68, 2457–2465, 2004.

207. Murakami, S., Hayashi, T., Maeda, T., Takenaka, S., and Aoki, K., Cloning and fu analysis of aniline dioxygenase gene cluster, from *Frateuria* species ANA-18, that me aniline via an *ortho*-cleavage pathway of catechol, *Biosci. Biotechnol. Biochem.*, 67, 235 2003.

208. Chang, H.K., Mohseni, P., and Zylstra, G.J., Characterization and regulation of the ger novel anthranilate 1,2-dioxygenase from *Burkholderia cepacia* DBO1, *J. Bacteriol.*, 185 5881, 2003.

209. Neidle, E.L., Hartnett, C., Ornston, L.N., Bairock, A., Rekik, M., and Harayama, S., Nu sequences of the *Acinetobacter calcoaceticus benABC* genes for benzoate 1,2-dioxygenas evolutionary relationships among multicomponent oxygenases, *J. Bacteriol.*, 173, 538 1991.

210. Haddad, S., Eby, D.M., and Neidle, E.L., Cloning and expression of the benzoate dio genes from *Rhodococcus* sp. strain 19070, *Appl. Environ. Microbiol.*, 67, 2507–2514, 2001.

211. Jeffrey, W.H., Cuskey, S.M., Chapman, P.J., Resnick, S., and Olsen, R.H., Characte of *Pseudomonas putida* mutants unable to catabolize benzoate: cloning and characteriz *Pseudomonas* genes involved in benzoate catabolism and isolation of a chromosoma fragment able to substitute for *xylS* in activation of the TOL lower-pathway pr *J. Bacteriol.*, 174, 4986–4996, 1992.

212. Tsoi, T.V., Plotnikova, E.G., Cole, J.R., Guerin, W.F., Bagdasarian, M., and Tiedj Cloning, expression, and nucleotide sequence of the *Pseudomonas aeruginosa* 142 *ohb* gene for oxygenolytic *ortho* dehalogenation of halobenzoates, *Appl. Environ. Microbiol.*, 65, 215 1999.

213. Suzuki, K., Ogawa, N., and Miyashita, K., Expression of 2-halobenzoate dioxygenas (*cbdSABC*) involved in the degradation of benzoate and 2-halobenzoate in *Burkhola* TH2, *Gene*, 262, 137–145, 2001.

214. Haak, B., Fetzner, S., and Lingens, F., Cloning, nucleotide sequence, and expressior plasmid-encoded genes for the two-component 2-halobenzoate 1,2-dioxygenase from *Pseu cepacia* 2CBS, *J. Bacteriol.*, 177, 667–675, 1995.

215. Francisco, P.J., Ogawa, N., Suzuki, K., and Miyashita, K., The chlorobenzoate dioxygena of *Burkholderia* sp. strain NK8 involved in the catabolism of chlorobenzoates, *Microbiolo* 121–133, 2001.

216. Nakatsu, C.H., Straus, N.A., and Wyndham, R.C., The nucleotide sequence of the 3-chlorobenzoate 3,4-dioxygenase genes (*cbaAB*) unites the class IA oxygenases in a single *Microbiology*, 141, 485–495, 1995.

217. Eaton, R.W., *p*-Cumate catabolic pathway in *Pseudomonas putida* F1: cloning and characte of DNA carrying the *cmt* operon, *J. Bacteriol.*, 178, 1351–1362, 1996.

218. Díaz, E., Ferrández, A., and García, J.L., Characterization of the *hca* cluster encoc dioxygenolytic pathway for initial catabolism of 3-phenylpropionic acid in *Escherichia co J. Bacteriol.*, 180, 2915–2923, 1998.

219. Hudlicky, T., Stabile, M.R., Gibson, D.T., and Whited, G.M., 1-Chloro-(2*S*, 3*S*)-dihydro hexa-4,6-diene, *Organic Synth.*, 76, 77–85, 1999.

220. van der Meer, J.R., Van Neerven, A.R.W., De Vries, E.J., De Vos, W.M., and Zehnder, Cloning and characterization of plasmid-encoded genes for the degradation of 1,2-di 1,4-dichloro, and 1,2,4-trichlorobenzene of *Pseudomonas* sp. strain P51, *J. Bacteriol.*, 17 1991.

221. Beil, S., Happe, B., Timmis, K.N., and Pieper, D.H., Genetic and biochemical characteriz the broad spectrum chlorobenzene dioxygenase from *Burkholderia* sp. strain PS12—dechlo of 1,2,4,5-tetrachlorobenzene, *Eur. J. Biochem.*, 247, 190–199, 1997.

222. Boyd, D.R., Sharma, N.D., Barr, S.A., Dalton, H., Chima, J., Whited, G., and Seema Chemoenzymatic synthesis of the 2,3- and 3,4-*cis*-dihydrodiol enantiomers of monosub benzenes, *J. Am. Chem. Soc.*, 116, 1147–1148, 1994.

223. Seto, M., Kimbara, K., Shimura, M., Hatta, T., Fukuda, M., and Yano, K., A novel tra ation of polychlorinated biphenyls by *Rhodococcus* sp. strain RHA1, *Appl. Environ. Microt* 3353–3358, 1995.

224. Barriault, D., Pelletier, C., Hurtubise, Y., and Sylvestre, M., Substrate selectivity pattern of *Comamonas testosteroni* strain B-356 towards dichlorobiphenyls, *Int. Biodeterior. Biodegrad.*, 39, 311–316, 1997.

225. Gibson, D.T., Cruden, D.L., Haddock, J.D., Zylstra, G.J., and Brand, J.M., Oxidation of poly-chlorinated biphenyls by *Pseudomonas* sp. strain LB400 and *Pseudomonas pseudoalcaligenes* KF707, *J. Bacteriol.*, 175, 4561–4564, 1993.

226. Ohtsubo, Y., Shimura, M., Delaware, M., Kimbara, K., Takagi, M., Kudo, T., Ohta, A., and Nagata, Y., A novel approach to the improvement of biphenyl and polychlorinated biphenyl degradation activity: promoter implantation by homologous recombination, *Appl. Environ. Microbiol.*, 69, 146–153, 2003.

227. Bedard, D.L. and Haberl, M.L., Influence of chlorine substitution pattern on the degradation of polychlorinated biphenyls by eight bacterial strains, *Microb. Ecol.*, 20, 87–102, 1990.

228. Habe, H., Miyakoshi, M., Chung, J., Kasuga, K., Yoshida, T., Nojiri, H., and Omori, T., Phthalate catabolic gene cluster is linked to the angular dioxygenase gene in *Terrabacter* sp. strain DBF63, *Appl. Microbiol. Biotechnol.*, 61, 44–54, 2003.

229. Ouchiyama, N., Miyachi, S., and Omori, T., Cloning and nucleotide sequence of carbazole catabolic genes from *Pseudomonas stutzeri* strain OM1, isolated from activated sludge, *J. Gen. Appl. Microbiol.*, 44, 57–63, 1998.

230. Nam, J.-W., Nojiri, H., Noguchi, H., Uchimura, H., Yoshida, T., Habe, H., Yamane, H., and Omori, T., Purification and characterization of carbazole 1,9a-dioxygenase, a three component dioxygenase system of *Pseudomonas resinovorans* strain CA10, *Appl. Environ. Microbiol.*, 68, 5882–5890, 2002.

231. Chang, H.-K. and Zylstra, G.J., Novel organization of the genes for phthalate degradation from *Burkholderia cepacia* DBO1, *J. Bacteriol.*, 180, 6529–6537, 1998.

232. Eaton, R.W., Plasmid-encoded phthalate catabolic pathway in *Arthrobacter keyseri* 12B, *J. Bacteriol.*, 183, 3689–3703, 2001.

233. Wang, Y.Z., Zhou, Y., and Zylstra, G.J., Molecular analysis of isophthalate and terephthalate degradation by *Comamonas testosteroni* YZW-D, *Environ. Health Perspect.*, 103(Suppl. 5), 9–12, 1995.

234. Schläfli, H.R., Weiss, M.A., Leisinger, T., and Cook, A.M., Terephthalate 1,2-dioxygenase system from *Comamonas testosteroni* T-2: purification and some properties of the oxygenase component, *J. Bacteriol.*, 176, 6644–6652, 1994.

235. Shigematsu, T., Yumihara, K., Ueda, Y., Morimura, S., and Kida, K., Purification and gene cloning of the oxygenase component of the terephthalate 1,2-dioxygenase system from *Delftia tsuruhatensis* strain T7, *FEMS Microbiol. Lett.*, 220, 255–260, 2003.

236. Harayama, S., Rekik, M., Bairoch, A., Neidle, E.L., and Ornston, L.N., Potential DNA slippage structure acquired during evolutionary divergence of *Acinetobacter calcoaceticus* chromosomal *benABC* and *Pseudomonas putida* TOL pWWO plasmid *xylXYZ*, genes encoding benzoate dioxy-genases, *J. Bacteriol.*, 173, 7540–7548, 1991.

237. Bernhardt, F.-H., Bill, E., Trautwein, A.X., and Twilfer, H., 4-Methoxybenzoate monooxygenase from *Pseudomonas putida*: isolation, biochemical properties, substrate specificity, and reaction mechanisms of the enzyme components, *Methods Enzymol.*, 161, 281–294, 1988.

238. Locher, H.H., Leisinger, T., and Cook, A.M., 4-Toluene sulfonate methyl-monooxygenase from *Comamonas testosteroni* T-2: purification and some properties of the oxygenase component, *J. Bacteriol.*, 173, 3741–3748, 1991.

239. Mampel, J., Ruff, J., Junker, F., and Cook, A.M., The oxygenase component of the 2-aminoben-zenesulfonate dioxygenase system from *Alcaligenes* sp. strain O-1, *Microbiology*, 145, 3255–3264, 1999.

240. Lessner, D.J., Johnson, G.R., Parales, R.E., Spain, J.C., and Gibson, D.T., Molecular character-ization and substrate specificity of nitrobenzene dioxygenase from *Comamonas* sp. strain JS765, *Appl. Environ. Microbiol.*, 68, 634–641, 2002.

241. Robertson, J.B., Spain, J.C., Haddock, J.D., and Gibson, D.T., Oxidation of nitrotoluene by toluene dioxygenase: evidence for a monooxygenase reaction, *Appl. Environ. Microbiol.*, 58, 2643–2648, 1992.

242. Parales, J.V., Kumar, A., Parales, R.E., and Gibson, D.T., Cloning and sequencing of ▮ encoding 2-nitrotoluene dioxygenase from *Pseudomonas* sp. JS42, *Gene*, 181, 57–61, 199▮

243. Parales, R.E., Emig, M.D., Lynch, N.A., and Gibson, D.T., Substrate specificities of hyb▮ thalene and 2,4-dinitrotoluene dioxygenase enzyme systems, *J. Bacteriol.*, 180, 2337–2344,

244. Suen, W.-C. and Spain, J.C., Cloning and characterization of *Pseudomonas* sp. strain D▮ for 2,4-dinitrotoluene degradation, *J. Bacteriol.*, 175, 1831–1837, 1993.

245. Nishino, S.F., Paoli, G.C., and Spain, J.C., Aerobic degradation of dinitrotoluenes pathway for bacterial degradation of 2,6-dinitrotoluene, *Appl. Environ. Microbiol.*, 6▮ 2147, 2000.

246. Seeger, M., Camara, B., and Hofer, B., Dehalogenation, denitration, dehydrogenat▮ angular attack on substituted biphenyls and related compounds by biphenyl dio▮ *J. Bacteriol.*, 183, 3548–3555, 2001.

247. Bestetti, G., Bianchi, D., Bosetti, A., Di Gennaro, P., Galli, E., Leoni, B., Pelizzoni, F., a▮ G., Bioconversion of substituted naphthalenes to the corresponding 1,2-dihydro-1,2-d▮ derivatives. Determination of the regio- and stereochemistry of the oxidation reactio▮ *Microbiol. Biotechnol.*, 44, 306–313, 1995.

248. Deluca, M.E. and Hudlicky, T., Microbial oxidation of naphthalene derivatives. Absolu▮ uration of metabolites, *Tetrahedron Lett.*, 31, 13–16, 1990.

249. Whited, G.M., Downie, J.C., Hudlicky, T., Fearnley, S.P., Dudding, T.C., Olivo, H▮ Parker, D., Oxidation of 2-methoxynaphthalene by toluene, naphthalene and biphen▮ genases: structure and absolute stereochemistry of metabolites, *Bioorg. Med. Chem.*, 2, 1994.

250. Brilon, C., Beckmann, W., and Knackmuss, H.-J., Catabolism of naphthalenesulfonic *Pseudomonas* sp. A3 and *Pseudomonas* sp. C22, *Appl. Environ. Microbiol.*, 42, 44–55, 19▮

251. Knackmuss, H.-J., Beckmann, W., and Otting, W., Microbiological synthesis of (+)-*cis*-droxy-1,2-dihydronaphthalene-2-carboxylic acid, *Angew. Chem. Int. Ed. Engl.*, 15, 549, ▮

252. Martin, V.J.J. and Mohn, W.W., A novel aromatic-ring-hydroxylating dioxygenase ▮ diterpenoid-degrading bacterium *Pseudomonas abietaniphila* BKME-9, *J. Bacteriol.*, 18▮ 2682, 1999.

253. Habe, H., Ashikawa, Y., Saiki, Y., Yoshida, T., Nojiri, H., and Omori, T., *Sphingomonas* ▮ KA1, carrying a carbazole dioxygenase gene homologue, degrades chlorinated dibenzo-▮ in soil, *FEMS Microbiol. Lett.*, 211, 43–49, 2002.

254. Shepherd, J.M. and Lloyd-Jones, G., Novel carbazole degradation genes of *Sphingomo*▮ sequence analysis, transcription, and molecular ecology, *Biochem. Biophys. Res. Comm*▮ 129–135, 1998.

255. Resnick, S.M., Torok, D.S., and Gibson, D.T., Oxidation of carbazole to 3-hydroxycarb▮ naphthalene 1,2-dioxygenase and biphenyl 2,3-dioxygenase, *FEMS Microbiol. Lett.*, 113, 1993.

256. Iida, T., Mukouzaka, Y., Nakamura, K., and Kudo, T., Plasmid-borne genes code for ar▮ dioxygenase involved in dibenzofuran degradation by *Terrabacter* sp. strain YK3, *Appl.* *Microbiol.*, 68, 3716–3723, 2002.

257. Kasuga, K., Habe, H., Chung, J.S., Yoshida, T., Nojiri, H., Yamane, H., and Omori, T., and characterization of the genes encoding a novel oxygenase component of angular dio▮ from the gram-positive dibenzofuran-degrader *Terrabacter* sp. strain DBF63, *Biochem.* *Res. Commun.*, 283, 195–204, 2001.

258. Takagi, T., Habe, H., Yoshida, T., Yamane, H., Omori, T., and Nojiri, H., Characteri▮ [3Fe-4S] ferredoxin DbfA3, which functions in the angular dioxygenase system of *Terral*▮ strain DBF63, *Appl. Microbiol. Biotechnol.*, 68, 336–345, 2005.

259. Armengaud, J., Gaillard, J., and Timmis, K.N., A second [2Fe-2S] ferredoxin from *Sphin*▮ sp. strain RW1 can function as an electron donor for the dioxin dioxygenase, *J. Bacter*▮ 2238–2244, 2000.

260. Armengaud, J., Happe, B., and Timmis, K.N., Genetic analysis of dioxin dioxyg▮ *Sphingomonas* sp. strain RW1: catabolic genes dispersed on the genome, *J. Bacteriol.*, 18▮ 3966, 1998.

261. Armengaud, J. and Timmis, K.N., Molecular characterization of Fdx1, a putidaredoxin-type [2Fe-2S] ferredoxin able to transfer electrons to the dioxin dioxygenase of *Sphingomonas* sp. RW1, *Eur. J. Biochem.*, 247, 833–842, 1997.

262. Armengaud, J. and Timmis, K.N., The reductase RedA2 of the multi-component dioxin dioxygenase system of *Sphingomonas* sp. RW1 is related to class-I cytochrome P450-type reductases, *Eur. J. Biochem.*, 253, 437–444, 1998.

263. Rosche, B., Tshisuaka, B., Hauer, B., Lingens, F., and Fetzner, S., 2-Oxo-1,2-dihydroquinoline 8-monooxygenase: phylogenetic relationships to other multicomponent nonheme iron oxygenases, *J. Bacteriol.*, 179, 3549–3554, 1997.

264. Sauber, K., Fröhner, C., Rosenberg, G., Eberspächer, J., and Lingens, F., Purification and properties of pyrazon dioxygenase from pyrazon-degrading bacteria, *Eur. J. Biochem.*, 74, 89–97, 1977.

265. Treadway, S.L., Yanagimachi, K.S., Lankenau, E., Lessard, P.A., Stephanopoulos, G., and Sinskey, A.J., Isolation and characterization of indene bioconversion genes from *Rhodococcus* strain I24, *Appl. Microbiol. Biotechnol.*, 51, 786–793, 1999.

266. Eaton, S.E., Resnick, S.M., and Gibson, D.T., Initial reactions in the oxidation of 1,2-dihydronaphthalene metabolism by *Sphingomonas yanoikuyae* strains, *Appl. Environ. Microbiol.*, 62, 4388–4394, 1996.

267. Sikkema, J. and De Bont, J.A.M., Metabolism of tetralin (1,2,3,4-tetrahydronaphthalene) in *Corynebacterium* sp. strain C125, *Appl. Environ. Microbiol.*, 59, 567–572, 1993.

268. Moreno-Ruiz, E., Hernaez, M.J., Martinez-Perez, O., and Santero, E., Identification and functional characterization of *Sphingomonas macrogolitabida* strain TFA genes involved in the first two steps of the tetralin catabolic pathway, *J. Bacteriol.*, 185, 2026–2030, 2003.

269. Dehmel, U., Engesser, K.-H., Timmis, K.N., and Dwyer, D.F., Cloning, nucleotide sequence, and expression of the gene encoding a novel dioxygenase involved in metabolism of carboxydiphenyl ethers in *Pseudomonas pseudoalcaligenes* POB310, *Arch. Microbiol.*, 163, 35–41, 1995.

270. Boyd, D.R., Sharma, N.D., and Allen, C.C.R., Aromatic dioxygenases: molecular biocatalysis and applications, *Curr. Opin. Biotechnol.*, 12, 564–573, 2001.

11 A Genomic Approach to Investigating Baker's Yeast Reductions

Jon D. Stewart

CONTENTS

Alcohols are one of the most common functional groups in organic chemistry, either as constituents of the final products or as "handles" for further synthetic transformations. While all primary alcohols are achiral, this is not necessarily true of secondary and tertiary alcohols. For these cases, controlling the absolute stereochemistry is always essential in modern synthesis. Solutions to this problem include kinetic resolutions by acylation or deacylation, stereoselective oxidations, and asymmetric reductions [1–3]. The last method—asymmetric ketone reduction—is particularly powerful as it potentially allows 100% of the starting material to be transformed into the desired enantiomer, as compared with kinetic resolution strategies that have maximal 50% yields.

11.1 INTRODUCTION TO YEAST-MEDIATED REDUCTIONS

Among the many biocatalytic approaches to asymmetric ketone reductions that ha
explored [4], the use of commercial baker's yeast has been by far the most popular o►
Whole cells of this organism are readily available, and inexpensive, from the brew:
baking industries. The experimental protocols are very simple: dried yeast cells are reh
in the presence of a carbohydrate (usually sucrose), the substrate is added and allo
react, then the alcohol product is recovered by solvent extraction [7]. Yeast-mediatec
tions are also applicable to a very wide range of aldehydes and ketone structures. In f.
difficult to find a substrate that is not reduced by baker's yeast.

Unfortunately, several problems counterbalance the many advantages of baker'
Most critically, reductions using whole yeast cells often afford stereoisomeric mixt
alcohols. In principle, this could be the result of either a single reductase with
enantioselectivity or multiple enzymes with conflicting stereopreferences. In fact, th
situation prevails in baker's yeast. This understanding has been the basis of most app►
to improve the stereoselectivities of yeast-mediated reductions, while still maintaining
mental simplicity. Before yeast reductions can be improved rationally, however, the
and identity of the participating enzymes must be established. Even early suggestions t
than half-a-dozen enzymes played key roles turned out to be incorrect, and it
recognized that several dozens of yeast reductases may be involved. Because the
leading to this conclusion have had a direct impact on the development of our genome
strategy (described below), these results are summarized briefly below.

11.1.1 EVIDENCE FOR MULTIPLE YEAST REDUCTASES

Alcohols with the general structure (R)-2 serve as precursors for pharmaceuticals an
tional supplement L-carnitine (Scheme 11.1). Because it is commercially available, eth
1b is the most convenient substrate; unfortunately, employing baker's yeast under s►
conditions affords a 55% ee favoring the (S)-enantiomer [8]. Sih and coworkers exp
number of approaches to improving the stereoselectivity of this process, and their
revealed a great deal about the biochemistry of yeast-mediated reductions. Most impo
they showed that enantioselectivity varied with the concentration of 1b. If a single
tase were responsible for the conversion, the stereoselectivity would be concen►
independent; however, if two or more enzymes with different stereopreferences accep
then decreasing the substrate concentration would favor reduction by the enzyme(s) v
lower K_M value(s). Sih's study provided unequivocal evidence that more than on
enzyme was involved in reducing exogenous ketones.

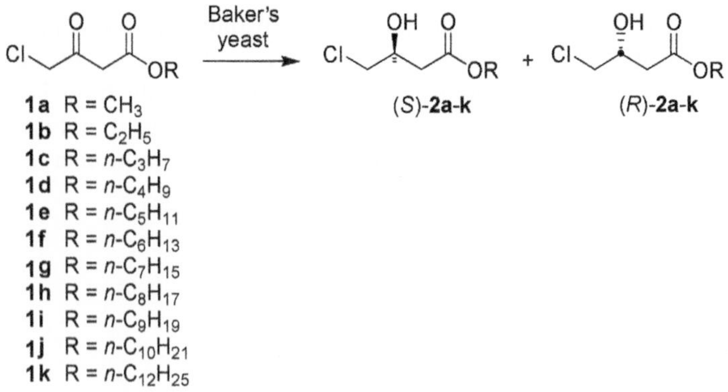

1a R = CH$_3$
1b R = C$_2$H$_5$
1c R = n-C$_3$H$_7$
1d R = n-C$_4$H$_9$
1e R = n-C$_5$H$_{11}$
1f R = n-C$_6$H$_{13}$
1g R = n-C$_7$H$_{15}$
1h R = n-C$_8$H$_{17}$
1i R = n-C$_9$H$_{19}$
1j R = n-C$_{10}$H$_{21}$
1k R = n-C$_{12}$H$_{25}$

(S)-2a-k   (R)-2a-k

SCHEME 11.1

To define the number and the properties of yeast enzymes that reduced 4-chloroacetoacetate esters, independent groups led by Sih and Nakamura purified yeast protein extracts on standard chromatographic supports [9,10]. When fractions were assayed for the ability to reduce **1b**, four peaks of enzyme activity were obtained, and each was purified to apparent homogeneity. Significantly, each purified protein reduced **1b** with >98% ee, although two were (R)-selective and two (S)-selective. The simultaneous participation of these four reductases neatly explained the modest enantioselectivities observed for reactions involving whole yeast cells. At least for β-keto ester substrates such as **1b**, it was believed that this collection of four yeast reductases was complete. Later work showed that this was not the case, however.

11.1.2 IMPROVING THE STEREOSELECTIVITIES OF YEAST-MEDIATED REDUCTIONS

Knowing the properties of the four purified yeast reductases opened the door to improving the stereoselectivities of whole-cell reductions by selectively depressing the catalytic activities of specific enzymes. One approach—already mentioned above—is to use low substrate concentrations, which encourages reduction by the enzyme(s) that binds the ketone most tightly. In the case of **1b**, however, incomplete stereoselectivity was still observed even at the lowest practical substrate concentration [8]. Moreover, this approach yields dilute product streams and diminished volumetric productivities that are undesirable in large-scale operations. Altering the substrate structure to restrict acceptance to a limited subset of reductases proved to be a more fruitful approach. Yeast-mediated reductions of a homologous series of 4-chloroacetoacetates differing in their alkoxy moieties revealed that essentially complete (R)-selectivity was obtained for side chains longer than C_6 (Figure 11.1) [8]. While such a strategy was successful in this particular case, it is not always possible (or desirable) to modify the substrate structure. This limits the applicability of the approach.

Eliminating the catalytic activities of one or more competing enzymes with undesirable stereoselectivities is a more direct approach to improving yeast-mediated reductions, and several methods for achieving this goal have been reported. Studies of the four purified yeast

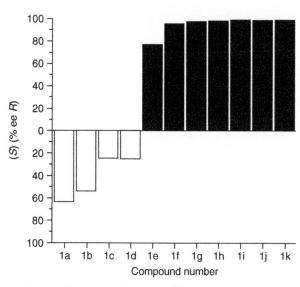

FIGURE 11.1 Yeast-mediated reductions of alkyl 4-chloroacetoacetates. Compound structures are depicted in Scheme 11.1. White bars indicate reactions that proceed with predominant (S)-selectivity, while those with predominant (R)-selectivity are represented by black bars.

reductases accepting **1b** revealed that reagents such as allyl alcohol (an acrolein precurs
methyl vinyl ketone, and ethyl chloroacetate inactivated particular yeast enzymes [10
vitro, these were highly selective inactivators, and they could also be used *in vivo* (Figure 1
In favorable cases, these additives afforded very high stereoselectivities in whole cell-medi
reductions. However, because the additives are reactive compounds, they are also toxic tov
the cells and this always requires the increasing cell mass:substrate to compensate.
increased biomass and the presence of additives complicate downstream processing
purification. Despite these drawbacks, the inclusion of reductase poisons has great prac
utility and remains the most common strategy for improving stereoselectivities in y
reductions [11]. In the case of ethyl 4-chloroacetoacetate **1b**, the (*S*)-alcohol was availab
90% ee and the (*R*)-enantiomer in 80% ee by using methyl vinyl ketone and ethyl chloroace
respectively (Figure 11.2).

Knocking out the genes encoding competing reductases provides an even more d
route to eliminating unwanted enzymatic activities. Knowing that the fatty acid synt
complex was one of the D-selective reductases for **1b**, Sih substituted whole cells of a fatty
synthase mutant strain [12,13] and observed improved stereoselectivity. We extended
strategy to include the other two L-specific reductases isolated by Sih and Nakamura (p
ucts of the *YPR1* and *GRE2* genes), and also added gene overexpression to improve st
selectivity even further. We created a complete set of modified yeast strains in which o
these three proteins was overexpressed, while the other two were ablated by gene knoc
[14,15]. The resulting strains were screened for the reduction of a panel of β-keto e
Knocking out fatty acid synthase virtually eliminated the formation of D-alcohols,
its overexpression enhanced their production, in line with our expectations. By com
more complex results were obtained from strains with altered levels of proteins encode
the *YPR1* and the *GRE2* genes. While their overexpression slightly increased the leve
the corresponding alcohol products, knocking these genes out had almost no effect o

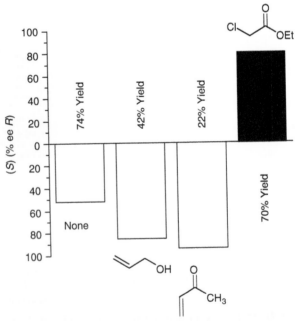

FIGURE 11.2 Yeast-mediated reductions of ethyl 4-chloroacetoacetate in the presence of ad
Inhibitor structures and the corresponding stereoselectivities are indicated.

reduction stereoselectivities. These observations provided the first clues that the baker's yeast genome actually encoded more than three reductases that accepted exogenous β-keto ester substrates such as **1b**.

11.1.3 DEFINING THE FULL COMPLEMENT OF *SACCHAROMYCES CEREVISIAE* REDUCTASES

The surprising results from reductase gene knockout studies made it clear that the complete set of enzymes accepting exogenous ketones had to be identified before we could take full advantage of the biocatalytic opportunities offered by baker's yeast. As previous biochemical approaches had yielded only a subset of the participating enzymes, we chose a bioinformatics approach to answering the question, "How many reductases can potentially be produced by *S. cerevisiae*?" We analyzed the complete genome sequence using a variety of criteria and probe sequences drawn from all major classes of reductases (lactate dehydrogenases, alcohol dehydrogenases, aldose reductases, D-hydroxyacid dehydrogenases, short- and medium-chain dehydrogenases) and discovered that approximately 50 genes likely encode reductases (Table 11.1) [16]. While the properties of a few of these—the well-known yeast alcohol dehydrogenases and the gene products of *YPR1* and *GRE2*, for example—had been delineated prior to our work, the rest were almost completely unknown.

Given the large number of reductases that could be expressed in a yeast cell, it was clear that attempting to create mutant strains with dozens of genes knocked out simultaneously was not feasible. It also seemed difficult to find inhibitors that would be sufficiently selective to eliminate the catalytic activities of only a few of the 50 potential reductases. Rather than continuing to use whole baker's yeast cells for asymmetric reductions, we decided instead to focus on the individual reductases. In this approach, yeast cells are treated as a genetic resource rather than a self-contained biocatalytic reagent. In addition to solving practical problems in asymmetric ketone reductions, we were also drawn to this strategy because it might be generally useful. Our methods can be applied to any genome or chemical conversion. All that is required is one or more amino acid sequences from proteins known to carry out the reaction of interest and the readily available computer software.

There were two key problems in moving from whole baker's yeast cells to the use of individual yeast reductases. First, each purified protein must be available, preferably by simple standard method that can be applied to large numbers of enzymes in parallel. Second, methods for carrying out reactions on preparative scales must be available. Our approaches to solving these two problems are described below.

11.2 A LIBRARY OF INDIVIDUAL YEAST REDUCTASES

11.2.1 CREATING THE LIBRARY

At the start of our work, almost no yeast reductases were available commercially in purified form. We, therefore, turned to a recombinant collection developed by Martzen and coworkers in which every open reading frame (ORF) from the baker's yeast genome was overexpressed in individual yeast strains (Figure 11.3) [17]. Each protein was fused to a common tag (glutathione *S*-transferase) that allowed them to be purified by a common, one-step method using a glutathione-containing solid support. The purified fusion proteins could then be assayed for catalytic activity by conventional methods. This collection is ideally suited to analyzing the substrate- and stereoselectivities of yeast reductases rapidly. A given ketone can be tested as a substrate for each protein in small-scale reactions. This defines the subset of yeast reductases that accept the substrate and the stereoselectivities of the reactions, and allows an informed choice of the best reductase for large-scale bioconversions.

TABLE 11.1
Saccharomyces cerevisiae Genes Encoding Potential Ketone Reductases

Protein	Gene(s)	ORF(s)
Lactate dehydrogenases		
∟-Lactate cytochrome *c* oxidoreductase	*CYB2*	Yml054c
ᴅ-Lactate dehydrogenase	*DLD1*	Ydl174c
ᴅ-Lactate dehydrogenase	*DLD2*	Ydl178w
ᴅ-Lactate dehydrogenase	*DLD3*	Yel071w
Alcohol dehydrogenases		
Long-chain alcohol dehydrogenase	*SFA1*	Ydl168w
Alcohol dehydrogenase I	*ADH1*	Yol086c
Alcohol dehydrogenase II	*ADH2*	Ymr303c
Alcohol dehydrogenase III	*ADH3*	Ymr083w
Alcohol dehydrogenase V	*ADH5*	Ybr145w
Alcohol dehydrogenase IV	*ADH4*	Ygl256w
Medium-chain alcohol dehydrogenases		
Sorbitol dehydrogenase	*SOR1*	Yjr159w
Xylitol dehydrogenase	*XYL2*	Ylr070c
Putative polyol dehydrogenase	—	Yal061w
(2*R*,3*R*)-2,3-Butanediol dehydrogenase	*BDH1*	Yal060w
NADPH-dependent dehydrogenase	*ADH6*	Ymr318c
NADPH-dependent dehydrogenase	*ADH7*	Ycr105w
ζ-Crystallin homolog	*ZTA1*	Ybr046c
Short-chain alcohol dehydrogenases		
NADPH-dependent methyl glyoxal reductase	*GRE2*	Yol151w
Putative oxidoreductase	—	Ydr541c
Oxidoreductase	—	Ygl157w
Oxidoreductase	—	Ygl039w
Microsomal β-keto reductase	—	Ylr426w
Putative oxidoreductase	—	Ybr159w
Putative oxidoreductase	—	Ydl114w
NADP⁺-dependent dehydrogenase	—	Ymr226c
Putative oxidoreductase	—	Yir035c
Putative oxidoreductase	—	Yir036c
NADPH-dependent 1-acyl dihydroxyacetone phosphate reductase	*AYR1*	Yil124w
Putative oxidoreductase	—	Ykl107w
3-Oxoacyl-(acyl-carrier-protein) reductase	*OAR1*	Ykl055c
Putative aryl alcohol dehydrogenase	*AAD14*	Ynl331c
Putative aryl alcohol dehydrogenase	*AAD3*	Ycr107w
Putative aryl alcohol dehydrogenase	*AAD4*	Ydl243c
Putative aryl alcohol dehydrogenase	*AAD10*	Yjr155w
Putative aryl alcohol dehydrogenase	*AAD16*	Yfl057c
Putative aryl alcohol dehydrogenase	*AAD15*	Yol165c
Aldose reductase family		
2-Methylbutyraldehyde reductase	*YPR1*	Ydr368w
Putative NADP⁺ coupled glycerol dehydrogenase	*GCY1*	Yor120w
Large subunit of NADP⁺-dependent arabinose dehydrogenase	*ARA1*	Ybr149w
Aldose reductase	*GRE3*	Yhr104w
Putative oxidoreductase	—	Yjr096w
NADPH-dependent α-keto amide reductase	—	Ydl124w

TABLE 11.1 (continued)
Saccharomyces cerevisiae Genes Encoding Potential Ketone Reductases

Protein	Gene(s)	ORF(s)	Plasmid
D-Hydroxyacid dehydrogenase family			
Putative dehydrogenase	—	Ypl113c	pIK15
Putative hydroxyacid dehydrogenase	—	Ygl185c	pAKS1
Putative hydroxyisocaproate dehydrogenase	—	Ynl274c	pIK13
NAD$^+$-dependent formate dehydrogenase	*FDH1*	Yor388c	pIK14
NAD$^+$-dependent formate dehydrogenase	*FDH2*	Ypl275w	pIK18
Multifunctional			
Fatty acid synthase	*FAS1,*	Ykl182w,	
	FAS2	Ypl231w	

The original Martzen library encompasses all 6144 ORFs in the yeast genome, and it was designed to allow the enzyme(s) catalyzing specific reactions to be identified directly, with no additional information required. We deliberately restricted ourselves to a subset of this library, since our prior genome analysis had focused attention on approximately 50 ORFs most likely to participate in ketone reductions. We narrowed this list further by eliminating proteins known to have narrow substrate specificities, for example lactate and alcohol dehydrogenases, and those with weak sequence similarity to authentic reductases. This left 23 proteins for our first-pass library (Table 11.1). Clearly, the number can be increased in the future, by including additional proteins either from the *S. cerevisiae* genome or from those of other organisms.

Applying yeast reductase GST-fusions to nonnatural substrates and millimolar substrate concentrations required relatively higher protein concentrations in the screening reactions. Originally, the GST-fusion proteins were overproduced in *S. cerevisiae*. While this provided

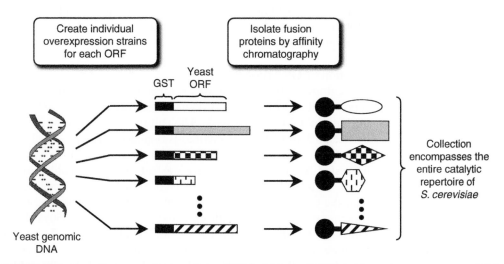

FIGURE 11.3 Overall strategy for creating a GST-fusion protein library from the yeast genome. DNA encoding the GST portion is shown in black and yeast ORFs (and their protein products) are indicated by different patterns.

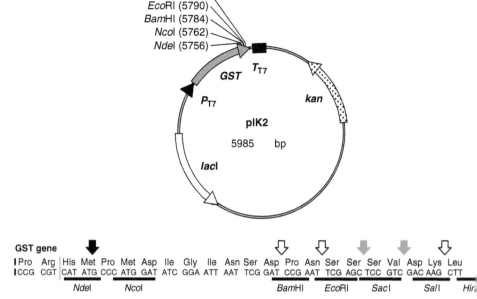

FIGURE 11.4 Structure of plasmid pIK2 is used to overproduce GST-fusion proteins. The loca
key restriction sites are indicated along with important genetic elements. The DNA at the 3′-en
GST portion is shown below. Arrows indicate the locations of restriction sites, those shown in
not unique.

the most "natural" environment for yeast proteins, it also severely limited the quantiti
could be obtained. We therefore expressed these the proteins in *Escherichia coli*. P
pIK2 was developed for this purpose, and it includes the strong T7 promoter to
overproduction of the mRNA corresponding to the GST-fusion protein (Figure 11.4
restriction sites (*Nde*I and *Nco*I) are provided for linking the 5′-end of the yeast ORF
3′-end of the glutathione *S*-transferase gene as well as several downstream restriction s
incorporating the yeast ORF. After construction, each plasmid was used to transform
strain BL21(DE3). Detailed experimental protocols for isolating and employing these
proteins are given below:

1. Isolation of GST-fusion proteins
    A culture of the appropriate overexpression strain grown overnight in LB m
    supplemented with 25 μg/mL kanamycin was diluted 1:100 into 500 mL of the
    medium in a 2 L baffled flask. The culture was shaken at 37°C until the O.D.
    between 0.5 and 1.0, isopropylthio-β-D-galactoside was added to final concentra
    100 μM and the culture was shaken for an additional 6 h at room temperature
    harvested by centrifugation were washed twice with cold water, then resuspen
    25 mL of 100 mM KP$_i$, pH 7.0. A cold French pressure cell was used to lyse th
    then the extract was clarified by centrifugation at 10,000 × g for 10 min at 4°
    supernatant was mixed with an equal volume of cold 50 mM Tris-Cl, 4 mM
    1 mM DTT, 10% glycerol, pH 7.5 and loaded onto a 2.4 × 5.0 cm column
    tathione resin (Clontech) at a flow rate of 0.5 mL/min. The column had been
    brated with 50 mM Tris-Cl, 4 mM MgCl$_2$, 1 mM DTT, 500 mM NaCl, 10% gl
    pH 7.5 (wash buffer). The flow-through was discarded and the resin was washe
    with 20 mL of the starting buffer. This material was also discarded. Essentiall

GST-fusion proteins were eluted with 40 mL of freshly prepared elution buffer (starting buffer (39.6 mL) plus 2 M NaOH (0.40 mL), and solid glutathione (0.31 g)). The eluant was dialyzed against several changes of 20 mM Tris-Cl, 2 mM EDTA, 4 mM $MgCl_2$, 1 mM DTT, 55 mM NaCl, 50% glycerol, pH 7.5 prior to storage at $-20°C$. The glutathione agarose resin was regenerated by washing with 20 column volumes of phosphate-buffered saline supplemented with 3 M NaCl (10 mM Na_2HPO_4, 1.8 mM KH_2PO_4, 2.7 mM KCl, 3.14 M NaCl, pH 7.4) followed by 10 column volumes of wash buffer. We have used a single column for purifying all GST-fusion proteins and have observed no cross-contamination problems.

2. Small-scale reactions involving GST-fusion proteins

$NADP^+$ (0.20 μmol, 0.15 mg), glucose-6-phosphate (14 μmol, 4.3 mg), glucose-6-phosphate dehydrogenase (5 μg), and ketone substrate (5 mM) were mixed with the appropriate purified GST-fusion protein (5–50 μg) in 1.0 mL of 100 mM KP_i, pH 7.0. Reaction mixtures were incubated at 30°C and periodically sampled for analysis by normal- and chiral-phase GC (DB-17 (J&W Scientific) and Chirasil-Dex CB (Varian) or Chirasil-Val (Varian) columns, respectively). Reactions were scaled up by 10- or 20-fold to allow product isolation and spectroscopic characterization. At the conclusion of the bioconversion, the reaction mixture was extracted with Et_2O (3 × (5 × reaction volume)). The combined organic extracts were washed with brine (1 volume) and water (1 volume), then dried with $MgSO_4$ and concentrated in vacuo. If required, the alcohol product was purified by flash column chromatography.

11.2.2 APPLICATIONS OF YEAST GST-FUSION PROTEINS

Access to a large fraction of baker's yeast reductases in purified form allowed us to probe several important issues. For example, if a substrate is known to be reduced by whole yeast cells, the collection makes it possible to define which one(s) accept the substrate. In principle, this knowledge can be combined with amino acid sequence and protein structural information to understand better the interplay between protein structure and stereoselectivity, although this has been proven difficult in practice (*vide infra*). The collection of individual enzymes also opens the possibility of finding a biocatalyst with high stereoselectivity to meet the needs of a specific synthetic route. Since only a single reductase is used, products formed by competing yeast enzymes are eliminated, and this maximizes the chances of success. Moreover, keeping purified GST-fusion proteins on hand allows the screening tasks to be accomplished rapidly, of particular importance is the pharmaceutical industry, where time-to-market is a key driver in process development.

We have used a number of ketones to probe our collection of yeast GST-fusion proteins [18–21], and four representative examples are summarized below. Collectively, they illustrate the strengths and limitations of our current collection. In each case, it was possible to identify a yeast reductase that provided the desired stereoisomer; in three of the four cases, the enantiomer was also available by employing a different yeast enzyme. This is an important achievement, because difficulties in supplying both product enantiomers have traditionally been an Achilles heel of biocatalysis. Whereas, chemical catalysts can be reengineered to produce the enantiomeric product simply by inverting the stereochemistry of the chiral ligand field, this option is not practically available in biocatalysis. Instead, one must identify pairs of enantiocomplementary enzymes. Since the degree of enantiocomplementary will depend greatly on the substrate structure, it is necessary to screen the entire collection of enzymes for each new application, and a larger collection and simple assay methodologies maximize the chances of success.

11.2.2.1 Ethyl 4-Chloroacetoacetate

As noted above, the (*R*)-alcohol derived from ethyl 4-chloroacetoacetate is an importar
building block in pharmaceutical and nutritional supplement synthesis. We screened 18 in◄
yeast reductases against this substrate, and the results are summarized in Figure 11
enzymes afforded only the (*S*)-alcohol within the limits of our detection, whereas nine
the homochiral (*R*)-alcohol. The remaining four enzymes displayed limited stereoselecti◄

Several interesting points emerged from these results. First, at least for our
collection of yeast reductases, **1b** is a "universal substrate" accepted by every enzyn
would, therefore, expect that using this ketone to assess reductase activity in yeast
fractions would reveal all 18 dehydrogenases in our collection. In fact, only four wer
[9,10]. Presumably, the "missing" reductases are produced at low levels under the
conditions used to produce yeast cells commercially. Using a genomic, rather thar
chemical, approach to identifying potentially useful enzymes side-steps problems of
expression levels under physiological conditions.

The second general lesson from the studies of **1b** is that most yeast reductases,
accept a given substrate, catalyze reactions with very high enantioselectivities. This re◄
the notion that the mixtures of products observed from reductions using whole ye◄
are primarily due to simultaneous participation by multiple reductases with o◄
stereoselectivities. This demonstrates the very important practical benefits of ex◄
individual reductases, rather than relying on whole organisms.

It is also noteworthy that we observed no simple correlation between stereoselecti◄
amino acid sequence. For example, the high sequence similarity between the six yeas◄

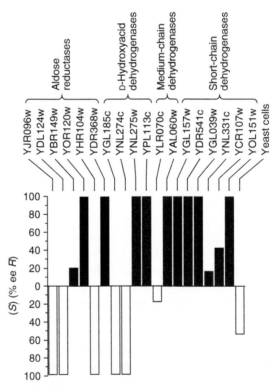

FIGURE 11.5 Reduction of ethyl 4-chloroacetoacetate by GST-fusion proteins. Systematic gene n◄
shown above, and enzymes are grouped by superfamily. White bars indicate reactions that proc◄
predominant (*S*)-selectivity while those with predominant (*R*)-selectivity are represented by black◄

reductases likely indicates that all share the same overall structure. Despite this similarity, three yield only the (S)-alcohol and two produce the (R)-product. This underscores the difficulties in predicting the outcomes of enzymatic reactions involving unnatural substrates, and it is one of the major reasons that we favor simple, rapid screening approaches to determine these outcomes empirically. Our results also make it clear that when choosing whether or not to add a new protein to the dehydrogenase collection, high sequence similarity to existing members should not automatically disqualify a candidate.

11.2.2.2 Ethyl 4-Phenyl-2-Oxobutyrate

Alcohol (R)-**4** is an important intermediate in commercial routes to a number of angiotensin converting enzyme (ACE) inhibitors such as Enalapril and Cilazepril, used clinically to regulate blood pressure (Scheme 11.2). A large number of synthetic approaches to (R)-**4** have been published [18]. In an effort to find a direct route to (R)-**4** from commercially available **3**, we screened our collection of isolated yeast dehydrogenases (as GST-fusion proteins) and discovered that the YDR368w enzyme afforded (R)-**4** in 97% ee [18]. By contrast, the short-chain dehydrogenase encoded by the YOL151w yielded (S)-**4** in 90% ee. While commercial interest in this enantiomer is much lower, the result further illustrates the stereochemical diversity of individual enzymes found in a single organism. These bioconversions were also successfully carried out with engineered E. coli strains overexpressing the yeast dehydrogenases.

11.2.2.3 Atorvastatin Building Block

The cholesterol-lowering drug, Atorvastatin, has been a top-selling pharmaceutical product over the past several years (Scheme 11.3). Not surprisingly, its synthesis in homochiral form has attracted a great deal of attention. The pyrrole and most of the flanking groups are achiral; however, the heptanoate side-chain contains two secondary hydroxyl groups whose stereochemistry must be controlled. This was the portion that attracted our interest. Our efforts were not aimed at developing a process to unseat the current manufacturing route, but instead to learn whether a chemoenzymatic approach could deliver all possible stereoisomer for this type of structure. Beyond its intrinsic academic appeal, synthesizing all potential

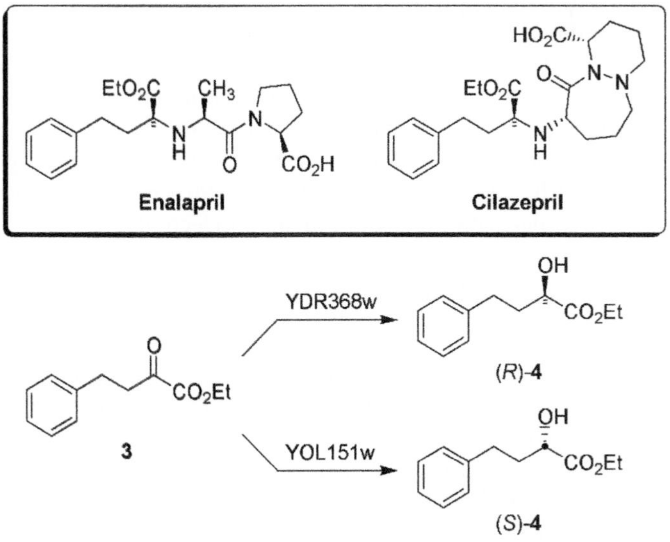

SCHEME 11.2

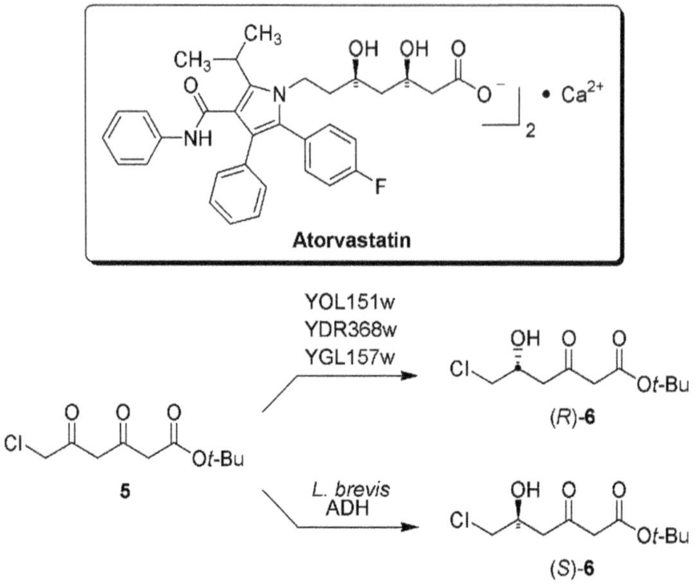

SCHEME 11.3

isomers of a target is required in early drug development and toxicology studies so th
biological activities can be profiled completely. In these applications, high volumetri
uctivity is not necessary; instead, high stereoselectivity and rapid enzyme discovery are
success.

We identified (R)- and (S)-6 as our targets for producing all four stereoisomers of
Atorvastatin building block since diastereoselective chemical transformations can
configuration of the δ-hydroxyl to afford either the syn- or anti-diol (Scheme 11.3). Th
to introduce the first asymmetric center. Müller and coworkers identified an alcoho
drogenase from *Lactobacillus brevis* that enantio- and regioselectively reduced β,δ
ester 5 to (S)-6 [22,23]. Whole cells of baker's yeast were originally described as t
biocatalyst for producing (R)-6; however, the modest enantioselectivity observed (4
prompted a search for better alternatives. In collaboration with the Müller gro
explored two complementary strategies. By systematically altering reaction condit
was possible to increase the enantiomeric purity of (R)-6 to 94% ee when whole yea
were used as the biocatalyst [20]. Alternatively, screening individual yeast dehydrogen
GST-fusion proteins revealed that three (encoded by the YOL151w, YDR368
YGL157w genes) produced (R)-6 in >96% ee. These reactions could also be carried
gram scales with no erosion of stereoselectivity.

11.2.2.4   Taxol Side-Chain Precursor

Taxol has proven highly effective against certain ovarian and breast cancers (Schem
The key step in the commercial route to this compound involves coupling an ad
precursor of the terpene core isolated from natural sources with the N-benzoyl-phenylis
side-chain prepared chemically [24]. Because of its small size, dense functionality and p
economic value, the Taxol side-chain has attracted a great deal of synthetic attenti
While few of these are likely to compete directly with the currently practiced commercia
the molecule has provided an important test-bed for methods in asymmetric synthesis

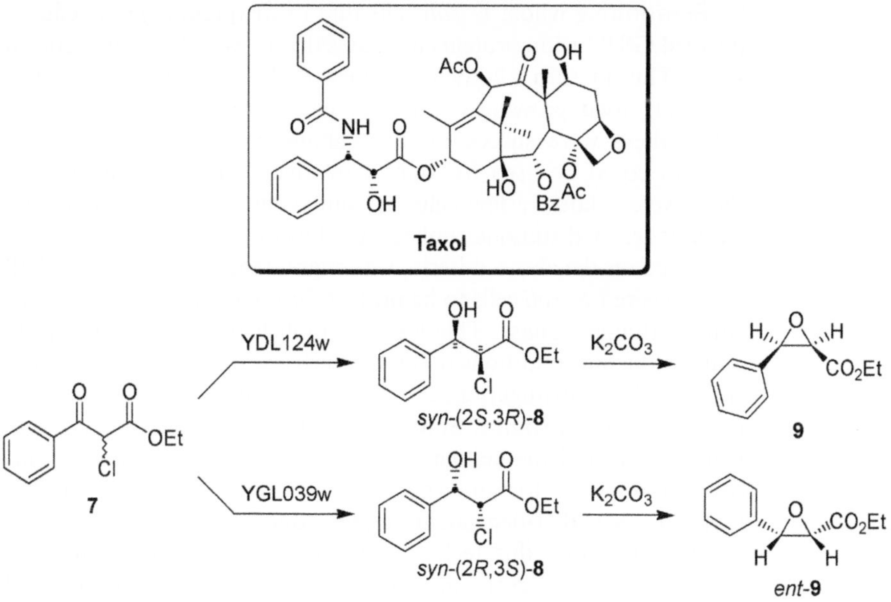

SCHEME 11.4

We envisioned a concise approach to the Taxol side-chain ethyl ester that would rely on an asymmetric reduction of α-chloro-β-keto ester **7** as the key step (Scheme 11.4) [26]. The highly acidic α-proton allows facile enol formation and reprotonation to occur, even at neutral pH values, so that **7** is racemic under the reduction conditions. This allows a dynamic kinetic resolution to be carried out, provided that the reduction methodology is both diastereo- and enantioselective. An early attempt to use whole yeast cells for stereoselective reduction of **7** yielded a disappointing mixture of diastereomers; we hoped that substituting individual yeast reductases might be more successful. Screening our collection of yeast GST-fusion proteins uncovered two with the desired properties. The dehydrogenase encoded by the YDL124w gene gave *syn*-(2S,3R)-**8** as the only observable product. This is the stereoisomer required to complete the synthesis of the natural enantiomer of the Taxol side-chain, and its absolute configuration was confirmed by conversion to the known *cis*-epoxide **9**. We also discovered that the protein product of the YGL039w gene afforded the enantiomeric *syn*-alcohol **8** as the major product (90% of the total). Unfortunately, it was not possible to separate the diastereomeric chlorohydrins from this reduction chromatographically; however, after cyclization to epoxide *ent*-**9**, the diastereomers could be separated cleanly. We have used both **9** and *ent*-**9** to produce the natural and unnatural enantiomers of the Taxol side-chain ethyl ester, respectively. This sequence illustrates how judicious choices of biocatalysts allow a common starting material to be converted into both target enantiomers without the need for resolution.

11.2.3 LARGE-SCALE REDUCTIONS

All of the efforts described above were focused on identifying yeast reductases with suitable substrate- and stereoselectivities toward specific ketones. Discovering the most appropriate enzyme(s), however, solves only half of the problem. Employing these conversions in synthetic routes demands that grams (or kilograms) of product can be prepared. One advantage of our strategy is that all reductases are produced by a common promoter in the same host cell. This simplifies process development.

Substituting whole *E. coli* cells that overexpress a yeast reductase for the corres
purified GST-fusion protein eliminates the time- and resource-consuming protein puri
steps. This in turn allows reductions to be carried out on larger (multiliter) sca
originally used growing cultures of engineered *E. coli* cells to carry out reductio
While these were successful, the volumetric productivities were relatively low si
cultures could be used for only a relatively short period of time. Early in the re
there were relatively few cells per unit volume, and this limited catalytic activity.
cultures reached stationary phase, the bioconversions slowly ceased.

Based on the above criteria, we sought conditions that would allow a high concer
of engineered *E. coli* cells to be present during the reduction while not allowing the cu
enter stationary phase. This was accomplished by first growing the engineered *E. c*
under inducing conditions to load the cells with the reductase of interest. After they h
collected by centrifugation, the cells were resuspended in a minimal salts medium la
reduced nitrogen source. Solutions of glucose (to provide a source of NADPH by
metabolism) and the ketone substrate were added by pumps. This two-step approa
originally devised for an enzymatic Baeyer–Villiger oxidation [28], but it also prov
useful for NADPH-dependent ketone reductions [29].

We have used this technique for preparing the chlorohydrins (**8** and its enan
required for our Taxol side-chain synthesis. As we have observed for many hydro
substrates, high concentrations of **7** were toxic and dramatically retarded *E. coli* cell
However, using fed-batch conditions, several grams of optically pure chlorohydrin c
obtained from a lab-scale 5 L fermenter [26].

11.3 SUMMARY AND FUTURE OUTLOOK

Our experience in cloning, expressing and screening a large fraction of the dehydro
from baker's yeast has yielded several useful lessons. First, it has demonstrated I
performance of yeast reductions can be increased dramatically by focusing on ine
enzymes. In contrast to earlier strategies that attempted to minimize the catalytic acti
undesirable yeast enzymes by including inhibitors in reductions by whole yeast cell
substituting strains with single gene knockouts, employing pure yeast reductases ve
yields stereochemically pure alcohols. In addition, we have shown that even a
collection of approximately 20 reductases culled from a single genome is sufficient
to address many synthetic needs. It was gratifying to discover that in most cases, e
complementary enzymes could be identified. Finally, our experience has shown that I
lyst and discovery and process development can be accelerated significantly by so
cloned proteins in pure form. Even without the use of robotic assistance, screening r
can be carried out rapidly and answers obtained in less than 48 h. Once the desired |
have been identified, the corresponding genes are also immediately known, which fa
further reaction engineering by altering overexpression levels, by moving the genes 1
process-friendly hosts, etc.

While it might be supposed that proteins with high sequence similarities wou
similar substrate- and stereoselectivities, our results argue that this is not necessa
case. This is an important reason that we favor empirical screening to find useful enzy
a given substrate, rather than trying to predict behavior on the basis of computer me
For nearly all of the enzymes investigated here, neither the physiological substrates
three-dimensional structures of the yeast reductases are known, so that little exper
data is available to guide computer modeling approaches. We are also reluctant to us
acid similarity to existing library members as a basis for rejecting potential library a
since their properties may actually be quite novel.

11.3.1 Judging Completeness in a Collection

There are two ways to judge whether a given collection of enzymes is "complete." On the one hand, if a library provides a solution to every synthetic problem investigated, it can be considered sufficiently large. Given the infinite number of potential substrates, this is obviously an impossible goal to achieve in practice. It might therefore be more reasonable to ask how many enzymes are required to fulfill a significant fraction, for example 80%, of the synthetic demands. This is a key question when deciding an adequate level of resources devoted to discovering, cloning and isolating the fusion proteins that make up a library. In the case of yeast dehydrogenases, even our modest collection of approximately 20 members has generally been up to the task. There are exceptions, however, and we are currently addressing these needs by adding additional proteins (*vide infra*).

The second way to judge "completeness" in an enzyme library is to ask whether they collectively yield all of the products observed when a substrate is tested against whole cells of the organism from which the genes were cloned. In our example of baker's yeast dehydrogenases, it is clear that we have not yet met this criterion for completeness. For example, none of the yeast enzymes in our current collection produced (*S*)-**6**, even though this was a major product from reactions involving whole yeast cells. Reductions of α-chloro-β-keto ester **7** provide another example, where whole yeast cells produce an alcohol diastereomer that is not observed when **7** is screened against our 20 library members [30]. This problem also highlights an important challenge in using a bioinformatics approach to identifying all of the relevant enzymes in a given organism. While it is probable that (*S*)-**6** originates from one or more of the enzymes listed in Table 11.1 that are not yet members of our yeast dehydrogenase library, we cannot discount the possibility that an enzyme with little or no sequence similarity to well-characterized dehydrogenases and carbonyl reductases is responsible. Such an enzyme would not appear in Table 11.1. The only recourse would be to examine every ORF in the yeast genome until every protein involved has been identified, the original purpose for which the library of yeast GST-fusion proteins was developed [17].

11.3.2 Expanding the Dehydrogenase Collection

In practice, the most common problem with the existing GST-fusion protein library was that not every alcohol stereoisomer could be prepared, particularly in the case of α-substituted β-keto esters. How should the collection be expanded to solve these problems? Given limited resources, the key is to focus the search for new enzymes on the subset most likely to broaden the range of available products. One possibility is to include the additional *S. cerevisiae* reductase candidates listed in Table 11.1. When "missing" products are known to be produced by reductions with whole yeast cells, this strategy has a good chance of succeeding. Because only a subset of genes are expressed in commercial yeast cells, this approach is also likely to provide additional products beyond those anticipated from whole-cell results. It is also possible to expand our current dehydrogenase collection by incorporating genes from other organisms. In cases, where gene sequences are known (either by cloning individual proteins or by sequencing whole genomes), these can be expressed as GST-fusion proteins in the same manner as used for the existing collection. On the other hand, when whole cells of an unsequenced organism are known to carry out the reaction of interest, more laborious approaches (protein isolation, amino acid microsequencing, etc.) are required. We favor this strategy only for the most difficult cases. Our first choice would be to examine already-known genes. Given the diversity in substrate- and stereoselectivities exhibited by proteins with highly similar sequences, we believe that this will provide synthetic solutions in most cases.

11.3.3   WIDER APPLICABILITY AND LIMITATIONS

We have used the *S. cerevisiae* genome and the asymmetric ketone reductions to prov
proof-of-principle for our approach. We deliberately designed the strategy so that it cc
applied to any genome and reaction of interest. All that is required is to have sequence
one or more proteins known to catalyze a particular reaction, which can be used to
sequence databases and provide library candidates. Creating expression clones for
fusion proteins is the most labor-intensive phase of the project. While our original
of *E. coli* overexpression plasmids was constructed by stepwise "cut-and-paste" mo
biology techniques, we have more recently developed recombination-based cloning m
that allow direct incorporation of PCR products into the fusion protein expression
[31]. This has dramatically shortened the time required to isolate purified protei
expedites screening. It should be noted that a variety of other expression systen
purification "handles" are available; using *E. coli* and fusion with GST was an ex
choice based on the methods used to construct the original genome-wide library
S. cerevisiae.

While we believe that our approach has a number of advantages, it is also important
several limitations in mind. First, success in identifying candidate proteins by genome
depends on access to sequences of proteins known to carry out the desired chemical con
("bait" sequences). Including members of multiple sequence families as sequence
increases both the number and the variety of candidates that may exhibit the desired c
characteristics. In some cases, however, no suitable "bait" sequences exist, either beca
enzyme catalyst for the reaction of interest has been identified or because the proteins ha
yet been sequenced. This precludes genome mining. The solution will be to screen who
for the reaction of interest, then to identify the enzyme(s) and the gene(s) involved. Gi
very large number of whole genomes sequences available, having even one "bait" se
from such an approach will likely result in a number of library candidates.

A second limitation on phenotypic screening of proteins is that they must be cataly
active. Problems with low expression, including body formation and misfolding
minimized by expressing library proteins in suitable hosts. To date, we have relied on
because of its rapid growth in inexpensive media coupled with efficient overexp
systems. Other hosts might be more appropriate for different classes of enzymes, ho
particularly those requiring specific posttranslational modifications or cofactors that
available in the *E. coli* cytoplasm.

Finally, protein structure may also limit the breadth of enzymes amenable to our
approach; membrane-bound enzymes must be purified and screened in the presence of
gents and those that are only active as part of multienzyme complexes are handled i
forms. To date, we have not made significant efforts to include these types of "di
proteins in our library. We always observe several enzyme "hits" for a given substrate, i
or poorly expressed proteins are simply dropped from further consideration. In ou
enlarging the library by adding well-expressed and stable proteins is more resource-e
than spending inordinate efforts to make even poorly behaved proteins available. Hamp
these problems, such enzymes are unlikely to be used as the basis for a scaleable biopro

ACKNOWLEDGMENTS

The author's research in this area has been generously supported by the National
Foundation (CHE-0130315) and Great Lakes Fine Chemicals. We are also grateful to
Martzen for sharing the original yeast GST-fusion protein clones. The real credit for th
belongs to the students and coworkers whose hard work has reduced the ideas to pra

REFERENCES

1. Patel, R.N., Enzymatic synthesis of chiral intermediates for drug development, *Adv. Synth. Catal.*, 343, 527–546, 2001.
2. Breuer, M., Ditrich, K., Habicher, T., et al., Industrial methods for the production of optically active intermediates, *Angew. Chem. Int. Ed. Engl.*, 43, 788–824, 2004.
3. Grasa, G.A., Singh, R., and Nolan, S.P., Transesterification/acylation reactions catalyzed by molecular catalysts, *Synthesis*, 971–985, 2004.
4. Nakamura, K., Yamanaka, R., Matsuda, T., et al., Recent developments in asymmetric reduction of ketones with biocatalysts, *Tetrahedron Asymmetry*, 14, 2659–2681, 2003.
5. Santaniello, E., Ferraboschi, P., Grisenti, P., et al., The biocatalytic approach to the preparation of enantiomerically pure chiral building blocks, *Chem. Rev.*, 92, 1071–1140, 1992.
6. Santaniello, E., Ferraboschi, P., and Manzocchi, A., Recent advances on bioreductions mediated by baker's yeast and other microorganisms, in *Enzymes in Action. Green Solutions for Chemical Problems*, Zwanenburg, B., Mikolajczyk, M, Kielbasinski, P., Eds., Kluwer Academic, Dordrecht, The Netherlands, 2000, pp. 95–115.
7. Seebach, D., Sutter, M.A., Weber, R.H., et al., Yeast reduction of ethyl acetoacetate: (*S*)-(+)-3-hydroxybutanoate, *Org. Synth.*, 63, 1–9, 1985.
8. Zhou, B.-N., Gopalin, A.S., VanMiddlesworth, F., et al., Stereochemical control of yeast reductions. 1. Asymmetric synthesis of L-carnitine, *J. Am. Chem. Soc.*, 105, 5925–5926, 1983.
9. Shieh, W.-R., Gopalin, A.S., and Sih, C.J., Stereochemical control of yeast reductions. 5. Characterization of the oxidoreductases involved in the reduction of β-keto esters, *J. Am. Chem. Soc.*, 107, 2993–2994, 1985.
10. Nakamura, K., Kawai, Y., Nakajima, N., et al., Stereocontrol of microbial reduction. 17. A method for controlling the enantioselectivity of reductions with bakers' yeast, *J. Org. Chem.*, 56, 4778–4783, 1991.
11. Baraldi, P.T., Zarbin, P.H.G., Vieira, P.C., et al., Enantioselective synthesis of (*R*)- and (*S*)-2-methyl-4-octanol, the male-produced aggregation pheromone of *Curculionidae* species, *Tetrahedron Asymmetry*, 13, 621–624, 2002.
12. Schweizer, E. and Bolling, H., A *Saccharomyces cerevisiae* mutant defective in saturated fatty acid biosynthesis, *Proc. Natl. Acad. Sci. USA*, 67, 660–666, 1970.
13. Kühn, L., Castorph, H., and Schweizer, E., Gene linkage and gene-enzyme relations in the fatty-acid-synthetase system of *Saccharomyces cerevisiae*, *Eur. J. Biochem.*, 24, 492–497, 1972.
14. Rodriguez, S., Kayser, M.M., and Stewart, J. D., Improving the stereoselectivity of bakers' yeast reductions by genetic engineering, *Org. Lett.*, 1, 1153–1155, 1999.
15. Rodriguez, S., Kayser, M.M., and Stewart, J.D., Highly stereoselective reagents for β-keto ester reductions by genetic engineering of bakers' yeast, *J. Am. Chem. Soc.*, 123, 1547–1555, 2001.
16. Stewart, J.D., Rodriguez, S., and Kayser, M.M., Cloning, structure and activity of ketone reductases from bakers' yeast, in *Enzyme Technology for Pharmaceutical and Biotechnological Applications*, Kirst, H.A., Yeh, W.-K., and Zmijewski, M.J., Eds., Marcel Dekker, New York, 2001, pp. 1–208.
17. Martzen, M.R., McCraith, S.M., Spinelli, S.L., et al., A biochemical genomics approach for identifying genes by the activity of their products, *Science*, 286, 1153–1155, 1999.
18. Kaluzna, I., Andrew, A.A., Bonilla, M., et al., Enantioselective reductions of ethyl 2-oxo-4-phenylbutyrate by *Saccharomyces cerevisiae* dehydrogenases, *J. Mol. Catal., B Enzym.*, 17, 101–105, 2002.
19. Kaluzna, I.A., Matsuda, T., Sewell, A.K., et al., A systematic investigation of *Saccharomyces cerevisiae* enzymes catalyzing carbonyl reductions, *J. Am. Chem. Soc.* 126, 12827–12832, 2004.
20. Wolberg, M., Kaluzna, I.A., Müller, M., et al., Regio and enantioselective reduction of *t*-butyl 6-chloro-3,5-dioxohexanoate with baker's yeast, *Tetrahedron Asymmetry*, 15, 2825–2828, 2004.
21. Kaluzna, I.A., Feske, B.D., Wittayanan, W., et al., Stereoselective, biocatalytic reductions of α-chloro-β-keto esters, *J. Org. Chem.*, 70, 342–345, 2005.
22. Wolberg, M., Hummel, W., and Müller, M., Biocatalytic reduction of β,δ-diketo esters: a highly stereoselective approach to all four stereoisomers of a chlorinated β,δ-dihydroxy hexanoate, *Chem. Eur. J.*, 7, 4562–4571, 2001.

23. Wolberg, M., Ji, A., Hummel, W., and Müller, M., Enzymatic reduction of hydrophobic β esters, *Synthesis*, 937–942, 2001.
24. Wall, M.E. and Wani, M.C., Paclitaxel—from discovery to clinic, *ACS Symp. Ser.*, 58. 1995.
25. Borah, J.C., Gogoi, S., Boruwa, J., et al., A highly efficient synthesis of the C-13 side-chain using Shibasaki's asymmetric henry reaction, *Tetrahedron Lett.*, 45, 3689–3691, 2004.
26. Feske, B.D., Kaluzna, I.A., and Stewart, J.D., Enantiodivergent, biocatalytic routes to be C-13 side-chain enantiomers, *J. Org. Chem.*, 70, 9654–9657, 2005.
27. Rodriguez, S., Schroeder, K.T., Kayser, M.M., et al., Asymmetric synthesis of β-hydroxy e α-alkyl-β-hydroxy esters by recombinant *Escherichia coli* expressing enzymes from bake: *J. Org. Chem.*, 65, 2586–2587, 2000.
28. Walton, A.Z. and Stewart, J.D., An efficient enzymatic Baeyer–Villiger oxidation by en *Escherichia coli* cells under non-growing conditions, *Biotechnol. Prog.*, 18, 262–268, 2002.
29. Walton, A.Z. and Stewart, J.D., Understanding and improving NADPH-dependent reac non-growing *Escherichia coli* cells, *Biotechnol. Prog.*, 20, 403–411, 2004.
30. Cabon, O., Buisson, D., Larcheveque, M., et al., The microbial reduction of 2-chloro-3-o *Tetrahedron Asymmetry*, 6, 2199–2210, 1995.
31. Bougioukou, D. and Stewart, J.D., unpublished results, 2005.

12 Immobilization of Enzymes as Cross-Linked Enzyme Aggregates: A Simple Method for Improving Performance

Roger A. Sheldon

CONTENTS

12.1 INTRODUCTION

Currently much attention is focused on the application of atom efficient catalytic methodologies—heterogeneous, homogeneous, and enzymatic—in organic synthesis, both in industry and in academe [1]. The ultimate goal is the development of green, sustainable technologies for the manufacture of (fine) chemicals. In this context, biocatalytic methodologies have many potential benefits compared with traditional organic syntheses. They generally employ mild reaction conditions (ambient temperature and pressure at physiological pH) and afford high chemo-, regio-, and stereoselectivities. Furthermore, enzymatic syntheses generally involve few steps by obviating the need for protection and deprotection steps, and by avoiding the use of environmentally unattractive organic solvents. This affords syntheses that are short, less energy intensive, and generates less waste; hence are both environmentally and economically more attractive.

The time is ripe for the widespread application of biocatalysis, using whole cells or isolated enzymes in industrial organic synthesis [2–16]. According to a 2002 review [2], there are more than 130 biotransformations currently performed in an industrial scale, mostly in the manufacture of pharmaceutical intermediates and other fine chemicals. Advances in recombinant

DNA techniques [17] have made it, in principle, possible to produce any enzyme
commercially acceptable price and widespread sourcing of enzymes from extreme er
ments and genomic DNA data has significantly expanded the pool of available enzyme
Furthermore, advances in protein engineering have made it possible, using techniques s
site-directed mutagenesis and *in vitro* evolution by means of DNA shuffling [19–
manipulate enzymes such that they exhibit the desired properties: substrate spec
activity, selectivity, productivity, stability, pH profile, and so forth. Nonetheless, the
mercialization of many enzymatic processes is often hampered by the lack of opera
stability of many enzymes, coupled with their relatively high price. This impedime
generally be overcome if one can find an effective method for their immobilization [7,2
If successful, this not only results in improved operational stability but also allows fo
separation and reuse of the enzyme, and simplifies downstream processing.

Conceptually, immobilization methods can be conveniently divided into three
(a) binding to a support (carrier), (b) encapsulating in an inorganic or organic pol
gel, (c) cross-linking of the protein molecules [23]. These three types can be further sube
on the basis of the technique used for binding, encapsulating, or cross-linking as sh
Figure 12.1.

A distinct disadvantage of carrier-bound enzymes, whether they involve binding
encapsulation in, a carrier, is the dilution of catalytic activity resulting from the int
tion of a large proportion of noncatalytic mass, generally ranging from 90 to >99%
total mass [23]. This inevitably leads to lower volumetric and space-time yields, and
catalyst productivities. Attempts to achieve high enzyme loadings usually lead to
activity due to leaching. The third type of immobilization, cross-linking of enzym
ecules with a bifunctional cross-linking agent, most commonly glutaraldehyde, do
suffer from this disadvantage. The molecular weight of the cross-linking agent is neg
compared with that of the enzyme and the resulting biocatalyst essentially comprises
active enzyme.

Whichever method is used, an ideal industrial immobilized enzyme has to meet
criteria, such as being recyclable, broadly applicable, cost-effective, and safe for use. F
more, deactivation is not always avoidable. Consequently, the disposal of deactivated
bilized enzymes also has to be taken into account, especially for a large-scale inc
production that might require multihundred tons of a carrier-bound immobilized e
on an annual basis, for example, the immobilized penicillin G acylase needed for the p
tion of semisynthetic β-lactam antibiotics.

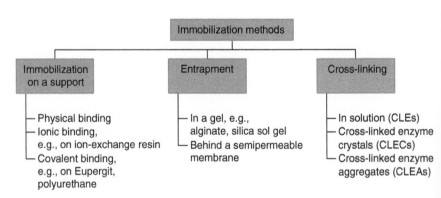

FIGURE 12.1 Methods for immobilization of enzymes.

12.2   IMMOBILIZATION BY CROSS-LINKING

The technique of protein cross-linking by the reaction of glutaraldehyde with reactive NH_2 groups on the protein surface was initially developed in the 1960s [24]. However, this method of producing cross-linked enzymes (CLEs) had several drawbacks, such as low activity retention, poor reproducibility, low mechanical stability, and difficulties in handling the gelatinous CLEs. Mechanical stability and ease of handling could be improved by cross-linking the enzyme in a gel matrix or a carrier, but this led to the disadvantageous dilution of catalytic activity. Consequently, in the late 1960s, emphasis switched to carrier-bound enzymes, which became the most widely used industrial methodology for enzyme immobilization for the next three decades.

The cross-linking of a crystalline enzyme by glutaraldehyde was first described by Quiocho and Richards in 1964 [25]. Their main objective was to stabilize enzyme crystals for x-ray diffraction studies, but they also showed that catalytic activity was retained. The use of cross-linked enzyme crystals (CLECs) as industrial biocatalysts was pioneered by scientists at Vertex Pharmaceuticals in the early 1990s [26], and subsequently commercialized by Altus Biologics [27–30]. The initial studies were performed with CLECs of thermolysin, of interest in the manufacture of aspartame, but the method was subsequently shown to be applicable to a broad range of enzymes. CLECs proved significantly more stable to denaturation by heat, organic solvents, and proteolysis than the corresponding soluble enzyme or lyophilized (freeze-dried) powder. CLECs are robust, highly active immobilized enzymes of controllable particle size, varying from 1 to 100 µm. Their operational stability and ease of recycling, coupled with their high catalyst and volumetric productivities, render them ideally suited for industrial biotransformations.

12.3   CROSS-LINKED ENZYME AGGREGATES

An inherent disadvantage of CLECs is the need to crystallize the enzyme, which is often a laborious procedure requiring enzymes of high purity. Consequently, we reasoned that comparable results could possibly be achieved by simply precipitating the enzyme from aqueous solution, using standard techniques, and cross-linking the resulting physical aggregates of enzyme molecules (Figure 12.2). This indeed proved to be the case and led to the development of a new family of cross-linked enzymes, which we have called cross-linked enzyme aggregates (CLEA). Protein purification by precipitation from aqueous solution with, e.g., ammonium sulfate or polyethylene glycol, is the most frequently used primary method of protein purification, and the preparation of CLEAs can be easily integrated into a purification protocol as it does not require a highly purified enzyme.

It is well known [31] that the addition of salts, organic solvents, or nonionic polymers to aqueous solutions of proteins leads to their precipitation as physical aggregates of protein molecules without perturbation of their tertiary structure, that is without denaturation. These

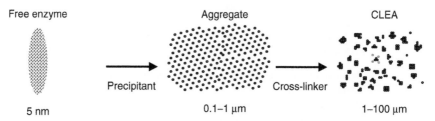

Free enzyme   Aggregate   CLEA

Precipitant   Cross-linker

5 nm   0.1–1 µm   1–100 µm

FIGURE 12.2 Preparation of a CLEA.

solid aggregates are held together by noncovalent bonding, and readily collapse an solve when dispersed in water. Cross-linking of these physical aggregates produces that are rendered permanently insoluble while maintaining the preorganized superstru the aggregates, and, hence their catalytic activity. Initial studies [32,33] focused synthesis of CLEAs from penicillin G acylase (penicillin amidohydrolase, E.C. 3.5.1 industrially important enzyme used in the synthesis of semisynthetic penicillin and ce porin antibiotics [34]. The free enzyme has limited thermal stability and low toler organic solvents, making it an ideal candidate for stabilization as a CLEA. Indeed, p G acylase CLEAs, prepared by precipitation with, for example, ammonium su *tert*-butanol, proved to be effective catalysts for the synthesis of ampicillin acco Reaction 12.1 [35].

D-Phenylglycine amide   6-APA   Ampicillin

Reaction 12.1

The CLEA exhibited a synthesis/hydrolysis ratio (S/H) comparable with that of the c cial catalyst, PGA-450 (penicillin G acylase immobilized on poly-acrylamide), and subs higher than that of the penicillin G acylase CLEC (Table 12.1). This suggests the sional limitations are more severe in the CLEC than in the CLEA or the PGA-450. Rem the productivity of the CLEA was even higher than that of the free enzyme that it was ma and substantially higher than that of the CLEC. Not surprisingly, the productivit commercial catalyst was very low, a reflection of the fact that it mainly consists of non ballast in the form of polyacrylamide carrier. Analogous to the corresponding CLE penicillin G acylase CLEAs also maintained their high activity in organic solvents [35–

12.4 LIPASE CLEAS: HYPERACTIVATION AND THE EFFECT OF ADDITIV

We next turned our attention to an investigation of the effect of various parameters, the precipitant and the addition of additives, and surfactants and crown ethers,

TABLE 12.1
Ampicillin Synthesis Catalyzed by Different Penicillin G Acylase Preparations

Biocatalyst	Conversion (%)	S/H[a]	V_{syn} (μmol/U/h)[b]	Relative Prc
Free enzyme	88	2.0	25.5	100
CLEC	72	0.71	39.6	39
T-CLEA[d]	85	1.58	38.2	151
PGA-450	86	1.56	15.6	0

[a]Synthesis and hydrolysis molar ratio at the conversion listed.
[b]Initial reaction rate.
[c]Relative productivity at maximum conversion, free enzyme set at 100.
[d]CLEA produced using *tert*-butanol as precipitant.

TABLE 12.2
Hyperactive Lipase CLEAs in Organic Medium

Lipase	Precipitant/Additive	Relative Reactivity (%)
C. antarctica A	$(NH_4)_2SO_4$/SDS	900
C. antarctica B	DME	75
P. alcaligenes	$(NH_4)_2SO_4$	1200
T. lanuginosus	DME	82
Rh. miehei	$(NH_4)_2SO_4$/SDS	250
A. niger	$(NH_4)_2SO_4$/SDS	200
C. rugosa	$(NH_4)_2SO_4$/TR	420

activities of CLEAs prepared from seven commercially available lipases [38]. The activation of lipases by additives, such as surfactants, crown ethers, and amines, is well documented and is generally attributed to the lipase induced to adopt a more active conformation [39]. We, therefore, reasoned that cross-linking of enzyme aggregates, resulting from precipitation in the presence of such an additive, would "lock" the enzyme in this more favorable conformation. Moreover, since the additive is not covalently bonded to the enzyme, the additive can subsequently be washed from the CLEA using, for example, an appropriate organic solvent.

Using this procedure, we succeeded in preparing a variety of hyperactive lipase CLEAs exhibiting activities even higher than the corresponding free enzyme, that is up to 12 times the activity of the free enzyme in the hydrolysis of ethyl octanoate in a 1,2-dimethoxyethane/water (95/5: v/v) mixture (Table 12.2) [39]. We also demonstrated that the experimental procedure for CLEA preparation could be further simplified by combining precipitation, either in the presence of additives or in the absence of additives, with cross-linking in a single operation [39].

Hence, the potential of the CLEA technology for preparing immobilized enzymes with high catalyst and volumetric productivities, in some cases exceeding those of the native enzymes that they were derived from, was firmly established. The method is exquisitely simple and can be performed with relatively impure enzyme preparations. Subsequent studies have been aimed at optimizing protocols with regard to parameters, such as temperature, pH, concentration, stirring rate, precipitant, additives, and cross-linking agent and exploring the scope of the technology (see later). The relative simplicity of the operation ideally lends itself to automation, e.g., using 96-well plates.

12.5 CLEA ACTIVITIES IN "ANHYDROUS" ORGANIC MEDIA

It is noteworthy that the CLEA derived from the most popular lipase, *Candida antarctica* lipase B (CaLB), showed only very moderate hyperactivation. Subsequent studies of CLEA preparations revealed that the optimum performance of lipase CLEAs observed in aqueous media could not be directly translated to organic media. Hence, the preparation of the CLEA was modified to allow for a better diffusion of substrates into the particles and to decrease the repulsion of organic solvents. Application of this modified "organic media technology" for

TABLE 12.3
Activities of Various Lipase B Formulations in Water and Organic Media

CaLB Preparation	Aqueous Activity[a]	Organic Activity[b]	Ratio
Native (lyophilized NOVO preparation)	22,000	—	
Novozym 435	7,300	250	
CLEA-AM[c]	38,000	50	
CLEA-OM[c]	31,000	1,500	

[a]Tributyrin: 5 vol% in 40 mM Tris buffer, pH 7.5, 40°C.
[b]Phenylethylamine 41 mM, n-butyl methoxyacetate 34 mM, 12 mg/ml, molecular sieve 4A, 40°C.
[c]AM = Developed for aqueous media, OM for organic media.

CLEA preparation dramatically improved the activity of CaLB CLEA in the enantiose
acylation of 1-phenethylamine in diisopropyl ether as solvent (Reaction 12.2). Tab
shows a comparison of the results obtained with Novozym 435, the immobilized f
CaLB available from Novozymes and with CaLB CLEAs optimized for operation i
or organic solvents. Clearly, the optimized CaLB CLEAs have activities surpassing t
Novozym 435 in both aqueous and organic media. It is also worth noting that Novoz
cannot be recycled in water as the enzyme is leached from the surface (immobilizatic
not involve covalent bonding to the support).

Reaction 12.2

We have also shown that CaLB CLEAs exhibit superior activities, compared with st
organic solvents, in ionic liquids and supercritical carbon dioxide (unpublished results)

12.6 PARTICLE SIZE AND MASS TRANSFER EFFECTS

One major property of CLEAs, that was up to now relatively unexplored, is their
shape and size, which obviously can have a direct effect on mass transfer limitations
operational conditions. The number of enzyme molecules and the way they are
together in an aggregate can be expected to have a crucial influence on the activity
aggregate as a whole. Hence, an understanding of the parameters which influence parti
and how to control them will pave the way for changing the CLEA from an inte
phenomenon into a mature, well-defined catalytic particle. Scanning electron micr
showed a very uniform structure of the aggregates (see Figure 12.3 for *C. antarctica* lip
The diameter of the CLEA is about 1 μm with a small deviation. Taking an enzyme
CaLB of $5 \times 5 \times 5$ nm, a single CLEA particle contains a maximum 8×10^6 enzyme mo
CLEAs can form larger clusters that do have mass-transport limitations, especially
UV-based assays. The size of these clusters can be up to 100 μm (Figure 12.3), makin
visible to the naked eye. The number of CLEAs in a cluster is less uniform than the e
in an aggregate. It can vary from a few to a hundred thousand. Our previous findin

FIGURE 12.3 *Candida antarctica* lipase A/B CLEA, 1 CLEA particle can contain up to 8 million enzyme molecules (magnification 3500×).

laser scattering (data not shown) suggested that a variable number of enzyme molecules per aggregate are now rationalized as variable number of aggregates per cluster.

When a dispersed CLEA is assayed for activity, directly after dilution of the cross-link medium, a higher activity is observed than when the sample is centrifuged and redispersed. This treatment presumably does not disturb the individual CLEA and thereby the enzyme structure, but it does squeeze the CLEA particles close together, elevating mass-transport limitations to a noticeable level. Some differences between enzymes were observed: with CaLB very large and hydrophobic clusters were obtained, whereas with β-galactosidase well-dispersible suspensions were common. The most noticeable structural difference between these two enzymes is that β-galactosidase is extensively glycosylated while CaLB is not. Comparing activities of CaLB CLEA, found in the relatively fast hydrolysis of *p*-nitrophenyl propionate measured by UV/Vis absorption, with slower hydrolysis of triacetin (monitored by titration) the mass-transport limitation was obvious. Compared with free enzyme, the first showed 35% activity recovery and the latter 177%. For β-galactosidase, however, activity recovery found in the hydrolysis of *p*-nitrophenyl-β-D-galactopyranoside and in lactose hydrolysis was the same. These two CLEAs emphasize the important effect of cluster formation on the apparent activity.

12.7 CROSS-LINKING AGENTS

Glutaraldehyde is generally the cross-linking agent of choice as it is inexpensive and readily available in commercial quantities. However, we found that good results were not always obtained when using glutaraldehyde as the cross-linker. With some enzymes, e.g., nitrilases,

we observed low or no retention of activity. We were surmized that this total loss of
might be caused by the reaction of cross-linker with amino acid residues, which are cr
the activity of the enzyme. If this was so, then, we expected that inactivation w
particularly severe with glutaraldehyde owing to its high reactivity and small size
allows it to penetrate the internal structure of the protein where it can react with ami
that are essential for activity. If this was the problem then the solution was eminently
use a bulky cross-linking agent, which cannot access the internal surface of the pro
this end, we employed bulky polyaldehydes obtained by periodate oxidation of dextr
followed by reduction of the Schiff's base moieties with sodium borohydride,
irreversible linkages. The activity retention of these CLEAs was generally much high
that observed with CLEAs prepared using glutaraldehyde. Dramatic results were c
with two nitrilases, one from *Pseudomonas fluorescens* and the other from the co
Biocatalytics (Biocatalytics 1004): it was observed that cross-linking with glutara
produced a completely inactive CLEA, while cross-linking with dextran polyaldehy
duced 50 to 60% activity retention (not optimized).

12.8 SCOPE OF THE METHODOLOGY

Having established the potential of the CLEA technology for preparing immobilized
with high catalyst and volumetric activities, in some cases with activities significantly
ing those of the native enzymes they were derived from, we set about investigating i
As noted earlier, the methodology essentially combines enzyme preparation (by preci;
and immobilization into one step. We note, however, that if an impure sample cont
mixture of enzymes is used, this can lead to a CLEA containing more than one enzy
have used this to our advantage in the deliberate preparation of combi CLEAs, co
two or more enzymes, for use in multistep, biocatalytic cascade processes (see later).

We have shown that, by a suitable optimization of the procedure, which may dif
one enzyme to another, the CLEA methodology is applicable to essentially any
including cofactor dependent oxidoreductases and lyases in addition to a wide va
hydrolases (Table 12.4).

For example, we have shown that a CLEA of β-galactosidase, which cataly
hydrolysis of lactose and is administered as "tolerase" to people suffering from
intolerance, was recycled with no loss of activity [41]. We have successfully prepare
lable CLEAs from glucose oxidase and galactose oxidase (Schoevaart, in press). Gui

TABLE 12.4
Examples of Enzymes That Have Been Successfully Cleated

Hydrolases	Oxidoreductases	Lyases
Penicillin acylases (2)	Glucose oxidase	*R*- and *S*-Oxynitrilase
Aminoacylase	Galactose oxidase	Pyruvate decarboxylase
Pig liver esterase	Laccase	*R*-Deoxyribose aldolase
Lipases (7)	Catalase	
Nitrilases (2)	Alcohol dehydrogenase	
Galactosidase	Formate dehydrogenase	
Proteases (2)		
Phytase		

coworkers [42] have recently pointed out that the CLEA technology has an additional benefit: it can stabilize the quaternary structures of multimeric enzymes. To demonstrate this, they prepared CLEAs from two tetrameric catalases [42]. The enzyme stability, which for the soluble enzyme is dependent on concentration, became independent of this parameter in the CLEA, which allowed for the use of low "concentrations" of catalase.

Another development is the inclusion of CLEAs in nonprotein material. Since CLEAs are an average factor 1000 bigger than free enzyme, enclosure into materials that would retain none or only low amounts of enzyme is now feasible [37,43].

12.9 COMBI CLEAS AND CATALYTIC CASCADE PROCESSES

Fine chemical syntheses generally involve multistep syntheses and the ultimate inefficiency is to combine these, preferably catalytic, steps into a one-pot multistep catalytic cascade process [44]. Indeed, this is truly emulating the metabolic pathways conducted in living cells by an elegant orchestration of a series of biocatalytic steps into an exquisite multicatalyst cascade, without the need for separation of intermediates. Catalytic cascade processes have numerous potential benefits: fewer unit operations, less reactor volume, and higher volumetric and space-time yields, shorter cycle times and less waste generation. Furthermore, by coupling steps together unfavorable equilibria can be driven toward product (see later).

Notwithstanding the considerable benefits, catalytic cascade processes are fraught with several problems. Different catalysts, e.g., combinations of biocatalysts and chemocatalysts, are often incompatible. The rates may be very different and the optimum conditions for each catalyst may differ considerably and recycling of a complex mixture of catalysts will not be simple. It is worth noting, in this context, that biocatalytic processes generally proceed under the same conditions: in water at ambient temperature, pressure, and physiological pH. In the living cell interference between the different biocatalytic steps is circumvented by compartmentalization in/behind membranes. Following Nature's example, the key to compatibility would appear to be compartmentalization. This could be achieved, for example, by immobilizing two or more (bio)catalysts, thereby avoiding interference between them. We have achieved this, by immobilizing two or more enzymes in "combi CLEAs," by coprecipitation and cross-linking. For example, we have successfully prepared combi CLEAs containing catalase in combination with glucose oxidase or galactose oxidase, respectively (R. Schoevaart, manuscript in preparation). The catalase serves to catalyze the rapid degradation of the hydrogen peroxide formed in the aerobic oxidation of glucose and galactose, respectively, catalyzed by these enzymes, thus suppressing deactivation of the enzyme by the hydrogen peroxide.

We [45] have recently used a combi CLEA containing an *S*-selective oxynitrilase (from *Manihot esculenta*) and a nonselective nitrilase, in diisopropyl ether/water (9:1) at pH 5.5, 1 h for the one-pot conversion of benzaldehyde to *S*-mandelic acid (Figure 12.4), in high yield and enantioselectivity.

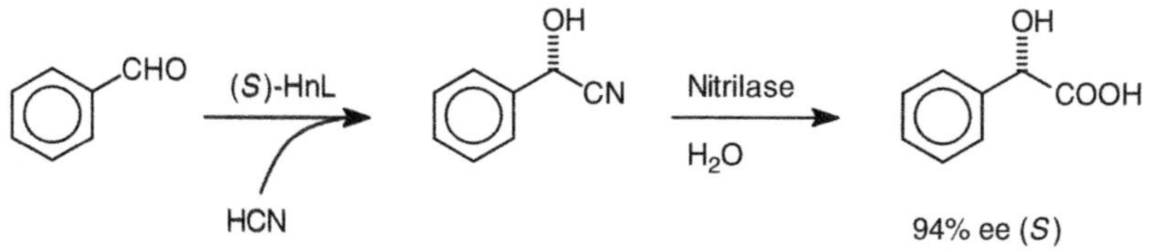

FIGURE 12.4 One-pot conversion of benzaldehyde to *S*-mandelic acid with a combi CLEA.

The enantioselectivity is provided by the oxynitrilase and *in situ* conversion by the nitr
serves to drive the equilibrium of the first step toward product. In principle, this could als
achieved by using an *S*-selective nitrilase in combination with nonenzymatic hydrocyana
(as we have previously shown with an *R*-nitrilase) but, unfortunately, there are no nitri
that exhibit *S*-selectivity with mandelonitriles. We also demonstrated that the combi C'
was more effective than a mixture of the two separate CLEAs.

12.10   CONCLUSIONS AND PROSPECTS

The CLEA technology has many advantages in the context of industrial applications.
method is exquisitely simple and amenable to rapid optimization, which translates tc
costs and short time-to-market. It is applicable to a wide variety of enzymes, including (
preparations, affording stable, recyclable catalysts with high retention of activity.

In contrast to CLECs, there is no need for the enzyme to be available in crystalline
and the technique can be applicable to the preparation of combi CLEAs containing tv
more enzymes. Synthesis of CLEAs in the presence of additives, such as crown ethe
surfactants, provides the possibility of "locking" the immobilized enzyme in a more favo
conformation, resulting in increased activity and/or (enantio)selectivities. We believe
CLEAs will, in the future, be widely applied in industrial biotransformations and other
requiring immobilized enzymes.

REFERENCES

1. Sheldon, R.A., Atom efficiency and catalysis in organic synthesis, *Pure Appl. Chem.*, 72, 1233
   2000.
2. Straathof, A.J.J., Panke, S., and Schmid, A., The production of fine chemicals by biotransl
   tions, *Curr. Opin. Biotechnol.*, 13, 548–556, 2002.
3. Patel, R.N., Biocatalytic synthesis of intermediates for the synthesis of chiral drug substances
   Opin. Biotechnol., 12, 587–604, 2001.
4. Patel, R.N., Microbial/enzymatic synthesis of chiral intermediates for pharmaceuticals, *E*
   Microb. Technol., 31, 804–826, 2002.
5. Liese, A. and Villela, F.M., Production of fine chemicals using biocatalysis, *Curr. Opin. Biote*
   10, 595–603, 1999.
6. Schmid, A., Dordick, J.S., Hauer, B., et al., Industrial biocatalysis today and tomorrow, *N*
   409, 258–268, 2001.
7. Liese, A., Seelbach, K., and Wandrey, C., *Industrial Biotransformations*, Wiley-VCH, Wei
   2000.
8. Thomas, S.M., DiCosimo, R., and Nagarajan, V., Biocatalysis: applications and potentials
   chemical industry, *TIBTECH*, 20, 238–242, 2002.
9. White Biotechnology: Gateway to a More Sustainable Future, EuropaBio Report, 2003,
   europabio.org.
10. Schoemaker, H.E., Mink, D., and Wubbolts, M.G., Dispelling the myths—biocatalysis in ind
    synthesis, *Science*, 299, 1694–1697, 2003.
11. Schmidt, A., Hollmann, F., Park, J.B., et al., The use of enzymes in the chemical industry in E
    Curr. Opin. Biotechnol, 13, 359–366, 2002.
12. Laumen, K., Kittelmann, M., and Ghisalba, O., Chemo-enzymatic approaches for the crea
    novel chiral building blocks and reagents for pharmaceutical applications, *J. Mol. Catal., B*
    19–20, 55–66, 2002.
13. Shaw, N.M., Robins, K.T., and Kiener, A., Lonza: 20 years of biotransformations, *Adv.*
    Catal., 345, 425–435, 2003.
14. Zaks, A., Industrial biocatalysis, *Curr. Opin. Chem. Biol.*, 5, 130–136, 2001.

15. Yazbek, D.R., Martinez, C.A., Hu, A., et al., Challenges in the development of an efficient enzymatic process in the pharmaceutical industry, *Tetrahedron Asymmetry*, 15, 2757–2763, 2004.
16. Huisman, G.W. and Gray, D., Towards novel processes for the fine-chemical and pharmaceutical industry, *Curr. Opin. Biotechnol.*, 13, 352–358, 2002.
17. Stewart, J.D., Dehydrogenases and transaminases in asymmetric synthesis, *Curr. Opin. Chem. Biol.*, 5, 120–129, 2001.
18. DeSantis, G., Burk, M.J., et al., An enzyme library approach to biocatalysis: development of nitrilases for enantioselective production of carboxylic acid derivatives, *J. Am. Chem. Soc.*, 124, 9024–9025, 2002.
19. Powell, K.A., Ramer, S.W., del Cardayré, S.B., et al., Directed evolution and biocatalysis, *Angew. Chem. Int. Ed. Engl.*, 40, 3948–3959, 2001.
20. Reetz, M.T. and Jaeger, K.-E., Enantioselective enzymes for organic synthesis created by directed evolution, *Chem. Eur. J.*, 6, 407–412, 2000.
21. Minshull, J. and Stemmer, W.P.C., Protein evolution by molecular breeding, *Curr. Opin. Chem. Biol.*, 3, 284–290, 1999.
22. May, O., Nguyen, P.T., and Arnold, F.H., Inverting enantioselectivity by directed evolution of hydantoinase for improved production of L-methionine, *Nature Biotechnol.*, 18, 317–320, 2000.
23. Cao, L., van Langen, L.M., and Sheldon, R.A., Immobilised enzymes: carrier-bound or carrier-free, *Curr. Opin. Biotechnol.*, 14, 387–394, 2003.
24. Doscher, M.S. and Richards, F.M., The activity of an enzyme in the crystalline state: ribonuclease S, *J. Biol. Chem.*, 238, 2399–2406, 1963.
25. Quiocho, F.A. and Richards, F.M., Intermolecular cross-linking of a protein in the crystalline state: carboxypeptidase A, *Proc. Natl. Acad. Soc. USA*, 114, 7314–7316, 1992.
26. St. Clair, N.L. and Navia, M.A., Cross-linked enzyme crystals as robust biocatalysts, *J. Am. Chem. Soc.*, 114, 7314–7316, 1992.
27. Margolin, A.L., Novel crystalline catalysts, *Tibtech.*, 14, 223–230, 1996.
28. Margolin, A.L. and Navia, M.A., Protein crystals as novel catalytic materials. *Angew. Chem. Int. Ed. Engl.*, 40, 2204–2222, 2001.
29. Lalonde, J., Practical catalysis with enzyme crystals, *Chemtech.*, 27(2), 38–45, 1997.
30. Haring, D. and Schreier, P., Cross-linked enzyme crystals, *Curr. Opin. Biotechnol.*, 3, 35–38, 1999.
31. Brown, D.L. and Glatz, C.E., Aggregate breakage in protein precipitation, *Chem. Eng. Sci.*, 47, 1831–1839, 1986.
32. Cao, L., van Rantwijk, F., and Sheldon, R.A., Cross-linked enzyme aggregates: a simple and effective method for the immobilization of penicillin acylase, *Org. Lett.*, 2, 1361–1364, 2000.
33. Cao, L., van Langen, L.M., van Rantwijk, F., et al., Cross-linked aggregates of penicillin acylase: robust catalysts for the synthesis of β-lactam antibiotics, *J. Mol. Catal., B Enzym.*, 11, 665–670, 2001.
34. Wegman, M.A., Janssen, M.H.A., van Rantwijk, F., et al., Cross-linked aggregates of penicillin acylase: robust catalysts for the synthesis of β-lactam antibiotics, *Adv. Synth. Catal.*, 343, 559–576, 2001.
35. van Langen, L.M., Oosthoek, N.H.P., van Rantwijk, F., et al., Pencillin acylase catalysed synthesis of ampicillin in hydrophilic organic solvents, *Adv. Synth. Catal.*, 345, 797–801, 2003.
36. Wilson, L., Illanes, A., Abián, O., et al., Co-aggregation of penicillin G acylase and polyionic polymers: an easy methodology to prepare enzyme biocatalysts stable in organic media, *Biomacromolecules*, 5, 852–857, 2004.
37. Wilson, L., Illanes, A., Pessela, B., et al., Encapsulation of crosslinked penicillin G acylase aggregates in lentikats: evaluation of a novel biocatalyst in organic media, *Biotechnol. Bioeng.*, 86, 558–562, 2004.
38. Lopez-Serrano, P., Cao, L., van Rantwijk, F., et al., Cross-linked enzyme aggregates with enhanced activity: application to lipases, *Biotechnol. Lett.*, 24, 1379–1383, 2002.
39. Theil, F., Enhancement of selectivity and reactivity of lipases by additives, *Tetrahedron*, 56, 2905–2909, 2000.
40. Mateo, C., Palomo, J.M., van Langen, L.M., et al., A new, mild cross-linking methodology to prepare cross-linked enzyme aggregates, *Biotechnol. Bioeng.*, 86, 273–276, 2004.

41. Schoevaart, R., Wolbers, M.W., Golubovic, M., et al., Preparation, optimization, and stru cross-linked enzyme aggregates (CLEAs), *Biotechnol. Bioeng.*, 87, 754–762, 2004.
42. Wilson, L., Betancor, L., Fernandez-Lorente, G., et al., Cross-linked aggregates of m enzymes: a simple and efficient methodology to stabilize their quaternary structure, *Biom lecules*, 5, 814–817, 2004.
43. Hilal, N., Nigmatullin, R., and Alpatova, A., Immobilization of cross-linked lipase a within microporous polymeric membranes, *J. Memb. Sci.*, 238, 131–141, 2004.
44. Bruggink, A., Schoevaart, R., and Kieboom, T., Concepts of nature in organic synthesis catalysis and multistep conversions in concert, *Org. Proc. Res. Dev.*, 7, 622–640, 2003.
45. Mateo, C. Chmura, A., Rustler, S., van Rantwijk, F., Stolz, A., Sheldon, R.A., Syn enantiomerically pure (S)-mandelic acid using an oxynitrilase-nitrilase bienzymatic ca nitrilase surprisingly shows nitrile hydratase activity, *Tetrahedron: Asymmetry*, 17, 320–32

Biotechnological
Applications of Aldolases

Wolf-Dieter Fessner and Stefan Jennewein

CONTENTS

13.1 INTRODUCTION

The demand for enantiomerically pure compounds, in particular for pharmaceutical application, has steered immense interest in the industrial asymmetric synthesis of chiral products
and intermediates [1]. As an alternative to classical chemical methodology, enzyme-catalyzed
reactions were used early on in chemical research and production, and are now gaining
increasing attention [2,3]. Indeed, processes using hydrolytic enzymes for biocatalytic resolution of racemates constitute the majority of current applications on industrial scale [4].
However, standard resolution approaches to enantiopure compounds can only achieve a
maximum molar yield of 50%.

Undoubtedly, the direct stereoselective synthesis of asymmetric molecules from prochiral
precursors using chiral catalysts represents an attractive alternative. In particular, stereoselective carbon–carbon bond forming reactions are among the most useful synthetic
methods in asymmetric synthesis because this allows the simultaneous creation of up to two
adjacent stereocenters [5]. Several classes of enzymes are able to catalyze stereoselective C–C
bond formation reactions (lyases) such as aldolases [6], oxynitrilases [7], or thiamine diphos-
phate-dependent enzymes [8]. Especially, the high stereospecificity of aldolases in
C–C bond forming reactions gives them substantial utility as synthetic biocatalysts, and
makes them an environmentally benign alternative to chiral transition metal catalysis for
the asymmetric aldol reaction [9,10]. In fact, the industrial process based on yeast pyruvate
decarboxylase for the production of (R)-phenylacetyl carbinol, a precursor to (–)-ephedrine,
is still utilized since its invention 70 years ago [11].

In vivo, aldolases are involved in the metabolism of amino acids and carbohydrates,
compounds that are important *per se*, or as chiral building blocks for more complex pharmaceuticals. Although this may seem to limit the scope of potential practical applications,

such enzymes open new windows of synthetic opportunity because these polyfuncti classes of compounds are difficult to prepare and handle by conventional organic chem methods and mandate the laborious manipulation of protective groups. Due to the levels of selectivity offered by enzymes under their mild operating conditions that compatible with most functional groups, biocatalytic conversions can usually be perfor on nonderivatized substrates, thus making tedious and costly protecting group manipulat superfluous [12].

13.2 MECHANISTIC DISTINCTIONS

Aldolases achieve the activation of an aldol donor substrate by two mechanistically diffe pathways (Figure 13.1). Class I aldolases (a) bind their substrates covalently at an activ lysine residue through an imine–enamine intermediate to initiate C–C bond formatic cleavage [13]. In contrast, class II aldolases (b) utilize transition metal ions as a Lewis cofactor, which facilitates the stereospecific deprotonation of the donor by bidentate co ination to stabilize the enediolate nucleophile (Figure 13.1) [14]. This effect is usually ach by means of a tightly bound Zn^{2+} ion. The mechanistic models for both classes of aldo have been substantiated by several crystal structures of enzymes liganded by substra inhibitor that altogether provide a detailed insight into the catalytic function and the vidual contribution of active site residues to the stereochemically determining event [6]

Apart from their mechanistic distinction, aldolases can be subdivided into cc families depending on their nucleophilic substrate: (1) dihydroxyacetone phosp

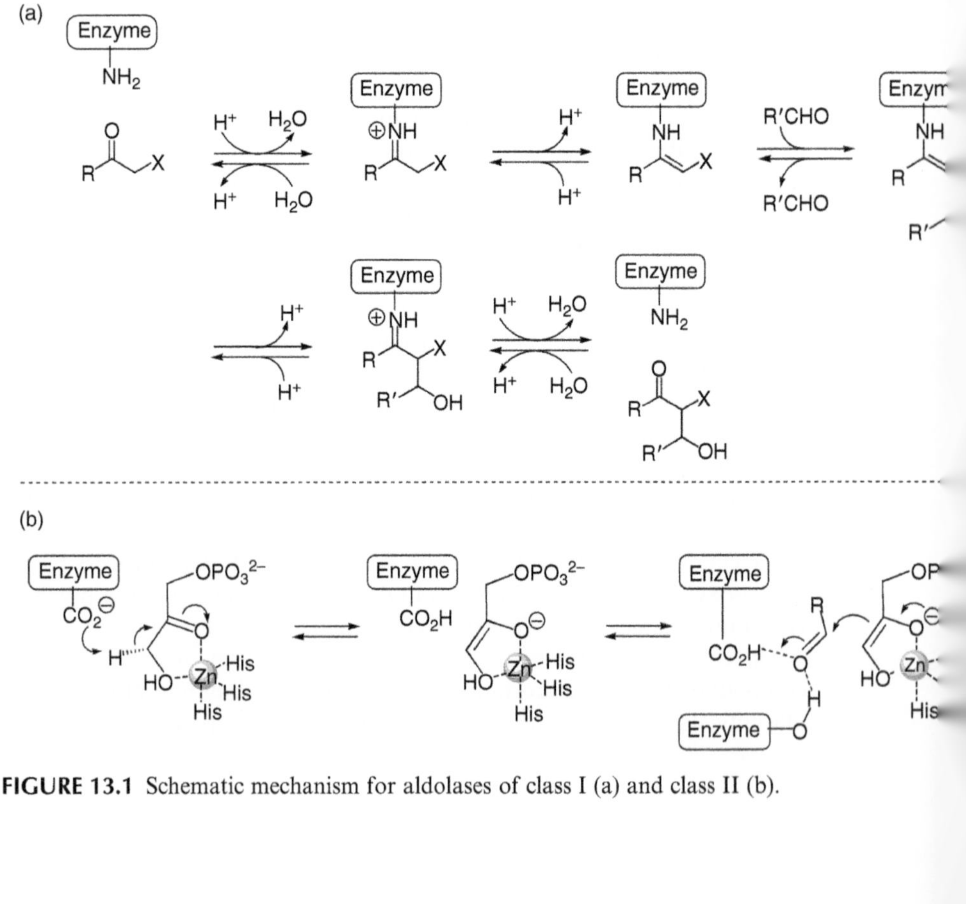

FIGURE 13.1 Schematic mechanism for aldolases of class I (a) and class II (b).

(DHAP)-dependent aldolases, (2) pyruvate- (and phosphoenolpyruvate-)dependent lyases, (3) aldehyde-dependent aldolases, and (4) glycine-dependent enzymes (Figure 13.2). Members of the first two families all add 3-carbon ketone fragments to the carbonyl group of an aldehyde, yielding 1,3,4-trihydroxylated methyl ketone derivatives or 3-deoxy-2-oxoacids, respectively. The latter two families use the carbon fragments acetaldehyde and glycine, respectively, as the nucleophilic component. Today, many useful synthetic applications have been demonstrated for all classes of aldolases and for related C–C bond forming enzymes. Due to the nature of the different substrates used and compound classes generated, the individual aldolase families will be treated separately with regard to their synthetic potential.

Because of the mechanistic complexity inherent to the C–C bond forming process, which requires the stereocontrolled binding and dipolar activation of two substrates, and because of some unique properties of the individual enzymes, still only a limited number of industrial lead processes have been successfully developed. Thus, this chapter will also deal with interesting synthetic examples, and prominent operating technology, to demonstrate preparative opportunities and to stimulate imagination toward future biotechnology developments.

13.3 DIHYDROXYACETONE PHOSPHATE-DEPENDENT ALDOLASES

While all aldolases typically represent catabolic enzymes, DHAP-dependent aldolases are involved in the degradation of phosphorylated ketosugars. A particular advantage for synthetic applications is the fact that nature has evolved a full set of four stereochemically complementary enzymes, which are termed as D-fructose 1,6-bisphosphate aldolase (FruA), D-tagatose 1,6-bisphosphate aldolase (TagA), L-fuculose 1-phosphate aldolase (FucA), and L-rhamnulose 1-phosphate aldolase (RhuA), based on their capacity to cleave the corresponding diastereoisomeric ketose 1-phosphates 6–9 in a retroaldol manner (Figure 13.3). In the direction of synthesis, this formally allows the preparation of any one of the four possible diastereomeric aldol adducts by simply choosing the appropriate enzyme and

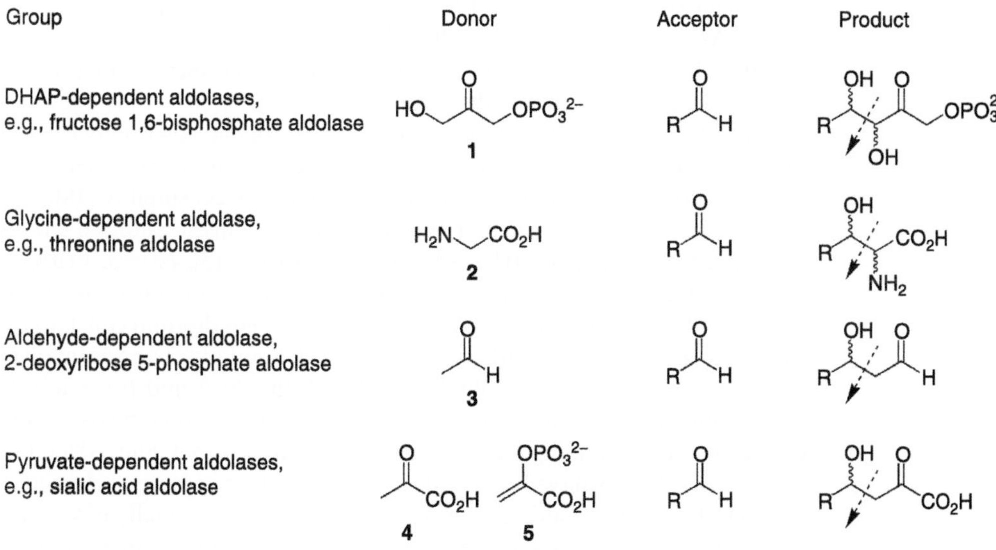

FIGURE 13.2 Subgrouping of aldolases according to donor type.

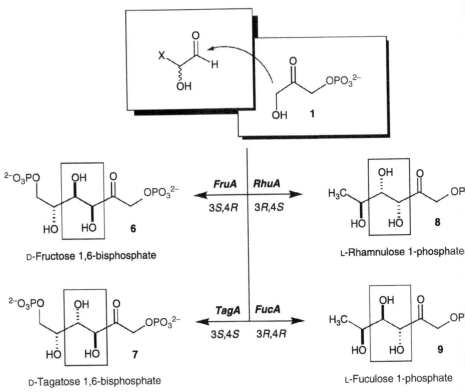

FIGURE 13.3 Stereochemically complementary set of dihydroxyacetone phosphate-dependent al

starting materials for a full control over constitution and absolute configuration
desired product [9].

All four members of the DHAP-dependent aldolase family are represented by
dependent types (class II), and Schiff-base-forming types (class I) are known for Fru
TagA. As a rule of thumb, metalloaldolases are more stable than their Schiff-base-f
relatives. In the presence of low concentrations of their metal cofactor, they possess ha
of several weeks or even months in solution, as compared to only a few days [15–17], a
tolerate the presence of significant portions of organic cosolvents ($\geq$30%), such as I
DMF, and ethanol [17]. Notable exceptions are the class I FruA members found in *St*
coccus strains, which show unusual thermal and process stability [18].

The FruA, and in particular the class I rabbit muscle aldolase enzyme (RAMA)
most extensively studied DHAP-dependent aldolase [15]. *In vivo*, FruA catalyzes the
reaction of glycolysis, which is the reversible cleavage of D-fructose 1,6-bisphosphat
D-glyceraldehyde 3-phosphate and DHAP (**1**). The equilibrium constant of the reactio
M^{-1} strongly favors the synthesis direction [19]. Similarly, TagA is involved in the cata
of D-*galacto*-configured carbohydrates [20]. Both RhuA and FucA are derived from
parallel microbial degradation pathways of L-rhamnose and L-fucose, respectively, wh
cleave the corresponding ketose 1-phosphate (**8,9**) into **1** and L-lactaldehyde [21–2
latter two aldolases have been studied for production at larger volumetric scale [24,2

DHAP-dependent aldolases have proved to be exceptionally powerful tools for as
ric synthesis, particularly for the stereocontrolled synthesis of polyoxygenated com
due to their relaxed acceptor specificity for aldehyde substrates and a generally high

stereocontrol. Class I FruA enzymes show a very broad tolerance for structurally diverse aldehyde substrates, and hundreds of aldehydes have so far been tested successfully to function as acceptor substrate in enzymatic assays or preparative experiments [26]. Table 13.1 lists a small compilation of aldehydes tested as replacement for D-glyceraldehyde 3-phosphate and the corresponding aldol products. Similar to FruA, also FucA and RhuA enzymes show broad acceptance of variously substituted aldehydes, compiled in Table 13.2, with conversion rates useful for synthetic applications [17]. Conversely, microbial FruA enzymes (class II) from *Escherichia coli* and yeast display high substrate specificity for phosphorylated substrates [9].

In direction of synthesis, stereospecificity for the natural configuration is somewhat substrate-dependent, in that small aliphatic aldehydes can give rise to a certain fraction of the diastereomer having an opposite configuration at C-4 [9,17]. In general, stereochemical fidelity is usually higher, and diastereospecific results are observed more often with FruA and FucA than with RhuA. While class I TagA enzymes examined from *Staphylococcus* and *Streptococcus* sp. have no stereochemical preference with regard to D-tagatose or D-fructose configuration [27], the class II enzyme from *E. coli* is highly stereoselective for its natural substrate in both cleavage and synthesis direction [28–30]. The aldolase also accepts a range of unphosphorylated aldehydes as electrophilic substrates, but unfortunately produces only mixtures of diastereomers [28].

In addition to the high-stereoselective formation of the C–C bond, both RhuA and FucA show strong kinetic preference for L-configurated enantiomers of 2-hydroxyaldehydes

TABLE 13.1
Substrate Tolerance of Fructose 1,6-Bisphosphate Aldolase

R–CHO $\xrightarrow[\text{DHAP}]{\text{FruA}}$ aldol product with OPO$_3^{2-}$

R	Relation Rate (%)	Yield (%)
D-CHOH-CH$_2$OPO$_3^{2-}$	100	95
H	105	—
CH$_3$	120	—
CH$_2$Cl	340	50
CH$_2$-CH$_3$	105	73
CH$_2$-CH$_2$-COOH	—	81
CH$_2$OCH$_2$C$_6$H$_5$	25	75
D-CH(OCH$_3$)-CH$_2$OH	22	56
CH$_2$OH	33	84
D-CHOH-CH$_3$	10	87
L-CHOH-CH$_3$	10	80
DL-CHOH-C$_2$H$_5$	10	82
CH$_2$-CH$_2$OH	—	83
CH$_2$-C(CH$_3$)$_2$OH	—	50
DL-CHOH-CH$_2$F	—	95
DL-CHOH-CH$_2$Cl	—	90
DL-CHOH-CH$_2$–CH=CH$_2$	—	85

TABLE 13.2

Substrates Accepted by L-Rhamnulose 1-Phosphate and L-Fuculose 1-Phosphate Ald■

$$R\text{-CHO} \xrightleftharpoons[\text{DHAP}]{\text{Aldolase}} R\text{-CH(OH)-CH(OH)-CO-CH}_2\text{-OPO}_3^{2-} + R\text{-CH(OH)-CH(OH)-CO-CH}_2\text{-OPO}_3^{2-}$$

R	RhuA Rel. Rate (%)	Selectivity threo/erythro	Yield (%)	FucA Rel. rate (%)	Selectivity threo/erythro
L-CH$_2$OH-CH$_3$	100	>97:3	95	100	<3:97
CH$_2$OH	43	>97:3	82	38	<3:97
D-CHOH-CH$_2$OH	42	>97:3	84	28	<3:97
L-CHOH-CH$_2$OH	41	>97:3	91	17	<3:97
CH$_2$-CH$_2$OH	29	>97:3	73	11	<3:97
CHOH-CH$_2$OCH$_3$	—	>97:3	77	—	<3:97
CHOH-CH$_2$N$_3$	—	>97:3	97	—	<3:97
CHOH-CH$_2$F	—	>97:3	95	—	<3:97
H	22	—	81	44	—
CH$_3$	32	69:31	84	14	5:95
CH(CH$_3$)$_2$	22	97:3	88	20	30:70

(Figure 13.4), which facilitates an effective racemate resolution [31,32]. This feature
the concurrent determination of three contiguous chiral centers in the final products (■
12) starting from readily accessible racemic aldehyde substrates. Enantiomer discrim■
by rabbit muscle FruA is limited to its natural substrate (20:1 preference for D-glyceral
3-phosphate over the L-antipode), but fails for nonphosphorylated aldehydes [15,33].

Due to the freely reversible nature of the aldol reactions, a high level of enant
differentiation can be gained alternatively by thermodynamic control under fully eq■
ing conditions, when diastereomeric products adopt cyclic half acetals that have ■
relative stability (Figure 13.5). Using racemic 2- or 3-hydroxylated aldehyde substra■
fold and 33-fold discrimination for the more stable *trans* and all-equatorial isomer h■
reported, respectively [15,34,35]. This thermodynamic product control (e.g., **17/23/27** ■
24/26) was utilized in the preparation of novel 4,6-deoxy sugars such as 4-deoxy-L-fuc■
from racemic 3-hydroxy-butanal (**14**) [9], and in the synthesis of unsaturated high■
starting from racemic unsaturated aldehydes [34,35].

Owing to their metabolic function, early utilization of aldolases concerned the pre■
synthesis of ketoses (e.g., **30**) through the corresponding ketose 1-phosphates, notably
isotopically labeled derivatives ^{13}C-enriched at defined positions [36–38]. Taking adva■
the kinetic selectivity of RhuA for L-configurated 2-hydroxyaldehydes, nonnatural L-f
(**31**) is accessible starting from DHAP to racemic glyceraldehyde (Figure 13.6) [39].

The scope and synthetic usefulness of FruA for the synthesis of novel monosacc■
and related molecules are further illustrated by the preparation of various unusual pr■
including branched-chain (**34**) and spiroannelated sugars (**37,38**) [40], natural product ■
such as nucleoside (**35**) [41] or oligosaccharide mimetics (**39**) [42], or a perfluoroa■
fructose surfactant (**36**) [43], each synthesized by FruA from the corresponding a■
precursors (Figure 13.6).

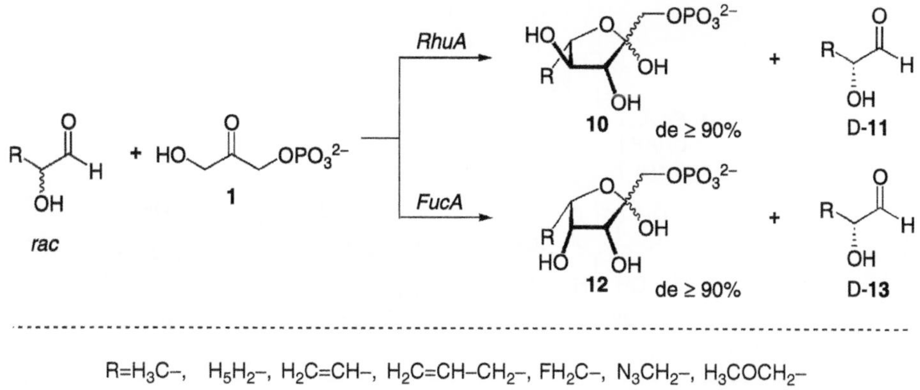

R=H$_3$C–, H$_5$H$_2$–, H$_2$C=CH–, H$_2$C=CH–CH$_2$–, FH$_2$C–, N$_3$CH$_2$–, H$_3$COCH$_2$–

FIGURE 13.4 Kinetic enantiopreference of class II DHAP-dependent aldolases useful for racemic resolution of 2- hydroxyaldehydes.

Complex 8- and 9-carbon monosaccharide derivatives, representing sialic acid or KDO analogs, could be obtained from pentose and hexose monophosphates by stereospecific chain extension using FruA from rabbit muscle (Figure 13.7) [44]. This approach provides a convenient route to novel α,ω-phosphorylated high-carbon sugars (e.g., **41**, **43**), which are difficult to obtain from either natural sources or chemical synthesis.

Twofold aldolase-catalyzed chain elongation of α,ω-dialdehydes ("tandem" aldolization) has been developed into an efficient method for the generation of high-carbon sugars (e.g., **45**)

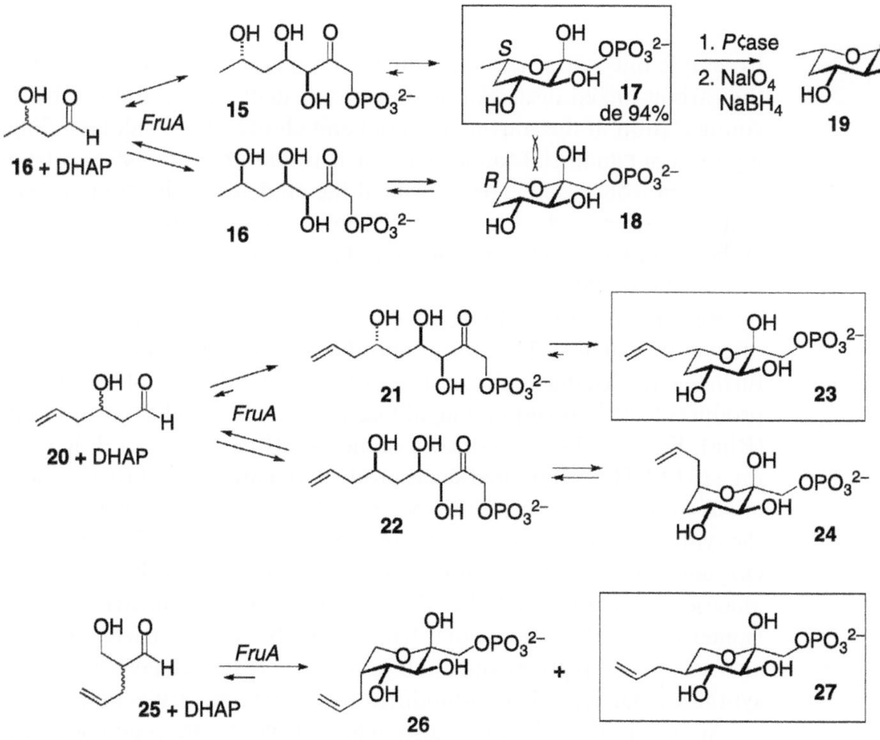

FIGURE 13.5 Thermodynamically controlled diastereoselectivity in FruA-catalyzed aldol additions under equilibrating conditions.

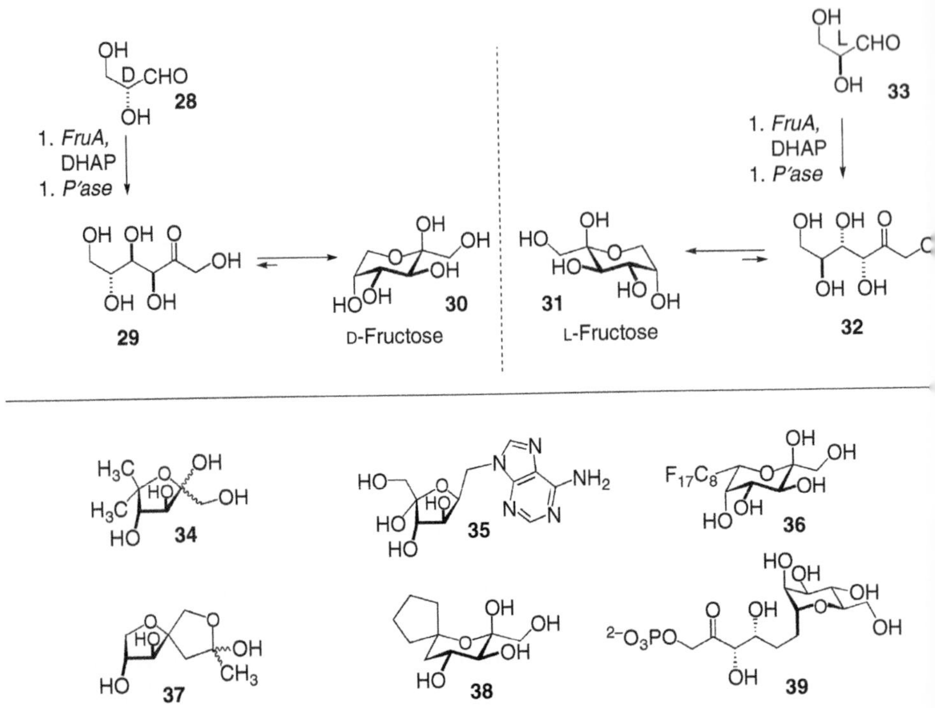

FIGURE 13.6 Enzyme-catalyzed asymmetric synthesis of D- and L-fructose by stereocomplem aldolases, and selected examples of carbohydrate-related products accessible by enzymatic aldoli

by simple one-pot operations (Figure 13.7) [45]. The overall specific substitution patt the carbon-linked disaccharide mimetics is deliberately addressable by the relative hy configuration in the starting material and choice of the aldolase. Single diastereomers ing *trans* or equatorial connectivity such as that present in **45** may be obtained in good yield even from racemic precursors, if the tandem aldolizations are conducted under th dynamic control (cf., Figure 13.5). Highly complex structures like anulated (**47**) and cyclic (**49**) carbohydrate mimics may be obtained from appropriately customized prec (Figure 13.7) [46].

Reaction products of DHAP aldolases are typically ketoses; however, isomeric aldo often more valuable. The utility of products from FruA, FucA, and RhuA catalyzed reac further extended due to the existence of corresponding ketol isomerases that convert the products to the corresponding aldose isomers. Stereochemically complementary L-rha (RhaI; EC 5.3.1.14) and L-fucose isomerases (FucI; EC 5.3.1.3) have been found to be s for a (3R)-OH configured sugar product, but are tolerant to modifications in stereoche or substitution pattern at other positions [9,31,47]. This strategy has been clearly illustra the synthesis of new L-fucose analogs (**55**) and other L-configured aldohexoses using d enzyme combinations (Figure 13.8) [47,48]. Similar results have been realized by util glucose isomerase (GlcI; EC 5.3.1.5), which is an industrially important enzyme isomerization of D-glucose to D-fructose but has a more narrow specificity. The latter also accepts derivatives and analogs of D-fructose and has been used in combined enz syntheses, particularly of 6-modified D-glucose derivatives [49].

Access to aldoses can be also achieved by an "inversion strategy" (Figure 13.8) utilizes a monoprotected dialdehyde (e.g., **56**) for aldolization and, after stereoselective reduction, provides free aldose (**58**) upon liberation of the masked aldehyde function

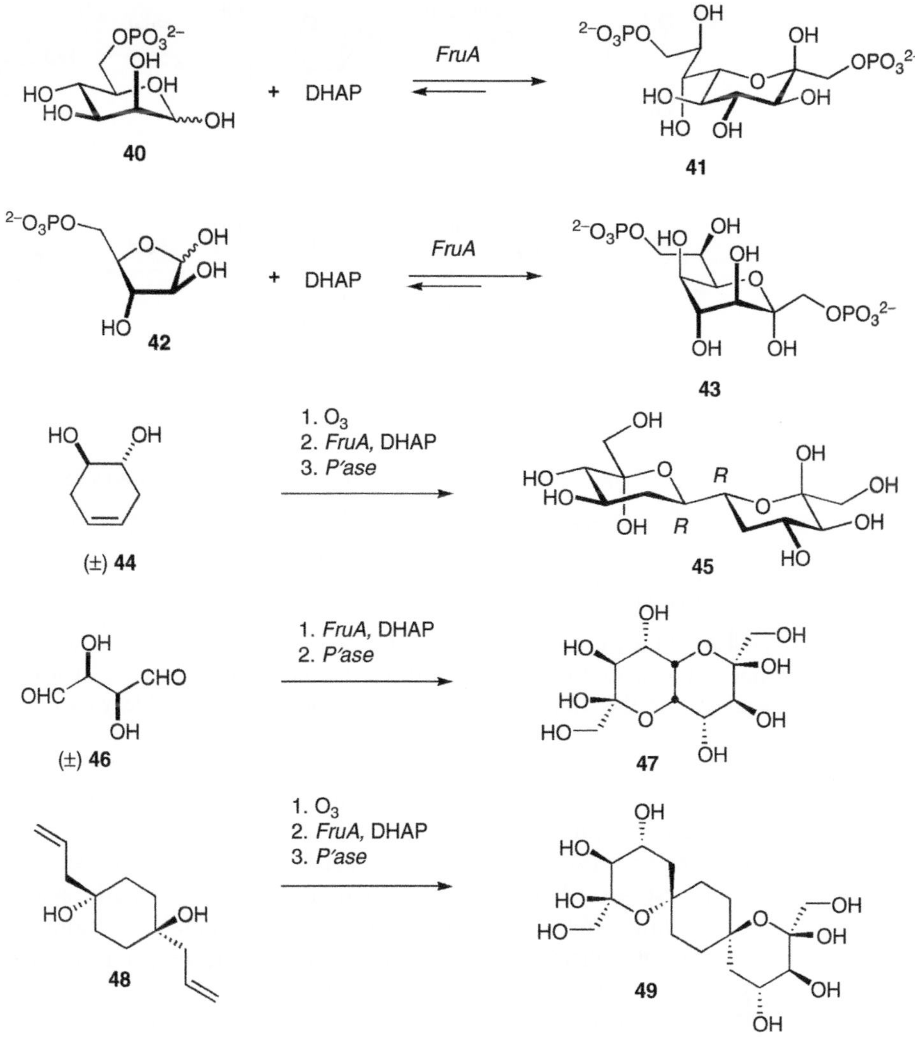

FIGURE 13.7 FruA-catalyzed enzymatic synthesis of α,ω-bisphosphate esters of higher carbon sugars, and bidirectional chain extension of dialdehyde substrates yielding potential disaccharide and oligosaccharide mimetics.

Stemming from the lessons learned in the synthesis of rare and novel monosaccharides, a major opportunity has been developed toward the preparation of "aza sugars," which represent powerful glycosidase inhibitors or show potent antiviral activity [51]. This rather flexible synthetic strategy consists of an aldol addition to an N-functionalized aldehyde, which usually contains an azide group, followed by hydrogenolytic intramolecular reductive amination of the keto functionality (including prior azide to amine reduction) to close an N-heterocyclic ring structure stereoselectively [52]. Prominent examples concern the FruA-catalyzed synthesis of deoxynojirimycin (**62**) and deoxymannojirimycin (**63**) from 3-azido-2-hydroxypropanal (**59**) (Figure 13.9) [16,53–56], which represent natural products found in plants or microorganisms. It is important to note that the stereodivergent preparation of a large variety of diastereomeric aza sugars of the nojirimycin type from any single azido aldehyde (such as **59**) can be achieved by choosing from the DHAP aldolases having different stereospecificity [9].

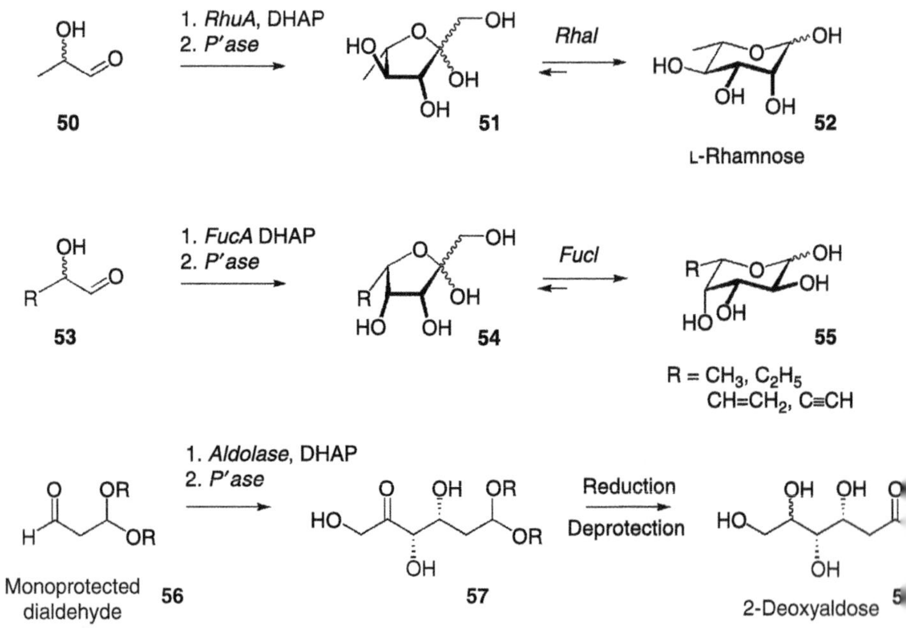

FIGURE 13.8 Enzymatic synthesis of L-rhamnose and L-fucose as well as hydrophobic ana aldolization-ketol isomerization with kinetic resolution of racemic hydroxyaldehydes, and "in strategy" for the synthesis of aldoses from monoprotected dialdehydes.

Starting from (S)- to (R)-3-azido-2-acetamidopropanal in FruA-catalyzed aldol rea aza sugar derivatives corresponding to N-acetyl-glucosamine and N-acetyl-manno. have been prepared [57]. Similarly, a FruA-mediated stereospecific DHAP addition to mediate **64** served as the key step in the chemoenzymatic synthesis of australin 3-epiaustraline (**68**), and 7-epialexine (Figure 13.9) [58]. Australine and alexine are na occurring pyrrolizidine alkaloids, which are of pharmacological interest due to their glycosidase inhibitory and antiviral activities.

Beyond the immediately obvious applications of this type of aldolases to the synth carbohydrates or carbohydrate derived compounds, the enzymes are highly valua the determination of asymmetric centers in the construction of stereochemically homo, fragments of complex noncarbohydrate natural products. An impressive illustra the FruA-catalyzed chemoenzymatic synthesis of (+)-exo-brevicomin (**72**), the aggre pheromone of the western pine bark beetle *Dendroctonus brevicomis* (Figure 13. two complementary routes, starting from either 5-oxohexanal (**69**) [59] or propan [44], which yields an intermediate **74** identical to that from a corresponding transke catalyzed transformation [60], the only independent two stereocenters of the natural p **72** are induced by the aldolase. Another elegant application employing FruA for the sy of a nonsugar molecule represents the synthesis of (−)-syringolide (**79**), a structurally c tricyclic microbial elicitor in plants (Figure 13.11) [61]. Again, the FruA-catalyzed re established the absolute and relative configuration of the vicinal diol defining the only centers that needed to be externally induced in the intermediates **77/78**; the config of all other stereocenters seemed to follow by kinetic preference during the subs cyclization.

Other remarkable applications of FruA for "noncarbohydrate" synthetic targets i the stereoselective generation of intermediates (e.g., **81**) for the synthesis of (+)-aspicill

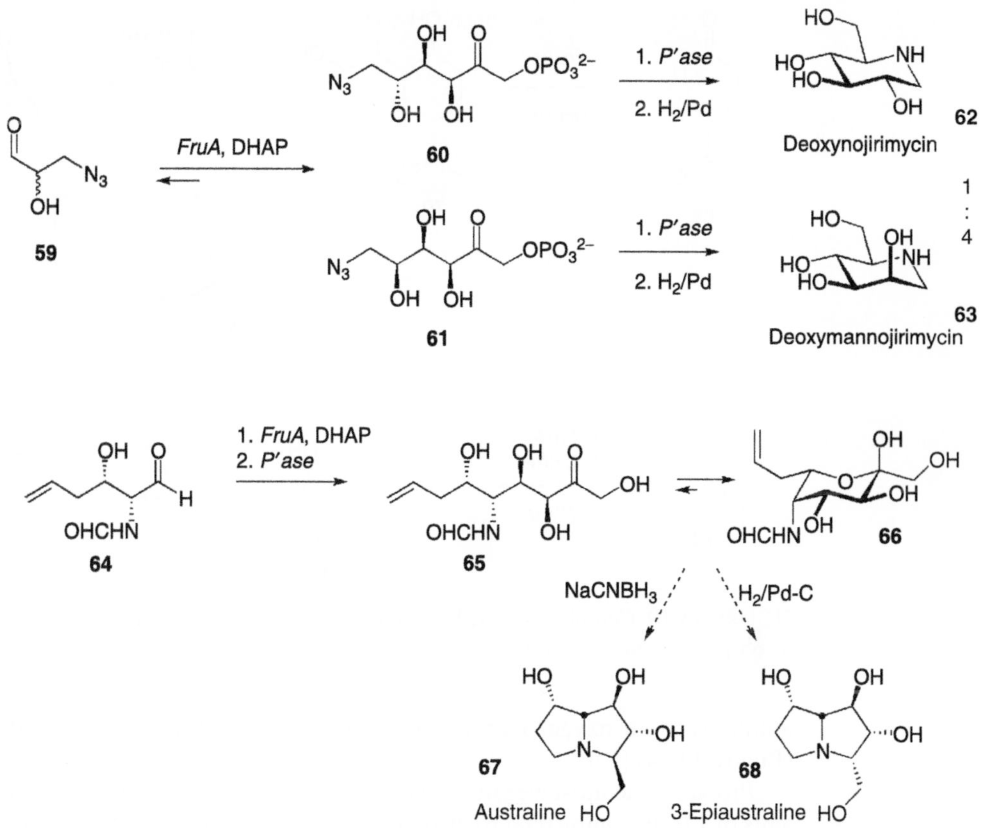

FIGURE 13.9 Chemoenzymatic synthesis of the potent glycosidase inhibitors deoxynojirimycin and deoxymannojirimycin, and of the pyrrolizidine alkaloids australine and 3-epiaustraline.

a lichen macrolactone [62], and of the skipped polyol C9–C16 chain fragment (84) of the macrolide antibiotic pentamycin (85) (Figure 13.12) [63]. A two-stage enzymatic sequence of arene dihydroxylation, using a naphthalene dioxygenase from *Pseudomonas putida*, followed by RhuA-catalyzed aldolization has been developed for the synthesis of novel analogs of the cytotoxic pancratistatin (91) pharmacophore (Figure 13.13) [64]. This strategy converts a naphthalene core (86) into a complex hybrid arene–carbohydrate structure (88), with simultaneous creation of four contiguous stereocenters, in just three steps.

It seems that all DHAP-dependent aldolases are highly specific for 1 as the nucleophilic component. However, the compound is chemically difficult to prepare, and in addition shows limited stability in solution, particularly at alkaline pH. Several protocols for the chemical synthesis of 1 have been developed [65–72]. However, due to its sensitive nature 1 is generated best enzymatically and consumed *in situ* by enzymatic aldol reaction, which avoids the buildup of high stationary concentrations. Dihydroxyacetone (96) can be enzymatically phosphorylated using a glycerol kinase with ATP regeneration [73–75], or by transphosphorylation from phosphatidyl choline using phospholipases [76]. The preparation from glycerol (93) by successive phosphorylation/dehydrogenation with an integrated double cofactor recycling scheme has also been developed [77]. A highly practical method for generation of 1 *in situ* is based on the oxidation of L-glycerol phosphate (91) catalyzed by a microbial flavine-dependent glycerol phosphate oxidase [78]. The process clearly generates 1 in practically quantitative yield and, due to the insensitivity of the DHAP-dependent

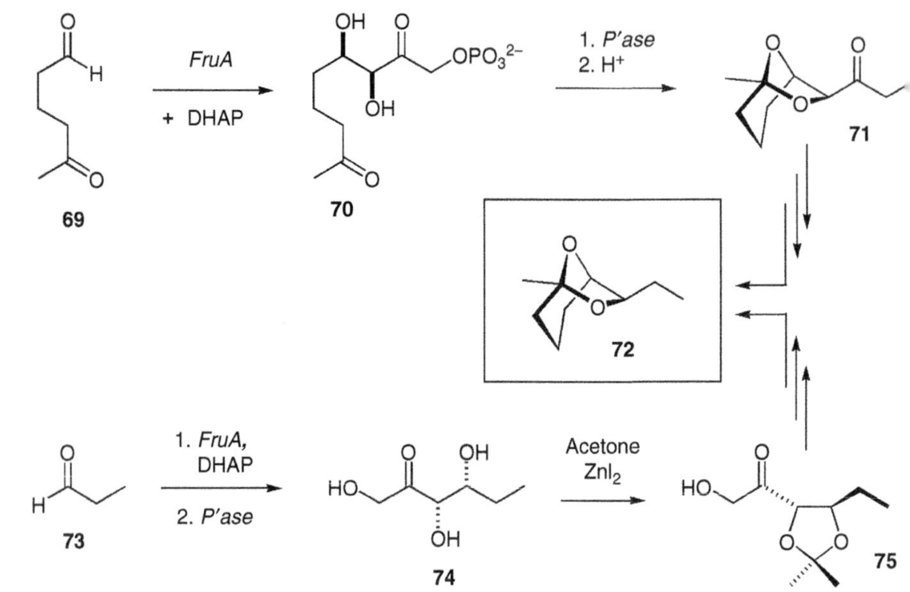

FIGURE 13.10 Complementary, backbone inverting approaches for the asymmetric synthes insect pheromone (+)-*exo*-brevicomin.

aldolases to an oxygenated solution, can be coupled directly to synthetic aldol re (Figure 13.14) [79].

Probably the most elegant and convenient method is the *in situ* formation of two lents of **1** from commercial fructose 1,6-bisphosphate (**6**) by a combination of FruA an phosphate isomerase [15]. This scheme has been extended into a highly integrated "a metabolism" for the efficacious *in situ* preparation of **1** from inexpensive feedstock glucose and fructose (two equivalents of **1** each), or sucrose (four equivalents of combination of up to seven inexpensive enzymes [40]. When employing the class II Fru *coli* for aldol cleavage of **6**, which displays high substrate specificity for glyceralde phosphate (**95**) and thus is inactive with other added aldehyde substrates, this system

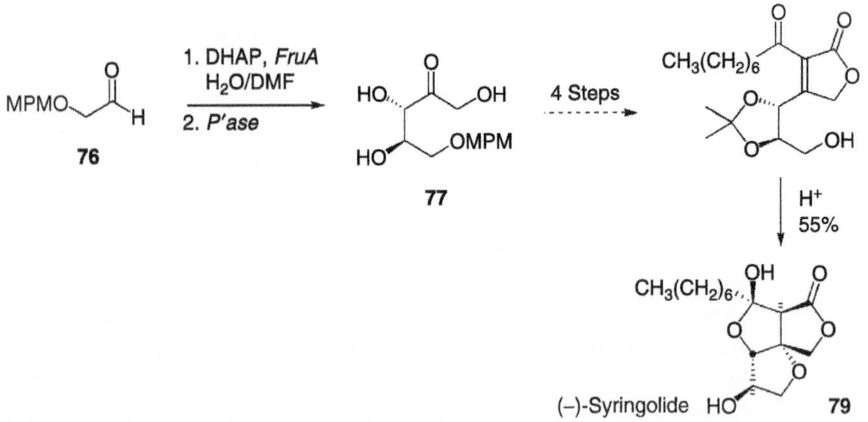

FIGURE 13.11 FruA-based creation of two independent chiral centers in the total synthes complex microbial plant defence elicitor (−)-syringolide.

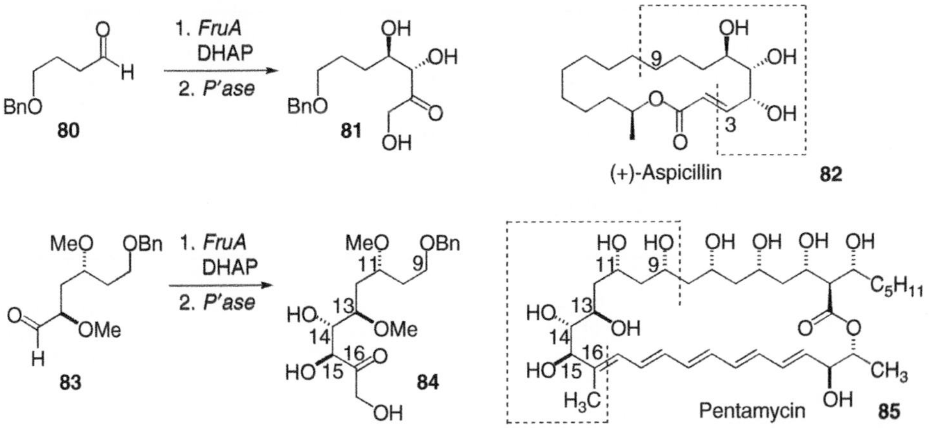

FIGURE 13.12 Stereoselective generation of chiral precursors for the synthesis of the macrolactone (+)-aspicillin and the macrolide antibiotic pentamycin using FruA catalysis.

"metabolically engineered" by adding another aldolase to furnish products having a stereo-configuration different from the starting material **6** (Figure 13.15) [9].

Due to the need of **1** as donor nucleophile, the generated aldol products will contain a phosphate ester moiety. The latter facilitates product isolation, for example, by barium salt precipitation or by use of ion-exchange techniques, but usually is undesired in the final product. The corresponding phosphate free compounds can easily be obtained by enzymatic hydrolysis using alkaline phosphatase at pH 8–9, whereas base labile compounds require acid

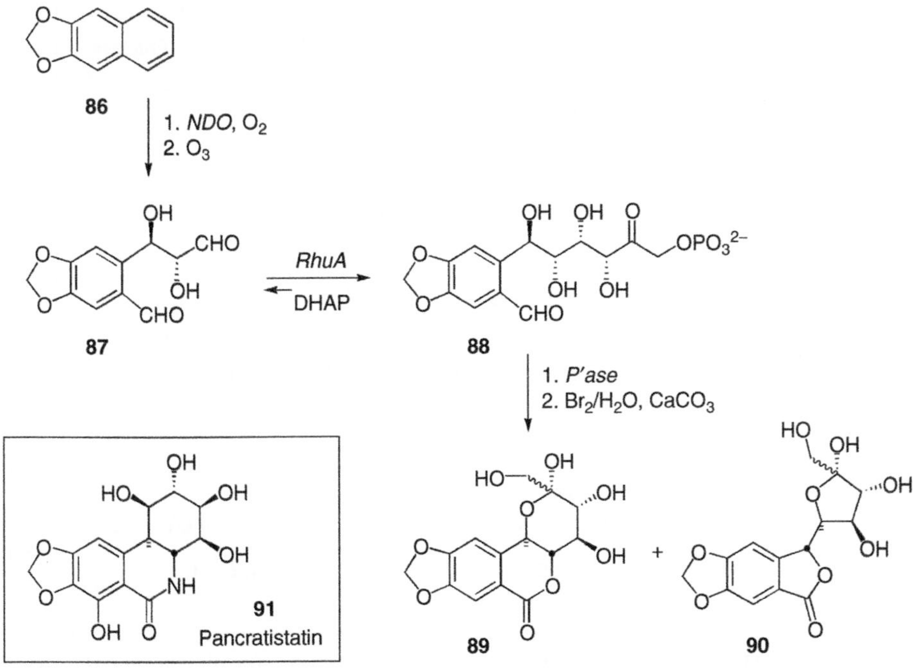

FIGURE 13.13 Chemoenzymatic synthesis of a pancratistatin analog using a RhuA-catalyzed aldolization reaction.

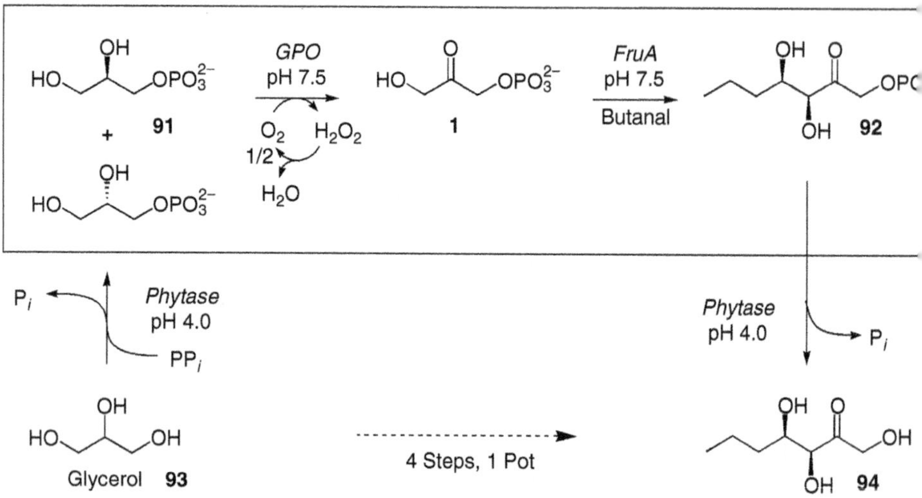

FIGURE 13.14 Enzymatic *in situ* generation of DHAP for stereoselective aldol reactions using aldolases (box), and extension by pH-controlled, integrated precursor preparation and liberation.

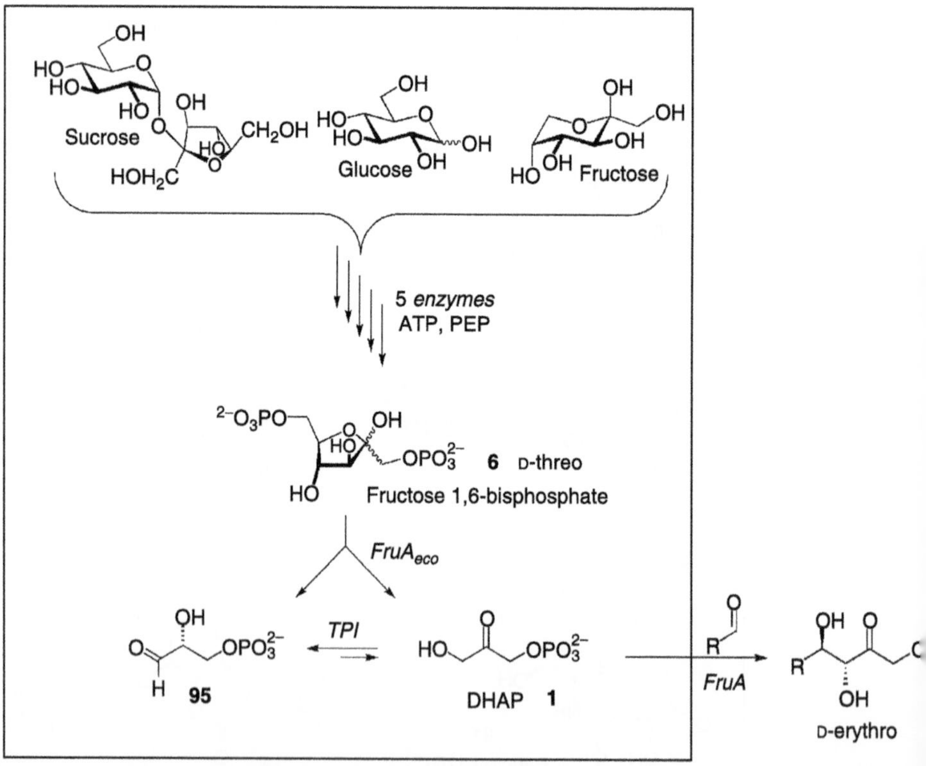

FIGURE 13.15 *In vitro* "artificial metabolism" for the *in situ* preparation of dihydroxya phosphate along the glycolysis cascade, and utilization for subsequent stereoselective carbon-bond formation using an aldolase with distinct stereoselectivity.

phosphatase treatment. The phosphate moiety also mandates an aqueous medium for solubility reasons and to prevent substrate/product inhibition of the enzymes; for organic substrates having only limited solubility in water, it could be shown that aldolases are compatible with water-in-oil microemulsion systems [80,81]. For industrial applications, the high specificity for **1** and the costs for its preparation severely limit the synthetic usefulness of this enzyme class. A partial solution was offered by coupling the GPO oxidative generation of **1** from **91** with a reversible phosphoryl transfer from inexpensive pyrophosphate to **93** catalyzed by phytase (an acid phosphatase), and staging the phosphorylation–aldolization product dephosphorylation sequence by appropriate pH shifts (Figure 13.14) [82].

Interestingly, dihydroxyacetone (**96**) in the presence of high concentration of inorganic arsenate ($\geq 0.5\,M$) reversibly forms the corresponding monoarsenate ester (**97a**) *in situ*, which can replace **1** in enzyme-catalyzed aldol reactions (Figure 13.16) [83,84]. The method suffers, however, from rather low-reaction rates in addition to the high toxicity of arsenate. Similarly, inorganic vanadate also spontaneously forms the corresponding vanadate ester (**97b**), but the oxidation potential has to be controlled by appropriate buffers to prevent oxidation of **96**. So far, only RhuA of *E. coli* has been shown to accept **97b** in place of **1** for preparative synthesis [9].

The GPO method for generation of **1** can also be used to generate analogs with isosteric replacements of the ester oxygen for sulfur (**98**), nitrogen (**99**), or methylene carbon (**100**) (Figure 13.17) [38], which are substrates of many aldolases of classes I and II [38,85,86]. Thus, sugar phosphonate analogs (e.g., **105**) can be rapidly prepared from **100** that mimic intermediates of carbohydrate metabolism but are stable to hydrolysis. The hydrogenolytic lability of the phosphorothioate analogs (e.g., **102**) makes terminally deoxygenated sugars accessible, as was demonstrated by a stereoselective synthesis of D-olivose (**103**) based on the "inversion strategy" (Figure 13.17) [87].

13.4 PYRUVATE- (AND PHOSPHOENOLPYRUVATE-) DEPENDENT LYASES

Pyruvate-dependent aldolases serve *in vivo* catabolic functions, whereas phosphoenolpyruvate (PEP)-dependent lyases are involved in biosynthetic reactions of 2-keto-3-deoxy acids. The most interesting enzymes for synthetic applications are those involved in the metabolism of sialic acids (e.g., **107**) or KDO (2-keto-3-deoxy-*manno*-octosonate, *ent*-**108**), which represent complex sugars typically found in mammalian or bacterial glycoconjugates [88–90], respectively. Due to thermodynamically often unfavorable equilibrium constants [19], the preparative aldol additions have to be driven by providing excess pyruvate (**4**) to achieve satisfactory

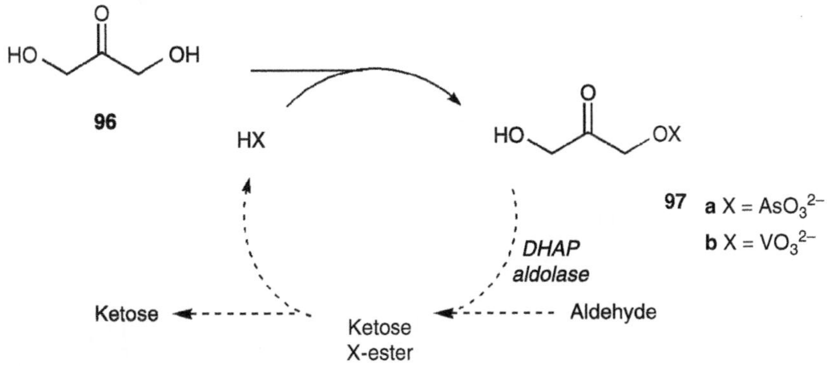

FIGURE 13.16 Spontaneous, reversible formation of arsenate and vanadate analogs of DHAP *in situ* for enzymatic aldol additions.

conversions. In contrast, PEP-dependent lyases (synthetases) use high-energy enol e
instead of **4**, which upon C–C bond formation liberates inorganic phosphate, a
renders the aldol addition essentially irreversible (Figure 13.18). Although attractive
synthetic point of view, PEP-dependent lyases have not yet been extensively stud
preparative applications [91].

Most extensively studied enzyme in the field of pyruvate-dependent lyases
N-acetylneuraminic acid aldolase (NeuA), also referred to as sialic acid aldolase [9]
found in both bacteria and animals mechanistically represent class I aldolases that
Schiff-base intermediate with **4** [92] to promote *si*-face attack on the sugar aldehyde c
group resulting in the formation of a (4*S*) configured stereocenter in the aldol product.
preparations from *Clostridium perfringens* and *E. coli* are commercially available, a
enzyme has broad pH optimum and useful stability at ambient temperature [93]. The fu
ally related 3-deoxy-D-*manno*-octulosonic acid aldolase (KdoA), which catalyzes *in*
degradation of the 8-carbon sugar D-KDO (*ent*-**108**) into D-arabinose and **4**, has b
investigated [94]. The enzyme, in principle, is interesting for synthetic development be
creates a (4*R*) configured chiral center, which is opposite to that produced by NeuA. A
2-keto-3-deoxy-6-phospho-D-gluconate aldolase (KDPG aldolase), which is produced b
bacteria for degradation of 6-phosphogluconate, has been probed for its synthetic utility

Due to the importance of sialic acids in many biological recognition processes, Ne
become popular for the chemoenzymatic synthesis of natural sialic acids and non
derivatives thereof. The specificity of NeuA for **4** as nucleophilic substrate, apa
fluoropyruvate [98], seems to be absolute. On the other hand, the enzyme displays
broad tolerance for analogs of the electrophilic substrate, such as a number of sug
their derivatives larger or equal to pentoses (Table 13.3) [93,99,100]. Permissible va
tolerated by the aldolase include replacement of the natural D-*manno* configurated s
with derivatives containing modifications such as epimerization, substitution, or del
positions C-2, C-4, or C-6. Epimerization at C-2, however, is restricted to smal
substituents at strongly reduced reaction rates [101,102]. In search for viral neuram
inhibitors, many sialic acid analogs have been prepared containing modifications at C
such as *N*-acylated derivatives (e.g., **111**, **113**) [103–107], or those having modification
(e.g., **112**, **114**) (Figure 13.19) [108–110]. 3-*N*-substutited mannosamine derivatives w
acceptable at all [111]. Examination of C-9 modified *N*-acetylneuraminic acid der

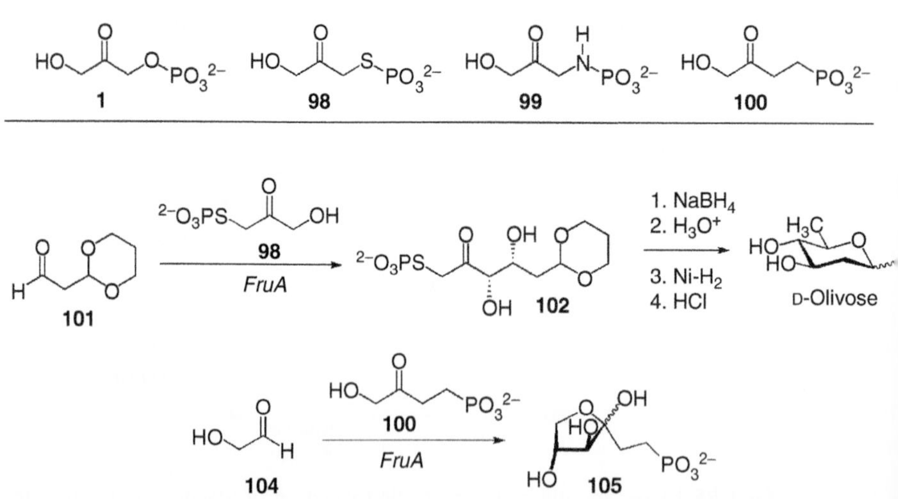

FIGURE 13.17 Substrate analogs of DHAP, and application in FruA-catalyzed aldolization re

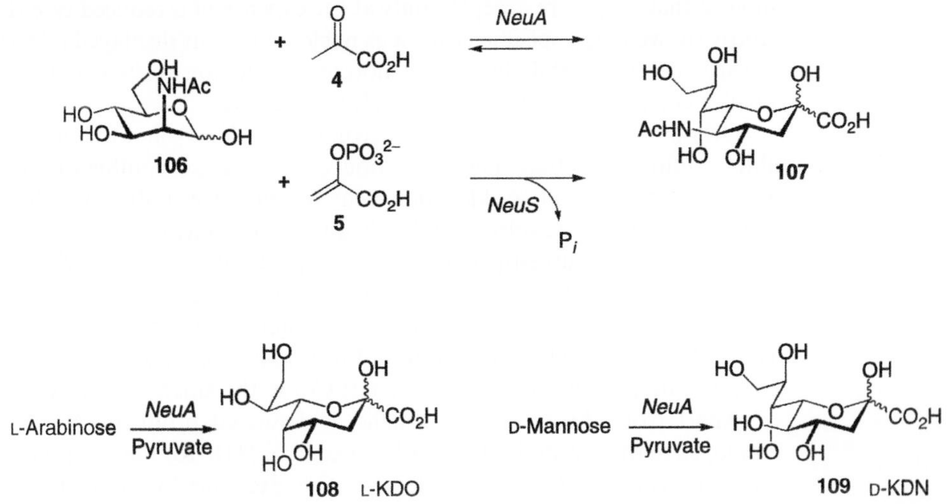

FIGURE 13.18 Synthesis of N-acetylneuraminic acid based on corresponding NeuA and NeuS catalysis, and preparation of other sialic acids on large scale by using NeuA.

TABLE 13.3
Substrate Tolerance of Neuraminic Acid Aldolase

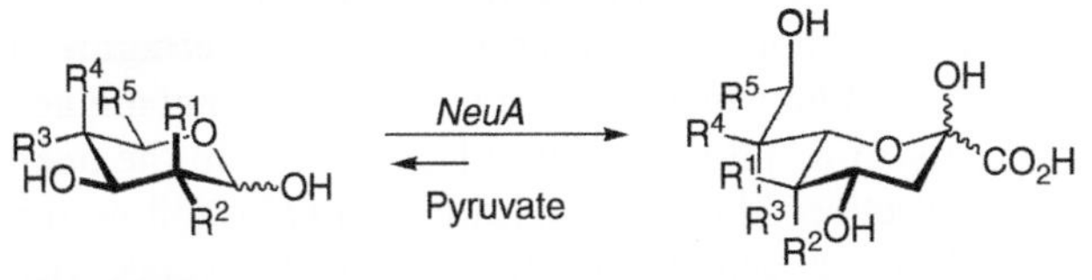

R_1	R_2	R_3	R_4	R_5	Yield (%)	Relation Rate (%)
NHAc	H	OH	H	CH_2OH	85	100
NHAc	H	OH	H	CH_2OAc	84	20
NHAc	H	OH	H	CH_2OMe	59	—
NHAc	H	OH	H	CH_2N_3	84	60
NHAc	H	OH	H	$CH_2OP(O)Me_2$	42	—
NHAc	H	OH	H	CH_2O(L-lactoyl)	53	—
NHAc	H	OH	H	CH_2O(Gly-N-Boc)	47	—
NHAc	H	OH	H	CH_2F	22	60
NHAc	H	OMe	H	CH_2OH	70	—
NHAc	H	H	H	CH_2OH	70	—
$NHC(O)CH_2OH$	H	OH	H	CH_2OH	61	—
NHCbz	H	OH	H	CH_2OH	75	—
OH	H	OH	H	CH_2OH	84	91
OH	H	H	H	CH_2OH	67	35
OH	H	H	F	CH_2F	40	—
OH	H	OH	H	H	66	10
H	F	OH	H	CH_2OH	30	—
H	H	OH	H	CH_2OH	36	130
Ph	H	OH	H	CH_2OH	76	—

showed that they were accepted only at the expense of a reduced rate [110]. Similar to
KdoA showed high specificity for **4** as nucleophile, but displayed a broad substrate to
toward the electrophile [94]. As a notable distinction to NeuA, KdoA was also act
smaller acceptor substrates, such as glyceraldehyde.

Owing to the broad substrate spectrum of the NeuA and the fully reversible nature
aldol addition reaction, the generation of a dynamic combinatorial library of sial
analogs from **4** and a few aldehyde precursors has been studied, which led to the amplif
of products that were selectively binding to added wheat germ agglutinin [112].

In most cases investigated so far, a high level of asymmetric induction by Net
the (4*S*) configuration is retained. However, a number of carbohydrates were also fo
be converted with random or even inverse stereoselectivity for the C-4 configuratio
115, 116) [100,113,114]. A critical and distinctive factor seems to be the recognition
configuration by the enzymic catalyst at C-3 in the aldehydic substrate [99,100].

Apart from the synthesis of sialic acid related products, NeuA was also succe
employed in the synthesis of a possible precursor (**118**) for the synthesis of macrolide an
amphotericin B (**119**) from 2-deoxy-2-*C*-hydroxymethyl-D-mannose (**117**) (Figure
[115,116].

Another elegant example for the use of NeuA in the synthesis of a nonsugar m
is represented by the synthesis of 3-(hydroxymethyl)-6-*epi*-castanospermine (**121**)
N-(carbobenzyloxy)-D-mannosamine (**120**) (Figure 13.21) [117]. The family of polyhy
indolizidine natural products, including castanospermine (**122**) and swainsonine
possess potent glycosidase inhibitor activity with possible applications in anticanc
anti-HIV therapy. It is remarkable that the sterically demanding *N*-substituent d
interfere with the enzyme catalytic function.

KDPG aldolase has been utilized for the highly stereoselective enzymatic synth
compound **126**, which represents the N-terminal amino acid part of the nucleoside ant
nikkomycin K_x and K_z (Figure 13.22) [118]. In the one-pot two-step enzymatic synthe
stereoselective aldol reaction leading to **125** is followed by a reductive amination, cataly
phenylalanine dehydrogenase, with high overall yield (75.9%) and enantiomeric
(>99.7%).

Zanamivir (**129**) was introduced by Glaxo, in 1999, for the treatment of influenza inf
which represented the first medicinally used sialic acid derivative with antiviral activit
required the cost-effective synthesis of the rare and expensive *N*-acetylneuraminic acid (**1**
which NeuA from *E. coli* was the first aldolase to find application in an industrial proce
multiton scale [119–121].

In a first approach, the expensive *N*-acetylmannosamine (**106**) was produced b
bining an enzymatic *in situ* isomerization of inexpensive *N*-acetylglucosamine (**128**
lyzed by *N*-acetylglucosamine 2-epimerase to the NeuA reaction in an enzyme mem
reactor (Figure 13.23) [122–124]. Unfortunately, both the steps in the sequence ar
reversible and are limited by unfavorable equilibrium constants (e.g., equilibrium cc
of $12.7\,M^{-1}$ in favor of the retroaldol reaction [19]), which mandates excess of **4** (
sevenfold to tenfold) to achieve a preparative useful conversion. Product isolation, ho
is complicated by the excess of **4**, and usually requires purification by ion ex
chromatography.

As an other example of this continuous process design, D-KDN (**109**) has been pr
on a 100-g scale from D-mannose and **4** using a pilot-scale enzyme membrane reactor
space-time yield of $375\,g\,L^{-1}\,d^{-1}$ and an overall crystallized yield of 75% [125]. In ana
the D-KDN process, nonnatural L-KDO (**108**) has been prepared using L-arabinose in p
D-mannose (Figure 13.18) [99].

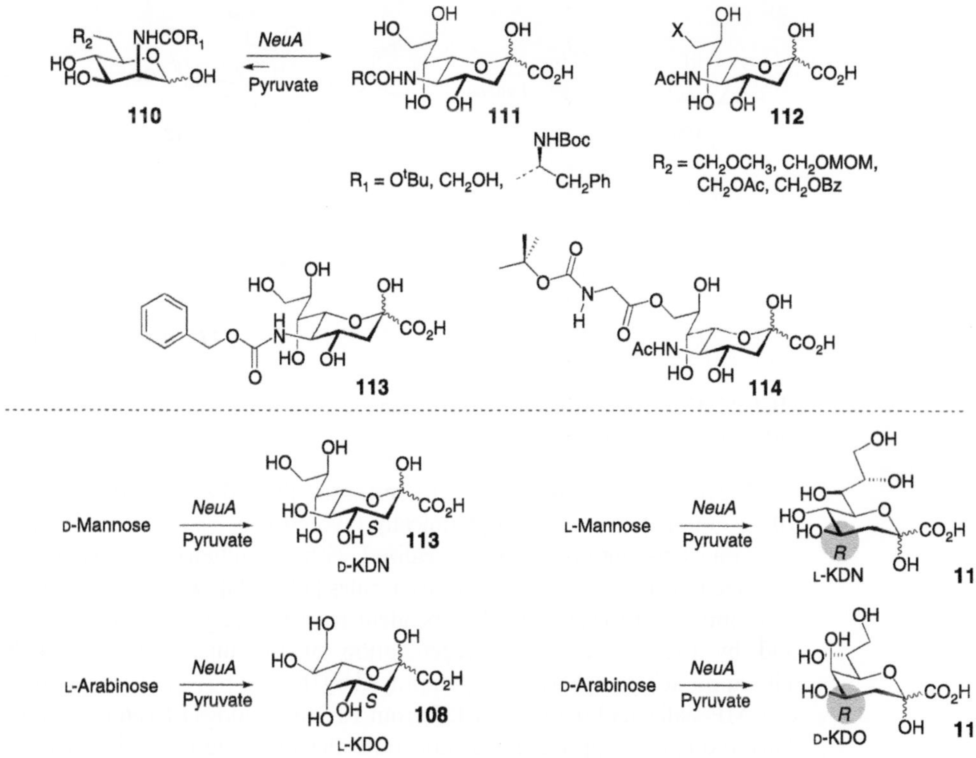

FIGURE 13.19 Examples of neuraminic acid derivatives accessible by NeuA catalysis.

In an alternative approach, serviceably high concentration of **106** may be achieved if the latter were to be produced separately by chemical base catalysis [119,121]. Also, it was found that product recovery can be facilitated by removing residual **4** through formation of a separable bisulfite adduct [121] or by decomposing **4** into volatile compounds using yeast pyruvate decarboxylase [113,119].

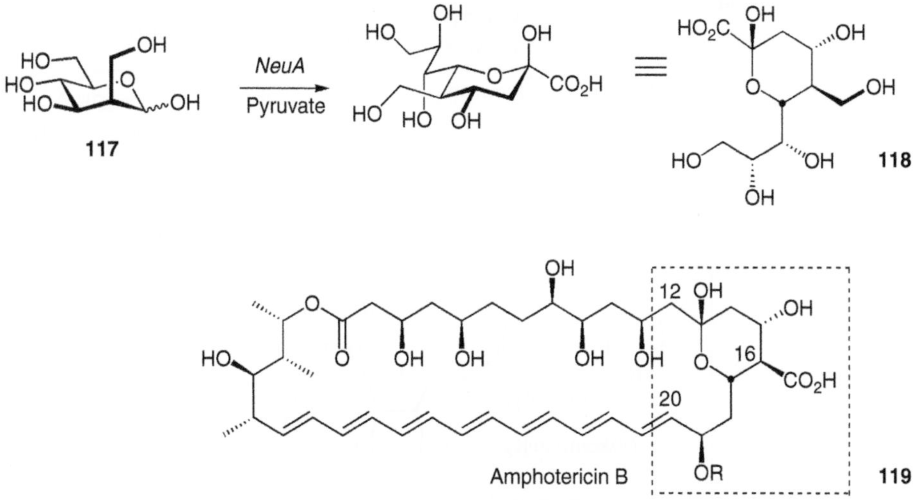

FIGURE 13.20 NeuA-catalyzed synthesis of a possible synthetic precursor for the macrolide antibiotic amphotericin B.

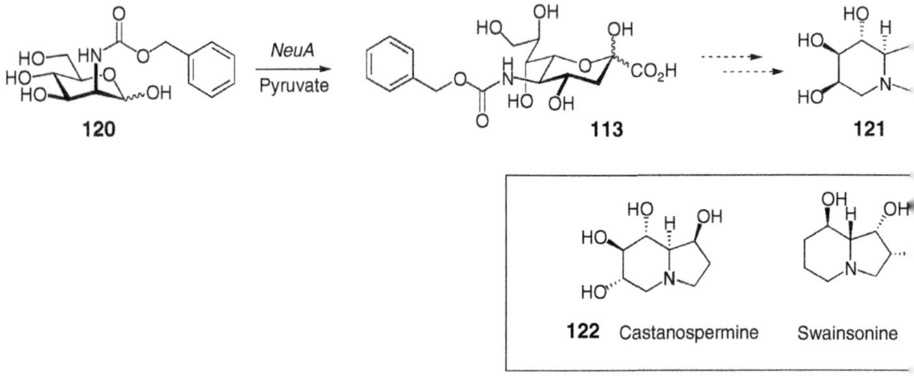

FIGURE 13.21 NeuA-based chemoenzymatic synthesis of the polyhydroxylated indolizidine 3-(hydroxymethyl)-6-epicastanospermine.

The need of excess **4** might be circumvented altogether when the NeuA-catalyzed syn **107** is coupled to a thermodynamically favored process, such as an irreversible sialyl reaction. This option has been realized with a multienzyme system for a sialyltran catalyzed synthesis of sialyloligosaccharides [126–128], where the driving force is pro consumption of **107** in the CTP-dependent activation to give the glycosyl donor CMP-Ne and by irreversible *in situ* regeneration of the nucleotide triphosphate (Figure This approach was also recently employed by the one-pot chemoenzymatic synthesis of a of CMP-sialic acid derivatives **132** from various analogs of **106** (e.g., **130**; Figure 13.2 This and earlier approaches demonstrate the substrate flexibility of not only the neu acid aldolase, but also the CMP-sialic acid synthases that were employed in the s reactions [130].

13.5 2-DEOXYRIBOSE 5-PHOSPHATE ALDOLASE

The family of aldehyde-dependent aldolases comprises only one member known so 2-deoxyribose 5-phosphate aldolase (DerA or RibA, EC 4.1.2.4). *In vivo* DerA cata

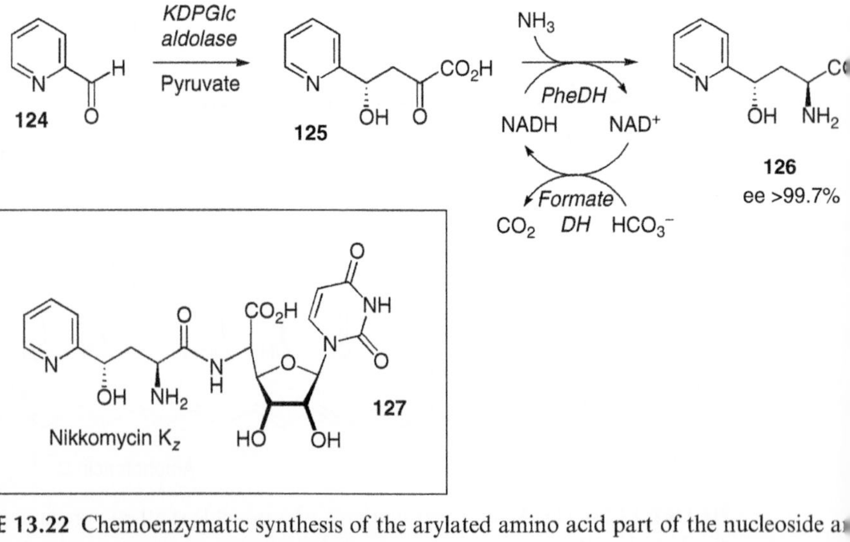

FIGURE 13.22 Chemoenzymatic synthesis of the arylated amino acid part of the nucleoside a nikkomycins K_x and K_z using KDPG aldolase.

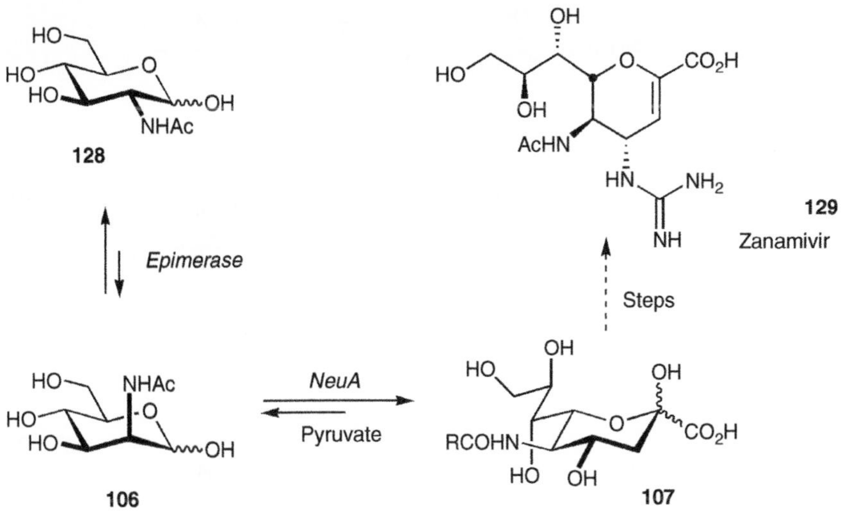

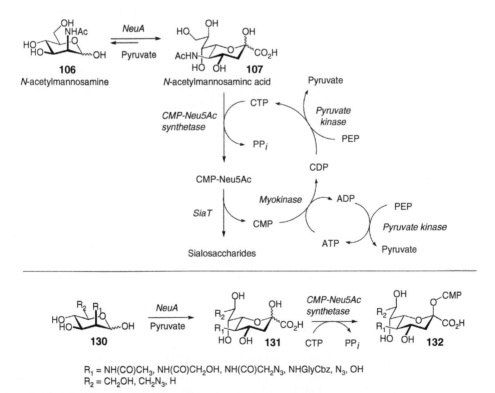

FIGURE 13.23 Industrial process for the production of *N*-acetyl neuraminic acid based on NeuA catalysis, as a precursor for the synthesis of the influenza drug Zanamivir.

reversible cleavage of 2-deoxyribose 5-phosphate (**134**) to D-glyceraldehyde 3-phosphate (**95**) and ethanal (**133**) (Figure 13.25). The equilibrium constant of the reaction with $4.2 \times 10^3 \text{ M}^{-1}$ favors product formation [19]. Mechanistically, DerA belongs to the class I aldolases, involving a covalent imine–enamine bond formation between the substrate and the enzyme

R_1 = NH(CO)CH$_3$, NH(CO)CH$_2$OH, NH(CO)CH$_2$N$_3$, NHGlyCbz, N$_3$, OH
R_2 = CH$_2$OH, CH$_2$N$_3$, H

FIGURE 13.24 One-pot multienzymatic synthesis of sialooligosaccharides with *in situ* cofactor regeneration, and one-pot synthesis of CMP-activated sialic acid derivatives based on NeuA catalysis.

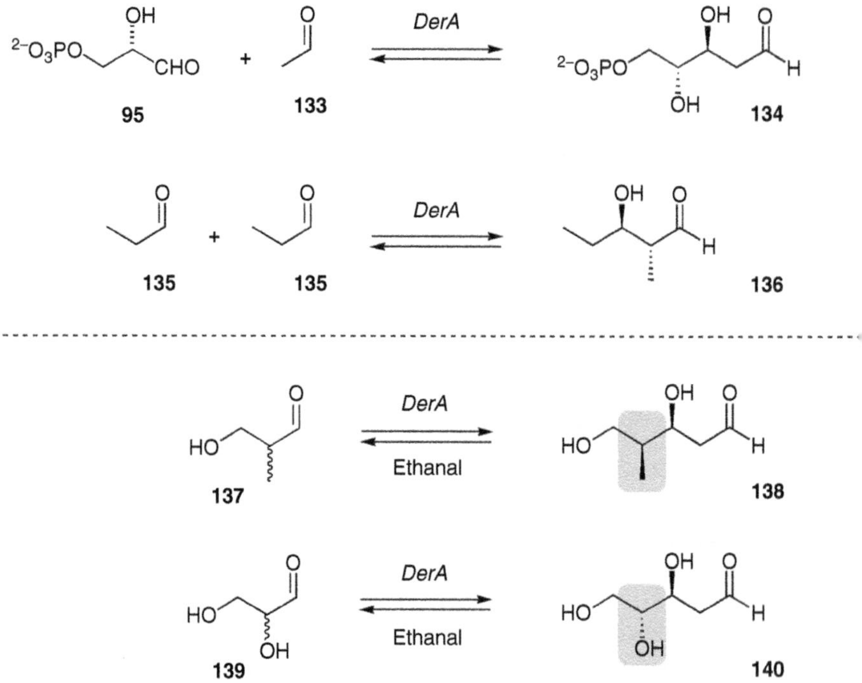

FIGURE 13.25 DerA-catalyzed stereoselective aldolization reactions.

lysine residue during the catalytic cycle [131]. DerA from *E. coli* can be easily overexp
by homologous overexpression in *E. coli*, and the crude cell-free extract often quali
preparative applications without further purification [132–134]. The enzyme is rather
under reaction conditions and at room temperature.

Like other aldolases, DerA is rather specific for its physiological donor substrate, but
of 133 also the structural closely related propanal (135), acetone, or fluoroacetone are acc
[132,133]. The substrate tolerance for acceptor aldehydes is rather relaxed, and many al
and variously substituted aldehydes can be used as a replacement for the natural subst
albeit at the expense of strongly reduced catalytic rates (Table 13.4) [132,133]. The enzyme
high stereoselectivity for the newly generated chiral center at C3; remarkably, 135 as a
substrate yielded only a single diastereomer of absolute (2R,3S) configuration [133]
13.25). This is indicative not only of the high level of asymmetric induction at the a
carbonyl, but also of the stereospecific deprotonation of the aldol donor substrate. Intere
acceptor aldehydes possessing a small hydrophobic group at C2 (e.g., 137) with L-config
were strongly preferred by DerA to yield *syn* adducts (138), whereas the D-isomers w
preferred substrate in case of polar groups, such as OH (139) or N_3 substituents [135,136

Corresponding to its natural function, DerA has been used for the synthesis of ^{13}C-
2-deoxy-D-ribose from labeled 133 and 95 [137], and in a commercial multienzyme pro
the synthesis of purine and pyrimidine containing deoxyribonucleosides (e.g., 142;
13.26) [138,139]. Synthetic applications of DerA also include the preparation of variou
derivatives, such as 2-deoxy, thio, and aza sugars [133]. Specifically, DerA was used
stereoselective synthesis of 2-deoxy-L-fucose, a constituent of several antibiotics, s
from (2R,3S)-dihydroxybutanal and 133 [134]. Recently, DerA has been used in the
controlled addition of 133 to *rac*-lactaldehyde and (S)-2-methyl-3-hydroxypropanal (
furnish chiral building blocks (e.g., 144–146) that are useful for the synthesis of ant
agents epothilone A (147) and C (Figure 13.27) [135].

TABLE 13.4

Substrate Spectrum Accepted by *E. coli* 2-Deoxy-D-Ribose-5-Phosphate Aldolase

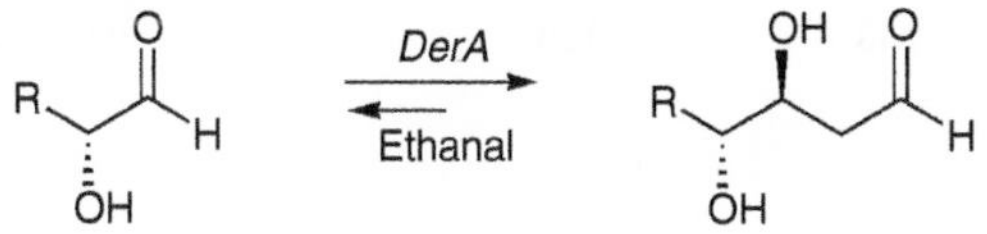

R	Yield (%)	Relation Rate (%)
$CH_2OPO_3^{2-}$	78	100
H	20	—
CH_2OH	65	0.4
CH_3	32	0.4
CH_2F	33	0.4
CH_2Cl	37	0.3
CH_2Br	30	—
CH_2SH	33	—
CH_2N_3	76	0.3
C_2H_5	18	0.3
$CH=CH_2$	12	—
$CHOH-CH_2OH$	62	0.3
CHN_3-CH_2OH	46	—
$CHOH-CH_3$	51	—
$CHOH-CH_2-C_6H_5$	46	—
$CH_2SCH_2-CHOH-CH_2OH$	27	—

An interesting transformation is the sequential asymmetric addition of two equivalents of donor **133** to a starting acceptor unit (Figure 13.28). In such a cascade, which is observed with simple α-substituted ethanal derivatives that contain a functionality unable to induce

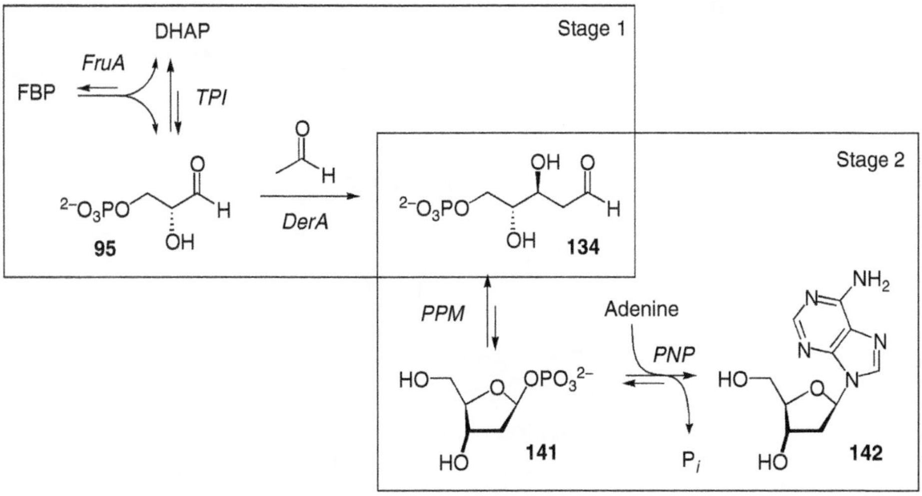

FIGURE 13.26 Two-stage aldolase-based process used for the preparative scale synthesis of deoxynucleosides.

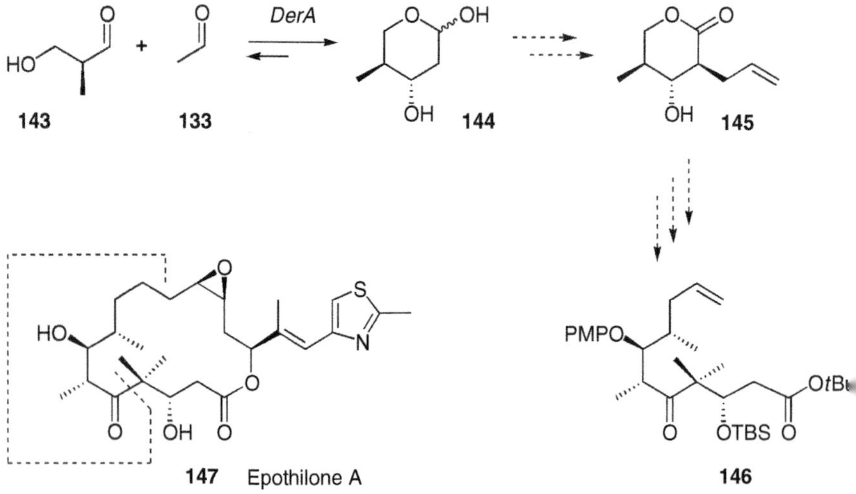

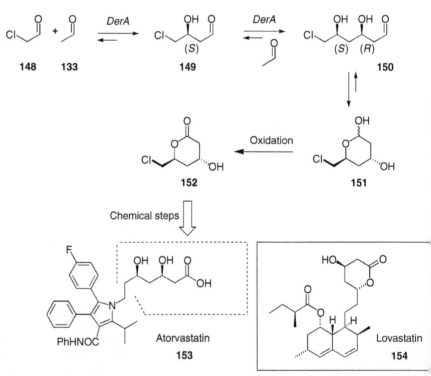

FIGURE 13.27 Chemoenzymatic synthesis of a chiral epothilone A building block using DerA-c aldolization reaction.

cyclization of the initial product (e.g., **149**) to a hemiacetal, the first aldol product an acceptor substrate for a second aldol addition. Only at this stage (e.g., **150**) a cyc can happen, which yields stable pyranoses and thereby prevents further aldol steps place [140]. A particular good acceptor substrate is chloroethanal (**148**), whicł

FIGURE 13.28 DerA-catalyzed sequential aldol reaction yielding a key chiral building block cholesterol lowering vastatin drugs.

($3R,5S$)-6-chloro-2,4,6-trideoxyhexapyranoside (**151**), a useful building block with two chiral centers for 1,3-polyol containing natural products. Recently, this reaction has been developed into an industrial process for the large-scale synthesis of the chiral side chain of the statins [141–143], which are marketed as important HMG-CoA reductase inhibitors. Although several elegant chemoenzymatic synthetic routes have been developed for such skipped polyols [144], the aldol route to this polyfunctional chiral building block (**152**) is economically most attractive because the asymmetric centers are created without cofactor demand directly from cheap bulk chemicals.

13.6 GLYCINE-DEPENDENT ALDOLASES

The metabolism of β-hydroxy-α-amino acids involves pyridoxal phosphate-dependent enzymes, classified as serine hydroxymethyltransferase (SHMT) and threonine aldolase (ThrA). Both enzymes catalyze reversible aldol-type cleavage reactions yielding glycine (**156**) and an aldehyde (Figure 13.29). Whereas SHMT *in vivo* has a biosynthetic function, ThrA catalyzes the degradation of threonine (**158, 159**) to **156** and **133**; both L- and D-specific ThrA enzymes are known [9,145]. Typically, ThrA enzymes show complete enantiopreference for their natural α-D- or α-L-amino configuration but, with few exceptions, have only low specificity for the relative *threo/erythro*-configuration (e.g., **158/159**) [146]. Likewise, SHMT is highly selective for L-configuration, but has poor *threo/erythro*-selectivity [147].

For biocatalytic applications, the known SHMT, D- and L-ThrA show broad substrate tolerance for various acceptor aldehydes (Table 13.5), including aromatic aldehydes [146–148]; however, α,β-unsaturated aldehydes are not accepted [149].

β-Hydroxyamino acids constitute an important class of compounds, representing natural products of their own right, like L-threonine (**158, 159**) or L-serine (**157**), or components of more complex structures, such as the β-hydroxytyrosine moiety present in vancomycin [150]. The synthesis of **157** has been developed on multimolar scale using SHMT from *Klebsiella aerogenes* or *E. coli* to furnish the product at high final concentration of >450 g/L [151–154]. SHMT has also been employed for the synthesis of L-*erythro*-2-amino-3-hydroxy-1,6-hexanedicarboxylic acid **160** [155], a potential precursor for carbocyclic β-lactams and nucleotides (Figure 13.30). The *erythro*-selective L-ThrA from the yeast *Candida humicola* has been used for the synthesis of (S,S,R)- and (S,S,S)-3,4-dihydroxyprolines (**167, 168**) [156], and for the preparation of a chiral building block (**162**) toward the synthesis of the immunosuppressive lipid mycestericin D (**163**) [157].

Due to the fully reversible equilibrium nature of the aldol addition process, enzymes with low diastereoselectivity will typically lead to a thermodynamically controlled mixture of

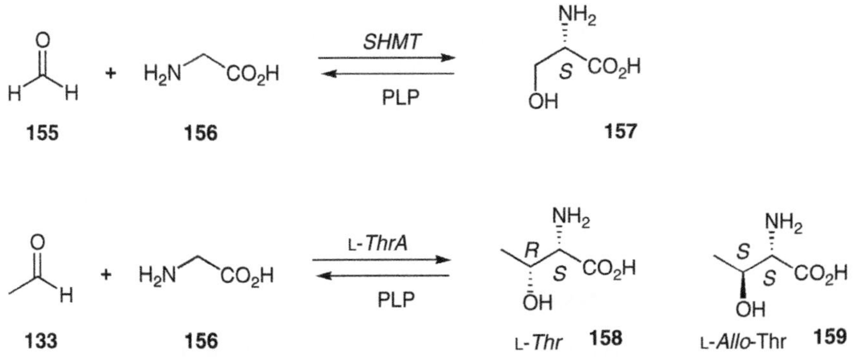

FIGURE 13.29 Aldol reactions catalyzed *in vivo* by SHMT and threonine aldolases.

TABLE 13.5
Substrate Tolerance of SHMT and Threonine Aldolases

R	SHMT Selectivity (threo/erythro)	Yield (%)	L-ThrA Selectivity (threo/erythro)	Yield (%)	D-ThrA Selectivity (threo/erythro)	Y
H	–	94				
CH_3	2:98	–	9 : 91	40	53 : 47	
C_3H_9			24 : 76	21	67 : 33	
C_5H_{11}	60:40	25	37 : 63	16	61 : 39	
$CH(CH_3)_2$			46 : 53	15	84 : 16	
CH_2OBn			2 : 98	88		
CH_2CH_2OBn			47 : 53	53		
$(CH_2)_2C_6H_5$			28 : 72	10	87 : 13	
C_6H_5	60:40	22	60 : 40	9	74 : 26	
p-$C_6H_4NO_2$			47 : 53	53	55 : 45	
o-$C_6H_4NO_2$			58 :42	93	72 : 28	
m-C_6H_4OH			73 : 27	43	74 : 26	
p-$C_6H_4CH_3$			55 : 45	17	43 : 57	
2-Tmidazolyl	33:67	10	66 : 34	40	61 : 39	
2-Turanyl	50:50	20				
2-Thienyl	56:44	11				

threo/erythro-isomers that are difficult to separate, e.g., by fractional crystallization
exchange chromatography. A further problem is that the overall equilibrium consta
not favor synthesis [19], which requires the reactions to de driven by an excess of **156** (w
difficult to separate from the product) or the aldehyde (which at high concentratio
engage in condensation reactions or inactivate the enzyme). Therefore, members of th
of aldolases have been more successfully applied in the kinetic resolution of racemic m
of β-hydroxy amino acids produced by chemical synthesis.

In this respect, SHMT has been used to resolve racemic *erythro* β-hydroxy amin
to produce pure D-*erythro* isomers [158,159]. An L-ThrA from *Streptomyces amaku*
was found to be particularly useful for the resolution of racemic *threo*-aryl serines
yielding enantiomerically pure D-amino acids (e.g., **170**; Figure 13.31) [155,160].
complementary example, the recombinant low-specificity D-ThrA of *Arthrobac*
DK-38 was used for the preparation of L-*threo*-β-(4-methylthiophenyl)serine (**173**
and L-*threo*-β-(3,4-methylenedioxyphenyl)serine (**176**) from DL-*threo*-isomeric m
[162]. The former represents an intermediate in the synthesis of the antibiotics flor
and thiamphenicol, while the latter serves as an intermediate for the productio
Parkinson drug.

13.7 TRANSKETOLASE

Transketolase (EC2.2.1.1) is an enzyme synthetically relevant to, but mechanistically
from, the aldolases for asymmetric C–C bond forming reactions. *In vivo*, the enzyme ca

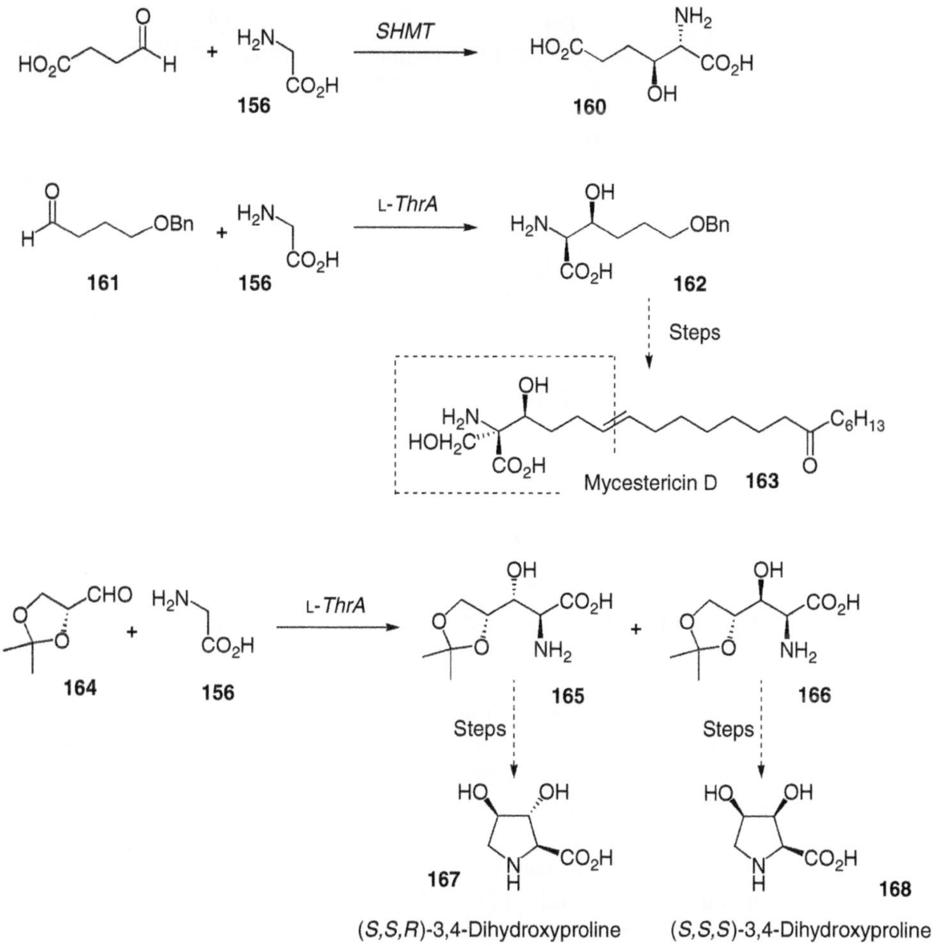

FIGURE 13.30 Synthetic application of ThrA for stereoselective synthesis of a potential chiral building block for the immunosuppressive lipid mycestericin D, and of dihydroxyprolines from glyceraldehyde.

the reversible transfer of a 2-carbon ketol unit between phosphorylated intermediates of the oxidative pentose phosphate pathway (Figure 13.32). Its mechanism resembles a classical benzoin addition, and the enzyme requires thiamine diphosphate and divalent Mg as cofactors [163]. For preparative reactions, the natural ketol donor D-xylulose 5-phosphate (178) [164] can be replaced by hydroxypyruvate (182) [165], which makes the enzyme more generally useful and renders synthetic reactions irreversible due to the spontaneous decarboxylation of the reactive intermediate (Figure 13.33). This feature compensates for the significant rate reduction with 182 as a nonphysiological substrate; also 182 causes no inhibition of the enzyme even at high concentrations [166].

Transketolase from different sources has been shown to possess a broad acceptor substrate spectrum, yielding products with complete (S)-stereospecificity for the newly formed chiral center [165]. Although generic aldehydes are converted with full stereocontrol and even α,β-unsaturated aldehydes are accepted to some degree, hydroxylated electrophiles are usually converted with higher rates [167]. In the latter case, transketolase discriminates 2-hydroxyaldehydes (181) with complete kinetic preference for the (R)-configuration to yield enantiopure diols of (3S,4R)-configuration (183; Figure 13.33) [168,169].

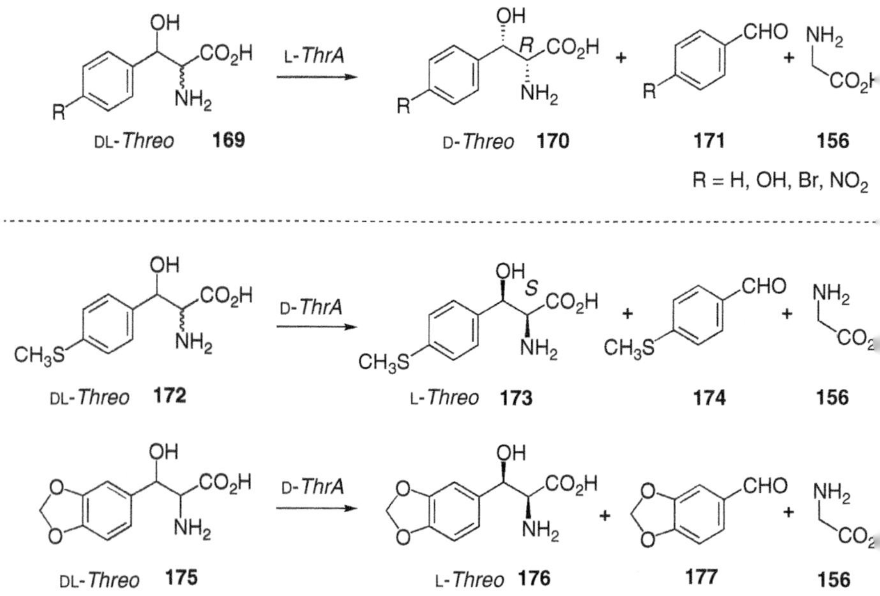

FIGURE 13.31 ThrA-catalyzed resolution of diastereomeric mixtures by retroaldolization unde control yielding enantiomerically pure arylserines.

Recently, the transketolase of *E. coli* has been developed for large-scale enzyme tion [170–172]. The enzyme is quite tolerant to organic cosolvents, and its immob stabilizes significantly against inactivation by aldehyde substrates [173]. Continuo reactions have been performed in a membrane reactor [172], and *in situ* product through borate complexation has been probed [174].

Synthetic applications of transketolase include the preparation of valuable ketose such as fructose analogs [165]. In addition, transketolase has been utilized for the k in the chemoenzymatic synthesis of (+)-exo-brevicomin (**72**; Figure 13.34) [60], az such as 1,4-dideoxy-1,4-imino-D-arabinol [53] and fagomine [175], or *N*-hydroxypyr (**188**) [176] from 3-azido, 3-cyano, and 3-O-benzyl (**186**) analogs and deriva glyceraldehyde, respectively. All syntheses take advantage of the intrinsic kinetic re

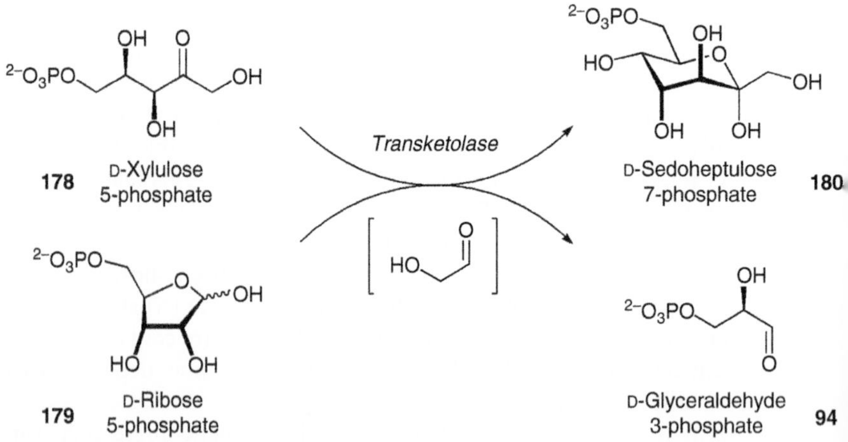

FIGURE 13.32 *In vivo* reaction catalyzed by transketolase.

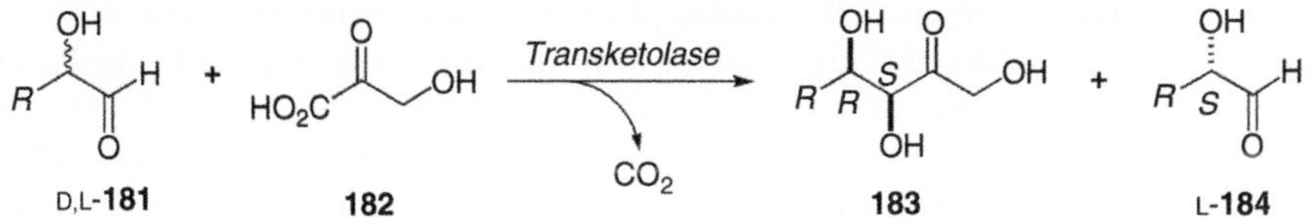

FIGURE 13.33 Kinetic resolution catalyzed by transketolase, and nonequilibrium C–C bond formation by decomposition of hydroxypyruvate.

of 2-hydroxyaldehydes by the enzyme and the thermodynamic driving force from the decarboxylation of **182**. Even the option to prepare (S)-configurated 2-hydroxyaldehydes (**184**) by consumption of the (R)-antipodes from the racemate in a transketolase reaction (Figure 13.33) has been verified [169]. More recently, transketolase has been utilized to prepare a fluorogenic substrate (**190**), useful for the high-throughput screening of stereoselective transaldolases (Figure 13.34) [177].

13.8 PERSPECTIVES

Aldol addition reactions are one among the most fundamental methodologies available to the synthetic chemist for the construction of new C–C bonds. Therefore, control of the stereochemical course of the aldol reaction for the preparation of enantiomerically and diastereomerically pure products has attracted considerable interest, and asymmetric aldol reactions are clearly some of the best-developed organic transformations today [5,178,179]. This chapter has attempted to demonstrate the remarkable progress of asymmetric enzymatic aldol reactions by highlighting the state of development of readily available enzymes, the most important synthetic examples, and the most efficient reaction techniques. This clearly proves that biocatalytic C–C bond formation is eminently useful, and highly predictable, for the asymmetric synthesis of complex multifunctional molecules. It is also evident that the

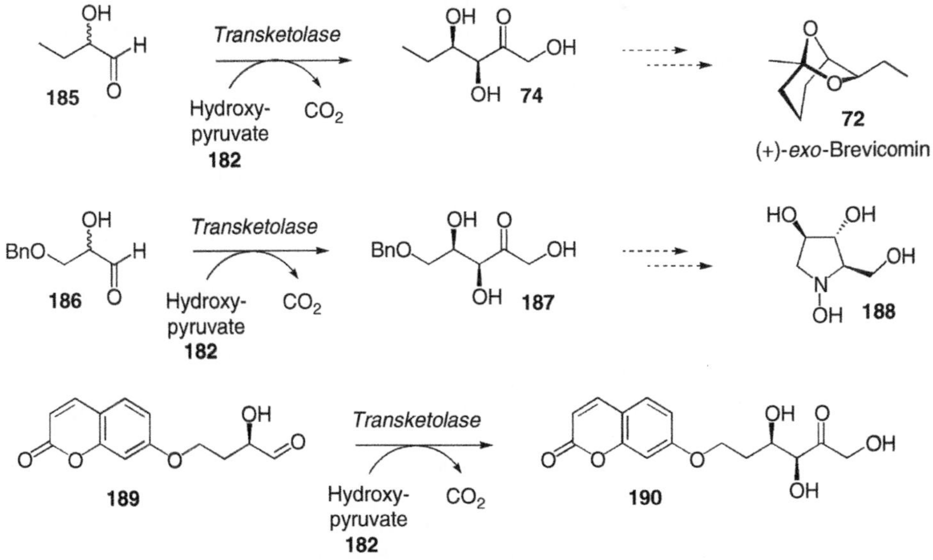

FIGURE 13.34 Chemoenzymatic synthesis of (+)-exo-brevicomin, an N-hydroxypyrolidine, and a fluorescence screening substrate based on stereospecific transketolase catalysis.

technology is well accepted in the chemical community and, with several examples fo: scale industrial processes now in operation, that the field is maturing rapidly.

Enzyme-catalyzed asymmetric aldol reactions are not only highly efficient, sin operate even at large scale, but can also be extremely cost-effective and environ friendly [143]. Clearly, in comparison with current chemical methods for asymmetri reactions, the ecological advantage of biocatalytic procedures stems from the fact that 1 not suffer from a need for harsh reaction conditions, corrosive reagents, costly pro groups, high catalyst loading, consumption of chiral auxiliaries, or heavy metal lea products. Despite a broad scope of synthetic opportunities, however, it becomes obvio enzymatic aldol reactions still suffer from restrictions that mostly derive from the flexibility of applicable donor components, or from the costs inherent in their preparat remain competitive with modern chemical developments such as proline-based orga lysis [180], current focus is therefore at lifting the restraints and tailoring aldolases for needs by different approaches, such as by rational protein engineering based on the available protein crystal structures [6] or by directed evolution through random muta [181,182]. Catalytic aldolase antibodies still suffer from inconveniently low catalyti [183,184], but they may guide in the construction of aldolases that act on more gene functionalized substrates. Prospecting for novel activities through whole-genome seqr efforts, mass screening of previously untapped natural [185,186], or panning of mar biodiversity may furnish attractive biocatalysts for new applications of C–C bond for in biotechnology [187].

REFERENCES

1. Collins, A.N., Sheldrake, G.N., and Crosby, J., *Chirality in Industry*, Wiley, Chichester. 1992, vol. 2, 1997.
2. Schmid, A., Dordick, J.S., Hauer, B., et al., Industrial biocatalysis today and tomorrow, 409, 258–268, 2001.
3. Schoemaker, H.E., Mink, D., and Wubbolts, M.G., Dispelling the myths—biocatalysis i trial synthesis, *Science*, 299, 1694–1697, 2003.
4. Liese, A., Seelbach, K., and Wandrey, C., *Industrial Biotransformations*, Wiley-VCH, Weinhei
5. Mahrwald, R., *Modern Aldol Reactions*, Wiley-VCH, Weinheim, Germany, 2004.
6. Fessner, W.-D., Enzyme-catalyzed aldol additions, in *Modern Aldol Reactions*, vol. 1, Mahrw Ed., Wiley-VCH, Weinheim, Germany, 2004, pp. 201–272.
7. Griengl, H., Schwab, H., and Fechter, M., The synthesis of chiral cyanohydrins by oxyni *Trends Biotechnol.*, 18, 252–256, 2000.
8. Pohl, M., Lingen, B., and Müller, M., Thiamin-diphosphate-dependent enzymes: new as asymmetric C–C bond formation, *Chem. Eur. J.*, 8, 5288–5295, 2002.
9. Fessner, W.-D. and Walter, C., Enzymatic C–C bond formation in asymmetric synthes *Curr. Chem.*, 184, 97–194, 1996.
10. Silvestri, M.G., Desantis, G., Mitchell, M., et al., Asymmetric aldol reactions using aldolas *Stereochem.*, 23, 267–342, 2003.
11. Hildebrandt, G. and Klavehn, W., L-1-Phenyl-2-methylamino-1-propanol, US 195695 (*Chem. Abstr.* 1934, 28, 40723).
12. Waldmann, H. and Sebastian, D., Enzymatic protecting group techniques, *Chem. R* 911–937, 1994.
13. Horecker, B.L., Tsolas, O., and Lai, C.Y., Aldolases, in *The Enzymes*, vol. VII, Boyer, P.l Academic Press, New York, 1972, pp. 213–258.
14. Fessner, W.-D., Schneider, A., Held, H., et al., The mechanism of class II, metal-de aldolases, *Angew. Chem. Int. Ed. Engl.*, 35, 2219–2221, 1996.
15. Bednarski, M.D., Simon, E.S., Bischofberger, N., et al., Rabbit muscle aldolase as a cat organic synthesis, *J. Am. Chem. Soc.*, 111, 627–635, 1989.

16. von der Osten, C.H., Sinskey, A.J., Barbas, C.F., et al., Use of a recombinant bacterial fructose-1,6-diphosphate aldolase in aldol reactions: preparative syntheses of 1-deoxynojirimycin, 1-deoxymannojirimycin, 1,4-dideoxy-1,4-imino-D-arabinitol, and fagomine, *J. Am. Chem. Soc.*, 111, 3924–3927, 1989.

17. Fessner, W.-D., Sinerius, G., Schneider, A., et al., Diastereoselective enzymatic aldol additions: L-rhamnulose and L-fuculose 1-phosphate aldolases from *E. coli.*, *Angew. Chem. Int. Ed. Engl.*, 30, 555–558, 1991.

18. Zannetti, M.T., Walter, C., Knorst, M., et al., Fructose 1,6-bisphosphate aldolase from *Staphylococcus carnosus*: overexpression, structure prediction, stereoselectivity, and application in the synthesis of bicyclic sugars, *Chem. Eur. J.*, 5, 1882–1890, 1999.

19. Goldberg, R.N. and Tewari, Y.B., Thermodynamics of enzyme-catalyzed reactions: Part 4. Lyases, *J. Phys. Chem. Ref. Data*, 24, 1669–1698, 1995.

20. Brinkkotter, A., Shakeri, G.A., and Lengeler, J.W., Two class II D-tagatose-bisphosphate aldolases from enteric bacteria, *Arch. Microbiol.*, 177, 410–419, 2002.

21. Ghalambor, M.A. and Heath, E.C., L-Fuculose 1-phosphate aldolase, *Meth. Enzymol.*, 9, 538–542, 1966.

22. Takagi, Y., L-Rhamnulose 1-phosphate aldolase, *Meth. Enzymol.*, 9, 542–545, 1966.

23. Chiu, T.H., Evans, K.L., and Feingold, D.S., L-Rhamnulose-1-phosphate aldolase, *Meth. Enzymol.*, 42, 264–269, 1975.

24. Durany, O., de Mas, C., and Lopez-Santin, J., Fed-batch production of recombinant fuculose-1-phosphate aldolase in *E. coli.*, *Process Biochem.*, 40, 707–716, 2005.

25. Vidal, L., Durany, O., Suau, T., et al., High-level production of recombinant his-tagged rhamnulose 1-phosphate aldolase in *Escherichia coli.*, *J. Chem. Technol. Biotechnol.*, 78, 1171–1179, 2003.

26. Toone, E.J., Simon, E.S., Bednarski, M.D., et al., Enzyme-catalyzed synthesis of carbohydrates, *Tetrahedron*, 45, 5365–5422, 1989.

27. Bissett, D.L. and Anderson, R.L., Lactose and D-galactose metabolism in *Staphylococcus aureus*. IV. Isolation and properties of a class I D-ketohexose-1,6-diphosphate aldolase that catalyzes the cleavage of D-tagatose 1,6-diphosphate, *J. Biol. Chem.*, 255, 8750–8755, 1980.

28. Fessner, W.-D. and Eyrisch, O., One-pot synthesis of tagatose 1,6-bisphosphate via diastereoselective enzymic aldol addition, *Angew. Chem. Int Ed. Engl.*, 31, 56–58, 1992.

29. Eyrisch, O., Sinerius, G., and Fessner, W.-D., Facile enzymic *de novo* synthesis and NMR spectroscopic characterization of D-tagatose 1,6-bisphosphate, *Carbohydr. Res.*, 238, 287–306, 1993.

30. Williams, G.J., Domann, S., Nelson, A., et al., Modifying the stereochemistry of an enzyme-catalyzed reaction by directed evolution, *Proc. Natl. Acad. Sci. USA*, 100, 3143–3148, 2003.

31. Fessner, W.-D., Badia, J., Eyrisch, O., et al., Enzymic syntheses of rare ketose 1-phosphates, *Tetrahedron Lett.*, 33, 5231–5234, 1992.

32. Fessner, W.-D., Schneider, A., Eyrisch, O., et al., 6-Deoxy-L-lyxo- and 6-deoxy-L-arabino-hexulose 1-phosphates. Enzymic syntheses by antagonistic metabolic pathways, *Tetrahedron Asymmetry*, 4, 1183–1192, 1993.

33. Lees, W.J. and Whitesides, G.M., Diastereoselectivity (enantioselectivity) of aldol condensations catalyzed by rabbit muscle aldolase at C-2 of RCHOHCHO if R has an appropriately placed negatively charged group, *J. Org. Chem.*, 58, 1887–1894, 1993.

34. Durrwachter, J.R., Wong, C.-H., Fructose 1,6-diphosphate aldolase-catalyzed stereoselective synthesis of C-alkyl and N-containing sugars: thermodynamically controlled C–C bond formations, *J. Org. Chem.*, 53, 4175–4181, 1988.

35. Kim, M.-J., Lim, I.T., Synthesis of unsaturated C8–C9 sugars by enzymatic chain elongation, *Synlett*, 138–140, 1996.

36. Serianni, A.S., Cadman, E., Pierce, J., et al., Enzymic synthesis of carbon-13-enriched aldoses, ketoses, and their phosphate esters, *Meth. Enzymol.*, 89, 83–92, 1982.

37. Wong, C.-H. and Whitesides, G.M., Synthesis of sugars by aldolase-catalyzed condensation reactions, *J. Org. Chem.*, 48, 3199–3205, 1983.

38. Fessner, W.-D. and Sinerius, G., Synthesis of dihydroxyacetone phosphate (and isosteric analogs) by enzymatic oxidation: sugars from glycerol, *Angew. Chem. Int. Ed. Engl.*, 33, 209–212, 1994.

39. Franke, D., Machajewski, T., Hsu, C.-C., et al., One-pot synthesis of L-fructose using multienzyme systems based on rhamnulose-1-phosphate aldolase, *J. Org. Chem.*, 68, 68 2003.

40. Fessner, W.-D. and Walter, C., "Artificial metabolisms" for the asymmetric one-pot syr branched-chain saccharides, *Angew. Chem. Int. Ed. Engl.*, 31, 614–616, 1992.

41. Liu, K.K.C. and Wong, C.-H., A new strategy for the synthesis of nucleoside analogues enzyme-catalyzed aldol reactions, *J. Org. Chem.*, 57, 4789–4791, 1992.

42. Wong, C.-H., Moris-Varas, F., Hung, S.- C., et al., Small molecules as structural and f mimics of sialyl Lewis X tetrasaccharide in selectin inhibition: a remarkable enhanc inhibition by additional negative charge and/or hydrophobic group, *J. Am. Chem. S* 8152–8158, 1997.

43. Zhu, W. and Li, Z., Synthesis of perfluoroalkylated sugars catalyzed by rabbit muscle (RAMA), *J. Chem. Soc. Perkin Trans. I*, 1105–1108, 2000.

44. Bednarski, M.D., Waldmann, H.J., and Whitesides, G.M., Aldolase-catalyzed synthesis plex C8 and C9 monosaccharides, *Tetrahedron Lett.*, 27, 5807–5810, 1986.

45. Eyrisch, O. and Fessner, W.D., Disaccharide mimetics by enzymatic tandem aldol a *Angew. Chem. Int. Ed. Engl.*, 34, 1639–1641, 1995.

46. Petersen, M., Zannetti, M.T., and Fessner, W.-D., Tandem asymmetric C–C bond form: enzyme catalysis, *Top. Curr. Chem.*, 186, 87–117, 1997.

47. Fessner, W.-D., Gosse, C., Jaeschke, G., et al., Enzymes in organic synthesis, 15. Short e synthesis of L-fucose analogs, *Eur. J. Org. Chem.*, 125–132, 2000.

48. Wong, C.-H., Alajarin, R., Moris-Varas, F., et al, Enzymatic synthesis of L-fucose and *J. Org. Chem.*, 60, 7360–7363, 1995.

49. Durrwachter, J.R., Drueckhammer, D.G., Nozaki, K., et al., Enzymic aldol condensatior ization as a route to unusual sugar derivatives, *J. Am. Chem. Soc.*, 108, 7812–7818, 1986

50. Borysenko, C.W., Spaltenstein, A., Straub, J.A., et al., The synthesis of aldose sugars fr protected dialdehydes using rabbit muscle aldolase, *J. Am. Chem. Soc.*, 111, 9275–9276,

51. Look, G.C., Fotsch, C.H., and Wong, C.-H., Enzyme-catalyzed organic synthesis— routes to aza sugars and their analogs for use as glycoprocessing inhibitors, *Acc. Chem.* 182–190, 1993.

52. Kajimoto, T., Chen, L., Liu, K.K.C., et al., Palladium-mediated stereocontrolled reductiv tion of azido sugars prepared from enzymic aldol condensation: a general approacl synthesis of deoxy aza sugars, *J. Am. Chem. Soc.*, 113, 6678–6680, 1991.

53. Ziegler, T., Straub, A., and Effenberger, F., Enzym-katalysierte synthese von 1-desoxyn imycin, 1-desoxynojirimycin und 1,4-didesoxy-1,4-imino-D-arabinitol, *Angew. Chem. Engl.*, 27, 716–717, 1988.

54. Liu, K.K.C., Kajimoto, T., Chen, L., et al., Use of dihydroxyacetone phosphate-d aldolases in the synthesis of deoxy aza sugars, *J. Org. Chem.*, 56, 6280–6289, 1991.

55. Lees, W.J. and Whitesides, G.M., The enzymatic synthesis of 1,5-dideoxy-1,5-diimino-D-ta 1-deoxygalactostatin using fuculuose-1-phosphate aldolase, *Bioorg. Chem.*, 20, 173–179,

56. Zhou, P.Z., Salleh, H.M., Chan, P.C.M., et al., A chemoenzymic synthesis of 1,5- 1,5-imino-L-mannitol and 1,5-dideoxy-1,5-imino-L-rhamnitol and investigation of their e glycosidases, *Carbohydr. Res.*, 239, 155–166, 1993.

57. Kajimoto, T., Liu, K.K.C., Pederson, R.L., et al., Enzyme-catalyzed aldol condensa asymmetric synthesis of azasugars: synthesis, evaluation, and modeling of glycosidase in *J. Am. Chem. Soc.*, 113, 6187–6196, 1991.

58. Romero, A. and Wong, C.-H., Chemo-enzymatic total synthesis of 3-epiaustraline, austra 7-epialexine, *J. Org. Chem.*, 65, 8264–8268, 2000.

59. Schultz, M., Waldmann, H., Kunz, H., et al., Chemoenzymatische "Chiral-Pool"-Synt (+)-*exo*-Brevicomin aus Kohlenhydraten mit Fructose-1,6-diphosphat-Aldolase, *Lieb. Chem.*, 1019–1024, 1990.

60. Myles, D.C., Andrulis, P.J.I., and Whitesides, G.M., A transketolase-based synthesis of brevicomin, *Tetrahedron Lett.*, 32, 4835–4838, 1991.

61. Chenevert, R. and Dasser, M., Chemoenzymatic synthesis of the microbial elicitor (−)-syringolide via a fructose 1,6-diphosphate aldolase-catalyzed condensation reaction, *J. Org. Chem.*, 65, 4529–4531, 2000.

62. Chenevert, R., Lavoie, M., and Dasser, M., Use of aldolases in the synthesis of non-carbohydrate natural products. Stereoselective synthesis of aspicilin C-3-C-9 fragment, *Can. J. Chem.*, 75, 68–73, 1997.

63. Shimagaki, M., Muneshima, H., Kubota, M., et al., Chemoenzymatic carbon–carbon bond formation leading to non-carbohydrate derivative—stereoselective synthesis of pentamycin C-11-C-16 fragment, *Chem. Pharm. Bull.*, 41, 282–286, 1993.

64. Phung, A.N., Zannetti, M.T., Whited, G., et al., Stereospecific biocatalytic synthesis of pancratistatin analogues, *Angew. Chem. Int. Ed. Engl.*, 42, 4821–4824, 2003.

65. Colbran, R.L., Jones, J.K.N., Matheson, N.K., et al., A synthesis of dihydroxyacetone phosphate from dihydroxyacetone, *Carbohydr. Res.*, 4, 355–358, 1967.

66. Effenberger, F. and Straub, A., A novel convenient preparation of dihydroxyacetone phosphate and its use in enzymatic aldol reactions, *Tetrahedron Lett.*, 28, 1641–1644, 1987.

67. Jung, S.H., Jeong, J.H., Miller, P., et al., An efficient multigram-scale preparation of dihydroxyacetone phosphate, *J. Org. Chem.*, 59, 7182–7184, 1994.

68. Pederson, R.L., Esker, J., and Wong, C.-H., An improved synthesis of dihydroxyacetone phosphate, *Tetrahedron*, 47, 2643–2648, 1991.

69. Valentin, M.L. and Bolte, J., A convenient synthesis of dihydroxyacetone phosphate from acetone, *Bull. Soc. Chim. Fr.*, 132, 1167–1171, 1995.

70. Gefflaut, T., Lemaire, M., Valentin, M.L., et al., A novel efficient synthesis of dihydroxyacetone phosphate and bromoacetol phosphate for use in enzymatic aldol syntheses, *J. Org. Chem.*, 62, 5920–5922, 1997.

71. Ferroni, E.L., DiTella, V., Ghanayem, N., et al., A three-step preparation of dihydroxyacetone phosphate dimethyl acetal, *J. Org. Chem.*, 64, 4943–4945, 1999.

72. Charmantray, F., El Blidi, L., Gefflaut, T., et al., Improved straightforward chemical synthesis of dihydroxyacetone phosphate through enzymatic desymmetrization of 2,2-dimethoxypropane-1,3-diol, *J. Org. Chem.*, 69, 9310–9312, 2004.

73. Crans, D.C., Kazlauskas, R.J., Hirschbein, B.L., et al., Enzymic regeneration of adenosine 5′-triphosphate: acetyl phosphate, phosphoenolpyruvate, methoxycarbonyl phosphate, dihydroxyacetone phosphate, 5-phospho-α-D-ribosyl pyrophosphate, uridine-5′-diphosphoglucose, *Meth. Enzymol.*, 136, 263–280, 1987.

74. Crans, D.C. and Whitesides, G.M., Glycerol kinase: synthesis of dihydroxyacetone phosphate, sn-glycerol-3-phosphate, and chiral analogs, *J. Am. Chem. Soc.*, 107, 7019–7027, 1985.

75. Sanchez-Moreno, I., Francisco G.-G.J., Bastida, A., et al., Multienzyme system for dihydroxyacetone phosphate-dependent aldolase catalyzed C–C bond formation from dihydroxyacetone, *Chem. Commun.*, 1634–1635, 2004.

76. D'Arrigo, P., Piergianni, V., Pedrocchi-Fantoni, G., et al., Indirect enzymatic phosphorylation: preparation of dihydroxyacetone phosphate, *J. Chem. Soc. Chem. Commun.*, 2505–2506, 1995.

77. Fessner, W.-D. and Sinerius, G., Phosphoenolpyruvate as a dual purpose reagent for integrated nucleotide/nicotinamide cofactor recycling, *Bioorg. Med. Chem.*, 2, 639–645, 1994.

78. Fessner, W.-D. and Sinerius, G., Synthesis of dihydroxyacetone phosphate (and isosteric analogues) by enzymatic oxidation: sugars from glycerol, *Angew. Chem. Int. Ed. Engl.*, 33, 209–212, 1994.

79. Eyrisch, O., Keller, M., and Fessner, W.- D., Higher-carbon sugars by enzymatic chain extension. Oxidative generation of aldol precursors *in situ*, *Tetrahedron Lett.*, 35, 9013–9016, 1994.

80. Espelt, L., Parella, T., Bujons, J., et al., Stereoselective aldol additions catalyzed by dihydroxyacetone phosphate-dependent aldolases in emulsion systems: preparation and structural characterization of linear and cyclic iminopolyols from aminoaldehydes, *Chemistry. Eur. J.*, 9, 4887–4899, 2003.

81. Espelt, L., Clapes, P., Esquena, J., et al., Enzymatic carbon–carbon bond formation in water-in-oil highly concentrated emulsions (gel emulsions), *Langmuir*, 19, 1337–1346, 2003.

82. Schoevaart, R., van Rantwijk, F., and Sheldon, R.A., A four-step enzymatic cascade for pot synthesis of non-natural carbohydrates from glycerol, *J. Org. Chem.*, 65, 6940–6943,

83. Schoevaart, R., van Rantwijk, F., and Sheldon, R.A., Facile enzymatic aldol reactic dihydroxyacetone in the presence of arsenate, *J. Org. Chem.*, 66, 4559–4562, 2001.

84. Drueckhammer, D.G., Durrwachter, J.R., Pederson, R.L., et al., Reversible and *in situ* fo of organic arsenates and vanadates as organic phosphate mimics in enzymatic reactions: i istic investigation of aldol reactions and synthetic applications, *J. Org. Chem.*, 54, 70–77,

85. Stribling, D., Properties of the phosponomethyl isosteres of two phosphate ester glycoly mediats, *Biochem. J.*, 141, 725–728, 1974.

86. Arth, H.L. and Fessner, W.-D., Practical synthesis of 4-hydroxy-3-oxobutylphosphonic a its evaluation as a bio-isosteric substrate of DHAP aldolase, *Carbohydr. Res.*, 305, 313–32

87. Duncan, R. and Drueckhammer, D.G., Preparation of deoxy sugars via aldolase-cataly thesis of 1-deoxy-1-thioketoses, *J. Org. Chem.*, 61, 438–439, 1996.

88. Raetz, C.R.H. and Dowhan, W., Biosynthesis and function of phospholipids in *Escheric J. Biol. Chem.*, 265, 1235–1238, 1990.

89. Varki, A., Diversity in the sialic acids, *Glycobiology*, 2, 25–40, 1992.

90. Varki, A., Biological roles of oligosaccharides—all of the theories are correct, *Glycobic* 97–130, 1993.

91. Fessner, W.-D. and Knorst, M., Manufacture of the *N*-acetylneuraminic acid synthase of *P meningitidis* for synthesis of modified sialic acids expression of the cloned gene, DE100345 (*Chem. Abstr.* 2002, 136, 166160).

92. Barbosa, J.A., Smith, B.J., DeGori, R., et al., Active site modulation in the N-acetylneur lyase sub-family as revealed by the structure of the inhibitor-complexed *Haemophilus in* enzyme, *J. Mol. Biol.*, 303, 405–421, 2000.

93. Kim, M.J., Hennen, W.J., Sweers, H.M., et al., Enzymes in carbohydrate s *N*-acetylneuraminic acid aldolase catalyzed reactions and preparation of *N*-acetyl-2-deoxy aminic acid derivatives, *J. Am. Chem. Soc.*, 110, 6481–6486, 1988.

94. Sugai, T., Shen, G.J., Ichikawa, Y., et al., Synthesis of 3-deoxy-D-*manno*-2-octulosonic acid and its analogs based on KDO aldolase-catalyzed reactions, *J. Am. Chem. Soc.*, 115, 4 1993.

95. Allen, S.T., Heintzelman, G.R., and Toone, E.J., Pyruvate aldolases as reagents for stere aldol condensation, *J. Org. Chem.*, 57, 426–427, 1992.

96. Shelton, M.C., Cotterill, I.C., Novak, S.T.A., et al., 2-Keto-3-deoxy-6-phosphogluconate a as catalysts for stereocontrolled carbon–carbon bond formation, *J. Am. Chem. Sc* 2117–2125, 1996.

97. Cotterill, I.C., Shelton, M.C., Machemer, D.E.W., et al., Effect of phosphorylation on the rate of unnatural electrophiles with 2-keto-3-deoxy-6-phosphogluconate aldolase, *J. Che Perkin Trans. I*, 1335–1341, 1998.

98. Beliczey, J., Kragl, U., Liese, A., et al., Method for making fluorinated sugars having a si and use thereof, US 6355453: 2002 (*Chem. Abstr.* 2002, 136, 231340).

99. Kragl, U., Gödde, A., Wandrey, C., et al., New synthetic applications of sialic acid ald useful catalyst for KDO synthesis—relation between substrate conformation and enzym selectivity, *J. Chem. Soc. Perkin Trans. I*, 119–124, 1994.

100. Fitz, W., Schwark, J.R., and Wong, C.H., Aldotetroses and C(3)-modified aldohexoses strates for *N*-acetylneuraminic acid aldolase: a model for the explanation of the normal inversed stereoselectivity, *J. Org. Chem.*, 60, 3663–3670, 1995.

101. Augé, C., Gautheron, C., David, S., et al., Sialyl aldolase in organic synthesis: from the t acid, 3-deoxy-D-glycero-D-galacto-2-nonulosonic acid (KDN), to branched-chain higher ke possible new chirons, *Tetrahedron*, 46, 201–214, 1990.

102. Liu, J.L.C., Shen, G.J., Ichikawa, Y., et al., Overproduction of CMP-sialic acid synthe organic synthesis, *J. Am. Chem. Soc.*, 114, 3901–3910, 1992.

103. Lin, C.-C., Lin, C.-H., and Wong, C.-H., Sialic acid aldolase-catalyzed condensation of p and *N*-substituted mannosamine: a useful method for the synthesis of *N*-substituted sial *Tetrahedron Lett.*, 38, 2649–2652, 1997.

104. Wu, W.-Y., Jin, B., Kong, D.C.M., et al., A facile synthesis of a useful 5-*N*-substituted-3,5-dideoxy-D-*glycero*-D-*galacto*-2-nonulosonic acid from 2-acetamido-2-deoxy-D-glucose, *Carbohydr. Res.*, 300, 171–174, 1997.

105. Kuboki, A., Okazaki, H., Sugai, T., et al., An expeditions route to *N*-glycolylneuraminic acid based on enzyme-catalyzed reaction, *Tetrahedron*, 53, 2387–2400, 1997.

106. Humphrey, A.J., Fremann, C., Critchley, P., et al., Biological properties of *N*-acyl and *N*-haloacetyl neuraminic acids: processing by enzymes of sialic acid metabolism, and interaction with influenza virus, *Bioorg. Med. Chem.*, 10, 3175–3185, 2002.

107. Pan, Y., Ayani, T., Nadas, J., et al., Accessibility of *N*-acyl-D-mannosamines to *N*-acetyl-D-neuraminic acid aldolase, *Carbohydr. Res.*, 339, 2091–2100, 2004.

108. Murakami, M., Ikeda, K., and Achiwa, K., Chemoenzymatic synthesis of neuraminic acid analogs structurally varied at C-5 and C-9 as potential inhibitors of the sialidase from influenza virus, *Carbohydr. Res.*, 280, 101–110, 1996.

109. Kong, D.C.M. and von Itzstein, M., The chemoenzymatic synthesis of 9-substituted 3,9-dideoxy-D-glycero-D-galacto-2-nonulosonic acids, *Carbohydr. Res.*, 305, 323–329, 1997.

110. Kiefel, M.J., Wilson, J.C., Bennett, S., et al., Synthesis and evaluation of C-9 modified *N*-acetylneuraminic acid derivatives as substrates for *N*-acetylneuraminic acid aldolase, *Bioorg. Med. Chem.*, 8, 657–664, 2000.

111. Kok, G.B., Campbell, M., Mackey, B.L., et al., Synthesis of C-3 nitrogen-containing derivatives of *N*-acetyl-alpha,beta-d-mannosamine as substrates for *N*-acetylneuraminic acid aldolase, *Carbohydr. Res.*, 332, 133–139, 2001.

112. Lins, R.J., Flitsch, S.L., Turner, N.J., et al., Generation of a dynamic combinatorial library using sialic acid aldolase and *in situ* screening against wheat germ agglutinin, *Tetrahedron*, 60, 771–780, 2004.

113. Lin, C.H., Sugai, T., Halcomb, R.L., et al., Unusual stereoselectivity in sialic acid aldolase-catalyzed aldol condensations—synthesis of both enantiomers of high-carbon monosaccharides, *J. Am. Chem. Soc.*, 114, 10138–10145, 1992.

114. David, S., Malleron, A., and Cavaye, B., Aldolases in organic synthesis: acylneuraminate-pyruvate lyase accepts furanoses as substrates, *New J. Chem.*, 16, 751–755, 1992.

115. Koppert, K. and Brossmer, R., Synthesis of the C-5 homologue of *N*-acetylneuraminic acid by enzymatic chain elongation of 2-C-acetamidomethyl-2-deoxy-D-mannose, *Tetrahedron Lett.*, 33, 8031–8034, 1992.

116. Malleron, A. and David, S., An enzymic approach to sequence 12–20 of Amphotericin B, *New J. Chem.*, 20, 153–159, 1996.

117. Zhou, P.Z., Salleh, H.M., and Honek, J.F., Facile chemoenzymatic synthesis of 3-(hydroxymethyl)-6-epicastanospermine, *J. Org. Chem.*, 58, 264–266, 1993.

118. Henderson, D.P., Shelton, M.C., Cotterill, I.C., et al., Stereospecific preparation of the N-terminal amino acid moiety of nikkomycins K-X and K-Z via a multiple enzyme synthesis, *J. Org. Chem.*, 62, 7910–7911, 1997.

119. Sugai, T., Kuboki, A., Hiramatsu, S., et al., Improved enzymatic procedure for a preparative-scale synthesis of sialic acid and KDN, *Bull. Chem. Soc. Jpn.*, 68, 3581–3589, 1995.

120. Blayer, S., Woodley, J.M., Lilly, M.D., et al., Characterization of the chemoenzymatic synthesis of *N*-acetyl-D-neuraminic acid (Neu5Ac), *Biotechnol. Prog.*, 12, 758–763, 1996.

121. Mahmoudian, M., Noble, D., Drake, C.S., et al., An efficient process for production of *N*-acetylneuraminic acid using *N*-acetylneuraminic acid aldolase, *Enzyme Microb. Technol.*, 20, 393–400, 1997.

122. Kragl, U., Gygax, D., Ghisalba, O., et al., Enzymatic, two-step synthesis of *N*-acetylneuraminic acid in a enzyme-membrane reactor, *Angew. Chem. Int. Ed. Engl.*, 30, 827–828, 1991.

123. Maru, I., Ohnishi, J., Ohta, Y., et al., Simple and large-scale production of *N*-acetylneuraminic acid from *N*-acetyl-D-glucosamine and pyruvate using *N*-acyl-D-glucosamine 2-epimerase and *N*-acetylneuraminate lyase, *Carbohydr. Res.*, 306, 575–578, 1998.

124. Lee, J.-O., Yi, J.-K., Lee, S.-G., et al., Production of *N*-acetylneuraminic acid from *N*-acetylglucosamine and pyruvate using recombinant human renin binding protein and sialic acid aldolase in one pot, *Enzyme Microb. Technol.*, 35, 121–125, 2004.

125. Salagnad, C., Godde, A., Ernst, B., et al., Enzymatic large-scale production of 2-keto-3-glycero-D-galacto-nonopyranulosonic acid in enzyme membrane reactors, *Biotechnol. F* 810–813, 1997.

126. Ichikawa, Y., Liu, J.L.C., Shen, G.J., et al., A highly efficient multienzyme system for the synthesis of a sialyl trisaccharide: *in situ* generation of sialic acid and N-acetyllactosamine with regeneration of UDP-glucose, UDP-galactose and CMP-sialic acid, *J. Am. Chem. S* 6300–6302, 1991.

127. Ichikawa, Y., Wang, R., and Wong, C.H., Regeneration of sugar nucleotide for e oligosaccharide synthesis, *Meth. Enzymol.*, 247, 107–127, 1994.

128. Blixt, O. and Paulson, J.C., Biocatalytic preparation of *N*-glycolylneuraminic acid, deam aminic acid (KDN) and 9-azido-9-deoxysialic acid oligosaccharides, *Adv. Synth. Ca* 687–690, 2003.

129. Yu, H., Yu, H., Karpel, R., et al., Chemoenzymatic synthesis of CMP-sialic acid derivat one-pot two-enzyme system: comparison of substrate flexibility of three microbial CMP-s synthetases, *Bioorg. Med. Chem.*, 12, 6427–6435, 2004.

130. Knorst, M. and Fessner, W.-D., CMP-sialate synthetase from Neisseria meningitidis—ov sion and application to the synthesis of oligosaccharides containing modified sialic ac *Synth. Catal.*, 343, 698–710, 2001.

131. Heine, A., Luz, J.G., Wong, C.-H., et al., Analysis of the class I aldolase binding site arc based on the crystal structure of 2-deoxyribose-5-phosphate aldolase at 0.99 Å resolution *Biol.*, 343, 1019–1034, 2004.

132. Barbas, C.F., Wang, Y.F., and Wong, C.-H., Deoxyribose-5-phosphate aldolase as a catalyst, *J. Am. Chem. Soc.*, 112, 2013–2014, 1990.

133. Chen, L., Dumas, D.P., and Wong, C.-H., Deoxyribose-5-phosphate aldolase as a ca asymmetric aldol condensation, *J. Am. Chem. Soc.*, 114, 741–748, 1992.

134. Wong, C.-H., Garcia-Junceda, E., Chen, L.R., et al., Recombinant 2-deoxyribose-5-p aldolase in organic synthesis: use of sequential two-substrate and three-substrate aldol r *J. Am. Chem. Soc.*, 117, 3333–3339, 1995.

135. Liu, J. and Wong, C.-H., Aldolase-catalyzed asymmetric synthesis of novel pyranose synt new entry to heterocycles and epothilones, *Angew. Chem. Int. Ed. Engl.*, 41, 1404–1407,

136. DeSantis, G., Liu, J., Clark, D.P., et al., Structure-based mutagenesis approaches toward ing the substrate specificity of D-2-deoxyribose-5-phosphate aldolase, *Bioorg. Med. C* 43–52, 2003.

137. Ouwerkerk, N., Van Boom, J.H., Lugtenburg, J., et al., Chemo-enzymatic synthesis of t 13C-labeled in the 2′-Deoxyribose moiety, *Eur. J. Org. Chem.*, 861–866, 2000.

138. Tischer, W., Ihlenfeldt, H.-G., Barzu, O., et al., Enzymatic synthesis of deoxyribonucleosi deoxyribose 1-phosphate and nucleobase, PCT WO0114566: 2001 (*Chem. Abstr.* 2C 208062).

139. Horinouchi, N., Ogawa, J., Sakai, T., et al., Construction of deoxyriboaldolase-overe> *Escherichia coli* and its application to 2-deoxyribose 5-phosphate synthesis from glu acetaldehyde for 2′-deoxyribonucleoside production, *Appl. Environ. Microbiol.*, 69, 37′ 2003.

140. Gijsen, H.J.M. and Wong, C.-H., Unprecedented asymmetric aldol reactions with three substrates catalyzed by 2-deoxyribose-5-phosphate aldolase, *J. Am. Chem. Soc.*, 116, 8422–84

141. Liu, J., Hsu, C.-C., and Wong, C.-H., Sequential aldol condensation catalyzed by DERA Ser238Asp and a formal total synthesis of atorvastatin, *Tetrahedron Lett.*, 45, 2439–2441

142. Greenberg, W.A., Varvak, A., Hanson, S.R., et al., Development of an efficient, scalable, catalyzed process for enantioselective synthesis of statin intermediates, *Proc. Natl. Acad. S* 101, 5788–5793, 2004.

143. Kierkels, J.G.T., Mink, D., Panke, S., et al., Process for the preparation of 2,4-dideoxyhex 2,4,6-trideoxyhexoses and therapeutic uses thereof, WO0306656: 2003 (*Chem. Abstr.* 2C 112523).

144. Müller, M. Chemoenzymatic synthesis of building blocks for statin side chains, *Angew. C Ed. Engl.*, 44, 362–365, 2005.

145. Liu, J.Q., Dairi, T., Itoh, N., et al., Diversity of microbial threonine aldolases and their application, *J. Mol. Catal., B Enzym.*, 10, 107–115, 2000.

146. Kimura, T., Vassilev, V.P., Shen, G.J., et al., Enzymatic synthesis of beta-hydroxy-alpha-amino acids based on recombinant D- and L-threonine aldolases, *J. Am. Chem. Soc.*, 119, 11734–11742, 1997.

147. Saeed, A. and Young, D.W., Synthesis of L-β-hydroxy amino acids using serine hydroxymethyltransferase, *Tetrahedron*, 48, 2507–2514, 1992.

148. Lotz, B.T., Gasparski, C.M., Peterson, K., et al., Substrate specificity studies of aldolase enzymes for use in organic synthesis, *J. Chem. Soc. Chem. Commun.*, 16, 1107–1109, 1990.

149. Vassilev, V.P., Uchiyama, T., Kajimoto, T., et al., L-Threonine aldolase in organic synthesis: preparation of novel beta-hydroxy-alpha-amino acids, *Tetrahedron Lett.*, 36, 4081–4084, 1995.

150. Genet, J.-P., New asymmetric syntheses of b-hydroxy a-amino acids and analogs. Components of biologically active cyclopeptides, *Pure Appl. Chem.*, 68, 593–596, 1996.

151. Hsiao, H.-Y. and Wei, T., Enzymatic production of L-serine with a feedback control system for formaldehyde addition, *Biotechnol. Bioeng.*, 28, 1510–1518, 1986.

152. Hsiao, H.-Y., Wei, T., and Campbell, K., Enzymatic production of L-serine, *Biotechnol. Bioeng.*, 28, 857–867, 1986.

153. Ura, D., Hashimukai, T., Matsumoto, T., et al., Enzymic preparation of L-serine, EP 421477: 1991 (*Chem. Abstr.* 1991, 115, 90657).

154. Anderson, D.M. and Hsiao, H.-H, Enzymatic method for L-serine production, in *Biocatalytic Production of Amino Acids and Derivatives*, Rozzell, D. and Wagner, F., Eds., Hanser, Munich, 1992, pp. 23–41.

155. Bycroft, M., Herbert, R.B., and Ellames, G.J., Heterocyclic β-hydroxy-α-amino acids as substrates for a novel aldolase from *Streptomyces amakusaensis*; preparation of (2R,3R)-3-(2-thienyl)serine and (2R,3R)-3-(2-furyl)serine from racemic *threo* material, *J. Chem. Soc. Perkin Trans. I*, 2439–2442, 1996.

156. Fujii, M., Miura, T., Kajimoto, T., et al., Facile synthesis of 3,4-dihydroxyprolines as an application of the L-threonine aldolase-catalyzed aldol reaction, *Synlett*, 1046–1048, 2000.

157. Shibata, K., Shingu, K., Vassilev, V.P., et al., Kinetic and thermodynamic control of L-threonine aldolase catalyzed reaction and its application to the synthesis of mycestericin D, *Tetrahedron Lett.*, 37, 2791–2794, 1996.

158. Morikawa, T., Ikemi, M., and Miyoshi, T., Manufacture of D-erythro-3-(3,4-dihydroxyphenyl)serine derivatives with microorganisms or enzymes, 1992 (*Chem. Abstr.* 1992, 118, 190107).

159. Morikawa, T., Ikemi, M., and Myoshi, T., D-Erythro-hydroxyamino acids enzymic preparation, 1994 (*Chem. Abstr.* 1994, 121, 155936).

160. Herbert, R.B., Wilkinson, B., and Ellames, G.J., Preparation of (2R,3S)-beta-hydroxy-alpha-amino acids by use of a novel streptomyces aldolase as a resolving agent for racemic material, *Can. J. Chem.*, 72, 114–117, 1994.

161. Liu, J.Q., Odani, M., Dairi, T., et al., A new route to L-threo-3-[4-(methylthio)phenylserine], a key intermediate for the synthesis of antibiotics: recombinant low-specificity D-threonine aldolase-catalyzed stereospecific resolution, *Appl. Microbiol. Biotechnol.*, 51, 586–591, 1999.

162. Liu, J.Q., Odani, M., Yasuoka, T., et al., Gene cloning and overproduction of low-specificity D-threonine aldolase from Alcaligenes xylosoxidans and its application for production of a key intermediate for parkinsonism drug, *Appl. Microbiol. Biotechnol.*, 54, 44–51, 2000.

163. Schörken, U. and Sprenger, G.A., Thiamin-dependent enzymes as catalysts in chemoenzymatic syntheses, *Biochim. Biophys. Acta*, 1385, 229–243, 1998.

164. Zimmermann, F.T., Schneider, A., Schörken, U., et al., Efficient multi-enzymatic synthesis of D-xylulose 5-phosphate, *Tetrahedron Asymmetry*, 10, 1643–1646, 1999.

165. Turner, N.J., Applications of transketolases in organic synthesis, *Curr. Opin. Biotechnol.*, 11, 527–531, 2000.

166. Bolte, J., Demuynck, C., and Samaki, H., Utilization of enzymes in organic chemistry: transketolase catalyzed synthesis of ketoses, *Tetrahedron Lett.*, 28, 5525–5528, 1987.

167. Morris, K.G., Smith, M.E.B., Turner, N.J., et al., Transketolase from *Escherichia coli*: a practical procedure for using the biocatalyst for asymmetric carbon–carbon bond synthesis, *Tetrahedron Asymmetry*, 7, 2185–2188, 1996.

168. Kobori, Y., Myles, D.C., and Whitesides, G.M., Substrate specifity and carbohydrate s using transketolase, *J. Org. Chem.*, 57, 5899–5907, 1992.

169. Effenberger, F., Null, V., and Ziegler, T., Enzyme catalyzed reactions. 12. Preparation of pure L-2-hydroxyaldehydes with yeast transketolase, *Tetrahedron Lett.*, 33, 5157–5160, 19

170. French, C. and Ward, J.M., Improved production and stability of *E. coli* recombinants ex transketolase for large scale biotransformation, *Biotechnol. Lett.*, 17, 247–252, 1995.

171. Hobbs, G.R., Mitra, R.K., Chauhan, R.P., et al., Enzyme-catalysed carbon–carbon bond tion: large-scale production of *Escherichia coli* transketolase, *J. Biotechnol.*, 45, 173–179,

172. Bongs, J., Hahn, D., Schörken, U., et al., Continuous production of erythrulose using tra lase in a membrane reactor, *Biotechnol. Lett.*, 19, 213–215, 1997.

173. Brocklebank, S., Woodley, J.M., and Lilly, M.D., Immobilised transketolase for carbon bond synthesis: biocatalyst stability, *J. Mol. Catal., B Enzym.*, 7, 223–231, 1999.

174. Chauhan, R.P., Woodley, J.M., and Powell, L.W., *In situ* product removal from *E. coli* tolase-catalyzed biotransformations, *Ann. N. Y. Acad. Sci.*, 799, 545–554, 1996.

175. Effenberger, F. and Null, V., Enzyme-catalyzed reactions. 13. A new, efficient synt fagomine, *Liebigs Ann. Chem.*, 1211–1212, 1992.

176. Humphrey, A.J., Parsons, S.F., Smith, M.E.B., et al., Synthesis of a novel *N*-hydroxypy using enzyme catalysed asymmetric carbon–carbon bond synthesis, *Tetrahedron L* 4481–4485, 2000.

177. Gonzalez-Garcia, E., Helaine, V., Klein, G., et al., Fluorogenic stereochemical probes fc aldolases, *Chemistry Eur. J.*, 9, 893–899, 2003.

178. Machajewski, T.D. and Wong, C.-H., The catalytic asymmetric aldol reaction, *Angew. Cl Ed. Engl.*, 39, 1352–1374, 2000.

179. Palomo, C., Oiarbide, M., and Garcia, J.M., Current progress in the asymmetric aldol a reaction, *Chem. Soc. Rev.*, 33, 65–75, 2004.

180. Kazmeier, U., Amino acids—valuable organocatalysts in carbohydrate synthesis, *Angew Int. Ed. Engl.*, 44, 2186–2188, 2005.

181. Powell, K.A., Ramer, S.W., del Cardayré, S.B., et al., Directed evolution and biocatalysis, *Chem. Int. Ed. Engl.*, 40, 3948–3959, 2001.

182. Williams, G.J., Nelson, A.S., and Berry, A., Directed evolution of enzymes for biocatalysis life sciences, *Cell. Mol. Life Sci.*, 61, 3034–3046, 2004.

183. Tanaka, F., Fuller, R., Shim, H., et al., Evolution of aldolase antibodies in vitro: correl catalytic activity and reaction-based selection, *J. Mol. Biol.*, 335, 1007–1018, 2004.

184. Zhu, X., Tanaka, F., Hu, Y., et al., The origin of enantioselectivity in aldolase antibodies structure, site-directed mutagenesis, and computational analysis, *J. Mol. Biol.*, 343, 126 2004.

185. Lorenz, P., Liebeton, K., Niehaus, F., et al., Screening for novel enzymes for biocataly cesses: accessing the metagenome as a resource of novel functional sequence space, *Cur Biotechnol.*, 13, 572–577, 2002.

186. Streit, W.R., Daniel, R., and Jaeger, K.-E., Prospecting for biocatalysts and drugs in the g of non-cultured microorganisms, *Curr. Opin. Biotechnol.*, 15, 285–290, 2004.

187. Wahler, D. and Reymond, J.-L., Novel methods for biocatalyst screening, *Curr. Opin. Chel* 5, 152–158, 2001.

14 Enzymatic Synthesis of Modified Nucleosides

Luís A. Condezo, Jesús Fernández-Lucas, Carlos A. García-Burgos, Andrés R. Alcántara, and José V. Sinisterra

CONTENTS

14.1 INTRODUCTION

Nucleosides are involved in many biochemical processes, notably the storage and transfer of genetic information. As a consequence, nucleoside analogs have been used in the treatment of anti-immunodeficiency syndrome (AIDS), and other viral infections, such as those caused by herpes viruses, and influenza A and B viruses [1]. Most of the approved antiviral drugs, such as 3′-azido-2′,3′-dideoxythymidine (AZT, **1**) or 2′,3′-dideoxyinosine (ddI, **2**), shown in Figure 14.1, are naturally occurring nucleoside analogs that act by interfering with the synthesis of viral nucleic acids.

In addition to the classical antiviral activity of nucleosides, there are some other therapeutical effects described for these nucleoside analogs, such as:

1. Nucleoside transport inhibitors: this strategy is used for cardio-protection and brain protection in ischemic heart disease and stroke, respectively, and in some human leukemia [2]
2. Treatment of inflammatory processes [3]
3. Antisense oligonucleotides, useful for preparing triple ADN helix, as potential and selective inhibitors of gene expression [4]
4. Antimicrobial drugs, such as oxetanocine (AXT-A) or oxanosine [5]

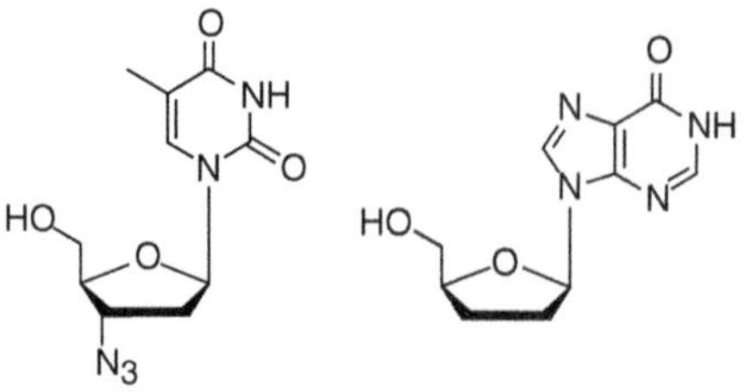

FIGURE 14.1 Approved antiviral drugs: 3′-azido-2′,3′-dideoxythymidine (AZT, 1) or 2′,3′-did sine (ddI, 2).

Therefore, much effort has been expended on the synthesis of nucleoside analogs. sense, the chemical syntheses of these compounds are often multistage processes, in several protection and deprotection steps in order to obtain the specific modifica certain groups of these polyfunctional molecules [6–11]. On the other hand, the en synthesis of nucleoside analogs offers several advantages over chemical methods:

1. Protecting groups are not usually required.
2. The enzymatic steps are highly stereospecific and/or regioselective.
3. Enzyme-catalyzed reactions are usually so efficient that the synthesized nucleos: be used in further processes without purification. Some reviews about this top been performed [12,13].

Four different types of enzymes can be used to carry out the chemoenzymatic synt nucleoside analogs:

1. Hydrolytic enzymes capable of modifying functional groups of the base; ad deaminase is the most interesting enzyme
2. Hydrolytic enzymes that can modify the functional groups of the sugar residue, penicillinacylase, lipases, proteases, and esterases
3. Oxygenases that can oxidize the CH_2OH group of the pentose ring to CO_2H
4. Enzymes that catalyze the transfer of glycosyl residues from a nucleoside don acceptor base. There are two main subclasses of these enzymes:

   • Nucleoside phosphorylases
   • N-2′-deoxyribosyl transferases

14.2   ENZYMATIC SYNTHESIS OF MODIFIED NUCLEOSIDES

14.2.1   Enzymes That Modify the Base

Intracellular adenosine deaminase (ADA) is widely used in the preparation of a structures by deamination of 6-aminopurines [14,15]. ADA specifically catalyzes the lytic deamination of 6-aminopurine nucleoside, leaving the amino group of the 2 sugar moiety intact (Figure 14.2).
    In addition, this enzyme displays broad substrate specificity and its use can be exte carbocyclonucleosides or acyclonucleosides [15]. Traditionally, it has been used in t thesis of guanosine or inosine nucleoside analogs with moderated to high yields (40

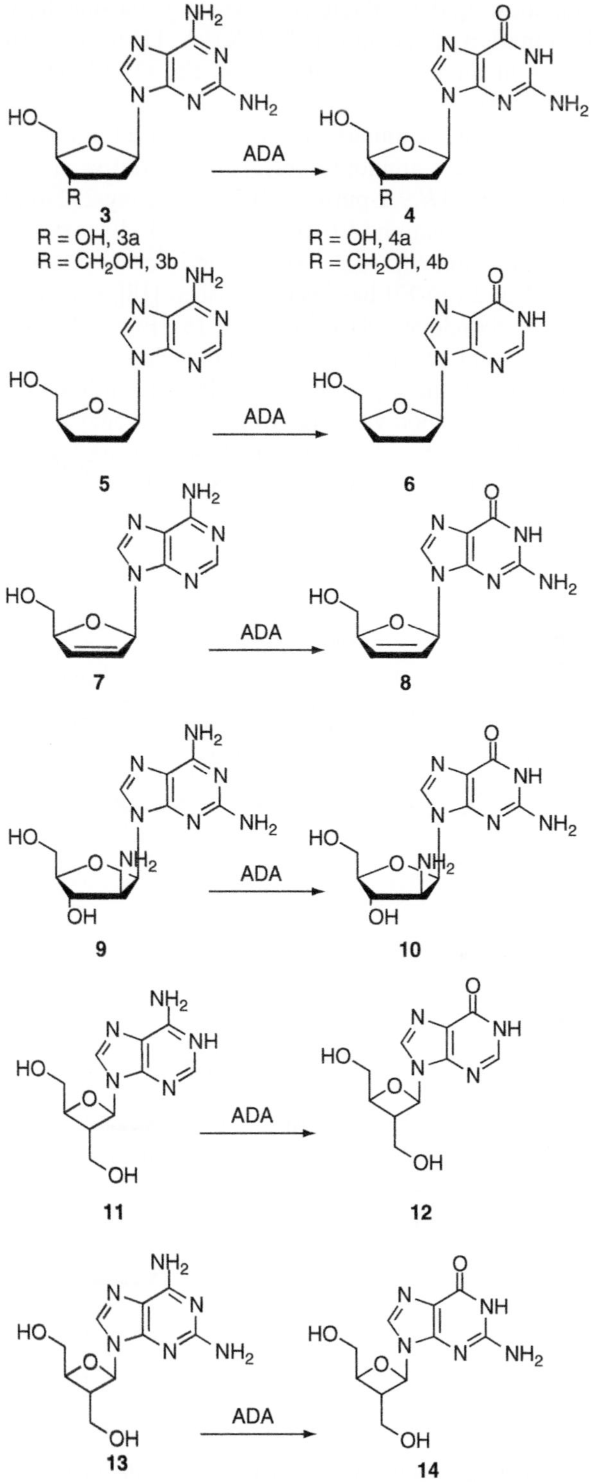

FIGURE 14.2 Applications of adenosine deaminase (ADA) to the synthesis of some nucleoside analogs. (From Rachakonda, S. and Cartee, L., *Curr. Med. Chem.*, 11(6), 775–793, 2004; Ferrero, M. and Gotor, V., *Monatsh. Chem.*, 131(6), 585–616, 2000; Kakefuda, A., Shuto, S., Nagahata, T., et al., *Tetrahedron*, 50(34), 10167–10182, 1994.)

[16], allowing the scalable synthesis of guanosine derivatives. ADA has also been used in deamination of oxetanocin A (OXT-A, **11**) to the corresponding hypoxanthine (OXT-H analog in quantitative yields, or in the specific transformation of 2-amino-OXT-A (**1.** OXT-G (**14**) (Figure 14.2) [5].

Adenosine deaminase has also been used in the resolution of racemic nucleoside anå derived from 6-aminopurines. Thus, Hertel et al. [17] reported the preparation of β-amino-6-oxo-1*H*,9*H*-purin-9-yl)-2′,3′-dideoxy-2′,2′-difluororibose (β-**15**) from a racemic ture of 1-(2,6-diamino-9*H*-purin-9-yl)-2′,3′-dideoxy-2′,2′-difluororibose (rac-**15**) (Figure 1 With the same strategy, the synthesis of both antipodes of the anti-HIV drug car [(−)-**17** and (+)-**17**] has been described [18], starting from racemic *cis*-[3-(2,6)-diamino purin-9-yl]cyclopentyl carbinol (rac-**16**) (Figure 14.3).

Margolin et al. [19] and Santaniello et al. [15] have used adenylic acid deam (AMPDA) from *Aspergillus niger* in the synthesis of 6-oxopurine nucleoside analogs moderated to good yields (40 to 70%). This enzyme shows much broader substrate speci than ADA, as depicted in Figure 14.4, and can lead to deamination of adenosine deriva including phosphorylated cyclic (**18**), carbocyclic (**20**), and acyclic analogs [15]. In addi

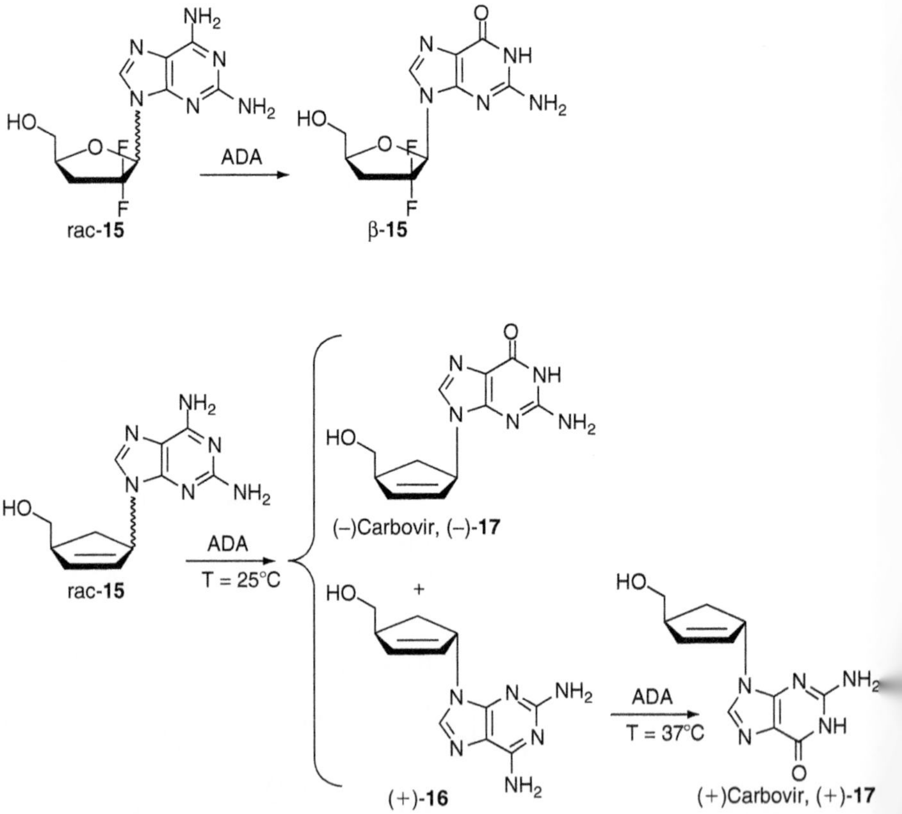

FIGURE 14.3 Resolution of racemic nucleoside analogs using adenosine deaminase (ADA). Hertel, L.W., Grossman, C.S., and Kroin, J.S., Eur. Pat. Appl. 0328345, 1989; Vince, R. and Br J., *Biochem. Biophys. p. Res. Co.*, 168(3), 912–916, 1990.)

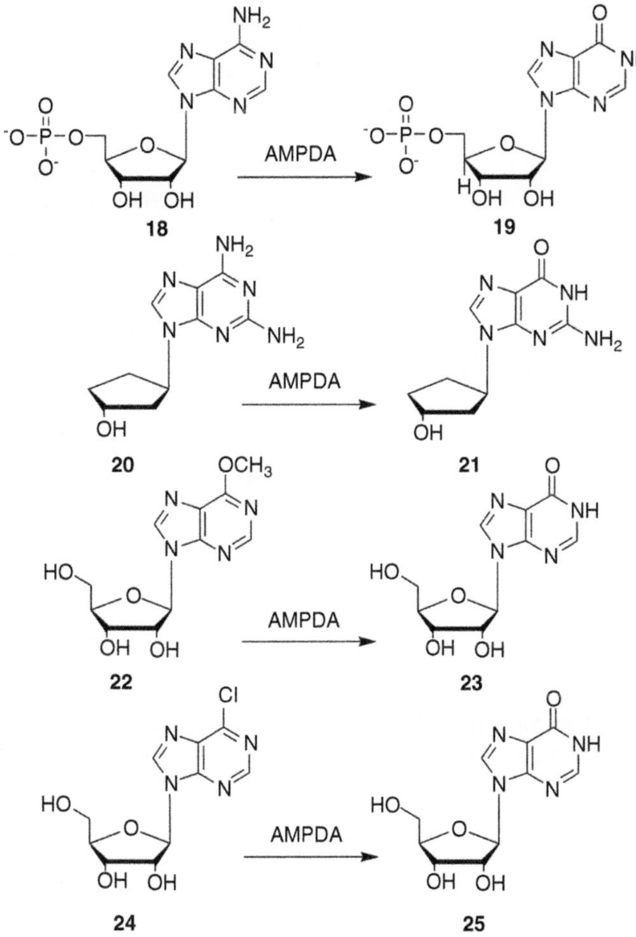

FIGURE 14.4 Synthesis of nucleoside analogs using adenylic acid deaminase (AMPDA). (From Margolin, A.L., Borcherding, D.R., Wolf-Kugel, et al., *J. Org. Chem.*, 59(24), 7214–7218, 1994.)

the enzyme can catalyze demethylation (**22**) and dechlorination (**24**) of purine ribonucleosides (Figure 14.4).

14.2.2 ENZYMES THAT MODIFY THE SUGAR

The direct protection–deprotection of hydroxyl groups of nucleosides is a key step in the synthesis of nucleoside analogs. Although several chemical methods are available for regioselective acylation of the nucleoside sugar moiety, enzymatic methods offer advantages considering yield, regioselectivity, and overall number of synthetic steps that have to be carried out. Gotor et al. [20–22] described oxime esters and oxime carbonates as suitable acyl donors in the lipase-catalyzed regioselective acylation of different −OH residues of sugars (Figure 14.5) with good yields. More concretely, *Pseudomonas cepacia* lipase regioselectively acylates in 3′-position when using the oxime carbonate (**27**), while *Candida antarctica* lipase B leads to the acylation exclusively in the 5′-position (**29**) (Figure 14.5). The regioselectivity is only associated to the enzyme rather than to the acyl-donor structure, as it was proved by the same research group [21] using vinyl benzoate and lipase B from *C. antarctica*

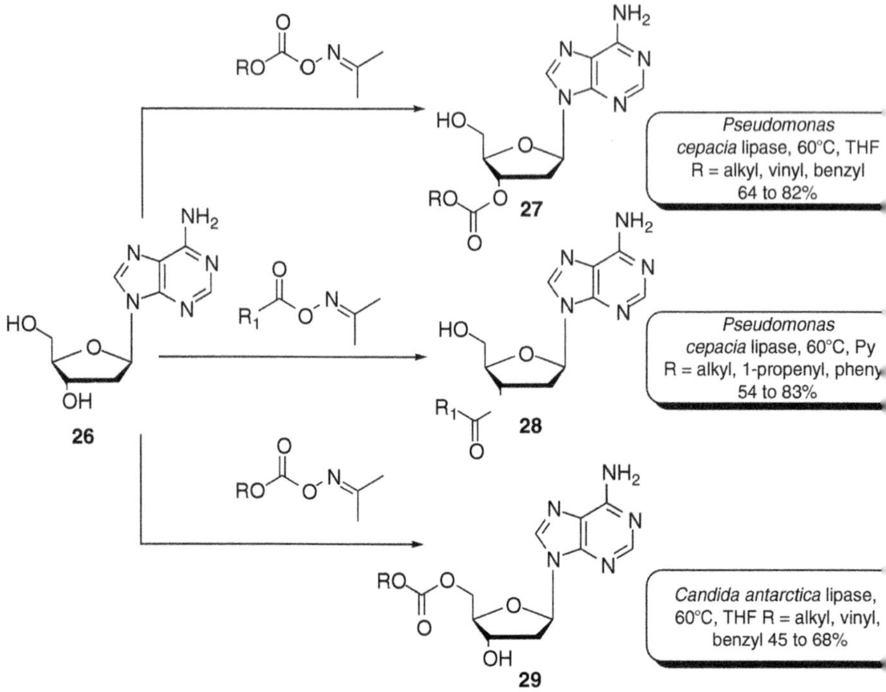

FIGURE 14.5 Regioselective acylation of sugar moiety using oxime esters or oxime carbonate▪ Moris, F. and Gotor, V., *J. Org. Chem.*, 57(8), 2490–2492, 1992; Moris, F. and Gotor, V., *J. Org* 58(3), 653–660, 1993; García, J., Fernández, S., Ferrero, M., et al., *Tetrahedron Lett.*, 45(8), 17▪ 2004.)

(**28**, Figure 14.5). Indeed, 2′-deoxyuridine, thymidine, 2′-deoxyadenosine, and 2′-deoxygu▪ were benzoylated in 5′-position with moderated yields (13 to 76%), although larger ▪ times were necessary to achieve the maximum yields (71 to 80 h).

Subtilisin leads to the same regioselectivity as *P. cepacia* lipase with similar yields, a higher amount (fivefold) of enzyme is required for this purpose. The lipase f▪ *fluorescens* catalyzes the acylation of nucleosides, using anhydrides in dimethylfor▪ (DMF) or dimethyl sulfoxide (DMSO), but poor regioselectivity is observed [23–25]. ▪ ing an opposite hydrolytic strategy, regioselective deacylations can be achieved either *fluorescens* lipase (3′-position) or with subtilisin and/or an alkaline protease from ▪ *subtillis* (5′-position) [23,24], as shown in Figure 14.6.

The highly regiospecific acylation of the OH group in C-2′ remains unsolved. On▪ from *Mucor javanicus* shows a moderated regioselectivity for this position (42% yield▪ *n*-octanoic anhydride as the acyl donor [12]. Another plausible alternative is the depr▪ of peracetylated nucleosides using two enzymatic steps [13]. Recently, wheat ger▪ (WGL) has been used to deacylate in one step C-5′ and C-3′ positions, but only mo▪ yields (26 to 29%) are reported [13].

Finally, Wang et al. [26] have described a highly diastereoselective thermophili▪ ESL-001-02 that afforded in phosphate buffer at 60°C α-L-talofuranosyluronamid▪ (**34**) with 100% diastereomeric excess (de) from 5′-*O*-acetyl-β-D-allofuranosyl-uro▪ uracil (**33**) (Figure 14.7).

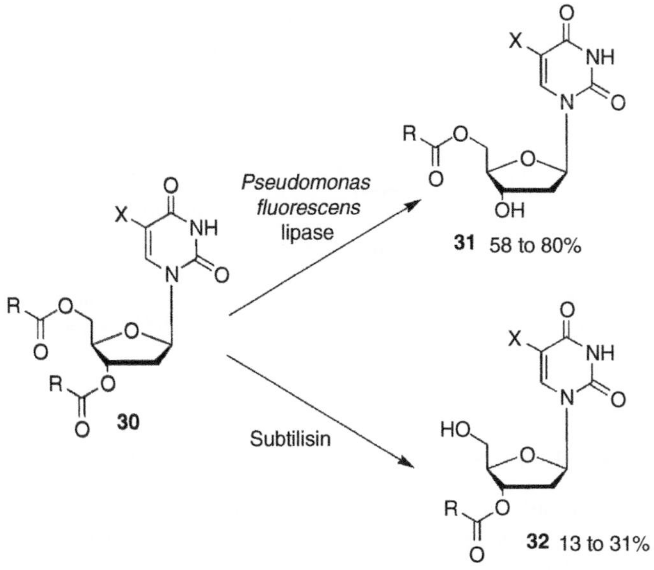

FIGURE 14.6 Regioselective deacylation of peracylated nucleosides using different enzymes. (From García, J., Fernández, S., Ferrero, M., et al., *Tetrahedron Lett.*, 45(8), 1709–1712, 2004.)

14.2.3 Enzymatic Oxidation of CH$_2$OH in Nucleoside Analogs

The enzymatic oxidation of these polyfunctional and labile compounds is more advantageous than the chemical process, because it is environmentally cleaner, and the avoiding of protection–deprotection steps of other groups makes it more attractive from an economical point of view. In fact, the preparation of carboxylic nucleoside derivatives in C-5′ is a powerful tool for obtaining a new family of anti-inflammatory drugs [3,12]. In this sense, nucleoside oxidase from *Stenotrophomonas maltophila* FERM BP-2252 has been used for this purpose [12,27]. This enzyme shows high tolerance for different functionalities in the base, especially in C-2 position, as shown in Figure 14.8. Furthermore, the *N*-1 can be modified either to *N*-oxide or *N*-methylated groups.

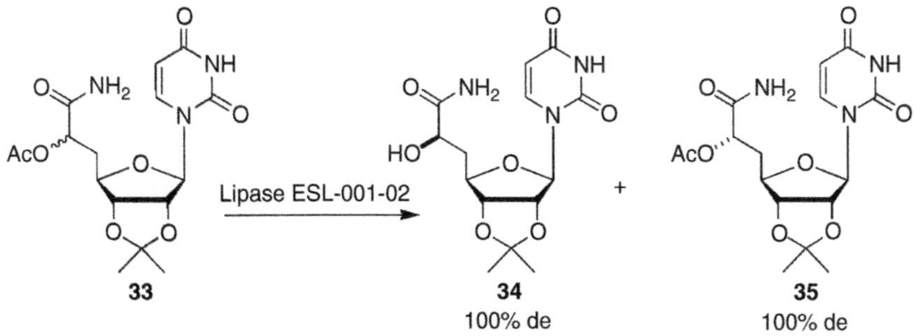

FIGURE 14.7 Selective hydrolysis of 5′-*O*-acetyl-β-D-allofuranosyluronamide uracil by lipase ESL-001-02. (From Wang, J-Q., Archer, C., Li, J., et al., *J. Org. Chem.*, 63(14), 4850–4853, 1998.)

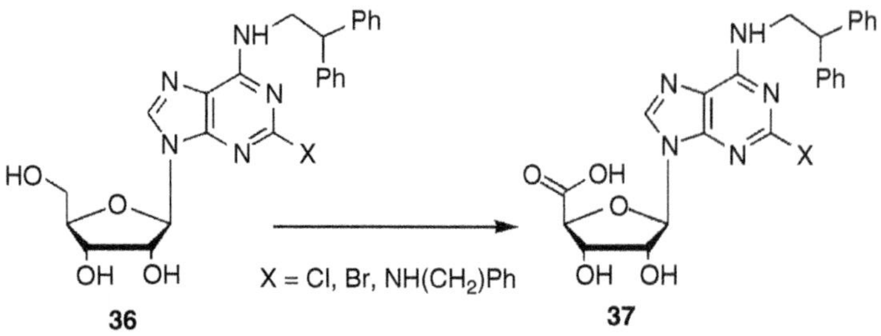

FIGURE 14.8 Selective oxidation of C-5 OH by oxidase from *Stenotrophomonas maltophila*
Zhang, F.J. and Sih, C.J., *Tetrahedron Lett.*, 36(51), 9289–9292, 1995.)

14.2.4 ENZYMES THAT CATALYZE THE SYNTHESIS OF NUCLEOSIDES BY TRANSGLYCOSYLAT

Nucleoside analogs can be prepared by base interchange using two different kinds o
cellular enzymes: nucleoside phosphorylases (NP) and *N*-2′-deoxyribosyl transferases

NPs catalyze the reversible phosphorolysis of nucleosides and the transferase r
involving purine or pyrimidine bases. Purine (E.C. 2.4.2.2) and pyrimidine (E.C. :
nucleoside phosphorylases have been isolated from a large number of bacteria [28].
enzymes display fairly broad substrate specificity. Contrarily, *N*-2′-deoxyribosyl trans
(E.C. 2.4.2.6.) specifically catalyze the exchange of the base from a 2′-deoxyribosyl nuc
with a free purine or pyrimidine [29], and especially those from *Lactobacilli* are wel
mented [30–33]. Both types of enzymes display a high regio-(*N*-1 glycosylation in pyri
and *N*-9 in purine) and stereoselectivity (β-anomers are exclusively formed). The
synthetic process could be pictured as shown in Figure 14.9.

In order to find new active strains that are able to catalyze this process, a taxe
screening of 147 microorganisms was performed in our laboratories, using strains bel
to different microbial groups: Bacillaceae, Enterobacteraceae, Lactobacillaceae, Phc
teraceae, Pseudomonaceae, Psychrobacter, and Vibrionaceae. Two reaction tests (
14.10) were used in the screening: reaction test I was the synthesis of adenosine from u
and reaction test II was the synthesis of 2′-deoxyadenosine from 2′-deoxyuridine.

The screening resulted in 41 microorganisms (28%) active in reaction test I, while 5:
microorganisms gave positive results in reaction test II, and 35 strains (24%) were pos
both reactions. Only six microorganisms were specific for the synthesis of ribonucleo:

The most interesting microorganisms were selected according to four main criteri:

1. Low production of hypoxanthine (low adenosine deaminase activity)
2. Nonpathogenic and easy to cultivate in standard culture media

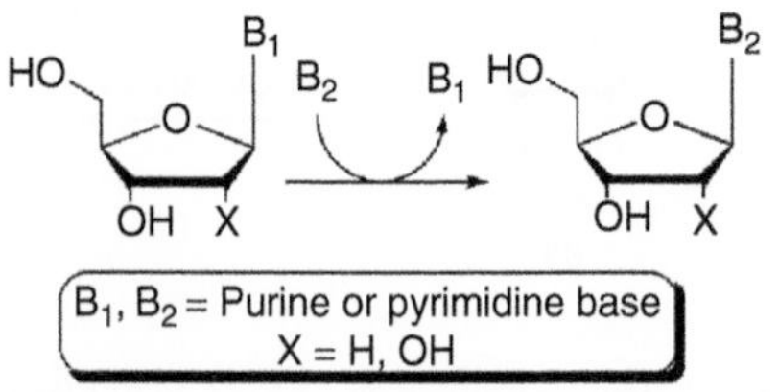

FIGURE 14.9 Synthetic activity of nucleoside phosphorilases (NPs).

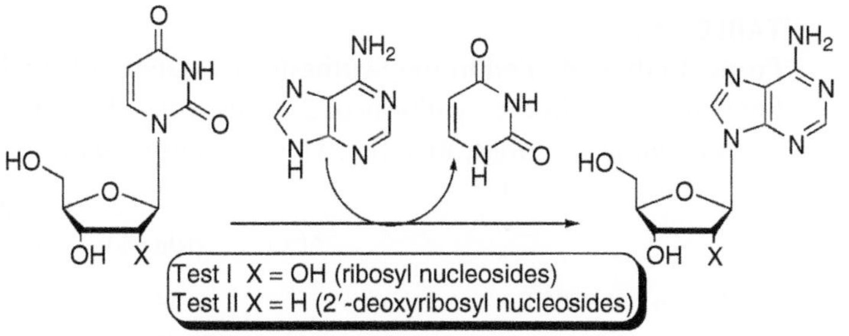

FIGURE 14.10 Test reactions used in the screening.

3. Ability to catalyze the reactions at low temperature
4. High productivity in adenosine or 2′-deoxyadenosine

Using these criteria, the microorganisms shown in Table 14.1 were selected as the most interesting strains for the synthesis of ribonucleosides, while those in Table 14.2 show the best strains for the synthesis of 2′-deoxyribosyl nucleosides. The assays were performed in duplicate experiments. In all cases, the reproducibility of the results was good, as we show in the case of *Xanthomonas translucens* (57°C, Table 14.1), and for *B. psychrosaccharolyticus* (57°C) and *Psychrobacter immobilis* (70°C) in Table 14.2. These two strains, and both *Photobacterium*, are described in this chapter for the first time as nucleoside phosphorylase producers.

TABLE 14.1

Microorganisms—Active in Reaction Test I and/or in Reaction Test II—Selected after the Screening Assays (Adenine = Uridine = 2′Deoxyuridine = 5 mM, Reaction Volume = 20 mL of 30 mM Phosphate Buffer (pH = 7), Time = 1 h)

Strain	T (°C)	Productivity [mM/(h×10^6 cells)]×10^5 Reaction Test I	Productivity (mM/(h×10^6 cells) ×10^5 Reaction Test II
Aeromonas salmonicida ssp. *achromogenes* CECT 895	57	10.0	0
Bacillus cereus CECT 131	57	3.3	0
	70	1.72	6.17
	57	3.3	0
B. subtillis CECT 4524	70	1.1	4.5
B. subtillis ssp.*niger* CECT 4071	57	6.7	0
Enterobacter amnigenus CECT 4078	70	9.0	1.25
E. sakazakii CECT 858	70	9.0	2.25
E. gergoviae CECT 857	70	18.0	12.25
	70	15.5	20.5
Erwinia amylovora CECT 222	70	20.8	0
Photobacterium leiognathi CECT 4191	57	12.1	2.2 (70°C)
P. phosphoreum CECT 4192	57	11.6	0
Serratia marcenses CECT 977	57	7.6	4.01
	70	6.6	0
S. rubidea CECT 868	70	19.7	0
Xanthomonas translucens CECT 4643	57	15.5	12.8
	57	18.8	13.4

TABLE 14.2
Productivity Obtained in the Synthesis of 2′-Deoxyadenosine from 2′-Deoxyuridine Specific Strains in the Synthesis of 2′-Deoxyribosyl Nucleosides (Adenine = 2′-Deoxyuridine = 5 mM, Reaction Time = 1 h; Reaction Volume = 50 mL)

Strain	T (°C)	Yield (%)	[Cells] (10^6 cells/mL)[a]	Productivi [mM/(h × 10^6 cell
Bacillus coagulans CECT 12	57	14	514	6.8
	70	24	1204	2.9
	70	13	1404	2.3
B. psychrosacharolyticus CECT 4074	57	71	554	32
	57	56	590	27
Enterobacter aerogenes CECT 684[b]	70	42	686	15
Lactobacillus alimentarius CECT 570[b]	57	19	1677	2.8
	70	6.5	750	2.2
L. pampilonensis CECT 4219[b]	57	42	2068	5.0
	70	7.7	1623	1.2
	70	8.8	2223	1.0
Psychrobacter immobilis CECT 4492	57	66	690	23
	57	53	713	18

[a]Cell cultured at 28°C.
[b]Hypoxanthine producer (yield > 10% referred to adenine).

Regarding the synthesis of ribonucleosides, some strains yielded adenosine detectable production of hypoxanthine at 57 ± 2°C: *B. cereus*, *B. subtilis*, *B. atro Serratia marcescens*, and *X. translucens* (Table 14.1). This temperature is lower th usually described for other strains in the synthesis of this nucleoside, which is n carried out at temperatures higher than 60°C in order to reduce the production of anthine [33,34]. The productivity values for 2′-deoxyadenosine (Table 14.1) are g lower than those obtained for adenosine in the case of nonspecific strains (Tabl Nevertheless, *B. cereus* (70°C), *B. subtilis* (70°C), *Enterobacter gergoviae*, *S. marcesce X. translucens* showed good productivity in both reactions. It has been previously r [35–37] that *B. subtillis*, *S. marcescens*, several *Enterobacter* strains, and *X. campestris* are nucleoside phosphorylase producers.

From the results presented in Table 14.2 we can deduce that the most interesting (considering the productivity obtained) are *B. coagulans*, *B. psychrosaccharolytic P. immobilis*. These strains displayed the best productivity values and the lowest anthine levels in reactions at 57°C. It must be considered that the presence of 2′-deoxy transferases in *Lactobacillus* spp. [30,31] is well documented, but that is not the c psychrotrophic strains [38].

14.2.5   Characterization of the Enzymes Involved in the Biotransformation of N-2′-Deoxyribosyl Nucleosides

As indicated in Table 14.2, six strains were selected as active and specific for modi of 2′-deoxyribosyl nucleosides. Two different enzymes can catalyze the synthesis of : yribonucleosides by means of base interchange. The mechanism proposed for 2′-deoxy transferases involves a covalent catalysis, by analogy with the glycoside hydrolases

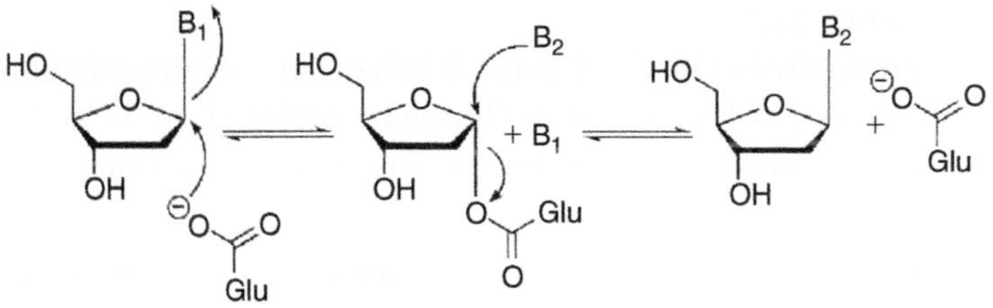

FIGURE 14.11 Mechanism proposed for 2'-deoxyribosyl transferases.

either a glutamyl or an aspartyl residue for catalytic activity. Indeed, several researchers [29,39,40] have demonstrated the presence of a glutamyl residue involved in the rate-controlling step in several N-2'-deoxyribosyl transferases, as depicted in Figure 14.11.

Another possibility involves two enzymes: thymidine nucleoside phosphorylase (TNP) and purine nucleoside phosphorylase [41,42]. TNP (E.C. 2.4.2.4) is specific for thymidine and the reaction takes place in two steps, through a 2'-deoxyribose-1-phosphate intermediate, as shown in Figure 14.12. TNP recognizes thymidine better than 2'-deoxyuridine [39,40].

The difference between both mechanisms is the presence of the ribose 1-α-phosphate intermediate in the case of TNP but not in the case of N-2'-deoxyribosyl transferases. Therefore, two parallel experiments were performed at the same time and with the same microbial culture. One reaction was performed using 30 mM NaH_2PO_4/Na_2HPO_4 buffer (pH = 7) and the other using 25 mM Tris/HCl buffer (pH = 7). When the microorganisms expressed a TNP, the yield and productivity of the reaction were higher in phosphate buffer than in the presence of Tris/HCl buffer. Contrarily, if the microorganisms expressed a nucleoside 2'-deoxyribosyl transferase, the yield were similar in both reactions.

From the results in Table 14.3 we can deduce that B. subtilis (nonspecific strain, Table 14.1) cannot give the test II in the presence of Tris buffer. E. aerogenes clearly express a TNP because the reaction does not take place in Tris/HCl buffer. On the contrary, B. psychrosaccharolyticus, B. coagulans, P. immobilis, and Lactobacillus pampilonensis gave similar yields in Tris/HCl buffer as in phosphate buffer. This confirms the presence of nucleoside 2'-deoxyribosyl transferases in these strains.

14.2.6 IMMOBILIZATION OF WHOLE CELLS

Nucleoside phosphorylases are extracellular and multimeric enzymes [43,44]. A synthesis with these enzymes at semi-industrial scale would demand a co-immobilization step after the overexpression of the enzymes in genetically modified microorganisms and disruption of cells by consecutive passage of biomass in a high-pressure homogenizer [45–47]. Therefore, the immobilization–stabilization of these enzymes is a rather complex, but not impossible, process. In fact, nucleoside phosphorylases from B. subtilis have been expressed in Escherichia

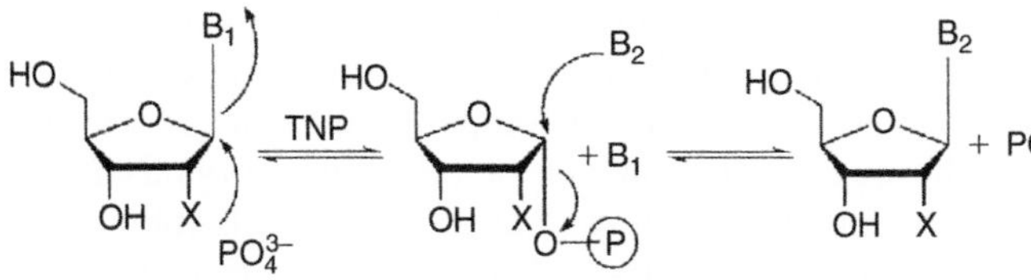

FIGURE 14.12 Mechanism involving tymidine and purine nucleoside phosphorylases.

TABLE 14.3

Productivity Obtained in the Synthesis of 2′-Deoxyadenosine from 2′-Deoxyuridine Adenine in Tris/HCl or in Phosphate Buffer (T = 57°C, Reaction Volume = 4 mL, 2′-Deoxiuridine = Adenine = 5 mM, Reaction Time = 20 min)

Strain	Buffer	Yield[a] (%)	Productivity [mM/ cells)] × 1C
Bacillus coagulans	Tris/HCl	15 (8)	3.7
CECT 12	Phosphate	14 (9)	3.8
B. subtillis	Tris/HCl	0	0
CECT 4524	Phosphate	35 (7)	31.5
B. psychrosaccharolyticus	Tris/HCl	48	31
CECT 4074	Phosphate	62	35
Enterobacter aerogenes[b]	Tris/HCl	0	0
CECT 4074	Phosphate	14 (6)	17
Lactobacillus pampilonensis	Tris/HCl	26 (8)	3.1
CECT 4219	Phosphate	23 (11)	2.7
Psycrhobacter immobilis	Tris/HCl	27	13
CECT 4492	Phosphate	24	10

[a]Yield in 2′-deoxyadenosine and (yield in hypoxanthine).
[b]Reaction time when it is not 1 h.

coli, purified, and immobilized in sepabeads with polyethyleneamine, yielding an enz derivative that is very active at pH = 10 and 45°C [46,47].

A cheap alternative is the immobilization of the whole cells [48,49]. With these b lysts, efficient syntheses of some purine nucleosides have been reported using *Erwinia cola* [34], *E. coli* BL21 [35,50], *B. stearothermophilus* [28], and *B. cereus T.* [51]. In o scale up the synthetic process using whole cells, two main conditions are necessary:

1. Absence of secondary reactions such as deamination of adenine in adenosine ri
2. Possibility of reusing the biocatalyst [34,52,53]

Thus, the most widely used technique for cell immobilization is the cell entrapment [54], in the living cells are enclosed in a polymeric porous matrix that allows the diffusion of sub and products. The main advantages of this methodology are the high operational stabil ease of cell handling and separation, and finally the feasibility of the scaling up [55–61].

Therefore, the best microorganisms as deduced by their performance shown in Tak and Table 14.2 were immobilized in different supports, in order to select the best bioc according to the maximum yield reached and the maximum number of reuse cycle immobilized derivatives in agars (with different methoxylation degree) or in agaros vided the best results. Different shapes of the biocatalysts—small beads, microbeads, o parallelepipeds—were prepared in our laboratories, while the immobilization in micr was rejected because the biocatalysts were not interesting from the reuse and scale-up a as we show in Figure 14.13 compared to Figure 14.14 (immobilization in small plates

The qualitative behavior of other microorganisms immobilized in these matrix similar. This finding shows that all thermogels (obtained at their optimum percenta with optimum shaking speed) do not exert dramatic restrictions on the diffusion of reagents or products for obtaining an optimum yield in adenosine. In Figure 14.15 we c

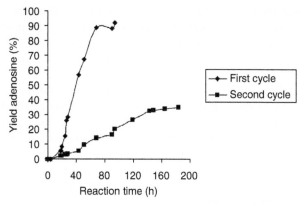

FIGURE 14.13 Synthesis of adenosine from uridine/adenine (30/10) mM using immobilized *Serratia marcenses* CECT 977 whole cells. Microbeads of highly methoxylated agar A28/03. Temperature = 57°C. Stirring speed = 250 r.p.m. Reaction time = 120 h. 40 g of wet biocatalyst.

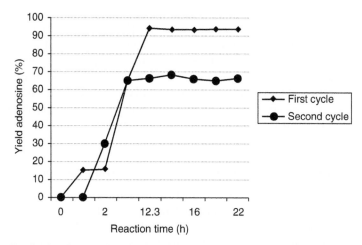

FIGURE 14.14 Synthesis of adenosine from uridine/adenine (30/10) mM using immobilized *Serratia marcenses* CECT 977 whole cells. Microplates (1 cm × 1 cm × 0.2 cm) of highly methoxylated agar A28/03. Temperature = 57°C. Stirring speed = 250 r.p.m. Reaction time = 120 h. 40 g of wet biocatalyst.

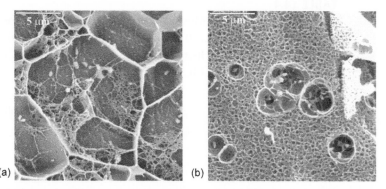

FIGURE 14.15 Electronic micrographs of immobilized whole cells of *Serratia marcenses* CECT 977 in (a) agar A28/03 or in (b) agarose.

that the bacteria are immobilized inside small cavities where only a few cells are ent
linked to the matrix through small linkers that probably could be mucopolysac
secreted by the cell. These microphotographs justify that in the washing of the bic
particles for reusing, no free bacteria were observed in the reaction medium with a
tional microscope.

The stability of the biocatalysts in standard storage conditions (4°C) was also teste
60 to 75 days all the immobilized biocatalysts retained the same activity in the synt
adenosine starting from uridine and adenine [38].

In Table 14.4 we compare the catalytic activity of several strains immobilized in
agarose microplates using resting cell conditions. We can observe that E. amnige
E. gergoviae gave similar yields in all cases, while S. marcenses showed a mo
diminution in the yield in adenosine after immobilization. So, these biocatalysts can
for a further scaling up.

In the case of the synthesis of 2′-deoxyribonucleosides, the immobilized whole cells
(A28/03) were as active in agar as in agarose, as we show for a nonspecific strain
E. gergoviae (Table 14.5).

Even though better yield was obtained using the immobilized biocatalysts comp
free resting cells, the productivity was lower (Table 14.5), due to the longer reactic
demanded for the immobilized cells. Finally the best biocatalyst—E. gergoviae immob
small plates of highly methoxylated agar—was reused several times, as shown in Figur
This biocatalyst could be reused at least seven times without diminution of the yie
immobilized biocatalysts in agarose gave good yield in the first cycle (Table 14.5), b
60% yield was achieved in the second cycle. The yields obtained with immobilized E. ge
were similar to those described by Yokozeki and Tsuji [53], using free cells of E. aeroge
11125, or better than those described by Prasad et al. [62].

14.2.7   SYNTHESIS OF DIFFERENT NUCLEOSIDE ANALOGS

The syntheses of several nucleosids analogs have been described in the literature usin
cells as biocatalysts [36,49,63–66] (Table 14.6).

Kulikowska et al. [64] showed that 7-methylguanosine and 7-methylinosine are su
of purine nucleoside phosphorylases. Taking into account this finding, Hennen and

---

TABLE 14.4
**Catalytic Activity of the Most Interesting Strains in the Synthesis of Adenosine from U
Uridine/Adenine (30 mL/10 mM). Resting Cells and Immobilized in Agar A28/03 an
Agarose D5 ($T = 70°C$, Reaction Time = 30 min (Resting Cells) or 1 h (Immobilizec
Reaction Volume = 15 ML, Wet Catalyst Weight = 20 g)**

Microorganisms	Adenosine (%) Resting Cells	Adenosine (%) Immobilized Agar	Adenosir Immobilizec
Enterobacter amnigenus CECT 4078	98.5	88	94
E. gergoviae CECT 857	100	97	95
E. sakazakii CECT 858	91	5.2	48
Serratia rubidea CECT 868	90	11	7
S. marcenses CECT 977	100	70	70

TABLE 14.5
Productivity Values in the Synthesis of 2'-Deoxyadenosine from 2'-Desoxyuridine/Adenine (30 mL/10 mM) Using *Enterobacter gergoviae* in Resting Cells Condition (1 h) or Immobilized in Agar A28/03 (3 h) or in Agarose D5 (3 h) (Reaction Time = 30 min (Resting Cells) and 2 h (Immobilized Biocatalysts), 40 g of Wet Biocatalyst, T = 70°C)

Catalyst	2'-Desoxyadenosine (%)	Cell Concentration (10^6 cell/ML)	Catalyst Load (10^6 cell/g cat)	Productivity [mM/(h × 10^6 cells)] × 10^5
Resting cells	42	1319	—	84
Immobilized agar A28/03	90	—	1319	23
Immobilized agarose	85	—	1110	25.5
Resting cells	43	1319	—	86

[65] proposed the use of 7-methylguanosinium or 7-methylinosinium as water-soluble glycosyl donor to obtain analogs of nucleosides in an irreversible and quantitative process, caused by the low solubility of guanine and inosine at pH = 7.0 to 7.4 (Table 14.7).

Shirae et al. [67] described the use of different *E. coli* strains (resting cells) for the preparation of different nucleoside analogs, such as 2',3'-dideoxyadenosine (**39**) or 2',3'-dideoxyinosine (**40**) (Figure 14.17). After a taxonomic screening of 436 microorganisms belonging to 39 genera, they selected one strain (optimal pH was ~6.5, and optimal temperature was 50°C). These authors indicated that 2',3'-dideoxypyrimidine nucleosides (**38 and 41**) could also be obtained from 2',3'-dideoxyadenosine (**39**), but with moderated yields, as shown in Figure 14.17, so this result indicates the great versatility of the strain.

Recombinant *E. coli* BL21 strain was used by Rogert et al. [50] to prepare some *ara*-nucleosides (**43 and 44**) from *ara*-uridine (**42**) using resting cells at 60°C and pH = 7.0 (Figure 14.17). On the other hand, Murakami et al. [68] described the use of *E. coli* JA-300

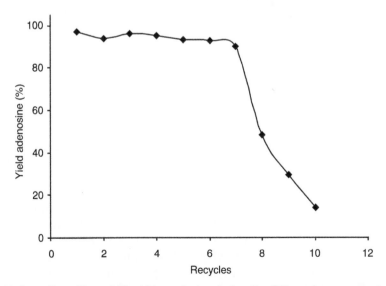

FIGURE 14.16 Recycling of immobilized biocatalysts: whole cells of *Enterobacter gergoviae* CECT 857 immobilized in small plates of agar A28/03, T = 57°C.

TABLE 14.6

Synthesis of Some Nucleoside Analogs by Base Interchange (Enzymes: Pyrimidine + Nucleoside Phosphorylase)

Pyridine Nucleoside	W	Z	Z'	Microorganism	Base	Yield in Nucleoside (%)	Time (h)
Uridine	OH	OH	H		6-Mercapto-purine	56	3
2'-Deoxyuridine	OH	H	H	E. gergoviae CECT 857	6-Mercapto-purine	18	3
Uridine	OH	OH	H		3-Carboxy-amido 1,2,4-triazol	45	3
Uridine	OH	H	H		Purine	80	3
Ara-uridine	OH	H	OH	E. coli BMT-ID/1A	Adenine	83	133
Uridine	OH	OH	H	E. coli free enzymes	4-Chloro-imidazo [4,5-c] pyrimidine	57	—

whole cells (50°C and pH = 6.5) in the synthesis of 6-halo-2',3'-dideoxypurine nucleosi and **46**), used as anti-HIV drugs, starting from 2',3'-dideoxyuridine (**38**) (Figure 14.17

Finally, recombinant *E. coli* BM-11 cells have been used by Zinchenko et al. [69] to 9-(β-D-arabinofuranosyl)guanine (**47**, Figure 14.18), active against viruses of herpes ty also for inhibiting T-lymphocyte proliferation. The synthesis was performed from arabinofuranosyl)uracil (**42**) and three different guanine donors: guanine (**48**), guanosi and 2'-deoxyguanosine (**49**), as shown in Figure 14.18.

TABLE 14.7
**Synthesis of Some Nucleoside Analogs Using 7-Methyl Guanidinium or
7-Methylinosinium Salts**

1 X = NH2
2 X = NH2
3 X = H

Initial Compound[a]	Base[b]	Yield in Nucleoside	Reaction Time (days)
1	adenine	100	4
2	Adenine	100	4
3	Adenine	100	2
2	3-Diazaedenine	70	2
1	3-Diazadenine	50	2
2	1,2,4-Triazol-3-carboxiamide	60	2
3	1,2,4-Triazol-3-carboxamide	57	2

[a]100 mol.
[b]25 mol.

Source: Hennen, W.J. and Wong, Ch-H., *J. Org. Chem.*, 54(19), 4692–4695, 1989.

The synthesis starting from guanine nucleosides leads to higher yields than using free base. The phenomenon is related to the poor solubility of guanine in the reaction conditions (~0.042 g/L) which leads to a low base concentration and thus the unfeasibility of an effective transglycosidation mediated by the cells. Contrarily, both nucleosides are used by the cells to render a high yield in *ara*-guanosine. As can be seen, the cross-linking with glutaraldehyde of whole cells increases both the stability of biocatalysts and the obtained yield.

Some other researchers have described the synthesis of nucleoside analogs using phosphorylases in combination with other enzymes. Thus, Pal and Nair [28] described how the addition of xanthineoxidase (xodase) to a synthesis of nucleosides catalyzed by whole cells of *B. stearothermophilus* ATCC 12980 leads to higher yields in the synthesis of thymidine (**51**) starting from 2'-deoxyinosine (**50**) than those described in the absence of xodase (Figure 14.19). This fact is explained because xodase shifts the equilibrium of the reversible transglycosylation

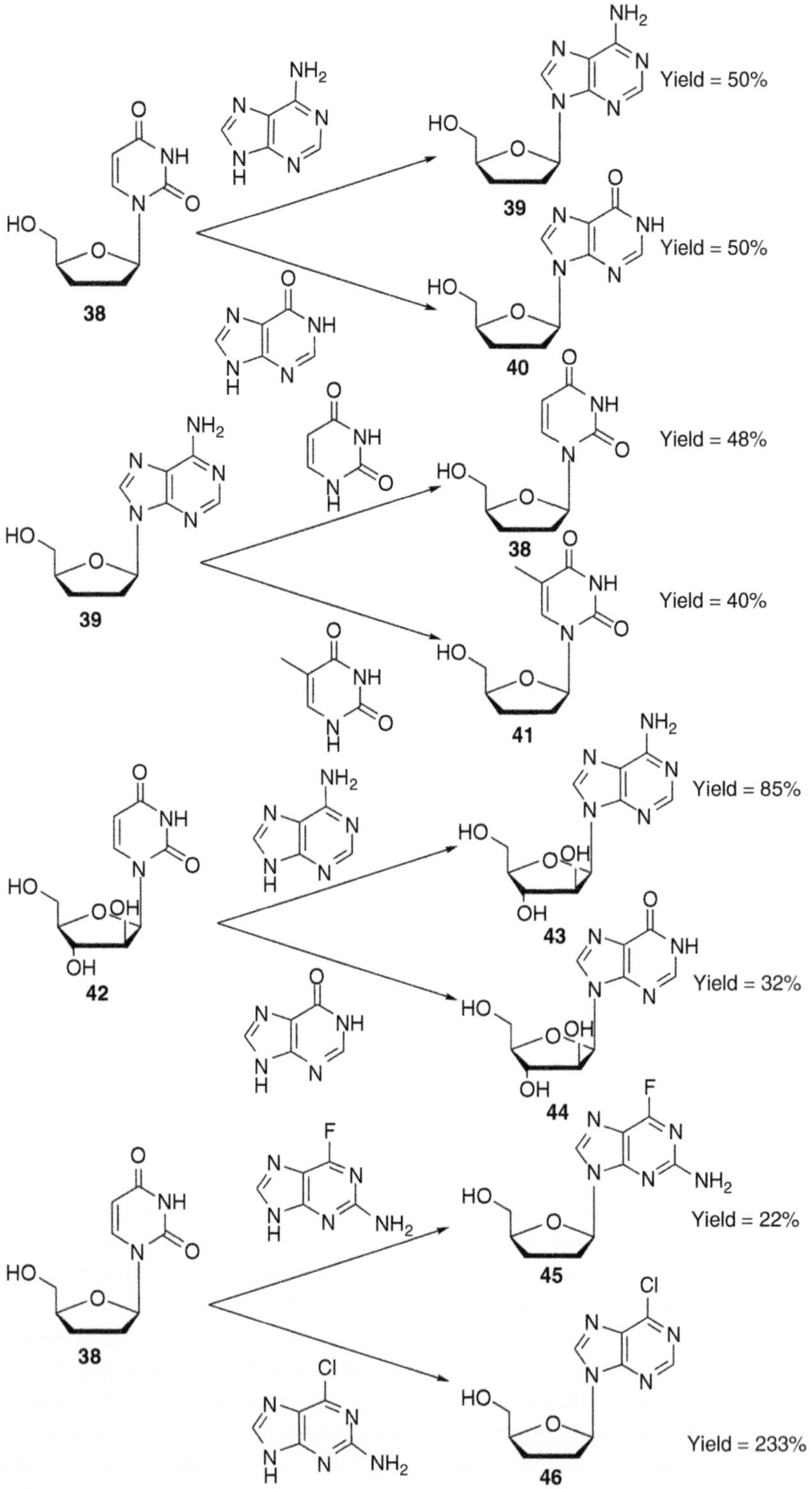

FIGURE 14.17 Synthesis of different nucleoside analogs using some *Escherichia coli* strains.

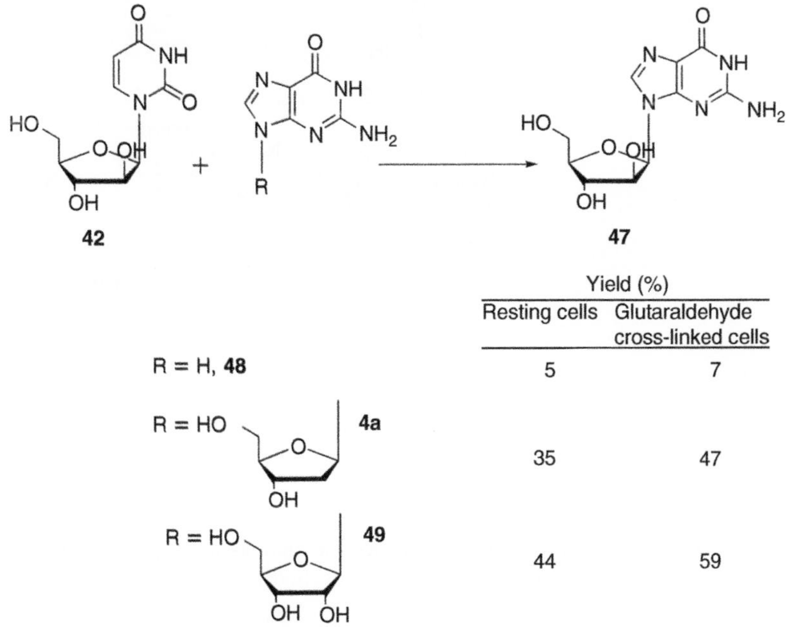

R	Yield (%) Resting cells	Yield (%) Glutaraldehyde cross-linked cells
R = H, **48**	5	7
4a	35	47
49	44	59

FIGURE 14.18 Synthesis of 9-(β-D-arabinofuranosyl) guanine using free cells and cross-linked cells of *Escherichia coli* BM-11. (From Zinchenko, A.I., Barai, V.N., Bokut, S.B., et al., *Appl. Microbiol. Biotechnol.*, 32(6), 658–661, 1990.)

reaction toward completion by the formation of uric acid (**52**, Figure 14.19), which is not recognized by phosphorylase.

Finally, Yokozeki and Tsuji [53] described the synthesis of 2′-deoxyguanosine (**4a**, Figure 14.20) with 100% yield at 25°C, using *E. gergoviae* AJ-11125 whole cells. This strain has a very active adenosine deaminase, so that the synthesis of **4a** could be described using

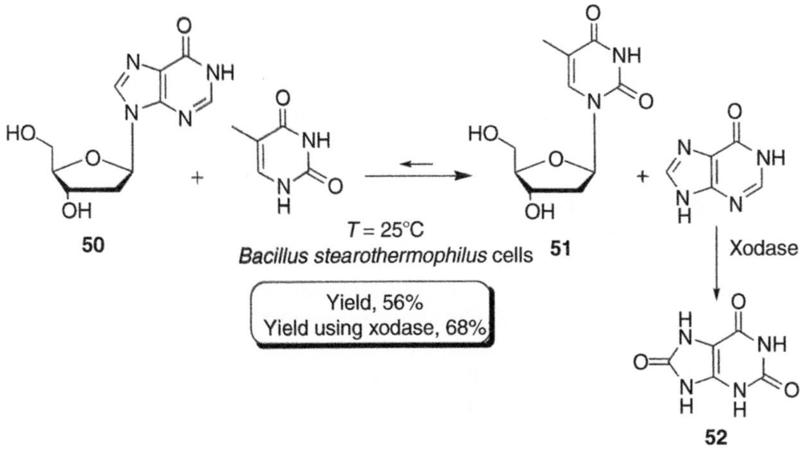

FIGURE 14.19 Synthesis of nucleoside analogs using phosphorylases in combination with xanthineoxidase. (From Pal, S. and Nair, V., *Biocatal. Biotransfor.*, 15(2), 147–158, 1997.)

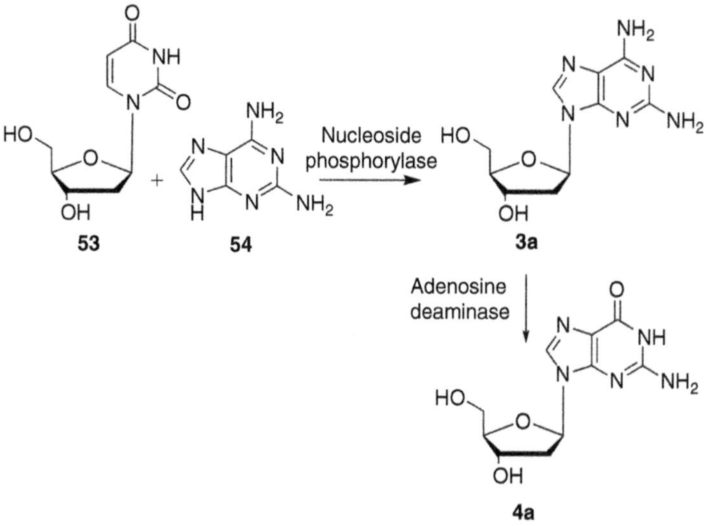

FIGURE 14.20 Synthesis of nucleoside analogs using phosphorylases in combination with ad
deaminase (ADA). (From Yokozeki, K. and Tsuji, T., *J. Mol. Catal. B Enzym.*, 10(1–3), 207–213

2,6-diaminopurine (**54**) and 2′-deoxyuracil (**53**) as base acceptor, through **3a** intern
(Figure 14.20). This indirect methodology overcomes the direct synthesis of **4a** from g
and **53** due to the low water solubility of the base in the medium.

14.3   CONCLUSIONS

The enzymatic synthesis of nucleoside analogs using either free enzymes or whole cells
advantageous over chemical synthesis due to the reduction in the number of steps a
high regio- and stereoselectivity observed. Different enzymes can be used in the synth
these compounds in order to selectively modify either the base or the sugar functional g
In addition, different expensive nucleoside analogs can be prepared starting from
nucleosides by base interchange using immobilized whole cells in processes that are e
scale up. The main problem for this synthetic methodology is to design a "homoge
reaction media" for both nucleosides (hydrophilic) and bases (hydrophobic). Therefo
development of new solvents displaying low toxicity for cells or enzymes and possessin
solving properties is the main work in the future for scaling up these syntheses.

REFERENCES

1. Garg, R., Gupta, S.P., Gao, H., et al., Comparative quantitative structure–activity relat
   studies on anti-HIV drugs, *Chem. Rev.*, 99(12), 3525–3601, 1999.
2. Buolamwimi, J.K., Nucleoside transport inhibitors: structure–activity relationships and p
   therapeutic applications, *Curr. Med. Chem.*, 4(1), 35–66, 1997.
3. Mahmoudian, M., Rudd, B., Cox, B., et al., A versatile procedure for the generation of nuc
   5′-carboxylic acids using nucleoside oxidase, *Tetrahedron*, 54(28), 8171–8182, 1998.
4. Fox, K.R., Targeting DNA with triplexes, *Curr. Med. Chem.*, 7(1), 17–37, 2000.
5. Rachakonda, S. and Cartee, L., Challenges in antimicrobial drug discovery and the pote
   nucleoside antibiotics, *Curr. Med. Chem.*, 11(6), 775–793, 2004.

6. Pankiewicz, K.W., Fluorinated nucleoside, *Carbohydr. Res.*, 327(1–2), 87–105, 2000.

7. Gómez, J.A., Trujillo, M.A., Campos, J., et al., Synthesis of novel 5-fluorouracil derivatives with 1,4-oxaheteroepine moieties, *Tetrahedron*, 54(43), 13295–13312, 1998.

8. Priego, E.M., Balzarini, J., Karlsson, A., et al., Synthesis and evaluation of thymidine carboxamides against mitochondrial thymidine kinase (TK-2) and related enzymes, *Bioorg. Med. Chem.*, 12(19), 5079–5090, 2004.

9. Raic-Malic, S., Herold-Brudic, A., Nagl, A., et al., Novel pyrimidine and purine derivatives of L-ascorbic acid: synthesis and biological evaluation, *J. Med. Chem.*, 42(14), 2673–2678, 1999.

10. Ramesh, N., Klunder, A.J.H., and Zwanenburg, B., Enantioselective synthesis of 4-acetylaminocyclopent-2-en-1-ols from tricycle [5,2,1,0(2,6)] decenylenaminone: precursor for 5′-norcarbocyclic nucleosides and related antiviral compounds, *J. Org. Chem.*, 64(10), 3635–3641, 1999.

11. Harword, E.A., Hopkins, P.B., and Sigurdsson, S.Th., Chemical synthesis of cross-link lesions found in nitrous acid treated DNA: a general method for the preparation of *N*-2-substituted-2′-deoxyguanosine, *J. Org. Chem.*, 65(10), 2959–2964, 2000.

12. Ferrero, M. and Gotor, V., Chemoenzymatic transformations in nucleoside chemistry, *Monatsh. Chem.*, 131(6), 585–616, 2000.

13. Ferrero, M. and Gotor, V., Biocatalytic modifications of conventional nucleosides, carbocyclic nucleosides and C-nucleosides, *Chem. Rev.*, 100(12) 4319–4347, 2000.

14. Discroll, J.S., Siddiqui, M.A., Ford, H., Jr., et al., Lipophilic acid-stable, adenosine deaminase-activated anti-HIV prodrugs for central nervous system delivery, 3: 6-amino prodrugs of 2′-β-fluoro-2′,3′-dideoxyinosine, *J. Med.Chem.*, 39(8), 1619–1625, 1996.

15. Santaniello, E., Ciuffreda, P., and Alessandrini, L., Synthesis of modified purine nucleosides and related compounds mediated by adenosine deaminase (ADA) and adenylate deaminase, *Synthesis*, 4, 509–526, 2005.

16. Kakefuda, A., Shuto, S., Nagahata, T., et al., Nucleosides and nucleotides 132. Synthesis and biological evaluation of ring-expanded oxetanocin analogues: purine and pyrimidine analogues of 1,4-anhydrido-2-deoxy-D-arabitol and 1,4-anhydro-2-deoxy-3-hydroxymethyl-D-arabitol, *Tetrahedron*, 50(34), 10167–10182, 1994.

17. Hertel, L.W., Grossman, C.S., and Kroin, J.S., Method of preparing β-2′,2′-difluoronucleosides, Eur. Pat. Appl. 0328345, 1989.

18. Vince, R. and Brownell, J., Resolution of racemic carbovir and selective inhibition of human immunodeficiency virus by the (−) enantiomer, *Biochem. Biophys. Res. Co.*, 168(3), 912–916, 1990.

19. Margolin, A.L., Borcherding, D.R., Wolf-Kugel, et al., AMP deaminase as a novel practical catalyst in the synthesis of 6-oxopurine ribosides and their analogs, *J. Org. Chem.*, 59(24), 7214–7218, 1994.

20. Moris, F. and Gotor, V., A novel and convenient route to 3′-carbonates from unprotected 2′-deoxynucleosides through an enzymatic reaction, *J. Org. Chem.*, 57(8), 2490–2492, 1992.

21. Moris, F. and Gotor, V., A useful and versatile procedure for the acylation of nucleosides through an enzymatic reaction, *J. Org. Chem.*, 58(3), 653–660, 1993.

22. García, J., Fernández, S., Ferrero, M., et al., A mild, efficient and regioselective enzymatic procedure for 5′-*O*-benzoylation of 2′-deoxynucleosides, *Tetrahedron Lett.*, 45(8), 1709–1712, 2004.

23. Uemura, A., Nozaki, K., Yamashita, J.-L., et al., Lipase-catalyzed regioselective acylation of sugar moieties of nucleosides, *Tetrahedron Lett.*, 30(29), 3817–3818, 1989.

24. Fan H., Kitawa, M., Raku, T., et al., Enzymatic synthesis of vinyl uridine esters catalyzed by Bioprase, alkaline protease from *Bacillus subtillis*, *Biochem. Eng. J.*, 21(3), 279–283, 2004.

25. Uemura, A., Nozaki, K., Yamashita, J.-L., et al., Regioselective deprotection of 3′,5′-*O*-acylated pyrimidine nucleosides by lipases and esterases, *Tetrahedron Lett.*, 30(29), 3819–3820, 1989.

26. Wang, J.-Q., Archer, C., Li, J., et al., Thermophilic esterases/lipases as an effective tool for the resolution of nucleoside diastereoisomers: convenient one-pot synthesis of α-L-taluronamide and β-D-alluronamide nucleosides, *J. Org. Chem.*, 63(14), 4850–4853, 1998.

27. Zhang, F.J. and Sih, C.J., Novel enzymatic cyclizations of pyrimidine nucleoside analogues: cyclic-GDP-ribose and cyclic-HDP ribose, *Tetrahedron Lett.*, 36(51), 9289–9292, 1995.

28. Pal, S. and Nair, V., Enzymatic synthesis of thymidine using bacterial whole cells and isolated purine nucleoside phosphorylase, *Biocatal. Biotransfor.*, 15(2), 147–158, 1997.

29. Short, S.A., Armstrong, S.R., Ealick, S.E., et al., Active site aminoacids that participa catalytic mechanism of nucleoside 2'-deoxyribosyltransferase, *J. Biol. Chem.*, 271(9), 49 1996.

30. Cardinaud, R. and Holguin, J., Nucleoside deoxyriboxyltransferase-II from *Lactobacillus h* Substrate specificity studied: pyrimidine bases as acceptors, *Biochim. Biophys. Acta.* 339–347, 1979.

31. Kaminski, P.A., Functional cloning, heterologous expression and purification of two *N*-deoxyrybosyltransferases, *J. Biol. Chem.*, 277(17), 14400–14407, 2002.

32. Carson, D.A. and Wasson, D.B., Synthesis of 2',3'-dideoxynucleosides by enzymatic trans-lation, *Biochem. Biophys. Res. Comm.*, 155(2), 829–834, 1988.

33. Huang, M.C., Montgomery, J.A., Thorpe, M.C., et al., Formation of 3'-(2'-deoxyribofu and 9-(2'-deoxyribofuranosyl) nucleosides of 8-substituted purines by nucleoside deoxyribo: ferases, *Arch. Biochim. Biophys.*, 222(1), 133–144, 1983.

34. Utagawa, T., Enzymatic preparation of nucleoside antibiotics, *J. Mol. Catal. B Enzy* 215–222, 1999.

35. Lewkowicz, E.S., Martínez, N., Rogert, M.C., et al., An improved microbial synthesis c nucleosides, *Biotechnol. Lett.*, 22(16), 1277–1280, 2000.

36. Hayashi, Y., Shiba, T., Katsuyama, A., et al., Method of manufacturing trifluorothyimi Pat. Appl. 0331853, 1989.

37. Trelles, J.A., Fernández, M., Lewkowicz, E.S., et al., Purine nucleoside synthesis from uridi immobilized *Enterobacter gergoviae* CECT 875 whole cells, *Tetrahedron Lett.*, 44(12), 26(2003.

38. Sinisterra, J.V., Condezo, L.A., and Fernández-Lucas, J., Procedimiento para la síntesis de sidos mediante la utlización de microorganismos psicrotrofos o psicrotolerantes, Spanish 200400817, PCT/ES/2005/00166, 2004.

39. Holguin, J., Cardinaud, R., and Salemnik, C.A., *trans*-*N*-deoxyriboxylase: substrate sp studies. Purine bases as acceptors, *Eur. J. Biochem.*, 54(2), 515–520, 1975.

40. Danzin, C. and Cardinaud, R., Deoxyribosyltransfer catalysis with *trans*-*N*-deoxyriboxylas study of purine (pyrimidine) to pyrimidine (purine) *trans*-*N*-deoxyriboxylase, *Eur. J. B* 62(2), 365–372, 1976.

41. Walter, M.R., Cook, W.J., Cole, L.B., et al., Three-dimensional structure of thymidine phosp from *E. coli* at 2.8 Å resolution, *J. Biol. Chem.*, 265(23), 14016–14022, 1990.

42. Zimmerman, M. and Seidenberg, J.I., Thymidine phosphorylase and nucleoside deoxyribo ferase in normal and malignal tissues, *J. Biol. Chem.*, 239(8), 2618–2621, 1964.

43. Pugmire, M.J. and Ealick, S.E., Structural analyses reveal two distinct families of nucleo: sphorylases, *Biochem. J.*, 361, 1–25, 2002.

44. Erion, M.D., Stoecker, J.D., Guida, W.C., et al., Purine nucleoside phosphorylase 2: mechanism, *Biochemistry* (USA), 36(39), 11735–11748, 1997.

45. Zuffi, G., Ghisotti, D., Oliva, I., et al., Immobilized biocatalysts for production of nucleos nucleoside analogues by enzymatic transglycosidation reactions, *Biocatal. Biotransfor.*, 22(1 2004.

46. Ubiali, D., Rocchietti, S., Scaramozzin, F., et al., Synthesis of 2'-deoxynucleosides by tra silation with new immobilized and stabilized uridine phosphorylase and purine nu phosphorylase, *Adv. Synth. Catal.*, 346(11), 1361–1366, 2004.

47. Rocchietti, S., Ubiali, D., Terreni, M., et al., Immobilization and stabilization of reco multimeric uridine and purine nucleoside phosphorylases from *Bacillus subtillis*, *Biomacrom* 5(6), 2195–2200, 2004.

48. Huang, J., Hooijmans, C.M., Briasco, C.A., et al., Effect of free-cell growth parameters or concentration profiles in gel-immobilized recombinant *E. coli.*, *Appl. Microbiol. Biotechno* 619–623, 1990.

49. Trelles, J.A., Fernández-Lucas, J., Condezo L.A., et al., Nucleoside synthesis by imm bacterial whole cells, *J. Mol. Catal. B Enzym.*, 35(5–6), 219–227, 2004.

50. Rogert, M.C., Trelles, J.A., Porro, S., et al., Microbial synthesis of antiviral nucleosid *Escherichia coli* BL21 as biocatalyst, *Biocatal. Biotransfor.*, 20(5), 347–351, 2002.

51. Gardner, R. and Kornberg, A., Biochemical studies of bacterial sporulation and germination, V: purine nucleoside phosphorylases of vegetative cells and spores of *Bacillus cereus, J. Biol. Chem.*, 242(10), 2383–2388, 1967.
52. Yokozeki, K., Shirae, H., Kobayashi, K., et al., Methods of producing 2′,3′-dideoxyinosine, US Patent No. 4970148, 1990.
53. Yokozeki, K. and Tsuji, T., A novel enzymatic method for the production of purine-2′-deoxyribonucleosides, *J. Mol. Catal. B Enzym.*, 10(1–3), 207–213, 2000.
54. Mazid, M.A., Biocatalysis and immobilized/cell bioreactors, *Biotechnology*, 11(6), 690–695. 1993.
55. Bickerstaff G.F., Immobilization of enzymes and cells: some practical considerations, in *Immobilization of Enzymes and Cells*, Bickerstaff, G.F., Ed., Humana Press, Totowa, NJ, 1987, pp. 1–3.
56. Hulst, A.C., Tramper, J., Riet, K., et al., A new technique for production of immobilized biocatalyst in large quantities, *Biotechnol. Bioeng.*, 27(6), 870–876, 1995.
57. Quintana, M.G. and Dalton, H., Production of toluene *cis*-diol by immobilized *Pseudomonas putida* UV4 in barium alginate beads, *Enzyme Microb. Technol.*, 22(8), 713–720, 1998.
58. Khattar, J.I.S., Sarma, T.A., and Singh, D.P., Removal of chromium ions by agar immobilized cells of the cyanobacterium *Anacystis nidulans* in a continuous flow reactor, *Enzyme Microb. Technol.*, 25(7), 564–568, 1999.
59. Lozinsky, V.I. and Plieva, F.M., Poly(vinylalcohol) cryogels employ matrix for cell immobilization, 3: overview of recent research developments, *Enzyme Microb. Technol.*, 23(3–4), 227–242, 1998.
60. Sinisterra, J.V. and Dalton, H., Influence of the immobilization methodology in the stability and activity of *P. putida* UV4 immobilized whole cells, in *Immobilized cells: Basics & Applications*, Wijffels, R.H., Buitlear, R.M., Bucke, C., and Tramper, J., Eds., Elsevier Science, Amsterdam, 1996, pp. 416–423.
61. Carballeira, J.D., Valmaseda, M., Alvarez, E., et al., *Gongronella butleri, Schizosaccharomyces octosporu, Diplogelasinospora grovesii*: novel microorganisms useful for the stereoselective reduction of ketones, *Enzyme Microb. Technol.*, 34(6), 611–623, 2004.
62. Prasad, A.K., Trikha, S., and Parmar, V.S., Nucleoside synthesis mediated by glycosyl transferring enzymes, *Bioorg. Chem.* 27(2), 135–154, 1999.
63. Trelles, J.A., Betancor, L., Schoijer, A., et al., Immobilized *Escherichia coli* BL21 as a catalyst for the synthesis of adenine and hypoxanthine nucleosides, *Chem. Biodivers.*, 1(2), 280–288, 2004.
64. Kulikowska, E., Bzowska, A., Wierzchowski, J., et al., Properties of two unusual and fluorescent substrates of purine-nucleoside phosphorylase 7-methylguanosine and 7-methylinosine, *Biochim. Biophys. Acta.*, 874(3) 355–363, 1986.
65. Hennen, W.J. and Wong, Ch.-H., A new method for the enzymatic synthesis of nucleosides using purine nucleoside phosphorylase, *J. Org. Chem.*, 54(19), 4692–4695, 1989.
66. Krenitsky, T.A., Freeman, G.A., Shaver, S.R., et al., Imidazo [4,5-c]pyrimidines (3-deazapurines) by glycosyl transferring and their nucleosides as immunosuppressive and anti-inflammatory agents, *J. Med. Chem.*, 29(1), 138–143, 1986.
67. Shirae, H., Kobayashi K., Shiragami, H., et al., Production of 2′,3′-dideoxyadenosine and 2′,3′-dideoxyinosine from 2′,3′-dideoxyuridine and the corresponding bases by resting cells of *Escherichia coli* AJ 2595, *Appl. Environ. Microbiol.*, 55(2), 419–424, 1989.
68. Murakami, K., Shirasaka, T., Yoshioka, H., et al., *Escherichia coli* mediated biosynthesis and *in vitro* anti-HIV activity of lipophilic 6-halo-2′,3′-dideoxynucleosides, *J. Med. Chem.*, 34(5), 1606–1612, 1991.
69. Zinchenko, A.I., Barai, V.N., Bokut, S.B., et al., Synthesis of 9-(β-D-arabinofuranosyl) guanine using whole cells of *Escherichia coli.*, *Appl. Microbiol. Biotechnol*, 32(6), 658–661, 1990.

15 Biocatalytic Reduction of Carboxylic Acids: Mechanism and Applications

Padmesh Venkitasubramanian, Lacy Daniels, and John P.N. Rosazza

CONTENTS

In nature, carboxylic acids are a ubiquitous class of organic compounds. They occur either as free acids like acetic, pyruvic, or citric or masked in the forms of esters, anhydrides, lactones amides, and lactams. Most free carboxylic acids occur as salts, leading to their chemical stability and relatively high solubilities in water. Carboxylic acids and their derivatives are often precursors for value-added chemicals for the pharmaceutical, agricultural, and food industries. Methods to harness this pool of valuable precursors have led to the development of several methods for chemical reductions and other manipulations of carboxylic acids.

Microorganisms and their enzymes are now widely used as classes of "biocatalytic reagents" in synthetic organic chemistry [1–12]. Biocatalysts intrinsically bind organic substrates, and catalyze highly specific and selective reactions under the mildest of reaction conditions. These selectivities and specificities are often realized because of highly rigid interactions occurring between enzyme active sites and substrate molecules. Microbial reductions of aromatic carboxylic acids, usually to their corresponding alcohols, have been observed with whole-cell biotransformations by a number of microorganisms including *Actinomyces* [13], *Clostridium thermoaceticum* [14], *Aspergillus niger* [15,16], *Corynespora melonis* [15], *Coriolus* [15], *Neurospora* [17], *Glomerella cingulata* [18,19], *Gloeosporium laeticolor* [19], and *Nocardia* sp. [20,21]. In all reported cases, substrates contain aromatic moiety, albeit not all directly attached to the carboxyl groups being reduced

Gross and Zenk first showed that *Neurospora crassa* gave an aldehyde intermediate carboxylic acid reductions [22,23]. The enzyme reducing aryl-carboxylic acids to aldehyd purified and identified as a monomeric protein with an apparent molecular m 120,000 Da [22]. The reduction reaction was adenosine triphosphate (ATP)-, Mg^{2+} NADPH-dependent [22]. The *N. crassa* aryl-aldehyde oxidoreductase (AAOR) (EC 1. or carboxylic acid reductase (CAR) initially catalyzed the condensation of carboxylic ac ATP to yield an acyl-adenosine monophosphate (AMP) intermediate, which was subsec reduced by NADPH to afford the aldehyde (Figure 15.1) [24]. Because the reaction is ir ible, AAOR (EC 1.2.1.30) is different from aldehyde oxidoreductases (EC 1.2.1.30), oxidize aldehydes to carboxylic acids using NAD^+ [24,25].

Kato et al. showed that *Nocardia asteroides* JCM 3016 could reduce benzoic acid to alcohol [21]. Similar to the CAR from *Neurospora*, the reductase from this or reduced benzoic acid to benzaldehyde using ATP, NADPH, and Mg^{2+} [26]. Report values for benzoic acid, NADPH, and ATP were extremely low at 260 nM, 6 nM, and respectively. This enzyme reduced a number of mono- and multisubstituted benzoic a corresponding aldehydes. This CAR also catalyzed the reduction of benzoyl-AMP. Al aryl-carboxylic acid reduction to aldehyde activities has been observed in many micro isms [13–20], enzymes catalyzing this reaction have been purified only from *N. crassa* [2 *N. asteroides* [26].

In our laboratory, a strain of *Nocardia* sp. NRRL 5646 efficiently reduced aryl-carl acids to aldehydes and alcohols. The substrate specificity for reducing carboxylic acids w organism appeared to be significantly different than that for the enzyme reported by Kat [26]. The availability of relatively large amounts of pure CAR was necessary for elucidatio properties and its catalytic mechanism. *Nocardia* sp. NRRL 5646 CAR was purified 196- homogeneity by a combination of Mono-Q, Reactive Green 19 agarose affinity, and hy apatite chromatographies. A key step was the binding and elution of CAR from Reactive 19 [27]. ATP, $NADP^+$, and NADPH were used to elute CAR from Reactive Gre suggesting that binding of the enzyme to this matrix was due to a CAR nucleotide bindi

At 128,000 Da, the molecular mass of the *Nocardia* sp.NRRL 5646 CAR by SDS- and gel filtration chromatography was similar to that for CAR from *N. asteroides* [27] *crassa* [22]. Apparent K_m values for benzoate, ATP, and NADPH for *Nocardia* sp.NRR CAR were more than 1000-fold higher than those reported for the enzyme from *N. ast*

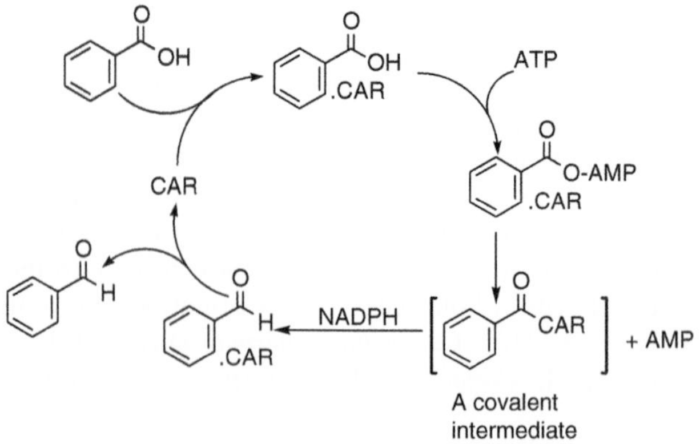

FIGURE 15.1 Enzyme catalyzed reduction of carboxylic acids.

[26]. The K_m of benzoate for *N. crassa* CAR was 63 µM and its activity was inhibited by 300 µM benzoate [23]. No inhibition was observed for the enzyme from *Nocardia* sp.NRRL 5646 even when benzoate concentrations were increased to 2 mM. This suggested that the *Nocardia* sp.NRRL 5646 CAR had different catalytic properties from those reported earlier [22,26].

15.1 MECHANISM OF CARBOXYLIC ACID REDUCTION BY CAR

Carboxylic acids can be chemically reduced to aldehydes, ketones, and alcohols by several methods [28]. Esters are reduced to aldehydes with diisobutylaluminum hydride (DIBAL-H). Reductions of acids and esters with lithium aluminum hydride give corresponding alcohols. Treatment of the lithium salts of acids with methyl Grignard or methyl lithium yields methyl ketones. In most of these cases, reaction conditions require temperatures below 0°C, using ether-based solvents and resulting in the aluminum salt by-products formed during reaction workup. Biocatalytic reduction of carboxylic acids would obviate many of these experimental difficulties, rendering such an important process more "green." However, to be fully utilized and optimized, understanding the mechanism of the biocatalytic reaction was essential.

Early studies by Gross and Zenk [22,24] implicated involvement of carbonyl-AMP intermediates during enzymatic reduction of carboxylic acids. This hypothesis was based primarily on the use of radiolabeled precursors and chromatographic [thin-layer chromatography (TLC), gas chromatography (GC)] analyses of reaction mixtures to identify reaction intermediates. With *N. crassa* CAR, ^{14}C-labeled benzoyl-AMP was trapped as a hydroxamate derivative that was identified by TLC [24]. ^{14}C-Labeled benzaldehyde was chromatographically identified as the product obtained when [14]COOH-labeled benzoate was used as substrate. Although Kato et al. [26] implicated benzoyl-AMP as an intermediate in benzoic acid reduction by an enzyme from *N. asteroides*, no experimental evidence was provided.

We designed a new one-pot synthesis of benzoyl-AMP under anhydrous conditions in *N,N*-dimethylformamide. Previous syntheses of such carbonyl-AMP derivatives were relatively crude, and products had never been well defined by spectral analysis. Reaction of benzoic acid with *N,N'*-carbonyldiimidazole and subsequently with 5'-adenosylmonophosphate gave the mixed anhydride in 76% isolated yield. The structure of this key intermediate in CAR reductions of benzoic acid was clearly confirmed by mass spectrometry and by proton, carbon, and phosphorous nuclear magnetic resonance (NMR) [29]. This method was also used to produce carboxy-^{13}C-benzoyl-AMP as a labeled intermediate for NMR analysis during the enzymatic reduction. With *Nocardia* sp. NRRL 5646 CAR, ^{13}C NMR analyses of enzymatic reductions of ^{13}C-carboxy-labeled benzoic acid and its derivatives provided clear evidence for the involvement of benzoyl-AMP in the reduction. Using ^{13}C-carbonyl-tagged benzoic acid afforded well-resolved ^{13}C NMR signals for COOH (169 ppm), CO-AMP (as a trapped hydroxamate at 176.4 ppm), and CH=O (200.8 ppm). *Nocardia* sp. NRRL 5646 CAR reduction of ^{13}C-carboxy-labeled benzoic acid with ATP only (Figure 15.2a), with ^{13}C-benzoyl-AMP plus NADPH (Figure 15.2b), or with ATP plus NADPH (Figure 15.2c) gave all expected signals [30].

Benzoyl-AMP had a significantly lower (67-fold) apparent K_m (9.66 $\pm$ 0.71 µM) and a higher (1.3-fold) V_{max} (7.50 $\pm$ 0.18 µmol/min/mg) than those for benzoic acid [30]. These results indicated that steady-state levels of benzoyl-AMP are likely to be low during the course of the reduction of benzoic acid to benzaldehyde. The results of this study, together with our previous work, in which benzaldehyde was isolated and characterized spectrally, clearly demonstrated that the mechanism for benzoic acid reduction to benzaldehyde by purified *Nocardia* sp. NRRL 5646 CAR involved benzoyl-AMP as an intermediate [29].

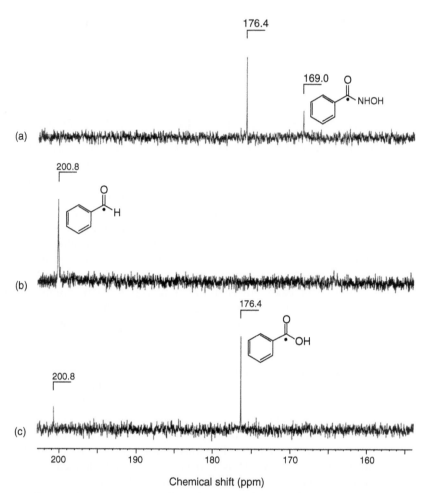

FIGURE 15.2 ^{13}C NMR study of the mechanism of benzoic acid reduction catalyzed by *Nocar* NRRL 5646 CAR. NMR spectra were obtained using incubations containing: (a) carboxy-^{13}C-b acid + ATP + NH$_2$OH; (b) carboxy-^{13}C-benzoyl-AMP + NADPH; and (c) carboxy-^{13}C-benzoic ATP + NADPH.

15.2 CAR CONTAINS A PHOSPHOPANTETHEINE ATTACHMENT SITE

CAR is produced in *Nocardia* sp., an organism that is relatively difficult to grow, an produces limited quantities of CAR. In order to obtain reproducibly large quantities of we sought to clone and express the enzyme in a common expression host such as *Esch coli*. *Nocardia* sp. NRRL 5646 genomic DNA was completely digested with *Sal*I or A and was then diluted fivefold, ligated, and used as the template for inverse Polymer reaction (PCR). Based on the N-terminal and internal sequences of purified CAR [3 entire *Nocardia car* gene sequence was derived by inverse PCR experiments to give a nucleotide of 6.9 kb, which included the entire *Nocardia car* gene and its flanking region DNA sequence and the deduced amino acid sequence of *Nocardia car* (accession m AY495697) indicated that *Nocardia car* consisted of 3525 base pairs (bp), correspondi protein with 1173 amino acid residues with a calculated molecular mass of 128.3 kDa a of 4.74. The N-terminal amino acid sequence of purified *Nocardia* CAR exactly match deduced amino acid sequence of the N terminus, with Ala as the first amino aci

The assignment of ATG as the start codon was supported by analysis of the 5' flanking region. At 6 bp upstream from the start codon ATG lies a conserved *Streptomyces* ribosomal binding site (GGGAGG) [31,32]. *Nocardia car* gene was amplified by PCR and then cloned into plasmid pHAT10 to give plasmid pHAT-305.

The deduced amino acid sequence of *Nocardia* CAR was 60% identical to the putative acyl-coenzyme A (CoA) synthase-substrate-CoA ligase *fadD9* of *Mycobacterium tuberculosis* and *M. bovis* and was 57% identical with its *M. leprae* homolog [33]. This finding suggested that these mycobacterial proteins might function in carboxylic acid reduction, a hypothesis that remains to be tested. BLAST analysis of CAR showed an N-terminal domain (amino acids 90 to 544) with high homology to known AMP-binding proteins. The C-terminal domain (amino acids 750 to 1094) had high homology with NADPH-binding proteins.

To express rCAR, a 100 ml culture of *E. coli* [BL21(DE3) or BL21-CodonPlus(DE3)-RP] harboring pHAT-305 was grown overnight in LB-ampicillin medium. This culture was diluted 20-fold in fresh medium, and then incubated at 170 rpm on a rotary shaker at 37°C to an optical density of 0.6 at 600 nm, at which time 1 mM IPTG was added, and the induced culture was further incubated for 4.5 h. Lysate from *E. coli* BL21(DE3) and CodonPlus cells carrying pHAT-305 had moderate CAR activity (0.003 and 0.009 U/mg, respectively) compared to that of *Nocardia* wild-type cells (0.03 U/mg of protein) [30]. SDS-PAGE of cell-free extracts (CFEs) gave a Coomassie blue-stained band with an apparent molecular mass of 132.4 kDa, which was confirmed to be HAT-CAR by Western blot analysis. The DHFR-positive control and pHAT10-negative control showed the absence of a 132.4 kDa band by SDS-PAGE and Western blot analysis. A typical 1 L culture of *E. coli* BL21(DE3) CodonPlus cells carrying pHAT-305 yielded approximately 50 to 70 mg of purified CAR. Recombinant CAR (rCAR) from *E. coli* BL21-CodonPlus-((DE3)RP)/pHAT-305 showed a specific activity of 0.1 U/mg, considerably less than that of wild-type CAR purified from *Nocardia* sp. NRRL 5646 (5.89 U/mg) [33].

To explain the difference in activity, we examined the CAR sequence. Buried within the deduced sequence (amino acids 683 to 693) was a serine residue (S689) and LGGxSxxA, which is the typical consensus sequence for a phosphopantetheine (Ppant) attachment site [34]. We then considered the possibility that CAR produced in wild-type *Nocardia* was actually a holo-enzyme produced by posttranslational phosphopantetheinylation of apo-CAR to afford maximal enzyme activity. A 4'-Ppant prosthetic group in active CAR could serve as a "swinging arm" that would react with acyl-AMP intermediates to form a covalently linked thioester to the C-terminal reductase domain for reduction by NADPH, aldehyde release, thiol regeneration, and a new catalytic cycle. This arrangement of the CAR protein would reflect a sequential catalytic mechanism wherein the N-terminal domain catalyzes substrate activation by formation of an initial acyl-AMP intermediate, while the C-terminal portion catalyzes the reduction of acyl-AMP by NADPH to finish a catalytic cycle (Figure 15.1). The 50-fold difference in activity between the rCAR and the wild-type CAR was attributed to the fact that the majority of the purified recombinant enzyme produced in *E. coli* was largely apo-CAR, while fully active wild-type CAR would be a phosphopantetheinylated holo-enzyme.

15.3 EVIDENCE FOR CAR PHOSPHOPANTETHEINYLATION

Phosphopantetheine transferases (Pptases) [34] are enzymes that catalyze the posttranslational modification of carrier proteins by covalently attaching the 4-Ppant moiety of CoA to a conserved serine residue (Figure 15.3). Evidence for such Pptases has been well documented in fatty acid and polyketide biosynthesis and in nonribosomal peptide synthesis [34–36]. By phosphopantethenylation, Pptases convert inactive apo-carrier proteins to active holo-carrier

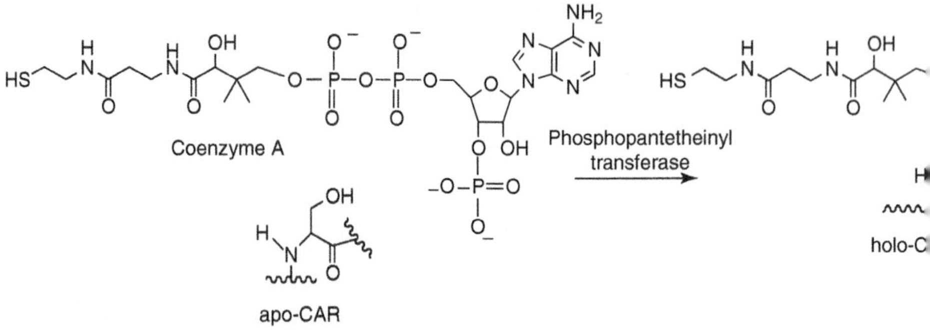

FIGURE 15.3 Phosphopantetheinylation mechanism catalyzed by Pptase.

proteins that participate in fatty acid biosynthesis, and in the biosynthesis of poly
like bleomycin and rifamycin, bacterial acyl oligosaccharides, and acylated proteins.
of the presence of a possible phosphopantethenylation site in the *car* ORF,
gested that *Nocardia* sp. NRRL 5646 contained a Pptase that converts apo-CAR t
holo-CAR [37].

Three hypotheses were thus postulated and examined:

1. Native CAR was more active because it was a phosphopantetheinylated holo-e
2. *Nocardia* contained a Pptase, which phosphopantetheinylated native apo-C
   produce active holo-CAR.
3. rCAR was predominantly an apo-enzyme because *E. coli* lacked a Pptase th
   ciently converts apo- to holo-enzyme.

If these ideas were correct, it would be possible to activate rCAR by incubat
purified rCAR "apo-enzyme" with crude *Nocardia* sp. NRRL 5646 CFE. In a
reaction, *Nocardia* CFE (~320 to 400 μg protein) was incubated with purified
(~2 nmol) in the presence of the Ppant donor CoA (1 mM) for 1 h at 28°C in a final
of 100 μL. Enzyme activity was measured by adding this enzyme mixture (50 μL) to a s
containing 50 mM Tris buffer (pH 7.5), 1 mM EDTA, 10 mM $MgCl_2$, 1 mM DTT, 1C
glycerol, 1 mM ATP, 0.2 mM NADPH, and 5 mM benzoate in a final volume of 1.4 n
rate of change of absorbance at 340 nm was measured as a means of assessing increase
specific activity of the enzyme preparation. The controls for this experiment were (i)
rCAR, (ii) rCAR + CoA, (iii) CFE + CoA, (iv) CAR, and (v) CFE. The various c
ations used with CFE, rCAR, and CoA are shown in Table 15.1 [37].

Complete reaction mixtures with *Nocardia* CFE increased rCAR specific activity f
The result suggests that *Nocardia* contains a Pptase that requires CoA for complete ac
of CAR. Addition of CoA (1 mM) to *E. coli* BL21-CodonPlus- ((DE3)RP)/pHAT-3
free lysate had no effect on the specific activity of CAR. This result confirmed our su
that *E. coli* expressing rCAR lacked, poorly expressed, or contained, a Pptase that
broadly functional.

In order to ensure that the activity increase observed was due to phosphopanteth
tion of CAR by a putative Pptase in *Nocardia* sp. NRRL 5646 and not due to an art
the CFE, we incubated rCAR with CFE from *E. coli* JM109/pUC8-*sfp*, which co
plasmid localized *sfp* gene encoding for the *Bacillus subtilis* Pptase Sfp. This Pptase h
used to convert apo carrier enzyme to holo form [34,36]. When rCAR was incubat

TABLE 15.1

Specific Activity Measurements of Recombinant Carboxylic Acid Reductase (CAR) When Incubated with *Nocardia* Cell-Free Extract (CFE) and Coenzyme A (CoA) for 1 h at 28°C

Conditions	Specific Activity (U/mg)
Nocardia CFE	0.01 ± 0.006
Recombinant CAR	0.3 ± 0.04
Recombinant CAR + CoA	0.3 ± 0.04
Nocardia CFE + CoA	0.01 ± 0.003
Nocardia CFE + recombinant CAR	0.57 ± 0.06
Nocardia CFE + recombinant CAR + CoA	1.6 ± 0.1

CFE from *E. coli* JM109/pUC8-*sfp* and CoA, a similar fivefold increase in specific activity was observed (data not shown) [37].

Benzoate induction increases the expression of wild-type CAR in *Nocardia* sp. *Nocardia* sp. NRRL 5646 CFEs obtained from cells with and without benzoate were equally effective at enhancing rCAR specific activity. This suggested that a putative *Nocardia* Pptase is constitutively expressed and may be involved in activation of other apo proteins in *Nocardia* sp. NRRL 5646 as are the "promiscuous" Pptase Sfp from *B. subtilis* [34,35] and Svp from *Streptomyces verticillus* [36].

15.4 A REVISED MECHANISM FOR HOLO-CAR REDUCTION OF CARBOXYLIC ACIDS

These results suggest that CAR reduces carboxylic acids by a more complex "swinging phosphopantetheinyl arm" process such as that illustrated in Figure 15.4 [37]. Not shown is the initial binding of both benzoic acid and ATP at the CAR N terminus where the formation of an acyl-adenylate occurs as shown in Figure 15.4a. The cystiene residue of the Ppant moiety of the holo-CAR reacts with benzoyl-AMP to form benzoyl thioester, which is covalently attached to the enzyme (Figure 15.4b). The thioester tethered to the enzyme is brought into the proximity of the C-terminal domain (Figure 15.4c) where hydride attack from NADPH reduces the thioester, yielding free benzaldehyde, $NADP^+$, and a free Ppant-sulfhydryl group for another catalytic cycle (Figure 15.4d).

15.5 APPLICATIONS OF CAR

Chemists continuously explore means for improving chemical processes to become both economically viable and environmentally sustainable. This involves selective catalysis, using more viable sources of raw material, and modifying synthetic procedures. The introduction of biological systems in traditional chemical synthesis provides the added advantages of catalyst versatility, regio-, chemo- and enantioselectivities, and catalysis at ambient temperature and pressure [38]. Highly selective biocatalytic reductions of carboxylic acids open new avenues for producing value-added chemicals, which are currently derived either from petroleum feedstocks or by chemical synthetic means. These range from vanillin to chiral aldehydes and alcohols obtained by kinetic resolution for the pharmaceutical, food, and agricultural industries.

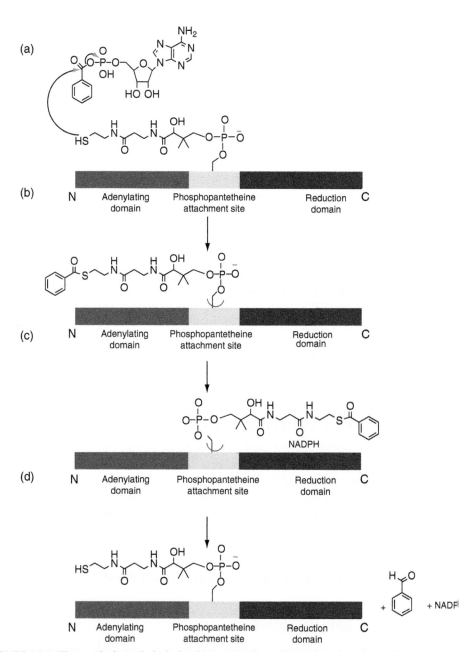

FIGURE 15.4 The catalytic cycle in holo-CAR reduction of benzoic acid to benzaldehyde.

15.6 CAR IS VERSATILE

Reduction of carboxylic acids can be accomplished by growing cultures of wild-type *Nc* sp. NRRL 5646 or heterologously expressed *car* in *E. coli* as well as by purified enzyme examined the reduction of a range of aromatic carboxylic acids and a variety of al acids. The range of oxidoreductase activities of crude *Nocardia* sp. NRRL 5646 CF different substrates is summarized in Table 15.2. Benzoates substituted with halogens, **r** methoxy, hydroxy, acetyl, nitro, benzoyl, phenyl, and phenoxyl groups as well as ar

TABLE 15.2
Substrate Specificity of Carboxylic Acid Reductase (CAR) Obtained from Different Sources

Substrate	*Nocardia* sp. NRRL 5646	*Nocardia asteroides* JCM 3016	*Neurospora crassa*
Benzoic acid	100	100	100
2-Anisic acid	3	—	—
3-Anisic acid	87	—	53
4-Anisic acid	36	—	—
2-Chlorobenzoic acid	3	0	—
3-Chlorobenzoic acid	124	97	—
4-Chlorobenzoic acid	100	78	—
2-Hydroxybenzoic acid	0	0	—
3-Hydroxybenzoic acid	77	47	24
4-Hydroxybenzoic acid	6	16	11
2-Nitrobenzoic acid	0	0	—
3-Nitrobenzoic acid	0	22	—
4-Nitrobenzoic acid	0	0	—
2-Toluic acid	6	6	—
3-Toluic acid	4	150	—
4-Toluic acid	82	103	—

Relative rates of substrate reductions were determined by measuring the ΔOD at 340 nm as NADPH was oxidized to $NADP^+$ during the reduction reaction.

systems containing two (naphthalene) and three (fluorene) rings, and heterocyclic aromatic acids including furoic, nicotinic, and indole carboxylic acids were examined [30]. In addition, cinnamate, phenylacetate, phenylmalonate, phenylsuccinate, and 2-phenylpropionic acids were examined. Except for fluoro- and methyl-substituted benzoic acids, 3-substituted benzoic acids (bromo, chloro, hydroxyl, methoxyl) were the best substrates within their respective aryl-carboxylic acid series. The range of aromatic substrates used by CAR is similar to that observed by Kato et al. [26]. In general, *ortho*-substituted benzoates were the poorest substrates from within any of the substrates compared. The CFE from *Nocardia* efficiently reduced naphthoic acids, but only reduced indole-3- and indole-5-carboxylic acids. In addition, CFEs reduced furoic acids, nicotinic acid and phenylmalonate, phenylsuccinate, phenylacetate and phenylpropionate, albeit at slower rates than benzoate itself. Of all the compounds examined, the best substrates were benzoic acid, 3-bromobenzoic acid, 3-chlorobenzoic acid, 4-fluorobenzoic acid, 4-methylbenzoic acid, 3-methoxy-benzoic acid, and 2-naphthoic acid. Nitrobenzoates were not reduced at all.

rCAR also reduced a number of natural mono-, di-, and tricarboxylic acids (Table 15.3) [37]. Rates of reduction of D-tartaric acid, α-ketoglutaric acid, and *cis*- and *trans*-aconitic, citric, and malic acids were higher than those for benzoic acid. Regiospecificity in carboxylic acid reduction remains to be established for citric, tartaric, and malic acids. rCAR in D-tartaric acid specifically offered the possibility of resolving tartaric acid racemates. Pyruvic, isocitric, fumaric, and maleic acids were not substrates, indicating that carboxyl moieties adjacent to sp^2 carbon centers might be less susceptible to rCAR reduction.

15.7   CAR IS STEREOSELECTIVE

The 2-arylpropionic acid derivative ibuprofen (**1**, Figure 15.5) is a potent, oral nonsteroidal antiinflammatory, antipyretic, and analgesic agent [39,40]. Pharmacological activities of

TABLE 15.3
Substrate Specificity of rCAR

Substrate	Relative A
Benzoic acid	10(
cis-Aconitic acid	13(
trans-Aconitic acid	13(
Citric acid	12(
Fumaric acid	(
3-Hydroxy-3-methylglutaric acid	4(
Isocitric acid	
2-Ketoglutaric acid	16(
DL-Lactic acid	5(
L-Lactic acid	5(
Maleic acid	
DL-Malic acid	10
D-Malic acid	10.
L-Malic acid	13(
Oxaloacetic acid	8(
Pyruvic acid	
D-Tartaric acid	12
L-Tartaric acid	
Tricarballylic acid	8.

Relative rates of substrate reductions were determined by measuring the ΔOD at 340 nm as NADPH was o:
to NADP$^+$ during the reduction reaction.

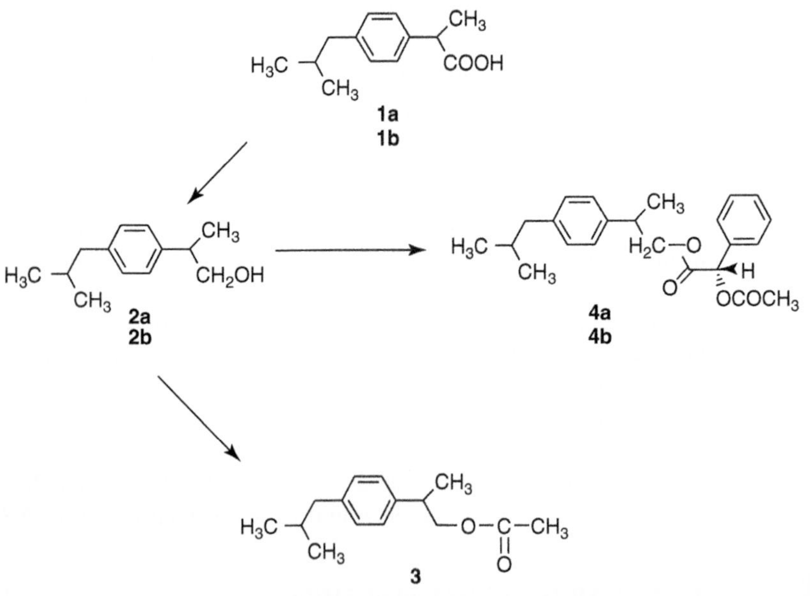

FIGURE 15.5 Schemes for ibuprofen, metabolites, and derivatives: **1a**, racemic ibuprofen; **1** ibuprofen; **2a**, racemic ibuprofenol; **2b**, S(+) ibuprofenol **3**, ibuprofen acetate; **4a**, S(+)-O-ace delate ester of compound **2a**; **4b**, S(+)-O-acetylmandelate ester of **2b**.

ibuprofen reside almost exclusively in its S-enantiomer [41]. Ibuprofen was an attractive substrate candidate for investigation because of its overall importance as a widely used drug, and the opportunity to examine the possible stereoselective properties of the CAR reaction.

With growing cultures of *Nocardia* sp. NRRL 5646 two major metabolites of racemic ibuprofen were formed, isolated, and identified as ibuprofenol (**2**) and the corresponding acetate derivative (**3**) by spectral methods (Figure 15.5) [20]. An unoptimized preparative scale reaction conducted with (+/−)-ibuprofen was stopped at 24 h when TLC analysis indicated that approximately 50% of the substrate had been converted into other products. Unreacted and recovered ibuprofen was largely $S(+)$-ibuprofen (**1b**, Figure 15.5). The enantiomeric excess (ee) of $R(+)$-ibuprofenol obtained from the biotransformation of racemic ibuprofen was determined to be 61.2% by NMR spectral analysis of the $S(+)$-O-acetylmandelate ester (Figure 15.6). This finding showed that whole-cell *Nocardia* sp. NRRL 5646 reactions were $R(−)$-selective with (+/−)-ibuprofen (**1a**, Figure 15.5), and offered the possibility that other racemic carboxylic acids can be resolved.

The availability of pure CAR from *Nocardia* sp. NRRL 5646 allowed the evaluation of enantioselectivity of the reduction of ibuprofen isomers [30]. The K_m and V_{max}, respectively, for (S)-$(+)$-ibuprofen were 155 ± 18 μM and 0.148 ± 0.003 μmol/min/mg. By comparing V_{max}/K_m ratios of carboxylic acid reduction for the two isomers, the enantiomeric ratio was calculated to be 40.4 for the (R)-$(−)$-isomer over the (S)-$(+)$-isomer [42]. Using these data, theoretically, the reduction of ibuprofen by the pure enzyme should give an ee of 86.9% at 50% conversion yield [42]. This compares favorably with the observed ee of 61.2% for the isolated and unoptimized reduction of racemic ibuprofen [20].

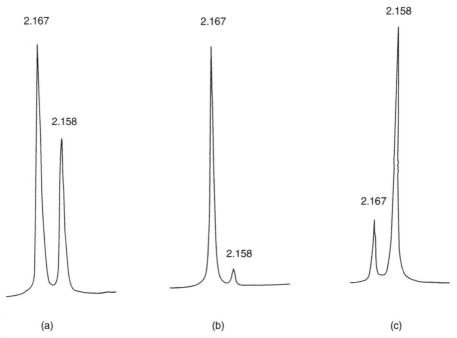

FIGURE 15.6 ^{1}H NMR signals of 600 Mhz, for the acetyl esters of **4a** $S(+)$-O-acetylmandelate ester of: (a) racemic ibuprofenol; (b) $S(−)$-ibuprofenol; and (c) $R(+)$-ibuprofenol from microbial transformation of racemic ibuprofen.

While it may appear disadvantageous for whole cells to reduce carboxylic a
aldehydes, alcohols, and further esterify them to the corresponding acetates, such a c
ation of reactions may be highly desirable, for example, in the conversion of ca
substrates into esters of value or fragrances and flavors.

15.8 SYNTHESIS OF VALUE-ADDED CHEMICALS

15.8.1 VANILLIN

Natural vanilla obtained from the dried pods of the orchid *Vanilla planifolia* accounts f
0.2% of the world flavor market (20 t/y out of 1.2×10^5 t/y) and the remainder is
mented by chemical synthesis of vanillin (6) from guaiacol (8) [43]. Vanillin ha
produced by microbial transformation from natural substrates, including phenolic st
[44,45], eugenol [46], and ferulic acid [47,48].

In whole-cell *Nocardia* biotransformation reactions, vanillic acid (5) decarboxyla
guaiacol (8) was the major complicating pathway [49,50]. The identification of g
(8) and vanillyl alcohol (7) as metabolites confirmed that *Nocardia* sp. strain NRR
possesses two different metabolic pathways for the biotransformation of vanillic a
These pathways are (i) decarboxylation to guaiacol (8) and (ii) reduction to vanillin (6)
subsequent reduction to vanillyl alcohol (7) (Figure 15.7).

With pure *Nocardia* CAR, the ATP- and NADPH-dependent reduction of vanill
was quantitative, yielding only vanillin and no complicating by-products [51]. In 24 h
reactions containing 0.1 mmol of substrate and rCAR, preparative TLC gave 7.5
vanillin (48% yield). The difference between the yield in the *in vitro* reaction of pure *N*
CAR and rCAR may be due to the differences in their specific activities [33]. With
biotransformations of vanillin with *E. coli* BL21-CodonPlus(DE3)-RP harboring p
pHAT-305, vanillin was observed at 2 h and was converted to the corresponding alco
24 h [33].

15.8.2 FERULIC ACID

Ferulic acid [3-(4-hydroxy-3-methoxyphenyl)-propenoic acid] (9) is an extremely ab
plant product available from corn kernel hulls obtained from wet milling [52,53]. Vanil

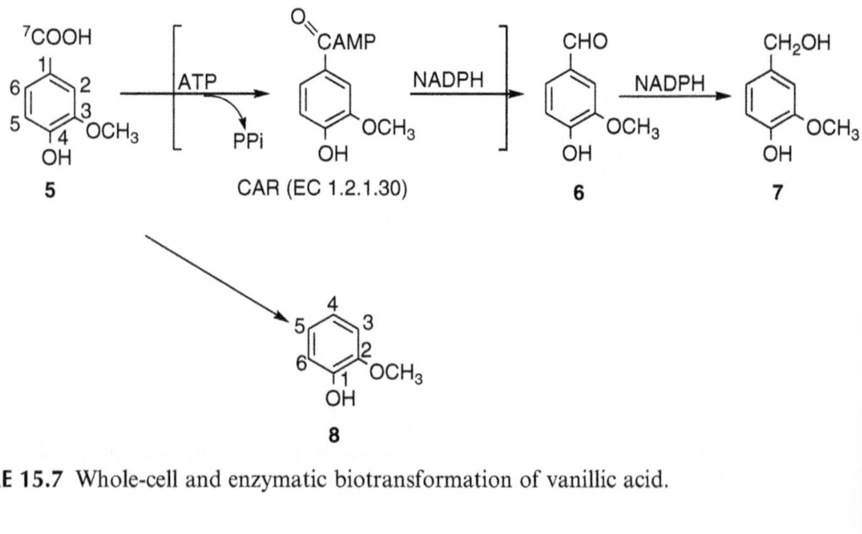

FIGURE 15.7 Whole-cell and enzymatic biotransformation of vanillic acid.

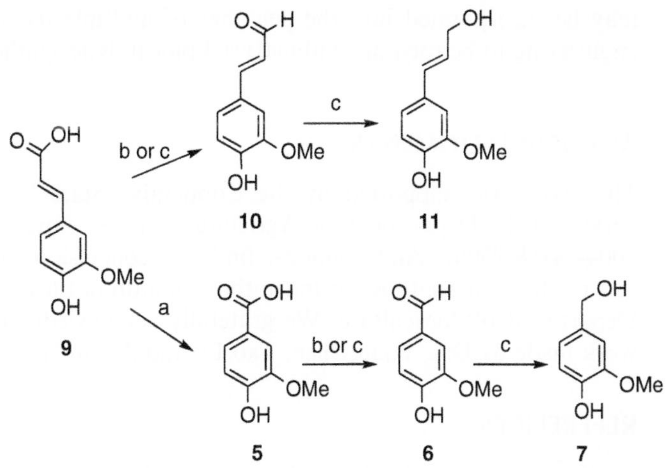

FIGURE 15.8 Reduction of ferulic acid to value-added chemicals: (a) microbial oxidation by strains of *Bacillus, Pseudomonas, Polyporus, Rhodotorula*, or *Streptomyces*; (b) Purified recombinant CAR; (c) *E. coli* BL21-CodonPlus-((DE3)RP)/pHAT-305.

is the major product obtained by oxidation of ferulic acid by species of *Bacillus* [54], *Pseudomonas* [55,56], *Polyporus* [57], *Rhodotorula* [58], and *Streptomyces* [59]. Thus, vanillic acid is an abundant, readily available precursor for the biocatalytic synthesis of vanillin using the AAOR. So in a two-step/single-pot process, one can envision a simple route to the preparation of vanillin from ferulic acid (Figure 15.8).

In vitro reaction of ferulic acid with rCAR afforded a smooth reduction of ferulic acid to coniferyl aldehyde (**10**) [33]. The current market price of coniferyl aldehyde is approximately $8,000/lb. Incubation of ferulic acid with growing cultures of *E. coli* BL21-CodonPlus-((DE3)RP)/pHAT-305 expressing CAR led to the production of coniferyl alcohol (**11**). The current market price of coniferyl alcohol is approximately $109,000/lb. Such reductions of the carboxyl moiety of ferulic acid to its aldehyde and alcohol were not observed when ferulic acid was incubated with resting cells of *Nocardia* sp. NRRL 5646.

15.9 CONCLUSIONS

The heterologous expression of CAR in *E. coli* provides a new avenue for the production of value-added chemical like vanillin, coniferyl aldehyde, and their alcohols that have a large interest in the pharmaceutical, food, and agricultural industries. Biotransformation reactions using IPTG-induced whole *E. coli* BL21(DE3)CodonPlus cells harboring plasmid pHAT-305 were simple to conduct and resulted in the smooth conversion of carboxylic acids to aldehydes and subsequently to alcohols. With whole cells, expensive cofactors are not necessary [33,51], rendering the biocatalytic reaction more practical on large scales. Reduction of aldehydes formed by CAR to alcohols by an endogenous *E. coli* alcohol dehydrogenase similar to that observed in *Nocardia* [51] is relatively slow. Biochemical engineering approaches with the recombinant organism can be exploited to diminish this unwanted side reaction.

The unique *car* sequence for the CAR enzyme may be used to produce recombinant *E. coli* cultures which can be grown easily for direct use in whole-cell biocatalytic conversions of natural or synthetic carboxylic acids [30,60]. Alternatively, this gene sequence or its homologs may be incorporated into the genomes of multiple recombinant strains through engineering to be used in combinatorial biocatalytic syntheses of useful compounds.

ACKNOWLEDGMENTS

This work was supported by the Cooperaive State Research, Education and Ex Service, U.S. Department of Agriculture, under Agreement Nos. 2002-34188-12(2004-34188-15067. Any opinions, findings, conclusions, or recommendations expre this publication are those of the authors and do not necessarily reflect the views of t Department of Agriculture. We gratefully acknowledge the scientific contributions work made by Drs. Yijun Chen, Tao Li, and Aimin He.

REFERENCES

1. Davies, H.G. et al., *Biotransformations in Preparative Organic Chemistry: the Use of Enzymes and Whole Cell Systems in Synthesis*, Academic Press, New York, 1989.
2. Hanc, O.F., The microbial transformation of organic compounds, *J. Ind. Chem. Soc.*, : 1975.
3. Jones, J.B., Perlman, D., and Sih, C.J., Eds., *Applications of Biochemical Systems in Pr Organic Chemistry*, vols. I and II, Wiley, New York, 1976.
4. Kieslich, K., Preparative anwendbare mikrobiologische reaktionen. zweiter teil. redu hydrolytische reaktionen, acylierungen, und kondensationen, aminierungen, decarboxyli *Synthesis*, 1, 147, 1969.
5. Leuenberger, H.G.W., Biotransformation: a useful tool in organic chemistry, *Pure Appl. Ci* 753, 1990.
6. Porter, R. and Clark, S., Eds., *Enzymes in Organic Synthesis*, CIBA Foundatation Sympo Pitman Publishers, London, 1985.
7. Roberts, S.M. and Wiggins, D.K., *Preparative Transformations: Whole Cell and Isolated En Organic Synthesis. A Core Manual*, Wiley, New York, 1992.
8. Liu, W.G. et al., Stereochemistry of microbiological hydroxylations of 1,4-cineoles, *J. Org* 53, 5700, 1988.
9. Sariaslani, F.S. and Rosazza, J.P.N., Biocatalysis in natural products chemistry, *Enzyme. Technol.*, 6, 242, 1984.
10. Sih, C.J. and Abushanab, E., Biochemical procedures in organic synthesis, *Ann. Rep. Mea* 12, 298, 1977.
11. Sih, C.J. and Rosazza. J.P.N., Microbial transformations in organic synthesis, Chap. III, in *tions of Biochemical Systems in Organic Chemistry*, Part I, Jones, J.B., Ed., Wiley, New Yo p. 69.
12. Ward, O.P. and Young, C.S., Reductive biotransformations of organic compounds by enzymes of yeast, *Enzyme. Microb. Technol.*, 12, 482, 1990.
13. Jezo, I. and Zemek, J., Enzymatische reduktion einiger aromatischer carboxysäuren, *Che* 40(2), 279, 1986.
14. White, H. et al., Carboxylic acid reductase, a new tungsten enzyme catalyses the reduction activated carboxylic acids to aldehydes, *Eur. J. Biochem.*, 184, 89, 1989.
15. Arfman, H.A. and Abraham, W.R., Microbial reduction of aromatic carboxylic acids, *Z. Natt* 48(c), 52, 1993.
16. Raman, T.S. and Shanmugasundaram, E.R.B., Metabolism of some aromatic acids by *As niger*, *J. Bacteriol.*, 84, 1340, 1962
17. Bachman, D.M., Dragoon, B., and John, S., Reduction of salicylate to saligenin by *Net Arch. Biochem. Biophys.*, 91, 326, 1960.
18. Tsuda, Y., Kawai, K., and Nakajima, S., Asymmetric reduction of 2-methyl-2-aryloxyace by *Glomerella cingulata*, *Agric. Biol. Chem.*, 48(5), 1373, 1984.

19. Tsuda, Y., Kawai, K., and Nakajima, S., Microbial reduction of 2-phenylpropionic acid, 2-benzyloxypropionic acid and 2-(2-furfuryl)propionic acid, *Chem. Pharm. Bull.*, 33(11), 4657, 1985.

20. Chen, Y. and Rosazza, J.P.N., Microbial transformation of ibuprofen by a *Nocardia* species, *Appl. Environ. Microbiol.*, 60(4), 1292, 1994.

21. Kato, N. et al., Microbial reduction of benzoate to benzyl alcohol, *Agric. Biol. Chem.*, 52(7), 1885, 1988.

22. Gross, G.G. and Zenk, M.H., Reduktion aromatischer säuren zu aldehyden und alkoholen im zellfreien system. 1. Reinigung und eigenschaften von aryl-aldehyde, NADP-oxidoreduktase aus *Neurospora crassa*, *Eur. J. Biochem.*, 8, 413, 1969.

23. Gross, G.G. and Zenk, M.H., Reduktion aromatischer säuren zu aldehyden und alkoholen im zellfreien system. 2. Reinigung und eigenschaften von aryl-alkohol, NADP-oxidoreduktase aus *Neurospora crassa*, *Eur. J. Biochem.*, 8, 420, 1969.

24. Gross, G.G., Formation and reduction of intermediate acyl-adenylate by aryl-aldehyde NADP oxidoreductase from *Neurospora crassa*, *Eur. J. Biochem.*, 31, 585, 1972.

25. Hempel,J., Nicholas, H. and Lindahl, R., Aldehyde dehydrogenases: widespread structural and functional diversity within a shared framework, *Protein Sci*; 2(11), 1890, 1993.

26. Kato, N. et al., Purification and characterization of aromatic acid reductase from *Nocardia asteroides* JCM 3016, *Agric. Biol. Chem.*, 55(3), 757, 1991.

27. Clonis, Y.D., Dye-ligand chromatography, in *Reactive Dyes in Protein and Enzyme Technology*, Clonis, Y.D. et. al., Eds., Stockton Press, New York, 1989, p. 33.

28. Smith, M.B. and March, J., *Advanced Organic Chemistry*, 5th ed., Wiley, New York, 2001, p. 532.

29. Li, T. and Rosazza, J.P.N., NMR identification of an acyl-adenylate intermediate in the aryl-aldehyde oxidoreductase catalyzed reaction, *J. Biol. Chem.*, 273, 34230, 1998.

30. Li, T. and Rosazza, J.P.N., Purification, characterization, and properties of an aryl aldehyde oxidoreductase from *Nocardia* sp. strain NRRL 5646, *J. Bacteriol.*, 179, 3482, 1997.

31. Mulder, M.A., Zappe, H., and Steyn, L.M., Mycobacterial promoters, *Tuber. Lung Dis.*, 78, 211, 1997.

32. Strohl, W.R., Compilation and analysis of DNA sequences associated with apparent streptomycete promoters, *Nucleic Acids Res.*, 20, 961, 1992.

33. He, A. et al., *Nocardia* sp. carboxylic acid reductase, cloning, expression, and characterization of a new aldehyde oxidoreductase family, *Appl. Environ. Microbiol.*, 70, 1874, 2004.

34. Lambalot, R.H. et.al., A new enzyme superfamily: the phosphopantetheinyl transferases, *Chem. Biol.*, 3, 923, 1996.

35. Sanchez, C. et al., Identification and characterization of a type II peptidyl carrier protein from the bleomycin producer *Streptomyces verticillus* ATCC 15003, *Chem. Biol.*, 8, 725, 2001.

36. Quadri, L.E. et al., Characterization of Sfp, a *Bacillus subtilis* phosphopantetheinyl transferase for peptidyl carrier protein domains in peptide synthetases, *Biochemistry*, 37, 1585, 1998.

37. Venkitasubramanian, P. and Rosazza, J.P.N., unpublished results, 2006.

38. Thomas S.M., DiCosimo, R., and Nagarajan, V., Biocatalysis, applications and potentials for the chemical industry, *Trends Biotechnol.*, 20(6), 238, 2002.

39. Simon, L. and Mills, J., Non-steroidal anti-inflammatory drugs, *New Engl. J. Med.*, 302, 1179, 1980.

40. Kantor, T.G., Ibuprofen, *Ann. Intern. Med.*, 91, 877, 1979.

41. Geissinger, G. et al., Pharmacological differences between $R(-)$ and $S(+)$ibuprofen, *Agents Actions*, 27, 455, 1989.

42. Chen, C.S. et al., Quantitative analysis of biochemical kinetic resolutions of enantiomers, *J. Am. Chem. Soc.*, 104, 7294, 1982.

43. Esposito, L. et al., in *Kirk–Othmer Encyclopedia of Chemical Technology*, 4th ed., Kroschwitz, J. I. and Howe-Grant, M., Eds., Wiley, New York, vol. 24, 1997, p. 812.

44. Clark, G.S., Vanillin, *Perfum. Flavor.*, 15, 45, 1990.

45. Priefert, H., Rabenhorst, J., and Steinbuchel, A., Biotechnological production of vanillin, *Appl. Microbiol. Biotechnol.*, 56, 296, 2001.

46. Yoshimoto, T. et al., Dioxygenase for styrene cleavage manufactured by *Pseudomonas*, Jap. Patent No. 2,195,871, 1990.

47. Washisu, Y. et al., Manufacture of vanillin and related compounds with *Pseudomonas*, Jap. Patent No. 5,227,980, 1993.

48. Rabenhorst, J. and Hopp, R., Process for the preparation of vanillin, U.S. Patent 5,017,38

49. Mulheim, A. and Lerch, K., Towards a high-yield bioconversion of ferulic acid to vanilli *Microbiol. Biotechnol.*, 51, 456, 1999.

50. Labuda, I.M., Goers, S.K., and Keon, K.A., Bioconversion process for the production of U.S. Patent 5,128,253, 1992.

51. Li, T. and Rosazza, J.P.N., Biocatalytic synthesis of vanillin, *Appl. Environ. Microbiol.*, 2000.

52. Huang, Z., Dostal, L., and Rosazza, J.P.N., Mechanisms of ferulic acid conversions to vani and guaiacol by *Rhodotorula rubra*, *J. Biol. Chem.*, 268, 23954, 1993.

53. Rosazza, J.P.N. et. al., Biocatalytic transformation of ferulic acid, an abundant aromatic product, *J. Ind. Microbiol.*, 15, 457, 1995.

54. Gurujeyalakshmi, G. and Mahadevan, A., Dissimilation of ferulic acid by *Bacillus subtili Microbiol.*, 16, 69, 1987.

55. Jurková, M. and Wurst, M., Biodegradation of aromatic carboxylic acids by *Pseudomon FEMS Microbiol. Lett.*, 111, 245, 1993.

56. Toms, A. and Wood, J.M., The degradation of *trans*-ferulic acid by *Pseudomonas acid Biochemistry*, 9, 337, 1970.

57. Ishikawa, H., Schubert, W.J., and Nord, F.F., The degradation by *Polyporus versicolor* and *fomentarius* of aromatic compounds structurally related to softwood lignin, *Archiv. B Biophys.*, 100, 140, 1970.

58. Mulheim, A. and Lerch, K., Towards a high-yield bioconversion of ferulic acid to vanilli *Microbiol. Biotechnol.*, 51, 456, 1999.

59. Sutherland, J.B., Crawford, D.L., and Pometto, A.L. III, Metabolism of cinnamic, *p*-couma ferulic acids by *Streptomyces setonii*, *Can. J. Microbiol.*, 29, 1253, 1983.

60. Rosazza, J.P.N. and Li, T., Purification characterization and properties of an aryl-aldehyde eductase from *Nocardia* sp. strain NRRL 5646. U.S. Patent 579,559, 1998.

16 Dehalogenases in Biodegradation and Biocatalysis

Dick B. Janssen

CONTENTS

16.1 INTRODUCTION

The first descriptions of dehalogenases date back to the 1960s. Davies and Evans [1] and Goldman [2] described enzymes that dehalogenate fluoroacetate, and Castro and Bartnicki [3] studied an enzyme that releases bromide from 2,3-dibromo-1-propanol. Dehalogenases were initially classified on the basis of their substrate range. However, genetic analysis has revealed that dehalogenases acting on similar substrates (e.g., fluoroacetate and chloroacetate) may belong to unrelated protein families, whereas dehalogenases that act on rather different substrates (e.g., fluoroacetate and 1,2-dibromoethane) may belong to the same phylogenetic superfamily. Fluoroacetate, 1,2-dichloroethane, 1,3-dichloropropene, and tetrachlorocyclohexadiene are dehalogenated by dehalogenases that have low sequence similarity, but all four belong to the α/β-hydrolase fold family. Most 2-chlorocarboxylic acid dehalogenases that

have been studied belong to the HAD superfamily of dehalogenases–phosphatase:
16.1, L-DEX, DhlB) [4], but there are also haloacid dehalogenases such as the DL-2-▮
dehalogenase from *Pseudomonas* sp. 113 [5], that are not related to the HAD supe
and form a separate group of enzymes, some of which are specific for D-chlorop▮
acid. The phylogenetic classification, as used in Table 16.1, is now replacing
substrate-based classification in most studies, and the majority of haloacid dehalc
can be grouped into two families [4].

The reaction types catalyzed by dehalogenases and the cofactors that these enzy
are highly diverse (Table 16.1). Several groups of dehalogenases catalyze a simple hy
reaction without the use of cofactors, prosthetic groups, or metal ions. This includes ▮
ane dehalogenases and haloacid dehalogenases. Other dehalogenases, especially thos
on halogens bound to an aromatic ring, have completely different mechanisms, often
ing cofactors such as glutathione or NADPH. For example, the dehalogena
4-chlorobenzoic acid occurs after activation by coupling of the compound to coen
(CoA) and is catalyzed by a dehalogenase (Table 16.1, CbzA) that produces 4-hydroxy
CoA through a covalent substrate–enzyme intermediate. The CoA is released by a thiol:
dechlorination. The first step in the pentachlorophenol biodegradation pathway is cata
an NADPH-dependent monooxygenase (Table 16.1, PcpC) that removes a chlorin
para-position, whereas a reductive glutathione-dependent enzyme is involved in the
dechlorination step, in which tetrachlorohydroquinone is the substrate.

Reductive dehalogenases (e.g., PceA in Table 16.1) use an external electron dor
replace a halogen substituent by hydrogen, releasing halide. The reductive dehalc
have drawn considerable attention because of their role in the reductive dechlorin.
notorious environmental pollutants such as PCBs, chlorobenzenes, and especiall
oethenes. Tetrachloroethene, trichloroethene, and other chloroethenes can be sequ
dechlorinated by dehalogenases that are produced by anaerobic bacteria, which
chlorinated compounds as their physiological electron acceptor. The bacterium *De*
coides ethenogenes, which can reductively dechlorinate tetrachloroethene all the
ethene, possesses 17 putative reductive dehalogenase genes. These genes show the ▮
of two iron–sulfur cluster binding motifs that also occur in ferredoxins, as well as co
tryptophan and histidine residues that may play a role in catalysis, but it is not kno
the cobalamin cofactor that is involved in dechlorination is bound to the enzyn
structures are known for these important proteins [6].

An intriguing characteristic of many dehalogenases is their capacity to bind and
xenobiotic compounds that have a structure which does not naturally occur in the bi
Although thousands of naturally occurring organohalogens have been identified, in
some structures that were rather unexpected, such as chloroethenes [7], it is likely th:
of the compounds that have been the subject of detailed biodegradation studies did n
on earth in biologically significant concentrations before the onset of their industrial
tion some 100 years ago. Thus, compounds such as 1,2-dichloroethane, tetrachloroe
1,3-dichloropropylene, epichlorohydrin, hexachlorocyclohexane, and pentachlor
only entered the biosphere and challenged microbial evolution after they were syn'
and applied by man. This raises the question whether the dehalogenases that are now f
convert these xenobiotics already existed in preindustrial times, in which case they p
have played a role in the metabolism of natural compounds that carry a halogen sub
or a substituent that is biochemically similar to a halogen group, or whether they ev
short evolutionary pathways from preindustrial enzymes that acted on unidentified
compounds. In the latter case, the evolutionary predecessor could again have been a
genase that converts a natural organohalogen, or an enzyme that acts on a nonhalo
compound and only acquired dehalogenase activity after some mutations.

TABLE 16.1
Diversity of Dehalogenases, Their Reaction Types, Phylogenetic Grouping and Key Mechanistic Features

Type (example)	Reaction with Xenobiotic Compound	Family/Cofactor	Mechanism
Haloalkane dehalogenase (DhlA, DhaA, LinB)	*(chemical reaction scheme)*	α,β-Hydrolase fold, similar to lipases	Catalytic triad, Asp as nucleophile, distinct halide-binding site, covalent intermediate [16, 17]
Halocarboxylic acid dehalogenase (L-DEX, DhlB)	*(chemical reaction scheme)*	Haloacid dehalogenase–phosphatase fold (HAD family)	Catalytic Asp close to N terminus as the nucleophile, covalent intermediate [47, 48]
Halohydrin dehalogenase (HheA, HheB, HheC)	*(chemical reaction scheme)*	Short-chain dehydrogenase/reductase family (SDR proteins)	Catalytic triad for H$^+$ abstraction, halide binding site, noncovalent catalysis [56]
Chloroacrylic acid dehalogenase (CaaD)	*(chemical reaction scheme)*	4-Oxalocrotonate tautomerase family (trimers or hexamers)	N-terminal nucleophilic Pro, no covalent intermediate, hydratase-like mechanism [58]
Hexachlorocyclohexane dechlorinase (LinA)	*(chemical reaction scheme)*	Scytalone dehydratase	Abstraction of an axial proton, with concomitant anti-elimination of a transaxial chloride from the adjacent carbon atom [80, 81]
Atrazine chlorohydrolase (AtzA)	*(chemical reaction scheme)*	Amidohydrolase superfamily	Activation by metal group [10]

continued

TABLE 16.1 (continued)
Diversity of Dehalogenases, Their Reaction Types, Phylogenetic Grouping and Key Mechanistic Features

Type (example)	Reaction with Xenobiotic Compound	Family/Cofactor	Mechanism
4-Chlorobenzoyl-CoA dehalogenase (CbzA)	*(structure: chlorobenzoyl-CoA + H$_2$O → HCl, hydroxybenzoyl-CoA)*	Enoyl hydratase superfamily	Asp as nucleophile, covalent intermediate [82]
Pentachlorophenol dehalogenase (PcpA)	*(structure: pentachlorophenol + O$_2$+NADPH → HCl+H$_2$O+ NADP$^+$, tetrachloroquinone)*	NADPH-dependent flavin monooxygenase, para-hydroxybenzoate hydroxylase group	Oxygen substitution at the para-position, elimination of halide from unstable product [83]
Tetrachlorohydroquinone reductive dehalogenase (PcpC)	*(structure: tetrachlorohydroquinone + 2GSH → HCl + GSS, trichlorohydroquinone)*	Zeta class of the glutathione S-transferase superfamily	Nucleophilic attack of GSH on substrate, displacement of GSH by Cys to form a GS-Cys heterodisulfide, release of GSH by second GSH [84, 85]
Tetrachloroethene reductive dehalogenase (PceA)	*(structure: tetrachloroethene + XH$_2$ → X, trichloroethene)*	Reductive dechlorination, cobalamin-containing enzyme	Either formation of organocobalt adduct or electron donation by corrinoid cofactor [6]
Methyltransferase I (CmuA), methyltransferase II (CmuB)	*(structure: H$_3$C–I + H$^+$+Co(I) → HI, Co(III)-CH$_3$ CmuA; THF Co(I) → THF=CH$_2$ CmuB)*	Methyltransferase domain fused to corrinoid-binding domain (CmuA), methyl transferase-like protein (CmuB)	Methyl group transfer, nucleophilic substitution of chlorine by Co [86]
Dichloromethane dehalogenase (DcmA)	*(structure: Cl–CH$_2$–Cl + GSH → HCl, Cl–SG DcmA; + H$_2$O → HCl, HO–SG DcmA; spont. → H$_2$C=O + GSH)*	Theta class of the glutathione S-transferase superfamily	Glutathione activation by Tyr, decomposition of adduct [87]

Evolutionary and mechanistic studies have indicated that both these options may occur, although data are very scarce. The *Xanthobacter autotrophicus* haloalkane dehalogenase (DhlA, Table 16.1) that converts 1,2-dichloroethane was proposed to have evolved from a debrominating enzyme by a few mutations in the cap domain that influence the activity, including generation of short tandem repeats that are found both in the wild-type enzyme and in derivatives with an enhanced rate of 1-chlorohexane hydrolysis [8,9]. Atrazine chlorohydrolase from *Pseudomonas* sp. strain adenosine diphosphate (ADP) (AtzA) may have evolved from a protein that acts on an amine-substituted triazine [10]. The enzyme, which acts as a hydrolytic dehalogenase and belongs to the amidohydrolase superfamily, differs by only nine amino acids in its sequence from melamine (2,4,6-triamino-1,3,5-triazine) deaminase (TriA), and just a few mutations suffice to switch the activity between deaminase and chlorohydrolase [11]. Another member of the amidohydrolase superfamily, AtzB, can catalyze both deamination and dechlorination of triazine derivatives, indicating that deamination and dechlorination indeed are mechanistically similar and can be catalyzed by the same protein scaffold.

Most dehalogenases appear to belong to protein superfamilies that mainly consist of enzymes that catalyze reactions with natural substrates. Examples are the HAD, the α/β-hydrolase fold, and the short-chain dehydrogenase/reductase (SDR) superfamilies, which are discussed in more detail later. A further example of such a superfamily that encompasses some dehalogenases is the 4-OT group of isomerases–tautomerases, of which 4-oxalocrotonate tautomerase is the classical example [12]. This group of trimeric or hexameric enzymes includes *cis*- and *trans*-3-chloroacrylic acid dehalogenases (CaaD, Table 16.1) [13]. *trans*-3-Chloroacrylic acid is remarkably easy to dehalogenate enzymatically in view of its extreme stability at high pH and temperature [14]. Apparently, the substrate and water are very effectively labilized along the reaction coordinate. This allows a reaction that, mechanistically, is addition of water to the double bond, which is catalyzed by the conserved N-terminal proline that deprotonates water to attack C3 and donates a proton to C2 [15].

Research on dehalogenases has been inspired by the possibility that they can play a role in the bioremediation of contaminated groundwater or wastewater and the certainty that they are important for biodegradation of halogenated pollutants in the natural environment. Dehalogenases appear to be attractive detoxification catalysts as they carry out the most critical reaction in the metabolism of halogenated organics: the cleavage of carbon–halogen bonds. Only a few dehalogenases have been explored for their use in biocatalysis, even though their capacity to recognize and convert synthetic chemicals obviously can have great potential for industrial biotechnology. This review will focus on the dehalogenases that can be applied in the production of fine chemicals, in particular on the chlorocarboxylic acid dehalogenases, which have been investigated because of their stereoselectivity, the haloalkane dehalogenases, which so far have mainly bioremediation potential, and the halohydrin dehalogenases, an emerging new class of highly interesting biocatalysts.

16.2 HALOALKANE DEHALOGENASES

16.2.1 PROPERTIES AND CATALYTIC MECHANISM

The biodegradation of chloro- and bromoalkanes often starts with a hydrolytic step in which the haloalkane is hydrolyzed to the corresponding alcohol. This occurs in the bacterial degradation of 1,2-dichloroethane, 1,3-dichloropropene, 1-chlorobutane, 1-chloro- and 1,6-dichlorohexane, chloroethylethers, and probably several other haloalkanes. All known haloalkane dehalogenases [16] are soluble enzymes of around 35 kDa and belong to the α/β-hydrolase fold superfamily of enzymes, which also includes lipases, esterases, hydrolases, carboxypeptidases, aminoester hydrolases, and epoxide hydrolases. The ha** dehalogenases are composed of two domains: the main domain, which has the co α/β-hydrolase fold topology, and a cap domain, which is comparable to the fle: of lipases.

Examples of well-studied haloalkane dehalogenases are DhlA from *X. autot*** [16,17], DhaA from *Rhodococcus erythropolis* [18], and LinB from *Sphingomonas pau*c [19,20]. The former two enzymes act mainly on simple haloalkanes such as 1,2-oethane, 1-chlorobutane, and 1,2-dibromoethane, whereas LinB converts tetrachlorocyclohexadiene, an intermediate in the catabolic pathway for the ins γ-hexachlorocyclohexane (lindane). The active site in these enzymes is located in a cavity between the two domains. The main domain contributes three catalytic resid* are positioned in the form of a triad (acid/His/Asp), ordered along the sequ Asp...acid...His. Structural and biochemical studies have revealed that the catalyti anism is based on the nucleophilicity of the conserved Asp residue, of which a car** oxygen displaces the halogen from the substrate by nucleophilic substitution, for covalent intermediate (Figure 16.1). This mechanistic information was originally o from an x-ray structure in which the covalent intermediate was trapped by soak crystals of DhlA with the substrate 1,2-dichloroethane at low pH [17]. The histidine is fully conserved in these dehalogenases, subsequently activates a water molecule that the carbonyl carbon of the ester intermediate, releasing the alcohol product (Figure 1€ other α/β-hydrolase fold enzymes, the nucleophilic Asp may be a Ser or Cys, like in e or lipases. The role of the acidic residue is to stabilize the positive charge developing histidine when the covalent intermediate is cleaved. It may be present at different top* positions and can be a Glu or an Asp. In DhlA, it is an Asp, located in the sequence € the cap domain, whereas in DhaA and LinB it is a Glu located on a strand that prec€ cap [21–23].

An intriguing feature of the haloalkane dehalogenases is the presence of a distinct binding site, which is essential for leaving group stabilization in the first half reactio**

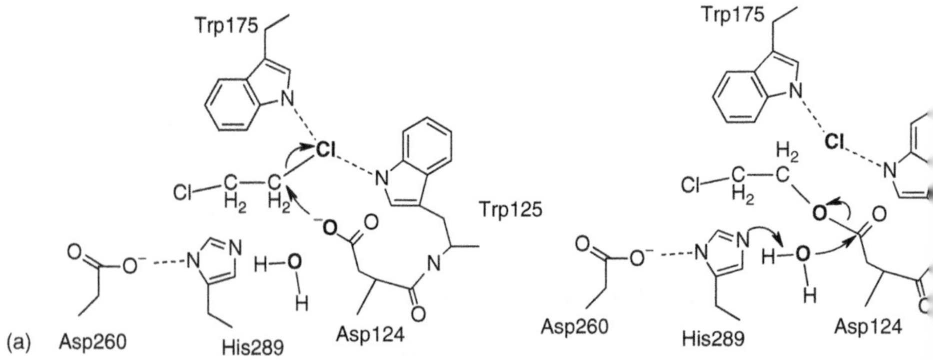

FIGURE 16.1 The catalytic mechanisms of three dehalogenases of different superfamilies. (a) Ha* dehalogenase (DhlA) from *X. autotrophicus* GJ10. (From Ref. [17,80].)

(c*

FIGURE 16.1 (continued) (b) Haloacid dehalogenase (HAD, DhlB) from *X. autotrophicus* GJ10. (From Ref. [47,48,51].) (c) Halohydrin dehalogenase (HheC) from *Agrobacterium radiobacter* AD1. (From Ref. [56,58,59].) (d) General mechanism of a short-chain dehydrogenase/reductase (SDR) enzyme.

the carbon–halogen bond is cleaved. The halide-binding site is composed of a Trp that flanks the nucleophilic Asp on the carboxy-terminal side and the side chain of a second residue, which is either contributed by the cap domain (Trp175 in DhlA) or by the main domain (Asn38 in LinB or Asn41 in DhaA). Sequence analysis has shown that catalytic residues are conserved in a much broader range of (putative) haloalkane dehalogenases [23]. In view of the conserved involvement in catalysis of five residues, the active site has been described as a catalytic pentad [16].

Several computational studies have confirmed that the catalytic mechanism that was revealed by x-ray crystallography for DhlA is energetically possible. They also provided further insight to the stabilizing interactions in the active site. For example, Bohac et al. [24]

concluded that electrostatic stabilization of the Trp residues is important for ca and emphasized the additional role of Phe172 in the stabilization of the halide in This was in agreement with a kinetic analysis of Phe172 mutants [25]. Such mutants reduced rate of carbon–halogen bond cleavage, but still show an elevated activit: for 1,2-dibromoethane conversion because halide release in the mutant has faster, and this step, at the end of the catalytic cycle, is the slowest step durii dibromoethane hydrolysis in the wild-type enzyme [26,27]. More recent studies were at describing the way in which haloalkane dehalogenase affects its remarkable ca power by comparing the reaction in the enzyme active site with that in water, and confirmed the importance of electrostatic stabilization [28] and positioning of the des reactant [29] close to the nucleophilic aspartate, but there is dispute about the most priate terminology and thermodynamic description of events that lead to transitic stabilization [28].

16.2.2 ENVIRONMENTAL RELEVANCE

The use of free or immobilized enzymes instead of live cells in bioreactors for the clea polluted water or soil does not seem to be attractive because of cost and limited stability under conditions that are difficult to control. Nevertheless, the general ca properties of dehalogenases have important environmental implications, since th influence the spectrum of halogenated compounds that is degraded during natural ation of soil and groundwater pollutants as well as the possibility to develop tre processes for the biological removal of halogenated compounds in reactors. Th environmental recalcitrance of compounds such as 1,2-dichloropropane and 1,1,2-tri ethane can be attributed to the fact that these organohalogens are hardly subject to mi dehalogenation. A treatment process for the removal of 1,2-dichloroethane has been oped and implemented successfully owing to the availability of pure bacterial cultur can rapidly degrade this compound and use it as a carbon source [30]. Thus, the disco new haloalkane dehalogenases with activity toward recalcitrant haloalkanes remains of considerable importance, as will be discussed for 1,2,3-trichloropropane.

16.2.3 PROTEIN ENGINEERING STUDIES

Several attempts have been made to improve the catalytic activity of haloalkane deha ase. Initial work, carried out with DhlA, showed that mutations in the cap doma enhance the activity for 1-chlorohexane [8]. The results showed that short tandem varying in length from 3 to 10 amino acids caused a 5- to 25-fold increase in activity (k with 1-chlorohexane, whereas the specificity constant for 1,2-dichloroethane decrease 60-fold. It was found that the activity for 1,2-dibromoethane and other bromo comp was much less influenced and the dehalogenation of bromoalkanes appears much less tive to mutations. Both with 1,2-dichloroethane and 1,2-dibromoethane, halide releas slowest step in the catalytic cycle, which is accompanied by large conformational rea ments in the cap domain [27]. For 1,2-dichloroethane, carbon–halogen bond cleavage i little bit faster than halide release. This causes loss of overall activity for 1,2-dichlor when halide release is made faster since the rates of carbon–halogen bond cleava halide release always appear to be inversely correlated. One can imagine that the requires strong binding in the active site with precise positioning of the substrate, v the latter requires flexibility, which allows the active site to open and become so A decrease of overall activity in mutants with a higher rate of halide release is not ob for 1,2-dibromoethane, since carbon–bromine bond cleavage is very rapid anyway and remains fast enough in most mutants. Several mutations that modify the selectivity of DhlA have been described [25,31,32].

Damborsky et al. [33] have used computational methods to explore dehalogenase selectivity. Using a comparative binding energy analysis of dehalogenases with a series of ligands that was automatically docked into the active site, the position of several substrates was calculated. It appeared that potentially reactive binding modes for most substrates could be found. With 1,2-dibromopropane, two binding modes were uncovered, one allowing attack on the α carbon, and the other on the β-bromine-substituted carbon. This showed that hydrolysis can take place at either carbon atom, which is in agreement with experimental data (D.B. Janssen and J. Kingma, unpublished data).

These engineering and modeling studies also provide information about possible targets for further mutagenesis. Chaloupkova et al. [34] compared the three available haloalkane dehalogenase structures and concluded that residue Leu177 of LinB, located at the entrance of the tunnel leading to the active site and partly blocking it, might be a good target for mutagenesis. Mutants in which a smaller residue was present generally showed an enhanced catalytic activity. Apparently, substrate access and binding significantly influence the selectivity constant (k_{cat}/K_m) of this enzyme for most substrates.

It has been possible to obtain mutants of haloalkane dehalogenase with enhanced temperature stability. This also holds for DhlA, where a computational approach was used in which molecular dynamics simulation identified a flexible region in the dehalogenase protein that likely is an early unfolding region. It was subsequently fixed by introducing a disulfide bond, yielding a more resistant mutant [35]. The temperature stability of DhaA was significantly improved by a directed evolution approach [36].

In spite of all the computational and modeling work, no haloalkane dehalogenase variant has yet been obtained that attacks really difficult compounds such as 1,1,2-trichloroethane or 1,2-dichloropropane. Also no DhlA variants have yet been described that exhibit significantly better conversion of the "natural" substrate 1,2-dichloroethane [31]. Attempts to engineer or select a dehalogenase for one such remaining problematic compound are described below.

16.2.4 CONVERSION OF 1,2,3-TRICHLOROPROPANE

The industrial synthesis of epichlorohydrin is accompanied by significant formation of side products of which 1,2,3-trichloropropane (TCP) is the most important one. Biodegradation and biotransformation studies with this chemical have been aimed at obtaining a dehalogenase enzyme or a whole-cells system that can be used for the hydrolytic conversion of 1,2,3-trichloropropane. Researchers at Dow Chemical, working with Diversa, and we at the University of Groningen, working with Ciba Fine Chemicals, have explored the development of dehalogenases that can degrade 1,2,3-trichloropropane to 2,3-dichloro-1-propanol, starting with almost identical dehalogenases and using a directed evolution approach. After random mutagenesis, either by site-saturation mutagenesis or error-prone polymerase chain reaction (PCR), mutants with increased activity toward TCP have been obtained by Bosma et al. [37], which allowed slow growth of a recombinant bacterium on TCP. One of the two mutations was a Cys176Tyr substitution and, surprisingly, a mutation at this position also occurs in a natural variant of DhaA that we have obtained from a 1,2-dibromoethane degrading *Mycobacterium* strain [38]. Position Cys176 of DhaA corresponds to Leu177 in LinB, and its role in determining the selectivity of that enzyme has already been mentioned [34].

Gray et al. [39] have attempted to obtain a new collection of dehalogenases by e≥ environmental gene libraries. An ingenious panning method based on sequence sim was used to enrich a library for clones possessing haloalkane dehalogenase activ several new enzymes were indeed obtained. However, no dehalogenase variants ha isolated so far, either by directed evolution, genetic engineering, or the use of meta screening, that have sufficient activity to develop a feasible full-scale process fe hydrolysis at the level at which it is present in industrial waste.

16.2.5   ENANTIOSELECTIVITY

Little work has been done on the enantioselectivity of haloalkane dehalogenases would potentially be interesting for the preparation of enantiopure alkylhalides e hols by kinetic resolution. Pieters et al. [40] have demonstrated that haloalkane genases in principle can be enantioselective, but the enantioselectivity factors w substrates that were tested were low (Figure 16.2, E = 4 to 9), both in kinetic res and in the conversion of prochiral compounds such as 1,2,3-tribromopropane. experiments showed that with methyl-3-bromo-methyl propionate, the large dif between the K_m values for the different enantiomers were compensated by large ences in k_{cat} that had the opposite effect. The enantioselectivity of the engineered haloalkane dehalogenase that showed enhanced 1,2,3-trichloropropane conversi also low [37].

16.2.6   IMMOBILIZED HALOALKANE DEHALOGENASES

Bioreactors containing immobilized enzyme for treating water or gas that contains chl anes as a contaminant have been described. Dravis et al. [41] studied conver 1-chlorobutane and 1,3-dichloropropane, supplied via the gas phase, by lyophilizec

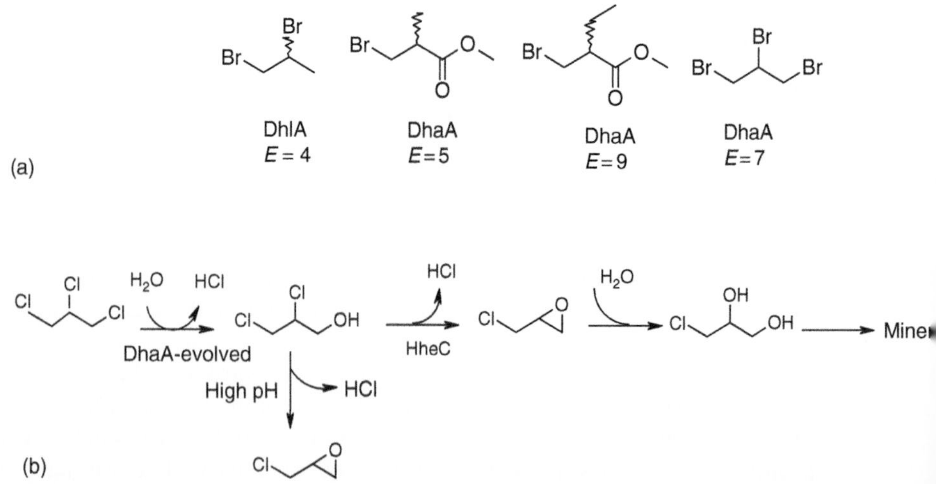

FIGURE 16.2 Biotechnologically relevant conversions catalyzed by haloalkane dehalc (a) Substrates tested for kinetic resolution of haloalkane dehalogenases. (From Ref. [9].) describe enantioselectivity [E = ($k_{cat,S}/K_{m,S}$)/($k_{cat,R}/K_{m,R}$)]. (b) A potential pathway for trichloro mineralization or conversion to epichlorohydrin for reuse in manufacturing. (From Ref. [37,39,4

from *R. rhodochrous*. Although it seems counterintuitive to perform a reaction that requires water and produces hydrochloric acid in the gas phase, significant conversion was obtained. The specific activity for 1,3-dichloropropane was about 3.5×10^{-3} µmol/min·mg protein using enzyme lyophilized from high pH buffer at optimal water activity in the presence of triethylamine as a buffer. This activity is about 170-fold lower than the k_{cat} of the enzyme [42]. Immobilized haloalkane dehalogenase (DhaA) has also been tested for TCP removal after covalent immobilization on a polyethyleneimine impregnated alumina support. The immobilized enzyme was more stable than the free protein, whereas the thermostability and resistance to organic solvent were improved after covalent attachment of the enzyme to the support [43].

Other attempts to use haloalkane dehalogenase in the gas phase were published recently and involved the use of DhlA for 1,2-dichloroethane removal [44,45]. With lyophilized DhlA, the V_{max} was about 0.08 µmol/min/g of cells, and it was found that activity and stability were dependent on the thermodynamic activity of the substrates and the water activity. As expected, the system displayed stability problems, which were attributed to accumulation of hydrochloric acid. If bioreactors must have a high space–time yield (or volumetric elimination rate), it is preferable to use packed bed systems with supply of buffer and removal of halide through the water phase. Trickling filters with immobilized whole cells seem to be a good option if rapid mass transfer of a compound that is sparingly soluble in water is required. They function well for dichloromethane removal from air [46], which is a more volatile organohalogen (in terms of partition coefficient) than 1,2-dichloroethane.

16.3 HALOACID DEHALOGENASES

16.3.1 PROPERTIES OF L-2-CHLOROPROPIONIC ACID DEHALOGENASES

A frequently studied group of aliphatic dehalogenases are the haloacid dehalogenases. Most of these enzymes can be classified as either group I or group II haloacid dehalogenases. The group II enzymes have been better characterized, and the x-ray structure of two of its members has been solved [47,48]. These proteins are somewhat smaller (about 29 kDa) than haloalkane dehalogenase, are active with the L-(S)-enantiomer of 2-chloropropionic acid, and convert this substrate with inversion of configuration at the chiral carbon atom. It is a diverse group of enzymes, members of which occur quite commonly, both in bacteria that grow on haloalkanoic acids and organisms that are not involved in haloalkanoic acid degradation. The structures confirmed the original proposal by Schneider et al. [49] that an aspartate residue located close to the N terminus of the protein acts as a nucleophile. Its carboxylate performs a nucleophilic displacement of the halogen, producing a covalent intermediate that is hydrolyzed by water in a similar way as the alkyl-enzyme intermediate in haloalkane dehalogenases (Figure 16.1b). The enzymes have a high pH optimum of around 9. The group II haloacid dehalogenases belong to the HAD superfamily of enzymes, which also includes proteins or protein domains that have phosphatase activity [50]. In these enzymes, the nucleophilic aspartate is also conserved.

The first haloacid dehalogenase whose structure was solved is the dimeric enzyme from *Pseudomonas* sp. YL was called L-DEX [47], and further information about the haloacid dehalogenase mechanism was obtained from structures in which the covalent intermediate was trapped by using an enzyme with a mutated oxyanion hole (Ser171Ala) [51] or by collecting data at low pH [48]. As in haloalkane dehalogenases, the active site is located between two domains. The haloacid dehalogenase main domain has an α/β structure, but the topology, in the sense of order and direction of strands in the β-sheet, is very different from that of haloalkane dehalogenases. In haloacid dehalogenase, the sheet of the main domain is composed of six parallel β-strands, and it is flanked by five α-helices. The second domain

is formed by an excursion in the sequence and has the topology of a four-helix bund
active site is lined by the nucleophilic Asp10 (in L-DEX), which is positioned on a lo
follows the N-terminal segment of the protein, and a series of other residues that influe
activity when mutated. Of the active site residues, Lys151 and Ser118 are also conse
other haloacid dehalogenase proteins, such as phosphatases and an *N*-carbamoylsa
amidohydrolase from *Arthrobacter* sp. In the dehalogenases, the serine contributes t
oxylate binding, as do main chain amides of two amino acids that follow the nucle
aspartate. The conserved lysine could possibly act as the catalytic base that activates wa
cleavage of the covalent intermediate. The halide-binding site in haloacid dehalogen
the HAD superfamily is composed of an arginine (Arg41 in L-DEX, Arg39 in DhlA)
tyrosine at a position +2 of the nucleophilic aspartate [48,51].

16.3.2   D-Specific Chloropropionic Acid Dehalogenase

Another important group of haloacid dehalogenases are the D-specific 2-chloropropior
dehalogenases and phylogenetically related nonenantioselective dehalogenases, classi
group I enzymes [4]. Perhaps the best-characterized member of this group of protei
nonstereoselective haloacid dehalogenase from *Pseudomonas* sp. 113, called DL-DEX [5
protein has a single active site for converting both enantiomers of 2-chloropropionic ac
sequence is similar to that of D-selective 2-haloacid dehalogenase from *Pseudomonas*
AJ1, which only converts D-2-haloalkanoic acids. The catalytic mechanism was prop
be fundamentally different from that of the HAD family of dehalogenases in the sen
the reaction does not involve a covalent intermediate. The conversion of 2-chloropro
acid proceeds with inversion of configuration. Mechanistic details are not known,
structure has been solved and sequence similarities to well-understood proteins are lo

16.3.3   Production of Optically Active Chloropropionic Acid

The excellent enantioselectivity of some 2-chloropropionic acid dehalogenases and th
for optically active chloropropionic acid as a building block in the production of pl
ceuticals have triggered the development of a biocatalytic process for producing
chloropropionic acid by kinetic resolution (Figure 16.3). This process has been commer
in England by Zeneca (ICI) and the product is used for the manufacture of the
xyherbicide fusilade. After hydrolysis of the (*R*)-chloropropionic acid, the remaini
chloropropionic acid is isolated by extraction and purified by distillation. The bioc
applied in this whole-cell process consisted of dried whole cells, which were stable eno
transport and storage [53].

Biotechnological processes for the production of enantiopure haloacids have been
oped, for example, by Ordaz et al. [54] using immobilized L-(*S*)-chloropropionic deha
ase, which was either His-tagged or coupled to an acrylic acid polymer. The L-2-ha
dehalogenase DehCI from *Pseudomonas* CBS3 was used and the acrylate-immo
enzyme was the more stable form. Immobilization of another 2-haloacid dehalogena
also been done with a DEAE Sephacel solid matrix, which yielded an enzyme that wa
temperature-stable and could be used in a plug-flow reactor [55].

16.4   HALOHYDRIN DEHALOGENASE

16.4.1   Properties

Halohydrin dehalogenases (also called haloalcohol hydrogen halide lyases) have been s
because of their environmental relevance and their potential application in biocatalys

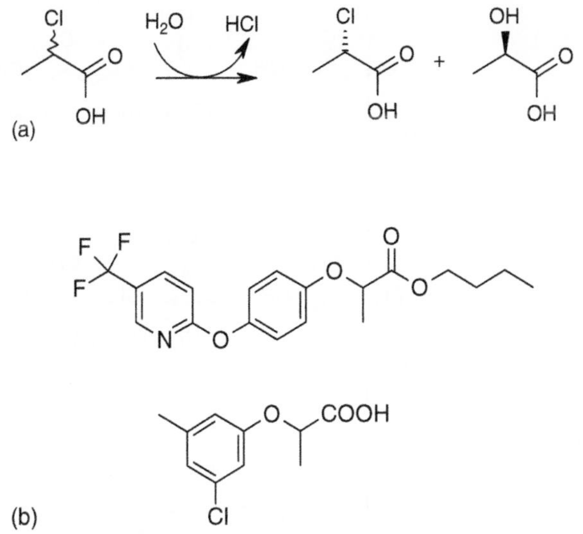

FIGURE 16.3 (a) Kinetic resolution of (R,S)-2-chloropropionic acid by a type I haloacid dehalogenase (HAD) to produce (b) (S)-2-chloropropionic acid, an intermediate for herbicides. (From Ref. [53].)

activity was first described by Castro and Bartnicki [3], who investigated the degradation of 2,3-dibromo-1-propanol. Later, halohydrin dehalogenases were purified and cloned from various organisms that degrade halohydrins [56,57], and the first x-ray structure was solved recently [58]. The latter work was done with HheC, a halohydrin dehalogenase from a strain of *Agrobacterium radiobacter* that grows on epichlorohydrin. The kinetics and substrate binding properties of this dehalogenase have been studied [59,60].

Sequence and structure analysis revealed that the catalytic mechanism of halohydrin dehalogenases is similar to that of members of the SDR superfamily of proteins. Both the dehalogenases and the SDR enzymes possess a conserved Ser/Tyr/Arg (or Ser/Tyr/Lys) catalytic triad for proton abstraction from the hydroxyl group of the substrate. In the case of SDR proteins, the negative charge developing on the hydroxyl oxygen is transferred by a hydride to the $NAD(P)^+$ cofactor (Figure 16.1c). In the dehalogenases, instead, it is passed on with the oxygen to the neighboring carbon atom and the halogen bound to this carbon atom is displaced as halide. This intramolecular substitution mechanism results in the formation of an epoxide. The halohydrin dehalogenases thus use noncovalent catalysis. The enzymes also possess a distinct halide-binding site.

Sequence analysis of the known SDR-type halohydrin dehalogenases indicates that they can be divided into three different phylogenetic subgroups, which in this case coincides well with the classification according to substrate range. The group A enzymes (HheA from *Corynebacterium* sp. strain N-1074 and HheA_AD2 from *Agrobacterium*) have a high sequence similarity (97%). HheB from *Corynebacterium* sp. strain N-1074 and HheB-GP1 from *Mycobacterium* sp. GP1 are also very similar (98%). Only HheB is enantioselective in the conversion of 1,3-dichloropropanol to epichlorohydrin, producing the (R)-enantiomer [61]. In group C, the sequence similarity between HheC from *A. radiobacter* and HalB from *A. tumefaciens* is only 80%. HheC of *A. radiobacter* is the best-characterized halohydrin dehalogenase, both from a mechanistic and biocatalystic point of view, and distinguishes itself from most of the other enzymes because of its high enantioselectivity [62]. Another source of halohydrin dehalogenases is *Arthrobacter erithii* H10A [63].

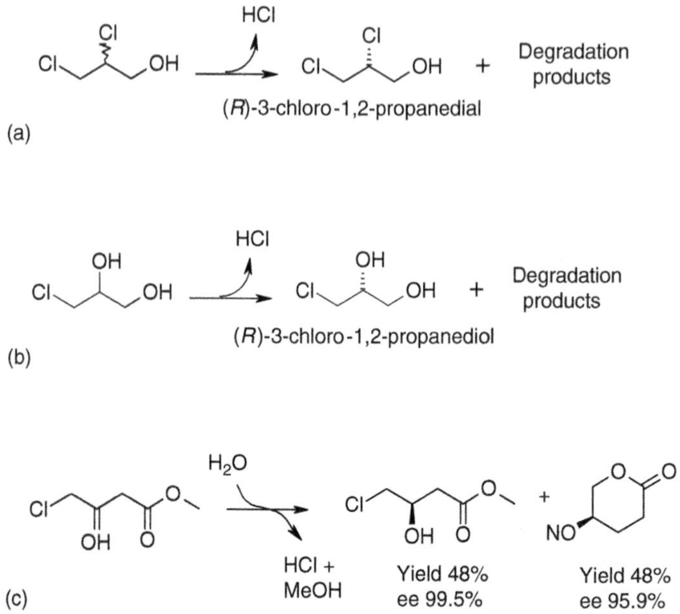

FIGURE 16.4 Dechlorination of chloropropanols and chlorinated C4 compounds cata[...] halohydrin dehalogenases and whole cells containing dehalogenases. (a) Resolution of [...] 3-dichloro-1-propanol by enantioselective degradation with *Alcaligenes* sp. DS-K-S38. (F[...] [64].). (b) Resolution of (*R,S*)-3-chloro-1,2-propanediol by enantioselective degradation wit[...] *monas* sp. DS-K-2DI. (From Ref. [65].) (c) Enantioselective conversion of a 3-hydroxy-4-chlo[...] acid methylester by an *Enterobacter* sp. (From Ref. [67].)

16.4.2 BIOCATALYTIC POTENTIAL OF HALOHYDRIN DEHALOGENATION

The enantioselectivity of halohydrin dehalogenases in the conversion of haloalcoho[...] them promising biocatalysts for the production of optically active epoxides and β-su[...] alcohols (Figure 16.4). Several bacterial cultures and enzymes that catalyze dechlori[...] bromo- and chloropropanols have been investigated for their potential in the produ[...] optically active C3 building blocks, and a historical account on the development of [...] lytic processes employing halohydrin dehalogenase containing microorganisms was p[...] by Kasai et al. [64] and Kasai and Sizuki [65].

The epichlorohydrin precursor 2,3-dichloro-1-propanol has been produced in enan[...] cally pure form by kinetic resolution using bacteria that possess halohydrin deha[...] activity. In this process, the unwanted isomer is degraded. The (*S*)-enantiomer [...] obtained using immobilized cells of *Pseudomonas* sp. OS-K-29, a strain that enantiose[...] degrades the (*R*)-enantiomer. Organisms that selectively degrade (*S*)-2,3-dichloro-1-[...] were also found and allowed the production of (*R*)-2,3-dichloro-1-propanol. The later [...] sions have been performed at pilot scale, mostly using whole cells of *Alcaligenes* sp. DS[...] The dehalogenase system in this organism is of a different type than the halohydrin[...] genases of the SDR family class, and will be briefly discussed below. The opticall[...] dichloropropanols that remain after the enantioselective degradation process can be [...] and used for the synthesis of a variety of products, including epichlorohydrin, whic[...] obtained from the halohydrins by simple chemical ring closure under basic condition[...]

The use of halohydrin dehalogenase producing bacteria for making optically active 3-chloro-1,2-propanediols has also been studied with complementary organisms that have opposite stereospecificities. Kasai et al. [64] have screened several bacterial strains for the enantioselective conversion of these chiral chloropropanols. Although the enzymes involved are not described in detail, it is most likely that they are halohydrin dehalogenases related to the SDR family proteins. (R)-3-chloro-1,2-propanediol was obtained by kinetic resolution using whole cells of *Pseudomonas* sp. DS-K-2D1, whereas (S)-3-chloro-1,2-propanediol was obtained by enantioselective transformation with *Alcaligenes* sp. DS-S-7G [66].

Suzuki et al. [67] employed the stereoselectivity of halohydrin degradation system present in *Pseudomonas* sp. OS-K-29 to perform a kinetic resolution of 4-chloro-3-hydroxybutyrate ethylester. The remaining (S)-enantiomer had a high enantiomeric excess (ee) (>98%), although the yield was modest (33%), indicating that the enzyme involved did not have a very high enantioselectivity. The enzyme also appeared to be very tolerant with respect to the halohydrin R-group: 1-halo-2-hydroxybutane was converted, as well as the corresponding nitrile and different 4-halo-3-hydroxy-butanoic acid esters.

A conversion with the opposite enantioselectivity toward 4-chloro-3-hydroxybutyrate was also described [65]. In this case, the catalyst consisted of whole cells of *Pseudomonas* sp. DS-K-NR818 (a mutant of *Pseudomonas* sp. OS-K-29). When exposed to 4-chloro-3-hydroxybutyric acid ethyl ester, the (S)-enantiomer was degraded and the remaining ethyl (R)-4-chloro-3-hydroxybutyrate was recovered in good ee (>95%).

The conversions mentioned above showed the potential of biocatalytic production of enantiomerically pure halohydrins and related compounds with the use of whole cells. Using free enzyme, the halohydrin dehalogenase from *A. radiobacter* AD1 (HheC) was found to catalyze the dehalogenation of phenyl-substituted halohydrins with good enantioselectivity, yielding enantiomerically enriched (R)-epoxides and (S)-halohydrins [62]. The enantioselectivity was further increased by an active-site mutation (Trp249Phe), which also increased the k_{cat} for some substrates [68], whereas other mutations improved the stability against oxidation [69]. In the Trp249Phe mutant, the affected residue is donated by an opposite subunit and sticks in the binding site for the phenyl ring of styrene epoxide.

16.4.3 BIOCATALYTIC RING OPENING OF EPOXIDES

The possibility to use halohydrin dehalogenase for ring-opening reactions (Figure 16.5, Figure 16.6) is in line with the early observation that the enzymes catalyze transhalogenation [3]. An early alternative nucleophile to be explored was cyanide. Nakamura et al. [70] explored the epoxide ring opening of epoxybutyronitrile, a conversion that yielded (R)-4-chloro-3-hydroxybutyronitrile, an intermediate in the synthesis of L-carnithine. The enzyme was originated from a *Corynebacterium* sp.

The scope of epoxide ring-opening reactions was further explored by Lutje Spelberg et al. [71], using halohydrin dehalogenase from *A. radiobacter* (HheC). This enzyme catalyzes epoxide ring-opening reactions with several alternative nucleophiles, most notably with azide, cyanide, and nitrite, which allows the production of a range of β-substituted derivatives. The enantioselective azidolysis of styrene oxide and substituted derivates yields 2-azido-1-phenylethanols of the (R) configuration, since the enzyme is enantioselective for (R)-epoxides in ring-opening reactions [72] (Figure 16.5). It was found that it is also possible to apply HheC in a dynamic kinetic resolution of epihalohydrins when using azide for ring opening [73]. This produces (S)-1-azidohaloalcohol. The racemization is based on the presence of catalytic amounts of halide, which opens the epoxide ring to produce nonchiral 1,3-dihalopropanol. With epichlorohydrin as the substrate, the rate of racemization was lower than the rate of ring opening by azide, resulting in a mixed kinetic resolution

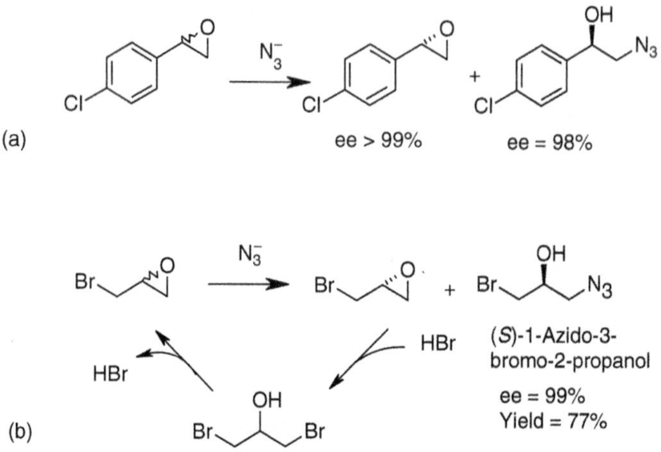

FIGURE 16.5 Epoxide ring-opening reaction catalyzed by HheC with azide as the nucle (a) Enantioselective ring opening of *para*-chlorostyrene epoxide with azide. (From Ref. [72].) (b) D resolution of epibromohydrin. All reactions shown are catalyzed by HheC. (From Ref. [73].)

and dynamic kinetic resolution and modest enantiopurity of the (*S*)-1-azido-3-chloro-2-pro With epibromohydrin as the substrate, the racemization rate was higher, which allov efficient dynamic resolution and production of (*S*)-1-azido-3-bromo-2-propanol at >9 and with a yield of 77%. This product can be used for preparing additional chiral comp

Another unexpected reaction of halohydrin dehalogenases was the opening of ep with nitrite as the nucleophile [74]. Nitrite predominantly attacks the epoxide throw nitrite oxygen, yielding an unstable nitrite ester intermediate that spontaneously hydrol form the corresponding diol (Figure 16.7). This reaction occurs with high regioselectiv the lesser substituted carbon atom of the oxirane ring and, depending on the e substrate, with high enantioselectivity by using the W249F mutant halohydrin dehalo with improved enantiodiscrimination. This conversion makes it possible to apply halo dehalogenases and nitrite as an alternative epoxide hydrolase to the kinetic resolut various racemic epoxides. In several cases, the *E*-values are higher than those foun epoxide hydrolases, which catalyze the same overall reaction.

A further alternative nucleophile is cyanide (Figure 16.6). The HheC-mediated conv of epoxides with this nucleophile was recently found to yield a variety of β-cyanoalcoho Again, the reactions proceed with varying enantioselectivity, depending on the epoxid conversion complements the reactions catalyzed by hydroxynitrile lyases. The later en mediate formation of α-hydroxynitriles from aldehydes, whereas halohydrin dehalog catalyze formation of β-hydroxynitriles from epoxides.

Another interesting application of halohydrin dehalogenases is in the manufact the statin side-chain building block 4-cyano-3-hydroxybutyric acid ester (Figure 16.6). ing from 4-bromo-3-ketobutyric acid ethylester, this product was obtained by a g dehydrogenase–mediated reduction to (3*S*)-hydroxybutyric acid, which was subseq converted by the epoxide to the product by an engineered HheC-type halohydrin deha ase [76]. This is one of the several possible routes to the statin side chain [77].

In conclusion, halohydrin dehalogenases are an important emerging new class of bioca They show unprecedented catalytic promiscuity by being able to catalyze carbon–chlorine, ca bromine, carbon–nitrogen, carbon–oxygen, and carbon–carbon bond formation. I hypo

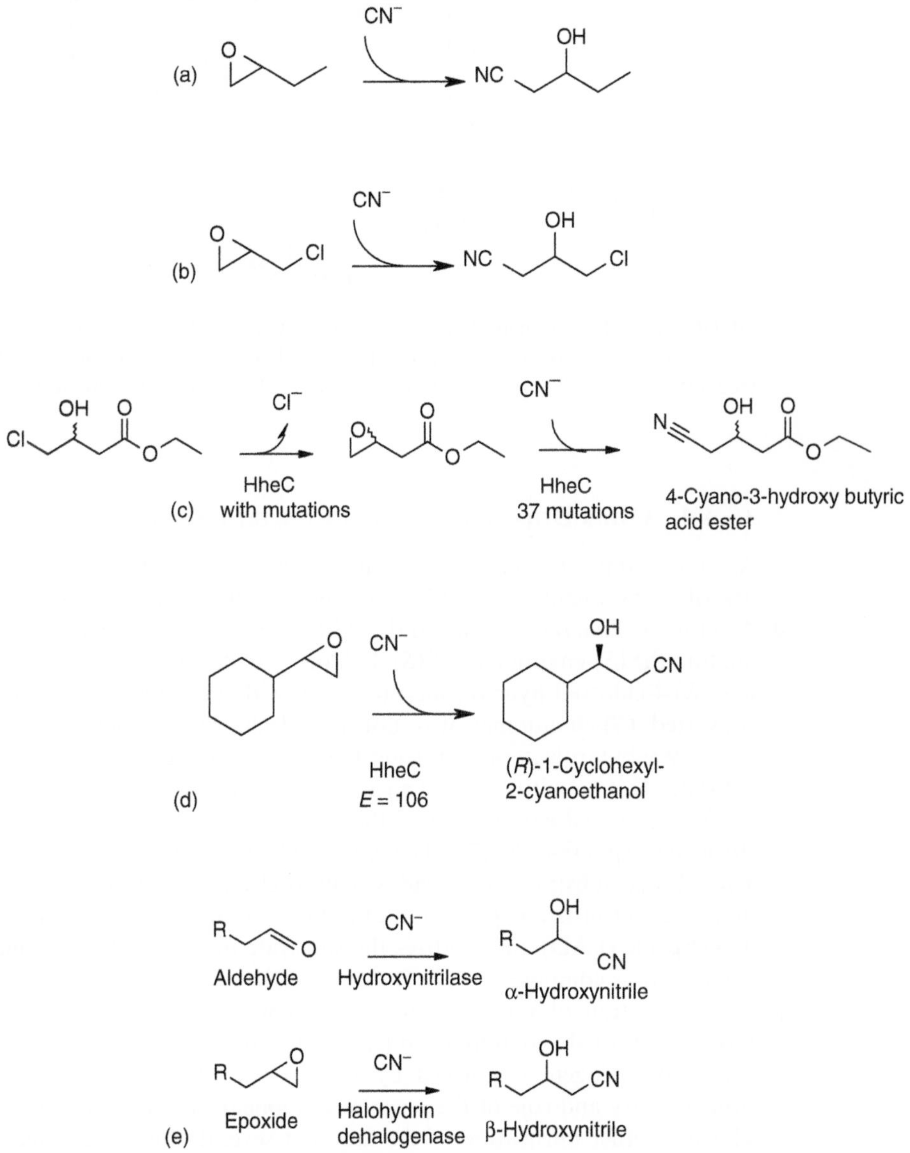

FIGURE 16.6 Epoxide ring opening with cyanide as catalyzed by halohydrin dehalogenases. (a) Ring opening of epoxybutane to produce hydroxyvaleronitrile. (From Ref. [70].) (b) Ring opening of epichlorohydrin to produce (R)-4-chloro-3-hydroxybutryonitrile. (From Ref. [81].) (c) Conversion of a 4-halo-3-hydroxybutryric acid ester to the corresponding nitrile for producing a statin side-chain building block. (From Ref. [76].) (d) Enantioselective conversion of cyclohexyloxirane to the corresponding cyanohydrin. (From Ref. [75].) (e) Comparison of halohydrin dehalogenase and hydroxynitrile lyase–mediated reactions.

that halohydrin dehalogenases are evolutionary primitive enzymes that have not evolved a high selectivity toward a specific substrate. The nonspecific active site can be used for a wide diversity of reactions. It is well possible that bacterial genomes harbor many such genes, which encode promiscuous enzymes that can be tailored to catalyze industrially important reactions.

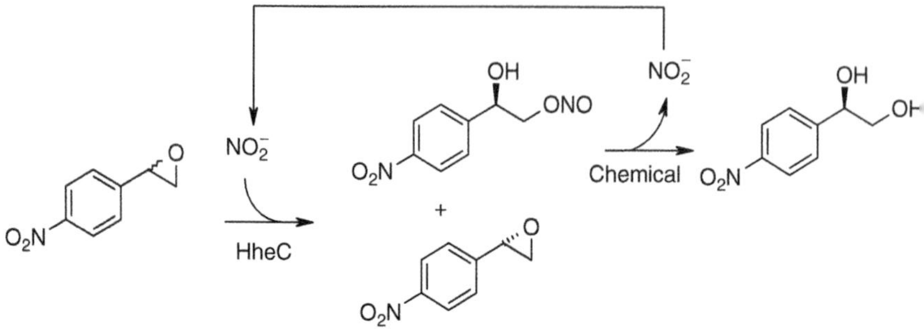

FIGURE 16.7 Nitrite-mediated ring opening of *para*-nitrostyrene epoxide catalyzed by Hh (*R*)-enantiomer is converted to a nitrite ester. This unstable product is hydrolyzed to the d recovery of nitrite. The net reaction is an epoxide hydrolase–like kinetic resolution. (From Re

16.4.4 OTHER DEHALOGENASES THAT CONVERT HALOHYDRINS

An interesting intramolecular transesterification that yields dehalogenation ha described by Suzuki et al. [78] although its biochemistry has not yet been studie A culture of *Enterobacter* sp. strain DS-S-75 could stereoselectively dechlorinate in a mixture the (*S*)-enantiomer of (*S*)-4-chloro-3-hydroxybutyrate ethylester to yield the ing (*R*)-4-chloro-3-hydroxybutyrate ester with excellent optical purity. Interesting converted (*S*)-enantiomer was not just dechlorinated but also transformed into hydroxy-γ-butyrolactone, releasing the corresponding alcohol moiety of the substr chloride. This activity required a carboxylate and a halogen group at terminal positi

As mentioned earlier, one of the bacteria known to degrade 2,3-dichloro-l-prop *Alcaligenes* sp. DS-S-7G [79]. The responsible two-component enzyme from this organ a very broad substrate range and is enantioselective for the (*R*)-enantiomer. The deha tion mechanism is complicated and involves a 70 kDa protein that contains flavin dinucleotide (FAD) and oxidizes the substrate in an NAD-dependent reaction. The component is composed of 33 kDa and 53 kDa subunits and seems to be involved in or regeneration of the electron transfer components. It was proposed that the r proceeds through 3-chloroacetone, which would decompose chemically in a reac which the halogen is replaced by a hydrogen. Details of this mechanism and th stoichiometry and role of the protein components and reactants have not been work This is another illustration of the fact that several unusual enzyme systems for hal dehalogenation still await exploration at the biochemical level.

16.5 CONCLUSIONS AND OUTLOOK

The current status of the research on dehalogenases has yielded fascinating insight i molecular mechanism of biological carbon–halogen bond cleavage and the diversity of e that can catalyze for such reactions. Development of whole-cell processes in which deh ase-producing organisms are applied for removing chlorinated compounds in biorem e schemes has been successful. Various enzyme systems for dehalogenation reactions have been explored, and it is likely that surprising new enzyme mechanisms can still be fou also foresee a growth in the range of reactions that are useful for biocatalysis, especial new enzymes that attack a broader range of organic substrates are found.

REFERENCES

1. Davies, J.I. and Evans, W.C., The elimination of halide ions from aliphatic halogen-substituted organic acids by an enzyme preparation from *Pseudomonas dehalogenans*, *Biochem. J.*, 82, 50–51, 1962.
2. Goldman, P., The enzymatic cleavage of the carbon–fluorine bond in fluoroacetate, *J. Biol. Chem.*, 240, 3434–3438, 1965.
3. Castro, C.E. and Bartnicki, E.W., Biodehalogenation, epoxidation of halohydrins, epoxide opening, and transhalogenation by a *Flavobacterium* sp., *Biochemistry*, 7, 3213–3218, 1968.
4. Hill, K.E., Marchesi, J.R., and Weightman, A.J., Investigation of two evolutionarily unrelated halocarboxylic acid dehalogenase gene families, *J. Bacteriol.*, 181, 2535–2547, 1999.
5. Nardi-Dei, V., Kurihara, T., Park, C., et al., DL-2-Haloacid dehalogenase from *Pseudomonas* sp. 113 is a new class of dehalogenase catalyzing hydrolytic dehalogenation not involving enzyme–substrate ester intermediate, *J. Biol. Chem.*, 274, 20977–20981, 1999.
6. Smidt, H., Akkermans, A.D.L., van der Oost, J., and de Vos, W.M., Halorespiring bacteria—molecular characterization and detection, *Enzyme Microb. Technol.*, 27, 812–820, 2000.
7. Gribble, G.W., Naturally occurring organohalogen compounds, *Acc. Chem. Res.*, 31, 141–152, 1998.
8. Pries, F., van den Wijngaard, A.J., Bos, R., et al., The role of spontaneous cap domain mutations in haloalkane dehalogenase specificity and evolution, *J. Biol. Chem.*, 269, 17490–17494, 1994.
9. Pikkemaat, M.G. and Janssen, D.B., Generating segmental mutations in haloalkane dehalogenase: a novel part in the directed evolution toolbox, *Nucleic Acids Res.*, 30, E35-5, 2002.
10. Seffernick, J.L., de Souza, M.L., Sadowsky, M.J., and Wackett, L.P., Melamine deaminase and atrazine chlorohydrolase: 98 percent identical but functionally different, *J. Bacteriol.*, 183, 2405–2410, 2001.
11. Wackett, L.P., Evolution of enzymes for the metabolism of new chemical inputs into the environment, *J. Biol. Chem.*, 279, 41259–41262, 2004.
12. Poelarends, G.J. and Whitman, C.P., Evolution of enzymatic activity in the tautomerase superfamily: mechanistic and structural studies of the 1,3-dichloropropene catabolic enzymes, *Bioorg. Chem.*, 32, 376–392, 2004.
13. Poelarends, G.J., Saunier, R., and Janssen, D.B., *trans*-3-chloroacrylic acid dehalogenase from *Pseudomonas pavonaceae* 170 shares structural and mechanistic similarities with 4-oxalocrotonate tautomerase, *J. Bacteriol.*, 183, 4269–4277, 2001.
14. Horvat, C.M. and Wolfenden, R.V., A persistent pesticide residue and the unusual catalytic proficiency of a dehalogenating enzyme, *Proc. Natl. Acad. Sci. USA*, 102, 16199–16202, 2005.
15. de Jong, R.M., Brugman, W., Poelarends, G.J., et al., The x-ray structure of *trans*-3-chloroacrylic acid dehalogenase reveals a novel hydration mechanism in the tautomerase superfamily, *J. Biol. Chem.*, 279, 11546–11552, 2004.
16. Janssen, D.B., Evolving haloalkane dehalogenases, *Curr. Opin. Chem. Biol.*, 8, 150–159, 2004.
17. Verschueren, K.H., Seljee, F., Rozeboom, H.J., et al., Crystallographic analysis of the catalytic mechanism of haloalkane dehalogenase, *Nature*, 363, 693–698, 1993.
18. Newman, J., Peat, T.S., Richard, R., et al., Haloalkane dehalogenases: structure of a *Rhodococcus* enzyme, *Biochemistry*, 38, 16105–16114, 1999.
19. Marek, J., Vevodova, J., Smatanova, I.K., et al., Crystal structure of the haloalkane dehalogenase from *Sphingomonas paucimobilis* UT26, *Biochemistry*, 39, 14082–14086, 2000.
20. Prokop, Z., Monincova, M., Chaloupkova, R., et al., Catalytic mechanism of the haloalkane dehalogenase LinB from *Sphingomonas paucimobilis* UT26, *J. Biol. Chem.*, 278, 45094–45100, 2003.
21. Krooshof, G.H., Ridder, I.S., Tepper, A.W., et al., Kinetic analysis and x-ray structure of haloalkane dehalogenase with a modified halide-binding site, *Biochemistry*, 37, 15013–15023, 1998.
22. Hynkova, K., Nagata, Y., Takagi, M., and Damborsky, J., Identification of the catalytic triad in the haloalkane dehalogenase from *Sphingomonas paucimobilis* UT26, *FEBS Lett*, 446, 177–181, 1999.
23. Damborsky, J. and Koca, J., Analysis of the reaction mechanism and substrate specificity of haloalkane dehalogenases by sequential and structural comparisons, *Protein Eng.*, 12, 989–998, 1999.

24. Bohac, M., Nagata, Y., Prokop, Z., et al., Halide-stabilizing residues of haloalkane dehalog
    studied by quantum mechanic calculations and site-directed mutagenesis, *Biochemistry*, 41,
    14280, 2002.
25. Schanstra, J.P., Ridder, I.S., Heimeriks, G.J., et al., Kinetic characterization and x-ray struct
    mutant of haloalkane dehalogenase with higher catalytic activity and modified substrate
    Biochemistry, 35, 13186–13195, 1996.
26. Schanstra, J.P., Kingma, J., and Janssen, D.B., Specificity and kinetics of haloalkane dehalo
    J. Biol. Chem., 271, 14747–14753, 1996.
27. Schanstra, J.P. and Janssen, D.B., Kinetics of halide release of haloalkane dehalogenase: e
    for a slow conformational change, *Biochemistry*, 35, 5624–5632, 1996.
28. Shurki, A., Strajbl, M., Villa, J., and Warshel, A., How much do enzymes really gain by rest
    their reacting fragments? *J. Am. Chem. Soc.*, 124, 4097–4107, 2002.
29. Lau, E.Y., Kahn, K., Bash, P.A., and Bruice, T.C., The importance of reactant positio
    enzyme catalysis: a hybrid quantum mechanics/molecular mechanics study of a haloalkane
    genase, *Proc. Natl. Acad. Sci. USA*, 97, 9937–9942, 2000.
30. Stucki, G. and Thüer, M., Experiences of a large-scale application of 1,2-dichloroethane de
    microorganisms for groundwater treatment, *Environ. Sci. Technol.*, 29, 2339–2345, 1995.
31. Schanstra, J.P., Ridder, A., Kingma, J., and Janssen, D.B., Influence of mutations of Val226
    catalytic rate of haloalkane dehalogenase, *Protein Eng.*, 10, 53–61, 1997.
32. Holloway, P., Knoke, K.L., Trevors, J.T., and Lee H., Alteration of the substrate ra
    haloalkane dehalogenase by site-directed mutagenesis, *Biotechnol. Bioeng.*, 59, 520–523, 199
33. Kmunicek, J., Bohac, M., Luengo, S., et al., Comparative binding energy analysis of halc
    dehalogenase substrates: modelling of enzyme–substrate complexes by molecular docki
    quantum mechanical calculations, *J. Comput. Aided Mol. Des.*, 17, 299–311, 2003.
34. Chaloupkova, R., Sykorova, J., Prokop, Z., et al., Modification of activity and specifi
    haloalkane dehalogenase from *Sphingomonas paucimobilis* UT26 by engineering of its e
    tunnel, *J. Biol. Chem.*, 278, 52622–52628, 2003.
35. Pikkemaat, M.G., Linssen, A.B., Berendsen, H.J., and Janssen, D.B., Molecular dynamics
    tions as a tool for improving protein stability, *Protein Eng.*, 15, 185–192, 2002.
36. Gray, K.A., Richardson, T.H., Kretz K., et al., Rapid evolution of reversible denaturati
    elevated melting temperature in a microbial haloalkane dehalogenase, *Adv. Synth. Catal.*, 34
    617, 2001.
37. Bosma, T., Damborsky, J., Stucki, G., and Janssen, D.B., Biodegradation of 1,2,3-trichlorop
    through directed evolution and heterologous expression of a haloalkane dehalogenase gene
    Environ. Microbiol., 68, 3582–3587, 2002.
38. Poelarends, G.J., van Hylckama Vlieg, J.E.T., Marchesi, J.R., et al., Degradation of 1,2
    moethane by *Mycobacterium* sp. strain GP1, *J. Bacteriol.*, 181, 2050–2058, 1999.
39. Gray, K.A., Richardson, T.H., Robertson, D.E., et al., Soil-based gene discovery: a new tech
    to accelerate and broaden biocatalytic applications, *Adv. Appl. Microbiol.*, 52, 1–27, 2003.
40. Pieters, R.J., Fennema, M., Kellogg, R.M., and Janssen, D.B., Design and synthesis of reage
    phage display screening of dehalogenases, *Bioorg. Med. Chem. Lett.*, 9, 161–166, 1999.
41. Dravis, B.C., LeJeune, K.E., Hetro, A.D., and Russell, A.J., Enzymatic dehalogenation
    phase substrates with haloalkane dehalogenase, *Biotechnol. Bioeng.*, 69, 235–241, 2000.
42. Bosma, T., Pikkemaat, M.G., Kingma, J., et al., Steady-state and pre-steady-state kinetic ana
    halopropane conversion by a *Rhodococcus* haloalkane dehalogenase, *Biochemistry*, 42,
    8053, 2003.
43. Dravis, B.C., Swanson, P.E., and Russell, A.J., Haloalkane hydrolysis with an immobilized h
    ane dehalogenase, *Biotechnol. Bioeng.*, 75, 416–423, 2001.
44. Erable, B., Goubet I., Lamare S., et al., Haloalkane hydrolysis by *Rhodococcus erythropol.*
    comparison of conventional aqueous phase dehalogenation and nonconventional gas phase
    logenation, *Biotechnol. Bioeng.*, 86, 47–54, 2004.
45. Erable, B., Goubet, I., Lamare, S., Seltana, A., et al., Nonconventional hydrolytic dehaloge
    of 1-chlorobutane by dehydrated bacteria in a continuous solid–gas biofilter, *Biotechnol. Bio
    304–313, 2005.

46. Diks, R.M.M. and Ottengraf, S.P.P., Verification studies of a simplified model for the removal of dichloromethane from waste gases using a biological trickling filter: part II, *Bioproc. Eng.*, 6, 131–140, 1991.

47. Hisano, T., Hata, Y., Fujii, T., et al., Crystal structure of L-2-haloacid dehalogenase from *Pseudomonas* sp. YL. An alpha/beta hydrolase structure that is different from the alpha/beta hydrolase fold, *J. Biol. Chem.*, 271, 20322–20330, 1996.

48. Ridder, I.S., Rozeboom, H.J., Kalk, K.H., and Dijkstra, B.W., Crystal structures of intermediates in the dehalogenation of haloalkanoates by L-2-haloacid dehalogenase, *J. Biol. Chem.*, 274, 30672–30678, 1999.

49. Schneider, B., Muller, R., Frank, R., and Lingens, F., Site-directed mutagenesis of the 2-haloalkanoic acid dehalogenase I gene from *Pseudomonas* sp. strain CBS3 and its effect on catalytic activity, *Biol. Chem. Hoppe Seyler*, 374, 489–496, 1993.

50. Aravind, L., Galperin, M.Y., and Koonin, E.V., The catalytic domain of the P-type ATPase has the haloacid dehalogenase fold, *Trends Biochem. Sci.*, 23, 127–129, 1998.

51. Li, Y.F., Hata, Y., Fujii, T., et al., Crystal structures of reaction intermediates of L-2-haloacid dehalogenase and implications for the reaction mechanism, *J. Biol. Chem.*, 273, 15035–15044, 1998.

52. Nardi-Dei, V., Kurihara, T., Park, C., et al., DL-2-Haloacid dehalogenase from *Pseudomonas* sp. 113 is a new class of dehalogenase catalyzing hydrolytic dehalogenation not involving enzyme–substrate ester intermediate, *J. Biol. Chem.*, 274, 20977–20981, 1999.

53. OECD, The application of biotechnology to industrial sustainability: manufacture of (S)-chloropropionic acid (Avecia, United Kingdom), OECD, Paris, 2001, pp. 67–90.

54. Ordaz, E., Garrido-Pertierra, A., Gallego, M., and Puyet, A., Covalent and metal–chelate immobilization of a modified 2-haloacid dehalogenase for the enzymatic resolution of optically active chloropropionic acid, *Biotechnol. Prog.*, 16, 287–291, 2000.

55. Diez, A., Prieto, M.I., Alvarez, M.J., et al., Improved catalytic performance of a 2-haloacid dehalogenase from *Azotobacter* sp. by ion-exchange immobilisation, *Biochem. Biophys. Res. Commun.*, 220, 828–833, 1996.

56. van Hylckama Vlieg, J.E.T., Tang, L., Lutje Spelberg, J.H., et al., Halohydrin dehalogenases are structurally and mechanistically related to short-chain dehydrogenases/reductases, *J. Bacteriol.*, 183, 5058–5066, 2001.

57. Yu, F., Nakamura, T., Mizunashi, W., and Watanabe, I., Cloning of two halohydrin hydrogen-halide-lyase genes of *Corynebacterium* sp. strain N-1074 and structural comparison of the genes and gene products, *Biosci. Biotechnol. Biochem.*, 58, 1451–1457, 1994.

58. de Jong, R.M., Tiesinga, J.J., Rozeboom, H.J., et al., Structure and mechanism of a bacterial haloalcohol dehalogenase: a new variation of the short-chain dehydrogenase/reductase fold without an NAD(P)H binding site, *EMBO J.*, 22, 4933–4944, 2003.

59. Tang, L., Lutje Spelberg, J.H., Fraaije, M.W., and Janssen, D.B., Kinetic mechanism and enantioselectivity of halohydrin dehalogenase from *Agrobacterium radiobacter*, *Biochemistry*, 42, 5378–5386, 2003.

60. Tang, L., van Merode, A.E., Lutje Spelberg, J.H., et al., Steady-state kinetics and tryptophan fluorescence properties of halohydrin dehalogenase from *Agrobacterium radiobacter*: roles of W139 and W249 in the active site and halide-induced conformational change, *Biochemistry*, 42, 14057–14065, 2003.

61. Nakamura, T., Nagasawa, T., Yu, F., et al., Resolution and some properties of enzymes involved in enantioselective transformation of 1,3-dichloro-2-propanol to (R)-3-chloro-1,2-propanediol by *Corynebacterium* sp. strain N-1074, *J. Bacteriol.*, 174, 7613–7619, 1992.

62. Lutje Spelberg, J.H.L., van Hylckama Vlieg, J.E.T., Bosma, T., et al., A tandem enzyme reaction to produce optically active halohydrins, epoxides and diols, *Tetrahedron Asymmetry*, 15, 2863–2870, 1999.

63. Assis, H.M, Sallis, P.J., Bull, A.T., and Hardman, D.J., Biochemical characterization of a haloalcohol dehalogenase from *Arthrobacter erithii* H10a, *Enzyme Microb. Technol.*, 22, 568–574, 1998.

64. Kasai, N., Suzuki, S., and Furukawa, Y., Optically active chlorohydrins as chiral C3 and C4 building units: microbial resolution and synthetic applications, *Chirality*, 10, 682–692, 1998.

65. Kasai, N. and Suzuki, T., Industrialization of the microbial resolution of chiral C3 and C4 synthetic units: from a small beginning to a major operation, a personal account, *Adv. Synth. Catal.*, 345, 437–455, 2003.

66. Suzuki, T. and Kasai, N., A novel method for the generation of (R)- and (S)-3-chloro-1, nediol by stereospecific dehalogenating bacteria and their use in the preparation of ((S)-glycidol, *Bioorg. Med. Chem. Lett.*, 1, 343–346, 1991.

67. Suzuki, T., Idogaki, H., and Kasai N., A novel generation of optically active ethyl-4-(hydroxybutyrate as a C4 chiral building unit using microbial dechlorination, *Tet Asymmetry*, 7, 3109–3112, 1996.

68. Tang, L., Torres Pazmino, D.E., Fraaije, M.W., et al., Improved catalytic properties of ha dehalogenase by modification of the halide-binding site, *Biochemistry*, 44, 6609–6618, 200

69. Tang, L., van Hylckama Vlieg, J.E.T., Fraaije, M., and Janssen, D.B., Improved sta halohydrin dehalogenase from *Agrobacterium radiobacter* AD1 by replacement of cysteine *Enzyme Microb. Technol.*, 30, 251–258, 2002.

70. Nakamura, T., Nagasawa, T., Yu, F., et al., A new catalytic function of halohydrin h halide-lyase, synthesis of beta-hydroxynitriles from epoxides and cyanide, *Biochem. Biop Commun.*, 180, 124–130, 1991.

71. Lutje Spelberg, J.H., Tang, L., van Gelder, M., et al., Exploration of the biocatalytic pote halohydrin dehalogenase using chromogenic substrates, *Tetrahedron Asymmetry*, 13, 1083–10

72. Lutje Spelberg, J.H., van Hylckama Vlieg, J.E., Tang, L., et al., Highly enantioselec regioselective biocatalytic azidolysis of aromatic epoxides, *Org. Lett.*, 3, 41–43, 2001.

73. Lutje Spelberg, J.H., Tang, L., Kellogg, R.M., and Janssen, D.B., Enzymatic dynami resolution of epihalohydrins, *Tetrahedron Asymmetry*, 15, 1095–1102, 2004.

74. Hasnaoui, G., Lutje Spelberg, J.H., de Vries, E., et al., Nitrite-mediated hydrolysis of catalyzed by halohydrin dehalogenase from *Agrobacterium radiobacter* AD1: a new too kinetic resolution of epoxides, *Tetrahedron Asymmetry*, 16, 1685–1692, 2005.

75. Majerić-Elenkov, M., Hauer, B., and Janssen, D.B., Enantioselective ring opening of epoxi cyanide catalysed by halohydrin dehalogenases: a new approach to enantiopure β-hydrox *Adv. Synth. Catal.*, 348, 579–585, 2006.

76. Davis, S.C., Grate, J.H., Gray, D.R., et al., Enzymatic processes for the production of 4-su 3-hydroxybutyric acid derivatives and vicinal cyano, hydroxy substituted carboxylic aci Patent WO 2004/015132.

77. Müller, M., Chemoenzymatic synthesis of building blocks for statin side chains, *Angew. Ch Ed.*, 44, 362–365, 1995.

78. Suzuki, T., Idogaki, H., and Kasai, N., Dual production of highly pure methyl (R)-4-c hydroxybutyrate and (S)-3-hydroxy-γ-butyrolactone with *Enterobacter* sp., *Enzyme Technol.*, 24, 13–20, 1999.

79. Suzuki, T., Kasai, N., Yamamoto, R., and Minamiura, N., A novel enzymatic dehalogenatio 3-chloro-1,2-propanediol in *Alcaligenes* sp. DS-S-7G, *Appl. Microbiol. Biotechnol.*, 42, 270–2

80. Pries, F., Kingma, J., Pentenga, M., et al., Site-directed mutagenesis and oxygen isotope incor studies of the nucleophilic aspartate of haloalkane dehalogenase, *Biochemistry*, 33, 1242–124

81. Nakamura, T., Nagasawa, T., Yu, F., et al., A new enzymatic synthesis of (R)-γ-chloro-(xybutyronitrile, *Tetrahedron*, 50, 11821–11826, 1994.

17 Enzymatic Synthesis of Sugar Esters and Oligosaccharides from Renewable Resources

A. Ballesteros, F.J. Plou, M. Alcalde, M. Ferrer,
H. García-Arellano, D. Reyes-Duarte, and I. Ghazi

CONTENTS

17.1 CARBOHYDRATES AND FATS AS RENEWABLE FEEDSTOCK

As fossil raw materials are diminishing, and as the pressure on our environment is strengthening, the progressive switch of industry toward renewable feedstock emerges as an unavoidable necessity [1]. In fact, one of the basic principles of Green Chemistry is to develop processes that use renewable starting materials as an alternative to fossil resources [2]. Plant biomass represents the dominant source of feedstock for biotechnological processes as well as the only predictable sustainable source of organic fuels, chemicals, and materials [3]. For the best utilization of biomass, it first needs to be separated into its principal components: cellulose, hemicellulose, lignin, xylooligomers, starch, nonstructural carbohydrates, vegetable and essential oils, etc.

Carbohydrates are considered, by far, the most important class of renewable compounds. About 200 billion MT/y of glucose are formed by photosynthesis, of which 95% remains in the carbohydrate stage as mono-, oligo-, and polysaccharides and glycoconjugates, whereas the remaining 5% is transformed through metabolic pathways into amino acids and later into proteins, nucleic acids, and steroids [4]. A minor fraction (~4%) of the total carbohydrates

produced by nature is used by man [1], and the rest decays and recycles along r
pathways [5].

Only a few carbohydrates fulfill the criteria of low price, quality, reactivity, and avail
to constitute interesting raw material sources. Polysaccharides constitute the bulk
renewable carbohydrate biomass (Table 17.1). Nowadays, corn is the dominant feedst(
biological productions, as its carbohydrate (starch) is in a form that is more homogeneo
more reactive than that found in cellulosic materials [3]. The disaccharide sucrose—fron
beet or sugar cane—is the most abundant pure organic molecule produced at the ind
scale ($>125 \times 10^6$ MT/y) [6]. Finally, the constituent repeating units of the main polysa
ides—glucose (cellulose, starch), fructose (inulin), and xylose (xylan)—are inexpensi
available on multiton scale. Thus, D-glucose is produced from starch in the form of syr
pure crystals [1]. Fructose is prepared by isomerization of starch-derived glucose or
drolysis of inulin. Sorbitol, the hydrogenated glucose derivative, is also widely availabl

Together with carbohydrates, plant oils and animal fats play an important r
renewable resources because of their availability and versatile applications [5,8]. Wh
largest share ($\sim 100 \times 10^6$ MT/y) of fats and oils is used as human foodstuff, 15% is av
for oleochemistry (soaps, detergents, cosmetics, biodiesel, lubricants, etc.) [5].

In general, carbohydrates and plant oils are attractive renewable raw materials i
production of a wide range of food and chemical products, including pharmace
agrochemicals, insecticides, and novel biodegradable surfactants [9]. The use of milc
energy-demanding biocatalytic processes for the transformations of these compounds
the lowest number of by-products and is the "greenest" alternative. Actually, one
principles of Green Chemistry is to use catalytic reagents as selective as possible [10].

In this chapter, by combining recent work of our group and literature data, we will r
(1) the applications of lipases and proteases in the synthesis of fatty acid esters of disa
ides and higher oligosaccharides; and (2) the use of transglycosidases for the synth
(prebiotic) oligosaccharides from different carbohydrate resources.

17.2   SYNTHESIS OF FATTY ACID SUGAR ESTERS BY LIPASES AND PROT

Sugar esters are nonionic surfactants formed by a carbohydrate moiety as hydrophilic
and one or more fatty acids as lipophilic component. By controlling the esterification
and the nature of fatty acid and sugar, it is possible to synthesize derivatives with di
hydrophile–lipophile balance [11]. Carbohydrate fatty acid esters, synthesized from ren
resources, have a vast number of applications in the food, polymers, cosmetics, ora
detergent, and pharmaceutical industries [12]. Their properties as antimicrobials fo
storage [13], antitumorals [14,15], anticaries [16], and insecticidals [17] indicate their
versatility. Among them, sucrose esters are the most developed carbohydrate esters a
produced at about 4000 MT/y [7].

TABLE 17.1
Carbohydrates Available as Feedstock in Multiton Scale

Polysaccharides	Disaccharides	Monosaccharides	Sugar Derivative
Starch	Sucrose	Glucose	Sorbitol
Cellulose		Fructose	
Inulin		Xylose	
Xylan			

As sucrose contains eight reactive hydroxyl groups, the degree of substitution and the acylation position on the disaccharide skeleton have a notable effect on its physicochemical properties. Accordingly, monosubstituted sucrose esters are oil-in-water emulsifiers, whereas diesters of long-chain fatty acids and higher esterified derivatives are water-in-oil ones [18]. Regioselective acylation of carbohydrates, i.e., the result of a chemospecificity of one alcohol function among all the others, is an arduous task. Thus, selective chemical acylation requires complex protecting-group methodologies. The regioselectivity can be controlled by structural and electronic factors [19]. It has been demonstrated that the classically accepted view of sucrose nucleophilic reactivity describing the primary 6-OH and 6'-OH as the more reactive positions is not valid anymore, as the reactions can be oriented toward different hydroxyls, depending on the nature of the electrophilic partner and on the reaction conditions and catalysts [19]. Thus, 2-OH selective reactions have been described based on electronic factors [20], primaries 6-OH and 6'-OH have been chemically acylated based on steric hindrance [19], and primary 1'-OH acylation is achieved using proteases as biocatalysts [21].

As stated, sugar esters can be produced using either chemical or biological catalysts. Although they are currently manufactured by base-catalyzed chemical synthesis, the high temperatures required, poor selectivity, and formation of furfurals have focused attention on the more selective enzymatic process [22,23]. Although remarkable work has been done on acylation of monosaccharides as is described elsewhere [24,25], we focus here basically on the acylation of disaccharides and other oligosaccharides, which offer a higher level of structural and functional complexity.

17.2.1 ENZYME REGIOSPECIFICITY

Lipases and carboxylesterases are the "natural" biocatalysts for reactions that involve fatty acids, including sugar ester synthesis. Both types of enzymes belong to the family of carboxylic ester hydrolases (EC 3.1.1), which catalyze the cleavage of ester bonds of lipids and other organic compounds, and are divided into 79 enzymatic families according to the specific bond, moiety, and substrate they hydrolyze [26]. These enzymes are very stable in water and nonpolar organic solvents, and posses regio- and stereospecificity. Unlike carboxylesterases (EC 3.1.1.1), which preferentially hydrolyze water-soluble "simple" esters and triglycerides bearing fatty acids shorter than C6, lipases (EC 3.1.1.3, also known as triacylglycerol lipases) prefer water-insoluble substrates, typically triglycerides composed of long-chain fatty acids. Apart from lipases and carboxylesterases, serine proteases (EC 3.4.21), specifically those of the subtilisin family, have also been successfully employed in sugar acylation processes [21,27], even with long-chain fatty acids [28].

For the enzyme-catalyzed transesterification of sugars, different regioisomers may be obtained with an appropriate selection of the biocatalyst (Figure 17.1). This is remarkable because, for example, the properties of different sugar monoesters have been reported to vary significantly [29]. While the lipases from *Pseudomonas* sp., *Mucor miehei*, and *Thermomyces lanuginosus* are regiospecific for the hydroxyl 6-OH [30,31], the lipase B from *Candida antarctica* yields a mixture of 6- and 6'-monoesters [13,32]. In contrast, several alkaline proteases catalyze selectively the acylation of the primary 1'-OH at the fructose ring [33–36]. This result is noteworthy because 1'-OH is sterically more hindered than the other primary hydroxyl groups at 6- and 6'-positions. Pedersen's group reported the synthesis of 2-O-lauroylsucrose catalyzed by the metalloprotein thermolysin [37]. Recently, several computational studies were performed to explain the differences in regiospecificity [34,38]. One large binding pocket was identified in *C. antarctica* lipase B that accommodates both the sucrose and the acyl moiety of the transition state, whereas in *T. lanuginosus* lipase the binding pocket was found to be smaller, leading to the localization of the two moieties in

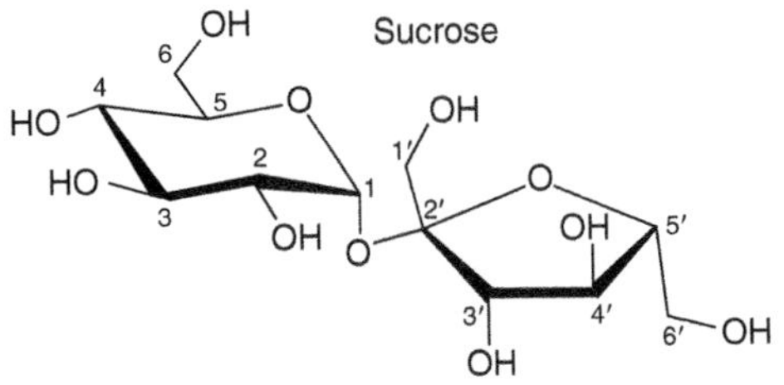

Position	Enzyme
6-OH	*Pseudomonas* sp. lipase [33] *M. miehei* lipase [31] *T. lanuginosus* lipase [30]
6-OH and 6'-OH	*C. antarctica* lipase B [13,32,49]
1'-OH	Subtilisin [21,27,28,33,36] Protease N [42] *B. licheniformis* protease [35]
2-OH	Thermolysin [37]

FIGURE 17.1 Regioselectivity displayed by lipases and proteases in the acylation of sucrose.

two distinct pockets (Figure 17.2). This partially explains the broader specific
C. antarctica lipase B [38].

The esterification of other carbohydrates has also been achieved. For example, the
group at the nonreducing end of maltose is enzymatically acylated by *T. lanuginosus*
and other biocatalysts [13,27,32]. Leucrose, an isomer of sucrose with $\alpha(1{\rightarrow}5)$-glu
fructose link, was also tested under similar conditions: the acylation gave rise to
monoesters, although 6-*O*-lauroyl-leucrose represented 92% of the total [39]. Trehalo
acylated at its primary 6-OH by a protease from *Bacillus subtilis* [40]. With the trisacc
maltotriose, the hydroxyl 6″-OH (at the nonreducing end) was selectively acylate
Subtilisin also esterified the C-1 position of the fructose moiety in the trisaccharide ra
using different vinyl donors [41].

17.2.2 Reaction Medium

Methodologies for enzymatic carbohydrate acylation (also applicable to other polyhyd
compounds) need to find a medium in which a polar reagent (carbohydrate) and a no
acyl donor are soluble and may react in the presence of an appropriate biocatalyst
solvents such as dimethylsulfoxide (DMSO), dimethylformamide (DMF), dimethy
mide, and pyridine were first reported as suitable solvents for sugar solubilization; ho
most lipases and carboxylesterases are readily inactivated by these solvents. An alterna
sugar esterification is based on the hydrophobization of the sugar moiety by comple
with phenylboronic acids or by formation of acetals, thus making the substrate more
in nonpolar solvents in which most enzymes are active [42].

Some proteases of the subtilisin family are able to catalyze the acylation of sugars i
and pyridine [21,28,43]. This fact has been applied to the synthesis of sucrose monome
lates, a group of interesting polymerizable compounds [44]. In this context, subtilisins ha
engineered by rational and evolution approaches to enhance their activity in polar solve

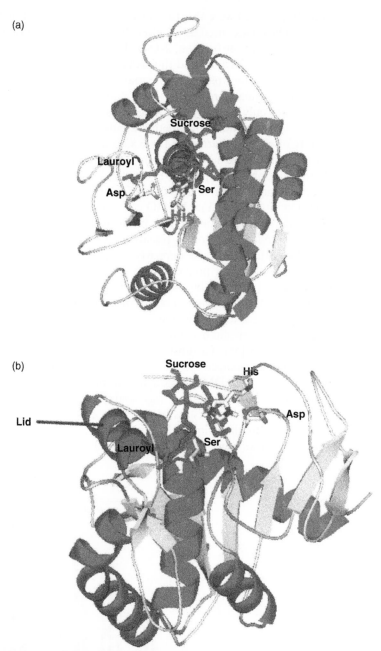

FIGURE 17.2 **(See color insert following page 526)** Structure of *C. antarctica* (a) and *T. lanuginosus* (b) lipases, docked with lauroylsucrose regioisomers. Both enzymes are characterized by similar overall folds and the same catalytic triad residues and oxyanion hole. (Adapted from Fuentes, G., et al., *Protein Sci.*, 13, 3092, 2004. With permission.)

Another drawback of dipolar aprotic solvents, such as DMSO or DMF, is that some acylating agents, e.g., acid anhydrides and vinyl esters, suffer activation processes that may result in spontaneous nonspecific acylation. Studying the transesterification of sucrose with vinyl laurate in DMSO using immobilized lipases, we unexpectedly found that some carriers

(Celite, Eupergit C) were able to catalyze the process [46]. Furthermore, the simple diso(
hydrogen phosphate, used as a buffer during enzyme immobilization, was an efficient (
lyst. In the monoester fraction, 2-*O*-acylsucrose was the major product ($\geq$60%), as hydr
2-OH has a special acidity and thus reactivity [19,20].

In an attempt to exploit the potential of lipases for sugar acylation, several processes
been reported using more benign solvents such as ketones or tertiary alcohols that dissolv
carbohydrate only partially. Although this strategy has been fruitful for monosaccha
using *tert*-butyl alcohol [24], 2-methyl-2-butanol [47], or acetone [48], the comparatively l
solubility of oligosaccharides in these solvents resulted in lower conversions [32].

In this context, we developed a simple process for the lipase-catalyzed acylation of sucr
and other carbohydrates [30,39]. The method was based on the presolubilization of sucr(
a polar solvent (DMSO) and its further mixing with a tertiary alcohol [2-methyl-2-buta
(2M2B)], adjusting the final DMSO content close to 20% (v/v). Our approach of using
miscible solvents as a compromise between sugar solubility and enzyme stability has
successfully followed by other researchers in the acylation of several polyhydroxilic
pounds [49–51]. We recently improved the reaction rate by modifying the substrate pr
ation protocol, namely preincubating sucrose overnight in 2M2B [52]. To emphasiz
importance of sugar solubilization, Halling et al. [53,54] observed that in the synthes
glucose esters, sugar dissolution rate in organic solvents is usually the main limiting para
for carbohydrate ester production.

Variations in medium polarity also influence the selectivity of enzyme-catalyzed
transfer reactions [55]. To substantiate it, we observed that the percentage of DMSO i
reaction medium changed the ratio of monoester/diester obtained [30,52,56]. In particul
concentrations $\leq$10% DMSO, the synthesis of diesters was favored, whereas at $\geq$15% D
the formation of monoesters was majoritary. This effect occurred with lipase
T. lanuginosus and vinyl esters as acylating agents [30], and with methyl palmitate as
donor and immobilized lipase B from *C. antarctica* as biocatalyst [52]. In the latter
Figure 17.3 shows how moving from 0 to 20% DMSO, the molar ratio of monopalm

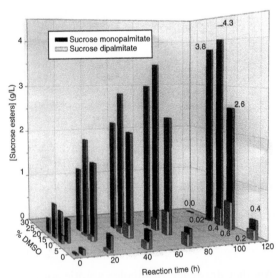

FIGURE 17.3 Effect of DMSO percentage on the ratio of monoester and diester in the transest
tion of methyl palmitate with sucrose. Conditions: 0.1 mol/L sucrose, 0.3 mol/L methyl palmitate
lipase B from *C. antarctica* (Novozym 435), 60°C, and 150 rpm. (Adapted from Reyes-Duarte, D
Biocatal. Biotransform., 23, 19, 2005. With permission.)

dipalmitate increased from 65:35 to 99.5:0.5. The effect of solvent polarity on reaction selectivity seems to be related with sucrose solubility [57]: the amount of dissolved sucrose is higher when increasing DMSO content, going from 0.2 g/L in pure 2M2B to 29.5 g/L in 30% DMSO. Consequently, at high DMSO percentage, sucrose competes more efficiently with the monoester formed, which is very soluble in these reaction media for the acyl-enzyme intermediate, resulting in a major presence of monoester. At $\geq 30\%$ DMSO, the reaction rate was negligible, probably due to the inactivation of the lipase by DMSO.

17.2.3 Acyl Donors

As acylations catalyzed by serine hydrolases take place through the formation of an acyl-enzyme intermediate, the nature of the acyl donor has a notable effect on reactivity [58]. In contrast to hydrolytic reactions, where the nucleophile water is always in great excess, the concentration of nucleophile in acylation processes is always limited; as a result, these reactions are generally reversible in contrast to the irreversible nature of a hydrolytic reaction. The ideal acyl donor should therefore be inexpensive, fast acylating, and completely non-reactive in the absence of the enzyme, but unfortunately such reagents do not exist yet.

For the transformation of inexpensive chemicals, cost is the most important parameter, so simple esters (methyl, ethyl, and glyceryl) are preferred as transesterification reagents. Acylations with these agents are often slow and reversible, and their utility is limited to a narrow range of lipases. This reversibility gives rise to low yields (the thermodynamic equilibrium is commonly not too far from the middle, i.e., $K_{eq} \approx 1$). For this reason, the simple way to increase the rate and yield is to use an excess of acyl donor—but its solubility is limited—or to remove *in situ* the alcohol formed.

A better solution, however, is offered by the use of special acyl donors, which ensure a more or less irreversible reaction. This can be achieved by the introduction of electron-withdrawing substituents (e.g., trihaloethyl esters and oxime esters) [21,59]. Furthermore, if the acyl donor is an enol ester (e.g., vinyl or isopropenyl ester), the reaction yields an enol as the protonated leaving group, which rapidly tautomerizes to acetaldehyde or acetone, respectively (Figure 17.4). These are not nucleophilic, therefore they cannot react with the intermediate acyl-enzyme. In addition, the aldehyde or the ketone formed can be removed by evaporation, thus displacing the equilibrium and making the reaction completely irreversible. Wang et al. [60] demonstrated that the rate of transesterification of hydroxyl-containing compounds with vinyl esters was about 20 to 100 times faster than with other activated esters, and up to 1000 times faster than using methyl or ethyl esters of the same fatty acid.

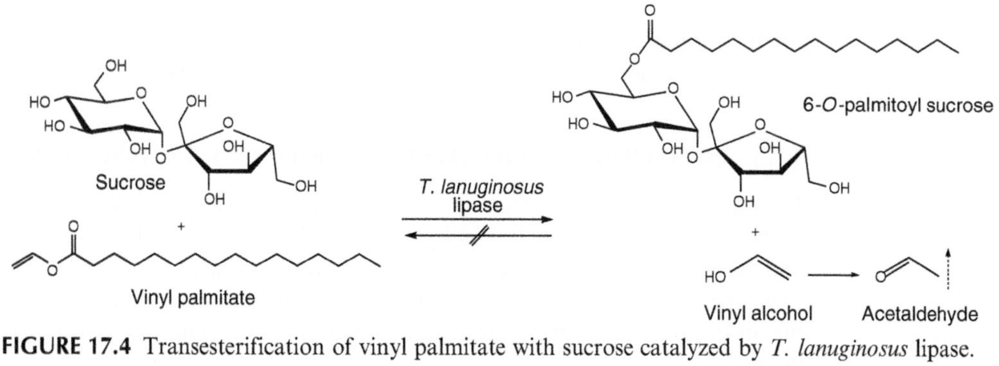

FIGURE 17.4 Transesterification of vinyl palmitate with sucrose catalyzed by *T. lanuginosus* lipase.

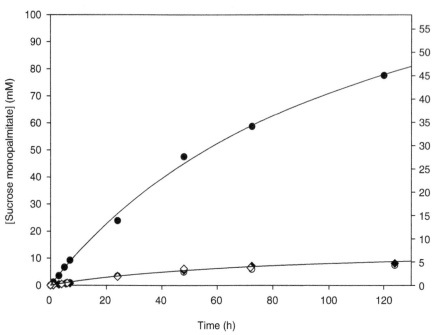

FIGURE 17.5 Effect of the nature of the acyl donor on transesterification. Conditions: 0 sucrose (34 g/L), 0.3 mol/L acyl donor, 2M2B:DMSO 85:15 (v/v), 25 g/L Novozym 435, 60 150 rpm. Acyl donors: (●) vinyl palmitate; (○) methyl palmitate; (▲) ethyl palmitate; and (◇) acid. (Adapted from Reyes-Duarte, D., et al., *Biocatal. Biotransform.*, 23, 19, 2005. With perm

Due to steric reasons, vinyl esters give higher reaction rates than isopropenyl este major disadvantage of vinyl esters is in the released acetaldehyde. Acetaldehyde may a alkylating agent of proteins by forming Schiff's bases in a Maillard-type reaction with groups of lysine residues, which can lead to inactivation. In fact, several lipases (*rugosa* or *Geotrichum candidum*) lose most of their activity when exposed to acetaldehyde The addition of molecular sieves (4 Å) to the medium to trap acetaldehyde seems to against inactivation [62]. However, considering the low boiling point of acetaldehyde it can be removed by simply performing the reaction in an open vessel. Vinyl ac routinely employed in different types of transesterifications and, for the past few year esters of longer saturated fatty acids (C_8–C_{18}) are also commercially available at a rea. price. Figure 17.5 shows that the use of vinyl palmitate gives rise to a productivity on of magnitude higher than that obtained with other activated (methyl palmitate c palmitate) or nonactivated (palmitic acid) acylating agents [52].

17.3 SYNTHESIS OF OLIGOSACCHARIDES BY TRANSGLYCOSIDASES

It is generally accepted that an oligosaccharide is a carbohydrate consisting of 2 monosaccharide residues linked by *O*-glycosidic bonds [63,64]. The development of e and scalable synthetic processes for addressing oligosaccharides in the food (e.g., pre sweeteners, stabilizers, bulking agents) and pharmaceutical industries (as therape prevention of infection, neutralization of toxins, and immunotherapies) is of great [65,66].

Considering the structural diversity of oligosaccharides (3 different amino acids would allow the synthesis of only 6 different peptides, while 3 different hexopyranose moieties would yield up to 720 trisaccharides), the stereo- and regioselectivity of enzymes are considered a valuable alternative to chemical synthesis, which needs complex protection and deprotection steps for the preparation of structurally well-defined oligosaccharides. Actually, enzymatic processes are preferred in the food industry for the production of most important oligosaccharides.

In vivo synthesis of glycosidic bonds is performed by glycosyltransferases (EC 2.4.) [67]. These enzymes catalyze the coupling (by a transfer reaction) of a glycosyl donor to an acceptor molecule forming a new glycosidic bond with regio- and stereoselectivity. According to the nature of the sugar residue transferred, glycosyltransferases are divided into hexosyltransferases (EC 2.4.1.), pentosyltransferases (EC 2.4.2.), and those transferring other glycosyl groups (EC 2.4.99.).

Depending on the nature of the donor molecule, glycosyltransferases are classified into three main mechanistic groups: (1) Leloir-type glycosyltransferases, which require sugar nucleotides (e.g., UDP-glucosyltransferase); (2) non-Leloir glycosyltransferases, which use sugar-1-phosphates (e.g., phosphorylases); and (3) transglycosidases, which employ nonactivated oligosaccharides (e.g., sucrose, starch) as glycosyl donors. A distinctive feature of transglycosidases, compared with Leloir and non-Leloir glycosyltranferases, is that they also display some hydrolytic activity, which can be regarded as a transfer of a glycosyl group from the donor to water. It is noteworthy that, in terms of reaction mechanism, transglycosidases belong to the same group as that of glycosidases (3.2.), a group of hydrolases that catalyze with exquisite stereoselectivity the hydrolysis of glycosidic bonds in oligo- and polysaccharides. According to the Henrissat classification, which is based on amino acid sequence comparisons, transglycosidases and glycosidases constitute the "glycoside hydrolase family (GH family)", with more than 2500 enzymes [68].

In vitro oligosaccharide synthesis can be performed with glycosyltransferases and glycosidases. There are several problems associated with the use of glycosyltransferases of Leloir and non-Leloir type: (1) the requirement of sugar nucleotides or sugar phosphates as substrates, whose synthesis is rather difficult and expensive; (2) the inhibitory effect of the nucleotide phosphate released; and (3) the limited availability of these enzymes [69]. Glycosidases are widely employed for oligosaccharide synthesis, as under appropriate conditions the normal hydrolytic reaction can be reversed toward glycosidic bond synthesis [70]. This can be achieved by thermodynamic control (using low-water concentrations) or by kinetic control (using activated glycosyl donors at high concentrations). Despite the broad specificity of glycosidases and their availability, the application of these catalysts is often limited by low yields and poor regioselectivity [4].

In this context, transglycosidases constitute the ideal biocatalysts for oligosaccharide synthesis *in vitro*, because they do not require special activated substrates, as they directly employ the free energy of cleavage of disaccharides (e.g., sucrose) or polysaccharides (e.g., starch) [71]. Transglycosidases present the same mechanism as retaining glycosidases, resulting in net retention of anomeric configuration. The active site contains two carboxylic acid residues, located at ~5.5 Å apart: one acting as a nucleophile and the other as an acid and base catalyst (Figure 17.6). The reaction proceeds by a double-displacement mechanism in which a covalent glycosyl-enzyme intermediate is formed by the attack of the deprotonated carboxylate to the anomeric center of the carbohydrate with concomitant C–O breaking of the scissile glycosidic bond [72]. The first step is assisted by the carboxylic residue acting as general acid. The second step is the attack of a nucleophile to the glycosyl-enzyme intermediate, which is assisted by the conjugate base of the second carboxyl residue.

The nucleophiles (H_2O) and the acceptor (carbohydrate) compete for the glycosyl-enzyme intermediate (Figure 17.6). When the nucleophile is H_2O, the enzyme acts as a hydrolase; when the sugar is the nucleophile, the enzyme acts as a transferase. The transferase to

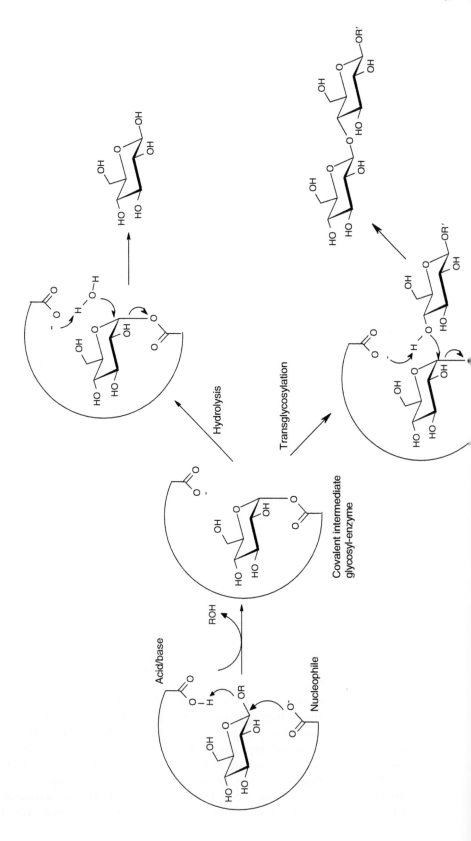

Hydrolysis

Transglycosylation

Acid/base

Covalent intermediate
glycosyl-enzyme

ROH

Nucleophile

hydrolase ratio depends on two main factors: (1) the concentration of acceptor (high concentrations must be used to enhance glycosyl transfer); and (2) the intrinsic enzyme properties. A transglycosidase will be considered efficient if it possesses significant ability to bind the acceptor and to exclude H_2O.

As a consequence of the progress in the understanding of the structures and catalytic mechanisms involved in the enzymatic synthesis of glycosidic bonds, a group of novel mutants—called glycosynthases—were developed by site-directed mutagenesis of glycosidases [73]. The glycosynthase concept was introduced in 1998 by Whiters' group on an exoglucosidase [74], and extended to endo-glycosidases by Planas' group [75]. A glycosynthase is a specifically mutated retaining glycosidase in which substitution of the catalytic carboxyl nucleophile by a noncatalytic residue (Ala, Gly, or Ser) renders a hydrolytically inactive enzyme, but is yet able to catalyze the transglycosylation of activated glycosyl fluoride donors (having the opposite anomeric configuration of that of normal substrates of the parental wild-type enzyme). The yield obtained with glycosynthases reaches 95 to 98% in some cases [69]. The impressive amount of glycosidases available clearly indicates that the potential biodiversity of glycosynthases is still largely unexplored, and new applications of these enzymes will emerge in the near future [76].

17.3.1 TRANSFRUCTOSIDASES

Many microbial organisms and about 12% (i.e., ~40,000 species) of the higher plants build carbohydrate storage based on fructans, polymers formed by β-D-fructofuranose units with a terminal D-glucose. The fructosyl moieties are β$(2\rightarrow6)$-linked in the case of levan or β$(2\rightarrow1)$-linked in the case of inulin, and the enzymes responsible for fructan synthesis are referred to as levansucrases and inulosucrases, respectively [77]. They belong to the glycansucrases family, a group of transglycosidases acting on sucrose and utilizing it as the sole energy source for oligo- and polysaccharide synthesis (Table 17.2). Glycansucrases are subdivided into glucansucrases and fructansucrases depending on the transferred monosaccharide.

An interesting feature of several glycansucrases is their ability to catalyze the synthesis of low-molecular-weight (LMW) oligosaccharides from sucrose when efficient acceptors are

---

TABLE 17.2
Polysaccharides Synthesized by Glycansucrases

Polysaccharide	Bond Formed	Branching	Enzyme	EC Number	Source Microorganism	Group Transferred
Dextran	α-$(1\rightarrow6)$	α-$(1\rightarrow2)$ α-$(1\rightarrow3)$	Dextransucrase (Glucansucrase)	2.4.1.5	*Leuconostoc mesenteroides Streptococcus strains*	Glucosyl
Mutan	α-$(1\rightarrow3)$	α-$(1\rightarrow6)$	Mutansucrase (Glucansucrase)	2.4.1.5	*Streptococcus strains*	Glucosyl
Amylose	α-$(1\rightarrow4)$	—	Amylosucrase	2.4.1.4	*Neisseria polysaccharea*	Glucosyl
Alternan	α-$(1\rightarrow3)$ alterning with α-$(1\rightarrow6)$	α-$(1\rightarrow3)$	Alternansucrase	2.4.1.140	*Leuconostoc mesenteroides*	Glucosyl
Inulin	β-$(2\rightarrow1)$	β-$(2\rightarrow6)$	Inulosucrase	2.4.1.9	*Leuconostoc citreum Lactobacillus reuteri*	Fructosyl
Levan	β-$(2\rightarrow6)$	β-$(2\rightarrow1)$	Levansucrase	2.4.1.10	*Bacillus subtilis Rahnella aquatilis*	Fructosyl

added to the reaction medium. The addition of acceptor causes a decrease in the am■
polymer formed, as both reactions are competitive [78].

17.3.1.1   Synthesis of Inulin-Type Fructooligosaccharides

Short-chain fructooligosaccharides (FOS) of the inulin type constitute one of the gr■
prebiotic oligosaccharides most established in the world [79]. Prebiotic agents a■
ingredients that are potentially beneficial to the health of the consumers. Prebiotics
enzymatic digestion in the upper gastrointestinal tract and enter the colon without any■
in their structure. None are excreted in the stools, indicating that they are ferme■
colonic flora so as to give a mixture of short-chain fatty acids (acetate, propiona■
butyrate), L-lactate, carbon dioxide, and hydrogen [80]. By selectively stimulating be
intestinal bacterial genera such as *Bifidobacterium* and *Lactobacillus*, they may h■
following implications for health: (1) potential protective effects against colorecta■
and infectious bowel diseases by inhibiting putrefactive (*Clostridium perfingens*) and p■
(*Escherichia coli*, *Salmonella*, *Listeria*, and *Shigella*) bacteria, respectively; (2) improve■
glucid and lipid metabolisms; (3) fiber-like properties by decreasing the renal ■
excretion; and (4) improvement in the bioavailability of essential minerals [81–85]. ■
also noncariogenic and have a sweet taste of about 40 to 60% that of sucrose.

FOS of the inulin type are fructose oligomers with a terminal glucose group, in wh■
fructosyl moieties are linked through $\beta(1\rightarrow2)$-glycosidic bonds [86]. Their structural ■
is α-D-glucopyranosyl-$(1\rightarrow2)$-$[\beta$-D-fructofuranosyl-$(1\rightarrow2)$-$]_n$ (GF_n). Commercial F■
mainly composed of 1-kestose (GF_2), nystose (GF_3), and 1^F-fructofuranosyl-nystose

FOS are industrially produced through fructosyl transfer from pure sucrose using ■
enzyme (Figure 17.7) [87]. FOS-synthesizing enzymes are produced by many highe■
(e.g., asparagus, chicory, onion, Jerusalem artichoke) and microorganisms, especial■
(e.g., *Aureobasidium pullulans*, *Aspergillus niger*, *A. oryzae*) [86,88–90]. Inulosucrases
not produce FOS but inulin polymers have also been isolated from plants and fungi

FOS-producing enzymes belong to families 32 and 68 in the Henrissat classificati■
The inclusion of FOS-producing enzymes in the group of glycosidases (β-fructofurano■
EC 3.2.1.26) or fructosyltransferases (transfructosidases, EC 2.4.1.9) still remains co■
sial. However, the mechanism displayed by both groups of enzymes is essentially t■
(Figure 17.6), and the assignation of a particular enzyme as β-fructofuranosidase o■
fructosidase should be based on the transferase to hydrolysis ratio. Only a few ■
enzymes have a significant level of transfructosylating activity to make them us■
industrial applications. Recently, several FOS-synthesizing enzymes from *Aspergil*■
have been purified and characterized [92], and the first three-dimensional structu■
β-fructofuranosidase, that of *Thermotoga maritima*, has been resolved [93].

The maximal FOS production for a particular enzyme depends on the relative ■
transfructosylation and hydrolysis [94], which means that the synthesis of oligosaccha■
transglycosidases is kinetically controlled. Consequently, the FOS concentration re■
maximum (Figure 17.8) that may be substantially higher than the equilibrium conce■
[95]. The time required to get the maximum FOS production depends inversely on the ■
of the enzyme; however, the FOS concentration at this maximum is not affected ■
amount of biocatalyst.

Ghazi et al. [95] obtained a maximum FOS production of 61.5% (w/w) in 24 h, ref■
total carbohydrates in the mixture, using an immobilized transfructosidase (Figure 17■
weight ratio of 1-kestose/nystose/1^F-fructofuranosylnystose was 6.2/3.7/0.1. At equi■
(150 h), the FOS production was slightly lower (57 to 58%), although the product dist■
changed notably. At this stage the rest of the carbohydrates were glucose (29 to 31%),

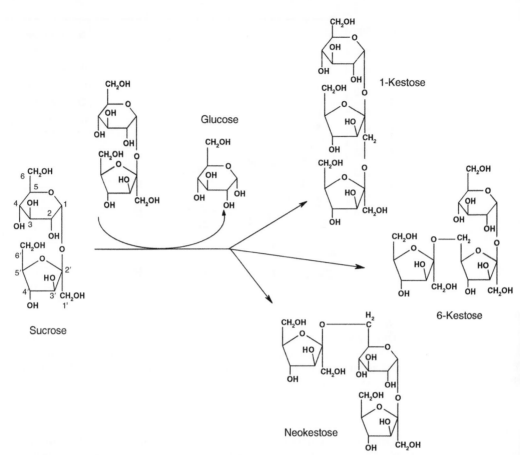

FIGURE 17.7 Reaction scheme for the synthesis of inulin-type fructooligosaccharides, neo-FOS, and ^{6}F-type FOS catalyzed by transfructosidases.

(9.5 to 10.5%) ^{6}F-type and fructose (2 to 3%). Similarly, yields of FOS have been reported with other immobilized transfructosidases [97,98].

Levansucrases catalyze the synthesis from sucrose of levan, a polymer with applications in food, cosmetic, and pharmaceutical industries. In addition to levan formation, levansucrase concomitantly produces FOS of the inulin type [99–101]. Levansucrases also catalyze other transfructosylation reactions in the presence of acceptors such as methanol [102], glycerol [103], and disaccharides [104]. Levansucrases fall into glycoside hydrolase (GH) family 68 in the "carbohydrate-active enzymes" database [68]. Several studies have focused on elucidating the structure and the function of these enzymes. The crystal structure of *B. subtilis* levansucrases was recently solved by Meng and Fütterer [105] at 1.5 Å resolution, and shows a rare fivefold propeller. Site-directed mutants of the three putative catalytic residues of the *Lactobacillus reuteri* 121 levansucrase and inulosucrase (the catalytic nucleophile, the general acid–base catalyst, and the transition state stabilizer) have been recently obtained [106].

17.3.1.2 Synthesis of ^{6}F-Type and Neo-Fructooligosaccharides

Neo-fructooligosaccharides (neo-FOS) consist mainly of neokestose (neo-GF2) and neo-nystose (neo-GF3), in which a fructosyl unit is β(2→6) bound to the glucose moiety of sucrose and 1-kestose, respectively (Figure 17.7).

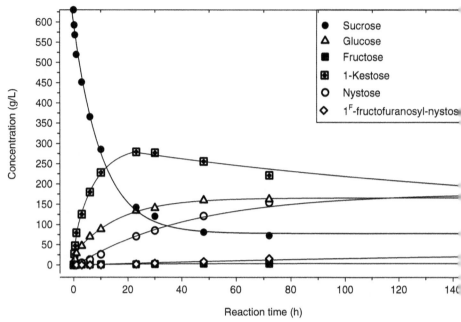

FIGURE 17.8 Time course of the fructooligosaccharides production catalyzed by a transfruct from *Aspergillus aculeatus* (Pectinex Ultra SP-L). Experimental conditions: 630 g/L sucrose, 0. 50 mM sodium acetate buffer (pH 5.4), and 60°C. (Adapted from Ghazi, I., et al., *J. Mol. C Enzym.*, 35, 19, 2005. With permission.)

Grizard and Barthomeuf [107] were the first in reporting the enzymatic synthesis FOS using a transfructosylating activity present in a commercial enzyme from *A. aw* The neo-FOS yield reached a maximum of 50% (w/w) based on the total weight of hydrates in the reaction mixture. Cultures of the astaxanthin-producing yeast *Xantho; myces dendrorhous* accumulated neokestose as a major transfructosylation product growing on sucrose [108,109]. In addition, neokestose occurs only as a minor transfru lating product of whole-cells or enzymes from various plants, yeasts (e.g., *Sacchar< cerevisiae*), and some filamentous fungi [110]. Investigations using human feces as inc *in vitro* have demonstrated that neokestose has prebiotic effects that surpass those o mercial FOS [111].

Short-chain ^{6}F-type FOS have also received some attention (Figure 17.7). In fact, and branched β-(2,6)-linked FOS (the first is 6-kestose) occur naturally in variou products [112]. However, the enzymatic synthesis of ^{6}F-type FOS has been scarcely re Bekers et al. [113] determined the presence of the trisaccharides 1-kestose, neokestos 6-kestose in the fructans syrup obtained with a levansucrase from the ethanol-pro bacteria *Zymomonas mobilis*.

17.3.2   TRANSGLUCOSIDASES

17.3.2.1   Glucansucrases

Several bacteria excrete a range of transglucosidases called glucansucrases that utilize s as the sole energy source to polymerize its glucosyl moiety (Table 17.2). Glucans belong to family 70 of the GH family in the Henrissat classification [68].

Dextransucrases and glucansucrases from the mutans streptococci are the most important members of this group. Glucansucrases from mutans streptococci (EC 2.4.1.5) are able to synthesize extracellular polysaccharides (glucans) from dietary sucrose *in situ,* which serve as adherence sites for streptococci [16,114]. In the protected environment conferred by the glucans, the mutans streptococci and other microorganisms form a stable and protected community (dental plaque) and release sufficient quantities of metabolic acids to demineralize tooth enamel and initiate dental caries. Mutans streptococci secrete at least three glucosyl-transferases: two synthesize α-1,6-linked water-soluble glucans (dextrans), which differ in glucan affinity and degree of α-1,3 branching, while the third synthesizes an α-1,3-linked water-insoluble glucan (mutan) [115].

Dextransucrases (sucrose: 1,6-α-D-glucan 6-α-D-glucosyltransferase, EC 2.4.1.5) are closely related to glucansucrases from mutans streptococci. Dextransucrases are produced by different *Leuconostoc mesenteroides* strains and catalyze the synthesis, from sucrose, of α(1→6)-linked glucose polymers called dextrans, releasing fructose [116]. In the presence of other compounds, generally mono-, di-, and short oligosaccharides, the synthesis of acceptor products may occur [117]. The three reactions catalyzed by dextransucrase are: (1) polymer-ization of the glucose moiety of sucrose; (2) glucose transfer to acceptors; and (3) sucrose hydrolysis, which are competitive, indicating that they take place at the same active site [118].

The regioselectivity displayed by dextransucrases is highly strain-dependent [119]. With dextransucrase from *L. mesenteroides* NRRL B-512F, the synthesis of oligosaccharides with α(1→6)-linked glucose moieties is observed [120]; however, dextransucrase from the strain B-1299 is also able to form α(1→2) linkages [121,122]. Due to the nature of the linkages formed between the glucose moieties, most of the products synthesized by dextransucrase display prebiotic properties.

Numerous sugars can act as acceptors of dextransucrase, which can be classified as (1) strong acceptors, e.g., maltose, which enhance the reaction rate (measured as fructose released) and strongly inhibit the synthesis of dextran; and (2) weak acceptors, e.g., fructose, which have an inhibitory effect on glucan formation, but yield a low amount of acceptor products [116]. Figure 17.9 shows how increasing concentrations of a strong acceptor (i.e., α-D-methyl glucoside) correlate with higher inhibition of dextran formation [123].

The acceptors can also be classified according to the acceptor products obtained: (1) those that give a homologous series of oligosaccharides, each differing one from the other by one glucose residue (e.g., isomaltose); and (2) those that only form a single acceptor product containing one glucose residue more than the acceptor [118]. The latter is the case of fructose, which is a major product in all dextransucrase-catalyzed reactions. Fructose yields leucrose (5-*O*-α-D-glucopyranosyl-D-fructopyranose) and a minor product, isomaltulose (4-*O*-α-D-glucopyranosyl-D-fructofuranose), in which fructose is in the furanose form. The leucrose synthesis process becomes particularly important at the final stages of the glucan synthesis reaction as the fructose concentration is high [124].

With several acceptors such as glucose, methyl 1-*O*-α-D-glucopyranoside, maltose or isomaltose, the glucose unit from sucrose is transferred to the C-6 hydroxyl of the monosac-charide or to the 6-OH of the nonreducing end glucose of the disaccharide. This yields a series of isomaltodextrins with degree of polymerization from 2 to 7 attached to the acceptor. Isomaltooligosaccharides constitute an important group of prebiotics that are also useful as immunostimulants and anticaries agents [125,126].

Using dextransucrase from *L. mesenteroides* B-1299, glucooligosaccharides containing α(1→2) are synthesized, which exhibit particular prebiotic properties [127,128]. Very recently, Boucher et al. [129] demonstrated that a diet supplemented with glucooligosaccharides containing α(1→2) linkages may regulate the carbohydrate metabolism, and it may be useful for patients exhibiting loss of insulin sensitivity. In addition, they are capable of promoting

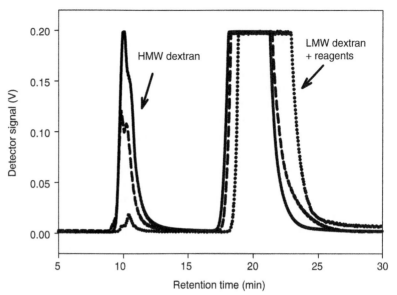

FIGURE 17.9 Size exclusion chromatography analysis of high-molecular-weight (HMW) dext low-molecular-weight (LMW) oligosaccharides synthesized by dextransucrase B-512F, at diffe▶ ratios sucrose:methyl α-D-glucopyranoside, 100:100 (g/L) (▬▬▬), 100:400 (g/L) (- - - - - -), an◀ sucrose in absence of acceptor (......). Reaction time: total consumption of sucrose (24 ▮ Experimental conditions: 100 g sucrose/L, 0.3 U/ml, 30°C, and pH 5.4. (Adapted from Gómez de Se et al., *Food Technol. Biotechnol.*, 42, 337, 2004. With permission.)

the development of the beneficial cutaneous flora in detriment of the undesirable micr◀ isms, either pathogenic or those associated with infections. Based on the acceptor ▮ with maltose, dextransucrase from *L. mesenteroides* B-1299 is being employed for pr▶ 50 MT/y of nondigestible glucooligosaccharides containing α(1→2) bonds for the de▮ metic industry [130].

Cellobiose gives an unusual series in which the first product is 2-α-D-glucopy◀ cellobiose with glucose attached to the C-2 of the reducing-end moiety. With la◀ cellobiose analog, only the first acceptor product is formed, but with similar regiose◀ [131]. Nonconventional acceptors such as D-sorbitol, α-D-glucosyl-1,6-D-sorbitol, α-D-g▮ 1,6-D-arabonic acid, and D-glucal can also be glucosylated with dextransucrase [132].

17.3.2.2 Cyclodextrin Glucosyltransferases

Cyclodextrin glucanotransferases (CGTases, EC 2.4.1.19) are a group of transglucosida◀ belong to the family 13 of GH [68], called α-amylase family. CGTases catalyze the form▶ cyclodextrins (CDs) from starch by way of an intramolecular transglucosylation reactio◀ zation), in which part of the α(1→4)-amylose chain is cyclized by formation of an ad◀ α(1→4)-glucosidic bond (Figure 17.10). CDs possess a hydrophilic exterior and a hydr▶ cavity, and have numerous applications in food, pharmaceutical, chemical, and cosmeti◀ tries [133]. CGTases usually produce a mixture of α, β, and γ-CDs (containing six, sev◀ eight α-D-glucose units, respectively), and are accordingly classified as α, β, and γ-CGTas◀ For example, the commercially available CGTase from *Thermoanaerobacter* sp., whose o◀ temperature is 85°C at pH 5.5 [135], is able to convert about 30% of a 25% (w/v) starch-sl▶ a mixture of CDs, with a ratio α:β:γ of 3:5:2 [136].

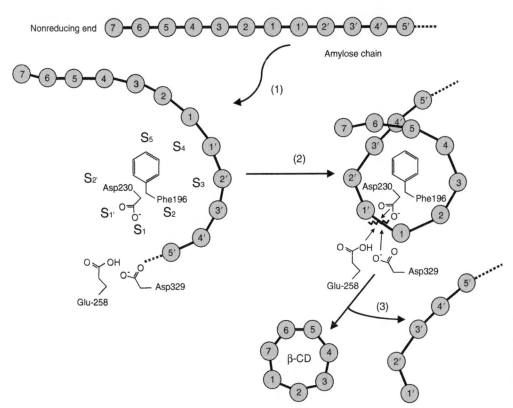

FIGURE 17.10 Cyclodextrin synthesis by CGTases. (1) The amylose chain is attached to the CGTase active site with the help of the cyclization axis (generally an aromatic residue, either Phe or Tyr). (2) The 3-carboxylic acid residues exert a combined attack between subsites S_1 and S_1'. (3) The reducing end of amylose is released from active site. Afterward, the amylose nonreducing end acts as an acceptor (placing at subsites S_1' and S_2') forming the new glycosidic bond and releasing the cyclodextrin.

Besides the cyclization process, CGTases catalyze intermolecular transglucosylation using CDs or oligosaccharides as glucosyl donors; these reactions are referred to as coupling and disproportionation, respectively. In addition, CGTases catalyze the hydrolysis of starch and maltooligosaccharides [137]. The three-dimensional structure elucidation and the biochemical characterization of site-directed mutants have yielded a detailed insight into the mechanism of the reactions catalyzed by CGTases [138]. CGTase can be changed into a starch hydrolase with high exospecificity by hampering substrate binding at the remote donor substrate binding subsites [139]. In addition, CGTase was transformed into a starch hydrolase by directed evolution [140]. Chemical modification of certain residues of CGTase has also allowed to increase transglycosylation [141] or hydrolysis [142] activities. When these acceptors are present in the reaction mixture, they inhibit the formation of cyclodextrins because the glucosyl moiety of the donor (starch) is deviated toward the acceptor reaction (Figure 17.11).

The acceptor specificity of CGTase is rather broad. It is able to use various carbohydrates and related compounds as acceptors by way of intermolecular transglucosylation reactions, coupling or disproportionation. The transglucosylation capability of CGTase seems to be very dependent on the enzyme source [143].

To act as a CGTase acceptor, a carbohydrate must have a D-glucopyranose structure (chair form) with equatorial hydroxyl groups at C-2, C-3, and C-4 [134]. Among monosaccharides, D-glucose, D-xylose, and L-sorbose are the most effective acceptors. CGTase

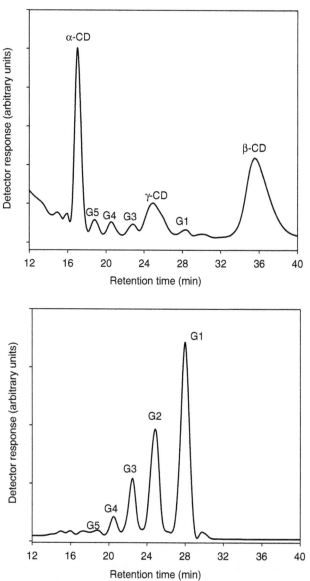

FIGURE 17.11 HPLC chromatograms of products synthesized by CGTase in the absence (top) the presence of D-glucose (bottom). Conditions: 0.7 μg/ml *Thermoanaerobacter* CGTase, 10% soluble starch, 20% (w/v) D-glucose, 60°C, 10 mM sodium citrate buffer (pH 5.5) containing 0. CaCl₂, and incubation for 48 h. (Adapted from Martin, et al., *Biocatal. Biotransform.*, 19, 21, 200 permission.)

transfers glucosyl residue to the hydroxyl group at the C-4 of D-glucose and D-xylose, the 3-OH of L-sorbose. On the contrary, D-galactose, D-ribose, D-mannose, D-arabinos D-fructose—which do not present the mentioned configuration of 2-OH, 3-OH, and 4- give rise to very low or even negligible yields of acceptor products.

The transglycosylation to disaccharides containing a nonreducing end glucose r occurs mainly or exclusively at the C-4 hydroxyl group of this glucose moiety. Nak

TABLE 17.3
Examples of Acceptors That Work with Transfructosidases and Transglucosidases

Type of compound	Acceptor	Enzyme	References
Carbohydrates	Monosaccharides, Disaccharides	CGTase, dextransucrase, transfructosidases	104,145,151–153
Sugar alcohols	Myo-inositol, Sorbitol, Lactitol, Xylitol, Maltitol	CGTase	147,154
Functionalized sugars	Alditols, Aldosuloses, Sugar acids, Glycals, Alkyl saccharides, Fructose dianhydride	Dextransucrase	155
Glycosides	Rutin, Salicin	CGTase	143,156
Vitamins	Ascorbic acid	CGTase	146,157
Simple alcohols	Methanol	Levansucrase	102
Polyols	Trimethylolpropane, Pentaerythritol	CGTase	158
	Glycerol	Levansucrase	103
Flavonoids	Hesperidin, Naringin, Catechin	CGTase	150,159,160

et al. [144] demonstrated that maltose and cellobiose were better acceptors than glucose for the transglycosylation reaction of CGTase. With maltose or glucose as acceptors and starch as donor, linear malto-oligosaccharides (MOS) composed of α-D-glucose residues linked by α(1→4)-glycosidic bonds are produced [145]. A homologous series of MOS is obtained, as shown in Figure 17.7. The degree of polymerization of the oligosaccharides formed can be modulated varying the starch to acceptor ratio. The higher the acceptor concentration, the higher will be the selectivity to LMW oligosaccharides.

It is well documented that CGTase has a higher affinity to disaccharides compared with monosaccharides [143]. This suggests that the acceptor-binding site can recognize at least two glucopyranose moieties. Disaccharides such as isomaltose, gentiobiose, turanose, maltulose, isomaltulose, cellobiose, and sucrose act as good CGTase acceptors, yielding the corresponding oligosaccharides. A steric factor possibly plays a major role in diminishing the acceptor capacity of maltooligosaccharides larger than maltose (e.g., maltotriose).

Other hydroxyl-compounds, such as glycosides, sugar alcohols, vitamins, flavonoids, etc., have been also reported to act as acceptors, in many cases with high efficiency [146–148]. Glucosylation often results in new stability and solubility properties [149,150]. Table 17.3 summarizes the wide nature of acceptors that can be "recognized" by transglycosidases.

ACKNOWLEDGMENTS

This work was supported by the European Union (Project MERG-CT-2004-505242) and the Spanish CICYT (Projects BIO2003-02473 and BIO2004-03773-C04-01). M.F. thanks the Spanish Ministerio de Ciencia y Tecnología for its support.

REFERENCES

1. Lichtenthaler, F.W. and Peters, S., Carbohydrates as green raw materials for the chemical industry, *C. R. Chim.*, 7, 65, 2004.
2. Jenck, J.F., Agterberg, F., and Droescher, M.J., Products and processes for a sustainable chemical industry: a review of achievements and prospects, *Green Chem.*, 6, 544, 2004.

3. Lynd, L.R., Wyman, C.E., and Gerngross, T.U., Biocommodity engineering, *Biotechnol. F* 777, 1999.

4. Thiem, J., Applications of enzymes in synthetic carbohydrate-chemistry, *FEMS Microbi* 16, 193, 1995.

5. Warwel, S. et al., Polymers and surfactants on the basis of renewable resources, *Chemosp* 39, 2001.

6. Descotes, G. et al., Preparation of esters, ethers and acetals from unprotected sucre *J. Chem.*, 73, 1069, 1999.

7. Hill, K. and Rhode, O., Sugar-based surfactants for consumer products and technical appl *Fett-Lipid*, 101, 25, 1999.

8. de Castro, H.F. et al., Modification of oils and fats by biotransformation, *Quim. Nova*, 2004.

9. Mulhaupt, R., The use of renewable resources: possibilities and limitations, *Chimia*, 50, 1⁹

10. Tundo, P. et al., Synthetic pathways and processes in green chemistry. Introductory overvi *Appl. Chem.*, 72, 1207, 2000.

11. Holmberg, K., Natural surfactants, *Curr. Opin. Colloid Interface Sci.*, 6, 148, 2001.

12. Nakayama, M. et al., Studies on growth characteristics of *Bacillus* strains isolated from seasoning, *J. Japan. Soc. Food Sci. Technol.*, 50, 537, 2003.

13. Ferrer, M. et al., Synthesis of sugar esters in solvent mixtures by lipases from *Ther lanuginosus* and *Candida antarctica* B, and their antimicrobial properties, *Enzyme Micro nol.*, 36, 391, 2005.

14. Okabe, S. et al., Disaccharide esters screened for inhibition of tumor necrosis factor-alph are new anti-cancer agents, *Jpn. J. Cancer Res.*, 90, 669, 1999.

15. Ferrer, M. et al., Antitumour activity of fatty acid maltotriose esters obtained by er synthesis, *Biotechnol. Appl. Biochem.*, 42, 35, 2005.

16. Devulapalle, K.S. et al., Effect of carbohydrate fatty acid esters on *Streptococcus sobr* glucosyltransferase activity, *Carbohydr. Res.*, 339, 1029, 2004.

17. Michaud, J.P. and Mckenzie, C.L., Safety of a novel insecticide, sucrose octanoate, to b insects in *Florida* citrus, *Fla. Entomol.*, 87, 6, 2004.

18. Ferrer, M. et al., Comparative surface activities of di- and trisaccharide fatty acid esters, *L* 18, 667, 2002.

19. Queneau, Y., Fitremann, J., and Trombotto, S., The chemistry of unprotected sucr selectivity issue, *C. R. Chim.*, 7, 177, 2004.

20. Cruces, M.A. et al., Improved synthesis of sucrose fatty acid monoesters, *J. Am. Oil Che* 78, 541, 2001.

21. Plou, F.J. et al., Enzymatic synthesis of partially acylated sucroses, *Ann. N. Y. Acad. Sci.*, ? 1995.

22. Polat, T. and Linhardt, R.J., Syntheses and applications of sucrose-based esters, *J. Surf. D* 415, 2001.

23. Plou, F.J. et al., Enzymatic acylation of di- and trisaccharides with fatty acids: choo appropriate enzyme, support and solvent, *J. Biotechnol.*, 96, 55, 2002.

24. Degn, P. et al., Lipase-catalysed synthesis of glucose fatty acid esters in tert-butanol, *Bi Lett.*, 21, 275, 1999.

25. Fregapane, G. et al., Enzymatic synthesis of monosaccharide fatty-acid esters and their c son with conventional products, *J. Am. Oil Chem. Soc.*, 71, 87, 1994.

26. Recommendations of the Nomenclature Committee of the International Union of Bioc and Molecular Biology on the Nomenclature and Classification of Enzyme-Catalysed Rea http://www.chem.qmw.ac.uk/iubmb/enzyme; http://us.expasy.org/enzyme, 2005.

27. Riva, S. et al., Protease-catalyzed regioselective esterification of sugars and related comp anhydrous dimethylformamide, *J. Am. Chem. Soc.*, 110, 584, 1988.

28. Polat, T., Bazin, H.G., and Linhardt, R.J., Enzyme catalyzed regioselective synthesis of fatty acid ester surfactants, *J. Carbohydr. Chem.*, 16, 1319, 1997.

29. Husband, F.A. et al., Comparison of foaming and interfacial properties of pure sucros laurates, dilaurate and commercial preparations, *Food Hydrocolloid.*, 12, 237, 1998.

30. Ferrer, M. et al., Lipase-catalyzed regioselective acylation of sucrose in two-solvent mixtures, *Biotechnol. Bioeng.*, 65, 10, 1999.
31. Kim, J.E. et al., Effect of salt hydrate pair on lipase-catalyzed regioselective monoacylation of sucrose, *Biotechnol. Bioeng.*, 57, 121, 1998.
32. Woudenberg van O.M., van Rantwijk, F., and Sheldon, R.A., Regioselective acylation of disaccharides in tert-butyl alcohol catalyzed by *Candida antarctica* lipase, *Biotechnol. Bioeng.*, 49, 328, 1996.
33. Rich, J.O., Bedell, B.A., and Dordick, J.S., Controlling enzyme-catalyzed regioselectivity in sugar ester synthesis, *Biotechnol. Bioeng.*, 45, 426, 1995.
34. Fuentes, G. et al., Computational studies of subtilisin-catalyzed transesterification of sucrose: importance of entropic effects, *Chembiochem*, 3, 907, 2002.
35. Park, O.J., Kim, D.Y., and Dordick, J.S., Enzyme-catalyzed synthesis of sugar-containing monomers and linear polymers, *Biotechnol. Bioeng.*, 70, 208, 2000.
36. Wang, X. et al., Chemo-enzymatic synthesis of disaccharide-branched copolymers with high molecular weight, *Carbohydr. Polym.*, 60, 357, 2005.
37. Pedersen, N.R. et al., Efficient transesterification of sucrose catalysed by the metalloprotease thermolysin in dimethylsulfoxide, *FEBS Lett.*, 519, 181, 2002.
38. Fuentes, G., Ballesteros, A., and Verma, C.S., Specificity in lipases: a computational study of transesterification of sucrose, *Protein Sci.*, 13, 3092, 2004.
39. Ferrer, M. et al., A simple procedure for the regioselective synthesis of fatty acid esters of maltose, leucrose, maltotriose and *N*-dodecyl maltosides, *Tetrahedron*, 56, 4053, 2000.
40. Raku, T. et al., Enzymatic synthesis of trehalose esters having lipophilicity, *J. Biotechnol.*, 100, 203, 2003.
41. Wu, Q. et al., Enzyme-catalyzed regioselective synthesis of novel raffinose vinyl esters, *Chem. J. Chin. Univ. Chin.*, 24, 1806, 2003.
42. Steverink de Zoete, M.C. et al., Enzymatic synthesis and NMR studies of acylated sucrose acetates, *Green Chem.*, 1, 153, 1999.
43. Carrea, G. et al., Enzymatic synthesis of various 1'-O-sucrose and 1-O-fructose esters, *J. Chem. Soc. Perkin Trans. I*, 1057, 1989.
44. Jhurry, D. et al., Sucrose-based polymers. 1. Linear-polymers with sucrose side-chains, *Macromol. Chem. Phys.*, 193, 2997, 1992.
45. Chen, K.Q. and Arnold, F.H., Tuning the activity of an enzyme for unusual environments—sequential random mutagenesis of subtilisin-E for catalysis in dimethylformamide, *Proc. Natl. Acad. Sci. USA*, 90, 5618, 1993.
46. Plou, F.J. et al., Acylation of sucrose with vinyl esters using immobilized hydrolases: demonstration that chemical catalysis may interfere with enzymatic catalysis, *Biotechnol. Lett.*, 21, 635, 1999.
47. Chamouleau, F. et al., Influence of water activity and water content on sugar esters lipase-catalyzed synthesis in organic media, *J. Mol. Catal., B Enzym.*, 11, 949, 2001.
48. Arcos, J.A., Hill, C.G., and Otero, C., Kinetics of the lipase-catalyzed synthesis of glucose esters in acetone, *Biotechnol. Bioeng.*, 73, 104, 2001.
49. Pedersen, N.R. et al., Effect of fatty acid chain length on initial reaction rates and regioselectivity of lipase-catalysed esterification of disaccharides, *Carbohydr. Res.*, 337, 1179, 2002.
50. Simerska, P. et al., Regioselective enzymatic acylation of *N*-acetylhexosamines, *J. Mol. Catal., B Enzym.*, 29, 219, 2004.
51. Castillo, E. et al., Lipase-catalyzed synthesis of xylitol monoesters: solvent engineering approach, *J. Biotechnol.*, 102, 251, 2003.
52. Reyes-Duarte, D. et al., Parameters affecting productivity in the lipase-catalysed synthesis of sucrose palmitate, *Biocatal. Biotransform.*, 23, 19, 2005.
53. Flores, M.V. and Halling, P.J., Full model for reversible kinetics of lipase-catalyzed sugar-ester synthesis in 2-methyl 2-butanol, *Biotechnol. Bioeng.*, 78, 794, 2002.
54. Flores, M.V. et al., Influence of glucose solubility and dissolution rate on the kinetics of lipase catalyzed synthesis of glucose laurate in 2-methyl 2-butanol, *Biotechnol. Bioeng.*, 78, 814, 2002.
55. Rendon, X., Lopez-Munguia, A., and Castillo, E., Solvent engineering applied to lipase-catalyzed glycerolysis of triolein, *J. Am. Oil Chem. Soc.*, 78, 1061, 2001.

56. Ferrer, M. et al., Effect of the immobilization method of lipase from *Thermomyces lanugin* sucrose acylation, *Biocatal. Biotransform.*, 20, 63, 2002.

57. Voutsas, E.C. et al., Solubility measurements of fatty acid glucose and sucrose esters in 2-1 2-butanol and mixtures of 2-methyl-2-butanol with dimethyl sulfoxide, *J. Chem. Eng. D* 1517, 2002.

58. Plou, F.J., Ferrer, M., and Ballesteros, A., Transesterification-Biological, in *Encyclop Catalysis*, Vol. 6, Horváth, I.T., Ed., Wiley-Interscience, New York, 2003, p. 483.

59. Pulido, R. and Gotor, V., Towards the selective acylation of secondary hydroxyl-grc carbohydrates using oxime esters in an enzyme-catalyzed process, *Carbohydr. Res.*, 252, 5:

60. Wang, Y.F. et al., Lipase-catalyzed irreversible transesterifications using enol esters as a reagents—preparative enantioselective and regioselective syntheses of alcohols, glycerol tives, sugars, and organometallics, *J. Am. Chem. Soc.*, 110, 7200, 1988.

61. Weber, H.K., Stecher, H., and Faber, K., Sensitivity of microbial lipases to acetaldehyde by acyl-transfer reactions from vinyl esters, *Biotechnol. Lett.*, 17, 803, 1995.

62. Weber, H.K. and Faber, K., Stabilization of lipases against deactivation by acetaldehyde fo acyl transfer reactions, *Meth. Enzymol.*, 286, 509, 1997.

63. Eggleston, G. and Cote, G.L., Oligosaccharides in food and agriculture, in *Oligosaccha Food and Agriculture*, Eggleston, G. and Cote, G.L., Eds., American Chemical Society, W ton, 2003, chap. 1.

64. McNaught, A.D., International union of pure and applied chemistry and international u biochemistry and molecular biology—joint commission on biochemical nomenclature— clature of carbohydrates (Recommendations 1996) (Reprinted from *Pure Appl. Chem.*, 68 2008, 1996), 43, 1997.

65. Macmillan, D. and Daines, A.M., Recent developments in the synthesis and discovery o saccharides and glycoconjugates for the treatment of disease, *Curr. Med. Chem.*, 10, 2733,

66. Kren, V. and Thiem, J., Glycosylation employing bio-systems: from enzymes to whole cells. *Soc. Rev.*, 26, 463, 1997.

67. Ichikawa, Y. et al., Synthesis of oligosaccharides using glycosyltransferases, *J. Synth. Org. Jpn.*, 50, 441, 1992.

68. Coutinho, P.M. and Henrissat, B., Carbohydrate-Active Enzymes Server at URL: afmb.cnrs-mrs.fr/CAZY/, 1999.

69. Planas, A. and Faijes, M., Glycosidases and glycosynthases in enzymatic synthesis of o charides. An overview, *Afinidad*, 59, 295, 2002.

70. Ajisaka, K. and Yamamoto, Y., Control of the regioselectivity in the enzymatic synth oligosaccharides using glycosidases, *Trends Glycosci. Glycotechnol.*, 14, 1, 2002.

71. Plou, F.J. et al., Glucosyltransferases acting on starch or sucrose for the synthesis of oligos ides, *Can. J. Chem.*, 80, 743, 2002.

72. Crout, D.H. and Vic, G., Glycosidases and glycosyl transferases in glycoside and oligosac synthesis, *Curr. Opin. Chem. Biol.*, 2, 98, 1998.

73. Davies, G.J., Charnock, S.J., and Henrissat, B., The enzymatic synthesis of glycosidic "glycosynthases" and glycosyltransferases, *Trends Glycosci. Glycotechnol.*, 13, 105, 2001.

74. Mackenzie, L.F. et al., Glycosynthases: mutant glycosidases for oligosaccharide synthesis, *Chem. Soc.*, 120, 5583, 1998.

75. Malet, C. and Planas, A., From beta-glucanase to beta-glucansynthase: glycosyl transfer to glycosyl fluorides catalyzed by a mutant endoglucanase lacking its catalytic nucleophile. *Lett.*, 440, 208, 1998.

76. Perugino, G. et al., Recent advances in the oligosaccharide synthesis promoted by catal engineered glycosidases, *Adv. Synth. Catal.*, 347, 941, 2005.

77. Tungland, B.C., Fructooligosaccharides and other fructans: Structures and occurrence, tion, regulatory aspects, food applications, and nutritional health significance, in *Oligosacc in Food and Agriculture*, Eggleston, G. and Cote, G.L., Eds., American Chemical S Washington, 2003, p. 135.

78. Monsan, P. and Paul, F., Enzymatic synthesis of oligosaccharides, *FEMS Microbiol. Rev.*, 1995.

79. Bornet, F.R.J. et al., Nutritional aspects of short-chain fructooligosaccharides: natural occurrence, chemistry, physiology and health implications, *Dig. Liver Dis.*, 34, S111, 2002.

80. Probert, H.M. and Gibson, G.R., Investigating the prebiotic and gas-generating effects of selected carbohydrates on the human colonic microflora, *Lett. Appl. Microbiol.*, 35, 473, 2002.

81. Fooks, L.J., Fuller, R., and Gibson, G.R., Prebiotics, probiotics and human gut microbiology, *Int. Dairy J.*, 9, 53, 1999.

82. Roberfroid, M., Functional food concept and its application to prebiotics, *Dig. Liver Dis.*, 34, S105, 2002.

83. Gibson, G.R. and Ottaway, R.A., *Prebiotics: New Developments in Functional Foods*, Chandos Publishing (Oxford), Oxford, 2000, chap. 2.

84. Tuohy, K.M. et al., Modulation of the human gut microflora towards improved health using prebiotics—assessment of efficacy, *Curr. Pharm. Design.*, 11, 75, 2005.

85. Grizard, D. and Barthomeuf, C., Non-digestible oligosaccharides used as prebiotic agents: mode of production and beneficial effects on animal and human health, *Reprod. Nutr. Dev.*, 39, 563, 1999.

86. Antosova, M. and Polakovic, M., Fructosyltransferases: the enzymes catalyzing production of fructooligosaccharides, *Chem. Pap. Chem. Zvesti*, 55, 350, 2001.

87. Hirayama, M., Sumi, N., and Hidaka, H., Purification and properties of a fructooligosaccharide-producing beta-fructofuranosidase from *Aspergillus niger* ATCC-20611, *Agric. Biol. Chem.*, 53, 667, 1989.

88. Sangeetha, P.T., Ramesh, M.N., and Prapulla, S.G., Fructooligosaccharide production using fructosyl transferase obtained from recycling culture of *Aspergillus oryzae* CFR 202, *Process Biochem.*, 40, 1085, 2005.

89. Shin, H.T. et al., Production of fructo-oligosaccharides from molasses by *Aureobasidium pullulans* cells, *Bioresour. Technol.*, 93, 59, 2004.

90. Fernandez, R.C. et al., Production of fructooligosaccharides by beta-fructofuranosidase from *Aspergillus* sp. 27H, *J. Chem. Technol. Biotechnol.*, 79, 268, 2004.

91. Olivares-Illana, V. et al., Characterization of a cell-associated inulosucrase from a novel source: a *Leuconostoc citreum* strain isolated from Pozol, a fermented corn beverage of Mayan origin, *J. Ind. Microbiol. Biotechnol.*, 28, 112, 2002.

92. Velasco, J. and Adrio, J.L., Microbial enzymes for food-grade oligosaccharide biosynthesis, in *Microorganisms for Health Care, Food and Enzyme Production*, Barredo, J.L., Ed., Research Signpost, Trivandrum, Kerala, India, 2002.

93. Alberto, F. et al., The three-dimensional structure of invertase (beta-fructosidase) from *Thermotoga maritima* reveals a bimodular arrangement and an evolutionary relationship between retaining and inverting glycosidases, *J. Biol Chem.*, 279, 18903, 2004.

94. Nguyen, Q.D. et al., Purification and some properties of beta-fructofuranosidase from *Aspergillus niger* IMI303386, *Process Biochem.*, 40, 2461, 2005.

95. Buchholz, K., Kasche, V., and Bornscheuer, U.T., *Biocatalysts and Enzyme Technology*, Wiley-VCH, Weinheim, Germany, 2005, chap. 2.

96. Ghazi, I. et al., Immobilisation of fructosyltransferase from *Aspergillus aculeatus* on epoxy-activated Sepabeads EC for the synthesis of fructo-oligosaccharides, *J. Mol. Catal., B Enzym.*, 35, 19, 2005.

97. Chiang, C.J. et al., Immobilization of beta-fructofuranosidases from *Aspergillus* on methacrylamide-based polymeric beads for production of fructooligosaccharides, *Biotechnol. Prog.*, 13, 577, 1997.

98. Tanriseven, A. and Aslan, Y., Immobilization of Pectinex Ultra SP-L to produce fructooligosaccharides, *Enzyme Microb. Technol.*, 36, 550, 2005.

99. Euzenat, O., Guibert, A., and Combes, D., Production of fructo-oligosaccharides by levansucrase from *Bacillus subtilis* C4, *Process Biochem.*, 32, 237, 1997.

100. Tambara, Y. et al., Structural analysis and optimised production of fructo-oligosaccharides by levansucrase from *Acetobacter diazotrophicus* SRT4, *Biotechnol. Lett.*, 21, 117, 1999.

101. Trujillo, L.E. et al., Fructo-oligosaccharides production by the *Gluconacetobacter diazotrophicus* levansucrase expressed in the methylotrophic yeast *Pichia pastoris*, *Enzyme Microb. Technol.*, 28, 139, 2001.

102. Kim, M.G. et al., Synthesis of methyl beta-D-fructoside catalyzed by levansucrase from *Rahnella aquatilis*, *Enzyme Microb. Technol.*, 27, 646, 2000.

103. Gonzalez-Muñoz, F. et al., Enzymatic synthesis of fructosyl glycerol, *J. Carbohydr. Chem.*, 1999.

104. Park, H.E. et al., Enzymatic synthesis of fructosyl oligosaccharides by levansucrase from *bacterium laevaniformans* ATCC 15953, *Enzyme Microb. Technol.*, 32, 820, 2003.

105. Meng, G.Y. and Fütterer, K., Structural framework of fructosyl transfer in *Bacillu* levansucrase, *Nat. Struct. Biol.*, 10, 935, 2003.

106. Ozimek, L.K. et al., Site-directed mutagenesis study of the three catalytic residues of the f transferases of *Lactobacillus reuteri* 121, *FEBS Lett.*, 560, 131, 2004.

107. Grizard, D. and Barthomeuf, C., Enzymatic synthesis and structure determination of NF *Food Biotechnol.*, 13, 93, 1999.

108. Kilian, S.G. et al., Transport-limited sucrose utilization and neokestose production by *rhodozyma*, *Biotechnol. Lett.*, 18, 975, 1996.

109. Kritzinger, S.M. et al., The effect of production parameters on the synthesis of the trisaccharide, neokestose, by *Xanthophyllomyces dendrorhous* (*Phaffia rhodozyma*), *Microb. Technol.*, 32, 728, 2003.

110. Hayashi, S. et al., Production of a novel syrup containing neofructo-oligosaccharides by th *Penicillium citrinum*, *Biotechnol. Lett.*, 22, 1465, 2000.

111. Kilian, S. et al., The effects of the novel bifidogenic trisaccharide, neokestose, on the colonic microbiota, *World J. Microbiol. Biotechnol.*, 18, 637, 2002.

112. Marx, S.P., Winkler, S., and Hartmeier, W., Metabolization of beta-(2,6)-linked fructo saccharides by different bifidobacteria, *FEMS Microbiol. Lett.*, 182, 163, 2000.

113. Bekers, M. et al., Fructooligosaccharide and levan producing activity of *Zymomonas* extracellular levansucrase, *Process Biochem.*, 38, 701, 2002.

114. Devulapalle, K.S. and Mooser, G., Glucosyltransferase inactivation reduces dental caries, *Res.*, 80, 466, 2001.

115. Mooser, G. and Iwaoka, K.R., Sucrose 6-alpha-D-glucosyltransferase from *Streptococcus sc* characterization of a glucosyl-enzyme complex, *Biochemistry*, 28, 443, 1989.

116. Monchois, V., Willemot, R.M., and Monsan, P., Glucansucrases: mechanism of act structure-function relationships, *FEMS Microbiol. Rev.*, 23, 131, 1999.

117. Robyt, J.F. and Walseth, T.F., Mechanism of dextransucrase action. 4. Mechanism of reactions of *Leuconostoc mesenteroides* B-512F dextransucrase, *Carbohydr. Res.*, 61, 433,

118. Robyt, J.F., Mechanism and action of glucansucrases, in *Enzymes for Carbohydrate Eng* Park, K.H., Robyt, J.F., and Choi, Y.D., Eds., Elsevier Science, Amsterdam, 1996, p. 1.

119. Jeanes, A. et al., Characterization and classification of dextrans from ninety-six strains of *J. Am. Chem. Soc.*, 76, 5041, 1954.

120. Robyt, J.F. and Eklund, S.H., Relative, quantitative effects of acceptors in the rea *Leuconostoc mesenteroides* B-512F dextransucrase, *Carbohydr. Res.*, 121, 279, 1983.

121. Dols-Lafargue, M. et al., Factors affecting alpha,-1,2 glucooligosaccharide synthesis by *L toc mesenteroides* NRRL B-1299 dextransucrase, *Biotechnol. Bioeng.*, 74, 498, 2001.

122. Gómez de Segura, A. et al., Encapsulation in LentiKats of dextransucrase from *Leu mesenteroides* NRRL B-1299, and its effect on product selectivity, *Biocatal. Biotransfe* 325, 2003.

123. Gómez de Segura, A. et al., Modulating the synthesis of dextran with the acceptor reacti native and encapsulated dextransucrases, *Food Technol. Biotechnol.*, 42, 337, 2004.

124. Buchholz, K., Noll-Borchers, M., and Schwengers, D., Production of leucrose by dextra *Starch-Starke*, 50, 164, 1998.

125. Goulas, A.K. et al., Synthesis of isomaltooligosaccharides and oligodextrans in a recyc brane bioreactor by the combined use of dextransucrase and dextranase, *Biotechnol. Bio* 778, 2004.

126. Buchholz, K. and Seibel, J., Isomaltooligosaccharides, in *Oligosaccharides in Food and Ag* Eggleston, G. and Coté, G.L., Eds., American Chemical Society, Washington, 2003, p. 6

127. Djouzi, Z. et al., Degradation and fermentation of alpha-gluco-oligosaccharides by strains from human colon: *in vitro* and *in vivo* studies in gnotobiotic rats, *J. Appl. Bacte* 117, 1995.

128. Simmering, R. and Blaut, M., Pro- and prebiotics—the tasty guardian angels? *Appl. Microbiol. Biotechnol.*, 55, 19, 2001.
129. Boucher, J. et al., Effect of non-digestible gluco-oligosaccharides on glucose sensitivity in high fat diet fed mice, *J. Physiol. Biochem.*, 59, 169, 2003.
130. Dols, M. et al., Structural characterization of the maltose acceptor-products synthesized by *Leuconostoc mesenteroides* NRRL B-1299 dextransucrase, *Carbohydr. Res.*, 305, 549, 1998.
131. Yamauchi, F. and Ohwada, Y., Synthesis of oligosaccharides by growing culture of *Leuconostoc mesenteroides*. 4. Oligosaccharide formation in presence of various types of glucobioses as acceptors, *Agric. Biol. Chem.*, 33, 1295, 1969.
132. Heincke, K. et al., Glucosylation by dextransucrase—modeling of reaction kinetics and unconventional products, *Ann. N. Y. Acad. Sci.*, 864, 203, 1998.
133. Qi, Q.S. and Zimmermann, W., Cyclodextrin glucanotransferase: from gene to applications, *Appl. Microbiol. Biotechnol.*, 66, 475, 2005.
134. Tonkova, A., Bacterial cyclodextrin glucanotransferase, *Enzyme Microb. Technol.*, 22, 678, 1998.
135. Alcalde, M. et al., Effect of chemical modification of cyclodextrin glycosyltransferase (CGTase) from *Thermoanaerobacter* sp. on its activity and product selectivity, *Ann. N. Y. Acad. Sci.*, 864, 183, 1998.
136. Zamost, B.L., Nielsen, H.K., and Starnes, R.L., Thermostable enzymes for industrial applications, *J. Ind. Microbiol.*, 8, 71, 1991.
137. van der Veen, B.A. et al., The three transglycosylation reactions catalyzed by cyclodextrin glycosyltransferase from *Bacillus circulans* (strain 251) proceed via different kinetic mechanisms, *Eur. J. Biochem.*, 267, 658, 2000.
138. Leemhuis, H. and Dijkhuizen, L., Engineering of hydrolysis reaction specificity in the transglycosylase cyclodextrin glycosyltransferase, *Biocatal. Biotransform.*, 21, 261, 2003.
139. Leemhuis, H. et al., Engineering cyclodextrin glycosyltransferase into a starch hydrolase with a high exo-specificity, *J. Biotechnol.*, 103, 203, 2003.
140. Leemhuis, H. et al., Conversion of cyclodextrin glycosyltransferase into a starch hydrolase by directed evolution: the role of alanine 230 in acceptor subsite+1, *Biochemistry*, 42, 7518, 2003.
141. Alcalde, M. et al., Succinylation of cyclodextrin glycosyltransferase from *Thermoanaerobacter* sp. 501 enhances its transferase activity using starch as donor, *J. Biotechnol.*, 86, 71, 2001.
142. Alcalde, M. et al., Chemical modification of lysine side chains of cyclodextrin glycosyltransferase from *Thermoanaerobacter* causes a shift from cyclodextrin glycosyltransferase to alpha-amylase specificity, *FEBS Lett.*, 445, 333, 1999.
143. Park, D.C., Kim, T.K., and Lee, Y.H., Characteristics of transglycosylation reaction of cyclodextrin glucanotransferase in the heterogeneous enzyme reaction system using extrusion starch as a glucosyl donor, *Enzyme Microb. Technol.*, 22, 217, 1998.
144. Nakamura, A., Haga, K., and Yamane, K., The transglycosylation reaction of cyclodextrin glucanotransferase is operated by a ping-pong mechanism, *FEBS Lett.*, 337, 66, 1994.
145. Martin, M.T. et al., Synthesis of malto-oligosaccharides via the acceptor reaction catalyzed by cyclodextrin glycosyltransferases, *Biocatal. Biotransform.*, 19, 21, 2001.
146. Aga, H. et al., Synthesis of 2-*O*-alpha-D-glucopyranosyl L-ascorbic-acid by cyclomaltodextrin glucanotransferase from *Bacillus stearothermophilus*, *Agric. Biol. Chem.*, 55, 1751, 1991.
147. Kim, T.K., Park, D.C., and Lee, Y.H., Synthesis of glucosyl-sugar alcohols using glycosyltransferases and structural identification of glucosyl-maltitol, *J. Microbiol. Biotechnol.*, 7, 310, 1997.
148. Ohara, S. and Hishiyama, S., Utilization of triterpenoids. 1. Synthesis of betulin glycosides by cyclodextrin glycosyltransferase, *Mokuzai Gakkaishi*, 40, 444, 1994.
149. Kometani, T. et al., Transglycosylation to hesperidin by cyclodextrin glucanotransferase from an alkalophilic *Bacillus* species in alkaline pH and properties of hesperidin glycosides, *Biosci. Biotechnol. Biochem.*, 58, 1990, 1994.
150. Kometani, T. et al., Synthesis of hesperidin glycosides by cyclodextrin glucanotransferase and stabilization of the natural pigments, *J. Jpn. Soc. Food Sci. Technol.*, 42, 376, 1995.
151. Robyt, J.F., Mechanisms in the glucansucrase synthesis of polysaccharides and oligosaccharides from sucrose, *Adv. Carbohydr. Chem. Biochem.*, 51, 133, 1995.
152. Canedo, M. et al., Production of maltosylfructose (erlose) with levansucrase from *Bacillus subtilis*, *Biocatal. Biotransform.*, 16, 475, 1999.

153. Baciu, L.E. et al., Investigations of the transfructosylation reaction by fructosyltransfera *B. subtilis* NCIMB 11871 for the synthesis of the sucrose analogue galactosyl-fru *J. Biotechnol.*, 116, 347, 2005.
154. Sato, M. et al., Synthesis of novel sugars, oligoglucosyl-inositols, and their growth stim effect for *Bifidobacterium*, *Biotechnol. Lett.*, 13, 69, 1991.
155. Demuth, K., Jordening, H.J., and Buchholz, K., Oligosaccharide synthesis by dextransucra unconventional acceptors, *Carbohydr. Res.*, 337, 1811, 2002.
156. Tachibana, Y. et al., Purification and characterization of an extremely thermostable cycl dextrin glucanotransferase from a newly isolated hyperthermophilic archaeon, a *Thermococ Appl. Environ. Microbiol.*, 65, 1991, 1999.
157. Jun, H.K., Bae, K.M., and Kim, S.K., Production of 2-*O*-alpha-D -glucopyranosyl L-ascor using cyclodextrin glucanotransferase from *Paenibacillus* sp., *Biotechnol. Lett.*, 23, 1793, 2
158. Nakano, H. et al., Syntheses of glucosides with trimethylolpropane and 2 related polyol r by cyclodextrin glucanotransferase and their esterification by lipase, *J. Ferment. Bioeng.*, 1992.
159. Kometani, T. et al., Synthesis of neohesperidin glycosides and naringin glycosides by cyclc glucanotransferase from an alkalophilic *Bacillus* species, *Biosci. Biotechnol. Biochem.*, 1996.
160. Funayama, M. et al., Enzymatic-synthesis of (+)catechin-alpha-glucoside and its effect on nase activity, *Biosci. Biotechnol. Biochem.*, 57, 1666, 1993.

18 Efficient Methods and Instrumentation for Engineering Custom Enzymes

Steven J. Robles, William J. Coleman, and Mary M. Yang

CONTENTS

18.1 NATURE'S DIVERSITY

The quest to identify highly active and stable biocatalysts generally requires sampling a highly diverse pool of enzyme variants in order to retrieve rare mutants with desirable properties. One major source of "natural" diversity is contained in the genomes of environmental microorganisms, the vast majority of which have never been cultivated or characterized [1,2]. Methods for screening such libraries for a number of different enzyme activities have been described [3]. However, despite the advantages of screening naturally occurring enzymes, the properties of these biocatalysts are often not optimal for industrial chemistry, as these processes typically involve reaction conditions that are rarely or never encountered in the natural environment [4]. In addition, the intrinsic properties of the enzyme (e.g., substrate affinity, catalytic efficiency, stability) may not be appropriate for the desired substrate. Properties that typically require optimization include:

- Regio- and enantioselectivity
- Activity on a novel substrate
- Resistance to inhibitors, extreme pH, or organic solvents
- Thermostability
- Thermoactivity

Methods have been developed to overcome the natural limitations of enzymes by using mutagenesis to create artificial diversity. Most of these methods are actually adapt natural evolutionary processes. Several of these directed evolution methods are re genetic algorithms (GAs), and will be briefly described in Section 18.2.

18.2   DIRECTED EVOLUTION: REVERSING THE METAPHOR OF THE GENETIC ALGORITHM

GAs were conceived by computer scientists several decades ago as a search techn quickly find solutions to combinatorial optimization problems by mimicking the bi processes of mutation and recombination [5]. In this metaphor on biological evolutic equate biological chromosomes with digital strings of alphanumeric characters. These exist within populations, and each can be scored for fitness. Thus, the basic eleme digital computer's GA are as follows:

1. Population of strings
2. Random mutation along each string
3. Random recombination among the strings
4. Scoring mechanism for fitness
5. Method of recursion

This metaphor was "reversed" in the early 1990s by molecular biologists who beg GAs to engineer proteins in an effort to solve the "protein folding problem," wherein structure and function are calculated from the amino acid sequence. These initial stud performed by computer simulations [6–8], followed by *in vitro* evolution of protein the laboratory [9–12].

In the laboratory, each of the five elements of a digital GA were reembodied in bi material and procedures:

1. Population of DNA molecules
2. Mutation (e.g., using an error-prone polymerase chain reaction [PCR] or a deg cassette)
3. Recombination (production of DNA strands having random combinations o tions)
4. Selection or screening criteria based on the properties of the encoded protein
5. Propagating "fit" clones

In what later became known as sequential random mutagenesis (SRM) for evolution, step 3 was not employed. This noncanonical GA utilized error-prone PCF genesis to create a pool of point mutations whose phenotypes could be screened, resu the isolation of improved variants after several iterations [13]. However, method: on introducing point mutations without thorough recombination require a large of screening because many of the point mutations will be either silent or dele and beneficial combinations of mutations are unlikely to occur within the sam strand [14,15].

DNA shuffling [11,12] and recursive ensemble mutagenesis (REM) [9,10] are ca GAs, using both mutation and random recombination as part of all five steps listed REM is based on combinatorial cassette mutagenesis and predates DNA shuffling. A but important difference between the two methods is that the recombination ope

DNA shuffling is only useful if the point mutations found in seminal strings are independently favorable. This requirement limits its ability to solve hard combinatorial problems where simultaneous mutations are necessary. However, as DNA shuffling uses PCR to amplify families of gene fragments and thereby reconstitute chimeric genes containing various combinations of point mutations, the advantage is that beneficial mutations scattered throughout various DNA fragments can be brought together into the same construct so that their cumulative and synergistic effects can be evaluated [14]. Likewise, deleterious mutations can be eliminated. The disadvantages of PCR-based reassembly are that (1) proper annealing of the fragments requires significant sequence similarity, and recombination is strongly favored in regions of high homology [16,17]; (2) truncated and multimeric sequences are generated by the mutagenesis procedure [18]; and (3) potentially desirable point mutations on different fragments cannot be located too close together in the sequence because their rate of recombination will be low.

The REM procedure (Figure 18.1) seeks to find a set of "optimal" amino acids at preselected positions in a given protein so that its functional behavior can be enhanced or appropriately altered. REM exploits the relationship between genotype (DNA) and phenotype (protein) by incorporating a computer program called CyberDope. This is an interactive program that restricts the sequence complexity of mixed bases in the synthesis of DNA oligonucleotides such that the combinatorial complexity can be reduced. This is accomplished by favoring functional and/or unique mutations in the library. The program calculates

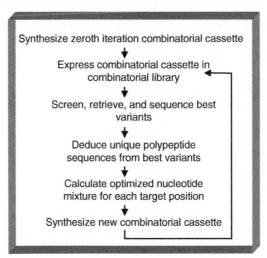

FIGURE 18.1 Flowchart for recursive ensemble mutagenesis (REM), which involves the iterative use of combinatorial cassette mutagenesis. Typical steps in the process are shown here. For the zeroth iteration, a combinatorial cassette can be synthesized using phylogenetic data, randomized codons [NN(G,C) or NNN], sequence information from prior screening (e.g., from error-prone polymerase chain reaction [PCR]), or elements of rational design. The starting cassette is expressed in a library, which is then screened for functional or positive clones. The positive clones are sequenced, and the unique polypeptide sequences are identified. These unique sequences form a target set of amino acids at a particular residue. Based on the target set at each position, the CyberDope program calculates a nucleotide mixture biased toward amino acids in the target set. The output of the program is used to synthesize a new cassette, which is incorporated into another round of mutagenesis and screening. A number of algorithms for modifying or refining this process are included in the CyberDope program. Constraints based on amino acid physicochemical properties can also be built into the design of the combinatorial cassette. Therefore, the program provides improved control over the population diversity.

optimized nucleotide mixtures using algorithms to maximize the occurrence of a speci
of amino acid residues. Information from phylogeny, PCR mutagenesis, or other p
rounds of combinatorial cassette mutagenesis can be used in an iterative manner to
sequence space more efficiently. Multiple REM segments can also be combined to pot
mutagenize entire proteins. With a 16-site library, a 10 million–fold enrichment of fun
and unique mutants, compared to random cassette mutagenesis, has been achieved [1

REM is especially advantageous for targeting particular regions of the protein sec
such as an antibody-binding site or a motif that comprises a portion of the enzyme act
[19]. This kind of targeted mutagenesis may be more advantageous than previously r
For cases in which the desired improvement involves enantioselectivity, substrate sele
or new catalytic activity, an analysis of directed evolution experiments by Morl
Kazlauskas [20] suggests that mutations close to the enzyme active site have a
probability of generating enhanced phenotypes than do more distant mutations.

REM may be seen as a GA constrained by experimental requirements or physicoch
properties. The repertoire of substitutions can be restricted to a certain class—for ex
charged or aromatic amino acids. Nucleotide mixtures can also be determined usin
physicochemical criteria to change the average molar volume, hydropathy, pK_a, o
parameters known to affect protein structure and function. These constraints give ri
powerful new optimization and diversification component not found in traditional G

18.3   THE REALITY OF THE SCREENING BOTTLENECK

Although the GA-based mutagenesis techniques described in Section 18.2 make it pos
explore sequence space with greater thoroughness and efficiency, finding a truly novel
or enzyme can still be fundamentally a numbers game. For example, a relatively small
(e.g., length = 100) may be represented by 20^{100} possible sequences. Randomly incorpe
all 20 amino acids at just six different positions in a given protein will produce a thec
complexity of 64 million unique sequences. The need to screen large libraries is also
part to the scarcity of positive phenotypes. Thus, the demands placed on the av
activity-based screening procedures to process such large libraries efficiently and acc
has created the "screening bottleneck."

To deal with this challenge, various liquid-phase platforms have been used to
libraries at high throughput [21]. Commercial microplate systems employ 96-, 3
1536-well plates with individual reaction volumes ranging from about 10 to 200 μ
advantage of using these systems is that they are versatile and adaptable. However, du
limited well density, screening a somewhat modest library of 1 million variants r
between 700 and 10,000 plates, with a total substrate volume of 10 to 200 L. The inve
required in terms of robotics, reagents, and laboratory space for handling, analyzir
washing large numbers of microplates can therefore be substantial. This means tha
experiments utilize a total throughput of fewer than 10,000 variants.

Fluorescence-activated cell sorting (FACS) has also been used to screen microbi
expressing enzymes [22], and this has been extended to mutagenized enzyme libraries [2
chief advantage of FACS is its extremely high throughput, which is typically 10,000 to
cells/s. However, the applicability of the method is limited by several factors: (1) the pro
be detected must be fluorescent; (2) the substrate must freely enter the cell; (3) the p
must not leak out of the cell; (4) only a single cell can be measured at a time; and (
end-point assays can be performed. The penetration and/or leakage problems can be r
by using cell-surface display of the enzyme and trapping the products [23], or by encaps
the cells in microdroplets [24].

However, despite their proven utility for screening many types of enzymes, liquid-phase techniques are not the only assay format that can be applied to the problem of the screening bottleneck. In order to pursue the goals of greater efficiency, versatility, information content, and cost savings (without sacrificing accuracy or performance), a platform of solid-phase methods, software, and instrumentation known commercially as Kcat technology was therefore developed [25]. This technology is designed to reduce or eliminate some of the drawbacks inherent in conventional liquid-phase screening. By providing very high throughput, the problem of the screening bottleneck can be reduced. Assay design can be more flexible, and sample handling can also be vastly simplified. Application of this method to enzyme engineering will be described in the following sections.

18.4 SOLID-PHASE ENZYME SCREENING TECHNOLOGY: MAXIMIZING DENSITY FOR HIGH THROUGHPUT

A key objective of solid-phase screening is to achieve very high throughput by maximizing the cell density. This can be accomplished in a straightforward manner by using microcolonies randomly distributed on a microporous surface. The basic solid-phase assay for high-throughput enzyme screening is shown in Figure 18.2. Typically, the enzyme gene is expressed using a plasmid vector in a microbial host, such as *Escherichia coli*. The transformed cells containing the mutagenized plasmid library are deposited randomly on an assay disk, which consists of a very thin, microporous membrane. The following steps are performed:

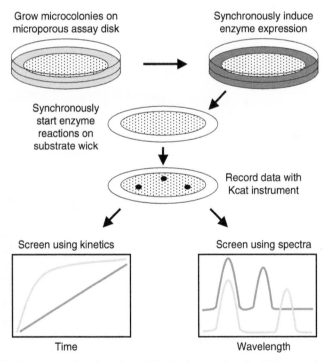

FIGURE 18.2 Solid-phase screening flowchart. The basic steps involved in preparing and analyzing an assay disk using Kcat technology are shown. The kinetics mode can be used to identify variants with enhanced activity, while the spectral mode can be used to compare activity on different substrates or detect novel colored products.

1. Place assay disk with transformed cells on nutrient agar containing the approp antibiotics, and incubate the disk until microcolonies develop on the surface.
2. (Optional) Transfer the assay disk to nutrient agar containing an inducer to sync ously start expression of the enzyme in all of the microcolonies.
3. Transfer the assay disk to a "wick" impregnated with substrate inside the instrument so that the enzyme reactions are synchronously initiated in all c microcolonies.
4. Collect data with the Kcat instrumentation as a function of wavelength or time.
5. Perform massively parallel data analysis and sorting using spectral and/or time- information.

Enzymatic activity can be detected by using chromogenic or fluorogenic substrates, c metric indicators, or coupled enzyme reactions. The Kcat instrument acquires images assay disk over time at the appropriate wavelength(s) to determine the enzymatic activi the microcolonies on the disk. The data obtained consist of full spectra and kinetics for pixel in the image. Kcat software analyzes the disk to identify the desired microcolonie then the "positive" plasmids are retrieved and purified.

Analysis is performed using software that processes and classifies tens to hundr thousands of spectra or kinetics traces. The data can be acquired in any of several c modes (absorption, fluorescence, or reflectance) with the additional dimension of time. are sorted based on criteria selected by the user and displayed as color-coded lines in a "cc plot," which makes them easier to view (see Figure 18.3). Sorting criteria include, for exa

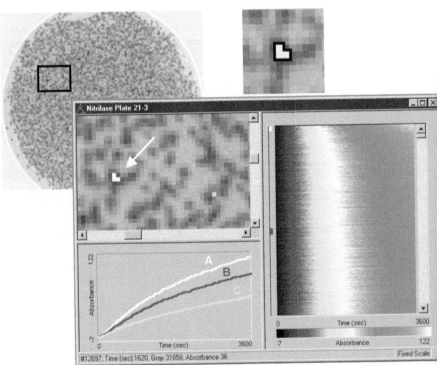

FIGURE 18.3 (See color insert following page 526) Operation of the Kcat graphical user interface Data were obtained during screening of a nitrilase library. In the background is an image of the assay disk, with the region appearing in the GUI enclosed in a rectangle. A single "high-a microcolony is also magnified.

wavelength of maximum absorbance, maximum absorbance at a given wavelength, ratio of absorbance at two wavelengths, and maximum absorbance at a given time. Similarly, derivatives or logarithmic plots of the data can also be generated. When the user selects the appropriate elements from the contour plot by clicking and dragging a computer mouse, the corresponding microcolonies are simultaneously highlighted in an image of the assay disk to facilitate retrieval.

Due to the very high density of microcolonies that can be attained on the disks, a throughput of more than 1 million variants per instrument per day is possible. The flexibility of the format makes it feasible to use cells that are intact or lysed, or that secrete the enzyme. The solid-phase format also provides additional operational advantages. For example, because the number of microcolonies per disk can be up to 50,000 or more and the cells that comprise each microcolony occupy a tiny volume, the amount of substrate solution required per microcolony is only about 50 nL. In addition, since the cells are randomly deposited and the substrate is delivered by diffusion, the sample handling is vastly simplified: pipetting is virtually unnecessary and robotics are not required. Likewise, since the microcolonies are placed in contact with the substrate by simply laying the assay disk onto the wick, high–molecular weight substrates, such as viscous polymer suspensions that are incompatible with automated dispensing, can be used. This type of screening cannot be performed using FACS because the polymers do not penetrate the cells.

Additional levels of complexity can be incorporated into solid-phase assays in order to perform more sophisticated screening procedures. For example, the assay disk can be transferred from one environment or chemical pretreatment to another without disturbing the microcolonies. This makes it possible to evolve multiple properties simultaneously by exposing the cells to heat treatments, pH changes, and inhibitors. On the detection side, the ability to obtain complete spectra during the reaction makes it possible to multiplex the substrates, a feature that is useful for changing the enzyme specificity [19,25]. This can be combined with the time-based information in order to identify variants with enhanced activity as well as altered substrate specificity [19,25,29].

18.5 SYNTHESIS OF PHARMACEUTICAL INTERMEDIATES: NITRILASE

The application of enzymes to pharmaceutical synthesis has been an active area of development for many years. Enzymes and whole-cell biocatalysts provide an attractive alternative to conventional chemical synthesis in situations where high regio- or enantioselectivity is required, or where elimination of chemical waste is a major concern [26].

Nitrilases have attracted attention as catalysts for pharmaceutical production because several important drug compounds are carboxylic acids with chiral centers. Examples of these compounds include Lipitor (atorvastatin), a cholesterol-lowering drug, and (S)-$(+)$-ibuprofen, an analgesic. Although carboxylic acids can be produced using conventional chemistry by hydrolyzing their corresponding nitriles, this process requires harsh chemicals. The chemical waste products must also be processed and disposed of. Nitrilase-catalyzed synthesis also offers the possibility of selectively generating either the (R)- or (S)-enantiomer from the nitrile precursors, and this enantioselectivity could be advantageous for producing a high yield of pure product [26,27].

Naturally occurring nitrilases, however, are not necessarily optimized for pharmaceutical processing conditions. For example, they might be inactive on unusual substrates or in concentrated solutions, and they might not be stable over long periods of time. Introducing large-scale changes to the structure and function of the enzyme is likely to require screening of large libraries of mutants, and in order to perform these directed evolution procedures efficiently, a high-throughput screening system is also required. We have demonstrate this could be accomplished in solid phase with some preliminary measurements on a genized bacterial nitrilase gene library.

The basic experiment was performed as follows. Error-prone PCR mutagenesis wa to create a random library of nitrilase variants expressed from a pUC vector in *A* Microcolonies were grown on assay disks as described in Section 18.4, and a color assay to detect ammonia (a by-product of nitrilase activity) was performed at 37°C increase in the amount of colored product in each microcolony was detected by mon the absorbance at 580 nm over time (60 s intervals).

Figure 18.3 shows the Kcat graphical user interface (GUI) after performing a comp analysis on microcolonies displaying three different levels of nitrilase activity. Approxi 9,000 microcolonies were screened in this experiment on a 47 mm-diameter assay disl the Kcat instrument and software, and approximately 15,000 kinetics traces were sort displayed. The Kcat software was used to sort the kinetics traces pixel by pixel and them as thin horizontal lines in the contour plot window (right). The horizontal axis window is time. In the actual GUI, the intensity of the absorbance at each time point is coded from black or blue (low) to pink or white (high), based on the color scale at the b Hence, some of the details are not discernible in this black-and-white figure. See colo on page xxx.

Pixels showing the greatest change in absorbance over time are sorted to the top. By c and dragging the computer mouse over three different regions of the contour plot w pixels representing high, medium, and low activity were selected. These pixels are identi color-coded markings (grouping bars) to the left of the contour plot window. (The grouping bar is visible after scrolling.) Several different microcolonies corresponding t selected pixels are automatically highlighted in the image window (top left). The a kinetics for each of the three classes are also automatically displayed in conventional f the plot window (lower left). By assigning a measurable activity to particular microcolon variants having the highest apparent activity can then be retrieved from the assay d further characterization. Figure 18.4 shows in greater detail how the sorting process o on kinetics traces to identify the most active microcolonies on the assay disk.

18.6 MODIFICATION OF BIOPOLYMERS: CELLULASE AND OTHER CARBOHYDRATE ENZYMES

Biopolymers encompass an enormous variety of different molecules produced by enz synthesis, including nucleic acids, proteins, polysaccharides, and polyhydroxyalkanoi (PHAs) [28]. The versatility of these compounds can be further extended by chem enzymatic modification to enhance their functional properties. This makes them valuable in industrial, medical, and biotechnological applications. Examples of imp applications include paper and construction (cellulose), biopharmaceuticals (anti and hormones), biodegradable packaging (PHAs), drug delivery (polylactic acid), engineering (hyaluronan), and ethanol (lignocellulosic biomass).

The challenge for researchers who are screening enzymes for the purpose of moc large biopolymers is that the substrates may be insoluble in aqueous buffer or they ma highly viscous suspensions. These properties can make them difficult to handle in c tional liquid-phase systems. For example, in screening for variants of galactose oxida enhanced activity on guar (a galactomannan polymer), it was desirable in some assays t suspension of 2% guar [29]. This material is very difficult to pipette due to its high vis However, using the solid-phase format, it was possible to mix the suspension with 2% a

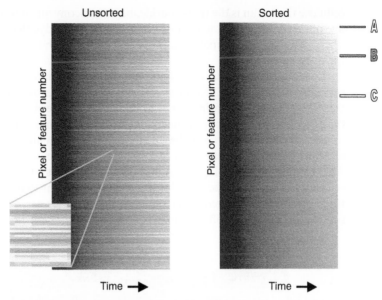

FIGURE 18.4 (See color insert following page 526) Sorting nitrilase kinetics using the contour plot window. All 15,000 kinetics traces are visible in each view. The data are the same as that of Figure 18.3. The *x*-axis corresponds to elapsed time, and the *y*-axis to pixel or feature number. In the actual window, the intensity of the absorption at a given time is color-coded. The unsorted traces on the left display a random color pattern. The traces on the right have been sorted based on their change in absorbance over time. The most "active" pixels are thereby moved to the top. Using the computer mouse, it is possible to select various groups of pixels (marked by the three grouping bars to the right of the window) to identify the corresponding microcolonies.

and then overlay an assay disk containing the microcolonies. This made it feasible to screen for activity directly on the target substrate. A variant with 16-fold higher activity was ultimately identified. Variants with improved thermostability were also isolated using solid-phase screening [19].

Another key enzyme class involved in modifying biopolymers is cellulase, which includes cellobiohydrolase and endoglucanase. Cellobiohydrolase hydrolyzes crystalline cellulose by sequentially removing cellobiose units from the ends of the cellulose chains. Endoglucanase randomly cleaves cellulose within the amorphous regions of the polymer to create chains of various lengths [30]. These enzymes have important applications in pulp and paper processing, textile finishing, and converting agricultural waste into fermentable sugars. Enhanced thermostability is one of the desired goals for cellulase engineering.

Quantitative, high-throughput assays for pulp and paper cellulases are difficult to design because of the high molecular weight and insolubility of the cellulose. In some cases, the actual substrate is wood pulp, which contains other polymeric constituents, such as hemi-celluloses and lignins. For these reasons, many laboratories screening for endoglucanase activity employ surrogate substrates, such as *p*-nitrophenyl cellotrioside (PNP-cellotrioside, which releases a yellow chromophore upon hydrolysis), or azurine cross-linked hydroxyethyl-cellulose (AZCL-HE cellulose, an insoluble fibrous material that releases a blue chromophore upon hydrolysis). In the case of PNP-cellotrioside, the reaction turns from colorless to yellow. However, in the case of AZCL-HE cellulose, the starting material is already colored, and the progress of the reaction is monitored by measuring release of the dye from the cellulose matrix. The latter substrate has the advantage of being more similar to the natural

cellulose fiber than is the trisaccharide, but it is inconvenient to use in liquid phase bec
dye released into solution must be separated from the dye that is still attached to the c
fibers. An analogous separation procedure must also be performed on the soluble ve
the dyed substrate.

In solid phase, the task of identifying the active microcolonies is somewhat easie
presence of AZCL-HE cellulose (dispersed in an agarose matrix), activity is indic
the development of clearing zones in the agarose underneath the microcolonies. By
the assay disk at 590 nm over time, it is possible to monitor the growth of these cleari
to find variants with the highest activity. Variants with improved thermostability
screened by subjecting the assay disk to heating for a fixed time prior to performing th
An example of screening on AZCL-HE cellulose is shown in Figure 18.5. Prel
screening of an endoglucanase library yielded a variant that showed a 15°C improve
thermostability compared to the parental enzyme after pretreatment at various temp
for 15 min. Ongoing software improvements will make it possible to quantitatively

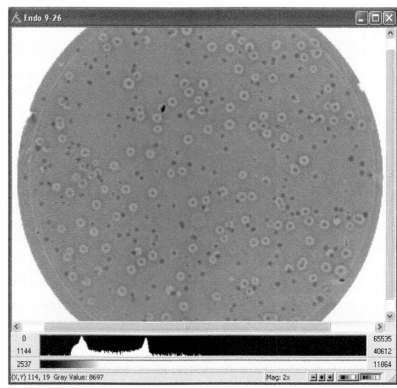

FIGURE 18.5 Thermostability screening of a library of endoglucanase mutants assayed on
cross-linked hydroxyethylcellulose (AZCL-HE cellulose) dispersed in agarose. Microcolonie
assay disk were exposed to 57°C heat treatment for 15 min. The assay for thermal stability v
performed on the substrate at 37°C to detect those mutants with the highest residual activity. Tl
was obtained at 550 nm in the Kcat instrument. Microcolonies containing active endoglucanase
hydrolyze the labeled substrate and release the bound dye. Combined diffusion of enzyme
creates the characteristic "donut" shape around the active microcolonies in this image. The c
intensity loss at this time point is approximately proportional to the enzyme activity remaining
heat treatment. The kinetics of clearing zone formation can be used to identify microcolonies ex
the most active enzyme variants.

the clearing zones and increase the microcolony density. This will benefit all types of assays on carbohydrate polymers that employ the dye release method, including assays for xylanase, mannanase, pullulanase, and amylase.

18.7 FUTURE PROSPECTS

18.7.1 ADVANTAGES OF CUSTOMIZED ENZYMES

The ability to perform very high-throughput screening with a user-friendly, bench-top device will make it possible for individual laboratories to create their own custom enzymes for specific applications. From the industrial process perspective, this means that companies can develop biocatalysts that are highly compatible with their own proprietary chemistry, rather than having to rely on commercial enzyme products that were designed for a different purpose. This new capability will hopefully broaden the range of applications for industrial enzymes.

18.7.2 GREEN CHEMISTRY FOR CHEAPER, CLEANER PHARMACEUTICAL MANUFACTURING

In the area of pharmaceutical and specialty chemical manufacturing, introducing custom enzyme catalysts will provide more options to reduce hazardous waste, streamline complex synthetic processes, and create new chemical entities. Toolboxes of enzymes will be available to provide alternative routes to conventional organic synthesis. Production cost savings and added functional value will enable the users to remain profitable despite increasing competition.

18.7.3 VALUE OF MODIFIED POLYMERS

In the area of polymer chemistry, the availability of customized enzymes will enable chemists to generate novel functionalized polymers. These new compounds will provide exotic materials that are both biocompatible and biodegradable. Moreover, their performance characteristics can be fine-tuned to specific applications. More efficient production of these products will also save energy, reduce pollution and greenhouse gas emissions, and decrease our reliance on petrochemical feedstocks.

REFERENCES

1. Hugenholtz, P., Goebel, B.M., and Pace, N.R., Impact of culture-independent studies on the emerging phylogenetic view of bacterial diversity, *J. Bacteriol.*, 180, 4765, 1998.
2. Streit, W.R., Daniel, R., and Jaeger, K.E., Prospecting for biocatalysts and drugs in the genomes of non-cultured microorganisms, *Curr. Opin. Biotechnol.*, 15, 285, 2004.
3. Brennan, Y. et al., Unusual microbial xylanases from insect guts, *Appl. Environ. Microbiol.*, 70, 3609, 2004.
4. Bommarius, A.S. and Riebel, B.R., Eds., *Biocatalysis*, Wiley-VCH, Weinheim, Germany, 2004.
5. Holland, J.H., Genetic algorithms, *Sci. Amer.*, 267, 66, 1992.
6. Arkin, A.P. and Youvan, D.C., An algorithm for protein engineering: simulations of recursive ensemble mutagenesis, *Proc. Natl. Acad. Sci. USA*, 89, 7811, 1992.
7. Arkin, A.P. and Youvan, D.C., Optimizing nucleotide mixtures to encode specific subsets of amino acids for semi-random mutagenesis, *Biotechnology* (New York), 10, 297, 1992.
8. Youvan, D.C., Arkin, A.P., and Yang, M.M., Recursive ensemble mutagenesis: a combinatorial optimization technique for protein engineering, in *Parallel Problem Solving from Nature II. Proceedings of the Second International Conference on Parallel Problem Solving from Nature*, Maenner, R. and Manderick, B., Eds., Elsevier Publishing, Amsterdam, 1992, p. 401.
9. Delagrave, S., Goldman, E.R., and Youvan, D.C., Recursive ensemble mutagenesis, *Protein Eng.*, 6, 327, 1993.

10. Delagrave, S. and Youvan, D.C., Searching sequence space to engineer proteins: exp(ensemble mutagenesis, *Biotechnology* (New York), 11, 1548, 1993.
11. Stemmer, W.P.C., DNA shuffling by random fragmentation and reassembly: *in vitro* recomt for molecular evolution, *Proc. Natl. Acad. Sci. USA*, 91, 10747, 1994.
12. Stemmer, W.P.C., Rapid evolution of a protein *in vitro* by DNA shuffling, *Nature*, 370, 38!
13. Chen, K.Q. and Arnold, F.H., Enzyme engineering for nonaqueous solvents: random muta to enhance activity of subtilisin E in polar organic media, *Biotechnology* (New York), 9, 107.
14. Tobin, M.B., Gustafsson, C., and Huisman, G.W., Directed evolution: the 'rational' b. 'irrational' design, *Curr. Opin. Struct. Biol.*, 10, 421, 2000.
15. Peng, W. et al., Analytical study of the effect of recombination on evolution via DNA sh *Phys. Rev. E Stat. Nonlin. Soft Matter Phys.*, 69, 051911, 2004.
16. Moore, G. et al., Predicting crossover generation in DNA shuffling, *Proc. Natl. Acad. Sci. L* 3226, 2001.
17. Joern, J.M., Meinhold, P., and Arnold, F.H., Analysis of shuffled gene libraries, *J. Mol. Bi.* 643, 2002.
18. Maheshri, N. and Schaffer, D.V., Computational and experimental analysis of DNA sh *Proc. Natl. Acad. Sci. USA.*, 100, 3071, 2003.
19. Delagrave, S. et al., Combinatorial mutagenesis algorithms, digital imaging spectroscopy, an phase assays for directed evolution, in *Enzyme Functionality: Design, Engineering and Sc.* Svendsen, A., Ed., Marcel Dekker, New York, 2004, p. 507.
20. Morley, K.L. and Kazlauskas, R.J., Improving enzyme properties: when are closer mt better? *Trends Biotechnol.*, 23, 231, 2005.
21. Cohen, N. et al., *In vitro* enzyme evolution: the screening challenge of isolating the one in a : *Trends Biotechnol.*, 19, 507, 2001.
22. Wittrup, K.D. and Bailey, J.E., A single-cell assay of beta-galactosidase activity in *Sacchaw cerevisiae*, *Cytometry*, 9, 394, 1988.
23. Olsen, M.J. et al., Function-based isolation of novel enzymes from a large library, *Nat. Biol* 18, 1071, 2000.
24. Griffiths, A.D. and Tawfik, D.S., Directed evolution of an extremely fast phosphotrieste *in vitro* compartmentalization, *EMBO J.*, 22, 24, 2003.
25. Bylina, E.J. et al., Solid-phase enzyme screening, *ASM News*, 66, 211, 2000.
26. Sheldon, R.A. and van Rantwijk, F., Biocatalysis for sustainable organic synthesis, *J. Chem.*, 57, 281, 2004.
27. Robertson, D.E. et al., Exploring nitrilase sequence space for enantioselective catalysis *Environ. Microbiol.*, 70, 2429, 2004.
28. Gross, R.A. and Cheng, H.N., Eds., *Biocatalysis in Polymer Science*, vol. 840, ACS Sym Series, American Chemical Society, Washington, DC, 2003.
29. Delagrave, S. et al., Application of a very high-throughput digital imaging screen to ev(enzyme galactose oxidase, *Protein Eng.*, 14, 261, 2001.
30. Hilden, L. and Johansson, G., Recent developments on cellulases and carbohydrate-bindin ules with cellulose affinity, *Biotechnol. Lett.*, 26, 1683, 2004.

19 Deaminating Enzymes of the Purine Cycle as Biocatalysts for Chemoenzymatic Synthesis and Transformation of Antiviral Agents Structurally Related to Purine Nucleosides

Enzo Santaniello, Pierangela Ciuffreda, and Laura Alessandrini

CONTENTS

19.1   INTRODUCTION

Many synthetic compounds structurally related to natural nucleosides are characterize
biopharmacological effects that include anticancer [1,2] or antiviral [3,4] activities. Se
compounds are effective against viruses that are causative agents of severe infections, su
the human immunodeficiency virus (HIV) [5,6]. Synthetic antiviral compounds that are s
turally related to nucleosides may be characterized by modifications in the carbohydrate o
moiety, while substitution of the furanose ring with an analogous cyclopentane system gen
carbocyclic nucleosides. In these compounds, the problem associated with the presence
β-N-glycosidic bond, making nucleoside structure vulnerable to chemical or enzymatic hydr
and favoring loss of pharmacological activity in a nucleosidic drug, is eliminated. Only
carbocyclonucleosides are naturally occurring, such as aristeromycin and neplanocin that
been isolated from microbial fermentation [7,8]. In general, carbocyclonucleosides are pre
by chemical total synthesis that sometimes uses chemoenzymatic methodologies [9–11].
synthetic acyclic analogs of nucleosides (acyclonucleosides) also present important an
activity and for the synthesis of two leading products such as acyclovir and ganciclovir che
and chemoenzymatic procedures have already been reviewed [12].

Due to the importance of modified nucleosides and analogs in medicinal chemistr
development of synthetic methodologies for the selective modification of natural nucle
or for the preparation of related compounds has been extensively investigated and
efficient procedures are currently available. A special place is occupied by the method
take advantage of the great potential that enzymes possess as biocatalysts for reactic
synthetic relevance and in this well-established field of research a few reviews deal
enzyme-catalyzed modifications of nucleosides and related compounds [13–15]. The sp
topic of the enzymatic synthesis of antiviral agents was reviewed a few years ago [16] an
work includes also nucleosides and structurally related compounds. Adenosine dean
(adenosine aminohydrolase, ADA, EC 3.5.4.4) and adenylate deaminase (5'-adenyli
deaminase, AMP deaminase, AMPDA, EC 3.5.4.6) are enzymes of the purine cycl
catalyze the hydrolytic deamination of purine nucleosides and nucleotides [17]. Recer
has been shown that these deaminating enzymes are able to transform a great num
modified nucleosides or related analogs and their potential as biocatalysts in nucl
chemistry has been recently reviewed [18].

This chapter will cover applications of the above deaminases to the preparatio
transformation of compounds structurally related to nucleosides that are characteriz
antiviral activity. It should be pointed out that in the majority of articles dealing with mc
nucleosides, positions of the purine ring are indicated by a conventional, rather than syst
numbering that has to be used for different heterocyclic systems related to purine. For th
of clarity, in Figure 19.1 we have reported both the above criteria for adenosine.

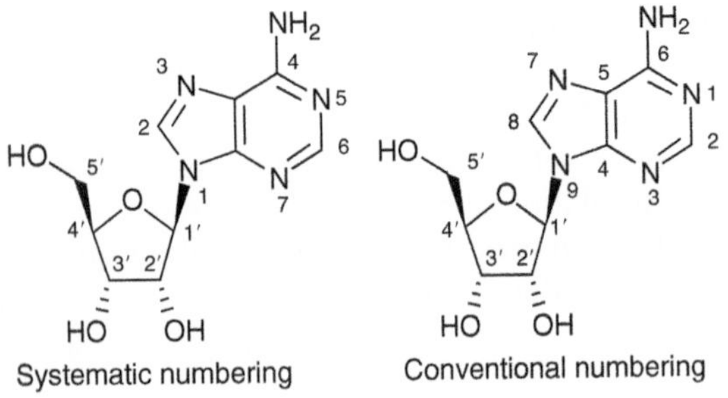

FIGURE 19.1 Systematic and conventional numbering of adenosine.

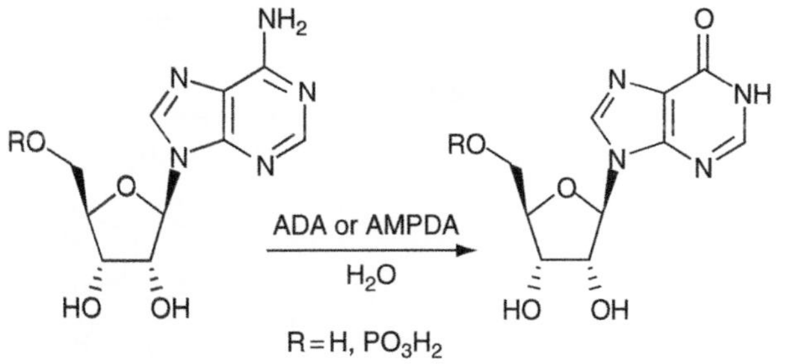

FIGURE 19.2 ADA- and AMPDA-catalyzed hydrolytic deamination of adenosine and adenylic acid.

19.2 DEAMINATING ENZYMES OF THE PURINE CYCLE

ADA and AMPDA belong to the class of hydrolases and are metalloenzymes whose activity depends on the presence of a Zn^{2+} cation [17]. ADA catalyzes the deamination of adenosine to inosine and AMPDA converts 5′-phosphate adenosine (adenylic acid, AMP) to inosine 5′-phosphate (Figure 19.2).

ADA is present in microorganism, invertebrate, and mammalian cells, including human, where its role at physiological and pathological level has been thoroughly investigated, also with the aim of establishing a gene therapy for specific diseases and designing of enzyme inhibitors [19]. The role of AMPDA has been mainly investigated at muscular level and it is now well established that it is one of the principal enzymes of the purine nucleotide cycle [20] that concurs to the regulation of the level of adenosine.

Both enzymes are commercially available at a considerable level of purity and adequate activity, thus ADA and AMPDA are potential biocatalysts for chemoenzymatic transformations in nucleoside chemistry. Calf intestinal mucosa is the main source of commercial ADA (Sigma, type II, 1 to 5 U/mg protein) and AMPDA from *Aspergillus* sp. is the enzymatic preparation used in food industry for large-scale production of inosine 5′-phosphate (inosinic acid), an important meat-like flavor [21]. The fact that these enzymes are hydrolases constitutes an additional advantage in their use as biocatalysts, because they do not require additional coenzymes for their catalytic action. Consequently, these deaminating enzymes do not present the inconvenience related to recycling of the cofactor, a major limitation in the use of coenzyme-dependent biocatalysts [22].

19.3 STRUCTURE AND CATALYTIC MECHANISM OF PURINE NUCLEOSIDE AND NUCLEOTIDE DEAMINASES

No studies on three-dimensional structure of AMPDA have been reported yet, whereas, since the first report on crystallized murine enzyme, a well-established set of structural information is currently available for ADA [23]. All x-ray studies have been carried out on ADA and inhibitor complexes [23–26] and, from the available structural data, a well-defined enzyme-catalyzed mechanism can be drawn for hydration of adenosine, as shown in Figure 19.3. Water attack on position 6 of the purine ring is assisted by His-238, Asp-295, and Zn^{2+} to form a tetrahedral intermediate that irreversibly eliminates ammonia with the formation of inosine.

Although structural data for AMPDA are not available, it is generally accepted that its deaminating activity should follow the same mechanism shown in Figure 19.3 for ADA, as evidenced by the fact that nucleoside-like inhibitors of ADA, after *in vivo* phosphorylation of the 5′-hydroxy group, tend to inhibit AMPDA as well [27].

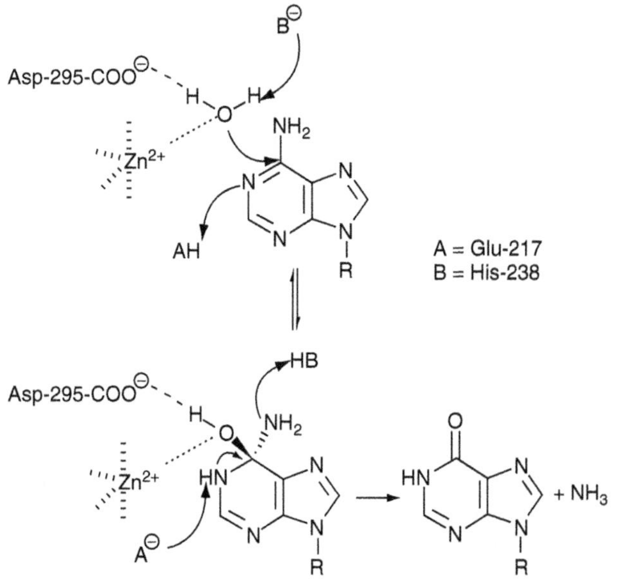

FIGURE 19.3 Mechanism of ADA-catalyzed hydrolytic deamination.

19.4 ADENOSINE DEAMINASE AS BIOCATALYST

Purified ADA from calf intestinal mucosa has been used for the biotransformation of a range of structurally modified purine nucleosides. The enzyme is relatively stable also presence of polar cosolvents [28], which may overcome problems related to the solubi modified nucleosides into the reaction medium constituted by water or buffer solu Works on kinetic determination of ADA activity with different substrates have outl clear picture of the enzyme specificity. In addition to the physiological hydrolytic dea tion of adenosine, ADA is able to transform into inosine a few other 6-substituted ribosides shown in Figure 19.4, although at a slower rate [29–31].

Also ADA from bovine placenta confirmed the above-described specificity with resp the hydrolysis of various 6-substituted adenosines [32]. Introduction of additional gro position 2, such as alkylamino, alkylthio, and halogen [32] as well as nitro [33] cau competitive inhibition of the enzyme. The activity of ADA on adenosines modified furanose moiety has been studied on kinetic basis [34] and interesting conclusions c relationship between substrate structure and enzyme activity were drawn [35]. Stu

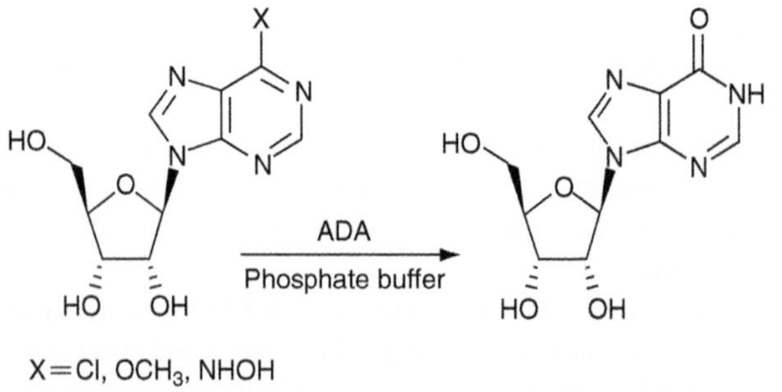

X=Cl, OCH₃, NHOH

FIGURE 19.4 ADA-catalyzed transformation of 6-substituted purine nucleosides.

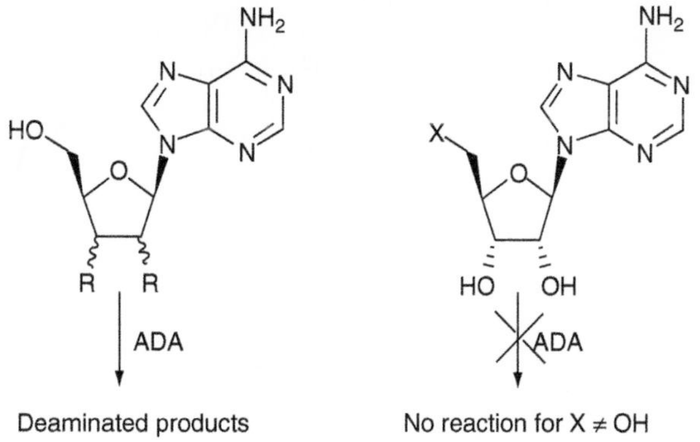

FIGURE 19.5 ADA activity on adenosines bearing modifications in the furanose moiety.

several structurally different substrates, including ribose epimers, it was demonstrated that protection, substitution, or elimination of 2'- and 3'-hydroxy groups were compatible with the enzyme activity. 5'-Protected or 5'-deoxy-5'-substituted nucleosides were not substrates (Figure 19.5) and the overall results suggested that the presence of the 5'-hydroxy group in any modified adenosine or 6-substituted purine nucleoside is essential for ADA activity.

Further, extensive investigation on another series of adenine nucleoside derivatives confirmed the crucial role of the 5'-hydroxymethyl function [36]. Apparent exception to the above statement could be represented by the deamination of 5'-deoxyadenosine [37], but long reaction time was required and yields were not clearly indicated. In the claimed conversion of adenosine 5'-deoxy-5'-thioether to the corresponding inosine derivative with ADA from *Aspergillus orizae* [38], probably the enzyme was not adenosine but adenylate deaminase, as it has recently been suggested [39]. The fundamental importance of the 5'-hydroxy group was evidenced by the results obtained from the ADA-catalyzed deamination of acetates of adenosine and 2'- or 3'-deoxyadenosine [40].

In a reinvestigation of ADA-catalyzed deamination of 5'-protected or 5'-deoxy-5'-substituted 2',3'-isopropylidene adenosines (Figure 19.6), the crucial role of the 5'-hydroxy group was confirmed with the only exception of slow reacting 5'-deoxy-5'-amino derivative [41]. It was also shown that the previously reported use of DMSO as cosolvent [42] did not significantly influence the deamination rate.

19.5 ADENYLATE DEAMINASE AS BIOCATALYST

As mentioned in the introduction, an AMPDA from the microorganism *Aspergillus* sp. is commercially available in bulk quantity for its use in large-scale production of inosine 5'-phosphate for food industry [21]. In spite of this availability, the enzyme has been applied only recently to the synthesis of 6-oxopurine nucleosides and their analogs [43]. In this report, Margolin and coworkers determined the relative rates of deamination of various modified nucleosides, including 6-substituted ribonucleosides, in the presence of AMPDA, comparing their data with those obtained using ADA as reference deaminating enzyme. Interestingly, AMPDA was able to deaminate also 3',5'-cyclic monophosphate adenosine and compounds such as (4'E)- and (4'Z)-4',5'-didehydro-5'-deoxy-5'-fluoroadenosine, 4',5'-didehydro-5'-deoxyadenosine, and 5'-deoxy-2',3'-O-(1-methylethylidene)-5'-(phenylsulfonyl) adenosine. These compounds were not substrates for ADA, thus indicating that for AMPDA the structural restriction related to the presence of the 5'-OH in the substrate was not as mandatory as for ADA (Figure 19.7).

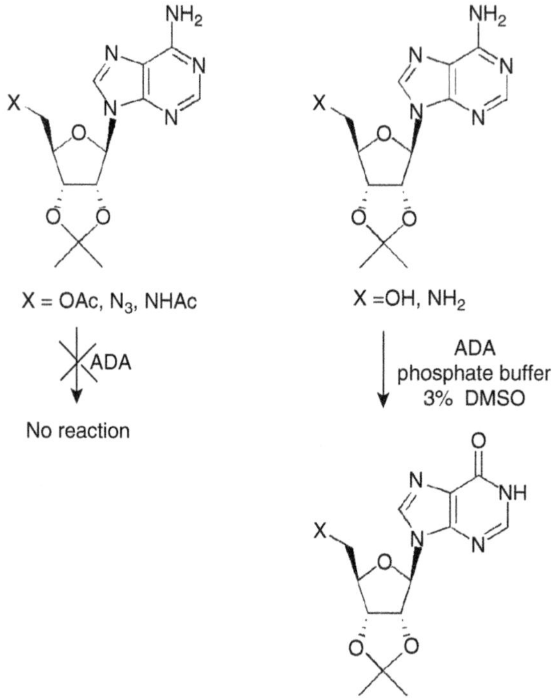

FIGURE 19.6 ADA-catalyzed deamination of 5'-protected or 5'-deoxy-5'-substituted 2',3'-iso dene of the adenosines.

This was later confirmed by the observation that a few 5'-protected or 5'-de substituted adenosines (Figure 19.8), whose ADA was not able to transform or tha only poorly deaminated, were completely converted into inosine derivatives with AMI room temperature in 3 to 150 min [39].

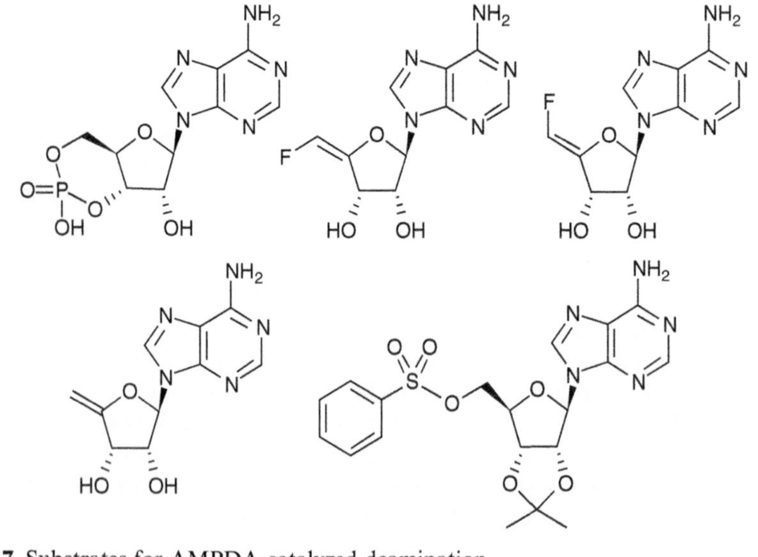

FIGURE 19.7 Substrates for AMPDA-catalyzed deamination.

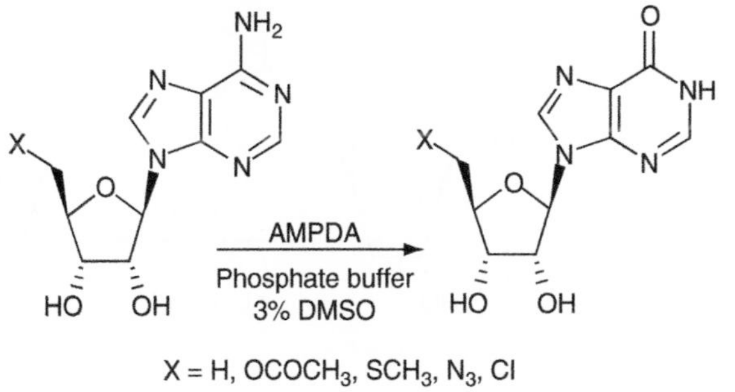

X = H, OCOCH$_3$, SCH$_3$, N$_3$, Cl

FIGURE 19.8 AMPDA-catalyzed deamination of 5′-protected and 5′-deoxy-5′-substituted adenosines.

19.6 ADA-ASSISTED CHEMOENZYMATIC SYNTHESIS AND TRANSFORMATION OF ANTIVIRAL NUCLEOSIDES

The above reported data on the activity of the enzymes constitute an important background for the rational use of ADA and AMPDA as biocatalyst in the preparation of modified nucleosides. Here, we will review the application of ADA to the synthesis and elaboration of nucleosides containing modified purine systems, adenosines with modification of the ribose (or 2′-deoxyribose) moiety, and synthetic nucleosides containing structural variations on the part of both purine and furanose. Substrates of the deaminating enzyme will be selected within the general framework of a recognized or potential antiviral activity.

19.6.1 2′,3′-Dideoxygenated Purine Nucleosides

Purine nucleosides that are deoxygenated at 2′ and 3′ positions are known inhibitors of HIV reverse transcriptase, a key enzyme for the retrovirus replication [44] and a few of them have been introduced as chemotherapeutic agents with antiretroviral activity [5]. For instance, 2′,3′-dideoxyinosine (ddI) is the drug Didanosine used for the treatment of HIV advanced infection [45], and this modified nucleoside can be prepared by a few synthetic approaches [46,47]. A chemoenzymatic route to the synthesis of ddI that relies on ADA-catalyzed deamination of 2′,3′-dideoxyadenosine (ddA) has been reported [48] and shown in Figure 19.9.

Another chemoenzymatic synthesis of ddI has been recently proposed [49] and consisted in ADA-catalyzed deamination of 2′-deoxyadenosine to 2′-deoxyinosine and in a selective 5′-acetylation of this compound catalyzed by the lipase from *Candida cylindracea* (CAL). Deoxygenation of the unprotected 3′-hydroxy group was chemically achieved by a classical approach [50] that consisted in the preparation of the 3′-*O*-phenoxythiocarbonyl derivative and in further reaction with tributyltin hydride (Figure 19.10).

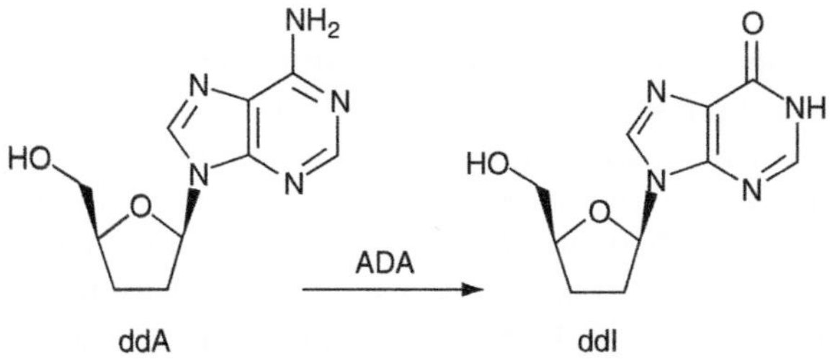

ddA   ddI

FIGURE 19.9 ADA-catalyzed deamination of 2′,3′-dideoxyadenosine to 2′,3′-dideoxyinosine.

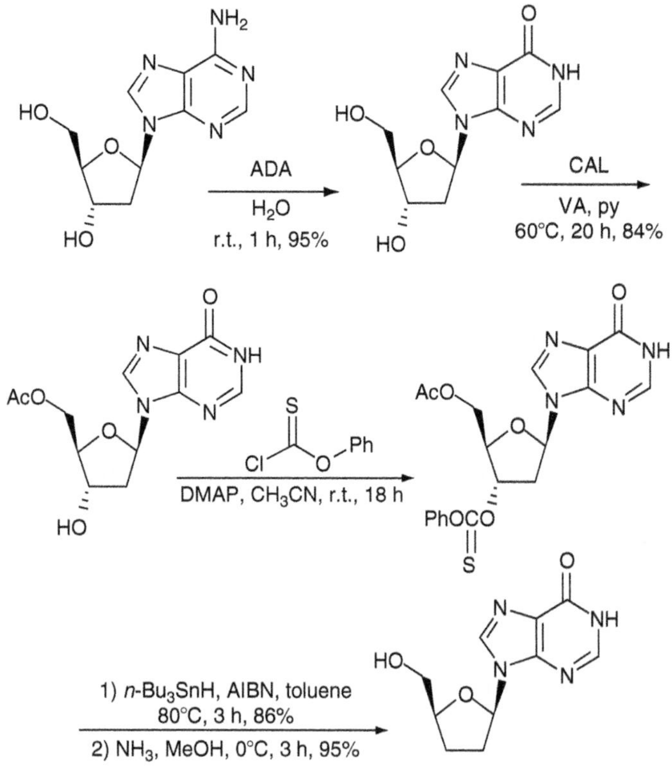

FIGURE 19.10 Chemoenzymatic synthesis of 2′,3′-dideoxyinosine from 2′-deoxyadenosine.

A series of 8-substituted 2′,3′-dideoxyadenosines, synthesized as potential inhibit HIV reverse transcriptase, were characterized by higher stability than 8-unsubstituted a [51]. The deaminating activity of ADA that had previously been evaluated on 8-substituted adenosines [52] was applied to 8-amino and 8-hydroxy derivatives tha converted into the corresponding inosines (4 and 12 h, respectively). 8-Methylthio, 8 oxy, and 8-benzyloxy derivatives were resistant to the enzymatic deamination (Figure

In addition, 8-aza-7-deaza-2′,3′-dideoxyadenosine, synthesized as a potential ant viral agent, was deaminated by ADA [53], as shown in Figure 19.12.

Although 2-substituted adenosines have been described as inhibitors of ADA [3 presence of an amino group at position 2 is compatible with the activity of the enzy shown by the slow deamination of 2-aminoadenosine to guanosine [54]. Therefore, diaminopurine can be considered a masked guanosine system and its ADA-catalyzed o nation may offer an alternative route to direct chemical transformation of guanosine s that is plagued by experimental problems, such as poor solubility, instability, and for of gel [55,56]. Following the above biocatalytic approach, Robins and coworkers [42 described the preparation of a few ribose-modified guanine nucleosides and it is mentioning that for a substrate with amino groups in positions 2, 6, and 2′ as, for ins 2,6-diamino-9-(2′-amino-2′-deoxy-β-D-arabinofuranosyl)purine, only the selective dea tion of the 6-amino group occurs in the presence of ADA (Figure 19.13). This chem matic approach has been applied to the preparation of 2′,3′-dideoxyguanosine, an ef inhibitor of hepatitis B virus [57,58] that can be enzymatically prepared from the corres ing 2,6-diaminopurine riboside, which is the inhibitor of the enzyme HIV and hepa viruses [59–61]. 2′,3′-Dideoxyguanosine can be also prepared by ADA-catalyzed deami

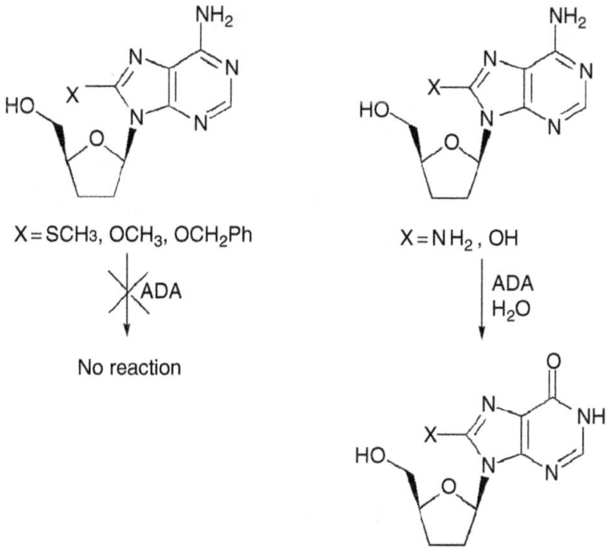

FIGURE 19.11 ADA-catalyzed deamination of 8-substituted 2′,3′-dideoxyadenosines.

of 2-amino-6-chloro-2′,3′-dideoxyfuranoside [62]. Within a work aimed to select cytotoxic or antihepatitis B modified nucleosides [63], other 2-amino-6-substituted 2′,3′-dideoxy nucleosides were deaminated by ADA to the corresponding guanosines with a decreasing efficiency of hydrolysis going from 6-amino to 6-methoxy and to 6-ethoxy nucleosides (Figure 19.13).

19.7 FLUORINE-CONTAINING NUCLEOSIDES

Introduction of a fluorine atom in the carbohydrate ring of nucleosides has been a means to improve the stability of the compound toward enzymatic or chemical hydrolysis of the β-N-glycosidic bond. The resistance to acidic pH of 2′-fluoro purine dideoxygenated nucleosides has been, in fact, verified [64]. The 2′,3′-dideoxy-2′-ara-fluoroadenosine, conveniently synthesized from 6-chloropurine, was converted to the corresponding inosine derivative by ADA-catalyzed deamination (room temperature, overnight, 77%). In this case, yield of enzymatic reaction was lower than the treatment with sodium nitrite in acetic acid (room temperature, 20 h, 95%; Figure 19.14).

A series of 6-substituted 2′-β-fluoro-2′,3′-dideoxypurine nucleosides (Figure 19.15) have been prepared with the aim of finding lipophilic, ADA-activated anti-HIV-prodrugs for the specific delivery in central nervous system [65,66]. The activity of ADA with these substrates has been evaluated and it turned out that the 6-fluoro nucleoside was hydrolyzed to the corresponding inosine with the highest relative rate.

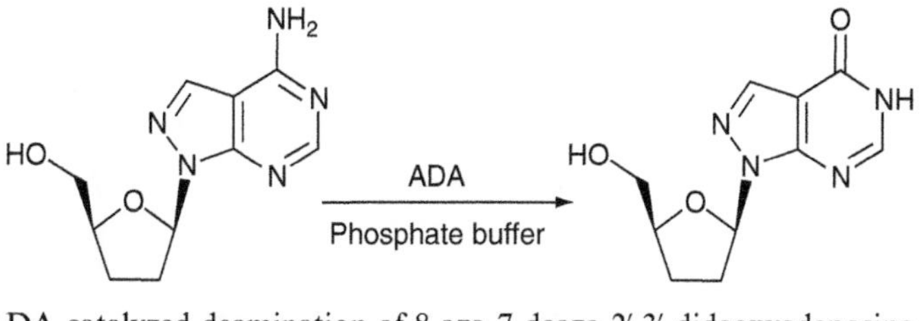

FIGURE 19.12 ADA-catalyzed deamination of 8-aza-7-deaza-2′,3′-dideoxyadenosine.

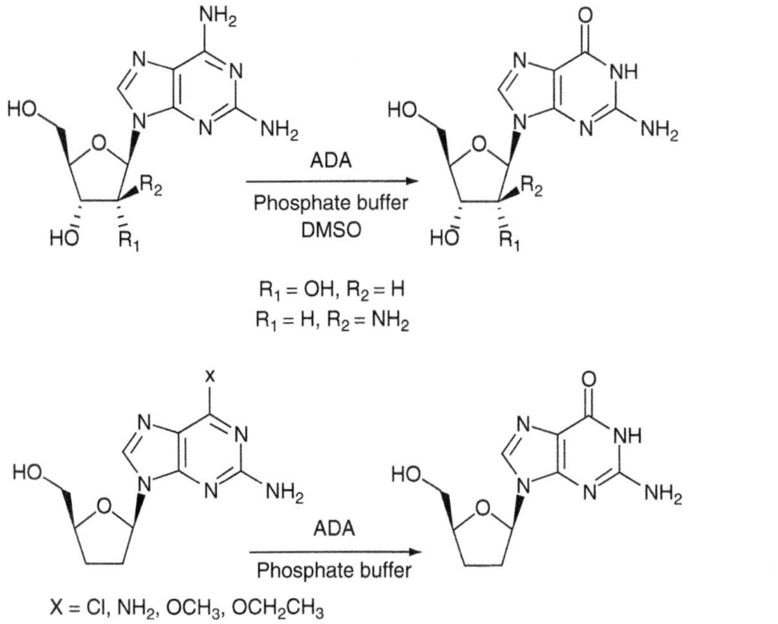
ADA
Phosphate buffer
DMSO
R_1 = OH, R_2 = H
R_1 = H, R_2 = NH_2
ADA
Phosphate buffer
X = Cl, NH_2, OCH_3, OCH_2CH_3

FIGURE 19.13 Selective ADA-catalyzed deamination of 2,6-substituted diaminopurine nucleos

A fluoro-*arabino* analog of 8-aza-7-deaza-2′,3′-dideoxyadenosine, i.e., 4-amin deoxy-2′-fluoro-β-D-arabinofurano)-1*H*-pyrazolo[3,4-*d*]pyrimidine, showed activity ag human herpes virus [67]. The enzymatic deamination with ADA is similar to that repo Figure 19.12 and readily proceeds to afford the corresponding 4-oxo compound (Figure

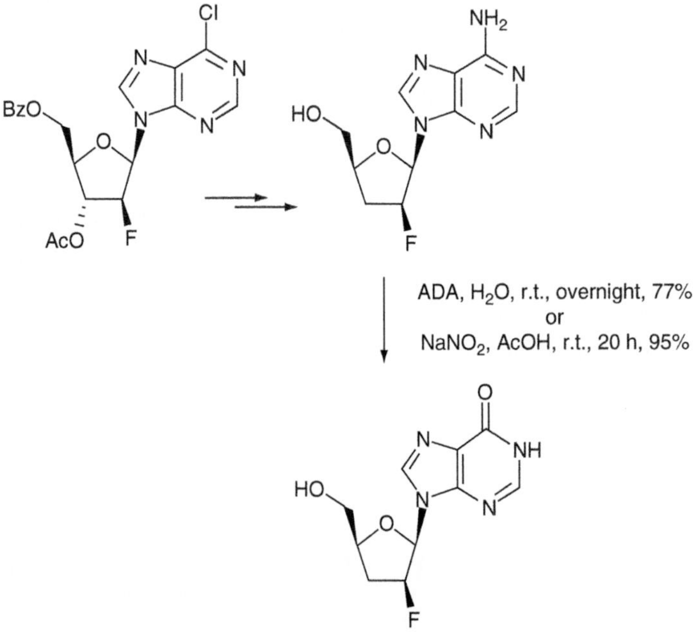
ADA, H_2O, r.t., overnight, 77%
or
$NaNO_2$, AcOH, r.t., 20 h, 95%

FIGURE 19.14 ADA-catalyzed deamination of 2′,3′-dideoxy-2′-ara-fluoroadenosine.

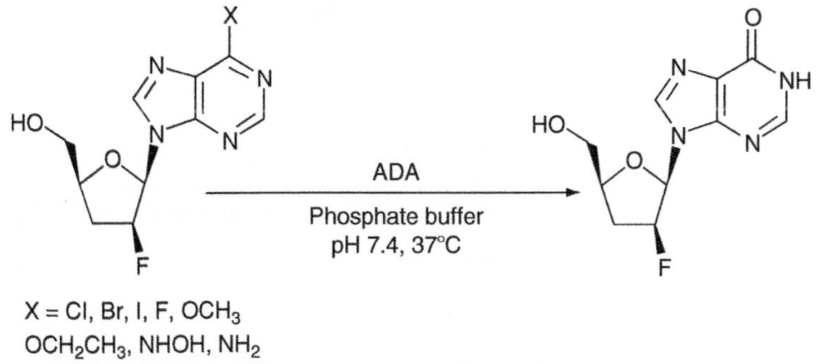

X = Cl, Br, I, F, OCH₃
OCH₂CH₃, NHOH, NH₂

FIGURE 19.15 ADA-catalyzed deamination of 6-substituted 2′-β-fluoro-2′,3′-dideoxypurine nucleosides.

19.8 NUCLEOSIDES CONTAINING SUBSTITUTIONS IN THE FURANOSE MOIETY

Numerous nucleosides and 2′-deoxynucleosides substituted at various positions of the furanose moiety show marked activities as antiviral agents, for instance, C-3′-branched deoxynucleosides. For C-3′-ethynyl nucleosides, it has been demonstrated that transformation into the corresponding nucleotides is required to become active on HIV reverse transcriptase [68]. 3′-β-Ethynyladenosine and the corresponding 2′-deoxynucleoside were deaminated by ADA, differently from 3′-vinyl and ethyl substituted compounds that were not substrates (Figure 19.17) [69].

4′-C-Substituted 2′-deoxynucleosides constitute a group of antiviral agents effective against HIV variants that are resistant to other antiretroviral drugs [70]. Previous studies on structure–activity of 4′-C-substituted nucleosides had directed the investigation toward compounds bearing small substituents at 4′-position as the most active compounds [71–73]. The synthesis of 4′-cyano- and 4′-ethynyl-2′-deoxypurine nucleosides was reported [74] and these compounds were shown to be substrate of ADA for the preparation of the corresponding 4′-substituted-2′-deoxyinosine and guanosine, respectively (Figure 19.18).

19.9 OXETANOCIN A

Oxetanocin A is a naturally occurring nucleoside characterized by a peculiar 4-member ring carbohydrate moiety that has been isolated from *Bacillus megaterium* [75] and was soon recognized as a HIV inhibitor [76]. Strictly related to oxetanocin A, the cyclobutyl guanine nucleoside analog, Lobucavir, has been synthesized as an antiviral agent for the treatment

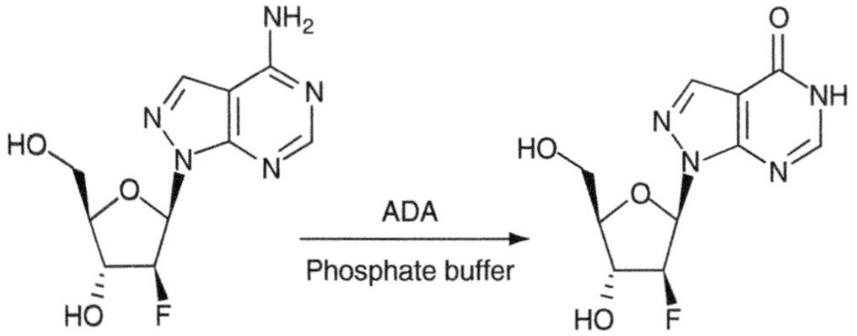

FIGURE 19.16 ADA-catalyzed deamination of 4-amino-1-(2′-deoxy-2′-fluoro-β-D-arabino furanosyl)-1*H*-pyrazolo [3,4-*d*]pyrimidine.

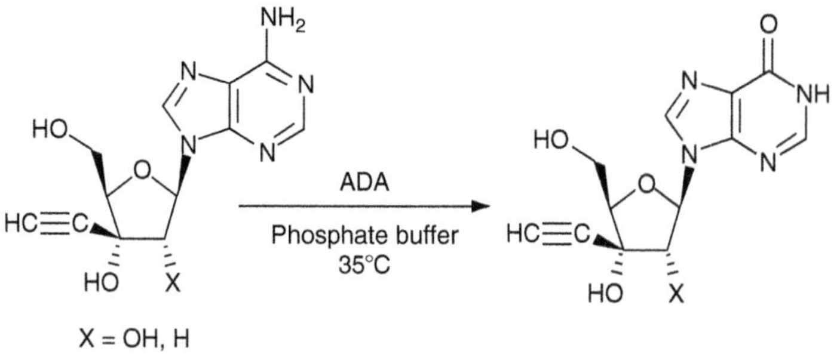

FIGURE 19.17 ADA-catalyzed deamination of 3'-β-ethynyladen osines.

of herpes viruses and hepatitis B [77]. Oxetanocin A and the 2-amino analog are read
quantitatively converted (Figure 19.19) to the corresponding hypoxanthine and gua
derivatives by an ADA-catalyzed reaction [78].

19.10   ADA-ASSISTED CHEMOENZYMATIC SYNTHESIS OF ANTIVIRAL CARBOCYCLONUCLEOSIDES

Replacement of the sugar moiety of nucleosides with a cyclopentyl ring leads to carbc
nucleosides, a class of compounds endowed with interesting antiviral activity, as
evidenced for the first naturally occurring carbocyclonucleosides, aristeromycin and
nocin [79,80]. ADA activity on carbocyclonucleosides is different with respect to nucle
and, in fact, complete deamination of adenosine was accomplished in 2 h, while for a
mycin, the carbocyclic analog of adenosine, 32 h, was required [81]. 2',3'-Dideoxyade
was completely deaminated at a similar rate to adenosine (2 h), while complete transfor

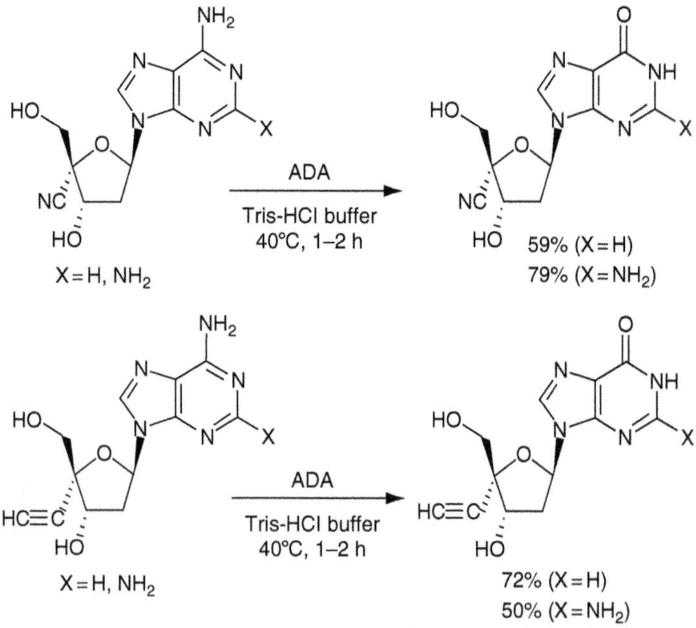

FIGURE 19.18 ADA-catalyzed deamination of 4'-substituted 2'-deoxypurine nucleosides.

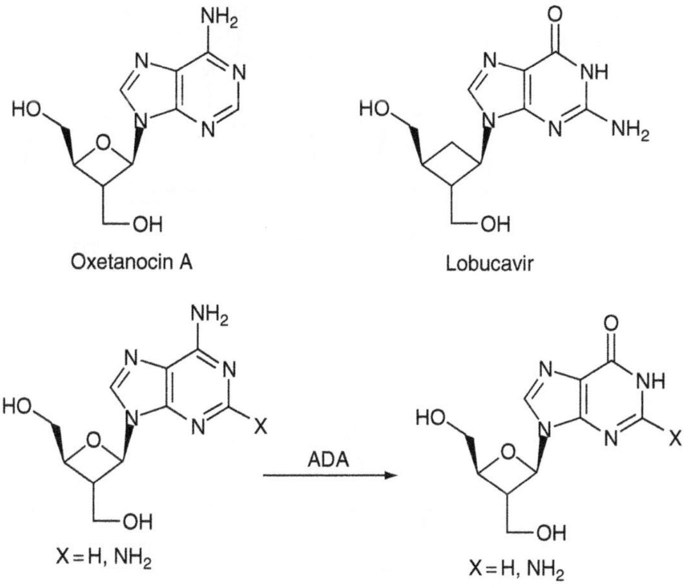

Oxetanocin A

Lobucavir

ADA

X = H, NH₂

X = H, NH₂

FIGURE 19.19 Structures of oxetanocin A, lobucavir, and ADA-catalyzed deamination of oxetanocin A and the 2-amino analog.

of 2′,3′-dideoxyaristeromycin required 72 h (Figure 19.20). The utility of the chemoenzymatic approach for a preparative synthesis of carbocyclic inosines was also demonstrated.

Carbovir is a synthetic carbocyclonucleoside that found application as a chemotherapeutic agent for the treatment of AIDS and the selective inhibition of HIV resides in the (−) enantiomer [82]. (−)-Carbovir could be synthesized from structurally related (−)-aristeromycin

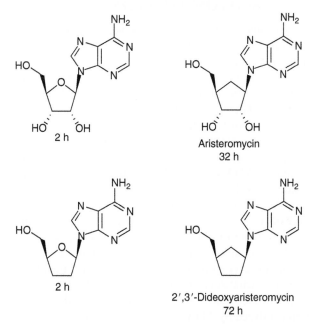

2 h

Aristeromycin
32 h

2 h

2′,3′-Dideoxyaristeromycin
72 h

FIGURE 19.20 Time for complete ADA-catalyzed deamination of nucleosides and carbocyclonucleosides.

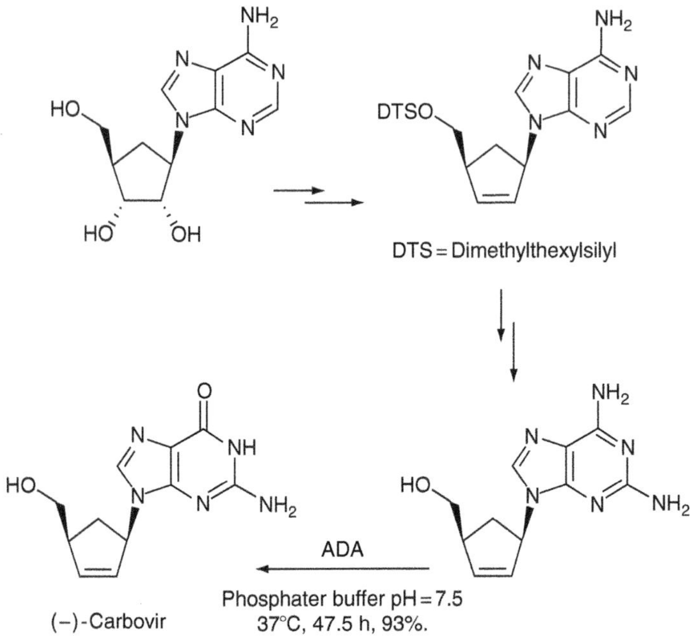

FIGURE 19.21 Chemoenzymatic synthesis of (−)-carbovir.

[83] and the guanine system was enzymatically prepared by quantitative deamina 2,6-diaminopurine derivative in the presence of ADA (Figure 19.21).

19.11 ADA-ASSISTED CHEMOENZYMATIC SYNTHESIS OF ANTIVIRAL ACYCLONUCLEOSIDES

Acyclic analogs of nucleosides (acyclonucleosides) often present important antiviral a as shown by representative compounds such as acyclovir, 9-[(2-hydroxyethoxy)metl nine and ganciclovir, 9-[(1,3-dihydroxy-2-propoxy)methyl]guanine, that are at prese established antiherpetic agents [84,85]. A few structurally related acyclonucleosides penciclovir [86], or its prodrug famciclovir [87] were introduced as antiviral agents du past years (Figure 19.22).

In the above acyclonucleosides, the purine system is constituted by a guanine alky the nitrogen atom indicated as N^9 and synthetic methods for the preparation of compounds acyclovir or ganciclovir and their prodrugs have already been reviewed [1 action of ADA on suitable precursors may constitute a chemoenzymatic approach synthesis of these antiviral agents, therefore extending the potentiality of ADA as a l lyst. As previously reported, 2,6-diaminopurine nucleosides can be deaminated in t ence of ADA to the corresponding guanosines and this biocatalytic approach was exte 2-amino-6-chloro- or 2,6-diamino-9-[(2-hydroxyethoxy)methyl]purines [88]. Complete catalyzed deamination of above compounds was observed in phosphate buffer a temperature (Figure 19.23), while for the quantitative transformation of a suital diaminopurine acyclonucleoside into ganciclovir an excess of ADA was needed (18 l [89]. Another synthesis of both acyclovir and ganciclovir relies on the hydrolytic defi tion catalyzed by ADA of a 6-fluoropurine derivative that took place quantitatively ov at room temperature [90].

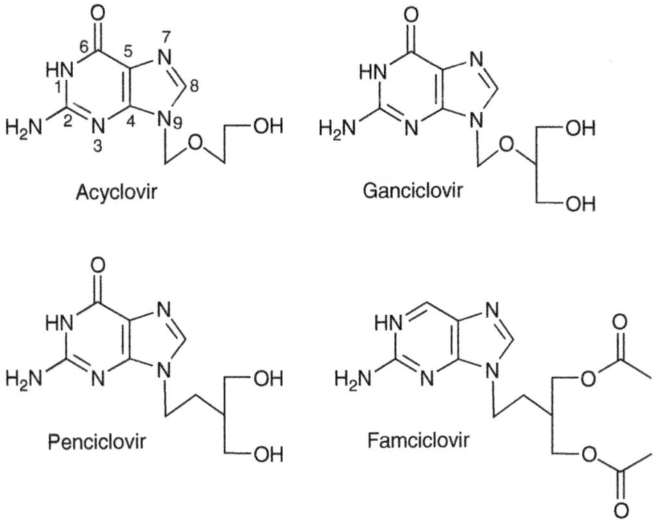

FIGURE 19.22 Structures of antiviral acyclonucleosides.

Unsaturated acyclonucleosides containing one or more double bonds that separate the nucleic base from an hydroxymethyl group are characterized by potent antiviral activity [91,92]. A few of the above compounds reported in Figure 19.24 are also substrates of ADA [92–97] and, as a general feature, the (E)-isomer reacts much faster than the (Z)-analog.

The same stereopreference has been shown by the enzyme for 2-[(hydroxymethyl)cyclopropylidene]methyladenine (Figure 19.25), characterized by a broad-spectrum antiviral activity [97]. ADA-catalyzed deamination proceeds more efficiently on the (E)-isomer and this can contribute to explain the higher activity of the (Z)-isomer (Synadenol).

19.12 AMPDA-ASSISTED SYNTHESIS AND TRANSFORMATION OF ANTIVIRAL NUCLEOSIDES AND ANALOGS

Compared to ADA, application of AMPDA to biocatalytic transformation of nucleosides and related analogs has been studied only recently [43]. The broader activity of AMPDA vs.

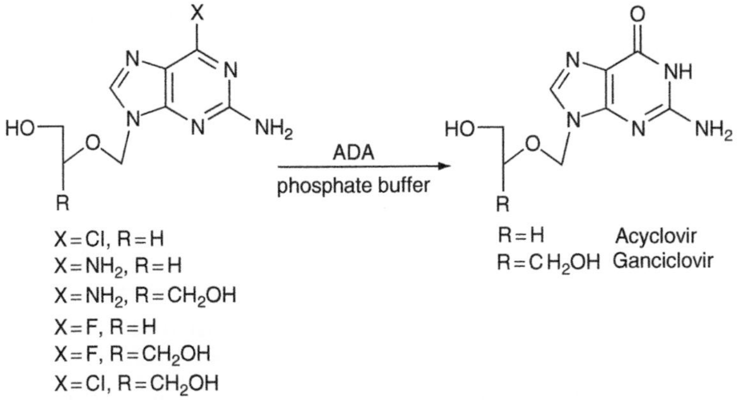

X = Cl, R = H
X = NH₂, R = H
X = NH₂, R = CH₂OH
X = F, R = H
X = F, R = CH₂OH
X = Cl, R = CH₂OH

R = H   Acyclovir
R = CH₂OH Ganciclovir

FIGURE 19.23 ADA-catalyzed deamination of precursors of acyclovir and ganciclovir.

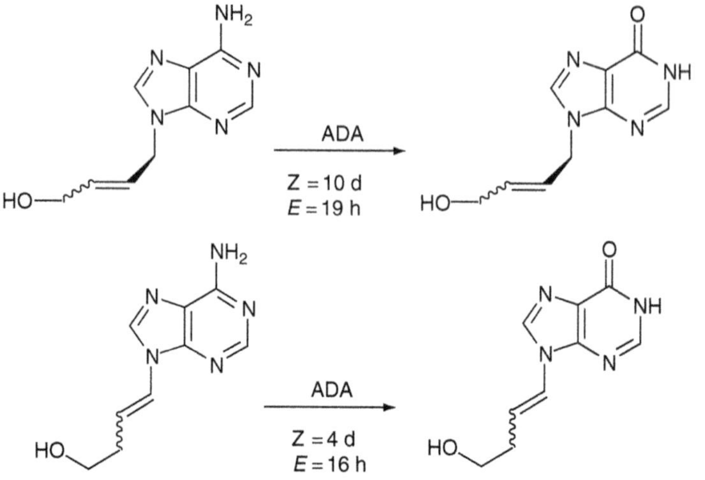

FIGURE 19.24 ADA-catalyzed selective deamination of unsaturated acyclonucleosides.

ADA, already evidenced in the above report, has been confirmed by more recent resul
Among compounds that were not substrate for ADA and were deaminated by AM
(1R-cis)-bis(1-methylethyl) [2-[4-(2,6-diamino-9H-purin-9-yl)-2-cyclopenten-1-yl]ethyl]-1
phonate (Figure 19.26) was enzymatically converted into the corresponding guanosi
rivative [43], i.e., carbovir phosphonate.

Another difference for AMPDA with respect to ADA has been observed in previous
on the antiviral activity of the 2-[(hydroxymethyl)cyclopropylidene]methyladenines [
fact, the more active (Z)-isomer, Synadenol, that was only slowly deaminated by ADA
efficiently transformed in the presence of AMPDA (>95%, 4.5 h, Figure 19.27).

19.13   STEREOSELECTIVITY OF DEAMINATING ENZYMES

The chemical synthesis of nucleosides invariably yields isomeric mixtures and the ster
mical specificity of deaminating enzymes (ADA and AMPDA) may represent an add
advantage of the biocatalysts. Therefore, applications of the enzymes might be sp
appreciated to resolve problems of enantio- and diasteroselectivity in asymmetric sy
of nucleosides. The stereochemical preference of ADA toward different isomers of sy
nucleosides can be evidenced through the examples that will now be examined and disc
AMPDA-catalyzed stereoselection of isomeric nucleosides or analogs will be also des

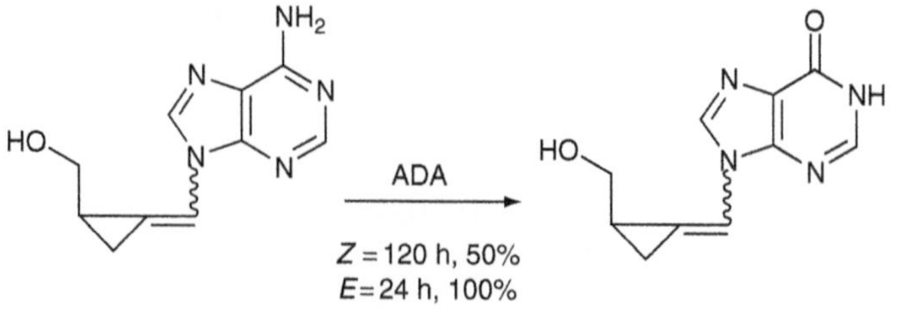

FIGURE 19.25 ADA-catalyzed selective deamination of (E)/(Z)-2-[(hydroxymethyl)cyclopropy
methyladenine.

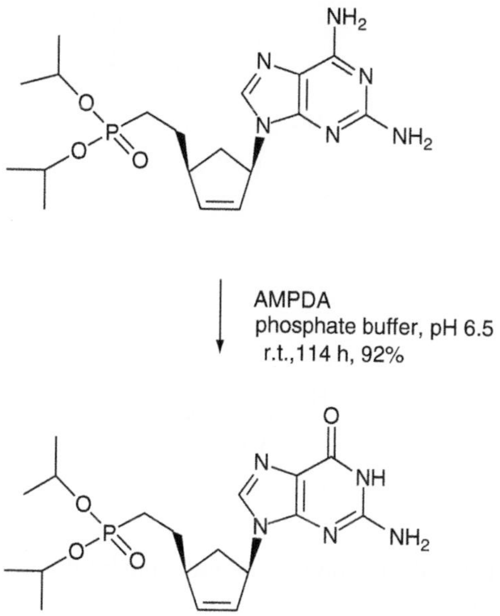

FIGURE 19.26 AMPDA-catalyzed deamination of a carbocyclonucleoside phosphonate.

but examples are rather limited in number, as the application of this enzyme in nucleoside chemistry has been only recently reported.

19.14 RESOLUTION OF SYNTHETIC NUCLEOSIDES AND ANALOGS

Since the introduction of Lamivudine (β-L-2′,3′-dideoxy-3′-thiacytidine), there has been a considerable synthetic effort directed to the preparation of β-L-nucleosides as specific anti-viral agents [98]. Based on the consideration that ADA is able to catalyze the deamination of natural nucleosides, it was expected that L-nucleosides should not be substrates of the enzyme or would react much slower than the naturally occurring D-isomer. This has been verified by kinetic studies on β-D- and L- 2′,3′-dideoxyadenosines [99] or D- and L-2′,3′-dideoxy-2′,3′-*endo*-methylene nucleosides, that were synthesized as potential antiviral agents [100]. Specifically, in the presence of ADA from calf intestinal mucosa and compared with adenosine ($t_{1/2}$ = 30 s), the D-2′,3′-dideoxy-2′,3′-*endo*-methylene nucleoside was a good substrate ($t_{1/2}$ = 90 s), while the L-isomer reacted slower ($t_{1/2}$ = 22 h) (Figure 19.28). Synthesis of carbocyclic analogs of natural nucleosides has to rely mainly, if not exclusively, on chemical methods that afford isomeric mixtures from which each stereoisomer has

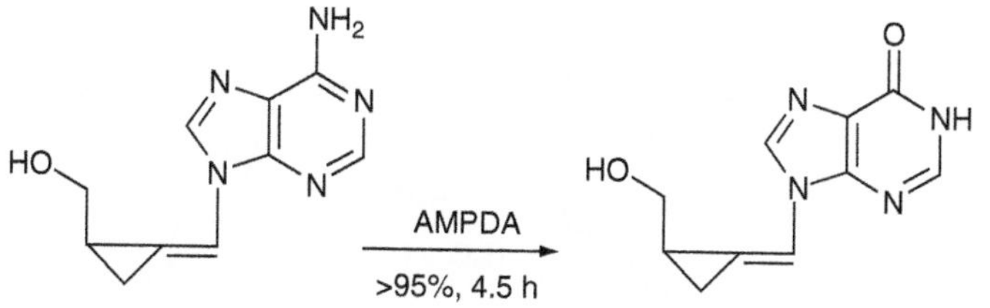

FIGURE 19.27 Selective deamination of (Z)-2-[(hydroxymethyl)cyclopropylidene]methyladenine by AMPDA.

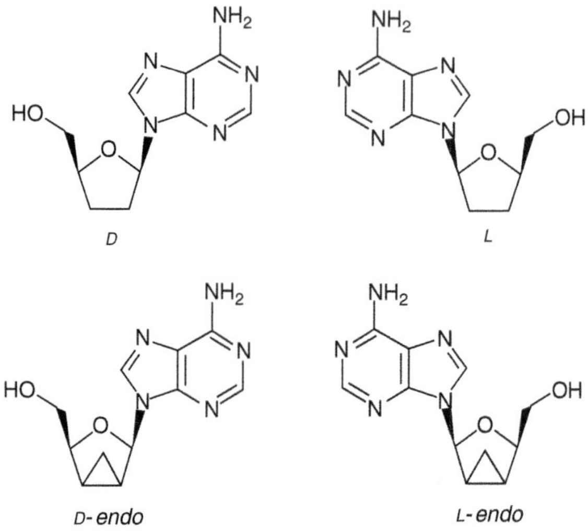

FIGURE 19.28 Structures of D- and L-modified adenosines, substrates for D-selective ADA-c deamination.

to be separated. Secrist and coworkers have described the first resolution of a racemic of synthetic carbocyclic nucleosides [101]. Specifically, (±)-aristeromycin was reso ADA that deaminated the (−)-isomer into to the corresponding hypoxanthine derivati room temperature), so that the stereochemically pure unnatural (+)-aristeromycin c recovered (Figure 19.29).

The same authors report another interesting application of ADA-catalyzed deamina the resolution of carbocyclic 2′-deoxyguanosine starting from the racemic 2,6-diamin analog [101]. The enzyme and substrate ratio and the temperature of the reaction pl essential role, since the unreacted (−)-compound was transformed into (−)-guanine de using ADA only at a lower temperature (37 instead of 50°C), but raising the enzy substrate ratio (from 0.5 to 4 U/μmol) (Figure 19.30).

Analogously, the same synthetic strategy was applied to obtain both enantior carbovir starting from 9-(4-hydroxymethylcyclopent-2-enyl)-9H-purine-2,6-diamir was deaminated to (−)-carbovir at 25°C in 72 h. Tranformation of the unreacted (+)-d compound under different experimental conditions (37°C, 2 d) allowed the prepara (+)-carbovir that was shown to be much less active as HSV-1 inhibitor (Figure 19.31

The resolution of racemic 2-amino-6-chloro pyrazolopyrimidine bound to a fluc cyclopentane moiety (Figure 19.32) is a further example of ADA versatility [102]. A ■

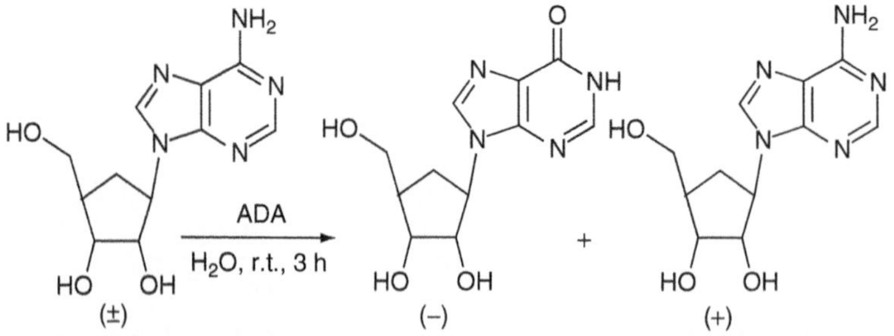

FIGURE 19.29 ADA-catalyzed resolution of synthetic racemic aristeromycin.

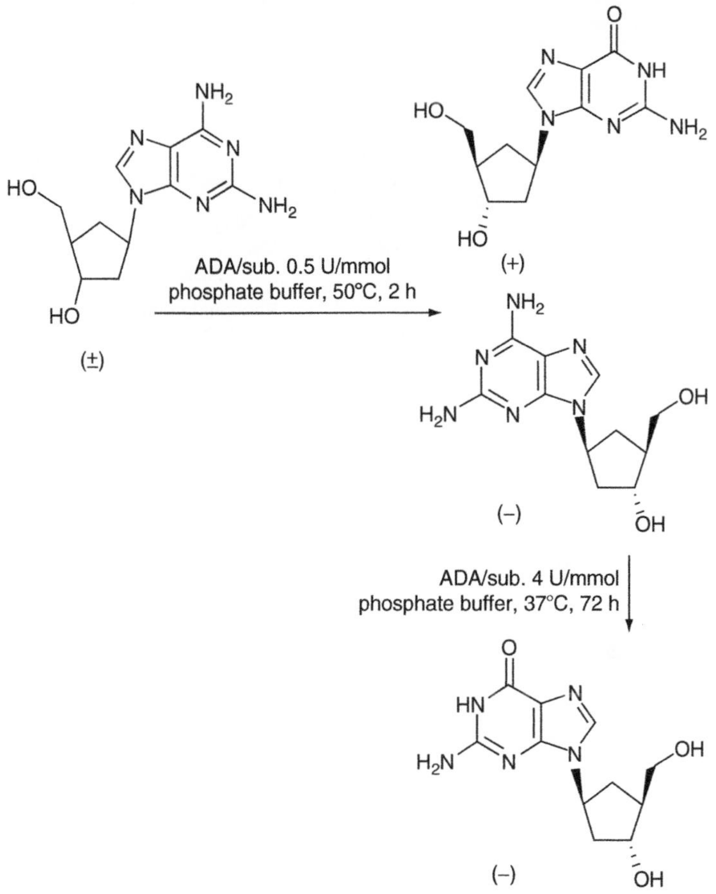

FIGURE 19.30 ADA-catalyzed selective deamination of a synthetic 2,6-diaminopurine carbocyclonucleoside.

program has been developed in recent years aimed to the synthesis and biological evaluation of conformationally restricted carbocyclic nucleosides, as exemplified by one of the latest reports [103]. Within this line of research, a racemic cyclopropyl analog of neplanocin (Figure 19.32) has been synthesized and an anti-HIV screening indicated that this adenosine analog possessed a level of activity that could be better evaluated by preparing individual enantiomers [104]. The resolution of the synthetic racemic mixture was successfully achieved by an ADA-catalyzed deamination.

Zemlicka and coworkers have prepared a series of antiviral acyclonucleosides characterized by the presence of unsaturations and cyclopropane ring in the acyclic moiety of the structure [105]. The compound containing the cyclopropyl moiety directly bound to adenine, 9-[(2-hydroxyethylidene)cyclopropyl]adenine, was incubated as a mixture (E)/(Z) with ADA. After 45 h at room temperature, 82% of the (E)-isomer was deaminated to the hypoxanthine analog, while the (Z)-isomer remained essentially unchanged (Figure 19.33) [106].

The stereopreference exhibited by ADA (Figure 19.33) is in agreement with the observation on single isomers of similar compounds reported by the same authors [97] and illustrated in Figure 19.25. It should also be recalled that (Z)-2-[(hydroxymethyl)cyclopropylidene]methyladenine has been deaminated in the presence of AMPDA, as shown in Figure 19.27 [97].

R,S-Adenallene is a strong inhibitor of the replication and cytopathic effect of HIV-1 and HIV-2 [107], the antiretroviral activity being higher for R-adenallene [108]. For the study of

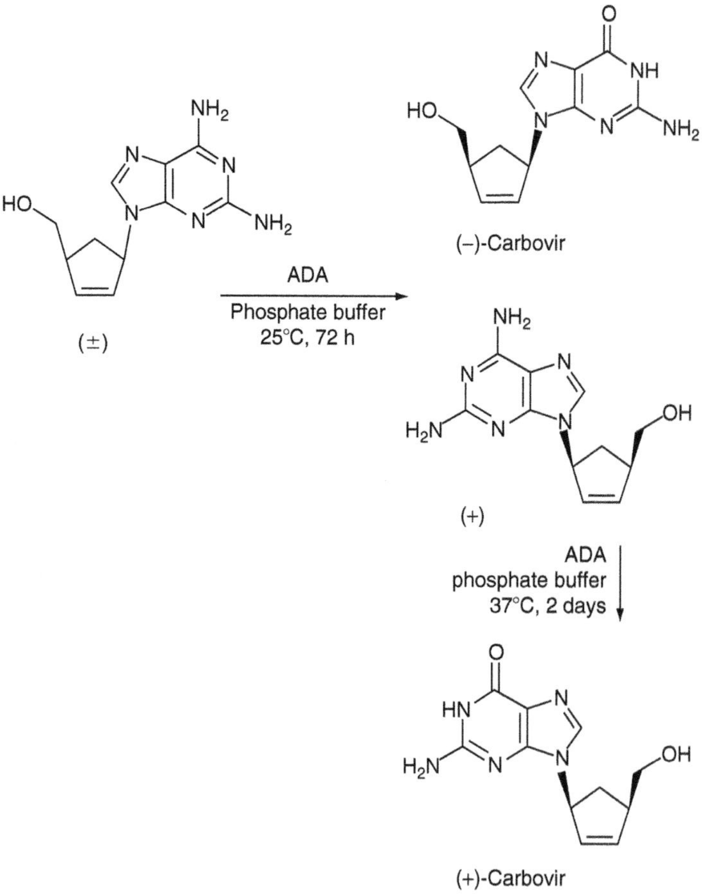

FIGURE 19.31 ADA-catalyzed selective deamination of 2,6-diaminocarbocyclonucleosides.

single-isomer action on enzymes of nucleic acid metabolism, *R*- and *S*-adenallene phos
were required. *R,S*-Adenallene 4′-phosphate is deaminated with AMPDA with no en.
electivity (Figure 19.34), whereas a partial resolution was achieved by slow enzymatic
sphorylation catalyzed by 5′-nucleotidase (*R* 26%; *S* 54%) [109].

    Other carbocyclonucleosides containing bicyclic systems characterized by small ri
spiroheptane and spiropentane mimics of nucleosides were studied as potential at
agents [110,111]. The spiroheptane analog of adenallene with axial dissymmetry w

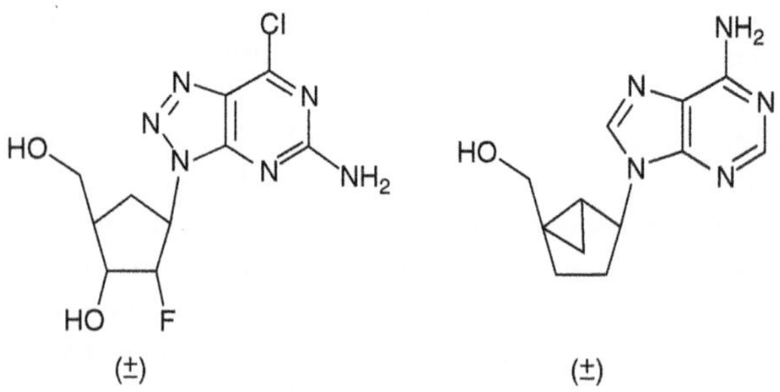

FIGURE 19.32 Racemic carbocyclic substrates resolved by ADA-catalyzed deamination.

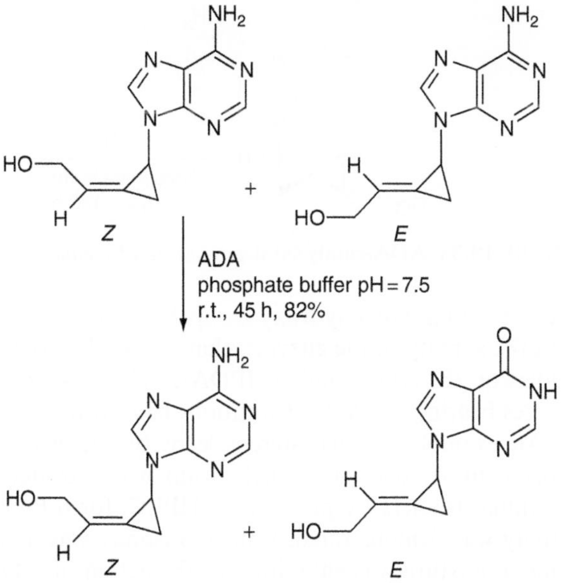

FIGURE 19.33 ADA-catalyzed selective deamination of (E)-9-[(2-hydroxyethylidene)-cyclopropyl]adenine.

substrate for ADA [110]. Among the synthesized stereoisomers of spiropentane mimics (Figure 19.35), only the *medial-anti*-isomer was slowly transformed by the deaminating enzyme (activity expressed as $t_{1/2} > 120$ h) [111].

19.15 DIASTEREOSELECTIVITY OF DEAMINATING ENZYMES

Finally, a particular type of diastereoselectivity was revealed for the enzymatic deamination catalyzed by ADA when a methyl group was introduced in the position 5′ of modified adenosine and related analogs. An additional stereogenic center is generated at that position

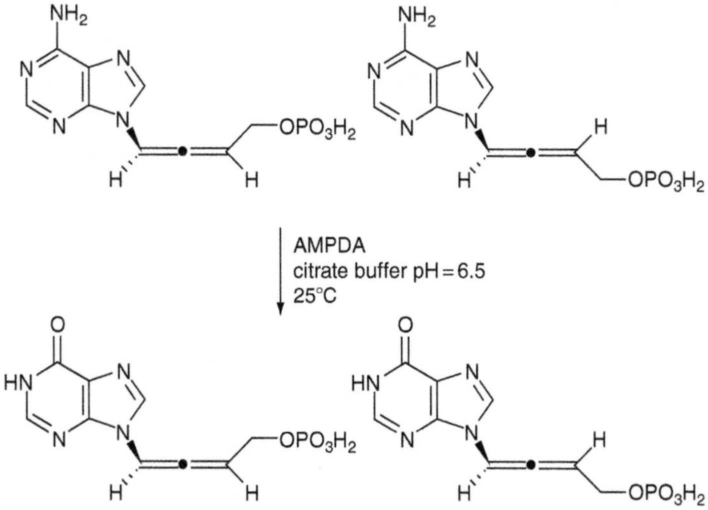

FIGURE 19.34 AMPDA-catalyzed nonselective deamination of R,S-adenallene-4′-phosphate.

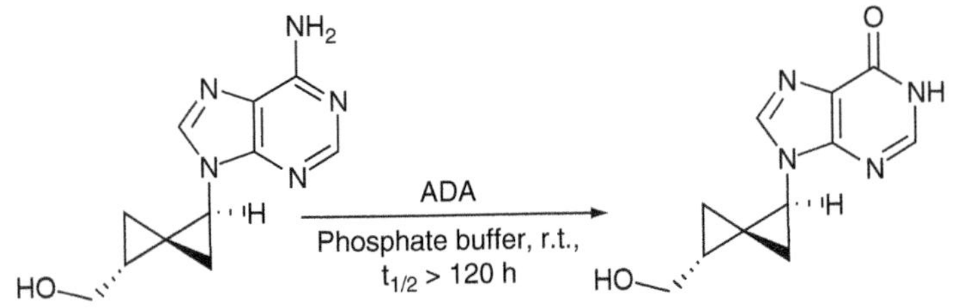

FIGURE 19.35 ADA-catalyzed deamination of *medial-anti*-spiropentane acyclonucleoside.

and the presence of a hydroxy group in the newly formed secondary alcohol is still comp
with the activity of the enzyme, that stereoselectively deaminates the 5'S-isomer [112].
studies on the ADA- and AMPDA-catalyzed selective deamination of a few modified
nosines bearing 5'-alkyl substitutions have also been recently reported [113].

An example of such stereoselectivity had been reported for the carbocyclonucle
neplanocin A, once a methyl group was introduced at 6'-position. After separati
individual isomers by preparative HPLC, for 6'R-methyl neplanocin a significant an
activity was demonstrated, while 6'S-isomer was inactive [114]. When the synthetic d
eomeric mixture was submitted to the action of ADA, a selective hydrolysis of 6'S-i
neplanocin to the corresponding hypoxanthine analog was observed and the
6'R-isomer could be recovered (Figure 19.36) [115].

19.16   CONCLUDING REMARKS

The synthesis of modified purine nucleosides and related systems that present ar
activity may rely on a vast array of selective chemical and biocatalytic procedures. Ade

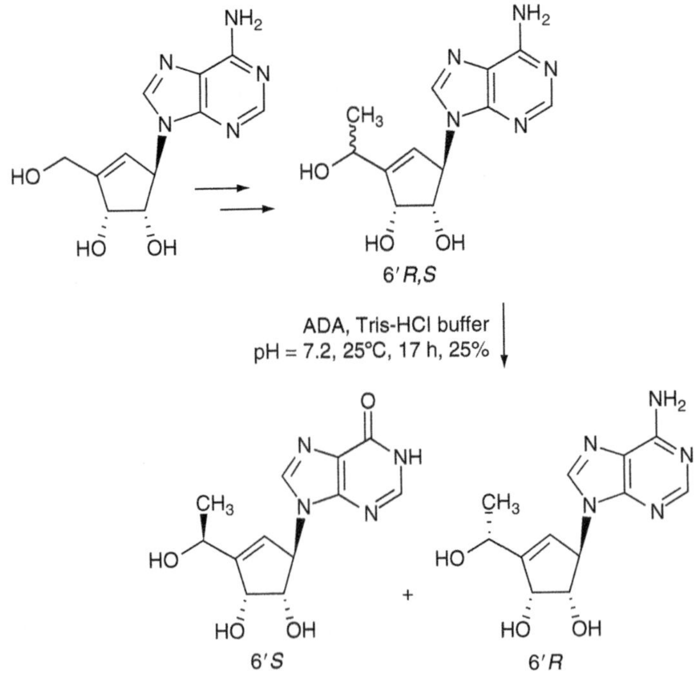

FIGURE 19.36 Diastereoselectivity of ADA-catalyzed deamination of 6'R,S-methyl neplanoci

or adenylate deaminase (ADA or AMPDA) are hydrolytic deaminating enzymes of the purine recycling or salvatage pathway that selectively converts 6-substituted purine nucleosides into the corresponding 6-oxo analogs. Both enzymes can accept substrates with a good degree of structural variations either in the base ring or in the furanose moiety. For this reason, they can be considered as valuable biocatalysts for application to synthetic chemistry of nucleosides and related carbocyclic or acyclic analogs. First applications of ADA date back to nearly 40 years ago and the well-established knowledge about its structure and mechanism makes this enzyme a relatively well-defined biocatalyst. Several examples of chemoenzymatic syntheses where ADA is required for selective transformation are already available. The substrates are nucleosides modified in the furanose and/or purine portion, carbocyclic or acyclic nucleosides and in the present review selected examples of antiviral compounds have been reported. Although AMPDA is commercially utilized for the synthesis of inosine monophosphate, the enzyme is less studied than ADA and only recent works have disclosed its capability. AMPDA appears to be less selective toward the substrate structure and is capable to offer a broader range of applications, thus further expanding potentiality of deaminating enzymes in the synthesis of modified nucleosides and analogs.

REFERENCES

1. Cheson, B.D., Keating, M.J., and Plunkett, W., (Eds.), *Nucleoside Analogs in Cancer Therapy*, Marcel Dekker, New York, 1997.
2. Galmarini, C.M., Mackey, J.R., and Dumontet, C., Nucleoside analogues and nucleobases in cancer treatment, *Lancet Oncol.*, 3(7), 415–424, 2002.
3. Marquez, V.E., Design, synthesis, and antiviral activity of nucleoside and nucleotide analogs, *ACS Symp. Ser.*, 401, 140–155, 1989.
4. Mackman, R.L. and Cihlar, T., Prodrug strategies in the design of nucleoside and nucleotide antiviral therapeutics, *Annu. Rep. Med. Chem.*, 39, 305–321, 2004.
5. Jonckheere, H., Annè, J., and De Clercq, E., The HIV-1 reverse transcription (RT) process as target for RT inhibitors, *Med. Res. Rev.*, 20(2), 129–154, 2000.
6. Macchi, B. and Mastino, A., Pharmacological and biological aspects of basic research on nucleoside-based reverse transcriptase inhibitors, *Pharmacol. Res.*, 46(6), 473–482, 2002.
7. Kusaka, T., Yamamoto, H., Shibata, M., et al., *Streptomyces citricolor* nov. sp. and a new antibiotic, aristeromycin, *J. Antibiot.*, 21(4), 255–263, 1968.
8. Yaginuma, S., Muto, N., Tsujino, M., et al., Studies on neplanocin A, new antitumor antibiotic. I. Producing organism, isolation and characterization, *J. Antibiot.*, 34(4), 359–366, 1981.
9. Borthwick, A.D. and Biggadike, K., Synthesis of chiral carbocyclic nucleosides, *Tetrahedron*, 48(4), 517–623, 1992.
10. Agrofoglio, L., Suhas, E., Farese, A., et al., Synthesis of carbocyclic nucleosides, *Tetrahedron*, 50(36), 10611–10670, 1994.
11. Crimmins, M.T., New developments in the enantioselective synthesis of cyclopentyl carbocyclic nucleosides, *Tetrahedron*, 54(32), 9229–9272, 1998.
12. Gao, H. and Mitra, A.K., Synthesis of acyclovir, ganciclovir and their prodrugs: a review, *Synthesis*, 3, 329–351, 2000.
13. Prasad, A.K. and Wengel, J., Enzyme-mediated protecting group chemistry on the hydroxyl groups of nucleosides, *Nucleosides Nucleotides*, 15(7–8), 1347–1359, 1996.
14. Prasad, A.K., Trikha, S., and Parmar, V.S., Nucleoside synthesis mediated by glycosyl transferring enzymes, *Bioorg. Chem.*, 27(2), 135–154, 1999.
15. Ferrero, M. and Gotor, V., Biocatalytic selective modifications of conventional nucleosides, carbocyclic nucleosides, and C-nucleosides, *Chem. Rev.*, 100(12), 4319–4347, 2000.
16. Hanrahan, J.R. and Hutchinson, D.W., The enzymatic synthesis of antiviral agents, *J. Biotechnol.*, 23(2), 193–210, 1992.

17. Zielke, C.L. and Suelter, C.H., Purine, purine nucleoside, and purine nucleotide aminohyd: in *Enzymes*, 3rd ed., vol. 4, Boyer, P.D., Ed., Academic Press, New York, 1971, pp. 47–78

18. Santaniello, E., Ciuffreda, P., and Alessandrini, L., Synthesis of modified purine nucleosi• related compounds mediated by adenosine deaminase (ADA) and adenylate dea (AMPDA), *Synthesis*, 4, 509–526, 2005.

19. Cristalli, G., Costanzi, S., Lambertucci, C., et al., Adenosine deaminase: functional impl• and different classes of inhibitors, *Med. Res. Rev.*, 21(2), 105–128, 2001.

20. Lushchak, V.I., Functional role and properties of AMP-deaminase, *Biochemistry (Mosc)*• 143–154, 1996.

21. Samejima, H., Kimura, K., Noguchi, S., et al., Production of 5'-mononucleotides using in• ized 5'-phosphodiesterase and 5'-AMP deaminase, in *Enzyme Engineering*, vol. 3, Pye, E• Weetall, H.H., Eds., Plenum Press, New York, 1978, pp. 469–475.

22. Jones, J.B. and Beck, J.F., Asymmetric syntheses and resolutions using enzymes, in *Applic• Biochemical Systems in Organic Chemistry*, Jones, J.B., Sih, C.J., and Perlman, D., Eds., New York, 1976, part I, pp. 107–401.

23. Wilson, D.K., Rudolph, F.B., and Quiocho, F.A., Atomic-structure of adenosine-dea complexed with a transition-state analog: understanding catalysis and immunodeficiency• tions, *Science*, 252(5010), 1278–1284, 1991.

24. Sharff, A.J., Wilson, D.K., Chang, Z., et al., Refined 2.5 Å structure of murine ad• deaminase at pH 6.0, *J. Mol. Biol.*, 226(4), 917–921, 1992.

25. Wang, Z. and Quiocho, F.A., Complexes of adenosine deaminase with two potent inhibitor• structures in four independent molecules at pH of maximum activity, *Biochemistry*, 37(23)• 8324, 1998.

26. Kinoshita, T., Nishio, N., Nakanishi, I., et al., Structure of bovine adenosine deaminase com• with 6-hydroxy-1,6-dihydropurine riboside, *Acta Cryst.*, D59, 299–303, 2003.

27. Frieden, C., Kurz, L.C., and Gilbert, H.R., Adenosine deaminase and adenylate dea• comparative kinetic studies with transition state and ground state analogue inhibitors, *Bio• try*, 19(23), 5303–5309, 1980.

28. Bolen, D.W. and Fisher, J.R., Kinetic properties of adenosine deaminase in mixed a• solvents, *Biochemistry*, 8(11), 4239–4246, 1969.

29. Cory, J.G. and Suhadolnik, R.J., Dechloronase activity of adenosine deaminase, *Bioche* 4(9), 1733–1735, 1965.

30. Wolfenden, R., Enzymatic hydrolysis of 6-substituents on purine ribosides, *J. Am. Chem* 88(13), 3157–3158, 1966.

31. Chassy, B.M. and Suhadolnik, R.J., Adenosine aminohydrolase. Binding and hydrolysis o• 6-substituted purine ribonucleosides and 9-substituted adenine nucleosides, *J. Biol.* 242(16), 3655–3658, 1967.

32. Maguire, M.H. and Sim, M.K., Studies on adenosine deaminase. 2. Specificity and mecha• action of bovine placental adenosine deaminase, *Eur. J. Biochem.*, 23(1), 22–29, 1971.

33. Wanner, M.J., Deghati, P.Y.F., Rodenko, B., et al., New nucleoside analogs, synthes• biological properties, *Pure Appl. Chem.*, 72(9), 1705–1708, 2000.

34. Cory, J.G. and Suhadolnik, R.J., Structural requirements of nucleosides for binding by ad• deaminase, *Biochemistry*, 4(9), 1729–1732, 1965.

35. Bloch, A., Robins, M.J., and McCarthy, J.R., Jr., The role of the 5'-hydroxyl group of aden• determining substrate specificity for adenosine deaminase, *J. Med. Chem.*, 10(5), 908–912,

36. Maury, G., Daiboun, T., Elalaoui, A., et al., Inhibition and substrate specificity of ad• deaminase. Interaction with 2',3'- and/or 5'-substituted adenine nucleoside derivatives, *Nuc• Nucleotides*, 10(8), 1677–1692, 1991.

37. Chae, W.-G., Chan, T.C.K., and Chang, C., Facile synthesis of 5'-deoxy- and 2',5'-did• thiopurine nucleosides by nucleoside phosphorylases, *Tetrahedron*, 54(30), 8661–8670, 199•

38. Wnuk, S.F., Stoeckler, J.D., and Robins, M.J., Nucleic acid related compounds. 82. Conv• of adenosine to inosine 5'-thioether derivatives with *Aspergillus oryzae* adenosine deami• alkyl nitrites. Substrate and inhibitory activities of inosine 5'-thioether derivatives with nucleoside phosphorylase, *Nucleosides Nucleotides*, 13(1–3), 389–403, 1994.

39. Ciuffreda, P., Loseto, A., Alessandrini, L., et al., Adenylate deaminase (5′-adenylic acid deaminase, AMPDA)-catalyzed deamination of 5′-deoxy-5′-substituted and 5′-protected adenosines: a comparison with the catalytic activity of adenosine deaminase (ADA), *Eur. J. Org. Chem.*, (24), 4748–4751, 2003.

40. Ciuffreda, P., Casati, S., and Santaniello, E., The action of adenosine deaminase (EC 3.5.4.4.) on adenosine and deoxyadenosine acetates: the crucial role of the 5′-hydroxy group for the enzyme activity, *Tetrahedron*, 56(20), 3239–3243, 2000.

41. Ciuffreda, P., Loseto, A., and Santaniello, E., Deamination of 5′-substituted-2′, 3′-isopropylidene adenosine derivatives catalyzed by adenosine deaminase (ADA, EC 3.5.4.4) and complementary enzymatic biotransformations catalyzed by adenylate deaminase (AMPDA, EC 3.5.4.6): a viable route for the preparation of 5′-substituted inosine derivatives, *Tetrahedron*, 58(29), 5767–5771, 2002.

42. Robins, M.J., Zou, R., Hansske, F., et al., Synthesis of sugar-modified 2,6-diaminopurine and guanine nucleosides from guanosine via transformations of 2-aminoadenosine and enzymatic deamination with adenosine deaminase, *Can. J. Chem.*, 75(6), 762–767, 1997.

43. Margolin, A.L., Borcherding, D.R., Wolf-Kugel, D., et al., AMP-deaminase as a novel practical catalyst in the synthesis of 6-oxopurine ribosides and their analogs, *J. Org. Chem.*, 59(24), 7214–7218, 1994.

44. Mitsuya, H. and Broder, S., Inhibition of the *in vitro* infectivity and cytopathic effect of human T-lymphotrophic virus type III/lymphadenopathy-associated virus (HTLV-III/LAV) by 2′,3′-dideoxynucleosides, *Proc. Natl. Acad. Sci. USA*, 83(6), 1911–1915, 1986.

45. Shelton, M.J., O'Donnell, A.M., Morse, G.D., *Didanosine. Annal. Pharmacother.*, 26(5), 660–670, 1992.

46. Bhat, V., Stocker, E., and Ugarkar, B.G., A new synthesis of 2′,3′-dideoxyinosine, *Synth. Commun.*, 22(10), 1481–1486, 1992.

47. Chu, C.K., Bhadti, V.S., Doboszewski, B., et al., General syntheses of 2′,3′-dideoxynucleosides and 2′,3′-didehydro-2′,3′-dideoxynucleosides, *J. Org. Chem.*, 70(14), 2217–2225, 1989.

48. Webb, R.R, Wos, J.A., Martin, J.C., et al., Synthesis of 2′,3′-dideoxyinosine, *Nucleosides Nucleotides*, 7(2), 147–155, 1988.

49. Ciuffreda, P., Casati, S., and Santaniello, E., Lipase-catalyzed protection of the hydroxy groups of the nucleosides inosine and 2′-deoxyinosine: a new chemo-enzymatic synthesis of the antiviral drug 2′,3′-dideoxyinosine, *Bioorg. Med. Chem. Lett.*, 9(11), 1577–1582, 1999.

50. Robins, M.J., Wilson, J.S., and Hansske, F., Nucleic acid related compounds. 42. A general procedure for the efficient deoxygenation of secondary alcohols. Regiospecific and stereoselective conversion of ribonucleosides to 2′-deoxynucleosides, *J. Am. Chem. Soc.*, 105(12), 4059–4065, 1983.

51. Buenger, G.S. and Nair, V., Dideoxygenated purine nucleosides substituted at the 8-position: chemical synthesis and stability, *Synthesis*, 10, 962–966, 1990.

52. Ikehara, M. and Fukui, T., Studies of nucleosides and nucleotides. LVIII. Deamination of adenosine analogs with calf intestine adenosine deaminase, *Biochim. Biophys. Acta*, 338(2), 512–519, 1974.

53. Seela, F. and Kaiser, K., 8-Aza-7-deaza-2′,3′-dideoxyadenosine: synthesis and conversion into allopurinol 2′,3′-dideoxyribofuranoside, *Chem. Pharm. Bull.*, 36(10), 4153–4156, 1988.

54. Baer, H.P., Drummond, G.I., and Duncan, E.L., Formation and deamination of adenosine by cardiac muscle enzymes, *Mol. Pharmacol.*, 2(1), 67–76, 1966.

55. Mengel, R. and Muhs, W., Nucleoside transformations. V. Transformation of guanosine into 2′-deoxy, 3′-deoxy-, 2′,3′-anhydro- and xylo-guanosine, *Chem. Ber.*, 112(2), 625–639, 1979.

56. Jain, T.C., Jenkins, D.I., Russell, A.F., et al., Reactions of 2-acyloxyisobutyryl halides with nucleosides. IV. A facile synthesis of 2′,3′-unsaturated nucleosides using chromous acetate, *J. Org. Chem.*, 39(1), 30–38, 1974.

57. Suzuki, S., Lee, B., Luo, W., et al., Inhibition of duck hepatitis B virus replication by purine 2′,3′-dideoxynucleosides, *Biochem. Biophys. Res. Commun.*, 156(3), 1144–1151, 1988.

58. Lee, B., Luo, W.-X., Suzuki, S., et al., *In vitro* and *in vivo* comparison of the abilities of purine and pyrimidine 2′,3′-dideoxynucleosides to inhibit duck hepadnavirus, *Antimicrob. Agents Chemother.*, 33(3), 336–339, 1989.

59. Balzarini, J., Pauwels, R., Baba, M., et al., The 2′,3′-dideoxyriboside of 2,6-diaminopurin ively inhibits human immunodeficiency virus (HIV) replication *in vitro*, *Biochem. Bioph Commun.*, 145(1), 269–276, 1987.

60. Balzarini, J., Robins, M.J., Zou, R.M., et al., The 2′,3′-dideoxyriboside of 2,6-diaminopu its 2′,3′-didehydro derivative inhibit the deamination of 2′,3′-dideoxyadenosine, an inhi human immunodeficiency virus (HIV) replication, *Biochem. Biophys. Res. Commun.*, 145(283, 1987.

61. Pauwels, R., Baba, M., Balzarini, J., et al., Investigations on the anti-HIV activity dideoxyadenosine analogues with modifications in either the pentose or purine moiety and selective anti-HIV activity of 2,6-diaminopurine 2′,3′-dideoxyriboside, *Biochem. Pha* 37(7), 1317–1325, 1988.

62. Nair, V. and Sells, T.B., Enzymatic synthesis of 2′,3′-dideoxyguanosine, *Synlett*, 10, 753–7ʃ

63. Robins, M.J., Wilson, J.S., Madej, D., et al., Nucleic acid related compounds. 114. Syn 2,6-(disubstituted)purine 2′,3′-dideoxynucleosides and selected cytotoxic, anti-hepatiti adenosine deaminase substrate activities, *J. Heterocycl. Chem.*, 38(6), 1297–1306, 2001.

64. Marquez, V.E., Tseng, C.K.H., Mitsuya, H., et al., Acid-stable 2′-fluoro purine dideox sides as active agents against HIV, *J. Med. Chem.*, 33(3), 978–985, 1990.

65. Ford, H., Jr., Siddiqui, M.A., Driscoll, J.S., et al., Lipophilic, acid-stable, adenosine dea activated anti-HIV prodrugs for central nervous system delivery. 2. 6-Halo and 6-alkoxy ɾ of 2′-β-fluoro -2′,3′-dideoxyinosine, *J. Med. Chem.*, 38(7), 1189–1195, 1995.

66. Driscoll, J.S., Siddiqui, M.A., Ford H., Jr., et al., Lipophilic, acid-stable, adenosine dea activated anti-HIV prodrugs for central nervous system delivery. 3. 6-Amino prodrugs fluoro-2′,3′-dideoxyinosine, *J. Med. Chem.*, 39(8), 1619–1625, 1996.

67. Shortnacy-Fowler, A.T., Tiwari, K.N., Montgomery, J.A., et al., Synthesis and biologica of 2′-fluoro-ᴅ-arabinofuranosylpyrazolo[3,4-*d*]pyridine nucleosides, *Helv. Chim. Acta*, 2240–2245, 1999.

68. Jung, P.M.J., Burger, A., and Biellmann, J.-F., Diastereofacial selective addition of ethyny reagent and Barton-McCombie reaction as the key steps for the synthesis of C-3′-ethy nucleosides and of C-3′-ethynyl-2′ -deoxyribonucleosides, *J. Org. Chem.*, 62(24), 8309–831

69. Tritsch, D., Jung, P.M., Burger, A., et al., 3′-β-ethynyl and 2′-deoxy-3′-β-ethynyl adenosiɪ 3′-β-branched-adenosines substrates of adenosine deaminase, *Bioorg. Med. Chem. Letɪ* 139–141, 2000.

70. Ohrui, H. and Mitsuya, H., 4′-C-substituted-2 ′-deoxynucleosides: a family of antiretrovirɑ which are potent against drug-resistant HIV variants, *Curr. Drug Targets Infect. Disoɪ* 1–10, 2001.

71. Nomura, M., Shuto, S., Tanaka, M., et al., Nucleosides and nucleotides. 185. Synth biological activities of 4′α-C-branched-chain sugar pyrimidine nucleosides, *J. Med.* 42(15), 2901–2908, 1999.

72. Ohrui, H., Kohgo, S., Kitano, K., et al. Syntheses of 4′-C-ethynyl-β -ᴅ-arabino- and 4′-C- 2′-deoxy-β-ᴅ-ribo-pentofuranosylpyrimidines and -purines and evaluation of their anti-Hɪ ity, *J. Med. Chem.*, 43(23), 4516–4525, 2000.

73. Kodama, E.-I., Kohgo, S., Kitano, K., et al., 4′-Ethynyl nucleoside analogs: potent inhiɪ multidrug-resistant human immunodeficiency virus variants *in vitro*, *Antimicrob. Ageɪ mother.*, 45(5), 1539–1546, 2001.

74. Kohgo, S., Yamada, K., Kitano, K., et al., Design, efficient synthesis, and anti-HIV ac 4′-C-cyano- and 4′-C-ethynyl-2′-deoxy purine nucleosides, *Nucleosides Nucleotides Nucleɪ* 23(4), 671–690, 2004.

75. Shimada, N., Hasegawa, S., Harada, T., et al., Oxetanocin, a novel nucleoside from ɪ *J. Antibiot.*, 39(11), 1623–1625, 1986.

76. Hoshino, H., Shimizu, N., Shimada, N., et al., Inhibition of infectivity of human imm ciency virus by oxetanocin, *J. Antibiot.*, 40(7), 1077–1078, 1987.

77. Ireland, C., Leeson, P.A., and Castaner, J., Lobucavir, *Drugs Future*, 22(4), 359–370, 199

78. Shimada, N., Hasegawa, S., Saito, S., et al., Derivatives of oxetanocin: oxetanocins H, X and 2-aminooxetanocin A, *J. Antibiot.*, 40(12), 1788–1790, 1987.

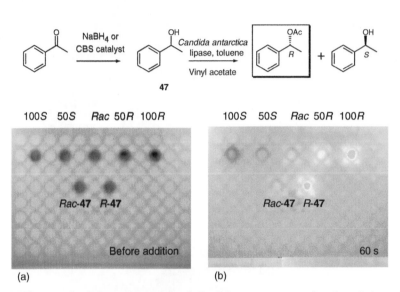

FIGURE 3.38 Time-resolved thermal imaging of *Candida antarctica*–catalyzed acylation of 1-phenylethanol **47**. NaBH₄ reduction of acetophenone afforded *Rac*-**47** (ee 1%, chiral HPLC), whereas Corey–Bakshi–Shibata (CBS) reduction gave *R*-**47** (ee 86%, chiral HPLC). Colors correspond to the following temperatures: (a) at time zero—black (12.5 to 13.5°C), purple (15 to 16°C), pink (17 to 18°C), orange (18.5 to 19.5°C); (b) after 60 s—orange (18.5 to 19°C), yellow (19.5 to 20°C), white (20.5°C), gray (21.4 to 21.6°C).

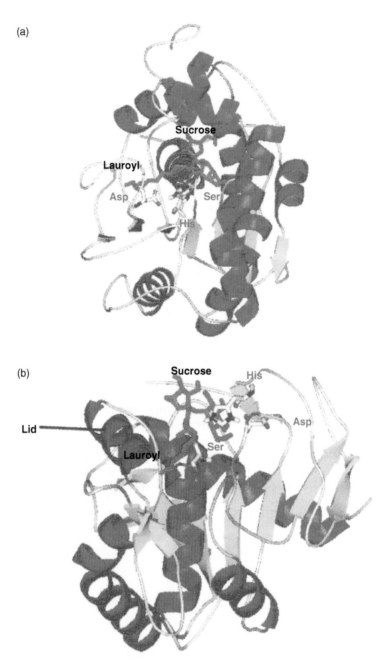

FIGURE 17.2 Structure of *C. antarctica* (a) and *T. lanuginosus* (b) lipases, docked with lauroyl regioisomers. Both enzymes are characterized by similar overall folds and the same catalytic residues and oxyanion hole. (Adapted from Fuentes, G., et al., *Protein Sci.*, 13, 3092, 2004, permission.)

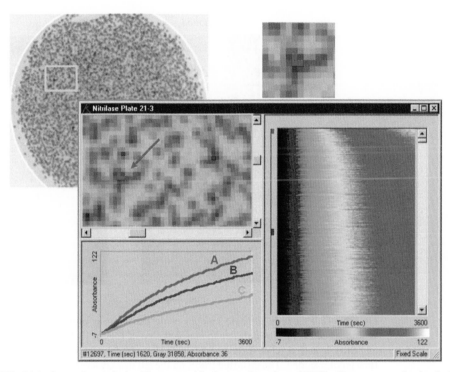

FIGURE 18.3 Operation of the Kcat graphical user interface (GUI). Data were obtained during screening of a nitrilase library. In the background is an image of the original assay disk, with the region appearing in the GUI enclosed in a rectangle. A single "high-activity" microcolony is also magnified.

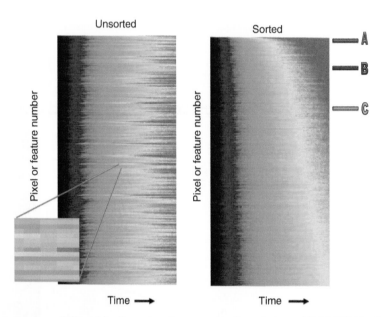

FIGURE 18.4 Sorting nitrilase kinetics using the contour plot window. All 15,000 kinetics tr
visible in each view. The data are the same as that of Figure 18.3. The x-axis corresponds to elaps
and the y-axis to pixel or feature number. In the actual window, the intensity of the absorption a
time is color-coded. The unsorted traces on the left display a random color pattern. The traces on
have been sorted based on their change in absorbance over time. The most "active" pixels are
moved to the top. Using the computer mouse, it is possible to select various groups of pixels (mark
three grouping bars to the right of the window) to identify the corresponding microcolonies.

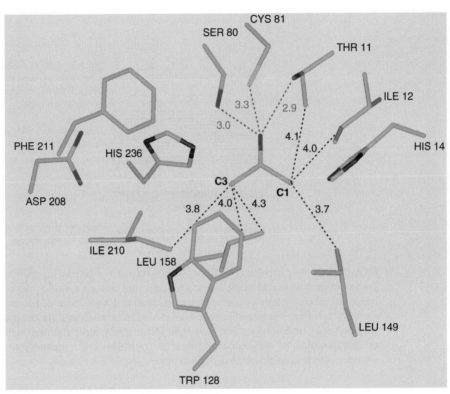

FIGURE 28.2 Structure of selected active-site residues of hydroxynitrile lyase from *Manihot eseulenta* (MeHNL) complexed with acetone.

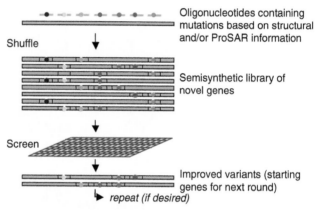

FIGURE 30.5 Semisynthetic shuffling. The nature of the genetic diversity that is recombined w. gene of interest is biased using oligonucleotides that contain specific mutations that are expecte. beneficial. Such mutations can be based on structural knowledge of the enzyme, hot spots in th. found to be of interest in earlier rounds of evolution (even from different projects that utilized th. starting gene), or mutations that have been identified by computational methods as desirable. Th. of incorporation of the oligonucleotides is controlled experimentally to create semisyntheti. libraries with a variable number of mutations.

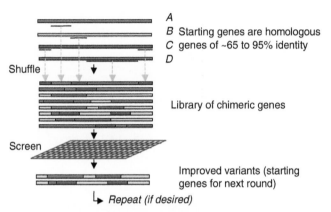

FIGURE 30.6 Family shuffling. Various genes encoding enzymes with identical (or at least similar) function are recombined to provide libraries of chimeric genes. The diversity introduced in this method is functional since it has been preselected in nature. In addition to full-length genes, partial genes and oligonucleotides can be added to the library generation procedure, thereby improving the quality of the libraries that are generated.

79. Marquez, V.E. and Lim M.-I., Carbocyclic nucleosides, *Med. Res. Rev.*, 6(1), 1–40, 1986.

80. Marquez, V.E., Carbocyclic nucleosides, in *Advances in Antiviral Drug Design*, vol. 2, De Clercq, E., Ed., JAI Press, Greenwich, CT, 1996, pp. 89–146.

81. Gala, D. and Schumacher, D.P., Deamination of carbocyclic adenosine analogs with adenosine deaminase and its synthetic utility, *Synlett*, 1, 61–63, 1992.

82. Vince, R. and Brownell, J., Resolution of racemic carbovir and selective inhibition of human immunodeficiency virus by the (−) enantiomer, *Biochem. Biophys. Res. Commun.*, 168(3), 912–916, 1990.

83. Exall, A.M., Jones, M.F., Mo, C.-L., et al., Synthesis from (−)-aristeromycin and x-ray structure of (−)-carbovir, *J. Chem. Soc. Perkin Trans. I*, 10, 2467–2477, 1991.

84. Schaeffer, H.J., Beauchamp, L., de Miranda, P., et al., 9-(2-Hydroxyethoxymethyl) guanine activity against viruses of the herpes group, *Nature*, 272(5654), 583–585, 1978.

85. Martin, J.C., Dvorak, C.A., Smee, D.F., et al., 9-[(1,3-Dihydroxy-2-propoxymethyl)]guanine: a new potent and selective antiherpes agent, *J. Med. Chem.*, 26(5), 759–761, 1983.

86. Harnden, M.R., Jarvest, R.L., Bacon T.H., et al., Synthesis and antiviral activity of 9-[4-hydroxy-3-(hydroxymethyl)but-1-yl]purines, *J. Med. Chem.*, 30(9), 1636–1642, 1987.

87. Vere Hodge, R.A., Sutton, D., Boyd, M.R., et al., Selection of an oral prodrug (BRL 42810; famciclovir) for the antiherpesvirus agent BRL 39123 [9-(4-hydroxy-3-hydroxymethylbut-l-yl)guanine; penciclovir], *Antimicrob. Agents Chemother.*, 33(10), 1765–1773, 1989.

88. Robins, M.J. and Hatfield, P.W., Nucleic acid related compounds, 37, convenient and high-yield syntheses of N-[(2-hydroxyethoxy)methyl]heterocycles as "acyclic nucleoside" analogues, *Can. J. Chem.*, 60(5), 547–553, 1982.

89. Ogilvie, K.K., Nguyen-ba, N., Gillen, M.F., et al., Synthesis of a purine acyclonucleoside series having pronounced antiviral activity, The glyceropurines, *Can. J. Chem.*, 62(2), 241–252, 1984.

90. Kim, D.K., Kim, H.K., and Chae, Y.-B., Design and synthesis of 6-fluoropurine acyclonucleosides: potential prodrugs of acyclovir and ganciclovir, *Bioorg. Med. Chem. Lett.*, 4(11), 1309–1312, 1994.

91. Haines, D., Tseng, C., and Marquez, V.E., Synthesis and biological activity of unsaturated carboacyclic purine nucleoside analogues, *J. Med. Chem.*, 30(5), 943–947, 1987.

92. Phadtare, S. and Zemlicka, J., Nucleic acid derived allenols. Unusual analogs of nucleosides with antiretroviral activity, *J. Am. Chem. Soc.*, 111(15), 5925–5931, 1989.

93. Phadtare, S., Kessel, D., Corbett, T.H., et al., Unsaturated and carbocyclic nucleoside analogues: synthesis, antitumor and antiviral activity, *J. Med. Chem.*, 34(1), 421–429, 1991.

94. Xu, Z.-Q., Qiu, Y.-L., Chokekijchai, S., et al., Unsaturated acyclic analogues of 2′-deoxyadenosine and thymidine containing fluorine: synthesis and biological activity, *J. Med. Chem.*, 38(6), 875–882, 1995.

95. Phadtare, S. and Zemlicka, J., Synthesis of (Z)- and (E)-N⁹-(4-hydroxy-1-buten-1-yl)adenine. New unsaturated analogs of adenosine, *Tetrahedron Lett.*, 31(1), 43–46, 1990.

96. Phadtare, S. and Zemlicka, J., Synthesis and biological properties of 9-(trans-4-hydroxy-2-buten-1-yl)adenine and guanine: open-chain analogues of neplanocin A, *J. Med. Chem.*, 30(2), 437–440, 1987.

97. Qiu, Y.-L., Ksebati, M.B., Ptak, R.G., et al., (Z)- and (E)-2-((Hydroxymethyl) cyclopropylidene)methyladenine and -guanine. New nucleoside analogues with a broad-spectrum antiviral activity, *J. Med. Chem.*, 41(1), 10–23, 1998.

98. Standring, D.N., Bridges, E.G., Placidi, L., et al., Antiviral β-L-nucleosides specific for hepatitis B virus infection, *Antivir. Chem. Chemother.*, 12(Suppl. 1), 119–121, 2001.

99. Pelicano, H., Pierra, C., Eriksson, S., et al., Enzymatic properties of the unnatural β-L-enantiomers of 2′,3′-dideoxyadenosine and 2′,3′-didehydro-2′,3′-dideoxyadenosine, *J. Med. Chem.*, 40(24), 3969–3973, 1997.

100. Chun, B.K., Olgen, S., Hong, J.H., et al., Enantiomeric syntheses of conformationally restricted D- and L-2′,3′-dideoxy-2′,3′-endo-methylene nucleosides from carbohydrate chiral templates, *J. Org. Chem.*, 65(3), 685–693, 2000.

101. Secrist III, J.A., Montgomery, J.A., Shealy, Y.F., et al., Resolution of racemic carbocyclic analogues of purine nucleosides through the action of adenosine deaminase. Antiviral activity of the carbocyclic 2′-deoxyguanosine enantiomers, *J. Med. Chem.*, 30(4), 746–749, 1987.

102. Borthwich, A.D., Biggadike, K., and Kirk, B.E., et al., Preparation of cyclopentyl triazolo dimones as antivirals and pharmaceutical compositions containing them, Eur. Pat. Appl. 660, 1991.

103. Moon, H.R., Ford, H., and Marquez, V.E., A remarkably simple chemicoenzymatic appro structurally complex bicyclo[3.1.0]hexane carbocyclic nucleosides, *Org. Lett.*, 2(24), 3793 2000.

104. Rodriguez, J.B., Marquez, V.E., Nicklaus, M.C., et al., Conformationally locked nuc analogues. Synthesis of dideoxycarbocyclic nucleoside analogues structurally related to nepl C, *J. Med. Chem.*, 37(20), 3389–3399, 1994.

105. Zemlicka, J., Unusual analogues of nucleosides: chemistry and biological activity, in *Advances in Nucleosides: Chemistry and Chemotherapy*, Chu, C.K., Ed., Elsevier Science, A dam, 2002, pp. 327–357.

106. Qiu, Y.-L., Ksebati, M.B., and Zemlicka, J., Synthesis of (Z)- and (E)-9-[(2-hydroxyethyl cyclopropyl]adenine. New methylenecyclopropane analogues of adenosine and their su activity for adenosine deaminase, *Nucleosides Nucleotides Nucleic Acids*, 19(1–2), 31–37, 2C

107. Zemlicka, J., Allenols derived from nucleic acid bases: a new class of anti-HIV agents: che and biological activity, in *Nucleosides and Nucleotides as Antitumor and Antiviral agents*, Chu and Baker, D.C., Eds., Plenum, New York, 1993, pp. 73–100.

108. Megati, S., Goren, Z., Silverton, J.V., et al., (R)-(−)- and (S)-(+)-Adenallene: synthesis, ab configuration, enantioselectivity of antiretroviral effect, and enzymic deamination, *J. Med.* 35(22), 4089–4104, 1992.

109. Jones, B.C.N.M. and Zemlicka, J., R,S-Adenallene 4'-phosphate: substrate activity and en electivity toward AMP deaminase and 5'-nucleotidase, *Biorg. Med. Chem. Lett.*, 5(15), 1633 1995.

110. Jones, B.C.N.M., Drach, J.C., Corbett, T.H., et al., (±)-N⁹-(2-(Hydroxymethyl) spiro[3.3]hep adenine. The first biologically active saturated analogue of adenallene with axial dissym *J. Org. Chem.*, 60(20), 6277–6280, 1995.

111. Guan, H.-P., Ksebati, M.B., Cheng, Y-C., et al., Spiropentane mimics of nucleosides: analog 2'-deoxyadenosine and 2'-deoxyguanosine. Synthesis of all stereoisomers, isomeric assignmen biological activity, *J. Org. Chem.*, 65(5), 1280–1290, 2000.

112. Ciuffreda, P., Loseto, A., and Santaniello, E., Stereoselective deamination of (5'RS)-5'-m 2',3'-isopropylidene adenosine catalyzed by adenosine deaminase: preparation of diastere cally pure 5'-methyl adenosine and inosine derivatives, *Tetrahedron Asymmetry*, 13(3), 23 2002.

113. Ciuffreda, P., Loseto, A., and Santaniello, E., Stereoselective adenylate deaminase (5'-adenyl deaminase, AMPDA)-catalyzed deamination of 5'-alkyl substituted adenosines: a compariso the action of adenosine deaminase (ADA), *Tetrahedron Asymmetry*, 15(2), 203–206, 2004.

114. Shuto, S., Obara, T., Toriya, M., et al., New neplanocin analogues. 1. Synthesis of 6'-m neplanocin A derivatives as broad-spectrum antiviral agents, *J. Med. Chem.*, 35(2), 324–331

115. Shuto, S., Obara, T., Yaginuma, S., et al., New neplanocin analogues. IX. A practical prepa of (6'R)-6'-C-methylneplanocin A (RMNPA), a potent antiviral agent, and the determinatio 6'-configuration. Diastereoselective deamination by adenosine deaminase, *Chem. Pharm.* 45(1), 138–142, 1997.

20 Microbial and Enzymatic Processes for the Production of Chiral Compounds

Kohsuke Honda, Takeru Ishige, Michihiko Kataoka, and Sakayu Shimizu

CONTENTS

20.1 INTRODUCTION

The global market of the chemical industry amounts to US $1.8 trillion, and the contribution to this of biotechnology has been increasing year by year. According to McKinsey & Company's report, white biotechnology accounted for only 5% of the market in 2000, but th mate that it will increase to 10 to 20% by 2010. Biotechnology will be a significa innovation driver in the next 10 years. As for the synthesis of various chemical comp biosynthesis has some advantages over conventional chemical synthesis. Reactions ca by biocatalysts proceed under modest conditions with remarkable rate acceleration 10^{18}), and thus applying them to industrial syntheses prevents the wasting of fossil fuels r for preparing high-temperature and high-pressure conditions that are commonly requ conventional chemical syntheses. Above all, the excellent stereoselectivities of bioc enabling the formation of enantiomerically pure products are a particularly attractive because the chirality of a molecule is a very important factor in the pharmaceutical I fact, over half of the top 100 pharmaceuticals are chiral molecules, and this chir. market netted over US $145 billion in 2003. The industrial production methods for o active compounds, regardless of the catalysts, have been reviewed by Breuer et al. [1] are two major strategies for the syntheses of optically active compounds: optical resolu racemic mixtures and stereoselective conversion of prochiral structures. In this chap describe recent successful examples of industrialized enzymatic chiral compound sy involving these two strategies.

20.2 A NOVEL TOOL FOR ENZYMATIC OPTICAL RESOLUTION—LACTC

20.2.1 WHAT IS A LACTONASE?

Carboxyl ester-hydrolyzing enzymes (EC 3.1.1.x) such as lipases and esterases exhib diversity and are the most widely used group of enzymes for industrial organic sy [2–4]. Reactions catalyzed by them are often stereo- and/or regioselective. In addi hydrolysis, they catalyze several other reactions, esterification (reverse reaction of hydr and transesterification (alcoholysis of ester bonds). Some of them even exhibit c activity in organic solvents, allowing the use of water-insoluble compounds as sub [5,6]. These features make carboxyl ester hydrolases convenient to use as synthetic ca

Lactonases, catalyzing hydrolysis of the intramolecular ester bonds of lactone belong to the esterase family. Lactones are widely found in nature as biologically compounds and metabolic intermediates. Lactonases comprise some of the enzymes ir in the synthesis and degradation of these compounds. For example, a lactonase from sp. hydrolyzes and inactivates N-acyl-homoserine lactones, which are produced by a r. gram-negative bacterial species as quorum-sensing signals and are involved in t density–dependent regulation of specific sets of genes [7–9]. Aldonate lactone hyd have been suggested to participate in the metabolism of aldoses [10–13]. Thus, a nun lactonases have already been found in various organisms. The reactions catalyzed by nases, as well as those by lipases and other esterases, are sometimes regio- and/or ste cific, and should be applicable to the chemoenzymatic synthesis of optically active la and the corresponding hydroxyacids. This section describes the development of a lac mediated production system. Some other potential lactonases are also dealt with.

20.2.2 OPTICAL RESOLUTION OF DL-PANTOYL LACTONE BY A FUNGAL LACTONASE

20.2.2.1 Discovery of a Novel Lactonase

D-Pantoyl lactone (D-PL) is a chiral building block for the production of c D-pantothenate and its derivatives, which are sold as vitamin supplements, feed ad and cosmetics. The commercial production of D-PL relies on a chemical synthesis proc includes a complicated resolution process for DL-PL. In order to simplify the res process, Shimizu et al. evaluated many microbial enzymes as potential catalysts [14–17]. At the beginning, they focused on the asymmetric reduction of a prochiral compound, ketopantoyl lactone, to yield D-PL, and screened a number of microbes for reductase activity. During the screening, they serendipitously found that some microorganisms degrade the product and accumulate D-pantoic acid. The degrading reactions, i.e., the hydrolysis of lactone, catalyzed by these microorganisms proceeded quite effectively and sometimes stereoselectively. This finding prompted them to perform intensive exploration of a novel industrial enzyme, lactonase.

PL-hydrolyzing activity is widely distributed in various microorganisms [18,19]. While bacterial strains tend to hydrolyze the L-enantiomer of PL, fungal strains such as *Fusarium*, *Gibberella*, *Penicillium*, and *Schizophyllum* preferentially hydrolyze the D-enantiomer. When an L-specific lactonase is used for the resolution of racemic PL, the optical purity of the remaining D-PL might be low, except when the hydrolysis of L-PL is completed. In the case of D-specific lactonase, D-pantoic acid of high optical purity can be constantly obtained regardless of the hydrolysis yield. After the asymmetric hydrolysis of DL-PL, the remaining L-enantiomer and D-pantoic acid formed can be separately extracted by altering the pH of the reaction mixture. The recovered L-PL is easily racemized by heating under acidic conditions and thus can be reused as a substrate (Figure 20.1).

Finally, a filamentous fungus, *Fusarium oxysporum* AKU3702, which showed the highest hydrolysis activity and stereoselectivity toward D-PL, was selected as a potential enzyme source.

20.2.2.2   Properties of the Lactonase from *Fusarium Oxysporum*

The D-PL hydrolyzing lactonase of *F. oxysporum* has been isolated and characterized in some detail [20]. The relative molecular mass of the lactonase is 125 kDa and the subunit molecular mass is 60 kDa, suggesting that the enzyme is a dimer of identical subunits. One molecule of calcium and 15.4% (w/w) glucose equivalent of carbohydrate are included in the enzyme. Calcium is necessary for both the enzyme activity and stability. The enzyme hydrolyzes aldonate lactones, such as D-galactono-γ-lactone, L-mannono-γ-lactone, and D-gulono-γ-lactone, stereospecifcally. The corresponding enantiomers not only is inert as substrates but also competitively inhibit the enzyme activity. For every substrate, the reverse reaction, i.e., the lactonization of aldonic acids and D-pantoic acid, takes place under acidic pH conditions.

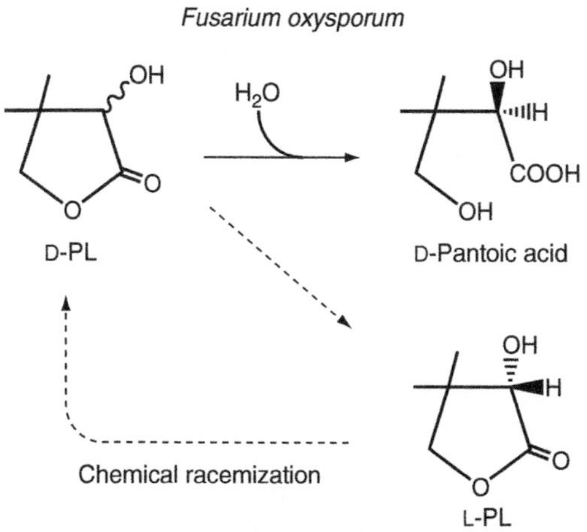

FIGURE 20.1 Principle of the optical resolution of DL-PL using the lactonase-producing fungus, *Fusarium oxysporum*.

20.2.2.3   Development of a Practical Resolution System

Under the optimum conditions, the asymmetric hydrolysis of DL-PL proceeded quite effe
with wet mycelia of *F. oxysporum* as a catalyst [18,21]. Both the concentration (>300 g/
the optical purity (>96% ee) of the product were sufficient for practical use. However, wh
mycelia were recovered after the reaction and reused for further reactions, the re
enzyme activity was <80% of the initial level. Such a decrease in enzyme activity is a
disadvantage for a commercial process that requires retention of higher activity for lon
repeated reactions. In order to improve the enzyme stability, the *Fusarium* mycelia
immobilized by calcium alginate gel [22]. Immobilization could protect the mycelia
the damage due to stirring and the resulting leakage of the enzyme from the mycelia in
reaction mixture. It had another advantage in that the immobilized mycelia could be
recovered after each reaction. When the calcium alginate gel–entrapped mycelia were
bated with a 300 g/L DL-PL solution with automatic control of the pH of the reaction m
(pH 6.0 to 7.0), 90 to 95% of the D-enantiomer in the racemic mixture was sele
hydrolyzed to D-pantoic acid of high optical purity (90 to 97% ee). The immobilized m
retained 70% of the initial lactonase activity, even after 180 reactions (Figure 20.2
reaction time = 3780 h). The estimated half-life of the lactonase activity of the immo
mycelia was 6000 h, which is 35 times higher than that of the free mycelia. The improven
the enzyme stability is partly due to calcium, as a stabilizer of the enzyme, contained in t

This enzymatic resolution process was scaled up and has been in commercial ope
since 1999. This process is highly satisfactory not only from economic aspects but als
environmental ones (water −49%, CO_2 −30%, and BOD −60%, compared with the
chemical resolution process). Nowadays, about 30% of the world production of ca
D-pantothenate (~6000 t/y) occurs through this chemoenzymatic process.

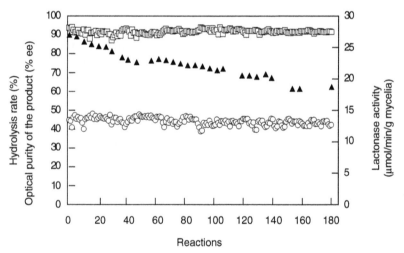

FIGURE 20.2 Asymmetric hydrolysis of DL-PL by immobilized *Fusarium oxysporum*. Batchwis
tion was repeated 180 times. Each reaction was performed as follows. Immobilized *F. oxysporu*
containing 4.6 g wet mycelia) was incubated in 100 mL of a 30% (w/v) DL-PL solution supplement
90 mM $CaCl_2$ at 30°C for 21 h. The pH of the mixture was automatically kept in the range of 6.5
After the reaction, immobilized cells were collected by filtration and used for the next reaction v
being washed. Hydrolysis rate for DL-PL (○), optical purity of D-pantoic acid formed (□), and la
activity of the immobilized cells (▲). PL, pantoyl lactone.

20.2.2.4 Recent Progress

Recently, the *Fusarium* lactonase gene was overexpressed in a heterologous fungus, *Aspergillus oryzae*, under the control of an artificially modified promoter [23]. The lactonase gene includes five introns and a presumed endoplasmic reticulum (ER)–transporting signal peptide in its NH_2-terminal region. *A. oryzae* can recognize the signal peptide of the *Fusarium* lactonase and the splice junctions in the gene. Posttranslational modifications, such as transportation to the ER, cleavage of the signal peptide, and glycosylation, occurred in the same manner as in the case of *F. oxysporum*, and the recombinant enzyme was produced as a mature protein like the wild-type enzyme.

When mycelia of the recombinant *A. oryzae* were used as a catalyst for asymmetric hydrolysis of DL-PL, the initial velocity of the reaction was 30 times higher than that with *F. oxysporum*, suggesting that the transformant is promising for industrial application.

20.2.3 OTHER MICROBIAL LACTONASES

20.2.3.1 L-Pantoyl Lactone Hydrolase from *Agrobacterium Tumefaciens*

As described above, L-PL hydrolyzing activity (Figure 20.3a) was largely found in bacterial strains. Two similar but distinct L-PL hydrolases have been isolated from *Agrobacterium tumefaciens* AKU316 and *A. tumefaciens* Lu681 by Kataoka et al. [19] and Kesseler et al. [24], respectively. Kesseler et al. also demonstrated the availability of L-PL hydrolase for an alternative racemic PL-resolution process. The asymmetric hydrolysis of DL-PL [30% (w/v)] by the Lu681 enzyme gave D-PL, the yield being 50 to 53% and the optical purity 90 to 95% ee. Covalent immobilization in a carrier material, EupergitC, made the enzyme more stable and easier to handle in repeated batch reactions. Furthermore, mutant enzymes with improved L-PL hydrolyzing activity were generated through directed evolution. The specific activities of mutants F62S, K197D, and F100L were 2.3, 1.7, and 1.5 times higher than that of the wild-type enzyme, respectively.

20.2.3.2 Dihydrocoumarin Hydrolase from *Acinetobacter Calcoaceticus*

The dihydrocoumarin hydrolase from *Acinetobacter calcoaceticus* is a lactonase specific for aromatic lactones such as dihydrocoumarin, homogentistic acid-γ-lactone, and 2-coumaranone (Figure 20.3b) [25]. Interestingly, this enzyme is bifunctional, being capable of both the hydrolysis of ester bonds and the hydrolytic degradation of peroxoacids [26].

This enzyme also catalyzes the stereo- and/or regiospecific hydrolysis of linear esters and thus is applicable to industrial processes (Figure 20.3b) [27]. Dihydrocoumarin hydrolase catalyzes the enantiospecific hydrolysis of methyl β-acetylthioisobutyrate. The reaction product, D-β-acetylthioisobutyrate, is an important chiral building block for the synthesis of a series of angiotensin-converting enzyme inhibitors. The same enzyme is also useful for regioselective deblocking (specific for the terminal ester bond) of methyl cetraxate to yield cetraxate, which is widely used as an antiulcer agent.

20.2.3.3 (*R*)-δ-Decanolactone Hydrolase from *Pseudomonas* sp. 3–1

Recently, a lactonase catalyzing the enantiospecific hydrolysis of (*R*)-δ-decanolactone ((*R*)-δ-decanolactone hydrolase (DLHase), Figure 20.3c) was found in the membrane fraction of *Pseudomonas* sp. 3–1, which was isolated as a δ-decanolactone-assimilating bacterium (Honda et al., unpublished data). This enzyme was the first example of the membrane-associated lactonase. δ-Decanolactone is known as a naturally occurring aroma in meat and

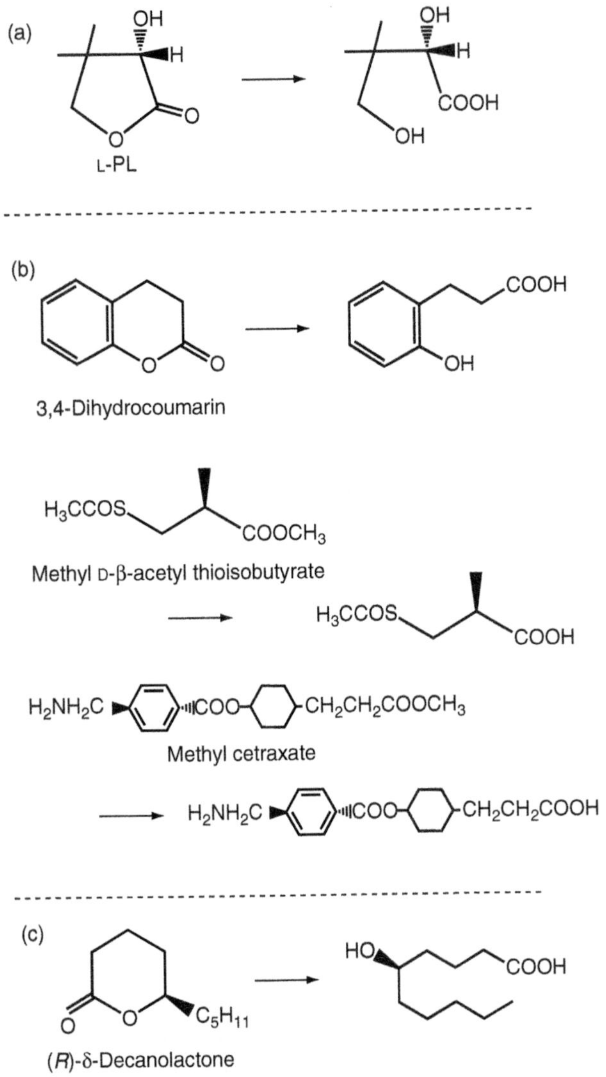

FIGURE 20.3 Hydrolysis reactions catalyzed by (a) L-pantoyl lactone hydrolase of *Agroba tumefaciens*, (b) dihydrocoumarin hydrolase of *Acinetobacter calcoaceticus*, and (c) (R)-δ-decan hydrolase of *Pseudomonas* sp. 3–1.

dairy products, and has been in great commercial demand as a food additive [28]. Bes decanolactone, the enzyme hydrolyzes lactones with both 5- and 6-membered rings, su caprolactone, γ-heptalactone, γ-decanolactone, δ-valerolactone, and δ-octanolactor deduced amino acid sequence of DLHase exhibits low but significant similarity to tl the *Fusarium* lactonase and the δ-valerolactone hydrolase from *Comamonas* sp. NCIM [29,30]. Although they all act on intramolecular ester bonds, their substrate specificit physiological roles are quite different from each other. This suggests that a lactonase p tor diverged into these specific lactonases with some parallel functions independe different evolutionary lines. Until this finding, lactonases had been classified into tl family according to their phenotype alone, i.e., catalysis of the hydrolysis of lacton no structural relationship had been shown among them. Structural information o homologous lactonases might enable the use of novel strategies, such as metagenome screening with specific probes and gene-shuffling among family proteins, to obtain potential lactonases.

20.3 DIRECT SYNTHESIS OF CHIRAL ALCOHOLS WITH STEREOSELECTIVE CARBONYL REDUCTASES

20.3.1 METHODLOGICAL ADVANTAGES OF DIRECT SYNTHESIS

Although the optical resolution of a racemic substrate is a versatile strategy for obtaining enantiomerically pure compounds, one major disadvantage of this strategy is that the theoretical maximum yield is only 50%, because only one enantiomer in the racemic mixture can serve as the substrate.

Direct synthesis is a simple and promising method. A high efficiency, i.e., a 100% theoretical yield, can be expected. Asymmetric reduction of prochiral carbonyl substrates is one of the most competitive methodologies for chiral alcohol production. Chiral alcohols are useful starting materials as building blocks for many compounds. Among 205 patents in the area of biotransformations issued after the year 2000 by companies producing pharmaceuticals, fine chemicals, and flavours and fragrances, 57 are for alcohols, 35 of which are for technologies concerning ketone reduction [31]. This section reviews successful examples of chiral alcohol production utilizing enantioselective carbonyl reductases.

20.3.2 DIVERSITY OF CARBONYL REDUCTASES

Carbonyl compounds, such as ketones or aldehydes, are ubiquitous compounds in cells, and are important intermediates in metabolic and synthetic pathways. However, the reactivity of the carbonyl groups can cause damage to cells. The enzymes that reduce these carbonyl compounds belong to one of three enzyme superfamilies: the short-chain dehydrogenase/reductase superfamily, the medium-chain dehydrogenase/reductase superfamily, and the aldo-keto reductase superfamily. Each enzyme can be classified into one of these superfamilies based on DNA/amino acid sequences, structures, catalytic mechanisms, and so on [32–35].

As for the applicability of carbonyl reductases to chemical syntheses, it has long been well known that baker's yeast (*Saccharomyces cerevisiae*) catalyzes the asymmetric reduction of prochiral carbonyl substrates in the presence of glucose (energy source). *S. cerevisiae* cells exhibit a broad substrate specificity toward various carbonyl compounds due to the existence of many oxidoreductases. Kaluzna et al. systematically investigated 18 key reductases from baker's yeast to a total 11 α- and β-keto ester substrates [36]. Such diversity in carbonyl-reducing enzymes is convenient for cells to survive under diverse environmental conditions, but is inconvenient for practical use of the cells for chiral alcohol production. For example, the actions of more than one enzyme on a prochiral carbonyl substrate often result in a low enantiopurity of the product. Purified enzymes and heterologous microorganism cells, in which a desired reductase gene is expressed, are usually used for industrial processes.

20.3.3 COFACTOR REGENERATING SYSTEMS

20.3.3.1 Single Enzyme Systems

The reduction of carbonyl groups requires a cofactor, NAD(P)H, and a continuous supply is necessary for an efficient reaction and high conversion rate. Considering their high cost, it is not economically feasible to supply them externally in stoichiometric amounts, and thus it is desirable to construct an *in situ* regeneration system involving a second redox reaction.

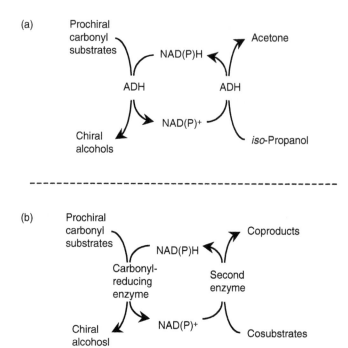

FIGURE 20.4 Cofactor regeneration systems. (a) Single enzyme (coupled substrate) system in ADH and *iso*-propanol. ADH, alcohol dehydrogenase. (b) Coupled enzyme system.

The combination of alcohol dehydrogenase (ADH) and a small aliphatic alcohol, e. ADH-*iso*-propanol system, is classic and well studied (Figure 20.4a). In this system, a enzyme (ADH) catalyzes both the asymmetric reduction of prochiral carbonyl substrate the regeneration of cofactors in the presence of a second substrate (*iso*-propanol). To dr reaction equilibrium in the desired direction, a large supply of *iso*-propanol is essential. Alt acetone is a problem for the thermodynamics and kinetics of the reaction, it can be easily re from the reaction mixture. This means that the equilibrium limitations of cosubstrat be ignored, and an increase in the conversion yield can be expected. A high concentrat *iso*-propanol increases the solubility of lipophilic substrates, but it also demands that the be chemostable. Stampfer et al. reported that *Rhodococcus ruber* whole cells showed ca activity toward various ketones with high enantioselectivity in a 50% (v/v) *iso*-pro containing reaction mixture [37]. Furthermore, the ADH partially purified from this stra retained high activity in the presence of 80% (v/v) *iso*-propanol [38].

20.3.3.2 Coupled Enzyme Systems

A two-enzyme two-substrate system means that reduction of the substrate and regene of the cofactor are catalyzed by different enzymes (Figure 20.4b). In this case, there is n to consider the limitations arising from thermodynamic equilibrium described above thermore, various carbonyl-reducing enzymes, not necessarily having ADH activity, used in this system. One good example of a second enzyme–substrate combination formate dehydrogenase (FDH)–formate system. Regarding this reaction, formate is a and safe hydrogen donor, the coproduct is innocuous CO_2, and the reaction is irreve These features have allowed this system to be applied to many commercialized prod systems, e.g., L-*tert*-leucine production in combination with leucine dehydrogena reductive amination of 2-oxo-3,3-dimethylbutanoic acid [39]. Although one disadvantage of this system is that FDHs from methylotrophs substantially only use NAD$^+$ as a cofactor, Tishkov et al. reported pilot-scale production of a mutant FDH from *Pseudomonas* sp. 101 that showed high specificity for NADP$^+$ and NAD$^+$ [40].

The pairing of glucose dehydrogenase (GDH)–glucose is another example of a coupled enzyme system. Glucose is a cheaper cosubstrate than *iso*-propanol and formate. GDH from *Bacillus megaterium* can regenerate both NADH and NADPH [41], so it can be used together with any carbonyl-reducing enzymes independent of their cofactor dependencies. The addition of an alkaline solution is required for continuous reactions, because coproduct gluconolactone is spontaneously converted to gluconate, which decreases the pH of the reaction mixture.

Novel candidates for cofactor regeneration systems have been reported recently. Hydrogenase I from *Pyrococcus furiosus* uses molecular hydrogen as a hydrogen donor for NADPH regeneration, and forms protons as the only coproduct [42]. The enzyme retains its activity at as high as 80°C because it is from a hyperthermophilic archaeon, but the instability of NADPH becomes a limitation of reactions at such high temperature conditions.

Nakamura and Yamanaka reported a unique NADPH regeneration system involving cyanobacteria that uses light energy as reducing power [43]. Cyanobacteria are characteristic phototrophs and microbes, so they possess the features of both direct utilization of light energy and a high growth rate. This is a promising system since light energy is ultimately an economical resource.

20.3.4 PRACTICAL APPLICATIONS OF CARBONYL REDUCTASES

20.3.4.1 Production of 4-Chloro-3-Hydroxybutanoate Ethyl Ester through Alternative Processes

(*R*)- and (*S*)-4-chloro-3-hydroxybutanoate ethylesters (CHBEs) are useful chiral building blocks applicable to the synthesis of pharmaceuticals: the (*R*)-enantiomer for a precursor of L-carnitine [44] and the (*S*)-enantiomer for hydroxymethylglutaryl-CoA (HMG-CoA) reductase inhibitors [45]. Both enantiomers can be synthesized through asymmetric reduction of 4-chloro-3-oxobutanoate ethyl ester (COBE) (Figure 20.5). Many microorganisms and enzymes are known to possess the ability to catalyze such reduction [46–48], and some successful examples of large-scale production are listed in Table 20.1. Among them, *Candida magnoliae* showed multiple COBE-reducing enzyme activities in a cell-free extract. Three of the enzymes were purified, characterized, and named S1, S4, and R, respectively. These enzymes showed different specific activities (13.5, 7.4, and 12.2 U/mg, respectively), K_m (4.6, 0.11, and 2.9 mM), and ee [100% for (*S*), 51.1% for (*S*), and 100% for (*R*)] toward COBE [49,50]. Recombinant *Escherichia coli* coexpressing the S1- and GDH-coding genes showed (*S*)-CHBE productivity of 208 g/L in 13 h in the presence of glucose and a catalytic amount of NADP$^+$ [51–53]. In this mono-aqueous phase reaction, the total turnover number of NADP$^+$ was as high as 21,600 mol/mol. The substrate, COBE, was continuously fed at the rate of 0.02 g/min because of its instability in an aqueous environment. When an *n*-butyl acetate/water diphasic system is applied, the one-shot supply of a large amount of COBE

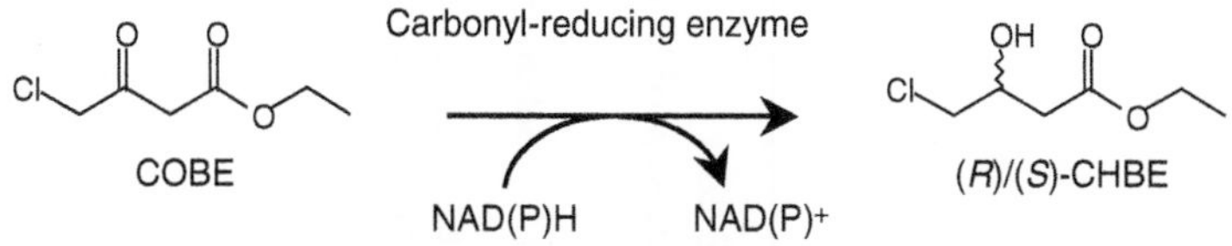

FIGURE 20.5 CHBE synthesis by asymmetric reduction of COBE. COBE, 4-chloro-3-oxobutanoate ethyl ester; CHBE, 4-chloro-3-hydroxybutanoate ethyl ester.

TABLE 20.1
Comparison of Various Systems for CHBE Production

Product	Substrate	Productivity	ee (%)	Yield (%)	Reaction System	Enzyme Source	Catalyst	Cofactor	Cofactor Regeneration	Reference
(S)-CHBE	COBE	100 g/L/d	>99	90.0	Aqueous	*Lactobacillus brevis*	Isolated enzyme	NADPH	*iso*-Propanol	[60]
(S)-CHBE	COBE	208 g/L	100	96.0	Aqueous	*Candida magnoliae*	Recombinant *Escherichia coli*	NADPH	GDH-Glucose	[53]
(S)-CHBE	COBE	430 g/L	100	85.0	*n*-Butyl acetate/water	*C. magnoliae*	Recombinant *E. coli*	NADPH	GDH-Glucose	[53]
(S)-CHBE	COBE	48.7 g/L	>99	99.8	Aqueous	*Ralstonia eutropha*	Recombinant *E. coli*	NADPH	GDH-Glucose	[68]
(S)-CHBE	COBE	45.6 g/L	>99	91.2	Aqueous	*Kluyveromyces aestuarii*	Recombinant *E. coli*	NADH	GDH-Glucose	[69]
(S)-CHBE	COBE	32.2 g/L	>99	98.5	Aqueous	*Pichia finlandica*	Recombinant *E. coli*	NADH	FDH-Formate	[55]
(R)-CHBE	COBE	268 g/L	92	94.0	*n*-Butyl acetate/water	*Sporobolomyces salmonicolor*	Recombinant *E. coli*	NADPH	GDH-Glucose	[54]
(R)-CHBE	COBE	36.6 g/L	>99	95.2	Aqueous	*C. parapsilosis*	Recombinant *E. coli*	NADH	*iso*-Propanol	[55]

CHBE, 4-chloro-3-hydroxybutanoate ethyl ester; COBE, 4-chloro-3-oxobutanoate ethyl ester; GDH, glucose dehydrogenase; FDH, formate dehydrogenase.

becomes possible. In this case, 430 g/L of CHBE was produced from 500 g/L COBE in 34 h with a total turnover number of NADP$^+$ of 15,400 mol/mol [53]. The other enantiomer, (R)-CHBE, can be produced as well if the carbonyl-reducing enzyme module is substituted with that from *Sporobolomyces salmonicolor* [54].

COBE is synthesized from acetic acid by diketene, from which 5-chloro-2-pentanone (CPON) can be synthesized as well. The same recombinant *E. coli* cells as listed in Table 20.1 overexpressing the ADH-coding gene from *Pichia finlandica* could asymmetrically reduce CPON to (R)-5-chloro-2-pentanol (CPOL) (26.1 g/L, 99% ee), and the cells over-expressing the genes of *C. parapsilosis*–derived ADH and *Mycobacterium* sp.–derived FDH could produce (S)-CPOL (33.8 g/L, 98% ee) (Figure 20.6) [55].

20.3.4.2 Chiral Alcohols from Nonaromatic Ketones

Table 20.2 shows other recent examples of chiral alcohols synthesized through asymmetric reduction of nonaromatic ketones. (R)-1,2-Propanediol (PDO) is a major commodity chemical, and there are two biosynthetic routes for it: fermentation and asymmetric reduction of acetol. *Thermoanaerobacterium thermosaccharolyticum* (previously known as *Clostridium thermosaccharolyticum*) is one of the best strains for the fermentation route. This strain can produce PDO through the fermentation of various sugars, such as D-glucose, D-xylose, D-mannose, L-arabinose, cellobiose, and galactose [56,57]. The highest productivity was obtained by Sánchez-Riera et al., i.e., a PDO titer of 9.05 g/L with a yield with glucose of 0.2 g/g, although the major product was lactate (11.1 g/L, 0.24 g/g glucose) [58].

Concerning hexanediol production, the substrate diketone is subjected to two-step asymmetric reduction. The selectivity regarding hydroxyketones produced in the first reduction step and diols in the second one is an important factor that affects downstream processing. In addition to the continuous production process for (2R,5R)-hexanediol described in Table 20.2, (2S,5S)-hexanediol production by fed batch reduction with *S. cerevisiae* has been achieved by Jülich Fine Chemicals GmbH on a 10 kg scale (conversion 95%, selectivity 68%, ee > 99%, de > 90%) [59].

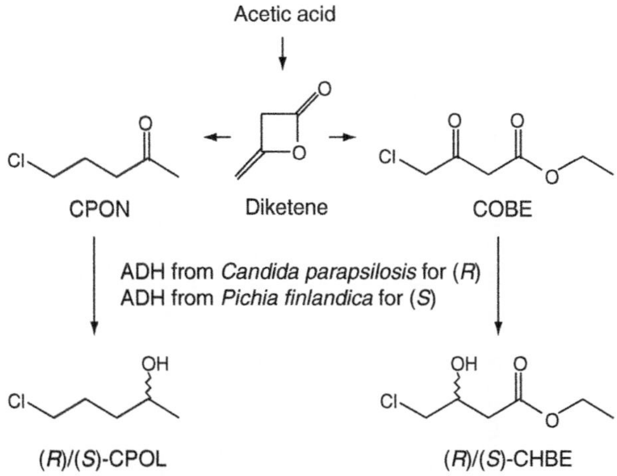

FIGURE 20.6 Production of chiral secondary alcohols with both configurations. ADH, alcohol dehydrogenase; CPON, 5-chloro-2-pentanone; CPOL, 5-chloro-2-pentanol; COBE, 4-chloro-3-oxobutanoate ethyl ester; CHBE, 4-chloro-3-hydroxybutanoate ethyl ester.

TABLE 20.2
Chiral Alcohols from Nonaromatic Ketones

Product	Substrate	Productivity	ee (%)	Yield (%)	Enzyme Source	Catalyst	Cofactor	Cofactor Regeneration	Reference
(R)-1,2-Propanediol	Acetol	>520 mM	>98	>95	Saccharomyces cerevisiae	Whole cells	—	Ethanol	[71]
(R)-1,2-Propanediol	Acetol	550 mM	99.9	71	Hansenula polymorpha	Recombinant Escherichia coli	NADH	GDH-Glucose	[72]
(2R,3R)-2,3-Butanediol	Acetoin (R:S = 3:4)	308 mM	99	39	H. polymorpha	Recombinant E. coli	NADH	GDH-Glucose	[72]
(S)-5-Chloro-2-pentanol	5-Chloro-2-pentanone	33.8 g/L	98	—	Candida parapsilosis	Recombinant E. coli	NADH	iso-Propanol	[55]
(R)-5-Chloro-2-pentanol	5-Chloro-2-pentanone	26.1 g/L	99	—	Pichia finlandica	Recombinant E. coli	NADH	FDH-Formate	[55]
(2R,5R)-Hexanediol	(2,5)-Hexanedione	64 g/L/d	>99 ee >99 de	Selectivity 78 Conversion 98	Lactobacillus kefir	Whole cells	—	Glucose	[73]
(S)-2-Heptanol	2-Heptanone	1.4 g/L	>99	54	S. cerevisiae	Whole cells	—	Part of substrate	[74]
(S)-2-Octanol	2-Octanone	57 g/L/d	>99	87	C. parapsilosis	Isolated enzyme	NADH	FDH-Formate	[59]
(S)-Cyclopropylethanol	Cyclopropyl methyl ketone	18.6 g/L	98	—	C. parapsilosis	Recombinant E. coli	NADH	iso-Propanol	[55]
Ethyl (S)-3-hydroxybutyrate	Ethyl acetoacetate	270 g/L/d	99	90	Rhodococcus erythropolis	Isolated enzyme	NADH	FDH-Formate	[59]
(S)-Ethyl 3-hydroxybutanoate	Ethyl acetoacetate	40.5 g/L	99.3	67	S. cerevisiae	Whole cells	—	Ethanol	[75]
tert-Butyl-6-chloro-(3R,5S)-dehydroxyhexanoate	tert-Butyl-6-chloro-3,5-dioxohexanoate	55.7 mM	99.5	68	L. kefir	Whole cells	—	Glucose	[76]
tert-Butyl-(S)-6-chloro-5-hydroxy-3-oxohexanoate	tert-Butyl-6-chloro-3,5-dioxohexanoate	10 g/L/d	>99	75	L. brevis	Cell-free extract of recombinant E. coli	NADP	iso-Propanol	[59,77]

GDH, glucose dehydrogenase; FDH, formate dehydrogenase.

Eckstein et al. recently reported (R)-2-octanol production from 2-octanone using ADH from *Lactobacillus brevis* as a catalyst and *iso*-propanol as a cosubstrate (88% conversion, >99% ee) [60]. The reaction proceeded in a biphasic system comprising buffer and an ionic liquid [BMIM and $(CF_3SO_2)_2N$]. As the ionic liquid removed acetone from the aqueous buffer phase, the reaction equilibrium shifted to cofactor regeneration, and thus resulted in a high reaction rate.

20.3.4.3 Chiral Alcohols from Aromatic Ketones

Chiral alcohols that are synthesized from aromatic ketones are listed in Table 20.3. These compounds with complex structures are important chiral synthons for optically active drugs that are synthesized through multistep enzyme reactions (not fermentation). The time-to-market factor is often given priority over productivity in pharmaceuticals. Several good reviews have been published on the synthesis of chiral drug intermediates by means of biocatalyst-involved methods [61–63].

TABLE 20.3
Chiral Alcohols from Aromatic Ketones

Product	Productivity	ee%	Yield%	Enzyme Source	Catalyst	Reference
(structure)	75 g/L/d	>99.9	>95	*Zygosaccharomyces rouxii*	Whole cells	[78]
(structure)	2.25 g/L	99	75	*Saccharomyces cerevisiae*	Whole cells	[79]
(structure)	17.5 g/L	97	99	*Candida maris*	Whole cells	
	91.5 g/L	99	89.5	*C. maris*	Cell-free extract	[80]
(structure)	80 g/L	99	80	*S. cerevisiae*	Whole cells	[81]
(structure)	8.6 g/L/d	93	85	*S. montanus*	Whole cells	[82]
(structure)	0.5 g/L	>95	67	*Microbacterium* sp.	Whole cells	[83]
(structure)	118 g/L/d	99	71	*C. parapsilosis*	Isolated enzyme	[84]
(structure)	19.7 g/L	99	63	*Rhodococcus erythropolis*	Isolated enzyme	[85]
(structure)	20.8 g/L	99	65	*R. ruber*	Whole cells	[86]
(structure)	0.64 g/L	99.8 ee 99 de	80	*Streptomyces nodosus*	Whole cells	[87]
(structure)	130 g/L	>99	87	*Rhodotorula glutinis*	Cell-free extract	[88]

20.4   CONCLUSION

Whereas the excellent specificities of enzymes (substrate-, stereo-, and/or regiospeci make them attractive catalysts for chiral syntheses, the narrow substrate tolera enzymes due to their strict specificities are a major disadvantage and, thus, a doubl sword from the industrial point of view. Various novel methodologies enhanced by ac molecular biological techniques have been proposed to overcome such limitations. M ome analysis in conjunction with bioinformatics and high-throughput screening is a p approach for exploring potential enzymes in biological niches [64,65]. Another app expansion of the "reaction-specificities" of known enzymes. There have been a nur reports on alteration of the catalytic properties of enzymes by means of protein engi techniques, and on the application of "old" enzymes with or without such artificial c zation to "new" processes [66,67]. These new technologies together with traditional sc should allow further applicability of enzymes to chiral syntheses.

ACKNOWLEDGMENTS

The works in our laboratory were supported in part by a Grant-in-Aid for Scientific R (No. 15658024, to S.S.) from the Ministry of Education, Science, Sports, and Culture, the COE for Microbial-Process Development Pioneering Future Production System program of the Ministry of Education, Science, Sports, and Culture, Japan, to S.S.), Project for the Development of a Technological Infrastructure for Industrial Bioproce R&D of New Industrial Science and Technology Frontiers (to S.S.) of the New Ene Industrial Technology Development Organization (NEDO), Japan. We wish to thank Fine Chemical (Takaoka), Kaneka (Osaka), Lonza (Visp), and DSM Vitamins (K gust) for their collaboration.

REFERENCES

1. Breuer, M., Ditrich, K., Habicher, T., et al., Industrial methods for the produciton of active intermediates, *Angew. Chem. Int. Ed.*, 43(7), 788–824, 2004.
2. Bornscheuer, U.T. and Kazlauskas, R.J., Hydrolases, in *Organic Synthesis: Regio- and Ste tive Biotransformations*, Wiley-VCH, Weinheim, Germany, 1999.
3. Bornscherer, U.T., Microbial carboxyl esterases: classification, properties and applicatio catalysis, *FEMS Microbiol. Rev.*, 26(1), 73–81, 2002.
4. Reetz, M.T., Lipase as practical biocatalysts, *Curr. Opin. Chem. Biol*, 6(2), 145–150, 2002.
5. Klibanov, A.M., Enzymatic catalysis in anhydrous organic solvents, *Trends Biochem. Sc* 141–144, 1989.
6. Klibanov, A.M., Improving enzymes by using them in organic solvents, *Nature*, 409(6817), 2001.
7. Dong, Y.H., Wang, L.H., Xu, J.L., et al., Quenching quorum-sensing-dependent bacterial i by an *N*-acyl homoserine lactonase, *Nature*, 411(6839), 813–817, 2001.
8. Dong, Y.H., Gusti, A.R., Zhang, Q., Xu, J.L., et al., Identification of quorum-quenchin; homoserine lactonases from *Bacillus* species, *Appl. Environ. Microbiol.*, 68(4), 1754–1759,
9. Lee, S.J., Park, S.Y., Lee, J.J., et al., Genes encoding the *N*-acyl homoserine lactone-d enzyme are widespread in many subspecies of *Bacillus thuringiensis*, *Appl. Environ. Microbio* 3919–3923, 2002.
10. Bublitz, C. and Lehninger, A.L., The role of aldonolactonase in the conversion of L-gul L-ascorbate, *Biochim. Biophys. Acta*, 47(2), 288–297, 1961.
11. Dilworth, M.J., Arwas, R., McKay, I.A., et al., Pentose metabolism in *Rhizobium legum* MNF300 and in cowpea *Rhizobium* NGR234, *J. Gen. Microbiol.*, 132(10), 2733–2742, 198

12. Novick, N.J. and Tyler, M.E., L-Arabinose metabolism in *Azospirillum brasiliense, J. Bacteriol.*, 149(1), 364–367, 1982.

13. Rigo, L.U., Marechal, L.R., Vieira, M.M., et al., Oxidative pathway for L-rhamnose degradation in *Pullularia pullulans, Can. J. Microbiol.*, 31(9), 817–822, 1985.

14. Shimizu, S., Hata, H., and Yamada, H., Reduction of ketopantoyl lactone to D-(−)-pantoyl lactone by microorganisms, *Agric. Biol. Chem.*, 48(9), 2285–2291, 1984.

15. Shimizu, S., Hattori, S., Hata, H., et al., One-step microbial conversion of a racemic mixture of pantoyl lactone to optically active D-(−)-pantoyl lactone, *Appl. Environ. Microbiol.*, 53(3), 519–522, 1987.

16. Hata, H., Shimizu, S., and Yamada, H., Enzymatic production of D-(−)-pantoyl lactone from ketopantoyl lactone, *Agric. Biol. Chem.*, 51(11), 3011–3016, 1987.

17. Kataoka, M., Shimizu, S., and Yamada, H., Novel enzymatic production of D-(−)-pantoyl lactone through the stereo-specific reduction of keto pantoic acid, *Agric. Biol. Chem.*, 54(1), 177–182, 1990.

18. Kataoka, M., Shimizu, K., Sakamoto, K., et al., Lactonohydrolase-catalyzed optical resolution of pantoyl lactone: selection of a potent enzyme producer and optimization of culture and reaction conditions for practical resolution, *Appl. Microbiol. Biotechnol.*, 44(3–4), 333–338, 1995.

19. Kataoka, M., Nomura, J., and Shinohara, M., Purification and characterization of a novel lactono-hydrolase from *Agrobacterium tumefaciens, Biosci. Biotechnol. Biochem.*, 64(6), 1255–1262, 2000.

20. Shimizu, S., Kataoka, M., Shimizu, K., et al., Purification and characterization of a novel lactono-hydrolase, catalyzing the hydrolysis of aldonate lactones and aromatic lactones, from *Fusarium oxysporum, Eur. J. Biochem.*, 209(1), 383–390, 1992.

21. Kataoka, M., Shimizu, K., Sakamoto, K., et al., Optical resolution of racemic pantolactone with a novel fungal enzyme, lactonohydrolase, *Appl. Microbiol. Biotechnol.*, 43(6), 974–977, 1995.

22. Sakamoto, K., Honda, K., Wada, K., et al., Practical resolution system for DL-pantoyl lactone using the lactonase from *Fusarium oxysporum, J. Biotechnol.*, 118, 99–106, 2005.

23. Honda, K., Tsuboi, H., Minetoki, T., et al., Expression of the *Fusarium oxysporum* lactonase gene in *Aspergillus oryzae*—molecular properties of the recombinant enzyme and its application, *Appl. Microbiol. Biotechnol.*, 66(5), 520–526, 2005.

24. Kesseler, M., Friedrich, T., Höffken, H.W., et al., Development of a novel biocatalyst for the resolution of *rac*-pantolactone, *Adv. Synth. Catal.*, 344(10), 1103–1110, 2002.

25. Kataoka, M., Honda, K., and Shimizu, S., 3,4-Dihydrocoumarin hydrolase with haloperoxidase activity from *Acinetobacter calcoaceticus, Eur. J. Biochem.*, 267(1), 3–10, 2000.

26. Honda, K., Kataoka, M., Sakuradani, E., et al., Role of *Acinetobacter calcoaceticus* 3,4-dihydrocou-marin hydrolase in oxidative stress defence against peroxoacids, *Eur. J. Biochem.*, 270(3), 486–494, 2003.

27. Honda, K., Kataoka, M., and Shimizu, S., Enzymatic preparation of D-β-acetylthioisobutyric acid and cetraxate hydrochloride using a stereo- and/or regioselective hydrolase, 3,4-dihydrocoumarin hydrolase from *Acinetobacter calcoaceticus, Appl. Microbiol. Biotechnol.*, 60(3), 288–292, 2002.

28. Adams, T.B., Greer, D.B., Doull, J., et al., The FEMA GRAS assessment of lactones used as flavour ingredients, *Food Chem.Toxicol.*, 36(4), 249–278, 1998.

29. Onakunle, O.A., Knowles, C.J., and Bunch, A.W., The formation and substrate specificity of bacterial lactonases capable of enantioselective resolution of racemic lactones, *Enzyme Microb. Technol.*, 21(4), 245–251, 1997.

30. Iwaki, H., Hasegawa, Y., Wang, S., et al., Cloning and characterization of a gene cluster involved in cyclopentanol metabolism in *Comamonas* sp. strain NCIMB 9872 and biotransformations effected by *Escherichia coli*–expressed cyclopentanone 1,2-monooxygenase, *Appl. Environ. Microbiol.*, 68(11), 5671–5684, 2002.

31. Panke, S., Held, M., and Wubbolts, M., Trends and innovations in industrial biocatalysis for the production of fine chemicals, *Curr. Opin. Biotechnol.*, 15(4), 272–279, 2004.

32. Jörnvall, H., Höog, J.-O., and Persson, B., SDR and MDR: completed genome sequences show these protein families to be large, of old origin, and of complex nature, *FEBS Lett.*, 445(2–3), 261–264, 1999.

33. Forrest, G.L. and Gonzalez, B., Carbonyl reductase, *Chem. Biol. Interact.*, 129(1–2), 21–40, 2000.

34. Ellis, E.M., Microbial aldo-keto reductases, *FEMS Microbiol. Lett.*, 216(2), 123–131, 2002.

35. Hyndman, D., Bauman, D.R., Heredia, V.V., et al., The aldo-keto reductase superfamily home-page, *Chem. Biol. Interact.*, 143–144, 621–631, 2003.

36. Kaluzna, I.A., Matsuda, T., Sewell, A.K., et al., Systematic investigation of *Saccharomyces* siae enzymes catalyzing carbonyl reductions, *J. Am Chem. Soc.*, 126(40), 12827–12832, 200

37. Stampfer, W., Kosjek, B., Moitzi, C., et al., Biocatalytic asymmetric hydrogen transfer, *Chem. Int. Ed.*, 41(6), 1014–1017, 2002.

38. Kroutil, W., Mang, H., Edegger, K., et al., Recent advances in the biocatalytic reduction of and oxidation of *sec*-alcohols, *Curr. Opin. Chem. Biol.*, 8(2), 120–126, 2004.

39. Kula, M.-R. and Kragl, U., Dehydrogenases in the synthesis of chiral compounds, in *Stereos Biocatalysis*, Patel, R.N., Ed., Marcel Dekker, New York, 2000, pp. 839–866.

40. Tishkov, V.I., Galkin, A.G., Fedorchuk, V.V., et al., Pilot scale production and isolation of rec ant NAD$^+$- and NADP$^+$-specific formate dehydrogenases, *Biotechnol. Bioeng.*, 64(2), 187–193

41. Makino, Y., Ding, J.-Y., Negoro, S., et al., Purification and characterization of a new dehydrogenase from vegetative cells of *Bacillus megaterium*, *J. Fermen. Bioeng.*, 67(6), 374–379

42. Mertens, R., Greiner, L., van den Banb, E.C.D., et al., Practical applications of hydrog from *Pyrococcus furiosus* for NADPH generation and regeneration, *J. Mol. Catal. B Enzym.* 39–52, 2003.

43. Nakamura, K. and Yamanaka, R., Light mediated cofactor recycling system in biocatalyti metric reduction of ketone, *Chem. Commun.*, (16), 1782–1783, 2002.

44. Zhou, B., Gopalan, A.S., VanMiddlesworth, F., et al., Stereochemical control of yeast reduct Asymmetric synthesis of L-carnitine, *J. Am. Chem. Soc.*, 105(18), 5925–5926, 1983.

45. Karanewsky, D.S., Badia, M.C., Ciosek, C.P.J., et al., Phosphorus-containing inhibitors of CoA reductase. 1. 4-[2-arylethyl]-hydroxyphos phinyl]-3-hydroxybutanoic acids: a new class selective inhibitors of cholesterol biosynthesis, *J. Med. Chem.*, 33(10), 2925–2956, 1990.

46. Yasohara, Y., Kizaki, N., Hasegawa, J., et al., Synthesis of optically active ethyl 4-ch hydroxybutanoate by microbial reduction, *Appl. Microbiol. Biotechnol.*, 51(6), 847–851, 199

47. Shimizu, S., Kataoka, M., and Kita, K., Chiral alcohol synthesis with microbial carbonyl red in a water-organic solvent two-phase system, *Ann. NY Acad. Sci.*, 864(1), 87–95, 1998.

48. Kataoka, M., Kita, K., Wada, M., et al., Novel bioreduction system for the production o alcohols, *Appl. Microbiol. Biotechnol.*, 62(5–6), 437–445, 2003.

49. Wada, M., Kawabata, H., Kataoka, M., et al., Purification and characterization of an al reductase from *Candida magnoliae*, *J. Mol. Catal. B Enzym.*, 6(3), 333–339, 1999.

50. Wada, M., Kataoka, M., Kawabata, H., et al., Purification and characterization of N dependent carbonyl reductase, involved in stereoselective reduction of ethyl 4-chloro-3-oxob ate, from *Candida magnoliae*, *Biosci. Biotechnol. Biochem.*, 62(2), 280–285, 1998.

51. Yasohara, Y., Kizaki, N., Hasegawa, J., et al., Molecular cloning and overexpression of t encoding an NADPH-dependent carbonyl reductase from *Candida magnoliae*, involved in ster tive reduction of ethyl 4-chloro-3-oxobutanoate, *Biosci. Biotechnol. Biochem.*, 64(7), 1430–143

52. Yasohara, Y., Kizaki, N., Hasegawa, J., et al., Stereoselective reduction of alkyl 3-oxibutan carbonyl reductase from *Candida magnoliae*, *Tetrahedron Asymmetry*, 12(12), 1713–1718, 2

53. Kizaki, N., Yasohara, Y., Hasegawa, J., et al., Synthesis of optically pure ethyl (*S*)-4-ch hydroxybutanoate by *Escherichia coli* transformant cells coexpressing the carbonyl reduct glucose dehydrogenase genes, *Appl. Microbiol. Biotechnol.*, 55(5), 590–595, 2001.

54. Kataoka, M., Yamamoto, K., Kawabata, H., et al., Stereoselective reduction of ethyl 4- 3-oxobutanoate by *Escherichia coli* transformant cells coexpressing the aldehyde reducta glucose dehydrogenase genes, *Appl. Microbiol. Biotechnol.*, 51(4), 486–490, 1999.

55. Matsuyama, A., Yamamoto, H., and Kobayashi, Y., Practical application of recombinant cell biocatalysts for the manufacturing of pharmaceutical intermediates such as chiral alcoho *Process Res. Dev.*, 6(4), 558–561, 2002.

56. Cameron, D.C., Altaras, N.E., Hoffman, M.L., et al., Metabolic engineering of prop pathways, *Biotechnol. Prog.*, 14(1), 116–125, 1998.

57. Cameron, D.C. and Cooney, C.L., A novel fermentation: the production of *R*(−)-1,2-prop and acetol by *Clostridium thermosaccharolyticum*, *Bio/Technology*, 4, 651–654, 1986.

58. Sánchez-Riera, F., Cameron, D.C., and Cooney, C.L., Influence of environmental factor production of *R*-(−)-1,2-propanediol by *Clostridium thermosaccharolyticum*, *Biotechnol. Let* 449–454, 1987.

59. http://www.juelich-chemicals.de/
60. Eckstein, M., Filho, M.V., Liesec, A., et al., Use of an ionic liquid in a two-phase system to improve an alcohol dehydrogenase catalysed reduction, *Chem. Commun.*, (9), 1084–1085, 2004.
61. Patel, R.N., Microbial/enzymatic synthesis of chiral drug intermediates, *Adv. Appl. Microbiol.*, 47, 33–78, 2000.
62. Node, M., Development of new methods in asymmetric reactions and their applications, *Yakugaku Zasshi*, 122(1), 71–88, 2002.
63. Patel, R.N., Microbial/enzymatic synthesis of chiral pharmaceutical intermediates, *Curr. Opin. Drug Dis. Devel.*, 6(6), 902–920, 2003.
64. Lorenz, P., Lieveton, K., Niehaus, F., et al., Screening for novel enzymes for biocatalytic processes: accessing the metagenome as a resource of novel functional sequence space, *Curr. Opin. Biotechnol.*, 13(6), 572–577, 2002.
65. Streit, W.R. and Schmitz, R.A., Metagenomics — the key to the uncultured microbes, *Curr. Opin. Microbiol.*, 7(5), 492–498, 2004.
66. Bornscheuer, U.T. and Kazlauskas, R.J., Catalytic promiscuity in biocatalysis: using old enzymes to form new bonds and follow new pathway, *Angew. Chem. Int. Ed.*, 43(45), 6032–6040, 2004.
67. Faber, K. and Kroutil, W., New enzymes for biotransformations, *Curr. Opin. Chem. Biol.*, 9, 174–180, 2005.
68. Yamamoto, H., Matsuyama, A., and Kobayashi, Y., Synthesis of ethyl (*S*)-4-chloro-3-hydroxybutanoate using *fabG*-homologues, *Appl. Microbiol. Biotechnol.*, 61(2), 133–139, 2003.
69. Yamamoto, H., Mitsuhashi, K., Kimoto, N., et al., A novel NADH-dependent carbonyl reductase from *Kluyveromyces aestuarii* and comparison of NADH-regeneration system for the synthesis of ethyl (*S*)-4-chloro-3-hydroxybutanoate, *Biosci. Biotechnol. Biochem.*, 68(3), 638–649, 2004.
70. Wada, M., Kawabata, H., Yoshizumi, A., et al., Occurrence of multiple ethyl 4-chloro-3-oxobutanoate-reducing enzymes in *Candida magnoliae*, *J. Biosci. Bioeng.*, 87(2), 144–148, 1999.
71. Kometani, T., Yoshii, H., Takeuchi, Y., et al., Large-scale preparation of (*R*)-1,2-propanediol through baker's yeast–mediated bioreduction, *J. Ferment. Bioeng.*, 76(5), 414–415, 1993.
72. Yamada-Onodera, K., Kawahara, N., Tani, Y., et al., Synthesis of optically active diols by *Escherichia coli* transformant cells that express the glycerol dehydrogenase gene of *Hansenula polymorpha* DL-1, *Eng. Life Sci.*, 4(5), 413–417, 2004.
73. Haberland, J., Hummel, W., Daussmann, T., et al., New continuous production process for enantiopure (2*R*,5*R*)-hexanediol, *Org. Process Res. Dev.*, 6(4), 458–462, 2002.
74. Cappaert, L. and Larroche, C., Behaviour of dehydrated baker's yeast during reduction reactions in a biphasic medium, *Appl. Microbiol. Biotechnol.*, 64(5), 686–690, 2004.
75. Kometani, T., Yoshii, H., Kitatsuji, E., et al., Large-scale preparation of (*S*)-ethyl 3-hydroxybutanoate with a high enantiomeric excess through baker's yeast–mediated bioreduction, *J. Ferment. Bioeng.*, 76(1), 33–37, 1993.
76. Amidjojo, M., Nowak, A., Franco-Lara, E., et al., Abstracts of papers, 2nd International Congress on Biocatalysis, Hamburg, Germany, Aug 29–Sept 1, 2004, p. 157.
77. Wolberg, M., Hummel, W., Wandrey, C., et al., Highly regio- and enantioselective reduction of 3,5-dioxocarboxylates, *Angew. Chem.*, 112(23), 4476–4478, 2000.
78. Vicenzi, J.T., Zmijewski, M.J., Reinhard, M.R., et al., Large-scale stereoselective enzymatic ketone reduction with in situ product removal via polymeric adsorbent resins, *Enzyme Microbial. Technol.*, 20(7), 494–499, 1997.
79. Aleu, J., Fronza, G., Fuganti, C., et al., On the baker's yeast mediated transformation of α-bromoenones: synthesis of (1*S*,2*R*)-2-bromoindan-1-ol and (2*S*,3*S*)-3-bromo-4-phenylbutan-2-ol, *Tetrahedron Asymmetry*, 9(9), 1589–1596, 1998.
80. Kawano, S., Horikawa, M., Yasohara, Y., et al., Microbial enantioselective reduction of acetylpyridine derivatives, *Biosci. Biotechnol. Biochem.*, 67(4), 809–814, 2003.
81. Kometani, T., Sakai, Y., Matsumae, H., et al., Production of (2*S*,3*S*)-2,3-dihydro-3-hydroxy-2-(4-methoxyphenyl)-1,5-benzothiazepin-4(W)-one, a key intermediate for diltiazem synthesis, by baker's yeast–mediated reduction, *J. Ferment. Bioeng.*, 84(3), 195–199, 1997.
82. Dehli, J.R. and Gotor, V., Dynamic kinetic resolution of 2-oxocycloalkanecarbonitriles: chemo-enzymatic syntheses of optically active cyclic β- and γ-amino alcohols, *J. Org. Chem.*, 67(19), 6816–6819, 2002.

83. Roberge, C., King, A., Pecore, V., et al., Asymmetric bioreduction of a keto ester to its coring (*S*)-hydroxy ester by *Microbacterium* sp. MB 5614, *J. Ferment. Bioeng.*, 81(6), 530–533

84. Zelinski, T., Liese, A., Wandrey, C., et al., Asymmetric reductions in aqueous media: e
synthesis in cyclodextrin containing buffers, *Tetrahedron Asymmetry*, 10(9), 1681–1687, 19

85. Groeger, H., Hummel, W., Buchholz, S., et al., Practical asymmetric enzymatic reduction
discovery of a dehydrogenase-compatible biphasic reaction media-176, *Org. Lett.*, 5(2),
2003.

86. Stampfer, W., Edegger, K., Kosjek, B., et al., Simple biocatalytic access to enantiopu
heteroarylethanols employing a microbial hydrogen transfer reaction, *Adv. Synth. Cat*
57–62, 2004.

87. Patel, R.N., Banerjee, A., McNamee, C.G., et al., Preparation of chiral synthon for HIV
inhibitor: stereoselective microbial reduction of *N*-protected α-aminochloroketone, *Tet*
Asymmetry, 8(15), 2547–2552, 1997.

88. Kizaki, N., Sawa, I., Yano, M., et al., Purification and characterization of a yeast
reductase for synthesis of optically active (*R*)-styrene oxide derivatives, *Biosci. Biotechnol. B*
69(1), 79–86, 2005.

21 Discovery of Arylmalonate Decarboxylase and Conversion of the Function by Rational Design

Kenji Miyamoto and Hiromichi Ohta

CONTENTS

21.1 INTRODUCTION

Decarboxylation of malonic acid is a well-known process in the biosynthesis of long-chain fatty acids starting from acetic acid. If this kind of enzymes exhibits enantioselectivity for α-substituted malonates, it will be useful as a novel method for preparing optically active carboxylic acids. Indeed malonyl-CoA decarboxylase from uropygial gland is enantioselective to the substrate and the product. Racemate of methylmalonyl-CoA (**1**) was incubated with the above decarboxylase in ^3H_2O. (*S*)-Enantiomer was smoothly decarboxylated to result in 2-(^3H)-propionyl-CoA (**2**), while (*R*)-isomer of the substrate remained intact (Equation 21.1). The absolute configuration of α-carbon of resulting **2** was revealed to be (*R*). Thus, the decarboxylase distinguishes the chirality of **1** and gives enantiomerically pure (*R*)-isomer by retention of configuration [1].

$$(21.1)$$

Serine hydroxymethyltransferase (SHMT) catalyzes a similar reaction. This e
decarboxylases α-amino-α-methylmalonate by the aid of pyridoxal-5′-phosphate (PLP
is unusual in that it promotes a wide range of different types of reactions of α-amino ac
It is capable of catalyzing aldol/retro-aldol reaction and transamination in addit
decarboxylation reaction. This is due to the similarity of the reaction pattern. The
step of each reaction is the decomposition of the Schiff base formed between the substra
pyridoxal coenzyme. The decarboxylation of α-amino-α-methylmalonate (3) cataly
SHMT is enantioselective. Thomas et al. synthesized chiral 3, containing ^{13}C in one of t
carboxyl groups. Both enantiomers of 3 were incubated with SHMT in the presence o
When (R)-isomer was used, no ^{13}C was found in the product. On the contrary, ^{13}
retained in the resulting alanine (4) when (S)-enantiomer was employed as the st
material. Apparently, pro-(R) carboxyl group of α-amino-α-methylmalonate was rel
as carbon dioxide. The absolute configuration of resulting alanine is (R), indicating th
stereochemical course of the reaction is retention of configuration, as illustrated in Eq
21.2 [2]. α-Aminomalonate decarboxylase utilized a similar mechanism. In this case, ho
the starting material is prone to racemize under reaction conditions, and careful experi
to evaluate the stereochemistry of the reaction are required [3].

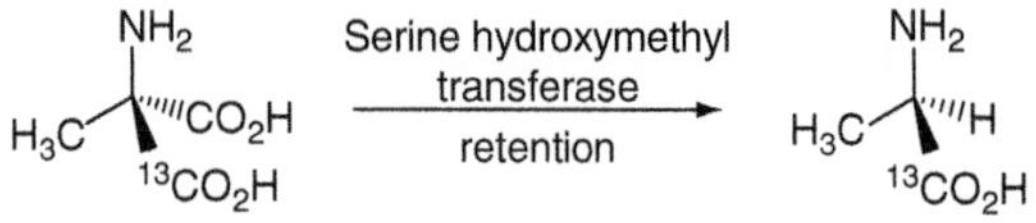

3   4

21.2   ASYMMETRIC DECARBOXYLATION OF α-ARYL-α-METHYLMALONA

Optically active α-arylpropionic acids are useful compounds as anti-inflammatory
(Figure 21.1) [4,5]. We intended to synthesize such optically active compounds by asym
decarboxylation of α-aryl-α-methylmalonates with the aid of biocatalysis, and have
that *Alcaligenes bronchisepticus* was the most active strain. Unfortunately the ab
configuration of the product was opposite to that of the above-mentioned active ph
ceuticals. Thus, this reaction is considered to be of little value for obtaining such comp
However, as this is a new type of biotransformation that can be carried out in a prepa
scale and the optical and chemical yields were high, it is expected that this decarboxy
reaction would be utilized to other objectives and be a good model for examini
interaction between an enzyme and synthetic substrates. Therefore, the enzyme was is
and studied in this direction.

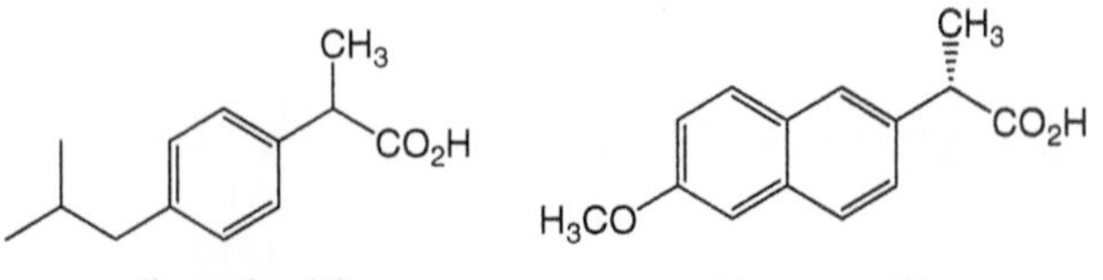

Ibuprofen (5)   Naproxen (6)

FIGURE 21.1 Nonsteroidal anti-inflammatory 2-arylpropionates.

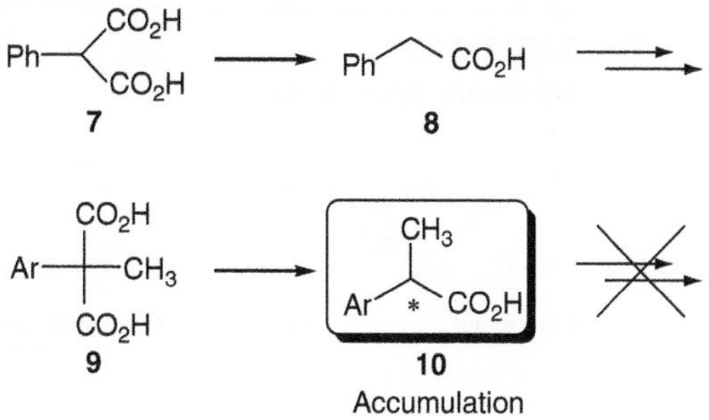

FIGURE 21.2 Screening of microorganisms.

21.2.1 SCREENING OF MICROORGANISM

Screening of a microorganism that is capable of decarboxylating α-aryl-α-methylmalonates was carried out using a medium containing phenylmalonate (**7**) as the sole source of carbon, because we supposed that the first step of the metabolic path of the acid would be decarboxylation to give phenylacetate (**8**), which would further be metabolized by oxidation of α-position (Figure 21.2). Accordingly, the microorganisms that assimilate the added carbon source and grow on this medium are expected to have decarboxylase, which is active to phenylmalonate (**7**). At least a few of the enzymes that catalyze decarboxylation of **7** can be expected to be active to α-aryl-α-methylmalonates (**9**), as the difference in the structure between two molecules is only the presence or the absence of a methyl group on the α-position. If the presence of a methyl group conveniently inhibits the following metabolic degradation, the expected monoacid would be obtained. A wide variety of soil samples and type cultures were tested and we found a few strains that were capable of growing on the medium. We selected a bacterium identified as *A. bronchisepticus* KU1201, because this strain gave optically active α-arylpropionate (**10**) starting from α-aryl-α-methylmalonate (**9**) [6,7]. This enzyme was revealed to be an inducible enzyme, the inducer being phenylmalonate (**7**).

21.2.2 SUBSTRATE SPECIFICITY

A medium (50 mL) containing phenylmalonate (250 mg) and peptone (50 mg), yeast extract, and some inorganic salts was inoculated with *A. bronchisepticus* and shaken for 4 d at 30°C. The substrate, 250 mg of α-methyl-α-phenylmalonate (**9a**), was added to the resulting suspension and the incubation was continued for an additional 5 d. The mixture was acidified, saturated with NaCl, and extracted with ether. After removal of the solvent, the organic residue was treated with an excess amount of diazomethane and the expected product was isolated as its methyl ester. Purification with preparative TLC afforded optically active methyl α-phenylpropionate. The absolute configuration was proved to be (*R*) by the specific rotation and the enantiomeric excess (ee) was determined to be 98% by HPLC using a column with an optically active solid phase (Table 21.1). The enzyme system of *A. bronchisepticus* was also effective to other compounds with a substituent on the phenyl ring. As is clear from Table 21.1, an electron-withdrawing substituent (**9b**) is preferable to promote the reaction. The enzyme also accepted β-naphthyl ring (**9d**) and thienyl ring (**9e**) as good substrates. No decarboxylation was observed when the malonate had a substituent on the *ortho*-position of the aryl ring (**9f**), or in the case when the alkyl group of α-position was ethyl instead of methyl (Figure 21.3). The fact that the compound with an α-naphthyl ring (**9h**) is inactive to the enzyme system is a marked contrast to the fact that **9d** is a good substrate. The inactivity of all these compounds may be due to the steric effect rather than electronic effect. One possible key

TABLE 21.1
Substrate Specificity

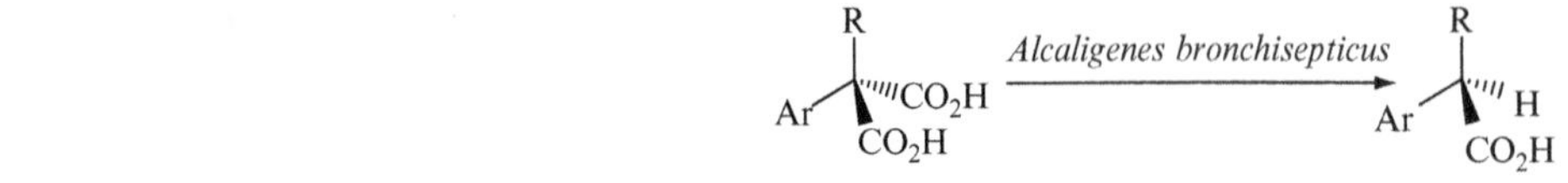

Compound	Ar	R	Sub. conc. (%)	Yield (%)	ee (%)
9a	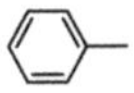 (phenyl)	CH$_3$	0.5	80	98
9b	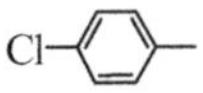 Cl—(phenyl)	CH$_3$	0.5	95	98
9c	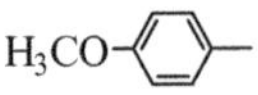 H$_3$CO—(phenyl)	CH$_3$	0.1	48	99
9d	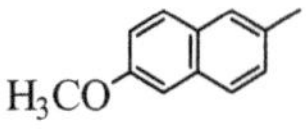 H$_3$CO—(naphthyl)	CH$_3$	0.5	96	>95
9e	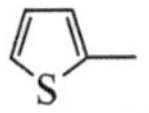 (thienyl)	CH$_3$	0.3	98	95

to an appropriate interpretation is that all of the inactive compounds have a substit
the *ortho*-position of the phenyl ring (the naphthyl ring of **9h** can be regarded as
m-substituted phenyl ring) or a bulkier alkyl group on the α-carbon. Insertion of a me
group or a hetero atom between the α-carbon and the phenyl ring made the compoun
inactive (**9i–k** in Figure 21.3). This is apparently due to the electronic effect. The π-el
of the aromatic ring have an essential effect to promote the reaction.

Putting all the results obtained together, it is expected that introduction of fluorin
in the substrate would bring about a favorable effect on the reaction, because flu
strongly electron-withdrawing and not so bulky. The results are summarized in Table
α-Fluorinated malonate (**9l**) gave the corresponding optically active monobasic ac
moderate yield. The effect of substitution of ring hydrogens showed again a marked
ence between *ortho*- and *meta*- or *para*-positions. Even the steric bulkiness of a fluorin
has a serious effect on the rate of reaction. The yield of the product after a 5 d rea
o-fluorinated compound **9m** was as low as 12%. The low ee value of the product is p
due to nonenzymatic decarboxylation, which will give a racemic monobasic acid. As
from Table 21.2, the *m*- and *p*-trifluoromethyl derivatives (**9p,q**) were good substra
chemical and optical yields of the expected products being very high. This can be attrib
the strong electron-withdrawing effect of these substituents.

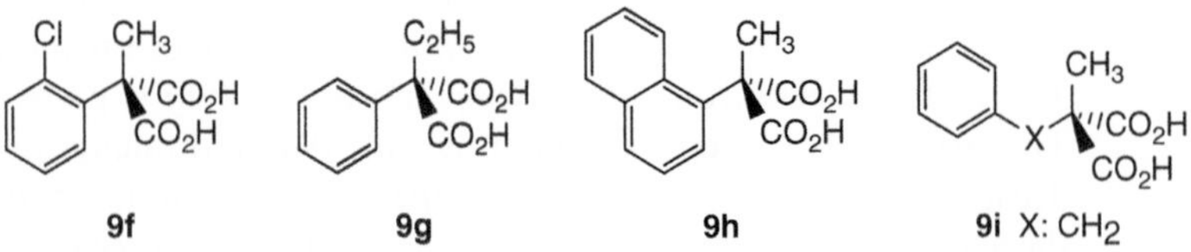

9f   9g   9h   9i X: CH$_2$
                                                                     9j X: S
                                                                     9k X: O

FIGURE 21.3 Inactive compounds to *Alcaligenes bronchisepticus*.

TABLE 21.2
Reaction of Fluorine-Containing Compounds

Compound	Ar	R	Sub. conc. (%)	Yield (%)	ee (%)
9l	phenyl	F	0.1	64	95
9m	2-fluorophenyl	CH_3	0.3	12	54
9n	3-fluorophenyl	CH_3	0.5	75	97
9o	4-fluorophenyl	CH_3	0.5	54	97
9p	3-(trifluoromethyl)phenyl	CH_3	0.5	99	>95
9q	4-(trifluoromethyl)phenyl	CH_3	0.5	91	95

21.2.3 PURIFICATION AND CHARACTERIZATION OF THE ENZYME

Although the steric bulkiness around the reaction center is rather restricted, the enzyme system of *A. bronchisepticus* was proved to have a unique reactivity. Thus, detailed studies on the isolated enzyme were expected to elucidate some new interesting mechanism of the new type of decarboxylation.

The bacterium was grown in a medium containing phenylmalonate as the enzyme inducer. The decarboxylation enzyme was purified according to the standard protocol. Column chromatography was performed using three kinds of gels after fractionation by ammonium sulfate. The enzyme was purified to about 300-fold to 377 U/mg protein (Table 21.3). SDS-PAGE and HPLC analysis showed that this enzyme was monomeric, its molecular mass being about 24 kDa (Table 21.4). The enzyme was named arylmalonate decarboxylase (AMDase), as the reaction rate of enzyme-catalyzed decarboxylation of phenylmalonate is faster than that of the α-methyl derivative [9].

To clarify the characteristics of the enzyme, the effects of additives were examined using phenylmalonate as the representative substrate [9]. The addition of ATP and coenzyme

TABLE 21.3
Purification of the Enzyme

Purification	Total Protein (mg)	Total Activity	Specific Activity (U/mg)	Yield (%)
Cell-free extract	8630	10950	1.26	100
Heat treatment	4280	9540	2.22	87
Ammonium sulfate	2840	10350	3.64	95
DEAE-Toyopearl	244	5868	24.1	54
Butyl-Toyopearl	14.7	3391	231	31
QAE-Toyopearl	4.32	1627	377	15

TABLE 21.4
Property of the Enzyme

Molecular Weight

Gel filtration	22,000
SDS-PAGE	24,000
Number of subunits	1
Stability	
pH	7.0–8.5
Temperature	<40°C
Optimum	
pH	8.5 (Tris-HCl Buffer)
Temperature	45°C

A (CoA) to the enzyme reaction mixtures did not enhance the rate of reaction. In the c
malonyl-CoA decarboxylase and others, ATP and substrate acid form a mixed anhy
which in turn reacts with CoA to form a thiol ester of the substrate. In the present case, a
ATP and CoA-SH had no effect, the normal mechanism is very unlikely to operate. It
established that avidin is a potent inhibitor of the formation of the biotin–enzyme comple
13]. In this case, addition of avidin had no influence on the decarboxylase activity, indi
that AMDase is not a biotin enzyme (Table 21.5). Thus, the cofactor requirements of AM
are entirely different from those of known analogous enzymes, such as acyl-CoA carbox
[14], methylmalonyl-CoA decarboxylases [10], and transcarboxylases [14,15].

A strong inhibitory effect on AMDase activity was found for sulfhydryl rea
(at 1 mM), such as $HgCl_2$ (relative activity, 0%), HgCl (8%), $AgNO3$ (3%), iodoa
(3%), and p-chloromercuribenzoate (PCMB) (0%) (Table 21.5). N-Ethylmaleimi
10 mM) causes 72% inhibition of the decarboxylase activity. Thus, AMDase was re
to be a thiol decarboxylase, i.e., at least one of the cysteine residues is present as a fr
that plays an essential role in the active site of the enzyme. The activity of the enzym
not lost upon incubation with the following reagents: several divalent metal cation

TABLE 21.5
Effect of Additives

Inhibitor	Concentration (mM)	Relative Activ
None	—	10
$NiCl_2$, $MnCl_2$, $CoCl_2$	10	90–9
$BaCl_2$, $MgCl_2$, $CaCl_2$	1	17–1
$SnCl_2$, $PbCl_2$	1	
$HgCl_2$		
HgCl	1	
$AgNO_3$	1	
DNB, PCMB, Indoacetate	1	0–
N-ethylmaleimide	1	9
PMSF	10	8
NaN_3, NH_2OH, KCN	10	103–10
EDTA, Bipyridil	10	103–10
Avidin	2.5 mg/ml	9

as Ni^{2+}, Co^{2+}, Ba^{2+}, Mg^{2+}, and Ca^{2+}; carbonyl reagents such as NaN_3, NH_2OH, and KCN; metal-chelating agents such as EDTA, 8-quinolinol, bipyridil, and 1,10-phenanthroline; serine inhibitors such as phenylmethanesulfonyl fluoride (at 10 mM); and others. It is concluded that AMDase is an unusual enzyme which does not contain metal ions or coenzymes that are present in decarboxylases and transcarboxylases.

21.2.4 Cloning of AMDase Gene

In order to obtain large amounts of the enzyme, we evaluated the cloning and overexpression of AMDase gene using the direct expression method [16]. The genomic DNA isolated from *A. bronchisepticus* was completely digested by *Pst*I and cloned into pUC19 *Pst*I site. *Escherichia coli* DH5α-MCR was transformed by the plasmids, and the genomic libraries were constructed. The transformants were screened by the change of color of the plate containing bromothymol blue and phenylmalonic acid, due to the local change of pH of the medium around the active colony as a result of the formation of monobasic acid from the starting dibasic acid. One in approximately 700 transformants exhibited AMDase activity. The plasmid (pAMD100) contained an insert of about 2.8 kbp and was sequenced. The gene was revealed to be consisting of 720 bp, which means that the enzyme consists of 240 amino acids (Figure 21.4). A *Pst*I–*Hind*III (1.2 kbp) fragment was subcloned in pUC19 to generate

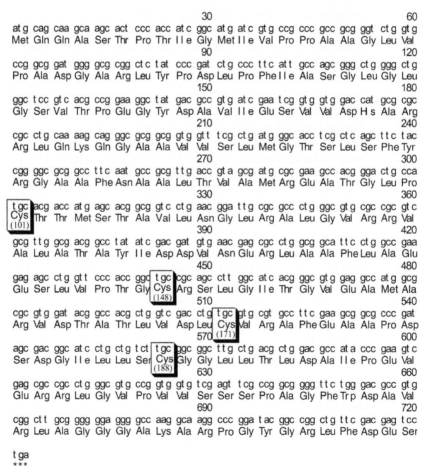

FIGURE 21.4 Nucleotide and deduced amino acid sequences of the AMDase.

pAMD101. The amount of the enzyme in a cell-free extract of the transformant, JM109/pAMD101, was elevated to 37,250 units/L culture broth. It was calculated t enzyme comprised over 25% of the total extractable cellular proteins. The enzyme pr by the E. coli transformant was purified to homogeneity and shown to be identical to the original strain. Both enzymes had the same enzymological properties and N-t amino acid sequences.

21.2.5 SITE-DIRECTED MUTAGENESIS OF AMDASE GENE

DNA sequence indicated that AMDase contains four cysteine residues located at 1(171, and 188 from the amino terminal (Figrue 21.4) [16]. At least one of these four is es to play a crucial role in the decarboxylation of disubstituted malonic acid. The mos method to locate which Cys is responsible for enzyme activity is site-directed muta; Then, we have to ascertain which mutation is effective for this purpose. To determ amino acid that should be introduced in place of Cys, we have to consider the rol cysteine residue. One possibility is that it works as a nucleophile. Attack on the c carbon will result in the thiol ester, which will stabilize the enolate-type transition place of CoA. Enantioselective protonation followed by hydrolysis at the final step v (R)-α-arylpropionate. SH group can work as an acid. Partial protonation to α-carb facilitate the C–C bond fission and accelerate the reaction. Formation of the C–H bor the observed optically active product. Then, from the standpoint of organic reactior anism, substitution of the sulfur atom with an oxygen atom would greatly decrease the reaction, because nucleophilicity, anion-stabilizing effect, and proton-donating abili hydroxyl group are far smaller than those of an SH group. Still, a hydroxyl group is of more or less keeping the hydrogen bondings in which the cystein residues are incorp Accordingly, the activity of enzyme was expected to remain partly, whichever Cys is r by Ser, different from the cases in which alanine or some other totally different ami residue is introduced in place of Cys. Thus, four mutant genes in which one of the four of Cys is replaced by that of Ser were prepared and expressed in E. coli. Four mutant e were isolated, purified, and incubated with phenylmalonate (7) [17]. Kinetic data mutant enzymes and the wild enzyme are summarized in Table 21.6. Among t mutants, C188S showed a drastic decrease in activity (k_{cat}/K_m). This low activity was a decrease in the catalytic turnover number (k_{cat}) rather than an affinity to the substra

The CD spectrum of the C188S mutant is essentially the same as that of the wi enzyme, which reflects that the tertiary structure of this mutant changed little comp that of the wild-type enzyme. Calculation of the content of the secondary structure of enzymes based on J-600S Secondary Structure Estimation system (JASCO) also show there is no significant change in the tertiary structure of the C188S mutant. The fact t k_{cat} value of this mutant is extremely small despite little change in conformation

TABLE 21.6
Reactivities of Wild Type and Mutant Enzymes

	K_m (mM)	k_{cat} (S^{-1})
Wild type	13.3	366
C101S	4.3	248
C148S	11.5	100
C171S	9.1	62.3
C188S	4.9	0.62

indicates that Cys188 is located in the active site. The catalytic activity of mutants C148S and C171S also decreased in spite of the smaller K_m values compared to that of the wild-type enzyme. It can be assumed that the decrease in α-helix structure caused a decrease in k_{cat} value. The distance between the catalytic amino acid and the binding substrate would become longer because of the change in conformation. It was thus concluded that cysteine188 is located in the catalytic site of the enzyme.

21.2.6   STEREOCHEMICAL COURSE OF THE REACTION

One can ask whether the enzyme distinguishes between two prochiral carboxyl groups? The clue to elucidation of this question is to prepare both enantiomers of α-methyl-α-phenylmalonate that have ^{13}C on one of the two carboxyl groups. Starting from ^{13}C-phenylacetate ([13]C-11), by optical resolution of an intermediate 14, both enantiomers of chiral ^{13}C-containing α-methyl-α-phenylmalonate were prepared (Figure 21.5). The absolute configuration of the chiral substrate was unambiguously determined by the optical rotation of the resolved hydroxyl acid 14.

The result of enzymatic decarboxylation was very clear [18]. While (S)-15 resulted in ^{13}C containing product (12), (R)-15 gave the product with ^{13}C no more than natural abundance. Apparently the reaction proceeds with net inversion of configuration. Thus, the presence of a planar intermediate, such as 21, can be postulated. Enantioface-differentiating protonation to 21 will give the optically active final product. One evidence supporting this intermediate is the

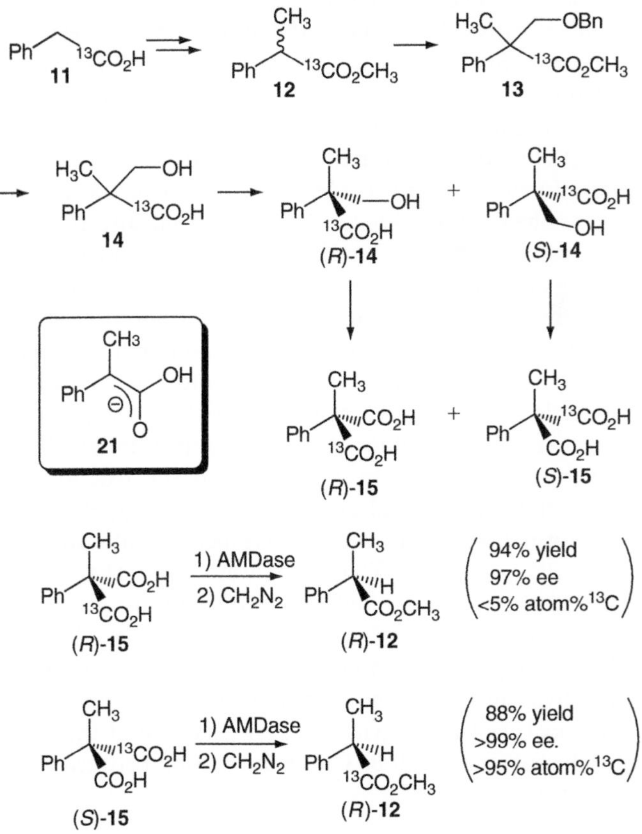

FIGURE 21.5 Stereochemistry of AMDase-catalyzed decarboxylation.

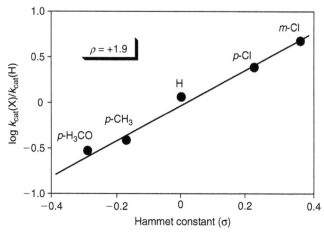

FIGURE 21.6 Hammett plot of the k_{cat} for the AMDase-catalyzed decarboxylation of se
X-phenylmalonates.

electronic effect of the substituents on the rate of decarboxylation of substituted (p-
p-Me, p-Cl, m-Cl, and H) phenylmalonic acid. The logarithm of $k_{cat}(X)/k_{cat}(H)$ c
correlated in a linear fashion to Hammett σ-values (Figure 21.6). The ρ-value was re
to be $+1.9$. The fact that the sign of the ρ-value is positive means that the transition sta
some negative charge. Thus, the supposition of a negatively charged intermediate such a
rationalized. In this case, the aromatic ring should occupy the same plane as that
olefinic part. It is then estimated that the conformation of the substrate is already rest
when it binds to the active site of the enzyme.

As the conformation of the intermediate **21** reflects the one of the starting materials,
substituents (X) and α-substituent (CH_3) should be arranged in *syn*- or *anti*-periplan
mentioned earlier (Figure 21.7), *ortho*-substituted compounds (X = Cl, CH_3) and o
derivative are inactive to this enzyme in contrast to substrates, in which either methy
is replaced by a hydrogen, that is readily decarboxylated by the enzyme. The differe
reactivities of these compounds is well deduced by supposing that the substrates are re
to take the *syn*-periplanar conformation to bind to the enzyme or to undergo decarboxy
reaction. Accordingly, if the loss of potential energy to take the *syn*-periplanar conforn
exceeds the binding energy between the substrate and the enzyme, the compound v
inactive. The K_m values of ordinary substrates are 10 to 20 mM, indicating that the b
energy is about 2.5 to 3 kcal/mol at the reaction temperature. Then, how much
difference of potential energies between *syn*-periplanar conformation and the most
conformation? As there are no experimental data available, the values were obtain
ab initio calculation. The results were clear. The difference in potential energy betwee
and *anti*-peripalanar conformation (the most stable conformation) of α-methyl-α-(o-
ophenyl)malonate (**9f**) is calculated to be 5.5 kcal/mol, whereas that of the nonr
lated compound is about 0.7 kcal/mol. These values together with the binding ene
reactive substrates demonstrate the essential importance of conformation of the subst
the pocket of the enzyme [19]. At least one of the attracting forces to fix the conforr
of the substrates will be CH–π interaction between the aromatic rings of the enzyme a
substrate [20].

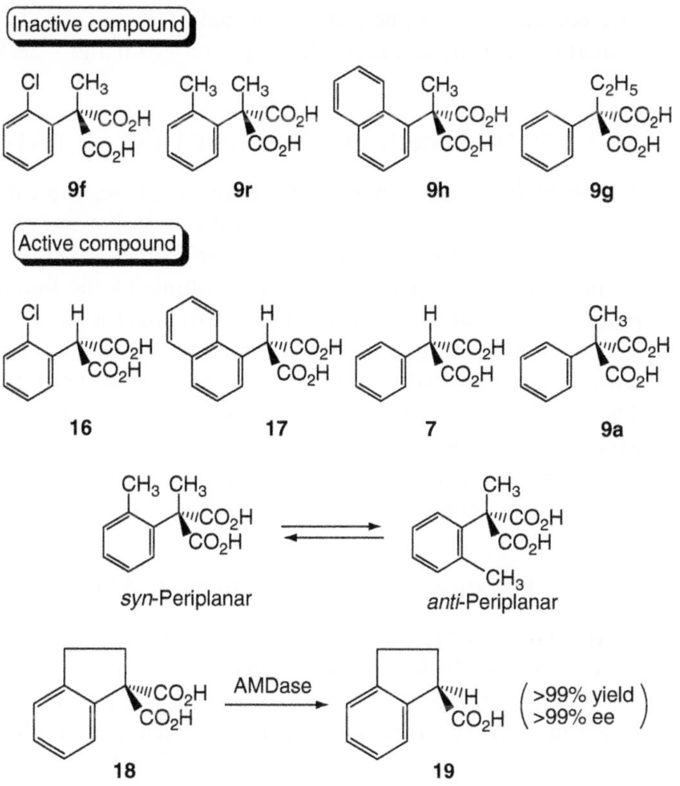

FIGURE 21.7 Effect of the conformation of the substrates.

The nonreactivity of the o-substituted compound, i.e., dimethyl derivative **9r**, is due to the high potential energy of *syn*-periplanar conformation, and the decarboxylation reaction will be realized if the conformation of **9r** could be fixed to *syn*-periplanar. Indane dicarboxylic acid (**18**) is considered to be a good model of *syn*-periplanar conformation of **9r**. As expected, cyclic substrate **18** was smoothly decarboxylated to give the corresponding monobasic acid (*R*)-**19** in high chemical and optical yield. It is noteworthy that the K_m value of this substrate is smaller by one order compared to hitherto mentioned noncyclic compounds (Table 21.7). According to the discussions proposed above, this is considered to be due to its conformation already being fixed to the one that fits to the binding site of the enzyme. This estimation was demonstrated to be true by kinetic studies at various temperatures. Arrhenius plots of the

TABLE 21.7
Reactivities of Representative Substrates

Entry	Substrate	K_m (mM)	V_{max} (U/mg)	Relative Activity (%)
1	7	13.9	883	100
2	16	12.6	2713	338
3	17	6.1	1440	372
4	9a	25.5	74	4.6
5	18	1.1	3.9	5.8

rate constants of indanedicarboxylic acid (18) and phenylmalonic acid (7) showed
activation entropy of these substrates are −27.6 and −38.5 cal/mol/K, respectively [2

21.3 CONVERSION OF THE FUNCTION BY RATIONAL DESIGN

The inversion of the enantioselectivity of the reaction might be possible by chang
binding mode of the substrate or by shifting the key cysteine residue from the *si*-face t
face of the enolate intermediate. In the former case, to exchange the binding site
aromatic ring and methyl group, the volume of the binding sites and the location
amino acid residues should be changed to effect a strong binding between the substr
the enzyme. The latter strategy seems more practical, because replacing the Cys188
with some appropriate amino acid that has little or no proton-donating ability is not
Then, the introduction of a new proton donor at a suitable position would bring ab
expected inversion of enantioselectivity. Herein, we report that the introduction of o
mutations, G74C and C188S, led to the formation of the opposite enantiomer, altho
activity of the mutant was lower than that of the native enzyme.

21.3.1 INVERSION OF ENANTIOSELECTIVITY OF AMDASE

As the tertiary structure of AMDase is not yet known, the most serious problem is to
the position at which the new proton donor should be introduced. The only clue that n
effective is a homology search as well as a comparison of the amino acid sequence
function of the enzymes. The characteristic features of AMDase are: (1) the reaction p
through an enolate-type transition state [9]; (2) the cysteine residue plays an essen
[9,17]; and (3) the reaction involves an inversion of configuration on the α-carbo
carboxyl group [18].

Some enzymes were found that had about 30% homology and some common fu
through multiple alignments using the PSI-BLAST program. These were glutamate ra
from *Lactobacillus fermenti* [22], aspartate racemase from *Streptococcus thermophi*
hydantoine racemase from *Pseudomanas* sp. strain NS671 [24], and maleate isomera
A. faecalis [25]. The important feature that is consistent for all these enzymes is the pre
Cys188. While all the racemases have another cysteine residue at around 74, AMDase
corresponding cysteine residue around this region as shown in Figure 21.8.

The reaction mechanism for glutamate racemase has been studied extensively
It has been proposed that the key for the racemization activity is that the two
residues of the enzyme are located on both sides of the substrate bound to the act

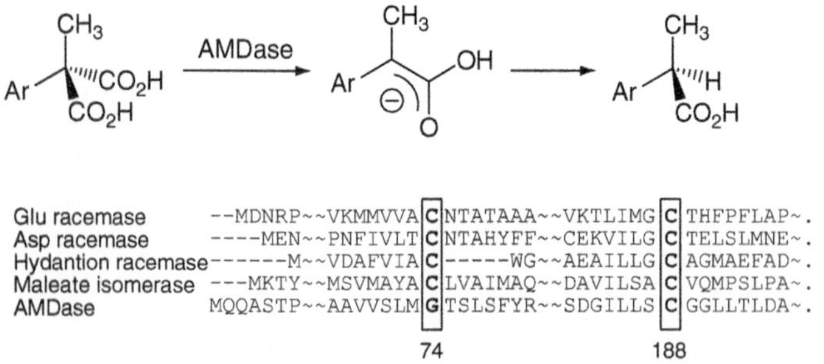

```
Glu racemase        --MDNRP~~VKMMVVA C NTATAAA~~VKTLIMG C THFPFLAP~.
Asp racemase        ----MEN~~PNFIVLT C NTAHYFF~~CEKVILG C TELSLMNE~.
Hydantion racemase  ------M~~VDAFVIA C -----WG~~AEAILLG C AGMAEFAD~.
Maleate isomerase   ---MKTY~~MSVMAYA C LVAIMAQ~~DAVILSA C VQMPSLPA~.
AMDase              MQQASTP~~AAVVSLM G TSLSFYR~~SDGILLS C GGLLTLDA~.
                                        74                       188
```

FIGURE 21.8 Amino acid homology between some racemases and AMDase.

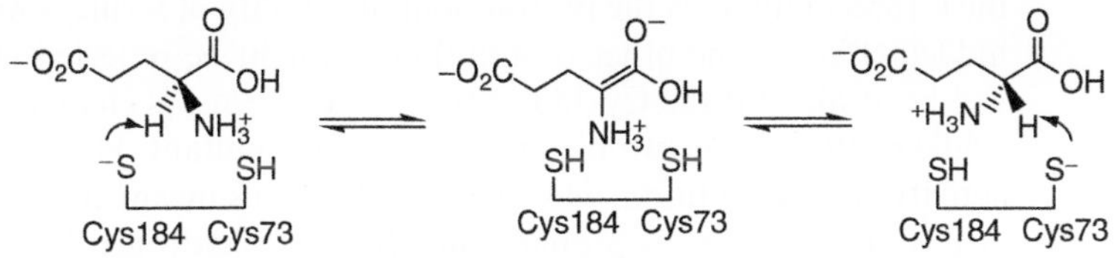

FIGURE 21.9 Reaction mechanism of glutamate racemase.

Thus, one cysteine residue abstracts the α-proton from the substrate, while the other delivers a proton from the opposite side of the intermediate enolate of the amino acid. In this way, the racemase catalyzes the racemization of glutamic acid through a so-called two-base mechanism (Figure 21.9).

The tertiary structure of glutamate racemase has already been resolved, and it has also been clarified that a substrate analog glutamine binds between two cysteine residues [29]. These data enabled us to predict that the new proton-donating amino acid residue should be introduced at position 74, replacing Gly for the inversion of enantioselectivity of the decarboxylation reaction.

First, we examined the enantioselectivity of a C188S mutant enzyme purified from *E. coli* JM109/pAMD101 (Table 21.8). Although, the proton-donating ability of Ser is weaker than Cys, the location of the proton donor does not change in this case. Thus, we presumed that the configuration of the product would be the same as in the case of the reaction by the wild-type enzyme. To our surprise, however, the results were entirely different: α-methyl-α-thienyl (**9e**) and α-methyl-α-naphthylmalonate (**20**) gave the corresponding monobasic acids with the configuration opposite to that given by the native enzyme (Table 21.8). This fact suggests that there are some other proton donors on the opposite side of the enantiomeric face of the intermediate enolate, although their effect is far smaller compared with Cys188. In the case of

TABLE 21.8
Enantioselectivities of Mutant Enzymes

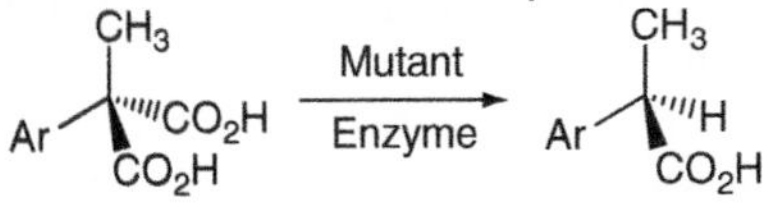

Enzyme	Ar	Yield (%)	ee (%)
Wild type		quant	99 (*S*)
C188S	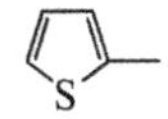 **9e**	17	50 (*R*)
G74C		37	0 (−)
G74C/C188S		60	94 (*R*)
Wild type		96	97 (*R*)
C188S	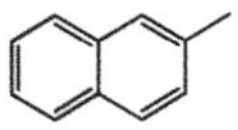 **20**	6	70 (*S*)
G74C		13	6 (*R*)
G74C/C188S		17	96 (*S*)

the C188S mutant, as the proton-donating ability of serine is weaker than that of cystei
hidden effect of the other proton donors might be reflected in the product. Then, hig
will be attained if the Cys188 of the native enzyme is changed by an amino acid that
acidic proton. Thus, we prepared the C188A mutant. However, this mutant had no e
activity and, in addition, was very unstable. This means that a Cys or Ser residue is ine
at position 188, and is probably involved in a hydrogen bonding. Therefore, we deci
introduce a Cys residue as a proton donor instead of the Gly74 of the native enzyme.

The G74C mutant was prepared by polymerase chain reaction (PCR) using the p
that contains the gene coding native AMDase (pAMD101). The amplified gene was di
by *Hind*III and *Pst*I followed by ligation with pUC19. The mutant enzyme was purifiec
the transformed *E. coli* cells. Although the change in amino acid is drastic, the muta
exhibited some activity. As expected, the products were nearly racemic, if not entirely,
case of both **9e** and **20** (Table 21.8). These results demonstrate that this position is effec
give a proton to the intermediate of the reaction.

If the proton-donating ability of the amino acid at 188 is weaker, the enantioselecti
the reaction is expected to become the opposite. So we prepared a double mutant
C188S gene starting from the one that already contains the codon for C188S mutatic
shown in Table 21.8, the absolute configuration of the products is opposite to that
products obtained by the native enzyme, and the ee of the products dramatically increa
94% and 96% for **9e** and **20**, respectively (Table 21.8). This inversion of the enantiosele
of the reaction supports the reaction mechanism where the Cys188 of the native enz
working as the proton donor.

REFERENCES

1. Kim, Y.S. and Kolattukudy, P.E., Stereospecificity of malonyl-CoA decarboxylase, acet
   carboxylase, and fatty acid synthetase from the uropygial gland of goose, *J. Biol. Chem.*,
   686–689, 1980.
2. Thomas, N.R., Schirch, V., and Gani, D., Synthesis of (2*R*)- and (2*S*)-[1–13C]-2-amino-2-
   malonic acid, probes for the serine hydroxymethyltransferase reaction: stereospecific decart
   tion of the 2-pro-*R* carboxy group with the retention of configuration, *J. Chem. Soc.
   Commun.*, 5, 400–402, 1990.
3. Thomas, N.R., Rose, J.E., and Gani, D., Decarboxylation of 2-aminomalonic acid cataly
   serine hydroxymethyltransferase is, in fact, a stereospecific process, *J. Chem. Soc. Chem. Co*
   14, 908–909, 1991.
4. Harrison, I.T., Lewis, B., Nelson, P., et al., Nonsteroidal anti-inflammatory agents. I. 6-Subs
   2-naphthylacetic acids, *J. Med. Chem.*, 13(2), 203–205, 1970.
5. Shen, T.Y., Perspectives in nonsteroidal anti-inflammatory agents, *Angew. Chem. Int. Ed.
   11(6), 460–472, 1972.
6. Miyamoto, K. and Ohta, H., Enzyme-mediated asymmetric decarboxylation of disubstitutec
   nic acids, *J. Am. Chem. Soc.*, 112, 4077–4078, 1990.
7. Miyamoto, K. and Ohta, H., Asymmetric decarboxylation of disubstituted malonic acid b
   ligenes bronchisepticus KU1201, *Biocatalysis*, 5, 49–60, 1991.
8. Miyamoto, K., Tsuchiya, S., and Ohta, H., Microbial asymmetric decarboxylation of fl
   containing arylmalonic acid derivatives, *J. Fluorine Chem.*, 59, 225–232, 1992.
9. Miyamoto, K. and Ohta, H., Purification and properties of a novel arylmalonate decarbc
   from *Alcaligenes bronchisepticus* KU1201, *Eur. J. Biochem.*, 210, 475–481, 1992.
10. Galivan, J.H. and Allen S.H.G., Methylmalonyl-CoA decarboxylase: partial purificatic
    enzymatic properties, *Arch. Biochem. Biophys.*, 126(3), 838–847, 1968.
11. Hoffmann, A., Hilpert, W., and Dimroth, P., The carboxyltransferase activity
    sodium-ion-translocating methylmalonyl-CoA decarboxylase of *Veillonella alcalescens*
    J. Biochem., 179(3), 645–650, 1989.

12. Green, N.M., A spectrophotometric assay for avidin and biotin based on binding of dyes by avidin, *Biochem. J.*, 94(3), 23c–24c, 1965.

13. Green N.M. and Toms E.J., Purification and crystallization of avidin, *Biochem. J.*, 118(1), 67–70, 1970.

14. Boyer, P.D., Ed., *The Enzymes*, vol. 6: *Carboxylation and Decarboxylation (Nonoxidative)*, Academic Press, New York, 1972, pp. 39–115.

15. Wood, H.G., Lochmueller, H., Riepertinger, C., et al., Transcarboxylase. IV. Function of biotin and the structure and properties of the carboxylated enzyme, *Biochem. Z.*, 337, 247–266, 1963.

16. Miyamoto, K. and Ohta, H., Cloning and heterologous expression of a novel arylmalonate decarboxylase gene from *Alcaligenes bronchisepticus* KU1201, *Appl. Microbiol. Biotechnol.*, 38, 234–238, 1992.

17. Miyazaki, M., Kakidani, H., Hanzawa, S., et al., Cysteine188 revealed as being critical for the enzyme activity of arylmalonate decarboxylase by site-directed mutagenesis, *Bull. Chem. Soc. Jpn.*, 70(11), 2765–2769, 1997.

18. Miyamoto, K., Tsuchiya, S., and Ohta, H., Stereochemistry of enzyme-catalyzed decarboxylation of α-methyl-α-phenylmalonic acids, *J. Am. Chem. Soc.*, 114, 6256–6257, 1992.

19. Miyamoto, K., Ohta, H., and Osamura, Y., Effect of conformation of the substrate on enzymatic decarboxylation of α-arylmalonic acid, *Bioorg. Med. Chem.*, 2, 469–475, 1994.

20. Kawasaki, T., Saito, K., and Ohta, H., The mode of substrate-recognition mechanism of arylmalonate decarboxylase, *Chem. Lett.*, 26(4), 351–352, 1997.

21. Kawasaki, T., Horimai, E., and Ohta, H., On the conformation of the substrate binding to the active site during the course of enzymic decarboxylation, *Bull. Chem. Soc. Jpn.*, 69(12), 3591–3594, 1996.

22. Gallo, K.A. and Knowles, J.R., Purification, cloning, and cofactor independence of glutamate racemase from *Lactobacillus*, *Biochemistry*, 32(15), 3981–3990, 1993.

23. Yohda, M., Okada, H., and Kumagai, H., Molecular cloning and nucleotide sequencing of the aspartate racemase gene from lactic acid bacteria *Streptococcus thermophilus*, *Biochim. Biophys. Acta*, 1089(2), 234–240, 1991.

24. Watabe, K., Ishikawa, T., Mukohara, Y., et al., Identification and sequencing of a gene encoding a hydantoin racemase from the native plasmid of *Pseudomonas* sp. strain NS671, *J. Bacteriol.*, 174, 3461–3466, 1992.

25. Hatakeyama, K., Asai, Y., Uchida, Y., et al., Gene cloning and characterization of maleate *cis–trans* isomerase from *Alcaligenes faecalis*, *Biochem. Biophys. Res. Commun.*, 239(1), 74–79, 1997.

26. Glavas, S. and Tanner, M.E., Catalytic acid/base residues of glutamate racemase, *Biochemistry*, 38(13), 4106–4113, 1999.

27. Gallo, K.A., Tanner, M.E., and Knowles, J.R., Mechanism of the reaction catalyzed by glutamate racemase, *Biochemistry*, 32(15), 3991–3997, 1993.

28. Tanner, M.E., Gallo, K.A., and Knowles, J.R., Isotope effects and the identification of catalytic residues in the reaction catalyzed by glutamate racemase, *Biochemistry*, 32(15), 3998–4006, 1993.

29. Hwang, K.Y., Cho, C.-S., Kim, S.S., et al., Structure and mechanism of glutamate racemase from *Aquifex pyrophilus*, *Nat. Struct. Biol.*, 6, 422–426, 1999.

22 Chemoenzymatic Preparation of Enantiopure Building Blocks of Synthetic Utility

Kenji Mori

CONTENTS

22.1 INTRODUCTION

Since 1978 we have been employing hydrolytic enzymes and yeasts as biocatalysts to p
enantiopure building blocks for our laboratory-scale synthesis of pheromones and ter
In this chapter, preparative methods of some enantiopure building blocks will be des
together with the experimental details for the preparation of especially useful building b
Unlike the traditional industrial use of biocatalysts ("fermentation") to provide comp
with complicated structures such as antibiotics starting from cane molasses or other o
sources, the approach detailed in this chapter is to prepare enantiopure building block
can eventually be converted by organic reactions to more value-added products like
mones and terpenes (Figure 22.1). Design of versatile building blocks can be best carri
by synthetic chemists, who know the scope and limitations of various organic reaction:
used in conversion of the building blocks to the target molecules.

There are two choices for the biocatalytic preparation of the enantiopure building b
One is the desymmetrization of *meso*-compounds and prochiral compounds, while the o
the kinetic resolution of racemic samples. These two methods will be discussed in de
must be added that desymmetrization is more efficient than kinetic resolution. The f
converts all of the starting material to the desired product, while the latter gives the o
enantiomer in 50% yield at most. In special cases, such as in the cases of chiral and eno
β-keto esters with an α-substituent, their enantiomers can interconvert to each other th

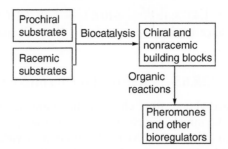

FIGURE 22.1 Roles of biocatalysis and organic reactions in the synthesis of pheromones and other bioregulators.

keto-enol tautomerism, and more readily reducible ones can be reduced with reductases in yeast to give optically active β-hydroxy esters in more than 50% yield. This is an example of dynamic kinetic resolution.

22.2 DESYMMETRIZATION OF ACYCLIC *MESO*-DIOLS WITH ESTERASES AND LIPASES

22.2.1 (2*R*,4*S*)-5-ACETOXY-2,4-DIMETHYL-1-PENTANOL (2)

The compound **2** is a useful building block for the synthesis of 1,3-dimethylated alkyl chains of natural products, such as (2*R*,4*R*)-supellapyrone, the female-produced sex pheromone of the broad-banded cockroach, *Supella longipalpa*. Asymmetric acetylation of diol **1** with vinyl acetate and lipase AK (Amano) gives (2*R*,4*S*)-**2** (98.0% ee) in 72% yield (Figure 22.2) [1]. In order to find out the optimal reaction conditions, it is necessary to screen lipases and esterases, solvents [usually tetrahydrofuran (THF), diisopropyl ether, or diethyl ether], acyl donors (usually vinyl acetate or isopropenyl acetate), and reaction temperature.

22.2.1.1 Preparation of (2*R*,4*S*)-2 [1]

Lipase AK (250 mg) was added to a cooled and stirred solution of **1** (5.18 g, 39.2 mmol) in THF (50 mL) at 0°C. Vinyl acetate (3.7 mL, 43 mmol) was added to the mixture, which was stirred for 3 d at 5°C. Vinyl acetate (1.5 mL, 17 mmol) was added again, and the mixture was stirred for 2 d, filtered through Celite, and the filtrate was concentrated *in vacuo*. The residue was chromatographed on SiO₂ (50 g, hexane/EtOAc, 20:1) to give (2*R*,4*S*)-2 (4.95 g, 72%) as an

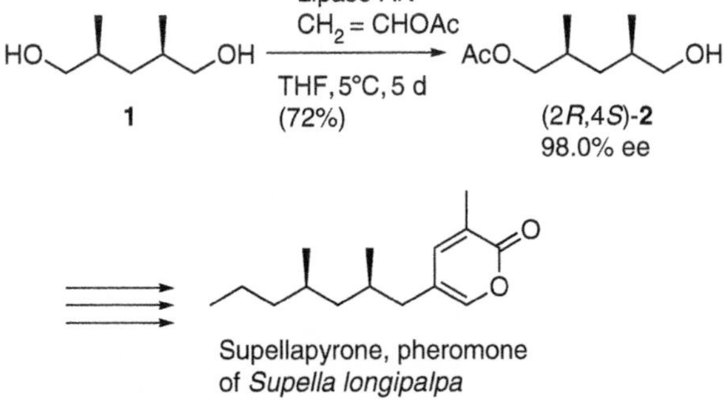

Lipase AK
CH₂ = CHOAc
HO⤳⤳⤳OH ──────────→ AcO⤳⤳⤳OH
THF, 5°C, 5 d
1   (72%)   (2*R*,4*S*)-2
                             98.0% ee

Supellapyrone, pheromone
of *Supella longipalpa*

FIGURE 22.2 Preparation and utilization of half-acetate (2*R*,4*S*)-**2**.

oil, $n_D^{25} = 1.4378$; $[\alpha]_D^{25} = +10.6$ ($c = 1.88$, CHCl$_3$). Enantiomeric purity determination performance liquid chromatography (HPLC) analysis on Chiralcel OD (hexane/EtOH 98.0% ee.

22.2.2 (2R,6S)-7-ACETOXY-2,6-DIMETHYL-1-HEPTANOL (4)

1,5-Dimethylated alkyl chain is the characteristic feature of isoprenoids and some mones. Asymmetric acetylation of diol 3 with isopropenyl acetate and immobilized li (Amano) affords (2R,6S)-4 (95% ee) in 57% yield (Figure 22.3) [2]. This acetylation v reported by Chênevert and Desjardins [3]. Several pheromones, as shown in Figure 22 tribolure [3] were synthesized from (2R,6S)-4 [2,4,5].

22.2.2.1 Preparation of (2R,6S)-4 [2]

Immobilization of Lipase PS: Lipase PS (Amano, 3.0 g) was mixed thoroughly with Super Cel (10.0 g). Then, 0.1 M potassium phosphate buffer (pH 7, 10 mL) was add mixture was shaken vigorously and dried *in vacuo*.

Acetylation: Immobilized lipase PS (2.20 g) and isopropenyl acetate (6 mL) were to a solution of 3 (5.30 g, 33.1 mmol) in THF (350 mL). The mixture was stirred at 0

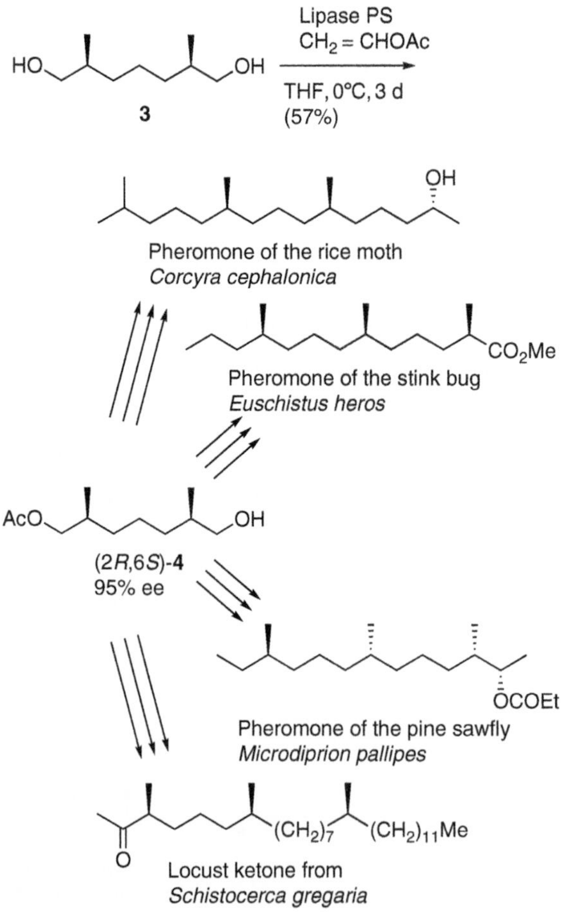

FIGURE 22.3 Preparation and utilization of half-acetate (2R,6S)-4.

monitored by TLC (3 d). The solid was filtered off and washed with THF, and the filtrate was concentrated *in vacuo*. The residue was chromatographed on SiO_2 (140 g, hexane/EtOAc 20:1 to 10:1) to give 3.8 g (57%) of **4** as a colorless oil along with diacetate (2.1 g, 26%) and diol (0.7 g, 13%). Properties of **4**: $n_D^{24} = 1.4426$; $[\alpha]_D^{24} = +8.8$ ($c = 0.98$, CHCl₃). Enantiomeric purity determination: derivatization followed by HPLC analysis, 95% ee.

22.2.3 (1*S*,2*R*)-1-ACETOXYMETHYL-2-HYDROXYMETHYLCYCLOPROPANE (6)

Cyclopropane-containing fatty acids are occasionally found as natural products. Plakoside A (Figure 22.4) is an immunosuppressive galactosphingolipid isolated from the Caribbean sponge *Plakortis simplex*, and contains two cyclopropane rings. Synthesis and stereochemical assignment of plakoside A were achieved by employing (1*S*,2*R*)-**6** as the key building block [6,7]. As shown in Figure 22.4, acetylation of diol **5** with vinyl acetate and lipase AK (Amano) gives (1*S*,2*R*)-**6** (>99.9% ee) in 86% yield [6]. Asymmetric hydrolysis of a diester of **5** was also reported by others ([8] and references therein), and the resulting optically active mono ester served as a starting material in pheromone synthesis [8].

22.2.3.1 Preparation of (1*S*,2*R*)-6 [6]

Lipase AK (Amano, 1.06 g) was added to a solution of **5** (21.3 g, 208 mmol) in THF (110 mL) and vinyl acetate (130 mL), and the mixture was stirred for 3.5 h at room temperature. The mixture was filtered through Celite and the filtrate was concentrated *in vacuo*. The residue was chromatographed on SiO_2 to give (1*S*,2*R*)-**6** as a colorless oil (25.8 g, 86%), $n_D^{25} = 1.4558$; $[\alpha]_D^{21} = -19.9$ ($c = 1.65$, CHCl₃). Enantiomeric purity determination: HPLC analysis on Chiralcel OD-H, >99.9% ee.

22.2.4 (2*S*,3*R*)-4-ACETOXY-2,3-EPOXY-1-BUTANOL (8)

Epoxides are frequently found among insect pheromones. Asymmetric hydrolysis of *meso*-diacetate **7** with pig pancreatic lipase (PPL) gives (2*S*,3*R*)-**8** (90% ee) in 71% yield

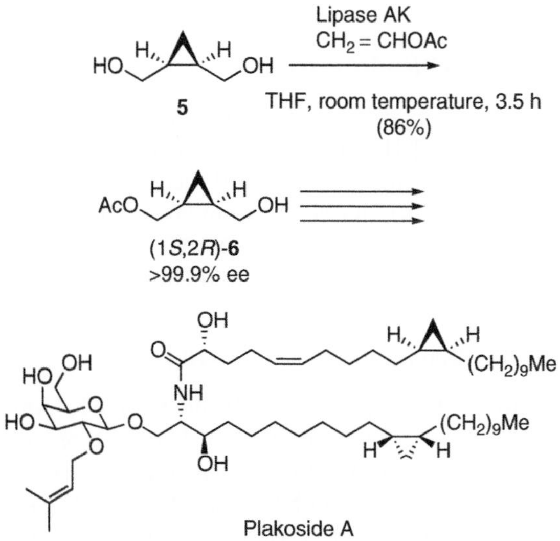

FIGURE 22.4 Preparation and utilization of half-acetate (1*S*,2*R*)-6.

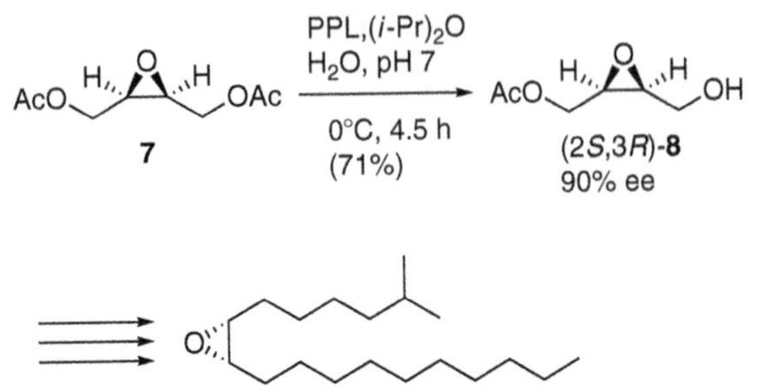

FIGURE 22.5 Preparation and utilization of half-acetate (2S,3R)-8.

(Figure 22.5) [9]. This half-acetate 8 could be converted to various pheromones as sl in Figure 22.6. Alkylative ring-cleavage of epoxides afforded other kinds of pheromones Chuche and coworkers reported the preparation of a similar building block [10]. In 20 became clear that enzymatic kinetic resolution of (1S*,2R*)-(±)-4-t-butyldiphenylsily 2,3-epoxy-1-butanol gives enantiomerically pure (98 to 99% ee) and therefore more ω building block (see Section 22.4.1)

22.2.4.1   Preparation of (2S,3R)-8 [9]

PPL (Sigma, 1.2 g) was added to a solution of 7 (2.82 g, 15 mmol) in (i-Pr)$_2$O(67 mL 0.1 M phosphate buffer (pH 7, 140 mL) at 0°C. The mixture was stirred vigorously a for 4.5 h, and the pH was kept constant by addition of 1 M NaOH using an autobu After addition of NaOH (1 equivalent), the (i-Pr)$_2$O layer was separated and the aq layer was extracted with Et$_2$O. The combined organic solution was dried (MgSO$_4$ concentrated in vacuo. The residue was chromatographed on SiO$_2$. Elution with he Et$_2$O (2:1 to 1:1) afforded 8 as a slightly yellow oil (1.55 g, 71%), $n_D^{20} = 1$ $[\alpha]_D^{22} = +17.4$ (c = 0.9, CH$_2$Cl$_2$). Enantiomeric purity determination: derivatizatioɪ lowed by HPLC analysis, 90% ee.

22.2.5   (2S,3R)-4-ACETOXY-2,3-ISOPROPYLIDENEDIOXY-1-BUTANOL (10)

Stereochemically defined vic-diols can be starting materials for stereochemically define dioxabicyclo[3.2.1]octanes, a common structure-type among insect pheromones. As shc Figure 22.7, erythritol derivative 9 was acetylated with vinyl acetate using lipase AK (Aɪ to furnish (2S,3R)-10 (98.5% ee) in 96% yield [11]. Vandewalle and coworkers also ob (2S,3R)-10 by employing lipase SAM II (Amano-Fluka) [12]. (+)-endo-Brevicomin, a beetle pheromone, can be prepared from (2S,3R)-10 [11].

22.2.5.1   Preparation of (2S,3R)-10 [11]

A mixture of 9 (5.60 g, 34.5 mmol) and lipase AK (Amano, 800 mg) in vinyl acetate (7.0 m stirred at room temperature for 5 h. The mixture was filtered through a Celite pad, whi washed with EtOAc. The filtrate was concentrated in vacuo. The residue was chromatogɪ on SiO$_2$ (80 g). Elution with hexane/EtOAc (2:1) gave 10 (6.75 g, 96%) as an oil, $n_D^{21} = 1$ $[\alpha]_D^{21} = +17.3$ (c = 1.08, CHCl$_3$). Enantiomeric purity determination: derivatization benzoate and HPLC analysis on Chiralcel OG), 98.5% ee.

22.2.6 (R)-2-Acetoxymethyl-3-Buten-1-ol (12)

To synthesize the β-lactone moiety of an antibiotic 1233A, (R)-12 was designed and prepared by asymmetric hydrolysis of prochiral diacetate 11 with lipase P (Amano) to give (R)-12 (90% ee) in 86% yield (Figure 22.8) [13]. Asymmetric acetylation or hydrolysis serves as a tool to convert prochiral diacetate or diol to an optically active building block.

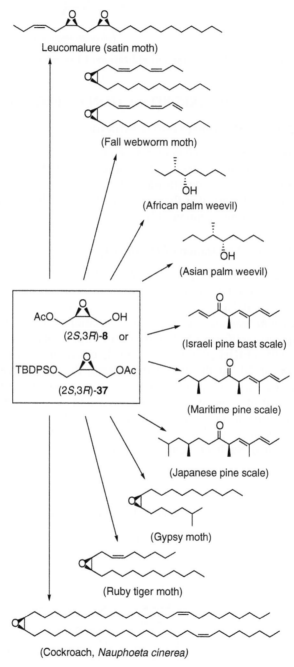

FIGURE 22.6 Pheromones synthesized from (2S,3R)-8 and (2S,3R)-37.

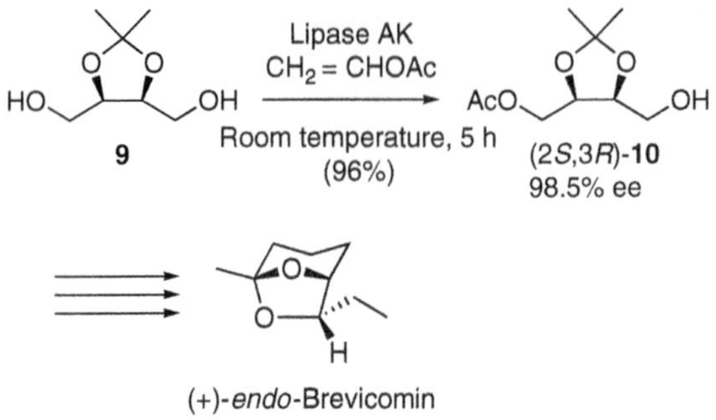

FIGURE 22.7 Preparation and utilization of half-acetate (2S,3R)-**10**

22.3 ASYMMETRIC REDUCTION OF PROCHIRAL KETONES WITH YEAST

22.3.1 (S)-1-PHENYLTHIO-2-BUTANOL (14)

Baker's yeast (*Saccharomyces cerevisiae*) reduced 1-thiophenyl-2-butanone (**13**) to
(91% ee) using ethanol as the energy source (Figure 22.9) [14]. Oxidation of **14** wi
butyl hydroperoxide afforded crystalline **15**, which could be purified by recrystall
Sulfoxide **15** served as the starting material for the synthesis of (+)-juvabione, the well
juvenile hormone mimic [14].

22.3.1.1 Preparation of (S)-14 [14]

A suspension of dry baker's yeast (Oriental Yeast Co., 300 g) in H_2O (3 L) was sti
10 min. A solution of **13** (9.03 g, 50.1 mmol) in 95% EtOH (20.0 mL) was divided i
portions and added every 30 to 150 min to the vigorously stirred and aerated yeast sus
at room temperature. When the evolution of CO_2 gas was too vigorous, a few mL
aqueous antifoam (AF) emulsion or 95% EtOH was added. After 19 h, baker's yeas
and 95% EtOH (10 mL) were added, and vigorous stirring and aeration were continue
total of 2 d. At the end of this period, the total volume of EtOH added was 200 m
reaction was stopped by the addition of Et_2O, and the mixture was filtered through
The filtrate was saturated with NaCl, and extracted three times with EtOAc. The fil
was suspended in acetone, sonicated, and filtered. The filtrate was concentrated *in vac*
residue was extracted three times with EtOAc. The combined extracts were washe
saturated $NaHCO_3$ aqueous solution and brine, dried ($MgSO_4$), and concentrated *i*

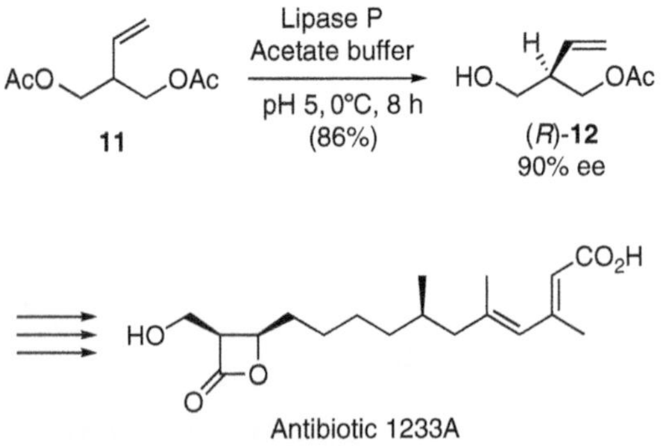

FIGURE 22.8 Preparation and utilization of half-acetate (R)-**12**.

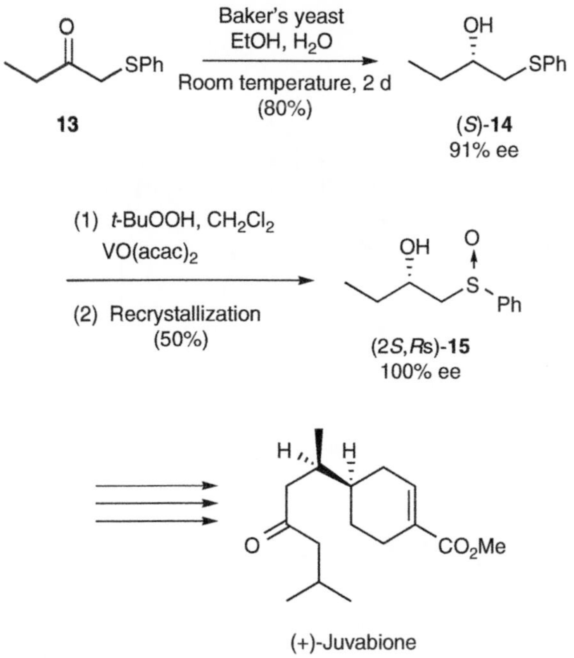

FIGURE 22.9 Preparation and utilization of (S)-14.

The residue (12.9 g) was chromatographed on SiO$_2$ (400 g, hexane/EtOAc 10:1) to give 8.50 g (80%) of (S)-14 as an oil, $n_D^{18} = 1.5589$; $[\alpha]_D^{22} = +52.5$ ($c = 1.66$, CHCl$_3$). Enantiomeric purity determination: HPLC analysis after derivatization to (R)-MTPA ester, 91% ee. After oxidation of (S)-14 to (2s, Rs)-15, its enantiomeric purity was improved to ~100% ee by recrystallization.

22.3.2 (S)-3-HYDROXY-2,2-DIMETHYLCYCLOHEXANONE (17)

Reduction of 2,2-dimethylcyclohexane-1,3-dione (16) with baker's yeast in a sucrose solution gave (S)-17 (99% ee) in 70 to 80% yield (Figure 22.10) [15,16]. Many terpenoids as shown in Figure 22.11 were synthesized from (S)-17, including glycinoeclepin A [17] and insect juvenile hormone III [18]. This hydroxy ketone (S)-17 is one of the most useful chiral building blocks of microbial origin, and affords not only cyclic terpenoids but also acyclic terpenoids like juvenile hormone III after Baeyer–Villiger oxidation.

22.3.2.1 Preparation of (S)-17 [15,16]

A solution of 16 (15.0 g, 107 mmol) in 95% EtOH (30 mL) and 0.2% Triton X-100 solution (150 mL) was added to a suspension of dry baker's yeast (Oriental Yeast Co., 200 g) in a 15%

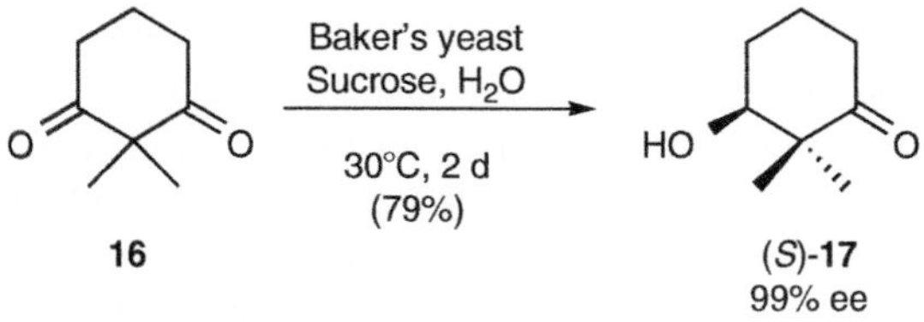

FIGURE 22.10 Preparation of hydroxy ketone (S)-17.

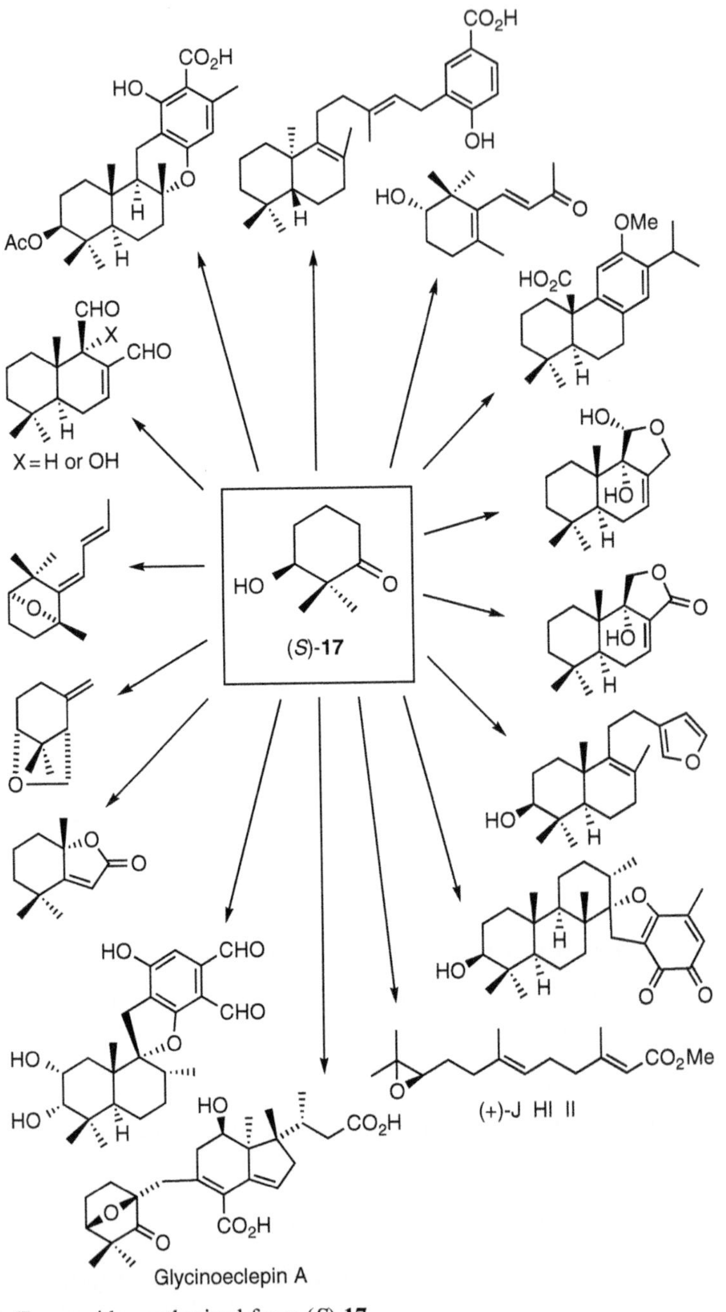

FIGURE 22.11 Terpenoids synthesized from (S)-17.

sucrose solution in H_2O (3 L) at 30°C with stirring. The mixture was stirred at
with aeration (16 L/min) for 48 h. Et_2O (200 mL) was added to the mixture, and the m
was left to stand overnight. After the flocculated yeast cells had precipitated, the mixtu
filtered through Celite. The filtrate was saturated with NaCl, and extracted with EtOA
filter-cake was washed with EtOAc. The extract and washings were combined, washe
brine, dried ($MgSO_4$), and concentrated *in vacuo*. The residual oil (27 g) was chrc
graphed on SiO_2 (220 g). Elution with hexane/EtOAc (5:1) recovered **16** (4.7 g). F

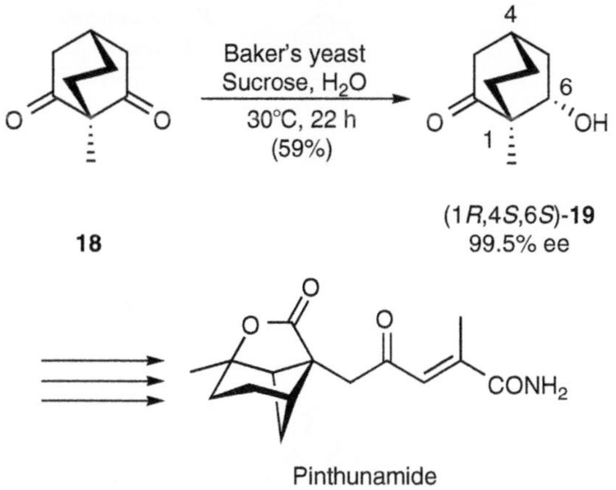

FIGURE 22.12 Preparation and utilization of hydroxy ketone (1R,4S,6S)-19.

elution with EtOAc gave 8.8 g (79% based on the consumed **16**) of (*S*)-**17**, bp 85 to 87°C/83.7 Torr; $n_D^{21} = 1.4747$; $[\alpha]_D^{21} = +24.1$ ($c = 1.12$, CHCl$_3$). Enantiomeric purity determination: HPLC analysis after derivatization to (*S*)-MTPA ester, 99% ee.

22.3.3 (1R,4S,6S)-6-Hydroxy-1-Methylbicyclo[2.2.2]octan-2-one (19)

Reduction of bridged bicyclic ketone **18** with baker's yeast in a sucrose solution furnished crystalline (1R,4S,6S)-**19** (99.5% ee) in 59% yield (Figure 22.12) [19]. This hydroxy ketone **19** served as the starting material for the synthesis of pinthunamide [20] and other sesquiterpenes.

22.3.3.1 Preparation of (1R,4S,6S)-19 [19]

A solution of **18** (0.79 g, 5.2 mmol) in 99% EtOH (6 mL) was reduced with baker's yeast (7 g) and sucrose (16.8 g) in H$_2$O (80 mL) in the presence of Triton S-100 (0.2% aqueous solution, 6 mL) at 30°C for 22 h. Subsequent work-up followed by SiO$_2$ chromatography as described for (*S*)-**17** gave 0.6 g of crude but crystalline **19**. This was recrystallized from hexane/EtOAc 85:1) to give **19** (0.47 g, 59%) as needles, mp 117.5 to 118.0°C; $[\alpha]_D^{16} = -5.6$ ($c = 1.0$, CHCl$_3$), 98.8% de [as determined by gas chromatography (GC)]. Determination of enantiomeric purity: HPLC analysis after derivatization to MTPA ester, 99.5% ee.

22.3.4 Other 3-Hydroxy Ketones

Figure 22.13 illustrates the preparation of some other 3-hydroxy ketones. Reduction of **20** was not stereoselective with baker's yeast. However, by employing another yeast, *Pichia terricola* KI 0117, **20** was stereoselectively reduced to (2S,3S)-**21** (99% ee), which was converted to the naturally occurring (+)-juvenile hormone I [21]. 1,3-Diketone **22** was reduced with baker's yeast to give (1R,4S,6S)-**23** [19], which was converted to (+)-juvabione, a juvenile hormone mimic [22].

4-Methylbicyclo[2.2.1]heptane-2,6-dione (**24**) was unstable in H$_2$O even at pH 7 due to the ring strain, and its reduction was only successful by employing a large amount of baker's yeast in the presence of phosphate buffer (pH 7) to give (1R,4S,6S)-6-hydroxy-4-methylbicyclo[2.2.1]heptan-2-one (**25**, 82.5% ee) in 42% yield, [17]. The right-hand portion of glycinoeclepin A was constructed from **25** [17].

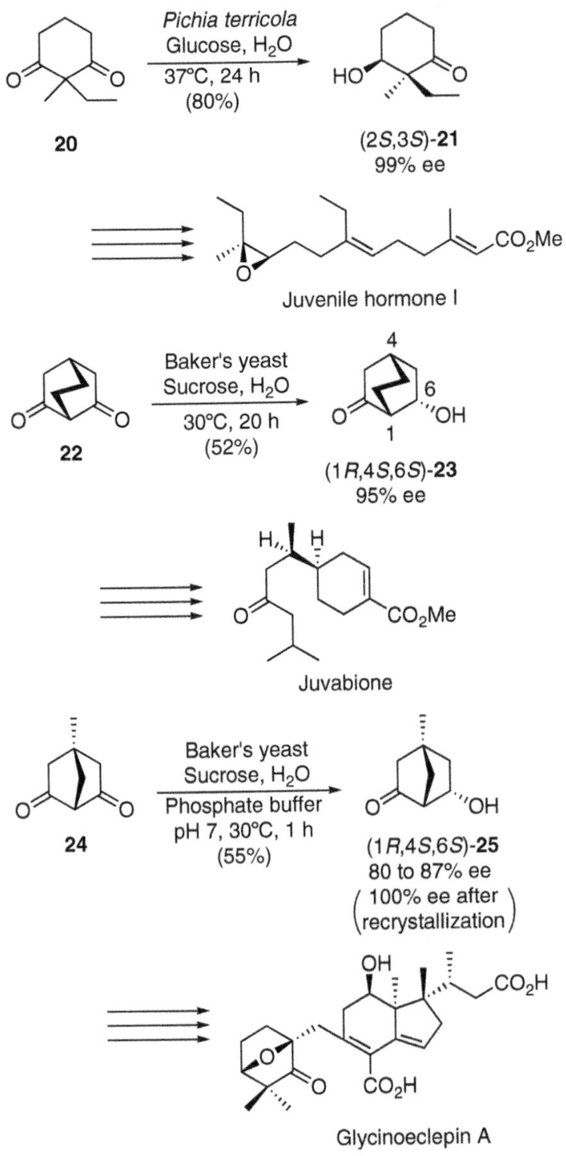

FIGURE 22.13 Preparation and utilization of hydroxy ketone (2S,3S)-21, (1R,4S,6S)-2
(1R,4S,6S)-25.

22.3.4.1 Preparation of (1R,4S,6S)-25 [17]

A suspension of dry baker's yeast (Oriental Yeast Co., 100 g), sucrose (100 g), K
(5.5 g), and Na₂HPO₄·12H₂O (21.5 g) in H₂O (1 L) was kept at 30°C on a rotary pl
shaker for 10 min. A solution of 24 (1.0 g, 7.2 mmol) in 95% EtOH (3 mL) was then ad
the mixture, and the incubation was continued at 30°C. After 10 min, sucrose (10 g) a
baker's yeast (10 g) were again added to the mixture, and after a further 2 min of incu
dione 24 (1.0 g, 7.2 mmol) was also added to the mixture. These additions of sucrose,
yeast, and 24 were repeated twice more. After the completion of the additions, the incu
was continued for 30 min. The mixture was then filtered through Celite, and the filt
was washed with acetone. The filtrate was made slightly alkaline (pH ~8, universal inc

by addition of $NaHCO_3$, and was then saturated with NaCl and extracted six times with EtOAc. The washings (acetone solution) were evaporated, and the residue was diluted with EtOAc and washed with saturated aqueous $NaHCO_3$. The combined organic phase was dried ($MgSO_4$), filtered, and concentrated *in vacuo*. The residue was purified by SiO_2 chromatography to give **25** as crystals (2.04 g, 55%), mp 56.5 to 57.5°C (prisms from hexane/ Et_2O); $[\alpha]_D^{23} = -16.6$ ($c = 1.06$, $CHCl_3$). Enantiomeric purity determination: HPLC analysis after derivatization to MTPA ester, 82.5% ee.

22.3.5 Octahydronaphthalenone (28) and Octahydroindenone (31)

Sugai and coworkers selected a yeast strain, *Torulaspora delbrueckii* IFO 10921, for the selective reduction of a carbonyl group of triketones **26** and **29** (Figure 22.14) [23]. Structure of the reduction product of **26** was determined as (1S,3S,6S)-**27** by x-ray analysis. The hemiacetal ketone **27** was converted to (1S,8aR)-**28**, a useful building block in isoprenoid synthesis [23]. Reduction of **29** gave **30**, which was not analyzed by x-ray [24]. The hemiacetal ketone **30** yielded (1S,7aR)-**31**, another useful building block in steroid synthesis [24].

22.3.6 Cyclic and Enolizable β-Keto Esters with a Single Stereogenic Center at α-Position

Yeast reduction of a cyclic and enolizable β-keto ester with a single stereogenic center at α-position is known to give a single optically active β-hydroxy ester in more than 50% yield, because enolization of the starting β-keto ester destroys the chirality at the α-position before

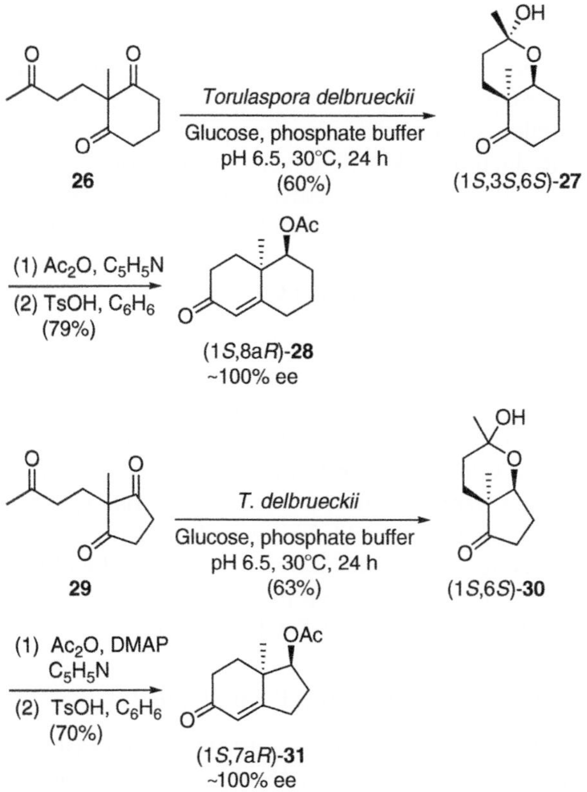

FIGURE 22.14 Preparation of octahydronaphthalenone **28** and octahydroindenone **31**.

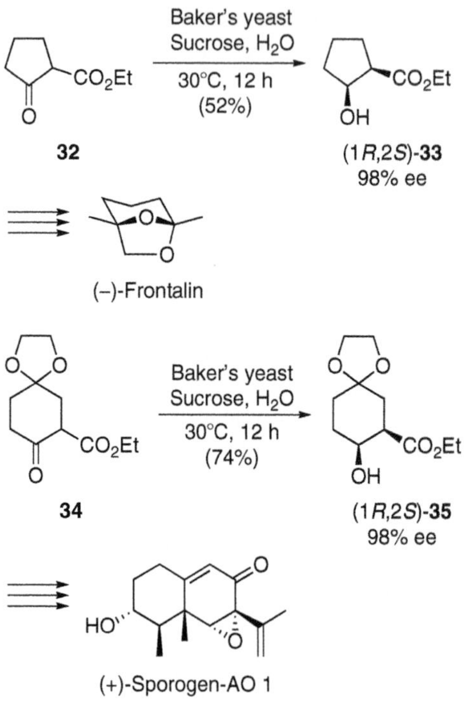

FIGURE 22.15 Preparation and utilization of cyclic β-hydroxy esters (1R,2S)-33 and (1R,2S)-3

reduction. As shown in Figure 22.15, reduction of **32** with baker's yeast gave ethyl (1
2-hydroxycyclopentane-1-carboxylate (**33**, 98% ee) in 53% yield, and **33** was converted
frontalin, a bark beetle pheromone [25].

Reduction of **34** with baker's yeast is also an efficient method to obtain a useful bu
block, ethyl (1R,2S)-5,5-ethylenedioxy-2-hydroxycyclohexane-1-carboxylate (**35**, 98%
74% yield [26], which eventually afforded (+)-sporogen-AO 1, a microbial bioregulate

22.3.6.1 Preparation of (1R,2S)-33 [25]

A solution of **32** (27.2 g, 174 mmol) in 95% EtOH (10 mL) was added dropwise with stir
a suspension of dry baker's yeast (Oriental Yeast Co., 140 g) in a solution of sucrose (3(
H$_2$O (2.5 L). The mixture was stirred and aerated at 30°C for 12 h. Occasionally n-C$_8$F
(a few drops) was added to the suspension to suppress the foaming. The mixture was f
through Celite, and the filtration bed was washed with EtOAc. The combined filtra
washings were washed with brine, dried (MgSO$_4$), and concentrated *in vacuo*. The resid
distilled to give crude **33**. This was chromatographed on SiO$_2$ (300 g, hexane/EtOAc =
The fractions containing **33** were further purified by distillation to give 14.4 g (52%) of
76°C/4 Torr; n_D^{25} = 1.4534; $[\alpha]_D^{22}$ = +14.5 (c = 2.83, CHCl$_3$). Enantiomeric puriy det
ation: GC on Chirasil DEX-CB, 97.8% ee.

22.3.6.2 Preparation of (1R,2S)-35 [26,27]

A suspension of dry baker's yeast (Oriental Yeast Co., 200 g) in a solution of sucrose
in H$_2$O (2 L) was stirred for 15 min at 30°C with aeration. A solution of **34** (15 g, 66 mr
95% EtOH (30 mL) was added to this, and the fermentation was continued overnight a

with stirring and aeration. Then NaHCO$_3$ (50 g) and EtOAc (500 mL) were added, and the mixture was filtered through Celite. The filtrate was saturated with NaCl, and extracted with EtOAc five times. The filter-cake was washed thoroughly with EtOAc. The combined EtOAc solution was washed with brine, dried (MgSO$_4$), and concentrated *in vacuo*. The residue was chromatographed on SiO$_2$ (400 g, hexane/EtOAc 4:1 to 1:1) to give 1 to 2 g of recovered **34** and 9 to 10 g (60 to 67%) of **35**, bp 117 to 118°C/0.35 Torr; $n_D^{25} = 1.4695$; $[\alpha]_D^{23} = +51.1$ ($c = 1.02$, CHCl$_3$). Enantiomeric puriy determination: HPLC analysis of the corresponding MTPA ester, 98.4% ee.

22.4 KINETIC RESOLUTION OF RACEMIC ALCOHOLS, HYDROXY ACIDS, AND AMINO ACIDS WITH HYDROLYTIC ENZYMES

22.4.1 (2R,3R)-1-Acetoxy-4-(*t*-Butyldiphenylsilyloxy)-2,3-Epoxybutane (37)

Asymmetric acetylation of (±)-epoxy alcohol **36** with lipase PS-C (Amano) and vinyl acetate gave acetate (2S,3R)-**37** (98.1% ee) and the unreacted alcohol (2R,3S)-**36** (99.5% ee) in 48 and 49% yields, respectively (Figure 22.16) [28]. These products could be obtained with satisfactory enantiomeric purities through a single enzymatic reaction. This asymmetric acetylation procedure is therefore more efficient than the asymmetric hydrolysis of *meso*-epoxy diacetate **7** (see Section 22.2.4) with regard to the optical yield (98 to 99% ee vs. 90% ee). The epoxide (2S,3R)-**37** was converted to (3Z,6R,7S,9R,10S)-leucomalure, the sex pheromone of the satin moth [28].

22.4.1.1 Preparation of (2S,3R)-37 and (2R,3S)-36 [28]

Lipase PS-C (Amano, 50 mg) was added to a solution of (±)-**36** (2.00 g, 5.84 mmol) in Et$_2$O (20 mL) and vinyl acetate (1.0 mL) at room temperature. After stirring for 5 h at room temperature, the enzyme was filtered off and the filtrate was concentrated *in vacuo*. The residue was purified by SiO$_2$ chromatography (hexane/EtOAc, 5:1) to give less polar (2S,3R)-**37** (1.07 g, 48%), $n_D^{22} = 1.5359$; $[\alpha]_D^{22} = -3.66$ ($c = 1.25$, CH$_2$Cl$_2$), and more polar (2R,3S)-**36** (971 mg, 49%), $n_D^{22} = 1.5161$; $[\alpha]_D^{22} = +6.5$ ($c = 1.0$, CH$_2$Cl$_2$). Enantiomeric purity determination of **36**: HPLC analysis on Chiralcel OD, 99.5% ee for (2R,3S)-isomer and 98.1% ee for (2S,3R)-isomer.

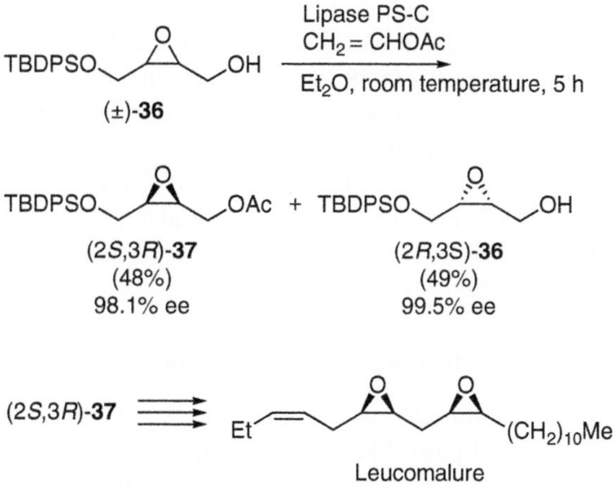

FIGURE 22.16 Preparation and utilization of epoxide (2S,3R)-**37**.

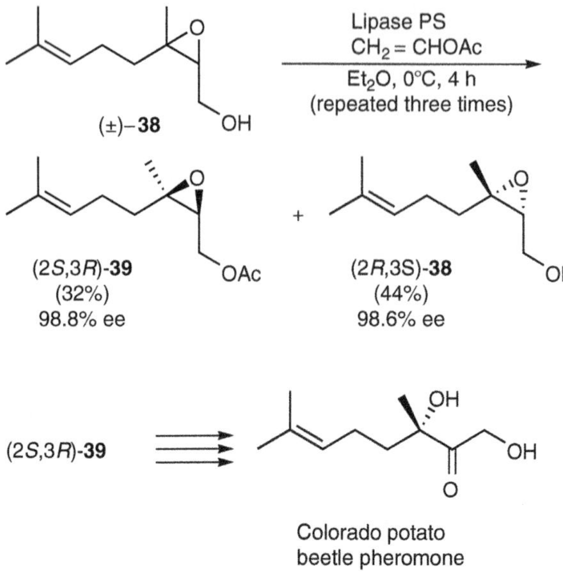

FIGURE 22.17 Preparation and utilization of epoxy acetate (2S,3R)-**39**.

22.4.2   (2S,3R)-3,7-Dimethyl-2,3-Epoxy-6-Octenyl Acetate (39) and (2R,3S)-3,7-Dimethyl- 2,3-Epoxy-6-Octen-1-ol (38)

Asymmetric acetylation of (±)-2,3-epoxynerol (**38**) with vinyl acetate using lip□ (Amano) was repeated three times to give (2S,3R)-**39** (98.8% ee) in 32% yield a□ (2R,3S)-**38** (98.6% ee) in 44% yield (Figure 22.17) [29]. The former acetate (2S,3R)□ converted to the pheromone of the Colorado potato beetle [29]. Since a similar asy■ acetylation is observed with (±)-2,3-epoxygeraniol to give (1R,3R)-acetate, lipase P□ nizes only the chirality at C-3 of **38** and 2,3-epoxygeraniol [29].

22.4.2.1   Preparation of (2S,3R)-39 and (2R,3S)-38 [29]

Lipase PS (Amano, 4.6 g) was added to a stirred solution of (±)-**38** (23.6 g, 162 mm■ vinyl acetate (60 mL) in Et₂O (230 mL) at 0°C. The mixture was stirred at 0°C for 4 h□ then filtered through Celite and the filtrate was concentrated *in vacuo*. The resid□ chromatographed on SiO₂ (1.1 kg, hexane/EtOAc = 5:1 to 2:1) to give (2S,3R)-**39**□ 79 mmol) and (2R,3S)-**38** (14.1 g, 83 mmol). The above (2S,3R)-**39** (16.8 g, 79 mmol) w□ methanolized with K₂CO₃ (1.1 g) in MeOH (200 mL) at room temperature for 4□ procedure—enzymatic resolution followed by methanolysis—was repeated with the r□ (2S,3R)-**38** twice more to give (2S,3R)-**39** (5.49 g, 98.8% ee) in 32% yield, n_D^{20} = $[\alpha]_D^{23}$ = −25.7 (c = 0.58, CHCl₃), and also (2R,3S)-**38** (5.19 g, 98.6% ee) in 44% yield■ 1.4702; $[\alpha]_D^{28}$ = +13.3 (c = 1.09, CHCl₃). Enantiomeric purity determination: HPLC □ on Chiralcel OD.

22.4.3   Enantiomers of 2-(2′,2′-Dimethyl-6′-Methylenecyclohexyl)ethanol (40)

(±)-γ-Cyclohomogeraniol (**40**, Figure 22.18) was resolved by asymmetric acetylation w■ acetate using lipase AK (Amano) to give (R)-**41** and (S)-**40** [30]. Lipase AK recogn■ stereochemistry at C-1′ of the ring-system. Ancistrodial, a defense sesquiterpene of a term■ prepared from (R)-**41** [31], while (S)-**40** was converted to γ-coronal, an ambergris odor■

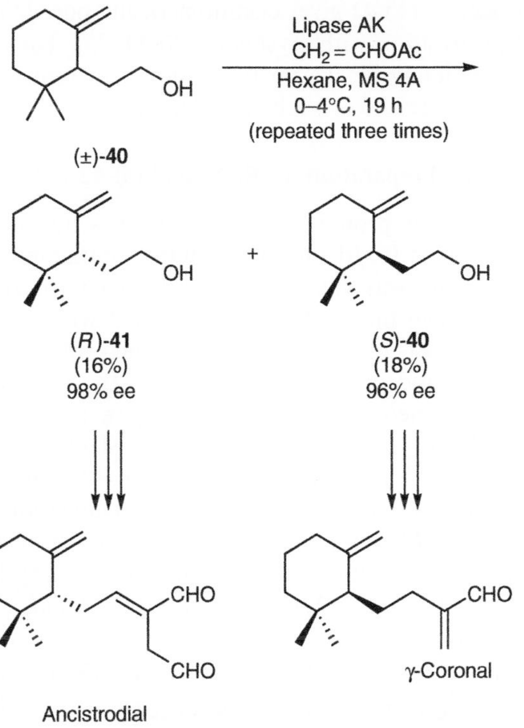

FIGURE 22.18 Preparation and utilization of the enantiomers of γ-cyclohomogeraniol **40**.

22.4.4 ENANTIOMERS OF 2,4,4-TRIMETHYL-2-CYCLOHEXEN-1-OL (43)

Asymmetric hydrolysis of ($\pm$)-2,4,4-trimethyl-2-cyclohexenyl acetate (**42**) with pig liver esterase (PLE) in phosphate buffer (pH 7.5) yielded (*R*)-2,4,4-trimethyl-2-cyclohexen-1-ol (**43**, ~100% ee) in 26% yield together with 67% of the recovered (*S*)-**42** (41% ee) (Figure 22.19) [32].

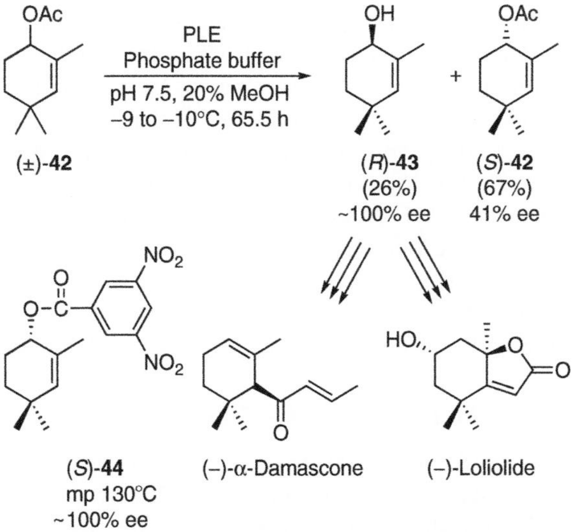

FIGURE 22.19 Preparation and utilization of the enantiomers of **43**.

The acetate (S)-**42** gives enantiomerically pure (S)-**42** after the second hydrolysis wit] and purification with crystalline (S)-**44** [32]. The enantiomers of **43** proved to be extr useful building blocks for the synthesis of terpenes and degraded carotenoids such a α-damascone and (−)-loliolide [33,34].

22.4.4.1   Preparation of (R)-43 and (S)-42 [32]

PLE (150 mg protein, 50,250 units) was added to a stirred mixture of (±)-**42** (145 mmol) in 0.1 M phosphate buffer containing 20% of MeOH (1.11 L) at −9 to − The mixture was vigorously stirred for 65.5 h, keeping the reaction temperature at −10°C. Then the mixture was saturated with NaCl and $(NH_4)_2SO_4$, and extracted s times with Et_2O. The extract was washed with saturated $NaHCO_3$ solution and brine $(MgSO_4)$, and concentrated *in vacuo*. The residue (27.5 g) was chromatographed or (550 g, hexane/Et_2O 30:1 to 10:1) to give 17.7 g (67%) of (S)-**42**, bp 57 to 57.5°C/2.5 $n_D^{22} = 1.4561$; $[\alpha]_D^{21} = -39.5$ ($c = 1.10$, MeOH), 41% ee. Further elution gave 5.32 g (2 (R)-**43**, bp 63 to 64°C/3 Torr; $n_D^{22} = 1.4751$; $[\alpha]_D^{21} = +95.7$ ($c = 1.13$, MeOH), ~10(Enantiomeric purity determination: derivatization to MTPA ester and HPLC analysi acetate (S)-**42** was treated with PLE once more to give (S)-**42** of 95.8% ee in 67% Further purification could be achieved by recrystallization of (S)-**44**, mp 130 to 13 $[\alpha]_D^{24} = -118.5$ ($c = 1.04$, $CHCl_3$), ~100% ee, in 77% recovery.

22.4.5   (3aR,5S,8bS)-5-Hʏᴅʀᴏxʏ-8,8-Dɪᴍᴇᴛʜʏʟ-3,3a,4,5,6,7,8,
8b-Oᴄᴛᴀʜʏᴅʀᴏɪɴᴅᴇɴᴏ[1.2-b]ꜰᴜʀᴀɴ-2-ᴏɴᴇ (45)

Our synthesis of (+)-strigol, a potent stimulant for the germination of parasitic ' necessitated an efficient synthesis of (+)-**45** (Figure 22.20). Asymmetric acetylation o **45** afforded (+)-**45** (99% ee) and (+)-**46** (87% ee) [35]. The acetate (+)-**46** could be f purified by recrystallization. This is a very simple method to prepare (+)-**45**, whic converted to (+)-strigol [35].

22.4.6   (R)-3-Hʏᴅʀᴏxʏ-15-Mᴇᴛʜʏʟʜᴇxᴀᴅᴇᴄᴀɴᴏɪᴄ Aᴄɪᴅ (47) ᴀɴᴅ (R)-2-Hʏᴅʀᴏxʏ-
21-ᴍᴇᴛʜʏʟᴅᴏᴄᴏsᴀɴᴏɪᴄ Aᴄɪᴅ (48)

These long-chain α-or β-hydroxy acids **47** and **48** are known as the acyl side-chain porti sphingolipids. The natural hydroxy acids are (R)-isomers and are highly crystalline therefore possible to prepare (R)-acids by treating (±)-acids with lipase PS and vinyl a and collecting the crystalline (R)-acids that have remained intact after the acylation. A ingly, (±)-**47** gave (R)-3-hydroxy-15-methylhexadecanoic acid (**47**, 100% ee) in 28% (Figure 22.21) [36]. Due to the low reactivity of **47**, the reaction was carried out at a temperature (60°C), and BHT (2,6-di-*t*-butyl-4-methylphenol, a polymerization inhibito added to the reaction mixture to prevent polymerization of vinyl acetate. (−)-Sulfobacin inhibitor of DNA polymerase α, was synthesized employing (R)-**47** [36]. Similarly, (furnished (R)-2-hydroxy-21-methyldocosanoic acid (**48**, 98% ee) in 25% yield, which wa for the synthesis of the sex pheromone of the female hair crab [37].

22.4.6.1   Preparation of (R)-47 [36]

Lipase PS (Amano, 1.00 g) was added to a stirred solution of (±)-**47** (2.00 g, 6.98 mmc 2,6-di-*t*-butyl-4-methylphenol (BHT, 20 mg, a polymerization inhibitor) in vinyl a (30 mL), and the mixture was stirred at 60°C for 48 h. After cooling, the mixture was f and the filtrate was concentrated *in vacuo*. The residue was chromatographed on SiO

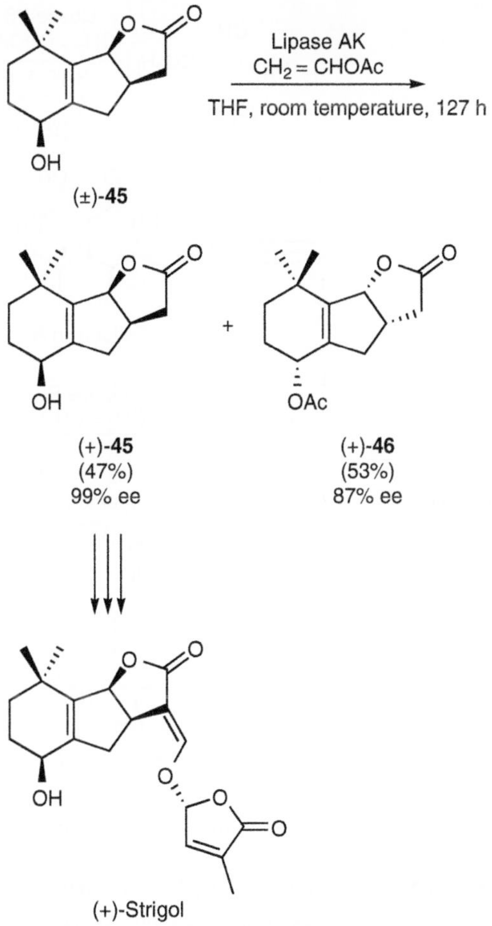

FIGURE 22.20 Preparation and utilization of hydroxy lactone (+)-**45**.

the resulting solid was recrystallized as colorless plates of (R)-**47** (562 mg, 28%, mp 55 to 57°C $[\alpha]_D^{23} = -12.7$ ($c = 1.02$, CHCl₃). Enantiomeric purity determination: gas–liquid chromatography (GLC) analysis of the corresponding methyl ester on Chirasil-DEXCB, ~100% ee.

22.4.7 (2R,3S)- AND (2S,3R)-3-ACETOXY-2-(OCTADECANOYLAMINO)-1-HEXADECANOL (50)

Sphingolipids are important biofunctional molecules. Sugai and coworkers developed an enzymatic method to prepare both the natural (2S,3R)- and unnatural (2R,3S)-isomers of 3-acetylated ceramide **50** (Figure 22.22) [38]. By employing an immobilized form of SC lipase A from *Burkholdeia cepacia* (Sumitomo Chemical Co.), (±)-**49** was regio- and enantioselectively hydrolyzed to give (2R,3S)-**50** (41% yield, 98% ee) belonging to the unnatural stereochemical series. The unreacted (2S,3R)-diacetate **49** (58% yield, 69% ee) could be further purified by repeating once more the enzymatic hydrolysis to give (2S,3R)-**49** of 96% ee. Curiously, when nonimmobilized SC lipase A was employed, (2S,3R)-**49** (96% ee) could be hydrolyzed to give (2S,3R)-**50** (~100% ee) in 85% yield. Accordingly, both the enantiomers of **50** became available. Due to the total insolubility of **49** in aqueous solvents, a biphasic system (decane/0.1 M phosphate buffer) was used as the reaction medium of enzymatic hydrolysis [38].

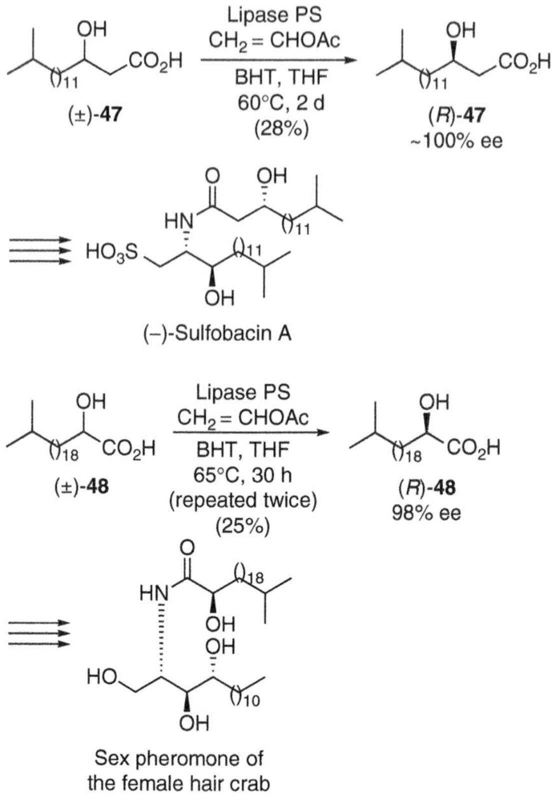

FIGURE 22.21 Preparation and utilization of hydroxy acids (R)-47 and (R)-48.

22.4.8 (4S,6S)-4-Hydroxy-4,9-Dioxaspiro[5.5]undecane (52) and (R)-Tetrahydropyranyl Alcohol 53

Heterocyclic alcohols can also be substrates for esterases and lipases. Acetate (±)-51 spiroacetal ring system was asymmetrically hydrolyzed to give alcohol (4S,6S)-52 (and the unreacted acetate (4R,6R)-51 (~100% ee) in 22 and 13% yield, respectivel repeating the enzymatic hydrolysis for three times (Figure 22.23) [39]. The alcohol (4S is a component of the sex pheromone of the olive fruit fly [39].

Asymmetric acetylation of tetrahydropyranyl alcohol (±)-53 afforded (S)-acetate unreacted (R)-alcohol 53, the latter of which was oxidized to (R)-(+)-hippospongic ac metabolite of the marine sponge *Hippospongia* sp. (Figure 22.23) [40].

22.4.9 (R)- and (S)-2-Aminohexadecanoic Acid (56)

Amino acylase of *Aspergillus* origin is known to possess broad substrate spe N-Chloroacetyl derivative 55 of (±)-2-aminohexadecanoic acid (56) was treated with the to give (S)-56 and (R)-55 in good yield (Figure 22.24)[41]. Removal of the N-chloroacetyl g (R)-55 afforded (R)-56, which was treated with nitrous acid to furnish (R)-2-hydroxyhe noic acid. This α-hydroxy acid is a useful building block in sphingolipid synthesis [41].

22.4.9.1 Preparation of the Enantiomers of 56 [41]

Amino acylase (from *Aspergillus*, 10,000 unit/g, Tokyo Kasei Co., 5 g) and CoCl$_2$ were added to a solution of (±)-55 (36.0 g, 103 mmol) in H$_2$O (4 L) adjusted to pH 7.3

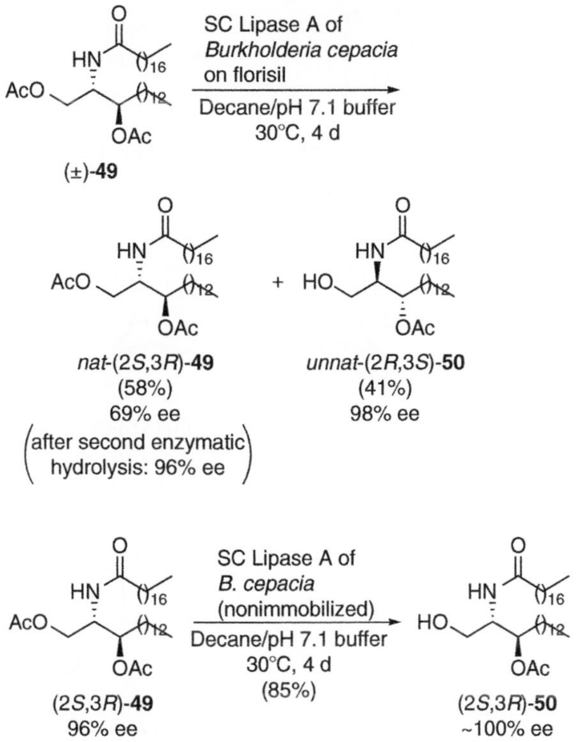

FIGURE 22.22 Preparation of natural and unnatural ceramides **50**.

addition of NaOH. The solution was left to stand for 44 h at 37°C. The precipitated crystalline (*S*)-**56** was collected on a filter, washed with MeOH and Et$_2$O, and dried over P$_2$O$_5$ to give 14.0 g (quantitative) of (*S*)-**56**, mp 234 to 236°C, $[\alpha]_D^{26} = +21.8$ ($c = 0.1$, AcOH). The filtrate obtained after removal of (*S*)-**56** was acidified with 3 M HCl. The precipitated solid was collected on a filter, and dissolved in EtOAc (1 L). The insoluble material was filtered off and the filtrate was concentrated *in vacuo*. The residue was recrystallized from hexane to give (*R*)-**55** (15.5 g, 86%), mp 87 to 88°C, $[\alpha]_D^{21} = -28.0$ ($c = 0.5$, CHCl$_3$). This acid (*R*)-**55** (15.5 g) was mixed with 4 M HCl (150 mL), and the mixture was stirred and heated under reflux for 3 h. After cooling, the mixture was neutralized with 28% NH$_3$ aqueous solution. The precipitated solid was collected on a filter, washed with H$_2$O, MeOH, and Et$_2$O, and dried over P$_2$O$_5$ to give (*R*)-**56** (12.0 g, 99.5%), mp 233 to 236°C, $[\alpha]_D^{26} = -21.0$ ($c = 0.1$, AcOH).

22.5 KINETIC RESOLUTION OF RACEMIC KETONES BY ASYMMETRIC REDUCTION WITH YEASTS

22.5.1 (1*S*,2*R*,5*R*)-2-Ethoxycarbonyl-7,7-Ethylenedioxybicyclo [3.3.0]octan-3-one (57) and (1*S*,5*R*,6*R*,7*S*)-6-Ethoxycarbonyl-3,3-Ethylenedioxy-7-Hydroxybicyclo[3.3.0]octane (58)

β-Keto ester (1*S*,2*R*,5*R*)-**57** was required as the starting material for the synthesis of (+)-carbaprostaglandin I$_2$ (Figure 22.25). Reduction of (±)-**57** with *S. bailli* KI 0116 afforded (1*S*,2*R*,5*R*)-**57** (93.6% ee) and (1*S*,5*R*,6*R*,7*S*)-**58** (99.2% ee) in 40 and 36% yield, respectively [42]. When baker's yeast was employed, the resulting (1*S*,2*R*,5*R*)-**57** was only of 62% ee [42]. It is important to select an appropriate microorganism for the desired reaction.

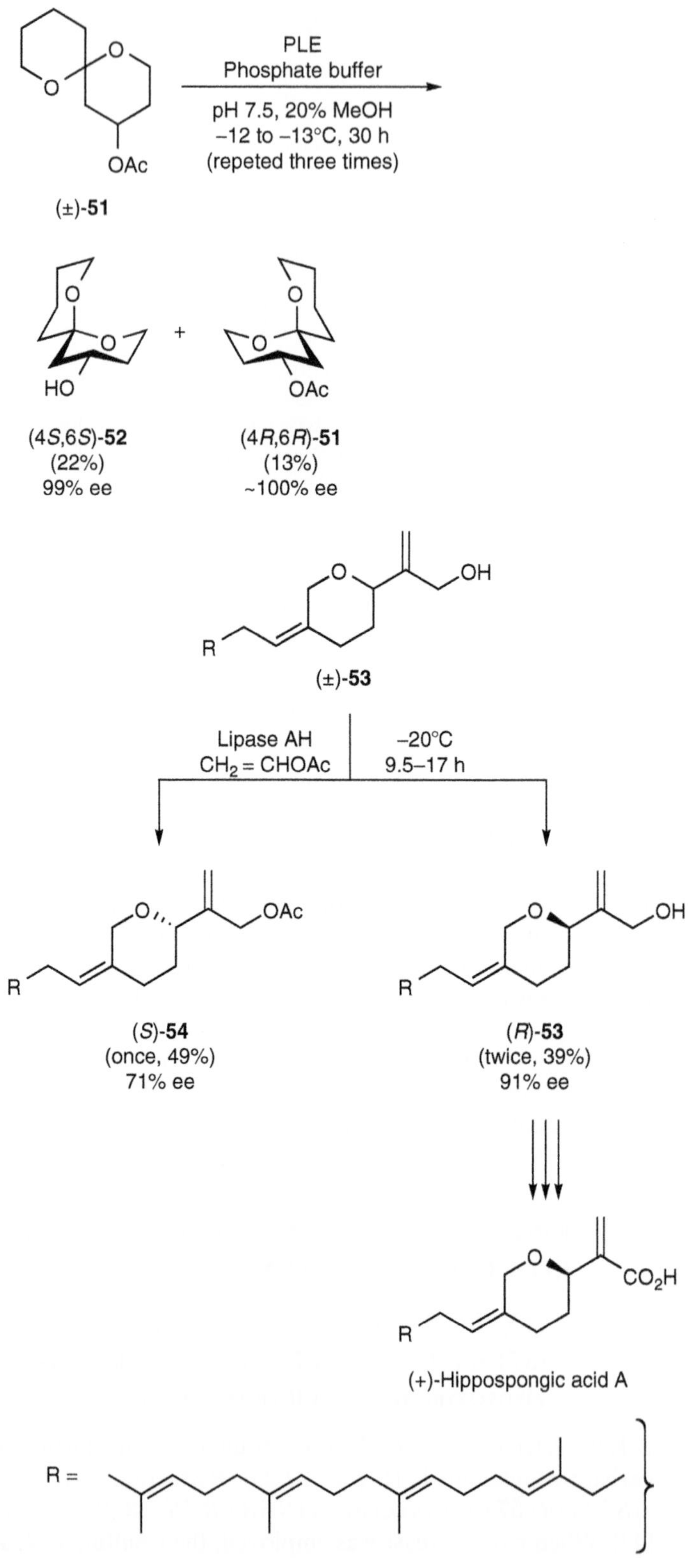

FIGURE 22.23 Preparation of oxygen heterocycles (4*S*,6*S*)-**52** and (*R*)-**53**.

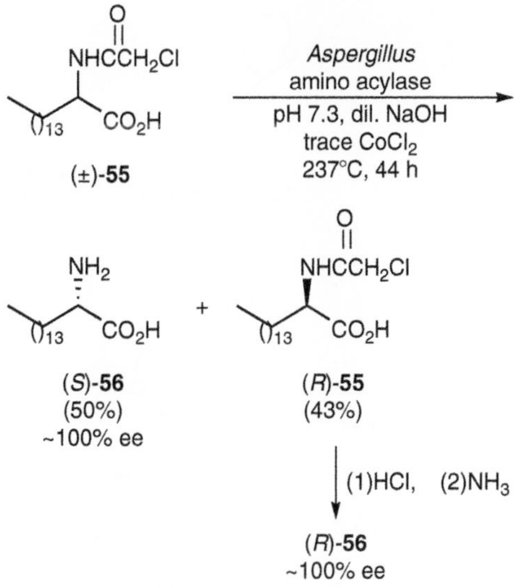

$$\underset{(\pm)\text{-}\textbf{55}}{\overset{\displaystyle \overset{O}{\parallel}}{\text{NHCCH}_2\text{Cl}}}$$

Aspergillus
amino acylase

pH 7.3, dil. NaOH
trace CoCl₂
237°C, 44 h

(S)-56
(50%)
~100% ee

+

(R)-55
(43%)

(1)HCl,   (2)NH₃

(R)-56
~100% ee

FIGURE 22.24 Preparation of the enantiomers of amino acid **56**.

22.5.2   (R)- AND (S)-WIELAND–MIESCHER KETONE (59)

Wieland–Miescher ketone (**59**, Figure 22.26) is a well-known and versatile starting material in isoprenoid synthesis. Both the enantiomers of **59** were prepared by proline-catalyzed asymmetric annelation reaction. The products are unfortunately enantiomerically impure (~70% ee). Although they can be purified by careful fractional recrystallization, this purification is very tedious. Sugai and coworkers devised a simple method for the purification of **59** by kinetic resolution through yeast reduction with *Torulaspora delbrueckii* IFO 10921 (Figure 22.26) [23]. This yeast can reduce only (S)-**59** at C-5. Therefore, treatment of (R)-**59** of 70% ee with *T. delbrueckii* leaves (R)-**59** (97.6% ee) in 82% yield, while (S)-**59** of 70% ee gives (4a*S*, 5*S*)-**60** of 94.4% de in 78% yield. Jones chromic acid converts (4a*S*,5*S*)-**60** to (S)-**59**, which can be purified readily to 100% ee by recrystallization.

22.5.2.1   Preparation of (R)-59 [23]

The wet cells of *T. delbrueckii* IFO 10921 (60 g) were suspended in a solution of glucose (20 g), KH₂PO₄ (3.0 g), and K₂HPO₄ (2.0 g) in H₂O (1 L). To this was added (R)-**59** of 70% ee (10.0 g, 56.1 mmol), and the mixture was shaken for 6 h at 30°C. The mixture was then filtered through Celite and the filtrate was extracted with EtOAc. The extract was washed with brine, dried (Na₂SO₄), and concentrated *in vacuo*. The residue was chromatographed on SiO₂ (450 g, hexane/EtOAc 3:1 to 1:1) to give (R)-**59** (8.2 g, 82%) as a solid. This was recrystallized from hexane/EtOAc to give (R)-**59** as needles, mp 48.6 to 49.4°C, $[\alpha]_D^{27} = -95.2$ ($c = 1.03$, toluene). Enantiomeric purity determination: HPLC analysis, 99.5% ee.

22.5.2.2   Preparation of (4a*S*,5*S*)-60 [23]

The wet cells of *T. delbrueckii* IFO 10921 (60 g) were suspended in a solution of glucose (20 g), KH₂PO₄ (3.0 g), and K₂HPO₄ (2.0 g) in H₂O (1 L). To this was added (S)-**59** of 70% ee (10.0 g, 56.1 mmol), and the mixture was shaken for 14 h at 30°C. The mixture was then filtered through Celite and the filtrate was extracted with EtOAc. The extract was washed with brine, dried (Na₂SO₄), and concentrated *in vacuo*. The residue was chromatographed on SiO₂ (450 g,

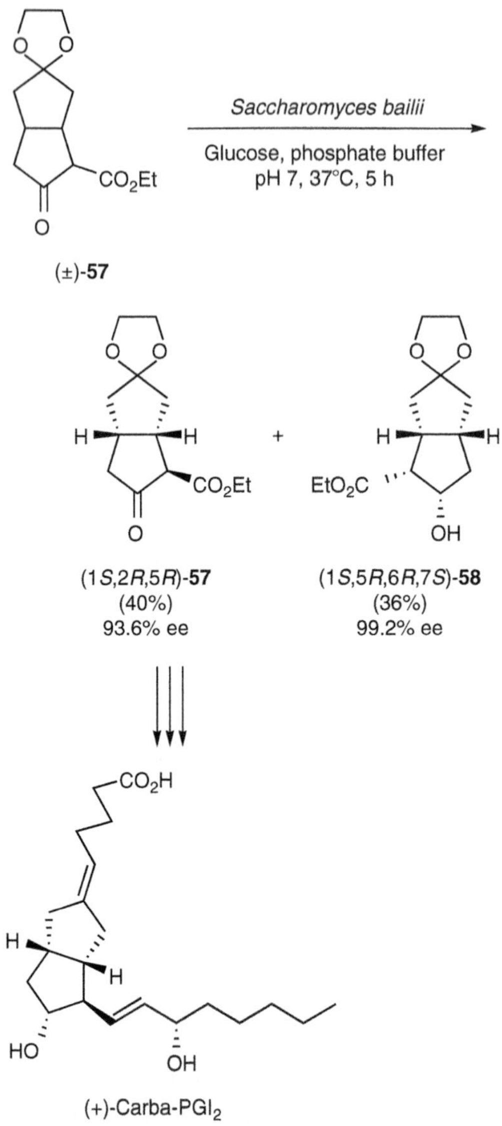

FIGURE 22.25 Preparation of β-keto ester (1*S*,2*R*,5*R*)-**57**.

hexane/EtOAc 3:1 to 1:1) to give (*R*)-**59** (2.2 g, 22%) first, and then (4a*S*,5*S*)-**60** (7.9 g, 78 solid. This was recrystallized from hexane/EtOAc to give needles, mp 44.0 to $[\alpha]_D^{24} = +191.6$ (*c* = 1.05, benzene). Enantiomeric purity determination: HPLC analy oxidation to (*S*)-**59**, >99.9% ee.

22.6 CONCLUSION

We have discussed various biocatalytic processes for preparations of enan building blocks for laboratory organic synthesis. Many new building blocks will be in future to facilitate enantioselective synthesis. Three recent reviews are available area [43–45].

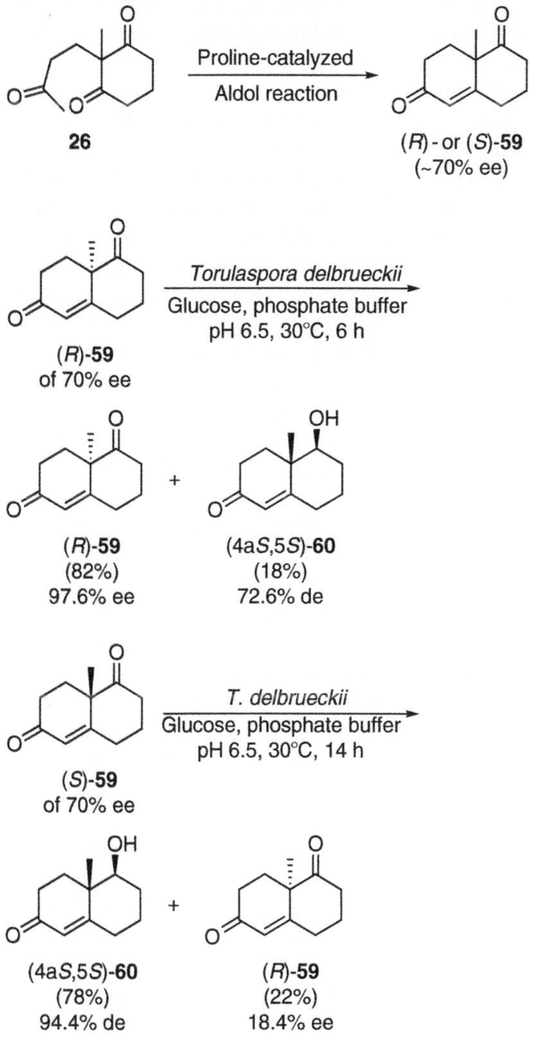

FIGURE 22.26 Purification of Wieland–Miescher ketone **59**.

ACKNOWLEDGMENT

I thank my past and present coworkers whose names appear in References for their experimental efforts and contributions. My thanks are due to Drs. T. Tashiro and K. Fuhshuku for their help in preparing the figures.

REFERENCES

1. Fujita, K. and Mori, K., Synthesis of (2R,4R)-supellapyrone, the sex pheromone of the brownbanded cockroach, *Supella longipalpa*, and its three stereoisomers, *Eur. J. Org. Chem.*, 493–502, 2001.
2. Nakamura, Y. and Mori, K., Pheromone synthesis, CXCVIII: synthesis of (1S,2S,6S,10R)- and (1S,2R,6R,10R)-1,2,6,10-tetramethyldodecyl propanoate, the component of the sex pheromone of the pine sawfly, *Microdiprion pallipes*, *Eur J. Org. Chem.*, 2175–2182, 1999.
3. Chênevert, R. and Desjardins, M., Enzymatic desymmetrization of *meso*-2,6-dimethyl-1,7-heptanediol: enantioselective formal synthesis of the vitamin E side chain and the insect pheromone tribolure, *J. Org. Chem.*, 61, 1219–1222, 1996.

4. Nakamura, Y. and Mori, K., Pheromone synthesis, CCI: synthesis of (3S,7S)- and (3S,7,
   stereoisomers of 3,7-dimethyl-2-heptacosanone and 3,7,15-trimethyl-2-heptacosanone, the k
   identified from the locust *Schistocerca gregaria*, *Eur. J. Org. Chem.*, 1309–1312, 2000.

5. Nakamura, Y. and Mori, K., New synthesis of the rice moth and stink moth pheromo
   employing (2R,6S)-7-acetoxy-2,6-dimethyl-1-heptanol as a building block, *Biosci. Biotechn.*
   chem., 64, 1713–1721, 2000.

6. Seki, M. and Mori, K., Synthesis of a prenylated and immunosuppressive marine galactosphin
   with cyclopropane-containing alkyl chains: (2S,3R,11S,12R,2'''R,5'''Z,11'''S,12'''R)-plakoside
   its (2S,3R,11R,12S,2'''R,5'''Z, 11'''R,12'''S)-isomer, *Eur. J. Org. Chem.*, 3797–3809, 2001.

7. Tashiro, T., Akasaka, K., Ohrui, H., Fattorusso, E., and Mori, K., Determination of the al
   configuration at the two cyclopropane moieties of plakoside A, an immunosuppressive
   galacosphingolipid, *Eur. J. Org. Chem.*, 3659–3665, 2002.

8. Grandjean, D., Pale, P., and Chuche, J., Enzymatic hydrolysis of cyclopropanes: total synth
   optically pure dictyopterenes A and C, *Tetrahedron*, 47, 1215–1230, 1991.

9. Brevet, J.-L. and Mori, K., Pheromone synthesis, CXXXIX: enzymatic preparation of (2S
   acetoxy-2,3-epoxybutan-1-ol and its conversion to the epoxy pheromones of the gypsy moth a
   ruby tiger moth, *Synthesis*, 1007–1012, 1992.

10. Grandjean, D., Pale, P., and Chuche, J., Synthesis of optically pure *cis* epoxyalcohols
    enzymatic route: an alternative to the Sharpless asymmetric epoxidation, *Tetrahedron Le*
    3043–3046, 1991.

11. Mori, K. and Kiyota, H., Pheromone synthesis, CXLII: a new synthesis of (3S,4R)-8-nonene-3
    the key intermediate for the synthesis of (+)-*endo*-brevicomin, *Liebigs Ann. Chem.*, 989–992, 1

12. Pottie, M., Van der Eycken, J., Vandewalle, M., and Röper, H., The synthesis of optically
    derivative of erythritol, *Tetrahedron: Asymmetry*, 2, 329–330, 1991.

13. Mori, K. and Takahashi, Y., Synthetic microbial chemistry, XXIV: synthesis of antibiotic
    an inhibitor of cholesterol biosynthesis, *Liebigs Ann. Chem.*, 1057–1065, 1991.

14. Watanabe, H., Shimizu, H., and Mori, K., Synthesis of compounds with juvenile hc
    activity, XXXI: stereocontrolled synthesis of (+)-juvabione from a chiral sulfoxide, *Sy.*
    1249–1254, 1994.

15. Yanai, M., Sugai, T., and Mori, K., Synthesis of (S)-2-hydroxy-β-ionone employing (S)-3-hy
    2,2-dimethylcyclohexanone as the chiral starting material, *Agric. Biol. Chem.*, 49, 2373–2377

16. Mori, K. and Mori, H., Yeast reduction of 2,2-dimethylcyclohexane-1,3-dione: (S)-(+)-3-hy
    2,2-dimethylcyclohexanone, *Org. Synth. Col. Vol.*, 8, 312–315, 1993.

17. Watanabe, H. and Mori, K., Triterpenoid total synthesis, Part 2: synthesis of glycinoeclepin A, a
    hatching stimulus for the soybean cyst nemadode, *J. Chem. Soc. Perkin Trans. 1*, 2919–2934, 1

18. Mori, K. and Mori, H., Synthesis of both the enantiomers of juvenile hormone III, *Tetrahed.*
    4097–4106, 1987.

19. Mori, K. and Nagano, E., Preparative bioorganic chemistry, Part 10: asymmetric reduc
    bicyclo[2.2.2]octane-2,6-diones with baker's yeast, *Biocatalysis*, 3, 25–36, 1990.

20. Mori, K. and Matsushima, Y., Synthesis of mono- and sesquiterpenoids, XXII: synthesis
    pinthunamide, a sesquiterpene metabolite of a fungi, *Ampulliferina* sp., *Synthesis*, 406–410,

21. Mori, K. and Fujiwara, M., Synthesis of enantiomerically pure (10R,11S)-(+)-juvenile ho.
    I and II, *Tetrahedron*, 44, 343–354, 1988.

22. Nagano, E. and Mori, K., Synthesis of (+)-juvabione, a compound with juvenile hormone a
    Biosci. Biotechnol. Biochem., 56, 1589–1591, 1992.

23. Fuhshuku, K., Funa, N., Akeboshi, T., Ohta, H., Hosomi, H., Ohba, S., and Sugai, T., Ac
    Wieland–Miescher ketone in an enantiomerically pure form by a kinetic resolution with
    mediated reaction, *J. Org. Chem.*, 65, 129–135, 2000.

24. Fuhshuku, K., Tomita, M., and Sugai, T., Enantiomerically pure octahydronaphthaleno
    octahydroindenone: elaboration of the substrate overcame the specificity of yeast-mediated
    tion, *Adv. Synth. Catal.*, 345, 766–774, 2003.

25. Nishimura, Y. and Mori, K., Pheromone synthesis, 188: a new synthesis of (−)-frontalin, th
    beetle pheromone, *Eur. J. Org. Chem.*, 233–236, 1998.

26. Kitahara, T. and Mori, K., Preparation of chiral cyclohexanol derivative with high optical purity by yeast reduction, *Tetrahedron Lett.*, 26, 451–452, 1985.
27. Kitahara, T., Kurata, H., and Mori, K., Efficient synthesis of the natural enantiomer of sporogen-AO 1 (13-desoxyphomenone), a sporogenic sesquiterpene from *Aspergillus oryzae*, *Tetrahedron*, 44, 4339–4349, 1988.
28. Muto, S. and Mori, K., Synthesis of all four stereoisomers of leucomalure, components of the female sex pheromone of the satin moth, *Leucoma salicis*, *Eur. J. Org. Chem.*, 1300–1307, 2003.
29. Tashiro, T. and Mori, K., Enzyme-assisted synthesis of (S)-1,2-dihydroxy-3,7-dimethyl-6-octen-2-one, the male-produced aggregation pheromone of the Colorado potato beetle, and its (R)-enantiomer, *Tetrahedron: Asymmetry*, 16, 1801–1806, 2005.
30. Horiuchi, S., Takikawa, H., and Mori, K., Enzyme resolution of (±)-γ-cyclohomogeraniol and conversion of its (S)-isomer to (S)-γ-coronal, the ambergris odorant, *Bioorg. Med. Chem.*, 7, 723–726, 1999.
31. Horiuchi, S., Takikawa, H., and Mori, K., Synthesis of (6S,7S)-7-hydroxy-6,11-cyclofarnes-3(15)-en-2-one, the opposite enantiomer of the antibacterial sesquiterpene from *Premna oligotricha*, and the (R)-enantiomer of ancistrodial, the defensive sesquiterpene from *Ancistrotermes cavithorax*, *J. Org. Chem.*, 2851–2854, 1998.
32. Mori, K. and Puapoomcharoen, P., Preparative bioorganic chemistry, XV: preparation of optically pure 2,4,4-trimethyl-2-cyclohexen-1-ol, a new and versatile chiral building block in terpene synthesis, *Liebigs Ann. Chem.*, 1053–1056, 1991.
33. Mori, K., Amaike, M., and Itou, M., Synthesis of (S)-α-damascone, *Tetrahedron*, 49, 1053–1056, 1993.
34. Mori, K. and Khlebnikov, V., Carotenoids and degraded carotenoids, VIII: synthesis of (+)-dihydroactinidiolide, (+)- and (−)-actinidiolide, (+)- and (−)-loliolide as well as (+)- and (−)-epiloliolide, *Liebigs Ann. Chem.*, 77–82, 1993.
35. Hirayama, K. and Mori, K., Plant bioregulators, 5: synthesis of (+)-strigol and (+)-orobanchol, the germination stimulants, and their stereoisomers by employing lipase-catalyzed asymmetric acetylation as the key step, *Eur. J. Org. Chem.*, 2211–2217, 1999.
36. Takikawa, H., Nozawa, D., Kayo, A., Muto, S., and Mori, K., Synthesis of sphingosine relatives, Part 22: synthesis of sulfobacin A,B and flavocristamide A, new sulfolipids isolated from *Chryseobacterium* sp., *J. Chem. Soc. Perkin Trans. 1*, 2467–2477, 1999.
37. Masuda, Y., Yoshida, M., and Mori, K., Synthesis of (2S,2′R,3S,4R)-2-(2′-hydroxy-21′-methyldocosanoylamino)-1,3,4-pentadecanetriol, the ceramide sex pheromone of the female hair crab, *Erimacrus isenbeckii*, *Biosci. Biotechnol. Biochem.*, 66, 1531–1537, 2002.
38. Bakke, M., Takizawa, M., Sugai, T, and Ohta, H., Lipase-catalyzed enantiomer resolution of ceramides, *J. Org. Chem.*, 63, 6929–6938, 1998.
39. Yokoyama, Y., Takikawa, H., and Mori, K., Enzymatic preparation of (4R,6R)-4-hydroxy-1,7-dioxaspiro[5.5]undecane and its antipode, the minor component of the olive fruit fly pheromone, *Bioorg. Med. Chem.*, 4, 409–412, 1996.
40. Ichihashi, M., Takikawa, H., and Mori, K., Synthesis of (R)-(+)-hippospongic acid A, a triterpene isolated from the marine sponge, *Hippospongia*, sp., *Biosci. Biotechnol. Biochem.*, 65, 2569–2572, 2001.
41. Mori, K. and Funaki, Y., Synthesis of (4E,8E,2S,3R,2′R)-N-2′-hydroxyhexadecanoyl-9-methyl-4,8-sphingadienine, the ceramide portion of the fruiting-inducing cerebroside in a basidiomycete *Schizophyllum commune*, *Tetrahedron*, 41, 2369–2377, 1985.
42. Mori, K. and Tsuji, M., A new synthesis of (+)-6a-carbaprostaglandin I₂ employing yeast reduction of a β-keto ester derived from *cis*-bicyclo[3.3.0]octane-3,7-dione as the key step, *Tetrahedron*, 42, 435–444, 1986
43. Fuhshuku, K., Oda, S., and Sugai, T., Enzyme reactions as the key steps in the synthesis of terpenoids, degraded carotenoids, steroids, and related substances, *Recent Res. Devel. Org. Chem.*, 6, 57–75, 2002.
44. Brenna, E., Enzyme-mediated synthesis of chiral communication substances: fragrances for perfumery applications, *Curr. Org. Chem.*, 7, 1347–1367, 2003.
45. Ghanem, A. and Aboul-Enein, H.Y., Lipase-mediated chiral resolution of racemates in organic solvents, *Tetrahedron: Asymmetry*, 15, 3331–3351, 2004.

23 Stereoselective Modifications of Polyhydroxylated Steroids

Elena Fossati and Sergio Riva

CONTENTS

23.1 INTRODUCTION

Nature is a tireless architect of complex chemical structures that can quite often be exploited to improve human health and well-being. Among them, steroids are probably the most widely investigated family of bioactive compounds. Specifically, their typical tetracyclic rigid structure decorated with a bunch of sensitive functional groups has attracted the synthetic interest of organic chemists, who have always regarded these molecules as ideal targets to develop or to apply new selective reactions [1–4]. Similarly, biocatalyzed transformations of steroids—by whole-cell processes [5–6] or by isolated enzyme-catalyzed reactions [7]—have been widely investigated, making available an enormous number of scientific papers, published in both chemical and biological journals.

Among these data, two kinds of stereoselective transformations of polyhydroxylated steroids will be discussed:

(1) The regio- and stereoselective oxidoreductions catalyzed by hydroxysteroid dehydrogenases (HSDH)
(2) The regioselective acylation catalyzed by lipases and proteases

An important parameter to be considered for these efficient enzymatic transformations is the reaction solvent. On this respect, different approaches, such as water, biphasic systems, and organic solvents, will be discussed.

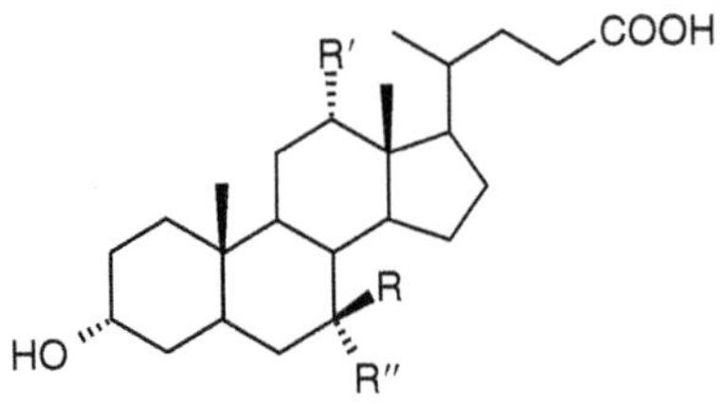

1   R = H; R'= R''= OH

2   R = R' = H; R'' = OH

3   R' = R'' = H; R = OH

FIGURE 23.1 Compounds 1–3

23.2   REGIO- AND STEREOSELECTIVE OXIDOREDUCTIONS CATALYZED BY HSDH

In nature, the regio- and stereoselective interconversions of secondary alcohols to keton
the steroid nucleus and side chain are catalyzed by HSDH, a group of NAD(P)-deper
oxidoreductases [8].

The synthetic exploitation of these enzymes was initially described with bile acid
These steroids are derivatives of 5β-cholan-24-oic acid and represent the major quantit
pathway by which cholesterol is metabolized in the human body. Cholic acid (**1**, Figure
and chenodeoxycholic acid (**2**) are the primary bile acids in human bile; the latter comp
and its 7β-hydroxy epimer, ursodeoxycholic acid (**3**), have important pharmaceutical a
cations related to their ability to solubilize cholesterol gallstones. Thanks to their carbc
group, bile acids are water-soluble at mild alkaline pH and therefore enzymatic oxidor
tions on these compounds are mainly performed in aqueous solutions. In these example
in situ regeneration of the nicotinamide cofactors was obtained by coupled enzymatic
tions that allowed the use of catalytic amounts of NAD(P)(H) [10].

As a first example, Figure 23.2 shows the regio- and stereospecific reductions of deh
cholic acid (**4**) with different HSDH [11,12]. The reduced products were obtained quantita
and, by using a suitable deuterated substrate donor for cofactor regeneration (i.e., 1-D-gl
and glucose dehydrogenase), it was also possible to synthesize selectively deuterated st
[13]. Similarly, quantitative regioselective oxidation of cholic acid (**1**) was also described [1
and, in a recent example, 12α-HSDH was used for the preparative-scale oxidation of G
complexes of ligands containing derivatives of 7,12-dihydroxy or 12-hydroxy chol
acids [15].

In a more general approach, the hydroxyl α/β inversion at position 3 and 7 of the s
skeleton was obtained in a two-step enzymatic process (Figure 23.3) [16–18].

HSDHs are also active on neutral steroids. To overcome the very low water solubi
these compounds, new and innovative (by those times) two-phase systems [19] were desc
in the late 1970s, the reaction conditions being optimized with a series of androstan
pregnane derivatives [20–23].

Figure 23.4 summarizes the industrial chemical route for the transformation of choli
(**1**) into the bioactive derivative ursodeoxycholic acid (**3**). Compound **5**, 12-chetochenoc
cholic acid, is a key intermediate and can be prepared enzymatically from **1** in just one-s
using 12α-HSDH [11,24,25]. Following optimization of the reaction conditions, compo

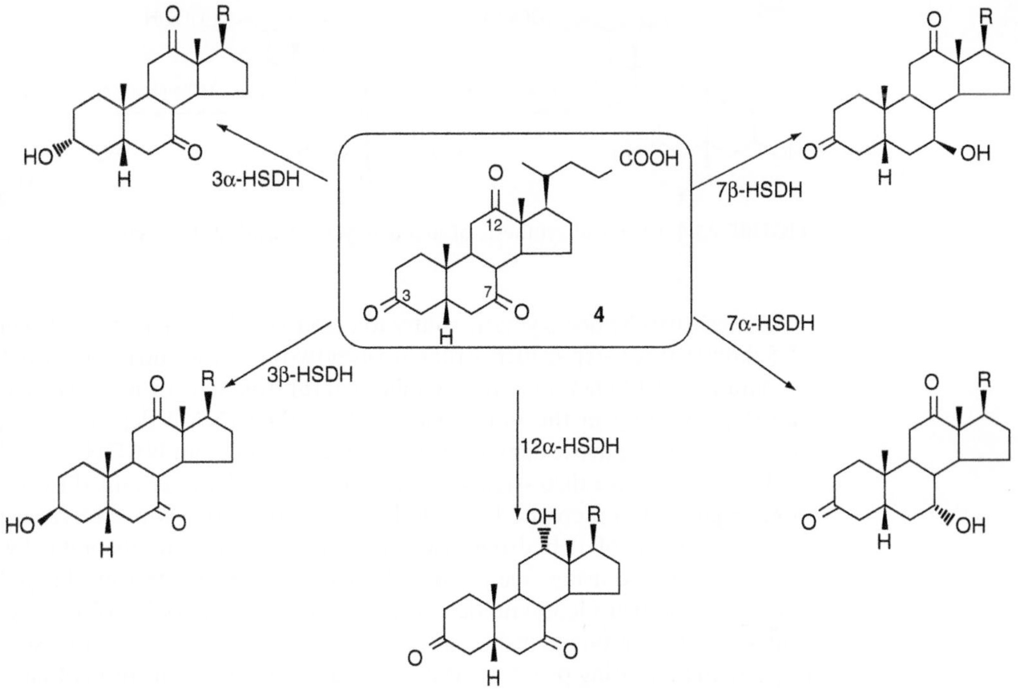

FIGURE 23.2 Regio- and stereospecific reduction of dehydrocholic acid (**4**) catalyzed by different hydroxystreoid dehydrogenases.

was obtained on a multikilogram scale in a membrane reactor containing an initial 4% w/v solution of cholic acid (**1**). Free 12α-HSDH and glutamate dehydrogenase, the enzyme used to regenerate NADP by reduction of α-ketoglutarate to glutamate, were retained inside the reactor and were reused for several cycles (up to 50) with satisfactory retention of their catalytic activity.

Recently, a different NADP-regeneration system, based on the use of alcohol dehydrogenases together with a large excess of acetone, has been discussed [26].

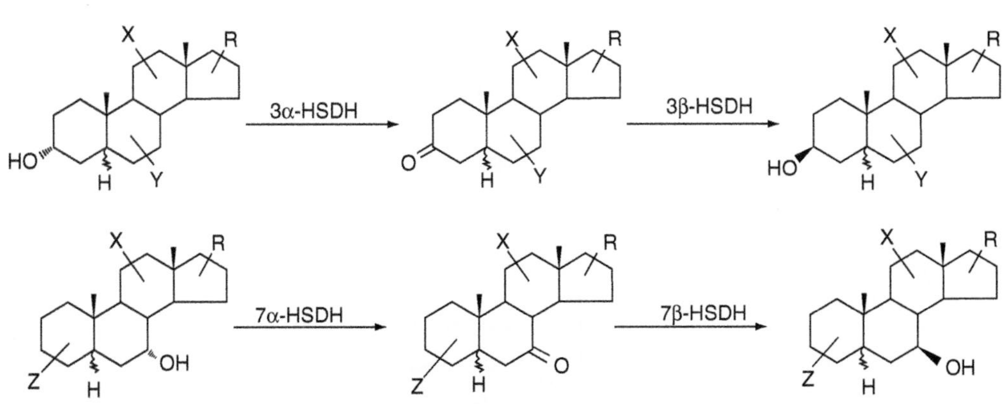

FIGURE 23.3 3α/3β-OH and 7α/7β-OH inversions catalyzed by hydroxysteroid dehydrogenases.

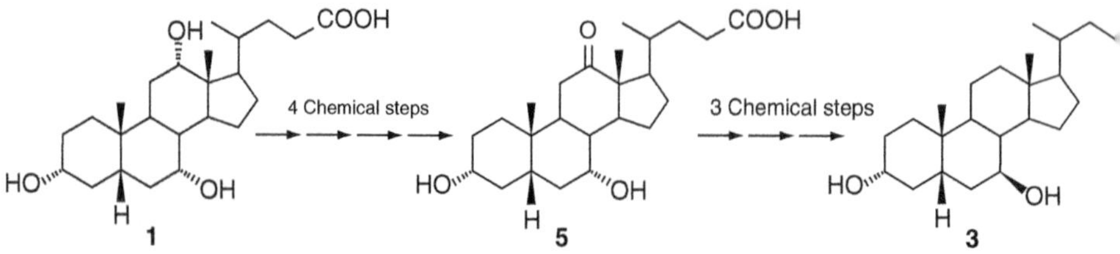

FIGURE 23.4 Chemical synthesis of ursodeoxycholic acid (3) from cholic acid (1).

Alternative chemoenzymatic routes to ursodeoxycholic acid were also investigated.. 23.5 shows a three-step sequence that involves two chemical and one enzymatic transform Preparation of 12-chetoursodeoxycholic acid (6) from dehydrocholic acid (4) was obta one step by using in the same reactors 3α-HSDH (NAD dependent), 7β-HSDH (dependent), and glucose dehydrogenase (to regenerate both NADH and NADPH) [27

In an alternative three-step sequence, compound 3 was obtained by a chemical re of 6, (Figure 23.6) prepared from cholic acid through two consecutive enzymatic step

The scale-up of dehydrogenase-catalyzed reactions can theoretically be perfor different reactors, using free or immobilized enzymes, neutral or charged membran native or modified nicotinamide cofactors. The general goal is obviously to recycle zymes—that are retained inside the reactor—and, if possible, the expensive coenzyme. respect, an interesting process was developed by the immobilization of native dehydrog

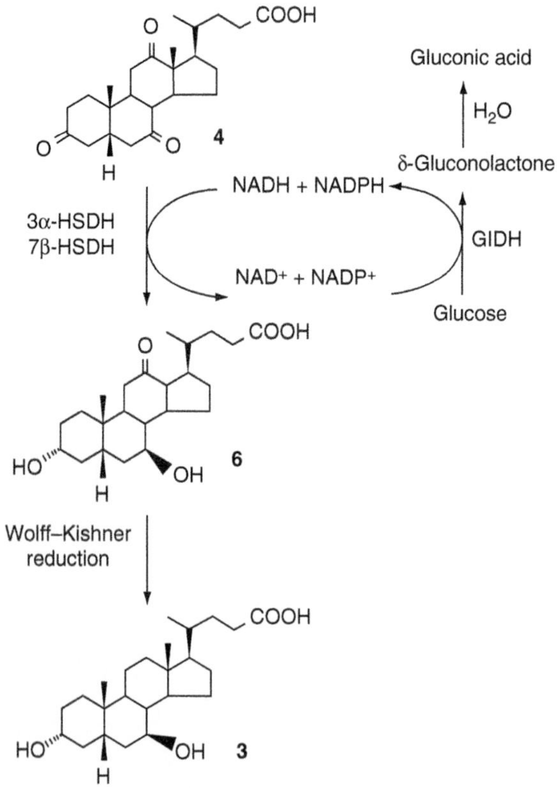

FIGURE 23.5 Proposed three-step chemoenzymatic route to ursodeoxycholic acid (3).

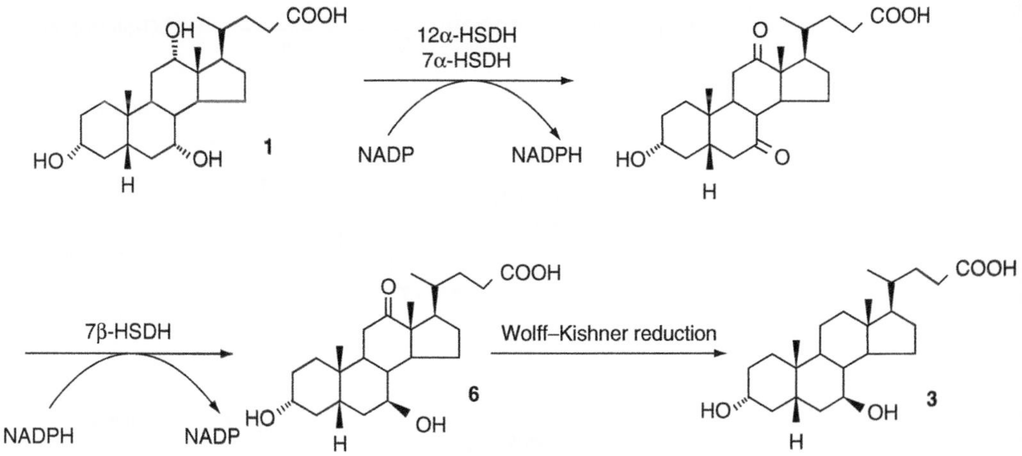

FIGURE 23.6 Alternative three-step chemoenzymatic route to ursodeoxycholic acid (**3**).

into isoelectric traps, formed by pairs of isoelectric membranes encompassing their pI values. The enzyme (3β-HSDH) was forced to perform its catalytic activity in an electric field coupled orthogonally to a hydraulic flow, which allowed the continuous transportation and harvesting of the charged bile acid product, by voltage gradient, into neighboring chambers [29].

The analytical exploitation of HSDHs for the quantitative enzymatic determination of neutral steroids and bile acid derivatives in serum or other physiological fluids has been reported [30–33]. Evaluation of steroid concentration was performed by spectrophotometric, fluorometric, and bioluminescence assays. The high sensitivity of the latter method allowed the detection of as little as picomoles of steroids using enzymes coimmobilized on nylon tubes [33].

23.3 REGIOSELECTIVE MODIFICATION OF POLYHYDROXYLATED STEROIDS CATALYZED BY LIPASES AND PROTEASES

Enzyme catalysis in organic solvents is a well-assessed methodology [34,35]. Specifically, when suspended in organic solvents hydrolytic enzymes, i.e., lipases and proteases, can efficiently catalyze stereoselective transesterification, esterification, aminolysis, acyl exchange, and thio-transesterification reactions [36], transformations that in water or biphasic systems are suppressed by the "natural" hydrolytic activity and therefore, do not occur to appreciable extent.

Lipase- and protease-catalyzed regioselective esterifications of sugars were initially achieved [37,38], but the methodology was soon extended to other polyhydroxylated natural compounds, including hydroxysteroids.

In a first report [39], two hydrolases—the lipase from *Chromobacterium viscosum* (*Chv*) and the protease subtilisin—were found to acylate the model dihydroxysteroid 5α-androstane-3β,17β-diol (**7**) in acetone with opposite regioselectivity; while *Chv* lipase reacted exclusively with the C-3 OH, subtilisin displayed a preference for the C-17 OH (Figure 23.7, the chemical reactivities of C-3 OH and C-17 OH were comparable). The selectivity of these two enzymes was examined with other hydroxysteroids and the results are reported in Table 23.1. Only C-3 OHs in the equatorial (β) position of steroids having a *trans* A/B ring fusion (or a C-5 to C-6 double bond) were acylated by *Chv* lipase, while changes in the A or B rings did not dramatically affect the reactivity of the steroid with subtilisin.

Later, two other lipases were found to be able to acylate hydroxysteroids, the lipase from *Candida cylindracea* (*Cac*, subsequently identified as *C. rugosa* lipase, *Car*, Figure 23.7) [40]

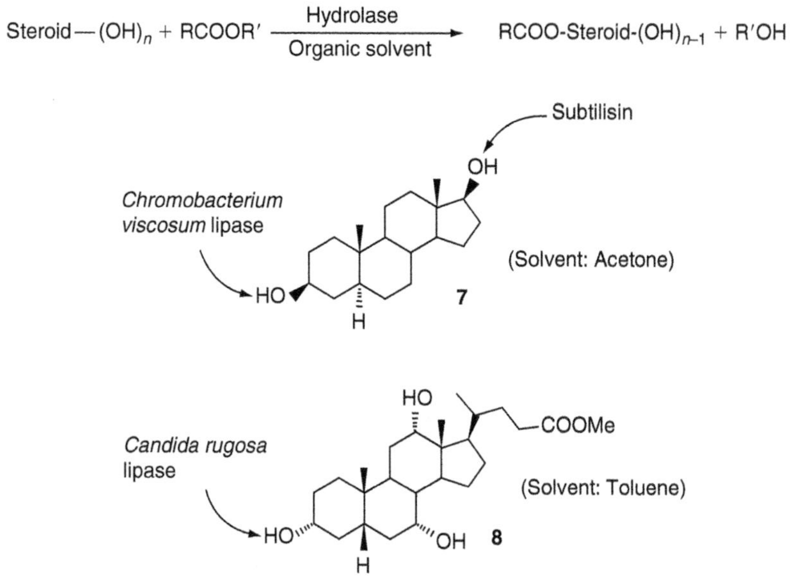

$$\text{Steroid} - (\text{OH})_n + \text{RCOOR}' \xrightarrow[\text{Organic solvent}]{\text{Hydrolase}} \text{RCOO-Steroid-}(\text{OH})_{n-1} + \text{R}'\text{OH}$$

FIGURE 23.7 Regioselective lipase-catalyzed acylation of polyhydroxylated steroids.

and the lipase B from *C. antarctica* (*CaaB*, also known as Novozym 435) [41], and
selectivities were evaluated. The former enzyme was effective in the esterification of C-3
of bile acid methyl esters (*cis* A/B ring fusion, i.e., **8**), while the latter one was active be
C-3 OHs and on side chain OHs. These reports, as well as further data published re
indicate that no enzymes are able to acylate hydroxyl groups located on the inner B
rings, while esterification of OHs on the A and D rings, as well as on the side chain,
obtained by a proper choice of the right hydrolase.

TABLE 23.1

Reactivities of Various Hydroxysteroids in the Acylation Catalyzed by
***Chromobacterium viscosum* Subtilisin and Lipase Suspended in Acetone**

Steroid	Initial Rate (μmol/h)	
	Subtilisin	Lipas
3β,17β-Dihydroxy-5α-androstane	0.63	3.30
3α,17β-Dihydroxy-5α-androstane	0.53	0
3β,17β-Dihydroxy-5β-androstane	0.41	0
3α,17β-Dihydroxy-5β-androstane	0.63	0
3α,17β-Dihydroxy-5-androstene	0.67	1.72
3α,17β-Dihydroxy-4-androstene	0.32	0
17β-Estradiol	0.63	0
3β-Hydroxy-5α-pregnane	0.06	6.26
3β-Hydroxy-5α-cholanic acid methyl ester	0	8.32
3β-Hydroxy-5-cholenic acid methyl ester	0	2.75
3β-Hydroxy-5-cholestene (cholesterol)	0	3.00
3β,20β-Dihydroxy-5α-pregnane	0.10	4.52
3β,20α-Dihydroxy-5α-pregnane	0.55	4.34

23.3.1   REGIOSELECTIVE ESTERIFICATION OF HYDROXYL GROUPS LOCATED ON THE "A" RING

Monoacylated derivatives of a complete set of 2,3- and 3,4-vicinal diols of steroids were recently prepared by regioselective lipase-catalyzed transesterifications [42]. As expected, different lipases (*CaaB, Car, Chv*, along with the lipase PS from a *Pseudomonas* strain) displayed different selectivities toward the vicinal diols depending on the OHs configuration. The results are shown in Figure 23.8, which shows that the acylation mainly took place at the C-3 position (as expected), except by lipase PS (esterification of C-4α OH in **14** and of C-2β OH in **15**).

In another paper, the same authors obtained stereoisomerically pure 3β-hydroxy-5,6-epoxysteroids by combining selective chemical methods for α- and β-epoxidation of Δ^5-unsaturated steroids with enzymatic stereoselective esterification of their 3β-OHs. Specifically, 3β-OH-5β,6β-epoxysteroids were acylated by *CaaB* and by the lipase AK (from a *Pseudomonas* strain), while 3β-OH-5α,6α-epoxysteroids were good substrates for *Car* (Figure 23.9) [43].

A similar approach was used by Santaniello and coworkers for the chemoenzymatic synthesis of the 3-hydroxy metabolites of the estrogen tibolone (**22**, Figure 23.10): *CaaB* catalyzed the selective acetylation of the 3β-OH epimer of **22** [44]. The peculiar regioselectivity of this lipase was also exploited by the same authors in an elegant preparation of oxandrolone (**23**), an anabolic synthetic hormone that was capable of improving the quality of life for patients with HIV-infections [45].

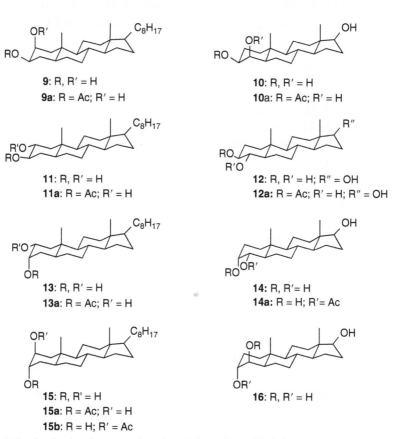

9: R, R′ = H
9a: R = Ac; R′ = H

10: R, R′ = H
10a: R = Ac; R′ = H

11: R, R′ = H
11a: R = Ac; R′ = H

12: R, R′ = H; R″ = OH
12a: R = Ac; R′ = H; R″ = OH

13: R, R′ = H
13a: R = Ac; R′ = H

14: R, R′= H
14a: R = H; R′= Ac

15: R, R′ = H
15a: R = Ac; R′ = H
15b: R = H; R′ = Ac

16: R, R′ = H

FIGURE 23.8 Regioselective lipase-catalyzed acylation of steroid diols.

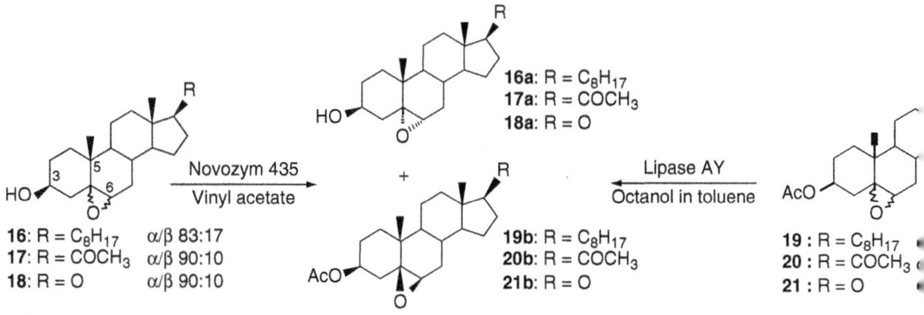

FIGURE 23.9 *CaaB* lipase-catalyzed stereoselective esterification of 3β-acetoxy-5,6-epoxyster Car lipase-catalyzed stereoselective alcoholysis of 3β-acetoxy-5,6-epoxysteroids.

Regioselective esterification of a series of ecdysteroids (like 20R-hydroxyecdisone, reported in the late 1990s [46]; despite the numerous OH groups present on these mo *CaaB* directed its action exclusively on the C-2 OH.

The lipase-catalyzed regioselective preparation of 3-hemisuccinates of polyhydro steroids [47] and of biologically active esters of dehydroepiandrosterone deserve als reported [48].

23.3.2 REGIOSELECTIVE ESTERIFICATION OF HYDROXYL GROUPS LOCATED ON THE "D" RING OR ON THE SIDE CHAIN

As discussed previously, the most interesting results reported in the first paper enzymatic acylation of steroids [39] was the complementary regioselectivity of th hydrolases, with a clear indication of the preference of the protease subtilisin for t located on the D ring or on the side chain. Additionally, the data reported in the last tv of Table 23.1 show that this protease was also sensitive to the stereochemistry nucleophile, such as in cases 20-OH epimers belonging to the pregnane family (t epimer being the preferred one).

Lipases do not acylate hydroxyl moieties on the D ring, but can act on side chain. For instance, a primary C-21 OH of a pregnene derivative was an excellent st for *CaaB* [41], and the same enzyme also showed a stereopreference for C-2(the 20β-epimer was acylated while the 20α-OH was not (a result complementary to obtained with subtilisin. It is known that lipases and proteases usually wor opposite enantioselectivity).

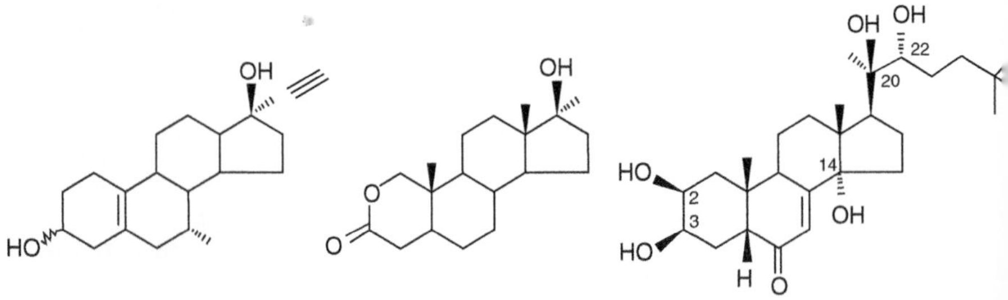

FIGURE 23.10 Compounds 22–24.

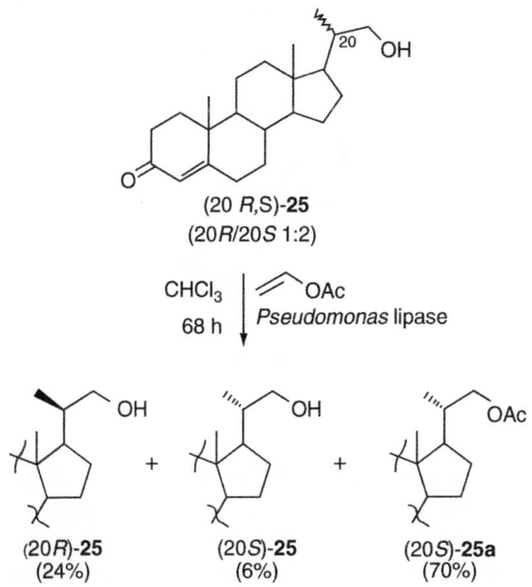

FIGURE 23.11 Stereoselective acylation of **22**.

In addition to the previously reported scant data, acylation of hydroxyl groups in steroid side chains was investigated in detail by Santaniello and coworkers and their results are reported in a series of papers [49–53]. For instance, Figure 23.11 shows the stereoselective acylation of the (20S)-isomer of the C-22 OH in compound **25** by action of a lipase from *Pseudomonas* [49].

Recently, the steroidal cyanohydrin derivative **26** could be isolated with a 89% d.e. through the subtilisin-catalyzed acylation of an epimeric mixture of the parent cyanohydrin [54].

23.3.3 Regioselective Enzymatic Alcoholysis of Polyacylated Hydroxysteroids

Hydrolases regioselectivity can also be exploited to get the mild removal of a protecting ester moiety, thus avoiding the negative drawbacks of classical chemical saponifications: lack of selectivity and promotion of undesirable side reactions. The application of this methodology to steroids was pioneered by Njar and Caspi in the late 1980s [55]. They showed that *Car* lipase suspended in isopropyl ether or in acetonitrile in the presence of *n*-octanol catalyzed the alcoholysis of different diacetylated steroids to give octyl acetate and monoacetylated steroids.

These transformations were later investigated in more detail by Baldessari and coworkers [56–58]. In a first report, they screened several lipases and proteases for the mild alcoholysis of the 3β-acetoxy derivatives **27** (a model compound) and **28**, evaluating the influence of the nature of the organic solvent and of the alcohol on the efficiency of the reaction. *Car* and *CaaB* lipases were the only active catalysts, toluene and acetonitrile were the best solvents, respectively, and octanol was the best nucleophile with both enzymes [56]. Under the best conditions the diol **28a** was isolated in 68% yields, without altering the highly sensitive 20β-ketol system that might easily undergo cleavage to give the 20-ketopregnene derivative (Figure 23.12).

The performances of the same enzymes were evaluated with the di- and triacetylated derivatives **29–34**, and the results were quite surprising [57]. While *Car* lipase gave the expected 3β-OH derivatives **28a–34a**, the lipase from *C. antarctica* directed its action to the

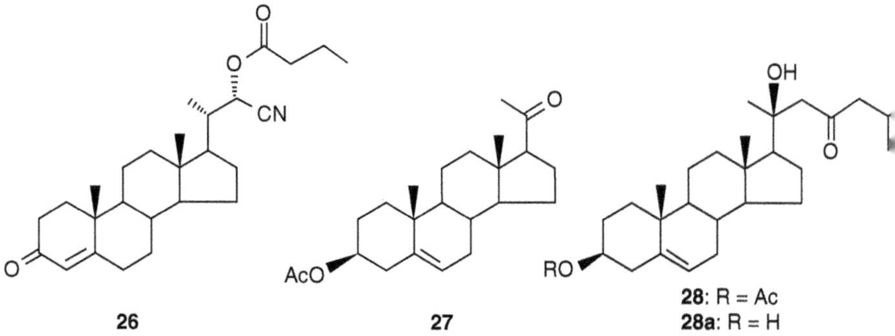

FIGURE 23.12 Compounds **26–28**.

16β-OAc of compounds **28–30** and to the 17-OAc of **31**, to give the corresponding acetates **29b–31b**. Compounds **32–34** were poor substrates for the same enzyme, an these compounds its action was directed toward the 3β-OAc. A similar behavior of the enzymes was observed with another series of androstane and pregnane polyace derivatives (i.e., **35**), and in the publication, additional information on the regiosele of the lipase from *Pseudomonas* on these new substrates was reported (Figure 23.13) [

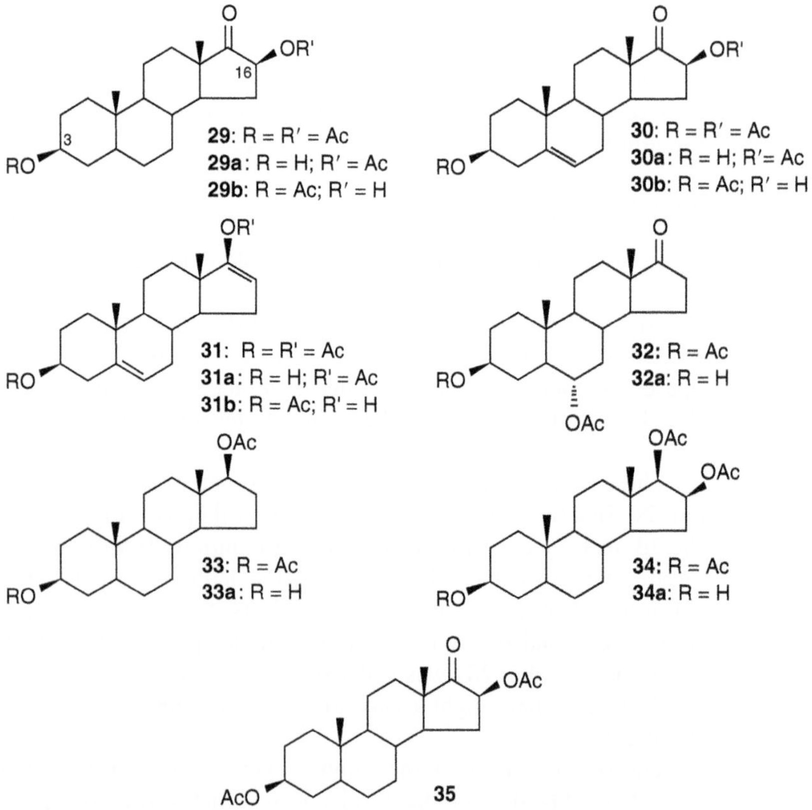

FIGURE 23.13 Compounds **29–35**.

Synthetic exploitation of enzymatic alcoholysis reactions was also described in two of the previously cited reports, for the preparation of oxandrolone [45] and diastereomeric 3β-5,6-epoxysteroids [43].

23.4 CONCLUSION

It has been shown that oxidoreductases and hydrolases are efficient catalysts for the selective modification of polyfunctionalized steroids. Needless to say, their action can be combined to increase the number of available derivatives. This point has been well-exemplified by Secundo and coworkers, who obtained a library of cholic acid methyl ester derivatives by alternating a stereo- and regioselective oxidative step, catalyzed by HSDH, and an acylation step with a series of different acyl donors (Figure 23.14), catalyzed by *C. antarctica* lipase B [59]. In the cited example, only about one third of the enzymatic reactions tested were productive, nevertheless a 39-member collection of bile acid derivatives (some of which are reported in Figure 23.15) was obtained in high purity and in good yields.

In conclusion, even when focusing on a small group of natural compounds (in this specific case polyhydroxylated steroids), it is possible to highlight the enormous opportunities that biocatalysis offers to organic chemists. Specifically, it has been shown that the highly

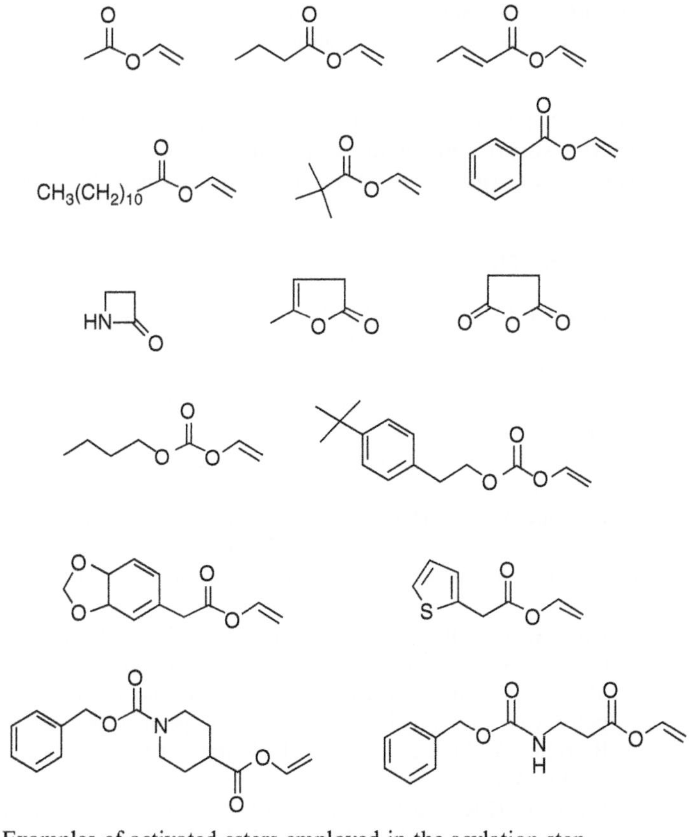

FIGURE 23.14 Examples of activated esters employed in the acylation step.

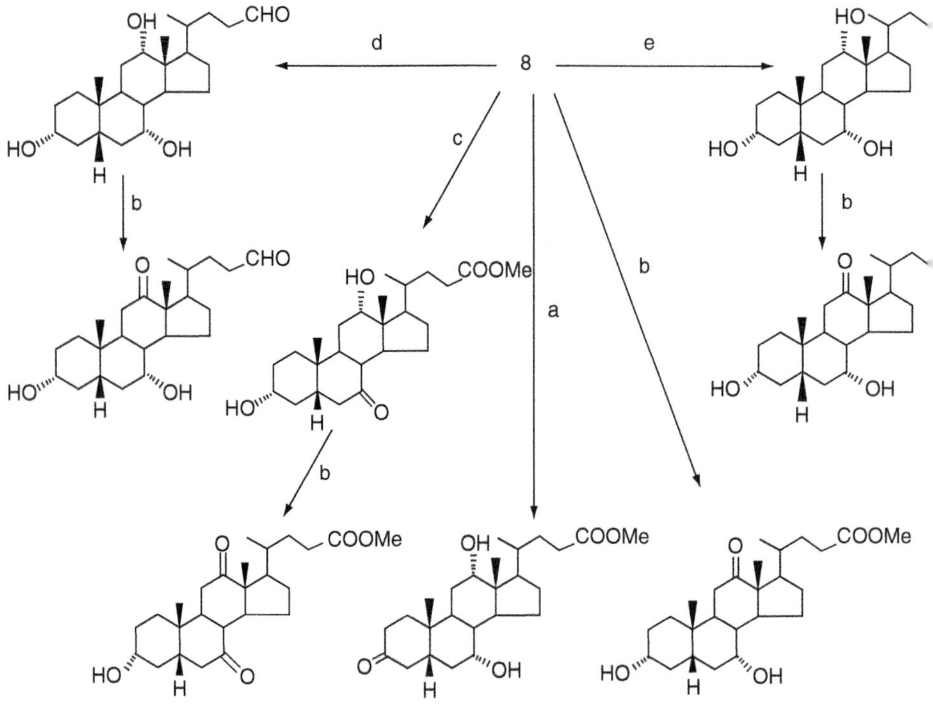

FIGURE 23.15 Example of combinatorial biocatalysis on polyhydroxylated steroids. (a–c) catalyzed oxidations; (d–e) chemical reductions.

substrate-specific enzymes HSDH are very close to industrial-scale exploitation and it p out the surprisingly wide, apparently never-ending, synthetic solutions that lipas proteases in organic solvents might offer to scientific investigators.

REFERENCES

1. Fieser, L.F. and Fieser, M., Eds., *Steroids*, Reinhold, New York, 1959.
2. Djerassi, C., Ed., *Steroid Reactions, An Outline for Organic Chemists*, Holden-Day, San Fr CA, 1963.
3. Fried, J. and Edwards, J.A., Eds., *Organic Reactions in Steroid Chemistry*, Reinhold, Ne 1972.
4. Blickenstaff, R.T., Ghosh, A.C., and Wolf, G.C., Eds., *Total Synthesis of Steroids*, Academ New York, 1974.
5. Charney, W. and Herzog, H.L., *Microbial Transformation of Steroids*, Charney, W. and H.L., Eds., Academic Press, New York, 1967.
6. Iisuka, K. and Naito, A., Eds., *Microbial Conversion of Steroids and Alkaloids*, Springer, Ne 1981.
7. Riva, S., Enzymatic modification of steroids, in *Applied Biocatalysis*, vol. 1, Blanch, H Clark, D.S., Eds., Marcel Dekker, New York, pp. 179–220, 1991.
8. Maser, E., Xenobiotic carbonyl reduction and physiological steroid oxidoreduction: th potency of several hydroxysteroid dehydrogenases, *Biochem. Pharmacol.*, 49(4), 421–440, 1
9. Bortolini, A., Medici, A., and Poli, S., Biotransformations on steroid nucleus of bile acids, 62, 564–577, 1997.
10. Chenault, H.K. and Whitesides, G.M., Regeneration of nicotinamide cofactors for use in synthesis, *Appl. Biochem. Biotechnol.*, 14, 147–197, 1987.

11. Riva, S., Bovara, R., Pasta, P., et al., Preparative-scale regio- and stereospecific oxidoreduction of cholic acid and dehydrocholic acid catalyzed by hydroxysteroid dehydrogenase, *J. Org. Chem.*, 51, 2902–2906, 1986.

12. Carrea, G., Bovara, R., Longhi, R., et al., Enzymatic reduction of dehydrocholic acid to 12-keto-chenodeoxycholic acid with NADH regeneration, *Enzyme Microb. Technol.*, 6, 307–311, 1984.

13. Riva, S., Ottolina, G., Carrea, G., et al., Efficient preparative-scale enzymatic synthesis of specifically deuterated bile acids, *J. Chem. Soc. Perkin Trans. I*, 2073–2074, 1989.

14. Bianchini, E., Chinaglia, N., Dean, M., et al., Regiospecific oxidoreductions catalyzed by a new *Pseudomonas paucimobilis* hydroxysteroid dehydrogenase, *Tetrahedron*, 55, 1391–1398, 1999.

15. Anelli, P.L., Brocchetta, M., Morosini, P., et al., Conjugates of Gd(III) complexes to di- and tri-hydroxy substituted cholanoic acids: regioselective oxidation with hydroxysteroid dehydrogenases, *Biocatalysis Biotrans.*, 20, 29–34, 2002.

16. Riva, S., Bovara, R., Zetta, L., et al., Enzymatic α/β inversion of C-3 OH of bile acids and study of the effects of organic solvents on reaction rates, *J. Org. Chem.*, 53, 88–92, 1988.

17. Bovara, R., Canzi, E., Carrea, G., et al., Enzymatic α/β inversion of the C-7-hydroxyl of steroids, *J. Org. Chem.*, 58, 499–501, 1993.

18. Medici, A., Pedrini, P., Bianchini, E., et al., 7α-Epimerization of bile acids via oxido-reduction with *Xanthomonas maltophilia, Steroids*, 67, 51–56, 2002.

19. Carrea, G., Biocatalysis in water–organic solvent two-phase systems, *Trends Biotech.*, 2(4), 102–106, 1984.

20. Carrea, G. and Cremonesi, P., Enzyme catalyzed transformations in water–organic solvent two-phase systems, in *Methods in Enzymology*, vol. 136, Mosbach, K., Ed., Academic Press, New York, pp. 150–157, 1987.

21. Cremonesi, P., Carrea, G., Sportoletti, G., et al., Enzymatic dehydrogenation of steroids by β-hydroxysteroid dehydrogenase in a two-phase system, *Arch. Biochem. Biophys.*, 159, 7–10, 1973.

22. Cremonesi, P., Carrea, G., Ferrara, L., et al., Enzymatic preparation of 20β-hydroxysteroids in a two-phase system, *Biotechnol. Bioeng.*, 17, 1101–1108, 1975.

23. Carrea, G., Riva, S., Bovara, R., et al., Enzymatic oxidoreduction of steroids in two-phase systems: effects of organic solvents on enzyme kinetics and evaluation of the performance of different reactors, *Enzyme Microb. Technol.*, 11, 333–340, 1988.

24. Carrea, G., Bovara, R., Cremonesi, P., et al., Enzymatic preparation of 12-ketochenodeoxycholic acid with NADP regeneration, *Biotechnol. Bioeng.*, 26, 560–563, 1984.

25. Carrea, G., Bovara, R., Longhi, R., et al., Preparation of 12-ketochenodeoxycholic acid using coimmobilized 12α-hydroxysteroid dehydrogenase and glutamate dehydrogenase with NADP cycling at high efficiency, *Enzyme Microb. Technol.*, 7, 597–600, 1985.

26. Fossati, E., Polentini, F., Carrea, G., et al., Exploitation of the alcohol dehydrogenase-acetone NADP regeneration system for the enzymatic preparative-scale production of 12-ketochenodeoxycholic acid,. *Biotechnol. Bioeng.*, 93, 1216–1220, 2006.

27. Carrea, G., Pilotti, A., Riva, S., et al., Enzymatic synthesis of 12-ketoursodeoxycholic acid from dehydrocholic acid in a membrane reactor, *Biotechnol. Lett.*, 14, 1131–1135, 1992.

28. Bovara, R., Carrea, G., Riva, S., et al., A new enzymatic route to the synthesis of 12-ketoursodeoxycholic acid, *Biotechnol. Lett.*, 18, 305–308, 1996.

29. Chiari, M., Dell'Orto, N., Mendozza, M., et al., Enzyme reactions in a multicompartment electrolyzer with isoelectronically trapped enzymes, *J. Biochem. Biophys. Methods*, 31, 93–104, 1996.

30. Mashige, F., Kazuhiro, I., and Osuga, T., A simple and sensitive assay of total serum bile acids, *Clin. Chim. Acta*, 70, 79–86, 1976.

31. Mashige, F., Tanaka, N., Maki, A., et al., Direct spectrophotometry of total bile acids in serum, *Clin. Chem.*, 27, 1352–1356, 1981.

32. Styrelius, I., Thore, A., and Bjorkhem, I., Bioluminescent assay for total bile acids in serum with use of bacterial luciferase, *Clin. Chem.*, 29, 1123–1127, 1983.

33. Roda, A., Girotti, S., Carrea, G., et al., Continuous-flow assays with nylon tube-immobilized bioluminescent enzymes, in *Methods in Enzymology*, vol. 137, Mosbach, K., Ed., Academic Press, New York, pp. 161–171, 1988.

34. Klibanov, A.M., Improving enzymes by using them in organic solvents, *Nature*, 409, 241–246, 2001.

35. Carrea, G. and Riva, S., Properties and synthetic applications of enzymes in organic so *Angew. Chem. Int. Ed. Engl.*, 33, 2226–2254, 2000.

36. Zaks, A. and Klibanov, A.M., Enzyme catalyzed processes in organic solvents, *Proc. Natl. Sci. USA*, 82, 3192–3196, 1985.

37. Therisod, M. and Klibanov, A.M., Facile enzymatic preparation of monoacylated sugars i dine, *J. Am. Chem. Soc.*, 108, 5638–5640, 1986.

38. Riva, S., Chopineau, J., Kieboom, A.P.G., et al., Protease-catalyzed regioselective esterification of and related compounds in anhydrous dimethylformamide, *J. Am. Chem. Soc.*, 110, 584–589, 198

39. Riva, S. and Klibanov, A.M., Enzymo-chemical regioselective oxidation of steroids without eductases, *J. Am. Chem. Soc.*, 110, 3291–3295, 1988.

40. Riva, S., Bovara, R., Ottolina, G., et al., Regioselective acylation of bile acids derivative *Candida cylindracea* lipase in organic solvents, *J. Org. Chem.*, 54, 3161–3164, 1989.

41. Bertinotti, A., Carrea, G., Ottolina, G., et al., Regioselective esterification of polyhydro: steroids by *Candida antarctica* lipase, *Tetrahedron*, 50, 13165–13172, 1994.

42. Cruz, S.M.M., Riva, S., and Sa e Melo, M.L., Regioselective enzymatic acylation of vicinal steroids, *Tetrahedron*, 61, 3065–3073, 2005.

43. Cruz, S.M.M., Riva, S., and Sa e Melo, M.L., Highly selective lipase-mediated discrimina diasteromeric 5,6-epoxysteroids, *Tetrahedron Asymmetry*, 15, 1173–1179, 2004.

44. Ferraboschi, P., Colombo, D., and Reza-Elahi, S., A practical chemoenzymatic approach synthesis of 3-hydroxy metabolites of tibolone, *Tetrahedron Asymmetry*, 13, 2583–2586, 200

45. Ferraboschi, P., Colombo, D., and Prestileo, P., A convenient synthesis of oxandrolone thr regioselective *Candida antarctica* lipase-catalyzed transformation, *Tetrahedron Asymmet* 2781–2785, 2003.

46. Danieli, B., Lesma, G., Luisetti, M., et al., *Candida antarctica* lipase B catalyzes the regios esterification of ecdysteroids at the C-2 OH, *Tetrahedron*, 53, 5855–5862, 1997.

47. Ottolina, G., Carrea, G., and Riva, S., Regioselective enzymatic preparation of hemisuccin polyhydroxylated steroids, *Biocatalysis*, 5, 131–136, 1991.

48. Bruttomesso, A.C., Tiscornia, A., and Baldessari, A., Lipase-catalyzed preparation of biolc active esters of dehydroepiandrosterone, *Biocatal. Biotransformation*, 22, 215–220, 2004.

49. Ferraboschi, P., Molatore, A., Verza, E., et al., The first example of lipase-catalyzed resoluti stereogenic center in steroid side chains by transesterification in organic solvent, *Tetra Asymmetry*, 6, 1551–1554, 1996.

50. Ferraboschi, P., Molatore, A., Verza, E., et al., The first example of lipase-catalyzed resoluti stereogenic center in steroid side chains by transesterification in organic solvent, *Addendum. hedron Asymmetry*, 6, 2160, 1996.

51. Ferraboschi, P., Reza-Elahi, S., Verza, E., et al., Lipase-catalyzed resolution of stereogenic in steroid side chains by transesterification in organic solvents: the case of a 26-hydroxychol *Tetrahedron Asymmetry*, 9, 2193–2196, 1998.

52. Ferraboschi, P., Pecora, F., Reza-Elahi, S., et al., Chemoenzymatic syntheses of (25*R*)- and 25-hydroxy-27-nor-cholesterol, a steroid bearing a secondary hydroxyl group in the side *Tetrahedron Asymmetry*, 10, 2497–2500, 1999.

53. Santaniello, E., Ferraboschi, P., and Reza-Elahi, S., Lipase-catalyzed regio- and stereos acylation of hydroxy groups in steroid side chains, *Mon. Chemie*, 131, 617–622, 2000.

54. Cruz, S.M.M., Sa e Melo, M.L., Parolin, M., et al., The biocatalyzed stereoselective prepara polycyclic cyanohydrins, *Tetrahedron Asymmetry*, 15, 21–27, 2004.

55. Njar, C.O. and Caspi, E., Enzymatic transesterification of steroid esters in organic so *Tetrahedron Lett.*, 28, 6549–6552, 1987.

56. Baldessari, A., Maier, M.S., and Gros, E.G., Enzymatic deacetylation of steroids bearing functions, *Tetrahedron Lett.*, 36, 4349–4352, 1995.

57. Baldessari, A., Bruttomesso, A.C., and Gross, E.G., Lipase-catalyzed regioselective deacyla androstane derivatives, *Helv. Chim. Acta*, 79, 999–1004, 1996.

58. Bruttomesso, A.C. and Baldessari, A., Lipase-catalysed deacetylation of androstane and pr derivatives: influence of ring D substitution, *J. Mol. Cat., B Enzym.*, 29, 149–153, 2004.

59. Secundo, F., Carrea, G., De Amici, M., et al., A combinatorial biocatalysis approach to an a cholic acid derivatives, *Biotechnol. Bioeng.*, 81, 391–396, 2003.

24 Recent Developments in Enzymatic Acyloin Condensations

Owen Ward and Ajay Singh

CONTENTS

24.1 INTRODUCTION

The production of (*R*)-phenylacetylcarbinol (*R*-PAC), as a precursor intermediate in the synthesis of L-ephedrine, was one of the first industrial biotransformations to be exploited. The reaction is catalyzed by fermenting baker's yeast, *Saccharomyces cerevisiae*, as a side reaction of pyruvate decarboxylase (PDC), a thiamine pyrophosphate (TPP)-linked α-ketoacid decarboxylases. *R*-PAC can be produced by chemical synthesis from cyanohydrins, but biotransformation represents the preferred industrial synthesis route. Other α-ketoacid decarboxylases include benzoylformate decarboxylase, phenylpyruvate decarboxylase, phosphonopyruvate decarboxylase, indole-3-pyruvate decarboxylase, and 3-sulfopyruvate decarboxylase. Bacterial and yeast decarboxylases have been evaluated as biocatalysts not only for the synthesis of acyloin-type compounds but also for the production of hydroxyl ketones [1], α-arylacetate [2], D-amino acids [3], dopamine [4], and organic acids [5]. Some

other TPP-dependent enzymes, which are not α-ketoacid decarboxylases, also have pot
roles in bioorganic synthesis.

Decarboxylases have various roles in nature: in the nonoxidative decarboxylation
and β-ketoacids and in the synthesis of carbohydrates [6]. Their application as biocataly
industrial biotransformation reactions derives from their ability to synthesize enanti
compounds or precursors with existing or potential applications in the pharmace
industry. These enzymes may be used in the form of whole cells or extracted enzym
both free and immobilized states. Use of biocatalysts, such as α-ketoacid decarboxyla
chemoorganic synthesis, avoids the problem of isomerization, racemization, epimeriz
and rearrangement, typically associated with organic synthesis reactions, as biotransf
tions are implemented under mild reaction conditions [7]. This chapter discusses
developments in the applications of enzymes in acyloin-type condensations.

24.2   R-PAC PRODUCTION

24.2.1   REACTION DESCRIPTION

The main reaction (Figure 24.1) in which PDC participates in metabolism is the nonoxi
decarboxylation of α-ketoacids to the corresponding aldehydes [8–10]. The enzyme i
vated by its natural substrate, pyruvate. The catalytic mechanism, involving the en
bound reaction intermediate, 2-α-hydroxyethylthiamine pyrophosphate (HETPP), ha
extensively investigated [11–13]. The ability of the TPP coenzyme to bind to the ca
group and act as an electron sink is what makes decarboxylation of the α-ketoacid p
[14]. Acetaldehyde inhibits the decarboxylation reaction, but has been reported

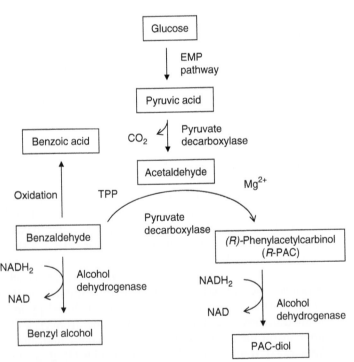

FIGURE 24.1 Pathway for the biosynthesis of (R)-phenylacetylcarbinol (R-PAC) and as
products.

inactivating the enzyme [9], and the inhibition decreases with increasing chain length of the substrate. PDCs (EC 4.1.1.1) have been characterized from many yeast, bacterial, and plant sources [15,16]. All appear to be made up of subunits consisting of around 560 to 610 amino acids, but with different quaternary structures. The enzyme from baker's yeast is a tetramer of similar subunits, while the brewer's yeast enzyme is an $\alpha 2\beta 2$ tetramer [8,17]. The holoenzyme contains as cofactors two to four molecules of TPP and magnesium. The structures of a number of PDCs have been fully elucidated [18–20]. With the exception of PDC from *Zymomonas mobilis*, all enzymes that have been characterized in detail exhibit a sigmoidal plot. Important microbial sources of decarboxylases are shown in Table 24.1.

The "carboligase" side reaction leads to the production of hydroxyl ketones or acyloins [14,21,22] and involves a two-site mechanism of formation of the acyloin compound by PDC. At the first site, pyruvate is decarboxylated to the aldehyde–diphosphatamine complex called "active acetaldehyde." Active acetaldehyde is irreversibly transferred to the second site, where reversible dissociation of free aldehyde may occur. However, in addition to forming the free aldehyde, PDC also catalyzes the formation of C–C bonds through the acyloin reaction in which free aldehyde competes with a proton for bond formation with the α-carbanion of HETPP. The addition of the C2 unit produces (*R*)-hydroxy ketone [23]. Acyclic and aromatic unsaturated aldehydes have been used in these acyloin condensations in addition to acetaldehyde.

24.2.2 STRAINS

A large number of microbes in addition to *S. cerevisiae* have been shown to produce *R*-PAC by biotransformation. Shukla and Kulkarni [24] noted the capacities of the yeasts *Hansenula*

TABLE 24.1
Microbial Sources of α-Ketoacid Decarboxylases

Enzyme	Microbial Sources		
	Bacteria	**Yeast**	**Filamentous Fungi**
Pyruvate decarboxylases (EC 4.1.1.1)	*Zymomonas mobilis, Zymobacter palmae, Acetobacter pasteurianus, Sarcina*	*Candida flareri, Kluyveromyces marxianus, Saccharomyces cerevisiae, S. carlsbergensis, S. ellipsoideus, S. delbrueckii*	*Neurospora crassa, Rhizopus javanicus, Aspergillus niger, A. parasiticus*
	Ventriculi	*Zygosaccharomyces rouxii, Hansenula anomala, Brettanomyces vini, Torula utilis, Torulaspora delbrueckii, Schizosaccharomyces pombe*	*A. nidulans, R. oryzae, Mucor rouxii Fusarium* sp., *M. circinelloides*
Benzoylformate decarboxylases (EC 4.1.1.7)	*Pseudomonas putida, P. aeruginosa, Acinetobacter calcoaceticus*		
Phenylpyruvate decarboxylases (EC 4.1.1.43)	*Achromobacter eurydice, A. calcoaceticus, P. putida, Thauera aromatica*	*S. cerevisiae*	

anomala, Brettanomyces vini, S. carlsbergensis, S. ellipsoideus, S. delbrueckii, Torul delbrueckii, and *Torula utilis* to produce significant amounts of *R*-PAC. Rosche et a evaluated 105 yeast strains from 10 genera (*Candida, Saccharomyces, Nadsonia, Kloe Kluyveromyces, Hansenula, Torulopsis, Pichia, Schizosaccharomyces,* and *Cryptococcu* 40 species for *R*-PAC production using a cell-free carboligase assay with benzaldehy(pyruvate as substrates, and detected activity in all but seven of the strains. Highest ac were found in strains of *Candida utilis, C. tropicalis,* and *C. albicans.* In other s *Torulopsis glabrata* was identified as a very high *R*-PAC producer [26]. PAC was not f from benzaldehyde and acetaldehyde by these yeasts. Most of these yeast strains we found [27] to have carboligase activity and were capable of fermenting glucose. Rosch [28] described a screening program of fungi, covering a broad taxonomic spectr Ascomycota, Zygomycota, and Basidiomycota, in which benzaldehyde and pyruvat transformed into *R*-PAC by all 16 strains tested. Species evaluated included *Rhizopus cus, R. oryzae, Aspergillus oryzae, A. tamari, Neurospora crassa, Polyporus eucalyp Fusarium lateritium,* another *Fusarium* sp., *Monilia sitophila, Paecilomyces lilacus,* and *rouxii.* In 12 of the fungal strains tested, enantiomeric excesses (ees) of *R*-PAC were similar to the values that were obtained with strains of *C. utilis* and *S. cerevisiae* whic tested in the same experiment. Highest *R*-PAC yields were observed with *F. javanic* another *Fusarium* sp.

Silk and Macaulay [29] observed that many metabolites, including PAC and PAC (tives, are produced by culture extracts of *Bjerkandera adjusta,* incubated with phenylal with or without cosubstrates such as glycerol, pyruvate, or glucose.

Some species such as *Saccharomyces, Candida, Hansenula, Rhizopus, Aspergillus, ium, Monilia, Paecilomyces,* and *Mucor* can synthesize *R*-PAC in the range of 4.5 to [28,30,31].

In contrast to the carboligase activity exhibited by yeasts and fungi, which cannot acetaldehyde as substrate in place of pyruvate, PDCs from the bacteria *Z. mobil Zymobacter palmae* can use acetaldehyde as substrate [32]. This has significance co cially, as acetaldehyde is considerably less expensive than pyruvate [33].

24.2.3 BIOTRANSFORMATIONS IN AQUEOUS MEDIA

The conventional commercial production of *R*-PAC consists of a yeast growth stage tating biomass production, accumulation of an intracellular pyruvate pool, and induc PDC, followed by a benzaldehyde-fed whole-cell biotransformation stage with benzal(concentration maintained below inhibiting levels. With *S. cerevisiae,* optimal concentra benzaldehyde for *R*-PAC production was indicated to be 10 mM [22]. The yeast who biotransformation process is limited by the denaturing effects of benzaldehyde on the talytic system, by insufficient accumulation of pyruvate, and by reduction in *R*-PAC yi a result of diversion of substrates into by-products, mediated by other yeast enzymes

Increased benzaldehyde concentrations decreased yeast cell viability [35] and ca cessation in *R*-PAC production. While cells grew at a reduced rate of 0.5 g/L benzald concentrations of 1 to 2 g/L completely inhibited growth and 3 g/L reduced cell vi: indicating that benzaldehyde concentrations for *R*-PAC production should be maintai or below 1 g/L. Higher benzaldehyde concentrations were found to damage yeast ce meability, with associated losses of cofactors, rather than by inactivating PDC activity. et al. [36] exploited a benzaldehyde feed strategy to accumulate *R*-PAC to a concentra 22 g/L. Other strategies involved development of aldehyde-resistant mutants. Addition cyclodextrin caused entrapment and controlled release of benzaldehyde, reduction of t toward the cells, and increased *R*-PAC yield [37]. Yields of more than 22 g/L *R*-PAC obtained using a wild-type strain of *C. utilis* through optimal control of metabolism by microprocessor-control of respiratory quotient to enhance pyruvate production and induction of PDC activity [38].

Quantitative conversion of benzaldehyde to *R*-PAC is never achieved due to the formation of by-products. Although a variety of by-products have been observed including acetyl benzoyl, *trans*-cinnamaldehyde, 2-hydroxy-1,1-phenylpropane, and benzyl alcohol, only the latter is produced in amounts substantial enough to influence the efficiency of *R*-PAC production [39–41]. Benzyl alcohol is formed from benzaldehyde in a reaction mediated by yeast alcohol dehydrogenase [42,43]. Ose and Hironaka [44] demonstrated that addition of acetaldehyde to the biotransformation increased and reduced formation of *R*-PAC and benzyl alcohol, respectively, and this may be attributed to the observation that alcohol dehydrogenase is more susceptible to denaturation by acetaldehyde than is PDC [35].

Miyata [26] addressed the problem of pyruvate limitation using a strain of *T. glabrata* and succeeded in producing a conversion to *R*-PAC of 70%, based on benzaldehyde, and volumetric yields of up to 30 g/L *R*-PAC.

Yeast cell immobilization reduced the toxic effect of benzaldehyde likely due to diffusion limitations within the immobilizing matrix [45–47]. By increasing benzaldehyde concentrations in the range of 2 to 6 g/L in an immobilized system, *R*-PAC concentrations were increased up to 7.5-fold the amount observed in free-cell controls. A high *R*-PAC concentration was produced by *S. cerevisiae* cells immobilized in an ENT-4000 matrix (containing polyethylene glycol). The nature of the support material also affected the ratio of product *R*-PAC to by-product benzyl alcohol [48]. Mahmoud et al. [47] exploited immobilized yeast cell systems to prolong semicontinuous transformation reactions and to reuse the cells through multiple [7–9] cycles. Shukla and Kulkarni [49] used *T. delbrueckii*, immobilized in alginate gels, for *R*-PAC synthesis and achieved superior yields with barium alginate compared with calcium alginate.

However, immobilized cell systems also limit rates of transformation, with diffusion limitations of substrates into the matrix and products from the matrix likely being the dominant causative factor responsible for the low transformation rates.

24.2.4 REACTION SPECIFICITY

PDC has been shown to be able to decarboxylate not only pyruvate but also hydroxyl pyruvate, acetaldehyde, aliphatic α-ketoacids, *p*-substituted α-ketoacids, and a number of higher homologs of pyruvate up to 2-oxohexanoate [17,50]. *S. cerevisiae* was found to be able to produce acyloin condensations of pyruvate by replacing benzaldehyde with substituted aromatic aldehydes, although acyloin yields were lower than those observed for benzaldehyde as substrate [42]. Aldehydes with substituents in the *ortho*-position were demonstrated consistently to be poor substrates for carbinol production. Aromatic aldehydes with –CH3, –CF3, and –Cl substituents located in the *para*-position produced higher carbinol yields than their *meta*-counterparts.

Acetohydroxyacid synthase, whose normal physiological function is the synthesis of (*S*)-acetohydroxyacids from pyruvate and a second ketoacid, also has the ability to catalyze chiral synthesis of *R*-PAC from pyruvate and benzaldehyde [51]. Some *Escherichia coli* acetohydroxyacid synthase isoenzymes, especially isoenzyme I, have advantages over PDC in this biotransformation in that high conversion yields of *R*-PAC are obtained and there is negligible acetaldehyde formation. Acetohydroxyacid synthase appears to be able to convert pyruvate and a range of substituted benzaldehydes and other aromatic aldehydes into the corresponding PAC analogs. In addition, pyridine, thiophene, furan, and naphthyl aldehydes are acceptable substrates in the biotransformation.

Culture extracts of *B. adjusta*, produce PAC from phenylalanine, with and wi
cosubstrate, and generate PACs substituted differently at the aromatic and/or aliphati
depending on the nature of the cosubstrate used in the reaction [29]. When PAC
incubated as substrate with the culture extracts, eight different biotransformation pro
were observed, indicating the potential to use enzymes from this strain to produce subst
PAC products. Many of the reactions described are stereoselective.

24.2.5 Aqueous–Organic Systems

Nikolova and Ward [52,53] investigated *R*-PAC production by yeast in a two-phase aqu
organic biotransformation system. When the effect of moisture content on *R*-PAC pr
tion by yeast cells, immobilized to celite, was investigated using hexane as organic solv
maximum biotransformation activity was observed at a moisture level of 10%. Better
formation activities were observed in hydrophobic solvents like hexane and hexac
compared to more hydrophilic solvents like chloroform, ethyl-, and butyl acetate.
latter solvents also caused puncturing of the yeast cells. Laurence at al. [54] used lacti
as cosubstrate and a pyruvate precursor with benzaldehyde in yeast *R*-PAC biotra
mations in petroleum spirit, but rates of transformation were inferior to those ob
with pyruvate.

With monophasic biotransformation systems, containing water-miscible organic so
for *R*-PAC production using aldehyde-resistant yeast mutants, *R*-PAC concentrations
to 15 g/L have been observed, with associated reduced production of benzyl alcoh
When the yeast-mediated condensation of benzaldehyde and pyruvate was carried ov
petroleum spirit solvent system, addition of a small amount of ethanol (0.5%) was fo
inhibit benzyl alcohol production and to increase *R*-PAC yield [55]. Solvents such as et
other monohydric solvents, and long- and short-chain polyols, such as ethylene glyc
glycerol, appeared to stimulate *R*-PAC production by immobilized yeast cell mass [56

24.2.6 Use of Mutants

Mutant strains of *S. cerevisiae* and *C. flareri*, resistant to acetaldehyde and *R*-PAC, pr
more of the condensation product than the wild-type parent strains, but not more than
[57]. Mutant strains of *S. cerevisiae*, lacking some or all of the alcohol dehydro
isoenzymes I, II, and III, manifested similar rates of conversion of benzaldehyde
undesirable by-product, benzyl alcohol [58].

Site-directed mutagenesis and molecular approaches have recently been used to
understand enzyme structure–function relationships of biocatalytic effects in the PA
transformation. Site-directed mutagenesis, replacing a Trp392 with Ala, enhanced th
ally low carboligase activity of *Z. mobilis* PDC, but enzyme conformational stabil
reduced [59]. Replacement of Trp with Ile or Met increased biotransformation activity
six times) as well as conformational stability. A number of other hydrophobic amino ac
specific mutations in or around the active site (Leu112, Ile472, and Ile476) of the *Z.*
PDC have been probed with respect to their effects on relative decarboxylation/carbo
activity, substrate specificity, and enantiospecificity [15].

Acetohydroxyacid synthase from *E. coli* can catalyze the stereospecific conver
pyruvate and benzaldehyde to *R*-PAC. Studies on the effects of mutation of four d
amino acids at the active site on biotransformation reaction rates and substrate speci
the ligation reaction demonstrated the crucial role of Arg276 in stabilizing enzyr
substrate transition states for ligation of the incoming second substrate [60]. These
demonstrated the potential to engineer this particular enzyme to enable it to catalyz
reactions of possible industrial utility.

24.2.7 USE OF PURIFIED ENZYMES

Because use of purified PDC enzymes avoids the problem of by-product formation due to the presence of alcohol dehydrogenases, purified PDCs from *S. cerevisiae*, *C. utilis*, and *Z. mobilis* have been evaluated for synthesis of *R*-PAC (Figure 24.2) [56,61]. Both *C. utilis* and *S. cerevisiae* enzymes were effective in producing *R*-PAC and were stable only at low temperatures. Shin and Rogers [61] used a partially purified PDC from *C. utilis* to achieve final *R*-PAC concentrations of 28 g/L. In contrast to *C. utilis*, the *Z. mobilis* PDC exhibited better stability and high catalytic activity at room temperature. However, this enzyme is disadvantageous in that it has low affinity for benzaldehyde and exhibits considerable substrate inhibitory effects.

Iwan et al. [62] exploited the high stability toward acetaldehyde of mutated PDC enzymes, PDC*W392I* and PDC*W392*, from *Z. mobilis* mutants, in continuous *R*-PAC production systems using enzyme-membrane bioreactors. With pyruvate (90 mM) and benzaldehyde (30 mM) as substrates, *R*-PAC was produced in a space–time yield of 27.4 g/L/d using purified PDC*W392I*. With acetaldehyde (50 mM) and benzaldehyde (50 mM) as substrates, a space–time yield of 81 g/L/d was obtained with PDC*W392M*. This approach also overcomes the problem of by-product formation in current fermentation processes [63].

It was noted earlier that high concentrations of *R*-PAC were obtained in fungal screening studies when cell-free extracts were used as biocatalyst in the biotransformation reaction [28]. Indeed, much lower PAC concentrations (19 mM, 2.9 g/L) were obtained in biotransformations with fermenting mycelia, and this appeared to be due to rapid reduction of the benzaldehyde to the by-product, benzyl alcohol. Using initial substrate concentrations of 100 mM benzaldehyde and 150 mM pyruvate, concentrations of 78 to 84 mM PAC (11.7 to 12.6 g/L) were achieved in reactions with cell-free extracts of *R. javanicus* and *Fusarium* sp. in 20 h incubations. Under the same conditions extracts of *S. cerevisiae* and *C. utilis* produced

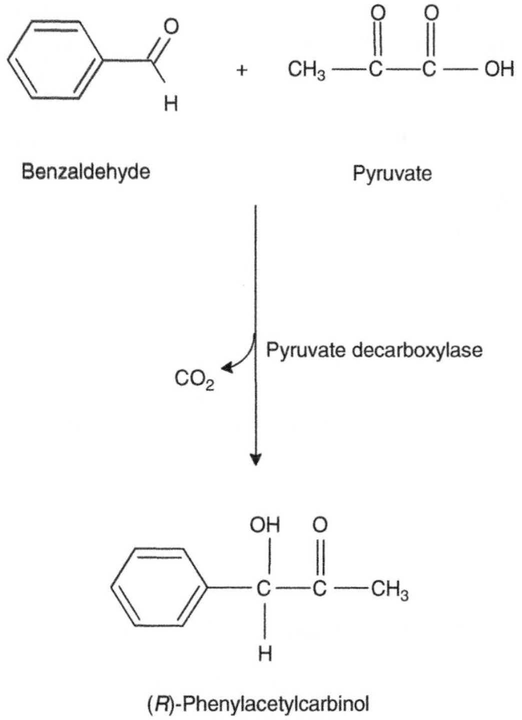

FIGURE 24.2 Enzymatic synthesis of (*R*)-phenylacetylcarbinol by pyruvate decarboxylase (PDC).

11.3 g/L PAC. Under optimized conditions a partially purified PDC from C. produced 14.6 g PAC/L from the same substrate concentrations [61].

Rosche et al. [28] reported significant increases in the production of R-PAC with the enzyme biotransformations in place of whole cells, noting that limitations on use of cells arise as a result of benzaldehyde toxicity, insufficient accumulation of pyruvate reduction in product yield through production of benzyl alcohol. Pyruvate and benzald were used as substrates in these biotransformations. Benzaldehyde was present in the fc an emulsion. Reduction in magnesium concentration from 20 to 0.5 mM and implemen of the biotransformation at 6°C resulted in high accumulation of PAC by PDC R. javanicus and C. utilis. Lowering the temperature from 23 to 6°C decreased initial re. rates but increased final R-PAC concentrations [28]. Loss of pyruvate has been attribu temperature and concentration-related impacts of magnesium [64]. The initial optimu was 6.5 and the tendency for the pH to rise during the reaction due to proton uptak countered through judicious buffer selection and use. The C. utilis enzyme formed le product from pyruvate and was more stable under the biotransformation conditions. PDC from yeast and from the fungus, R. javanicus, produced concentrations of R-P. excess of 50 g/L with no formation of benzyl alcohol by-product.

The emulsion process using C. utilis PDC was limited by inactivation of the enzy benzaldehyde, R-PAC, and the by-product acetoin [28]. Even higher concentratic product (>100 g/L) were achieved in two-phase aqueous–organic systems, containing o and nonanol designed to partition the enzyme-inactivating substrate, benzaldehyde, in organic phase. The product R-PAC was also substantially partitioned into the organic In a rapidly stirred two-phased system, buffered at pH 6.5 and incubated at 4°C, cont 160 g/L benzaldehyde and 123 g/L pyruvate, levels of benzaldehyde in the aqueous m were less than 50 mM. After 49 h, when all pyruvate was used up, final R-PAC concent was 140 g/L, representing a 91% molar conversion from benzaldehyde. In a similar s not rapidly stirred so as to maintain phase separation, a molar conversion of 93% ben hyde to R-PAC was achieved but in a more prolonged incubation phase of 395 h. In bc rapidly stirred and phase-separated system ee of R-PAC was >99%.

These studies indicated how substantial improvements can be achieved in the R biotransformation: by using isolated enzymes instead of whole cells, thereby elimi problems of pyruvate limitation and alcohol dehydrogenase-mediated benzyl alcohol f tion; and by employing a two-phase aqueous–organic system to partition the sub benzaldehyde predominantly in the organic phase, thereby reducing its inactivating on PDC located in the aqueous phase.

Leksawasdi et al. [65] implemented initial rate and biotransformation investigati mathematical modeling studies on C. utilis PDC-mediated production of R-PAC from vate and benzaldehyde. The model was validated as being capable of predicting time cour substrates, R-PAC, by-products acetaldehyde and acetoin, and enzyme activity level.

Rosche et al. [66] screened organic solvents for use in the biphasic biotransformat acetaldehyde and benzaldehyde to R-PAC by PDC isolated from Z. mobilis PDCW. Best PAC formation was observed with alcohols, in particular 1-pentanol, 1-hexano isobutanol, which, in contrast to other solvents, produced lower concentrations of acetaldehyde substrate in the aqueous phase.

24.3   BENZOYLFORMATE DECARBOXYLASE

Benzoylformate decarboxylase (EC 4.1.1.7) catalyses conversion of benzoylformate t zaldehyde [67–69], and is an integral part of the mandelic acid pathway for catabol aromatic compounds by different Pseudomonas and Acinetobacter sp. The purified en from *Acinetobacter calcoaceticus* [67] and *P. putida* [70,71] have tetrameric structures like PDC and are likewise TPP- and Mg-dependent enzymes. Benzoylformate decarboxylase has higher substrate specificity than PDC, using only benzoylformate and *para*-substituted benzoylformates as substrates [70–72].

24.3.1 THE LIGATION REACTION

Wilcocks et al. [73] demonstrated the capacity of cells of *P. putida* containing benzoylformate decarboxylase to form the acyloin compound (*S*)-2-hydroxypropiophenone or (*R*)-benzoin, when incubated with benzoylformate and acetaldehyde [21]. Figure 24.3 shows a benzoylformate decarboxylase catalyzed reaction. Cell extract or purified enzyme was also able to carry out the same biotransformation. Accumulation of benzyl alcohol was observed in reactions catalyzed by whole cells or cell extracts. The product was formed with a high degree of enantiospecificity, having an ee of 91 to 92% as 2-(*S*)-hydroxypropiophenone [74]. Under optimized conditions, in the presence of 100 mM benzoylformate and 1600 mM acetaldehyde, 61.76 mM acyloin product was produced in 1 h. Stereoselectivity of the reaction was improved by the use of *A. calcoaceticus* cells instead of *P. putida*, where optical purity of the (*S*)-enantiomer was >98% [75–77]. Optimization of reaction conditions resulted in the achievement of extremely high productivities, namely 6.95 g product/L/h.

Demir et al. [78] described the first general synthesis of benzoin and substituted benzoins, mediated by benzoylformate decarboxylase. By increasing the benzaldehyde concentration or decreasing the acetaldehyde concentration in the reaction mixture, (*R*)-benzoin formation was observed. Optimization of biotransformation conditions with respect to time, enzyme, cofactor, and medium produced a yield of 70% of (*R*)-benzoin, having a >99% ee. Addition of dimethylsulfoxide had no effect on ee but increased biotransformation reaction rate. Park and Jung [79] found that encapsulation of whole cells of *P. putida* in calcium alginate liquid core capsules resulted in the biotransformation of benzaldehyde and acetaldehyde to 2-hydroxypropiophenone, proceeding without the associated production of the undesirable

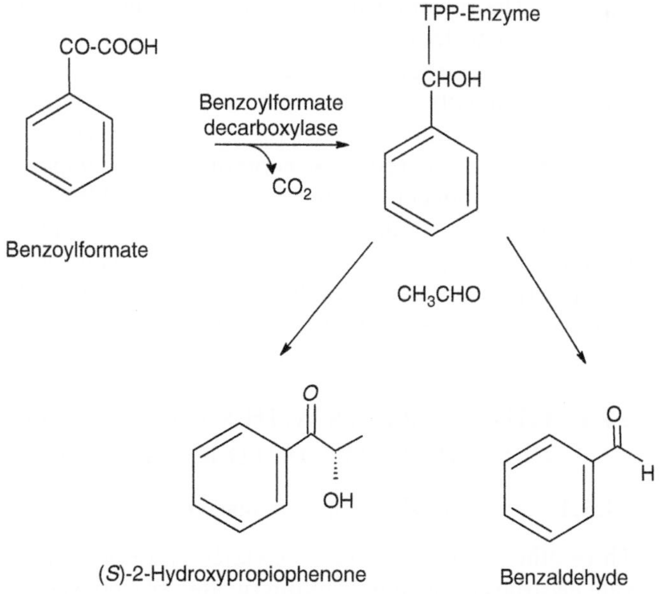

FIGURE 24.3 Synthesis of (*S*)-2-hydroxypropiophenone by benzoylformate decarboxylase.

by-product, benzyl alcohol. In addition, the cells retained their biotransformation a
with minimum loss by recycling many times.

24.3.2 Development of Mutants

The gene encoding the *P. putida* enzyme and the crystal structure of the protein hav
characterized [80,81]. Lingen et al. [82] used error-prone polymerase chain reaction
to develop benzoylformate decarboxylase mutant enzymes with altered substrate sp
ties. Mutant enzymes L476Q and M365L-L461S accepted *ortho*-substituted benza
derivatives such as 2-chloro-benzaldehyde or 2-bromo-benzaldehyde as donor sub
leading to production of 2-hydroxyketones, and produced enantiopure (*S*)-2-hydrox
methylphenyl)propan-1-one with excellent yields. The wild-type enzyme shows ver
catalytic activity toward this reaction. In aqueous buffer benzoylformate decarb
L476Q also exhibited a fivefold increase in carboligase activity compared to the wi
enzyme producing (*S*)-2-hydroxy-1-phenyl-propanone with a high ee. From m
investigations of the structure of the wild-type enzyme, it has been postulated that rec
of the side chain size at L461 can alter the substrate specificity of the enzyme by faci
accommodation and binding the enlarged *ortho*-substituted donor substrates to th
mine diphosphate.

24.4 PHENYLPYRUVATE DECARBOXYLASE

Phenylpyruvate decarboxylase (EC 4.1.1.43), another TPP-dependent α-ketoacid dec
ylase, has been observed in *Acinetobacter auridice*, *A. calcoaceticus* [68,83], *Thauera ar*
[84], and *S. cerevisiae* [85], but there are few studies on the properties of this e
Phenylpyruvate decarboxylase activity appears to be induced when cells are grc
substrates like mandelate [67,86], phenylalanine, or tryptophan [84]. The enzyme part
in conversion of phenylpyruvate to phenylacetate, through phenylacetaldehyde. Lik
and benzoylformate decarboxylase, this enzyme is also a tetramer, and has four ic
subunits. The substrate specificity of phenylpyruvate decarboxylase is broader than
benzoylformate decarboxylase, with compounds like indolepyruvate and α-ketoacids,
more than six carbon atoms in a straight chain, acting as substrates. Phenylpyruva
arboxylase activity was present in *S. cerevisiae* cultures grown with phenylalanine
nitrogen source, and the genes for this enzyme have been identified and characterize
Phenylpyruvate decarboxylase is presumed to be important in the production of fusel
yeast, one component of which is phenylethanol [87].

Asymmetric acyloin condensations catalyzed by phenylpyruvate decarboxylase ha
been reported [88]. *Achromobacter eurydice*, *Pseudomonas aromatica*, and *P. putida* cells
on L-phenylalanine catalyzed the acyloin condensation of phenylpyruvate and acetalde
produce 3-hydroxy-1-phenyl-2-butanone (Figure 24.4) with high degrees of enantiosele

24.5 OTHER ENZYMES WITH KNOWN OR POTENTIAL ROLES
IN LIGATION-MEDIATED BIOTRANSFORMATIONS

24.5.1 Other α-Ketoacid Decarboxylases

Three other α-ketoacid decarboxylases—phosphonopyruvate decarboxylase, indole-
vate decarboxylase, and 3-sulfopyruvate decarboxylase—have been characterized. Ph
nopyruvate decarboxylase [89] participates in the biosynthetic pathways for produc

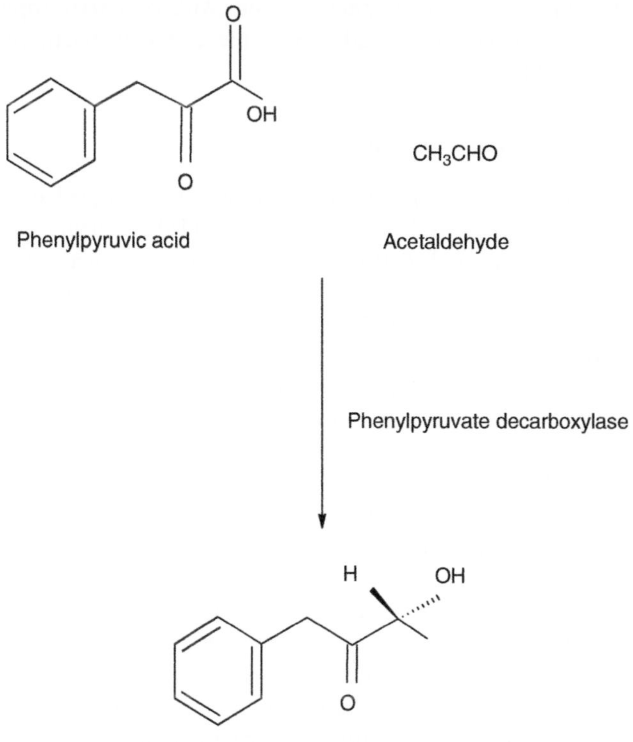

FIGURE 24.4 Asymmetric acyloin condensation catalyzed by phenylpyruvate decarboxylase.

bialaphos, fosfomycin, phosphinothricin, and tripeptide antibiotics. This enzyme is produced by *Bacteroides fragilis* and certain *Streptomyces* spp. Indole-3-pyruvate decarboxylase, involved in tryptophan catabolism of *Enterobacter cloacae* and many other bacteria [90,91], is a tetrameric enzyme with four active sites and has structural homology with the above decarboxylases. *Archea* spp. produce 3-sulfopyruvate decarboxylases that participate in coenzyme M biosynthesis [92].

Sequence similarities among PDCs, phenylpyruvate decarboxylase, and indole-3-pyruvate decarboxylase are greater than 50%. Benzoylformate decarboxylase shows less than 30% sequence similarity with the above group [93]. Phosphonopyruvate decarboxylase and 3-sulfopyruvate decarboxylases appear to be more distantly related both in terms of structure and amino acid sequence. Structural and mutagenesis investigations have established that differences in the amino acid sequences near the active sites of α-ketoacid decarboxylases strongly influence substrate selectivity and enantiospecificity of this group of enzymes [94].

24.5.2 OTHER THIAMINE DIPHOSPHATE–DEPENDENT ENZYMES

Apart from the α-ketoacid decarboxylases, a variety of other TPP-dependent enzymes have the potential to participate in ligation reactions, because they also produce active aldehyde intermediates. Acetohydroxyacid synthase from *E. coli* [95] and acetolactate synthase [96] catalyze decarboxylation of pyruvate, followed by ligation of the active acetaldehyde to the C2 of another α-ketoacid. As indicated above, acetohydroxyacid synthase I from *E. coli* was also shown to catalyze *R*-PAC synthesis [97]. Benzaldehyde lyase catalyzes ligation of aldehydes to chiral 2-hydroxyketones and may be exploited for synthesis of various 2-hydroxyketones

[98,99]. A number of other enzymes with potential applications in C–C ligation react
bioorganic synthesis and also some enzymes participating in C–N ligation reactio
described by Pohl et al. [100].

24.6 CONCLUSIONS

Recent developments have substantially advanced our scientific knowledge of α-ke
decarboxylases and, indeed, of other TPP-requiring enzymes [100,101]. Extensive scr
programs to find R-PAC producers have shown a widespread presence of ligase a
capable of synthesizing R-PAC from pyruvate and benzaldehyde among fungi and
and have resulted in the identification of some strains with very high biotransforr
activity. While carboligase activity of fungi and yeasts cannot use acetaldehyde as su
instead of pyruvate, PDCs from bacteria such as Z. mobilis and Zymobacter can use t
expensive acetaldehyde substrate.

While whole-cell yeast biotransformations have been manipulated to produce up t
L R-PAC, the biotransformation efficiency is limited by the availability of pyruvate
reaction, by toxic effects of benzaldehyde and R-PAC, and by participation of other c
enzymes in the production of by-products. While whole-cell immobilization appr
could be exploited to prolong the biotransformation or to reduce benzaldehyde toxi
manipulate R-PAC/by-product ratios, rates and extents of transformation were ty
limited by diffusion of reactants through the immobilization matrix. Mutation str
have been used with limited success to make yeast strains more resistant to aldehyd
ADH-defective mutants still form benzyl alcohol by-product from benzaldehyde. Whe
biotransformations in aqueous–organic reaction media could be manipulated to inhi
product formation but such systems did not improve biotransformation rates obser
aqueous media.

Purified PDCs have been used to overcome the limitations outlined above for who
R-PAC biotransformations. Reduction of incubation temperature from 23 to 4°C min
losses of pyruvate through nonbiological side reactions, and adequate buffering
reaction countered the tendency for pH rises due to proton uptake. Use of a cont
enzyme–membrane reactor in the Zymomonas-acetaldehyde–benzaldehyde biotransi
tion produced a space–time yield of 81 g/L/d and was free of undesirable by-pro
Incorporation of nonmiscible organic solvents, notably octanol, in the reaction substa
portioned benzaldehyde and R-PAC into the octanol, thereby minimizing the dena
effects of these reactants on the enzyme. Optimizing these various parameters resu
production of final concentrations of 100 to 140 g/L R-PAC, representing molar conve
of >90% based on benzaldehyde and a product ee of 99%.

The capacities of benzoylformate decarboxylase and phenylpyruvate decarboxyl
carry out efficient and enantiospecific carboligase biotransformations have also be
scribed, and it is assumed that these systems would be likewise amenable to the ki
reaction optimization strategies that have been applied to PDC.

In the past few years, applications of mutagenesis and molecular methods, combine
advances in our ability to analyze and characterize the structures of these enzymes, have a
researchers to understand the key structure–functional relationships that are important i
biocatalytic reactions. These approaches have been successfully exploited to understand th
of amino acids near the active site of PDC and benzoylformate decarboxylase with resp
relative decarboxylation/ligation activity, substrate- and enantiospecificity, and to genera
tants with improved activity and conformational stability. These studies demonstrat
improved biocatalysts may be developed for some of the more established biotransform
resulting in higher volumetric productivities and yields of desired product.

A number of other α-ketoacid decarboxylases and other thiamine diphosphate–dependent enzymes having potential to participate in ligation reactions have been identified. It is clear that many new enzymes from this family will be discovered and/or engineered in the coming years. In addition, the interesting capacity of *B. adjusta* to produce PAC from phenylalanine and to transform PAC substrates into a variety of PAC-substituted analogs has been noted. These various developments will undoubtedly lead to a diversity of new biocatalytic opportunities for bioorganic synthesis.

REFERENCES

1. Iding, H., Siegert, P., Mesch, K., and Pohl, M., Application of alpha-keto acid decarboxylases in biotransformations, *Biochim. Biophys. Acta*, 1385, 307–322, 1998.
2. Ohta, H. and Sugai, T., Enzyme-mediated decarboxylation reactions in organic synthesis, in *Stereoselective Biocatalysis*, Patel, R.N., Ed., Marcel Dekker, New York, 2000, pp. 487–526.
3. Yagasaki, M. and Ozaki, A., Industrial biotransformations for the production of amino acids, *J. Mol. Catal., B Enzym.*, 4, 1–11, 1998.
4. Lee, S.-G., Hong, S.-P., and Sung, M.-H., Development of an enzymatic system for production of dopamine from catechol, pyruvate and ammonia, *Enzyme Microb. Technol.*, 25, 298–302, 1999.
5. Abelyan, V.A. and Abelyan, L.A., Production of L-lactic acid and L-malic acid by using intact and immobilized cells of *Lactobacillus casei*, *Appl. Biochem. Microbiol.*, 34, 578–580, 1998.
6. Schulze, B. and Wubbolts, M.G., Biocatalysis for industrial production of fine chemicals, *Curr. Opin. Biotechnol.*, 10, 609–615, 1999.
7. Ward, O.P. and Singh, A., Enzymatic asymmetric synthesis by decarboxylases, *Curr. Opin. Biotechnol.*, 11, 520–526, 2000.
8. Huebner, G., Koenig, S., Schellenberger, A., and Koch, M.H.J., An x-ray solution scattering study of the cofactor and activator induced structural changes in yeast pyruvate decarboxylase (PDC), *FEBS Lett.*, 266, 17, 1990.
9. Juni, E., Evidence for a two-site mechanism for decarboxylation of alpha-carboxylase, *J. Biol. Chem.*, 236, 2302–2308, 1961.
10. Ulrich, J., Wittorf, J.H., and Gubber, C.J., Molecular weight and coenzyme content of pyruvate decarboxylase from brewer's yeast, *Biochim. Biophys. Acta*, 113, 595–604, 1966.
11. Utter, M.F.M., Non-oxidative carboxylation and decarboxylation, in *The Enzymes*, vol. 5, Boyer, P.D., Lardy, H., and Myrbak, K., Eds., Academic Press, New York, 1961, p. 319.
12. Schellenberger, A., Structur and wirkungsmechanismus des aktiven zentrums der hefe-pyruvate decarboxylase, *Angew. Chem. Int. Ed. Engl.*, 79, 1050–1061, 1967.
13. Ulrich, J., Structure–function relationship in pyruvate decarboxylases of yeast and wheat germ, *Ann. N.Y. Acad. Sci.*, 378, 287–305, 1982.
14. Kern, D., Kern, G., Neef, H., Tittman, K., Killanberg-Jabs, M., Wikner, C., Schneider, G., and Hubner, G., How thiamin diphosphate is activated in enzymes, *Science*, 275, 67–70, 1997.
15. Pohl, M., Siegert, P., Mesch, K., Bruhn, H., and Grotzinger, J., Active site mutants of pyruvate decarboxylase from *Zymomonas mobilis*—a site-directed mutagenesis study of Lii2, I472, I476, E473 and N482, *Eur. J. Biochem.*, 257, 538–546, 1998.
16. Candy, J.M. and Duggleby, R.G., Structure and properties of pyruvate decarboxylase and site-directed mutagenesis of the *Zymomonas mobilis* enzyme, *Biochim. Biophys. Acta*, 1385, 323–338, 1998.
17. Schomburg, D. and Salzmann, N., Lyases. Pyruvate decarboxylase, in *Enzyme Handbook*, vol. 1, Springer-Verlag, Berlin, 1990, p. 1.
18. Dyda, F., Furey, W., Swaminathan, S., Sax, M., Farrenkopf, B.C., and Jordan, F., Catalytic centers in the thiamin diphosphate dependent enzyme pyruvate decarboxylase at 2.4-Å6170, 1993.
19. Arjunan, P., Umland, T., Dyda, F., Swaminathan, S., Furey, W., Sax, M., Farrenkopf, B., Gao, Y., Zhang, D., and Jordan, F., Crystal structure of the thiamine diphosphate–dependent enzyme pyruvate decarboxylase from the yeast *Saccharomyces cerevisiae* at 2.3 angstrom resolution, *J. Mol. Biol.*, 256, 590–600, 1966.

20. Konig, S., Subunit structure, function and organization of pyruvate decarboxylases from va organisms, *Biochim. Biophys. Acta*, 1385, 271–286, 1998.

21. Sprenger, G.A. and Pohl, M., Synthetic potential of thiamin diphosphate–dependent enz *J. Mol. Catal. B Enzym.*, 6, 145–159, 1999.

22. Tripathi, C.M., Agarwal, S.C., and Basu, S.K., Production of L-phenylacetylcarbinol by fe tation, *J. Ferment Bioeng.*, 84, 487–492, 1997.

23. Csuk, R. and Glanzer, B.I., Baker's yeast mediated transformations in organic chemistry, ◀ *Rev.*, 91, 49, 1991.

24. Shukla, V.B. and Kulkarni, P.R., L-Phenylacetylcarbinol (L-PAC) biosynthesis and ind◀ applications, *World J. Microbiol. Biotechnol.*, 16, 499–506, 2000.

25. Rosche, B., Breuer, M., Hauer, B., and Rogers, P.L., Screening of yeast for enzy (R)-phenylacetylcarbinol production, *Biotechnol. Lett.*, 28, 841–845, 2003.

26. Miyata, R., Igarashi, K., and Yonehara, T., Production of pyruvic acid, Jap. Patent 2,000,07 2000.

27. Barnett, J.A., Payne, R.W., and Yarrow, D., *Yeasts: Characteristics and Identification*, 3◀ Cambridge University Press, Cambridge, 2000.

28. Rosche, B., Sandford, V., and Breuer, M., Biotransformation of benzaldehyde int◀ phenylacetylcarbinol by filamentous fungi or their extracts, *Appl. Microbiol. Biotechnc* 309–315, 2001.

29. Silk, P.J. and Macauley, J.B., Stereoselective biosynthesis of chloroarylpropane diols ▮ basidiomycete *Bjerkandera adjusta*: exploring the roles of amino acids, pyruvate, glycer◀ phenylacetylcarbinol, *FEMS Microbiol. Lett.*, 228, 11–19, 2003.

30. Baev, M.V. and Ward, O.P., Decarboxylases in stereoselective catalysis, in *Stereoselective talysis*, Patel, R.N., Ed., Marcel Dekker, New York, 2000, pp. 267–287.

31. Oliver, A.L., Anderson, F.A., and Roddick, F.A., Factors affecting the producti◀ L-phenylacetylcarbinol by yeast: a case study, *Adv. Microb. Physiol.*, 41, 1–45, 1999.

32. Breur, M., Hauer, B., and Mesch, L., Method for producing enantiomer-free phenylacety◀ nols from acetaldehyde and benzaldehyde in the presence of pyruvate decarboxylase from *monas*, Ger. Patent 19,736,104 A1, 1997.

33. Hauer, B., Breuer, M., and Rogers, P., Process for production of R-phenylacetylcarbinol enzymatic process in a two-phase system, Ger. Patent Application, DE 10,142,574 (W 020942), 2001.

34. Rosche, B., Sandford, V., Breuer, M., Hauer, B., and Rogers, P.L., Enhanced produc◀ R-phenylacetylcarbinol (R-PAC) through enzymatic biotransformation, *J. Mol. Catal. B I* 19–20, 109–115, 2002.

35. Long, A. and Ward, O.P., Biotransformation of benzaldehyde by *Saccharomyces cer*◀ Characterization of the fermentation and toxicity effects of substrates and products, *Bio*◀ *Bioeng.*, 34, 933–941, 1989.

36. Wang, B., Shin, H.S., and Rogers, P.L., in *Better Living Through Innovative Biochemical* *eering*, Teo, W.K., Yap, M.G.S., and Oh, S.W.K., Eds., Continental Press, Singapore, 1994,◀

37. Coughlin, R.W., Mahmoud, W.M., and El-Sayed, A.H., Enhanced bioconversion of to◀ stances, U.S. Patent 663,828, 1992.

38. Rogers, P.L., Shin, H.S., and Wang, B., Biotransformation for ephedrine productio◀ *Biochem. Eng. Biotechnol.*, 56, 33–59, 1997.

39. Voets, J.P., Vandamme, E.J., and Vlerick, C., Some aspects of the phenylacetylcarbinol thesis by *Saccharomyces cerevisiae*, *Z. Allg. Mikrobiol.*, 13, 355–366, 1973.

40. Smith, P.F. and Hendlin, J., Mechanism of phenylacetylcarbinol synthesis by yeast, *J. Ba* 65, 440–445, 1953.

41. Nikolova, P. and Ward, O.P., Production of L-phenylacetylcarbinol by biotransformation: and by-product formation and activities of the key enzymes in wild-type and ADH iso mutants of *Saccharomyces cerevisiae*, *Biotechnol. Bioeng.*, 38, 493–498, 1991.

42. Long, A., James, P., and Ward, O.P., Aromatic aldehydes as substrates for yeast and yeast dehydrogenase, *Biotechnol. Bioeng.*, 33, 657–660, 1989.

43. Smith, P.F. and Hendlin, D., Further studies on phenylacetylcarbinol synthesis by yeast, *Appl. Microbiol.*, 2, 294–296, 1954.

44. Ose, S. and Hironaka, J., Studies on production of phenylacetylcarbinol by fermentation, *Proc. Int. Symp. Enzyme Chem.* (Tokyo), 2, 457–460, 1979.

45. Nikolova, P. and Ward, O.P., Effect of organic solvent on biotransformation of benzaldehyde to benzyl alcohol by free and silicone-alginate entrapped cells, *Biotech. Tech.*, 7, 897–902, 1993.

46. Mahmoud, W.M., El-Sayed, A.-H.M.M., and Caughlin, R.W., Production of L-phenylacetyl-carbinol by immobilized yeast cells: I. Batch fermentation, *Biotechnol. Bioeng.*, 36, 47–54, 1990.

47. Mahmoud, W.M., El-Sayed, A.-H.M.M., and Caughlin, R.W., Production of L-phenylacetylcarbinol by immobilized yeast cells: II. Semi-continuous fermentation, *Biotechnol. Bioeng.*, 36, 55–60, 1990.

48. Nikolova, P. and Ward, O.P., Effect of support matrix on ratio of product to byproduct formation in L-phenylacetylcarbinol synthesis, *Biotechnol. Lett.*, 16, 7–10, 1994.

49. Shukla, V.B. and Kulkarni, P.R., Comparative studies on bioconversion of benzaldehyde to L-phenylacetylcarbinol (L-PAC) using calcium alginate- and barium alginate-immobilized cells of *Torulaspora delbrueckii*, *J. Chem. Technol. Biotechnol.*, 78, 949–951, 2003.

50. Barman, T.E., Pyruvate dehydrogenase, in *Enzyme Handbook*, vol. 2, Springer-Verlag, New York, 1969, p. 701.

51. Heider, J., Boll, M., Breese, K., Breinig, S., Ebenau-jehle, C., Feil, U., Gadon, N., Laempe, D., Leuthner, B., and Mohamed, M.E., Differential induction of enzymes involved in metabolism of aromatic compounds in the denitrifying bacterium *Thauera aromatica*, *Arch. Microbiol.*, 170, 120–131, 1998.

52. Nikolova, P. and Ward, O.P., Production of phenylacetylcarbinol by biotransformation using baker's yeast in two-phase systems, in *Biocatalysis in Non-Conventional Media*, Tramper, J., Ed., Elsevier, Amsterdam, 1992, pp. 675–680.

53. Nikolova, P. and Ward, O.P., Whole cell yeast biotransformations in two-phase systems: effect of solvent on product formation and cell structure, *J. Indust. Microbiol.*, 10, 169–177, 1992.

54. Laurence, G., Smallridge, A.J., and Trewhella, M.A., Lactate as an alternative to pyruvate in the yeast mediated preparation of PAC, *J. Mol. Catal., B Enzym.*, 19–20, 399–403, 2002.

55. Seely, R.J., Hageman, R.V., Yarus, M.J., and O'Sullivan, S.A., USP WO 90/04639; USP WO 90/04631, 1990.

56. Kostraby, M.M., Smallridge, A.J., and Trewhella, M.A., Yeast-mediated preparation of L-PAC in an organic solvent. *Biotechnol. Bioeng.*, 77, 827–831, 2002.

57. Seely, R.J., Heefner, D.L., Hageman, R.V., Yarus, M.J., and Sullivan, S.A., Process for making L-phenylacetylcarbinol (PAC), microorganisms for use in the process, and a method of preparing the microorganisms, U.S. Patent 5,312,742, 1994.

58. Nikolova, P. and Ward, O.P., Reductive biotransformation by wild type and mutant strains of *Saccharomyces cerevisiae* in aqueous–organic solvent systems, *Biotechnol. Bioeng.*, 39, 870–876, 1992.

59. Bruhn, H., Pohl, M., Grotzinger, J., and Kula, M.R., The replacement of Trp392 by alanine influences the decarboxylase/carboligase activity and stability of pyruvate decarboxylase from *Zymomonas mobilis*, *Eur. J. Biochem.*, 234, 650–655, 1995.

60. Engel, S., Vyazmensky, M., Vinogradov, M., Berkovich, D., Bar-Ilan, A., Qimron, U., Rosiansky, Y., Baral, Z., and Chipman, D., Role of conserved arginine in the mechanism of acetohydroxyacid synthase, *J. Biol. Chem.*, 279, 24803–24812, 2004.

61. Shin, H.S. and Rogers, P.L., Production of L-phenylacetylcarbinol (L-PAC) from benzaldehyde using partially purified pyruvate decarboxylase (PDC), *Biotechnol. Bioeng.*, 49, 52, 1996.

62. Iwan, P., Goetz, G., Schmitz, S., Hauer, B., Breur, M., and Pohl, M., Studies on the continuous production of (R)-(−)-phenylacetylcarbinol in an enzyme-membrane bioreactor, *J. Mol. Catal. B Enzym.*, 11, 387–396, 2001.

63. Goetz, G., Iwan, P., Hauer, B., Breuer, M., and Pohl, M., Continuous production of (R)-phenylacetylcarbinol in an enzyme-membrane bioreactor using a potent mutant of pyruvate decarboxylase from *Zymomonas mobilis*, *Biotechnol. Bioeng.*, 74, 317–325, 2001.

64. Rosche, B., Breuer, M., Hauer, B., et al., Increased pyruvate efficiency in enzymatic production of (R)-phenylacetylcarbinol, *Biotechnol. Lett.*, 25, 847–851, 2003.

65. Leksawasdi, N., Chow, Y.Y.S., Breuer, M., Hauer, B., Rosche, B., and Rogers, P.L., analysis and modeling of enzymatic (R)-phenylacetylcarbinol batch biotransformation p *J. Biotechnol.*, 111, 179–189, 2004.

66. Rosche, B., Breuer, M., Hauer, B., and Rogers, P.L., Biphasic aqueous/organic biotransfor of acetaldehyde and benzaldehyde by *Zymomonas mobilis* pyruvate decarboxylase, *Biog* *Bioeng.*, 86, 788–794, 2004.

67. Barrowman, M.M. and Fewson, C.A., Phenylglyoxylate decarboxylase and pyruvate dec ylase from *Acinetobacter calcoaceticus*, *Curr. Microbiol.*, 12, 235–240, 1985.

68. Barrowman, M.M., Harnett, W., Scott, A.J., Fewson, C.A., and Kusel, J.R., Immun comparison of microbial TPP-dependent non-oxidative alpha-keto acid decarboxylases, *Microbiol. Lett.*, 34, 57–60, 1986.

69. Hegeman, G.D., Synthesis of the enzymes of the mandelate pathway by *Pseudomonas* *J. Bacteriol.*, 91, 1140–1159, 1966.

70. Hegeman, G.D., Benzoylformate decarboxylase (*Pseudomonas putida*), *Meth. Enzymol.*, 17. 678, 1970.

71. Reynolds, L.J., Garcia, G.A., Kozarich, J.W., and Kenyon, G.L., Differential reactivity processing of [*para*-(halomethyl)benzoyl]formates by benzoylformate decarboxylase, a t pyrophosphate dependent enzyme, *Biochemistry*, 27, 5530–5538, 1988.

72. Weiss, P.M., Garcia, G.A., Kenyon, G.L., and Clealand, W.N., Kinetics and mechan benzoylformate decarboxylase using C-13 and solvent deuterium isotope effects on benzoylf and benzoylformate analogs, *Biochemistry*, 27, 2197–2205, 1988.

73. Wilcocks, R., Ward, O.P., Collins, S., Dewdney, N.J., Hong, Y., and Prosen, E., Acyloin tion by benzoylformate decarboxylase from *Pseudomonas putida*: a novel biotransformation *Environ. Microbiol.*, 58, 1699–1704, 1992.

74. Wilcocks, R. and Ward, O.P., Factors affecting 2-hydroxypropiophenone formation by b formate decarboxylases from *Pseudomonas putida*, *Biotechnol. Bioeng.*, 39, 1058–1063, 199

75. Ward, O.P., Wilcocks, R., Prosen, E., Collins, S., Dewdney, N.J., and Hong, Y., Acyloin tion mediated by benzoylformate decarboxylases, in *Microbial Reagents in Organic Sy* Servi, S., Ed., Kluwer Publishers, Amsterdam, 1992, pp. 67–75.

76. Prosen, E., Ward, O.P., Collins, S., Dewdney, N.J., Hong, Y., and Wilcocks, R., Enantio production of S-(–)-2-hydroxypropiophenone mediated by benzoylformate decarboxylas *Acinetobacter calcoaceticus*, *Biocatalysis*, 8, 21–29, 1993.

77. Prosen, E. and Ward, O.P., Optimisation of reaction conditions for production of S hydroxypropiophenone by *Acinetobacter calcoaceticus*, *J. Indust. Microbiol.*, 13, 287–291,

78. Demir, A.S., Dunwald, T., Iding, H., Pohl, M., and Muller, M., Asymmetric benzoin r catalyzed by benzoylformate decarboxylase, *Tetrahedron Asymmetry*, 10, 4769–4774, 1999

79. Park, J.K. and Jung, J.Y., Production of benzaldehyde by encapsulated whole-cell benzoylf decarboxylase, *Enzyme Microb. Technol.*, 30, 726–733, 2002.

80. Tsou, A.Y., Ransom, S.C., Geret, G.A., Buechter, B.D., Babbitt, P.C., and Kenyon Mandelate pathway of *Pseudomonas putida* involving mandelate racemase, (S)-mandelat drogenase and benzoylformate decarboxylase, and expression of benzoylformate decarbox *Escherichia coli*, *Biochemistry*, 29, 9856–9867, 1990.

81. Hasson, M.S., Muscate, A., McLeish, M.J., Polovnikova, L.S., Gerlt, J.A., Kenyon, G.L., Petsk and Ringe, D., The crystal structure of benzoylformate decarboxylase at 1.6Å res diversity of catalytic residues in thiamine diphosphate–dependent enzymes, *Biochemistry*, 37 9930, 1998.

82. Lingen, B., Grotzinger, J., Kolter, D., Kula, M.-R., and Pohl, M., Improving the carl activity of benzoylformate decarboxylase from *Pseudomonas putida* by a combination of e evolution and site-directed mutagensis, *Prot. Eng.*, 15, 585–593, 2002.

83. Asakawa, T., Wada, H., and Yamano, T., Enzyme conversion of phenylpyruvate to phenyl. *Biochim. Biophys. Acta*, 170, 375–384, 1968.

84. Schneider, S., Mohamed, M.E., and Fuchs, G., Anaerobic metabolism of L-phenylalar benzoyl-CoA in the denitrifying bacterium *Thauera aromatica*, *Arch. Mikrobiol.*, 168, 310–32

85. Vulalhan, Z., Morais, M.A., Tai, S.-L., Piper, M.D.W., and Pronk, J.T., Identification and characterization of phenylpyruvate decarboxylase genes in *Saccharomyces cerevisiae*, *Appl. Environ. Microbiol.*, 69, 4534, 2003.

86. Heider, J., Boll, M., Breese, K., Breinig, S., Ebenau-Jehle, C., Feil, U., Gadon, N., Laempe, D., Leuthner, B., and Mohamed, M.E., Differential induction of enzymes involved in metabolism of aromatic compounds in the denitrifying bacterium *Thauera aromatica*, *Arch. Microbiol.*, 170, 120–131, 1998.

87. Sentheshanmuganathan, S., The mechanism of the formation of higher alcohols from amino acids by *Saccharomyces cerevisiae*, *Biochem. J.*, 74, 568–576, 1960.

88. Guo, Z., Goswami, A., Mirfakhrae, D., and Patel, R.N., Assymetric acyloin condensation catalysed by phenylpyruvate decarboxylase, *Tetrahedron Asymmetry*, 10, 4667–4675, 1999.

89. Schutz, A., Golbik, R., Tittmann, K., Svergun, D.I., Koch, M.H., Hubner, G., and Konig, S., Studies on structure–function relationships of indolepyruvate decarboxylase from Enterobacter cloacae, a key enzyme of the indole acetic acid pathway, *Eur. J. Biochem.*, 270, 2322–2331, 2003.

90. Schutz, A., Sandalova, T., Ricagno, S., Hubner, G., Konig, S., and Schneider, G., Crystal structure of thiamindiphosphate–dependent indolepyruvate decarboxylase from *Enterobacter cloacae*, an enzyme involved in the biosynthesis of the plant hormone indole-3-acetic acid, *Eur. J. Biochem.*, 270, 2312–2321, 2003.

91. Pang, S.S., Duggleby, R.G., Schowen, R.L., and Guddat, L.W., The crystal structure of *Klebsiella pneumoniae* acetolactate synthase with enzyme-bound cofactor and with an unusual intermediate, *J. Biol. Chem.*, 279, 2242–2253, 2004.

92. Zhang, G., Dai, J., Lu, Z., and Dunaway-Mariano, D., The phosphonopyruvate decarboxylase from *Bacteroides fragilis*, *J. Biol. Chem.*, 278, 41302–41308, 2003.

93. Graupner, M., Xu, H., and White, R.H., Identification of the gene encoding sulfopyruvate decarboxylase, an enzyme involved in biosynthesis of coenzyme, *M. J. Bacteriol.*, 182, 4862–4867, 2000.

94. Polnikova, E.S., McLeish, M.J., Sergienko, E.A., Burgner, J.T., Anderson, N.L., Bera, A.K., Jordan, G.L., and Hasson, M.S., Structural and kinetic analysis of catalysis of a thiamine diphosphate–dependent enzyme, benzoylformate decarboxylase, *Biochemistry*, 42, 1820–1830, 2003.

95. Duggleby, R.G., Pang, S.S., Yu, H., and Guddat, L.W., Systematic characterization of mutations in yeast acetohydroxyacid synthase: interpretation of herbicide resistance data, *Eur. J. Biochem.*, 270, 2895–2904, 2003.

96. Siegert, P., Pohl, M., Keen, M.M., Pogozheva, I.D., Kenyon, G.L., and McLeish, M.J., Exploring the substrate specificity of benzoylformate decarboxylase, pyruvate decarboxylase and benzaldehyde lyase, in *Thiamine: Catalytic Mechanisms in Normal and Disease States*, Jordan, F. and Patel, M.S., Eds., Marcel Dekker, New York, 2004, pp. 275–290.

97. Engel, S., Vyazmensky, M., Geresh, S., Barak, Z., and Chipman, D.M., Acetohydroxyacid synthase: a new enzyme for chiral synthesis of *R*-phenylacetylcarbinol, *Biotechnol. Bioeng.*, 83, 833–840, 2003.

98. Demir, A.S., Sesenoglu, O., Dunkelmann, P., and Muller, M., Benzaldehyde lyase-catalyzed enantioselective carboligation of aromatic aldehydes with mono- and dimethoxy acetaldehyde, *Org. Lett.*, 5, 2047–2050, 2003.

99. Sanchez-Gonzalez, M. and Rosazza, J.P.N., Mixed aromatic acyloin condensations with recombinant benzaldehyde lyase: synthesis of alpha-hydroxydihydrochalcones and related alpha-hydroxyketones, *Synthetic. Catal.*, 345, 819–823, 2003.

100. Pohl, M., Sprenger, G.A., and Muller, M., A new perspective on thiamine catalysis, *Curr. Opin. Biotechnol.*, 15, 335, 2004.

101. Lingen, B., Kolter-Jung, D., Dunkelmann, P., Feldmann, R., Grotzinger, J., Pohl, M., and Muller, M., Alteration of the substrate specificity of benzoylformate decarboxylase from *Pseudomonas putida* by directed evolution, *Chem. Bio. Chem.*, 4, 721–726, 2003.

25 Synthesis of Chiral Alcohols with Carbonyl Reductase Library and Robust NAD(P)H Regenerating System

Hiroaki Yamamoto and Akinobu Matsuyama

CONTENTS

25.1 INTRODUCTION

Optically pure compounds are useful as building blocks and intermediates of pharmaceuticals and fragrances. Chiral compounds can be synthesized mainly by the following three methods:

1. A chiral pool method to obtain them by chemical conversion of naturally occurring chiral compounds such as L-amino acids, D-sugars, and organic acids [1]
2. An optical resolution of racemic mixture to obtain them by preferential crystallization [2], diastereomeric resolution [3], kinetic resolution [4], and chromatographic separation with chiral stationary phases [5]
3. An asymmetric synthesis method with chiral auxiliaries [6], metal-catalysts [7], and biocatalysts [8]

The optimal method to obtain chiral compounds is selected according to the quantity needed, allowable lead time, and cost. The biocatalytic process has several advantages: (1) it is generally highly regio- and stereoselective; (2) it is performed at a moderate temperature in aqueous solution; (3) biocatalysts can be reproduced easily and inexpensively; and (4) it is low environmental burden.

We have been investigating the synthesis of chiral alcohols by asymmetric reductio biocatalysts because its theoretical yield is 100% without any particular recycle process. have been, however, several problems in the manufacture of chiral alcohols by biol asymmetric reduction: (1) low enzyme activity; (2) sometimes low optical purity a variance; (3) cost of coenzyme and its regeneration; (4) low concentration of product mulated; (5) time-consuming for development, from enzyme screening to process de ment; and so on. As a strategy to overcome these problems, we have constructed the enzyme library consisting of recombinant whole-cell biocatalysts. Low enzyme activi inadequate optical purity have been overcome by high expression of excellent enzym have strict substrate- and enatioselectivity as well as toughness, in a heterologous especially *Escherichia coli. E. coli* is a suitable host for asymmetric reduction because it produces carbonyl reductases, which lead to decrease in optical purity. Subjects coenzymes have been significantly, although not perfectly, overcome using whole reco ant biocatalysts coexpressing glucose dehydrogenase (GDH) [9] and formate dehydro (FDH) [10] as NAD(P)H-regenerating systems with a carbonyl reductase. Whole-cell talysis enabled an efficient use of intracellular coenzymes without addition of extra cc [11]. A cloned enzyme library could shorten development time. If an appropriate enzym obtained by cloned enzyme library screening, we could advance to process develo without improvements of microorganism and culture conditions, enzyme purificatio cloning, optimization of gene expression, and so on. In this chapter, the proced constructing a cloned enzyme library and its application are described.

25.2 (*R*)-1,3-BUTANEDIOL AND (*S*)-SPECIFIC ALCOHOL DEHYDROGEN

(*R*)-1,3-butanediol (BDO) is used as a raw material for the production of azeti derivatives [12,13], which are intermediates for the syntheses of penem and carba antibiotics [14]. It may be produced by kinetic resolution, asymmetric reducti 4-hydroxy-2-butanone [15], and enantioselective oxidation of racemic 1,3-BDC (*R*)-1,3-BDO has been manufactured by enantioselective oxidation on an industrial sc 4-hydroxy-2-butanone is more expensive and more toxic to microorganisms than comr racemic 1,3-BDO for cosmetic use.

The enzyme responsible for the enantioselective oxidation of (*S*)-1,3-BDO was p from *Candida parapsilosis* and characterized [17]. The enzyme was NAD$^+$-dep (*S*)-specific alcohol dehydrogenase (ADH) and designated as CpSADH. It catalyz enantioselective oxidation of several (*S*)-alcohols such as (*S*)-2-butanol, (*S*)-2-octanc (*S*)-1-phenylethanol, and (*S*)-β-hydroxy esters, such as ethyl (*S*)-3-hydroxybutanoate, a asymmetric reduction of several ketones and β-oxo esters. A gene encoding CpSADI cloned and highly expressed in *E. coli* [18]. The gene consists of 1009 nucleotides enco protein with a molecular weight of 35,964. CpSADH is assumed to belong to group I long-chain, zinc-dependent ADH [19]. The properties of CpSADH and compariso several (*S*)-specific ADHs are summarized in Table 25.1 and its substrate specifi shown in Table 25.2.

The enzymatic synthesis of (*R*)-1,3-BDO was developed with whole recombinant cells expressing CpSADH as shown in Figure 25.1 [27]. There are only a few repc efficient production systems of optically active alcohols by enantioselective oxidatio recombinant *E. coli* cells expressing an ADH. The optimal pH for the reaction w which suggested that NAD$^+$ regeneration is a rate-limiting step. The optimal pH a stable pH for the regeneration of the coenzyme NAD$^+$ were around 6.8 becau optimal pH of CpSADH for oxidation of (*S*)-1,3-BDO and its stable pH were arou

TABLE 25.1
Properties of (S)-Specific Secondary Alcohol Dehydrogenases (ADHs)

Name	CpSADH	CpCR	CPAR	NfSADH	ReCR	RrSADH
Origin	Candida parapsilosis IFO 1396	C. parapsilosis DSM 70125	Corynebacterium sp. ST-10	Nocardia fusca AKU 2123	Rhodococcus erythropolis DSM 743	R. ruber DSM 44551
Mr						
Native	140,000	136,000	155,000	150,000	161,000	62,000
Subunit	40,000	67,000	42,000	39,000	40,000	38,000
ORF	336 aa	Not cloned	385 aa	Not cloned	348 aa	Not cloned
Optimum for oxidation						
Temperature (°C)	50	52–56	ND	60	ND	54
pH	9.0	7.8–8.6	10.5	8.5–9.5	9.5	9.0
Optimum for reduction						
Temperature (°C)	ND	36–40	ND	65	40	54
pH	6.0	6.5–7.2	7.0	5.5–6.5	5.5	6.5–7.5
Cofactor	NAD^+/NADH	NAD^+/NADH	NAD^+/NADH	NAD^+/NADH	NAD^+/NADH	NAD^+/NADH
Activity						
Substrate	(S)-1,3-Butanediol	Ethyl 5-oxohexanoate	Phenylacetaldehyde	2-Hexanol	Ethyl 3-oxobutanoate	1-Phenylethanol
Specific activity (U/mg)	244	1,855	64	92	269	17.5
Superfamily	SDR	Unknown	MDR	Unknown	MDR	Unknown
References	[17,18]	[20]	[21,22]	[23]	[24,25]	[26]

Mr, relative molecular mass; ND, not determined; SDR, short-chain ADH; MDR, medium-chain ADH.

TABLE 25.2
Substrate Specificity of CpSADH

Substrate	Concentration (mM)	Cofactor	pH	Relative Activi
Oxidation				
(S)-1,3-Butanediol	50	NAD$^+$	9.0	100
	50	NADP$^+$	9.0	3.4
(R)-1,3-Butanediol	50	NAD$^+$	9.0	1.7
Methanol	100	NAD$^+$	9.0	1.0
Ethanol	100	NAD$^+$	9.0	5.4
1-Propanol	100	NAD$^+$	9.0	8.5
2-Propanol	100	NAD$^+$	9.0	337
2-Butanol	100	NAD$^+$	9.0	244
(S)-2-Butanol	50	NAD$^+$	9.0	562
(R)-2-Butanol	50	NAD$^+$	9.0	18.8
2-Pentanol	100	NAD$^+$	9.0	191
3-Pentanol	100	NAD$^+$	9.0	58.3
2-Hexanol	50	NAD$^+$	9.0	156
2-Octanol	5	NAD$^+$	9.0	220
(S)-2-Octanol	5	NAD$^+$	9.0	381
(R)-2-Octanol	5	NAD$^+$	9.0	0
Cyclohexanol	20	NAD$^+$	9.0	297
(S)-1-Phenylethanol	50	NAD$^+$	9.0	502
(R)-1-Phenylethanol	50	NAD$^+$	9.0	6.4
Methyl (S)-3-hydroxybutanoate	50	NAD$^+$	9.0	266.9
Methyl (R)-3-hydroxybutanoate	50	NAD$^+$	9.0	9.9
Ethyl (S)-3-hydroxybutanoate	20	NAD$^+$	9.0	736.5
Ethyl (R)-3-hydroxybutanoate	20	NAD$^+$	9.0	3.2
Reduction				
4-Hydroxy-2-butanone	100	NADH	6.0	100
Acetone	100	NADH	6.0	299
2-Butanone	100	NADH	6.0	243
Ethyl acetoacetate	100	NADH	6.0	465
Ethyl 4-chloroacetoacetate	100	NADH	6.0	341
Acetophenone	20	NADH	6.0	296
Phenacyl chloride	10	NADH	6.5	18.8
3-Quinuclidinone	20	NADH	6.0	13.0

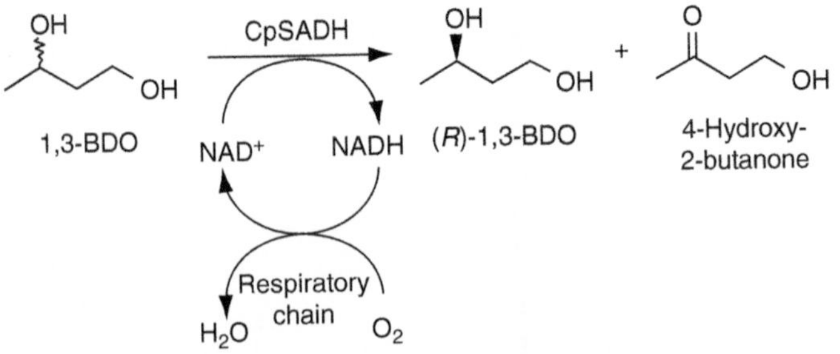

FIGURE 25.1 Enzymatic synthesis of (R)-1,3-butanediol using recombinant whole cells of *Esc coli* expressing CpSADH. 1,3-BDO, 1,3-butanediol.

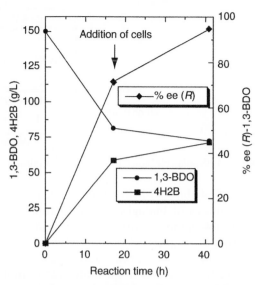

FIGURE 25.2 Synthesis of (*R*)-1,3-butanediol with whole *E. coli* cells expressing CpSADH. The reaction mixture (25 mL) containing 15% 1,3-butanediol and cells obtained from 25 mL culture in YT medium (bacto-tryptone 10 g, bacto-yeast extract 5 g, and NaCl 5 g) and 100 mM potassium phosphate (pH 6.8) at 30°C with shaking; after 17 h incubation cells obtained from 25 mL culture were added and further incubated. 1,3-**BDO**, 1,3-butanediol; 4H2B, 4-hydroxy-2-butanone.

NAD^+ was thought to be regenerated by aerobic respiratory chain in *E. coli*. *E. coli* has no cytochrome *c* and no equivalent to the mitochondrial Complex III (*bc*1 complex) or Complex IV (cytochrome *c* oxidase). Instead, two terminal oxidases in the *E. coli* cytoplasmic membrane, cytochrome *o* and cytochrome *d* complexes, oxidize ubiquinol and directly reduce molecular oxygen to water, concomitantly generating an electrochemical proton gradient across the membrane [28]. The optimal pH of NADH oxidase of *E. coli* membrane vesicles was reported to be 7.5 to 8.0 [29]. From these reports, the optimal pH for the synthesis of (*R*)-1,3-BDO using whole recombinant *E. coli* cells seems to result from the stability of NAD^+ and/or the NADH oxidase complex. Furthermore, in a reaction containing more than 7% 1,3-BDO, the cessation of enantioselective oxidation was observed. The cessation of the reaction resulted not from the inactivation of CpSADH but from the loss of NAD^+ and/or NAD^+-regeneration activity. The cessation was overcome by the addition of a fresh YT medium or its component, a yeast extract, to the reaction mixture. The addition of a YT-medium afforded cell growth during the reaction and is thought to supply fresh biocatalysts. Moreover, the yield of (*R*)-1,3-BDO reached 72.6 g/L, in 48.4% reaction yield (maximum theoretical yield is 50%) and 95% ee at 15% racemate by the addition of fresh cells after the 17 h incubation, as shown in Figure 25.2.

25.3   SYNTHESIS OF SEVERAL CHIRAL ALCOHOLS WITH CpSADH

Several optically active alcohols were synthesized with *E. coli* cells expressing CpSADH. Two NADH-regeneration systems were used for the asymmetric reduction in *E. coli*, as shown in Figure 25.3. One is NAD^+ reduction coupled with the oxidation of 2-propanol by CpSADH itself, and the other is that coupled with the oxidation of formate by formate dehydrogenase (McFDH) from *Mycobacterium vaccae* [30,31]. In both systems only intracellular NAD^+ was used.

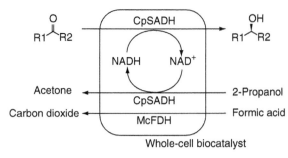

FIGURE 25.3 Whole-cell biocatalyst system containing CpSADH and NADH regenerator.

Ethyl (R)-4-chloro-3-hydroxybutanoate [(R)-ECHB] is a chiral compound that is for the synthesis of biologically and pharmacologically important materials: (R)-ca [32,33], (R)-4-amino-3-hydroxybutyric acid [34], and (R)-4-hydroxy-2-pyrrolidon (R)-ECHB was synthesized from ethyl 4-chloroacetoacetate (ECAA) and 2-propan whole recombinant E. coli cells expressing CpSADH only [11]. The ratio of 2-propan cosubstrate to ECAA as a substrate was optimized in the reaction mixture. Two excesses of 2-propanol over ECAA gave the best result, and 1.2-fold excess also gave results. Surprisingly, CpSADH had little dehydrogenase activities for (R)-ECH (S)-ECHB, although it showed a high activity for ethyl (S)-3-hydroxybutanoate w chlorine substitution at the 4-position. The extremely efficient regeneration of NAD due to the unique properties of CpSADH, which functioned as an exclusive reduct ECAA. The yield of (R)-ECHB reached 36.6 g/L in 95.2% conversion yield with a >5 under optimized reaction conditions comprising 3.8% ECAA, 2 molar excesses of 2-pr over ECAA, 200 mM potassium phosphate (pH 6.5), and recombinant E. coli cells exp CpSADH, at 20°C.

Several optically active alcohols, furthermore, were synthesized with whole E. co expressing CpSADH only or CpSADH and McFDH as an NADH regenerator, an examples are shown in Figure 25.4. CpSADH produced (S)-1,3-BDO from 4-hydr butanone efficiently by asymmetric reduction, and (R)-1,3-BDO from the racem enantioselective oxidation. Optical purity of alcohols obtained was more than 9 2-Acetylbutyrolactone, which had a chiral center at α-position, was reduced to (1'S,2R hydroxyethyl)butyrolactone (HEBL) in 99% ee and 95% de by syn-reduction accomp dynamic resolution. These results show that CpSADH is a powerful and versatile t produce optically active (S)-alcohols. CpSADH, also, enantioselectively catalyzed asy ric reduction of α-haloketones such as phenacyl chloride and m-chlorophenacyl ch Although the reactions were catalyzed at low concentrations of α-haloketones in a b system, they could not accumulate (R)-α-haloalcohols at an industrial level. α-Halok were found to be inhibitory and suicide substrates to CpSADH. Further investigation t directed evolution and protein engineering is needed to manufacture (R)-α-haloal with CpSADH.

25.4 INNOVATIVE SCREENING OF ETHYL 4-CHLOROACETOACETATE REDUCTASES AND SYNTHESIS OF ETHYL (S)-4-CHLORO-3-HYDROXYBUTANOATE

Ethyl (S)-4-chloro-3-hydroxybutanoate [(S)-ECHB] is a chiral synthon that is useful synthesis of pharmacologically active compounds, such as hydroxymethylglutaryl coe A (HMG-CoA) reductase inhibitors [36] and 4-hydroxypyrrolidone [37]. The asym

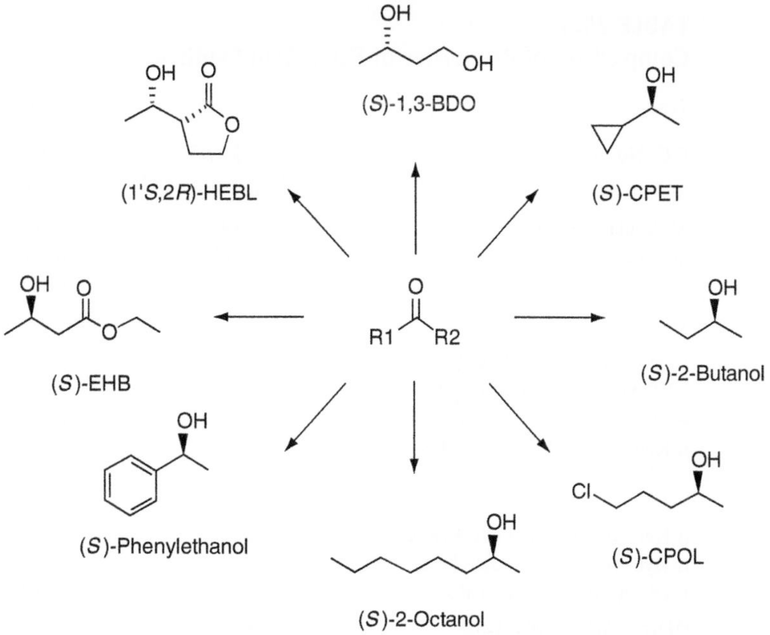

FIGURE 25.4 Synthesis of optically active alcohols with CpSADH. (*S*)-1,3-BDO, (*S*)-1,3-butanediol; (*S*)-CPET, (*S*)-cyclopropylethanol; (*S*)-CPOL, (*S*)-5-chloro-2-pentanol; (*S*)-EHB, ethyl (*S*)-3-hydroxy-butanoate; (1′*S*,2*R*)-HEBL, (1′*S*,2*R*)-2-(1′-hydroxyethyl)butyrolactone.

reduction using enantioselective oxidoreductases is a practical method for the production of (*S*)-ECHB. Several enzymes reducing ECAA to (*S*)-ECHB have been found and purified from *Saccharomyces cerevisiae* [38], *Geotrichum candidum* [39], *Sporobolomyces salmonicolor* [40], *C. macedoniensis* [41], and *C. magnoliae* [42].

We found several classes of ECAA reductases by innovative screening methods. First, we screened microorganisms that reduced ECAA asymmetrically to synthesize (*S*)-ECHB by a conventional method and selected *Kluyveromyces lactis* NRIC 1329 as the best producer of (*S*)-ECHB. This strain, however, was unsuitable for the production of (*S*)-ECHB on an industrial scale from the viewpoints of optical purity of ECHB produced and productivity. We found that the enzyme which participated in the asymmetric reduction of ECAA in *K. lactis* was a type-IB fatty acid synthase (FAS-IB) [43]. Type FAS-IB was not appropriate for the overexpression in heterologous hosts, such as *E. coli*, as it is a complex protein comprising α6β6 subunits and has a multifunctional enzyme that catalyzes eight kinds of reactions for fatty acids synthesis [44].

Second, we screened ECAA reductases from functional homology. We noticed that there were several types of FAS (IA, IB, IC, and II) among living organisms, as shown in Table 25.3 [45]. Type II FAS, in particular, consisted of several small monofunctional enzymes, and some bacteria, viruses, and higher plants were reported to have type II FAS components. The reduction of ECAA by type IB FAS seemed to be analogous to that of β-ketoacyl-ACP by general FAS. In type II FAS, the reduction of β-ketoacyl-ACP was catalyzed by β-ketoacyl-ACP reductase (KR, E.C.1.1.1.100) [46] encoded by *fabG* [47]. We examined the capabilities of β-ketoacyl-ACP reductases for synthesis of (*S*)-ECHB from ECAA.

Third, we screened ECAA reductases from amino acid sequence homology. A similarity search for the amino acid sequence of KR from *E. coli* with public databases showed significant similarity to acetoacetyl-CoA reductases (ARs, E.C.1.1.1.36) encoded by *phbB*,

TABLE 25.3
Comparison of Properties of Fatty Acid Synthase

Type	IA	IB	IC
E.C. No.	2.3.1.85	2.3.1.86	2.3.1.86
Structure	α2	α6β6	α6
Molecular weight	500K	2400K	1390K
Subunit size	α: 250K	α: 213K	α: 250K
	—	β: 203K	—
Function	*Localization*		
Acyl-carrier protein (ACP)	α	α	α
ACP *S*-acetyltransferase (AT)	α	β	α
ACP *S*-malonyltransferase (MT)	α	β	α
β-Ketoacyl-ACP synthase (KS)	α	α	α
β-Ketoacyl-ACP reductase (KR)	α	α	α
β-Hydroxyacyl-ACP dehydratase (DH)	α	β	α
Enoyl-ACP reductase (ER)	α	β (FMN)	α (FMN)
Acyl-ACP hydrolase (TE)	α	—	—
Palmitoyl transferase (PT)	—	β	α

which play a physiological role for poly-β-hydroxybutyric acid (PHB) [48]. We also exa
the capabilities of AR to synthesize (*S*)-ECHB. Two KR-genes from *E. coli* (EcKR1)
and *Bacillus subtilis* (BsKR1) [50], and two AR-genes from *Ralstonia eutropha* (Re
[51,52] and *Zoogloea ramigera* (ZrAR1) [53,54] were cloned and expressed in *E. co*
their substrate specificities and enantioselectivities were examined. All four enzymes cat
the asymmetric reduction of ECAA to synthesize (*S*)-ECHB as expected, but the
purity of ECHB obtained varied as shown in Table 25.4 [55]. ReAR1 was selected
best enzyme and a plasmid coexpressing ReAR1, and a GDH from *B. subtilis* (BsGDH

TABLE 25.4
Substrate Specificities and Enantioselectivities of β-Ketoacyl-ACP Reductases and Acetoacetyl-CoA Reductases

	Relative Activity (%)								
	ECAA		AcAc-SCoA		EAA		(*R*)-ECHB	(*S*)-ECHB	(*S*)-E
Enzyme	NADPH	NADH	NADPH	NADH	NADPH	NADH	NADP⁺		ee
EcKR1	100	<1	116	ND	1.9	ND	<1	<1	93
BsKR1	100	<1	22.3	ND	4.8	ND	1.3	1.3	98
ReAR1	100	2.3	39.3	ND	1.3	ND	<1	<1	99
ZrAR1	100	<1	284	ND	38.4	ND	<1	<1	99

The relative activities for substrates were expressed as percentage to the activity for ECAA.
AcAc-SCoA, acetoacetyl-CoA; EAA, ethyl acetoacetate; ND, not determined.

pSG-AER1, was constructed. (S)-ECHB was produced using whole E. coli cells harboring pSG-AER1 on a 1.2 L jar scale. The synthesis of (S)-ECHB was performed at 25°C and pH 6.5 in a working volume of 750 mL, as the substrate, ECAA, was unstable at more than pH 7.0 in an aqueous solution and at a higher temperature. To keep the concentration of ECAA in the reaction mixture at a low level, the initial concentration of ECAA was 2% (w/v), and ECAA was fed into the reaction mixture for 4 h at the rate of 7.5 g/L/h after 2 h of initiation. After 18 h, the amount of ECHB reached 48.7 g/L with 99.8% ee (S).

Furthermore, we screened and discovered two kinds of NADH-dependent ECAA reductases that had not been found. NADH-dependent ADH/carbonyl reductases have several advantages for manufacturing chiral compounds: (1) NAD(H) was chemically more stable than NADP(H); (2) the intracellular concentrations of NAD(H) were reported to be higher than those of NADPH in several microorganisms [57,58]; (3) FDH can be utilized for NADH regeneration instead of GDH. One was an NADH-dependent carbonyl reductase from K. aestuarii (KaCR1). KaCR1 was purified to homogeneity and its apparent subunit molecular mass was 32,000 on an SDS-PAGE [59]. The enzyme mainly used NADH as an electron donor, and NADHPH could replace NADH with only 6.4% activity. The enzyme had maximal activity at pH 5.0 to 5.5 and 45°C for the reduction of ECAA but no oxidative activity of (S)- or (R)-ECHB. These findings suggested that the enzyme functioned as an exclusive reductase. KaCR1 had restricted substrate specificity: it showed high activities for 4-chloroacetoacetate esters such as ethyl ester and methyl ester, and α,β-diketones such as 2,3-butanedione and 2,3-pentanedione; however, it showed none or little activity for ethyl acetoacetate without chlorine substitution at the 4-position, simple ketones, or aldehydes such as pyridine-3-aldehyde, a typical substrate for the aldo-keto reductase superfamily enzymes. The KaCR1 gene was cloned and was shown to contain an open reading frame of 876 nucleotides encoding a polypeptide of 292 amino acid residues with a calculated molecular weight of 31,687. The other was a novel (R)-specific ADH, (R)-2-octanol dehydrogenase (PfODH), from a methanol-assimilating yeast, Pichia finlandica. It was difficult to discover PfODH by conventional screening because it was a minor enzyme similar to ECAA reductases in P. finlandica. We screened using native polyacrylamide gel electrophoresis and differential activity staining. We found PfODH as an enzyme that had activities both for the reduction of ECAA and for the oxidation of (R)-2-octanol, and no activity for the oxidation of (S)-2-octanol in gels. PfODH was purified to homogeneity and characterized, and its apparent subunit molecular mass was 30,000 on an SDS-PAGE [60]. The enzyme was reversible and had maximal activity at pH 10.5 and 50°C for the oxidation of (R)-2-octanol, and at pH 6.0 and 55 to 60°C for the reduction of ECAA. The enzyme had very broad substrate specificity. The PfODH gene was cloned and was shown to contain an open reading frame of 765 nucleotides encoding a polypeptide of 254 amino acid residues with a calculated molecular weight of 27,143.

We obtained several kinds of ECAA reductases by innovative screening methods such as functional homology search, structural homology search, and differential activity staining. Properties of several ECAA reductases containing KaCR1 and PfODH are summarized in Table 25.5.

Both KaCR1 and PfODH reduced ECAA asymmetrically to synthesize (S)-ECHB in >99% ee. Coexpression plasmids of KaCR1 and PfODH with McFDH-26, pSFR426, and pSF-PFO3, respectively, were constructed for the use of whole-cell biocatalysts. McFHD-26 is a mutant FDH from M. vaccae having three amino acid substitutions [61], and the construction McFDH-26 is described in detail in Section 25.8. The synthesis of (S)-ECHB was examined with E. coli harboring pSFR426 and pSF-PFO3. Reaction conditions at a mini-jar scale were the same as those for E. coli cells harboring pSG-AER1. Using E. coli cells harboring pSFR426 and pSF-PFO3, the amount of ECHB reached 49.9 g/L in >99%

TABLE 25.5
Comparison of Properties of (S)-ECHB-Forming ECAA Reductases

Name	PfODH	KaCR1	ReAR1	KlCR1 (FASα)	CmCR-S1	LbR
Origin	Pichia finlandica	Kluyveromyces aestuarii	Ralstonia eutropha	K. lactis	Candida magnoliae	Lactob brev
E.C.	1.1.1.1	—	1.1.1.36	2.3.1.86	—	1.1.1.2
Cofactor	NADH	NADH	NADPH	NADPH	NADPH	NADH
Mr						
Native	83,000	85,000	84,000	>800,000	76,000	104,00
Subunit	30,000	32,000	23,000	190,000	32,000	27,000
ORF	254 aa (27,100)	292 aa (31,700)	246 aa (26,400)	—	283 aa (30,400)	252 aa (26,80
Optimum for reduction						
Temperature	55–60	45	—	45	50–55	50
pH	6.0	5.0–5.5	—	6.5	5.5–6.5	7.0
Activity						
Substrate	(R)-2-Octanol	ECAA	Acetoacetyl-CoA	ECAA	ECAA	Acetoa
Specific activity (U/mg)	91.1	28.6	44.0	50.8	13.5	489
Substrate specificity	Broad	Narrow	Narrow	Narrow	Broad	Broad
Reversibility	Reversible	Irreversible	Irreversible	Irreversible	Irreversible	Revers
Superfamily	SDR	SDR	SDR	SDR	SDR	SDR
References	[60]	[59]	[51]	[43]	[42]	[24]

Mr, relative molecular mass; ND, not determined; SDR, short-chan ADH; MDR, medium-chain ADH.

ee (S) and 49.4 g/L in >99% ee, respectively. The discovery of NADH-dependent reductases, such as KaCR1 and PfODH, and the use of McFDH-26 potentially the manufacturing of (S)-ECHB with low environmental burden.

25.5 SYNTHESIS OF SEVERAL CHIRAL ALCOHOLS WITH PfODH

Application of KaCR1was limited because of its narrow substrate specificity compare that of PfODH, which has very broad substrate specificity and high stability. Several alcohols such as (R)-alcohol, (S)-α-haloalcohol, and (R)-3-hydroxy esters, were synth in >99% ee from corresponding ketones with recombinant E. coli cells harboring PFO3 as shown in Figure 25.5. 2,3-Difluoro-6-nitro-[[(R)-2-hydroxypropyl]oxy]be (R)-FNHB, was a key intermediate for the synthesis of an anitibacterial agent, Levoflc (R)-FNHB was efficiently synthesized with 86% yield and >99% ee from the correspo ketone at 100 g/L input. 2-Acetylbutyrolactone, which had a chiral center at α-position carbonyl group, was reduced to (1'R,2S)- 2-(1-HEBL) in 99% ee and 90% de by syn-red accompanying dynamic resolution. As the PfODH had a high tolerance to α-haloketo was used to synthesize several α-haloalcohols such as (S)-2-chloro-1-phenylethanol, chloro-1-(4'-fluorophenyl)ethanol, and (S)-2-chloromethyl-3-pyridinemethanol in >9!

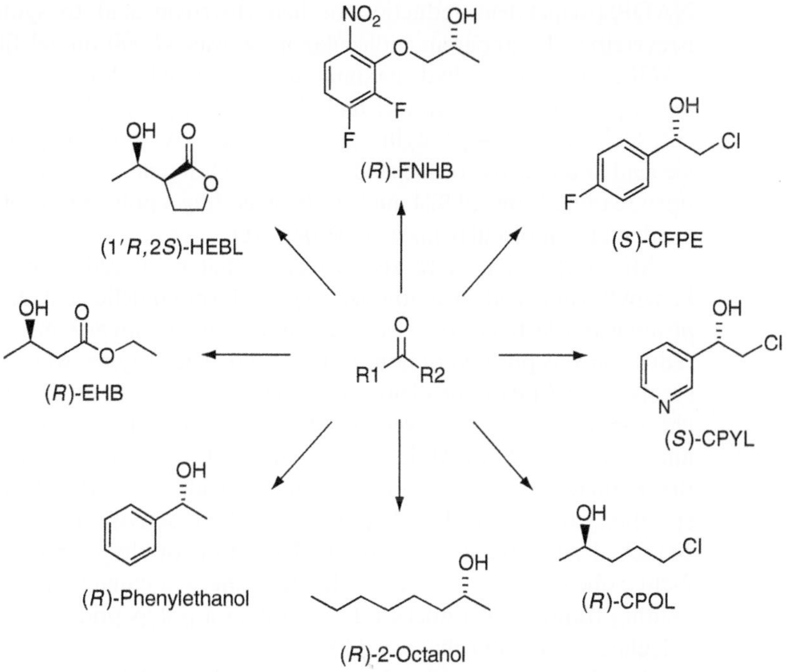

FIGURE 25.5 Synthesis of optically active alcohols with PfODH. (*R*)-FNHB, 2,3-difluoro-6-nitro-[[(*R*)-2-hydroxypropyl]oxy]benzene; (*S*)-CFPE, (*S*)-2-chloro-1-(4′-fluorophenyl)ethanol; (*S*)-CPYL, (*S*)-2-chloromethyl-3-pyridinemethanol; (*R*)-CPOL, (*R*)-5-chloro-2-pentanol; (*R*)-EHB, ethyl (*R*)-3-hydroxybutanoate; (1′*R*,2*S*)-HEBL, (1′*R*,2*S*)-2-(1′-hydroxyethyl)butyrolactone.

These compounds were easily transformed to corresponding chiral epoxides for versatile chemical synthesis. These results show that PfODH is a powerful enzyme for synthesis of optically active alcohols, especially α-haloalochols.

25.6 α-KETO ACID REDUCTASES FOR SYNTHESIS OF CHIRAL α-HYDROXY ACIDS

Chiral α-hydroxy acids are useful as building blocks and intermediates for the synthesis of several pharmaceuticals. (*R*)-benzyllactate ethyl ester is useful as an intermediate for the synthesis of angiotensin-converting enzyme (ACE) inhibitor [62,63]. (*R*)-Phenyllactic acid is useful as a precursor of anthelmintic cyclodepsipeptide PF1022A produced by *Mycelia sterilia* [64]. (*R*)-2-Chloromandelic acid, designated as (*R*)-2-CMA, is useful as an intermediate for the synthesis of an inhibitor of platelet aggregation and a fungicide [65,66]. (*R*)-2-CMA could be synthesized by diastereomeric resolution with chiral amines [67–69], enzymatic resolution [70], and asymmetrically enzymatic synthesis with hydroxynitrile lyases [71], D-lactate dehydrogenase [72,73], and D-hydroxyisocaproate dehydrogenases [74].

We found unique α-keto acid reductases that had wide substrate specificity by screening with a variety of substrates such as benzylpyruvic acid and 2-chlorobenzoylformic acid. At first, we screened microorganisms that catalyzed asymmetric reduction of benzylpyruvic acid to synthesize (*R*)-benzyllactic acid. *Leuconostoc oenos* (*L. dextranicum* subsp. *vinerium*) was selected as the best microorganism, and an enzyme responsible for the asymmetric reduction was purified to homogeneity [75]. The enzyme, which was designated as LoKAR1, catalyzed

NADPH-dependent reduction of benzylpyruvic acid to synthesize (R)-benzyllactic
irreversibly. Its apparent molecular mass was 47,000 on gel filtration and 33,000 on
PAGE. The enzyme had maximal activity at pH 6.5 to 7.0 and 40°C for the reducti
benzylpyruvic acid. LoKAR1 had higher activity for substrates having a long-chain sub
ent, such as 2-oxo-4-phenylbutyric acid and 2-oxo-5-phenylpentanoic acid, than phenyl
vic acid and benzoylformic acid. The LoKAR1 gene was cloned and was shown to cont
open reading frame of 924 nucleotides encoding a polypeptide of 307 amino acid residue
a calculated molecular mass of 33,049 [76].

Microorganisms were also screened that catalyzed asymmetric reduction of 2-cl
benzoylformic acid to synthesize (R)-2-chloromandelic acid. L. mesenteroides subsp. d
nicum was selected as the best microorganism, and an enzyme responsible for the asymm
reduction was purified to homogeneity [77]. The enzyme, which was designated as LmK
catalyzed NADH-dependent reduction of 2-chlorobenzoylformic acid to synthesize
chloromandelic acid, irreversibly. Its apparent molecular mass was 62,600 on gel filt
and 34,100 on SDS-PAGE. The enzyme had maximal activity at pH 5.5 to 6.0 and 50
the reduction of 2-chlorobenzoylformic acid. LmKAR1 had quite contrasting sul
specificity to LoKAR1. It had extremely high activity for substrates having a bulky a
branched substituent at a proximal position of the carbonyl group, e.g., 2,620 U/n
2-chlorobenzoylformic acid. LmKAR1 gene was cloned and was shown to contain ar
reading frame of 954 nucleotides encoding a polypeptide of 317 amino acid residues
calculated molecular mass of 33,872.

Another α-keto acid reductase, D-mandelate dehydrogenase, was purified from L
coccus faecalis and designated as EfDMDH [78]. Its apparent molecular mass was 130,0
gel filtration and 40,000 on SDS-PAGE. The enzyme had maximal activity at pH 4.5 a
to 45°C for the reduction of benzoylformic acid. EfDMDH had high activity for benze
mate, phenylpyruvate, and 2-oxoisocaproate. The substrate specificity of EfDMDl
complementary to those of LoKAR1 and LmKAR1. Enzymatic properties and sul
specificity of α-keto acid reductases are summarized in Table 25.6.

25.7 SYNTHESIS OF (R)-α-HYDROXY ACIDS WITH α-KETO ACID REDUCTASES

Various (R)-α-hydroxy acids were synthesized with whole-cell biocatalysts express
appropriate α-keto acid reductase for a target compound, as shown in Figure 25.
LoKAR1, a coexpression plasmid—pSG-LOR1—was constructed which coexp
BsGDH for the regeneration of NADPH. For EfDMDH and LmKAR1, coexp
plasmids—pSF-EFM2 and pSF-LMK1 respectively—with McFDH-26 were cons
for the regeneration of NADH.

Ethyl (R)-benzyllactic acid is an intermediate for the synthesis of ACE inhibi
could be synthesized by several biological reduction processes: (1) asymmetric reduc
ethyl benzylpyruvate; (2) asymmetric reduction of benzylpyruvic acid followed by ch
esterification; (3) asymmetric reduction of 2-oxo-4-phenyl-3-butenoic acid follow
chemical hydrogenation and esterification, as shown in Figure 25.7. Ethyl benzylpy
could be reduced to ethyl (R)-benzyllactate with some of our library enzymes, e.g.,
(a carbonyl reductase from Torulaspora delbrueckii). Benzylpuruvic acid
2-oxo-4-phenyl-3-butenoic acid, alternatively, could be reduced to (R)-benzyllact
and (R)-2-hydroxy-4-phenyl-3-butenoic acid, respectively, by LoKAR1. All enz
reductions were very effectively performed and the reductive process of benzyl
acid with LoKAR1 was selected further because of productivity and easy purifica

TABLE 25.6
Properties of α-Keto Acid Reductases and α-Hydroxy Acid Dehydrogenases

Name	EfDMDH	LmKAR1	LoKAR1	LmDLDH	LcDHicDH
Origin	*Enterococcus faecalis*	*Leuconostoc mesenteroides* subsp.*dextranicum*	*L. oenos*	*L. mesenteroides* subsp. *cremoris*	*Lactobacillus casei* subsp. *pseudo-plantarum*
E.C.	—	—	—	—	—
Cofactor	NADH	NADH	NADPH	NADH	NADH
Mr					
Native	130,000	62,600	46,000	80,000	74,000
Subunit	40,000	34,100	46,000	40,000	38,000
ORF	301 aa (33,100)	317 aa (33,900)	307 aa (33,049)	331 aa (36,316)	333 aa (36,893)
Optimum for reduction					
Temperature(°C)	35–45	50	40	ND	50
pH	4.5	5.5–6.0	6.5–7.0	ND	5.5–7.0
Activity					
Substrate	Benzoylformic acid	2-Chloro benzoylformic acid	Benzylpyruvic acid	Pyruvic acid	α-Ketoiso caproic acid
Specific activity (U/mg)	378	2,620	3.45	544	24.9
Substrate specificity (relative activity, %)					
Pyruvic acid	<1	<1	ND	100	ND
α-Ketoisoca proic acid	48.8	14.7	12.0	2.2	100
Benzoylfor mic acid	100	12.9	2.0	1.5	ND
2-Chloro-benzoyl formic acid	23.7	100	<1	ND	ND
Phenylpyruvic acid	36.0	<1	12.0	3.2	251
Benzylpyruvic acid	<1	<1	100	ND	ND
Reversibility	Significantly irreversible	Irreversible	Irreversible	Reversible	Reversible
References	[78]	[77]	[75,76]	[79]	[80,81]

Mr, relative molecular mass; aa, amino acids; ND, not determined.

product. (*R*)-Benzyllactic acid was synthesized from 2% benzylpyruvic acid and D-glucose with *E. coli* cells coexpressing LoKAR1 and BsGDH with 17.2 g/L concentration in >99% ee. By a similar process, (*R*)-2-hydroxy-5-phenylpentanoic acid and their derivatives could effectively be synthesized in >99% ee from their corresponding α-keto acids.

(*R*)-2-CMA is useful as an intermediate for the synthesis of several pharmaceuticals, such as an inhibitor of platelet aggregation and a reagent for diastereomeric resolution of several chiral amines. D-Lactate dehydrogenases from *Lactobacillus*, D-mandelate dehydrogenase

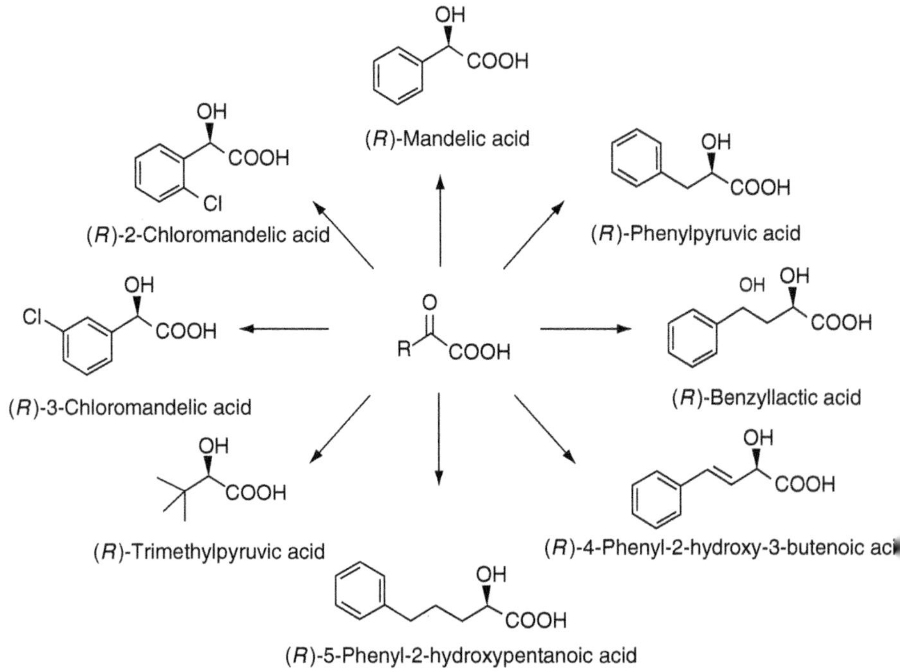

FIGURE 25.6 Synthesis of optically active α-hydroxy acids using several α-keto acid reductases

such as EfDMDH, and LoKAR1 could not efficiently synthesize (R)-2-CMA fr
chlorobenzoylformic acid because the substitution of chlorine on *o*-position, in g
decreased enzyme activity and stability. LmKAR1 is preferred to 2-chlorobenzoyl
acid as a substrate over benzoylformic acid and 3-chlorobenzoylformic acid, and i
efficiently synthesized (R)-2-CMA. The parent strain of LmKAR1, *L. mesenteroides*
dextranicum synthesized (R)-2-CMA with 26.2 g/L in >99% ee. The reaction, ho
required excess amount of glucose for the regeneration of NADPH and D-gluconat
duced, as the by-product of the regeneration system caused an extra environmental b

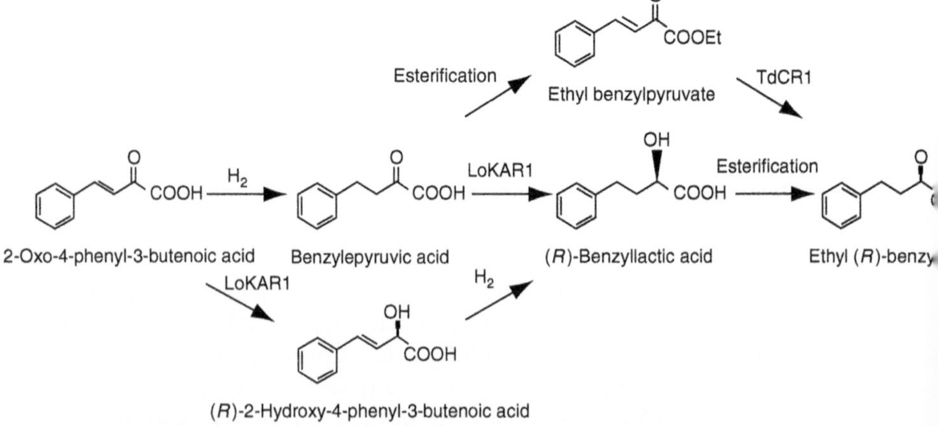

FIGURE 25.7 Synthetic routes of ethyl (R)-benzyllactate LoKAR1, α-keto acid reductase from
nostoc oenos. TdCR1, carbonyl reductase from *Torulaspora delbrueckii*.

The use of recombinant *E. coli* cells coexpressing LmKAR1 and McFDH-26 as a biocatalyst enabled more efficient production of (*R*)-2-CMA with 68.0 g/L in >99% ee from 7% 2-chlorobenzoylformic acid, and formic acid with carbon dioxide as a by-product caused lesser environmental burden. By a similar process, (*R*)-trimethyllactic acid was synthesized quantitively from 5% trimethylpyruvic acid in >99% ee.

(*R*)-Phenyllactic acid and D-leucinic acid [(*R*)-2-hydroxyisocaproic acid] are also useful as raw materials and intermediates for the synthesis of pharmaceuticals. These compounds and their derivatives were efficiently synthesized with *E. coli* cells coexpressing EfDMDH and McFDH-26. As described earlier, our α-keto acid reductases library having distinct substrate specificity enabled extremely broad application in reduction processes.

25.8 ROBUST NADH REGENERATOR, McFDH-26

Biological asymmetric reduction requires an efficient regeneration system of the coenzyme NAD(P)H, especially in recombinant *E. coli* cells that highly express ADHs and carbonyl reductases, as *E. coli* cells have poor coenzyme-regeneration activities [11,82]. To overcome this problem, a GDH from *B. megaterium* as an NADPH-regeneration system was coexpressed in *E. coli* with an aldehyde reductase from *S. salmonicolor* [9]. Currently, the coenzyme-regeneration method using GDH is the most frequently used because it can regenerate not only NADPH but also NADH, and has high activity in broad pH range as well as high stability against several organic compounds. However, GDHs have several problems: gluconate is generated as a by-product in an amount equimolar to the desired optically active alcohol, which interferes with purification and is not environment-friendly.

NAD$^+$-dependent FDHs (E.C. 1.2.1.2), have also been used for the regeneration of NADH *in vitro* [83]. The coenzyme-regeneration method using FDH has several advantages: (1) its reaction is irreversible; (2) formic acid as its substrate is a small molecule and is inexpensive; (3) carbon dioxide as its product does not inhibit and inactivate carbonyl reductases involved and is easily removed from a target product; and (4) its environmental burden is relatively low. FDH, however, has a low activity and is labile to heat, several metals, oxidation by air, and hydrophobic compounds, especially α-haloketones.

Using an FDH from *M. vaccae* [30], which was one of the highest specific-activity and stable FDHs reported previously, we compared FDH with GDH as an NADH regenerator to produce (*S*)-ECHB from ECAA with *E. coli* cells expressing KaCR1 [59]. Using BsGDH, the amount of ECHB reached 45.6 g/L in >99% ee (*S*), as shown in Figure 25.8. The use of BsGDH yielded 45.6 g/L of (*S*)-ECHB from 50 g/L of ECAA, but the use of McFDH yielded

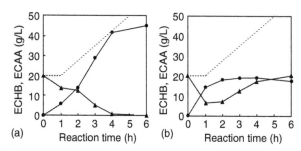

FIGURE 25.8 Synthesis of (*S*)-ECHB with *E. coli* cells haboring pSG-KAR1 and pSFR415. (a) pSG-KAR1, a coexpression plasmid of KaCR1 and BsGDH; (b) pSFR415, a coexpression plasmid of KaCR1 and McFDH. Symbols: ●, ECHB; ▲, ECAA. The total amount of ECAA added is shown by a dotted line.

only 19.0 g/L of ECHB in >99% ee (S). The low productivity in the case of FD suggested to result from the low activity and instability of FDH against ECAA.

On the assumption that the modification of some Cys residues by ECAA cau inactivation of FDH, we examined the substitution of five Cys residues by site-o mutagenesis (Cys-6, -146, -249, -256, and -355) among seven Cys residues that w conserved in several FDHs [84]. Mutant FDHs were evaluated by enzyme activ amounts of (S)-ECHB produced from 3% ECAA with E. coli cells coexpressing the l gene without NAD$^+$, which had a protective effect against the inactivation by ECAA. and C256S mutants yielded a higher amount (19.3 g/L and 30.5 g/L, respectiv (S)-ECHB than wild-type FDH (17.5 g/L). Cys-146 mutations (C146S and C6S C256S) did not show as remarkable effects as Cys-256 mutations, but yielded larger a of (S)-ECHB in spite of lower FDH activity. Cys-6, Cys-146, and Cys-256 residu substituted by Ala, Ser, and Val in several combinations, and in consequence a C6A C256V mutant, designated as McFDH-26, was selected as the best, as shown in Tab E. coli cells coexpressing an McFDH-26 gene and a KaCR1 gene had 8.15 U/mL-c broth of FDH activity and produced 32.2 g/L of (S)-ECHB from 30 g/L of ECAA mutants were characterized from the viewpoint of the tolerance to ECAA. Substitu Cys-256 by Ala, Ser, and Val gave higher tolerance to ECAA, as shown in Table 25.8 course of experiments on the inactivation of the enzyme, ECAA, ethyl acetoacetate, al acetate were found to activate FDH, though some mutant FDHs were inactivated less by incubation with ECAA for 20 min. C146S and C146A mutants, especiall activated to 140 to 220%, while FDHs that had Cys-146 were activated only slightly to 120%. These findings strongly suggested that Cys-256 mutations caused the to against ECAA, and Cys-146 mutations caused the activation by several organic comp McFDH-26 was evaluated from the productivity of (S)-ECHB on a mini-jar scale E. coli cells coexpressing McFDH-26 (C6A/C146S/C256V) and KaCR1, the amc ECHB reached 49.1 g/L in >99% ee (S), as shown in Figure 25.9. These results show McFDH-26 had comparative capacity to GDH as an NADH regenerator. Recently haloketone-resistant formate dehydrogenase (TsFDH) was obtained from Thiobaci

---

TABLE 25.7
Effects of Substitutions of Cys-Residues by Ala, Ser, and Val in McFDH

| ORF | Amino Acid Residues | | | FDH |
No.	6-Cys	146-Cys	256-Cys	(U/mg)
Wild	—	—	—	3.24
11	Ser	—	Ser	0.285
16	Ala	—	Ser	1.98
17	Val	—	Ser	1.62
18	Ser	—	Ala	0.635
19	Ser	—	Val	0.504
25	Ala	—	Val	1.83
23	—	—	Val	2.58
24	—	Ser	Val	2.29
26	Ala	Ser	Val	2.65
27	Ala	Ala	Val	1.15
28	Ala	Val	Val	1.49

(S)-ECHB, the amount of (S)-ECHB synthesized from 30 g/L ECAA with each recombinant whole-cell bio without NAD$^+$.

TABLE 25.8
Effects of Mutant FDHs on the Resistance against ECAA and the Activation by Ethyl Acetate

ORF No.	6-Cys	146-Cys	256-Cys	(S)-ECHB[a] (g/L)	Activation[b] by AcOEt (%)	Tolerance[c] against ECAA (%)
15	—	—	—	17.5	121	7.43
9	Ser	—	—	9.4	ND	6.58
23	—	—	Val	30.6	111	100
11	Ser	—	Ser	15.9	105	125
16	Ala	—	Ser	25.8	109	97.4
17	Val	—	Ser	25.2	105	92.1
18	Ser	—	Ala	22.4	125	120
19	Ser	—	Val	25.5	110	94.0
21	—	—	Ser	29.1	125	99.3
20	—	Ser	—	19.3	181	7.67
12	Ser	Ser	Ser	24.5	225	116
22	—	Ser	Ser	25.9	224	118
24	—	Ser	Val	31.0	185	104
25	Ala	—	Val	29.6	111	94.5
26	Ala	Ser	Val	32.2	137	108
27	Ala	Ala	Val	27.7	219	106
28	Ala	Val	Val	26.9	84.5	100

[a](S)-ECHB was synthesized from 3% ECAA without NAD$^+$ using whole cells in a flask.
[b]FDH was assayed with 5% AcOEt. To calculate the relative activity, the activity without AcOEt was taken as 100%.
[c]FDH was assayed after the incubation for 20 min with 20 mM ECAA at 25°C. To calculate the relative activity, the activity without incubation was taken as 100%.

strain KNK65MA by screening [85]. The corresponding residue of TsFDH to 256-Cys of McFDH is 256-Val, similarly to McFDH-26, and 256-Val of TsFDH was thought to cause the resistance of TsFDH against α-haloketone. 145-Cys residue of PsFDH corresponding to 146-Cys of McFDH was reported to be located in the region of the intersubunit contact [86] and not modified by DTNB even in the presence of 8 M urea to dissociate the enzyme into individual subunits [87]. C146A, C146S, and C146V mutations may strengthen an intersubunit binding. The relationship between this speculation and the activation by organic solvent remains to be unclear. The tolerance to organic solvent and activation of enzyme by organic solvent is very useful for bioconversion, because most artificial substrates for pharmaceutical intermediates are hydrophobic.

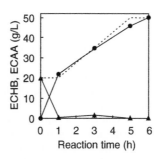

FIGURE 25.9 Synthesis of (S)-ECHB with *E. coli* cells coexpressing KaCR1 and McFDH-26. Symbols: ●, ECHB; ▲, ECAA. The total amount of ECAA added is shown by a dotted line.

25.9   REMARKS

The cloned enzyme library has several advantages: (1) highly active and regio- and selective; (2) speedy enzyme screening because their characteristics, such as su specificity and stereoselectivity, are well known; (3) easy scale-up of process becaus host is common *E. coli* and culture, induction, bioconversion, and purification m may be quite similar; (4) safe and green (white) process as the host *E. coli* K-12 st nonpathogenic to humans and animals, and is not viable in the environment. W constructed a cloned enzyme library comprising 140 unique enzymes having di substrate specificity and enatioselectivity, and 70 ADHs and carbonyl reductases. C tional enzyme screening has been significantly replaced by enzyme library screening company. Not only conventional screening but also the use of environmental DNA ing will continue to be necessary for broader applicability of biocatalytic process enrichment of the enzyme library. As shown in this chapter, an McFDH mutant, McFI has solved some of the problems associated with NADH cofactor regeneration. Incap of NADPH regeneration by FDH, however, remains to be solved on an industrial s biphasic system, such as *n*-butyl acetate and water, improved enzyme stability, solub substrates and products, and space yields [88,89]. These innovations have increas number of industrial manufacturing by asymmetric reduction. For continuous im ment in commercial production of chiral compounds by asymmetric reduction, innovations are necessary. One is an artificial design of a biocatalyst including the alte of substrate specificity and some properties of enzymes by directed evolution ar engineering of an expression host. Improvement in process engineering and down processing is required as the regulation of impurities in a product has become increa important in the pharmaceutical application.

REFERENCES

1. Hollingsworth, R.I. and Wang, G., General three carbon chiral synthons from carbohydrate pool and chiral auxiliary approaches, *ACS Symp. Ser.*, 841, 85, 2003.
2. Yamada, S. et al., Combined effect of polymorphism and process on preferential crystall example with ($\pm$)-5(4′-methylphenyl)-5-methylhydantoin, *Tetrahedron Asymmetry*, 9, 1713
3. Kinbara, K. et al., Effect of a substituent on an aromatic group in diastereomeric res *Tetrahedron*, 56, 6651, 2000.
4. Ghanem, A. and Aboul-Enein, H.Y., Application of lipases in kinetic resolution of rac *Chirality*, 17, 1, 2004.
5. Franco, P. et al., The impact of productivity in the preparative separation of enantio chromatography, *Chimica Oggi*, 22, 28, 2004.
6. Camps, P. and Munoz-Torrero, D., Synthesis and applications of (*R*)- and (*S*)-pantolac chiral auxiliaries, *Curr. Org. Chem.*, 8, 1339, 2004.
7. Ohkuma, T. and Noyori, R., *Comprehensive Asymmetric Catalysis*, Suppl. 1, Springer Berlin, 2004, p. 1.
8. Patel, R.N., Microbial/enzymatic synthesis of chiral intermediates for pharmaceuticals, *Microb. Technol.*, 31, 804, 2002.
9. Kataoka, M. et al., Stereoselective reduction of ethyl 4-chloro-3-oxobutanoate by *Escheri* transformant cells coexpressing the aldehyde reductase and glucose dehydrogenase gene *Microbiol. Biotechnol.*, 51, 486, 1999.
10. Galkin, A. et al., Synthesis of optically active amino acids from α-keto acids with *Escheri* cells expressing heterologous genes, *Appl. Environ. Microbiol.*, 63, 4651, 1997.
11. Yamamoto, H., Matsuyama, A., and Kobayashi, Y., Synthesis of ethyl (*R*)-4-cl hydroxybutanoate with recombinant *Escherichia coli* cells expressing (*S*)-specific secondary dehydrogenase, *Biosci. Biotechnol. Biochem.*, 66, 481, 2002.

12. Tanaka, R., Iwata, H., and Ishiguro, M., Synthesis of 5,6-*cis*-penems, *J. Antibiot.*, 43, 1608, 1990.

13. Nakatsuka, T. et al., A facile conversion of the phenylthio group to acetoxy by copper reagents for a practical synthesis of 4-acetoxyazetidin-2-one derivatives from (*R*)-butane-1,3-diol, *J. Chem. Soc. Chem. Commun.*, 662, 1991.

14. Hart, D.J. and Ha, D.C., An enantioselective approach to carbapenem antibiotics: formal synthesis of (+)-thienamycin, *Tetrahedron Lett.*, 26, 5493, 1985.

15. Matsuyama, A., Kobayashi, Y., and Ohnishi, H., Microbial production of optically active 1,3-butanediol from 4-hydroxy-2-butanone, *Biosci. Biotech. Biochem.*, 57, 348, 1993.

16. Matsuyama, A., Kobayashi, Y., and Ohnishi, H., Microbial production of optically active 1,3-butanediol from the racemate, *Biosci. Biotechnol. Biochem.*, 57, 685, 1993.

17. Yamamoto, H. et al., Purification and characterization of (*S*)-1,3-butanediol dehydrogenase from *Candida parapsilosis*, *Biosci. Biotechnol. Biochem.*, 59, 1769, 1995.

18. Yamamoto, H. et al., Cloning and expression in *Escherichia coli* of a gene coding for a secondary alcohol dehydrogenase from *Candida parapsilosis*, *Biosci. Biotechnol. Biochem.*, 63, 1051, 1999.

19. Reid, M.F. and Fewson, C.A., Molecular characterization of microbial alcohol dehydrogenases, *Crit. Rev. Microbiol.*, 20, 13, 1994.

20. Peters, J., Minuch, T., and Kula, M.-R., A novel NADH-dependent carbonyl reductase with an extremely broad substrate range from *Candida parapsilosis*: purification and characterization, *Enzyme Microb. Technol.*, 15, 950, 1993.

21. Itoh, N. et al., Purification and characterization of phenylacetaldehyde reductase from a styrene-assimilating *Corynebacterium* strain ST-10. *Appl. Environ. Microbiol.*, 63, 3783, 1997.

22. Wang, J.-C. et al., Cloning, sequence analysis, and expression in *Escherichia coli* of the gene encoding phenylacetaldehyde reductase from styrene-assimilating *Corynebacterium* sp. strain ST-10, *Appl. Microbiol. Biotechnol.*, 52, 386, 1999.

23. Xie, S.-X., Ogawa, J., and Shimizu, S., NAD$^+$-dependent (*S*)-specific secondary alcohol dehydrogenase involved in stereoinversion of 3-pentyn-2-ol catalyzed by *Nocardia fusca* AKU 2123, *Biosci. Biotech. Biochem.*, 63, 1721, 1999.

24. Hummel, W., New alcohol dehydrogenases for the synthesis of chiral compounds, *Adv. Appl. Microbiol.*, 58, 145, 1997.

25. Abokitsel, K. and Hummel, W., Cloning, sequence analysis, and heterologous expression of the gene encoding an (*S*)-specific alcohol dehydrogenase from *Rhodococcus etythropolis* DSM 43297, *Appl. Microbiol. Biotechnol.*, 62, 380, 2003.

26. Stampfer, W. et al., Biocatalytic asymmetric hydrogen transfer, *Angew. Chem. Int. Ed. Eng.*, 41, 1014, 2002.

27. Yamamoto, H., Matsuyama, A., and Kobayashi, Y., Synthesis of (*R*)-1,3-butanediol by enantioselective oxidation using whole recombinant *Escherichia coli* cells expressing (*S*)-specific secondary alcohol dehydrogenase, *Biosci. Biotechnol. Biochem.*, 66, 925, 2002.

28. Anraku, Y. and Gennis, R.B., The aerobic respiratory chain of *Escherichia coli*, *Trends Biochem. Sci.*, 12, 262, 1987.

29. Kasahara, M. and Anraku, Y., Succinate- and NADH oxidase systems of *Escherichia coli* membrane vesicles: mechanism of selective inhibition of the systems by zinc ions, *J. Biochem.*, 76, 967, 1974.

30. Karzanov, V.V. et al., An alternative NAD$^+$-dependent formate dehydrogenases in the facultive methylotroph *Mycobacterium vaccae* 10, *FEMS Microbiol. Lett.*, 81, 95, 1991.

31. Galkin, A. et al., Cloning of formate dehydrogenase gene from a methanol-utilizing bacterium *Mycobacterium vaccae* N10, *Appl. Microbiol. Biotechnol.*, 44, 479, 1995.

32. Zhou, B. et al., Stereochemical control of yeast reductions. 1. Asymmetric synthesis of L-carnitine, *J. Am. Chem. Soc.*, 105, 5925, 1983.

33. Kitamura, M. et al., A Practical asymmetric synthesis of carnitine, *Tetrahedron Lett.*, 29, 1555, 1988.

34. Jung, M.E. and Shaw, T.J., Total synthesis of (*R*)-glycerol acetonide and the antiepileptic and hypotensive drug (−)-γ-amino-β-hydroxybutyric acid (GABOB): use of vitamin C as a chiral starting material, *J. Am. Chem. Soc.*, 102, 6304, 1980.

35. Pifferi, G. and Pinza, M., Cyclic GABA-GABOB analogues. I. Synthesis of new 4-hy(pyrrolidone derivatives, *Farmaco. Ed. Sci.*, 32, 602, 1977.

36. Karanewsky, D.S. et al., Phosphorous-containing inhibitors of HMG-CoA reductase. (arylethyl)hydroxyphosphinyl]-3-hydroxy-butanoic acids: a new class of cell-sensitive inhit cholesterol biosynthesis, *J. Med. Chem.*, 33, 2925, 1990.

37. Santaniello, E., Casati, R., and Milani, F., Chiral synthesis of a component of *Amanita m* (−)-4-hydroxypyrrolidin-2-one, and assessment of its absolute configuration, *J. Chem. Re* opsis), 132, 1984.

38. Shieh, W., Gopalan, A.S., and Sih, C.J., Stereochemical control of yeast reductions. 5. Ch ization of the oxidoreductases involved in the reduction of β-keto esters, *J. Am. Chem. S* 2993, 1985.

39. Patel, R.N. et al., Stereoselective reduction of β-keto esters by *Geotrichum candidum*, *Microb. Technol.*, 14, 731, 1992.

40. Kita, K. et al., Purification and characterization of new aldehyde reductases from *Sporobo salmonicolor* AKU4429, *J. Mol. Catal. B Enzym.*, 6, 305, 1999.

41. Kataoka, M. et al., A novel NADPH-dependent carbonyl reductase of *Candida maced* purification and characterization, *Arch. Biochem. Biophys.*, 294, 469, 1992.

42. Wada, M. et al., Purification and characterization of NADPH-dependent carbonyl re involved in stereoselective reduction of ethyl 4-chloro-3-oxobutanoate, from *Candida m* *Biosci. Biotech. Biochem.*, 62, 280, 1998.

43. Yamamoto, H. et al., Purification and properties of a carbonyl reductase that is useful production of ethyl (*S*)-4-chloro-3-hydroxybutanoate from *Kluyveromyces lactis*, *Biosci. Bi* *Biochem.*, 66, 1775, 2002.

44. Stoops, J.K. et al., Studies on the yeast fatty acid synthetase, *J. Biol. Chem.*, 253, 4464, 19

45. Wakil, S.J., Stoops, J.K., and Joshi, V.C., Fatty acid synthesis and its regulation, *A Biochem.*, 52, 537, 1983.

46. Toomey, R.E. and Walkil, S.J., Studies on the mechanism of fatty acid synthesis. XV. Pre and general properties of β-ketoacyl acyl carrier protein reductase from *Escherichia coli*, *Biophys. Acta*, 116, 189, 1966.

47. Rawlings, M. and Cronan, J.E. Jr., The gene encoding *Escherichia coli* acyl carrier pro within a cluster of fatty acid biosynthetic genes, *J. Biol. Chem.*, 267, 5751, 1992.

48. Schubert, P., Steinbuchel, A., and Schlegel, H.G., Cloning of the *Alcaligenes eutrophus* g synthesis of poly-β-hydroxybutyric acid (PHB) and synthesis of PHB in *Escherich J. Bacteriol.*, 170, 5837, 1988.

49. Vagelos, P.R., Alberts, A.W., and Majerus, P.W., β-Ketoacyl acyl carrier protein reductas *Enzymol.*, 14, 60, 1969.

50. Morbidoni, H.R., de Mendoza, D., and Cronan, J.E. Jr., *Bacillus subtilis* acyl carrier p encoded in a cluster of lipid biosynthesis genes, *J. Bacteriol.*, 178, 4794, 1996.

51. Haywood, G.W. et al., The role of NADH- and NADPH-linked acetoacetyl-CoA reductas poly-3-hydroxybutyrate synthesizing organism *Alcaligenes eutrophus*, *FEMS Microbiol. L* 259, 1988.

52. Peoples, O.P. and Sinskey, A.J., Poly-β-hydroxybutyrate biosynthesis in *Alcaligenes eutrop J. Biol. Chem.*, 264, 15293, 1989.

53. Fukui, T. et al., Purification and characterization of NADP-linked acetoacetyl-CoA reduct *Zoogloea ramigera* I-16-M, *Biochim. Biophys. Acta*, 917, 365, 1987.

54. Peoples, O.P. and Sinskey, A.J., Fine structural analysis of the *Zoogloea ramigera phbA-ph* encoding β-ketothiokase and acetoacetyl-CoA reductase: nucleotide sequence of *phbB*, *Mol biol.*, 3, 349, 1989.

55. Yamamoto, H., Matsuyama, A., and Kobayashi, Y., Synthesis of ethyl (*S*)-4-chloro-3-hyd tanoate using *fabG*-homologues, *Appl. Microbiol. Biotechnol.*, 61, 133, 2003.

56. Lampel, K.A. et al., Characterization of the developmentally regulated *Bacillus subtilis* dehydrogenase gene, *J. Bacteriol.*, 166, 238, 1986.

57. Takebe, I. and Kitahara, K., Levels of nicotinamide nucleotide coenzymes in lactic acid *J. Gen. Appl. Microbiol.*, 9, 31, 1963.

58. London, J. and Knight, M., Concentration of nicotinamide nucleotide coenzymes in micro-organism, *J. Gen. Microbiol.*, 44, 241, 1966.

59. Yamamoto, H. et al., A novel NADH-dependent carbonyl reductase from *Kluyveromyces aestuarii* and comparison of NADH-regeneration systems for the synthesis of ethyl (*S*)-4-chloro-3-hydroxybutanoate, *Biosci. Biotechnol. Biochem.*, 68, 638, 2004.

60. Kudoh, M. and Yamamoto, H., (*R*)-2-Octanol dehydrogenase, process for producing the enzyme, DNA encoding the enzyme and process for producing alcohol by using the same, WO Patent Appl. 2001061014, 2001.

61. Yamamoto, H. et al., Robust NADH-regenerator: improved α-haloketone-resistant formate dehydrogenase, *Appl. Microbiol. Biotechnol.*, 67, 33, 2005.

62. Urbach, H. and Henning, R., A favourable diastereoselective synthesis of *N*-(1-*S*-ethoxycarbonyl-3-phenylpropyl)-9*S*-alanine, *Tetrahedron Lett.*, 25, 1143, 1984.

63. Boyer, S.K. et al., Note on the synthesis of an optically active ACE inhibitor with amino-oxo-benzazepine-1-alkanoic-acid structure by means of an enantioconvergent crystallization-based resolution, *Helv. Chim. Acta*, 71, 337, 1988.

64. Sasaki, T. et al., A new anthelmintic cyclodepsipeptide, PF1022A, *J. Antibiot.*, 45, 692, 1992.

65. Badorc, A. and Frehel, D., Dextro-rotatory enantiomer of methyl alpha-5 (4,5,6,7-tetrahydro(3,2-c) thieno pyridyl) (2-chlorophenyl)-acetate and the pharmaceutical compositions containing it, U.S. Patent Appl. 4,847,265, 1989.

66. Seo, A., Hiraga, K., and Omi, T., Preparation of optically active ketene dithioketal derivative and its use as medicinal fungicide (in Japanese), Jap. Patent Appl. 02–275877, 1990.

67. Biquard, D., Absorption of certain classes of organic molecules, *Ann. Chim.*, 20, 97, 1933.

68. Collet, A. and Jacques, J., Optical antipode mixtures. V. Substituted mandelic acids, *Bull. Soc. Chim. France*, 12, 3330, 1973.

69. Valente, E. et al., Discrimination in resolving systems. II. Ephedrine-substituted mandelic acids, *Chirality*, 7, 652, 1995.

70. Yamamoto, K. et al., Production of *R*-(−)-mandelic acid from mandelonitrile by *Alcaligenes faecalis* ATCC 8750, *Appl. Environ. Microbiol.*, 57, 3028, 1991.

71. Glieder, A. et al., Comprehensive step-by-step engineering of an (*R*)-hydroxynitrile lyase for large-scale asymmetric synthesis, *Angew. Chem. Int. Ed. Eng.*, 42, 4815, 2003.

72. Simon, E.S., Plante, R., and Whitesides, G.M., D-Lactate dehydrogenase substrate specificity and use as a catalyst in the synthesis of homochiral 2-hydroxy acids, *Appl. Biochem. Biotechnol.*, 22, 169, 1989.

73. Hummel, W., Schutte, H., and Kula, M.R., Large-scale production of D-lactate dehydrogenase for the stereospecific reduction of pyruvate and phenylpyruvate, *Eur. J. Appl. Microbiol. Biotechnol.*, 18, 75, 1983.

74. Hummel, W. et al., New NADH-dependent dehydrogenases from *Lactobacillus* strains for synthesis and analysis L- and D-hydroxyisocaproate dehydrogenase, D-mandelate dehydrogenase and (+)-acetoin dehydrogenase large-scale purification, *Biocatalysis*, 4, 79–80, 1990.

75. Nakajima, N. et al., Purification and characterization of aldehyde reductase from *Leuconostoc dextranicum*, *Biosci. Biotech. Biochem.*, 57, 160, 1993.

76. Nakajima, T. et al., unpublished data, 2005.

77. Yamamoto, H. and Kimoto, N., α-Keto acid reductase, method for producing the same, and method for producing optically active α-hydroxy acids using the same, U.S. Patent Appl. 2004086993, 2004.

78. Tamura, Y. et al., Two forms of NAD-dependent D-mandelate dehydrogenase in *Enterococcus faecalis* IAM 10071, *Appl. Environ. Microbiol.*, 68, 947, 2002.

79. Dartois, V. et al., Purification, properties and DNA sequence of the D-lactate dehydrogenase from *Leuconostoc mesenteroides* subsp. *cremoris*, *Res. Microbiol.*, 146, 291, 1995.

80. Hummel, W., Schutte, H., and Kula, M.R., D-2-Hydroxyisocaproate dehydrogenase from *Lactobacillus casei*: a new enzyme suitable for stereospecific reduction of 2-ketocarboxylic acids, *Appl. Microbiol. Biotechnol.*, 21, 7, 1985.

81. Lerch, H.P. et al., Cloning, sequencing and expression in *Escherichia coli* of D-2-hydroxyisocaproate dehydrogenase gene of *Lactobacillus casei*, *Gene*, 78, 47, 1989.

82. Kataoka, M. et al., Enzymatic production of ethyl (*R*)-4-chloro-3-hydroxybutanoate: asymm reduction of ethyl 4-chloro-3-oxobutanoate by an *Escherichia coli* transformant expressing aldehyde reductase gene from yeast, *Appl. Microbiol. Biotechnol.*, 48, 699, 1997.

83. Wichmann, R. et al., Continuous enzymatic transformation in an enzyme membrane reactor simultaneous NAD(H) regeneration, *Biotechnol. Bioeng.*, 23, 2789, 1981.

84. Popov, V.O. and Lamzin, V.S., NAD$^+$-dependent formate dehydrogenase, *Biochem. J.*, 301, 1994.

85. Nanba, H., Takaoka, Y., and Hasegawa, J., Purification and characterization of an α-haloke resistant formate dehydrogenase from *Thiobacillus* sp. strain KNK65MA, and cloning of the *Biosci. Biotech. Biochem.*, 67, 2145, 2003.

86. Lamzin, V.S. et al., Crystal structure of NAD-dependent formate dehydrogenase, *Eur. J. Bioc.* 206, 441, 1992.

87. Fedorchuk, V.V. et al., Effect of interactions between amino acid residues 43 and 61 of the stability of bacterial formate dehydrogenases, *Biochemistry* (Moscow), 67, 1385, 2002.

88. Shimizu, S. et al., Stereoselective reduction of ethyl 4-chloro-3-oxobutanoate by a micr aldehyde reductase in an oganic solvent–water biphasic system, *Appl. Environ. Microbiol* 2374, 1990.

89. Kizaki, N. et al., Synthesis of optically pure ethyl (*S*)-4-chloro-3-hydroxybutanoate by *Esche coli* transformant cells coexpressing the carbonyl reductase and glucose dehydrogenase genes, *Microbiol. Biotechnol.*, 55, 590, 2001.

26 Comparative Analysis of Chemical and Biocatalytic Syntheses of Drug Intermediates

Michael J. Homann, Wen-Chen Suen, Ningyan Zhang, and Aleksey Zaks

CONTENTS

26.1 INTRODUCTION

In the last decade, biocatalysis has been firmly established as a powerful synthetic tool. While the majority of synthetic targets can be attained by purely chemical means, chemistry alone often does not provide the most elegant solutions. Numerous examples of biosynthetic approaches incorporating chemistry and biology have resulted in a significant simplification of synthetic strategies to generate complex chiral molecules including drugs and drug intermediates. Indeed, the number of biocatalytic approaches to chiral intermediates has grown significantly in recent years in support of a continuously expanding market for single enantiomer drugs [1,2]. Despite numerous examples of integrated biochemical approaches reported in the literature, some argue that the area of biocatalysis has not fulfilled earlier expectations based on the limited number of commercial biological processes. However, this position overlooks the fact that, while large-scale commercial bioprocesses garner most of the attention, they represent only a fraction of the overall contribution of biocatalysis to drug discovery and development.

Today biocatalysts are commonly employed in initial discovery-type approa
biologically active compounds and in creating targeted libraries [3–6]. They are also
used for generation of metabolites in support of drug safety and metabolism program
and continue to play a critical role in the area of process development. Enzym
approaches are routinely incorporated into initial drug supply routes due to their si
as they rarely require protection or deprotection steps and often enable rapid ac
kilogram quantities of drug substance or key intermediates. While the cost of thes
routes is rarely critical, the selection criterion for later-stage syntheses shifts from
access to drug supplies, to providing a reliable, safe, and economical process. He
definition of the "best" process depends largely on the stage of the drug devel
program and, consequently, on the selection criteria. As a drug candidate progress
early development into commercialization the initial synthetic route is often mod
completely redesigned to satisfy the commercial process needs. Therefore, as the com
process evolves, biocatalytic steps may be incorporated, modified, or removed altoge

In this review, we will describe three case studies involving development c
candidates at Schering-Plough that at one point or another utilized a biocatalytic ap
We will describe criteria for selecting a particular approach as it relates to a specific i
development of the drug candidate. We will also present examples of new strate,
enhancing the functional properties of biological catalysts that should ultimately p
their synthetic utility.

26.2 PROCESS DEVELOPMENT

26.2.1 Posaconazole

Posaconazole is a broad-spectrum orally active azole-antifungal discovered and devel
Schering-Plough (Figure 26.1). This compound exhibits enhanced activity against s
Candida and Aspergillus infections, thus providing greater potency for resolving
infections when compared with standard therapies. Development of this antifungal rep
a case in which an initial chemical approach designed to provide speedy delivery of t
substance for phase I clinical studies was later replaced with a new route that inc
biocatalytic step. This second generation synthesis being significantly more efficient t
original one was successfully scaled up to a commercial process.

The initial supply route to the key 2R, 4S-phenylsulfonate intermediate, 8, used SI
epoxidation to form a chiral epoxy alcohol 2 [13,14] (Figure 26.1). Although a pile
campaign based on this approach generated 10 kg of drug substance to support initial
trials, the route was deemed inadequate for scale-up. Indeed, the synthesis was lc
provided poor diastereomeric control in cyclization, resulting in predominant formati
of trans-8, and thus required chromatography to separate the isomers. Clearly, a
approach was needed to support increasing drug substance requirements.

Further process improvement efforts resulted in the discovery of a highly efficie
step three-component coupling giving the desired cis-triazole, albeit in the racemic
was thought that the desired stereochemistry at the 4-position of the tetrahydrofuran
the intermediate 15 (Figure 26.2) could be achieved by selectively blocking one of
primary hydroxyls of the diol 12 by enzymatic transesterification. To that end, ab
enzymes were screened for selective acylation of the diol in various organic solvent:
number of acyl donors. Candida antarctica lipase B (CALB) showed the highest selecti
the pro-S acylation of 12. Following reaction optimization including examination
binations of solvents, acylating agents, temperatures, substrate concentrations, and ac
the first-generation biocatalytic step was established [15,16] (Table 26.1).

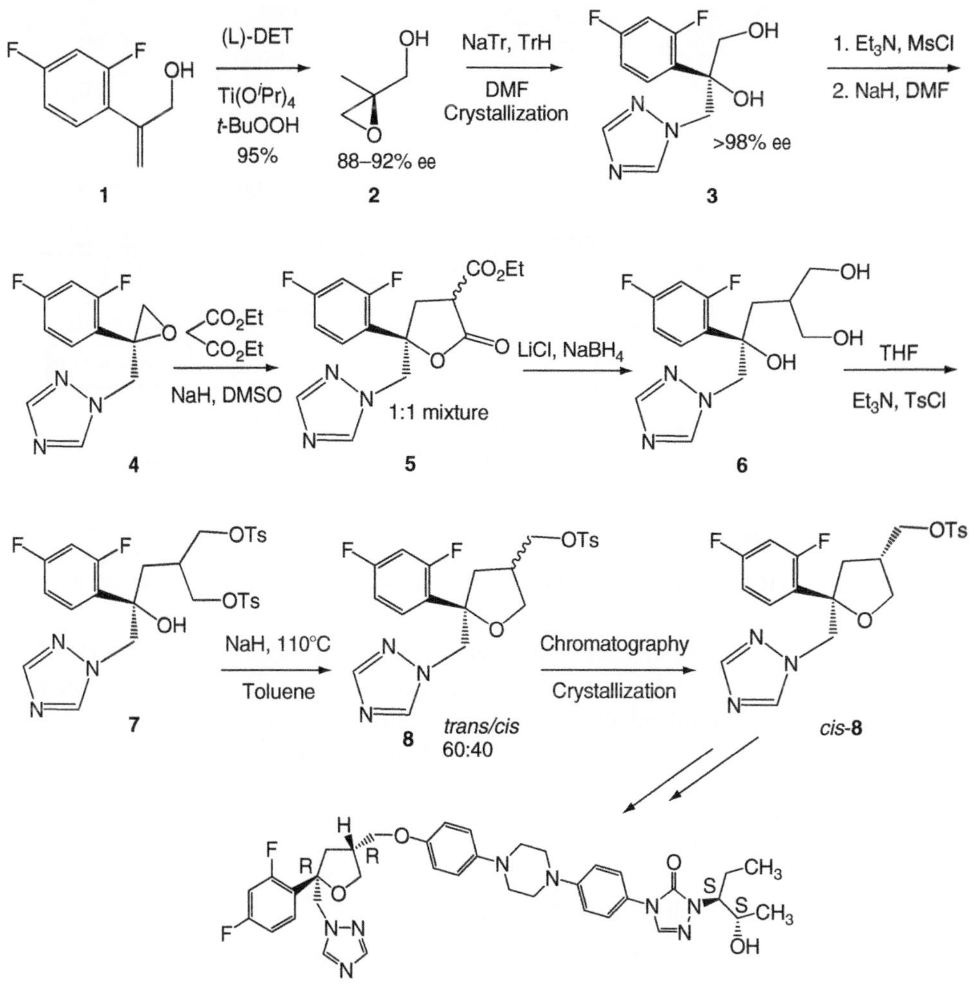

FIGURE 26.1 Initial chemical route to posaconazole.

Although >100 kg of **13a** were generated under conditions described in Table 26.1 the outcome of the biocatalytic step utilizing vinyl acetate as an acylating agent was somewhat unpredictable. The yield of monoester product after reaching 80 to 85% decreased to 70 to 75% if the reaction was allowed to progress. Moreover, both the enantiomeric purity of **13a** and the level of the diacetate by-product varied from batch to batch. Since lipase-catalyzed esterifications proceed through the formation of an acyl–enzyme intermediate [17], it was hypothesized that an acylating agent with a steric volume larger than vinyl acetate would increase the steric hindrance within the enzyme's active site. This was expected to decrease the rate of acyl transfer to the undesired pro-R hydroxyl, thereby improving the enzyme's enantioselectivity. To test this hypothesis, 11 acylating agents were examined in 8 solvents to determine an optimal solvent–acylating agent combination. As predicted, isobutyric anhydride was found to be a significantly superior acylating agent, reducing the maximum level of diester formation to no more than 7%. To minimize the possibility of nonenzymatic acylation with a highly reactive anhydride, the reaction temperature was lowered to −15°C. Lowering reaction temperature and introducing solid NaHCO$_3$ also eliminated the risk of

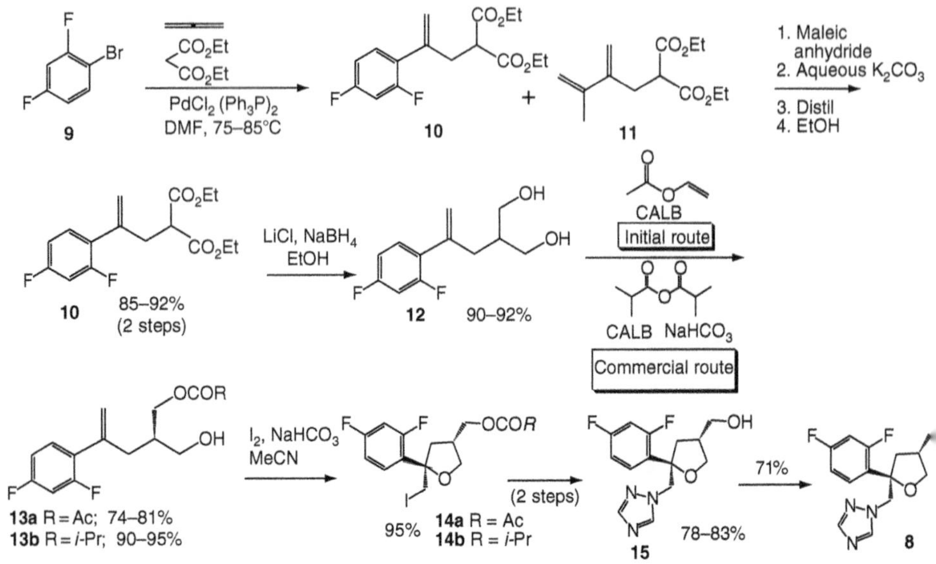

FIGURE 26.2 Chemoenzymatic commercial route to posaconazole key intermediate **8**.

$1 \rightarrow 3$ acyl migration. This second-generation biocatalytic process [15,18] has been scaled provide multiple tons of the key (2R, 4S)-phenylsulfonate intermediate **8**.

The introduction of a new chemical route coupled with enzymatic desymmetrizat the diol resulted in the development of a highly efficient seven-step synthesis that inexpensive raw materials and features 35% overall yield with only one isolated intern (diol **12**). This represents a striking improvement when compared with the original approach using Sharpless epoxidation which provided only a 6% overall yield.

26.2.2   RIBAVARIN ALANINE ESTER

The supply route to a new antiviral candidate, ribavirin alanine ester (RAE), represe example of establishing a biocatalytic route primarily for expedited deliveries of supplies for phase I studies. This strategy provided an ample opportunity for develo lower-cost chemical route to support large-scale deliveries of the drug substance.

TABLE 26.1
Comparison of the First- and Second-Generation Biocatalytic Routes to 13

	First Generation	Second Gen
Solvent	Acetonitrile	Acetonitrile
Diol concentration (g/L)	200	200
Enzyme loading (g/L)	10	10
Acylating agent	Vinyl acetate	Isobutyric at
Temperature (°C)	0	−10 to −15
Reaction time (h)	6–8	6–8
Isolated yield (%)	74–81	90–95
ee (%)	97–99	98

Ribavirin is a potent antiviral agent used in combination with alfa-2β interferon for the treatment of hepatitis C. Unfortunately, variations in the intersubject bioavailability of this drug have resulted in an increased risk of hemolysis and anemia in patients with high bioavailability, and low response in patients exhibiting low bioavailability. A series of preclinical evaluations indicated that ribavirin administered in the form of a prodrug, such as the alanine ester, afforded an improved pharmacokinetic profile. In order to provide support for toxicological and formulation development studies and early clinical trials, rapid access to 50 to 100 kg of the prodrug was required.

Synthesis of RAE centered on the formation of intermediate **19**, which was converted into the product **20** by hydrogenation in acetic acid (Figure 26.3). Attempts to obtain **19** by direct chemical acylation of the unprotected ribavirin with Cbz-Ala were not successful as they resulted in the formation of a mixture of mono-, di-, and triacylated products. A chemical strategy based on protection of the two secondary hydroxyls in the form of acetonide showed some promise. However, the projected 6-month timeline to scale up this three-step approach (Figure 26.4) would delay initiation of the clinical trial. To circumvent this problem, an alternative strategy based on a direct regioselective enzymatic acylation of ribavirin was proposed which obviated the need for protection/deprotection steps.

The use of CALB for the selective acylation of several 5′-hydroxy nucleosides with amino acid derivatives [19,20] prompted us to evaluate this enzyme for acylation of ribavirin with Cbz-Ala. The enzyme was found to be active and regioselective. Reaction conditions were optimized with respect to the solvent, temperature, and nature of the acylating agent used, enabling a fivefold improvement in yield when compared with initial conditions. Typically a yield of 85% of the 5′-acylated ribavirin was achieved in dry tetrahydrofuran (THF) at 60°C using the acetone oxime ester of Cbz-Ala as an acylating agent. As the coupling of **16** with acetone oxime proceeded in a nearly quantitative yield, we investigated

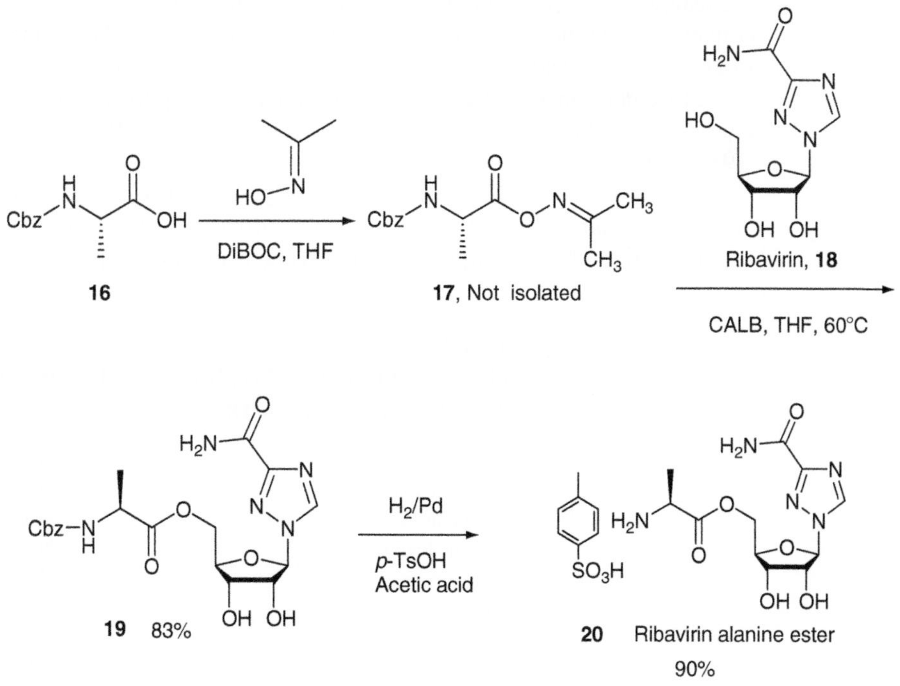

FIGURE 26.3 Chemoenzymatic supply route to ribavirin alanine ester (RAE) **20**.

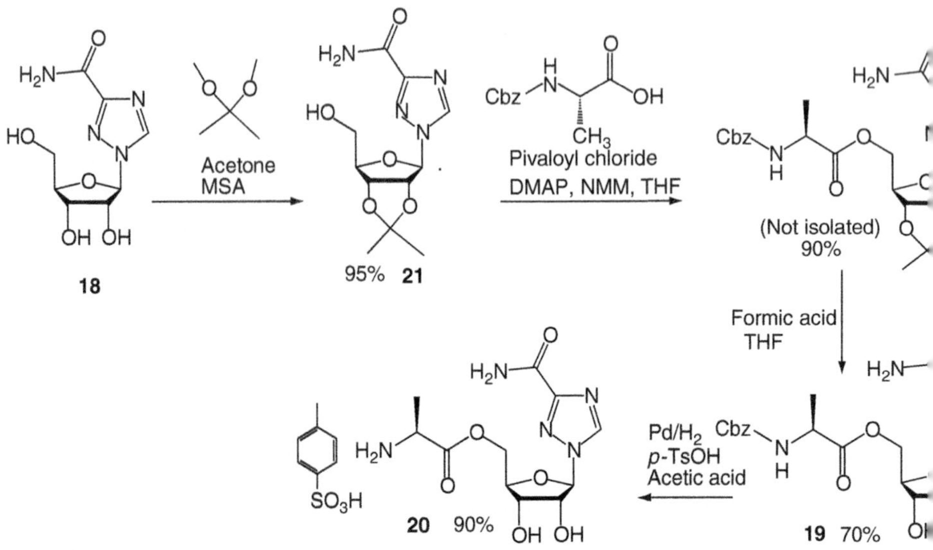

FIGURE 26.4 Chemical route to ribavirin alanine ester (RAE) **20**.

the option of running the acylation reaction without isolating intermediate **17** (Figur
To that end, **16** was esterified with acetone oxime, the reaction mixture was diluted th
with THF, ribavirin was added, and acylation was initiated with the addition of the
Minimizing moisture (<0.05%) in the solvent was found to be critical for achievi
product yield by minimizing lipase-catalyzed hydrolysis. The immobilized catalyst re
completely insoluble in THF at 60°C, forming a uniform free-flowing suspensior
though most of the ribavirin also remained in a suspension (due to poor solubility in
the reaction rate was found to be independent of the rate of mixing, indicating that the
conversion was not controlled by diffusion. Following 24 h of incubation at 60°C,
isolated in 85% yield, which was comparable to the yield obtained in the two-step pro
The aforementioned process was transferred to the pilot plant where it was run un
conditions outlined in Table 26.2 to produce ~80 kg of **19** in 80 to 85% isolated yi
>98% purity.

The above supply strategy not only provided drug substance for form
development and clinical trials, but also afforded ample opportunity to explore and
an alternative chemical approach involving protection/deprotection of the two sec

---

TABLE 26.2
Pilot Scale Reaction Conditions for Enzymatic
Acylation of Ribavirin 18

Solvent	THF
Substrate loading	25 g/L
Temperature	55–60°C
Enzyme loading	20 g/L
Reaction time	24 h
Average yield	83%
Purity	99.3%
de	>99.9%

TABLE 26.3
Comparison of Chemical and Enzymatic Approaches to 20

Route	Chemical	Enzymatic
Steps	4	2
Isolated intermediates	2	1
Concentration of ribavirin (g/L)	100	25
Reagents cost/kg product ($)	600–700	600–700
Enzyme cost/kg product ($)	0	700

hydroxyls (Figure 26.4). Although both approaches were found to be commercially viable, the chemical route having a fourfold higher volumetric productivity proved to be more efficient and less costly (Table 26.3). Despite the fact that the biocatalytic approach was not utilized beyond the pilot scale, its rapid implementation accelerated the clinical program by about 6 months, providing a significant economic benefit to the company.

26.2.3 NK₁/NK₂ Receptor Antagonist

A new class of nonpeptide oxime-based antagonists of NK_1/NK_2 receptors have drawn much attention in recent years as potential drugs for treating a variety of chronic diseases including asthma, bronchospasm, arthritis, and migraine. The biological evaluation of one such antagonist, SCH 206272 (Figure 26.5), revealed that the affinity for the NK_2 receptor resided predominately in the R,R-diastereomer. The proposed approach to establishing the R-stereochemistry of the carbon adjacent to the oxime centered on desymmetrization of prochiral diethyl 3-[3′,4′-dichlorophenyl] glutarate, **22**. The monoglutarate product, **23**, was then elaborated into the desired drug substance following the route illustrated in Figure 26.5. Unlike the two examples discussed earlier, in which the biological and the chemical approaches were investigated sequentially or in parallel, synthesis of SCH 206272 represents a paradigm in which the biological approach was the only one considered.

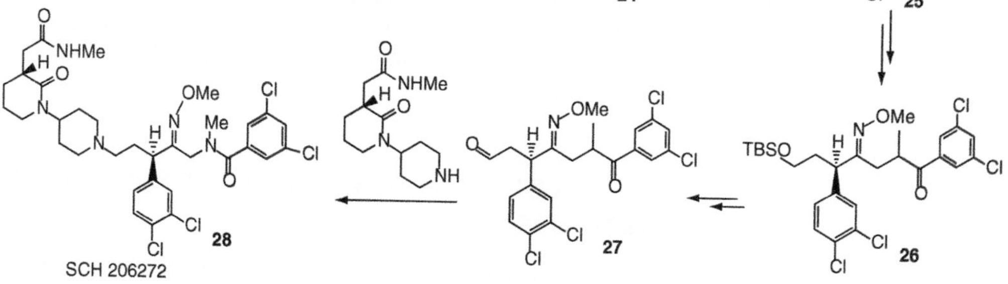

FIGURE 26.5 Chemoenzymatic supply route to NK_1/NK_2 receptor antagonist SCH 206272.

TABLE 26.4
Pilot-Scale Reaction Conditions for Selective Hydrolysis of 22

Substrate loading	100 g/L
Immobilized enzyme (CALB)	12.5 g/L
pH	7.5
Temperature	38–40°C
Reaction time	18–24 h
Conversion yield	92–96%
ee	>99%
Enzyme cost/kg monoester	~$200

Several enzymes capable of hydrolyzing **22** to both *S*- and *R*-**23** with good to ex
enantioselectivity were identified from a screen of ~200 commercial hydrolases [21]. Ou
candidates with *pro-S* selectivity, CALB was selected for further development due to t
zyme's excellent selectivity, moderate cost, and broad commercial availability. More tha
conversion of 100 g/L of **22** to *S*-**23** was achieved in the presence of 20 g/L of immobilized
in 24 h at 40°C. Operating at temperatures above 37°C (the melting point of **22**) was criti
completing the conversion as the reaction rate increased sevenfold between 36 and
coinciding with the transition of **22** from the suspension to the emulsion state. No p
inhibition was observed and the initial rate, as expected, was proportional to the concentra
the catalyst. Reducing the enzyme concentration by half resulted in a significant extension
required to complete the reaction due to progressive inactivation of the catalyst. In fa
immobilized lipase preparation lost ~30% of its original activity within the first 18 h. This
was somewhat unexpected in light of the commonly accepted view that CALB is a therm
enzyme. The rate of inactivation was found to decrease at more neutral pH, which allowe
reduce enzyme loading from 20 to 12.5 g/L. A 97% conversion of 100 g/L of **22** was achieve
50 to 70 kg scale in 18 to 24 h at 38 to 40°C as outlined in Table 26.4.

Given the low solubility of the substrate, the high reaction yield, and the ex
solubility of the product in aqueous medium at basic pH, product isolation was straightfo
and easy to scale up. The enzyme preparation was not recycled due to partial inactivat
the catalyst.

Further investigations were conducted in pursuit of greater process efficienci
commercial scale-up. The use of soluble enzyme as catalyst was examined to elimina
need for the enzyme filtration step, thereby simplifying the purification procedu
possibly reducing the cost contribution of the immobilized enzyme. However, des
slightly higher recovery yield of the isolated product, greater enzyme inactivation was ob
with the soluble form of the enzyme. Consequently, the overall cost contribution of the s
catalyst was about twice that of the immobilized enzyme preparation. In order to in
the process economics and reduce the catalyst cost, modification of CALB activity
stability through protein engineering was carried out.

26.3 DIRECTED EVOLUTION TO IMPROVE THE FUNCTIONAL PROPERTIES OF CALB

Various targeted protein-engineering techniques have been used to generate CALB va
with improved activity, thermostability, and enantioselectivity [22–24]. However, n

these approaches has matched the success achieved by directed evolution technology [25–29]. For this reason, we attempted to improve two functional properties of CALB by utilizing the two most commonly used directed evolution strategies: error-prone polymerase chain reaction (PCR) for improving thermostability, and family shuffling for increasing specific activity for the selective hydrolysis of **22**.

26.3.1 APPLICATION OF ERROR-PRONE PCR FOR IMPROVING THERMOSTABILITY

Error-prone PCR is the most common approach to creating mutants of a single gene. This approach coupled with proper screening and/or selection often gives mutants with significantly altered functional characteristics [30–33]. To construct an error-prone PCR mutation library, we first cloned and expressed CALB in *Saccharomyces cerevesiae* to obtain secreted active N-terminal FLAG-fused protein. This strategy allowed for simple purification of recombinant CALB using anti-FLAG monoclonal antibodies. The cloned CALB gene was then subjected to error-prone PCR under conditions generating one to two amino acid changes in a protein molecule for each cycle of mutation, followed by screening for mutants with improved thermostability. Two high-throughput assays using commercially available 6,8-difluoro-4-methylumbelliferyl octanoate and *p*-nitrophenyl butyrate were developed and utilized for screening the library. The screening protocol involved incubating the mutant libraries for 1 h at 60 or 70°C and then determining their remaining activities at 20°C. Mutants having higher residual activity compared with wild-type CALB (WT-CALB) were selected for further characterization. This approach was expected to identify CALB variants with either higher melting point (T_m) or those with an increased propensity for refolding. Two rounds of error-prone PCR mutagenesis resulted in the isolation of two mutants, 23G5 and 195F1, having more than a 20-fold increase in half-life at 70°C but lower T_m compared with WT-CALB (Table 26.5). Circular dichroism (CD) and protein precipitation studies suggested that the increase in half-life of these two mutants was due to their diminished capacity to aggregate in the unfolded state and also to improved refolding efficiency. Sequence analysis revealed that the first-generation mutant, 23G5, had two amino acid mutations, V210I and A281E. The second-generation mutant, 195F1, derived from 23G5, had one additional mutation, V221D (Table 26.5). On the basis of structure modeling and CD analysis of the three individual variants at each mutation site, we concluded that A281E and V221D, but not V210I, were critical to improving the refolding efficiency of both mutants [34]. As expected, no improvement in activity was observed when these mutants were evaluated for hydrolysis of **22**.

26.3.2 APPLICATION OF DNA FAMILY SHUFFLING FOR IMPROVING ACTIVITY

Two homologous lipase B genes derived from *Hyphozyma* (designated P60) and *Crytococcus* (designated P57) were cloned and expressed in *S. cerevisiae*. FLAG was fused to the N-termini of the enzymes to facilitate their secretion and to simplify isolation. The two aforementioned

TABLE 26.5
Amino Acid Substitutions, T_m, and $T_{1/2}$ of Wild-Type and Mutant CALB

Enzyme	Altered Amino Acid	T_m (°C)	$T_{1/2}$ (min)
WT-CALB	—	57.7	8
23G5	V210I, A281E	52.1	211
195F1	V210I, A281E, V221D	50.8	232

genes and the one coding for WT-CALB (designated P52) showed identities of 67 to
the DNA level and 73 to 81% at the protein level. The three genes were used as par
create a diverse library based on a DNA family shuffling procedure of *in vitro* and
recombination of homologous genes in yeast [28]. Sequence and functional character
of randomly picked clones revealed that the library was highly diverse at the molecu
functional levels. A pH indicator-based high-throughput screen was developed and
identify chimeras having improved hydrolytic activity with **22**. Out of approximate
screened variants, 69 demonstrated over a twofold improvement in the rate of hyc
when compared with the most active parent P57. Sixteen variants produced the
(*S*)-monoglutarate **23** with >99% ee. Of these 16 clones, 7 had a 3- to 13-fold higher
activity than P57, and up to a 20-fold higher activity than CALB (P52) as summar
Table 26.6. Sequence analysis revealed that seven clones had no point mutatio
contained DNA elements derived from two to three parents resulting from two
crossover events [35].

CALB is a moderately thermostable enzyme, retaining activity for an extended pe
time between 30 and 40°C. Above 40°C and especially in the presence of a high concen
of a water-insoluble substrate the enzyme becomes progressively unstable [21]. Since
parent was reported to be more thermostable than CALB [36], we set out to investiga
clones selected for increased activity toward the hydrolysis of **22** also acquired im
thermostability. Determination of T_m revealed that all seven variants had higher T_m
than that of CALB (Table 26.6). One of the two most thermostable variants, 3A4, exh
11-fold longer half-life at 45°C and a 6.4°C higher T_m compared with CALB.

Having created several CALB variants with significantly improved thermostabil
activity toward the hydrolysis of **22**, we then set out to develop a heterologous exp
system to produce enzyme. Variant 3A4 was chosen for development for its combina
high specific activity (16-fold more than CALB) and thermostability (6.4°C mor
CALB) [35].

TABLE 26.6
Specific Activity, $T_{1/2}$, and T_m of Parental and Chimeric Enzymes

Enzyme	Activity[a]	$T_{1/2}$ (h)
P52 (WT-CALB)	1.2	2
P57(*Crytococus*)	1.8	11
P60 (*Hyphozyma*)	1.1	24
2A10	11	ND[b]
2B3	5.7	ND
3A4	20	24
4D6	24	ND
5D10	9.3	ND
8C7	16	ND
12E10	9.0	ND

[a]Average values (nmol/min/μg) were determined for the hydrolysis of **22** based on five assays having a star
deviation ≤10%.
[b]Not determined.
[c]Based on two separate transition phases in the unfolding profile.

26.4 PRODUCTION OF CHIMERIC LIPASE B BY RECOMBINANT FERMENTATION

Three microbial hosts were initially evaluated for CALB expression including *Aspergillus niger*, *S. cerevisiae*, and *Pichia pastoris*. Following the optimization of expression in flask fermentations, *P. pastoris* was selected for all further studies because it exhibited up to fourfold higher titers than the other two host systems (Table 26.7).

Numerous publications describe the use of *Pichia* systems to express proteins from a diverse array of sources [37–39]. Protein expression in *P. pastoris* is mediated by the highly inducible AOXI promoter derived from the alcohol oxidase I gene. As recombinant expression is repressed by glycerol but induced by methanol, tight regulation of protein expression can be readily accomplished through manipulation of the carbon source. *Pichia* has a great propensity for respiratory growth, and therefore *Pichia*-based expression can be exploited to achieve high cell density affording high recombinant protein production.

As indicated in Section 26.3, chimera 3A4 expressed in *S. cerevisiae* exhibited the best improvement in thermostability and activity among the enzyme variants studied and was therefore chosen for pilot-scale development in 30 L fermentors. The 3A4 chimera gene was inserted into shuttle vector pPIC9K for expression in *P. pastoris*. Transformants selected on His⁻ plates were screened for growth on medium plates containing the aminoglycoside G418. Prolonged exposure of transformants to increasing levels of G418 (0.25 to 4 mg/L) through sequential plate transfers employing selective and complex media enabled isolation of transformants with single and multicopy chromosomal inserts of the 3A4 chimera gene as confirmed by colony PCR. Interestingly, multicopy gene integration did not necessarily lead to higher enzyme production in complex media. Two transformants, one of the methanol-tolerant strain GS115 harboring a single gene copy, and the other of the methanol-sensitive strain KM71 with multiple gene copies, exhibited a similar level of hydrolytic activity. Both transformants provided a notable 50-fold improvement in expression compared with the original *S. cerevisiae*–derived 3A4. Further development of these transformants was pursued in minimal salts medium as recommended by Invitrogen.

TABLE 26.7
Secreted Expression Levels of CALB by Microbial Hosts

Host	Vector	Expression	Activity[a]
Saccharomyces cerevisiae (BJ505-protease deficient)	YEpFLAG-1 ADH2 Promoter TRP1 Selectable marker α-Factor secretion leader N-terminal FLAG tag	Mature sequence regulated by glucose repression	45 U
Aspergillus niger (ATCC 20739)	pBARGPE1 gpdA Promoter argB Selectable marker	Full-length sequence constitutive expression	39 U
Pichia pastoris (KM71)	pPIC9K AOX1 Promoter α-Factor secretion leader HIS4 Selectable marker	Mature sequence regulated by methanol induction	160 U

[a]CALB Activity 1 Unit = 1 μmol tributyrin hydrolyzed/min/mL supernatant.

In contrast to complex medium where both transformants provided similar lev
expression, in minimal salts medium the level of 3A4 expression in GS115 was re
by a factor of 5. Consequently, the KM71 transformant was chosen for fermentation
up. Initial development was patterned after the approach developed by Jahic et al
wherein recombinant CALB expression was regulated by either a methanol-limit
temperature-limited fed-batch fermentation strategy. The rationale for controlling e:
sion by using growth-restricting temperatures was based on the observation that ce
lysis and concomitant extracellular proteolysis of CALB diminished at lower temper
In this study, however, the difference in the degree of proteolysis occurring in the fe
tation maintained at 28°C, and the one having a temperature transition from 28° to
was negligible. Notably, decreasing the temperature from 28 to 20°C during glycerol fe
and maintaining the lower temperature throughout methanol feeding provided
control of DO ($\geq$5% saturation), growth kinetics, and methanol concentration.

The growth rate in minimal salts medium was enhanced by initiating the fermentat
pH 6.3, allowing natural acidification to pH 5.0, and maintaining the pH by auto
addition of ammonium hydroxide (7N). Glycerol feeding (60% solution) admini
for about 9 h at 2 to 3 mL/min followed by an overnight feed at half the rate al
significant increases in cell density ($A_{600 nm} > 300$) while minimizing the level of re
glycerol prior to induction with methanol. Residual glycerol proved to be a mitigating 1
in predisposing the culture to methanol toxicity. As methanol catabolism is repress
glycerol, methanol feeding in the presence of glycerol was likely leading to transient elev
of methanol to toxic levels. Hence, methanol feed rates were adjusted based on freqι
performed enzymatic analysis of glycerol and methanol content using an oxygen p
based analyzer (Analox GM8, Lunenburg, Massachusetts). In the course of ferment
the culture was progressively adapted to methanol catabolism by feeding a 25% met
solution (at 0.5 to 1.0 mL/min) for the first 7 h, followed by a 50% methanol solution f
next 18 h, and finally feeding 100% methanol for the duration of the fermentation. Met
feed rate, agitation, and temperature were adjusted to support cell growth and eι
expression while maintaining a nontoxic methanol concentration (<0.5%). Particular
tion was also given to controlling foam formation throughout growth. As depicted in F
26.6, with proper control of methanol feeding, methanol concentrations were maint
below the toxic level (<0.5% v/v), resulting in high culture densities ($A_{600 nm} \geq 500$
chimera 3A4 lipase production levels of 1.2 to 1.5 g/L following 10 d of fermentation.

Secreted enzyme was isolated from the fermentation by first removing ce
centrifugation (10,000 × g, 15 min), followed by concentration of the remaining superr
~15-fold by ultrafiltration using 10 kD exclusion membrane and lyophilization
lyophilized material was purified by passing the crude suspension over a mono-Q
exchange column that retained most of the contaminating proteins. Over 10 g of pι
chimeric 3A4 lipase was obtained using 20 L of clarified supernatant. When com
with purified wild-type CALB expressed in *P. pastoris* (KM71), chimera 3A4
exhibited a tenfold higher specific activity (8.6 nmol/min/μg protein) in the sel
desymmetrization of **22**.

26.5   CONCLUSIONS

Today biocatalysis serves a multitude of functions ranging from discovery-type syn
of leading drug candidates and elaboration of bioactive compound libraries, to gene
of drug metabolites. As revealed by the case studies presented in this chapter, biocatalys
been used successfully as a means to expedite drug substance supplies for toxicologic;
clinical studies, as a primary approach to key drug intermediates, and as enabling techn

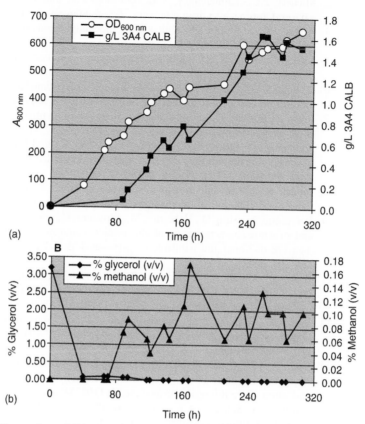

FIGURE 26.6 Expression of CALB 3A4 chimera in *Pichia pastoris* in 30 L fermentors. (a) Cell density and CALB expression level. (b) Residual glycerol and methanol concentrations.

in commercial production displacing initial chemical approaches. By incorporating protein engineering and recombinant expression technologies, biocatalysts can now be tailored to meet continuously evolving process criteria. As such, future developments will likely reveal greater reliance on biocatalysts to provide the "best" process route for production of pharmaceuticals.

ACKNOWLEDGMENTS

The authors would like to thank G. Wong, W. Tong, R. Raghavan, and J. Park for their contribution to the chemical synthesis of ribavirin alanine ester, as well as all the chemists, engineers, and operators in Chemical Development who supported process development of the three developmental candidates described in this chapter.

REFERENCES

1. Breuer, M., Ditrich, K., Habicher, T., Hauer, B., Kesseler, M., Sturmer, R., and Zelinski, T., Industrial methods for the production of optically active intermediates, *Angew. Chem. Int. Ed. Eng.*, 43(7), 788–824, 2004.
2. Panke, S. and Wubbolts, M., Advances in biocatalytic synthesis of pharmaceutical intermediates, *Curr. Opin. Chem. Biol.*, 9(2), 188–194, 2005.

3. Michels, P.C., Khmelnitsky, Y.L., Dordick, J.S., and Clark, D.S., Combinatorial bioca a natural approach to drug discovery, *Trends Biotechnol.*, 16(5), 210–215, 1998.

4. Krstenansky, J.L. and Khmelnitsky, Y., Biocatalytic combinatorial synthesis, *Bioorg. Med.* 7(10), 2157–2162, 1999.

5. Rich, J.O., Michels, P.C., and Khmelnitsky, Y.L., Combinatorial biocatalysis, *Curr. Opin. Biol.*, 6(2), 161–167, 2002.

6. Secundo, F., Carrea, G., De Amici, M., Joppolo Di Ventimiglia, S., and Dordick, J.S., A combi biocatalysis approach to an array of cholic acid derivatives, *Biotechnol. Bioeng.*, 81(4), 391–39•

7. Reiss, P., Burnett, D.A., and Zaks, A., An enzymatic synthesis of glucuronides of azetidinon cholesterol absorption inhibitors, *Bioorg. Med. Chem.*, 7(10), 2199–2222, 1999.

8. Kamimori, H., Ozaki, Y., Okabayashi, Y., Ueno, K., and Narita, S., Synthesis of acylglucu of drugs using immobilized dog liver microsomes octadecylsilica particles coated with phospl *Anal. Biochem.*, 317(1), 99–106, 2003.

9. Linnet, K., Glucuronidation of olanzapine by cDNA-expressed human UDP–glucuronos ferases and human liver microsomes, *Hum. Psychopharmacol.*, 17(5), 233–238, 2002.

10. Kuuranne, T., Aitio, O., Vahermo, M., Elovaara, E., and Kostiainen, R., Enzyme-assisted sy and structure characterization of glucuronide conjugates of methyltestosterone (17α-methyla 4-en-17β-ol-3-one) and nandrolone (estr-4-en-17β-ol-3-one) metabolites, *Bioconjug. Chem.* 194–199, 2002.

11. Kren, V., Ulrichova, J., Kosina, P., Stevenson, D., Sedmera, P., Prikrylova, V., Hala and Simanek, V., Chemoenzymatic preparation of silybin β-glucuronides and their bic evaluation, *Drug Metab. Dispos.*, 28(12), 1513–1517, 2000.

12. Vail, R.B., Homann, M.J., Hanna, I., and Zaks, A., Preparative synthesis of drug metabolit human cytochrome P450s 3A4, 2C9 and 1A2 with NADPH-P450 reductase expressed in *Esch coli*, *J. Ind. Microbiol. Biotechnol.*, 32(2), 67–74, 2005.

13. Saksena, A.K., Girijavallabhan, V.M., Pike, R.E., Wang, H., Lovey, R.G., Liu, Y.-T., Gangul Morgan, W.B., and Zaks, A., Preparation of chiral 2-azolylmethyl-2-phenyl-4-sulfonyloxymeth hydrofurans as antifungal intermediates, WO 9425452, 1994.

14. Lovey, R.G., Saksena, A.K., and Girijavallabhan, V.M., PPL-catalyzed enzymic asymmet of a 2-substituted prochiral 1,3-diol with remote chiral functionality: improvements synthesis of the eutomers of SCH 45012, *Tetrahedron Lett.*, 35(33), 6047–6050, 1994.

15. Morgan, B., Dodds, D.R., Homann, M.J., Zaks, A., and Vail, R., Biocatalysis in pharma process development: SCH56592, a case study, *Methods Biotechnol.*, 15(Enzymes in Nona Solvents), 423–467, 2001.

16. Morgan, B., Dodds, D.R., Zaks, A., Andrews, D.R., and Klesse, R., Enzymic desymmetriz prochiral 2-substituted-1,3-propanediols: a practical chemoenzymic synthesis of a key precu SCH 51048, a broad-spectrum orally active antifungal agent, *J. Org. Chem.*, 62(22), 7736–774•

17. Martinelle, M. and Hult, K., Kinetics of acyl transfer reactions in organic media cataly *Candida antarctica* lipase B, *Biochim. Biophys. Acta*, 1251(2), 191–197, 1995.

18. Nielsen, C.M. and Sudhakar, A., Process for preparing intermediates for the synthesis of an• agents, WO 9722710, 1997.

19. Moris, F. and Gotor, V., Selective aminoacylation of nucleosides through an enzymatic r with oxime aminoacyl esters, *Tetrahedron*, 50(23), 6927–6934, 1994.

20. Mahmoudian, M., Eaddy, J., and Dawson, M., Enzymatic acylation of 506U78 (2-amino arabinofuranosyl-6-methoxy-9H-purine), a powerful new anti-leukemic agent, *Biotechno. Biochem.*, 29(3), 229–233, 1999.

21. Homann, M.J., Vail, R., Morgan, B., Sabesan, V., Levy, C., Dodds, D.R., and Zaks, A., En: hydrolysis of a prochiral 3-substituted glutarate ester, an intermediate in the synthesis of a NK2 dual antagonist, *Adv. Synth. Catal.*, 343(6&7), 744–749, 2001.

22. Patkar, S.A., Svendsen, A., Kirk, O., Clausen, I.G., and Borch, K., Effect of mutation • consensus sequence Thr-X-Ser-Gly of *Candida antarctica* lipase B on lipase specificity, activity and thermostability, *J. Mol. Catal. B Enzym.*, 3, 51–54, 1997.

23. Patkar, S., Vind, J., kelstrup, E., Christensen, M.W., Svendsen, A., Borch, K., and Kirk, O. of mutations in *Candida antarctica* B lipase, *Chem. Phys. Lipids*, 93, 95–101, 1998.

24. Rotticci, D., Rotticci-Mulder, J.C., Denman, S., Norin, T., and Hult, K., Improved enantioselectivity of a lipase by rational protein engineering, *Chem. Biochem.*, 2, 766–770, 2001.

25. May, O., Nguyen, P.T., and Arnold, F.H., Inverting enantioselectivity by directed evolution of hydantoinase for improved production of L-methionine, *Nat. Biotechnol.*, 18, 317–320, 2000.

26. Ness, J.E., Welch, M., Giver, L., Bueno, M., Cherry, J.R., Borchert, T.V., Stemmer, W.P. C., and Minshull, J., DNA shuffling of subgenomic sequences of subtilisin, *Nat. Biotechnol.*, 17, 893–896, 1999.

27. Ness, J.E., Kim, S., Gottman, A., Pak, R., Krebber, A., Borchert, T.V., Govindarajan, S., Mundorff, E.C., and Minshull, J., Synthetic shuffling expands functional protein diversity by allowing amino acids to recombine independently, *Nat. Biotechnol.*, 20, 1251–1255, 2002.

28. Abecassis, V., Pompon, D., and Truan, G., High efficiency family shuffling based on multi-step PCR and *in vivo* DNA recombination in yeast: statistical and functional analysis of a combinatorial library between human cytochrome p450 1A1 and 1A2, *Nucleic Acids Res.*, 28, 1–10, 2000.

29. Christians, F.C., Scapozza, L., Crameri, A., Folkers, G., and Stemmer, W.P.C., Directed evolution of thymidine kinase for AZT phosphorylation using DNA family shuffling, *Nat. Biotechnol.*, 17, 259–264, 1999.

30. Chen, Z. and Zhao, H., Rapid creation of a novel protein function by *in vitro* coevolution, *J. Mol. Biol.*, 348(5), 1273–1282, 2005.

31. Henke, E. and Bornscheuer, U.T., Directed evolution of an esterase from *Pseudomonas fluorescens*: random mutagenesis by error-prone PCR or a mutator strain and identification of mutants showing enhanced enantioselectivity by a resorufin-based fluorescence assay, *Biol. Chem.*, 380, 1029–1033, 1999.

32. Song, J.K. and Rhee, J.S., Enhancement of stability and activity of phospholipase A_1 in organic solvents by directed evolution, *Biochim. Biophys. Acta*, 1547, 370–378, 2001.

33. Zha, D., Wilense, S., Hermes, M., K.-E. Jaeger, and Reetz, M.T., Complete reversal of enantioselectivity of an enzyme-catalyzed reaction by directed evolution, *Chem. Commun.*, 2664–2665, 2001.

34. Zhang, N., Suen, W.-C., Windsor, W., Xiao, L., Madison, V., and Zaks, A., Improving tolerance of *Candida antarctica* lipase B towards irreversible thermal inactivation through directed evolution, *Protein Eng.*, 16(8), 599–605, 2003.

35. Suen, W.-C., Zhang, N., Xiao, L., Madison, V., and Zaks, A., Improved activity and thermostability of *Candida antarctica* lipase B by DNA family shuffling, *Protein Eng. Des. Sel.*, 17(2), 133–140, 2004.

36. Hashida, M., Abo, M., Takamura, Y., Kirk, O., Halkier, T., Pedersen, S., Patkar, S.A., and Hansen, M.T., U.S. Patent 5,856,163, 1999, Lipases from *Hyphozyma*.

37. Cregg, J., Cereghino, J., Shi, J., and Higgins, D., Recombinant protein expression in *Pichia pastoris*, *Mol. Biotechnol.*, 16(1), 23–52, 2000.

38. Zhang, W., Inan, M., and Meagher, M., Fermentation strategies for recombinant protein expression in the methylotrophic yeast *Pichia pastoris*, *Biotechnol. Bioprocess Eng.*, 5(4), 275–287, 2000.

39. Lin Cereghino, G., Sunga, A., Lin Cereghino, J., and Cregg, J., Expression of foreign genes in the yeast *Pichia pastoris*, *Genet. Eng.* (*New York*), 23, 157–169, 2001.

40. Jahic, M., Wallberg, F., Bollok, M., Garcia, P., and Enfors, S.-O., Temperature limited fed-batch techniques for control of proteolysis in *Pichia pastoris* bioreactor cultures, *Microb. Cell Factories*, 2(6), 1–11, 2003.

27 Industrial Processes Using Lyases for C–C, C–N, and C–O Bond Formation

Martina Pohl and Andreas Liese

CONTENTS

27.1 INTRODUCTION

Industrial biocatalysis is becoming a standard tool in the synthesis of fine chemicals [1,2]. Although a major number of industrially applied biotransformations is still catalyzed by hydrolases (E.C. 3.x.x.x) [3], oxidoreductases (E.C. 1.x.x.x), and lyases (E.C. 4.x.x.x), become more and more important. This chapter gives an overview of the application of lyases in industrial processes, either as isolated enzymes or as immobilized enzymes, and by whole cell biocatalysis. In contrast to hydrolases, which in general catalyze kinetic resolutions, lyases are usually employed for asymmetric syntheses, enabling 100% conversion and 100% (enantio) selectivity under ideal conditions [4].

27.2   C–C BOND FORMATION

The first asymmetric C–C bond formation catalyzed by hydroxynitrile lyase, from *P.* *amygdalins*, was described in 1909 using bitter almond meal [5]. In addition to hydroxyr lyases, aldolases, ketolases, and a certain type of thiamine-diphosphate dependent lyase important roles in industrial biotransformations.

27.2.1   PYRUVATE DECARBOXYLASE

Pyruvate decarboxylases (E.C. 4.1.1.1) (synonym: PDC; systematic name: 2-oxo carboxy-lyase) are tetrameric enzymes requiring thiamine diphosphate and magnesiun as cofactors. They are found in many types of yeast, fungi, plants, and some bacteria. *A* from their physiological role, the decarboxylation of pyruvate to acetaldehyde, these enz have the potential to ligate two aldehyde molecules enantioselectively to 2-hydroxy ket The carboligation of acetaldehyde **1** and benzaldehyde **2** to (*R*)-phenylacetylcarbinol **3** (was one of the first biotransformations applied in industrial scale for the producti intermediates in the synthesis of fine chemicals (Figure 27.1).

PAC is not a marketed product, but is produced as a chiral precursor for various having α- and β-adrenergic properties, such as (1*R*,2*S*)-ephedrine, (1*R*,2*R*)-pseudophe norephedrine, and norpseudoephedrine. Although (1*R*,2*S*)-ephedrine is a natural pr found in various plant species of genus Ephedra, extraction from the natural sou not competitive to its biocatalytic or chemical synthesis. The fermentative process performed, by fermenting *Saccharomyces cerevisiae* cells in a fed-batch reactor, by BAS (Germany) and Krebs Biochemicals Ltd./Malladi Drug (India). Typical concentratic (*R*)-PAC produced during yeast fermentation are 4 to 12 g/L. The flood of recent public shows that process and strain optimization is still a matter of strong interest [6–9]. Most parameters that have been investigated so far are summarized [10,11]. Product concentr were improved up to 12 to 20 g/L (*R*)-PAC by optimization of the fermentative proc and the benzaldehyde feeding process [12]. Apart from *S. cerevisiae*, PDC from *C* *utilis* and *Rhizopus javanicus* have proved their potential for the production of PAC [In addition to optimization of the whole-cell biotransformation, recent work fc on the development of processes with isolated PDC to circumvent problems o product formation caused by various enzymes in whole-cell systems. In such enz biotransformations product concentrations up to 100 g/L PAC were obtained [7,12].

A similar carboligase reaction has also been described by other thiamine-dipho dependent enzymes like benzoylformate decarboxylase (BFD), phenylpyruvate decarbo: and benzaldehyde lyase. Although these enzymes are not yet commercially applied, the the potential for the enantioselective synthesis of a broad range of 2-hydroxy keto synthons for organic chemistry [14–16]. The differences and similarities between PD BFD were recently demonstrated by site-directed mutagenesis [17].

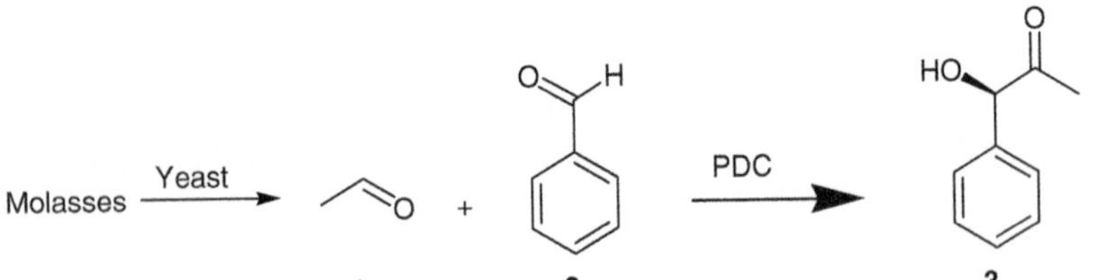

FIGURE 27.1 PDC catalyzed synthesis of (*R*)-phenylacetylcarbinol (PAC) **3**.

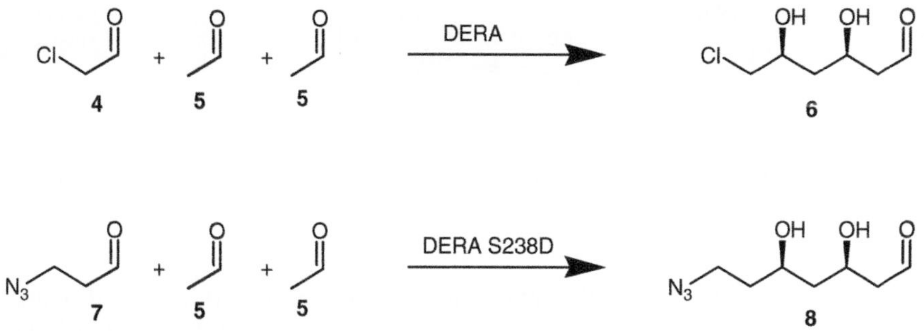

FIGURE 27.2 Deoxyribose-5-phosphate aldolase (DERA) as biocatalyst in the synthesis of intermediates **6** and **8** for (HMG-CoA) reductase inhibitors (statins).

27.2.2 Deoxyribose-5-Phosphate Aldolase

Deoxyribose-5-phosphate aldolase (E.C. 4.1.2.4) (synonyms: DERA, phosphodeoxyriboaldolase, deoxyriboaldolase, deoxyribose-5-phosphate aldolase, 2-deoxyribose-5-phosphate aldolase; systematic name: 2-deoxy-D-ribose-5-phosphate acetaldehyde-lyase).

Wong et al. first described the potential of DERA to catalyze the aldol condensation of chloroacetaldehyde **4** with two molecules of acetaldehyde **5** yielding (3*R*,5*S*)-6-chloro-3,5-dihydroxyhexanal **6** (Figure 27.2). This chiral compound is an important precursor in the synthesis of 3-hydroxy-3-methylglutaryl CoA (HMG-CoA) reductase inhibitors (statins), hypolipidaemic agents, which are multibillion-dollar drugs [4]. The need for high enantioselectivity at both stereo centers has led to the development of six different routes involving biocatalysis [18,19]. The product **6** (97% de) of the DERA biotransformation is stabilized as hemiacetal under optimized process conditions at DSM. Independently, a similar approach utilizing DERA aldolase was also developed by the Wong group together with DIVERSA Coop. (USA) [20]. To broaden the range of accepted substrates they isolated a DERA variant (S238A), which accepts azidopropionaldehyde **7** alternatively to chloroacetaldehyde enabling access to 7-azido-(3*R*,5*S*)-dihydroxy-heptanal **8**, the key intermediate for Atorvastatin [21] (Figure 27.2).

27.2.3 N-Acetyl-D-Neuraminic Acid Aldolase

N-Acetyl-D-neuraminic acid aldolases (E.C. 4.1.3.3) (synonyms: Neu5Ac-aldolase, *N*-acetylneuraminate lyase, *N*-acetylneuraminate lyase, *N*-acetylneuraminate pyruvate-lyase, sialic acid aldolase; systematic name: *N*-acetylneuraminate pyruvate-lyase) have been characterized from many organisms. Many genes are cloned and three-dimensional (3D) structures of the enzymes from *Escherichia coli* and *Haemolphilus influenzae* are available [22–23]. *In vivo* Neu5Ac-aldolase catalyzes the reversible aldol reaction of *N*-acetyl-D-mannosamine (ManNAc) **10** and pyruvate **11** to *N*-acetyl-5-amino-3,5-dideoxy-D-glycero-D-galacto-2-nonulosonic acid (Neu5Ac) **12**. Neuraminic acid, a C-9 amino sugar, is the aldol condensation product of pyruvic acid and *N*-acetyl-D-mannosamine, mostly important as its nitrogen- and oxygen-substituted *N*-acyl derivatives (sialic acids) that are found as the terminal sugar of cell surface glycoproteins, especially in animal tissue and blood cells. Sialic acids are components of lipids, polysaccharides, and mucoproteins, as cell surface glycoproteins that play an important role in cell adhesion, recognition, and interaction. The Neu5Ac analogs and Neu5Ac-containing oligosaccharides are precursors for inhibitors of neuraminidase, hemagglutinin, and selectin-mediated leucocyte adhesion [24]. For industrial biotransfomation, the enzyme from

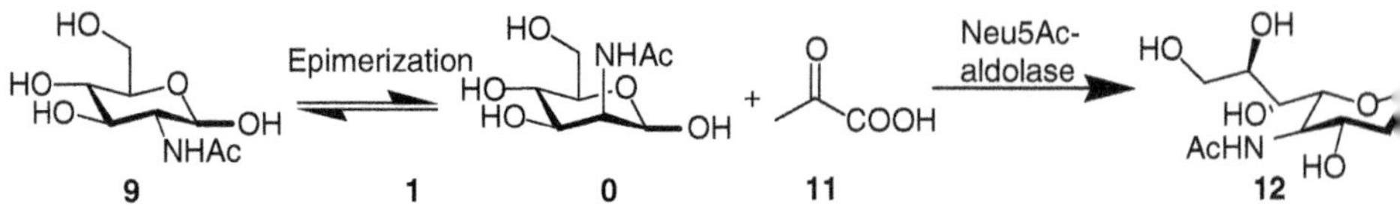

FIGURE 27.3 Synthesis of Neu5Ac **12** in a reaction sequence starting from *N*-acetyl-D-glucosam epimerization at C2.

E. coli is used either immobilized on Eupergit-C (GlaxoSmithKline, UK) or in its solu form (Marukin Shoyu Co. Ltd., Japan, Research Center Jülich, Germany).

As *N*-acetyl-D-mannosamine **10** is very expensive, it is synthesized from *N*-ac glucosamine **9** by epimerization at C2. The equilibrium of the epimerization is on tl of *N*-acetyl-D-glucosamine (GlcNAc:ManNAc = 4:1). After neutralization and addi isopropanol, GlcNAc precipitates leaving behind a ratio of GlcNAc:ManNAc = 1:4 remaining solution. This chemical epimerization is used in the multiton production pro Neu5Ac in a repetitive batch process by GlaxoSmithKline [25,26]. By contrast, M Shoyu Co. Ltd. and the Research Center Juelich integrated enzymatic epimerizatior *N*-acetyl-D-glucosamine-2-epimerase from *E. coli* in an one-pot biotransformation (27.3). Downstream processing of Neu5Ac can easily be achieved by crystallizatio acidification with acetic acid [27,28]. The fed-batch process of Marukin Shoyu Co which is operated in a multikilogram scale, requires an additional thermal denaturatic (80°C, 5 min) to remove the enzymes from the product.

27.2.4 TRYPTOPHAN SYNTHASE

Tryptophan synthase (E.C. 4.2.1.20) (synomyms: tryptophan desmolase, L-trypt synthetase, indoleglycerol phosphate aldolase, systematic name: L-serine hydro-lyas dimeric pyridoxal-5'-phosphate-dependent enzyme, which has been found in a ra organisms [22]. The alpha-subunit catalyzes the conversion of 1-(indol-3-yl)glyc phosphate to indole and glyceraldehydes-3-phosphate. The indole then migrates beta-subunit where, with serine in the presence of pyridoxal-5'-phosphate, it is conve tryptophane [29]. For commercial purposes, the ability of the enzyme to catalyze the c sion of serine **13** and indole **14** to tryptophan **15** and water is used for the produc L-tryptophan as a pharmaceutical active ingredient for parenteral nutrition (i solution) as well as active ingredient in sedativa, neuroleptica, antidepressiva, an additives. In the commercial process run at the Amino GmbH (Germany), the e from *E. coli* is used in the form of suspended whole cells (Figure 27.4) [30]. The su L-serine is separated from molasse by ion exchange chromatography. The fed-batch

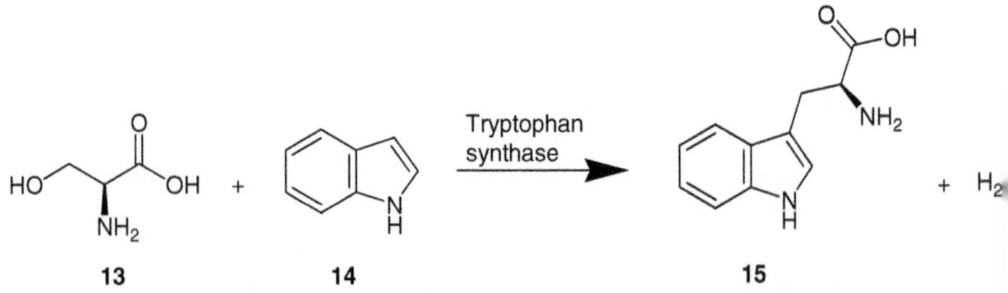

FIGURE 27.4 Synthesis of L-tryptophan **15** starting from L-serine **13** and indole **14**.

regulated and the indole dosage is controlled through online HPLC analysis of the product/educt ratio. Based on indole >95% yield is obtained in a fed-batch reactor. The process is run with 30 t/y.

27.2.5 HYDROXYNITRILE LYASE

Hydroxynitrile lyases (E.C. 4.1.2.X) (synonyms: HNL, oxynitrilase, mandelonitrile lyase, (R)-hydroxynitrile lyase, (S)-hydroxynitrile lyase, hydroxymandelonitrile lyase; systematic names: mandelonitrile benzaldehyde-lyase (E.C. 4.1.2.10), (S)-4-hydroxymandelonitrile hydroxybenzaldehyde-lyase (E.C. 4.1.2.11), acetone-cyanohydrin acetone-lyase (E.C. 4.1.2.39)) are enzymes that are originated in plants. This class of enzymes is a good example for convergent evolution, because hydroxynitrile lyase activity was introduced in different frameworks of oxidoreductases and α/β-hydrolases yielding hydroxynitrile lyases differing in enantioselectivity and substrate range [31]. A detailed study of 3D structures and mutant enzymes led to good understanding of the reaction mechanisms of (R)-specific mandelonitrile lyase from *Prunus* sp. (E.C. 4.1.2.10), acetone-cyanohydrin lyase from *Linum usitatissimum* (E.C. 4.1.2.37), (S)-specific hydroxymandelonitrile lyase from *Sorghum bicolor* (E.C. 4.1.2.11), and (S)-hydroxynitrile lyases (acetone-cyanohydrin acetone-lyases, E.C. 4.1.2.39) from *Hevea brasiliensis* and from *Manihot esculenta* [32]. Their ability to catalyze the cleavage of HCN from natural cyanide compounds, which is probably part of a plant defence system against herbivores and microbial attack, is synthetically applied for the synthesis of chiral hydroxynitriles [33–38]. The (R)-specific HNL from almond (*P. amygdalus*, *Pa*HNL) [39] and two (S)-specific HNLs from rubber tree (*H. brasiliensis*, *Hb*HNL), and cassava (*M. esculenta*, *Me*HNL) are nowadays available by overexpression in *Pichia pastoris*, *S. cerevisiae*, and *E. coli*, respectively [40–42]. This paved the way for the introduction of HNL-based industrial processes for the production of chiral hydroxynitriles and 2-hydroxycarboxylic acid [2,31].

*Pa*HNL has recently been implemented in industrial syntheses of some chiral aromatic 2-hydroxycarboxylic acids, such as (R)-2-chloromandelic acid **18**, an intermediate for the synthesis of the antidepressant and platelet-aggregation inhibitor clopidogrel by DSM Fine Chemicals Austria, Nippon Shokubai, and Clariant (Figure 27.5) [2,43–46]. For the enantioselective addition of HCN **17** to 2-chlorobenzaldehyde **16**, *Pa*HNL is applied in the form of almond-flour extract or is immobilized on Avicel microcrystalline cellulose. Depending on the solvent employed the enzyme can be used for several months. The (R)-2-choromandelonitrile formed is converted into the corresponding carboxylic acid by hydrolysis with concentrated HCl, without racemization. Thus, 100% theoretical yield is possible. One recent example is the production of (S)-3-phenoxybenzaldehyde cyanohydrin **20** in a biphasic process [47] by DSM Fine Chemicals Austria and Nippon Shokubai [47–50]. The cyanohydrin is a chiral intermediate in the production of pyrethroids [2]. Recently, the carving of the active site of almond R-HNL by means of site-directed mutations for increased enantioselectivity was demonstrated in respect to (R)-2-hydroxy-4-phenylbutyronitrile **22**, the key intermediate in the synthesis of different angiotensin-converting enzyme inhibitors [51].

27.3 C–N BOND FORMATION

27.3.1 TYROSINPHENOL LYASE

Tyrosinphenol lyase (E.C. 4.1.99.2) (synonyms: beta-tyrosinase, TLP; systematic name: L-tyrosine phenol-lyase (deaminating)) is a pyridoxal-5′-dependent multifunctional enzyme that catalyzes degradation of tyrosine to phenol, pyruvate, and ammonia. The reversibility of

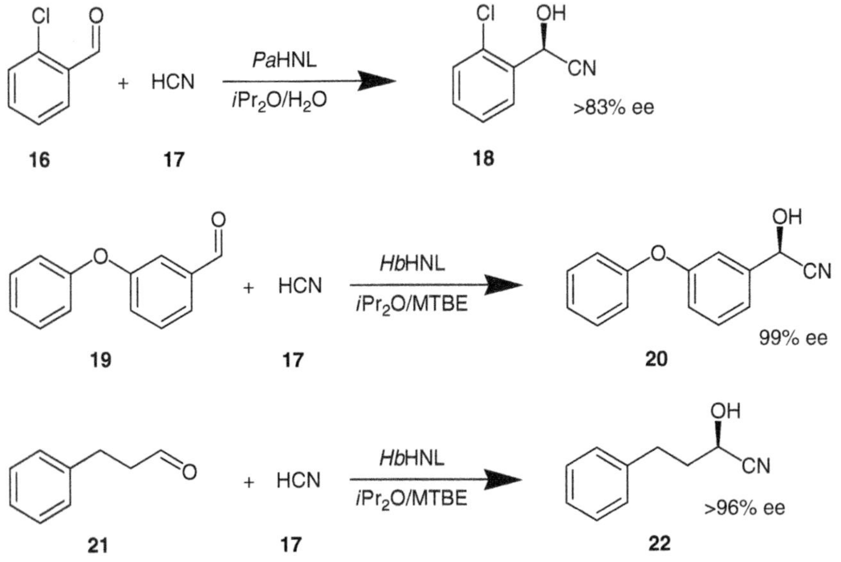

FIGURE 27.5 Asymmetric HCN addition to aldehydes catalyzed by hydroxynitrile lyases. *
hydroxynitrile lyase from *Prunus amygdalus* and *Hb*HNL, hydroxynitrile lyase from *Manihot e*

this reaction is used for the production of L-DOPA using catechol instead of phen
substrate (Figure 27.6).

From a thorough strain screening, bacterium *Erwinia herbicola* was selected for L-
production and the commercially applied process using suspended whole cells was op
intensively [52]. Ajinomoto Co. Ltd. (Japan) produces L-DOPA **26** by this
biotransformation using suspended *E. herbicola* cells with extremely high TLP activ
fed-batch reactor [53]. First, cells are prepared by cultivation in a medium con
L-tyrosine as an inducer of TLP. The intact cells are then harvested by centrifugati
transferred to the reactor together with the substrate catechol **23**. The one-step biotr
mation is more economic than the established chemical route that involves eight cl
reaction steps including optical resolution [52,54]. L-DOPA **26** is insoluble in the r
medium, so it precipitates during the reaction from a final concentration of 110 g
About 250 t of L-DOPA are supplied per year, and more than half of it are produced t
biotransformation.

The product is applied for the treatment of Parkinsonism that is caused by a
L-dopamine and its receptors in the brain. *In vivo* L-dopamine is synthesized by deca
lation of L-3,4-dihydroxyphenylalanine (L-DOPA). Since L-dopamine cannot p
blood–brain barrier L-DOPA is applied in combination with dopadecarboxylase-inl
to avoid formation of L-dopamine outside the brain.

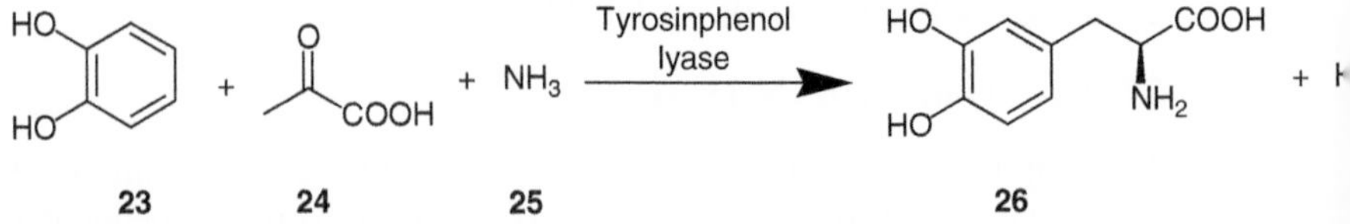

FIGURE 27.6 Tyrosinphenol lyase catalyzed L-DOPA **26** synthesis.

27.3.2 Aspartase

Aspartase (E.C. 4.3.1.1) (synonyms: L-aspartase, aspartate ammonia lyase, fumaric aminase; systematic name: L-aspartate ammonia lyase) enzyme was identified and cloned from various bacteria and the 3D structure from the tetrameric enzyme from *E. coli* was solved [22]. Aspartase does not require cofactors to catalyze the reversible cleavage of ammonia **28** from aspartate **29** to fumarate **27**. The reverse reaction is used for the production of L-aspartate **29**, since 1953 (Figure 27.7) [55], which is used in parenteral nutrition, as food additives and as a starting material for the low-calorie sweetener aspartame. Different procedures using whole cells and isolated aspartase, either mobilized or immobilized, have been developed for already four decades.

The industrial L-aspartate production by biotransformation was started using *E. coli* cells with high enzyme activity in a batch fermentation of fumaric acid and ammonia. This procedure had the disadvantage that cells with active enzyme had to be discarded at the end of each batch [55]. On the one hand, to overcome this problem, Tanabe Seiyaku Co. (Japan) produced L-aspartate, since 1973, using immobilized aspartase that was isolated from *E. coli* cells. The enzyme was bound to a weak ion exchange resin and since 1974, this procedure was used for L-aspartate production at Kyowa Hakko Kogyo Co. Ltd. On the other hand, Tanabe Seikyaku Co. Ltd. tested a broad range of entrapping matrices and found best results with polyacrylamide. In the plug-flow process, it was found that suspending the immobilized cells in their substrate for 24 to 28 h increased their activity tenfold. Once immobilized, the cell column was very stable, with a half-life of 120 d at 37°C. After acidification of the effluent from the column, the L-aspartic acid crystallizes and can be collected by centrifugation or filtration. This system has been operated since 1973 [55]. Overall costs compared with the fermentative batch production were reduced by ~40% due to a significant increase in productivity and due to the reduction of labor and waste-water costs. In 1978, the trapping matrix changed to κ-carrageenan, a polysaccharide obtained from seaweed that increased the relative productivity 15-fold. Approximately 100 t/month of L-aspartic acid can be manufactured using a 1000 L reactor. The process was further improved by developing *E. coli* strains with increased aspartase activity. These improved *E. coli* strains are used, since 1982, in the production process at Tanabe Seikyaku Co. Ltd. [56].

BioCatalytics (USA) uses a plug-flow reactor (75 L vol.) with immobilized *E. coli* cells with high aspartase activity on polyurethane and polyazitidine. Space-time yields of 3 kg/L/h are obtained and downstream processing of L-aspartate is performed by acid induced precipitation. A new production system was established in the 1980s by using intact resting cells of coryneform bacteria without immobilization [52]. This genetically engineered bacterial strain possesses high maleate isomerase and aspartase activites. A repetitive batch process using suspended *Corynebacterium glutamicum* (synonym: *Brevibacterium flavum*) cells was started at Mitsubishi Petrochemical Co. Ltd. (Japan), in 1986 [57,58]. The bacterial cells are retained by ultrafiltration.

FIGURE 27.7 Synthesis of L-aspartate **29** catalyzed by aspartase.

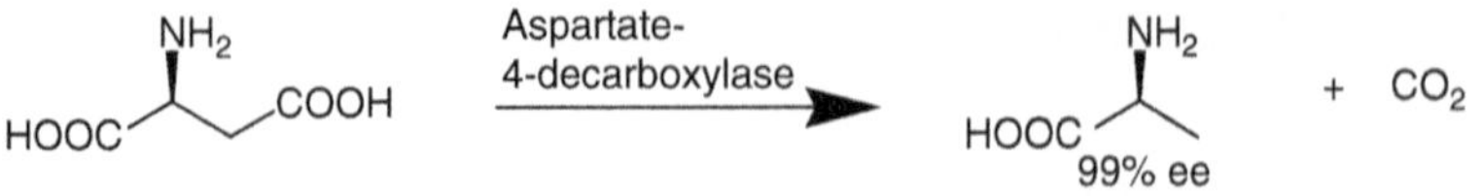

FIGURE 27.8 Decarboxylation of L-aspartate **30** to L-alanine **31** catalyzed by aspartate-4-decarbox

27.3.3   ASPARTATE-4-DECARBOXYLASE

Aspartate-4-decarboxylase (E.C. 4.1.1.12) (synonym: aspartate beta-decarbox
systematic name: L-aspartate 4-carboxy-lyase) is a pyridoxal-5'-phosphate dependent en
that catalyzes the nonoxidative decarboxylation of L-aspartate **30** to L-alanine **31** [59].
reaction is used for a one-step batch process using immobilized whole cells of *Pseudor*
dacunhae at Tanabe Seikyaku Co. Ltd. (Japan) since 1965 (Figure 27.8). The setup
continuous plug-flow reactor using immobilized cells was first complicated by the evol
of gaseous CO_2 from the decarboxylation reaction, which made plug flow of the subst
difficult and caused significant pH changes. This problem was overcome with the dev
ment of a pressurized fixed-bed reactor operating at 10 bar [60]. To improve the yie
L-alanine, the alanine racemase, and fumarase activities can be destroyed in the cells by
treatment (pH 4.75, 30°C). During this treatment aspartate-4-decarboxylase is stabiliz
the addition of pyruvate and pyridoxal-5'-phosphate.

Tanabe Seikyaku Co. Ltd. combined the process with the aspartase catalyzed synthe
L-aspartic acid from fumarate in a two-step biotransformation [61]. The main reason fo
separation of the reaction in two reactors is the difference in pH-optima of both enz
(aspartase from *E. coli*: pH 8.5, and aspartate-4-decarboxylase from *P. dacunhae*: pH
Using a 1000 L reactor for immobilized *E. coli* cells and a 2000 L pressurized reacto
P. dacunhae cells, about 100 t of L-aspartic acid and 100 t of L-alanine can be produce
month, respectively. Tanabe commercialized this system of a sequential enzyme rea
using two kinds of immobilized microbial cells in 1982 [56].

27.3.4   L-PHENYLALANINE AMMONIA LYASE

L-Phenylalanine ammonia lyase (E.C. 4.3.1.5) (synonyms: PAL, L-phenylalanine deami
tyrase; systematic name: L-phenylalanine ammonia lyase) catalyzes the reversible cleava
ammonia **25** from L-phenylalanine **34** yielding *trans*-cinnamic acid **33** without additional c
tor. The reverse reaction is used in industrial biotransformations to produce L-phenylalanin
(Figure 27.9). This amino acid is used in the synthesis of the sweetener aspartame a
parenteral nutrition. Further it is the chiral building block in the synthesis of the maci
antibiotic rutamycin B [63]. PALs from a broad range of bacteria and eukaryotes have
characterized and the genes were cloned [22].

The Genex Corporation (USA) established a fed-batch process, in 1986, with suspe
Rhodotorula rubra cells in aqueous medium at pH 10.6 [64]. As PAL from *R. rut*

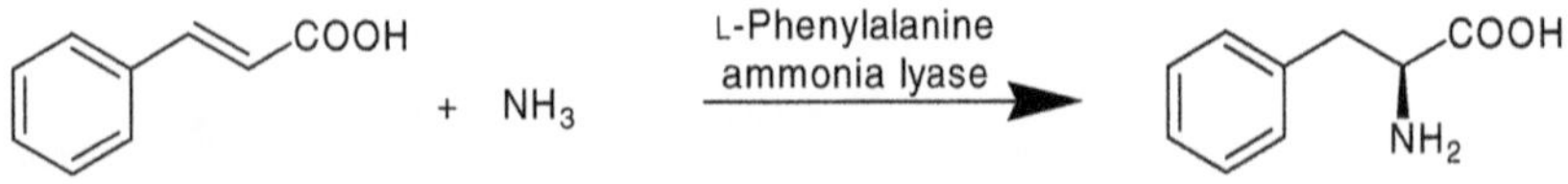

FIGURE 27.9 L-Phenylalanine **34** synthesis catalyzed by L-phenylalanine ammonia lyase.

sensitive to oxygen, the biotransformation is performed under anaerobic, static conditions. The reaction is performed in fed-batch mode with periodical addition of concentrated ammonium cinnamate solution and pH is adjusted by the addition of CO_2 [65]. The fermentative process was improved by starting the process directly from glucose as a substrate and L-phenylalanine is now manufactured by fermentation using overproducing cells on a scale of 8 to 10,000 t/y [16].

27.4   C–O BOND FORMATION

27.4.1   Fumarase

Fumarase (E.C. 4.2.1.2) (synonyms: fumarate hydratase; systematic name: (S)-maleate hydro-lyase) catalyzes the reversible cleavage of water from (S)-malate **36** without a cofactor. Fumarases have been identified in a broad range of bacteria and eukaryotes [22]. For industrial production of (S)-malate—which is used as an acidulant, in competition with citric acid, in fruit, in vegetable juices, carbonated soft drinks, jams and candies, in amino acid infusions, and in the treatment of hepatic malfunctioning—the enzymes from *C. glutamicum* and *C. ammoniagenes* are applied (Figure 27.10) [66,67]. Approximately 40,000 t of malic acid are used worldwide annually. The process developed by the Amino GmbH uses suspended cells of *C. glutamicum* in a batch reactor [68,69]. The production was started in the year 1988, and (S)-malate **36** was produced with an annual productivity of 2000 t. The process is highly enantioselective with no (R)-malate detected. The reaction is carried out in a slurry of crystalline calcium fumarate and calcium malate. The process is forced to quantitative conversion of fumarate **35** by *in situ* precipitation of the product shifting the equilibrium toward calcium malate.

27.4.2   Malease

Malease (E.C. 4.2.1.31) (synonym: maleate hydratase; systematic name: (R)-maleate hydro-lyase). Maleases have been found not only in many bacteria, yeasts, and fungi but also in plants and animals. However, the protein chemical and enzymological data known so far suggest that malease activity is catalyzed by different types of enzymes So far, there are no sequence data, neither of the protein nor of the DNA-sequence available [22]. *P. pseudoalcaligenes*, which is used in a commercial process of DSM Ltd. (NL) for the production of (R)-malic acid **39** (Figure 27.11) was identified from a thorough microbial screening [70]. (R)-Malic acid is an optically active 2-hydroxy acid that can be used as a chiral synthon, as a resolving agent, or as a ligand in asymmetric synthesis [71,72]. Malease from *P. pseudoalcaligenes* is a hetero dimeric enzyme that requires no additional cofactors for catalysis. The enzyme is used in a batch process with immobilized whole cells in aqueous medium. The cheaper maleic anhydride **37** can be used instead of maleic acid **38**, as it hydrolyzes *in situ* to the substrate. Optimization of the process is still a matter of research [73].

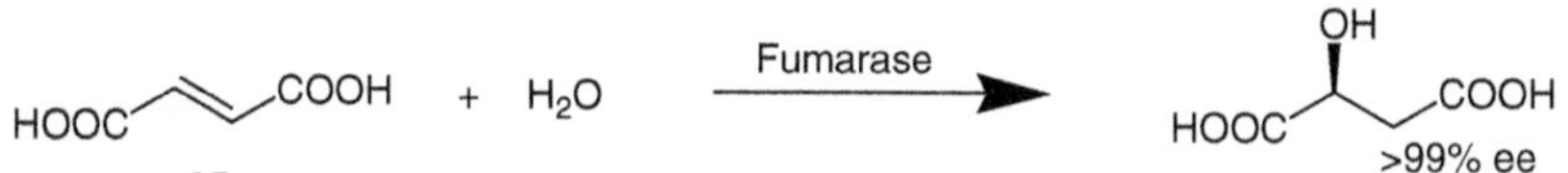

FIGURE 27.10 Addition of water to fumarate **35** yielding L-malate **36** catalyzed by fumarase.

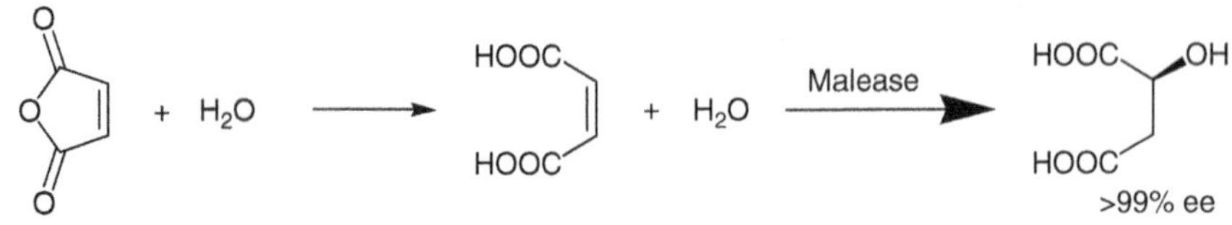

FIGURE 27.11 Production of (R)-malic **39** acid catalyzed by malease.

27.4.3 NITRILE HYDRATASE

Nitrile hydratases (E.C. 4.2.1.84) (systematic name: nitrile hydro-lyase) catalyze the rev cleavage of water from aliphatic amides. The enzyme has been found in different bacter structural data are available [22].

Nitrile hydratases belonging to the enzyme class of lyases (E.C. 4) should not be co with the nitrilases belonging to the class of hydrolases (E.C. 3) that hydrolyze nitriles corresponding carboxylic acids. For technical purposes, the enzymes from *Pseudo chloroaphis* and *Rhodococcus rhodochrous* are used as immobilized whole cells to p various amides by the addition of water to the respective nitriles.

DuPont uses the strain *P. chlororaphis B23* to synthesize 5-cyano valeramide **41** ure 27.12), which is used as an intermediate for the synthesis of the herbicide azafenidir Whole cells from *P. chlororaphis* are immobilized in calcium alginate beads. For selection, it was important that the cells did not show any amidase activity that would hydrolyze the amide to the carboxylic acid. The biotransformation is carried ou two-phase system with pure adiponitrile **40** forming the organic phase [74]. A re temperature of 5°C is chosen, as the solubility of the by-product adipodiamide is only 42 mM in 1 to 1.5 M 5-cyanovaleramide **41**. A batch reactor is preferred over a fix reactor, because of the lower selectivity toward 5-cyanovaleramide that was observe possibility of precipitation of adipodiamide, and plugging of the column. Excess is removed by distillation at the end of the reaction. The by-product adipodiam precipitated by dissolution of the resulting oil in methanol at >65°C. The crude p solution is directly transferred to the herbicide synthesis.

Using this method, 13.6 t have been produced in 58 repetitive batch cycles wit conversion and 96% selectivity. This biotransformation was chosen over the chemical formation because of the higher conversion and selectivity, production of more prod catalyst weight (3150 kg/kg dry cell weight), and less waste. The catalyst consump 0.006 kg/kg product.

Acrylamide **43** (Figure 27.13) is an important commodity monomer used in coagu soil conditioners, and stock additives for paper treatment and paper sizing, for adh paints, and petroleum recovering agents. It is biocatalytically produced by Nitto Ch Industry Co. Ltd. (Japan), using *R. rhodochrous* as a production strain that was select of 1000 microbial strains [75–79].

Acrylamide **43** is unstable and polymerizes easily; therefore the process is carri at low temperature (5°C). Although the cells are immobilized on polyacrylamide g

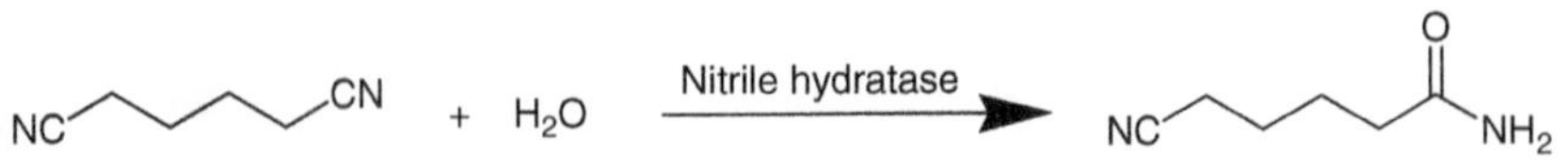

FIGURE 27.12 Synthesis of 5-cyano valeramide **41** catalyzed by nitrile hydratase.

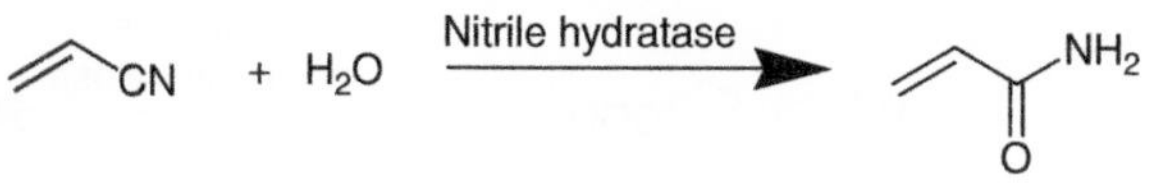

FIGURE 27.13 Synthesis of acrylamide **43** catalyzed by nitrile hydratase.

the containing enzymes are very stable toward acrylnitrile **42**, the starting material has to be fed continuously to the reaction mixture because of the inhibition effects at higher concentrations. The biotransformation is started with an acrylnitrile concentration of 0.11 M and is stopped at 5.6 M. The process is operated at a capacity of 30,000 t/y.

The chemical synthesis uses copper salt as a catalyst for the hydration of acrylnitrile and has the following disadvantages:

- The rate of acrylamide formation is lower than that of acrylic acid formation.
- The double bond of the starting material and the product causes the formation of by-products such as ethylene, cyanohydrin, and nitrilotrispropionamide.
- Polymerization occurs.
- Copper needs to be separated from the product (an extra step in the chemical synthesis).

Biotransformation has the advantage of 100% conversion of the nitrile, thereby avoiding recovering nonreacted nitrile and removing the copper catalyst required in the chemical process. This is also the first case of a biocatalytic conversion of a bulk fiber monomer.

Nitrile hydratase from *R. rhodochrous* also acts on other nitriles with yields of 100%. The most impressive example is the conversion of 3-cyanopyridine **44** to nicotinamide **45**, a biotransformation by Lonza AG (production site China) [80]. Nicotinamide (vitamin B3) is used as a vitamin supplement for food and animal feed. The biotransformation (a continuous process) is operated at low temperature and atmospheric pressure and is carried out on a scale of 3000 t/y (Figure 27.14).

The product concentration is about 1465 g/L. This conversion (1.17 g/L dry cell mass) can be named "pseudocrystal enzymation," because the substrate is solid at the beginning of the reaction and dissolved with ongoing reaction [80]. In contrast to the chemical alkaline hydrolysis of 3-cyanopyridine where nicotinic acid (4%) is formed as a by-product, the biotransformation is absolutely selective and no acid or base is required. Compared to the old synthetic route of nicotinamide at Lonza, the new one is environmentally friendly and safe. There is only one organic solvent used throughout the whole process in four highly selective continuous and catalytic reaction steps.

27.4.4 CARNITINE DEHYDRATASE

Carnitine dehydratases (E.C. 4.2.1.89) (systematic name: L-carnitine hydrolyase) have been described in only a few microorganisms. The enzyme catalyzes the reversible cleavage

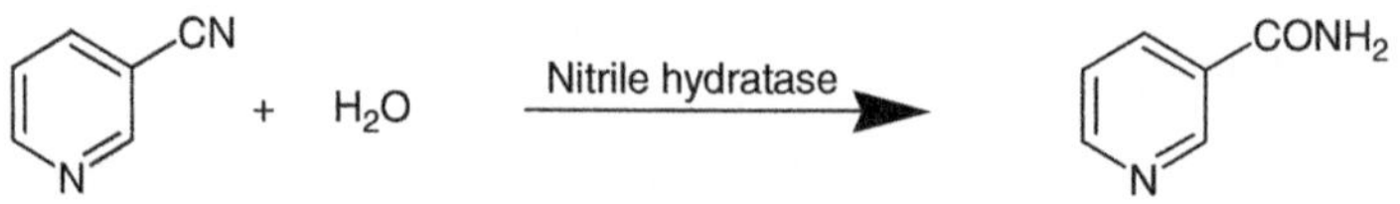

FIGURE 27.14 Synthesis of nicotinamide **45** catalyzed by nitrile hydratase.

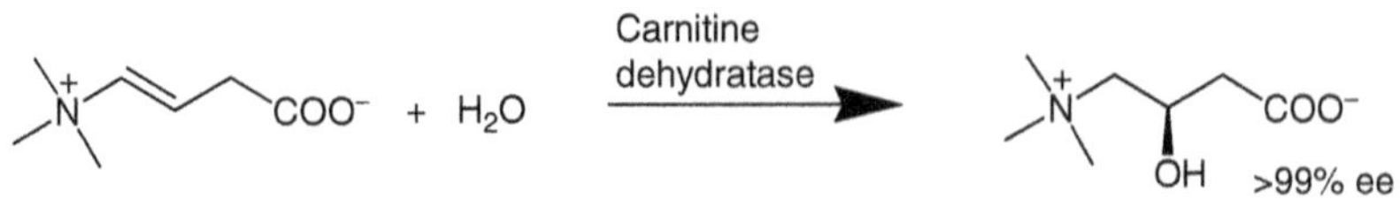

FIGURE 27.15 Synthesis of L-carnitine **47** catalyzed by carnitine dehydratase.

of water from L-carnitine **47** to 4-(trimethylamino)butanoate **46** and the reverse re
is used for the production of L-carnitine (Figure 27.15). L-Carnitine facilitates the tra:
of fatty acids across mitochondrial membrane, and thus is used as a nutritional suppl
ingredient, especially in sports and health beverages. Only the L-isomer is biolo;
active, which gives stereoselective enzyme-based processes an advantage over ch
synthesis [81]. Carnitine dehydratase from *E. coli* was thoroughly characterized a:
gene was cloned [22].

A good example is the manufacture of L-carnitine from butyrobetaine throu
butyrobetaine CoA by Lonza AG (Visp, Switzerland) [82,83]. This process in
site-selective reaction of a nonactivated carbon atom. What is interesting is that L-ca:
has been a product, for a sufficiently long period, for the butyrobetaine process t
second-generation bioprocess, replacing the original bioprocess that was based on the
chloroacetoacetic acid ester as the precursor. Furthermore, Lonza has described how
took them to develop a high-yielding process (300 g/L) for butyrobetaine conversion
further improvements becoming increasingly difficult to achieve as the concentrati
the product became gradually higher. This involved improving the original low-y
microbial strain to a production strain that produces about 2000-fold higher concent
of carnitine. This was achieved by avoiding the metabolism of L-carnitine with strains l:
in carnitine dehydrogenase [82]. 4-Butyrobetainyl coenzyme A is the key intermediate
metabolism. An aspect of the microorganisms physiology that greatly im
the productivity of the process is that butyrobetaine uptake is mediated by a perip
adenosine triphosphate-linked protein, which is closely coupled to carnitine excretion
means that more butyrobetaine substrate is taken up by the cell for every molec
carnitine exported. Therefore, alternative mutants of the original strain are now us
production and Lonza has achieved a 103-fold increase in activity from the original r
after 1.5 years of work.

27.5 CONCLUSION

Presently, biocatalytic C–C, C–N, and C–O formations play an important role in ind
syntheses of advanced intermediates. Often, new chiral centers are formed starting
achiral compounds like aldehydes. As there is no kinetic resolution involved, a ma:
theoretical yield of 100%, is possible. In future, new biocatalytic activities will be availa
molecular engineering and directed evolution. Lyases have a bright future in ind
applications for preparation of fine chemicals and chiral intermediates.

REFERENCES

1. Liese, A., Seelbach, K., and Wandrey, C., *Industrial Biotransformations*, 2nd ed., Wiley
   Weinheim, Germany, 2006.
2. Breuer, M., Ditrich, K., Habicher, T., et al., Industrial methods for the production of o:
   active intermediates, *Angew. Chem. Int. Ed. Engl.*, 43, 788–824, 2004.

3. Straathoff, A.J.J., Panke, S., Schmid, A., The production of fine chemicals by biotransformations, *Curr. Opin. Biotechnol.*, 13, 548–556, 2002.

4. Panke, S. and Wubbolts, M., Advances in biocatalytic synthesis of pharmaceutical intermediates, *Curr. Opin. Chem. Biol.*, 9, 188–194, 2005.

5. Rosenthaler, L., Asymmetric syntheses produced by enzymes, *Biochem. Z.*, 14, 238–253, 1909.

6. Rosche, B., Breuer, M., Hauer, B., et al., Screening of yeasts for cell-free production of (*R*)-phenylacetylcarbinol, *Biotechnol. Lett.*, 25, 841–845, 2003.

7. Goetz, G., Iwan, P., Hauer, B., et al., Continuous production of (*R*)-phenylacetylcarbinol in an enzyme-membrane reactor using a potent mutant of pyruvate decarboxylase from *Zymomonas mobilis*, *Biotechnol. Bioeng.*, 74, 317–325, 2001.

8. Leksawasdi, N., Chow, Y.Y.S., Breuer, M., et al., Kinetic analysis and modelling of enzymatic (*R*)-phenylacetylcarbinol batch biotransformation process, *J. Biotechnol.*, 111, 179–189, 2004.

9. Shukla, V.B. and Kulkarni, P.R., Comparative studies on bioconversion of benzaldehyde to L-phenylacetylcarbinol (L-PAC) using calcium alginate- and barium alginate-immobilized cells of Torulaspora delbrueckii, *J. Chem. Technol. Biotechnol.*, 78, 949–951, 2003.

10. Shukla, V.B. and Kulkarni, P.R., L-Phenylacetylcarbinol (L-PAC): biosynthesis and industrial applications, *World J. Microbiol. Biotechnol.*, 16, 499–506, 2000.

11. Rosche, B., Leksawasdi, N., Sandford, V., et al., Enzymatic (*R*)-phenylacetylcarbinol production in benzaldehyde emulsions, *Appl. Microbiol. Biotechnol.*, 60, 94–100, 2002.

12. Rosche, B., Sandford, V., Breuer, M., et al., Enhanced production of R-phenylacetylcarbinol (R-PAC) through enzymatic biotransformation, *J. Mol. Catal., B Enzym.*, 19, 109–115, 2002.

13. Oliver, A.L., Anderson, B.N., Roddick, F.A., Factors affecting the production of L-phenylacetyl-carbinol by yeast: a case study, *Adv. Microb. Physiol.*, 41, 1–45, 1999.

14. Pohl, M., Sprenger, G.A., Müller, M., A new perspective on thiamine catalysis, *Curr. Opin. Biotechnol.*, 15, 335–342, 2004.

15. Ward, O.P. and Singh, A., Enzymatic asymmetric synthesis by decarboxylases, *Curr. Opin. Biotechnol.*, 11, 520–526, 2000.

16. Breuer, M. and Hauer, B., Carbon–carbon coupling in biotransformation, *Curr. Opin. Biotechnol.*, 14, 570–576, 2003.

17. Siegert, P., McLeish, M.J., Baumann, M., et al., Exchanging the substrate specificities of pyruvate decarboxylase from *Zymomonas mobilis* and benzoylformate decarboxylase from *Pseudomonas putida*, *Protein Eng. Des Sel.* 18, 345–357, 2005.

18. Müller, M., Chemoenzymatic synthesis of building blocks for statine side chains, *Angew. Chem. Int. Ed. Engl.*, 44, 362–365, 2005.

19. Kierkels, J.G.T., Mink, D., Panke, S., et al., Process for the preparation of 2,4-dideoxyhexoses and therapeutic uses thereof, WO 2003/006656 (DSM).

20. DeSantis, G., Liu, J., Clark, D.P., et al., Structure based mutagenesis approaches towards expanding the substrate specifity of D-2-deoxyribose-5-phosphate aldolase, *Bioorg. Med. Chem.*, 11, 43–52, 2003.

21. Liu, J., Hsu, C.-C., and Wong, C.-H., Sequential aldol condensation catalyzed by DERA mutant Ser238Asp and a formal total synthesis of atorvastatin, *Tetrahedron Lett.*, 45, 2439–2441, 2004.

22. BRENDA_Enzyme_Database at http://www.brenda.uni-koeln.de/.

23. Ohta, Y., Tsukada, Y., Sugimori, T., et al., Isolation of a constitutive *N*-acetylneuraminate lyase-producing mutant of *Escherichia coli* and its use for NPL production, *Agric. Biol. Chem.*, 53, 477–481, 1989.

24. Maru, I., Ohnishi, J., Ohta, Y., et al., Simple and large-scale production of *N*-acetylneuraminic acid from *N*-acety-D-glucosamine and pyruvate using *N*-acyl-D-glucosamine 2-epimerase and *N*-acetylneuraminate lyase, *Carbohydr. Res.*, 306, 575–578, 1998.

25. Dawson, M., Noble, D., and Mahmoudia, D.S., Process for the preparation of *N*-acetyl-D-neuraminic acid, PCT WO 9429476, 1994.

26. Mahmoudian, M., Noble, D., Drake, C.S., et al., An efficient process for production of *N*-acetylneuraminic acid using *N*-acetyneuraminic acid aldolase, *Enzyme Microb. Technol.*, 45, 393–400, 1997.

27. Ghisalba, O., Gygax, D., Kragl, U., et al., Enzymatic method for *N*-acetyneuramin acid pro Novartis, EP 0428947, 1991.

28. Kragl, U., Gygax, D., Ghisalba, O., et al., Enzymatic process for preparing *N*-acetyneurami. *Angew. Chem. Int. Ed. Engl.*, 30, 827–828, 1991.

29. Bang, W., Lang, S., Sahm, H., et al., Production of L-tryptophan by *Escherichia cc Biotechnol. Bioeng.*, 25, 999–1011, 1983.

30. Plischke, H. and Steinmetzer, W., Verfahren zur Herstellung von L-Tryptophan und D AMINO GmbH, DE 3630878 C1, 1988.

31. Sharma, M., Sharma, N.N., and Bhalla, T.C., Hydroxynitrile lyases: at the interface of biol chemistry, *Enzyme Microb. Technol.*, 37, 279–294, 2005.

32. Gruber, K. and Kratky, Ch., Biopolymers for biocatalysis: structure and catalytic mecha hydroxynitrile lyases, *J. Polym. Sci. [A]*, 42, 479–486, 2004.

33. Effenberger, F., Synthesis and reactions of optically active cyanohydrins, *Angew. Chem. Engl.*, 33, 1555, 1994.

34. Effenberger, F., Optically active cyanohydrins—important sources for chiral drugs, *Enant* 359, 1996.

35. Effenberger, F., Enzyme-catalyzed preparation and synthetic applications of opticall cyanohydrins, *Chimia.*, 53, 3, 1999.

36. Griengl, H., Hickel, A., Johnson, D.V., et al., Enzymatic cleavage and formation of cyano a reaction of biological and synthetic relevance, *Chem. Commun.*, 20, 1933–1940, 1997.

37. Johnson, D.V., Zabelinskaja-Mackova, A.A., and Griengl, H., Oxynitrilases for asymmet bond formation, *Curr. Opin. Chem. Biol.*, 4, 103, 2004.

38. Griengl, H., Schwab, H., and Fechter, M., The synthesis of chiral cyanohydrins by oxyn *Ophtalmic Genet.*, 18, 252, 2000.

39. Glieder, A., Weis, R., Skranc, W., et al., Comprehensive step-by-step engineerinɡ (*R*)-hydroxynitrile lyase for large-scale asymmetric synthesis, *Angew. Chem. Int. Ed. E* 4815–4818, 2003.

40. Chemie Linz GmbH: DNA encoding *Hevea brasiliensis* (*S*)-hydroxy:nitrilase, DE 19522911(*Abstr.*, 126, 208953, 1997.

41. Semba, H. (Nippon Shokubai): Process for producing *S*-hydroxynitrile lyase, EP 1016712 *Abstr.*, 133, 85120, 2000.

42. Effenberger, F., Wajant, H., Lauble, P., et al. (Fine Chemicals Austria): (*S*)-Hydroxynitri with improved substrate acceptance and uses thereof, EP 969095, *Chem. Abstr.*, 132, 19122

43. Bousquet, A. and Musolino, A. (Sanofi-Synthelabo): Hydroxyacetic ester derivatives, prej method and use as synthesis intermediates, EP 1021449, *Chem. Abstr.*, 130, 296510, 1999.

44. Pöchlauer, P. and Mayrhofer, H. (DSM Fine Chemicals Austria): Method for producing optic chemically pure (*R*)- and (*S*)-hydroxycarboxylic acids, EP 1148042, *Chem Abstr.*, 135, 30360

45. Okuda, N., Semba, H., and Dobashi, Y. (Nippon Shokubai): A method for producinɡ hydroxycarboxylic acid, EP 1160235, *Chem. Abstr.*, 136, 5721, 2002.

46. Semba, H. and Dobashi, Y. (Nippon Shokubai): An enzyme reaction method and a me enzymatically producing an optically active cyanohydrin, EP 1160329, *Chem. Abstr.*, 136, 48(

47. Hasslacher, M., Schall, M., Schwab, H., et al. (DSM Chemie Linz): (*S*)-Hydroxynitrilyɑ *Hevea brasiliensis*, WO 97/03204, *Chem. Abstr.*, 126, 196118, 2000.

48. Griengl, H., Klempier, N., Pöchlauer, P., et al., Enzyme catalyzed formation of (*S*)-cyanɑ derived from aldehydes and ketones in a biphasic solvent system, *Tetrahedron*, 54, 1447, 1

49. Semba, H. (Nippon Shokubai): Method for the production of optically active cyanohyd 1026256, *Chem. Abstr.*, 133, 134248, 2000.

50. Pöchlauer, P., Schmidt, M., Wirth, I., et al. (DSM Fine Chemicals Austria): Enzymatic pre the preparation of (*S*)-cyanohydrins, EP 927766, *Chem. Abstr.*, 131, 72766, 2000.

51. Weis, R., Gaisberger, R., Skranc, W., et al., Carving the active site of almond *R*-HNL for i enantioselectivity, *Angew. Chem. Int. Ed. Engl.*, 44, 4700–4704, 2005.

52. Kumagai, H., Microbial production of amino acids in Japan, in *Advances in Biochemical Enɡ and Biotechnology*, vol. 69, Schepers, T., Ed., Springer, Berlin/Heidelberg, 2000, pp. 71–85.

53. Tsuchida, T., Nishimoto, Y., Kotani, T., et al., Production of L-3,4-dihydroxyphenylalanine, Ajinomoto Co. Ltd., JP 5123177A, 1993.
54. Yamada, H., Screening of novel enzymes for the production of useful compounds, in *New Frontiers in Screening for Microbial Biocatalysis*, Kieslich, K., van der Beek, C.P., de Bont, J.A.M., et al., Eds., *Studies in Organic Chemistry*, Elsevier, Amsterdam, 1998, 53, pp. 13–17.
55. Tanaka, A., Tosa, T., and Kobayashi, T., *Industrial Application of Immobilized Biocatalysts*, Marcel Dekker, New York, 1993.
56. OECD. The application of biotransformation to industrial sustainability, ISBN 92-64-1946-7 No 52184 200, OECD Publications, France, 2001.
57. Terasawa, M., Yukawa, H., and Takayama, Y., Production of L-aspartic acid from *Brevibacterium* by cell re-using process, *Process Biochem.*, 20, 124–128, 1985.
58. Yamagata, H., Teresawa, M., and Yukawa, H., A novel industrial process for L-aspartic acid production using an ultrafiltration-membrane, *Catal. Today*, 22, 621–627, 1994.
59. Schmidt-Kastner, G. and Egerer, P., Amino acids and peptides, in *Biotechnology*, vol. 6a, Kieslich, K., Ed., Verlag Chemie, Weinheim, 1984, pp. 387–419.
60. Furui, M. and Yamashita, K., Pressurized reaction method for continuous production of L-alanine by immobilized *Pseudomonas dacunhae* cells, *J. Ferment. Technol.*, 61, 587–591, 1983.
61. Takamatsu, S., Umemura, J., Yamamoto, K., et al., Production of L-alanine from ammonium fumarate using two immobilized microorganisms, *Eur. J. Appl. Biotechnol.*, 15, 147–152, 1982.
62. Sheldon, R.A., *Chirotechnology*, Marcel Dekker, New York, 1993.
63. Cheetham, P.S.J., Bioprocess for the manufacture of ingredients for food and cosmetics, *Adv. Biochem. Eng. Biotechnol.*, 86, 83–158, 2004.
64. Vollmer, P.J., Montgomery, J.P., Schruber, J.J., et al., Method for stabilizing the enzymatic activity of phenylalanine ammonia lyase during L-phenylalanine production, Genex Corporation, US 4584269, 1985.
65. Crosby, J., Synthesis of optically active compounds: a large scale perspective, *Tetrahedron*, 47, 4789–4846, 1991.
66. Mattey, M., The production of organic acids, *Crit. Rev. Biotechnol.*, 12, 87–132, 1988.
67. Tosa, T. and Shibatani, T., Industrial applications of immobilized biocatalysts in Japan, *Ann. N. Y. Acad. Sci.*, 750, 364–375, 1995.
68. Daneel, H.J., Busse, M., and Faurie, R., Pharmaceutical grade L-malic acid from fumaric acid— development of an integrated biotransformation and product purification process, *Med. Fac. Landbouww. Univ. Gent.*, 60(4a), 2093–2096, 1995.
69. Daneel, H.J., Busse, M., and Faurie, R., Fumarate hydratase from *Corynebacterium glutamicum*— process related optimization of enzyme productivity for biotechnical L-malic acid synthesis, *Med. Fac. Landbouww. Univ. Gent*, 61(4a), 1333–1340, 1996.
70. Van der Werf, M., van der Tweel, W., and Hermans, S., Screening for microorganisms producing D-malate from maleate, *Appl. Environ. Microbiol.*, 58, 2854–2860, 1992.
71. Chibata, I., Tosa, T., and Shibatani, T., The industrial production of optically active compounds by immobilized biocatalysts, in *Chirality in Industry*, Collins, A.N., Sheldrake, G., and Crosby, J., Eds., Wiley, New York, 1992, pp. 351–370.
72. Submarmarian, S.S. and Rao, R.M.R., Purification and properties of citraconase, *J. Biol. Chem.*, 243, 2367–2372, 1968.
73. Michielsen, M., Frielink, C., Wijffels, R., et al., D-Malate production by permeabilized *Pseudomonas speudoalcaligenes*: optimization of conversion and biocatalyst productivity, *J. Biotechnol.*, 79, 13–26, 2000.
74. Hann, E.C., Eisenberg, A., Fager, S.K., et al., 5-Cyanovaleramide production using immobilized *Pseudomonas chlororaphis* B23, *Bioorg. Med. Chem.*, 7, 2239–2245, 1999.
75. Nagasawa, T., Shimizu, H., and Yamada, H., The superiority of the third-generation catalyst, *Rhodococcus rhodochrous* J1 nitrile hydratase, for industrial production of acrylamide, *Appl. Microb. Biotechnol.*, 40, 189–195, 1993.
76. Shimizu, H., Ogawa, J., Kataoka, M., et al., Screening of novel microbial enzymes for the production of biologically and chemically useful compounds, in *New Enzymes for Organic Synthesis*,

Ghose, T.K., Fiechter, A., and Blakebrough, N., Eds., Springer, Berlin/Heidelberg, 199'
Biochem. Eng. Biotechnol., 58, 56–59, 1997.

77. Yamada, H. and Tani, Y., Process for biologically producing amide, EP 093782, 1982.
78. Yamada, H. and Kobayashi, M., Nitrile hydratase and its application to industrial produc
acrylamide, *Biosci. Biotech. Biochem.*, 60(9), 1391–1400, 1996.
79. Yamada, H. and Tani, Y. (Nitto Chemical Industry Co. Ltd.): Process for biological prepara
amides. US 4637982, 1987.
80. Petersen, M. and Kiener, A., Biocatalysis—preparation and functionalization of *N*-hetero
Green Chem., 4, 99–106, 1999.
81. Kulla, H., Enzymatic hydroxylations in industrial application, *Chimica* 45, 81–85, 1991.
82. Zimmermann, Th.P., Robins, K.T., Werlen, J., et al., Bio-transformation in the produc
L-carnitine, in *Chirality in Industry*, Collins, A.N., Sheldrake, G.N., and Crosby, J., Eds.,
New York, 1997, pp. 287–305.
83. Hoeks, F.W.J.M.M., Verfahren zur diskontinuierlichen Herstellung von L-Carnitin auf mi
logischem Weg. Lonza AG, EP 410 430 A2, 1991.

28 State of the Art and Applications in Stereoselective Synthesis of Chiral Cyanohydrins

*Franz Effenberger, Siegfried Förster,
and Christoph Kobler*

CONTENTS

28.1   INTRODUCTION

Michael North, guest editor of the *Tetrahedron Symposium in Print* Number 1C
summarized in his preface the most important actual research activities on the sy
and applications of nonracemic (chiral) cyanohydrins [1]. Since the beginning of rema
research activities on chiral cyanohydrins in 1987 [2], several excellent review artic
this topic have been published [3–12]. For this reason, this chapter only discuss
developments of the last 5 y in synthesis and applications of nonracemic cyanohydrin
special emphasis on applications of hydroxynitrile lyases (HNLs) as chiral catalysts.

28.2   METAL-CATALYZED SYNTHESES OF CHIRAL CYANOHYDRINS

Although HNL-catalyzed preparations of nonracemic cyanohydrins are still the
important synthetic approach to chiral cyanohydrins, metal-catalyzed reactions have re
considerable interest during the last decade [13,14]. The discovery and developm
new efficient catalysts, containing a metal and chiral ligands, has resulted in much
enantioselectivities of the cyanohydrins prepared. Since enzyme-catalyzed reactions
strongly on structural requirements of the reacting substrates, metal-catalyzed cyano
formations could become important and interesting in cases of sterically dem
aldehydes and especially in the reactions of ketones [13,14]. North classifies the
metal catalysts according to the nature of the chiral ligands [13].

28.2.1   Application of Chiral Bidentate and Tridentate Ligands

Corey and Wang have generated a catalytically active system for the asymmetric a
of trimethylsilyl cyanide (TMSCN) to aldehydes (Scheme 28.1) by using a combina
magnesium bisoxazoline complex **1** as a chiral Lewis acid and uncomplexed bisoxazoli
a chiral Lewis base [15]. The best results (63 to 95% enantiomeric excesses (ee)) were o
with aliphatic and α,β-unsaturated aldehydes [15]. In a recent publication, chiral oxa
olidinium salts were efficiently applied to the enantioselective cyanosilylation of keton

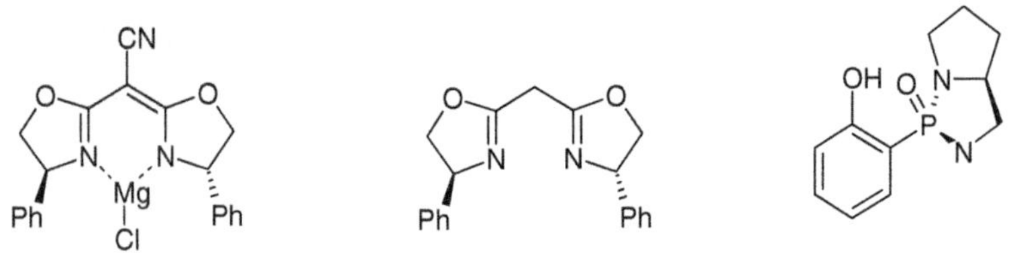

SCHEME 28.1 Bidentate ligands for asymmetric cyanosilylation.

For the addition of TMSCN to aromatic aldehydes, Brunel et al. were able to obtain
98% by using a titanium complex of diastereomerically pure phosphine oxide **3** (Scheme 28
BiNOL and its derivatives have also been used successfully as bidentate chiral liga
the enantioselective addition of TMSCN to aldehydes and ketones, respectively [1
most efficient BiNOL complex in asymmetric cyanohydrin synthesis is the Al com
BiNOL derivative **4** (Scheme 28.2) [18]. The best results (up to 98% ee) are obtained
presence of an additional phosphine oxide additive [18].

Hamashima et al. successfully developed tridentate ligands **5–7** for the enantios
addition of TMSCN to carbonyl compounds (Scheme 28.2) [19,20]. Although the Al c
of ligand **5** shows no catalytic activity, the corresponding titanium complex was found
good catalyst especially for the synthesis of (*R*)-ketone cyanohydrins [19,20].

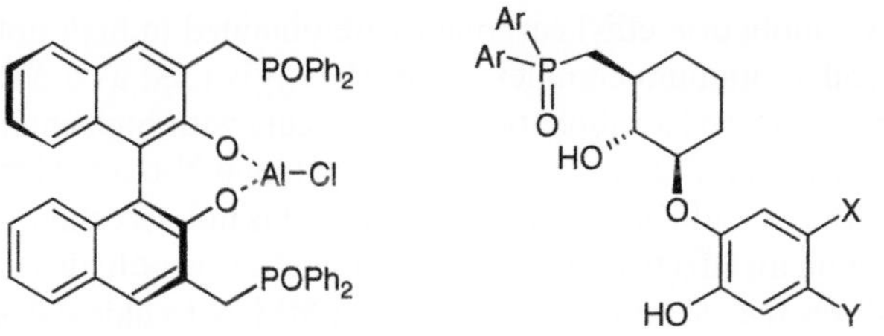

5: X = Y = H; Ar = Ph
6: X = COPh; Y = H; Ar = Ph
7: X = Y = F; Ar = Ph

SCHEME 28.2 Bidentate BiNOL derivatives and tridentate ligands for asymmetric cyanosilylation

Interestingly, it was possible to invert the stereochemistry of the TMSCN addition by substitution of titanium by gadolinium [21–23]. Several methyl and ethyl ketones were converted into (*S*)-cyanohydrin trimethylsilyl ethers with 62 to 92% ee. Aryl cycloalkyl ketones also react with high chemical (100%) and optical (95% ee) yields to form the corresponding (*S*)-cyanohydrin trimethylsilyl ethers [21–23].

28.2.2 APPLICATION OF CHIRAL TETRADENTATE LIGANDS

Schiff bases derived from salicylic aldehydes are widely and successfully used chiral tetradentate salen ligands with chiral 1,2-diamines. Interactive studies by the groups of Belokon and North showed ligand **8** to be one of the most promising compounds (Scheme 28.3) [24–29]. The titanium complex **9** is isolated as crystalline solid by the treatment of ligand **8** with TiCl$_4$ (Scheme 28.3).

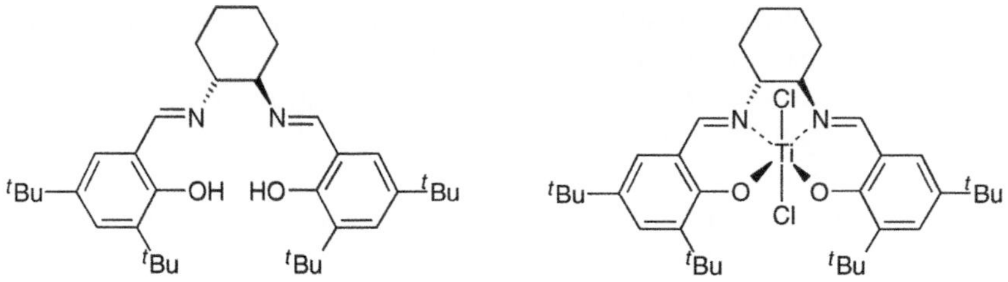

8   **9**

SCHEME 28.3 Tetradentate ligands for cyanohydrin synthesis. (From Tararov, V.I., et al., *Chem Commun.*, 387, 1998; Belokon, Y.N., et al., *J. Am. Chem. Soc.*, 121, 3968, 1999; Belokon, Y.N., et al., *Chem Commun*, 244, 2002; Belokon, Y.N., et al., *Tetrahedran*, 60, 10433, 2004; Chen, F.-X., et al., *Chem Eur. J.*, 10, 4790, 2004.)

Only 0.1 mol% of **9** was needed to convert benzaldehyde to (*S*)-mandelonitrile trimethylsilyl ether with 86% ee [25]. Under the reaction conditions, with traces of water present, complex **9** reacts to give a bimetallic complex that can be isolated and characterized [26]. The bimetallic complex allows the use of potassium cyanide instead of TMSCN as a cyanide source, which is highly advantageous for practical applications. Reactions using KCN are carried out in the presence of an acid anhydride and lead to (*S*)-cyanohydrin carboxylates with 85 to 93% ee [27]. Recently, it could be shown that ethyl cyanoformate can also serve as a cyanide source.

(R)-Cyanohydrin ethyl carbonates are obtained in high optical yields (up to 99% ee) when bimetallic titanium complex [(salen)TiO$_2$]$_2$ is used as a chiral catalyst [28]. Cyanosilylatic aromatic as well as aliphatic ketones occurs with high enantioselectivity when a chiral sale complex (2 mol%) and a chiral tertiary amine N-oxide (1 mol%) are combined [29].

Holmes and Kagan have shown that it is not necessary for complex 8 with a transition r to obtain an effective catalyst for asymmetric cyanohydrin formation: the monolithium salt catalyzes the asymmetric addition of TMSCN to aldehydes, giving (R)-cyanohydrin trime silyl ethers with up to 97% ee [30]. The amount of water present has a significant effect on enantioselectivity of the Li-salts of complex 8 [31]. Recently, a highly enantioselective cy silylation of aldehydes that was catalyzed by novel alcohol–titanium complexes was reported

Polymeric salen–Ti (IV) or –V (V) complexes were employed in the enantiosele O-acetyl cyanation of aldehydes with KCN and acetic anhydride [33].

28.2.3 SUMMARY OF METAL-CATALYZED SYNTHESIS OF CHIRAL CYANOHDRINS

The metal-catalyzed asymmetric addition of TMSCN to aldehydes and ketones has enormous developments during the last decade [13,14]. In particular, the possibility of potassium cyanide instead of TMSCN as a cyanide source is an important improvemen practical applications of this methodology. Simultaneously, enormous progress has been in the enzyme-catalyzed synthesis of nonracemic cyanohydrins, especially in HNL-cata reactions. Metal-catalyzed additions will be superior to enzyme catalysis only when the sta carbonyl compounds are not accepted as substrates by the enzymes. This will be the cas sterically demanding aldehydes and especially, ketones.

28.3 ENZYME-CATALYZED PREPARATIONS OF CHIRAL CYANOHYDRINS

Although considerable progress has been made in metal-catalyzed synthesis of nonra cyanohydrins during the last 5 y (see Chapter 27), the HNL-catalyzed preparation of cyanohydrins is still the most important method, especially for large-scale reactions. Pro in lipase-catalyzed kinetic resolution of racemic cyanohydrins and cyanohydrin deriv will also be dicussed, respectively. The kinetic resolution of racemic cyanohydrins by nitr and amidases [34] will not be discussed in this chapter.

28.3.1 OPTIMIZATION OF REACTION CONDITIONS FOR HNL-CATALYZED
CYANOHYDRIN FORMATION

All applications of HNLs in catalysis of nonracemic cyanohydrin formation aim to a the following objectives: first, it is crucial to get high enantioselectivity [2], suppressi chemical addition of HCN to the carbonyl substrates. Second, for industrial application kinetics are necessary for efficient and economical large-scale production. Third, formation of HCN is desirable to avoid free handling of toxic HCN.

All the three objectives can be achieved by using the enzymes on a solid support (i cellulose) in an organic solvent (diisopropylether and methyl-tert-butylether), by ap aqueous media at low pH, or by working in a biphasic system (methyl-tert-butylether/v A comparison of advantages and disadvantages of the various reaction conditions men here can be found in the comprehensive reviews by North [4] and Gröger [35]. For larg productions using inexpensive enzymes, the two-phase system is usually the method of c In contrast, reactions in organic solvents, with HNLs adsorbed on a support, considerably smaller amounts of enzymes, and product recovery is facilitated compa the two-phase system. An interesting alternative to the adsorption of HNLs on ce

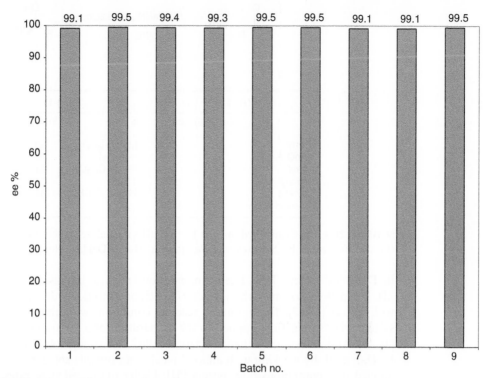

FIGURE 28.1 Long-term stability of PVA-entrapped PaHNL in aqueous media. (From Gröger, H. et al, Bundesforschungsanstalt für Landwirtschaft, Braunschweig, SI 241, 87, 2002.)

is the encapsulation of the enzymes in polyvinyl alcohol (PVA) [36–38]. The lens-shaped HNL-containing catalyst shows a well-defined particle of diameter 3 to 5 mm and thickness 0.3 to 0.4 mm. PVA-entrapped (R)-PaHNL can be applied in buffered aqueous media (pH 3.75) or in a biphasic system [36–38]. The catalyst can be reused without any loss of enantioselectivity (Figure 28.1) [36–38]. (R)-PaHNL has also been immobilized by cross-linking enzyme aggregates (CLEA) using glutaraldehyde [38]. The CLEA of (R)-PaHNL are stable and can be recycled many times [38].

Recently, the use of ionic liquids instead of organic solvents was published for the biphasic system [39]. For PaHNL and HbHNL, the reaction rates were increased in comparison to organic solvents, while the enantioselectivity was identical [39].

28.3.2 (R)-HNL CATALYSIS

(R)-PaHNL from almonds (*Prunus amygdalus*), the first enzyme ever used in asymmetric synthesis [40], is not only easily available in large amounts from almonds but is also surprisingly enantioselective when accepting a broad spectrum of substrates (Scheme 28.4) [1–12]. (R)-HNLs isolated from other plants do not show any improvements as catalysts for the preparations of (R)-cyanohydrins, and therefore cannot compete with (R)-PaHNL for applications [41–43].

Recently, cloning of the gene of HNL isoenzyme 5 (PaHNL5) and overexpression in the methylotrophic yeast *Pichia pastoris* was reported [44,45]. The authors stated that the recombinant PaHNL5 shows considerably higher reactivity than PaHNL for reactions of nonnatural substrates like 2-chlorobenzaldehyde [44,45]. However, comparable results (89% yield, 91% ee) can be obtained by working with wild-type enzyme adsorbed on cellulose

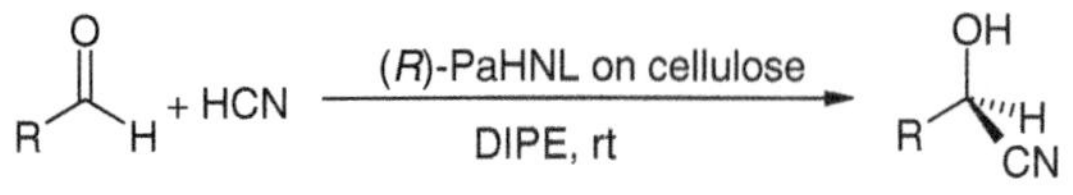

SCHEME 28.4 (*R*)-Cyanyohydrins (*R*)-**11** by PaHNL-catalyzed addition of HCN to Aldeh (From Effenberger, F., *Seteoselective Biocatalysis*, Marcel Dekker, New York, 2000, P. 321.)

(*R*)-11			(*R*)-11		
R	Yield (%)	ee (%)	*R*	Yield (%)	ee (%)
Phenyl	96	>99	MeSCH$_2$CH$_2$—	98	96
3-PhO—C$_6$H$_4$	99	98	*n*-Propyl	99	98
2-furyl	96	99	PhCH$_2$CH$_2$CH$_2$—	94	90
3-thienyl	95	>99	*t*-Butyl	84	83
3-pyridyl	97	82	HOCH$_2$C(CH$_3$)$_2$—	84	89
Me—CH=CH—	94	95			

in DIPE as solvent [46]. Even almond powder in a biphasic system (Na-citrate pH 3.3/methyl-*tert*-butylether) gives (*R*)-2-chlorobenzaldehyde cyanohydrin in 94% with 83% ee [46,47]. The optimized reaction conditions with almond powder in a two system were successfully applied to the preparation of a large number of (*R*)-cyano [48,49]. An interesting synthetic variation called "micro-aqueous conditions" was dev by Han et al. by working with almond meal in an organic solvent [50]. This method applied in a continuous-flow system [51]. Under optimized flow rates, both aroma heteroaromatic aldehydes [51] could be converted to cyanohydrins with 84 to 100% two-phase reactions described here were optimized empirically. Straathof has devel mathematical model, which demonstrates that the reaction rate will be influenced substrate concentration, by the initial enzyme concentration, and by the volume-interfacial area [52].

28.3.3 (*S*)-HNL CATALYSIS

The first (*S*)-HNL used in the preparation of (*S*)-cyanohydrins was isolated from *S bicolor* (SbHNL). SbHNL was applied in organic [53] as well as in aqueous [54] so Although very high ee result with SbHNL as catalyst, this enzyme has two disadva First, only aromatic and heteroaromatic aldehydes are accepted as substrates and the availability from natural sources is far more limited than the availability of F from almonds. Until now, SbHNL could not be obtained in recombinant form.

Two other (*S*)-HNLs, however, which do not show these disadvantages have beer oped in the last few years and are now widely used for the synthesis of (*S*)-cyanohydri first is (*S*)-MeHNL, which is isolated from *Manhiot esculenta* [55], cloned and overexpr *Escherichia coli* [56]; the second is (*S*)-HbHNL from *Hevea brasiliensis*, overexpresse yeast *P. pastoris* [57,58]. Since, MeHNL is highly homologous to HbHNL, both enzym similar catalytic behavior. Like (*R*)-PaHNL from almonds, (*S*)-MeHNL and (*S*)-H respectively, accept a wide range of aldehydes (aromatic, heteroaromatic, aliphat saturated as well as unsaturated) and even ketones as substrates (Scheme 28.5) [10].

During the last years, the overexpression of both enzymes, MeHNL and H has been improved considerably [59], so that large quantities are now cheaply av Jülich Fine Chemicals, GmbH, Germany, for example, is selling (*S*)-MeHNL units [59].

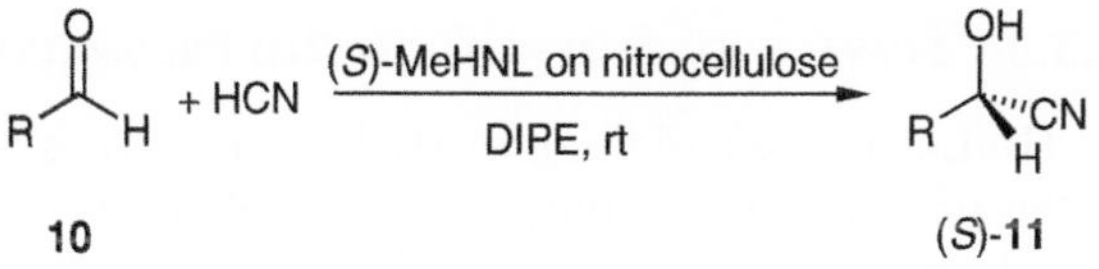

	(S)-11			(S)-11	
R	Yield (%)	ee (%)	R	Yield (%)	ee (%)
Phenyl	100	98	CH_3CH_2-	86	91
2-ClC_6H_4-	100	92	n-Butyl	100	91
4-MeOC_6H_4	82	98	$(CH_3)_2CH-$	91	95
2-Thienyl	85	96	$(CH_3)_3C-$	80	94
3-Furyl	98	92	$E\text{-CH}_3CH{=}CH-$	100	92

SCHEME 28.5 (S)-Cyanohydrins(S)-**10** by MeHNL-catalyzed addition of HCN to aldehydes **11** (From Effenberger, F., *Stereoselective Biocatalysis*, Marcel Dekker, New York, 2000, p. 321.)

28.3.4 LIPASE-CATALYZED PREPARATION OF CHIRAL CYANOHYDRINS

The kinetic resolution of racemic cyanohydrins and cyanohydrin derivatives, respectively by using lipases, has been summarized comprehensively [5,13]. Disadvantages of this method lie not only in the maximum yield of 50%, but also in the racemization tendency of unprotected chiral cyanohydrins under the conditions (pH$\geq$6) used for lipase-catalyzed reactions. This problem was avoided in the synthesis of (1R,cis,αS)-cypermethrine through lipase-catalyzed kinetic resolution of racemic m-phenoxybenzaldehyde cyanohydrin acetate by direct acylation of the obtained (S)-m-phenoxybenzaldehyde cyanohydrin with the corresponding cyclopropane carboxylic acid chloride [60]. The unreacted (R)-cyanohydrin is removed and racemized with triethylamine to give the racemic cyanohydrin acetate again, which is returned into the process. The total yield of the optically pure pyrethroide (1R,cis,αS)-cypermethrine referred to the racemic cyanohydrin acetate is more than 80% [60]. The kinetic resolution of racemic cyanohydrin acetates and protection of the formed (S)-cyanohydrins was performed with several racemic aromatic cyanohydrin acetates. In this case, both enantiomers are obtained in high optical purity [61]. Recently, the lipase-catalyzed resolution of protected hydroxycyanohydrins was published by Gotor et al. [62] The dynamic kinetic resolution using lipase has been successfully developed as a new methodology for the preparation of only one enantiomer, starting from racemic cyanohydrins or cyanohydrin derivatives [63,64,65,66]. When choosing optimized experimental conditions, the preparation of only one enantiomer in almost quantitative yield is possible starting from racemic cyanohydrins [63,64,65,66]. (Scheme 28.6)

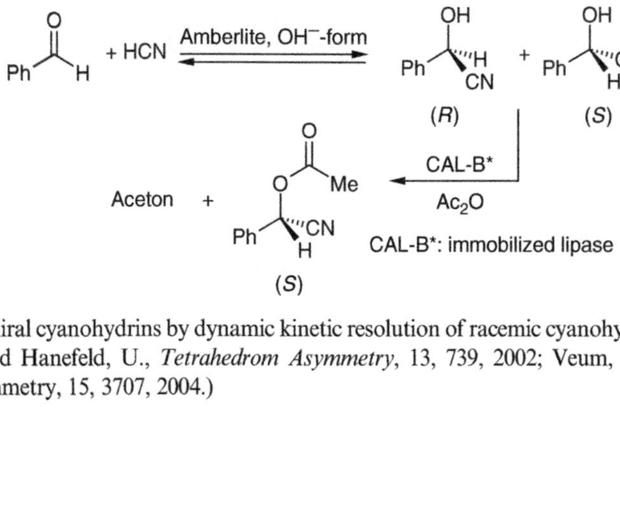

SCHEME 28.6 Chiral cyanohydrins by dynamic kinetic resolution of racemic cyanohydrins (From Li, Y-K., Stratthof, A.J., and Hanefeld, U., *Tetrahedrom Asymmetry*, 13, 739, 2002; Veum, L. and Hanefeld, U., Tetrahedron Asymmetry, 15, 3707, 2004.)

28.3.5 SUMMARY OF ENZYME-CATALYZED PREPARATIONS OF CHIRAL CYANOHYDRINS

The HNL-catalyzed addition of HCN to aldehydes is still the single most important n for the preparation of nonracemic cyanohydrins. The usefulness of these enzymes h creased substantially since (R)-PaHNL from almonds and two recombinant (S)-(MeHNL and SbHNL) are easily available in large amounts. Many (R)- as well a cyanohydrins are manufactured nowadays as intermediates for the production of ph ceuticals and plant protecting compounds, respectively (see Chapter 5). Further im ments of the dynamic kinetic resolution using lipases could be an enzymatic alterna HNLs in cases where the starting carbonyl compounds (aldehydes or ketones) a accepted as substrates by HNLs.

28.4 CRYSTAL STRUCTURES OF HNLs AND MECHANISM OF CYANOGEI

Several structurally different types of HNLs occur in nature, which likely origina convergent evolution from different ancestral proteins [67]. The properties and characte of the five most important HNLs for applications in the preparation of chiral cyanoh are listed in Table 28.1 [10].

The crystal structures of these five HNLs have only been determined during t decade, which allowed researchers to establish the mechanistic pathways of cyanogen

28.4.1 CRYSTAL STRUCTURES OF HNLs

A summary of 3D structures of HNLs known so far was recently published by Kratk [68] The enzyme from almonds (PaHNL) was already crystallized in 1994, and the str was solved by multiple wavelength anomalous dispersion of a mercury derivative [6 first 3D-structure analysis of PaHNL to a crystallographic resolution of 1.5 Å was perf in 2001 [70]. (R)-PaHNL from almonds uses FAD as cofactor and is related to reductases [71,72]. It is interesting to note that the enzyme exhibits HNL activity c the oxidized form of the cofactor FAD [73]. From indirect evidence, it was inferred tl active site of PaHNL is located close to the isoalloxazine of the cofactor [68].

TABLE 28.1
Properties and Characteristics of HNLs

Enzyme Source	Natural Substrate (Cyanogenic Glycoside)	R/S-Selectivity	Molecular Weight (kD	
			Native	S
Prunus amygdalus (Rosacea)	(R)-Mandelonitrile (amygdalin, prunasin)	R	55–80	5!
Sorghum bicolor (Gramineae)	(S)-p-Hydroxy-mandelonitrile (dhurrin)	S	105	3(
Manihot esculenta (Euphorbiaceae)	Acetone cyanohydrin (linamarin, lotaustralin)	S	92–124	2!
Hevea brasiliensis (Euphobiaceae)	Acetone cyanohydrin (linamarin)	S	58	3(
Linum usitatissimum (Linaceae)	Acetone cyanohydrin (linamarin)	R	82	4:

The crystal structure of the HNL from *S. bicolor* was determined in a complex with the inhibitor benzoic acid [74]. The folding pattern of SbHNL is similar to that of the closely related wheat serine carboxy peptidase (CPD-WII) [75] and alcohol dehydrogenase [76]; however, a unique two-amino acid deletion in SbHNL is forcing the putative active-site residues away from the hydrolase binding site toward a small hydrophobic cleft, thereby defining a completely different SbHNL active-site architecture where the classic triad of a carboxy peptidase is missing.

The most extensively investigated HNL structures are those of the highly homologous (76% sequence identity) enzymes from *H. brasiliensis* (HbHNL) and *M. esculenta* (MeHNL), both belonging to the family of α,β-hydrolases [77,78]. The 3D structure of HbHNL was first published in 1996 [79], and an ultrahigh resolution was reported in 1999 [80]. For MeHNL, the crystal structures of the wild-type enzyme complexed with acetone and chloroacetone, respectively, have been reported in 2001 [81,82]. The 3D structures of both enzymes clearly confirm that they belong to the family of α,β-hydrolases. The active site of both enzymes is located inside the protein and is connected to the outside through a small channel, which is covered by a bulky amino acid (tryptophane 128) [81,82]. It was possible to obtain the crystal structure of the complex with the natural substrate acetoncyanohydrin with the active-site mutant Ser80Al of MeHNL [81,82]. This complex allowed determination of the mode of subtrate binding in the active site of the enzyme [81,82].

28.4.2 REACTION MECHANISM OF CYANOGENESIS

Although mechanisms for cyanogenesis have been published based on the crystal structure investigations for HNL from *S. bicolor* [74] as well as for PaHNL from almonds [83], here, we will only dicuss the thoroughly proven mechanism of the closely related enzymes HbHNL and MeHNL. Originally, a mechanism involving a general acid/base catalysis was proposed for MeHNL on the basis of sequence and mutational analysis [84]. The first suggested mechanism for HbHNL, based on its crystal structure, was a covalently bound reaction intermediate in analogy to the accepted mechanism for hydrolases [79]. Later, this possibility was dismissed on the basis of structural data of HbHNL-inhibitor complexes [85]. Combining x-ray crystallography and site-directed mutagenesis results in a mechanism of cyanogenesis in which His236 acts as a general base and abstracts a proton from Ser80, thereby allowing proton transfer from the hydroxy group of acetoncyanohydrin to Ser80. The His236 imidazolium cation then facilitates the leaving of the nitrile group by proton transfer (Scheme 28.7) [81,82]. In a recent publication [86], it was demonstrated that the active-site lysine residue (Lys237 in MeHNL and Lys236 in HbHNL) is also necessary for the catalytic activity of these enzymes. The structure of selected active site residues complexed with acetoncyanohydrin is depicted in Figure 28.2 to explain the stereoselectivity.

The active site of MeHNL is located in a cavity burried inside the enzyme and is only accessible through a narrow tunnel that is covered by the large amino acid Trp128. As shown, the ketone carbonyl is hydrogen bonded to Ser80 and Thr11. One methyl group of acetone (labeled C1) is held in position by van der Waals contacts to Leu149, Thr11, and Ile12. The side chains of these three residues define a small hydrophobic site S_1. The second methyl group (labeled C3) points in the opposite direction toward the active-site tunnel, defining the putative second subsite, S_2 in the binding cavity. Asymmetrical carbonyl compounds, for instance aldehydes, are fixed in a way that the smaller substituent (hydrogen) is situated in S_1 and the larger group (R) in S_2. This binding mode suggests that the incoming cyanide exclusively attacks the *Si*-face of the carbonyl compound when bound to the acitve site of the enzyme. Thus implies that HCN has to be deprotonated by His236, which is located on the *Si*-face of the carbonyl compound to be consistent with the (S)-stereospecifity of the addition [81,82].

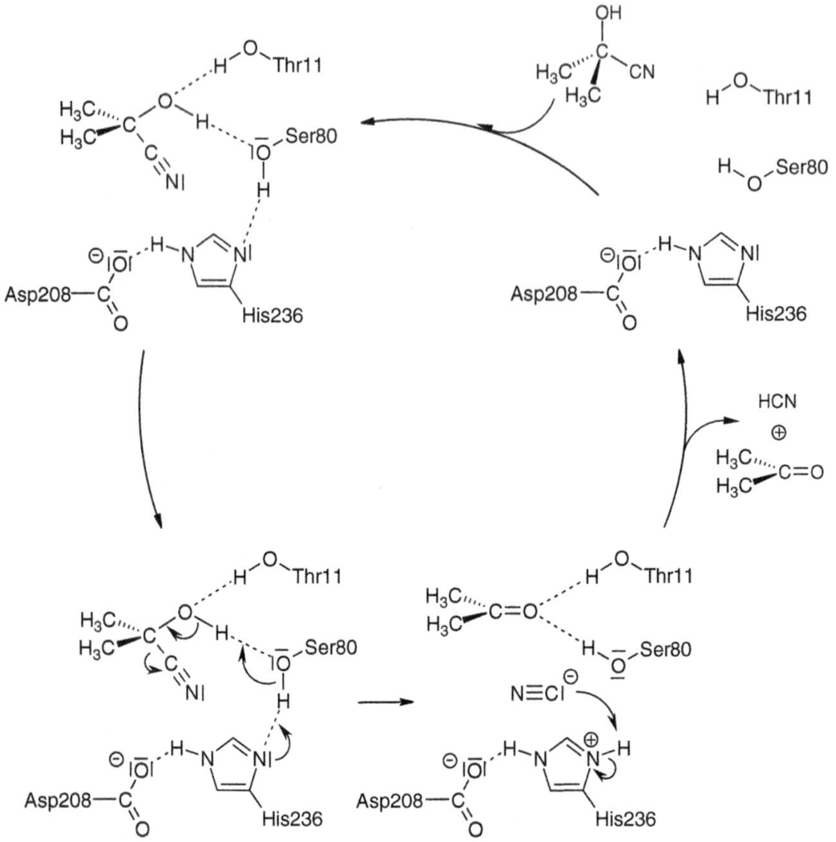

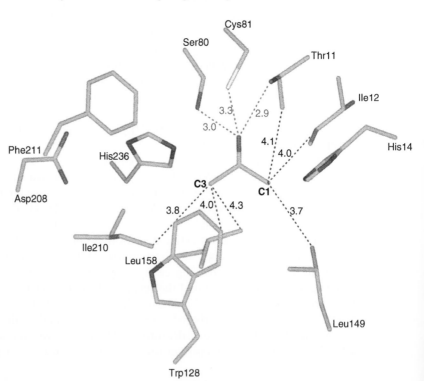

SCHEME 28.7 Catalytic mechanism of cyanogenesis by MeHNL.

FIGURE 28.2 (See color insert following page 526) Structure of selected active-site residues o
xynitrile lyase from *Manihot eseulenta* (MeHNL) complexed with acetone.

28.4.3 CHANGING SUBSTRATE SPECIFICITY AND STEREOSELECTIVITY BY APPLYING MUTANTS OF MeHHNL

Although MeHNL accepts a wide range of carbonyl compounds as substrates [9–11], the catalytic activity for bulky substrates is limited with respect to conversion and ee, as demonstrated for the preparation of (S)-3-phenoxybenzaldehyde cyanohydrin, an important starting compound for pyrethroids [9–11]. As mentioned earlier, the x-ray crystal structure of MeHNL has revealed that the active site of the enzyme is accessible by a narrow channel, where the channel entrance is capped by tryptophan (Trp128). Substitution of Trp128 by amino acids with decreasing size (alanine, cysteine, leucine, and tyrosine) gives the corresponding wt-MeHNL mutants. MeHNL-W128A, for example, could be prepared and overexpressed [87]. It is obvious from the crystal structure of MeHNL-W128A that the entrance to the active site of the enzyme is significantly less hindered, which facilitates access of sterically demanding substrates to the active site. Comparable reactions of 3-phenoxybenzaldehyde with wt-MeHNL and MeHNL-W128A in both aqueous citrate buffer and biphasic system of water/MTBE reveal the superiority of the W128A mutant. The aldehyde was converted quantitatively with high ee using the mutant W128A, nearly independent of the amount of enzyme present, with a reaction rate of 57 g/L/h [87]. Channel mutants differ not only in rates and ee-values, but can even invert the configuration of the cyanohydrins prepared [88]. In the MeHNL-catalyzed addition of HCN to *rac*-2-phenylpropionaldehyde, the (R)-enantiomer reacts with wt-MeHNL and with all MeHNL Trp128 mutants highly (S)-selective to give exclusively the (2S,3R)-diastereoisomers. The (S)-enantiomer of *rac*-3-phenylpropionaldehyde, however, reacts (S)-selectively only with wt-MeHNL to give the (2S,3S)-diastereoisomer, while the MeHNL-W128A mutant is exclusively (R)-selective resulting in the (2R,3S)-diastereoisomer [88]. The inversion of stereoselectivity can be explained and rationalized through crystal structure-based molecular modeling [88].

28.4.4 SUMMARY OF CRYSTAL STRUCTURES OF HNLs AND MECHANISM OF CYANOGENESIS

Since HNLs have originated by convergent evolution from different ancestral proteins, the 3D structure of the various HNLs differ considerably. In all cases, however, a general base catalysis is responsible for the catalytic activity of HNLs. During the last decade, the 3D structures of the most important HNLs (PaHNL, HbHNL, MeHNL, LuHNL, and SbHNL) have been determined. It is possible to formulate mechanisms for cyanogenesis by combining these structural informations with kinetics of certain mutants of the active site of HNLs. During the last 5 y, the mechanisms of cyanogenesis for both MeHNL and HbHNL was unambiguously elucidated.

28.5 IMPORTANT APPLICATIONS OF HNLs IN ORGANIC SYNTHESIS

The fundamentals of HNL-catalyzed syntheses of nonracemic cyanohydrins have been well-established [3–12]. Some applications of HNLs in organic synthesis are summarized in several recent review articles [89–92]. Most of the possible stereoselective follow-up reactions of chiral cyanohydrins have also been conceptually investigated [3–12]. Therefore, the application of chiral cyanohydrins will play a major role in the preparation of biologically active compounds with stereogenic centers, for the synthesis of pharmaceuticals and agrochemicals. For technical applications, the enzymes used must be available in sufficient amounts and reasonably priced. For the preparation of (R)-cyanohydrins, the wild-type PaHNL from almond fulfills both requirements. For the synthesis of (S)-cyanohydrins, the recombinant HNLs from rubber tree

(HbHNL) [93,94] and cassava (MeHNL) [95] have been optimized to high efficiencies. Th
produced in giga units [95].

28.5.1 BIOLOGICALLY ACTIVE CYANOHYDRIN ESTERS

The insecticidal activities of cyanohydrin esters are well known and investigated [96
esters of chiral cyanohydrins with chiral pyrethrum acids are the most important insect
[93]. The majority of the commercially important pyrethroids contain (S)-3-phenoxyb
dehyde cyanohydrin as alcohol component (Scheme 28.8). Nowadays, the technical p
tion of this (S)-cyanohydrin is performed almost exclusively through HNL-catalyzed a
of HCN to 3-phenoxybenzaldehyde using HbHNL or MeHNL, as catalysts. As men
earlier (Chapter 4.3), the channel mutant MeHNL-W128A is especially advantageous f
reaction of sterically demanding substrates like 3-phenoxybenzaldehyde [87].

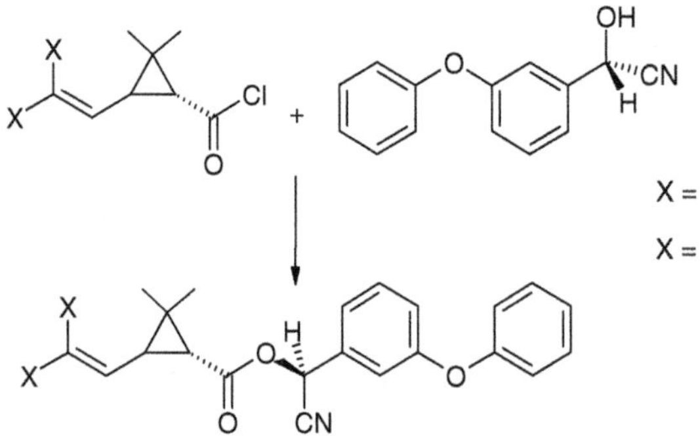

X = Cl: cypermetrin

X = Br: deltamethrin

SCHEME 28.8 Pyrethroids containing cyanohydrin moieties.

28.5.2 CHIRAL 2-HYDROXY ACIDS

In contrast to the abundant 2-amino acids, only a few nonracemic 2-hydroxy acids are
in nature. They have considerable synthetic promise [97]. Hydrolysis of chiral cyanoh
offers a general route to (R)- as well as (S)-2-hydroxy carboxylic acids. (R)- and (S)-C
hydrins are easily hydrolyzed in concentrated hydrochloric acid to give the correspo
(R)- and (S)-2-hydroxy acids, respectively, in excellent yields and with complete retent
configuration (Scheme 28.9) [5]. Under mild conditions (e.g., low temperature and
reaction times), the carboxylic acid amides can be obtained selectively (Scheme 28.9) [

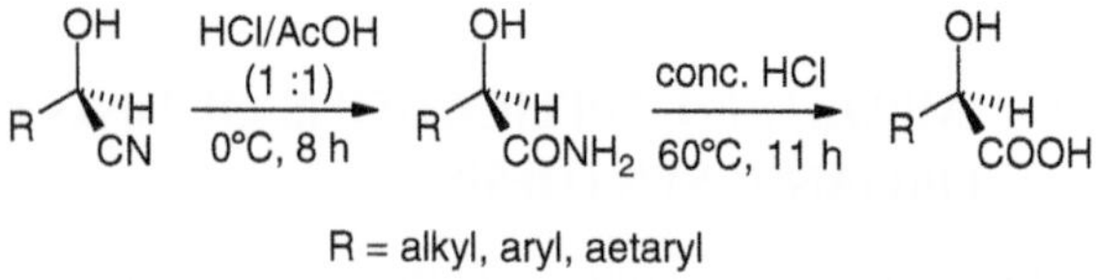

R = alkyl, aryl, aetaryl

SCHEME 28.9 Selective hydrolysis of (R)-cyanohydrins under complete retention of configurati

Aromatic α-hydroxy carboxylic acids are of special interest [35]. Among them, en
merically pure mandelic acid and substituted derivatives thereof are regarded as the
important representatives from a commercial point of view [35]. The great synthetic po
of optically active α-hydroxy acids lies in the possibility to activate the hydroxyl functic
example, by sulfonylation [99–101]. The activated derivatives react readily with any k
nucleophile under complete inversion of configuration [99–101].

The production of the blockbuster Clopidogrel (Plavix) is a convincing example for the synthetic potential of follow-up reactions of optically active cyanohydrins [102]. In the first step, (R)-2-chlorobenzaldehyde cyanohydrin is prepared by almond meal catalyzed by the addition of HCN to 2-chlorobenzaldehyde. The cyanohydrin is then transformed into the corresponding α-hydroxy carboxylic ester, which can be reacted with tetrahydrothieno[3,2-c]pyridine after activation with phenyl sulfonylchloride to give clopidrogel (Scheme 28.10).

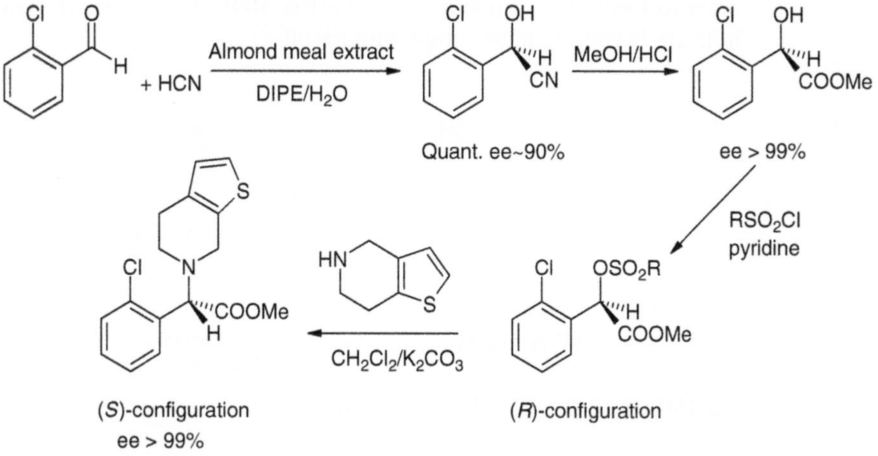

SCHEME 28.10 Stereoselective synthesis of clopidogrel (From Bousquet, A. and Musolino, A., PCT Int. Appl. 1999, Wo 9918110 A1, CAN 130: 296510, 1999.)

Another class of pharmaceuticals that can be synthesized through optically active cyanohydrins are angiotensin-converting enzyme (ACE) inhibitors [103]. (R)-3-Phenylpropionaldehyde cyanohydrin is an important starting cyanohydrin for the preparation of ACE inhibitors. After transformation of the cyanohydrin into the corresponding carboxylic ester and activation of the hydroxyl function by sulfonylation, the reaction with dipeptides, for example, yields under inversion of configuration stereoselectively ACE inhibitors known as enalapril and lisinopril (Scheme 28.11) [103,104].

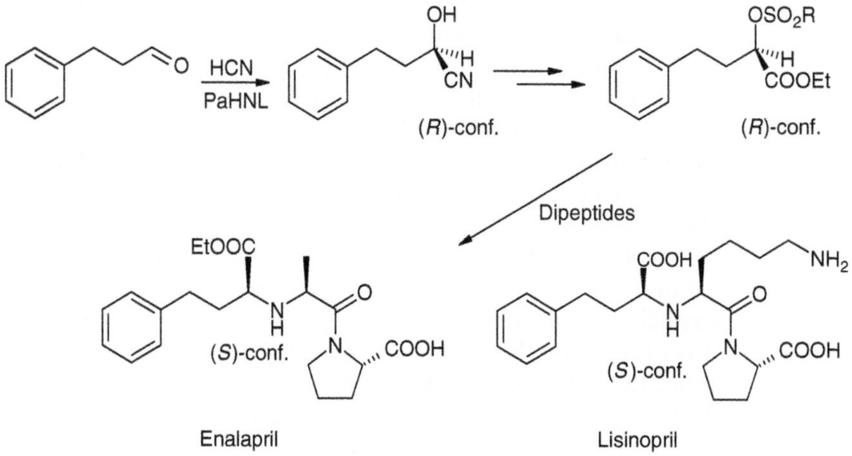

SCHEME 28.11 Stereoselective synthesis of ACE-inhibitors.

28.5.3 NONRACEMIC 1,2-AMINO ALCOHOLS

1,2-Amino alcohols have a broad spectrum of biological activity [105]. They can be cat
ized as adrenalin-like with one chiral center at C-1 or as ephedrine-like with two chiral ce
at C-1 and C-2 (Scheme 28.12). Although it is known that only the compounds with (1R)
(1R,2S)-configuration, respectively, are responsible for the desired biological activity, in
cases racemates are still applied as pharmaceuticals. Since some of the other enantic
seem to have undesirable side effects [106–109], stereoselective syntheses of 1,2-amino
hols are becoming increasingly important.

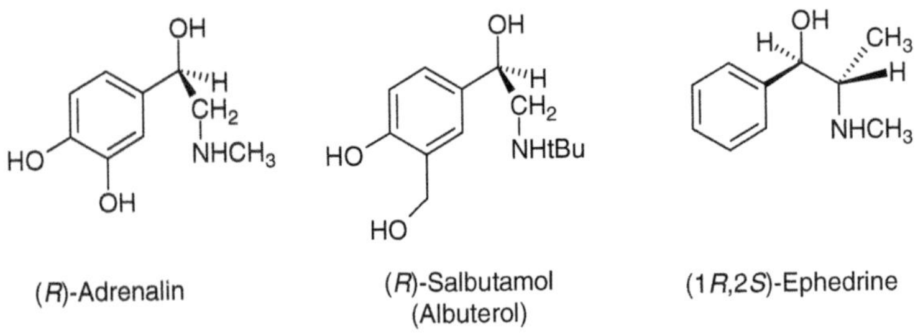

(R)-Adrenalin   (R)-Salbutamol   (1R,2S)-Ephedrine
                       (Albuterol)

SCHEME 28.12 Biologically active 1,2-amino alcohols.

A variety of methods have been developed for the stereoselective preparation c
amino alcohols [106–109]. However, the stereoselective preparation of amino al
starting from chiral cyanohydrins is in many cases easier and more efficient than
methods. Both free and O-protected chiral cyanohydrins can be hydrogenated
LiAlH$_4$ without any racemization to give adrenalin-type 1,2-amino alcohols. The
tert-butyldimethylsilyl(TBDMS)-protecting groups allows partial hydrogenation wit
sobutylaluminium hydride (DIBALH) and transformation of the formed imino inte
ates, for example: exchanging = NH against = NR. Subsequent hydrogenation giv
pharmacologically important N-alkyl-substituted (1R)-2-amino alcohols (Scheme
[110,111]. Ephedrine-like 2-amino alcohols can be prepared stereoselectively from
tected chiral cyanohydrins by the addition of a Grignard reagent and by subs
hydrogenation with NaBH$_4$ (Scheme 28.13) [112,113]. Again, NH$_3$ in the imino
mediate can be exchanged by primary amines to get after hydrogenation the N-
tuted 1,2-amino alcohols. The hydrogenation of the imino intermediates is
diastereoselective due to a chelate-controlled reaction; erythro products are formed
exclusively [113]. (1R,2S)-Ephedrine is commercially produced by a fermentation
that is usually limited to specific substrates. In contrast, almost any structural varia
the 1,2-amino alcohol is possible when starting from chiral cyanohydrins. The
selective preparation of heteroatomic analogs of L-ephedrine is an interesting exan
this synthetic methodology [114]. In the case of furyl-1,2-amino alcohols, it is p
to transform the furyl group into a carboxyl group by ozonization, which opens
possibility of a stereoselective route to 2-hydroxy-3-amino carboxylic acids (5
28.14) [115]. Another synthetic possibility is the addition of allyl Grignard to the
group. Ozonolysis of the acetylated amino alcohols affords the corresponding alde
which can be oxidized to give the optically active β-hydroxy-α-amino acids [49].

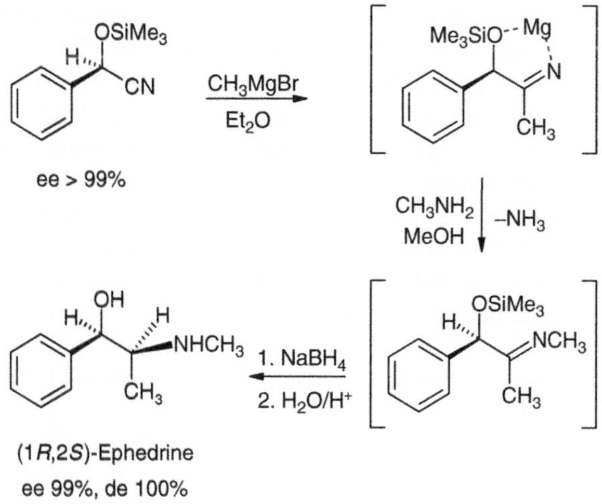

SCHEME 28.13 Stereoselective synthesis of R(2S)-Ephedrine.

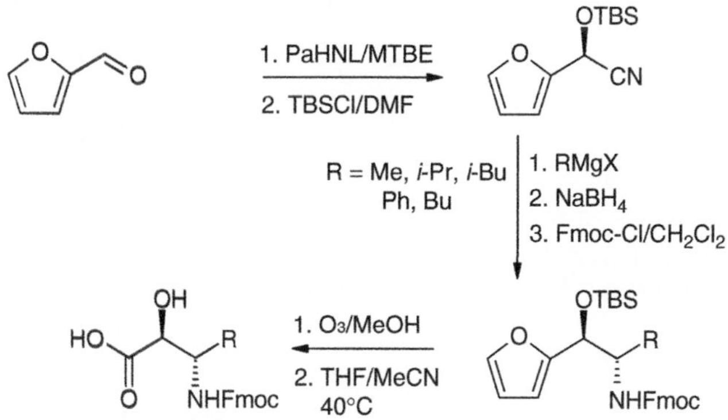

SCHEME 28.14 Synthesis of (3S)-2-hydroxy-3-amino acids (From Tromp, R.A., et al., *Tetrahedron Asymmetry*, 14, 1645, 2003.)

28.5.4 CYANOHYDRINS OF SUBSTITUTED CYCLOHEXANONES

Monosubstituted cyclohexanone cyanohydrins, which can easily be hydrolyzed to the corresponding α-hydroxy carboxylic acids, are important as both pharmaceuticals and plant protective agents [116–118]. This structural unit is also common in several natural products [119–122]. Some examples are given in Scheme 28.15. Although each of the compounds **12**, **13**, and **14** has two stereogenic centers, stereoselective syntheses of these compounds have been published only recently [123].

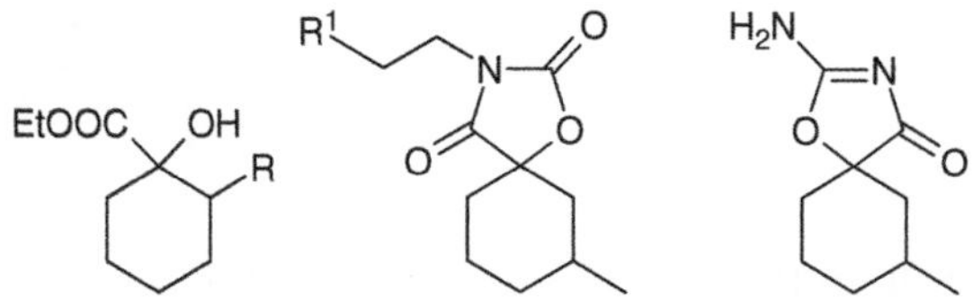

12   **13**   **14**

SCHEME 28.15 Examples of substituted hydroxy cyclohexanone carboxylic acids.

The HNL-catalyzed addition of HCN to 4-substituted cyclohexanones, which c
contain a prochiral center, is very interesting and completely unexpected [124,125].
hexanones are excellent subtrates for both PaHNL and MeHNL, respectively,
surprisingly exhibit high *cis/trans*-selectivity. The (*R*)-PaHNL-catalyzed reaction a
almost exclusively the *trans*-isomers, while with (*S*)-MeHNL the *cis*-addition is fa
(Scheme 28.16) [124,125]. Since the chemical addition of HCN to 4-substituted
hexanones always results in mixtures of *cis/trans*-isomers, a stereoselective additio
great importance. For example, in the chemical addition of HCN to 4-alkyl subst
cyclohexanones, the *cis/trans*-ratio is ~13:87. Important natural products that can be
obtained from corresponding cyclohexanone cyanohydrins; however, have exclusive
configuration [119–122]. Therefore, the chemical addition of HCN would be high
favorable because of the large *trans*-ratio. However, the *cis/trans*-ratio in many case
high as 98:2 with MeHNL as catalyst [124,125]. The synthesis of the natural monote
cis-p-menth-8-ene-1,7-diol and *cis-p*-menthane-1,7,8-triol by this new methodol
summarized in Scheme 28.17 [120].

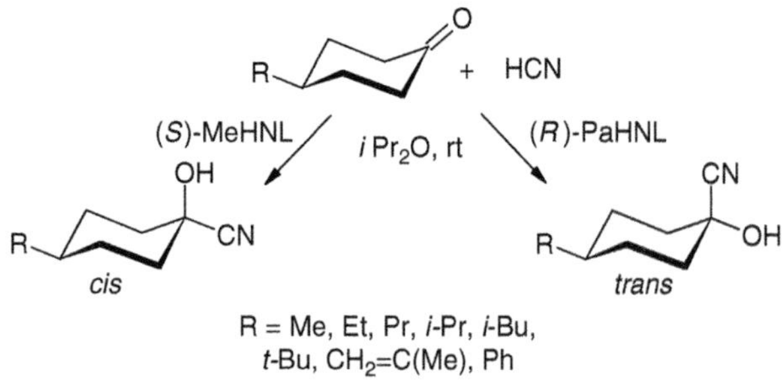

SCHEME 28.16 *Cis/trans*-selectivity of HNL-catalyzed additions of HCN to 4-substituted cycl
nones (From Effenberger, F., Kobler, C., and Roos, J., *Angew. Chem. Int. Ed.*, 41, 1876
Effenberger, F., et al., *Can. J. Chem.*, 80, 671, 2002.)

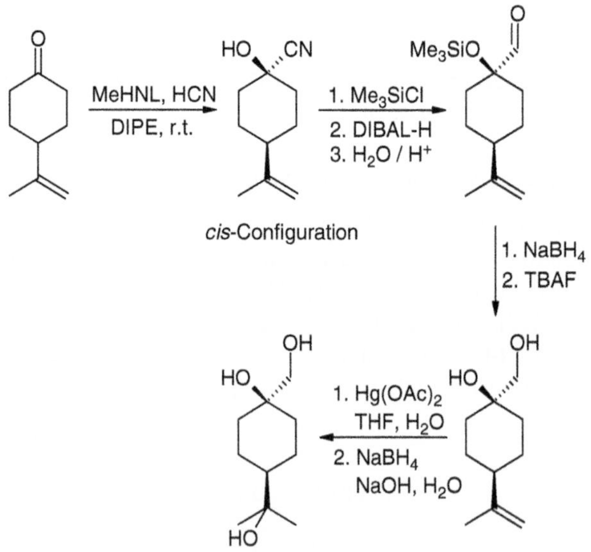

SCHEME 28.17 Synthesis of biologically active monoterpenes (From Kobler, C. and Effenber
Chem. Eur. J., 11, 2783, 2005.)

28.6 CONCLUSION

The synthetic potential of chiral cyanohydrins for the stereoselective preparation of biologically active compounds (pharmaceuticals and plant protective agents) has been known for more than 10 y [3–12]. The main objective was therefore to summarize the progress achieved during the last 5 y, with special emphasis on improvements in chemical and optical yields of the prepared chiral cyanohydrins, on simplification of the reaction procedures, and on developed conditions for large-scale productions of commercially interesting chiral cyanohydrins. The HNL-catalyzed mechanism of cyanogenesis has been elucidated unambiguously from x-ray structures of the most important HNLs and site-directed mutagenesis. It was possible to improve yields and stereoselectivity in a straightforward manner through knowledge of the active site and the reaction mechanism of the enzymes. Now not only (*R*)-PaHNL from almonds is available in more or less unlimited amounts, but the recombinant (*S*)-HNLs from cassava (MeHNL) and rubber tree (HbHNL) are also produced in giga units, the application of HNLs in large-scale productions of chiral cyanohydrins as intermediates for the stereoselective synthesis of pharmaceuticals and agrochemicals has become possible. Some of the most important technical applications of chiral cyanohydrins have been summarized in Section 28.5.

REFERENCES

1. North, M., Synthesis and applications of non-racemic cyanohydrins and α-amino-nitriles, *Tetrahedron*, 60, 10371, 2004.
2. Effenberger, F., Ziegler, T., and Förster, S., Enzyme-catalyzed cyanohydrin synthesis in organic solvents, *Angew. Chem. Int. Ed. Engl.*, 26, 458, 1987.
3. North, M., Catalytic asymmetric cyanohydrin synthesis, *Synlett*, 807, 1993.
4. North, M., Introduction of the cyano group by addition to a carbonyl group, *Sci. Synth.*, 19, 235, 2004.
5. Effenberger, F., Synthesis and reactions of optically active cyanohydrins, *Angew. Chem. Int. Ed. Engl.*, 33, 1555, 1994.
6. Kanerva, L.T., Biocatalytic ways to optically active 2-amino-1-phenylethanols, *Acta Chem. Scand.*, 50, 234, 1996.
7. Schmidt, M. and Griengl, H., Oxynitrilases: from cyanogenesis to asymmetric synthesis, *Top. Curr. Chem.*, 200, 193, 1999.
8. Gregory, R.J.H., Cyanohydrins in nature and the laboratory: biology, preparations, and synthetic applications, *Chem. Rev.*, 99, 3649, 1999.
9. Effenberger, F., Enzyme-catalyzed preparation and synthetic applications of optically active cyanohydrins, *Chimia*, 53, 3, 1999.
10. Effenberger, F., Hydroxynitrile lyases in stereoselective synthesis, in *Stereoselective Biocatalysis*, Patel, R.N., Ed., Marcel Dekker, New York, 2000, p. 321.
11. Effenberger, F., Förster, S., and Wajant, H., Hydroxynitrile lyases in stereoselective catalysis, *Curr. Opin. Biotechnol.*, 11, 532, 2000.
12. Brussee, J. and van der Gen, A., Biocatalysis in the enantioselective formation of chiral cyanohydrins, valuable building blocks in organic synthesis, in *Stereoselective Biocatalysis*, Patel, R.N., Ed., Marcel Dekker, New York, 2000, pp. 289–320.
13. North, M., Synthesis and application of non-racemic cyanohydrins, *Tetrahedron Asymmetry*, 14, 147, 2003.
14. Brunel, J.-M. and Holmes, I.P., Chemically catalyzed asymmetric cyanohydrin syntheses, *Angew. Chem. Int. Ed. Engl.*, 43, 2752, 2004.
15. Corey, E.J. and Wang, Z., Enantioselective conversion of aldehydes to cyanohydrins by a catalytic system with separate chiral binding sites for aldehydes and cyanide components, *Tetrahedron Lett.*, 34, 4001, 1993.

16. Ryu, D.H. and Corey, E.J., Enantioselective cyanosilylation of ketones catalyzed by oxazaborolidiniumion, *J. Am. Chem. Soc.*, 127, 5384, 2005.
17. Brunel, J.-M., Legrand, O., and Buono, G., Enantioselective trimethylsilyl cyanation of a aldehydes catalyzed by titanium alkoxide-chiral-*o*-hydroxyarylphosphine oxides, *Tetr Asymmetry*, 10, 1979, 1999.
18. Hamashima, Y., et. al., Highly enantioselective cyanosilylation of aldehydes catalyzed by acid–Lewis-base bifunctional catalyst, *Tetrahedron*, 57, 805, 2001.
19. Hamashima, Y., Kanai, M., and Shibasaki, M., Catalytic enantioselective cyanosilyl ketones, *J. Am. Chem. Soc.*, 122, 7412, 2000.
20. Hamashima, Y., Kanai, M., and Shibasaki, M., Catalytic enantioselective cyanosilyl ketones: improvement of enantioselectivity and catalyst turn-over by ligand tuning, *Tetr Lett.*, 42, 691, 2001.
21. Yabu, K., et al., Switching enantiofacial selectivities using one chiral source: catalytic en lective synthesis of the key intermediate for (20*S*)-camptothecin family by (*S*)-selective cya tion of ketones, *J. Am. Chem. Soc.*, 123, 9908, 2001.
22. Masumoto, S., et al., A practical synthesis of (*S*)-oxybutynin, *Tetrahedron Lett.*, 43, 8647
23. Yabu, K., et al., Studies toward practical synthesis of (20*S*)-camptothecin family through enantioselective cyanosilylation of ketones: improved catalyst efficiency by ligand-tuning *hedron Lett.*, 43, 2923, 2002.
24. Belokon, Y., et al., Asymmetric addition of trimethylsilyl cyanide to aldehydes catalyzed (salen)TiIV complexes, *J. Chem. Soc. Perkin Trans. 1*, 1293, 1997.
25. Tararov, V.I., et al., First structurally defined catalyst for the asymmetric addition of trime cyanide to benzaldehyde, *Chem. Commun.*, 387, 1998.
26. Belokon, Y.N., et al., The asymmetric addition of trimethylsilyl cyanide to aldehydes cata chiral (salen)titanium complexes, *J. Am. Chem. Soc.*, 121, 3968, 1999.
27. Belokon, Y.N., et al., Catalytic asymmetric synthesis of *O*-acetyl cyanohydrins from KCl and aldehydes, *Chem. Commun.*, 244, 2002.
28. Belokon, Y.N., et al., Synthetic and mechanistic studies on asymmetric cyanohydrin s using a titanium(salen)bimetallic catalyst, *Tetrahedron*, 60, 10433, 2004.
29. Chen, F.-X., et al., Enantioselective cyanosilylation of ketones by a double-activation metl an aluminium complex and an N-oxide, *Chem Eur. J.*, 10, 4790, 2004.
30. Holmes, I.P. and Kagan, H.B., The asymmetric addition of trimethylsilycyanide to al catalyzed by anionic chiral nucleophiles, Part 2, *Tetrahedron Lett.*, 41, 7457, 2000.
31. Hatano, M., et al., Chiral lithium binaphtolate aqua complex as a highly effective asy catalyst for cyanohydrin synthesis, *J. Am. Chem. Soc.*, 127, 10776, 2005.
32. Li, Y., et al., Highly enantioselective cyanosilylation of aldehydes catalyzed by novel alcohol–titanium complexes, *J. Org. Chem.*, 69, 7910, 2004.
33. Huang, W., et al., Polymeric salen–Ti(IV) complex catalyzed asymmetric synthesis of *O*-a nohydrins from KCN, Ac$_2$O and aldehydes, *Tetrahedron*, 60, 10469, 2004.
34. Breuer, M., et al., Industrial methods for the production of optically active intermediates *Chem. Int. Ed. Engl.*, 43, 788, 2004.
35. Gröger, H., Enzymatic routes to enantiomerically pure aromatic α-hydroxy carboxylic further example for the diversity of biocatalysis, *Adv. Synth. Catal.*, 343, 547, 2001.
36. Gröger, H., et al., Asymmetric synthesis of an (*R*)-cyanohydrin using enzymes entrapped shaped gels, *Org. Lett.*, 3, 1969, 2001.
37. Gröger, H., et al., Practical aspects of encapsulation technologies, in *Landbauforschung rode*, Prüae, U. and Vorlop, K.-D., Eds., Bundesforschungsanstalt für Landwirtschaft, schweig, SI 241, 87, 2002.
38. van Langen, L.M., et al., Cross-linked aggregates of (*R*)-oxynitrilase: a stable, recyclable lyst for enantioselective hydrocyanation, *Org. Lett.*, 7, 327, 2005.
39. Gaisberger, R.P., Fechter, M.H., and Griengl, H., The first hydroxynitrile lyase catalyze hydrin formation in ionic liquids, *Tetrahedron Asymmetry*, 15, 2959, 2004.
40. Rosenthaler, L., Enzyme-effected asymmetric synthesis, *Biochem. Z.*, 14, 238, 1908.

41. Kiljunen, E. and Kanerva, L., Novel (*R*)-oxynitrilase sources for the synthesis of (*R*)-cyanohydrins in diisopropylether, *Tetrahedron Asymmetry*, 8, 1225, 1997.
42. Luna, H., et al., New potential sources of oxynitrilase enzymes, *Rev. Soc. Chim. Mex.*, 41, 111, 1997.
43. Solis, A., et al., New sources of (*R*)-oxynitrilase: capulin (*Prunus capuli*) and Mamey (*Mammea americana*), *Biotechnol. Lett.*, 20, 1183, 1998.
44. Glieder, A., et al., Comprehensive step-by-step engineering of an (*R*)-hydroxynitrile lyase for large-scale asymmetric synthesis, *Angew. Chem. Int. Ed. Engl.*, 42, 4815, 2003.
45. Weis, R., et al., Biocatalytic conversion of unnatural substrates by recombinant almond R-HNL isoenzyme 5, *J. Mol. Cat., B Enzym.*, 29, 211, 2004.
46. Bühler, H., Optimierung der (S)-Hydroxynitril lyase aus *Manihot esculenta* durch gezielte Muta-tionen–Anwendungen von optisch aktiven Cyanhydrinen in der Synthese, Dissertation, Universi-tät Stuttgart, Stuttgart, Germany, 2000.
47. Kirschbaum, B., Wilbert, G., and Effenberger, F., Preparation of optically active cyanohydrins using (*R*)-oxynitrile lyase, U.S. Pat. Appl. Publ. (2002) US 202052523 A1, CAN 136:355075.
48. Effenberger, F. and Gaupp, S., Enzyme-catalyzed reactions. 35. Stereoselective substitution of (*R*)-2-(sulfonyloxy)nitriles with sulfur nucleophiles, *Tetrahedron Asymmetry*, 10, 1765, 1999.
49. Roos, J. and Effenberger, F., Stereoselective synthesis of β-amino-γ-butyrolactones, *Tetrahedron Asymmetry*, 13, 1855, 2002.
50. Han, S., Lin, G., and Li, Z., Synthesis of (*R*)-cyanohydrins by crude (*R*)-oxynitrilase-catalyzed reactions in micro-aqeuous medium, *Tetrahedron Asymmetry* 9, 1835, 1998.
51. Chen, P.-R., et al., A practical high throughput continuous process for the synthesis of chiral cyanohydrins, *J. Org. Chem.*, 67, 8251, 2002.
52. Straathof, A.J., Enzymatic catalysis via liquid–liquid interfaces, *J. Biotechnol. Bioeng.*, 83, 371, 2003.
53. Effenberger, F., et al., Enzyme-catalyzed reactions. 5. Enzyme-catalyzed synthesis of (*S*)-cyanohydrins and subsequent hydrolysis to (*S*)-α-hydroxy-carboxylic acids, *Tetrahedron Lett.*, 31, 1249, 1990.
54. Niedermeyer, U. and Kula, M.-R., Enzyme-catalyzed synthesis of (*S*)-cyanohydrins, *Angew. Chem. Int. Ed. Engl.*, 29, 386, 1990.
55. Wajant, H., et al., Acetone cyanohydrin lyase from *Manihot esculenta* (cassava) is serologically distinct from other hydroxynitrile lyases, *Plant Sci.*, 108, 1, 1995.
56. Foerster, S., et al., The first recombinant hydroxynitrile lyase and its application in the synthesis of (*S*)-cyanohydrins, *Angew. Chem. Int. Ed. Engl.*, 35, 437, 1996.
57. Schmidt, M., et al., Preparation of optically active cyanohydrins using the (*S*)-hydroxynitrile lyase from *Hevea brasiliensis*, *Tetrahedron*, 52, 7833, 1996.
58. Hasslacher, M., et al., Molecular cloning of the full-length cDNA of (*S*)-hydroxynitrile lyase from *Hevea brasiliensis*, *J. Biol. Chem.*, 271, 5884, 1996.
59. Effenberger, F., Wajant, H., and Förster, S., Method for the production of hydroxynitrile lyases, PCT Int. Appl. (2001), WO 200148178 A1, CAN 135:45300.
60. Roos, J., Stelzer, U., and Effenberger, F., Enzyme-catalyzed reactions. 34. Synthesis of (1*R,cis*, α*S*)-cypermethrine via lipase catalyzed kinetic resolution of racemic *m*-phenoxybenzaldehyde cyanohydrin acetate, *Tetrahedron Asymmetry*, 9, 1043, 1998.
61. Veum, L., et al., Enantioselective synthesis of protected cyanohydrins, *Eur. J. Org. Chem.*, 1516, 2002.
62. de Gonzalo, G., et al., Enzymatic acylation reactions on ω-hydroxycyanohydrins, *Tetrahedron*, 60, 10525, 2004.
63. Paizs, C., et al., Candida antartica lipase A in the dynamic resolution of novel furylbenzothiazole-based cyanohydrin, *Tetrahedron Asymmetry*, 14, 619, 2003.
64. Paizs, C., Biocatalytic enantioselective preparation of phenothiazine-based cyanohydrin acetates: kinetic and dynamic kinetic resolution, *Tetrahedron*, 60, 10533, 2004.
65. Li, Y.-X., Stratthof, A.J., and Hanefeld, U., Enantioselective formation of mandelonitrile acetate: investigation of a dynamic kinetic resolution, *Tetrahedron Asymmetry*, 13, 739, 2002.

66. Veum, L. and Hanefeld, U., Enantioselective formation of mandelonitrile acetate: investiga a dynamic kinetic resolution II, *Tetrahedron Asymmetry*, 15, 3707, 2004.

67. Wajant, H. and Effenberger, F., Hydroxynitrile lyases of higher plants, *Biol. Chem. Hoppe-* 377, 611, 1996.

68. Gruber, K. and Kratky, C.J., Biopolymers for biocatalysis: structure and catalytic mechar hydroxynitrile lyases, *J. Polym. Sci. [A]*, 42, 479, 2004.

69. Lauble, H., et al., Crystallization and preliminary x-ray diffraction studies of mandelonitri of almonds, *Proteins Struct. Funct. Genet.*, 19, 343, 1994.

70. Dreveny, I., et al., The hydroxynitrile lyase from almond: a lyase that looks like an oxidored *Structure*, 9, 803, 2001.

71. Cavener, D.R., GMC oxidoreductases. A newly defined family of homologous proteir diverse catalytic activities, *J. Mol. Biol.*, 223, 811, 1992.

72. Kiess, M., Hecht, H.J., and Kalisz, H.M., Glucose oxidase from *Penicillium amagasa* Primary structure and comparison with other glucose–methanol–choline (GMC) oxidored *Eur. J. Biochem.*, 252, 90, 1998.

73. Baerwald, K.R. and Jaenicke, L., D-Hydroxynitrile lyase: involvement of the prosthetic adenine dinucleotide in enzyme activity, *FEBS Lett.*, 90, 255, 1978.

74. Lauble, H., et al., Crystal structure of hydroxynitrile lyase from *Sorghum bicolor* in compl the inhibitor benzoic acid: a novel cyanogenic enzyme, *Biochemistry*, 41, 12043, 2002.

75. Wajant, H., Mundry, K.-W., and Pfizenmaier, K., Molecular cloning of hydroxynitrile lyas *Sorghum bicolor* (L.). Homologies to serine carboxypeptidases, *Plant Mol. Biol.*, 26, 735, 1

76. Trummler, K. and Wajant, H., Molecular cloning of acetone cyanohydrin lyase from flax (usitatissimum). Definition of a novel class of hydroxynitrile lyases, *J. Biol. Chem.*, 272, 477C

77. Ollis, D.L., et al., The alpha/beta hydrolase fold, *Protein Eng.*, 5, 197, 1992.

78. Cygler, M., et al., Relationship between sequence conservation and three-dimensional struc a large family of esterases, lipases and related proteins, *Protein Sci.*, 2, 366, 1993.

79. Wagner, U.G., et al., Mechanism of cyanogenesis: the crystal structure of hydroxynitrile lyas *Hevea brasiliensis*, *Structure*, 4, 811, 1996.

80. Gruber, K., et al., Atomic resolution crystal structure of hydroxynitrile lyase from *Hevec liensis*, *Biol. Chem.*, 380, 993, 1999.

81. Lauble, H., et al., Structure of hydroxynitrile lyase from *Manihot esculenta* in comple acetone and chloroacetone: implications for the mechanism of cyanogenesis, *Acta Chrys D Biol. Crystallogr.*, D57, 194, 2001.

82. Lauble, H., et al., Mechanistic aspects of cyanogenesis from active-site mutant Ser80 hydroxynitrile lyase from *Manihot esculenta*, *Protein Sci.*, 10, 1015, 2001.

83. Dreveny, I, Kratky, C., and Gruber, K., The active site of hydroxynitrile lyase from amygdalus: modelling studies provide insights into the mechanism of cyanogenesis, *Prote* 11, 292, 2002.

84. Wajant, H. and Pfizenmaier, K., Identification of potential active-site residues in the hydr trile lyase from *Manihot esculenta* by site-directed mutagenesis, *J. Biol. Chem.*, 271, 25830,

85. Zuegg, J., et. al., Three-dimensional structures of enzyme-substrate complexes of the hy nitrile lyase from *Hevea brasiliensis*, *Protein Sci.*, 8, 1990, 1999.

86. Gruber, K., et. al., Reaction mechanism of hydroxynitrile lyases of the α/β-hydrolase super the three-dimensional structure of the transient enzyme-substrate complex certifies the cruc of LYS236, *J. Biol. Chem.*, 279, 20501, 2004.

87. Buehler, H., et. al., Enzyme-catalyzed reactions, Part 47. Substrate specifity of mutants hydroxynitrile lyase from *Manihot esculenta*, *Chem. Bio.*, 4, 211, 2003.

88. Buehler, H., Miehlich, B., and Effenberger, F., Inversion of selectivity by applying mutants hydroxynitrile lyase from *Manihot esculenta*, *Chem. Chem.*, 6, 711, 2005.

89. Fessner, W.-D. and Helaine, V., Biocatalytic synthesis of hydroxylated natural product aldolases and related enzymes, *Curr. Opin. Biotechnol.*, 12, 574, 2001.

90. Gotor, V., Bioctalysis applied to the preparation of pharmaceuticals, *Org. Process Res. I* 420, 2002.

91. Schoemaker, H.E., Mink, D., and Wubbolts, M.G., Dispelling the myths-biocatalysis in industrial synthesis, *Science*, 299, 1694, 2003.
92. Sukumaran, J. and Hanefeld, U., Enantioselective C–C bond synthesis catalysed by enzymes, *Chem. Soc. Rev.*, 34, 530, 2005.
93. Griengl, H., Schwab, H., and Fechter, M., The synthesis of chiral cyanohydrins by oxynitrilases, *Trends Biotechnol.*, 18, 252, 2000.
94. Poechlauer, P., Skranc, W., and Wubbolts, M., The large scale-biocatalytic synthesis of enantiopure cyanohydrins, in *Asymmetric Catalysis on Industrial Scale*, Blaser, H.U. and Schmidt, E., Eds., Wiley-VCH Verlag, Weinheim, 2004, p. 151.
95. Rosen, T.C. and Dauamann, T., (*S*)-Oxynitrilase: biocatalyst for asymmetric HCN addition, *Chim. Oggi/Chemistry today, Suppl. Chiral Catalysis*, 22(11–12), 21, 2004.
96. Peterson, C.J., et al., Insecticidal activity of cyanohydrin and monoterpenoid compounds, *Molecules*, 5, 648, 2000.
97. Coppola, G.M. and Schuster, H.F., *Chiral α-Hydroxy Acids in Enantioselective Synthesis*, Wiley-VCH Verlag, Weinheim, 1997.
98. Rochowski, A., Untersuchungen zur Säurekatalysierten Hydrolyse von Nitrilen zu Amiden, Dissertation, Universität Stuttgart, Stuttgart, Germany, 1997.
99. Burkard, U. and Effenberger, F., Enantioselective substitution of 2-(sulfonyloxy)carboxylates with oxygen and sulfur nucleophiles, *Chem. Ber.*, 119, 1594, 1986.
100. Effenberger, F., Burkard, U., and Willfahrt, J., Amino acids. 4. Enantioselective synthesis of N-substituted α-amino carboxylic acids from α-hydroxy carboxylic acids, *Justus Liebigs Ann. Chem.*, 314, 1986.
101. Fleming, P.R. and Sharpless, K.B., Selective transformation of threo-2,3-dihydroxy esters, *J. Org. Chem.*, 56, 2869, 1991.
102. Bousquet, A. and Musolino, A., Hydroxyacetic ester derivatives, namely (*R*)-methyl-2-(sulfonyloxy)-2-(chlorophenyl)acetates, preparation method, and use as synthesis intermediates for clopidogrel, PCT Int. Appl. 1999, WO 9918110 A1, CAN 130:296510, 1999.
103. Sheldon, R.A., et al., The synthesis of angiotensin-converting enzyme (ACE) inhibitors, *Chim. Oggi*, 9, 35, 1991.
104. Weis, R., et al., Erhöhung der Enantioselektivität von Mandel-*R*-HNL durch rationales design des aktiven Zentrums, *Angew. Chem. Int. Ed. Engl.*, 117, 4778, 2005.
105. Kleemann, A. and Engel, J., *Pharmaceutical Substances*, 3rd ed., Thieme, Stuttgart, Germany, 1999.
106. Corey, E.J. and Link, J.O., The first enantioselective syntheses of pure *R*- and *S*-isopreterenol, *Tetrahedron Lett.*, 31, 601, 1990.
107. Enders, D. and Reinhold, U., Diastereo- and enantioselective synthesis of 1,2-amino alcohols and protected α-hydroxy aldehydes from glycol aldehyde hydrazones, *Justus Liebigs Ann. Chem.*, 11, 1996.
108. Li, G., Angert, H.H., and Sharpless, K.B., N-Halocarbamate salts lead to more efficient catalytic asymmetric aminohydroxylation, *Angew. Chem. Int. Ed. Engl.*, 35, 2813, 1996.
109. Bakale, R.P., The development of routes to (*R*)-albuterol hydrochloride, *Spec. Chem.*, 15, 249, 1995.
110. Zandbergen, P., et al., A one-pot reduction–transimination–reduction synthesis of substituted β-ethanolamines from cyanohydrins, *Tetrahedron*, 48, 3977, 1992.
111. Effenberger, F. and Jäger, J., Synthesis of the adrenergic bronchodilators (*R*)-terbutalin and (*S*)-salbutamol from (*R*)-cyanohydrins, *J. Org. Chem.*, 62, 3867, 1997.
112. Brussee, J., et al., Synthesis of optically active ethanolamines, *Tetrahedron*, 46, 1653, 1990.
113. Effenberger, F., Gutterer, B., and Jäger, J., Stereoselective synthesis of (1*R*)- and (1*R*,2*S*)-1-aryl-2-alkylamino alcohols from (*R*)-cyanohydrins, *Tetrahedron Asymmetry*, 8, 459, 1997.
114. Effenberger, F. and Eichhorn, J., Stereoselective synthesis of thienyl and furyl analogues of ephedrine, *Tetrahedron Asymmetry*, 8, 469, 1997.
115. Tromp, R.A., et al., Synthesis of Fmoc-protected (2*S*,3*S*)-2-hydroxy-3-amino acids from furyl substituted chiral cyanohydrin, *Tetrahedron Asymmetry*, 14, 1645, 2003.

116. Skinner, W.A., et al., Topical mosquito repellents. XIII: cyclic analogs of lactic acid, *J.* *Sci.*, 69, 196, 1980.
117. Shapiro, S.L., et al., Pyridylethylated oxazolidinediones. II, *J. Am. Chem. Soc.*, 81, 386, 1
118. Fischer, R., et al., Preparation of 3-aryl-4-hydroxy-Δ3-dihydrofuranones as pesticides, Ger 1995, DE 4337853 A1(1995), CAN, 123, 32947, 1995.
119. Ishikawa, T., Kitajima, J., and Tanaka, Y., Water-soluble constituents of fennel. IV. Me type monoterpenoids and their glycosides, *Chem. Pharm. Bull.*, 46, 1603, 1998.
120. Kobler, C. and Effenberger, F., Enzyme catalyzed reactions. 51. Stereoselective synthesi *p*-menth-8-ene-1,7-diol, *cis-p*-menthane-1,7-diol and *cis-p*-menthane-1,7,8-triol, *Chem. Eur* 2783, 2005.
121. Endo, K. and Hikino, H., Structures of regyol, rengyoxide, and rengyolone, new cyclohexy derivatives from Forsythia suspensa fruits, *Can. J. Chem.*, 62, 2011, 1984.
122. Ozaki, Y., Rui, J., and Tang, Y.T., Antiinflammatory effect of Forsythia suspensa V(AHL active principle, *Biol. Pharm. Bull.*, 23, 365, 2000.
123. Kobler, C., Bohrer, A., and Effenberger, F., Hydroxynitrile lyase-catalyzed addition of I 2-and 3-substituted cyclohexanones, *Tetrahedron*, 60, 10397, 2004.
124. Effenberger, F., Kobler, C., and Roos, J., Enzyme-catalyzed reactions, Part 44. *cis–trans* se of enzyme-catalyzed additions to 4-substituted cyclohexanones-correlation with the Ringold model of enzymatic hydrogenation, *Angew. Chem. Int. Ed.*, 41, 1876, 2002.
125. Effenberger, F., et al., Hydroxynitrle lyase-catalyzed addition of HCN to 4-sub cyclohexanones; stereoselective preparation of tetronic acids, *Can. J. Chem.*, 80, 671, 200:

29 Chiral Switches: Problems, Strategies, Opportunities, and Experiences

René Csuk

CONTENTS

29.1 INTRODUCTION AND HISTORIC DEVELOPMENTS

Chiral molecules are ubiquitous in nature and have ever-increasing importance in the pharmaceutical industry. This process started with Cushny's observation that one enantiomer of hyoscamine possessed greater pharmacological activity than the other. Thus, chirality of drugs and pharmaceuticals has been in the focus of interest for several decades. The discovery or design and the development of new drugs have been influenced to a great extent by a new understanding of molecular recognition in many pharmacologically relevant events [1–5].

There are several reasons why chiral pharmaceuticals are important. Frequently, the biological activity of one enantiomer is much greater than that of the other, or the other stereomer has no activity at all. Although in some instances the activity of the pure isomer is lower than that of the mixture due to synergistic effects of the isomers, in the worst case, the second isomer can have distinct, even undesirable biological activity [6]. Potential advantages of single-enantiomer drugs [7] reside in a less complex pharmacokinetic profile, a less complex but more selective pharmacodynamic profile, an improved therapeutic index, a reduced potential for complex drug interactions, and finally in a less complex relationship between effects and plasma concentration of the drug.

As these scenarios emerged, it became clear that the failure to address issues raised by the stereochemistry of racemic drugs was the cause of some drug–drug interactions, severe adverse reactions, and withdrawals from the market [4]. From another viewpoint, putting at least nonessential material to patients or into the environment is undesirable and uneconomic. Thus, producing companies have the opportunity to increase productio. factor of 2 by switching from the generation of a racemic mixture to the production pure enantiomer.

Thus, single-enantiomer drugs continue to be a significant force in the global ph ceutical market. Although the difference in pharmacological activity of enantiomei known in the early 1900s, it was not until the late 1980s that chirality became a c issue in the manufacture of agrochemicals and drugs. Interestingly and contrary to co thinking, an enantiomerically pure thalidomide would not have prevented the well-kno tragedy. Although thalidomide is the best known example of pharmaceuticals for whi stereoisomer has the desired effect and the other isomer displays harmful properties, examples include tuberculostatic (S,S)-ethambutol where the (R,R)-enantiomer causes ness, or antiarthritic (S)-penicillamine, where the (R)-enantiomer is extremely toxi toxicity of the (R) = L-enantiomer to animals including weight loss and death ha known since the 1950s. Whereas in the initial clinical evaluation of the drug in the l States the use of the racemate resulted in severe side effects and was withdrawn, in the l Kingdom (S) = D-penicillamine was obtained by a hydrolysis of penicillin as a single tiomer, and thus the adverse side effects were not observed [8,9].

Similarly, the use of racemic DOPA [10] for the treatment of Parkinson's disease re in adverse side effects, whereas the use of L-DOPA resulted in halving the required c reduction of adverse effects, a higher compliance and thus an increased number of imp patients. It was also in the 1980s that the racemate of perhexiline, a drug used tc abnormal heart rhythms, killed a number of people who had accumulated gram qua [4] of the wrong stereoisomer impurity which was more slowly metabolized.

Thus, "chiral switches" that exploit one single isomer of an approved racemic mixtu an important feature of drug development [11–14].

In addition to pharmacological benefits, chiral switches–based pharmaceuticals hav limited application in the generic marketplace; they are often intended to allow "line sions" of blockbuster drugs. Thus, economical bridging strategies have been devised tc speedy regulatory approval of single enantiomers.

In this concept, chiral switches are chiral drugs that have already been claimed, app and marketed as mixtures of enantiomers (or as mixtures of diastereomers), but hav redeveloped as single enantiomers.

The basic definition of a chiral switch can be extended to include chiral drugs tha been sold already as a mixture of diastereomers, but have since been developed as enantiomers or single-enantiomers E, which have been redeveloped and launched paired enantiomer E* [4]. A summary of this approach is depicted in Figure 29.1.

Following standard U.S. laws, a single enantiomer of a previously approved ra mixture is not considered to be a new chemical entity (NCE) [4,15]. As none of the ap chiral switches was classified as a new molecular entity (defined as an active ingredie has never been manufactured in the United States), chiral switches are elegible [4,16] fc 3 y exclusivity and are barred from the 5 y of exclusivity that is granted to new drugs.

Currently, there is no single global standard for the development of enantiomericall active pharmaceutical compounds. Producers have to develop these compounds accor the regulatory requirements of each individual country.

Although at present there are only a few agrochemicals sold as single enanti environmental considerations will most likely bring changes. As agrochemicals are applied in relatively large amounts compared to drugs for human use, the econom environmental impacts of a reduced application rate can be significant [4]. Frost and S [4] forecast that global revenues from chiral technology would grow between 2000 an at an annual rate of 13% during that period. Agricultural chemicals comprised 14%

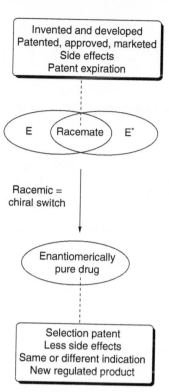

FIGURE 29.1 The chiral switch concept. (Adapted from Agranat, I., Caner, H., and Caldwell, J., *Nat. Rev. Drug Discovery*, 1, 735–768, 2002.)

revenues, while flavors and fragrances accounted for slightly less than 5%. Thus, the most important market for chiral molecules is pharmaceutical drugs.

Single-enantiomer drugs are commanding an ever-increasing presence in the global pharmaceutical market. The worldwide sales of single-enantiomer pharmaceutical products in 2001 was US$139 billion; the top ten single-enantiomer products resulted in sales of US$34.2 billion in 2002; in 2003, nine of the top ten drugs contained chiral as the active ingredient, and sales increased to a total of US$48.3 billion. The chiral switch process has resulted in a number of agents being remarketed as single enantiomers. A summary of those drugs is depicted in Figure 29.2.

In addition to the drugs depicted in Figure 29.2, a number of other compounds are in the pipeline of development [11] or introduction to the market. Among these are (*S*)-oxybutinin (for the treatment of urinary incontinence), (*R*,*R*)-formoterol (for an improved treatment of asthma), (+)-norcisapride (for the treatment of nocturnal heartburn with a reduction of cardiotoxicity), (*S*)-doxazosin (for the treatment of benign prostatic hyperplasia), (*S*)-amlodipine (for anti-hypertension treatment), (*S*)-fluoxetine (for migraine prophylaxis), as well as (*S*)-lansoprazole and (−)-pantoprazole for the treatment of gastroesophageal reflux.

However, there are some financial risks associated with the development of single enantiomers from racemates. Thus, the termination of the licensing agreement between Sepracor and Eli Lilly [17] for the development of (*R*)-fluoxetine resulted in an estimated loss of ~US$70 million for Sepracor.

Miconazole, an antifungal agent for the topical treatment of skin diseases, is another candidate for a chiral switch. Although a chiral switch is not needed at all for topical applications, for oral treatment of other diseases, e.g., tuberculosis, a chiral switch has to

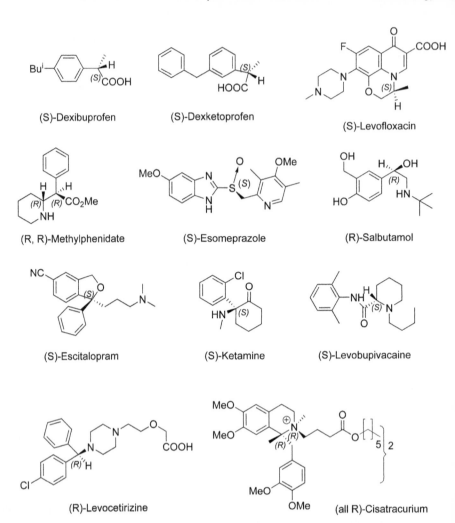

FIGURE 29.2 Marketed single enantiomers of drugs that have undergone the chiral switch. Hutt, A.J. and Valentova, J., *Acta Facult. Pharm. Univ. Comenianae*, 50, 7–23, 2003.)

be considered. In addition, several other structurally related antifungal drugs are a taken orally, and some of these are produced in an enantiomerically pure form. Both tir price are determining factors for a successful chiral switch. According to calculatio additional cost to make one enantiomer of miconazole at 99% enantiomeric excess (ee) be ~US$120/kg using advanced simulated-moving-bed (SMB) technology [18]. A undesired enantiomer cannot be re-racemized, it has to be discarded at the end (thus le to costs of US$25/kg).

Several technologies can be used for chiral switches in industrial-scale applica Among them traditional resolution of racemates still has a leading role, but one has t in mind that the cost of resolution rapidly escalates with scale-up. In some cases, technology is cost-effective on a commercial scale. To develop alternative routes to er pure materials beginning with racemic raw material will likely be more costly or consuming than a chiral-pool approach or the choice of a proper catalytic system consensus is that catalytic asymmetric routes are most desirable.

There are only two primary sources for pure enantiomers. The first (and probably oldest) technique involves the isolation of naturally occurring molecules from plants or microorganisms, and, more seldom, from animals. This approach also involves the techniques of *de novo* fermentation of readily available, inexpensive feedstocks. All of these compounds form a "chiral pool." This chiral pool is therefore a collection of relatively inexpensive, readily available natural products. The main types of chemicals included in the chiral pool are amino and hydroxyl acids, carbohydrates, alkaloids, and terpenes. An "extended chiral pool" also includes newly developed, synthetically derived stereoisomers formed as products, intermediates, or by-products in more or less large-scale commercial processes. The second technique is through synthesis, using optically active, prochiral, or achiral compounds (followed by a suitable process for the separation of the corresponding stereoisomers). These main pathways have been depicted in Figure 29.3.

Many pharmaceuticals have been prepared in an enantiomerically pure form by the crystallization of diastereomeric salts. Among them [19,20] are the antibiotics ampicillin and chloramphenicol (resolved by camphosulfonic acid), the tuberculostatic ethambutol and the anti-infective thiamphenicol (resolved using tartaric acid), or the antibiotic fosfomycin and the calcium antagonist diltiazem (using R-(+)-phenethylamine); the anti-inflammatory agent naproxen, however, has been subject to a classical resolution using the alkaloid cinchonidine.

Crystallization of diastereomeric salts can also involve a process of deracemization. One example [21] includes the conversion of the naturally occurring amino acid L-proline to D-proline using (+)-L-tartaric acid in the presence of a catalytic amount of an aldehyde, finally leading to D-proline (93–95% ee) in up to 95% yield (Figure 29.4). In addition, this methodology can be used for the synthesis of enantiomerically pure compounds starting from the corresponding racemates.

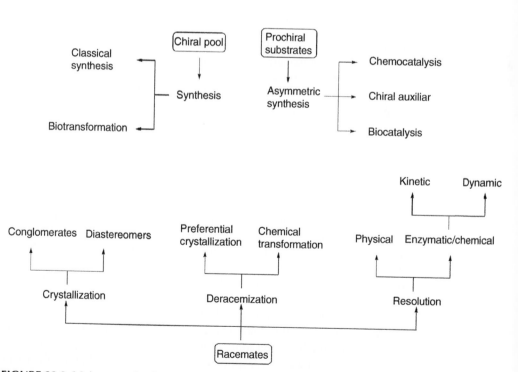

FIGURE 29.3 Main strategies for the synthesis of enantiomerically pure materials.

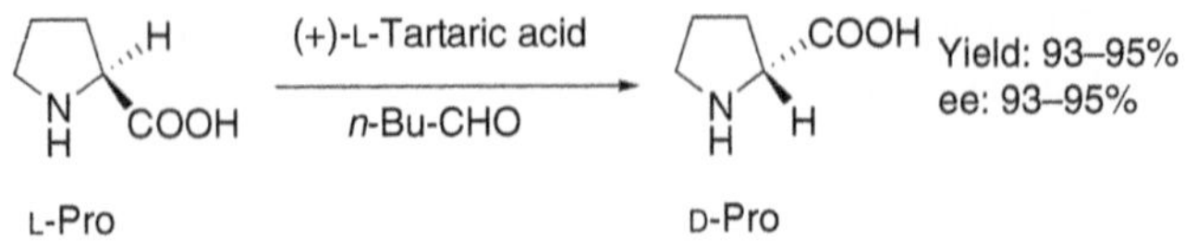

FIGURE 29.4 Deracemization of proline. (From Shiraiwa, T., Shinjo, K., and Kurokawa, H., *Lett.*, 1413–1414, 1989.)

Contrary to these almost 100% yields resulting from deracemization reactions, k resolutions allow the preparation of enantiomerically pure material, albeit in a maxi yield of 50%. The process of kinetic resolution relies on the differing reactivities o stereoisomers (preferentially enantiomers in a racemic mixture) with a chiral reagent of these stereomers reacts more readily than the other and, as a result, two produc obtained: one stereoisomer is converted into a different product and the other re unchanged. Kinetic resolutions can be achieved with enzymatic, microbial, or che methods; an example for the latter is the catalytic Sharpless epoxidation of allylic alc [22], as well as several rhodium BINAP [23] catalyzed oxidation processes.

29.2 CHIRAL SWITCH PROCESSES USING ENZYMES OR MICROORGANI!

Enzymatic kinetic resolutions have been in the focus of interest for the past several year use of enzymes (showing some advantages over the use of whole microorganisms) is attractive because the reactions are often highly selective, can be performed under conditions, are catalytic in nature, and, due to the use of water as a solvent, are enviror tally benign. No dynamic resolutions have been reported so far for a chiral switch.

29.2.1 DEXIBUPROFEN

α-Arylpropionic acid derivatives form a major class of nonsteroidal anti-inflammatory and household painkillers. Among these, naproxen [2-(4-methoxynaphthyl)-propanoic and ibuprofen [(2-(4-(2-methylpropyl)-phenyl]-propanoic acid] are probably the known and are sold in large quantities. The sole administration of the (*S*)-enantion ibuprofen is indicated as it has been demonstrated [24] that the (*R*)-enantiomer accum in fatty tissue as a glycerol ester whose long-term effects may be critical.

For the synthesis of enantiomerically pure dexibuprofen = (*S*)-ibuprofen, se approaches have been devised using either microorganisms or isolated enzymes. Severa steps have been identified for a straightforward synthesis of this anti-inflammatory Among them, the kinetic resolution of racemic 4-(isobutylphenyl)-propanoic acid enzyme membrane reactor led to enantiomerically pure products with an ee >99%. Th been accomplished [25] either by enzymic esterification or enzymic transesterification acid with vinyl acetate. The enzyme membrane consisted of a dense surface layer asymmetric polyamide capillary membrane combined with an immobilized lipase from *domonas* sp. It is interesting to note that the immobilized enzyme was significantly more than the native; however, its catalytic activity reached only 27% of that of the native The best resolution of the acid racemate was achieved in the sequential process, startin an esterification followed by a hydrolysis of the formed ester (Figure 29.5).

An enzyme-faciliated enantioselective transport of (*S*)-ibuprofen through a supp liquid membrane based on ionic liquids [26] has been investigated and ees up to 75% been obtained. Thus, a hydrophobic isotactic 1-propene homopolymer membrane wa to create a dived cell. One compartment is a feed phase containing racemic ibuprofen a

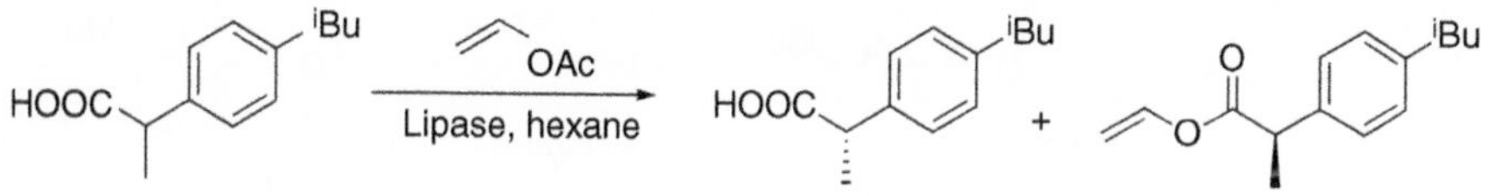

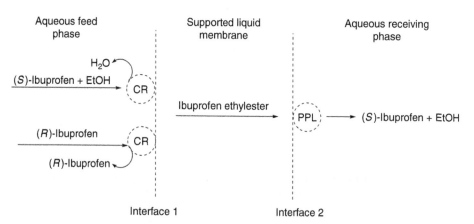

FIGURE 29.5 Dexibuprofen by transesterification. (From Ceynowa, J. and Rauchfleisz, H., *J. Mol. Catal. B Enzymatic*, 23(1), 43–51, 2003.)

lipase from *Candida rugosa* in a buffer–ionic liquid mixture. The receiving phase contained porcine pancreatic lipase (PPL) in a buffer. (*S*)-Ibuprofen is selectively esterified by the *C. rugosa* lipase in the feed phase, the ester dissolves in the ionic liquid phase and permeates across the membrane, and finally (*S*)-ibuprofen ethyl ester is hydrolyzed by the PPL in the receiving phase (Figure 29.6) [26].

The lipase from *C. rugosa* [27] has also been used for the resolution of the corresponding ibuprofen methyl esters (Figure 29.7).

It could be shown that higher conversions were obtained in a fed-batch reactor (compared to a batch reactor), using lower feed rates and high aqueous-phase hold-up. Effects of temperature and the pH value were established, and the reaction also showed some sensitivity to the shaking speed.

As an alternative to *C. rugosa* the use of the carboxypeptidase NP from *Bacillus subtilis 1–85* clones in sorbitan monoleate [28] has been suggested. The stability of the esterase under application conditions was improved by modification with glutaraldehyde, formaldehyde, or glyoxal. The methyl and ethyl esters of ibuprofen have been hydrolyzed by quite a number of different microorganisms [29] such as *Arthrobacter citeus thai, Pseudomonas mendocina thai, Streptomyces flavovirens*, as well as several strains of *Pseudomonas* and *B. subtilis*. The methyl ester, in addition, has been used [30] to access labeled ibuprofen.

Racemic ibuprofen has been used in another approach [31], but the resolution using PPL has been performed after reduction and acetylation, followed by a reoxidation, thus making this sequence rather lengthy (Figure 29.8).

The ee was high (94%) but both conversion rate (35%) and overall yield low. The same intermediates [32] could be obtained with a different approach using α-methylstyrene oxides (Figure 29.9).

Aqueous feed phase	Supported liquid membrane	Aqueous receiving phase

H_2O

(*S*)-Ibuprofen + EtOH

CR

Ibuprofen ethylester

PPL ⟶ (*S*)-Ibuprofen + EtOH

(*R*)-Ibuprofen

CR

(*R*)-Ibuprofen

Interface 1   Interface 2

FIGURE 29.6 Dexibuprofen by enantioselective transport. (Adapted from Miyako, E., Maruyama, T., Kamiya, N., and Goto, M., *Chem. Comm.*, 2926–2927, 2003.)

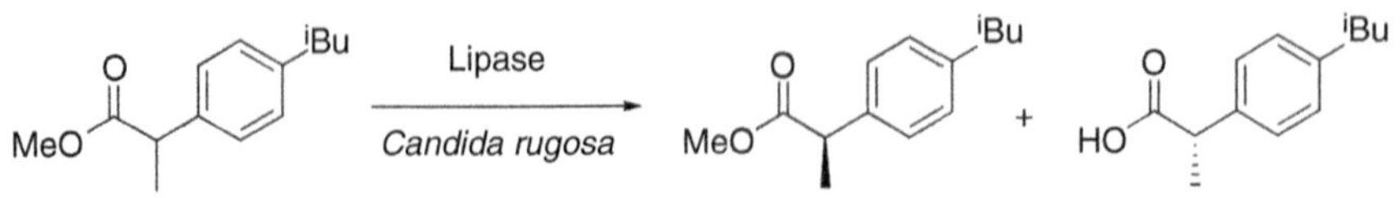

FIGURE 29.7 Dexibuprofen by hydrolysis with the lipase from *Candida rugosa*. (From Madha and Ching, C.B., *J. Chem. Technol. Biotechnol.*, 76, 941–948, 2001.)

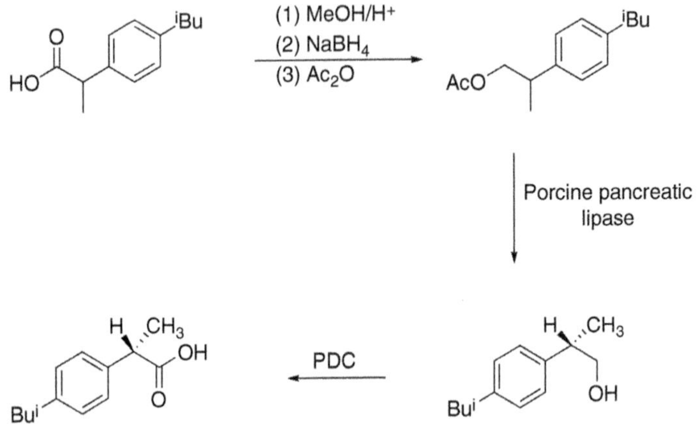

FIGURE 29.8 Dexibuprofen by hyrolysis with the lipase from porcine pancreas. (From Basak, A., Bhattacharya, G., Mandal, S., and Nag, S., *Tetrahedron Asymmetry*, 11, 2403–2407, 2000.)

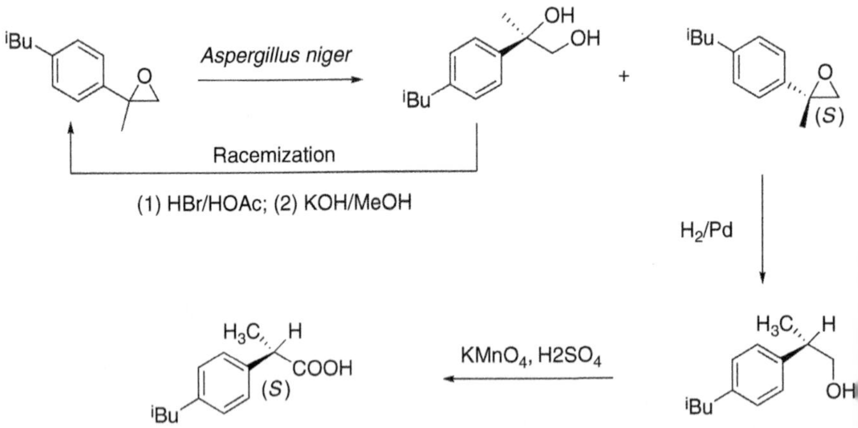

FIGURE 29.9 Dexibuprofen by hydrolysis with *Aspergillus niger*. (From Cleij, M., Archelan, Furstoss, R., *J. Org. Chem.*, 64, 5029–5035, 1999.)

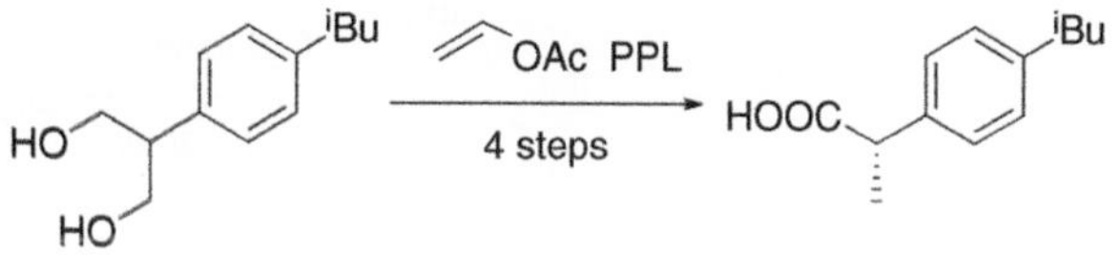

FIGURE 29.10 Dexibuprofen by transesterification. (From Bander, T., Namba, Y., and Shishido, K., *Tetrahedron Asymmetry*, 8, 2159–2165, 1997.)

Using this combined chemoenzymatic strategy (*S*)-ibuprofen was obtained in an optically pure form with a 47% overall yield. The enantioselectivity of the hydrolysis strongly depends on the nature of the enzyme. PPL-mediated transesterification of prochiral 2-aryl-1,3-propane-diols seems a rather lengthy [33] approach (Figure 29.10).

In search for higher enantioselectivity in the synthesis of (*S*)-ibuprofen several strategies were followed. New lipases from *Actinomycetes* have been isolated [34] and screened. Several of them displayed higher activity than commercial lipases from *C. rugosa* in the resolution of chiral secondary alcohols. Values of ee that were >99% were obtained using a biocatalyst from *Pseudomonas fluorescens MTCCB0015*, especially when hydrolyzing the 2,2,2-trifluor-oethyl ester [35] of ibuprofen. Thus, the (*S*)-acid could be obtained from the corresponding racemic ester after 20 h in 40% isolated yield at 47% conversion. The enzyme lost all of its enantioselectivity using substrates possessing bulky *ortho*-substituents. As far as the lipase from *C. rugosa* is concerned [36], the vinyl ester of racemic ibuprofen is more quickly hydrolyzed [37] than the methyl or chloroethyl ester.

Selectivity of the lipase from *C. rugosa* could be increased by using its cross-linked [38] enzyme crystal (CLEC). CRL-CLEC has been shown to be an attractive replacement for the crude enzyme due to its insolubility (and hence it is easy to recover from the reaction mixture), and these glutaraldehyde cross-linked enzymes are usually 2–3 orders of magnitude more stable than the soluble proteins. Thus, CRL-CLEC has been used for the enantioselective hydrolysis of racemic ibuprofen methyl ester to afford (*S*)-ibuprofen in 87% yield with an ee of 93%.

The influence of temperature [39] and of organic additives, e.g., polar solvents such as *N*,*N*-dimethylformamide, onto the selectivity of crude lipase from *C. rugosa* for the cleavage of racemic ibuprofen methyl ester has been investigated. As an alternative to the use of soluble enzyme or CLRCs the use of physically adsorbed lipase B from *C. antarctica* was investigated.

It was shown [40] that the reaction yield of the esterification of racemic ibuprofen using adsorbed *C. antarctica* lipase B on an anionic resin could be improved by the addition of benzo[18]crown-6 or metal-tetraphenylporphyrins. While the reaction rate of these reactions was increased, no increase in the enantioselectivity of the enzyme could be observed.

The corresponding nitrile [41] has also been transformed into ibuprofen using immobilized cells. Cells immobilized with both calcium alginate and κ-carrageenan gave a product with lower ee values than for free, intact cells; in addition, the reaction rate was decreased. However, the reaction of cells immobilized in cellulose porous beads with an average diameter of 200 μm containing pores of 10 to 30 μm diameter resulted in improved yields and higher optical purity than those from intact cells. The beads could be used repeatedly without decreasing the optical purity, but the conversion dropped to 32% (Figure 29.11).

Cells from *Rhodococcus butanica* have also been used for the hydrolysis of racemic nitriles, albeit in only 13% yield. The hydrolyses of these α-arylpropionitriles [42] afforded the (*R*)-amides and the corresponding (*S*)-carboxylic acids. A process using an ibuprofen-derived racemic aldehyde or a bisulfite adduct [43] has been described. These procedures utilize

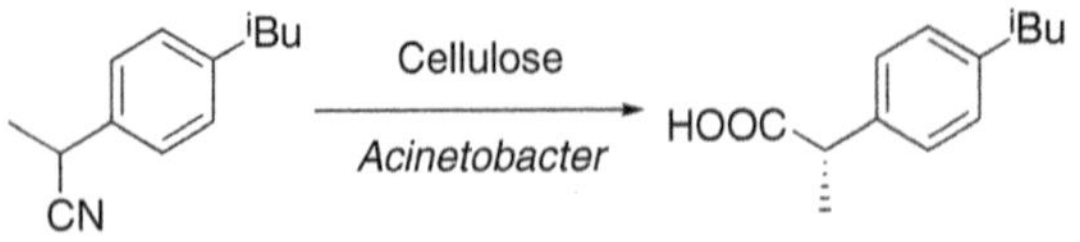

FIGURE 29.11 Dexibuprofen from nitriles. (From Takagi, M., Shirokaze, J.-I., Oishi, K., Otsub Yamamoto, K., Yoshida, N., and Fujimatsu, I., *J. Ferm. Bioeng.*, 78, 191–193, 1994.)

microorganisms, e.g., *Pseudomonas* sp., *Aeromonas* sp., *Escherichia* sp., *Alcaligenes* s enzyme preparations from these microorganisms containing oxidoreducases (Figure 29

29.2.2 DEXKETOPROFEN = (S)-KETOPROFEN

Lipase catalysis is a well-established method to obtain enantiomerically pure compo However, as most of the commercial lipases contain a bunch of competing (iso)enzymes also some additives), the enantiorecognition [44] of a lipase is lowered. The enanioselect however, can be increased either by using CLRCs or by increasing the purity of the en [45,46].

The use of crude lipase from *C. rugosa* for the hydrolysis of racemic ketop 2-chloroethyl ester [47] under extremely acidic conditions (pH 2.5) and in the presen Tween-80 as an emulsifier gave (S)-ketoprofen, possessing (after purification) an ee of S at 42.2% conversion (Figure 29.13).

A purification process has been used for improvement of this approach. Two process the partial purification and for the immobilization of the crude lipase preparation (*C. r* lipase OF) have been successfully integrated into one by simple adsorption of the en onto a cation exchange resin (Sephadex) at low pH.

The enantioselectivity was also improved [48] by a partial removal of isoenzymes a the addition of an emulsifier. Except for Tween-60, Tween-80, and nonylphenol polye neoxy ether, most of the surfactants tested had inhibitory influence on the lipase. Onl addition of Tween-80 [49] resulted in an increase of enantioselectivity. Interestingly addition of an enzyme inhibitor [50], e.g., dextromethorphan, enhanced the selectivi this transformation by a factor of 10. As an alternative, a purification of lipase OF mercurial affinity column can be performed, leading to three portions of these enzyme have remarkably different abilities to differentiate between the enantiomers of α-arylpro acids in the lipase-catylzed hydrolysis of the corresponding esters (Figure 29.14) [48].

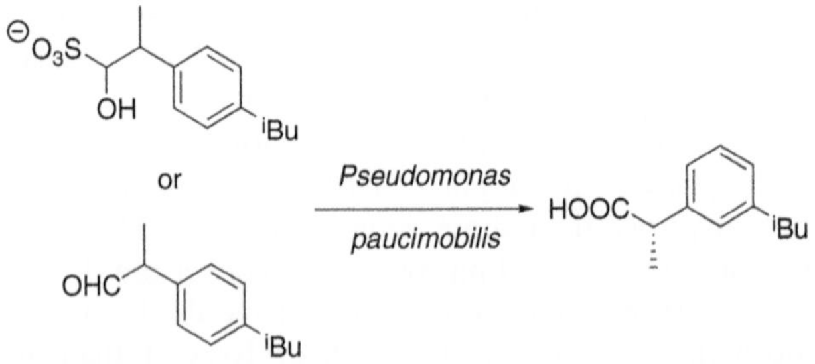

FIGURE 29.12 Dexibuprofen from an aldehyde or its bisulfite adduct. (From Rossi, R.F., H D.L., and Zepp, C.M., U.S. 5273895 (28.12.1993), *Chem. Abs.*, 120, 215480, 1994.)

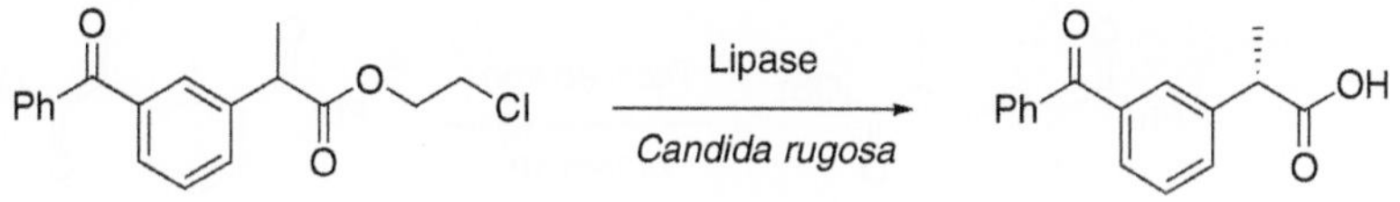

FIGURE 29.13 Dexketoprofen by hydrolysis of its 2-chloroethyl ester. (From Wu, H.-Y., Xu, J.-H., and Liu, Y.-Y., *Synth. Commun,* 31, 3491–3496, 2001.)

Improved production of (*S*)-ketoprofen has been reported using a mutant of *Trichosporon brassicae CGMCC0574* (Figure 29.15) [51–54].

The reverse process [55], i.e., the stereoselective esterification, has been accomplished using immobilized lipase from *C. antarctica* (Novozym 435). This process allows the large-scale preparation of (*S*)-ketoprofen with 96% ee as unreacted enantiomer, whereas the (*R*)-enantiomer is recovered as the ester that can easily be separated, racemized by hydrolysis, and be reused in the process.

Finally, some unusual microorganisms [56], e.g., *Aspergillus soje, Mycobacterium smegmatis, Mucor mihei* [57], *P. fluorescens MTCCB0015* [58], and acetone powders from liver [59] of rabbit, horse, sheep, dog, etc., have been used.

29.2.3 ESOMEPRAZOLE AND PANTOPRAZOLE

Esomeprazole and pantoprazole are proton pump inhibitors that are indicated for patients with gastroesophageal reflux disease.

Characteristic for this class of compounds are syntheses by microbial redox reactions. Thus, enantiomerically pure or enriched sulfoxides were prepared by the stereoselective biological reduction of racemic sulfoxides. Racemic omeprazole was reacted with *Proteus vulgaris*, reducing (−)-omeprazole [60] to leave (+)-omeprazole in >99% ee.

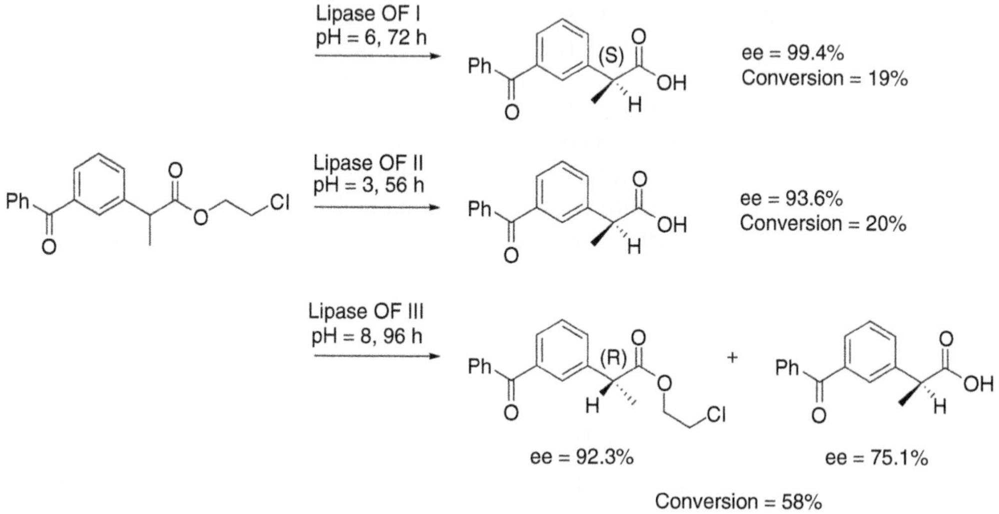

FIGURE 29.14 Dexketoprofen by hydrolysis using purified lipase OF from *Candida rugosa.* (From Liu, Y.-Y., Xu, J.-H., Wu, H.-Y., and Shen, D., *J. Biotechnol.*, 110, 209–217, 2004.)

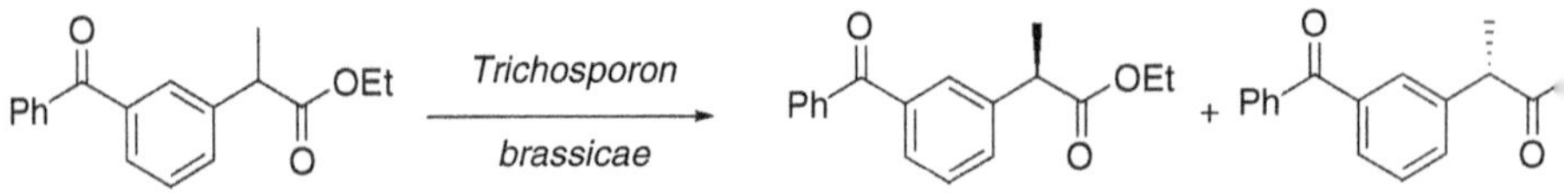

FIGURE 29.15 Dexketoprofen by hydrolysis with *Trichosporon brassicae*. (From Wu, H.-Y., X⌐ Shen, D., and Xin, Q., *J. Industrial Microbiol. Biotechnol.*, 30, 357–361, 2003; Shen, D., Xu, J. P., Liu, Y., and Wu, H., *Weishengwuxue Tongbao*, 29, 45–49, 2002; Shen, D., Xu, J.-H., Wu, H.- Liu, Y.-Y., *J. Mol. Catal. B Enzym.*, 18, 219–224, 2002; Shen, D., Xu, J.-H., Gong, P.-F., Wu, and Liu, Y.-Y., *Can. J. Microbiol.*, 47, 1101–1106, 2001.)

As an alternative, the sulfoxide can be accessed by the microbial oxidation corresponding sulfides. Thus (−)-omeprazole was produced in >99% ee by the oxida the sulfide [61] with *Penicillium frequentans* or *Ustilago maydis* (Figure 29.16).

29.2.4 METHYLPHENIDATE

(+/−)-threo-Methylphenidate is a mild nervous system–stimulating agent marke Ritalin for the treatment of children with attention deficit hyperactivity disorde (2*R*,2′*R*)-(+)-threo-derivative has been shown to be several times more active th corresponding (2*S*,2′*S*)-(−)-threo analog. The enantioselective hydrolysis [62] of the ra⌐ has successfully been performed using α-*chymotrypsin* in phosphate buffer to yield e⌐ merically pure material in ~16% yield. Subtilisin from Carlsberg gave 98% ee (14.5% (Figure 29.17).

29.2.5 LEVOFLOXACIN

Levofloxacin is an antimicrobial. Two strategies have been developed for the synth enantiomerically pure levofloxacin: a resolution at a very early step of the synthetic se⌐ and, as an alternative, a resolution at a very late step.

A representative for the first approach was the treatment of racemic lactate deri⌐ [63] with an esterase to achieve a resolution. In a series of steps the lactate is transform⌐ the product (Figure 29.18).

A late resolution utilizes immobilized porcine liver esterase (PLE) [64] to cleave of⌐ butyl ester enantioselectively to levofloxacin (Figure 29.19).

The immobilized esterase (calcium alginate/polyacrylamide gel) exhibited 58% im⌐ ization efficiency, and could be reused five times without severe loss of activity. Sig⌐

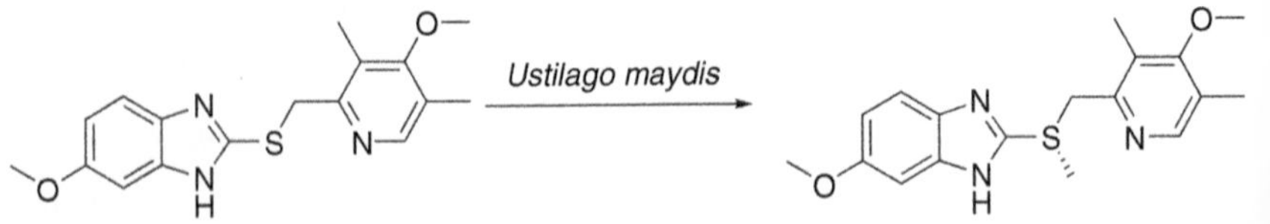

FIGURE 29.16 Esomeprazole by oxidation with *Ustilago maydis*. (From Graham, D., H⌐ Lindberg, P., and Taylor, S., PCT Int. Appl. 9617077 (28.11.1994), *Chem. Abs.*, 125, 112927, 1

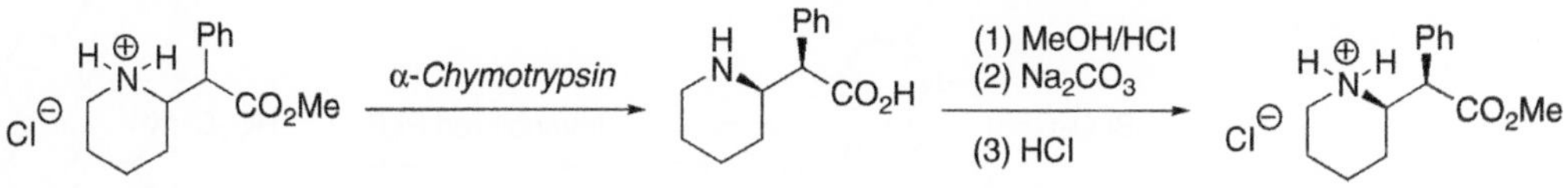

FIGURE 29.17 Methylphenidate by hydrolysis with α-*chymotrypsin*. (From Prashat, M., Har, D., Repic, O., Blacklock, T.J., and Giannousis, P., *Tetrahedron Asymmetry*, 9, 2133–2136, 1998.)

reduction of the enzyme activity was found, however, when the enzyme was physically adsorbed onto QAE-sephadex.

Similar results were obtained [65] when the enzyme was immobilized in polyacrylamide gel. It was found that the initial activity of the immobilized esterase is significantly affected by the gel composition. The activity of the immobilized enzyme was about 55% compared with that of the free enzyme; the ee was maintained at 60%, which corresponds well to the level obtained with the free enzyme.

In a quite different but rather lengthy approach [66] the use of microorganisms belonging to the genera *Bacillus, Micrococcus, Actinomycetes, Aspergillus, Rhizopus, Candida, Saccharomyces,* and *Zygoascus* has been the subject of a patent.

29.2.6 DOXYZOSIN

The quinazoline-derived compound (+/−)doxyzosin mesylate (Cardura) is indicated for the treatment of hypertension and has been proven effective in the treatment of benign prostatic hyperplasia. Doxazosin seems to be a selective inhibitor of the α1 adrenergic receptor. A very elegant and large-scale approach [67] utilizes a microbial esterase derived from *Serratia marcescens* for the stereoselective hydrolysis of racemic ethyl 1,4-benzodioxan-2-carboxylate to yield ethyl (S)-1,4-benzodioxan-2-carboxylate (41 to 43% yield, 95.6 to 98.4% ee) (Figure 29.20).

As an alternative to the use of the esterase from *S. marcescens* the use of the lipase from *C. antarctica* B (Novozyme A/S) has been suggested [68] to obtain the (S)-enantiomer with an ee > 95%.

29.2.7 OTHER COMPOUNDS

The development of the antidepressant (R)-fluoxetine has been stopped. Nevertheless, a few approaches [69–71] to the target molecule used biotransformations for the synthesis of intermediates. Among them, the stereoselective reduction of ethyl benzoyl acetate with an ee > 99% has been reported using *Geotrichum* sp. or the microorganism genera *Saccharomyces, Hansenula, Dekkera,* or *Klyveromyces marxianus*.

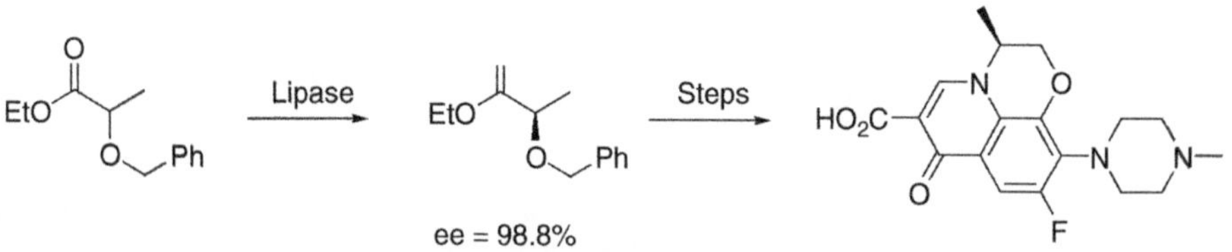

FIGURE 29.18 Levofloxacin from lactate ethers. (From Sato, K., Yagi, T., Kubota, K., and Imura, A., PCT Int. Appl. 2002070726 (12.09.2002), *Chem. Abs.*, 137, 232675, 2002.)

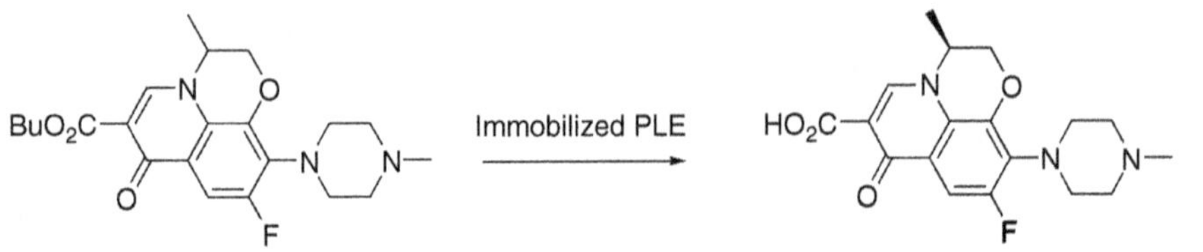

FIGURE 29.19 Levofloxacin by hydrolysis with immobilized porcine liver esterase (PLE). Lee, S.-Y., Min, B.-H., Hwang, S.-H., Koo, Y.-M., Lee, C.-K., Song, S.-W., Oh, S.-Y., Lim, Kim, S.-L., and Kim, D.-I., *Biotechnol. Lett.*, 23, 1033–1037, 2001.)

No racemic switches for the potent β2 agonists salbutamol and formoterol ut biocatalysis have been established so far. Both compounds are used as bronchodilat the therapy of asthma and chronic bronchitis. The enantiomers of the local anes bupivacaine exhibit stereoselectivity with respect to blockade of ion channels wi (*R*)-enantiomer being more potent. For this compound, for the selective serotonin re-u inhibitor escitalopram, for the anesthetic (*S*)-ketamine, for the H₁-antihistamine lev izine, and for the neuromuscular blocker cistracurium no enzyme- or microorganism- chiral switches have been reported so far.

29.3 CONCLUDING REMARKS

Reevaluation of the enantiomers of racemic drugs will continue and result in the introd of single enantiomers of established drugs into the market. Although this process is quite these costs can be regarded as modest compared to the costs associated with taking therapeutic agent to market. In addition, chiral switch approaches provide, at lower drugs of improved efficacy, higher safety, and lower risks due to cleaner pharmacol profiles.

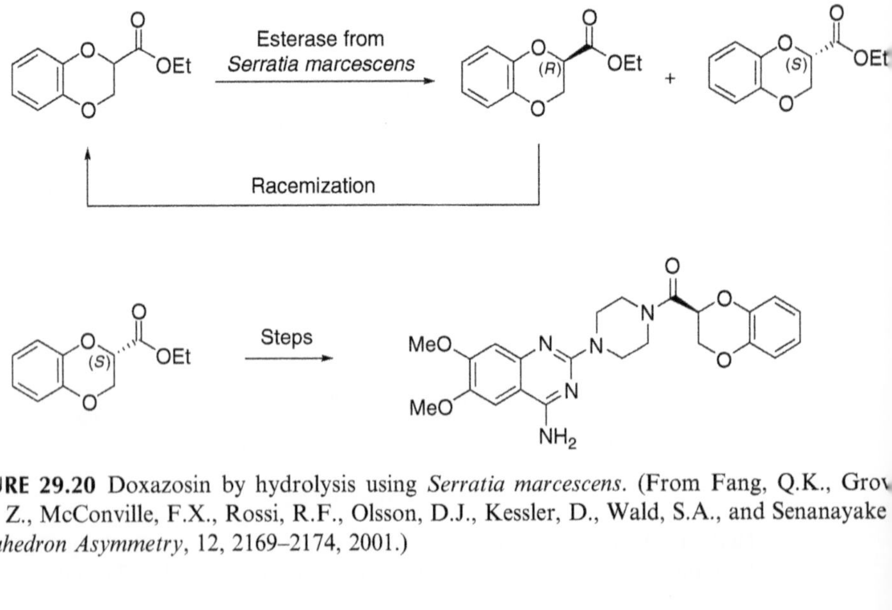

FIGURE 29.20 Doxazosin by hydrolysis using *Serratia marcescens*. (From Fang, Q.K., Gro Han, Z., McConville, F.X., Rossi, R.F., Olsson, D.J., Kessler, D., Wald, S.A., and Senanayake *Tetrahedron Asymmetry*, 12, 2169–2174, 2001.)

REFERENCES

1. Eichelbaum, M. and Gross, A., Stereochemical aspects of drug action and disposition, *Adv. Drug Res.*, 28, 1–64, 1996.
2. Triggle, D.J., Stereoselectivity of drug action, *Drug Dis. Today*, 2, 138–147, 1997.
3. Challener, C.A., Ed., *Chiral Drugs*, Ashgate, Burlington, Vermont, 2001.
4. Agranat, I., Caner, H., and Caldwell, J., Putting chirality to work: the strategy of chiral switches, *Nat. Rev. Drug Dis.*, 1, 735–768, 2002.
5. Eichelbaum, M., Testa, B., and Somogyi, A., Eds., *Stereochemical Aspects of Drug Action*, Springer, Heidelberg, 2002.
6. Szelenyi, I., Geisslinger, G., Polymeropoulos, E., Paul, W., Herbst, M., and Brune, K., The real gordian knot: racemic mixture versus pure enantiomers, *Drug News Perspect.*, 11, 139–160, 1998.
7. Hutt, A.J. and Valentova, J., The chiral switch: the development of single enantiomer drugs from racemates, *Acta Facult. Pharm. Univ. Comenianae*, 50, 7–23, 2003.
8. Williams, K.M., Enantiomers in arthritic disorders, *Pharmacol. Ther.*, 46, 273–295, 1990.
9. Walshe, J.M., Chirality of penicillamine, *Lancet*, 339, 254, 2002.
10. Cotzias, G.C., Papavasiliou, P.S., and Gellena, R., Modification of parkinsonism chronic treatment with L-DOPA, *N. Engl. J. Med.*, 280, 337–345, 1969.
11. Agranat, I. and Caner, H., Intellectual property and chirality of drugs, *Drug Dis. Today*, 4, 313–321, 1999.
12. Stinson, S.C., Chiral Switches, *Chem. Eng. News*, 79, 45–56, 2001.
13. Tucker, G.T., Chiral Switches, *Lancet*, 355, 1085–1087, 2000.
14. Rouhi, A.M., Chiral roundup, *Chem. Eng. News*, 80, 43–50, 2002.
15. Strong, M., FDA policy and regulation of stereoisomers: paradigm shift and the future of safer more efficient drugs, *Food Drug Law J.*, 54, 463–487, 1999.
16. Carlson, S.C., The case against market exclusivity for purified enantiomers of approved drugs, *Yale Symp. Law Technol.*, 6, 1999.
17. Thayer, A., Lilly pulls the plug on prozac isomer drug, *Chem. Eng. News*, October 30, p. 8, 2000.
18. Rouhi, A.M., Chiral Chemistry, *Chem. Eng. News*, 82, 47–62, 2004.
19. Bayley, C.R. and Vaidya, N.A., in *Chirality in Industry*, Collins, A.N., Sheldrake, G.N., and Crosby, J., Eds., Wiley, New York, 1992, p. 71.
20. Sheldon, R.A., *Chirotechnology*, Marcel Dekker, New York, 1993, p. 174.
21. Shiraiwa, T., Shinjo, K., and Kurokawa, H., Facile production of (*R*)-proline by asymmetric transformation of (*S*)-proline, *Chem. Lett.*, 1413–1414, 1989.
22. Martin, V.S., Woodard, S.S., Katsuki, T., Yamada, Y., Ikeda, M., and Sharpless, K.B., Kinetic resolution of racemic allylic alcohols by enantioselective epoxidation: a route to substances of absolute enantiomeric purity? *J. Am. Chem. Soc.*, 103, 6237–6240, 1981.
23. Kitamura, M., Manabe, K., and Noyori, R., Kinetic resolution of 4-hydroxy-2-cyclopentenone by rhodium-catalyzed asymmetric isomerization, *Tetrahedron Lett.*, 28, 4719–4720, 1987.
24. Williams, K., Day, R., Romualda, K., and Duffield, A., The stereoselective uptake of ibuprofen enantiomers into adipose tissue, *Biochem. Pharmacol.*, 35, 3403–3405, 1986.
25. Ceynowa, J. and Rauchfleisz, H., High enantioselective resolution of racemic 2-arylpropionic acids in an enzyme membrane reactor, *J. Mol. Catal. B Enzym.*, 23(1), 43–51, 2003.
26. Miyako, E., Maruyama, T., Kamiya, N., and Goto, M., Enzyme-facilitated enantioselective transport of (*S*)-ibuprofen through a supported liquid membrane based on ionic liquids, *Chem. Comm.*, 2926–2927, 2003.
27. Madhar, M.V. and Ching, C.B., Study on the enzymatic hydrolysis of racemic methyl ibuprofen ester, *J. Chem. Technol. Biotechnol.*, 76, 941–948, 2001.
28. Mutsaers, J.H.G.M. and Kooreman, H.J., Preparation of optically pure 2-aryl- and 2-aryloxypropionates by selective enzymatic hydrolysis, *Recl. Trav. Chim. Pays-Bas*, 110, 185–188, 1991.
29. Gist-Brocades, N.V., Manufacture of stereospecific 2-aryl-propionic acids by fermentation, Jpn. Kokai Tokkyo Koho 63045234 (26.02.1988), *Chem. Abs.*, 109, 168–975, 1988.

30. Chen, C.S., Copeland, D., Harriman, S., and Liu, Y.C., Preparation of enantiomericall deuterium-labeled ibuprofen, *J. Labelled Comp. Radiopharm.*, 28, 1017–1024, 1990.

31. Basak, A., Nag, A., Bhattacharya, G., Mandal, S., and Nag, S., Chemoenzymatic synt antiinflammatory drugs in enantiomerically pure form, *Tetrahedron Asymmetry*, 11, 240 2000.

32. Cleij, M., Archelan, A., and Furstoss, R., Microbiological transformations. 43. Epoxide hy as tools for the synthesis of enantiopure α-methylstyrene oxides: a new and efficient synthesi ibuprofen, *J. Org. Chem.*, 64, 5029–5035, 1999.

33. Bander, T., Namba, Y., and Shishido, K., Lipase-mediated asymmetric construction of 2-pionic acids: enantiocontrolled synthesis of (*S*)-naproxen and (*S*)-ibuprofen, *Tetrahedron metry*, 8, 2159–2165, 1997.

34. Cardenas, F., Alvarez, E., de Castro-Alvarez, M.S., Sanchez-Montero, J.M., Elson, S., an terra, J.V., Three new lipases from actinomycetes and their use in organic reactions, *E Biotrans.*, 19, 315–329, 2001.

35. Kumar, I., Manju, K., and Jolly, R.S., A new biocatalyst for the preparation of enantion pure 2-arylpropanoic acids, *Tetrahedron Asymmetry*, 12, 1431–1434, 2001.

36. Sih, C.J., Gu, Q.M., Fülling, G., Wu, S.H., and Reddy, D.R., The use of microbial enzymes synthesis of optically active pharmaceuticals, *Dev. Ind. Microbiol.*, 29, 221–229, 1988.

37. Fülling, G., Schlingmann, M., and Keller, R., Enzymic hydrolysis of 2-arylpropionic vinyl e manufacture of the optically active acids, Ger. Offen 3919029 (13.12.1990), *Chem. Ab* 245930, 1991.

38. Lalonde, J.J., Govardhan, C., Khalef, N., Martinez, A.G., Visuri, K., and Margolin, A.L. linked crystals of *Candida rugosa* lipase: highly efficient catalysts for the resolution of chira *J. Am. Chem. Soc.*, 117, 6845–6852, 1995.

39. Lee, H.W., Kim, K.-J., Kim, M.G., and Lee, S.B., Enzymic resolution of racemic ibuprofe effects of organic cosolvents and temperature, *J. Ferm. Bioeng.*, 80, 613–615, 1995.

40. Gradillas, A., del Campo, C., Sinisterra, J.V., and Llama, E.F., Alteration of the reaction ra esterification of (*R,S*)-ibuprofen by addition of crown ether or porphyrin, *Biotechnol. Lett.*, 90, 1996.

41. Takagi, M., Shirokaze, J.-I., Oishi, K., Otsubo, K., Yamamoto, K., Yoshida, N., and Fujim Production of (*S*)-(+)-ibuprofen with high optical purity from a nitrile compound by cells bilized on cellulose porous beads, *J. Ferm. Bioeng.*, 78, 191–193, 1994.

42. Kakeya, H., Sakai, N., Sugai, T., and Ohta, H., Microbial hydrolysis as a potent method preparation of optically active nitriles, amides and carboxylic acids, *Tetrahedron Lett.*, 32 1346, 1991.

43. Rossi, R.F., Heefner, D.L., and Zepp, C.M., Enantioselective production of chiral carboxyl U.S. 5273895 (28.12.1993), *Chem. Abs.* 120, 215480, 1994.

44 Chang, R.C., Chou, S.J., and Shaw, J.F., Multiple forms and functions of *Candida rugos Biotechnol. Appl. Biochem.*, 19, 93–97, 1994.

45. Chang, Y.-F. and Tai, D.-F., Enhancement of the enantioselectivity of lipase OF cataly drolysis, *Tetrahedron Asymmetry*, 12, 177–179, 2001.

46. Hernaiz, M.J., Sanchez-Montero, J.M., and Sinisterra, J.V., Comparison of the enzymatic of commercial and semi-purified lipase from *Candida cylindracea* in the hydrolysis of the (*R,S*)-2-arylpropionic acids, *Tetrahedron*, 50, 10749–10760, 1994.

47. Wu, H.-Y., Xu, J.-H., and Liu, Y.-Y., A practical enzymic method for preparation of (*S*)-ket with a crude *Candida rugosa* lipase, *Synth. Commun.*, 31, 3491–3496, 2001.

48. Liu, Y.-Y., Xu, J.-H., Wu, H.-Y., and Shen, D., Integration of purification with immobiliz *Candida rugosa* lipase for kinetic resolution of racemic ketoprofen, *J. Biotechnol.*, 110, 2 2004.

49. Liu, Y.-Y., Xu, J.-H., and Hu, Y., Enhancing effect of Tween-80 on lipase perfe in enantioselective hydrolysis of ketoprofen ester, *J. Mol. Catal. B Enzym.*, 20, 523–529, 20

50. Sih, C.J., Improving the enantioselectivity of biocatalytic resolution of racemic compounds, Eur. Pat. Appl. 387068 (12.09.1990), *Chem. Abs.*, 114, 245–954, 1991.
51. Wu, H.-Y., Xu, J.-H., Shen, D., and Xin, Q., Improved production of (*S*)-ketoprofen ester hydrolase by a mutant of *Trichosporon brassicae CGMCC0574*, *J. Ind. Microbiol. Biotechnol.*, 30, 357–361, 2003.
52. Shen, D., Xu, J., Gong, P., Liu, Y., and Wu, H., Isolation of the esterase producer *Trichosporon brassicae* and its catalytic performance in the kinetic resolution of ketoprofen, *Weishengwuxue Tongbao*, 29, 45–49, 2002.
53. Shen, D., Xu, J.-H., Wu, H.-Y., and Liu, Y.-Y., Significantly improved esterase activity of *Trichosporon brassicae* cells for ketoprofen resolution by 2-propanol treatment, *J. Mol. Catal. B Enzym.*, 18, 219–224, 2002.
54. Shen, D., Xu, J.-H., Gong, P.-F., Wu, H.-Y., and Liu, Y.-Y., Isolation of an esterase-producing *Trichosporon brassicae* and its catalytic performance in kinetic resolution of ketoprofen, *Can. J. Microbiol.*, 47, 1101–1106, 2001.
55. D'Antona, N., Lombardi, P., Nicolosi, G., and Salvo, G., Large-scale preparation of enantiopure (*S*)-ketoprofen by biocatalysed kinetic resolution, *Process Biochem.* (Oxford, UK), 38, 373–377, 2002.
56. Iriuchijima, S. and Keiyu, A., Asymmetric hydrolysis of esters by biochemical methods. I. Asymmetric hydrolysis of (+/−)-α-substituted carboxylic acid esters with microorganism, *Agric. Biol. Chem.*, 45, 1389–1392, 1981.
57. Carganico, G., Mauleon Casellas, D., and Palomer Benet, A., A process for the preparation of (*S*)-(+)-2-(3-benzoylphenyl)propionic acid by enzyme-catalyzed enantioselective hydrolysis, PCT Int. Appl. 9325703 (23.12.1993), *Chem. Abs.*, 120, 215–459, 1994.
58. Kumar, I., Manju, K., and Jolly, R.S., A new catalyst for the preparation of enantiomerically pure 2-aryl-propanoic acids. *Tetrahedron Asymmetry*, 12, 1431–1434, 2001.
59. Goswami, A., Stereospecific resolution by hydrolysis of esters of 2-aryl-propionic acids by liver enzymes, PCT Int. Appl. 91131632 (5.9.1991), *Chem. Abs.*, 115, 254–315, 1991.
60. Graham, D., Holt, R., Lindberg, P., and Taylor, S., Enantioselective preparation of pharmaceutically active sulfoxides by bioreduction, PCT Int. Appl. 9617077 (28.11.1994), *Chem. Abs.*, 125, 112927, 1996.
61. Holt, R., Lindberg, P., Reeves, C., and Taylor, S., Enantioselective preparation of pharmaceutically active sulfoxides by biooxidation, PCT Int. Appl. 9617076 (28.11.1994), *Chem. Abs.*, 115, 112–926, 1996.
62. Prashat, M., Har, D., Repic, O., Blacklock, T.J., and Giannousis, P., Enzymatic resolution of (+/−)-threo-methylphenidate, *Tetrahedron Asymmetry*, 9, 2133–2136, 1998.
63. Sato, K., Yagi, T., Kubota, K., and Imura, A., Process for preparation of optically active 2-hydroxypropoxyaniline derivatives as intermediates for levofloxacin via enzymic or microbial stereoselective hydrolysis of racemic lactic acid ester, PCT Int. Appl. 2002070726 (12.09.2002), *Chem. Abs.*, 137, 232675, 2002.
64. Lee, S.-Y., Min, B.-H., Hwang, S.-H., Koo, Y.-M., Lee, C.-K., Song, S.-W., Oh, S.-Y., Lim, S.-M., Kim, S.-L., and Kim, D.-I., Enantioselective production of levofloxacin by immobilized porcine liver esterase, *Biotechnol. Lett.*, 23, 1033–1037, 2001.
65. Lee, S.-Y., Min, B.-H., Song, S.-W., Oh, S.-Y., Lim, S.-M., Kim, S.-L., and Kim, D.-I., Polyacrylamide gel immobilization of porcine liver esterase for he enantioselective production of levofloxacin, *Biotechnol. Bioprocess Eng.*, 6, 179–182, 2001.
66. Sato, K., Takayanagi, Y., Okano, K., Nakayama, K., Imura, A., Itoh, M., Yagi, T., Kobayashi, Y., and Nagai, T., Process for the preparation of benzoxazine derivatives and intermediates therefrom, PCT Int. Appl. 2001018005 (15.03.2991), *Chem. Abs.*, 134, 222719, 2001.
67. Fang, Q.K., Grover, P., Han, Z., McConville, F.X., Rossi, R.F., Olsson, D.J., Kessler, D., Wald, S.A., and Senanayake, C.H., Practical chemical and enzymatic technologies for (*S*)-1,4-benzodioxan-2-carboxypiperizine intermediate in the synthesis of (*S*)-doxazosin mesylate, *Tetrahedron Asymmetry*, 12, 2169–2174, 2001.

68. Kasture, S.M., Varma, R., Kalkote, U.R., Nene, S., and Kulkarni, B.D., Novel enzymatic rout kinetic resolution of (+/−)-1,4-benzodioxan-2-carboxylic acid, *J. Biochem. Eng.*, 27, 66–71, 2

69. Schneider, M.P. and Görgens, U., An efficient route to enantiomerically pure antidepress Tomoxetine, Nisoxetine and Fluoxetine. *Tetrahedron Asymmetry*, 3, 525–528, 1992.

70. Kamal, A., Khanna, G.B., Rao, M.V., and Kondapuram, K.V., Stereoselective preparation hydroxy-3-phenyl-propionitrile, Int. Pat. Appl. 2002057475 (22.01.2001), *Chem. Abs.*, 137, 10 2002.

71. Master, H.E., Newadkar, R.V., Rane, R.A., and Kumar, A., Highly efficient enzymic resoluti homoallyl alcohols leading to a simple synthesis of optically pure fluoxetine and related compo *Tetrahedron Lett.*, 37, 9253–9254, 1996.

30 Enzyme Evolution for Chemical Process Applications

Gjalt W. Huisman and James J. Lalonde

CONTENTS

30.1 INTRODUCTION

The advent of advanced enzyme evolution technologies promises to be the breakt technology that will transform biocatalysis into a robust, economically attractive metho for the manufacture of complex chiral molecules. For many decades, enzymes have been as promising catalysts in chemical applications, yet the number of commercial bioc processes has grown only slowly. Recently developed enzyme evolution technologie for the rapid development of tailor-made catalysts for commercial processes. This discusses the reasons why such technologies were needed, provides an overview of the d tools that are available and the rationale that guides enzyme evolution, and sum recent successes.

30.2 THE BIOCATALYST ENVIRONMENT

30.2.1 THE NATURAL ENVIRONMENT

The "active site" of enzymes consists of amino acid ligands that influence the bin substrate, product, inhibitors, and activators. In nature, the sequence and type of the acids in the polypeptide strands of enzymes have evolved to optimally organize the act and surrounding ligands such that the function of the enzyme is appropriate for the substrates under the right conditions. The environment under which an enzyme fu exerts selective pressure on the structure of the enzyme, specifically the amino acid se providing a competitive advantage to organisms that express an improved enzyme. Wi context of overall fitness of the organism, there are intrinsic limitations on enzymes th consequence of the provision of optimal fitness to the host.

30.2.1.1 Specificity

The substrate specificity of an enzyme is a consequence of the role that the enzyme fu cellular metabolism and within the context of alternative substrates that may be pr within the cell. For instance, nucleic acid polymerases are either specific for deoxyribo tides [DNA polymerase] or ribonucleotides [RNA polymerase]. If this were not the case could be doped with ribonucleotides and RNA with deoxyribonucleotides with consequences for the accuracy of DNA replication, transcription, and RNA translat nature, DNA and RNA polymerases have ubiquitously evolved to each accept on nucleotides out of the eight naturally occurring possible substrates as differentiated presence of a hydroxyl substituent at the $2'$ position of the ribose moiety (ribonucleotide absence (deoxyribonucleotides). While the absence of this differentiating moiety is exq well recognized by DNA polymerases, the absence of a hydroxyl group at the $3'$ positio and $2',3'$-dideoxynucleotides are accepted by DNA polymerase as a substrate. This prom is not a problem for a living cell because dideoxynucleotides are not encountered *in vivo*, practically exploited for the development of DNA sequencing *in vitro*.

30.2.1.2 Chirality

Enantioselectivity or enantiospecificity of an enzyme is the key to ensure the reprodu of cellular events that involve chiral molecules, such as amino acids and proteins. If both L-amino acids were incorporated in a growing polypeptide chain, the population of enzyme would be heterogeneous and largely inactive. Naturally, only *S*-amino aci glycine) are incorporated into proteins and the whole cellular machinery has adapted so that the undesired amino acid enantiomers are typically not produced, besides (important) exceptions. Thus, natural transaminases, amino acid–ammonia lyases, amino acid synthases, amino acid dehydrogenases, etc. are strictly S-specific. Interestingly, the nonnatural R-amino acids are frequently encountered in pharmaceutical products, as humans are typically not equipped with enzymes that rapidly clear R-amino acid containing molecules. However, for enzymes in catabolic pathways, it may be advantageous not to be enantioselective so that both the enantiomers of an available substrate can be utilized as a source of elements and energy.

30.2.1.3 Stability

Controlled instability of enzymes is extremely important for an organism when it needs to adapt to rapidly changing environmental conditions that require the synthesis of a different set of polypeptides. Protein synthesis is an energy intensive process and synthesis of new enzymes is more efficient—if these building blocks can be salvaged from already produced but superfluous peptides rather than resynthesized from the basic cellular intermediates. Several natural processes have been identified that regulate the stability and hence the half-life of proteins. ppGpp is a global regulator of gene expression in bacteria and is suggested to cause a redirection of transcription so that genes important for starvation survival are favored at the expense of those required for growth and proliferation [1]; protein turnover is catalyzed by a range of different enzymes such as Clp proteases, proteasomes, and aminopeptidases [2–4].

Enzymes are susceptible to environmental conditions, particularly oxidation and certain chemical instabilities, for instance, aspartate isomerization. In nature, such degenerated polypeptides are removed from circulation by proteolytic system [5]. Although some enzymes from nature may be stable for days, the typical half-life of natural biocatalysts is in the range of minutes [6], which is typically insufficient for an industrial chemical process.

30.2.1.4 Regulation of Enzyme Activity

It is crucial for a cell to regulate the relative activity of enzymes, particularly as they operate in pathways or regulons. While randomly scavenging available nutrients may appear as an opportune approach for an organism in dealing with a plethora of substrates, such uncontrolled action would put a burden on the cellular machinery with the need to synthesize a broad range of catabolic enzymes. For efficiency reasons, utilization of preferred substrates over less-desired nutrient sources is facilitated in organisms by mechanisms such as induction and repression at the gene level (e.g., induction of the *lac* operon by lactose and repression by glucose), isozyme availability at the regulon level (e.g., three different isozymes for the common first-step in the synthesis of methionine, threonine, and lysine from aspartate), and substrate/product inhibitions at the enzyme level, (e.g., phosphofructokinase and fructose-1,6-bisphosphatase) [7,8]. Feedback control mechanisms, such as substrate and product inhibitions, are important in biological systems but limit the utility of enzymes in chemical manufacturing processes.

30.2.2 THE INDUSTRIAL ENVIRONMENT

The environment that an enzyme encounters during a chemical process is significantly different from that in a biological environment. Where metabolite concentrations in living cells are typically <10 mM [9] and are subjected to natural homeostasis mechanisms such as product inhibition of enzymes involved in metabolism, concentrations in chemical processes are orders of magnitude higher and as a consequence, the biocatalyst must function over a large concentration range. For a commercial chemical process, economic performance is

key and product concentrations of at least 10 to 15% (w/v) are generally required,
often correspond to concentrations in excess of 500 mM. As a result, substrate conver
product can lead to significant changes in the environment that the enzyme enco
(Figure 30.1). For instance, at 1 M substrate concentrations, enzyme-catalyzed rea
may produce 1 M of additional ionic strength in the form of an acid salt (in a l
catalyzed reaction), chloride (in a dehalogenase-catalyzed reaction), and gluconate
ketoreductase/glucose dehydrogenase cofacor regeneration system) with a po
destructive effect on the enzymes. Such changes in osmotic strength of the reaction m
are not encountered in nature and enzymes have not evolved naturally for such fluctua
The development of glucose dehydrogenase variants, which are less sensitive to hig
concentrations, is an example of how new protein engineering technologies have beg
address such issues [10].

The substrates for biocatalytic transformations are frequently hydrophobic, le
to the presence of a secondary phase in the reaction where the enzyme functic
the organic–aqueous interface. Usually, the substrate of interest is so poorly solubl
organic cosolvents can be required to afford appreciable reaction phase concentrati
the substrate. As enzymes in nature do not typically encounter large amounts of o
substrates or solvents, this adds physical stress to the enzyme. Given that the concent
of organics in water is dependent, among others, on the makeup of the aqueous phas
changing ionic strength in the medium can greatly alter the organic substrate availa
over the course of a reaction.

Natural regulation mechanisms for enzyme activity, such as substrate and pr
inhibition, are important to protect the cellular environment from toxic levels of metab
yet in a chemical process, the volumetric productivity and the ease of process ope
depend on fast and complete conversion of all substrate. With the need for high conve
in short periods of time, substrate and product inhibition are a major problem for ind
application of enzymes.

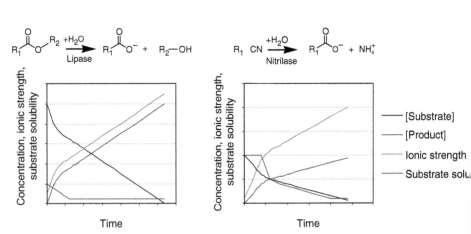

FIGURE 30.1 Environmental changes in a chemical process to which an enzyme may be ex
(a) Lipase reaction in 100 mM buffer at $t = 0$. With substrate available at 1 M, the ionic stre
the end of the reaction is >1 M. With the increase in osmotic strength, the substrate so
(accessibility) may diminish resulting in a slower reaction. (b) Nitrilase reaction in 100 mM b
$t = 0$. The ionic strength of the reaction medium increases quickly with two ionic products. In a
the organic substrate is converted to a fatty acid, which at the critical micelle concentration l
partitioning of substrate and product.

30.3   MERITS AND LIMITATIONS OF BIOCATALYTIC AND WHOLE-CELL TRANSFORMATIONS

Historically, both (semi) purified enzymes and whole cells (microbial/plant/mammalian) have been used in chemical conversions. Either system has advantages and disadvantages and can in principle be selected for process development. Before we examine the shortcomings of these systems, it is important to consider all aspects of a commercial biocatalytic process, including the conditions that the catalyst is subjected to prior and subsequent to chemical conversion.

1. The biocatalytic process starts with catalyst manufacture, generally by fermentation of a natural or recombinant microbial strain.
2. At the end of fermentation, the biomass is used directly or is harvested and formulated as whole-cell catalyst.
3. The enzyme is separated from other cell constituents in a down-stream process and formulated. In considering the different options for formulation, storage and transport are important factors.

   • If the catalyst is used at the same site as where it was produced, a catalyst suspension can directly be loaded into the chemical reactor or the reactants can be added to the fermenter where the catalyst was produced.
   • When catalyst manufacture is tolled to a site different from the chemical plant, the catalyst needs to be transported under controlled conditions and volume minimization (e.g., removal of water by lyophilization) may be required.
   • Upon arrival, the material may need to be stored for variable periods of time and prepared prior to use.

4. At the end of the biocatalytic reaction, the product is isolated and removal of all biologic material is important. Especially in the synthesis of pharmaceutical end products, the absence of proteinaceous material and endotoxins is key as these molecules carry the potential to cause significant side effects (such as allergies and other immunity related side effects).

When large catalyst-to-substrate ratios are needed to accomplish the desired reaction (low substrate-to-catalyst ratio), the separation of product and catalyst becomes increasingly difficult due to emulsion formation and isolated yield of product is compromised. This short overview indicates some general phenomena that are important for biocatalysts, whether the application is as whole cells or formulated enzyme: manufacturability, formulation options, stability during production, long-term storage stability, compatibility with the chemical reaction conditions, and separation from the chemical product.

30.3.1   Biocatalysts

In pure form, enzymes often exhibit longevity and are easy to store and transport. The early purification and removal of biological contaminants facilitate the work-up of the chemical process. Biocatalytic processes are much more adaptable than whole-cell processes, as the function of the catalyst is not deterred by the limitations that other components of a whole cell may exert. Consequently, more degrees of freedom in process design are available, including substrate concentration, pH, temperature, nature of organic solvent, and use of immobilization matrices if desired. However, the use of free enzymes in chemical reactions

is typically limited by the availability of commercial enzymes with the desired activity
substrate of interest. Many of these enzymes have been developed for use in the f
detergent industry and thus may not be suitable for chemical manufacture. Eve
commercial enzyme is found to have activity on a target substrate, frequently t
insufficient activity to make the process economically viable. In cases where an enz
identified that performs the chemistry well, application at industrial scale is frec
hampered by insufficient stability. Frequently applied engineering solutions such as
ization by cross-linking, immobilization (with or without recycling of the catalyst), su
engineering, solvent change, and continuous product extraction have sometimes been
mented at scale. In reactions that involve the stoichiometric use of cosubstrates, s
NAD(P)H, the efficient use of these molecules by regeneration is crucial for the econoi
the process. This puts additional requirements on the enzymes in terms of affinity
cosubstrate as well as an activity requirement, because the enzymatic reaction need:
completed before the intrinsic instability of cosubstrate impedes the reaction.

30.3.2 WHOLE CELLS

For whole-cell reductions, cosubstrate regeneration is not an issue as the cell is na
equipped for this. Microbes have adapted to the natural environment and produc
simple and complex metabolic products from their nutrient sources through co
integrated pathways. Whole cells can be used directly in chemical processes thereby
minimizing the formulation costs. Whole cells are cheap to produce and no knowle
genetic details is required.

However, processes that use whole organisms as catalyst are limited due to the ii
need to grow biomass, which results in the formation of by-products that need to be re
during the work-up after the chemical step. Both the cells themselves and the wide
of typical metabolites are potential process impurities. Although the use of filt
and liquid extractions can limit the contamination of the product, these procedu
significant cost. Additionally, if it is required that the catalyst is intact during the J
(e.g., during reactions that require cofactor regeneration), the survival boundaries of
limit the chemical process conditions. Another limitation is the short lifetime of wh
catalysts and the relatively long time required for the biomass production. Typicall
transformations or fermentative processes are slow and thus require significant proc
capacity. As in biocatalytic processes, engineering solutions have been sought for some
problems encountered with whole-cell processes with some success. The manufact
acrylamide at 1000s mT scale using immobilized whole cells of a *Rhodococcus* stra
great example [11]. However, when no engineering solutions exist, improving these cata
difficult due to the complexity of whole cells and the inherent problems with adapt
physiology of a cell.

30.4 METHODS TO OVERCOME LIMITATIONS, ENZYME IMPROVEMEN'

30.4.1 ENZYME ENGINEERING

With the advent of recombinant DNA technology in the early 1970s, the number of ei
available for biocatalytic studies and applications increased as enzymes no longer nee
be isolated from their natural source. Genes encoding enzymes of interest were free
overexpressed in *Escherichia coli*, with mixed success. As these tools became more estab
the number of recombinantly produced enzymes continued to climb. In the 1980s, to
advanced and hence knowledge of the three-dimensional structure of an enzyme cc

used to engineer enzymes with high precision at the molecular level. Although initial technical results were promising, the speculated increase in the potential of enzymatic conversion was not fully realized; primarily due to the expense of protein structure determination and the complexity of enzymes that limits the full understanding of how enzymes function at a molecular level, let alone under different environmental conditions. Recognizing the power or recombination in generating variety among antibodies, Stemmer invented what became the first of many recombination-based *in vitro* laboratory evolution methods, DNA shuffling [12]. His landmark paper initiated an added interest in the field of biocatalysis and led to the development of several *in vitro* and *in vivo* evolution methods.

30.4.2 ENZYME EVOLUTION

Given that the starting, natural enzyme is typically not sufficiently "fit" for a chemical process, several rounds or iterations of library generation and high-throughput (HTP) screening are generally required. "Fitness" is frequently used in the enzyme evolution field as a term that encompasses all the enzyme properties desired for large-scale application. Properties such as specific activity, substrate and/or product inhibition, tolerance to organic solvents, thermostability, oxidative stability, stability to high and low ionic strength, storage stability together are parameters in the fitness function. Accordingly, in enzyme evolution it is desirable to address as many of these parameters during the HTP screens as possible, so that no variants are taken forward that are deteriorated for any of these parameters. The essence of such an approach can be summarized in a relatively simple adage "you get what you screen for." While this indicates that knowing what the desired enzymatic properties are is key, oversimplification should be cautioned against as often you may "lose what you have, if you do not screen for it" as can be observed in the literature [13].

In vitro laboratory, enzyme evolution is comprised of three key components that form an iterative process until the desired enzyme has been identified (Figure 30.2):

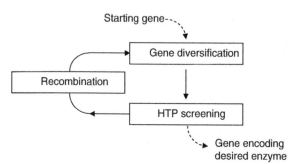

FIGURE 30.2 The evolution cycle. The gene of interest is diversified by procedures as described in the text, resulting in mutations that have been preselected in nature (e.g., homologous enzymes), or mutations that are novel, such as by error-prone PCR. Resulting libraries are screened and enzyme variants with improved properties are identified. If further improvement of the best variants is desired, the corresponding genes are recombined (optionally including additional diversity) and rescreened. It is desirable that the HTP screen resembles the ultimate process conditions as closely as possible so that the most attractive process can be enabled. The most informative screens tend to require larger reaction volumes than less-informative screens, and consequently, the throughput may be compromised. A reduced throughput can be tolerated, however, if the quality of the gene libraries is high, i.e., the chance of finding improved variants is high. The different gene evolution technologies described in this chapter provide gene libraries of different quality, each requiring different levels of throughput.

1. The generation of diverse enzyme libraries
2. The identification of improved variants in such libraries using HTP screening
3. The recombination of beneficial mutations to generate diverse enzyme libraries th further improved for desired function

All three components are of equal importance. In the absence of "high-quality" div "highly relevant" screens, or "efficient" recombination, the generation and identificat enzyme variants becomes difficult, thereby decreasing the overall pace of the *in vitro* evo process.

30.4.2.1 Enzyme Diversity

Large-scale genome sequencing efforts have provided a wealth of diverse er sequences. The number of publicly available microbial genomes for which the cor DNA sequence has been determined is approaching 250 [14]. Genes encoding enzyme have evolved for optimal fitness in the natural environment are thus abundantly ava However, as discussed above, such enzymes are rarely adept at functioning well in a che plant. Methods to generate nonnatural variants have been developed and include cla methods such as error-prone polymerase chain reaction (PCR), site directed mutage cassette mutagenesis, as well as more sophisticated methods that are specific [15] or rar such as "deletion-duplication" [16] and "sequence saturation muagenesis" [17].

30.4.2.2 High-Throughput Screening

HTP screening equipments for the identification of improved enzymes are widely ava as stand-alone units or are combined with other instruments in robotic setups. In instances, commercially available equipment can be rebuilt for specific applications in screening for improved enzymes.

The generally used approach in HTP screening for improved enzymes is a tiered pro where the number of variants under analysis is quickly reduced through successive scre decreasing throughput, providing information of increasing relevance. For instance, fir screens can be microbiological selections (TP: $\sim 10^9$), fluorescence-assisted cell s (TP: $\sim 10^9$), plate assays on bacterial colonies, e.g., for lipases, proteases, etc. (TP: second tier screens can be 96-well kinetic assays (TP: $\sim 10^3$), and third tier screens c chemical reactions at a 25 to 100 ml scale (TP: $\sim 10^0$)/day.

30.4.2.3 DNA Shuffling

The iterative process of gene diversification, library construction, and HTP scre comprise the enzyme evolution cycle depicted in Figure 30.2. In the present tradi approach [12], a number of gene variants that encode enzymes with improved fu would enter the next round of evolution to further improve fitness. The understand the power of such genetic algorithms was recognized in the engineering disciplines, 1950s and was applied almost universally in complicated construction and engin problems since then, but not in the protein engineering field. Recently, the importal iterative homologous recombination for sequence evolution was underscored by *in* simulations [18]. With the advent of bioinformatics tools [19], this paradigm has s in that the single best evolved variant of each round is designated as the "backbon further libraries, which are then generated by introducing a set of mutations that identified in other "hits." While the traditional approach advances genes and m into next rounds of evolution, these newer methods advance mutations into subse

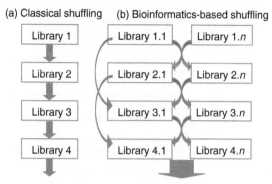

FIGURE 30.3 Evolution of enzyme evolution approaches. (a) Traditonal shuffling approach—a few "hits" are taken forward from each library in consecutive, serial rounds. Diversity is incorporated at the beginning and is derived from random mutagenesis, recombination, and homologous genes. (b) Bioinformatics aided shuffling approach—beneficial and neutral diversities are taken forward from parallel libraries ($n = 1$–20) spanning multiple rounds. Diversity is incorporated throughout and derived from random mutagenesis, saturation mutagenesis, homologous sequences, and structural information.

rounds (Figure 30.3). One of the additional advantages of this method is that mutations can be added more freely into several, parallel library designs and at different times during the evolution program. In fact, individual mutations can be introduced into library generation, even if they were identified much earlier in the program or in different projects that utilized the same genetic information.

The following section provides a synopsis of some evolution methods that involve a recombination step, but is by no means a complete summary. Several recent reviews are also available for the interested reader [20–23].

30.4.2.3.1 Gene Shuffling (Figure 30.4)

Method description: This is a method for *in vitro* homologous recombination of pools of selected mutant genes by random fragmentation and PCR reassembly.

Example: β-Lactamase was evolved in three cycles of mutagenic DNA shuffling and two cycles of backcrossing with wild-type DNA, each round followed by selection on increasing concentrations of the antibiotic cefotaxime. The minimum inhibitory concentration was increased 32,000-fold, while nonrecombinatory procedures such as cassette mutagenesis and error-prone PCR resulted in only a 16-fold increase. [12]

Note: The revolutionary method that changed the field of enzyme engineering.

Semisynthetic Shuffling (Figure 30.5)

Method description: Genes and plasmids were reassembled from DNA fragments 10 to 50 bp in size. Complete recombination was obtained between two markers on two genes separated by 75 bp. Oligonucleotides could be incorporated into the reassembled gene to target specific regions.

Example: A library of chimeras of the human and murine genes for interleukin 1 beta was prepared [24].

Note: The first account of the use of synthetic oligonucleotides in recombination-based enzyme engineering.

Low Mutageneses DNA Shuffling

Method description: An alternative DNA shuffling protocol for random recombination of homologous genes *in vitro* with a low rate of point mutagenesis (0.05%). The mutagenesis rate is controlled over a wide range by the inclusion of Mn^{2+} or Mg^{2+} during DNase I digestion,

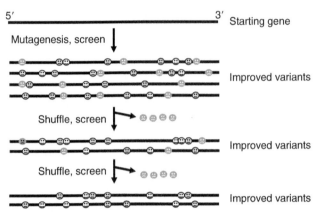

FIGURE 30.4 Single gene shuffling. A single gene is improved in iterative cycles of recombina
screening. Initial diversity is created by mutagenesis, such as error-prone PCR. This random
creates mutations that are desired (good mutations—happy faces), undesired (bad mutations—s
and those that are neutral. Evolution methods that involve iterative mutagenesis of improved
accumulate all three types of mutations and overall improvement is limited by the extent that the
mutations can offset the negative effects of the undesired mutations. In recombination-based n
the undesired mutations are removed so that the extent of possible improvement is not lim
undesired mutations.

by choice of DNA polymerase used during gene reassembly as well as how the ge
prepared for shuffling (PCR amplification vs. restriction enzyme digestion of plasmid

Example: Subtilisin E and a thermostable mutant were shuffled to demonstrate tha
shuffling can be achieved with high precision and that closely linked mutations
separated during the protocol [25].

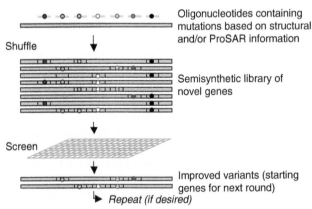

FIGURE 30.5 (See color insert following page 526) Semisynthetic shuffling. The nature of the
diversity that is recombined with the gene of interest is biased using oligonucleotides that contain
mutations that are expected to be beneficial. Such mutations can be based on structural knowledg
enzyme, hot spots in the gene found to be of interest in earlier rounds of evolution (even from
projects that utilized the same starting gene), or mutations that have been identified by compu
methods as desirable. The level of incorporation of the oligonucleotides is controlled experime
create semisynthetic gene libraries with a variable number of mutations.

Staggered Extension Process (StEP)

Method description: This method consists of priming the template sequence(s) followed by repeated cycles of denaturation and extremely abbreviated annealing/polymerase-catalyzed extension. In each cycle, the growing fragments anneal to different templates based on sequence complementarity and extend further. This is repeated until full-length sequences have formed. Due to template switching, most of the polynucleotides contain sequence information from different parental sequences.

Example: A set of five thermostabile subtilisin E variants were recombined and yielded an enzyme whose half-life at 65°C is 50 times that of wild-type subtilisin E [26].

Note: An alternative to the methods that use enzymatic, chemical or physical DNA fragmentation.

Heteroduplex Formation

Method description: Heteroduplexes prepared *in vitro* can be used to transform bacterial cell where they are repaired to form recombinant genes composed of elements from each parent.

Example: The method is demonstrated using GFP as model system [27].

Note: A laborious procedure that does not seem to be used frequently.

Direct Transformation of PCR Products

Method description: Libraries of shuffled gene PCR products are directly transformed into *Bacillus subtilis* or *Acinetobacter calcoaceticus*. Reconstitution of a selectable marker provides a simple selection for evolved genes [28].

Note: The transformation efficiency is rather low such that the library production may be problematic.

Functional Salvage Screen

Method description: The activity of nonfunctional enzymes is restored by random cloning of genomic DNA fragments in a predetermined region of the gene of interest.

Example: The method is demonstrated using GFP [29].

Note: An interesting approach that may be of value when structural elements need replacement, but due to the disruptive character of the method, ultra HTP screens or selections are needed, limiting the overall applicability of the method.

Synthetic Shuffling

Method description: The variety in functional libraries is increased by using degenerate oligonucleotides; starting genes are no longer necessary and codon usage criteria can be incorporated as well.

Example: Synthetically shuffled libraries of 15 subtilisin genes were generated, yielding active and highly chimeric enzymes with desirable combinations of properties that were not obtained by other directed-evolution methods [30].

Note: This paper shows the versatility of shuffling approaches varying from natural genes as starting materials to completely synthetic genes.

EndoV Fragmentation

Method description: Endonuclease V nicks uracil-containing DNA at the second or third phosphodiester bond 3' to uracil sites and is used to randomly fragment DNA. Cleavage occurs at random sites and the length of the fragments can be adjusted by varying the concentration of dUTP in the PCR. Unlike the DNase I methods, no partial digestion or gel separation of fragments is required.

Example: Two truncated GFP genes were recombined [31].

Note: An alternative method for DNA fragmentation.

Mutagenic and Unidirectional Reassembly (MURA)

Method description: This method generates libraries of DNA-shuffled and randomly truncated proteins. The method involves random fragmentation of the template gene(s) and

PCR reassembly with a unidirectional primer. The MURA products were treated with DNA polymerase and subsequently with a restriction enzyme whose site was located on region of the MURA primer.

Example: Phospholipase was converted into variants that exhibited absolute li activity by truncation at a region beginning with amino acids 61 to 71, together with ar acid substitutions [32].

Note: A variant library manufacturing technique.

Computational Prescreen

Method description: A combined computational and experimental method for the r optimization of proteins

Example: The active site of β-lactamase was redesigned using a computational screen. screen eliminated sequences that were incompatible with the protein fold, thereby redu the number of sequences that need to be screened experimentally. In a single round, 1280 improved variants were obtained from an antibiotic resistance selection. None of the n tions had been observed before in either natural or nonnatural β-lactamases [33].

Note: Even after size reduction, the experimentally screened library is very large requires ultra HTP sequencing or selections. This limits the application scope of the met

Gene Reassembly

Method description: Gene variants are digested using restriction enzymes and religat form hybrid genes encoding hybrid enzymes.

Example: A library of >20,000 hybrid genes was generated by gene reassembly of α-amylase wild-type genes that encoded enzymes with varying properties. Improved mu with combined optimal phenotypes of expression, temperature stability, and pH opti were obtained [34].

Note: The rational design element as well as the randomness of the religation proce limits the utility of these libraries, as additional genetic engineering to create and or re restriction sites may often be required and a large number of variants need to be screen

Assembly of Designed Oligonucleotides (ADO)

Method description: A synthetic shuffling approach based on oligonucleotides v design is guided by sequence information.

Example: From a library prepared with information from two *B. subtilis* lipase g several variants were obtained that displayed increased enantioselectivity in a model rea involving the hydrolysis of a *meso*-diacetate [35].

Note: A synthetic shuffling approach.

Biased Mutation Assembling

Method description: The library construction process is manipulated by prov different fragments, e.g., wild-type mutant, in different ratios to favor only a few or mu mutations in the progeny.

Example: The method was applied to the generation of more thermostable prolyl peptidases. A mutant with a 1200-fold longer activity, half-life at 60°C was identified.

NExT DNA Shuffling

Method description: Uridine triphosphate (dUTP) is used during the amplificatio gene library by PCR. The incorporated uracil bases were excised using uracil-DNA-glyco and the DNA backbone subsequently cleaved with piperidine. Polyacrylamide ure demonstrated adjustable fragmentation size over a wide range. The oligonucleotide po reassembled by internal primer extension to full length with a proofreading polymer. improve yield over Taq.

Example: Chloramphenicol acetyltransferase was shuffled using a 33% dUTP PCR, resulted in shuffled clones (average fragment size of 86 bases) and revealed a low mu rate (0.1%) [37].

Note: An alternative to Miyazaki's method [31].

Multiplex-PCR-Based Recombination (MUPREC)

Method description: A method for the *in vitro* recombination of single-point mutations that reduces the introduction of novel point mutations, which usually occur during recombination processes. A multiplex-PCR reaction generates gene fragments that contain preformed point mutations. These fragments are subsequently assembled into full-length genes by a recombination-PCR step. The process of MUPREC does not require DNase I digestion for gene-fragmentation. The low error rate resulted in high-quality variant libraries of true recombinants, thereby minimizing the screening efforts and saving time and money.

Example: The MUPREC method was used in the directed evolution of a *B. subtilis* lipase that can catalyze the enantioselective hydrolysis of a model *meso*-compound. The method was useful in producing a reliable second-generation library of true recombinants from which better performing variants were identified by using a high-throughput electrospray ionization mass spectrometry (ESI-MS) screening system [38].

Note: A method to reduce random mutagenesis during the recombination procedure.

Offset Recombinant PCR (OR-PCR)

Method description: A method that uses standard PCR to promote high recombination frequencies among compact heterologous domains.

Example: Pfu polymerase generated chimeric crossover events in 13% of the population when markers were separated by only 70 nt. The fraction of recombinant sequences reached 42% after six consecutive rounds of PCR, a value close to 50% expected from a fully shuffled population [39].

Note: This study describes how variations in the PCR process can influence the library quality.

30.4.2.3.2 Family Shuffling (Figure 30.6)

Method description: Libraries of hybrid genes were generated by random fragmentation of a pool of homologous genes, and reassembly of the fragments in a self-priming polymerase reaction.

Example: Moxalactamase activity was evolved when four different cephalosporinase genes were shuffled together instead of separately. The best clone was up to 540-fold improved

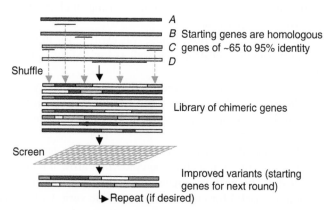

FIGURE 30.6 (See color insert following page 526) Family shuffling. Various genes encoding enzymes with identical (or at least similar) function are recombined to provide libraries of chimeric genes. The diversity introduced in this method is functional since it has been preselected in nature. In addition to full-length genes, partial genes and oligonucleotides can be added to the library generation procedure, thereby improving the quality of the libraries that are generated.

and contained eight segments from three of the four genes as well as 33 amino-acid mu [40].

Note: Homologous genes provide functional diversity that has been tested in na contrast to random diversity that is generated in the laboratory and that has no subjected to selection mechanisms.

Single-Stranded DNA Shuffling

Method description: Family shuffling efficiency is reportedly improved by using mentary strands of the two parents.

Example: ssDNAs of two catechol 2,3-dioxygenase genes were fragmented and re bled. The frequency of obtaining hybrid genes was 14-fold higher compared to stranded family shuffling and more thermostabile enzyme variants were obtained [41

Note: This method may provide advantages only when two genes are shuffled; increased number of parents some bias may slip in.

Combinatorial Libraries Enhanced by Recombination in Yeast (CLERY)

Method description: A family shuffling strategy involving both PCR-based and recombination and expression in yeast.

Example: Two human cytochrome P450s were shuffled and screened in yeast. S analysis of randomly picked and functionally selected clones confirmed the sh efficiency [42].

Note: A method that should be useful for the evolution of enzymes that are diff express in bacteria, but not in eukaryotic systems like yeast.

Degenerate Oligonucleotide Gene Shuffling (DOGS)

Method description: A shuffling method in which degenerate primers control th tive levels of recombination between the parents thereby increasing the recombi frequency.

Example: A diverse family of β-xylanase genes that differ widely in $G + C$ cont shuffled [43].

Note: This procedure avoids the use of endonucleases for gene fragmentation p shuffling and allows the use of random mutagenesis of selected segments of the gene as the procedure.

Rachitt

Method description: Randomly cleaved parental DNA fragments are anneale transient polynucleotide scaffold resulting in chimeric libraries that average a high of crossovers per gene.

Example: A monooxygenase was evolved for increased rate and extent of biodesu tion on complex substrates, as well as for 20-fold faster conversion of a nonnatural su [44].

Note: A laborious procedure that provides many crossovers between genes.

Degenerate Homoduplex Gene Family Recombination (DHR)

Method description: A synthetic DNA recombination method that provides more recombination

Example: A chimeric protein whose agonist activity was enhanced 123-fold was o from a human–mouse chimeric epidermal growth factor library [45].

Note: The authors suggest that this is a less-biased approach to DNA shuffli should be useful for engineering of a wide variety of proteins.

30.4.2.3.3 *"Distant" Family Shuffling*

Incremental Truncation for the Creation of Hybrid Enzymes (ITCHY)

Method description: ITCHY creates combinatorial fusion libraries between gen pendent of DNA homology.

Example: Fragments of the *E. coli* and human glycinamide ribonucleotide transformylase genes (50% identity at the DNA level) were fused by ITCHY and DNA shuffling. ITCHY identified a more diverse set of active fusion points including those in regions of nonhomology [46].

Note: The number of active clones in such libraries is small and very high-throughput or selection methods are needed to identify active variants.

SCRATCHY

Method description: The approach combines ITCHY and DNA shuffling. A library of hybrid enzymes created by ITCHY is subjected to a DNA-shuffling step to augment the number of crossovers.

Example: Functional hybrid enzymes containing multiple crossovers were obtained from SCRATCHY libraries that were created from the glycinamide-ribonucleotide formyltransferase genes from *E. coli* and human [47].

Sequence Homology–Independent Protein Recombination (SHIPREC)

Method description: A method to create single crossover libraries of distant (unrelated) genes at structurally related sites.

Example: A library of a human membrane-associated cytochrome P450 and the bacterial heme domain of a soluble P450 was generated and properly folded variants were selected with an antibiotic marker. Screening for the activity of the human enzyme identified two functional P450 hybrids with improved expression in *E. coli* [48].

Note: A method for chimaragenesis of unrelated enzymes with an interesting application.

Structure-Based Combinatorial Protein Engineering (SCOPE).

Method description: A semirational protein engineering approach that uses information from protein structure coupled with established DNA manipulation techniques to design and create multiple crossover libraries from nonhomologous genes.

Example: Libraries of chimeric genes of a rat and a African swine fever virus DNA polymerase with up to five crossovers were synthesized in a series of PCR reactions by employing hybrid oligonucleotides that code for variable connections between structural elements. Genetic complementation in *E. coli* enabled identification of several novel DNA polymerases with enhanced phenotypes [49].

Note: Rational protein engineering coupled with a recombination approach increases the functional diversity in the generated libraries.

Enhanced Crossover SCRATCHY

Method description: Gene variants prepared by ITCHY technology are shuffled so that multiple crossover points can be found in the progeny of the evolved library.

Example: Libraries were generated from rat and human glutathione transferase genes, while random sequencing indicated greater diversity than obtained by family shuffling or SCRATCHY itself. Several variants were isolated by HTP flow-cytometry that had retained rat-like substrate specificity [50].

Note: The best variants were primarily rat sequence with the only human DNA found at the 3′-end of the gene. The technology demonstrates that dramatic changes in the primary structure often result in loss of function.

Sequence-Independent Site-Directed Chimeragenesis (SISDC)

Method description: A computer-aided method using the SCHEMA algorithm [51] for the facile recombination of distantly related (or unrelated) genes at multiple discrete sites using endonucleases.

Example: Two β-lactamases were recombined at seven sites and screened for functionality. Among the active clones, 14 unique chimeras were identified [52].

Note: The use of computational methods increases the probability of generating hybrid enzymes with general activity.

30.4.2.3.4   Pathway Shuffling

Method description: DNA shuffling was used for the evolution of an operon.

Example: A three-gene arsenate resistance operon encoding two enzymes and a tran tional regulator was evolved by DNA shuffling, involving three rounds of *in vitro* recc ation, followed by selection in *E. coli* for increased resistance. The evolved cells grew in 0.5 M arsenate, a 40-fold increase in resistance. The resulting shuffled operon had spo eously integrated in the chromosome and contained 13 mutations [53].

Note: These results demonstrate that the fitness of proteins and cells is determin many parameters some of which are poorly understood.

30.4.2.3.5   Genome Shuffling

Method description: Whole-genome shuffling is an *in vivo* procedure that combin advantage of multiparental crossing allowed by DNA shuffling with the recombinat entire genomes normally associated with conventional breeding. Recursive genomic r bination within a population of bacteria efficiently generated combinatorial libraries (strains.

Example: Production of tylosin in *Streptomyces fradiae* was improved several fold

Note: This approach has the potential to facilitate cell and metabolic engineerin provides a nonrecombinant alternative to the rapid generation of improved organisms

30.5   EVOLVING ENZYMES FOR THE PROCESS

Given the potential utility of enzymes for the synthesis of complex chiral molecule the sensitivity of these biocatalysts under typical chemical processing conditions, it is standable that one of the most active applications of directed evolution has been for the enzymes in chemical synthesis. Section 30.4 provided an overview of laboratory methoc were developed over the last decade for the generation of large enzyme libraries which improved enzymes may be isolated. The applicability of natural enzymes was re summarized as "Natural selection has created a large variety of enzymes superbly adap catalyze an array of chemical reactions. However, there are still too few enzymes availa catalyzing reactions of interest and many of these have suboptimal properties for proc conditions." [55]. Before, chemical processes were developed to optimally utilize the po performance of the biocatalyst, optimal chemical processes can now be designed first principles, followed by the development of the biocatalyst to enable the designed p (Figure 30.7).

A robust manufacturing process is characterized by commercially important cons such as:

- High volumetric productivity with at least 10% substrate.
- Chirality introduced early in the process by chiral synthesis or resolution using reagents.
- Raw materials that are commercially available in large scale, rather than throu suppliers.
- Inexpensive solvents rather than expensive aprotic solvents.
- Reaction work-ups that are simple, involving filtration, extraction and distillatio evaporation to dryness, drying over magnesium sulfate, or chromatography.

Generally, these constraints are not an issue for biocatalytic processes, as enzyme exquisite enantioselectivity (or it can be evolved into the enzyme), are chemoselective, c venting the need for complicated separation processes in work-up, and can run wit

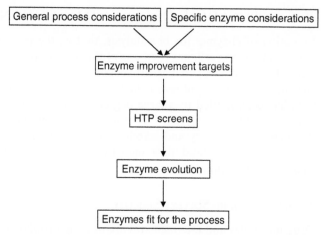

FIGURE 30.7 Generating the optimal enzyme for the process of interest. Before embarking on the enzyme improvement process, the process of interest is designed and process parameters are defined. An appropriate starting enzyme is identified and evaluated under the process conditions of choice to provide a set of enzyme improvement targets. Based on these targets, HTP screens are designed and developed so that enzyme evolution can be initiated. Iterative rounds of evolution will provide the enzyme of interest.

productivity. The extent to which all criteria can be met varies on an enzyme-by-enzyme and process-by-process basis. Enzyme manufacturing is a well-established technology that is available to (almost) every enzyme of interest and as a result, enzyme cost is predictable. Using more active variants and thereby decreasing enzyme loading and increasing the substrate-to-catalyst ratio can further lower the enzyme cost contribution to a process. Finally, expensive cosubstrates can be reused using efficient regeneration systems.

Based on these general guidelines and the known characteristics of the enzyme, a process is designed and the typically desired biocatalyst improvements are:

- Higher specific activity
- Increased or maintained enantioselectivity
- Increased solvent tolerance
- Operable at lower cofactor concentrations
- Increased in-process stability
- Reduced inhibition by substrate and product

The parameters that are determined to be key for the desired catalyst are then used to develop the tiered screening approach, addressing simultaneously as many of these parameters as possible in the earliest possible tier. Only in this way can the number of assumptions that have to be made when downscaling a chemical process to a HTP format, be kept at a minimum.

When this approach is well executed, enzymes will emerge that catalyze the desired conversion with a high volumetric productivity (high specific activity, stability, not inhibited) to give high quality product (purity and enantiomeric excess). The small amount of enzyme that is needed for these reactions simplifies the handling of the catalyst and facilitates the work-up of the reaction. Since the enzyme is biological in nature, it can easily be disposed of with the normal waste-treatment stream and no precautions are required in ensuring the absence of toxic heavy metals in the waste stream or in the product of the reaction.

30.6   APPLICATIONS OF DIRECTED EVOLUTION FOR BIOCATALYSIS

Evolution of enzymes for biocatalysis, that is, the preparation of chemicals using e
outside of a living cell, has been one of the most important applications of directed ev
technologies. While, it has been recognized for more than a century that the applica
enzymes to organic synthesis is highly attractive [56], practical application of these b
lysts has been limited to a large degree by the limitations of wild-type proteins evo
function within a living cell. Two primary application areas that have seen recent succ
the increase in activity and stability of enzymes in the presence of high concentrat
organic substrates (and thus increase of volumetric productivity) and the evolut
enantioselective enzymes.

30.6.1   ORGANIC SOLVENT STABILITY

Drug molecules and intermediates are often hydrophobic in nature and so, high activ
stability in the presence of high concentrations of organic substrates and solvent i
desirable. Virtually all enzymes evolved in nature to be active and stable in water. Za
Klibanov showed that while some lipases can be stable in very hydrophobic med
activity of most enzymes tends to be orders of magnitude lower in nonnatural med
in aqueous media [57]. Thus, it is surprising that relatively few evolution studies hav
directed toward producing biocatalysts that are tolerant of high concentrations of o
solvents and hydrophobic substrates. Woodley has pointed out that while one would
evolve enzymes for activity and stability in the presence of organic substrates and cos
there is a bias toward performing directed evolution screening in growth media
conditions that are amenable to cell growth and protein production [58]. Growth con
are unlikely to accurately mimic the environment that the biocatalyst would experienc
industrial setting. Separation of enzyme production from fitness of function testing, a:
examples given below, has allowed evolution of activity and stability toward organic

30.6.2   ENANTIOSELECTIVITY AND ENANTIOSPECIFICITY

Enantiopurity, the single most important driver for developing a biocatalytic process, i
one of the most difficult parameters to improve by directed evolution methods due
paucity of HTP methods for determination of enantiomeric purity. Frequently, the en
electivity and enantiospecificity of a wild-type enzyme for an unnatural substrate is l
cannot be improved by reaction engineering. A rational and systematic method to
enantiospecificity and enantioselectivity in enzymes will transform biocatalysis by imj
the ability to adapt enzymes to fit the target substrate and process [59,60]. Directed ev
by DNA shuffling, which does not require extensive screening of libraries appears
itself to complex problems, such as increasing enantioselectivity. A complementary met
limit library size is to use either a protein structural model of the parent enzyme to
regions in the protein, which may have a higher probability of impacting chiral intera
or the use of mutant library generation in which a small sampling of the library g
indication of overall library content.

30.6.3   EXAMPLES OF EVOLVED ENZYMES FOR BIOCATALYSIS APPLICATIONS

30.6.3.1   Amino Acid Oxidase

The dynamic kinetic "deracemization" of chiral amines was enabled by evolving an e
selective amino acid oxidase with broad substrate specificity that is coupled with a nons

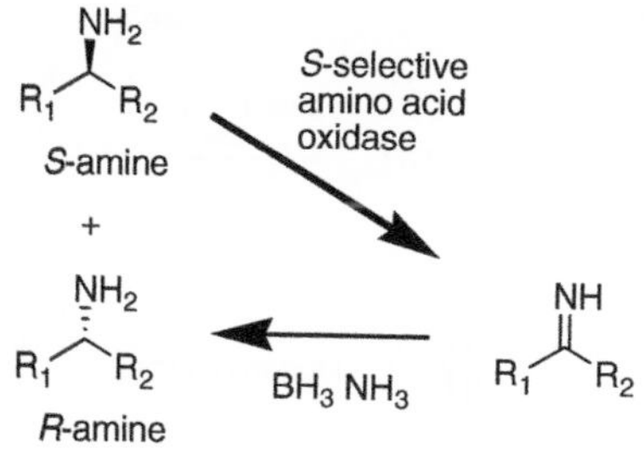

FIGURE 30.8 Amine "deracemization" through S-selective amino acid oxidase (R_1: alkyl, aryl; R_2: methyl).

imine reduction [61,62] (Figure 30.8). The enantioselective oxidase preferentially converts one enantiomer of the amine to an imine, (typically the S-amine), and then the nonspecific chemical reductant converts the imine back to the R,S-amine resulting in enrichment in the unoxidized (R)-amine. The process is run until the R,S-mixture is converted to the single enantiomer in high optical purity. Error-prone PCR was used to generate mutants of an *Aspergillus niger* derived amino acid oxidase. A colorimetric screen based on the generation of hydrogen peroxide enabled evaluation of ~150,000 clones for enantioselectivity in the oxidation of α-benzylamine. A single Asn336Ser mutation imparted a 47-fold increase in activity and a fivefold increase in enantioselectivity on this substrate. Surprisingly, the single mutant showed greatly improved selectivity and activity on a wide range of substrates. The results are unusual when compared with other directed evolution studies in that a single mutation resulted in not only an increase of activity and enantioselectivity, but also a significant broadening of substrate specificity.

30.6.3.2 Hydantoinase

One of the first demonstrations of the plasticity of enzymes for modification of enantiospecificity was the inversion of specificity of a hydantoinase for the production of L-methionine [63]. Inversion of enantiopreference from D to modest L specificity gave rise to increased throughput of a dynamic kinetic resolution system based on a racemase, hydantoinase, and L-carbamoylase combination. As outlined in Figure 30.9, the optical purity of L-methionine is determined by the specificity of the carbomoylase, but increasing the L-preference of the hydantoinase increases throughput by increasing the concentration of the L-carbomoyl methionine precursor rather than the nonproductive D-carbomoyl pathway. Following a now common strategy, an initial round of random mutagenesis was performed to identify sensitive residues, followed by saturation mutagenesis to screen all possible amino acid changes at these sensitive positions. Interestingly on scale-up the inverted selectivity found in high-throughput screening was only evident in the presence of the expression media component present in screening. A single amino acid change I95F was responsible for the inversion of selectivity. Subsequent coexpression of the three-enzyme system: the racemase, the moderately L-selective hydantoinase and the L-carbamoylase led to reaching 90% conversion five times faster than the parent hydantoinase system. The study indicates a number of important features common to directed evolution programs, that is, the sensitivity to screening conditions ("you get what you screen for"), the ability of proteins to make large changes in function with apparently minor amino acid substitutions, and the difficulty in predicting such changes *a priori*.

30.6.3.3 Epoxide Hydrolase

More recently, Janssen et al. evolved an epoxide hydrolase for the hydrolytic kinetic resolution of styrene epoxide (Figure 30.10) [64]. Analogous to the hydantoinase program,

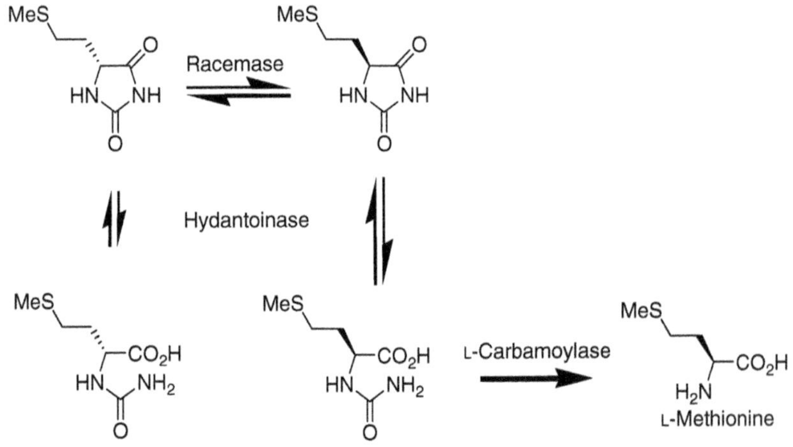

FIGURE 30.9 Dynamic kinetic resolution for the synthesis of L-Methionine.

a first round of screening of 40,000 clones from error-prone PCR with two to
mutations per mutant was used to identify mutants with improved enantiospecifici
these, 15 were selected for gene shuffling [12] to recombine beneficial mutations. A b
of 20,000 clones was screened for activity in an agar plate assay and of these, eight va
were identified with E values ranging from 20 to 44 (vs. 3.4 for the WT) on p-nitrop
glycidyl ether (NPGE). The most selective variants contained three or more muta
including at least one change in an active site tyrosine. Interestingly, kinetic charac
tion showed that the source of improved enantiospecificity on NPGE in the pr
differed for different mutants. Mutations that were responsible for increased spec
over the wild type gave rise to increased differentiation via either changes in the relati
or k_{cat} for the two enantiomers. Screening of the eight clones with high specific
NPGE on other epoxides showed that each had increased selectivity on at least
additional epoxides.

30.6.3.4 Lipase

Reetz et al. [65] demonstrated the utility of a "focused directed evolution approach"
enzyme model by targeting amino acid residues near the active site, which are likely to in
with the substrate to broaden the substrate selectivity of a lipase. Randomization of re
in the substrate-binding pocket of *Pseudomonas aeruginosa* lipase resulted in the accep
of α-substituted carboxylic acid esters by this enzyme. While α-methyl branching gav
rates of reaction with p-nitrophenyl esters with the natural enzyme, aryl substitution
alpha position was tolerated with moderate enantiospecificity in a single and double m
It was the supposition that pairs of amino acid changes would be required; however,
mutations were found to impart the broadest substrate acceptance and moderate sele
($E = 20$–25) on bulky esters that were not accepted by the wild-type lipase. Broadeni

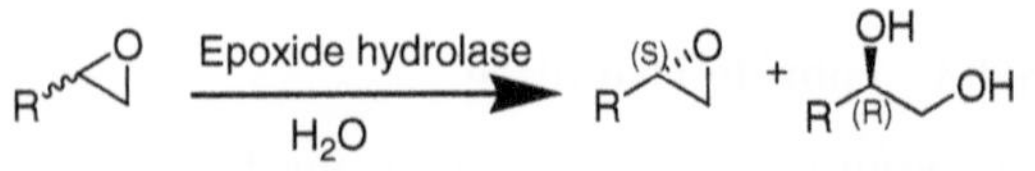

FIGURE 30.10 Hydrolytic resolution of epoxides (R = aryl, chloroethyl, butyl).

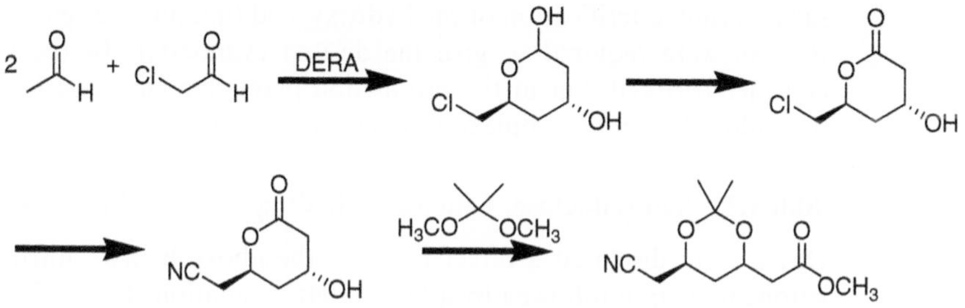

FIGURE 30.11 Potential route to key atorvastatin intermediate through aldolase coupling.

substrate range to bulkier side chains was not simply a result of decreasing steric bulk of residues in the active site, since L17F mutation resulted in some of the widest substrate tolerance.

30.6.3.5 Aldolase

In the evolution of a 2-deoxy-D-ribosephosphate aldolases (DERA) for construction of statin side chains, stability in the presence of high concentrations of the reactive substrates acetaldehyde and chloroacetaldehyde was necessary to make the process viable [66]. The installation of both chiral centers by DERA in an intermediate for the HMG CoA reductase inhibitor atorvastatin (Lipitor) is shown in Figure 30.11. Two rounds of random mutagenesis were performed on the aldolase to select for more stable mutants, followed by recombination of stability and activity hits. A double mutant with tenfold improved yield was identified. The route is attractive in that chirality was installed with excellent atom economy using relatively inexpensive reagents, although Mink notes that the reaction is made complex by the high reactivity of these aldehydes.

30.6.3.6 Nitrilase

Evolution of a biocatalyst for use in a potential industrial process often requires the evolution of several parameters at once. Desantis et al. [67] reported the directed evolution of a nitrilase for the synthesis of a chiral intermediate for atorvastatin from prochiral 3-hydroxyglutaronitrile (Figure 30.12). A screen of 200 nitrilases identified a wild-type enzyme that effectively carried out this reaction at an industrially relevant substrate concentration (3 M). However, the enantiomeric excess of the (R)-4-cyano-3-hydroxybutyric acid product was only 87.6%. Gene site saturation mutagenesis was used to generate libraries of each of the 20 amino acids at every residue of the enzyme. Using HTP mass spectrometry assay with ^{15}N-labeled R-dinitrile, >30,000 clones were evaluated. The libraries were screened for substrate tolerance, productivity, and enantioselectivity and the best variant was found to contain an Ala190His mutation that gave a product with 98.5% ee at 3 M substrate loading and with an overall volumetric productivity of 619 g/L/d. At first glance, this process appears to be extremely simple: a single enzymatic transformation on an inexpensive prochiral dinitrile. However, the process does not give the desired ester but a hydroxy acid. Thus, precipitation of the half-acid (e.g., calcium

FIGURE 30.12 Nitrilase desymmetrization for hydroxynitrile acid.

salt), drying, esterification of the hydroxy acid (minimizing self-condensation) and the
fication were required to give the desired cyanoester. In the laboratory, these st
considered trivial, but in the production plant they give rise to by-products and yi
that add substantial complexity to the overall process.

30.6.3.7 Ketoreductase, Glucose Dehydrogenase, and Halohydrin Dehalogenase

Davis et al. disclosed a direct route to the above hydroxynitrile ester using an en
ketone reduction followed by a biocatalytic cyanation (Figure 30.13) [68,69]. In the fi
of the three-enzyme process, ethyl 4-chloroacetoacetate was reduced by a ketone re
coupled with a glucose dehydrogenase NADPH recycling system to give the chiral
with very high enantiomeric excess. Both the enzymes were evolved for activity and s
in the presence of high concentrations of substrate (20%) and solvent (butyl acetat
evolved enzyme variants catalyzed this reaction with high volumetric productivity (
substrate was converted in 10 h with <1% (w/w) biocatalyst) and the high ee (>
product is recovered after a simple work-up. The second step consists of an en

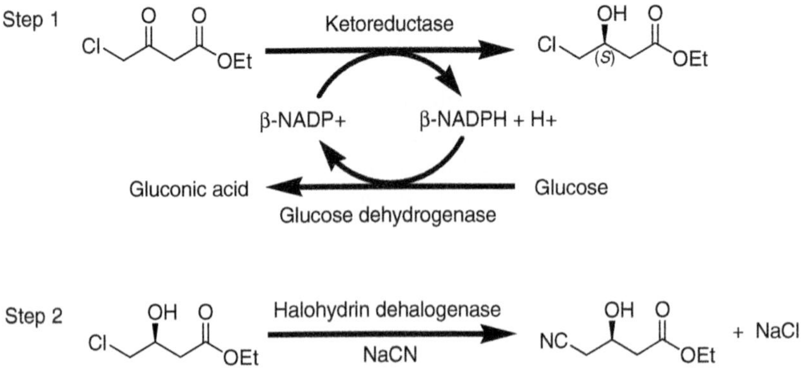

FIGURE 30.13 Three-enzyme process ethyl 3-hydroxybutyrate 4-nitrile.

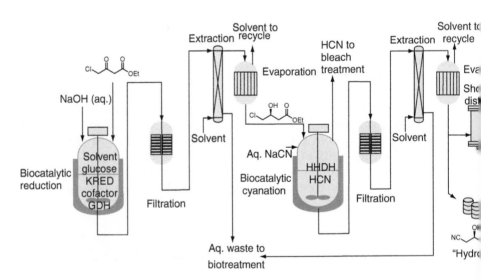

FIGURE 30.14 Process flow diagram for three-enzyme hydroxynitrile process.

cyanation using a halohydrin dehalogenase evolved to efficiently catalyze the displacement of chloride by cyanide. A combination of several directed evolution technologies including error-prone PCR, site directed mutagenesis, focused evolution, NNN randomization, semi-synthetic shuffling, and family shuffling was used to increase the volumetric productivity of the halohydrin dehalogenase more than 4000-fold by mutation of ~15% of the amino acids of the protein, including 4 of the 11 residues indicated to be directly in the active site [70]. The overall process is operationally very simple with direct extraction of the intermediate and product from the reaction(Figure 30.14).

30.7   CONCLUSION AND PROGNOSIS

The confluence of advancements in HTP screening and recombination technologies, and the rapid expansion of available genetic diversity has led to a renaissance in biocatalysis. For decades, the potential of biocatalysis for synthesis of chiral and complex molecules was great, but was largely limited by the inefficiency of these catalysts on unnatural substrates and sensitivity to typical processing conditions in a chemical plant. Examples described here provide proof that directed evolution of enzymes for chemical production is a powerful option for creating highly efficient processes that exceed the performance of processes based on traditional chemical catalysis. The malleability not only of catalytic efficiency and (stereo- and chemo-) selectivity, but of catalytic function as well, indicates that we are just beginning to explore the potential of creating new biocatalytic activities.

REFERENCES

1. Magnusson, L.U., Farewell, A., and Nystrom, T., ppGpp: a global regulator in *Escherichia coli*, *Trends Microbiol.*, 13, 236–242, 2005.
2. Porankiewicz, J., Wang, J., and Clarke, A.K., New insights into the ATP-dependent Clp protease: *Escherichia coli* and beyond, *Mol. Microbiol.*, 32, 449–458, 1999.
3. Bochtler, M., Ditzel, L., Groll, M., et al., The proteasome, *Annu. Rev. Biophys. Biomol. Struct.*, 28, 295–317, 1999.
4. Gonzales, T. and Robert-Baudouy, J., Bacterial aminopeptidases: properties and functions, *FEMS Microbiol. Rev.*, 18, 319–344, 1996.
5. Carrard, G., Bulteau, A.L., Petropoulos, I., et al., Impairment of proteasome structure and function in aging, *Int. J. Biochem. Cell. Biol.*, 34, 1461–1474, 2002.
6. Dice, J.F. and Goldberg, A.L., A statistical analysis of the relationship between degradative rates and molecular weight of proteins, *Arch. Biochem. Biophys.*, 170, 213–319, 1975.
7. Choe, J.Y., Nelson, S.W., Arienti, K.L., et al., Inhibition of fructose-1,6-bisphosphatase by a new class of allosteric effectors, *J. Biol. Chem.*, 278, 51176–51183, 2003.
8. Fenton, A.W., and Reinhart, G.D., Mechanism of substrate inhibition in *Escherichia coli* phospho-fructokinase, *Biochemistry*, 42, 12676–12681, 2003.
9. Chassagnole, C., Noisommit-Rizzi, N., Schmid, J.W., et al., Dynamic modeling of the central carbon metabolism of *Escherichia coli*, *Biotechnol. Bioeng.*, 79, 53–73, 2002.
10. Baik, S.H., Ide, T., Yoshida, H., et al., Significantly enhanced stability of glucose dehydrogenase by directed evolution, *Appl. Microbiol. Biotechnol.*, 61, 329–335, 2003.
11. Ashina, Y. and Suto, M., Development of an enzymatic process for manufacturing acrylamide and recent progress, *Bioprocess Technol.*, 16, 91–107, 1993.
12. Stemmer, W.P., Rapid evolution of a protein *in vitro* by DNA shuffling, *Nature*, 370, 389–391, 1994.
13. Zhang, N., Stewart, B.G., Moore, J.C., et al., Directed evolution of toluene dioxygenase from *Pseudomonas putida* for improved selectivity toward *cis*-indandiol during indene bioconversion, *Metab. Eng.*, 2, 339–348, 2000.
14. Entrez Genome Project. Available at: http://www.ncbi.nlm.nih.gov/genomes/lproks.cgi.

15. Murakami, H., Hohsaka, T., and Sisido, M., Random insertion and deletion of arbitrary nur bases for codon-based random mutation of DNAs, *Nat. Biotechnol.*, 20, 76–81, 2002.

16. Pikkemaat, M.G. and Janssen, D.B., Generating segmental mutations in haloalkane dehalo a novel part in the directed evolution toolbox, *Nucl. Acids Res.*, 30, e3s–5, 2002.

17. Wong, T.S., Tee, K.L., Hauer, B., et al., Sequence saturation mutagenesis (SeSaM): a novel r for directed evolution, *Nucl. Acids Res.*, 32, e26, 2004.

18. Fox, R., Roy, A., Govindarajan, S., et al., Optimizing the search algorithm for protein engi by directed evolution, *Protein Eng.*, 16, 589–597, 2003.

19. Fox, R., Directed molecular evolution by machine learning and the influence of nonlinea actions, *J. Theor. Biol.*, 234, 187–199, 2005.

20. Otten, L.G. and Quax, W.J., Directed evolution: selecting today's biocatalyst, *Biomol. E* 1–9, 2005.

21. Roodveldt, C., Aharoni, A., and Tawfik, D.S., Directed evolution of proteins for heter expression and stability, *Curr. Opin. Struct. Biol.*, 15, 50–56, 2005.

22. Lutz, S. and Patrick, W.M., Novel methods for directed evolution of enzymes: quality, not qu *Curr. Opin. Biotechnol.*, 15, 291–297, 2004.

23. Neylon, C., Chemical and biochemical strategies for the randomization of protein encoding sequences: library construction methods for directed evolution, *Nucl. Acids Res.*, 32, 1448 2004.

24. Stemmer, W.P., DNA shuffling by random fragmentation and reassembly: *in vitro* recomb for molecular evolution, *Proc. Natl. Acad. Sci. USA*, 91, 10747–10751, 1994.

25. Zhao, H. and Arnold, F.H., Optimization of DNA shuffling for high fidelity recombination *Acids Res.*, 25, 1307–1308, 1997.

26. Zhao, H., Giver, L., Shao, Z., et al., Molecular evolution by staggered extension process *in vitro* recombination, *Nat. Biotechnol.*, 16, 258–261, 1998.

27. Volkov, A.A., Shao, Z., and Arnold, F.H., Recombination and chimeragenesis by *in vitro* duplex formation and *in vivo* repair, *Nucl. Acids Res.*, 27, e18, 1999.

28. Melnikov, A. and Youngman, P.J., Random mutagenesis by recombinational capture c products in *Bacillus subtilis* and *Acinetobacter calcoaceticus*, *Nucl. Acids Res.*, 27, 1056 1999.

29. Kim, G.J., Cheon, Y.H., Park, M.S., et al., Generation of protein lineages with new sequence by functional salvage screen, *Protein Eng.*, 14, 647–654, 2001.

30. Ness, J.E., Kim, S., Gottman, A., et al., Synthetic shuffling expands functional protein diver allowing amino acids to recombine independently, *Nat. Biotechnol.*, 20, 1251–1255, 2002.

31. Miyazaki, K., Random DNA fragmentation with endonuclease V: application to DNA sh *Nucl. Acids Res.*, 30(24), e139, 2002.

32. Song, J.K., Chung, B., Oh, Y.H., et al., Construction of DNA-shuffled and incrementally tru libraries by a mutagenic and unidirectional reassembly method: changing from a substrate city of phospholipase to that of lipase, *Appl. Environ. Microbiol.*, 68, 6146–6151, 2002.

33. Hayes, R.J., Bentzien, J., Ary, M.L., et al., Combining computational and experimental sc for rapid optimization of protein properties, *Proc. Natl. Acad. Sci. USA*, 99, 15926–15931

34. Richardson, T.H., Tan, X., Frey, G., et al., A novel, high performance enzyme for starch li tion. Discovery and optimization of a low pH, thermostable alpha-amylase, *J. Biol. Chem* 26501–26507, 2002.

35. Zha, D., Eipper, A., and Reetz, M.T., Assembly of designed oligonucleotides as an efficient r for gene recombination: a new tool in directed evolution, *Chembiochem*, 4, 34–39, 2003.

36. Hamamatsu, N., Aita, T., Nomiya, Y., et al., Biased mutation-assembling: an efficient met rapid directed evolution through simultaneous mutation accumulation, *Protein Eng. Des. S* 265–271, 2005.

37. Muller, K.M., Stebel, S.C., Knall, S., et al., Nucleotide exchange and excision technology (DNA shuffling: a robust method for DNA fragmentation and directed evolution, *Nucl. Aci* 33, e117, 2005.

38. Eggert, T., Funke, S.A., Rao, N.M., et al., Multiplex-PCR-based recombination as a novel high method for directed evolution, *Chembiochem*, 6, 1062–1067, 2005.

39. Rozak, D.A. and Bryan, P.N., Offset recombinant PCR: a simple but effective method for shuffling compact heterologous domains, *Nucl. Acids Res.*, 33(9), e82, 2005.

40. Crameri, A., Raillard, S.A., Bermudez, E., et al., DNA shuffling of a family of genes from diverse species accelerates directed evolution, *Nature.*, 391, 288–291, 1998.

41. Kikuchi, M., Ohnishi, K., and Harayama, S., An effective family shuffling method using single-stranded DNA, *Gene*, 243, 133–137, 2000.

42. Abecassis, V., Pompon, D., and Truan, G., High efficiency family shuffling based on multi-step PCR and *in vivo* DNA recombination in yeast: statistical and functional analysis of a combinatorial library between human cytochrome P450 1A1 and 1A2, *Nucl. Acids Res.*, 28, E88, 2000.

43. Gibbs, M.D., Nevalainen, K.M., and Bergquist, P.L., Degenerate oligonucleotide gene shuffling (DOGS): a method for enhancing the frequency of recombination with family shuffling, *Gene*, 271, 13–20, 2001.

44. Coco, W.M., Levinson, W.E., Crist, M.J., et al., DNA shuffling method for generating highly recombined genes and evolved enzymes, *Nat. Biotechnol.*, 19, 354–359, 2001.

45. Coco, W.M., Encell, L.P., Levinson, W.E., et al., Growth factor engineering by degenerate homo-duplex gene family recombination, *Nat. Biotechnol.*, 20, 1246–1250, 2002.

46. Ostermeier, M., Shim, J.H., and Benkovic, S.J., A combinatorial approach to hybrid enzymes independent of DNA homology, *Nat. Biotechnol.*, 17, 1205–1209, 1999.

47. Lutz, S., Ostermeier, M., Moore, G.L., et al., Creating multiple-crossover DNA libraries independent of sequence identity, *Proc. Natl. Acad. Sci. USA*, 98, 11248–11253, 2001.

48. Sieber, V., Martinez, C.A., and Arnold, F.H., Libraries of hybrid proteins from distantly related sequences, *Nat. Biotechnol.*, 19, 456–460, 2001.

49. O'Maille, P.E., Bakhtina, M., and Tsai, M.D., Structure-based combinatorial protein engineering, *J. Mol. Biol.*, 321, 677–691, 2002.

50. Kawarasaki, Y., Griswold, K.E., Stevenson, J.D., et al., Enhanced crossover SCRATCHY: construction and high-throughput screening of a combinatorial library containing multiple non-homologous crossovers, *Nucl. Acids Res.*, 31, e126, 2003.

51. Voight, C.A., Martinez, C., Wang, Z.G., et al., Protein building blocks preserved by recombination, *Nat. Struct. Biol.*, 9, 553–558, 2002.

52. Hiraga, K. and Arnold, F.H., General method for sequence-independent site-directed chimeragenesis, *J. Mol. Biol.*, 330, 287–296, 2003.

53. Crameri, A., Dawes, G., Rodriguez, E., Jr., et al., Molecular evolution of an arsenate detoxification pathway by DNA shuffling, *Nat. Biotechnol.*, 15, 436–438, 1997.

54. Zhang, Y.X., Perry, K., Vinci, V.A., et al., Genome shuffling leads to rapid phenotypic improvement in bacteria, *Nature*, 415, 644–646, 2002.

55. Bommarius, A. and Bommarius-Riebel, B., *Biocatalysis: Fundamentals and Applications,* John Wiley, 2005.

56. Halling, P. and Kvittingen, L., Why did biocatalysis in organic media not take off in the 1930s? *Trends Biotechnol.*, 17, 334–344, 1999.

57. Zaks, A. and Klibanov, A., Enzymatic catalysis in organic media at 100 degrees C, *Science*, 224, 1249–1251, 1994.

58. Lye, G.J., Dalby, P.A., and Woodley, J.M., Better biocatalytic processes faster: new tools for the implementation of biocatalysis in organic synthesis, *Org. Proc. Res. Dev.*, 6, 434–440, 2002.

59. Reetz, M., Changing the enantioselectivity of enzymes by directed evolution, *Methods Enzymol.*, 388, 238–256, 2004.

60. Turner, N.J., Directed evolution of enzymes for applied biocatalysis, *Trends Biotechnol.*, 21, 474–478, 2003.

61. Alexeeva, M., Carr, R., and Turner, N.J., Directed evolution of enzymes: new biocatalysts for asymmetric synthesis, *Org. Biomol. Chem.*, 1, 4133–4137, 2003.

62. Carr, R., Alexeeva, M., Enright, A., et al., Directed evolution of an amine oxidase possessing both broad substrate specificity and high enantioselectivity, *Angew. Chem. Intl. Ed. Eng.*, 42, 4807–4810, 2003.

63. May, O., Nguyen, P.T., and Arnold, F.H., Inverting enantioselectivity by directed evolution of hydantoinase for improved production of L-methionine, *Nat. Biotechnol.*, 18, 317–320, 2000.

64. Van Loo, B., Spelberg, J.H., Kingma, J., et al., Directed evolution of epoxide hydrola A. *radiobacter* toward higher enantioselectivity by error-prone PCR and DNA shuffling *Biol.*, 11, 981–990, 2004.
65. Reetz, M.T., Bocola, M., Carballeira, J.D., et al., Expanding the range of substrate accep enzymes: combinatorial active-site saturation test, *Angew. Chem. Int. Edn.*, 44, 4192–4196,
66. Mink, D., Industrial enzymatic C–C bond formation, *Biotrans*, Delft, July 3–8, 2005.
67. Desantis, G., Wong, K., Farwell, B., et al., Creation of a productive, highly enantioselective through gene site saturation mutagenesis (GSSM), *J. Am. Chem. Soc.*, 125, 11476–11477,
68. Davis, C., Grate, J., Gray, D., et al., Enzymatic process for the production of 4-sub 3-hydroxybutyric acid derivatives, Codexis, Inc., WO04015132, 2005.
69. Davis, C., Grate, J., Gray, D., et al., Enzymatic process for the production of 4-sub 3-hydroxybutyric acid derivatives and vicinal cyano, hydroxyl substituted carboxylic aci Codexis, Inc., WO05018579, 2005.
70. Grate, J., A Green and Economic Biocatalytic Process for the Key Chiral Intermed Atorvastatin. Using Three Biocatalysts Evolved to Enable the Process, ACS Green Cl Conference, Washington, June 24, 2005.

31 Biocatalytic Routes to Nonracemic Chiral Amines

Nicholas J. Turner and Reuben Carr

CONTENTS

31.1 INTRODUCTION

Nonracemic chiral amines are an important, but not easily prepared, class of organic molecules. In addition to their value as building blocks for the preparation of pharmaceutical and agrochemical end products (Figure 31.1) [1], they are increasingly used as ligands for asymmetric catalysis [2] and resolving agents for crystallization. As with the manufacture on scale of many fine chemicals, the current trend is to develop catalytic processes for their production, thereby eliminating the need for the stoichiometric use of reagents that is both expensive and environmentally unattractive. Both chemo- and biocatalytic approaches to the preparation of nonracemic chiral amines are currently being explored, with the former relying to a large extent upon technologies based on the asymmetric hydrogenation of imines [3,4]. In terms of biocatalysis there are essentially three distinct routes by which optically active amines can be prepared: (i) kinetic resolutions of the corresponding racemates, including dynamic kinetic resolution processes; (ii) asymmetric transamination of ketone precursors; and (iii) deracemization reactions. The aim of this chapter is to highlight the various approaches that are being investigated and provide some examples of different routes by which optically active amines can be prepared. The application of some of these routes for the preparation of chiral amines at industrial scale is also discussed with relevant examples [5].

31.2 KINETIC RESOLUTION

While the use of hydrolytic enzymes (e.g., lipases, esterases, proteases) for the kinetic resolution of racemic chiral alcohols and carboxylic acids is well established, the corresponding transformation on racemic amines is less well studied, although it is now starting to receive considerably more attention [6,7]. By far the most commonly used hydrolase for this purpose is *Candida antarctica* lipase B (CAL-B) and, in addition, most of the reported examples involve acylation of racemic chiral amines under low water conditions. One issue that needs

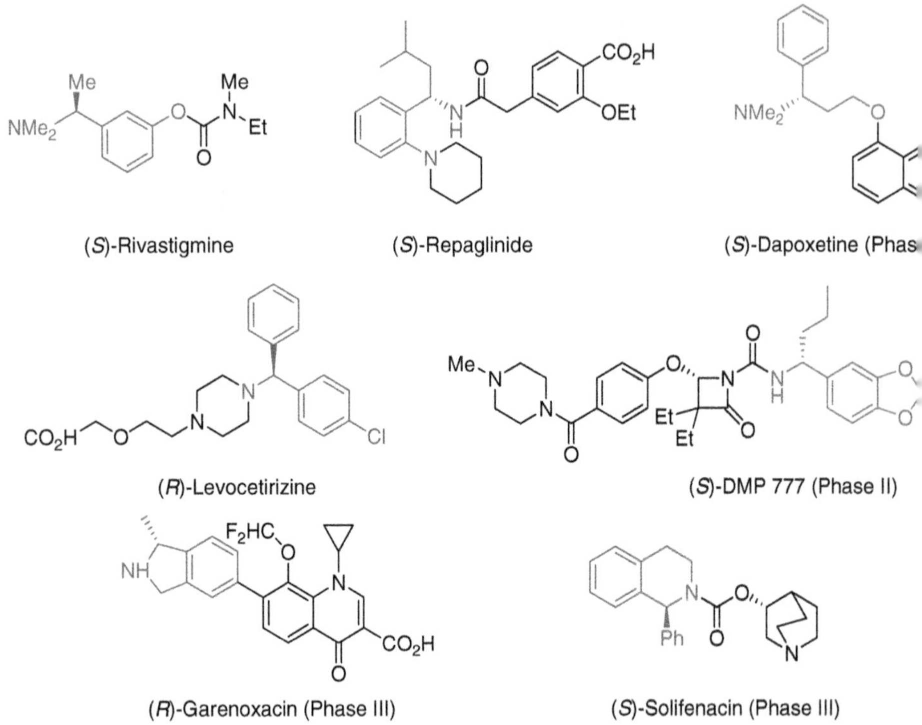

FIGURE 31.1 Pharmaceutical drugs containing chiral amine building blocks (grey).

to be addressed is that amines are more nucleophilic than alcohols and hence are more
to undergo background reaction with the acyl donor, leading to lower overall enantic
excesses (ees) for the product. The first example of this type of reaction was describ
Klibanov et al. in 1989, in which they screened a number of hydrolytic enzymes fo:
ability to catalyze the acylation of racemic α-methyl benzylamine and other chiral a
using trifluoroethyl butyrate as the acylating agent [8]. Interestingly they found that the
a strong dependence of the enantioselectivity on the solvent, with 3-methyl-3-pentanol
the highest selectivity. Using subtilisin as the catalyst they reported ees up to 99%.

Thereafter, several other groups began to examine the scope of this reaction an
investigate various enzyme/solvent/acyl donor combinations in order to optimize conve
and enantioselectivities. Schneider et al. [9] employed CAL-B for the resolution of v
aryl alkynyl amine derivatives **1**, obtaining very high selectivities ($E > 100$) in a num
cases to yield the corresponding (R)-amine **1** and (S)-amide **2**. The chiral (R)-amine pr
were subsequently used as building blocks for the preparation of a range of anti
aromatase inhibitors **3** (Figure 31.2).

Likewise, in another pharmaceutical application, researchers at Pfizer Inc. in San
demonstrated that CAL-B could be used for the resolution of a simple chiral building bl
which was used as an intermediate for the synthesis of cyclin-dependent kinase (C
inhibitors (Figure 31.3) [10].

1,1'-Binaphthylamine derivatives are useful chiral ligands for various asymmetric
tions: 2,2'-diamino-1,1'-binaphthyl has been used as the starting material for the synth
BINAP, which is a chiral ligand used in the asymmetric hydrogenation of ketones [11,12
lipase-catalyzed acylation of amine **5** was recently reported using LIP-300 (*Pseudo*
aeruginosa lipase immobilized on Hyflo Super-Cel) and LPL-311 (*P. aeruginosa* lipa

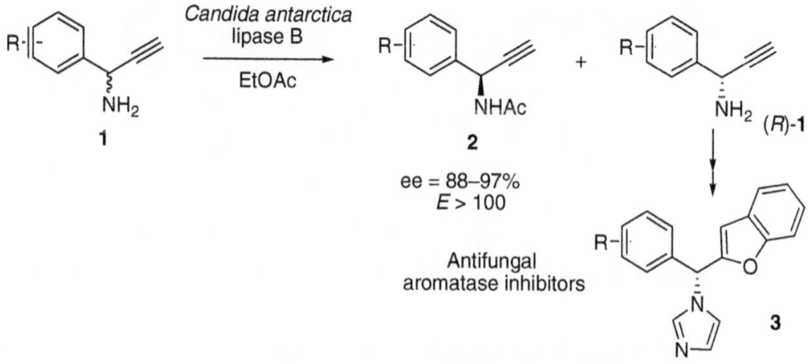

FIGURE 31.2 Resolution of amine building blocks for aromatase inhibitors.

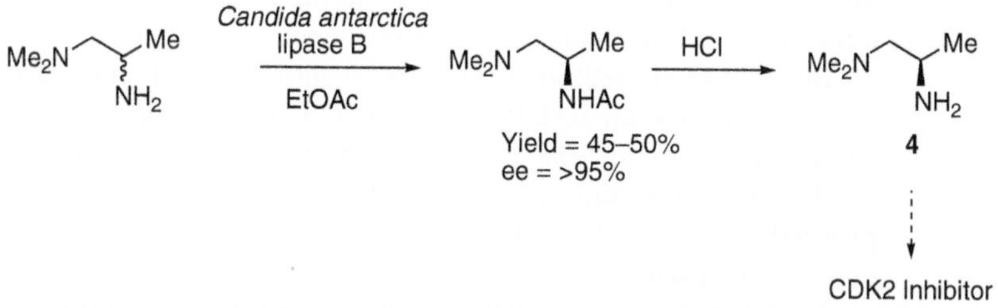

FIGURE 31.3 Chiral amine building blocks for CDK2 inhibitors obtained by CAL-B catalyzed kinetic resolution.

Toyonite 200-M) for the kinetic resolution of 1,1′-binaphthylamine derivatives (Figure 31.4). The amidation reaction was sensitive to the length of the alkyl chain between the binaphthyl ring and amino group, and improved enantioselectivities were observed for increasing alkyl chain length of the (R)-substituent on the acyl donor.

In 1993, BASF reported the results of an extensive screening of lipase-catalyzed acylations of amines, particularly with respect to the nature of the acyl donor. It was found that racemic α-methyl benzylamine 6 could be acylated with *Burkholderia plantarii* lipase with esters of methoxyacetic acid 7, giving optically pure (S)-amine 6 and (R)-amide 8 [13] (Figure 31.5). After separation by distillation, the (R)-amine was released by basic hydrolysis of the amide without racemization, in quantitative yield. The attractive feature of this process was the broad substrate tolerance of the catalyst and a wide variety of amines have been resolved, in some cases on a multiton scale [14]; recycling of the undesired enantiomer by racemization and recovery of the acylating agent help to lower the cost of the industrial process.

Optically active amines can also be accessed by exploiting alternative hydrolytic enzymes such as amidases and acylases. Compared to lipases, however, these enzymes are somewhat restricted in terms of commercial availability but many are found in nature and can be isolated as microbial cultures. Growing microorganisms in simple growth media containing an amide as the sole carbon and/or nitrogen source for growth selects active microorganisms over the remainder of the population for amidase. Following successive subculturing, the microbial population becomes enriched in microorganisms possessing the desired amidase activity. Using this approach with N-acetyl-2-butylamine as the amide substrate, researchers at Avecia Ltd. isolated ~60 microorganisms and found that *Arthrobacter* sp. predominated [15].

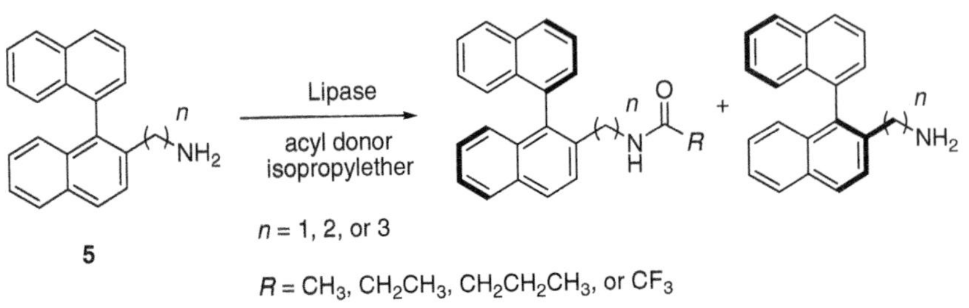

FIGURE 31.4 Lipase-catalyzed amidation of 1,1'-binaphthylamine derivatives 5.

A freeze-dried microbial sample (BH2-N1) was shown to exhibit (S)-selective am
activity in the hydrolysis of racemic amide substrates. For example, (1S,2R)-N-ace
aminoindanol 9, a key intermediate in the synthesis of the HIV protease inhibitor Indi
11 (Crixivan, Merck), was resolved by hydrolysis of the corresponding amide 10 usin
amidase approach (Figure 31.6). This amino alcohol has also found widespread applicati
a useful ligand in asymmetric catalysis using various transition metals [16].

Sheldon et al. have also exploited acylases in the kinetic resolution of racemic amir
this case through an acylation approach. Thus, they demonstrated the use of aminoacy
from *Aspergillus melleus* together with methyl methoxyacetate as the acyl donor fc
resolution of 1-amino indane 12 as shown in Figure 31.7 [17]. The reaction was fou
proceed with modest selectivity ($E = 9.3$), resulting in an ee of amine (S)-12 of 72% a
conversion (Figure 31.7).

Despite the general approaches described for the resolution of primary amines, there
few reported methods for the lipase-mediated enantioselective acylation of secondary a
Moreover, the resolutions that are reported tend to suffer from poor yields and/or
enantioselectivities. The cyclic secondary amine 1-methyltetrahydroisoquinoline 13 (M'
a building block for YH1885 14, a potential treatment for gastroesophageal reflux c

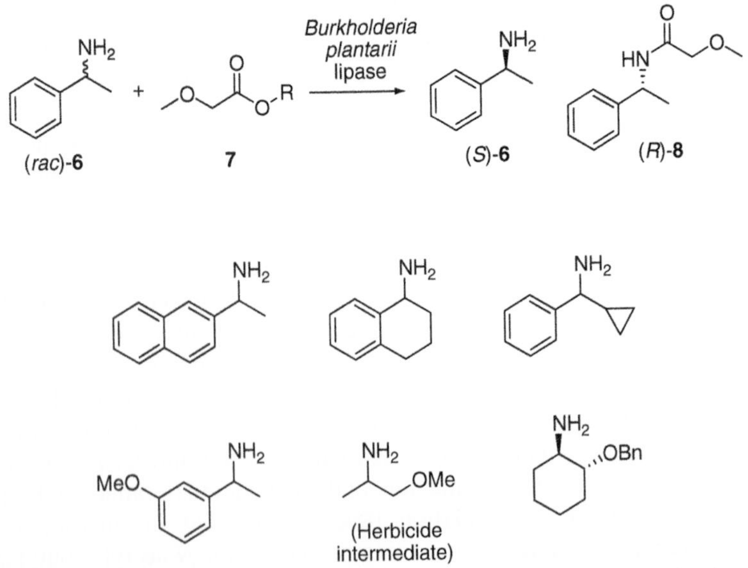

FIGURE 31.5 Lipase-catalyzed resolution of racemic 1-phenylethylamine 6 and other chiral
using methoxyacetic acid esters.

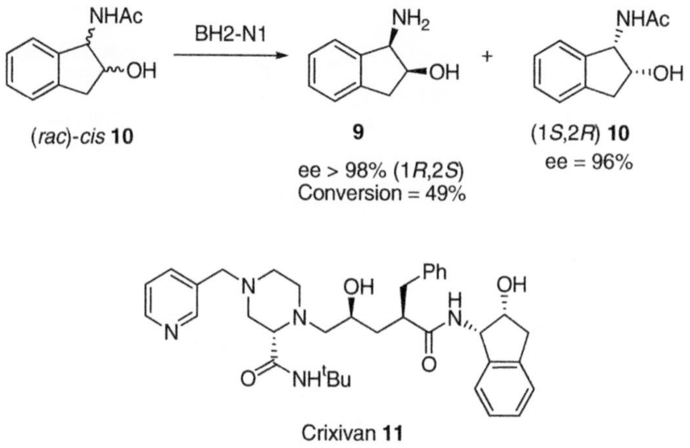

FIGURE 31.6 Kinetic resolution of *N*-acetyl-1-aminoindanol by BH2-N1 amidase. The (1*S*,2*R*)-enantiomer of 1-aminoindanol (shown in grey) is a key intermediate in Indinavir-**11** (Crixivan, Merck).

(GERD) and duodenal ulcers [18]. Researchers at GlaxoSmithKline required access to both enantiomers for the development of the drug. In this example, the kinetic resolution was achieved using substituted phenyl allyl carbonates **15** as the acyl donors. Enantioselective acylation of (*rac*)-MTQ **13** with *C. rugosa* lipase yielded (*S*)-**13** and the corresponding allylcarbamate derivative **16** (Figure 31.8) [19].

31.3   DYNAMIC KINETIC RESOLUTION

Enzyme-catalyzed dynamic kinetic resolution (DKR) reactions, in which the unreacted enantiomer is racemized *in situ*, have been successfully applied in the arena of chiral α-amino acids, secondary alcohols, and carboxylic acids [20]. The DKR of chiral amines has proved more problematic and challenging, principally because chiral amines require much higher temperature for racemization, leading to conditions that are generally incompatible with most hydrolases. The catalytic cycle for a DKR is shown in Figure 31.9, in which it is essential to identify methods for selective racemization of the substrate but not of the product.

The first reported example of DKR of an amine derives from Reetz and Schimossek in which they employed palladium as a catalyst for the racemization of α-methyl benzylamine in the presence of CAL-B [21]. The reaction was carried out in a solution of pyridine, with ethyl acetate as the acyl donor, at a temperature of 50 to 55°C. In order to achieve total conversion

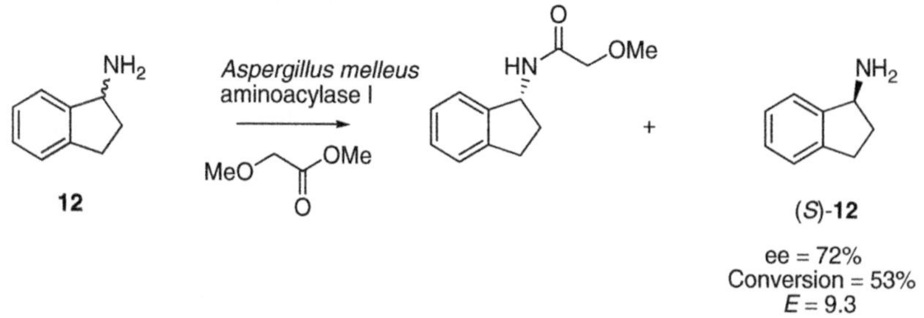

FIGURE 31.7 Resolution of rac-1-aminoindane **12** to (*S*)-1-aminoindane.

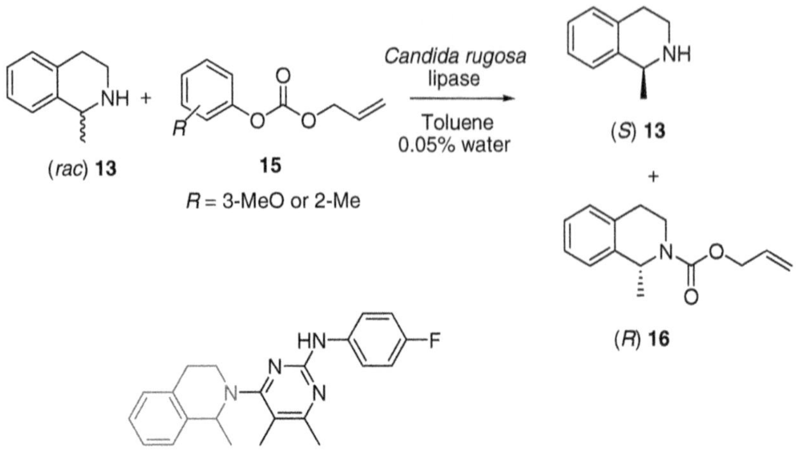

FIGURE 31.8 Lipase-catalyzed kinetic resolution of MTQ **13** with a phenylallyl carbonate acyl **15**. MTQ is a building block (shown in grey) in the drug candidate YH1885 **14**.

to the *N*-acylated product it was necessary to leave the reaction for 5 d giving the (*R*)-am 64% yield and 99% ee. Kim et al. [22] extended this approach by using the *in situ* reduct the corresponding oxime **17** to the racemic amine, thereby avoiding the presence of concentrations of the amine at the beginning of the reaction (Figure 31.10).

Backvall has recently taken this area off into a new direction by examining the ruthenium-based catalysts for the racemization of amines (Figure 31.11). Such catalyst previously been shown to be effective for the DKR of secondary alcohols in combination CAL-B and subtilisin as biocatalysts. However, as discussed above, the rates of racemi with chiral amines are substantially reduced; hence, the reactions require higher tempera

In fact, the initial work involving Shvo's catalyst **18** required a reaction temperat 110°C, which precluded the *in situ* use of the CAL-B [23]. However, very recently Backva shown that catalyst **19**, in which the phenyl substituents on the cyclopentadiene rir replaced by *para*-methoxyphenyl groups, is able to catalyze efficient amine racemizatio lower temperature (90°C). At this lower temperature, it is now possible to combir biocatalyst (CAL-B) with the transition metal catalyst to effect a true DKR of amines

31.4 ASYMMETRIC PROCESSES

Transaminases have also been used in the production of optically active amines. I process, a carbonyl compound such as a ketone **20** or an α-keto acid is converted di into an amine (Figure 31.12). The enantioselective transfer is dependent on the transan

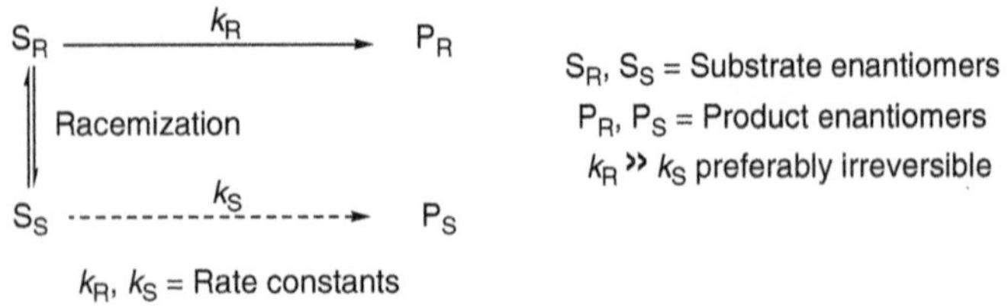

FIGURE 31.9 Reaction scheme for a DKR.

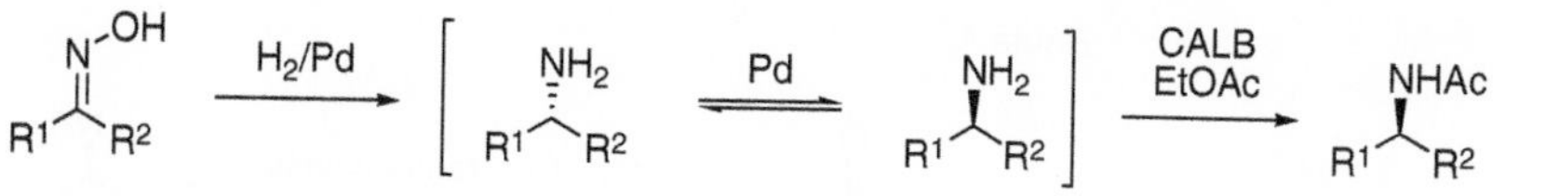

FIGURE 31.10 DKR of racemic chiral amines using palladium.

cofactor pyridoxal phosphate **23**. Celgene has developed both (S)- and (R)-selective transaminases, enabling both enantiomers to be accessed [25]. The reaction can be carried out either in the synthesis mode (route A) or as a kinetic resolution (route B) (Figure 31.12). In the synthetic procedure isopropyl amine **21** provides an amino group, which is transferred to a prochiral ketone enantioselectively to give the optically active amine product plus acetone **22**. The kinetic resolution is essentially the synthesis in reverse, where the racemic amine substrate (e.g., **6**) is enantioselectively converted to the ketone. The amino group is transferred to a low–molecular weight aldehyde such as propionaldehyde **24**. If α-keto acids are used in place of ketones, amino acids are obtained.

Stirling et al., [26] found that the wild-type enzymes possessed low activity and enantioselectivity but it was able to improve these properties by successive rounds of random mutagenesis and screening (directed evolution) such that the ee could be improved to >99%. It also employed directed evolution to improve the tolerance of the transaminase enzymes to high amine concentration, leading to improvement in the process development for large-scale production of amines. For example, it reported pilot-scale production of amines up to multi-100 kg scale and also fermentation of *Escherichia coli* cells expressing the enzymes from 20 L to several m³ capacity. An example of the range of chiral amines accessible with this technology is shown in Figure 31.13.

Kim et al. [27,28] have recently reported. the use of an ω-transaminase from *Vibrio fluvialis* as shown in Figure 31.14. In order to shift the equilibrium toward formation of the product

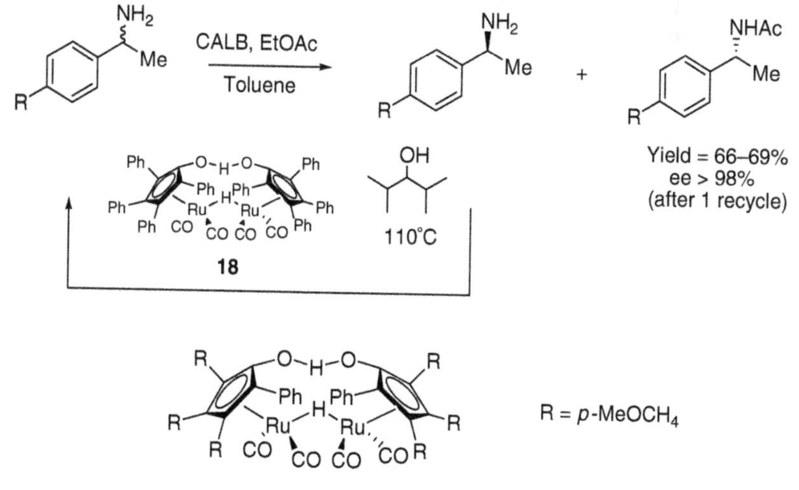

FIGURE 31.11 DKR of α-methyl benzylamine using Shvo's catalyst **18** and analog **19**.

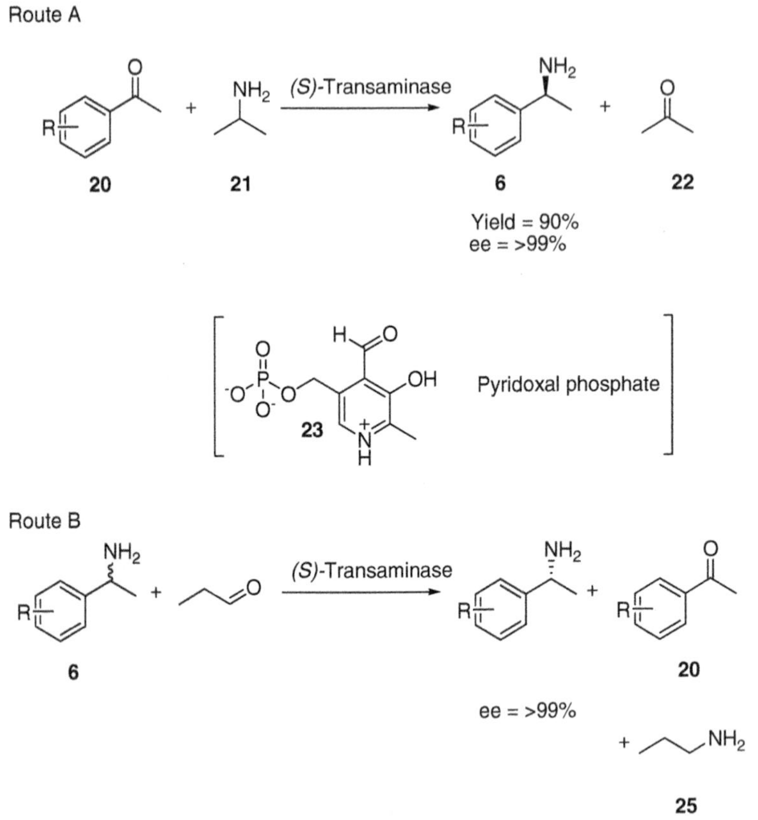

Route A

Yield = 90%
ee = >99%

Pyridoxal phosphate

Route B

ee = >99%

FIGURE 31.12 Route A is the synthesis procedure for the transamination of a prochiral ketone 2 an (*S*)-transaminase and isopropylamine. Route B is the kinetic resolution procedure of a (rac)- to the (*R*)-amine **6** (Celgene).

amine they employed excess L-alanine as the amino donor. However, they also found t reaction was inhibited by the formation of pyruvate. This problem could be overc the addition of lactate dehydrogenase, which converted the pyruvate to lactate. By using cells and a onefold excess of alanine they were able to obtain the product amine **6** in ~90 in less than 24 h.

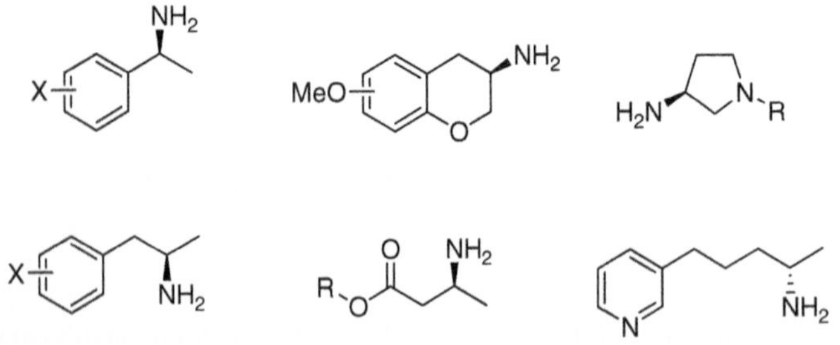

FIGURE 31.13 Chiral amines accessible with transaminase technology.

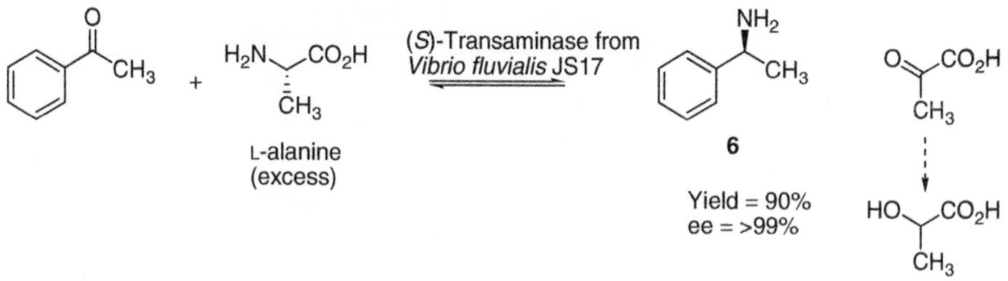

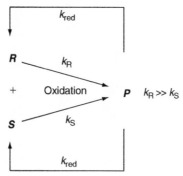

FIGURE 31.14 Transaminase from *Vibrio fluvialis*.

31.5 DERACEMIZATION

Deracemization is a process during which a racemate is converted into a nonracemic product in 100% theoretical yield without intermediate separation of materials. According to this definition, DKR, dynamic thermodynamic resolution, stereoinversion, and enantioconvergent transformations of a racemate are all classified as deracemization processes [29]. In principle, a cyclic oxidation and reduction sequence as shown in Figure 31.15 can lead to the deracemization of chiral alcohols and amines.

This process relies upon the enantioselective oxidation of the substrate ($k_R \gg k_S$) such that one enantiomer of the starting racemate is oxidized to the achiral intermediate, either a ketone or imine. The oxidized product is then converted back to the alcohol or amine by a chemical reduction in a nonselective manner.

The first example of such a process originates from Hafner and Wellner, who reported the generation of L-alanine and L-leucine from the corresponding D-enantiomers by the use of porcine kidney D-amino acid oxidase and sodium borohydride [30]. Subsequently, Soda et al. [31,32] extended this method for the deracemization of DL-proline and DL-pipecolic acid also using D-amino acid oxidase and sodium borohydride (Figure 31.16).

It is interesting to note that even after just four cycles of oxidation and reduction, the ee of L-amino acid is >93% from a starting racemate assuming a completely enantioselective oxidation and nonselective chemical reduction. By seven cycles the ee has risen to >99% (Figure 31.17).

Beard and Turner [33] have recently considerably expanded the scope and application of the deracemization of amino acids including the discovery of novel reducing agents such as sodium cyanoborohydride, amine-borane complexes, and catalytic transfer hydrogenation [34] that can be used in place of sodium borohydride. The use of catalytic transfer hydrogenation has also shown to be effective as the reducing agent in the deracemization of amino

FIGURE 31.15 Deracemization by a cyclic oxidation and reduction sequence.

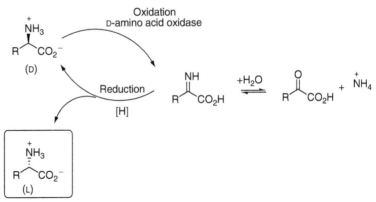

FIGURE 31.16 Deracemization of DL-amino acids to D-amino acids using D-amino acid c combined with a chemical reducing agent.

acids with L-amino acid oxidase giving D-amino acids [34]. Extension o deracemization approach of α-amino acids has been used in the stereoinversion of (γ-substituted α-amino acids using the chemo-enzymatic oxidation and reduction pro [35,36].

In a further recent development, the concept of deracemization by an oxidatio reduction sequence has been applied to chiral amines. In a key proof-of-principle exper Turner et al. [37] were able to demonstrate successfully the deracemization of α-» benzylamine **6** in 77% yield and 93% ee. The amine oxidase required for this proce obtained from *A. niger*. However, the wild-type enzyme had very narrow substrate spec typically reacting with simple achiral amines such as amyl amine, butyl amine, and amine. In order to improve the activity of the wild-type enzyme toward chiral amines suc methyl benzylamine **6**, the amine oxidase was subjected to an initial round of directed evo using a high-throughput colorimetric screen on solid phase. A single mutant (Asn336Se identified with considerably enhanced (~50-fold) activity toward α-methyl benzylamine. estingly this mutant was also found to have a much broader substrate specificity than the and catalyzed the oxidation of a wide range of structurally different chiral amines (Figure [38]. In all cases examined, the enzyme was found to have high (*S*)-selectivity. Subseq»

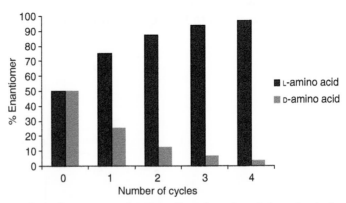

FIGURE 31.17 Enantiomeric excess as a function of number of catalytic cycles during the derac tion process.

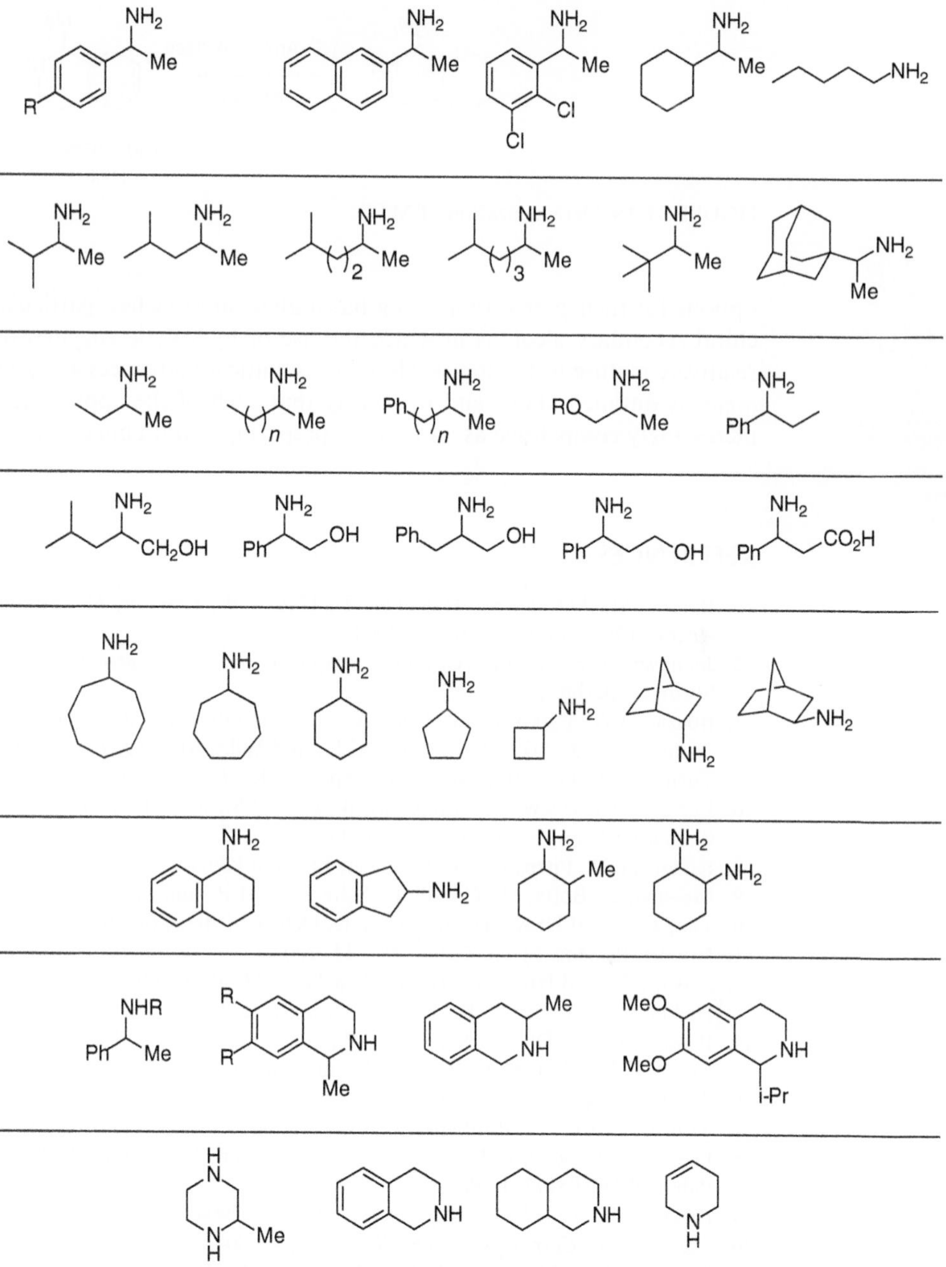

FIGURE 31.18 Chiral amines that are substrates for double mutants of amine oxidase.

a second mutation was also reported (Ile246Met) which further enhanced the activity of the amine oxidase toward chiral cyclic secondary amines as exemplified by the preparative deracemization of MTQ (Figure 31.19) [39].

31.6 CONCLUSIONS

Chiral, nonracemic amines are increasingly in demand, particularly as building blocks for the preparation of pharmaceutical and agrochemical products. Currently, there are relatively few

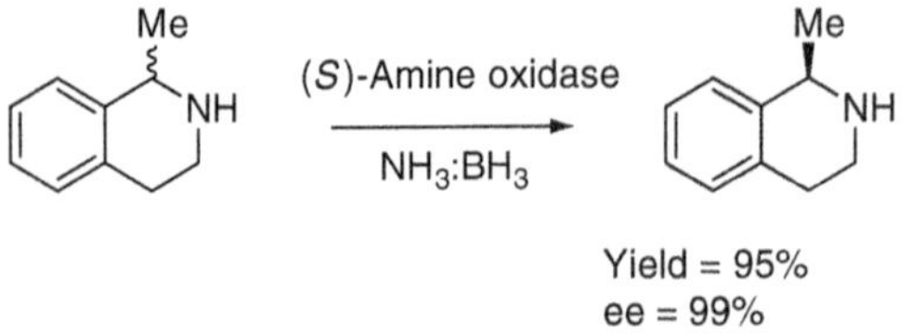

FIGURE 31.19 Deracemization of MTQ.

options for their preparation using biocatalytic approaches, particularly in comparisc
chiral secondary alcohols in which the use of hydrolytic enzymes or dehydrogenase
relatively mature technologies. However, significant advances are being made in a nur
areas as outlined above and it is likely that each of the approaches described will I
increasingly competitive as a means of preparing chiral amines on scale.

REFERENCES

1. Breuer, M., Ditrich, K., Habicher, T., Hauer, B., Kesseler, M., Stuermer, R., and Zelir
   Angew. Chem. Int. Ed., 43, 788, 2004.
2. Jacobsen, E.N., Pfaltz, A., and Yamamoto, H., in *Comprehensive Asymmetric Catalyst*
   Springer, Berlin, 1999.
3. Boezio, A.A., Pytkowicz, J., Côté, A., and Charette, A.B., *J. Am. Chem. Soc.*, 125, 14260,
4. Williams, G.D., Pike, R.A., Wade, C.E., and Wills, M., *Org. Lett.*, 5, 4227, 2003.
5. Turner, N.J., Fotheringham, I., and Speight, R., *Innov. Pharm. Technol.*, 4, 114, 2004.
6. For a recent review, see: van Rantwijk, F., and Sheldon, R.A., *Tetrahedron*, 60, 501, 2004.
7. García-Urdiales, E., Alfonso, I., and Gotor, V., *Chem Rev.*, 105, 313, 2005.
8. Kitaguchi, H., Fitzpatrick, P.A., Huber, J.E., and Klibanov, A.M., *J. Am. Chem. Soc.*, 111, 309
9. Messina, F., Botta, M., Corelli, F., Schneider, M.P., and Fazio, F.J., *J. Org. Chem.*, 64, 376
10. Tao, J., Chiral USA, Scientific Update, October 2003, Chicago.
11. Noyori, R., *Adv. Synth. Catal.*, 345, 15, 2003.
12. Aoyagi, N. and Izumi, T., *Tetrahedron Lett.*, 43, 5529, 2002.
13. Hieber, G. and Ditrich, K., *Chimica Oggi*, 16, 2001.
14. Balkenhohl, F., Ditrich, K., Hauer, B., and Ladner, W., *J. Prakt. Chem.*, 339, 381, 1997.
15. Reeve, C.D., Holt, R.A., Rigby, S.R., and Hazel, K., *Chimica Oggi*, 19, 31, 2001.
16. Senanayke, C.H., *Aldrichimica Acta*, 31, 3, 1998.
17. Youshko, M.I., van Rantwijk, F., and Sheldon, R.A., *Tetrahedron Asymmetry*, 12, 3267, 2
18. Lee, J.W., Chae, J.S., Kim, C.S., Kim, J.K., Lim, D.S., Shon, M.K., Choi, Y.S. and Le
   Patent WO9605177, 1996.
19. Breen, G.F., *Tetrahedron Asymmetry*, 15, 1427, 2004.
20. Turner, N.J., *Curr. Opin. Chem. Biol.*, 8, 114, 2004.
21. Reetz, M.T. and Schimossek, K., *Chimia*, 50, 668, 1996.
22. Choi, Y.K., Kim, M.-J., Ahn, Y., and Kim, M.J., *Org. Lett.*, 3, 4099, 2001.
23. Pamies, O., Ell, A.H., Samec, J.S.M., Hermanns, N., and Bäckvall, J.E., *Tetrahedron L*
   4699, 2002.
24. Paetzold, J. and Bäckvall, J.E., *J. Am. Chem. Soc.*, 127, 17620, 2005.
25. Stewart, J.D., *Curr. Opin. Chem. Biol.*, 5, 120, 2001.
26. Stirling, D.I., Zeitlin, A., and Matcham, G.W., 1990, US Patent 4,950,606; see also www.ca
   com.
27. Hwang, B.-Y., Cho, B.-K., Yun, H., Koteshwar, K., and Kim, B.-G., *J. Mol. Catal. B Enz*
   47, 2005.
28. Shin, J.-S. and Kim, B.-G., *Biotechnol. Bioeng.*, 65, 206, 1999.

29. Faber, K., *Chem. Eur. J.*, 7, 5005, 2001.
30. Hafner, E.W. and Wellner, D., *Proc. Nat. Acad. Sci.*, 68, 987, 1971.
31. Huh, J.W., Yokoigawa, K., Esaki, N., and Soda, K., *J. Ferment. Bioeng.*, 74, 189, 1992.
32. Huh, J.W., Yokoigawa, K., Esaki, N., and Soda, K., *Biosci. Biotech. Biochem.*, 56, 2081, 1992.
33. Beard, T. and Turner, N.J., *Chem. Commun.*, 246, 2002.
34. Alexandre, F.-R., Pantaleone, D.P., Taylor, P.P., Fotheringham, I.G., Ager, D.J., and Turner, N.J., *Tetrahedron Lett.*, 43, 707, 2002.
35. Enright, A., Alexandre, F.-R., Roff, G., Fotheringham, I.G., Dawson, M.J., and Turner, N.J., *Chem. Commun.*, 2636, 2003.
36. Roff, G.J., Lloyd, R.C., and Turner, N.J., *J. Am. Chem. Soc.*, 126, 4098, 2004.
37. Alexeeva, M., Enright, A., Dawson, M.J., Mahmoudian, M., and Turner, N.J., *Angew. Chem. Int. Ed.*, 41, 3177, 2002.
38. Carr, R., Alexeeva, M., Enright, A., Eve, T.S.C., Dawson, M.J., and Turner, N.J., *Angew. Chem. Int. Ed.*, 42, 4807, 2003.
39. Carr, R., Alexeeva, M., Dawson, M.J., Gotor-Fernández, V., Humphrey, C.E., and Turner, N.J., *Chem. Bio. Chem.*, 6, 637, 2005.

32 Enantioselective Biocatalytic Reduction of Ketones for the Synthesis of Optically Active Alcohols

Stefan Buchholz and Harald Gröger

CONTENTS

32.1 INTRODUCTION

The enantioselective reduction of ketones—according to Scheme 32.1—represents both a straightforward and an atom-economical approach toward optically active alcohols, which are important building blocks, for example, for the production of pharmaceuticals [1]. Often so-called blockbusters, drugs that are marketed with sales in the billion US$ range, are based on the

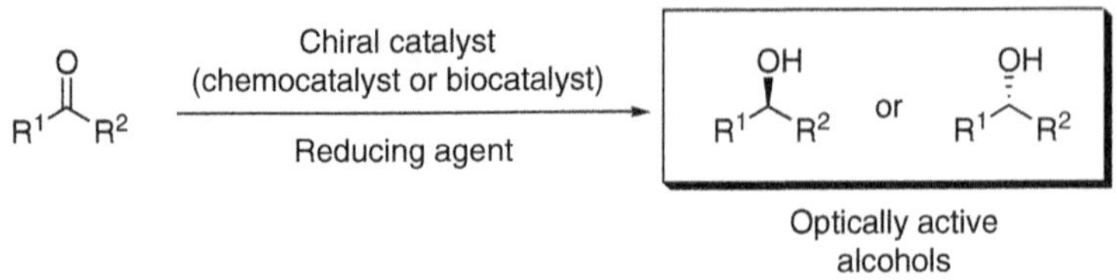

SCHEME 32.1

use of a chiral alcohol moiety. Thus, it is not surprising that numerous efficient asyn catalytic routes based on different types of concepts—from the fields of kinetic resolu and asymmetric synthesis [3–5]—have been developed up to date. Without any dou standing technologies in the latter field are the metal-catalyzed asymmetric hydrogena ketones [6] and the borane reduction [7]. Both technologies, which can be regarded as lan in industrial asymmetric catalysis in general, are applied widely on technical scale, and c represent the benchmark for any other type of alternative catalytic reduction methodo

However, biocatalysis [8] has turned out more and more to be an alternative, competitive technology for asymmetric ketone reductions. The recent increase in the of industrial applications of the biocatalytic asymmetric ketone reductions underli tremendous potential of this type of "white biotechnology" for large-scale manufac enantiomerically pure alcohols. Notably, the "biocatalytic reduction of ketones" is one type of reaction but rather consists of a range of different, often complementary, co

The goal of this review is to cover the state of the art in the field of enantios biocatalytic reductions of ketones for the synthesis of optically active alcohols (for p reviews, see references [9–14]). The major focus is on those methods that have demonstrated their feasibility on preparative and/or technical scale. In this connecti industrial impact of the different enzymatic approaches is discussed, and selected appli for large-scale manufacture of chiral alcohols are presented. The review is subdivided ing to the different concepts of cofactor regeneration, which is a key issue for ac economically attractive processes. In the scope of this review, typical yeast whole-c cesses (which have been reviewed extensively elsewhere; see, for example, [15–19]), " tation-type processes" using resting cells under metabolism of glucose for c regeneration [20], and diastereoselective reductions, for example, of water-soluble sug carbonyl compounds [21] are not included. With respect to technical applications it be noted that, for example, diastereoselective biocatalytic reduction is already on industrial scale at Avecia Ltd. for the synthesis of (4S,6S)-5,6-dihydro-4-hydroxy-6-4H-thieno[2,3b]thiopyran-7,7-dioxide, starting from the corresponding enantiom pure (6S)-ketone [22,23]. Therein this pharmaceutically important alcohol is manuf by the wild-type strain *Neurospora crassa*, obtaining high yield (>85%) and enantiosel (>98% ee) [24].

32.2   THE CONCEPTS OF BIOCATALYTIC REDUCTIONS OF KETONES

The basic principle of the enantioselective biocatalytic reduction of ketones (Scheme 32. use of an alcohol dehydrogenase (ADH) as a catalyst, and a cofactor as a reducing ag ADH is an enzyme capable of reducing carbonyl moieties under formation of (chiral) a and requires a specific "cofactor" as reducing agent. The most preferred cofactors ar NADH or NADPH, respectively. As the cofactors are very expensive reducing agents, costly to be applied in stoichiometric amount, a common key feature of all preparati technical) biocatalytic reductions is the use of cofactors in low, even catalytic, amounts, a

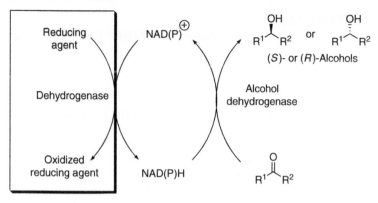

SCHEME 32.2

recycling *in situ* by coupling the ketone reduction process with a second process, in which the cofactor is regenerated (Scheme 32.2). Accordingly, the efficiency of regeneration of the cofactor contributes directly to the process economy of the whole reduction process and its synthetic applicability.

Notably, there are several options for an *in situ* cofactor recycling. In particular, substrate-coupled cofactor regeneration (Section 32.3), and enzyme-coupled cofactor regeneration (Section 32.4 through Section 32.6) have been developed. With respect to the latter, established cofactor regenerations based on formate dehydrogenase (FDH), glucose dehydrogenase (GDH), and glucose-6-phosphate dehydrogenase (G6PDH) are covered by this review. It should be added at this stage that alternative enzymatic cofactor regeneration concepts based on, for example, hydrogenase [25], phosphate dehydrogenase [26], and malic enzyme [27], have also been developed, which, albeit promising, are not covered by this review. This is in part because so far data from synthetic application of these methods are still rare, and also because there are no technical applications yet. In addition, electrochemical cofactor regeneration is known [28], but less commonly applied in asymmetric synthesis of alcohols compared to the above-mentioned methodologies. A common feature of all these types of cofactor regenerations is the use of a cheap and easily available source for the reduction of the oxidized form of the cofactor, namely NAD^+ or $NADP^+$. In case of the substrate- and enzyme-coupled cofactor regenerations, this source is a cheap and easily available reducing agent such as 2-propanol in the first case, and formate or glucose in the latter one.

The second key success factor is related to the properties of the ADHs. Although numerous ADHs are known, a multitude of biocatalytic reductions are based on the use of a relatively limited range of enzymes. Some selected examples of ADHs, which are widely applied in enzymatic reduction of ketones, are shown in Table 32.1. The synthetic applications of these enzymes are discussed in more detail in subsequent sections. It is noteworthy that the majority of these enzymes are already available in recombinant form by means of expression in *Escherichia coli*. These enzymes have already been often synthetically applied as they fulfil the prerequisites for synthetic applicability, for example, with respect to availability, high activity, and selectivity as well as a broad substrate tolerance. The variety of ADHs available is tremendous. This has been impressively demonstrated by, for example, Homann et al. in a broad ADH screening for about 30 ketones with a microbial library of about 300 microorganisms, identifying 60 cultures that contain highly enantioselective ADHs [29]. It should be added that several ADHs have already been characterized with respect to their stability in organic solvents. For example, a detailed study has been done by Müller et al. investigating the compatibility of solvents with ADHs from horse liver, *Thermoanaerobium brockii*, and *Lactobacillus brevis* [30].

TABLE 32.1
Selected Alcohol Dehydrogenases (ADHs) That Have Been Used in the Asymmetric Reduction of Ketones

Organism	Cofactor	Enantiospecificity	Source
Acinetobacter calcoaceticus	NADH	(*R*)	Wild-type organism
Baker's yeast (19 different ADHs)	NADPH	Both (*S*) and (*R*)	Recombinant in *Escheric*
Candida boidinii	NADH	(*S*)	Wild-type organism
C. magnoliae	NADPH	(*R*)	Recombinant in *E. coli*
C. parapsilosis	NADH	(*S*)	Recombinant in *E. coli*
Corynebacterium strain ST-10	NADH	(*S*)	Recombinant in *E. coli*
Geotrichum candidum	NADH	(*S*)	Wild-type organism
Gluconobacter oxydans	NADH	(*S*)	Wild-type organism
Hansenula polymorpha	NADPH	(*R*)	Recombinant in *E. coli*
Horse liver	NADH	(*S*)	Isolated from horse liver
Leifsonia sp.	NADH	(*R*)	Wild-type organism
Lactobacillus kefir	NADPH	(*R*)	Recombinant in *E. coli*
L. kefir	NADPH	(*R*)	Recombinant in *E. coli*
L. minor	NADPH	(*R*)	Recombinant in *E. coli*
Pichia finlandica	NADH	(*S*)	Recombinant in *E. coli*
P. methanolica	NADPH	(*S*)	Recombinant in *E. coli*
Pseudomonas sp.	NADH	(*R*)	Wild-type organism
Pseudomonas fluorescens	NADH	(*R*)	Recombinant in *E. coli*
Rhodococcus erythropolis	NADH	(*S*)	Recombinant in *E. coli*
R. ruber	NADH	(*S*)	Recombinant in *E. coli*
Sporobolomyces salmonicolor	NADPH	(*S*)	Recombinant in *E. coli*
Thermoanaerobium brockii	NADPH	Both (*S*) and (*R*)	Recombinant in *E. coli*

The ADHs can be used as isolated enzymes (in purified form or as a crude extra as whole cells. With respect to the latter approach, the use of wild-type cells or recomb whole-cell organisms, called "designer bugs," is conceivable. Recently, the tailor-made v cell catalysts, bearing the ADH and (in case of the enzyme-coupled cofactor regenera an additional enzyme, gained more and more interest due to their beneficial pr ties. Due to overexpression, the desired enzymes are available within the cells in amount, thus avoiding undesired side reactions by other dehydrogenases and allowi economically attractive access to them, particularly when using high-cell density fer tation for biocatalyst production.

32.3 PROCESSES BASED ON SUBSTRATE-COUPLED COFACTOR REGENERATION

32.3.1 THE CONCEPT

In the asymmetric reduction of ketones to alcohols according to the concept of subs coupled cofactor regeneration, a second alcohol is used as reducing agent for the ox cofactor. Typically the same enzyme, namely the same ADH, catalyzes both the format the product and the regeneration of the cofactor (Scheme 32.3). The substrate-co cofactor regeneration bears a special potential for the application of NAD(P)H-depe

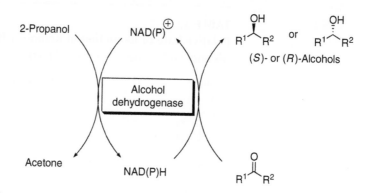

SCHEME 32.3

ADHs due to the possibility to use only one enzyme for both required transformations [31]. Another big advantage and the simplicity of this concept is that only one enzyme has to be produced, and often in the applied whole-cell systems not even the addition of an "external" cofactor, which represents an additional cost factor, is necessary. The reaction concept is the biotechnological analog to the classical Meerwein–Pondorf–Verley reduction that has been known since the 1920s. Simple alcohols like 2-propanol are cheap, environment-friendly, and also claim to be good examples for green chemistry [32]. A principal drawback of this concept, however, is also obvious: as we are dealing with an equilibrium reaction, a huge excess of reducing alcohol has to be added in order to reach high conversion rates. This high cosubstrate concentration might lead to an enzyme inhibition and/or inactivation, or only very low concentrations of the reducing alcohol and the substrate can be applied [33]. If industrially relevant substrate concentrations are being used, the necessary concentration of the reducing alcohol often becomes prohibitive. Although this hampered the application of the substrate-coupled cofactor regeneration for a long time, several successful examples of technical applications based on this concept have been reported.

32.3.2 PROCESSES WITH ISOLATED ENZYMES

One of the early works describing the enantioselective reduction of ketones using a substrate-coupled cofactor regeneration was published by Wong et al., who isolated and characterized the ADH from the *Pseudomonas* sp. strain PED [34]. The PED-ADH that utilizes NADH as cofactor has an activity of 36 U/mg with respect to 2-propanol under NAD^+ saturation conditions. As can be seen from Table 32.2 aromatic as well as cyclic and aliphatic ketones are accepted as substrates. Although substrates with a methyl group on one side of the keto-function are typically better substrates for the PED-ADH, other substrates such as ethyl phenylketone are also being accepted.

PED-ADH yields the (R)-alcohols in a low to good enantioselectivity (Table 32.3). All reactions have been carried out in a two-phase solvent system with n-hexane as organic phase.

Adlercreutz et al. [35] analyzed the influence of temperature and cosubstrate concentration on the enantiomeric excess (ee) of ketone reductions by the NADPH-dependent ADH of *T. brockii* with 2-propanol as cosubstrate. The reactions were carried out with the immobilized enzyme in hexane with 2.5% (v/v) water, allowing conversions to be carried out at temperatures as low as −20°C. The ee of the alcohol formed from the ketone decreased during the course of reaction (from 53 to 0% in the formation of (R)-2-butanol). This is interpreted as being a consequence of the reversibility of all reactions involved. By using a

TABLE 32.2
Relative Rate of Reduction of Ketones with PED Alcohol Dehydrogenase (PED-ADH)

Substrate	Relative Rate of Reaction
	1
	7
	<1
	6
	100
	5
	135

large excess of 2-propanol, this effect was suppressed. While in the reduction of 2-butar (R)-2-butanol the ee increased with increasing temperature, in the reduction of 2-pen to (S)-2-pentanol the ee decreased with increasing temperature. This effect, however, cc explained by thermodynamic effects, and one should keep in mind that in the reduc 2-butanone the (R)-alcohol is the predominant product and in the reduction of 2-pen (S)-alcohol is the main product. An increasing temperature thus favors the formation (R)-alcohol in both cases.

Specifically for commercial applications besides yield and enantioselectivity, c lifetime and total turnover number are crucial. Margolin et al. [36] reported on the suc stabilization of horse liver ADH–NADH complexes (HLADH–NADH–CLEC) by t mation of glutaraldehyde cross-linked enzyme crystals. While the soluble enzyme deactivated at 40°C in 25% 2-propanol ($t_{1/2}(40°C) = 4$ h; $t_{1/2}(25°C) = 24$ h), the HL NADH–CLEC had a half-life of more than 4 d at 40°C and still showed nearly full a after 48 h at 25°C. It has been shown that the cofactor is not leaching, which is explai the enzyme being cross-linked in the cofactor-binding conformation. However, by add external cofactor the activity of the crystallized enzyme for the reduction of 6-me heptene-2-one could be increased from 64 to 92% of the soluble HLADH activity. A

TABLE 32.3

Product	Enantiomeric Excess (%)	Yield (%)
	94	34
	98	79
	27	43

limitation of the HLADH, however, is the low specific activity and the nonavailability on a large scale [14].

Gupta et al. [37] reported on the recombinant expression of the NADPH-dependent *L. minor* ADH in *E. coli* and its application in a two-phase system. They were able to quantitatively convert 4-chloro-3-oxobutanoate to the corresponding ethyl (*S*)-chloro-3-hydroxy butanoate with an enantioselectivity of >99.9% ee at a substrate concentration of about 15% (v/v). The enzyme has broad pH stability and a remarkable stability against organic solvents. Some water-miscible solvents even increase stability: e.g., in a 10% solution of glycerol 62% of activity is left after 24 h, while only 3% of activity remains in *tris*-buffer.

The use of a two-phase system composed of a water-immiscible solvent like hexane and a PVA-hydrogel containing *L. kefir* ADH has been shown to be advantageous for many poorly water-soluble substrates by Ansorge-Schumacher et al. [38].

The sequence and cloning of the (*R*)-ADH from *Pseudomonas fluorescens* as well as its use in a cosubstrate-coupled NADH regeneration has been reported by Bornscheuer et al. [39,40]. For acetophenone as a model substrate, 95% conversion and 92% ee have been observed after a reaction time of 21 h using 2-propanol for cofactor regeneration. With methyl-3-oxo-butanoate, 83% conversion after 19 h and >99% ee have been obtained.

Recently Itoh et al. [41,42] characterized an alcohol dehydogenase from *Leifsonia* sp. S749 (LSADH) having a broad substrate specificity and showing a high enantioselectivity. All reactions had been carried out with the purified enzyme using 2-propanol for cofactor regeneration (Table 32.4).

The potential of ionic liquids for the biocatalytic reduction of ketones has been demonstrated by Kragl et al. [43]. Due to the favorable partitioning coefficients the highly enantioselective reduction of 2-octanone to (*R*)-2-octanol, catalyzed by an ADH from *L. brevis*, is faster in a two-phase system containing buffer and the ionic liquid [BMIM][(CF$_3$SO$_2$)N] compared to the reduction in a two-phase system containing buffer and methyl *tert*-butyl ether (MTBE). After a reaction time of 180 min, a conversion of 88% has been observed for the two-phase system with the ionic liquid in contrast to 61% in the presence of MTBE.

32.3.3 WHOLE-CELL REDUCTIONS

In the whole-cell bioreductions of ketones to chiral alcohols (Scheme 32.4), either wild-type cells or recombinant strains may be used. A significant advantage of this reaction concept is

TABLE 32.4

Product	Enantiomeric Excess (%)	Relative Acitivity	Conversion (%)	Yield (%
(structure: acetophenone)	(R) = 99	6	82	40
(structure: F₃C trifluoromethyl phenyl ketone)	(S) > 99	100	100	81
(structure: propiophenone)	(R) > 99	70	100	81
(structure: benzyl ketone)	(R) > 99	60	81	58
(structure: methyl ketone)	(S) > 99	29	100	79
(structure: methyl ketone chain)	(S) > 99	67	100	91
(structure: keto ester)	(R) 79	2	12	8
(structure: OH alcohol)	(R) > 99	17	79	17
(structure: OH alcohol)	(R) > 99	104	83	31
(structure: OH alcohol)	(R) > 99	229	87	38
(structure: OH ester Cl)	(S) > 99	809	100	54
(structure: N-Boc pyrrolidine, HO)	(R) > 99	4	35	24

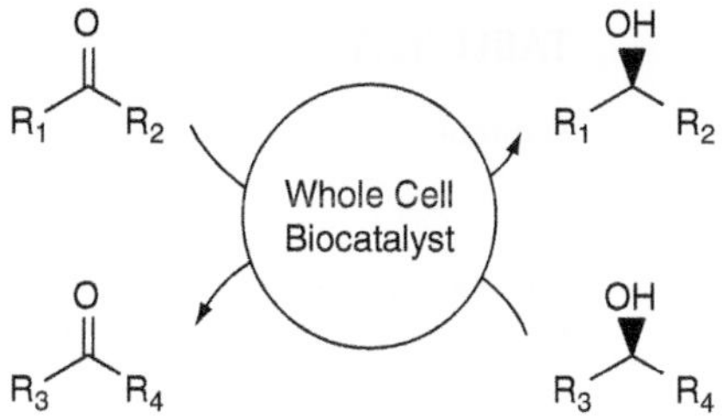

SCHEME 32.4

that no expensive external cofactor might be used. In some instances, however, the application of an additional cofactor proved to be useful to increase the reaction rate.

In an early work, Matsumura et al. [44] demonstrated the use of *Candida boidinii* wild-type cells grown on methanol in the production of (R)-1,3-butandiol. They could recycle the cells several times; however, extremely high biocatalyst/substrate ratios (5 g vs. 100 mg) were applied, limiting the technical potential of this system. Instead of methanol, glucose could also be used as a regenerating system, leading, however, to much poorer results with respect to yield and ee.

Itoh et al. [45] found and recombinantly expressed a phenylacetaldehyde reductase from a *Corynebacterium* strain. The enzyme has 29% identity with the ADH from *Sulfolobus solfotaricus* and 27% identity with the *C. parapsilosis* ADH. As can be seen from Table 32.5, the enzyme has broad substrate specificity.

As a model, the conversion of α,m-dichloroacetophenone was analyzed in more detail. The internal NADH-pool of the recombinant *E. coli* cells was insufficient to get high reaction rates. Therefore, 1 mM of NADH was added. The optimal 2-propanol concentration was 3 to 7%. The reaction rate was low, probably due to the inactivation of the enzyme. For all substrates analyzed, a very high enantioselectivity has been observed (Table 32.6).

For NADH regeneration, several secondary alcohols were tested at 5% v/v, including 2-propanol (relative activity at 3 mM: 6%), 2-butanol (15%), 2-pentanol (100%), 2-hexanol (240%), 2-heptanol (410%), and 2-octanol (422%). In the integrated reaction system, however, 2-propanol gave the highest conversion followed by 2-heptanol, 2-octanol, and 2-butanol.

In a major screening program, Matsuyama et al. [46] from Daicel found *C. parapsilosis* IFO 1396 exhibiting significant alcohol oxidizing activity when using (R)-1,3-butandiol as substrate. They also demonstrated that the enzyme could also be applied in the reductive mode using 2-propanol as reducing agent and without the addition of an external cofactor. They designated the enzyme as CpSADH. By recombinant expression of the enzyme in *E. coli* they were able to increase its activity by a factor of 78-fold. This recombinant biocatalyst reduces ethyl 4-chloroacetoacetate to ethyl (R)-4-chloro-3-hydroxybutanoate [(R)-ECHB] at 36.6 g/L with a yield of 95.2% and 99% ee (Scheme 32.5). (R)-ECHB is an important intermediate for the preparation of HMG-CoA-reductase inhibitors, a class of compounds that lowers the cholesterol level in human blood.

Amidjojo and Weuster-Botz [47] reported on the production of the corresponding product (S)-ECHB using *L. kefir* wild-type cells (analog to Scheme 32.5). The LKADH uses NADPH as cofactor, and again no additional cofactor was needed. The reaction was performed at high substrate concentration, which led to a two-phase system. With 5% (v/v) 2-propanol as cosubstrate, a final product concentration of 1.2 M, a yield of 97%, and an ee of 99.5% were achieved. The space–time yield was 85.7 mmol/L/h. Instead of 2-propanol, glucose can also be used as cosubstrate; however, the performance is much poorer.

TABLE 32.5

Substrate	Relative Rate of Reaction (%)
	100
	546
	258
	70
	188
	449
	868
	2247

Müller et al. [48] reported that by applying a recombinant *E. coli* strain expressing *L*
ADH (LBADH) 3,5-dioxocarboxylates **1** can be reduced to the corresponding 5-h█
3-oxocarboxylates **2** in a highly regio- and enantioselective manner, leaving the 3-oxo
untouched (Scheme 32.6). Thus, with 2-propanol as reducing agent **1** is reduced to
99.4% ee and complete regioselectivity. The corresponding chloro-derivative **3** is co█
with >99.5% ee to the expected *tert*-butyl (*S*)-6-chloro-5-hydroxy-3-oxohexanoate
isolated yield).

Applying a substrate-coupled cofactor regeneration, the limited stabilities of the
ADHs against organic reducing agents like 2-propanol were a significant hurdle
broader technical application as high cosubstrate concentrations are needed in orde█
high conversion rates. Two ways to overcome these limitations have been demon█
applying a clever reaction design and screening for a solvent-tolerant enzyme.

During the reduction of ethyl 5-oxohexanoate (Scheme 32.7), high acetone concen█
led to reducing the equilibrium concentration of ethyl (*S*)-5-hydroxyhexanoate. Lie█

TABLE 32.6

Product	Enantiomeric Excess (%)
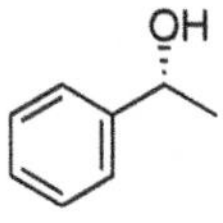	$R \gg 99$
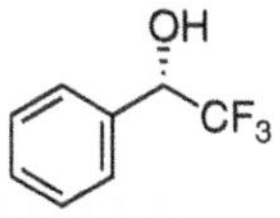	$S \gg 99$
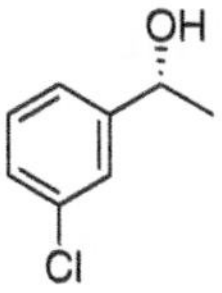	$S \gg 99$

[31] demonstrated that by pervaporation or stripping of the acetone the conversion can be increased from 75 to 95 and >97%, respectively.

The Faber group reported conversion of ketones very selectively and efficiently to the corresponding (*S*)-alcohol by *Rhodococcus ruber* DSM 4451 at 2-propanol concentrations of up to 50% (v/v) and acetone concentrations of up to 15% (v/v) [32,49]. Thus, the flexibility in applying substrate-coupled cofactor regeneration was significantly increased. Specifically, for commercial applications the high 2-propanol tolerance of the strain and enzyme considerably broadened the scope of this reaction scheme and thus represented a major breakthrough. From *R. ruber* at least four NAD$^+$/NADH-dependent active fractions could be isolated which showed the same enantioselectivity. However, only one showed the pronounced stability toward 2-propanol and acetone and could use 2-propanol as a substrate. As can be seen from Table 32.7, many methy alkyl ketones and methyl aryl ketones are converted with an enantioselectivity of more than 99%. Shifting the keto-function to the 3-position, however, resulted in a drop of the enantioselectivity to 97%.

The ADH from *R. ruber* not only exhibits a high solvent stability but is also stable at temperatures up to 60°C and active over a broad pH range from 5.5 up to 11 [50]. Meanwhile, the Faber group also reported the recombinant expression of the *R. ruber* ADH, thus further expanding the scope of this versatile enzyme [51].

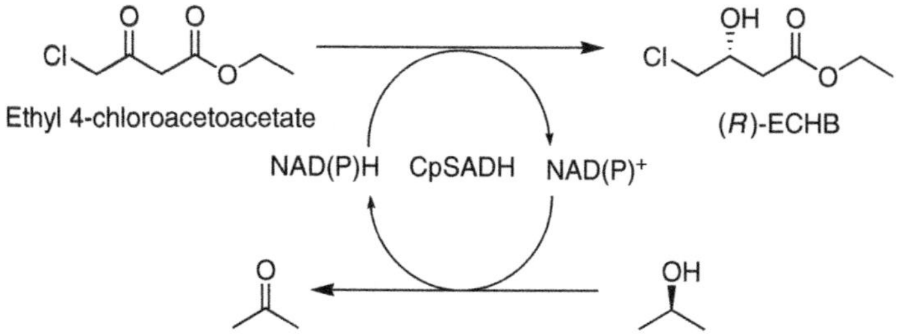

Ethyl 4-chloroacetoacetate   (*R*)-ECHB

NAD(P)H   CpSADH   NAD(P)$^+$

SCHEME 32.5

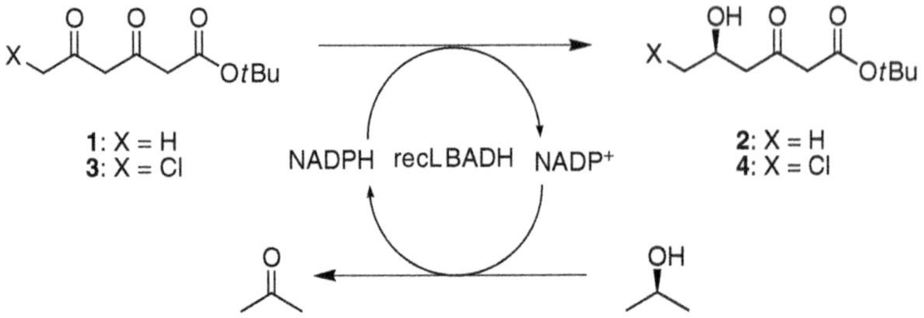
OtBu
OH O O
OtBu
1: X = H
3: X = Cl
NADPH recLBADH NADP+
2: X = H
4: X = Cl
OH

SCHEME 32.6

32.4 PROCESSES BASED ON ENZYME-COUPLED COFACTOR REGENERA WITH A FORMATE DEHYDROGENASE

32.4.1 The Concept

When applying enzyme-coupled cofactor regeneration for asymmetric biocatalytic red processes, the use of an FDH turned out to represent a very popular approach. The catalyzes the oxidation of formate to carbon dioxide, while reducing the oxidized form cofactor into its reduced form, NAD(P)H. The whole concept of an asymmetric reduct ketones in the presence of an ADH and FDH is shown in Scheme 32.8.

The most widely applied FDH is probably the FDH from *C. boidinii* and opt mutants thereof [52], developed in the Kula group, who are—jointly with Wandrey e the pioneers in the field of FDH-based cofactor regeneration [53,54], as well as the Whi group [55]. Although limited to the regeneration of NADH alone, the FDH from *C. boi* available in recombinant form on a large scale, and has already proved its suitabil industrial purpose in the reductive amination of trimethylpyruvate, a process wh running on industrial scale at Degussa AG for the manufacture of L-*tert*-leucine [56, addition to the FDH from *C. boidinii*, there are further FDHs known to be suital biotransformation such as that from the methylotrophic bacterium *Pseudomonas* sp. [5 from *Mycobacterium vaccae* N10 [59,60]. A key advantage when using an FDH for cc regeneration certainly is the irreversible step of carbon dioxide formation and remova shifting the equilibrium toward (complete) product formation. However, not only equilibrium shifted toward the product side, leading to high conversions, but the down processing is also simplified as (ideally) no organic by-product is generated in the re mixture. A general limitation of FDHs so far is the specific activity that does not exce range of 10 U/mg in spite of numerous attempts for improvement. For example, the from *C. boidinii* shows a specific activity of about 6 U/mg. Thus, these activities are

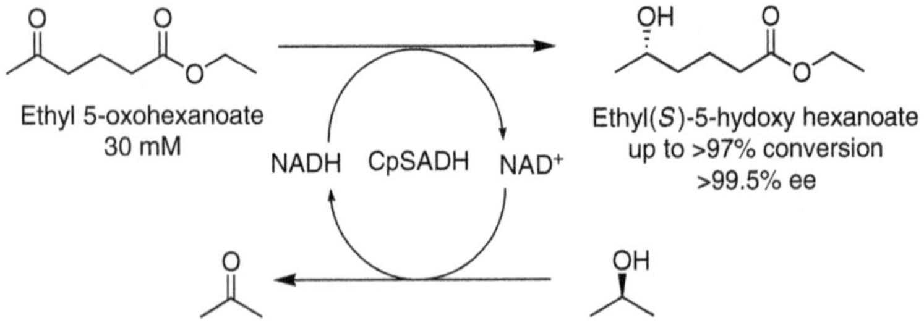
Ethyl 5-oxohexanoate
30 mM
NADH CpSADH NAD+
OH
Ethyl(S)-5-hydoxy hexanoate
up to >97% conversion
>99.5% ee
OH

SCHEME 32.7

TABLE 32.7

Substrate	Enantiomeric Excess (Product) (%)	Conversion (%)
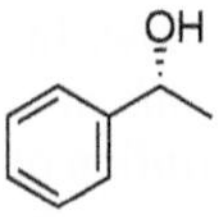	>99	81
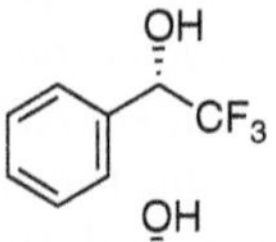	>99	82
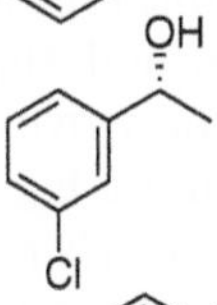	>99	92
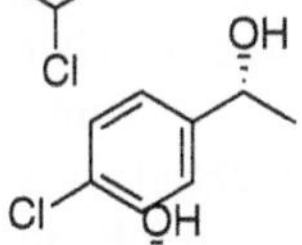	>99	91
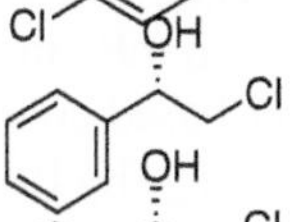	97	79
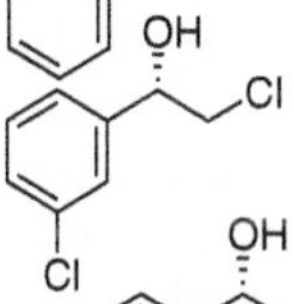	>99	70
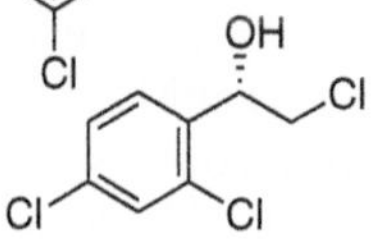	> 99	65

comparison with those for related cofactor-regenerating enzymes such as the GDH and malate dehydrogenase, which typically show activities of >100 U/mg. However, as can be seen from the subsequent examples, the FDH is an efficient and highly suitable enzyme for cofactor regeneration in synthetic reduction processes in spite of its limitations.

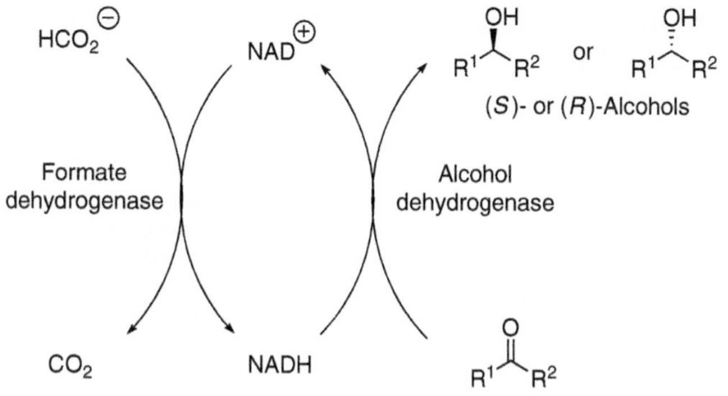

SCHEME 32.8

32.4.2 Processes with Isolated Enzymes

The initial work on the enzymatic reduction of ketones has been carried out based on
of isolated enzymes in homogeneous aqueous media. Due to the low solubility of the
phobic ketones in water, the reactions were carried out at low substrate concentration
long time, typically in the range of 5 to 20 mM or below. In 1993, Hummel and Go
reported the suitability of an ADH from *R. erythropolis* in combination with an FI
asymmetric reduction of ketones at a substrate concentration of 1.16 mM [61]. In additi
Kula and Wandrey groups studied in detail this cofactor regeneration methodology
preparation of a variety of alcohols using ADHs from *R. erythropolis* [62], *C. parapsilo*
and *C. boidinii*, respectively. Therein, the NADH-dependent ADH from *R. erythropolis*
out to be stable at 37°C for a couple of days, and suitable for the reduction of a broad
of ketones comprising keto esters, aromatic ketones, and aliphatic 2-alkanones [62
enzyme shows both a high enantioselectivity and specific activity with, for example, 26
for ethyl acetoacetate. However, a key feature of most highly suitable substrates
presence of an aceto-moiety, as in case of most of the ADHs reported so far.

Adlercreutz reported a crude enzyme preparation from permeabilized and lyop
Gluconobacter oxydans cells to contain suitable ADH(s) for enantioselective reductio
variety of ketones with high enantioselectivity of 93 to 99% in most cases [64].

Another example for the suitability of the FDH-based cofactor regeneration conc
asymmetric ketone reduction has been demonstrated by the Itoh group using an (*S*)-se
ADH isolated from styrene-assimilating *Corynebacterium* strain ST-10 [65]. An imp
range of ketones, comprising alkanones and aromatic ketones, has been reduced with
lent enantioselectivities of >99% in many cases. So far, however, reactions were carrie
a low substrate concentration of <10 mM, often <1 mM.

The Kula group also carried out preparative conversions based on these enzy
coupling the ADH reduction reactions with the FDH regeneration (Scheme 32.9)
As enzymes, ADHs from *R. erythopolis* and *C. parapsilosis* were used in combination w
FDH from *C. boidinii*. Carrying out the reductions of several keto esters and a keto
acetal at a substrate concentration of 100 mM in most cases furnished the desired a
with high conversion (up to 100%), and high enantioselectivities of >99%. A selected e
is given in Scheme 32.9 [66].

A modified methodology for enantioselective ADH- and FDH-coupled reduct
aqueous media has also been reported by Wandrey et al. carrying out enzymatic red
in cyclodextrin-containing buffers [68]. The corresponding alcohols were obtained i
yield and with high enantioselectivities when using an ADH from *C. parapsilosis*. N
high stability of the enzymes in media containing heptakis-(2,6-di-O-methyl)-β-cyclo
has been found. In the continuous reaction with the ADH and an FDH from *C. A*
a steady conversion of up to 79% for the synthesis of (*S*)-1-(2-naphthyl)ethan
been achieved.

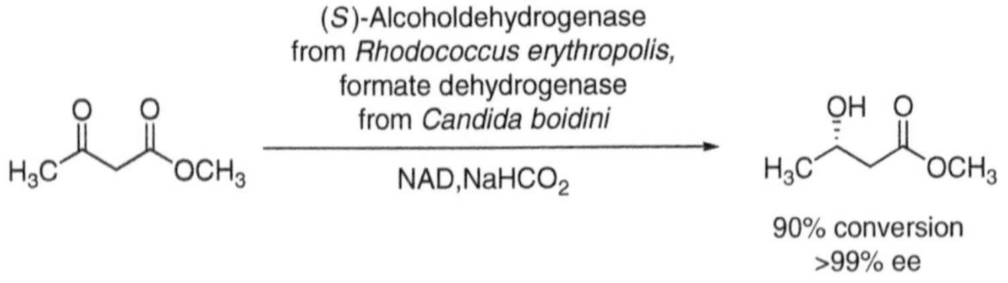

SCHEME 32.9

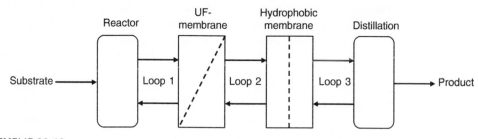

SCHEME 32.10

Also the issue of high space–time yields in spite of the limitation of low ketone solubility has been successfully addressed by the Wandrey group, who developed a very elegant engineering solution by means of a continuous process with an enzyme-membrane reactor (Scheme 32.10). An efficient "three-loops"-concept is based on an enzymatic reaction in pure aqueous medium, a separation of the aqueous phase from the enzyme by ultrafiltration, and a subsequent continuous extraction of the aqueous phase with an organic solvent. Organic and aqueous phases are separated by a hydrophobic membrane [69–71]. This is required as organic solvents can generally cause significant enzyme inactivation. In particular, this is known for the FDH from *C. boidinii* which is sensitive to organic solvents [72]. Although the reaction in this enzyme-membrane reactor is limited by the low solubility of the ketone in water (9 to 12 mM), good space–time yields in the range of 60 to 104 g/(L/d) have been obtained [69–71]. With respect to synthesized products, for example, (*S*)-1-phenylpropan-2-ol and (*S*)-4-phenylbutan-2-ol have been produced in enantiomerically pure form. Typical data for the consumption of the enzymes are ~3.0 to 3.5 kU/kg for the ADH and ~5 to 10 kU/kg for the FDH.

An extended, newly designed emulsion-membrane reactor concept has been successfully applied by Wandrey et al. for the asymmetric reduction of 2-octanone [73]. As enzymes, the ADH from *C. parapsilosis* (CP-ADH) and an FDH from *C. boidinii* have been used. The reactor consists of a stirred emulsion vessel, from which only the aqueous phase—separated by a hydrophilic ultrafiltration membrane—enters the enzyme-membrane reactor (Scheme 32.11).

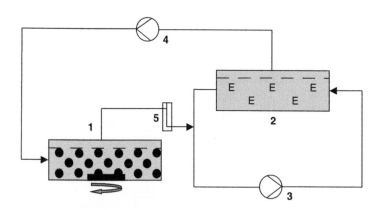

1: Stirred emulsion vessel with hydrophilic ultrafiltration membrane.
2: Enzyme–membrane reactor loop with ultrafiltration module.
3,4: Circulation pumps. 5: Bubble trap; E: Enzyme.

SCHEME 32.11

Therein, the desired enzymatic reduction takes place at a pH of 7, and the product outflc recirculated to the stirred emulsion reactor, where the alcohol is extracted into the org phase, and the water phase is recharged with substrate. A conversion of 87% (with a resid time of 1 h) has been achieved over a period of 200 h. The conversion itself has been fur increased to 97% at a residence time of 1 h when increasing the formate concentration fro: to 50 mM, corresponding to a space–time yield of 21.1 g/(L/d). Excellent half-life times been found with $\tau_{1/2} = 67$ d for the CP-ADH and $\tau_{1/2} = 88$ d for the FDH. Compared the "classic" enzyme-membrane reactor the total turnover number has been increase factor 9. Thus, this emulsion-membrane reactor, which has been operated over a peric >4 months, overcomes limitations of large volumes and low total turnover numbers.

An interesting comparative study of the ADH- and FDH-based reduction wit. oxazaborolidine-catalyzed borane reduction has been done by the Wandrey group Both reactions were carried out in continuously operating membrane reactors, whicl shown in Scheme 32.12. Notably, the stability of the enzymes is higher with a half-life ti: 31.1 d compared with 1.2 d for the chemical catalyst. Furthermore, enantioselectivity c enzymatic reaction was >99% ee, whereas a somewhat lower enantioselectivity of 90 to 9- was obtained in the chemical reduction method. However, space–time yield of the che: approach, which takes place in an organic solvent, was higher with 1.4 kg/(L/d) compared 0.088 kg/(L/d) for the enzymatic approach. The lower productivity when using enzyme been explained with the low solubility of substrate and product in water.

The search for suitable reaction media, which guarantee high ketone solubility, how has been ongoing. Although the presence of an organic solvent could improve the solubil poorly water-soluble ketones, the known instability of the FDH from *C. boidinii* toward : organic solvents remains a challenge. Addressing this issue, Gröger et al. developed a su: aqueous–organic two-phase solvent reaction medium based on the use of *n*-heptane *n*-hexane as organic phases (Scheme 32.13) [75,76]. This reaction medium fulfils the cr that both enzymes, namely the recombinant (*S*)-ADH from *R. erythropolis* and FDH *C. boidinii* (and mutants thereof), remain stable in the presence of the organic solve: addition, a good solubility of poorly water-soluble ketones led to substrate concentratic up to 200 mM with a simple reaction protocol. In contrast, the stability of the FDH fro *boidinii* (mutant C23S, C262A) [53] was low in the presence of only 10% (v/v) of many o: "standard" solvents such as MTBE, ethyl acetate, and toluene. With this enzyme-comp reaction medium at hand, preparative conversions were carried out that gave good co: sions accompanied by high enantioselectivities with a variety of aromatic ketone subs: (Scheme 32.13).

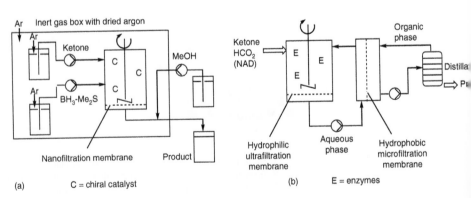

(a)   C = chiral catalyst   (b)   E = enzymes

SCHEME 32.12

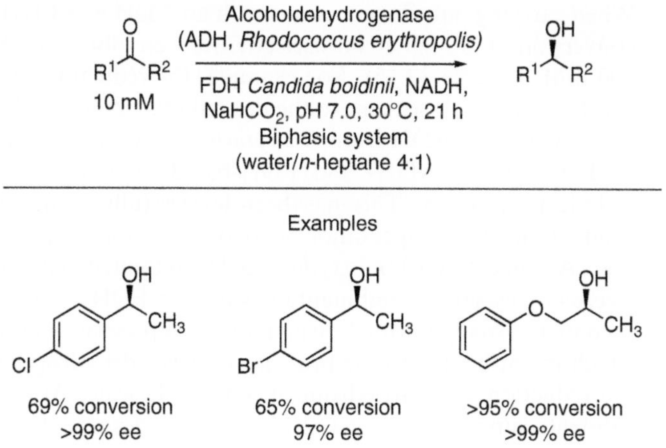

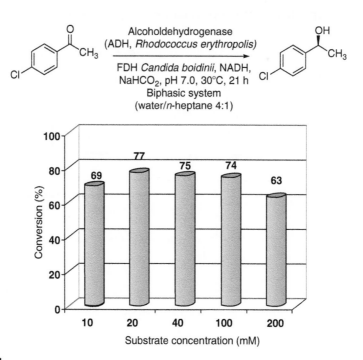

SCHEME 32.13

For example, in the presence of the (*S*)-ADH from *R. erythropolis*, *p*-chloroacetophenone was converted to the optically active (*S*)-enantiomer with >99% ee and a conversion of 69%. Furthermore, the reactions also proceed in water–*n*-heptane two-phase reaction media at higher substrate concentrations of up to 200 mM (Scheme 32.14) [75,76]. However, at >200 mM substrate concentration, lower conversions were obtained and prolonged reaction times were required.

A further improvement of the substrate concentrations up to 500 mM has been realized when using an "emulsion system" for the synthesis of the corresponding alcohols [77,78].

SCHEME 32.14

When carrying out the reductions with an ADH and FDH in pure aqueous media, bo conversions and high enantioselectivities were obtained at substrate concentrations o 500 mM (Scheme 32.15). For example, the reduction of 4-chloroacetophenone as a substrate on a 6 L scale gave the desired (S)-alcohol with >98% conversion and >99 As enzymes, the ADH from *R. erythropolis* and the FDH from *C. boidinii* have been

Furthermore, ADH- and FDH-based reduction processes have already prove technical feasibility. This has been successfully demonstrated by the Patel group production of (S)-2-pentanol on pilot scale using an ADH from *G. oxydans* (SC [79]. As biocatalyst, *G. oxydans* cells pretreated with Tritone X-100 were used (freeze/taw cycle) in combinantion with the FDH from *C. boidinii*. The latter enzy used in a partially purified form in order to prevent undesired formation of racemic p which might be due to the presence of other dehydrogenases in the crude extract. N the reduction was carried out at a 1.500 L scale. After a reaction time of 46 h a substrate input of 3.2 kg (~2.13 g/L), the desired (S)-(+)-2-pentanol was formed conversion of 32.2% and an enantioselectivity of >99% (Scheme 32.16).

Another elegant and efficient approach in process design has been reported Schomäcker group [80]. Therein, an enzymatic reduction with FDH-based NADH r ation of a less water-soluble ketone was carried out in reverse micelles. When using the of water-in-oil microemulsions as reaction media, the reduction of 2-heptanone un mation of (S)-2-heptanol proceeds with complete conversion and an excellent enant tivity of >99%. Furthermore, it turned out that the reaction rate of ADH in microem increases up to 12 times compared with the rate in water. The concept of this microe reaction system is shown in Scheme 32.17.

A further study by the same group focusing on the stability and activity of differe of ADHs in water-in-oil microemulsions showed significant changes of activity and s of these enzymes depending on the water and surfactant concentration of the emulsion, whereas—notably—the FDH from *C. boidinii* did not show any kineti depending on the microemulsion components [81].

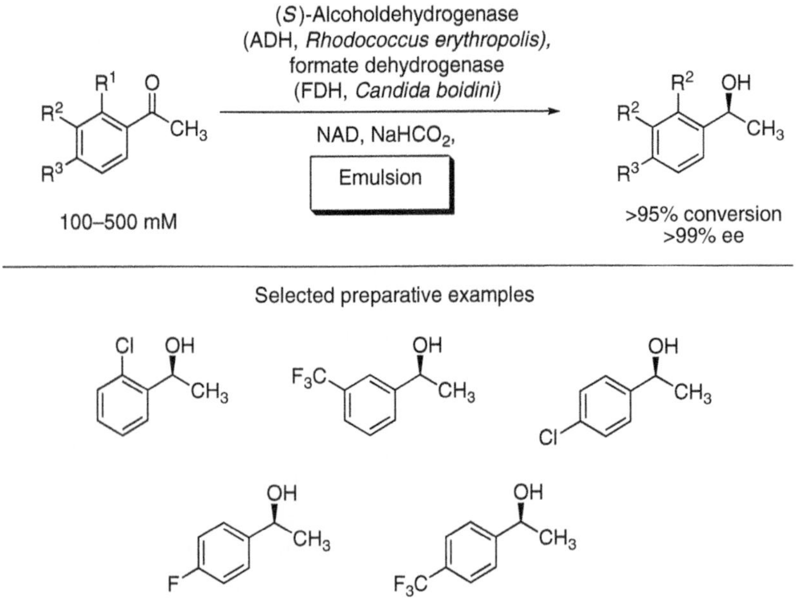

SCHEME 32.15

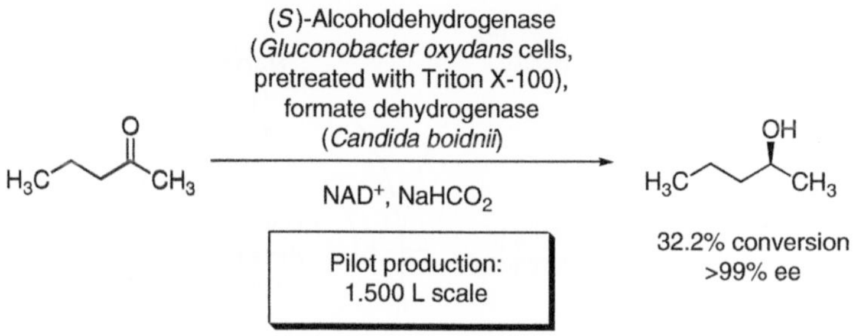

SCHEME 32.16

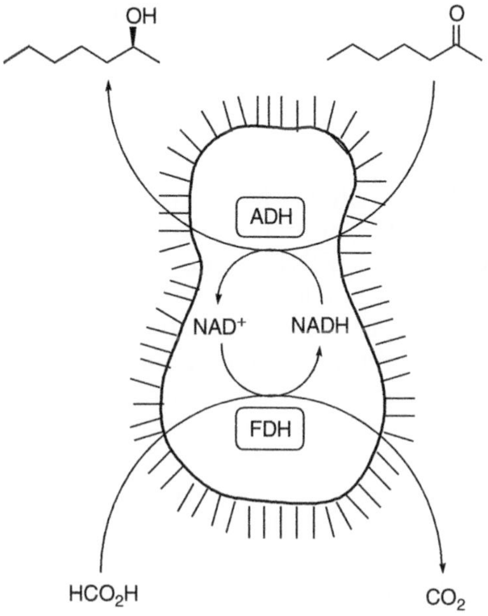

SCHEME 32.17

32.4.3 WHOLE-CELL REDUCTIONS

The potential of an FDH-based whole-cell catalyst for synthetic applications has been reported by Matsuyama et al., using a recombinant *E. coli* W3110 strain, which coexpresses an ADH from *Pichia finlandica* and an FDH from *Mycobacterium* [82]. As motivation for using the FDH, the authors mentioned the low affinity of the ADH for 2-propanol as a substrate, thus making the substrate-coupled cofactor regeneration method not attractive in this case. A "one-plasmid-strategy" has been used inserting the ADH- and FDH-genes in a single plasmid. The tailor-made whole-cell catalyst has subsequently been applied in the enantioselective reduction of ethyl 4-chloro-3-oxobutanoate to the corresponding (*S*)-alcohol at 32.2 g/L, with 98.5% yield and 99% ee (Scheme 32.18). A further successful example is the synthesis of (*R*)-1-chloro-4-pentanol from the corresponding ketone at 26.1 g/L with 99% ee.

Another ADH- and FDH-based whole-cell system, which contains an NADPH-dependent ADH from *L. brevis* with the FDH from *M. vaccae* N10, has been constructed by the Sahm group [83]. The FDH from *M. vaccae* N10, is suitable for the regeneration of both NADH and NADPH. Overexpression of both enzymes, ADH and FDH, led to an efficient

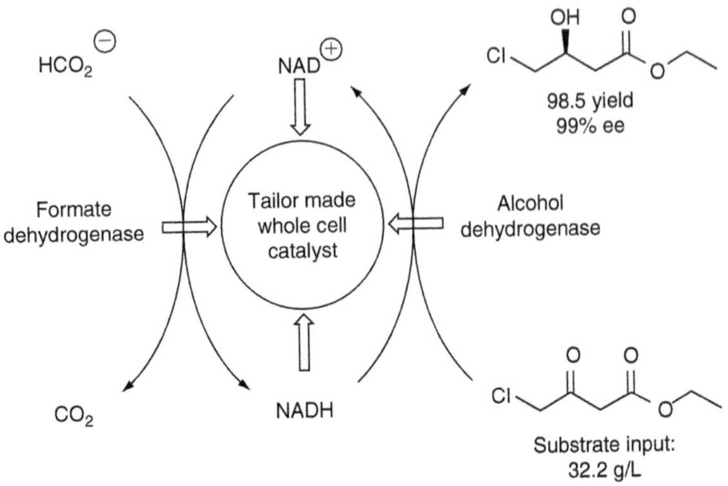

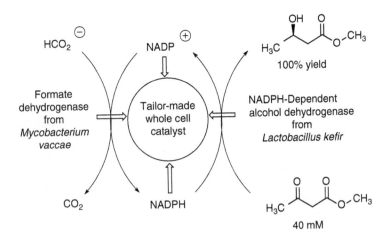

SCHEME 32.18

whole-cell biocatalyst, which gave a quantitative conversion in the reduction of m
acetoacetate at a substrate concentration of 40 mM (Scheme 32.19). It is noteworthy
the presence of the FDH led to a sevenfold increase of the intracellular NADH/NAD$^+$
in the recombinant cells.

 For a long time, a major limitation for applications using the FDH from *C. boidinii* w
unability to regenerate NADP$^+$, thus being limited to the regeneration of NAD$^+$ only
elegant solution for this problem has been recently found by the Hummel group,
expanding the application range of this FDH-based cofactor regeneration also to NA
dependent ADHs [84]. As an ADH, the ADH from *L. kefir* [85,86] was chosen. The key s
the integration of an additional enzymatic step within the cofactor regeneration cycle, na
the pyridine nucleotide transhydrogenase (PNT)–catalyzed regeneration of NADPH
NADP$^+$ under consumption of NADH forming NAD$^+$ [84]. The concept is graph
shown in Scheme 32.20, exemplified for the synthesis of phenylethanol. The Hummel
constructed a whole-cell system, coexpressing the ADH from *L. kefir*, PNT from *E. co*

SCHEME 32.19

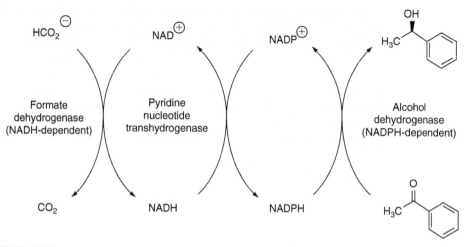

SCHEME 32.20

FDH from *C. boidinii*. In the presence of this catalyst, the desired product was formed in 66% yield at a substrate concentration of 10 mM in the presence of both cofactors.

32.5 PROCESSES BASED ON ENZYME-COUPLED COFACTOR REGENERATION USING A GLUCOSE DEHYDROGENASE

32.5.1 THE CONCEPT

An efficient option for recycling the cofactor NADH is based on the use of a GDH. Therein, glucose is oxidized to the corresponding gluconolactone, while the oxidized cofactor NAD(P)$^+$ is reduced to NAD(P)H required as reducing agent for the reduction process. This concept of cofactor regeneration is shown in Scheme 32.21. As the gluconate is subsequently hydrolyzed to gluconic acid (as its sodium salt at neutral pH), this reaction also can be regarded as an irreversible step, thus shifting the whole reaction into the direction of the desired alcohol product. Compared to the FDHs described in Section 32.4, the GDHs offer the advantage of higher specific activities, which are typically >100 U/mg (compared with FDH activity of ~10 U/mg).

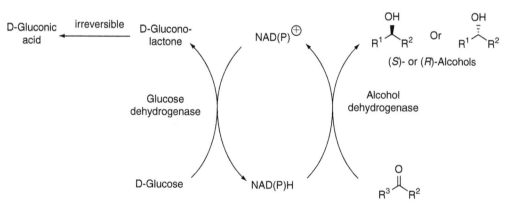

SCHEME 32.21

In addition, numerous GDHs are already known and available in recon form, which is a prerequisite for an efficient application on technical scale. Typical ex are the GDH from *Bacillus megaterium* [87,88], *B. subtilis* [89,90], *B. cereus* [91,9 *Thermosplasma acidophilum* [93]. Although some preparative synthetic applicati means of isolated enzymes are known, most of the reported applications of GDH-c cofactor regeneration in asymmetric reduction are based on the use of recombinant wh systems. Notably, industrial applications of this recombinant whole-cell technology ba an ADH and a GDH have already been reported by specialty chemicals compa: particular by Kaneka Corporation and Degussa AG.

32.5.2 Processes with Isolated Enzymes

The proof of principle and pioneering work for a biocatalytic reduction using a coupled cofactor regeneration process has been done by Wong et al., who also desc readable overview of the motivation to use this type of enzymes [94,95]. Notably, th from *B. cereus* is a thermostable enzyme and is beneficial in terms of stability. In add has a high specific activity of 250 U/mg. The corresponding enzymatic reduction of ket the presence of different types of ADHs, such as ADHs from horse liver, yeast, and *T.* (TADH), gave the desired alcohols with good to high enantioselectivities. Both AD GDH were used in an immobilized form. The conversions of these enzymatic biotrans tions were in the range of 72 to 90%. Although enantioselectivities varied, they exceed ee in many cases. Selected examples are shown in Scheme 32.22. The reactions are ty carried out at a substrate concentration of 125 mM.

This reduction method based on the use of isolated ADH and GDH enzymes h been applied for the synthesis of (S)-sulcatol [96]. In the presence of the NADPH-dep ADH of *T. brockii* and at a substrate concentration of 25 mM, sulcatone was reducec conversion of about 84%. Compared with the 2-propanol-based cofactor regen approach (conversion: 100%), conversion was somewhat lower when using a GDH.

A recent contribution to this field has been made by the Hua group focusing asymmetric reduction of α-chlorinated ketones in the presence of isolated ADHs and regeneration of the cofactor with a GDH [97]. The reactions have been carried o substrate concentration of ~4 g/L, and led to the formation of a range of α-chlo

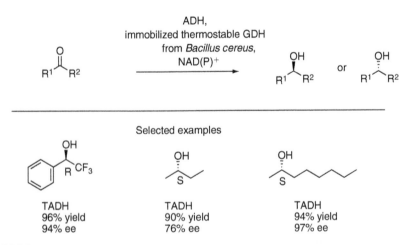

SCHEME 32.22

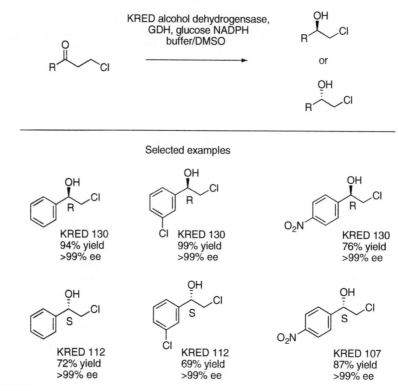

SCHEME 32.23

alcohols in high yields of 72 to 99%, with excellent enantioselectivities of typically >99% ee. Selected examples are shown in Scheme 32.23.

The feasibility of this methodology for an enantio- and diastereoselective reduction of a ketone, namely ethyl 6-benzyloxy-3,5-dioxohexanoate, has been demonstrated by the Patel group [98,99]. When using cell extracts of *Acinetobacter calcoaceticus* in combination with a GDH and glucose, the desired product ethyl (3*R*,5*S*)-6-benzyloxy-3,5-dihydroxyhexanoate was obtained with a conversion of 92% and an ee of 99% (Scheme 32.24). After product isolation, (3*R*,5*S*)-6-benzyloxy-3,5-dihydroxyhexanoate was obtained in 72% yield, with an ee of 99.5%.

In addition to screening for various ketones, the preparative asymmetric reduction of benzoyl hydroxyacetone and α-tetralone in the presence of isolated ADH and GDH enzymes has been described by BioCatalytics researchers [100]. Using the isolated ADH enzymes in an amount of 1 to 7% (w/w) compared with the amount of substrate and a catalytic amount of cofactor led to the synthesis of the optically active alcohols in high yields. Reductions have

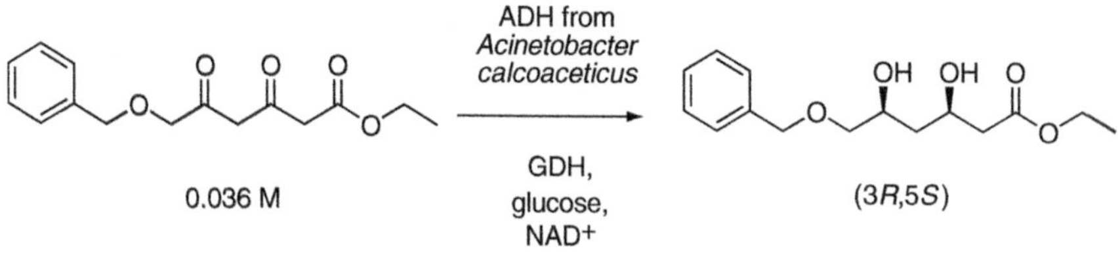

SCHEME 32.24

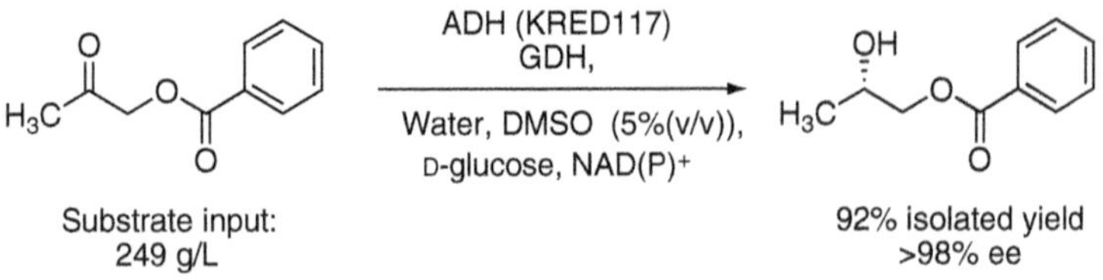

SCHEME 32.25

been carried out at high substrate concentrations of up to 0.75 to 1.4 M. A selected exam
shown in Scheme 32.25.

32.5.3 WHOLE-CELL REDUCTIONS

The design of recombinant whole cells is an elegant approach toward tailor-
(bio-)catalysts, called "designer bugs," which contain not only the cofactor "for free
also both of the desired enzymes, ADH and GDH, in overexpressed form. The correspo
reduction of ketones proceeds within the cell according to the concept shown in Scheme
thus converting glucose in a single step into gluconolactone, which itself is subseq
opened into its free acid form, gluconic acid. In parallel, the cell-internal ADH consum
formed NAD(P)H by reducing the prochiral ketone into the desired optically active al-
Advantages of such a recombinant whole-cell system over the wild-type ones are the
amount of the desired enzymes within the cell (due to overexpression) and its abil
produce these biocatalysts in a high-cell density fermentation process. Therein, wet bi
concentrations of >200 g/L of fermentation broth can be obtained, thus representing a
effective approach toward recombinant whole-cell catalysts. In addition, performance
cells with respect to enantioselectivity is better in the case of recombinant whole cells, a
the desired enzymes, namely ADH and GDH, are produced in large amounts. In cor
wild-type organisms contain a multitude of ADHs, which compete for the correspo
substrate. Due to the presence of substrate-tolerating (R)- and (S)-selective ADHs this
leads to low, nonsatisfactory enantioselectivities.

As synthetic efficiency depends on the performance of the (bio-)catalyst, the constr
of these catalysts is an issue of crucial importance for the later process. A well-k
approach is based on the use of E. coli as a host organism suitable for high cell d
fermentation and, thus, economical production.

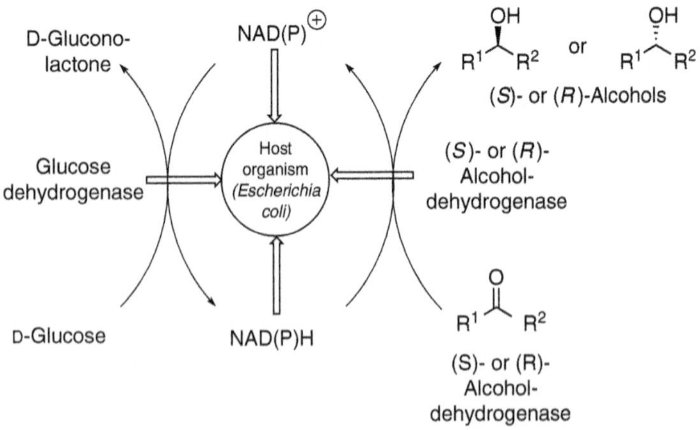

SCHEME 32.26

With respect to this recombinant whole-cell concept, the pioneers in the design and application of highly efficient recombinant whole-cell biocatalysts, consisting of an ADH and GDH, are Shimizu et al. [101]. As a GDH, the GDH from *B. megaterium*, which accepts both NADH and NADPH as a cofactor, was used. Already in the 1990s, Shimizu et al. had developed an effective *E. coli* catalyst, as well as a highly efficient reaction system for the reduction of 4-chloro-3-oxobutanoate. To start with the developed recombinant ADHs, several enzymes from *Sporobolomyces salmonicolor* [102–104] and *C. magnoliae* [105,106] have been screened, characterized, and successfully cloned in *E. coli* together with the GDH gene. Thus, a library of highly efficient biocatalysts, namely *E. coli* transformant cells coexpressing the corresponding ADH and GDH from *B. megaterium*, has been successfully established by Shimizu et al. [101].

The use of these efficient recombinant whole-cell catalysts in the asymmetric reduction of 4-chloro-3-oxobutanoate forming the corresponding pharmaceutically important alcohol has been intensively investigated and optimized by the Shimizu group. As a reaction media, an *n*-butyl acetate–water two-phase solvent system turned out to be very suitable [107]. When using the *E. coli* host organism overexpressing an NADP$^+$-dependent ADH from *S. salmonicolor*, and an isolated GDH enzyme or GDH-expressing cells as biocatalysts, the desired optically active (*R*)-alcohol was formed with up to 255 g/L in the organic phase under optimized conditions [108,109]. The conversion reached 91% and an ee of 91% was obtained [109]. In addition to glucose as a cosubstrate, a low amount of NADP$^+$ is required. A further improvement has been achieved when using *E. coli*, coexpressing both the ADH from *S. salmonicolor* and the GDH from *B. megaterium*, resulting in the formation of the desired optically active (*R*)-alcohol with 94.1% conversion and 91.7% ee at a substrate concentration of 300 g/L in the presence of a catalytic amount of the NADP$^+$ cofactor [110]. In Scheme 32.27, the concept of this application of a tailor-made whole-cell biocatalyst in a two-phase reaction media along with the experimental results are shown.

It is noteworthy that the Shimizu group also designed a whole-cell catalyst for the synthesis of the analog (*S*)-enantiomeric form of ethyl 4-chloro-3-hydroxybutanoate, which is pharmaceutically an important intermediate [111]. When using the recombinant *E. coli*

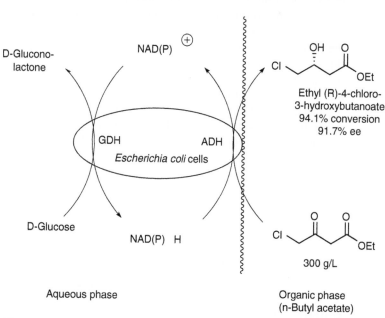

SCHEME 32.27

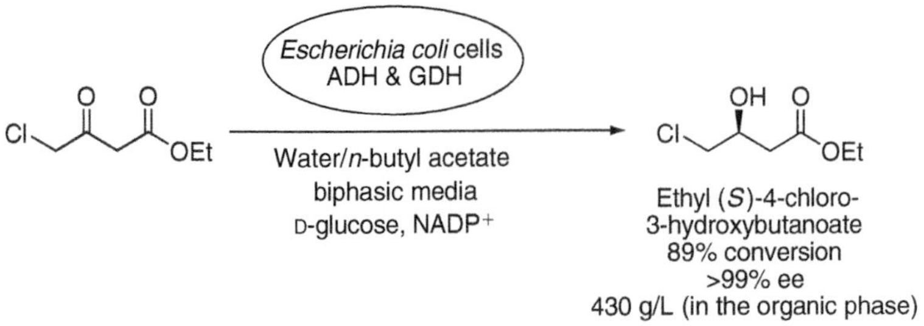

SCHEME 32.28

whole-cell catalyst, harboring an ADH from *C. magnoliae* and the GDH from *B. mega*
in an *n*-butyl acetate–water two-phase reaction media, the desired reduction proceeds
conversion of up to 89%, leading to 430 g/L of product in the organic phase (Scheme
The ee was >99%, and the turnover number (TON) of $NADP^+$ was high, with a TON
to 16.200. When feeding the substrate continuously, operating in aqueous solution ha
possible at a substrate concentration of 208 g/L, leading to 96% conversion and >9
With respect to the added $NADP^+$, a TON of 21.600 was achieved.

This impressive technology developed by the Shimizu group has already been co
cialized. Since 2000, Kaneka Corporation has applied this methodology to manufactur
(*S*)-4-chloro-3-hydroxybutanoate on an industrial scale [101].

In addition, Kaneka researchers jointly with the Shimizu group reported the exten
this reduction technology for the synthesis of other types of functionalized β-keto ester
For example, reduction of 4-bromo-3-oxobutanoate at a substrate concentration of 0.
the presence of an *E. coli* biocatalyst proceeds with >95% conversion, reaching a concen
of 0.38 M in the organic phase, within 20 h. The reaction course is shown in Scheme
The reaction was carried out in a two-phase system using *n*-butyl acetate as an organic
and a catalytic amount of the $NADP^+$ cofactor. In addition, the whole-cell rec
methodology has been further extended to a range of other substrates [101].

The application of a recombinant whole-cell catalyst, containing an ADH
P. methanolica SC 13825, has been developed by Patel et al., and successfully appl
the reduction of an acetophenone substituted with a keto ester–containing moiety [1
this case, however, an "external" amount of GDH has been added. The reaction pro
with a yield of 95% leading to an ee of 99.9% (Scheme 32.30). Notably, this biotransfor
has been scaled up to a 500 L scale.

The construction of an *E. coli* whole-cell catalyst, harboring the widely used (*R*)-se
ADH from *L. kefir* and a GDH from *B. subtilis*, has been successfully completed
Hummel group [114]. Interestingly, this GDH from *B. subtilis* also accepts both cof

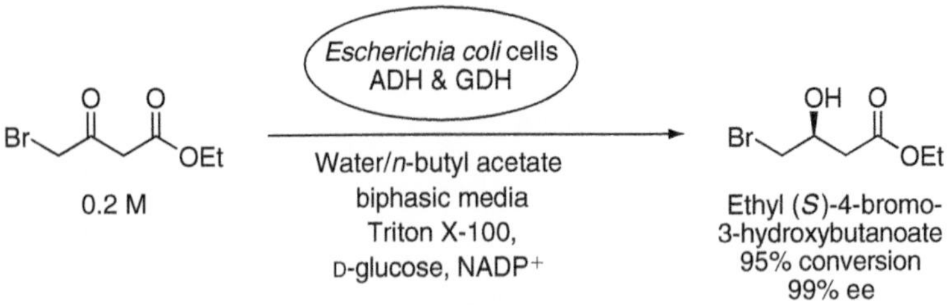

SCHEME 32.29

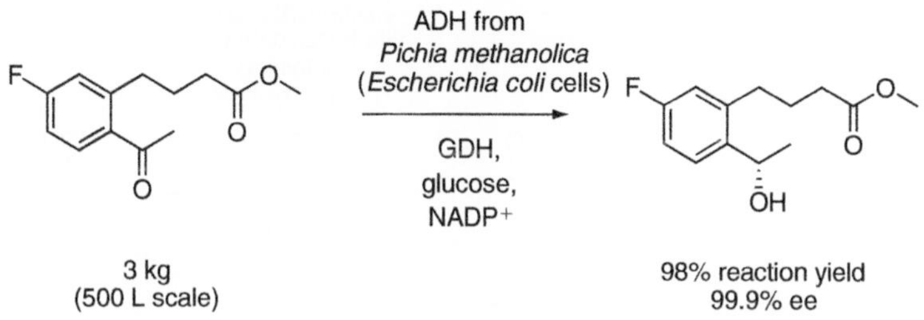

SCHEME 32.30

NAD$^+$ and NADP$^+$. Several reactions with acetophenone in many water–organic two-phase reaction media have been done using a substrate concentration of 20 mM. In general, the limited permeability of the membrane under the applied reaction conditions and limited stability of *E. coli* cells in organic solvents have been identified as limitations of the whole-cell approach.

In addition, Degussa AG researchers jointly with the Hummel group developed recombinant whole-cell biocatalysts and applied them in asymmetric reductions of a range of ketones at high substrate input, exceeding 150 g/L, in pure aqueous media, in general without the need of an external amount of cofactor [115]; in some exceptional cases a very low amount of cofactor had to be added with respect to a sufficient reaction course. Both types of enantiomers are available due to the use of (*S*)- and (*R*)-selective whole-cell biocatalysts. While the latter is based on the use of an ADH from *L. kefir*, an ADH from *R. erythropolis* has been used for the former. The suitability of both of these biocatalysts has been demonstrated at first by the synthesis of both enantiomeric forms of 4-chlorophenylethan-1-ol at a substrate concentration of 0.5 M (Scheme 32.31).

Subsequently, this methodology, which is both economical and simple to be carried out, has been used for the preparation of a wide range of optically active alcohols. Typically, the substrate concentrations have been in the range of 0.5 to 1 M, thus exceeding 100 g/L. An overview of examples is given in Scheme 32.32. Notably, in all of these examples the external addition of a cofactor was not required. In the presence of the tailor-made (*R*)- or (*S*)-selective whole-cell catalyst, the reduction proceeds with high conversions of up to 95% and an ee of up to 99.4%.

Furthermore, a 10 L scale reduction of 4-chloroacetophenone as a model substrate gave the (*R*)-4-chlorophenylethan-1-ol in 95% conversion and >99.8% ee [115]. After further process development and scale up, this recombinant whole-cell reduction technology platform has been applied on an industrial scale at Degussa AG.

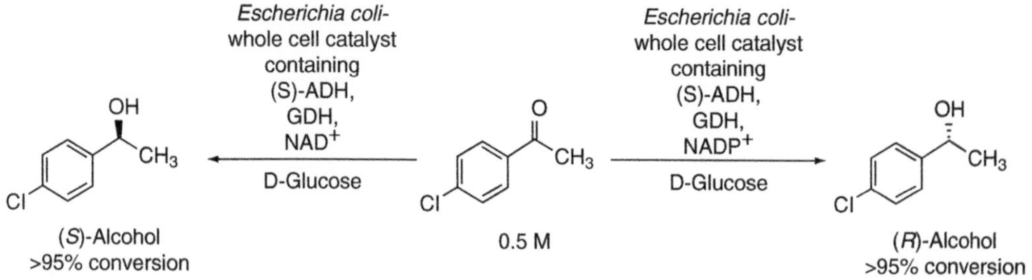

SCHEME 32.31

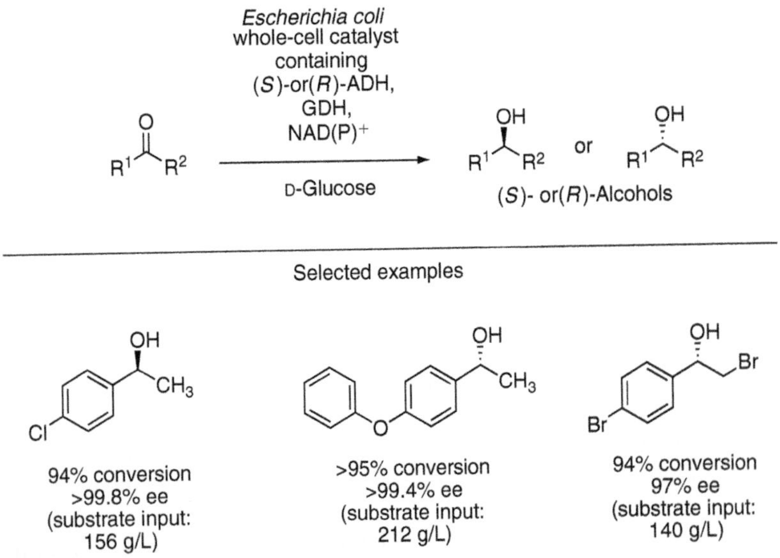

Escherichia coli
whole-cell catalyst
containing
(S)-or(R)-ADH,
GDH,
NAD(P)+

D-Glucose

(S)- or(R)-Alcohols

Selected examples

94% conversion
>99.8% ee
(substrate input:
156 g/L)

>95% conversion
>99.4% ee
(substrate input:
212 g/L)

94% conversion
97% ee
(substrate input:
140 g/L)

SCHEME 32.32

32.6 PROCESSES BASED ON COFACTOR REGENERATION WITH GLUCOSE-6-PHOSPHATE DEHYDROGENASE

32.6.1 THE CONCEPT

Another concept for cofactor regeneration related to the one with a GDH is based on th
of a G6PDH. The synthesis of chiral alcohols by means of this methodology has already
reported in 1981 by Wong and Whitesides [116]. These types of enzymes can also be use
the regeneration of cofactors NADH and NADPH. The concept based on the use of s
G6PDH in combination with an ADH for the ketone reduction for the synthesis of opt
active alcohols is shown in Scheme 32.33.

The high specific activity of G6PDHs, which is, for example, 676 U/mg for G6PDH
Saccharomyces carlsbergensis, contributes to their attractiveness as enzymes for co
regeneration. As a substrate, however, glucose-6-phosphate is needed, which can be fo
by the cells starting from glucose (in case of an whole-cell approach) or used directly (i
of isolated enzymes). The latter option, however, is less attractive as glucose-6-phos

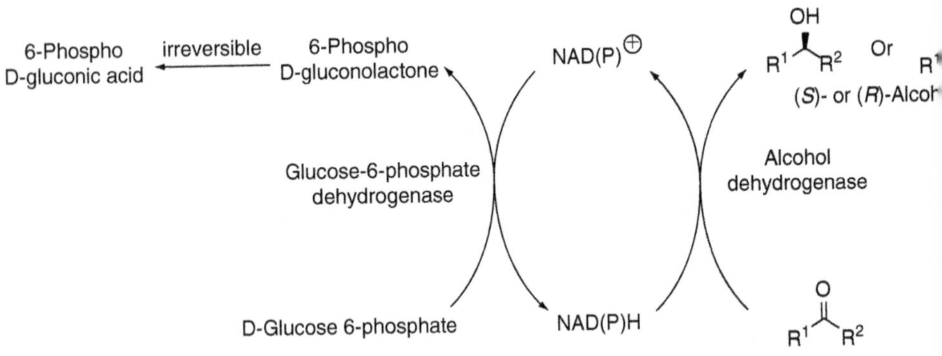

6-Phospho
D-gluconic acid   *irreversible*   6-Phospho
D-gluconolactone

$NAD(P)^{\oplus}$

(S)- or (R)-Alcoh

Glucose-6-phosphate
dehydrogenase

Alcohol
dehydrogenase

D-Glucose 6-phosphate   NAD(P)H

SCHEME 32.33

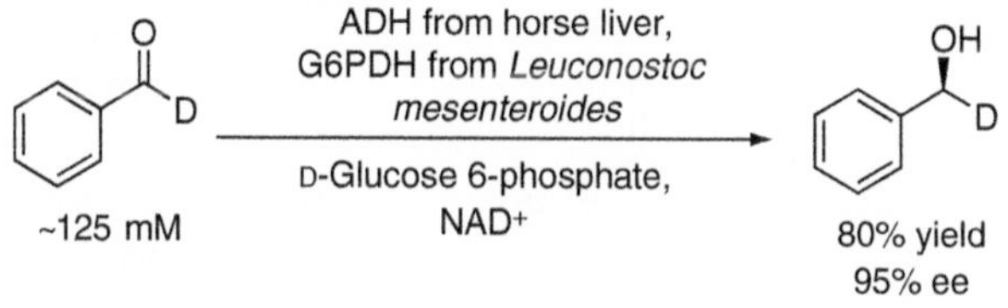

SCHEME 32.34

itself is an expensive compound. Accordingly, syntheses based on this type of cofactor regeneration with isolated enzymes have only been reported for small-scale applications. In contrast, the whole-cell approach has certainly the potential for large-scale applications, and promising process development has already been reported by Hanson et al. [117], which will be described in more detail in Section 32.6.2. An interesting option of this concept is the possibility to use intact cells, as glucose is transported to *E. coli* through a phosphotransferase system, which uses phosphoenol pyruvate as a phosphoryl donor, thus converting glucose into glucose-6-phosphate. Accordingly, a permeabilization of *E. coli* cells to ensure glucose supply (as in the case of the GDH-based cofactor regeneration) is not required.

32.6.2 ISOLATED ENZYMES

The initial studies on this G6PDH-based cofactor regeneration system and its proof of principle for synthetic application have been carried out by the Whitesides group using isolated enzymes (Scheme 32.34) [116]. In this pioneering work, G6PDH from *Leuconostoc mesenteroides* in combination with added glucose-6-phosphate turned out to be highly suitable for cofactor regeneration of both cofactors, NAD^+ and $NADP^+$. This G6PDH is stable and can be easily immobilized. The synthetic utility has already been shown in several examples, such as in the synthesis of (S)-1-d_1-phenylethanol which has been obtained in 80% yield with an ee of 95% at a substrate input of ~125 mM.

The combination of the G6PDH with the ADH from *L. kefir* for the synthesis of optically active (R)-phenylethan-1-ol has also been reported by Hummel [118]. The reductions, which were carried out at a 10 mM scale, gave the corresponding alcohol in 86% conversion.

The application of a two-phase solvent system for the reduction using an ADH and G6PDH has been described by Wong et al. for the synthesis of a variety of alcohols using an ADH from horse liver, *T. brockii*, and a hydroxyl steroid ADH [119]. In the presence of an ADH from horse liver, ethyl and butyl esters of 4-keto hexanoate were reduced in an aqueous buffer–hexane two-phase system to the corresponding (S)-alcohol in quantitative yield and >98% ee.

A range of G6PDH-based reductions have been reported by the Stewart group, who applied their impressive set of 19 recombinant ADHs from *S. cerevisia* in a screening of 11 α- and β-keto ester substrates [120]. Notably, in all cases ADHs have been identified which are complementary to each other, thus delivering both types of enantiomers of the corresponding alcohols in optically pure form. A selected example is shown in Scheme 32.35. The reactions, which have been carried out at a 5 mM substrate, have been done in combination with a commercial G6PDH for cofactor regeneration during the reduction process [117]. Very recently, the Stewart group extended the application range of their recombinant ADHs from *S. cerevisia* toward a highly stereoselective enzymatic reduction of an α-chloro-β-keto ester [121]. The corresponding alcohol represents a key intermediate of the α-hydroxy-β-amino acid moiety of $(-)$-bestatin.

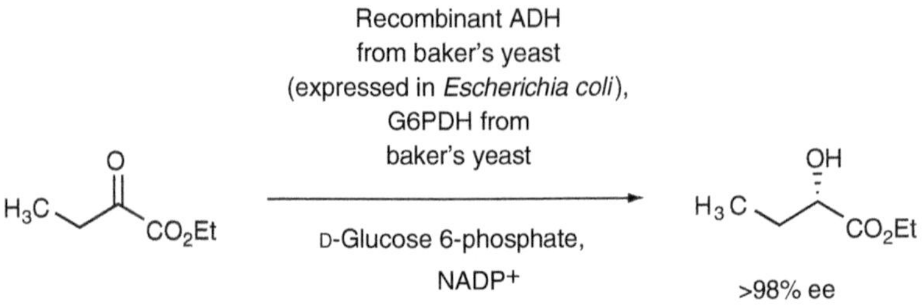

Recombinant ADH
from baker's yeast
(expressed in *Escherichia coli*),
G6PDH from
baker's yeast

D-Glucose 6-phosphate,
NADP+

>98% ee

SCHEME 32.35

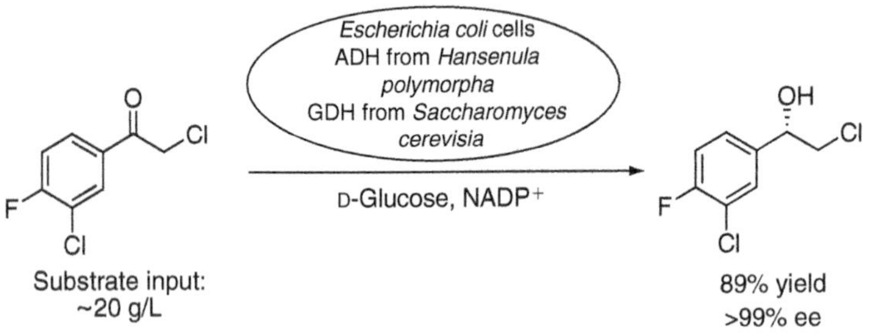

Escherichia coli cells
ADH from *Hansenula
polymorpha*
GDH from *Saccharomyces
cerevisia*

D-Glucose, NADP+

Substrate input:
~20 g/L

89% yield
>99% ee

SCHEME 32.36

32.6.3 WHOLE-CELL REDUCTIONS

For the asymmetric reductions with *E. coli* whole cells bearing a G6PDH from *S. cere*
recombinant whole cell was constructed recently by Hanson et al. [117]. Besides this G
the recombinant *E. coli* cells contained an NADPH-dependent ADH from *Ha
polymorpha*, which turned out to give an ee of >99% in the target reduction of
dichloro-4'-fluoroacetophenone. Subsequently, in the presence of a low amount of
NADP$^+$ whole-cell biotransformation was carried out leading to the desired produc
chloro-1-(3'-chloro-4'-fluorophenyl)-ethanol in 89% yield and >99% ee (Scheme 32.3
substrate input of this reduction was ~20 g/L, and the intact *E. coli* cells were provide
glucose directly.

32.7 SUMMARY AND OUTLOOK

In summary, numerous methodologies for asymmetric biocatalytic reductions of l
have been developed based on the use of isolated enzymes and whole-cell catalysts
respect to the latter, in particular, recombinant whole-cell systems very high impo
Notably, these reduction methods are based on different types of concepts for an
cofactor recycling, comprising substrate-coupled cofactor regeneration with 2-propan
enzyme-coupled cofactor regenerations with an FDH, GDH, and G6PDH. Although
ent from a conceptual point of view, high efficiency in organic synthetic transformat
ketones to optically active alcohols has been demonstrated by means of all these m
This is underlined by several applications on industrial scale that have been achie
addition to efficient cofactor regeneration, easy, cost-attractive, and large-scale acces
of the enzymes are another key criterion. Thus, availability of recombinant AI

combination with the production of the biocatalysts by means of a high cell density fermentation process plays a crucial role. In the future, we will certainly see further screening for new ADHs and the extension of the application range of biocatalytic reduction.

REFERENCES AND NOTES

1. Kleemann, A., Engels, J., Kutscher, B., and Reichert, D., *Pharmaceutical Substances: Synthesis, Patents, Applications*, 4th ed., Thieme, Stuttgart, Germany, 2001.
2. Breuer, M., Ditrich, K., Harbicher, T., Hauer, B., Keßeler, M., Stürmer, M., and Zelinski, T., *Angew. Chem.*, 116, 804–843, 2004.
3. Helmchen, G., Hoffmann, R.W., Mulzer, J., and Schaumann, E., Eds., *Houben-Weyl, Methods of Organic Chemistry, Stereoselective Synthesis*, vol. E21, Thieme, Stuttgart, Germany, 1995,
4. Ojima, I., Ed., *Catalytic Asymmetric Synthesis*, 2nd ed., Wiley-VCH, Weinheim, Germany, 2000.
5. Jacobsen, E.N., Pfaltz, A., and Yamamoto, H., Eds., *Comprehensive Asymmetric Catalysis I–III*, Springer, Berlin, 1999.
6. Noyori, R., *Angew. Chem. Int. Ed.*, 41, 2008–2022, 2002.
7. Corey, E.J. and Helal, C.J., *Angew. Chem. Int. Ed.*, 37, 1986–2012, 1998.
8. Schmid, A., Dordick, J.S., Hauer, B., Kiener, A., Wubbolts, M., and Witholt, B., *Nature*, 409, 258, 2001.
9. Nakamura, K. and Matsuda, T., in *Enzyme Catalysis in Organic Synthesis*, Drauz, K. and Waldmann, H., Eds., vol. 3, 2nd ed., Wiley-VCH, Weinheim, Germany, 2002, p. 991.
10. Hummel, W., *Adv. Biochem. Eng. Biotechnol.*, 58, 146, 1997.
11. Faber, K., *Biotransformations in Organic Chemistry*, vol. 5, Auflage, Springer, Berlin, 2004, Kapitel 2.2.3, p. 192.
12. Hummel, W., *TIBTECH*, 17, 487–492, 1999.
13. Eckstein, M., Daußmann, T., and Kragl, U., *Biocat. Biotransf.*, 22, 89, 2004.
14. Kula, M.-R. and Kragl, U., *Stereoselective Biocatalysis*, Patel, R.N., Ed., Marcel Dekker, New York, 2000, p. 839.
15. Csuk, R. and Glanzer, B.I., *Chem. Rev.*, 91, 49, 1991.
16. Stewart, J.D., *Curr. Op. Drug Dis. Dev.*, 1, 278, 1998.
17. Csuk, R. and Glanzer, B.I., *Stereoselective Biocatalysis*, Patel, R.N., Ed., Marcel Dekker, New York, 2000, p. 527.
18. Stewart, J.D., *Curr. Op. Biotechnol.*, 1, 363, 2000.
19. Bertau, M. and Burli, M., Industrial applications of Baker's yeast reduction, *Chimia*, 54, 503, 2000.
20. Haberland, J., Hummel, W., and Daussmann, T., *Org. Proc. Res. Dev.*, 6, 458–462, 2002.
21. Haltrich, D., Nidetzky, B., Miemietz, G., Gollhofer, D., Lutz, S., Stolz, P., and Kulbe, K.D., *Biocatal. Biotrans.*, 14, 31, 1996.
22. Holt, R.A., *Chimica Oggi*, 9, 17, 1996.
23. Holt, R.A. and Rigby, S.R., Avecia Limited, EP658211, 1995.
24. Liese, A., Seelbach, K., and Wandrey, C., *Industrial Biotransformations*, Wiley-VCH, Weinheim, Germany, 2000, p. 99.
25. Mertens, R., Greiner, L., van den Ban, E.C.D., Haaker, H.B.C.M., and Liese, A., *J. Mol. Catal. B Enzym.*, 24–25, 39, 2003.
26. Vrtis, J.M., White, A.K., Metcalf, W.M., and van der Donk, W.A., *Angew. Chem.*, 114, 3391, 2002.
27. Naahmniek, S., Hummel, W., and Gröger, H., Degussa AG, DE10240603, 2004.
28. Yuan, R., Watanabe, S., Kuwabata, S., and Yoneyama, H., *J. Org. Chem.*, 62, 2494, 1997.
29. Homann, M.J., Vail, R.B., Previte, E., Tamarez, M., Morgan, B., Dodds, D.R., and Zaks, A., *Tetrahedron*, 60, 789–797, 2004.
30. Villela Filho, M., Stillger, T., Müller, M., Liese, A., and Wandrey, C., *Angew. Chem. Int. Ed.*, 42, 2993–2996, 2003.
31. Stillger, T., Bönitz, M., Vilella, M., and Liese, A., *Chemie Ingenieur Technik*, 74, 1035, 2002.

32. Stampfer, W., Kosjek, B., Moitzi, C., Kroutil, W., and Faber, K., *Angew. Chem. Int. Ed.*, 4 2002.
33. Nakamura, K., Inoue, Y., Matsuda, T., and Misawa, I., *J. Chem. Soc. Perkin Trans.* 1, 239
34. Bradshaw, C.W., Fu, H., Shen, G.-J., and Wong, C.-H., *J. Org. Chem.*, 57, 1526, 1992.
35. Yang, H., Jönsson, A., Wehtje, E., Adlerkreutz, P., and Mattiasson, B., *Biochim. Biophy.* 1336, 51, 1997.
36. Clair, N. St, Wang, Y.-F., and Margolin, A.L., *Angew. Chem. Int. Ed.*, 39, 380, 2000.
37. Gupta, A., Breese, K., Bange, G., and Neubauer, P., Juelich Enzyme Products (WO 02086126, 2002.
38. De Temino, D.M.-R., Hartmeier, W., and Ansorge-Schumacher, M.B., *Enzyme Microb. T.* 36, 3, 2005.
39. Riermeier, T., Bornscheuer, U., Altenbuchner, J., and Hildebrandt, P., Degussa AG, EP 1. 2002.
40. Hildebrandt, P., Riermeier, T., Altenbuchner, J., and Bornscheuer, U., *Tetrahedron Asyr.* 12, 1207, 2001.
41. Inoue, K., Makino, Y., and Itoh, N., *Appl. Environ. Microbiol.*, 71, 3633, 2005.
42. Inoue, K., Makino, Y., and Itoh, N., *Tetrahedron Asymmetry*, 16, 2539, 2005.
43. Eckstein, M., Villela, M., Liese, A., and Kragl, U., *Chem. Commun.*, 1984, 2004.
44. Matsumura, S., Imafuku, H., Takahashi, Y., and Toshima, K., *Chem. Lett.*, 251, 1993.
45. Itoh, N., Matsuda, M., Mabuchi, M., Dairi, T., and Wang, J., *Eur. J. Biochem.*, 269, 239(
46. Matsuyama, A., Yamamoto, H., and Kobayashi, Y., *Org. Process Res. Dev.*, 6, 558, 2002
47. Amidjojo, M. and Weuster-Botz, D., *Tetrahedron Asymmetry*, 16, 899, 2005.
48. Wolberg, M., Hummel, W., Wandrey, C., and Müller, M., *Angew. Chem.*, 112, 4476, 200(
49. Stampfer, W., Kosjek, B., Faber, K., and Kroutil, W., *J. Org. Chem.*, 68, 402, 2003.
50. Stampfer, W., Kosjek, B., Kroutil, W., and Faber, K., *Biotechnol. Bioeng.*, 81, 865, 2003.
51. Stampfer, W., Kosjek, B., Kroutil, W., Faber, K., Niehaus, F., and Eck, J., Ciba S) Chemicals Holding Inc., WO2005026338, 2005.
52. Slusarczyk, H., Felber, S., Kula, M.-R., and Pohl, M., *Eur. J. Biochem.*, 267, 1280, 2000.
53. Kroner, K.H., Schütte, H., Stach, W., and Kula, M.R., *J. Chem. Technol. Biotechnol.*, ? 1982.
54. Wandrey, C., Wichmann, R., Bückmann, A.F., and Kula, M.-R., *Enzyme Eng.*, 5, 453, 19
55. Shaked, Z. and Whitesides, G.M., *J. Am. Chem. Soc.*, 102, 7104, 1980.
56. Bommarius, A.S., Drauz, K., Hummel, W., Kula, M.-R., and Wandrey, C., *Biocatalysis*, 1994.
57. Liese, A., Seelbach, K., and Wandrey, C., *Industrial Biotransformations*, Wiley-VCH, We Germany, 2000, p. 103.
58. Tishkov, V.I., Galkin, A.G., Fedorchuk, V.V., Savitsky, P.A., Rojkova, A.M., Gieren,) Kula, M.-R., *Biotechnol. Bioeng.*, 64, 187, 1999.
59. Galkin, A., Kulakova, L., Tishkov, V., Esaki, N., and Soda, K., *Appl. Microbiol. Biotech* 479, 1995.
60. Ernst, M., Kaup, B., Müller, M., Binger-Meyer, S., and Sahm, H., *Appl. Microbiol. Biotech* 629, 2005.
61. Hummel, W. and Gottwald, C., Bayer AG, DE4209022, 1992.
62. Zelinski, T., Peters, J., and Kula, M.-R., *J. Biotechnol.*, 33, 283, 1994.
63. Peters, J., Minuth, T., and Kula, M.-R., *Enzyme Microb. Technol.*, 15, 950, 1993.
64. Adlercreutz, P., *Enzyme Microb. Technol.*, 13, 9, 1991.
65. Itoh, N., Mizuguchi, N., and Mabuchi, M., *J. Mol. Catal. B Enzym.*, 6, 41, 1999.
66. Peters, J., Zelinski, T., Minuth, T., and Kula, M.-R., *Tetrahedron Asymmetry*, 4, 1683, 19
67. Peters, J., Zelinski, T., and Kula, M.-R., *Appl. Microbiol. Biotechnol.*, 38, 334, 1992.
68. Zelinski, T., Liese, A., Wandrey, C., and Kula, M.-R., *Tetrahedron Asymmetry*, 10, 1681,
69. Kruse, W., Hummel, W., and Kragl, U., *Recl. Trav. Chim. Pays-Bas*, 115, 239, 1996.
70. Kragl, U., Kruse, W., Hummel, W., and Wandrey, C., *Biotechnol. Bioeng.*, 52, 309, 1996.
71. Liese, A., Seelbach, K., and Wandrey, C., *Industrial Biotransformations*, Wiley-VCH, We Germany, 2000, p. 103.

72. Kruse, W., Kragl, U., and Wandrey, C., Forschungszentrum Jülich GmbH, DE 4436149, 1996.
73. Liese, A., Zelinski, T., Kula, M.-R., Kierkels, H., Karutz, M., Kragl, U., and Wandrey, C., *J. Mol. Catal. B Enzym.*, 4, 91, 1998.
74. Rissom, S., Beliczey, J., Giffels, G., Kragl, U., and Wandrey, C., *Tetrahedron Asymmetry*, 10, 923, 1999.
75. Gröger, H., Hummel, W., Buchholz, S., Drauz, K., Nguyen, T.V., Rollmann, C., Hüsken, H., and Abokitse, K., *Org. Lett.*, 5, 173, 2003.
76. Gröger, H., Hummel, W., Rollmann, C., Chamouleau, F., Hüsken, H., Werner, H., Wunderlich, C., Abokitse, K., Drauz, K., and Buchholz, S., *Tetrahedron*, 60, 633, 2004.
77. Gröger, H., Rollmann, C., Hüsken, H., Werner, H., Chamouleau, F., Hagedorn, C., Drauz, K., and Hummel, W., Degussa AG, WO 2004085662, 2004.
78. Gröger, H., *Chiral Europe Conference Proceedings*, 2004.
79. Nanduri, V.B., Banerjee, A., Howell, J.M., Brzozowski, B.B., Eiring, R.F., and Patel, R.N., *J. Ind. Microbiol. Biotechnol.*, 25, 171, 2000.
80. Orlich, B. and Schomäcker, R., *Biotechnol. Bioeng.*, 65, 357, 1999.
81. Orlich, B., Berger, H., Lade, M., and Schomäcker, R., *Biotechnol. Bioeng.*, 70, 638, 2000.
82. Matsuyama, A., Yamamoto, H., and Kobayashi, Y., *Org. Proc. Res. Dev.*, 6, 558, 2002.
83. Ernst, M., Kaup, B., Bringer-Meyer, S., and Sahm, H., *Appl. Microbiol. Biotechnol.*, 66, 629, 2005.
84. Weckbecker, A. and Hummel, W., *Biotechnol. Lett.*, 26, 1739, 2004.
85. Hummel, W. and Kula, M.-R., Forschungszentrum Jülich GmbH, EP 456107, 1991.
86. Bradshaw, C.W., Hummel, W., and Wong, C.-H., *J. Org. Chem.*, 57, 1532, 1992.
87. Haltrich, D., Nidetzky, B., Miemitz, D., Gollhofer, D., Lutz, S., Stolz, P., and Kulbe, K.D., *Biocatal. Biotrans.*, 14, 31, 1996.
88. Kataoka, M., Rohani, L.P.S., Wada, M., Kita, K., Yanase, H., Urabe, I., and Shimizu, S., *Biosci. Biotechnol. Biochem.*, 62, 167, 1998.
89. Hilt, W., Pfleiderer, G., and Fortnagel, P., *Biochim. Biophys. Acta*, 1076, 298, 1991.
90. Karmali, A. and Serralheiro, L., *Biochimie*, 70, 1401, 1988.
91. Bach, J.A. and Sadoff, H.L., *J. Bacteriol.*, 83, 699, 1962.
92. Wong, C.-H., Drueckhammer, D.G., and Sweers, H.M., *J. Am. Chem. Soc.*, 107, 4028, 1985.
93. Smith, L.D., Budgen, N., Bungard, S.J., Danson, M.J., and Jough, D.W., *Biochem. J.*, 261, 973, 1989.
94. Wong, C.-H., Drueckhammer, D.G., and Sweers, H.M., *J. Am. Chem. Soc.*, 107, 4028, 1985.
95. Wong, C.-H. and Drueckhammer, D.G., *Biol. Technol.*, 3, 649, 1995.
96. Bastos, F.M., dos Santos, A.G., Jones J. Jr., Oestreicher, E.G., Pinto, G.F., and Paiva, L.M.C., *Biotechnol. Technol.*, 13, 661, 1999.
97. Zhu, D., Mukherjee, C., and Hua, L., *Tetrahedron Asymmetry*, 16, 3275, 2005.
98. Patel, R.N., Banerjee, A., McNamee, C.G., Brzozowski, D., Hanson, R.L., and Szarka, L.J., *Enzyme Microb. Technol.*, 15, 1014, 1993.
99. Liese, A., Seelbach, K., and Wandrey, C., *Industrial Biotransformations*, Wiley-VCH, Weinheim, Germany, 2000, p.107.
100. Kaluzna, I.A., Rozzell, J.D., and Kambourakis, S., *Tetrahedron Asymmetry*, 16, 3682, 2005.
101. Kataoka, M., Kita, K., Wada, M., Yasohara, Y., Hasegawa, J., and Shimizu, S., *Appl. Microbiol. Biotechnol.*, 62, 437, 2003.
102. Yamada, H., Shimizu, S., Kataoka, M., Sakai, H., and Miyoshi, T., *FEMS Microbiol. Lett.*, 70, 45, 1990.
103. Kataoka, M., Sakai, H., Morikawa, T., Katoh, M., Miyoshi, T., Shimizu, S., and Yamada, H., *Biochim. Biphys. Acta*, 1122, 57, 1992.
104. Kita, K., Matuszaki, K., Hashimoto, T., Yanase, H., Kato, N., Chung, M.C.M., Kataoka, M., and Shimizu, S., *Appl. Environ. Microbiol.*, 62, 2303, 1996.
105. Wada, M., Kataoka, M., Kawabata, H., Yasohara, Y., Kizaki, N., Hasegawa, J., and Shimizu, S., *Biosci. Biotechnol. Biochem.*, 62, 280, 1998.
106. Yasohara, Y., Kizaki, N., Hasegawa, J., Wada, M., Kataoka, M., and Shimizu, S., *Biosci. Biotechnol. Biochem.*, 64, 1430, 2000.
107. Shimizu, S., Kataoka, M., Katoh, M., Morikawa, T., Miyoshi, T., and Yamada, H., *Appl. Environ. Microbiol.*, 56, 2374, 1990.

108. Kataoka, M., Rohani, L.P.S., Yamamoto, K., Wada, M., Kawabata, H., Kita, K., Yam and Shimizu, S., *Appl. Microbiol. Biotechnol.*, 48, 699, 1997.

109. Kataoka, M., Rohani, L.P.S., Wada, M., Kita, K., Yanase, H., Urabe, I., and Shimizu, S. *Biotechnol. Biochem.*, 62, 167, 1998.

110. Kataoka, M., Yamamoto, K., Kawabata, H., Wada, M., Kita, K., Yanase, H., and Shin *Appl. Microbiol. Biotechnol.*, 51, 486, 1999.

111. Kizaki, N., Yasohara, Y., Hasegawa, J., Wada, M., Kataoka, M., and Shimizu, S., *Appl. biol. Biotechnol.*, 55, 590, 2001.

112. Yasohara, Y., Kizaki, N., Hasegawa, J., Wada, M., Kataoka, M., and Shimizu, S., *Tetr Asymmetry*, 12, 1713, 2001.

113. Patel, R.N., Goswami, A., Chu, L., Donovan, M.J., Nanduri, V., Goldberg, S., Johns Siva, P.J., Nielsen, B., Fan, J., He, W., Shi, Z., Wang, K.Y., Eiring, R., Cazzulino, D., Si Mueller, R., *Tetrahedron Asymmetry*, 15, 1247–1258, 2004.

114. Weckbecker, A. and Hummel, W., *Methods in Biotechnology, Microbial Enzymes and Biot mations*, Barredo, J.L., Ed., vol. 17, Humana Press, Totowa, NJ, 2004, p. 225.

115. Gröger, H., Chamouleau, F., Orologas, N., Rollmann, C., Drauz, K., Hummel, W., Weckbec and May, O., *Angew. Chem.*, 2006, in press.

116. Wong, C.-H. and Whitesides, G.M., *J. Am. Chem. Soc.*, 103, 4890, 1981.

117. Hanson, R.L., Goldberg, S., Goswami, A., Tully, T.P., and Patel, R.N., *Adv. Synth. Cat* 1073, 2005.

118. Hummel, W., *Appl. Microbiol. Biotechnol.*, 34, 15, 1990.

119. Drueckhammer, D.J., Sadozai, S.K., Wong, C.-H., and Roberts, S.M., *Enzyme Microb. T* 9, 564–570, 1987.

120. Kaluzna, I.A., Matsuda, T., Sewell, A.K., and Stewart, J.D., *J. Am. Chem. Soc.*, 126, 12827–128.

121. Feske, B.D. and Stewart, J.D., *Tetrahedron Asymmetry*, 16, 3124–3127, 2005.

33 Enzyme Catalysis in Nonaqueous Media: Past, Present, and Future

Susanne Dreyer, Julia Lembrecht, Jan Schumacher, and Udo Kragl

CONTENTS

33.1 INTRODUCTION

Nature has designed its biocatalysts to perform best in an aqueous surrounding, neutral pH, and temperatures below 50°C. However, these conditions are often contrary to the requirements of the process engineer or chemist to optimize a reaction with respect to volumetric productivity or an easy downstream processing when substrates and/or products are not easily soluble in water. To overcome these bottlenecks addition of organic solvents is common practice. This has been used by various authors over the years, but a breakthrough was achieved during the second half of the 20th century [1–3].

The solvent can either be water-miscible or water-immiscible, resulting in one- o
phase systems. Depending on the solvent phase the enzyme will be dissolved o
suspended. Thus, the systems might be liquid, liquid–liquid, or liquid–solid, but even gas
reactions are gaining increasing attention. The possibilities are summarized in Figur
The aim of this chapter is to highlight some of the developments over the years. Thus
way can it be exhaustive.

In a pure aqueous system the solvents implied are the dissolved enzyme(s), cofacto
cosubstrates if needed, the substrate, and the product. In a pure organic phase most of
enzyme is not soluble and therefore applied in suspension either in its native form o
carrier. In a two-phase system formed by water and an organic solvent the enzyme is
dissolved in the bulk aqueous phase or can be localized at the interface. The organic
then can be seen as a reservoir for a poorly water-soluble substrate. During the reacti
substrate and the product formed will partition between the two phases. On the one
using two-phase systems for enzyme-catalyzed reactions might have several advai
Firstly, higher concentrations of hydrophobic substrates and therefore higher volu
productivities can be achieved. Secondly, no enzyme inhibition by high substrate con
tions will occur. Thirdly, an easy workup by phase separation is possible. Fourthly
immobilization" of the enzyme and cofactor in the aqueous phase allows reuse of the e
and the cofactor [4–6].

On the other hand, the influence of the organic phase on enzyme activity, stabilit
regio- and stereospecificity has to be evaluated. The choice of the organic solvent is
influenced by the enzyme behavior and by the partitioning behavior of substrate and p
[7–9]. It should be noted that despite all success there is no general rule as to which s
is "enzyme friendly." To a certain extent, the log P concept, based on the distri

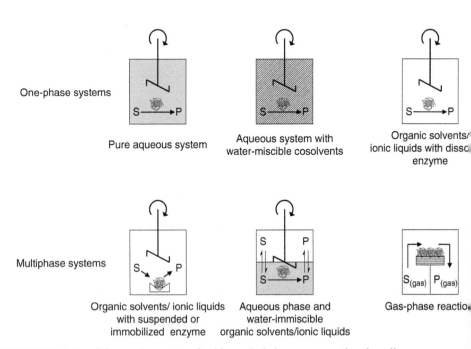

One-phase systems

Pure aqueous system

Aqueous system with
water-miscible cosolvents

Organic solvents/
ionic liquids with dissc
enzyme

Multiphase systems

Organic solvents/ ionic liquids
with suspended or
immobilized enzyme

Aqueous phase and
water-immiscible
organic solvents/ionic liquids

Gas-phase reactio

FIGURE 33.1 Possible reaction system for biocatalysis in nonconventional media.

coefficient between water and n-octanol, can be used as a guideline [10]. In general, solvents with a log P >3 such as xylene (3.1) or hexane (3.9) are less deactivating than those with a low log P such as ethanol (−0.24). Surprisingly, *tert*-butanol (0.35) stabilizes enzymes [11]. Certainly the hydrophilicity of the cosolvent is important, as it allows interaction and breaking of hydrogen bonds that are stabilizing the tertiary structure of the enzyme. However, not only have common organic solvents been used for biocatalysis, but supercritical CO_2 [12] and recently even ionic liquids (ILs) have also been shown to be compatible with enzymes or whole cells [13–16].

33.2   EC 1: OXIDOREDUCTASES

The oxidoreductases represent the first group of the classification formulated by the enzyme community (EC numbers). Enzymes that are part of this group reversely catalyze the oxidation and reduction of substrates by transferring two electrons. Mostly they need a cofactor, also called coenzyme, which activates the enzyme and acts as a mediator between the active site of the biocatalyst and the substrate [17]. These cofactors are small organic nonprotein molecules that can covalently or noncovalently bind to the inactive apoenzyme to build the active catalyst (haloenzyme). Such coenzymes are molecules like nicotinamide adenine dinucleotide (NAD), nicotinamide adenine dinucleotide phosphate (NADP), and flavine adenine dinucleotide (FAD). As can be seen in Table 33.1, oxidoreductases are divided into 22 subclasses.

TABLE 33.1
Subclasses of the Oxidoreductases

EC 1 Oxidoreductases

EC 1.1	Acting on the CH–OH group of donors
EC 1.2	Acting on the aldehyde or oxo group of donors
EC 1.3	Acting on the CH–CH group of donors
EC 1.4	Acting on the CH–NH_2 group of donors
EC 1.5	Acting on the CH–NH group of donors
EC 1.6	Acting on the NADH or NADPH
EC 1.7	Acting on the other nitrogenous compounds of donors
EC 1.8	Acting on a sulfur group of donors
EC 1.9	Acting on a heme group of donors
EC 1.10	Acting on diphenols and related substances of donors
EC 1.11	Acting on peroxide as acceptor
EC 1.12	Acting on hydrogen of donors
EC 1.13	Acting on single donors with incorporation of molecular oxygen (oxygenases)
EC 1.14	Acting on paired donors with incorporation or reduction of molecular oxygen
EC 1.15	Acting on superoxide as acceptor
EC 1.16	Oxidizing metal ions
EC 1.17	Acting on the CH or CH_2 groups
EC 1.18	Acting on iron–sulfur proteins as donor
EC 1.19	Acting on reduced flavodoxin as donor
EC 1.20	Acting on phosphorus or arsenic in donor
EC 1.21	Acting on X–H and Y–H to form an X–Y bond
EC 1.97	Other oxidoreductases

33.2.1 ALCOHOL DEHYDROGENASES

The biggest subclass of the oxidoreductases comprise alcohol dehydrogenases (AD 1.1). In the past, the ADHs were typically used in a buffered aqueous system because t their cofactors are sensitive against organic solvents. For example, Hummel pu the reduction of acetophenone to (R)-1-phenylethanol by the ADH from Lacto kefir in potassium phosphate buffer with NADPH regenerated by a glucose-6-ph dehydrogenase (G6PDH) [18].

An oft-applied enzyme is the yeast alcohol dehydrogenase (YADH). The devel started with the use of whole cells as biocatalysts to avoid the cofactor regeneratio additional reaction step. Since the end of the 20th century YADH has also been depl nonconventional media. Howarth et al. reported that it is possible to use immobilized yeast as whole cells in ILs mixed with water (10:1) to reduce prochiral ketones Organic solvents offer the chance to selectively form one of the enantiomers. It h reported that both enantiomers were produced when reducing 2-oxohexanolate by yeast in water, but when the biotransformation was conducted in benzene, the (R)- was formed in high yields [21]. Although the ADH can be successfully used under n ventional conditions, it should be noted that the catalytic activity strongly depends water activity in the system [22].

Further, Liao et al. performed a YADH covalently bonded onto Fe_3O_4 magneti particles, which is active in a water/AOT/isooctane microemulsion (Figure 33.2). The r activity immobilized YADH after 700 h was 78% and, in contrast to that, the free loses most of its activity in the same reaction system after 1 h [23]. Beyond this, Y highly stable and active in reverse micelles of AOT/isooctane depending on water c pH, and time [24,25].

For further research cross-linked enzyme aggregates (CLEAs), gas-phase reactio combinations of different organic solvents would be a promising field. However, there some open questions in the areas of one- and two-phase systems that have to be analyz better understanding of the mechanism.

Since the last 10 to 15 years there have appeared a number of reports about A nonconventional media. A new ADH from L. brevis (LBADH), first mentioned by

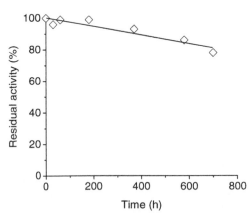

FIGURE 33.2 Storage stability of bound YADH at 25°C. The activity measurement was prefo 10 ml microemulsion solution at 0.1 M AOT, 0.2 mM NADH, 0.1 M 2-butanone, 25°C, and The concentration of bound YADH was 0.5 mg/ml. The initial 100% absolute values of act bound YADH was 8 nmol/(min·mg). (From Liao, M.H. and Chen, D.H., *J. Mol. Catal. B Enz* 81–87, 2002.)

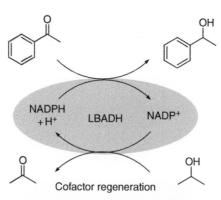

FIGURE 33.3 Model reaction in the gas phase: reduction of acetophenone with cofactor regeneration by 2-propanol oxidation. Enzyme and cofactor are immobilized on glass beads. (From Ferloni, C. et al., *Biokatalyse* (Transkript Sonderheft), 105–108, 2003.)

is able to catalyze the reduction of prochiral ketones [26]. This can be done not only in buffer but also in many different reaction systems. The advantage of all these systems is a higher stability of the **LBADH**. As published by Filho et al. the half-lives of ADHs are not directly related to the log P values of the solvent, but it is possible to appraise the miscibility with water and thus enable the contact between the enzyme and the organic solvent [27].

Ferloni et al. reported that the immobilized **LBADH** shows an enzyme activity of 100% for the reduction of substrates in the aqueous phase and also in the gas phase (Figure 33.3) [28]. This implies a high reactivity of the **ADH** to the reduction of acetophenone in the gas phase. The conversion of the substrate and the reactivity of the enzyme depend on relative moisture in the reaction system, on the pressure, and on the molecular ratio of the cosubstrate 2-propanol to the substrate.

The reduction of prochiral ketons catalyzed by **LBADH** including substrate-coupled regeneration with 2-propanol has been used for the production of chiral alcohol on a 10 to 100 kg scale, e.g., ethyl-(R)-3-hydroxybutyrate or (R)-2-octanol. The reaction was performed in a one-phase system with an extraction step at the end [29]. Schumacher et al. researched the **LBADH**-catalyzed reduction of short aliphatic ketones in a one-phase system with cosolvents [30]. They investigated an increasing enantiomeric excess (ee) of (R)-butan-2-ol (36.5 to 43.0%) by increasing the amount of acetonitrile from 0% (v/v) to 24.5% (v/v).

It is also possible to arrange the reaction in a two-phase system (Figure 33.4). Eckstein et al. investigated the reduction of 2-octanone to (R)-2-octanol with the cofactor regeneration by the same enzyme in two different binary systems [31]. In the first system of methyl *tert*-butyl ether (MTBE) and buffer the reduction catalyzed by the free **LBADH** achieved a conversion of 61% after 180 min. In contrast, the reaction is much faster when using an IL instead of an organic solvent as second phase. In the case of [BMIM][(CF$_3$SO$_2$)$_2$N]/buffer the conversion after 180 min reaches 88%.

Using the **LBADH** in a two-phase system has two advantages. Firstly, the concentration of poorly water-soluble substrates can be raised in the total system for even decoupling as reported by Kroutil et al. [32]. Secondly, there is an increase of stability of the **LBADH** in comparison to conversion in one-phase systems with cosolvents. It depends on the organic solvent in comparison to the conversion in buffer [27]. The half-life of the **LBADH** is effected by the nature of the organic solvent that is used as second phase, although the contact between the enzyme and the organic media should be marginal (Figure 33.5) [33].

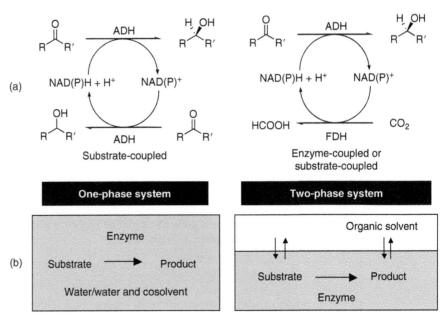

FIGURE 33.4 (a) Cofactor regeneration with substrate-coupled and enzyme-coupled ap
(b) Scheme of a one-phase system and a two-phase system for enzyme catalysis. (From Eckst
et al., *Chem. Comm.*, 1084–1085, 2004).

The enzymatic reduction by LBADH requires NADPH, an expensive cofact
configure this reaction economically an *in situ* regeneration is necessary. Eckstein
reviewed various methods for one- and two-phase systems to minimize the required co
concentration [29]. A common method is the use of a second substrate, mostly 2-propa
recycle the cofactor into its catalytically active form. This substrate-coupled regenera

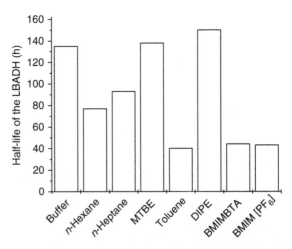

FIGURE 33.5 Half-life of the LBADH in buffer and different two-phase systems. Condition
volume 8 ml, organic solvent/water 1:1 ($\alpha = 1$), 3 U/ml (aq) LBADH, 0,1 mM (aq) $NADP^+$/N
30°C, 900 rpm, phosphate buffer (50 mM, 1 mM $MgCl_2$, pH 7,0). (From Lembrecht, J., unpu
data, 2006.)

mostly implemented by the same enzyme as that for the reduction of the ketone. Alternatively, an enzyme-coupled regeneration can be used. In this case, a second enzyme is deployed to restore NAD(P)H. Firstly, a regeneration by formate dehydrogenase (FDH) that is able to oxidize formate to CO_2 [34] has been reported, but also an application of glucose dehydrogenase (GDH), which oxidizes D-glucose, or D-glucose-6-phosphate dehydrogenase, which converts D-glucose-6-phosphate, was mentioned in literature. These are the two mainly used regeneration methods for the reduction of prochiral ketones in one- and two-phase systems [35].

Another possibility to permit the production of chiral hydrophobic alcohols catalyzed by ADHs in nonconventional media is the immobilization on a support. De Temiño et al. reported a higher stability of ADH from *L. kefir* by entrapping the enzyme and its cofactor in polyvinyl alcohol gel beads [36]. In the case of an immobilized enzyme it is possible to convert in pure organic solvents. An encapsulation in reverse micelles also causes a higher half-life time of ADHs in pure organic solvents [25]. To create a higher stability of the LBADH the preparation of CLEAs has been mentioned by Mateo et al., although the cross-linked ADH exhibits a recovered activity of 7 to 10% relative to the native enzyme cross-linked with dextran polyaldehyde [37].

To facilitate the catalysis in pure organic solvents or in the environment-friendly ILs genetic rearrangement of the enzyme would be a possibility.

33.2.2 LACCASES

Laccases are the second group of oxidoreductases that will be discussed in this chapter. These copper-containing enzymes oxidize substrates by reduction of molecular oxygen to water. Typically, laccases are deployed in textile, pulp, and paper industries, but they can be used as "green" catalysts in organic synthesis as well. Khmelnitsky studied the degeneration of laccase from *Polyporus versicolor* and other enzymes for the oxidation of a model substrate in many different organic solvents [38]. Laccases, which are very sensitive to nonconventional conditions, are only active in organic solvents till a limiting concentration of the cosolvent is exceeded. This barrier depends on the enzyme and also on the cosolvent; so the border where the enzyme is still native (catalytically active) must be determined for each protein.

Commonly, laccases are used in buffer solution, e.g., in the oxidative polymerization of 4-chloroguaiacol [39]. It has been found that a pH between 4 and 5 is the optimized condition for the described reaction.

As a result of the inactivation by organic solvents immobilization is a typical method for the laccase-catalyzed synthesis in organic solvents. Pilz et al. published the synthesis of coupling products of phenolic substrates (Figure 33.6) in different reactors [40]. They used

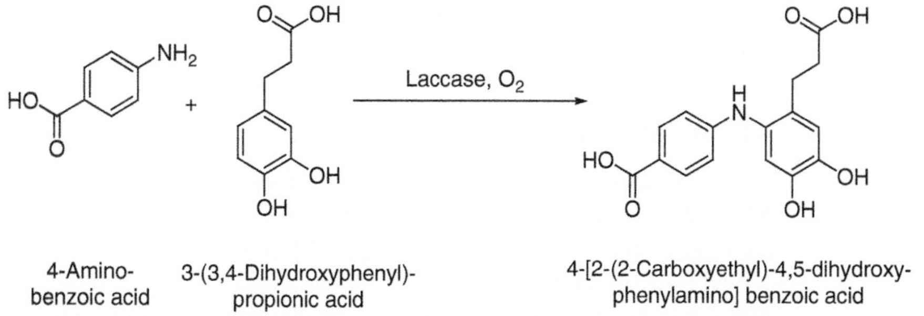

4-Amino-   3-(3,4-Dihydroxyphenyl)-   4-[2-(2-Carboxyethyl)-4,5-dihydroxy-
benzoic acid   propionic acid   phenylamino] benzoic acid

FIGURE 33.6 Cross-coupling reaction catalyzed by laccase from *Pycnoporus cinnabarinus*. (From Pilz, R. et al., *Appl. Microbiol. Biotechnol.*, 60, 708–712, 2003.)

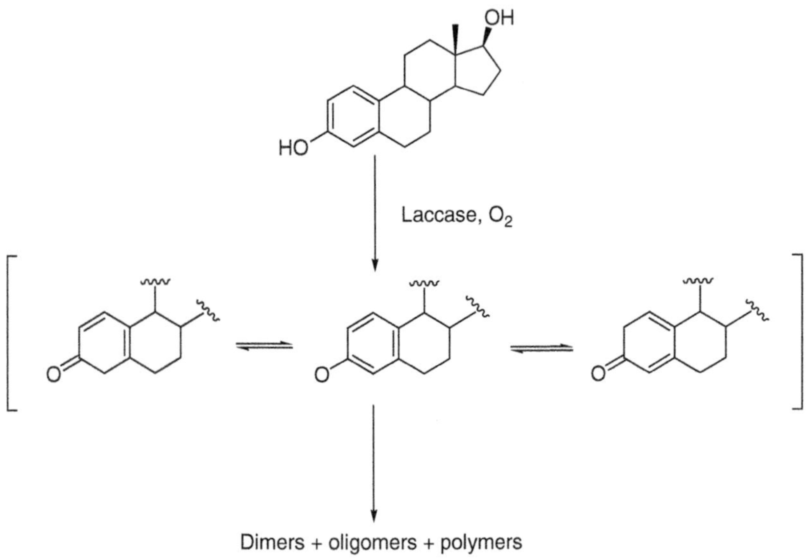

FIGURE 33.7 Laccase-mediated oxidation of β-estradiol. (From Nicotra, S. et al., *Tetrahedra metry*, 15, 2927–2931, 2004.)

an immobilized laccase prepared from the white rot fungus *Pycnoporus cinnabarinus* in a tank reactor (STR) and also in a continuously operated enzyme-membrane reactor in acetate buffer (pH 5). Immobilization of the enzyme on a support offers the possi reduce the amount of enzyme required for the reaction.

It has also been reported that the oxidation of the steroid hormone 17β-estra laccase from *Mycelypthora* and *Trametes pubescens* in organic media is feasible. Nicot accomplished the laccase-mediated oxidation in pure organic solvents and also in tw systems using the enzyme immobilized on glass beads (Figure 33.7) [41]. As pure so mixture of dioxin and water-saturated toluene has been applied and as two-phase AcOEt/buffer (pH 4.5) was utilized. The oxidation of β-estradiol by laccase from *P. ve* was described more than 30 years ago in one of the first papers on the use of enzymes phase systems [42].

In the future it might be possible to use laccases not only in the paper industry but produce new polymers that cannot be synthesized by organometallic catalysts o enzymes. Therefore, a further development of immobilization methods to generate active and stabile laccases in organic media is necessary. Perhaps it would be pos deploy immobilized laccases in a gas-phase reaction.

33.2.3 Monooxygenases

The third subclass of oxidoreductases that will be discussed is made up of monooxyg This subclass of the EC 1 catalyzes the oxidation of unsaturated substrates to oxir lactones by reducing molecular oxygen. Monooxygenases are dependent on cofactors, NADH and NADPH, so around 10 years ago the deployment of whole cells as bio was common. Schmid et al. availed whole cells containing styrene monooxygenase binant in *Escherichia coli*. [43]. The synthesis of (*S*)-styrenes from styrene and its deriva arranged in a two-phase system containing buffer and dioctyl phthalate. Because

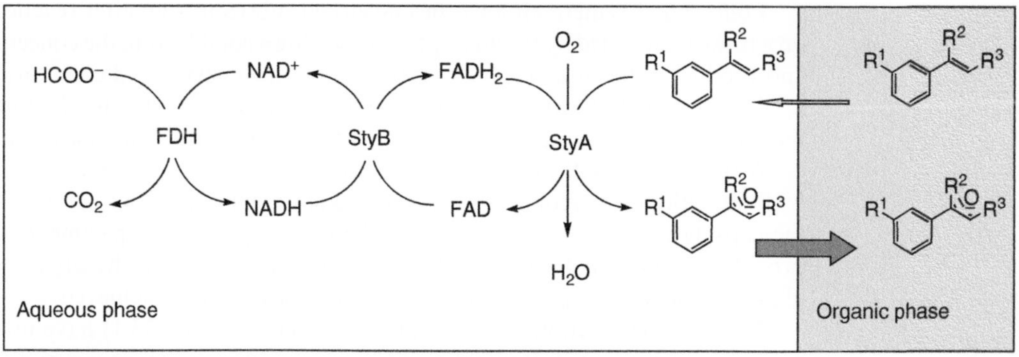

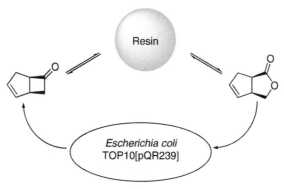

FIGURE 33.8 Reaction pathway during biocatalytic epoxidation in a two-phase system. The organic phase serves as a substrate reservoir and product sink. In the aqueous phase, formate dehydrogenase (FDH) and formate were used for regeneration of NADH. StyB transfers the reducing equivalents from NADH to flavine adenine dinucleotide (FAD). $FADH_2$ and oxygen are cosubstrates for olefin epoxidation by StyA. $R^1 = H$, Cl; $R^2 = R^3 = H$, CH_3. (From Hofstetter, K. et al., *Angew. Chem. Int. Ed.*, 43, 2163–2166, 2004.)

toxicity of the substrates the second phase could work as a reservoir to keep the substrate concentration low in the environment of the enzyme. The two-phase system also allows an *in situ* product removal (IPS) to isolate the toxic and water-labile product.

A biocatalytic asymmetric epoxidation with NADH regeneration in organic–aqueous emulsions has also been published [44]. As can be seen in Figure 33.8 there are two styrene monooxygenases involved, StyA and StrB (flavin- and NADH-dependent). As reported by Schmid et al. the second phase acts as reservoir of the substrate and also as an *in situ* extraction of the product [43]. The yield and also the ee of the epoxidation are comparable to the epoxidation by whole cells.

Alternatively, monooxygenases can transform racemic bicyclo[3.2.0]hept-2-en-6-one to chiral lactones and thioanisole (methyl phenyl sulphide) into its chiral (*R*)-sulphoxide [45,46]. In this case, coimmobilization of the enzyme and the cofactor is necessary to keep the biocatalyst active in the organic reaction media, e.g. on Eupergit C [45]. However, it is also possible to use a membrane reactor for removing the product continuously from the reaction media. Hilker et al. reported an *in situ* substrate feeding/product removal for the Baeyer–Villiger oxidation process catalyzed by cyclohexanone monooxygenase (CHMO) (Figure 33.9) [46].

FIGURE 33.9 Regiodivergent Baeyer–Villiger oxidation of *rac*-bicyclo[3.2.0]hept-2-en-6-one. (From Hilker, I. et al., *Org. Lett.*, 6, 1955–1958, 2004.)

Following this methodology they adsorbed substrate and, after reaction, also the pro on a resin. Controlled by the adsorption/desorption equilibrium, the concentration of subs and product in the aqueous bulk phase, containing enzyme and cofactor, is low. Theret inhibition caused by either the substrate or the product is minimized without stabilizatic the biocatalyst. This approach has been first described by Zmijewsky et al. [47,48].

A functional P450cam monooxygenase was created in water–oil (w/o) emulsion fo: with tetraethylene glycol dodecyl ether as a surfactant [49]. This can be used alternative the capsulation on a support, because the inner aqueous compartment of the w/o emu provides the monooxygenase with a cell-like environment in the organic bulk phase. higher efficiency and productivity of the biocatalyst can be achieved.

Many of the reaction systems that are possible (Figure 33.1) have not been reporte monooxygenases as yet. It might be effective to study the conversion of different substra the gas phase, because one of the reactants, oxygen, is gaseous.

In this chapter the possibilities of different reaction systems for the catalysis of o reductases have been shown by some examples. Oxidoreductases are, in addition t hydrolases, the most important EC group for industrial processes, because they are able in a large scale. It could be revealed in this chapter that the oxidoreductases are to catalyze conversions of many different substrates in various reaction systems jus two-phase systems. In the future is would be interesting to have a look at CLEAs promising immobilization method [37]. However, it should be kept in mind that o: eductases are often used in whole-cell processes adding different requirements t reaction system [168].

33.3   EC 2: TRANSFERASES

Transferases catalyze the transfers of functional residues of various substrates. The divided into nine subclasses, shown in Table 33.2 [50].

The utility in organic synthesis for transferase-catalyzed reactions in nonconven media is very low, compared to their catalytic importance in the living organism Nevertheless, several reports of organic solvent effects on transferases have been pub in the past. An early example deals with the effect of the organic solvents ethan 1,4-dioxane on a citrate synthase from pig heart [52]. For both solvents it has been

---

TABLE 33.2
Subclasses of the Transferases

EC 2 Transferases

2.1	Transferring one-carbon groups
2.2	Transferring aldehyde or carbon groups
2.3	Acyltransferases
2.4	Glycosyltransferases
2.5	Transferring alkyl or aryl groups, other than methyl groups
2.6	Transferring nitrogenous groups
2.7	Transferring phosphorous-containing groups
2.8	Transferring sulphur-containing groups
2.9	Transferring selenium-containing groups

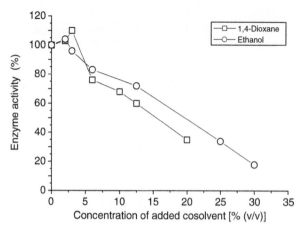

FIGURE 33.10 Effect of 1,4-dioxane and ethanol on citrate synthase from pig heart.

that with increasing amount of added solvent the enzyme activity decreases (Figure 33.10). The 50% inhibition was found to be at 16% dioxane and 19% ethanol.

In 1979 Singh and Wang reported the effects of organic solvents on a glycogen phosphorylase kinase from rabbit skeletal muscle [53]. It was observed that several organic solvents stimulate the enzyme up to a 28-fold enhancement (acetone). An overview is given in Table 33.3.

Screening the enzyme activity in correlation with increasing concentration of ethanol, a complex behavior has been observed. With an increasing amount of ethanol a relatively sharp maximum at 1.72 M was found, yielding in a kinase activity above 4 units/mg. This can be explained by ethanol-caused modification of the affinity of the protein phosphorylase b toward the kinase, which can also be done by a modification of the pH. At this point the pH-activity profile can also be modified by the organic solvent ethanol (Figure 33.10).

As shown in Figure 33.11 the activating effect is also transferable to other pH values, resulting in an activating effect. With an increase of pH the activation became negligible above a pH of 8.5; however, the causes for the behavior are not known. The authors were also able to show that all the observed effects caused by ethanol are mostly reversible, diluting out the ethanol. Therefore, the origin of the observed effects is not completely known, because no kinetic data were available.

TABLE 33.3
Enhancement of Phosphorylase Kinase–Catalyzed
Synthesis Effected by Various Organic Solvents at 1 M
Each (Unactivated Enzyme)

Solvent	Stimulation (-fold)
None (buffer)	1.0
Methanol	3.3
Ethanol	8.8
2-Propanol	18.6
Acetone	27.5
Dimethyl sulfoxide	17.8
Tetrahydrofuran	18.6

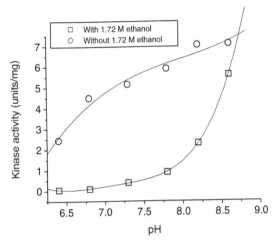

FIGURE 33.11 Ethanol-induced modification of pH dependence of (unactivated) kinase activity

Contrarily to this, an influence of ethanol on the enzyme kinetics has been investiga a myristoyl CoA:protein N-myristoyltransferase (NMT) [54]. For this transferase s results were obtained following addition of the organic solvents, in this case for et and acetonitrile (Figure 33.12).

For both added organic solvents it has been found that with increasing amount of o solvent the NMT activity rises. After a certain amount the activity decreases again, pro because of the inactivation of the enzyme through the organic solvent.

It is noteworthy that at 10% (v/v) ethanol the NMT activity increased nearly fiv but the cause of this activation has yet to be found in the altered enzyme kinetic constan has been observed at pp60^{scr} (a peptide substrate from NH$_2$-terminal sequences), showed nearly a fivefold rise in the V_{max} value during an additional ethanol concent of 10% (v/v) at a constant K_M value (Table 33.4).

It has been presumed that this activation is caused by a time-and-concentra dependent unfolding mechanism. So an interference of the organic solvent molecul arises from their different dielectric constants could lead to secondary effects on the co

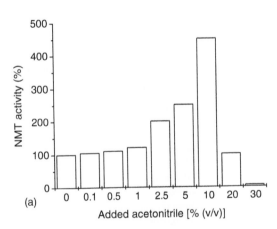

FIGURE 33.12 Effect of (a) acetonitrile and

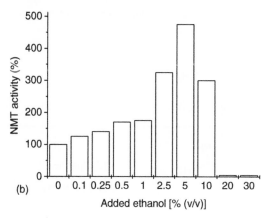

FIGURE 33.12 (continued) (b) ethanol on N-myristoyltransferase (NMT).

TABLE 33.4
Effect of Ethanol at Kinetic Constants

Substrate	Ethanol [% (v/v)]	K_M (μM)	Relative V_{max}[a]
pp60[src]	10	41.6	100
pp60[src]	0	41.0	21.4

[a] V_{max} values are reported as a percentage of velocity observed with pp60[scr]-derived peptide substrate at 100 to 0% (v/v) ethanol.

ion atmosphere and then to the binding of the substrate [55]. Final arguments for explaining the observed behavior have not been found yet.

In contrast, transaminases are very useful catalysts for amino acid synthesis, but are generally more complex and require special expertise compared to proteases and lipases [51].

The possibility to overcome product inhibition for a ω-transaminase has been investigated by applying one- and two-phase systems [56]. Several organic solvents were tested for the enzymatic resolution reaction of α-methylbenzylamine (α-MBA), but ethyl acetate and cyclohexanone as organic solvents yielded best results for enzyme activity and biocompatibility.

The usage of a two-phase system (Figure 33.13) for transaminase-catalyzed reaction has the major advantage of an *in situ* substrate and product removal from the organic phase. Due to this the acetophenone concentration is kept at a very low level, preventing product inhibition and leading to high reaction rates.

Additionally, the easy recovery of the chiral amine can be accomplished by adjusting the pH of the aqueous phase.

The phase ratio, expressed as the volume fraction of organic phase, can also lead to much higher reaction rates (Figure 33.14). Compared with the standard system (aqueous system), the reactivity increases ninefold at a value of 0.2, which is a further advantage of the two-phase system.

In analogy to hydrolytic enzymes like subtilisin that are also used for enzymatic resolution of racemic amines, transaminases are of minor interest in nonaqueous reaction systems [51].

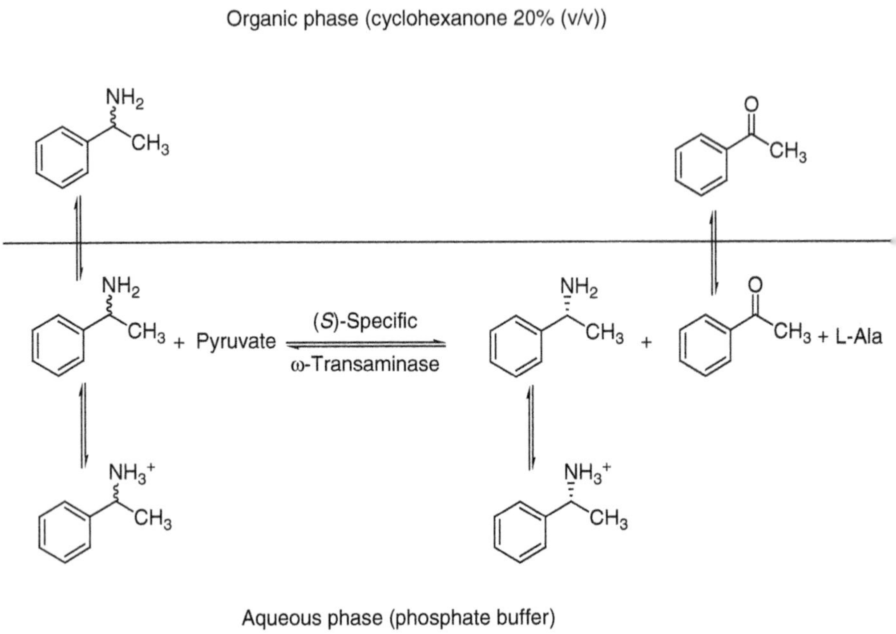

Organic phase (cyclohexanone 20% (v/v))

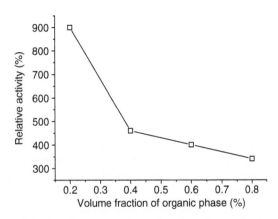

Aqueous phase (phosphate buffer)

FIGURE 33.13 Two-phase-catalyzed enzymatic cleavage of α-methylbenzylamine using a ω-transa.

33.4   EC 3: HYDROLASES

The hydrolase family is classified in EC 3 and represents a group of enzymes that ca
bond cleavage by reaction with water. Several advantageous characteristics of these l
lytic enzymes give them a high biotechnological potential and make them of special i
particularly with regard to their application in organic chemistry: (i) lack of se
cofactors; (ii) broad substrate specificity; (iii) high stereoselectivity; (iv) catalysis of s
related reactions, such as condensation and alcoholysis; and (v) commercial availabili
   Within the group of hydrolases, lipases (EC 3.1.1.3) stand amongst the most imp
biocatalysts, carrying out novel reactions in both aqueous and nonaqueous media. '

FIGURE 33.14 Enzyme activity in contrast to volume fraction of organic phase.

TABLE 33.5
Important Areas of Industrial Applications of Lipases

Industry	Examples of Use
Food industry	Bakery, diary: flavor improvement, transesterification of fats and oils, hydrolysis of milk fat [140,141]
Paper and wood industry	Bleaching of wood and recovered paper [142]
Pharmacy	Synthesis of chiral intermediates with high enantiomeric excess [143,144]
Medicine	Application in biosensors for identification of specific lipids (diagnostic of cardiovascular diseases) [145]
Detergent industry	Cleavage of fats in laundry [141,146]
Cosmetic industry	Surfactant synthesis and flavor synthesis [147]
Environment	Wastewater treatment, conditioning of waste fat and oils [148]
Agricultural economy	Pesticide synthesis [149]
Chemistry	Polyethylene terephthalate (PET) synthesis, bioconversions, seperation of enantiomers, oleochemistry [59,140,150–153]
Oleochemistry	Regioselective hydrolysis, transesterification, enantioselective processes; biodiesel, lubricants [141,154]

primarily due to their ability to utilize a wide spectrum of substrates, their high stability toward extremes of temperature, pH, and organic solvents, and their chemo-, regio-, and enantioselectivity. Lipases catalyze a wide range of reactions including hydrolysis, interesterification, alcoholysis, acidolysis, esterification, and aminolysis. They are used in a variety of biotechnological fields such as food and dairy, detergent, and pharmaceutical, agrochemical, and oleochemical industries (Table 33.5). Lipases can be further exploited in many newer areas where they can serve as potential biocatalysts. Due to the tremendous potential of lipases for exploitation in biotechnology this section will mainly review this group of hydrolases.

Lipases are ubiquitous in nature: they occur in plants, animals, and microorganisms, and are primarily responsible for the hydrolysis of triglycerides with concomitant production of free fatty acids and glycerol. Many lipases are excreted extracellularly by fungi and bacteria, which makes their large-scale production particularly easy. In addition to their specific and limited function in metabolism, lipases play an important role in biotechnology: about 40% of all biotransformations reported so far have been performed with lipases [51]. In general, lipases are characterized by their unique feature of acting at water–organic interfaces, which distinguishes them from esterases [58,59]. Research in this field suggests that a lipid-induced conformational change alters the orientation of a lid that covers the enzyme active site. This phenomenon is commonly known as interfacial activation [57]. Consequently, lipase-catalyzed reactions are preferably conducted in two-phase systems.

Bearing in mind that the natural substrates of lipases are esters of an alcohol—glycerol—with an achiral acid, it is understandable that lipases are particularly useful for the resolution or asymmetrization of esters bearing a chiral alcohol moiety. On the basis of a thorough survey of the literature on chiral resolutions with lipases from *Candida rugosa* (CRL) and *Pseudomonas cepacia* (PCL), Bornscheuer and Kazlauskas proposed rules for the enantiopreference of these two enzymes on the spatial requirements of the substituents on the reagent [57]. The basics of "Kazlauskas rules" are shown in Figure 33.15 and literature has shown this rule to be highly predictive for lipase action on secondary alcohols, but less accurate for lipase-catalyzed transformations of primary alcohols and acids.

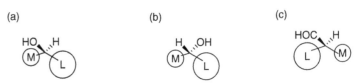

FIGURE 33.15 Scheme of the "Kazlauskas rule" to predict which enantiomer of (a) a se≡ alcohol, (b) a primary alcohol, and (c) a carboxylic acid reacts faster in lipase-catalyzed react≡ medium-sized substituent, L, large-sized substituent.

Commercial lipase from *C. rugosa* is probably the most often used biocatalyst. Thi≡ has been applied for several selective hydrolysis reactions of esters of cyclic sec≡ alcohols, because its active site accepts larger substrates than those of other lipas≡ One example to illustrate this point is the resolution of racemic 2,3-dihydroxy carbo≡ and cyclohexane-1,2,3-triol esters by CRL [60,61]. PCL possesses a "smaller" active si≡ CRL and can be extremely selective on "slim" counterparts as was shown for the des≡ trization of some prochiral dithioacetal esters [62–66].

During the past decades the general opinion about the use of enzymes in organic re≡ has changed. Conventional biocatalysis had mainly been performed in aqueous so≡ because enzymes were considered to be most active in water—their natural milieu. ≡ quent years have seen the replacement of this prejudice by a more refined position initi≡ the discovery that many enzymes are active in organic media.

Publications from the early 20th century and the 1930s already reported about b≡ lysis in organic media, but most of this work was forgotten [67]. About 60 ye≡ researchers published first results about lipase-catalyzed esterifications in organic s≡ containing approximately 8 to 12% of water [68,69]. However, it took more than 40 y≡ convince researchers that enzyme-catalyzed reactions are not only possible but also≡ times more convenient in organic solvents. Since the pioneering work of Camb≡ Klibanov on this subject many publications have appeared on enzyme-catalyzed≡ resolutions in organic solvents (especially lipase-catalyzed reactions) and on related≡ like the use of organic cosolvents in water, reactions in solvent-free media or in compre≡ supercritical gases, and the addition of salts or (thia)crown ethers to the lipase [70].

When performing lipase-catalyzed reactions in organic media, the ability to do an ≡ cation reaction instead of hydrolysis is one advantage. It has been found that although≡ favor the same prochiral group in both cases, the two reactions yield opposite enant≡ Figure 33.16 gives an example of the phenomenon: the acetylation of 2-benzylglycer≡ PCL yields the (*S*)-monoacetate, while hydrolysis of the diacetate with porcine par≡ lipase (PPL) yields the (*R*)-monoacetate.

Another advantage results from potential changes in the selectivity of lipases by≡ ing different solvents, sometimes called medium or solvent engineering. Changes≡

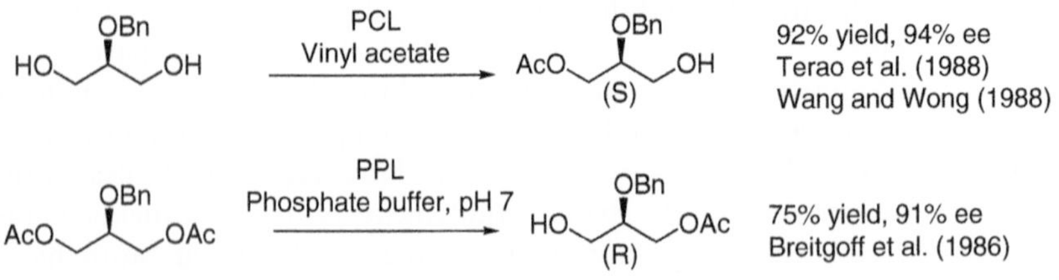

FIGURE 33.16 Acetylation of benzylglycerol with lipase of *Pseudomons cepacia* (PCL) and hydr≡ the diacetate with porcine pancreatic lipase (PPL).

enantioselectivity after varying the solvent could be shown for PPL. Mori et al. found no enantioselectivity for the hydrolysis of seudenol acetate, but Johnston et al. reported moderate enantioselectivity ($E = 17$) in the acetylation of seudenol with trifluoroethyl acetate in ethyl ether [71,72]. These are only a few examples from the large volume of literature on lipase-catalyzed reactions in organic solvents. In order to give an overview of other techniques applied for these reactions the following sections will briefly discuss recent developments in other reaction systems.

33.4.1 IMMOBILIZATION

Since the second half of the 20th century, investigations have focused on the development of immobilized enzymes in order to improve process economy by allowing the continuous or repetitive use and easy recovery of enzymes. Immobilization refers to the preparation of insoluble derivatives of enzymes, and has been performed by various methods including (i) noncovalent adsorption or deposition; (ii) covalent attachment; (iii) entrapment in a polymeric gel, membrane, or capsule; and (iv) cross-linking of the enzyme [73]. These techniques can dramatically affect enzyme properties such as pH dependence, temperature profile, and kinetics, and have often resulted in biocatalysts exhibiting significantly higher stability than the native enzyme.

Back in the 1950s, first immobilization methods for lipases have been reported. Although enzyme immobilization was dominated by physical methods during this time, first results of specific ionic adsorption like the binding of lipase on styrenepolyaminostyrene (Amberlite XE-97) have been published [74]. Unfortunately, those early-developed carriers were found to be not very suitable for enzyme immobilization. Since then many types of enzymes immobilized by different immobilization techniques have been found to exhibit higher stabilities, activities, and/or selectivities than native enzymes. Regarding lipases, entrapment in alkyl-substituted organic silane precursors has led to increased activity of these enzymes in organic solvents. The obtained sol–gels even show a significantly higher activity than freely dispersed enzymes [75]. The development of a novel technique for the immobilization of lipases by entrapping the enzymes within an aqueous solution in bead-shaped silicone elastomers ("static emulsion") led to an enhancement in enzyme activity by factor 31 for *C. antarctica* lipase A (CAL-A) and factor 250 for *Thermomyces lanuginosa* lipase in comparison to the native enzyme in hexane [76]. By entrapment of lipase–lipid complexes in *n*-vinyl-2-pyrrolidone gel matrix, Goto et al. succeeded in increasing the activity up to 50-fold in comparison to the native enzyme [77].

A variety of immobilization techniques like covalent bonding, entrapment, and adsorption can be applied to change the selectivity of lipases. For example, the *S*-selective lipase from *C. rugosa* presented a high enantioselectivity ($E = 400$) toward the *R*-isomer for the resolution of mandelic acid esters after covalent immobilization on glutaraldehyde supports [78].

In recent years, carrier-bound CLEAs have attracted increasing attention, due to their simplicity, broad applicability, high stability, and high volume activity. CLEAs of CRL with enhanced activity, stability, and defined particle size have been designed by impregnation of enzyme solution in a porous membrane of controlled pore size followed by subsequent aggregation and cross-linking [79].

33.4.2 IONIC LIQUIDS

Over the last decade ILs have emerged as alternative reaction media for performing all types of reactions with sometimes remarkable results [80–82]. In 2000, Lau et al. investigated the reactivity of *C. antarctica* lipase B (CAL-B) in ILs, such as [BMIM][PF$_6$] and [BMIM][BF$_4$],

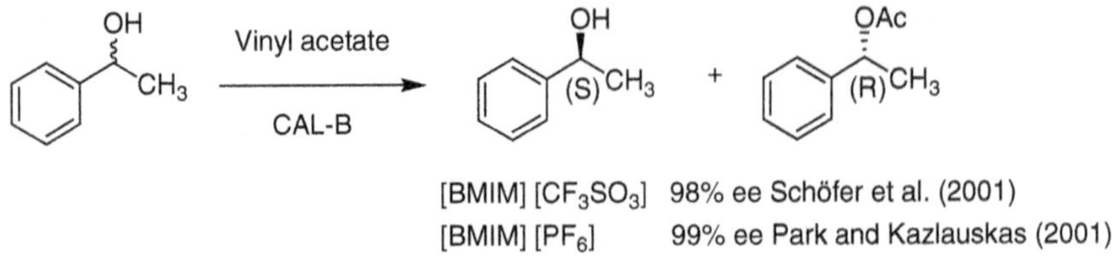

FIGURE 33.17 Lipase-catalyzed reaction in the presence of ionic liquids (ILs).

in comparison to conventional organic solvents [83]. They found similar reaction rates
the reactions investigated. This work represents the second publication to demonstra
potential use of ILs for enzyme catalysis and the first to show their use with lipases.

Since then several other lipases have been applied to catalyze reactions in systems w
as pure solvents, and as cosolvents with water or in two-phase systems. For example,
mediated kinetic resolution of racemates in ILs provides extremely high enantiopurities
products [84–86]. Kragl et al. investigated the kinetic resolution of 1-phenylethanol fo
of eight different lipases in ten different ILs with MTBE as reference [84]. To inve
transesterification vinylacetate was used as acetyl donor. The best results were obtair
CAL-B in [BMIM][CF$_3$SO$_3$], [BMIM][(CF$_3$SO$_2$)$_2$N], and [OMIM][PF$_6$]. Park and K
kas reported good activities for CAL-B by using the ILs [BMIM][BF$_4$] and [BMIM][
the same system (see Figure 33.17) [86].

Moreover, CAL-B has been found to exhibit impressive regioselectivities for t
monoacetylation of β-D-glucose in [MOEMIM][BF$_4$]. Due to their solvation properti
dissolve not only hydrophobic compounds but also hydrophilic compounds such as
hydrates. Park and Kazlauskas reported the regioselective acylation of glucose with 99
and 93% selectivity in [MOEMIM][BF$_4$] (Figure 33.18). These values are much highe
those observed in the organic solvents commonly used for this purpose (reaction carri
in acetone: 72% yield, 76% monoacetylation) [86].

33.4.3 REVERSE MICELLES

The low water content necessary to favor synthesis reactions in organic media by lipa
be achieved by microencapsulation of the biocatalyst within reverse micelles. Reverse n
consist of tiny aqueous droplets stabilized by surfactants in a bulk water-immiscible c
solvent. The biocatalyst remains soluble and active in the water phase, while reactin
water-insoluble or poorly soluble compounds present in the organic phase. The re
micellar system has been proven to be highly suitable for lipase-catalyzed reactions.
amount of aqueous phase is very small, lipases can catalyze transesterification and
synthesis. Moreover, the system provides a high interfacial area and thus eliminate

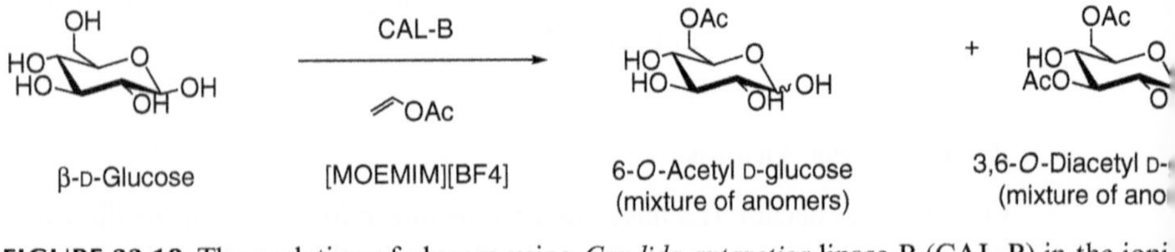

FIGURE 33.18 The acylation of glucose using *Candida antarctica* lipase B (CAL-B) in the ioni
(IL) [MOEMIM][BF$_4$]. (From Park, S. and Kazlauskas, R.J., *J. Org. Chem.*, 66, 8395–8401, 20

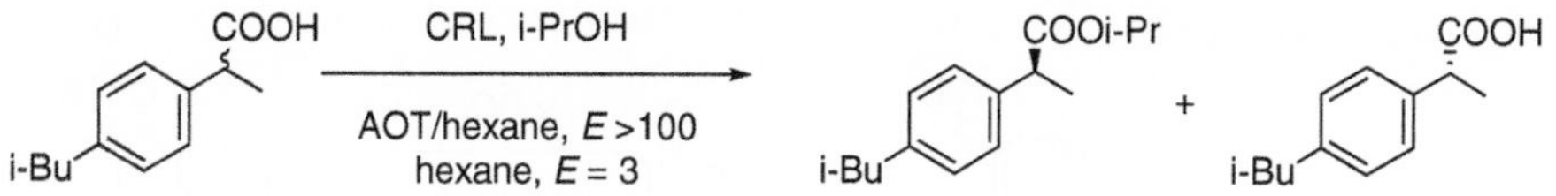

FIGURE 33.19 Esterification of ibuprofen by *Candida rugosa* lipase (CRL). (From Hedstrom, G., Backlund, M., and Slotte, J.P., *Biotechnol. Bioeng.*, 42, 618–624, 1993.)

transfer limitations. Many surfactants and solvents can be applied, but anionic surfactants, in particular AOT, have been proven to be best in lipase-catalyzed reactions [87].

The application of lipases in reverse micelles results in small changes in enzyme selectivity. In 1987 Bello et al. investigated the selectivity of CRL in the transesterification of triglycerides in reversed micelles and reported that CRL, which normally shows little fatty acid chain length selectivity, favored longer chain lengths in this system [88]. Moreover, the enantio-selectivity of CRL can be increased by applying the biocatalyst in reverse micelles. Hedström et al. reported enantioselectivities of $E > 100$ for the CRL-catalyzed esterification of ibuprofen in reversed micellar systems in comparison with enantioselectivites of $E = 3$ in hexane (Figure 33.19) [89].

Many other applications of lipases in reverse micellar systems can be found in literature. Table 33.6 shows selected examples of reactions catalyzed by lipases with potential applications in food, pharmaceutical, and chemical industries as well as in the environmental area.

However, the recovery of products from surfactant-containing organic solvents still represents a problem that must be overcome before the reverse micellar system can be effectively applied at industrial scale. One possible solution might be the continuous operation in membrane reactors: an ultrafiltration membrane can be used to retain the micelles while the small molecules of substrate and products pass freely [90].

33.4.4 SUPERCRITICAL FLUIDS

Lipase-catalyzed reactions have proven to be feasible in supercritical fluids (scF). These fluids represent substances heated above their critical temperature and compressed above their critical pressure. They exhibit properties similar to those of hydrophobic solvents, show

TABLE 33.6
Selected Examples of Reactions Catalyzed by Lipases in Reversed Micellar Systems

Enzyme/Source	System (Surfactant/Organic Solvent)	Reaction
Thermomyces lanugionsa	AOT Isooctane	Synthesis of ethyl-laurate [155]
Bacillus megaterium	AOT n-Heptane	Hydrolysis of pNPP [156]
Rhizopus delemar	AOT Isooctane	Hydrolysis of triolein [157]
Mucor javanicus	AOT Isooctane	Acylation of doxorubicin [158]
Candida lypolytica	AOT Isooctane	Esterification of octanoic acid with 1-octanol [159]

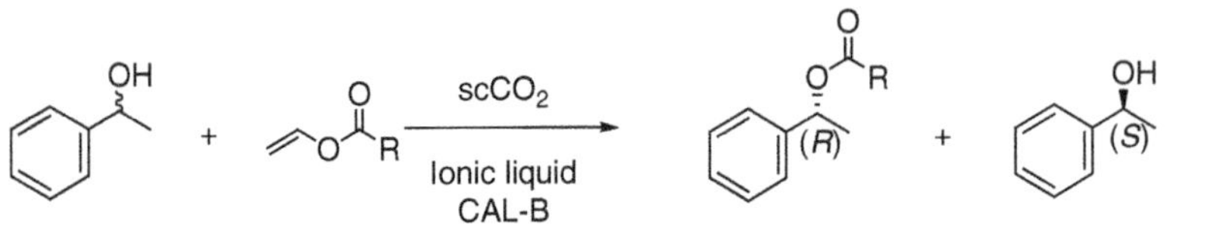

FIGURE 33.20 Lipase-catalyzed kinetic resolution in $scCO_2$. (From Reetz, M.T. et al., *Adv Catal.*, 345, 1221–1228, 2003.)

rapid mass transfer due to low viscosity, and allow a simple downstream process evaporation. Moreover, their solvation properties can be changed by changing the pa

The use of supercritical carbon dioxide is probably the most common way to c enzyme-catalyzed reactions in scFs, because it is nonflammable, nontoxic, cheap, and the supercritical state at 31.1°C. The first application of lipase-catalyzed reactions dates back to 1986. Nakamura et al. reported about the *Rhizopus oryzae* lipase catalyzed interesterification of triolein and stearic acid to 8% conversion in scCC Since then researchers have examined a wide range of reactions and observed cha conversion, enantioselectivity, and lipase stability similar to those in organic solver example, Reetz et al. found high enantioselectivities for the lipase-catalyzed kinetic res of chiral racemic secondary alcohols (Figure 33.20) [92]. Lozano et al. and Reet reported the immobilization of CAL-B in ILs (Section 34.4.5), whereas substra products are dissolved in a second phase formed by supercritical CO_2 [93,94]. examples for lipase-catalyzed reactions in scFs can be seen in Table 33.7.

To summarize, this chapter aims to deliver a short insight into the immense possibi lipase application in organic synthesis. Future research will focus on the developme new generation of lipases by extensive screening and genetic manipulations in order t task-specific lipases for special applications. Protein-engineering methods will help t

TABLE 33.7
Reactions in Supercritical Fluids (scFs) Catalyzed by Lipases

Enzyme/Source	System (Surfactant/Organic Solvent)	Reaction
Free and immobilized lipases: *Rhyzomucor meihei*, *Pseudomonas fluorescens*, *Rhizopus javanicus*, *R. niveus*, *Candida rugosa*	$scCO_2$ and scPropane	Ester synthesis: oleyl ol
Free and immobilized lipases: *C. antarctica*, *Mucor miehei*	$scCO_2$ and ionic liquids	Synthesis of glycidyl es kinetic resolution of *rac*-glycidol [161]
C. antarctica lipase (Novozyme 435)	scMethanol, scEthanol, and $scCO_2$	Synthesis of biodiesel [
Immobilized lipase: *M. miehei*	$scCO_2$	Hydrolysis of blackcurr
Immobilized lipase: *C. antarctica*	$scCO_2$	Butyl butyrate synthesi

TABLE 33.8
Selected Pig Liver Esterase (PLE)-Catalyzed Reactions

Reaction	Substrate	Enzyme Source
Asymmetrization of nonchiral substrates [98]	COOH / COOMe	PLE
Optical resolution of racemates [165]	HO, CO_2Et	PLE
Separation of endo/exoisomers [166]	CO_2Et	PLE
Regioselective hydrolysis of ester groups [167]	COOMe HO··· CH_2 COOMe	PLE

lipases versatile industrial biocatalysts by enabling the production of large amounts of recombinant enzymes and the improvement of their biochemical and catalytic features.

33.4.5 OTHER HYDROLYTIC ENZYMES

Esterases (carboxylester hydrolases, EC 3.1.1.1) catalyze, like lipases, the hydrolysis of carboxylic acid esters but, in contrast to lipases, only a few esterases have practical use in organic synthesis. Most of the esterase-catalyzed reactions in literature have been performed by the esterase isolated from pig liver (PLE) [95]. In contrast to lipases, PLE shows highest activity in aqueous buffered or two-phase systems, and usually does not accept highly hydrophobic substrates. Thus, selectivity tuning is more or less limited to the addition of up to 20% of polar protic water-miscible cosolvents like methanol, *tert*-butanol, DMSO, acetone, or acetonitrile [96]. In 1997 Ruppert and Gais succeeded to enhance the activity of PLE in organic solvents after colyophilization of the esterase with methoxypolyethylene glycol [97].

Esterases are in general very useful biocatalysts for the production of chiral intermediates through hydrolysis reactions. Examples include the asymmetrization of prochiral substrates and the optical resolution of racemates [98,99]. Esterases have also been used to separate endo/exo-mixture and for the regioselective hydrolysis of an ester group in the presence of a second ester function (Table 33.8).

Proteases are the last group of hydrolases to be mentioned in this review because they represent one of the three largest groups of industrial enzymes and find application in detergents, leather, food, and pharmaceutical industries, as well as bioremediation processes [100,101].

The most important commercial proteases are subtilisin, α-chymotrypsin, and—to a lesser extent—trypsin, pepsin, papain, and penicillin acylase. In the field of organic synthesis, two main applications of proteases can be found: (i) the enantioselective hydrolysis of carboxylic acid esters, where they seem to retain a preference for the hydrolysis of that enantiomer, which mimics the configuration of an L-amino acid more closely; and (ii) the synthesis of di- and oligopeptides by coupling of *N*-protected amino acids and peptide esters. The latter has been carried out on an industrial scale by Tosoh Corporation (Japan) for the thermolysin catalyzed synthesis of aspartame (Figure 33.21) [102,103].

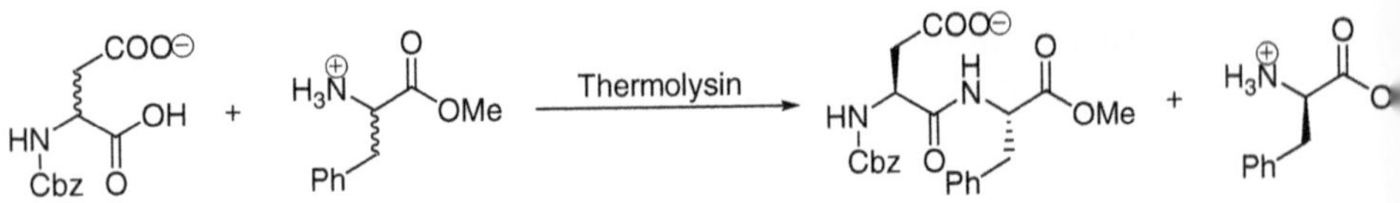

FIGURE 33.21 Commercial process for the production of aspartame by Tosoh Corporation (Japa

TABLE 33.9
Subclasses of the Lyases

	EC 4 Lyases
4.1	Carbon–Carbon lyases
4.2	Carbon–oxygen lyases
4.3	Carbon–nitrogen lyases
4.4	Carbon–sulfur lyases
4.5	Carbon–halide lyases
4.6	Phosphorous–oxygen lyases
4.99	Other lyases

33.5   EC 4: LYASES

Apart from oxidoreductases and hydrolases, lyases are the most frequently used enzyme
[51]. They are divided into several subgroups, which are shown in Table 33.9 [50].
    These enzymes catalyze several bond formation and cleavage reactions and have
extensive usage in several large-scale applications [104]. Two important examples, sho
Figure 33.22—(a) acrylamide synthesis and (b) N-acetylneuraminic acid—illustrate the
in the synthesis of bulk and fine chemicals.

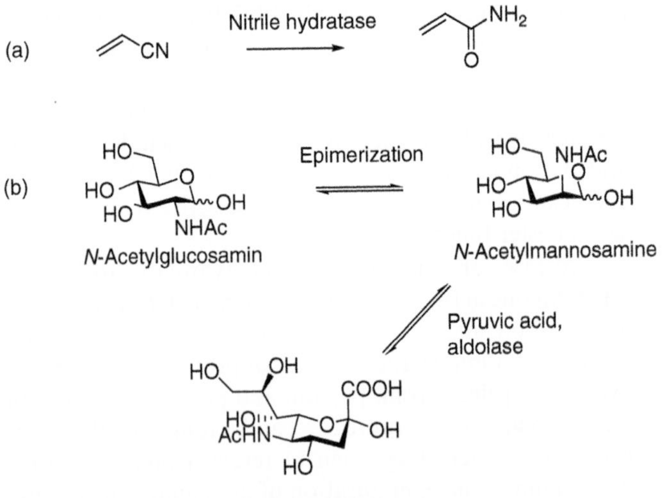

FIGURE 33.22 Large-scale applications of lyases. (From Wandrey, C., Liese, A., and Kihum
Org. Proc. Res. Dev., 4, 286–290, 2000.)

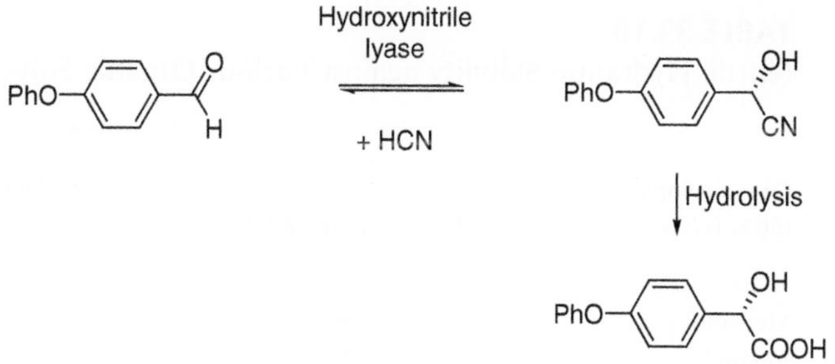

FIGURE 33.23 α-Hydroxy-carboxylic acids synthesis using hydroxynitrile lyases.

The cyanohydrin formation using hydroxynitrile lyases (HNLs) from several sources has produced enormous interest in the last decade. Important follow-up products are α-hydroxy-carboxylic acids, which are easily accessible using these enzymes (Figure 33.23), and have also been applied in large-scale operations [105].

33.5.1 Nitrile Hydratases

Nitrile hydratases hydrolyze nitriles selectively to the resulting amides. An important example is the enzymatic formation of acrylamide, see also Figure 33.24. The amide can be converted afterwards to the corresponding carboxylic acid. This two-step approach can also be simplified by using a nitrilase (Figure 33.24), yielding in one step the desired carboxylic acid [51].

In amide synthesis, using a nitrile hydratase, an important factor is the stability of the biocatalyst related to the rising product concentration [>50% (w/v) acrylamide], causing a rapid enzyme deactivation [106].

Several nitrile hydratases have been adapted to these challenges in acrylamide synthesis, e.g., the nitrile hydratase from *Rhodococcus* sp. N-774 in 1987 [107,108]. In 1988 a much more stable enzyme has been purified from *P. chloroaphis* B23 that exhibited a higher stability and reactivity in synthesis. An even more resistant nitrile hydratase from *Rhodococcus rhodochrous* J1 has been found in 1993. This enzyme exhibited an enormous stability against a variety of organic solvents (Table 33.10) [106].

As shown in Table 33.10, the variant from *R. rhodochrous* exhibits the highest stability against the organic solvents and also an activating effect for ethylene glycol. It has been assumed that the outstanding stability derives from the very high molecular mass of 505 kDa (20 subunits), by suppressing the flexibility of the protein.

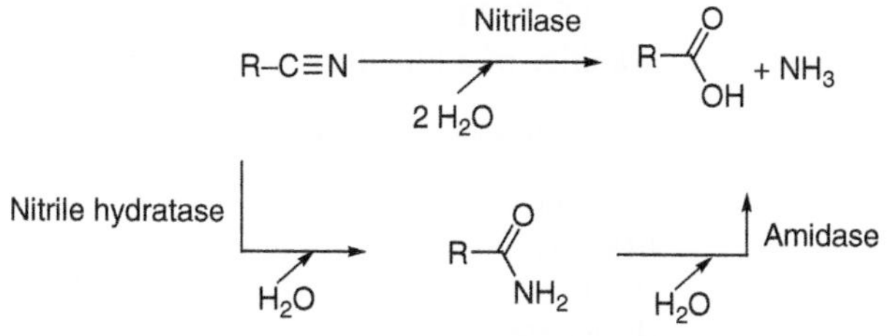

FIGURE 33.24 General pathway of the enzymatic hydrolysis of nitriles.

TABLE 33.10
Nitrile Hydratase Stability against Various Organic Solvents (Abstract)

Organic Solvent (50% (v/v))	Relative Activity (%)		
	Brevibacterium R312	*Pseudomonas chlororaphis* B23	*Rhodococcus rhodochrous* J1
None	100	100	100
Methanol	6	8	89
Ethanol	8	10	94
Acetone	10	12	66
Dimethyl sulfoxide	22	38	58
Ethylene glycol	64	55	136

In the past few years an enormous number of new nitrile-converting enzymes ha▾ discovered, possessing high enantioselectivities and broad substrate ranges [109]. Ni▸ have also been sucessfully applied to organic solvents, like *Pseudomonas* sp. DSM 11 shown in Table 33.11 [110].

It can be seen that the relative activity is rising with increasing log P value, an obse that is comparable with other enzymes [111].

In the future, new immobilization techniques will provide even more stable enzyme attempts using the technique of CLEAs have been successfully arranged for nitrilas Additionally, the search for new enzyme sources and the modification of established converting enzymes could lead to higher reactivities and selectivities. Potentially the c ation of HNLs for obtaining cyanohydrins and the usage of nitrile hydratases for hyc of the cyanohydrin in one step may also become an "interesting reaction" in the [112,113].

33.5.2 HYDROXYNITRILE LYASES

The HNLs reversely catalyze the enantioselective formation and cleavage of cyanoh Several HNLs are known that convert (in the synthesis reaction) a broad range of ald and ketones into the corresponding (*R*)- or (*S*)-cyanohydrins (Figure 33.25 gives exan enzyme sources) [114].

Since the early works of Rosenthaler, in which he uses the enzyme preparation ⦁ (from almond), several other enzyme sources have been intensely studied and reviewed cyanohydrin syntheses [114,115]. During the 1960s and 1970s HNLs were rediscovei used in several large-scale applications. In the last decade an intense study on these e₁ led to the enclosing overlook about mechanism and structure for nearly all known

TABLE 33.11
Stability of a Nitrilase in Organic/Aqueous Mixtures

Organic Solvent (50% (v/v))	log P	Relative Activity
No addition	—	100
1-Octanol	2.9	47
Octane	4.5	66
Hexadecane	8.8	97

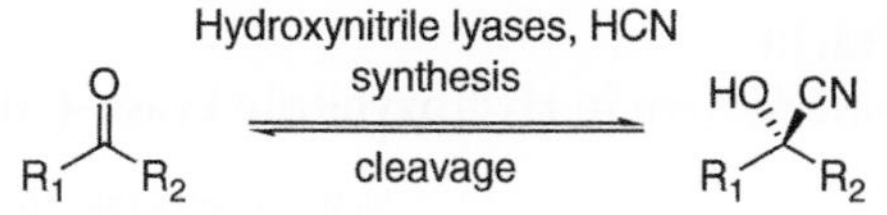

R-selective	*Prunus amygdalus* (almonds), *Prunus serotina* (black cherry), *Linum usitatissimum* (linseed), *Prunus domestica* (plum), *Phlebodium aureum* (goldfoot fern)
S-selective	*Manihot ecsculenta* (manioc), *Hevea brasiliensis* (rubber tree), *Sorghum bicolor* (millet), *Annonia muricata* (guanabana)

FIGURE 33.25 Hydroxynitrile lyase–catalyzed reactions.

They can be divided into FAD-dependent (e.g., *Prunus amygdalus*) and non-FAD-dependent (e.g., *Manihot esculenta*) HNLs [116].

Beginning with an emulsion of benzaldehyde in water, used by Rosenthaler, several different reaction systems have been developed. As an important example, the immobilized enzyme on various supports has to be mentioned and has found broad usability for various HNLs like the successful application to the HNL from *M. esculenta* (Table 33.12) [117–119]. The enzyme is adsorbed on nitrocellulose suspended in an organic solvent (e.g., diisopropyl-ether). Using this approach several aldehydes and ketones can be converted with high yields and very good ee [120].

Even though a macroscopic suspension (enzyme support suspended in an organic solvent) is formed, this system behaves like a one-phase system (see Section 33.1). Due to this restriction, limitations like substrate and product inhibition may occur. This disadvantage can be solved by using the very effective two-phase approach, using a buffer phase containing the enzyme and a water-immiscible organic solvent containing substrates. Applying this system the substrate and product concentration in the aqueous phase is very low, resulting from the partition coefficients between organic and aqueous phases.

Using this approach, several aldehyde and ketone cyanohydrins were easily accessible using the HNL from *Hevea brasiliensis* (Table 33.13) in a two-phase system consisting of a buffer and MTBE [121]. Recently, Lou et al. presented the first hydroxynitrile lyase–catalyzed reaction with encouraging results using the HNLs from *P. amygdalus* (*Pa*HNL) and *H. brasiliensis* (*Hb*HNL) in the IL–buffer two-phase system [122].

TABLE 33.12
Cyanohydrin Synthesis Using Adsorbed Wild-Type Hydroxynitrile Lyase from *Manihot esculenta*, Abstract

Substrate	Reaction Time (h)	Conversion (%)	ee (*S*) (%)
Benzaldehyde	0.5	97	99
2-Chlorobenzaldehyde	1	96	98
3-Phenoxybenzaldehyde	6.25	47	96
Decanal	17	65	78
Acetophenone	3	13	78
2-Thiophen	0.5	75	97

TABLE 33.13
Two-Phase System in Hydroxynitrile Lyase–Catalyzed Synthesis

Substrate	Water-Immiscible Solvent	Conversion (%)	ee
Benzaldehyde	MTBE	97	
3-Phenoxy-benzaldehyde	MTBE	99	
Propenal	MTBE	92	
Benzaldehyde (*Pa*HNL)	PMIM BF$_4$	99	
Benzaldehyde (*Hb*HNL)	PMIM BF$_4$	99	

Furthermore, with the HNL from *H. brasiliensis* it has been demonstrated tl enzymatic reaction is performed only at the interfacial area and not in the aqueo phase. This has been observed for the mandelonitrile cleavage by a blockade of the in between diisopopylether and buffer [123].

Due to several new discovered HNLs and large-scale applications for cyanohydrins established enzymes, the observation of this enzyme class will proceed. Additionally, few reports were published using ILs in lyase-catalyzed reactions, especially for HNL The research interest in the field of "enzymes in nonconventional media" will probably the next few years. Also the usage of new molecular-biological attempts will gair interest, with techniques like directed evolution. For example, a single residual repla improves the folding and stability of a HNL from *M. esculenta* [125].

33.5.3 ALDOLASES

The usability of directed evolution in aldolase synthesis has been shown for a f bisphosphate aldolase. After four rounds of directed evolution using DNA shuffling *fda* genes from *E. coli* and *Edwardsiella ictaluri* the resistance against various organic s and also the thermostability have been increased [126].

The wild-type enzyme (*E. coli*) lost 40 to 90% of activity, whereas the variant a remained nearly constant for an additional 20% (v/v) organic solvent (Figure 33.26)

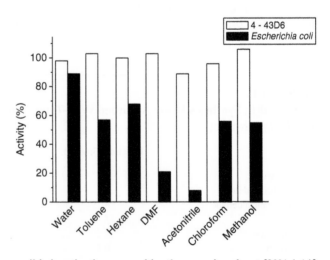

FIGURE 33.26 Irreversible inactivation, caused by the organic solvent [20% (v/v)].

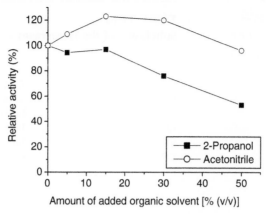

FIGURE 33.27 Influence of 2-propanol and acetonitrile on the aldolase from *Hyperthermophic archaea*.

heat-stable aldolase from *Methanococcus jannaschii* also a slight activating effect has been observed. By an addition of 15% (v/v) acetonitrile the relative activity increased to 23%, as shown in Figure 33.27 [127].

This stability against organic solvents may also be valuable for less-soluble substrates in enzyme catalysis in organic synthesis. A rabbit muscle aldolase (RAMA) showed an impressive stability with several water-miscible organic solvents (Figure 33.28).

Only for a few water-miscible organic solvents the enzyme activity was significantly reduced, whereas the water-immiscible organic solvents showed only a low decrease (maximum 50%).

As many substrates for aldolases are well soluble in water, the use of nonconventional media will be of limited interest.

33.6 EC 5: ISOMERASES

Isomerases catalyze isomerization reactions like racemization, epimerization, and rearrangement of different substrates. The utility of such enzymes is very low in industry because, on the one hand, there are only a few commercially available and, on the other hand, the

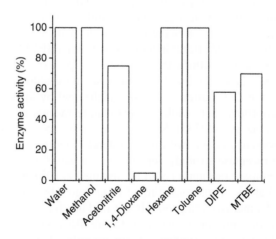

FIGURE 33.28 Rabbit muscle aldolase (RAMA) activity after 24 h of incubation at 20% (v/v), abstract.

TABLE 33.14
Subclasses of the Isomerases

	EC 5 Isomerases
EC 5.1	Racemases and epimerases
EC 5.2	*cis–trans*-Isomerases
EC 5.3	Intermolecular Oxidoreductases
EC 5.4	Intermolecular transferases
EC 5.5	Intermolecular lyases
EC 5.99	Other isomerases

exigency of isomerization in industry is marginal. Therefore, only about 2% of the tions in the period between 1987 and 2003 dealt with isomerase-catalyzed synthe. Isomerases are further divided into six subclasses (Table 33.14) [50].

In the last 25 years of the 20th century most reports were dealing with racema epimerases, which can be used for the dynamic kinetic resolution to get 100% conversi Yagasaki and Ozaki published the production of γ-aminobutyrat [128]. D-Glutamic a produced from L-glutamic acid. L-Glutamate was converted to the DL-form by the rec ant glutamate racemase of *L. brevis* ATCC8287. Then L-glutamate in the racemic was selectively decarboxylated to the product by L-glutamate decarboxylase of ATCC11246. This was successfully realized in a one-pot reaction. Before this, there few articles dealing with cofactor independence of glutamate racemase from *Lactobac* Some racemases like alanine racemase need a cofactor to catalyze the enzymatic zation. In the case of the alanine racemase, pyridoxal phosphate (PLP) is the n cofactor. In addition to the relatively well-studied alanine racemase, arginine ra from *Pseudomonas* sp. and two serine racemases from *Streptomyces* sp. require I their catalytic activity [129].

The epimerization of threonine could be observed by the new amino acid racema *P. putida*. As mentioned earlier this epimerization has been done in buffered aqueous as well, with attention on cofactor dependence—again the cofactor is pyridoxal 5′-ph —and on substrate specificity [130].

A typical enzyme to employ dynamic kinetic resolution is the mandelate racem combination with a second biocatalyst, it is possible to transfer a racemate into stereoisomer in 100% theoretical yield [131]. Mandelate racemase catalyzes the racem of different substrates in aqueous solution (HEPES pH 7.5 with 3.3 mM $MgCl_2 \cdot H_2O$) temperature, with further reaction by lipase from *Pseudomonas* sp. [132].

Syntheses catalyzed by isomerases are usually not investigated in nonconventiona But the *O*-acylation of (±)-mandelic acid by lipase from *Pseudomonas* sp. was ach diisopropylether, while the mandalate racemase–catalyzed racemization took plac aqueous phase [133]. Kaftzik et al. first reported the use of mandalate racemase ir cosolvent and in two-phase systems (Figure 33.29) [134]. It is shown that the act the racemase strongly depends on the water activity of the reaction system. Activit mandalate racemase could be obtained in [BMIM][OctSO$_4$] at water activity a_w:0.75 biphasic systems consisting of water and [OMIM][PF$_6$] in a ratio of 1:10.

Isomerases will probably not find much consideration in organic synthesis taking nonconventional media in the future. Because of their scarce availability and limited in biocatalytical synthesis, there are only a few applications.

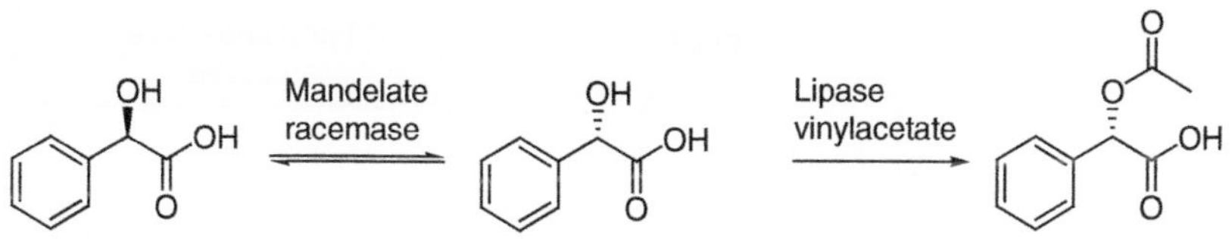

FIGURE 33.29 Deracemization of ($\pm$)-mandelic acid through a lipase-mandelate racemase two-enzyme system: *Pseudomonas* sp. lipase catalyzes *O*-acylation of ($\pm$)-mandelic acid and mandelate racemase–catalyzed racemization of remaining unreacted (*R*)-mandelic acid. (From Kaftzik, N. et al., *Mol. Catal. A Chem.*, 214, 107–112, 2004.)

33.7   EC 6: LIGASES

Ligases (synthetases) are classified in group 6 by the EC and represent a class of enzymes that catalyze the formation of bonds between two substrate molecules. They are further divided into subclasses according to the type of bond formed (Table 33.15).

The synthesis reaction catalyzed by ligases requires the hydrolysis of a nucleoside triphosphate such as adenosine triphosphate (ATP). In the field of ligases, examples of industrial biocatalysis are more or less missing [104]. This is further indicated by the fact that only about 1% of research from 1987 to 2003 has been performed with enzymes from the class of ligases [51]. One of the few examples for the biotechnological application of an enzyme from the group of ligases was described in 1998 by Aresta et al. [135]. They reported about a cheap, fast, and easy method for the phosphorylation of phenol at room temperature and subatmospheric pressure of CO_2: catalyzed by phenyl phosphate carboxylase, the synthesis of 4-OH benzoic acid from phenol and CO_2 with 100% selectivity was achieved (Figure 33.30).

Ligases also play an important role in the field of genetic engineering: DNA ligases (6.5.1.1) have become an indispensable tool for generating recombinant DNA sequences in modern molecular biology. By the development of a variety of nucleic acid–based detection systems for genetic disorders as well as for bacterial, viral, and other pathogens, ligases have been applied in a number of DNA amplification methods including polymerase chain reaction, self-sustained sequence replication, Q-beta replicase, and ligase chain reaction.

The discovery of DNA ligases and the biochemical studies of the ligase reaction by Lehman et al. were the first reports about host and bacteriophage-induced DNA ligases from eubacteria [136]. Subsequently, related enzymes from a wide range of organisms have been identified and studied. Today it is known that DNA ligases represent a large family of evolutionarily related proteins that play an important role in many essential reactions within the living cell including replication, recombination, and repair of DNA in all three kingdoms

TABLE 33.15
Subclasses of the Ligases

EC 6 Ligases

EC 6.1	Forming carbon–oxygen bonds
EC 6.2	Forming carbon–sulfur bonds
EC 6.3	Forming carbon–nitrogen bonds
EC 6.4	Forming carbon–carbon bonds
EC 6.5	Forming phosphoric ester bonds
EC 6.6	Forming nitrogen–metal bonds

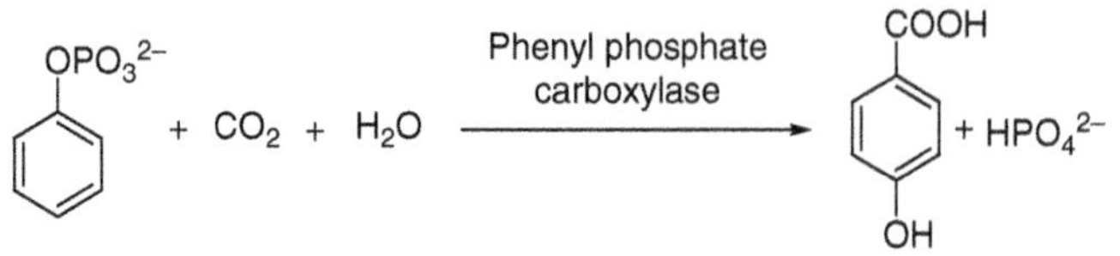

FIGURE 33.30 Synthesis of 4-OH benzoic acid from phenol and CO_2 by phenyl phosphate c ylase. (From Aresta, M. et al., *Tetrahedron*, 54, 8841–8846, 1998.)

of life. They catalyze the formation of phosphodiester bonds at single-stranded breaks (between adjacent 3′-hydroxyl and 5′-phosphate termini in double-stranded DNA by either ATP or NAD^+ as a cofactor [136,137].

As the presence of NAD^+-dependent DNA ligases is restricted to eubacteria, th makes them an attractive target for novel antibiotics. For example, bacterial N synthetase (EC 6.3.5.1) catalyzes the last step in both the *de novo* biosynthetic and s. pathways for NAD^+ and thus plays an essential role in the life cycle of bacteria. In or find novel inhibitors for this enzyme for future antibacterial drug development, an enzy assay was developed by Yang et al. for the purpose of screening compounds for e. inhibition [138]. To improve the solubility of the compounds screened, the water-m organic solvent dimethyl sulfoxide (DMSO) was added to the assay buffer as cosc Although no effects could be observed on behalf of the enzyme activity, concentrati 2.5% (v/v) DMSO led to changes in the stability of the dimer and its unfolding mecha

Although bacterial NAD^+-dependent DNA ligases have been studied for more 30 years, surprisingly few genetic and biochemical details are known about their regul In order to find new and interesting targets for future antibacterial drug development f investigations will be needed.

To summarize, ligases represent a class of enzymes found ubiquitously in nature. I their involvement in numerous essential reactions within the living cell and their co dependency, the catalytic activity of ligases is more or less restricted to aqueous r though some might act in the presence of small concentrations of cosolvents as While the subclass of DNA-ligases has already found numerous applications in mol biology, the discovery of other ligases, e.g., acetyl-coenzyme A carboxylase (ACC) (EC 6. will be useful for the investigation of future potential targets for drug discovery. ACC p crucial role in fatty acid metabolism in most living organisms. It catalyzes the carboxy of acetyl-CoA to produce malonyl-CoA and represents an attractive target for the thera intervention in the control of obesity and the treatment of metabolic syndrome [139].

REFERENCES

1. Zaks, A. and Klibanov, A.M., Enzyme-catalyzed processes in organic solvents, *Proc. Natl. Sci. USA.*, 82, 3192–3196, 1985.
2. Kilbanov, A.M., Enzymes that work in organic solvents, *Chemtech*, 16, 354–359, 1986.
3. Klibanov, A.M., Why are enzymes less active in organic solvents than in water? *Trends Biote* 15, 97–101, 1997.
4. Carrea, G., Biocatalysis in water–organic solvent 2-phase systems, *Trends Biotechn* 102–106, 1984.
5. Halling, P.J., Biocatalysis in multiphase reaction mixtures containing organic liquids, *Biot Adv.*, 5, 47–84, 1987.
6. Shimizu, S. and Yamada, H., Stereospecific reduction of 3-keto acid-esters by a novel al reductase of *Sporobolomyces salmonicolor* in a water organic solvent 2-phasic system, *A Acad. Sci.*, 613, 628–632, 1990.

7. Halling, P.J., Thermodynamic predictions for biocatalysis in nonconventional media—theory, tests, and recommendations for experimental design and analysis, *Enzyme Microb. Technol.*, 16, 178–206, 1994.

8. Heinemann, M. et al., Experimental and theoretical analysis of phase equilibria in a two-phase system used for biocatalytic esterifications, *Biocatal. Biotrans.*, 21, 115–121, 2003.

9. Martinek, K. et al., Preparative enzymatic synthesis in biphasic aqueous–organic system, *Bioorganicheskaya Khimiya*, 3, 696–702, 1977.

10. Laane, C., Tramper, J., and Lilly, M.D., *Biocatalysis in Organic Media*, Elsevier, Amsterdam, 1987.

11. van Deurzen, M.P.J. et al., Chloroperoxidase-catalyzed oxidations in *t*-butyl alcohol/water mixtures, *J. Mol.. Catal. A Chemical*, 117, 329–337, 1997.

12. Patel, R., *Stereoselective Biocatalysis*, Marcel Dekker, New York, 2000, p. 799.

13. Cull, S.G. et al., Room-temperature ionic liquids as replacements for organic solvents in multiphase bioprocess operations, *Biotechnol. Bioeng.*, 69, 227–233, 2000.

14. Erbeldinger, M., Mesiano, A.J., and Russell, A.J., Enzymatic catalysis of formation of Z-aspartame in ionic liquid—an alternative to enzymatic catalysis in organic solvents, *Biotechnol. Prog.*, 16, 1129–1131, 2000.

15. Kragl, U., Eckstein, M., and Kaftzik, N., Enzyme catalysis in ionic liquids, *Curr. Opin. Biotechnol.*, 13, 565–571, 2002.

16. van Rantwijk, F., Lau, R.M., and Sheldon, R.A., Biocatalytic transformations in ionic liquids, *Trends Biotechnol.*, 21, 131–138, 2003.

17. Wong, C.-H. and Whitesides, G.M., *Enzymes in Synthetic Organic Chemistry*, Pergamon Press, Oxford, 1994, pp. 12, 370.

18. Hummel, W., Reduction of acetophenone to $R(+)$-phenylethanol by a new alcohol dehydrogenase from *Lactobacillus kefir*, *Appl. Microbiol. Biotechnol.*, 34, 15–19, 1990.

19. Howarth, J., James, P., and Dai, J.F., Immobilized baker's yeast reduction of ketones in an ionic liquid, [bmim]PF$_6$ and water mix, *Tetrahedron Lett.*, 42, 7517–7519, 2001.

20. Kragl, U., Eckstein, M., and Kaftzik, N., Enzyme catalysis in ionic liquids, *Curr. Opin. Biotechnol.*, 13, 565–571, 2002.

21. Nakamura, K. et al., Recent developments in asymmetric reduction of ketones with biocatalysts, *Tetrahedron Asymmetry*, 14, 2659–2681, 2003.

22. Matsue, S. and Miyawaki, O., Influence of water activity and aqueous solvent ordering on enzyme kinetics of alcohol dehydrogenase, lysozyme, and beta-galactosidase, *Enzyme Microb. Technol.*, 26, 342–347, 2000.

23. Liao, M.H. and Chen, D.H., Characteristics of magnetic nanoparticles-bound YADH in water/AOT/isooctane microemulsions, *J. Mol. Catal. B Enzym.*, 18, 81–87, 2002.

24. Hirakawa, H. et al., Regioselective reduction of a steroid in a reversed micellar system with enzymatic NADH-regeneration, *Biochem. Eng. J.*, 16, 35–40, 2003.

25. Chen, D.H. and Liao, M.H., Effects of mixed reverse micellar structure on stability and activity of yeast alcohol dehydrogenase, *J. Mol. Catal. B Enzym.*, 18, 155–162, 2002.

26. Riebel, B., Biochemische und molekularbiologische Charakterisierung neuer mikrobieller NAD(P)-abhängiger Alkoholdehydrogenasen, University Düsseldorf, Düsseldorf, 1996.

27. Villela Filho, M. et al., Is the log P a convenient criterion to guide the choice of solvents for biphasic enzymatic reactions? *Angew. Chem.* 115, 3101–3104, 2003.

28. Ferloni, C. et al., Enzymatische Gasphasenkatalyse zur Produktion chiraler Substanzen, *Biokatalyse* (Transkript Sonsderheft), 105–108, 2003.

29. Eckstein, M., Daußmann, T., and Kragl, U., Recent development in NAD(P)H regeneration for enzymatic reductions in one- and two-phase systems, *Biocatal. Biotrans.*, 22, 89–96, 2004.

30. Schumacher, J., Eckstein, M., and Kragl, U., Influence of water-miscible organic solvents on kinetics and enantioselectivity of the (R)-specific alcohol dehydrogenase from *Lactobacillus brevis*, *Biotechnol. J.*, 1, 574–581, 2006.

31. Eckstein, M. et al., Use of an ionic liquid in a two-phase system to improve an alcohol dehydrogenase catalysed reduction, *Chem. Comm.*, 1084–1085, 2004.

32. Kroutil, W. et al., Recent advances in the biocatalytic reduction of ketones and oxidation of sec-alcohols, *Curr. Opin. Chem. Biol.*, 8, 120–126, 2004.

33. Eckstein, M. et al., Maximise your equilibrium conversion in biphasic catalysed reactions: H obtain reliable data for equilibrium constants? *Adv. Syn. Catal.*, in press.

34. Kragl, U. et al., Repetitive batch as an efficient method for preparative scale enzymic synthe 5-azido-neuroaminic acid and 15*N*-L-glutamic acid, *Tetrahedron Asymmetry*, 4, 1193–1202,

35. Adlercreutz, P., Cofactor regeneration in biocatalysis in organic media, *Biocatal. Biotrans* 1–30, 1996.

36. De Temino, D.M., Hartmeier, W., and Ansorge-Schumacher, M.B., Entrapment of the al dehydrogenase from *Lactobacillus kefir* in polyvinyl alcohol for the synthesis of chiral hydrop alcohols in organic solvents, *Enzyme Microb. Technol.*, 36, 3–9, 2005.

37. Mateo, C. et al., A new, mild cross-linking methodology to prepare cross-linked er aggregates, *Biotechnol. Bioeng.*, 86, 273–276, 2004.

38. Khmelnitsky, Y.L. et al., Denaturation capacity—a new quantitative criterion for selecti organic-solvents as reaction media in biocatalysis, *Eur. J. Biochem.* 198, 31–41, 1991.

39. Tanaka, T. et al., Enzymatic oxidative polymerization of 4-chloroguaiacol by laccase, *J.* *Eng. Jap.*, 36, 1101–1106, 2003.

40. Pilz, R. et al., Laccase-catalysed synthesis of coupling products of phenolic substrates in dif reactors, *Appl. Microbiol. Biotechnol.*, 60, 708–712, 2003.

41. Nicotra, S. et al., Laccase-mediated oxidation of the steroid hormone 17 beta-estradiol in o solvents, *Tetrahedron Asymmetry*, 15, 2927–2931, 2004.

42. Lugaro, G., Carrea, G., and Cremonesi, P., The oxidation of steroid hormones by fungal lac emulsion of water and organic solvents, *Arch. Biochem. Biophys.*, 159, 1–6, 1973.

43. Schmid, A. et al., Integrated biocatalytic synthesis on gramm scale: the highly en selective preparation of chiral oxiranes with styrene monooxygenase, *Adv. Synth. Catal* 732–737, 2001.

44. Hofstetter, K. et al., Coupling of biocatalytic asymmetric epoxidation with NADH regenera organic–aqueous emulsions, *Angew. Chem. Int. Ed.*, 43, 2163–2166, 2004.

45. Zambianchi, F. et al., Use of isolated cyclohexanone monooxygenase from recombinant *E ichia coli* as a biocatalyst for Baeyer–Villiger and sulfide oxidations, *Biotechnol. Bioeng.*, 78 496, 2002.

46. Hilker, I. et al., Microbiological transformations. 57. Facile and efficient resin-based *in situ* preparative-scale synthesis of an enantiopure "unexpected" lactone regioisomer via a B Villiger oxidation process, *Org. Lett.*, 6, 1955–1958, 2004.

47. Anderson, B.A. et al., Application of a practical biocatalytic reduction to an enantiose synthesis of the 5H-2,3-benzodiazepine Ly300164, *J. Am. Chem. Soc.*, 117, 12358–12359, 1

48. Vicenzi, J.T. et al., Large-scale stereoselective enzymatic ketone reduction with in situ p removal via polymeric adsorbent resins, *Enzyme Microb. Technol.*, 20, 494–499, 1997.

49. Michizoe, J. et al., Functionalization of the cytochrome P450cam monooxygenase system cell-like aqueous compartments of water-in-oil emulsions, *J. Biosci. Bioeng.*, 99, 12–17, 200

50. http://www.brenda.uni-koeln.de/.

51. Faber, K., *Biotransformations in Organic Chemistry*, 5th ed., Springer, Berlin, 2004,

52. Wu, J.Y. and Yang, J.T., Effects of salts and organic solvents on activity of citrate synthase, *Chem.*, 245, 3561, 1970.

53. Singh, T.J. and Wang, J.H., Stimulation of glycogen-phosphorylase kinase from rabbit s muscle by organic-solvents, *J. Biol. Chem.*, 254, 8466–8472, 1979.

54. Rajala, R.V.S. and Sharma, R.K., Myristoyl CoA-Protein *N*-myristoyltransferase—subc localization, activation and kinetic-behavior in the presence of organic-solvents, *Biochem. E Res. Commun.*, 208, 617–623, 1995.

55. Singer, S.J., The properties of proteins in nonaqueous solvents, *Adv. Prot. Chem.*, 17, 1–6

56. Shin, J.S. and Kim, B.G., Kinetic resolution of alpha-methylbenzylamine with transaminase screened from soil microorganisms: application of a biphasic system to o product inhibition, *Biotechnol. Bioeng.*, 55, 348–358, 1997.

57. Bornscheuer, U.T. and Kazlauskas, R.J., *Hydrolases in Organic Synthesis. Regio- and Ster tive Biotransformations*, Wiley-VCH, Weinheim, Germany, 1999.

58. Verger, R., 'Interfacial activation' of lipases: facts and artifacts, *Trends Biotechnol.*, 15, 32–38, 1997.
59. Schmid, R.D. and Verger, R., Lipases: interfacial enzymes with attractive applications, *Angew. Chem. Int. Ed.*, 37, 1609–1633, 1998.
60. Pottie, M., Vandereycken, J., and Vandewalle, M., Enzymatic enantioselective hydrolysis of 2,2-dimethyl-1,3-dioxolane-4-carboxylic esters, *Tetrahedron Lett.*, 30, 5319–5322, 1989.
61. Dumortier, L., Vandereycken, J., and Vandewalle, M., The synthesis of chiral isopropylidene derivatives of 1,2,3-cyclohexanetriols by enzymatic differentiation, *Tetrahedron Lett.*, 30, 3201–3204, 1989.
62. Xie, Z.F. et al., An insight into the enantioselective hydrolyses of cyclic acetates catalyzed by *Pseudomonas fluorescens* lipase, *J. Chem. Soc. Chem. Commun.*, 966–977, 1988.
63. Seemayer, R. and Schneider, M.P., Enzymatic preparation of tetrahydrofuran derivatives of high optical purity, *J. Chem. Soc. Perkin 1*, 2359–2360, 1990.
64. Laumen, K. and Schneider, M.P., A highly selective ester hydrolase from *Pseudomonas* sp. for the enzymatic preparation of enantiomerically pure secondary alcohols—chiral auxiliaries in organic-synthesis, *J. Chem. Soc. Chem. Commun.*, 598–600, 1988.
65. Kalaritis, P. et al., Kinetic resolution of 2-substituted esters catalyzed by a lipase ex *Pseudomonas fluorescens*, *J. Org. Chem.*, 55, 812–815, 1990.
66. Hughes, D.L. et al., Lipase-catalyzed asymmetric hydrolysis of esters having remote chiral prochiral centers, *J. Org. Chem.*, 55, 6252–6259, 1990.
67. Halling, P. and Kvittingen, L., Why did biocatalysis in organic media not take off in the 1930s? *Trends Biotechnol.*, 17, 343–344, 1999.
68. Sym, E.A., Action of esterase in the presence of organic solvents, *Biochem. J.*, 30, 609–617, 1936.
69. Sperry, W.M. and Brand, F.C., A study of cholesterol esterase in liver and brain, *J. Biol. Chem.*, 137, 377–387, 1941.
70. Cambou, B. and Klibanov, A.M., Unusual catalytic properties of usual enzymes, *Ann. NY Acad Sci.*, 434, 219–223, 1984.
71. Mori, K. et al., Pheromone synthesis. 101. Synthesis and bioactivity of optically active forms of 1-methyl-2-cyclohexen-1-ol, an aggregation pheromone *Dendroctonus-pseudotsugae*, *Tetrahedron*, 43, 2249–2254, 1987.
72. Johnston, B.D. et al., A convenient synthesis of both enantiomers of seudenol and their conversion to 1-methyl-2-cyclohexen-1-ol (Mcol), *Tetrahedron Asymmetry*, 2, 377–380, 1991.
73. Hartmeier, W., *Immobilized Biocatalysts: An Introduction.*, Springer, Berlin, 1988.
74. Brandenberg, H., Mehods for linking enzymes to insoluble carriers, *Angew. Chem. Int. Ed.*, 67, 244–246, 1955.
75. Reetz, M.T. et al., Characterization of hydrophobic sol–gel materials containing entrapped lipases, *J. Sol–Gel Sci. Technol.*, 7, 35–43, 1996.
76. Buthe, A. et al., Generation of lipase-containing static emulsions in silicone spheres for synthesis in organic media, *J. Mol. Catal. B Enzym.*, 35, 93–99, 2005.
77. Goto, M., Hatanaka, C., and Goto, M., Immobilization of surfactant–lipase complexes and their high heat resistance in organic media, *Biochem. Eng. J.*, 24, 91–94, 2005.
78. Palomo, J.M. et al., Modulation of the enantioselectivity of lipases via controlled immobilization and medium engineering: hydrolytic resolution of mandelic acid esters, *Enzyme Microb. Technol.*, 31, 775–783, 2002.
79. Hilal, N., Nigmatullin, R., and Alpatova, A., Immobilization of cross-linked lipase aggregates within microporous polymeric membranes, *J. Membrane Sci.*, 238, 131–141, 2004.
80. Wasserscheid, P. and Keim, W., Ionic liquids: new "solutions" for transition metal catalysis, *Angew. Chem. Int. Ed.*, 39, 3773–3789, 2000.
81. Welton, T., Room-temperature ionic liquids: solvents for synthesis and catalysis, *Chem. Rev.*, 99, 2071–2083, 1999.
82. Seddon, K.R., Ionic liquids for clean technology, *J. Chem. Technol. Biotechnol.*, 68, 351–356, 1997.
83. Lau, R.M. et al., Lipase-catalyzed reactions in ionic liquids, *Org. Lett.*, 2, 4189–4191, 2000.
84. Schofer, S.H. et al., Enzyme catalysis in ionic liquids: lipase catalysed kinetic resolution of 1-phenylethanol with improved enantioselectivity, *Chem. Commun.*, 425–426, 2001.

85. Itoh, T., Akasaki, E., and Nishimura, Y., Efficient lipase-catalyzed enantioselective acyla under reduced pressure conditions in an ionic liquid solvent system, *Chem. Lett.*, 154–155, 2

86. Park, S. and Kazlauskas, R.J., Improved preparation and use of room-temperature ionic liqui lipase-catalyzed enantio- and regioselective acylations, *J. Org. Chem.*, 66, 8395–8401, 2001.

87. Skagerlind, P., Jansson, M., and Hult, K., Surfactant interference on lipase-catalyzed reactio microemulsions, *J. Chem. Technol. Biotechnol.*, 54, 277–282, 1992.

88. Bello, M., Thomas, D., and Legoy, M.D., Interesterification and synthesis by *Candida cylindr* lipase in microemulsions, *Biochem. Biophys. Res. Commun.*, 146, 361–367, 1987.

89. Hedstrom, G., Backlund, M., and Slotte, J.P., Enantioselective synthesis of ibuprofen esters in isooctane microemulsions by *Candida cylindracea* lipase, *Biotechnol. Bioeng.*, 42, 618–624, 199

90. Krieger, N. et al., Non-aqueous biocatalysis in heterogeneous solvent systems, *Food Tec Biotechnol.*, 42, 279–286, 2004.

91. Nakamura, K. et al., Lipase activity and stability in supercritical carbon-dioxide, *Chem. Commun.*, 45, 207–212, 1986.

92. Reetz, M.T. et al., Continuous flow enzymatic kinetic resolution and enantiomer separation ionic liquid/supercritical carbon dioxide media, *Adv. Synth. Catal.*, 345, 1221–1228, 2003.

93. Lozano, P. et al., Continuous green biocatalytic processes using ionic liquids and supercr carbon dioxide, *Chem. Commun.*, 692–693, 2002.

94. Reetz, M.T. et al., Biocatalysis in ionic liquids: batchwise and continuous flow processes supercritical carbon dioxide as the mobile phase, *Chem. Commun.*, 992–993, 2002.

95. Ohno, M. and Otsuka, M., Chiral synthons by ester hydrolysis catalysed by pig liver esterase *React.*, 37, 1–55, 1990.

96. Faber, K., Ottolina, G., and Riva, S., Selectivity-enhancement of hydrolase reactions, *Biocat* 8, 91–132, 1993.

97. Ruppert, S. and Gais, H.J., Activity enhancement of pig liver esterase in organic solver colyophilization with methoxypolyethylene glycol: kinetic resolution of alcohols, *Tetrah Asymmetry*, 8, 3657–3664, 1997.

98. Mohr, P. et al., A study of stereoselective hydrolysis of symmetrical diesters with pig-liver est *Helv. Chim. Acta*, 66, 2501–2511, 1983.

99. Schneider, M. et al., Hydrolytic enzymes in organic-synthesis. 3. Enzymatic synthes chiral building-blocks from prochiral meso-substrates preparation of methyl(hydroger cycloalkanedicarboxylates, *Angew. Chem. Int. Ed. Engl.*, 23, 67–68, 1984.

100. Anwar, A. and Saleemuddin, M., Alkaline proteases: a review, *Bioresour. Technol.*, 64, 175–183

101. Gupta, R., Beg, Q.K., and Lorenz, P., Bacterial alkaline proteases: molecular approache industrial applications, *Appl. Microbiol. Biotechnol.*, 59, 15–32, 2002.

102. Isowa, Y. et al., Thermolysin-catalyzed condensation-reactions of N-substituted aspart glutamic acids with phenylalanine alkyl esters, *Tetrahedron Lett.*, 28, 2611–2612, 1979.

103. Oyama, K., *Industrial Production of Aspartame*, in: *Chirality in Industry*, Wiley, Chichester, pp. 237–247.

104. Wandrey, C., Liese, A., and Kihumbu, D., Industrial biocatalysis: past, present, and future *Proc. Res. Dev.*, 4, 286–290, 2000.

105. Glieder, A. et al., Comprehensive step-by-step engineering of an (R)-hydroxynitrile lyase for scale asymmetric synthesis, *Angew. Chem. Int. Ed.*, 42, 4815–4818, 2003.

106. Nagasawa, T., Shimizu, H., and Yamada, H., The superiority of the 3rd-generation ca *Rhodococcus rhodochrous* J1 nitrile hydratase, for industrial-production of acrylamide, *Microbiol. Biotechnol.*, 40, 189–195, 1993.

107. Watanabe, I., Satoh, Y., and Enomoto, K., Screening, isolation and taxonomical proper microorganisms having acrylonitrile-hydrating activity, *Agric. Biol. Chem.*, 51, 3193–3199,

108. Watanabe, I. et al., Optimal conditions for cultivation of *Rhodocossus* sp. *N*-774 and for c sion of acrylonitrile to acrylamide by resting cells, *Agric. Biol. Chem.*, 51, 3201–3206, 1987

109. Martinkova, L. and Mylerova, V., Synthetic applications of nitrile-converting enzymes, *Cur Chem.*, 7, 1279–1295, 2003.

110. Layh, N. and Willetts, A., Enzymatic nitrile hydrolysis in low water systems, *Biotechnol. L* 329–331, 1998.

111. Bauer, M., Griengl, H., and Steiner, W., Kinetic studies on the enzyme (S)-hydroxynitrile lyase from *Hevea brasiliensis* using initial rate methods and progress curve analysis, *Biotechnol. Bioeng.*, 62, 20–29, 1999.

112. Reisinger, C. et al., Enzymatic hydrolysis of cyanohydrins with recombinant nitrile hydratase and amidase from *Rhodococcus erythropolis*, *Biotechnol. Lett.*, 26, 1675–1680, 2004.

113. Sheldon, R.A., Green solvents for sustainable organic synthesis: state of the art, *Green Chem.*, 7, 267–278, 2005.

114. Fechter, M.H. and Griengl, H., Hydroxynitrile lyases: biological sources and application as biocatalysts, *Food Technol. Biotechnol.*, 42, 287–294, 2004.

115. Rosenthaler, L., Durch Enzyme bewirkte asymmetrische Synthesen, *Biochemische Zeitschrift: Beiträge zur chem. Physiologie u. Pathologie*, 14, 238–253, 1908.

116. Hickel, A., Hasslacher, M., and Griengl, H., Hydroxynitrile lyases: functions and properties, *Physiol. Plant.*, 98, 891–898, 1996.

117. Buhler, H., Bayer, A., and Effenberger, F., A convenient synthesis of optically active 5,5-disubstituted 4-amino- and 4-hydroxy-2(5H)-furanones from (S)-ketone cyanohydrins, *Chem. Eur. J.*, 6, 2564–2571, 2000.

118. Buhler, H. et al., Substrate specificity of mutants of the hydroxynitrile lyase from *Manihot esculenta*, *ChemBioChem*, 4, 211–216, 2003.

119. Buhler, H., Miehlich, B., and Effenberger, F., Inversion of stereoselectivity by applying mutants of the hydroxynitrile lyase from *Manihot esculenta*, *ChemBioChem*, 6, 711–717, 2005.

120. Buhler, H., Optimierung der (S)-Hydroxynitril Lyase aus *Manihot esculenta* durch gezielte Mutationen—Anwendungen von optisch aktiven Cyanhydrinen in der Synthese, University Stuttgart, Stuttgart, Germany, 2000.

121. Griengl, H. et al., Enzyme catalysed formation of (S)-cyanohydrins derived from aldehydes and ketones in a biphasic solvent system, *Tetrahedron*, 54, 14477–14486, 1998.

122. Lou, W.Y., Xu, R., and Zong, M.H., Hydroxynitrile lyase catalysis in ionic liquid-containing systems, *Biotechnol. Lett.*, 27, 1387–1390, 2005.

123. Hickel, A., Radke, C.J., and Blanch, H.W., Hydroxynitrile lyase at the diisopropyl ether/water interface: evidence for interfacial enzyme activity, *Biotechnol. Bioeng.*, 65, 425–436, 1999.

124. Trummler, K. and Wajant, H., Molecular cloning of acetone cyanohydrin lyase from flax (*Linum usitatissimum*): definition of a novel class of hydroxynitrile lyases, *J. Biol. Chem.*, 272, 4770–4774, 1997.

125. Yan, G.H. et al., A single residual replacement improves the folding and stability of recombinant cassava hydroxynitrile lyase in *E. coli*, *Biotechnol. Lett.*, 25, 1041–1047, 2003.

126. Hao, J.J. and Berry, A., A thermostable variant of fructose bisphosphate aldolase constructed by directed evolution also shows increased stability in organic solvents, *Prot. Eng. Des. Sel.*, 17, 689–697, 2004.

127. Choi, I.G. et al., Overproduction, purification, and characterization of heat-stable aldolase from *Methanococcus jannaschii*, a hyperthermophic archaea, *J. Biochem. Mol. Biol.*, 31, 130–134, 1998.

128. Yagasaki, M. and Ozaki, A., Industrial biotransformations for the production of D-amino acids, *J. Mol. Catal. B Enzym.*, 4, 1–11, 1998.

129. Gallo, K.A. and Knowles, J.R., Purification, cloning, and cofactor independence of glutamate racemase from *Lactobacillus*, *Biochemistry*, 32, 3981–3990, 1993.

130. Lim, Y.H. et al., A new amino-acid racemase with threonine alpha-epimerase activity from *Pseudomonas putida*: purification and characterization, *J. Bacteriol.*, 175, 4213–4217, 1993.

131. Asano, Y. and Yamaguchi, S., Discovery of amino acid amides as new substrates for alpha-amino-epsilon-caprolactam racemase from *Achromobacter obae*, *J. Mol. Catal. B Enzym.*, 36, 22–29, 2005.

132. Felfer, U. et al., Substrate spectrum of mandelate racemase—Part 2. (Hetero)-aryl-substituted mandelate derivatives and modulation of activity, *J. Mol. Catal. B Enzym.*, 15, 213–222, 2001.

133. Strauss, U.T. and Faber, K., Deracemization of (+/−)-mandelic acid using a lipase-mandelate racemase two-enzyme system, *Tetrahedron Asymmetry*, 10, 4079–4081, 1999.

134. Kaftzik, N. et al., Mandelate racemase activity in ionic liquids: scopes and limitations, *J. Mol. Catal. A Chem.*, 214, 107–112, 2004.

135. Aresta, M. et al., Enzymatic synthesis of 4-OH-benzoic acid from phenol and CO_2: the first of a biotechnological application of a carboxylase enzyme, *Tetrahedron*, 54, 8841–8846, 19!

136. Lehman, I.R., DNA ligase: structure, mechanism, and function, *Science*, 186, 790–797, 1S

137. Lindahl, T. and Barnes, D.E., Mammalian DNA ligases, *Annu. Rev. Biochem.*, 61, 251–28

138. Yang, Z.R.W. et al., Dimethyl sulfoxide at 2.5% (v/v) alters the structural cooperativ unfolding mechanism of dimeric bacterial NAD(+) synthetase, *Prot. Sci.*, 13, 830–841, 2C

139. Tong, L., Acetyl-coenzyme A carboxylase: crucial metabolic enzyme and attractive target f discovery, *Cell. Mol. Life Sci.*, 62, 1784–1803, 2005.

140. Jaeger, K.E. and Reetz, M.T., Microbial lipases form versatile tools for biotechnology, *Biotechnol.*, 16, 396–403, 1998.

141. Pandey, A. et al., The realm of microbial lipases in biotechnology, *Biotechnol. Appl. Bioch* 119–131, 1999.

142. Farrell, R.L., Hata, K., and Wall, M.B., Solving pitch problems in pulp and paper processe use of enzymes or fungi, *Adv. Biochem. Eng. Biotechnol.*, 57, 197–212, 1997.

143. Parmar, V.S. et al., Chiral discrimination by hydrolytic enzymes in the synthesis of optica materials, *Proc. Ind. Acad. Sci. Chem. Sci.*, 108, 575–583, 1996.

144. Rehm, H.J., Reed, G., and Stadler, P., *Biotransformations I—Water, Organic Solvents an Reaction Media*, VCH, Weinheim, Germany, 1998, 8a,

145. Shoemaker, M. et al., The lipodygenase sensor, a new approach in essential fatty acid de ation in foods, *Biosens. Bioelectron.*, 12, 1089–1099, 1997.

146. Godfrey, T. and West, S.H., *Industrial Enzymology: The Application of Enzymes in I* Stockton, New York, 1996,

147. Izumi, T. et al., Enzymatic transesterification of 3,7-dimethyl-4,7-octadien-1-ol using *J. Chem. Technol. Biotechnol.*, 68, 57–64, 1997.

148. Salleh, A.B. et al., Extracellular and intracellular lipases from a thermophilic *Rhizopus ory* factors affecting their production, *Can. J. Microbiol.*, 39, 978–981, 1993.

149. Kaneki, H. and Tanaka, M., Activation of *Rhizopus delemar* lipase-catalyzed hydrolysis o pionin by DDT and aldrin: tightly bound pesticide-lipase complexes, *Eisei Kagak J. Toxicol. Environ. Health*, 32, 1–12, 1986.

150. Boland, W., Frossl, C., and Lorenz, M., Esterolytic and lipolytic enzymes in organic-sy *Synth. Stutt.*, 1049–1072, 1991.

151. Otera, J., Transesterification, *Chem. Rev.*, 93, 1449–1470, 1993.

152. Theil, F., Lipase-supported synthesis of biologically active compounds, *Chem. Rev.*, 95 2227, 1995.

153. West, S.I., Enzymes in the food-processing industry, *Chem. Br.*, 24, 1220–1222, 1988.

154. Gotor, V., Non-conventional hydrolase chemistry: amide and carbamate bond formatio lyzed by lipases, *Bioorg. Med. Chem.*, 7, 2189–2197, 1999.

155. Fernandes, M.L.M. et al., Hydrolysis and synthesis reactions catalysed by *Thermomyc ginosa* lipase in the AOT/isooctane reversed micellar system, *J. Mol. Catal. B Enz* 43–49, 2004.

156. Falcone, R.D. et al., Effect of the addition of a nonaqueous polar solvent (Glycerol) on en catalysis in reverse micelles. Hydrolysis of 2-naphthyl acetate by alpha-chymotrypsin, *L* 20, 5732–5737, 2004.

157. Naoe, K. et al., Solvent condition in triolein hydrolysis by *Rhizopus delemar* lipase using reverse micellar system, *Biochem. Eng. J.*, 18, 49–55, 2004.

158. Altreuter, D.H., Dordick, J.S., and Clark, D.S., Optimization of ion-paired lipase for non-media: acylation of doxorubicin based on surface models of fatty acid esterification, *Microb. Technol.*, 31, 10–19, 2002.

159. Zhou, G.W. et al., Kinetic studies of lipase-catalyzed esterification in water-in-oil microem and the catalytic behavior of immobilized lipase in MBGs, *Colloids Surf. A Physicoche Aspects*, 194, 41–47, 2001.

160. Knez, Z. and Habulin, M., Compressed gases as alternative enzymatic-reaction solvents: review, *J. Supercrit. Fluids*, 23, 29–42, 2002.

161. Lozano, P. et al., Synthesis of glycidyl esters catalyzed by lipases in ionic liquids and supercritical carbon dioxide, *J. Mol. Catal. A Chem.*, 214, 113–119, 2004.
162. Madras, G., Kolluru, C., and Kumar, R., Synthesis of biodiesel in supercritical fluids, *Fuel*, 83, 2029–2033, 2004.
163. Sovova, H. and Zarevucka, M., Lipase-catalysed hydrolysis of blackcurrant oil in supercritical carbon dioxide, *Chem. Eng. Sci.*, 58, 2339–2350, 2003.
164. Lozano, P. et al., Membrane reactor with immobilized *Candida antarctica* lipase B for ester synthesis in supercritical carbon dioxide, *J. Supercrit. Fluids*, 29, 121–128, 2004.
165. Moorlag, H. et al., Pig-liver esterase catalyzed hydrolyzes of racemic alpha-substituted alpha-hydroxy esters, *J. Org. Chem.*, 55, 5878–5881, 1990.
166. Sicsic, S., Leroy, J., and Wakselman, C., Geometric selectivity of pig-liver esterase and its application to the separation of fluorinated bicyclic esters, *Synth. Stutt.*, 155–156, 1987.
167. Papageorgiou, C. and Benezra, C., Use of enzymatic-hydrolysis of dimethyl malates for a short synthesis of tulipalin-b and of its enantiomer, *J. Org. Chem.*, 50, 1144–1145, 1985.
168. Pfründer Weuster-Botz, H., et al., Effiziente Ganzzell—Biotransformation im zweiphasigen System ionische Flüssigkeit Wasser, *Angew. Chem.*, 116, 4629–4631, 2004.

34 Biocatalytic Concepts for the Synthesis of Optically Active Amines

Stefan Buchholz and Harald Gröger

CONTENTS

34.1 INTRODUCTION

A look into the structure of many drugs and agrochemicals often reveals a chiral amine subunit in the molecular framework [1]. Thus, there is a wide interest from the chemical and pharmaceutical industries for these types of optically active amines and manufacturing processes thereof. Many natural products, comprising highly complicated structures with various stereogenic centers, often contain chiral amine moieties (e.g., alkaloids and derivatives). Consequently, the synthesis of chiral amine building blocks is a field of intensive research in organic chemistry, and is also still challenging with respect to enzymatic approaches. In spite of the fact that numerous chiral amine structures exist in nature, most of the enzymatic synthetic approaches had not been developed until the 1990s [2]. On the other hand, biocatalysis offers a multitude of potential solutions covering a wide range of concepts based on resolution and asymmetric synthesis. Nevertheless, the "best solution" does not

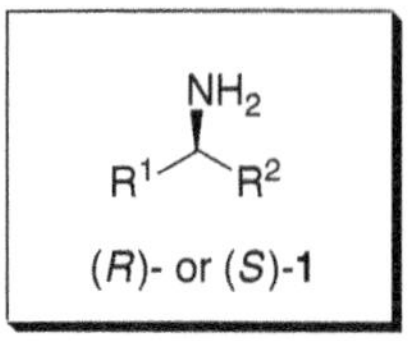

$$NH_2$$

$$R^1 \overset{|}{\diagup} R^2$$

(R)- or (S)-**1**

SCHEME 34.1

exist, and different concepts turn out to be the preferred solution depending on tl
studied.

This chapter covers the state of the art in the field of developed enantioselective sy
methods for optically active amines **1** (Scheme 34.1) by means of biocatalysis and foc
those reactions that lead to a chiral amine functionality itself. Thus, preparation of
functional compounds that bear an amino group but whose asymmetric synthesis is b;
the modification of functional groups other than the amino group and the synthesis of
acids in general are not included in this chapter as these are reviewed elsewhere
additional focus here will be on those biocatalytic routes that have already been pro
preparative, sometimes even technical, scale.

To start with the resolution concepts, a popular approach is the modification of the
functionality in the presence of an enzyme capable of carrying out an acylation proce
approach, shown as Route (A) in Scheme 34.2, is in most cases based on the use of lip
combination with an acyl donor, and the concept has been applied at industrial scale fo
some time already. Alternatively, amine oxidases can serve as suitable enzymes
resolution of amines through enantioselective oxidation (Route (B)). Moreover, this
resolution can be extended toward a deracemization process when combining it
chemical *in situ* reduction of the imine intermediate formed during the oxidatic
(Route (E)). Other types of enzymes suitable for conversions described in Route
transaminases. In Route (C) a similar way—also following the concept of a resolutio
consists of a transformation of the racemic amine into an amide, followed by an en;
hydrolysis of the desired enantiomer under formation of the optically active amine is
With the exception of the extension toward a dynamic kinetic resolution process (kno
lipase, see Route (D)) and deracemization (known for amine oxidase, see Route (
resolutions are limited by a 50% conversion and require subsequent separation
unwanted enantiomer and recycling.

A direct (theoretically) 100% approach toward optically active amines is p
when starting from prochiral compounds, namely ketones, as shown in Route (F
required enzymes, which make this asymmetric catalytic process possible, are transam
This Route (F) has also already proven its feasibility on technical scale. Route (A) t
Route (F) are subsequently described in more detail.

34.2 KINETIC RESOLUTION

34.2.1 OVERVIEW ABOUT TYPES OF ENZYMATIC RESOLUTIONS

Kinetic resolutions are a widely applied method for the synthesis of chiral building blc
spite of numerous examples for amino acids, carboxylic acids, alcohols, and many othe
of molecules, it took surprisingly long until the first efficient resolutions for racemic
were reported. Until now, kinetic resolutions according to Route (A) and Route
known, based on the use of hydrolases, in particular lipases (for both Route (A) and

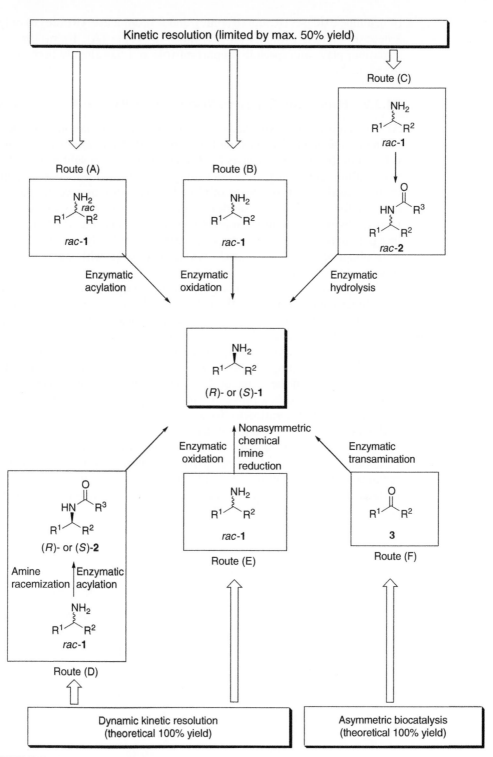

SCHEME 34.2

(C)), amine oxidases (Route (B) only), and transaminases (Route (B) only). Among the
kinetic resolution with lipases according to Route (A), by means of an enantiosel
acylation, plays an especially important role and has already become an industrially
lished method.

34.2.2 HYDROLASES-CATALYZED RESOLUTION OF RACEMIC AMINES THROUGH ACYLATION

Although most of the hydrolases applied so far in the acylation of amines belong to the
of lipases, the first efficient example for amines that was demonstrated by the Klibanov
is based on the use of proteases [4,5]. The protease from subtilisin Carlsberg in combi
with trifluoroethyl butyrate as acylating agent turned out to be a suitable enzyme
enantioselectivities in the range of 60 to 99% ee, whereas several lipases were inefficie
example for an efficient resolution in the presence of lipases by means of an acylatio
reported by Ito and Nemori, applying immobilized enzymes [6]. Immobilization turned
be a prerequisite for high enantioselectivities. Accordingly, enantioselectivities of 90 to 9
were obtained with immobilized enzymes compared with significantly lower 63 to 6
when using non-immobilized enzymes.

A wide range of work has been done using "standard" acyl donors such as ethyl acet
homologous alkyl derivatives thereof, and high enantioselectivities can be obtained in
cases when using, for example, ethyl acetate itself or butyl acetate. The first efficie
general example in this field was reported by the Reetz group. Ethyl acetate functions
acyl donor. As an enzyme, the lipase from *Candida antarctica* was used and gave high en
selectivities in the range of 90 to 98% ee for the formed amide, (*R*)-**4** (Scheme 34.3
conversions were in the range of 20 to 44%. However, high enzyme loadings were use
reaction times varied in a range of 7 to 60 h [7]. Selected examples are given in Scheme

This method of enantioselective acylation of racemic amines using ethyl acetate as a
acyl donor was further studied in detail and optimized by the Davis group (Scheme 34.
When carrying out the reaction in ethyl acetate as a solvent and in the presence of lipase
C. antarctica, a range of amines (*S*)-**1**, comprising aliphatic and aromatic products,
obtained in yields of up to 47%, and with excellent enantioselectivities of up to 99% e
preference for the (*S*)-enantiomer. A downstream processing for the remaining
cally active (*S*)-enantiomers, (*S*)-**1**—by means of precipitation of the corresponding o

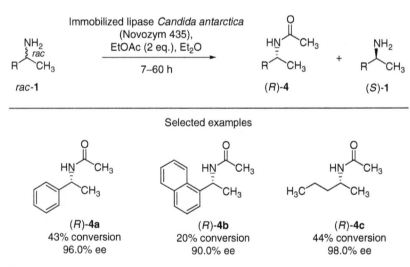

SCHEME 34.3

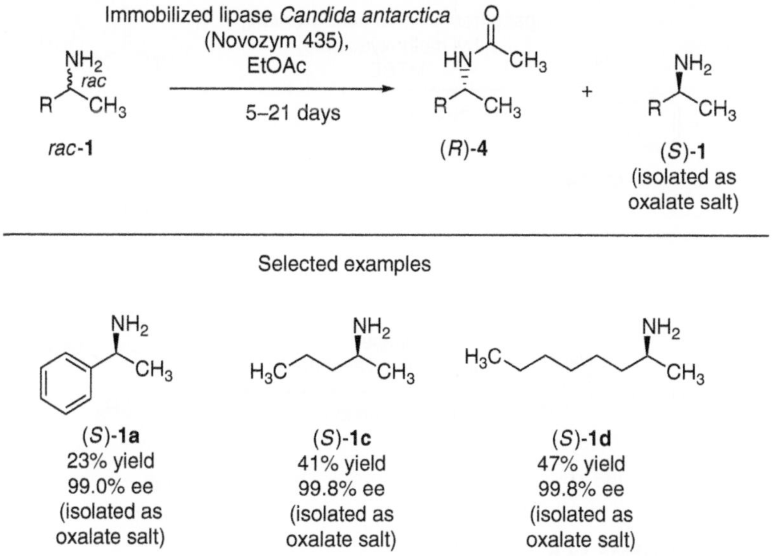

Immobilized lipase *Candida antarctica* (Novozym 435), EtOAc

5–21 days

rac-1

(R)-4

(S)-1
(isolated as oxalate salt)

Selected examples

(S)-1a
23% yield
99.0% ee
(isolated as oxalate salt)

(S)-1c
41% yield
99.8% ee
(isolated as oxalate salt)

(S)-1d
47% yield
99.8% ee
(isolated as oxalate salt)

SCHEME 34.4

salts—has also been developed. Reaction time, however, was long with 5 to 21 days. Compared with other solvents, reaction in ethyl acetate was faster, although enantioselectivity was somewhat lower in some cases.

The impact of the chain-length of the acyl donor was demonstrated by Patel et al. for the resolution of *rac*-2-butylamine, *rac*-**1e**, using the lipase from *C. antarctica* [9]. Due to low stereodifferentiation between the two similar substituents methyl and ethyl attached to the carbon atom bearing the amino group, low enantioselectivity was obtained when using ethyl butyrate ($E < 10$) as an acyl donor. However, the use of ethyl decanoate as acyl donor and methyl *tert*-butyl ether (MTBE) as a solvent led to an increase of the enantioselectivities of up to 99% at a yield of 25% for (S)-**1e** (Scheme 34.5) [9]. In addition, benzyl esters have been used as donors for the resolution of racemic amines [10].

A drawback of all types of alkyl ester-type acyl donors, however, is the low reaction rate in the acylation process, thus being the limiting factor with respect to commercial applications. The breakthrough with respect to industrial applications in the acylation was achieved by BASF researchers, using ethyl methoxyacetate as an acyl donor in the presence of lipases [11]. As a biocatalyst, the lipase from *Burkholderia plantarii*, which shows a high activity of 1000 U/mg for the crude enzyme product, turned out to be particularly useful. Preferred solvents are ethers, in particular MTBE. In the presence of methoxyacetate, the acylation reaction of *rac*-phenylethyl-1-amine, *rac*-**1a**, proceeds very fast, and the initial reaction rates

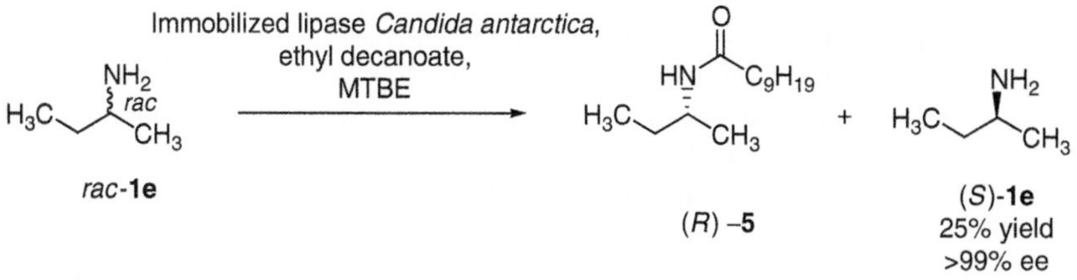

Immobilized lipase *Candida antarctica*, ethyl decanoate, MTBE

rac-1e

(R)-5

(S)-1e
25% yield
>99% ee

SCHEME 34.5

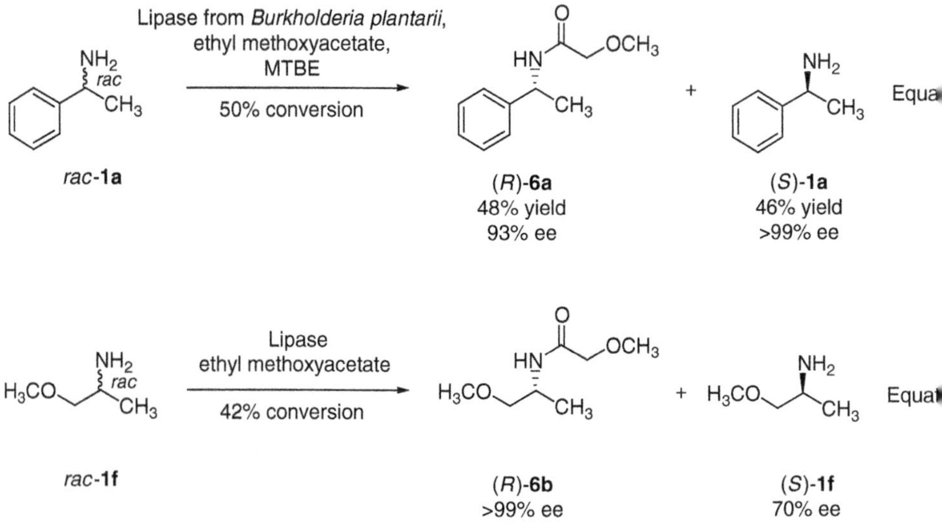

SCHEME 34.6

are 100 times higher compared with the one when using ethyl butyrate. After reac
quantitative conversion of 50%, downstream processing gave the remaining (S)-amine,
in 46% yield with a high enantioselectivity of >99% ee (Scheme 34.6, Equation (1)
corresponding amide, (R)-**6a**, is obtained in 48% yield, with 93% ee. The reactio
proceeds highly efficiently with other type of substrates, allowing an efficient acc
numerous (S)-amines (S)-**1**, as well as—after subsequent deacylation—(R)-amines. Nc
the enzymatic kinetic resolution of racemic amines is one of the most impressive exa
demonstrating how efficient a resolution process can be in spite of the limitation c
conversion. The major industrial application of this lipase resolution is the production
methoxyisopropylamine (S)-**1f** (Scheme 34.6, Equation (2)), which is an intermediate
production of the herbicide "Outlook," with an annual production capacity of 2500 t [
Furthermore, the resolution of several optically active amines under cGMP condit
carried out at an annual scale of 1500 t [13]. It should be added that the use of
methoxyacetate as an acyl donor was also reported by Bayer researchers. In their stu
lipase from *C. antarctica* was used as a biocatalyst [14]. In addition, the Reetz
developed a sol–gel encapsulated form of the lipase from *C. antarctica*, which is suita
resolve racemic 1-phenylethylamine with 50% conversion (within 29 h) and high er
selectivity (E > 100) [15]. This catalyst is more active than the commercial powder; it s
constant performance when recycled within five reaction cycles. Jaeger et al. also d
strated cloning and overexpression of a lipase from *Pseudomonas aeruginosa* in a
negative *P. aeruginosa* strain. The strain secreted the recombinant lipase, which turn
to be highly efficient in several resolution processes, including resolution of race
pentylamine (50% conversion, 96% ee) [16].

In addition to acetates and substituted derivatives thereof, other types of acyl donor
been applied as well. The kinetic resolution of racemic amines using dialkyl or
carbonates has been reported by the Wong group (see, e.g., Scheme 34.7, Equatio
[17]. The "nonsymmetrical" substituted phenyl allyl carbonate was found to be a new
acylation agent for the lipase-catalyzed resolution of 1-methyltetrahydroisochinolir
Alkoxycarbonylation of racemic amines by a lipase of *C. rugosa* was studied with di
lipase samples from *C. rugosa* that are based on cultivation with different inducers and c

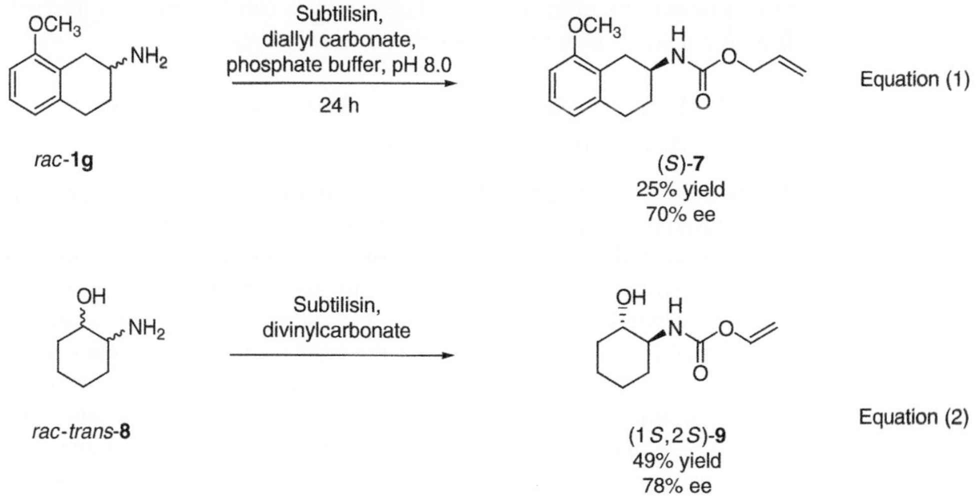

SCHEME 34.7

sources [19]. Enantioselective acylation was also done with divinyl carbonate as a donor in the presence of subtilisin as a hydrolase; functionalized amines were resolved as well. The subtilisin-catalyzed resolution of *rac-trans*-1-amino-1-hydroxycyclohexane, *rac-trans*-**8**, led to the formation of the corresponding enantiomerically enriched *N*-acylated (1*S*,2*S*)-enantiomer (1*S*,2*S*)-**9** in 49% yield with 78% ee (Scheme 34.7, Equation (2)) [20].

Furthermore, methyl acrylates have been used as acyl donors for the resolution of aliphatic amines [21]. Although enantioselectivities of up to 95% were obtained for the remaining optically active amines, reaction times were rather long with 7 to 11 days, and yields were in the range of 20 to 40%. Another acyl donor applied in the enzymatic acylation of racemic amines and amino alcohols is cyanomethyl pent-4-enoate, which was reported by the Wong group [22].

An elegant approach demonstrating that "nonactivated" acyl donors, namely carboxylic acids, can also be used was reported by Montet et al. (in organic solvents) [23], as well as Kato et al. (under neat conditions and in ionic liquids). Compared with neat reaction conditions, ionic liquids turned out to be superior [24]. Interestingly, the choice of preferred ionic liquid depends on the type of amine. The lipase-catalyzed kinetic resolution in 1-butyl-2,3-dimethylimidazolium trifluoromethanesulfate was also done on preparative scale.

In addition to intensive process development with respect to a high reaction rate, detailed study of the impact of the acyl group on the enzymatic process, and reaction media engineering, a major focus has been on the investigation of the substrate range [25,26]. Among "typical" nonfunctionalized amines, aliphatic as well as aromatic amines are accepted (see also section above). Lipases, however, have been used not only for resolution of nonfunctionalized amines but also for amines bearing other functionalized groups. For example, an enzymatic resolution of a variety of 1-heteroarylamines through enantioselective acylation proceeds very efficiently in the presence of the *C. antarctica* lipase B [27]. High yields and high enantioselectivities in the range of 90 to 99% are obtained. An example for amines that are further functionalized are β-amino esters such as ethyl-3-aminobutyrate. The Gotor group demonstrated that such types of amines could be efficiently resolved by means of a lipase-mediated acylation. In the presence of the lipase from *C. antarctica* and ethyl acetate as both acyl donor and solvent, the remaining amine was obtained in 45% yield with a high

enantioselectivity of 99% ee [28]. The products can be further converted into optically
β-amino acids, which are of pharmaceutical interest.

34.2.3 PENICILLIN ACYLASE-CATALYZED RESOLUTION OF RACEMIC AMINES THROUGH ACYLATION

The popularity of the penicillin acylase is not only because of its use as an efficient cata...
the industrial production of 6-aminopenicillanic acid (6-APA) [29], but also due to its...
utility range with respect to the substitution pattern at the amino functionality. In addi...
many amino acids and derivatives that are accepted as substrates in the acylation proc...
their corresponding amides in the hydrolytic resolution) [30], several groups demon...
that penicillin acylase is suitable for the resolution of racemic amines [31,32]. The pe...
acylase from *Alcaligenes faecalis* [33] was successfully applied for the enantioselective
tion of racemic amines using phenylacetamide as an acylating agent. The reaction was ...
out at a high pH of 11 in order to ensure deprotonation of the amines [32]. The use of c...
cosolvents turned out to be beneficial. Compared with pure aqueous media, a rema...
increase of the enantioselectivity was found when carrying out the resolution of ar...
amines in an aqueous solution with an acetonitrile content of 10 to 25%. Thus, an incr...
the *E* value from 110 to 220 was observed for the resolution of 4-phenylbutan-2-amin...
acetonitrile-containing solution (10%), while the reactivity was maintained in the same
A representative example of the penicillin acylase-based resolution is shown in Schem...
In contrast, enantioselectivities of analog aliphatic amines were rather low with *E*
below 8. It should be added that the penicillin acylase from *Escherichia coli* ga...
satisfactory results.

34.2.4 LIPASE-CATALYZED RESOLUTION OF RACEMIC AMIDES THROUGH HYDROLYSIS

Although most of the lipase-based resolutions for the syntheses of optically active ami...
related to the acylation of racemic amines, several examples for the reverse reaction, ...
the hydrolysis of racemic *N*-acetylated amines, are also known. There has been a consid...
interest from industry on this type of reaction, which is underlined by several contrib...
from different companies. At the beginning of the 1990s, Shell researchers report...
hydrolysis of *N*-acetylated *rac*-phenylethyl-1-amine in the presence of whole cells
Arthrobacter sp., giving access to the enantiomerically pure (*S*)-amine as well as t...
amide with a high enantioselectivity of >99.5% [34].

Another early example for this type of reaction is the use of an arylalkyl acylamid...
the preparation of both the (*S*)- and (*R*)-forms of the corresponding desired enanti...
forms of the amines [35]. The (*S*)-enantiomers of phenylethyl-1-amine and 4-phenylb...
amine were synthesized with high enantioselectivity with several microbes, e.g., *N*...

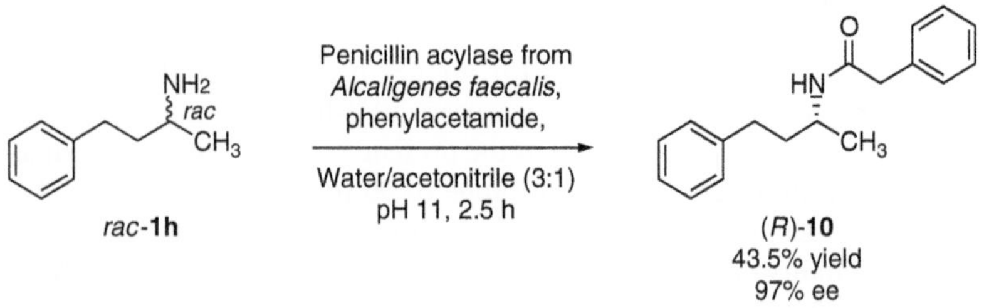

SCHEME 34.8

erythropolis. The corresponding (*R*)-enantiomer of 4-phenylbutyl-2-amine was obtained when using a crude amidase from *P. putida.* Microbial hydrolysis starting from the corresponding *N*-acyl derivative has also been reported for the production of D-aminobutanol [36].

Furthermore, three new, stable amidohydrolases and their successful application in the synthesis of several amines as pharmaceutical building blocks have recently been reported by a research group from Novartis [37]. After discovering these (*S*)-selective enzymes from *Rhodococcus equi* and *R. globerulus*, as well as the (*R*)-selective enzyme from *Arthrobacter aurescens* in a screening program, their production was optimized, demonstrating a high operational stability. All of these enzymes showed an excellent enantioselectivity with *E* values of >500 for the desired target reactions. Selected examples are given in Scheme 34.9. The feasibility of the synthesis of (*S*)-2-amino-1-(4-chlorophenyl)-4-pentene, (*S*)-**1j**, has already been successfully demonstrated on a laboratory pilot scale of 20 L.

A highly enantioselective hydrolytic process of *N*-acetylated amines based on the use of lipase B from *C. antarctica* has been developed by Bayer researchers [38]. A representative example is shown in Scheme 34.10. For the resolution of *N*-acetylated *rac*-4-chlorophenylethyl-1-amine, *rac*-**2j**, a very good yield of 43% with an excellent enantioselectivity of >99.5% was obtained for the product (*R*)-4-chlorophenylethyl-1-amine, (*R*)-**1j**. However, the reaction, which had been carried out in pure aqueous buffer media at pH 8 and 50°C, required a long reaction time of 7 days.

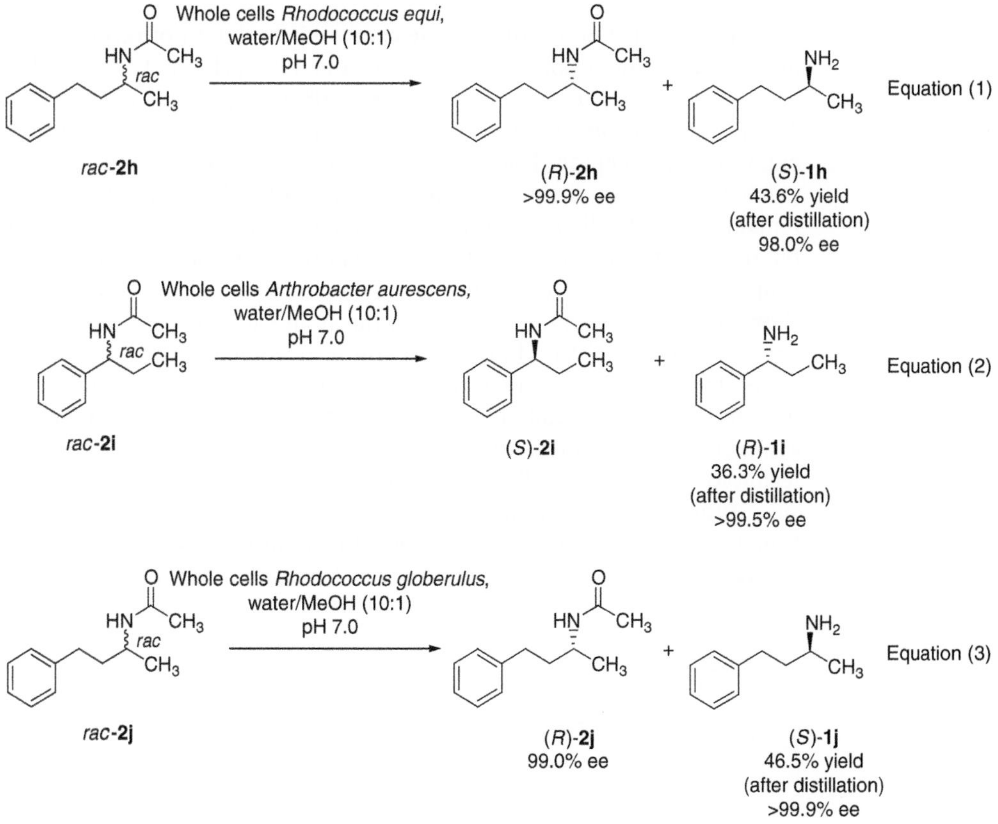

SCHEME 34.9

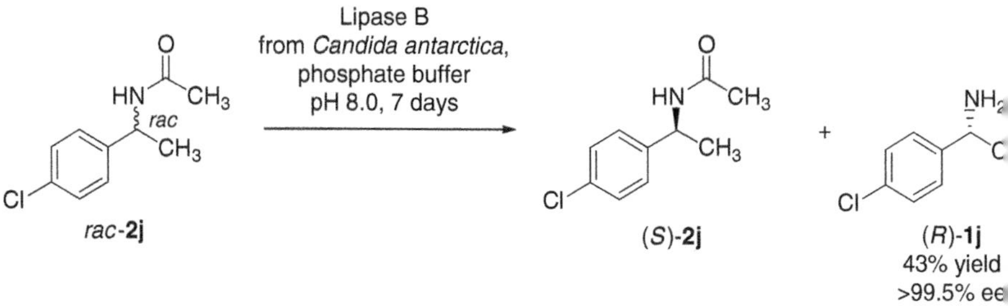

SCHEME 34.10

34.2.5 AMINE OXIDASE-CATALYZED RESOLUTION OF RACEMIC AMINES THROUGH OXIDA

In spite of acylation and hydrolytic processes, selective oxidation of one enantiomer amine oxidase can also represent a useful tool for the synthesis of optically active amin ability of these enzymes to selectively oxidize amine enantiomers has been known for time [39], although preparative applications for the synthesis of optically active ami this method have only recently been reported. In particular, this process has becc interesting key step in the deracemization of racemic amines, and will be discussed i detail in the section "Dynamic Kinetic Resolution and Deracemization."

34.2.6 TRANSAMINASE-CATALYZED RESOLUTION OF RACEMIC AMINES THROUGH OXIDAT

An early example of a transaminase-catalyzed resolution of racemic amines [40] is de in a Celgene patent application published in 1990 [41,42]. The optically active amine been prepared by using a pyridoxal phosphate-dependent ω-amino acid transami *Bacillus* or *Pseudomonas* in the presence of a carbonyl compound such as pyruvate **1** amino group acceptor. The (S)-amine enantiomer is oxidized to the corresponding k (while the pyruvate is converted into L-alanine **12**). Thus, the desired amine enantiome remained in optically active form. For example, the (S)-enantiomer of *rac*-1-phenyl-3 butane was transformed 20-fold faster than the corresponding (R)-enantiomer. The r principle, exemplified for *rac*-phenylethyl-1-amine as a donor and pyruvate as an acce shown in Scheme 34.11. The choice of the amino acceptor is of importance since pyru oxaloacetate gave fivefold faster reaction rates compared with butan-2-one.

By means of this scalable transaminase technology, a wide range of products are ible, and pharmaceutically interesting, substituted phenylethyl-1-amines have also bee available in developmental (kg) quantities [42]. Several examples of this efficient res technology by means of transaminases are given in Scheme 34.12. Independent substitution pattern, excellent enantioselectivities of up to 99% have been obtaine broad range of aryl-containing amines. For example, resolution of *rac*-phenylethyl-1 *rac*-**1a** proceeds with 50% conversion furnishing the optically active (S)-phen

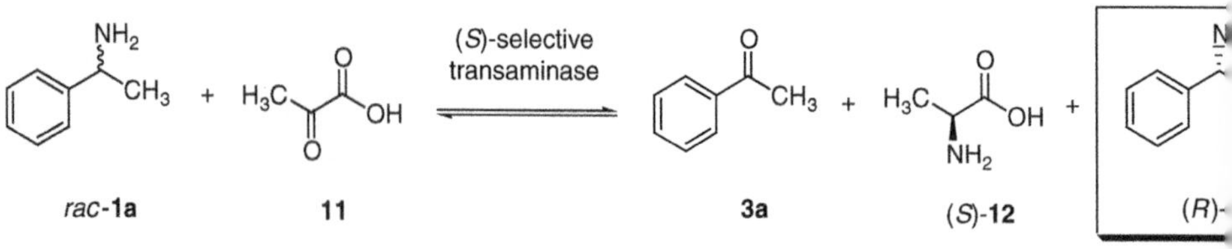

SCHEME 34.11

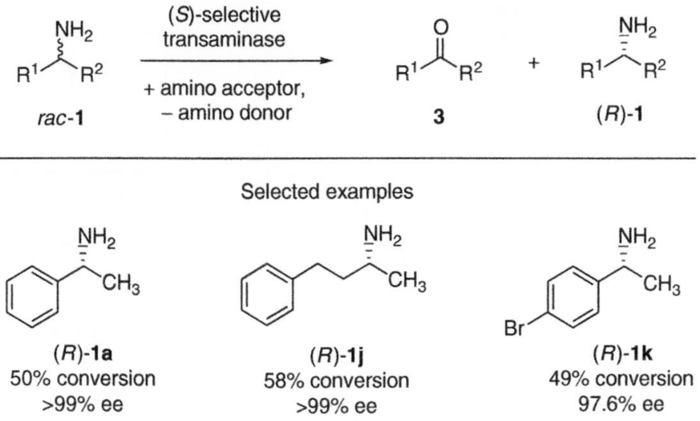

SCHEME 34.12

1-amine, (*S*)-**1a**, with a high enantioselectivity of >99%. In addition to (*S*)-selective transaminases, the analog (*R*)-selective enzymes have been developed as well.

A detailed study on the properties of three ω-transaminases from different microorganisms such as *Klebsiella pneumoniae*, *Bacillus thuringiensis*, and *Vibrio fluvialis* has been done by the Kim group [43]. All three transaminases showed high enantioselectivity when using *rac*-phenylethyl-1-amine as a substrate with *E* values of >50. Substrate tolerance is broad, accepting aryl as well as aliphatic amines. As an amino acceptor, alkyl aldehydes, e.g., propionaldehyde, can be used besides pyruvate. However, a limiting factor in synthetic applications is significant substrate inhibition by (*S*)-phenylethyl-1-amine at concentrations exceeding 200 mM. Acetophenone shows a significant product inhibition too. As transaminase, the one from *V. fluvialis* turned out to be preferred, and has been successfully used in the resolution of aliphatic amines. For example, resolution of *rac*-2-butylamine used at a substrate concentration of 20 mM gave an enantioselectivity of 94.7% after 12 h of reaction time (Scheme 34.13). The reaction was carried out under reduced pressure to selectively remove the by-product 2-butanone, thus suppressing the effect of product inhibitions. Another method to overcome the severe product inhibitions has also been reported by the Kim group for the syntheses of optically active aryl amines by means of a biphasic system [44]. Application of an enzyme-membrane reactor has also proven to be an efficient tool for the kinetic resolution of amines with a transaminase [45].

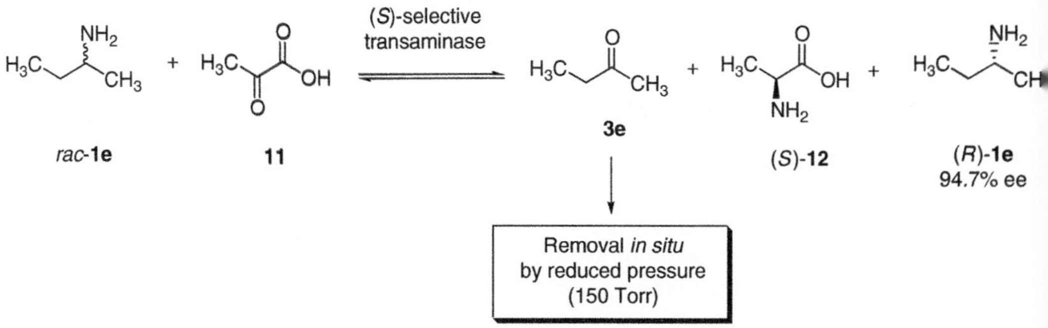

SCHEME 34.13

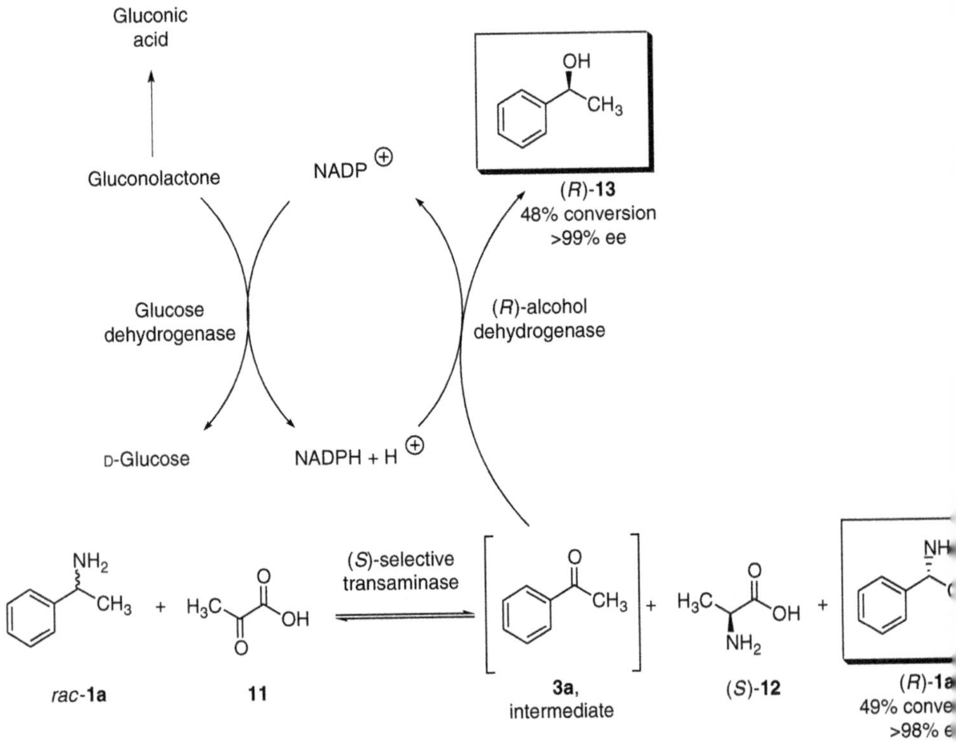

SCHEME 34.14

A unique resolution concept has recently been developed by the Kim group, coup resolution of a racemic amine with an enantioselective synthesis of an optically active a [46]. One enantiomer of the amine is oxidized to the ketone, and subsequently reduc means of an alcohol dehydrogenase-catalyzed reduction process under formation corresponding optically active alcohol. By means of this simultaneous reduction pr product inhibition of the formed ketone can be suppressed allowing an efficient sym of both the desired (R)-amine and (R)-phenylethan-1-ol. The cofactor regeneration ketone reduction process was carried out through a glucose dehydrogenase-catalyzed tion of D-glucose under recycling of the required reduced form of the cofactor, NA The concept—exemplified by the resolution of rac-phenylethyl-1-amine—is show Scheme 34.14. At a substrate concentration of 100 mM of rac-phenylethyl-1-amine, conversion was observed, leading to the corresponding (R)-phenylethyl-1-amine in 49 (R)-phenylethan-1-ol in 48% within a reaction time of 18 h.

34.3 DYNAMIC KINETIC RESOLUTION AND DERACEMIZATION

A prerequisite for dynamic kinetic resolutions is a sufficient *in situ* racemization proc combination with the applied resolution process. This has been achieved for lipase resolution processes by several groups. Deracemization processes also allow the ef transformation of a racemic compound into an enantiomerically pure form with (the ally) quantitative conversion. Both types of concepts, dynamic kinetic resolutions (Rou in Scheme 34.1) and deracemization (Route (E) in Scheme 34.1), are described in more in the following section.

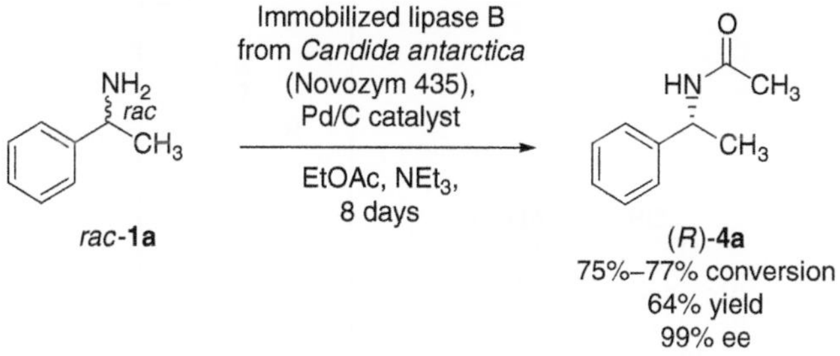

SCHEME 34.15

34.3.1 LIPASE-CATALYZED DYNAMIC KINETIC RESOLUTION OF RACEMIC AMINES

The extension of the lipase resolution toward a dynamic kinetic resolution process has been developed by several groups. The first example was reported by the Reetz group, applying Pd/ C as a metal catalyst for the racemization of the amine in combination with the enzymatic acylation of a racemic amine [47]. As an enzyme, the lipase Novozym 435 (*C. antarctica* lipase B) was used. After a reaction time of 8 days, the desired optically active amides were formed with conversions of 75 to 77%. The enantioselectivity was excellent, with 99% ee for the synthesis of *N*-acetylated (*R*)-phenylethyl-1-amine (Scheme 34.15).

The Bäckvall group studied mild racemization conditions for optically active amines, and developed a highly efficient metal catalytic procedure based on a ruthenium complex for this purpose [48,49]. The racemization protocol is very general with respect to substrate range, and also led to high recovery yields of the amine in the range of 95 to 99%. This racemization process was combined with a lipase-based esterification in a two-step manner for the synthesis of (*R*)-*N*-(1-phenylethyl)acetamide, (*R*)-**4a**. As an enzyme catalyst, the lipase B from *C. antarctica* was applied, and ethyl acetate used as a donor. After two resolutions and one racemization, the desired product was obtained in a yield of 69% with a high ee of >98% (Scheme 34.16).

Another dynamic kinetic resolution that is based on the use of ketoximes as precursor has been developed by the Kim group [50,51]. The one-pot synthesis concept, which is shown in Scheme 34.17, is based on the use of two catalysts, namely palladium and a lipase. Whereas the biocatalyst—an immobilized *C. antarctica* B lipase—selectively catalyzes the acylation of one amine enantiomer, the palladium catalyst is required for the initial conversion of the ketoxime into the racemic amine, as well as for the subsequent racemization of the amine. Typically the reactions were carried out in toluene at a reaction temperature of 60°C, with a

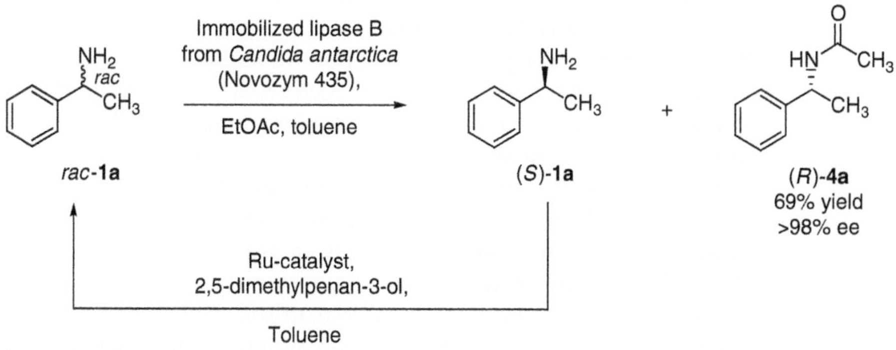

SCHEME 34.16

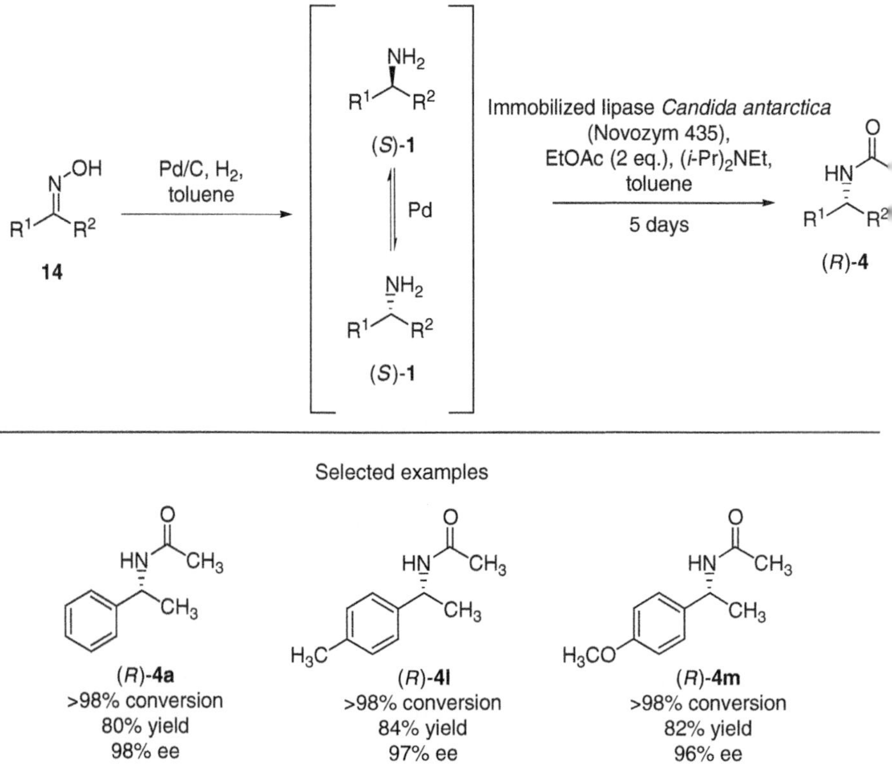

Selected examples

(R)-4a
>98% conversion
80% yield
98% ee

(R)-4l
>98% conversion
84% yield
97% ee

(R)-4m
>98% conversion
82% yield
96% ee

SCHEME 34.17

reaction time of 5 days. Several examples are shown in Scheme 34.17. In addition to ketoximes, cyclic analogs thereof can be used. In general, excellent conversions of accompanied by high enantioselectivities in the range of 94 to 99% ee were obtained purification, the desired (R)-amine products were obtained in yields of 70 to 98%.

Thus, highly enantioselective dynamic kinetic resolution methodologies of racemic are available. However, a challenge for the future certainly remains the improvement reaction time, which is still in the range of several days.

34.3.2 Amine Oxidase-Catalyzed Deracemization of Racemic Amines

Deracemization reactions play an important role in the synthesis of optically activ pounds, comprising those bearing a secondary alcohol as well as amino functionali Whereas several examples of applications of amine oxidases for deracemization proce α-amino acid synthesis have already been reported [53], their efficient extension tow synthesis of amines remained a challenge for a long time. The extension of the amine o processes (see section above) toward a deracemization process has been impressively c strated by the Turner group [54]. The deracemization reaction is based on the combina an enzymatic, highly enantioselective amine oxidation process with a nonasymmetric ically reductive amination reaction. Both reactions are compatible with each othe allowing a one-pot reaction. As a reducing agent, sodium cyanoborohydride was which turned out to be more suitable than sodium borohydride. For the synthetic d kinetic resolution process, an optimized mutant of the amine oxidase from Aspergillu was used. This mutant has been found by in vitro evolution technology, and turned

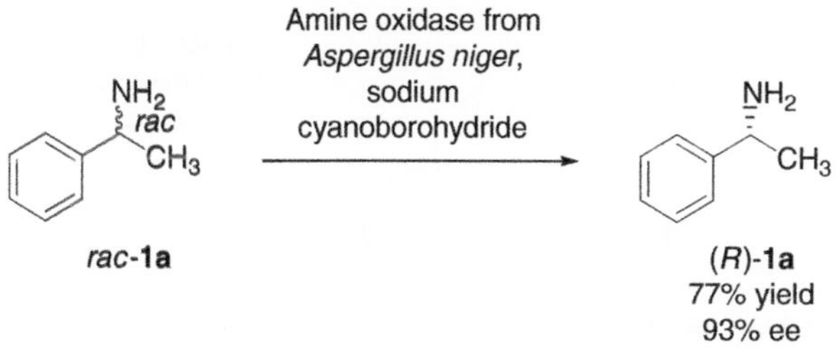

rac-**1a**   (R)-**1a**
77% yield
93% ee

SCHEME 34.18

give improved results with respect to both catalytic activity and enantioselectivity in the deracemization of rac-1-phenylethylamine. By means of this improved enzyme, the corresponding (R)-1-phenylethylamine was formed in 77% yield with an enantioselectivity of 93% ee (Scheme 34.18).

34.4 ASYMMETRIC SYNTHESIS

34.4.1 ASYMMETRIC TRANSAMINATION

The direct conversion of a ketone into the desired chiral amine—shown as Route (F) in Scheme 34.1—probably represents the most straightforward approach toward this class of compounds, as it represents an asymmetric catalytic reaction with (theoretically) 100% conversion. Such a process can be carried out by means of a transamination (originally known for the synthesis of enantiomerically pure α-amino acids starting from keto acids). Several transaminases have been identified that tolerate not only α-keto acids (as a precursor for amino acids) but also ketones, thus offering the potential to synthesize the corresponding optically active amines. In addition to the transaminase resolution of racemic amines with keto acids as an "amino acceptor," the first efficient syntheses of direct asymmetric transaminations have been reported in the early 1990s by Celgene researchers. The basic concept of the transamination process, which is shown in Scheme 34.19—exemplified for the synthesis of (S)-phenylethyl-1-amine, (S)-**1a**, using L-alanine as an amine donor—is based on the transformation of the ketone into the optically active amine under simultaneous conversion of an amino donor (L-amino acid in this case) into its carbonyl derivative (keto acid in this case) [41,42]. Alternatively, isopropylamine can also be used as an amine donor instead of an L-amino acid [55].

The key step, however, is not only the transamination process but, particularly, the issue of how to shift the equilibrium to the (thermodynamically nonpreferred) direction of amino and keto acids. Accordingly, the resolution of racemic amines using a keto acid as an "amino acceptor" is more favored and proceeds with high conversion without needing further process steps. In contrast, removing a formed product—preferably the formed carbonyl by-product—from the reaction mixture is essential for an asymmetric transamination process according to Scheme 34.20. Several options are conceivable, most of them addressing a way to decompose the oxidation product formed by deamination of the amine donor. Alternatively, removal of the formed keto product by evaporation has also been efficiently done, especially when isopropylamine has been used as an amine donor under formation of acetone.

A highly efficient application of the transamination concept based on isopropylamine as a donor was reported by the Matcham group for the synthesis of (S)-methoxyisopropylamine [56], which is of commercial relevance as it can be used as an intermediate for the production

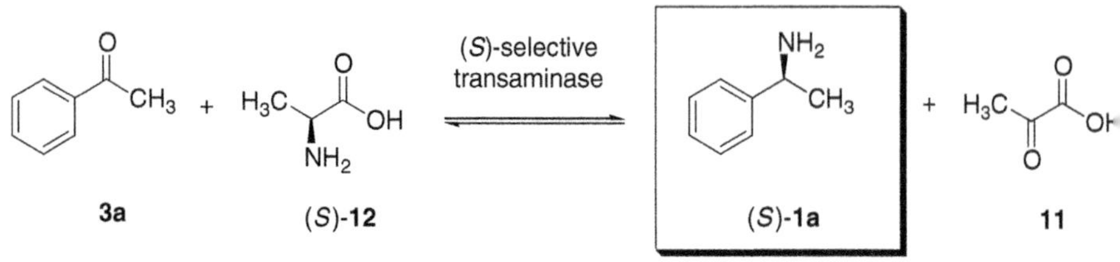

SCHEME 34.19

of herbicides such as (S)-dimethenamid and (S)-metolachlor. The latter is produc
quantities of 10,000 t/plant through a metal-catalyzed hydrogenation process, which,
ever, provides the product in only 79% ee. Choosing the transaminase route, Matcham
succeeded—after optimization of the enzyme catalyst through directed evolution as w
process conditions—in the formation of the desired product (S)-methoxyisopropylamine
93% conversion when removing the acetone during the reaction. The reactions were run ι
vacuum (100 mm Hg) at 50°C, and loss of methoxyacetone was avoided by means
overhead condenser. The substrate concentration was at an impressive 2.08 M, corresp
ing to a substrate input of 183 g/L. The enantioselectivity of this process is excellent
>99% ee (Scheme 34.20). As a biocatalyst, recombinant whole cells bearing an opti
transaminase mutant with respect to thermal and chemical stabilities were used at a cat
amount of 5 g/L.

Celgene also applied its transamination technology for numerous other types of opt
active amines, comprising both (R)- and (S)-enantiomers. This transamination techn
has already been scaled up to the production of optically active amines on a >500 kg
[55].

The use of L-alanine as an alternative α-amino acid donor has also been investigat
the Kim group [57]. In order to shift the reaction toward the product side, removal ι
formed keto acid as a side product turned out to be essential. When using crude extract
a transaminase, removal of pyruvate was achieved by means of additionally added la
dehydrogenase (LDH). The preferred way that was found, however, consisted of a whol
approach under consumption of the formed pyruvate within the cell metabolism as w
secretion of the synthesized (S)-amine. In the presence of whole cells, (S)-phenylethyl-1-a
was formed with a conversion of 90% and an enantioselectivity of >99% ee after a rea
time of 1 day. The substrate concentration of acetophenone, however, was 30 mM, and
requires further optimization with respect to a technical application.

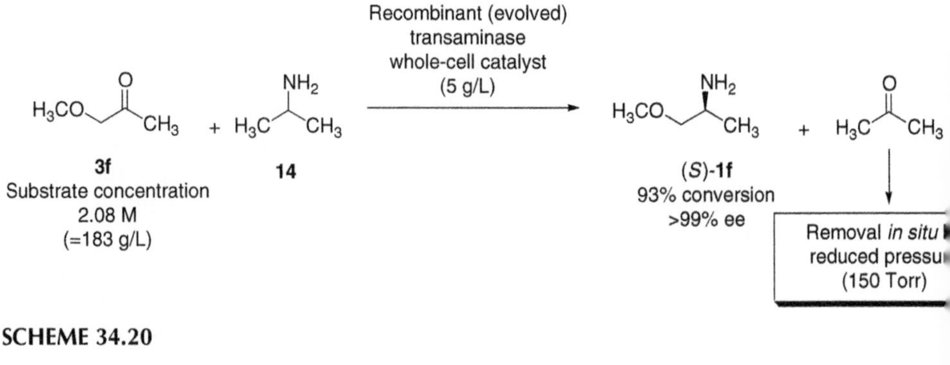

SCHEME 34.20

34.5   SUMMARY

For the development of synthetic methods for optically active amines, which are key building blocks for a variety of drugs and agrochemicals, numerous biocatalytic methods are available. Interestingly, these methods are based on various synthetic concepts such as (i) several types of kinetic resolutions using hydrolases, amine oxidases, and transaminases; (ii) dynamic kinetic resolutions with hydrolases; (iii) deracemization with amine oxidases; and (iv) asymmetric catalytic synthesis using transaminases. In addition to the broad variety of synthetic tools, it is further impressive that some technologies already made the "jump" from a laboratory scale application to a technically applied production process. In particular, the lipase-based acylation of amines, which is carried out at BASF on industrial scale, represents a highly efficient method for the large-scale manufacture of amines. Furthermore, this technology underlines that kinetic resolutions—although restricted in principle by a maximum yield of 50%—can be highly competitive on industrial scale, e.g., when achieving high volumetric productivities in combination with an efficient recycling of the undesired enantiomer. Another impressive technology already proven on a large scale is the transaminase technology, which has been developed at Celgene.

Thus, the biocatalytic syntheses of optically active amines are already valuable tools in the pharmaceutical and biotechnological industries for the synthesis of the desired amine target molecules. It can be expected that their importance will further increase in the future. Among the major challenges for the future might be the improvement of the volumetric productivity of the dynamic kinetic resolutions by combining enzymes (for resolution) and metal catalysts (for racemization).

REFERENCES

1. For a review, see Gotor, V., *Bioorg. Med. Chem.*, 7, 2189, 1999.
2. In addition to biocatalytic approaches, several highly efficient chemocatalytic syntheses have been developed. Selected recent excellent examples are: (a) asymmetric resolution of racemic amines: Arai, S., Bellemin-Laponnaz, S., and Fu, G., *Angew. Chem.*, 113, 242, 2001; (b) asymmetric hydrogenation of *N*-phosphinoyl imines: Spindler, F. and Blaser, H.-U., *Adv. Synth. Catal.*, 343, 68, 2001; (c) asymmetric hydrogenation of imines: Cobley, C.J. and Henschke, J.P., *Adv. Synth. Catal.*, 345, 195, 2001.
3. Reviews for enantioselective biocatalytic amino acid syntheses: (a) α-Amino acids: Gröger, H. and Drauz, K., in *Asymmetric Catalysis on Industrial Scale*, Blaser, H.-U. and Schmidt, E. Eds., Wiley-VCH, Weinheim, Germany, 2003, p. 131; (b) β-Amino acids: Liu, M. and Sibi, M.P., *Tetrahedron*, 58, 7991, 2002.
4. Kitaguchi, H., Fitzpatrick, P.A., Huber, J.E., and Klibanov, A.M., *J. Am. Chem. Soc.*, 111, 3094, 1989.
5. Notably, for some specific amino alcohols the porcine pancreatic lipase was found to be suitable already in 1988 by the Gotor group; see Gotor, V., Brieva, R., and Rebolledo, F., *J. Chem. Soc. Chem. Commun.*, 957, 1988.
6. Ito, I. and Nemori, R. (Fuji Photo), JP Pat. Appl. 03191797, 1991.
7. Reetz, M.T. and Dreisbach, C., *Chimia*, 48, 570, 1994.
8. Davis, B.A. and Durden, D.A., *Synth Commun.*, 31, 569, 2001.
9. Goswami, A., Guo, Z., Parker, W.L., and Patel, R.N., *Tetrahedron: Asymmetry*, 16, 1715, 2005.
10. Adamczyk, M. and Grote, J., *Tetrahedron Lett.*, 37, 7913, 1996.
11. Balkenhohl, F., Ditrich, K., Hauer, B., and Ladner, W., *J. Prakt. Chem.*, 339, 381, 1997.
12. Review: Schmid, A., Dordick, J.S., Hauer, B., Kieners, A., Wubbolts, M., and Witholt, B., *Nature*, 409, 258, 2001.
13. Review: Breuer, M., Ditrich, K., Habicher, T., Hauer, B., Keßeler, M., Stürmer, R., and Zelinski, T., *Angew. Chem.*, 116, 806, 2004.
14. Steltzer, U. and Dreisbach, C., DE Pat. 19637336, 1996.

15. Reetz, M.T., Tielmann, P., Wiesenhöfer, W., Könen, W., and Zonta, A., *Adv. Synth. Catal* 717, 2003.
16. Jaeger, K.-E., Liebeton, K., Zonta, A., Schimmosek, K., and Reetz, M.T., *Appl. Mic* *Biotechnol.*, 46, 99, 1996.
17. Wong, C.-H., Orsat, B., Moree, W.J., and Takayama, S., U.S. Pat. Appl. 5981267, 1999.
18. Breen, G.F., *Tetrahedron: Asymmetry*, 15, 1427, 2004.
19. Alcantara, A.R., Dominguez de Maria, P., Fernandez, M., Hernaiz, M.J., Sanchez-Montero, and Sinisterra, J.V., *Food Technol. Biotechnol.*, 42, 343, 2004.
20. Orsat, B., Alper, P.B., Moree, W., Mak, C.-P., and Wong, C.-H., *J. Am. Chem. Soc.*, 118, 712.
21. Puertas, S., Brieva, R., Rebolledo, F., and Gotor, V., *Tetrahedron*, 49, 4007, 1993.
22. Takayama, S., Moree, W.J., and Wong, C.-H., *Tetrahedron Lett.*, 37, 6287, 1996.
23. Montet, D., Pina, M., Graille, J., Renard, G., and Grimaud, J., *Fat. Sci. Technol.*, 91, 14, 1$
24. (a) Irimescu, R. and Kato, K., *Tetrahedron Lett.*, 45, 523, 2004; (b) Irimescu, R. and Ka$ *J. Mol. Catal. B.*, 30, 189, 2004.
25. van Rantwijk, F., Hacking, M.A.P.J., and Sheldon, R.A., *Monatsh. Chem.*, 131, 549, 2000.
26. (a) Garcia, M.J., Rebolledo, F., and Gotor, V., *Tetrahedron: Asymmetry*, 3, 1519, 1992; (b) Garcia Rebolledo, F., and Gotor, V., *Tetrahedron: Asymmetry*, 4, 2199, 1993; (c) Gotor, V., Menend Mouloungui, Z., and Gaset, A., *J. Chem. Soc. Perkin Trans.*, 1, 2453, 1993; (d) Quiros, M., Sa V.M., Brieva, R., Rebolledo, F., and Gotor, V., *Tetrahedron: Asymmetry*, 4, 1105, 1993.
27. Skupinska, K.A., McEachern, E.J., Baird, I.R., Skerlj, R.T., and Bridger, G.J., *J. Org. Chem* 3546, 2003.
28. Sanchez, V.M., Rebolledo, F., and Gotor, V., *Tetrahedron: Asymmetry*, 8, 37, 1997.
29. Verweij, J. and De Vroom, E., *Recl. Trav. Chim. Pays-Bas*, 112, 66, 1993.
30. α-Amino acids: (a) Lucente, G., Romeo, A., and Rossi, A., *Experiencia*, 21, 317, 1965; (b) Co *Biochem. J.*, 115, 741, 1969; (c) Kasche, V., Michaelis, G., and Wiesemann, T., *Biochem. B* *Acta*, 50, 38, 1991.
31. Rossi, D., Calcagni, A., and Romeo, A., *J. Org. Chem.*, 44, 2222, 1979.
32. van Langen, L.M., Oosthoek, N.H.P., Guranda, D.T., van Rantwijk, F., Svedas, V.K Sheldon, R.A., *Tetrahedron: Asymmetry*, 11, 4593, 2000.
33. Svedas, V., Guranda, D., van Langen, L., van Rantwijk, F., Sheldon, R.A., *FEBS Lett.*, 199 414, and references mentioned therein.
34. Phillips, J.T. and Shears, J.H. (Shell International Research Maatschappij), EP Pat. 399589,
35. (a) Shimizu, S., Ogawa, J., and Yamada, H., *Appl. Microbiol. Biotechnol.*, 37, 164, 1992; (b) Oga Shimizu, S., and Yamada, H., *Bioorg. Med. Chem.*, 2, 429, 1994; (c) Shimizu, S., Ogawa, J., (M.C.M., and Yamada, H., *Eur. J. Biochem.*, 209, 375, 1992.
36. Chisso Corp., JP Pat. 58198296; *Chem. Abstr.*, 100, 119354, 1997.
37. (a) Laumen, K., Kittelmann, M., and Gishalba, O., *J. Mol. Catal. B: Enzym.*, 19, 55, 20$ Laumen, K., Brunella, A., Graf, M., Kittelmann, M., Walser, P., and Gishalba, O., *Pharmace istry Library: Trends in Drug Research II*, Vol. 29, Elsevier, Amsterdam, p. 17, 1998.
38. Smidt, H., Fischer, A., Fischer, P., Schmidt, R.D., and Stelzer, U. (Bayer AG), EP Pat. 8 1995.
39. For example, see: Summers, M.C., Markovic, R., and Klinman, J.P., *Biochemistry*, 18, 1969
40. Review: Stirling, D.I., in *Chirality in Industry*, Collins, A.N., Sheldrake, G.N., and Crosby, J. Wiley, New York, 1992.
41. Stirling, D.I., Zeitlin, A.L., and Matcham, G.W. (Celgene Corp.), U.S. Pat. Appl. 4950606,
42. Matcham, G., *Specialty Chemicals*, 12, 178, 1992.
43. Shin, J.-S. and Kim, B.-G., *Biosci. Biotechnol. Biochem.*, 65, 1782, 2001.
44. Shin, J.-S. and Kim, B.-G., *Biotechnol. Bioeng.*, 55, 348, 1997.
45. Shin, J.-S., Kim, B.-G., Liese, A., and Wandrey, C., *Biotechnol. Bioeng.*, 73, 179, 2001.
46. Yun, H., Yang, Y.-H., Cho, B.-K., Hwang, B.-Y., and Kim, B.-G., *Biotechnol. Lett.*, 25, 809
47. Reetz, M.T. and Schimossek, K., *Chimia*, 50, 668, 1996.
48. Pamies, O., Ell, A.H., Samec, J.S.M., Hermanns, N., and Bäckvall, J.-E., *Tetrahedron Le* 4699, 2002.

49. Review about combination of enzymes and metal catalysts in asymmetric catalysis: Pamies, O. and Bäckvall, J.-E., *Chem. Rev.*, 103, 3247, 2003.
50. Choi, Y.K., Kim, M.J., Ahn, Y., and Kim, M.-J., *Org. Lett.*, 3, 4099, 2001.
51. Review: Kim, M.-J., Ahn, Y., and Park, J., *Curr. Opin. Biotechnol.*, 13, 578, 2002.
52. Kroutil, W. and Faber, K., *Tetrahedron: Asymmetry*, 9, 2901, 1998.
53. (a) Hafner, E.W. and Wellner, D., *Proc. Natl. Acad. Sci.*, 68, 987, 1971; (b) Huh, J.W., Yokoigawa, K., Esaki, N., and Soda, K., *J. Ferment. Bioeng.*, 74, 189, 1992; (c) Huh, J.W., Yokoigawa, K., Esaki, N., and Soda, K., *Biosci. Biotech. Biochem.*, 56, 2081, 1992; (d) Soda, K., Oikawa, T., and Yokoigawa, K., *J. Mol. Catal. B: Enzym.*, 11, 149, 2001; (e) Beard, T.M. and Turner, N.J., *Chem. Commun.*, 246, 2002.
54. Alexeeva, M., Enright, A., Dawson, M.J., Mahmoudian, M., and Turner, N.J., *Angew. Chem. Int. Ed.*, 41, 3177, 2002.
55. Matcham, G.G. and Bowen, A.R.S., *Chim. Oggi*, 6, 20, 1996.
56. Matcham, G., Bhatia, M., Lang, W., Lewis, C., Nelson, R., Wang, A., and Wu, W., *Chimia*, 53, 548, 1999.
57. Shin, J.-S. and Kim, B.-G., *Biotechnol. Bioeng.*, 65, 206, 1999.

Index

Printed and bound by CPI Group (UK) Ltd, Croydon, CR0 4YY
23/10/2024
01778226-0016